Handbook of Numerical Analysis

Volume 21

Geometric Partial Differential Equations - Part I

Handbook of Numerical Analysis

Volume 21

Geometric Partial Differential Equations - Part I

Edited by

Andrea Bonito

Department of Mathematics, Texas A&M University, College Station, TX, United States

Ricardo H. Nochetto

Department of Mathematics and Institute for Physical Science and Technology, University of Maryland, College Park, MD, Unites States

ELSEVIER

N·H North-Holland

An imprint of Elsevier

North-Holland is an imprint of Elsevier
Radarweg 29, PO Box 211, 1000 AE Amsterdam, The Netherlands
The Boulevard, Langford Lane, Kidlington, Oxford OX5 1GB, United Kingdom

ISBN: 978-0-444-64003-1
ISSN: 1570-8659

For information on all North-Holland publications visit our website at https://www.elsevier.com/

Publisher: Zoe Kruze
Acquisition Editor: Sam Mahfoudh
Editorial Project Manager: Joanna Collett
Production Project Manager: James Selvam
Cover Designer: Ricardo H. Nochetto

Typeset by SPi Global, India

Contents

Contributors

Numbers in Parentheses indicate the pages on which the author's contributions begin.

Eberhard Bänsch (555), Applied Mathematics III, Department Mathematik, Friedrich-Alexander-Universität Erlangen-Nürnberg, Erlangen, Germany

John W. Barrett (275), Department of Mathematics, Imperial College London, London, United Kingdom

Sören Bartels (221), Abteilung für Angewandte Mathematik, Albert-Ludwigs-Universität Freiburg, Freiburg im Breisgau, Germany

Andrea Bonito (1), Department of Mathematics, Texas A&M University, College Station, TX, United States

Alan Demlow (1), Department of Mathematics, Texas A&M University, College Station, TX, United States

Qiang Du (425), Department of Applied Physics and Applied Mathematics and the Data Science Institute, Columbia University, New York, NY, United States

Xiaobing Feng (425), Department of Mathematics, The University of Tennessee, Knoxville, TN, United States

Harald Garcke (275), Fakultät für Mathematik, Universität Regensburg, Regensburg, Germany

Behrend Heeren (621), Institute for Numerical Simulation, University of Bonn, Bonn, Germany

Michael Neilan (105), Department of Mathematics, University of Pittsburgh, Pittsburgh, PA, United States

Ricardo H. Nochetto (1), Department of Mathematics and Institute for Physical Science and Technology, University of Maryland, College Park, MD, Unites States

Robert Nürnberg (275), Department of Mathematics, Imperial College London, London, United Kingdom

Martin Rumpf (621), Institute for Numerical Simulation, University of Bonn, Bonn, Germany

Abner J. Salgado (105), Department of Mathematics, University of Tennessee, Knoxville, TN, United States

Robert I. Saye (509), Mathematics Group, Lawrence Berkeley National Laboratory, Berkeley, CA, United States

Alfred Schmidt (555), Center for Industrial Mathematics and MAPEX Center for Materials and Processes, Universität Bremen, Bremen, Germany

James A. Sethian (509), Mathematics Group, Lawrence Berkeley National Laboratory; Department of Mathematics, University of California, Berkeley, Berkeley, CA, United States

Max Wardetzky (621), Institute of Numerical and Applied Mathematics, University of Göttingen, Göttingen, Germany

Benedikt Wirth (621), Institute for Analysis and Numerics, University of Münster, Münster, Germany

Wujun Zhang (105), Department of Mathematics, Rutgers University, Piscataway, NJ, United States

Preface

This and the following volumes are devoted to the numerical approximation of *geometric partial differential equations* (GPDEs). Before describing the objectives and contents of this project, it is appropriate to explain the meaning of GPDEs and the reasons why they deserve two volumes of the *Handbook of Numerical Analysis*. GPDEs are governing equations of natural, social and economic phenomena where geometry plays a prominent to dominant role. Examples abound from interfaces and free boundaries in fluids and solids—such as modelling of surface tension and bending effects involving second fundamental forms—to the development of defects in director and line fields in liquid crystal modelling, motion, merging and splitting of droplets within incompressible fluids, total variation minimization in imaging science and shape morphing and extrapolation, just to name a few. These problems possess an intrinsic mathematical beauty and pose a formidable challenge both in analysis and computation. Besides their overwhelming mathematical richness, GPDEs are ubiquitous in many scientific, engineering and industrial applications, such as fluid and solid mechanics, materials science, biology, chemistry, astrophysics, plasma physics, imaging and computer animation.

The last three decades have witnessed the development of powerful algorithms and corresponding numerical analysis for the description and computation of interfaces. The level set and phase field methods have joined the more traditional front tracking techniques and, together with the advent of ever more powerful and versatile computers, have allowed for the simulation and understanding of rather complex phenomena involving interfaces. It is fair to assert that, besides theory and experimentation, mathematical modelling and computation have established themselves as the third pillar of scientific inquiry. A well-designed computational model can replace a very expensive or even unrealizable experimental setting and give new insight into the theoretical developments of a specific discipline. Because of their technical complexity and practical relevance, GPDEs are a chief example.

The purpose of this two-volume contribution is to provide a missing reference book that portrays the state of the art on basic algorithms for GPDEs and their analysis, along with their impacts in a wide variety of areas of science and engineering. Since this field has grown tremendously over the last few years, the selection of topics and authors is a formidable task. Our intention is to present different and complementary approaches ranging from fundamental numerical analysis of basic algorithms to scientific computation and exciting applications. The choice of authors reflects their expertise on various aspects of this ambitious and multifaceted project.

This first volume consists of eight chapters which encompass numerical analysis of GPDEs, algorithm design, analysis and simulation of interfaces and shape morphing. A brief description follows.

- *Numerical analysis of GPDEs.* Chapter 1, by Bonito, Demlow and Nochetto, reviews and extends three popular finite element methods to approximate the Laplace–Beltrami operator on a codimension one surface; they are the parametric, trace and narrow band methods. The discussion centres around the relationship between the (minimal) regularity of the surface and the manner it is represented and approximated. Chapter 2, by Neilan, Salgado and Zhang, describes and analyzes several approximation techniques for the fully nonlinear Monge–Ampère equation. Key notions such as stability, consistency and continuous dependence on data are developed in the max norm and used to derive rates of convergence to the viscosity solution.
- *Geometrically nonlinear plates and rods.* Chapter 3, by Bartels, discusses geometrically nonlinear bending models for plates and rods that allow large deformations. Plate deformations are constrained to be isometries, whereas rods are elastic and self-avoiding. Recent finite element algorithms are reviewed along with their numerical analysis.
- *Interfaces and free boundaries.* This is the focus of the next four chapters. Chapter 4, by Barrett, Garcke and Nürnberg, collects their work on the parametric finite element approximation of curvature driven interface evolutions. Finite element methods for surface geometric equations, coupling of surface geometric equations with bulk equations, and two phase flows are presented and analyzed. Chapter 5, by Du and Feng, overviews the phase-field modelling and corresponding diffuse interface approximation. The main numerical analysis techniques are described and applied. Chapter 6, by Saye and Sethian, reviews the level-set method starting from basic ideas and techniques and culminating with complex and intricate interface dynamics. A description of the inherent challenges to multiphase and sharp-interface physics is provided along with numerical algorithms to overcome them. These three chapters present variational front tracking, phase field and level set methods, three competing techniques for interface evolution. Chapter 7, by Bänsch and Schmidt, discusses different aspects of phase transitions in materials science and fluids along with efficient algorithms for their approximation.
- *Riemmanian calculus and applications.* Chapter 8, by Heeren, Rumpf, Wardetzky and Wirth, develops a Riemannian calculus on the space of discrete triangular shells along with its discretization. This leads to adequate algorithms for shape morphing, shape extrapolation, parallel transport, smooth interpolation and other geometric processes of interest. Computer-animated video sequences and movies are presented.

Acknowledgements

A.B. is partially supported by NSF grant DMS-1817691. R.H.N. is partially supported by NSF grants DMS-1411808 and DMS-1908267.

Andrea Bonito
Ricardo H. Nochetto

Chapter 1

Finite element methods for the Laplace–Beltrami operator

Andrea Bonito[a,*], Alan Demlow[a] and Ricardo H. Nochetto[b]

[a]*Department of Mathematics, Texas A&M University, College Station, TX, United States*

[b]*Department of Mathematics and Institute for Physical Science and Technology, University of Maryland, College Park, MD, Unites States*

[*]*Corresponding author: e-mail: bonito@math.tamu.edu*

Chapter Outline

Abstract

Partial differential equations posed on surfaces arise in a number of applications. In this survey we describe three popular finite element methods for approximating solutions to the Laplace–Beltrami problem posed on an n-dimensional surface γ embedded in $\mathbb{R}^{n+1}$: the *parametric*, *trace*, and *narrow band* methods. The parametric method entails constructing an approximating polyhedral surface Γ whose faces comprise the finite

Handbook of Numerical Analysis, Vol. 21. https://doi.org/10.1016/bs.hna.2019.06.002

element triangulation. The finite element method is then posed over the approximate surface Γ in a manner very similar to standard FEM on Euclidean domains. In the trace method it is assumed that the given surface γ is embedded in an $n + 1$-dimensional domain Ω which has itself been triangulated. An n-dimensional approximate surface Γ is then constructed roughly speaking by interpolating γ over the triangulation of Ω, and the finite element space over Γ consists of the trace (restriction) of a standard finite element space on Ω to Γ. In the narrow band method the PDE posed on the surface is extended to a triangulated $n + 1$-dimensional band about γ whose width is proportional to the diameter of elements in the triangulation. In all cases we provide optimal a priori error estimates for the lowest order finite element methods, and we also present a posteriori error estimates for the parametric and trace methods. Our presentation focuses especially on the relationship between the regularity of the surface γ, which is never assumed better than of class C^2, the manner in which γ is represented in theory and practice, and the properties of the resulting methods.

Keywords: Surface partial differential equations, Laplace–Beltrami operator, Surface finite element methods, Parametric finite element methods, Trace finite element methods, Narrow band finite element methods, A priori and a posteriori error estimates

AMS Classification Codes: 65N15, 65N30, 35A99, 53-01, 58J32

1 Introduction

Partial differential equations (PDEs) posed on surfaces play an important role in many domains of pure and applied mathematics, including geometry, modelling of materials, fluid flow, and image and shape processing. The numerical approximation of such surface PDEs is both practically important and the source of many mathematically rich problems.

We consider a closed, compact and orientable surface γ in $\mathbb{R}^{n+1}$ of co-dimension 1. The Laplace–Beltrami operator $-\Delta_\gamma$, which acts as a generalization of the standard Euclidean Laplace operator, plays a central role in both static and time-dependent surface PDE models arising in a wide range of applications. Because of this a wide variety of numerical methods have been developed for the Laplace–Beltrami equation

$$-\Delta_\gamma \widetilde{u} = \widetilde{f},$$

where $\widetilde{f}$ is a given forcing function satisfying $\int_\gamma \widetilde{f} = 0$. In this article we first lay out some important notions from differential geometry. We then describe three important classes of finite element methods (FEMs) for the Laplace–Beltrami problem: the parametric method, the trace method, and the narrow band method. In all three cases we focus on the simplest case of piecewise linear finite element spaces and give an in-depth discussion of the effects of geometry on error behaviour.

The *parametric finite element method* was introduced by Dziuk (1988), with some important related techniques appearing in earlier works on boundary element methods (Bendali, 1984; Nédélec, 1976). This method is the simplest of the many FEM that have been developed for solving the Laplace–Beltrami problem. The given PDE is first written in weak form as: Find $\widetilde{u} \in H^1(\gamma)$ such that $\int_\gamma \widetilde{u} = 0$ and

$$a(\widetilde{u}, \widetilde{v}) := \int_\gamma \nabla_\gamma \widetilde{u} \cdot \nabla_\gamma \widetilde{v} = \int_\gamma \widetilde{f}\widetilde{v} \quad \forall \widetilde{v} \in H^1(\gamma).$$

Here $H^1(\gamma)$ is the set of functions $\widetilde{v}$ in $L_2(\Omega)$ whose tangential gradient $\nabla_\gamma \widetilde{v} \in [L_2(\gamma)]^{n+1}$. The continuous surface γ is approximated by a polyhedral surface Γ whose faces serve as a finite element mesh, and the finite element space $\mathbb{V}$ is made of continuous piecewise linear functions over Γ. The finite element method then consists of finding $U \in \mathbb{V}$ such that

$$A(U,V) = \int_\Gamma \nabla_\Gamma U \cdot \nabla_\Gamma V = \int_\gamma FV \quad \forall V \in \mathbb{V},$$

where F is a suitable approximation (lift) of f defined on Γ. In its conception and implementation, the resulting method is very similar to canonical FEM for solving Poisson's problem on Euclidean domains. To quote Dziuk, "...the numerical scheme is just the same as in a plane-two dimensional problem. The only difference is that in our case the computer has to memorize three-dimensional nodes instead of two-dimensional ones" (Dziuk, 1988, p. 143). The strategy underlying parametric surface finite element methods—direct translation of FEM on Euclidean spaces to triangulated surfaces—has subsequently been applied to a variety of methods. These include higher order standard Lagrange methods (Demlow, 2009), various types of discontinuous Galerkin methods (Antonietti et al., 2015; Cockburn and Demlow, 2016; Dedner et al., 2013), and mixed methods in classical, hybridized, and finite element exterior calculus formulations (Bendali, 1984; Cockburn and Demlow, 2016; Ferroni et al., 2016; Holst and Stern, 2012). A posteriori error estimation and adaptivity have been studied in Demlow and Dziuk (2007), Wei et al. (2010), Bonito et al. (2013), Dedner and Madhavan (2016), Bonito et al. (2016), and Bonito and Demlow (2019). Finally, we refer to the survey article (Dziuk and Elliott, 2013).

In many applications in which surface PDEs are to be solved, a background volume (bulk) mesh is already present. A paradigm example is two-phase fluid flow, in which effects on the interface between the two phases such as surface tension are coupled with standard equations of fluid dynamics on the bulk. In these cases it is advantageous to utilize the background volume mesh to solve surface PDEs instead of independently meshing γ. This is especially the case when γ is evolving, since the meshes needed for the parametric method typically distort as γ changes and periodic remeshing is thus necessary. The trace and narrow band methods both employ background bulk meshes in order to solve surface PDEs.

Trace (or cut) FEMs for the Laplace–Beltrami problem were first introduced in Olshanskii et al. (2009). In this method an approximating surface is constructed as in the parametric method, but using a different approach. An *implicit* representation of γ as the level set of some function ϕ is used, that is, it is assumed that

$$\gamma = \{x \in \mathbb{R}^{n+1} \ : \ \phi(x) = 0\}.$$

A discrete surface Γ is then defined as the zero level set of an interpolant of ϕ on the background mesh, and the finite element space is taken to be the trace of the bulk finite element space on Γ. The FEM is posed and solved on Γ as in the parametric method. Note that the finite element space in the trace method consists of continuous piecewise linear functions over the faces of Γ. However, because the faces of Γ are arbitrary intersections of n-dimensional hyperplanes with $n+1$-simplices, they are not shape regular, and in particular may either fail to satisfy a minimum angle condition or be much smaller than the bulk simplices from which they are derived. Counter to natural intuition about the quality of a finite element method posed on such a mesh, the trace method satisfies optimal error bounds and works well in practice. In addition to the basic analysis of piecewise linear methods that we present below, the literature on trace methods for the Laplace–Beltrami problem includes study of matrix properties (Olshanskii and Reusken, 2010), adaptive versions (Chernyshenko and Olshanskii, 2015; Demlow and Olshanskii, 2012), and extensions to higher order (Grande and Reusken, 2016; Grande et al., 2018; Reusken, 2015), stabilized (Burman et al., 2015, 2016), and discontinuous Galerkin (Burman et al., 2017) methods. We refer to the recent survey article (Olshanskii and Reusken, 2017).

Narrow band methods also employ a bulk mesh in order to approximate surface PDEs, but extend a surface PDE to the bulk instead of restricting a bulk finite element space to a surface. This idea appeared first in Bertalmío et al. (2001) and is based on an extension of the PDE into a tubular neighbourhood $\mathcal{N}(\delta)$ of width 2δ about γ that reads

$$L(u_\delta) = -\mathrm{div}((I - \nabla d \otimes \nabla d)\nabla u_\delta) + u_\delta = f_\delta.$$

Here f_δ is an extension of $\widetilde{f}$ from γ to $\mathcal{N}(\delta)$ and d is the signed distance function to γ. The latter is chosen for simplicity over a generic level set function ϕ to represent γ throughout this article. Because ∇d is the unit outward normal to γ, the coefficient matrix $I - \nabla d \otimes \nabla d$ is degenerate in the direction normal to γ, and the operator L is thus elliptic but degenerate. We emphasize that in contrast to most previous literature on narrow band FEM we do not include a zero order term in our presentation, thereby adding extra difficulty due to the need to account for the nontrivial kernel of L on closed surfaces. In narrow band FEMs, the Galerkin approximations to u_δ are posed over a discrete approximation $\mathcal{N}_h(\delta)$ to the narrow band $\mathcal{N}(\delta)$. Related methods that involve extending surface PDEs to bulk domains include the closest point method (Ruuth and Merriman, 2008).

Narrow-band unfitted finite element methods have been proposed and analyzed by different authors. In Burger (2009), the aforementioned degenerate extension is shown to be well posed and error bounds in the weighted bulk energy norm are derived. Subsequently, error estimates in the $H^1(\gamma)$ norm are obtained in Deckelnick et al. (2010) for the lower order method. An alternate nondegenerate extension $L(u_\delta) = -\Delta u_\delta + u_\delta$ is then proposed in Deckelnick et al. (2014) leading to optimal $H^1(\gamma)$ and also $L^2(\gamma)$ error estimates for the lower order method when f_δ is (or is close to) the constant normal extension of $\widetilde{f}$. Independently, higher order methods are proposed and analyzed in Olshanskii and Safin (2016) using the extension

$$L(u_\delta) = -\mathrm{div}\left(\mu(I - dD^2d)^{-2}\nabla u_\delta\right) + u_\delta,$$

with $\mu := \det(I - dD^2d)$ and f_δ the constant normal extension of $\widetilde{f}$. Note also that the associated FEM requires a sufficiently accurate approximation of D^2d (if not known explicitly). For the case of lowest order (piecewise linear) finite element spaces, it is enough to approximate D^2d with zero and thereby retrieve the discrete formulation in Deckelnick et al. (2014).

In the construction of all three FEMs above, we incur variational crimes (consistency errors) due to the approximation of geometry. In the parametric and trace methods, these errors arise because the finite element method is posed over a discrete approximation Γ to γ, thereby leading to different bilinear forms (a and A) used to compute the continuous and finite element solutions ($\widetilde{u}$ and U). In the narrow band method the finite element equations are posed over a discrete narrow band $\mathcal{N}_h(\delta)$ instead of over the domain $\mathcal{N}(\delta)$ on which the extended solution u_δ is defined. This again entails the use of different bilinear forms in the definitions of the continuous and discrete solutions. A core problem in surface FEMs is understanding and controlling these errors, which are typically called *geometric consistency errors* or *geometric errors*. In order to analyze these errors, it is necessary to define a map $\mathbf{P}:\Gamma \to \gamma$ and then compare $a(\widetilde{v}, \widetilde{w})$ with $A(\widetilde{v} \circ \mathbf{P}, \widetilde{w} \circ \mathbf{P})$ for given functions $\widetilde{v}, \widetilde{w} \in H^1(\gamma)$. This is done via a change of variables argument for the map $\mathbf{P}$. There may be several competing demands of both theoretical and practical nature that come into play when choosing the map $\mathbf{P}$, and a main focus of this article is to elucidate how this choice affects analysis and implementation of surface FEMs.

The canonical choice of the map $\mathbf{P}$ is defined via the so-called *signed distance function* $d:\mathcal{N} \to \gamma$. The distance function is defined on a tubular neighbourhood $\mathcal{N}$ of γ and is of the same regularity class as γ provided that γ is at least C^2 and $\mathcal{N}$ is sufficiently narrow. In such a case, the map (also called distance lift or orthogonal closest point projection)

$$\mathbf{P}_d(\mathbf{x}) := \mathbf{x} - d(\mathbf{x})\nabla d(\mathbf{x}) \quad \forall \ \ \mathbf{x} \in \mathcal{N}$$

is well defined and is of class C^1. The maps d and $\mathbf{P}_d$ play a crucial role in analyzing and in some cases defining the numerical algorithms presented below. In particular, the distance function is a critical tool in proving error estimates that are of optimal order with respect to geometric consistency errors. When a generic map $\mathbf{P}: \Gamma \to \gamma$ is instead used to analyze surface FEMs, the predicted behaviour of geometric errors is of one order less than is seen in practice and also than may be proved using the closest point projection. More precisely, when quasi-uniform meshes of size h are used with affine surface approximations in the parametric and trace methods, arguments which use special properties of the closest point projection predict an $O(h^2)$ geometric errors, and these are in fact observed in practice. On the other hand, standard proofs employing a generic map $\mathbf{P}$ instead of the distance function map $\mathbf{P}_d$ predict only order h geometric errors. This increase in convergence order due to the properties of the closest point projection may be viewed as a *super-convergence* effect.

Reliance on $\mathbf{P}_d$ may however also constitute a serious drawback for several reasons. First, $\mathbf{P}_d$ has a closed form expression only for the sphere and torus, so it is in general not directly available to the user. We thus discuss how to use the distance function only as a theoretical tool for the parametric FEM and yet retain the superconvergence properties of $\mathbf{P}_d$. On a practical level, the user is still free to choose from a much more general class of lifts to implement an algorithm. Our presentation includes optimal a priori and a posteriori estimates in H^1 and optimal a priori estimates in L_2 for an algorithm whose implementation only requires access to a generic lift $\mathbf{P}$; the latter appear to be new in the literature even for smooth surfaces. Second, if γ is merely $C^{1,\alpha}$ for $\alpha < 1$, then the closest point projection $\mathbf{P}_d$ is not uniquely defined in any neighbourhood of γ. We thus also provide an analysis of parametric FEMs for γ of class $C^{1,\alpha}$ that instead makes use of a generic parametric map. The price we pay is a possible order reduction of the method due to the loss of superconvergence properties of $\mathbf{P}_d$. Finally, previous proofs of optimal-order error estimates employing $\mathbf{P}_d$ have required that $\mathbf{P}_d$ is of class C^2 and thus γ of class C^3 (cf. Dziuk, 1988). However, the solution u to the Laplace–Beltrami problem already possesses the H^2 regularity needed to ensure optimal convergence of piecewise linear finite element methods when γ is of class C^2. In this survey we bridge this gap by giving a novel error analysis for the three FEMs which is based exclusively on C^2 regularity of d and γ, but which also preserves the superconvergence property in the geometric error. In the case of the trace and narrow band methods we achieve this by a regularization argument.

This article is organized as follows. In Section 2 we introduce surface gradient, divergence and Laplace–Beltrami operators along with the signed distance function and its most relevant properties. In Section 3 we quantify the geometric effects of perturbing surfaces γ of class $C^{1,\alpha}$ and C^2. We also present H^2 extensions u to a tubular neigborhood $\mathcal{N}(\delta) \subset \mathcal{N}$ of width δ

$$\|u\|_{H^2(\mathcal{N}(\delta))} \lesssim \delta^{\frac{1}{2}} \|\widetilde{u}\|_{H^2(\gamma)}$$

of functions $\widetilde{u} \in H^2(\gamma)$ provided γ is of class C^2. This turns out to be essential for our later error analysis of the trace and narrow band methods for C^2 surfaces. In Section 4 we give a self-contained exposition of parametric FEMs for surfaces of class $C^{1,\alpha}$ and C^2, including a priori and a posteriori error analyses. In Section 5 we describe the trace method and conclude in Section 6 with the narrow band method. Both discussions assume C^2 regularity of γ.

2 Calculus on surfaces

In this section we discuss basic concepts of differential geometry. We start in Section 2.1 by describing the parametric representation of γ via charts. This classical point of view is critical to introduce the first fundamental form $\mathbf{g}$, the area element q, and the unit normal $\boldsymbol{\nu}$ of γ. We present in Section 2.2 the tangential operators (gradient ∇_γ, divergence div_γ, and Laplace–Beltrami Δ_γ) as well as the Weingarten map; we also discuss H^2-regularity for Δ_γ on surfaces γ of class C^2. We introduce the signed distance function d in Section 2.3 and derive several important properties of it; this intrinsic approach avoids parametrizations and allows for implicit representations of γ. We devote Section 2.4 to the second fundamental form of γ and its principal curvatures using both parametric and intrinsic approaches.

2.1 Parametric surfaces

We assume that γ is a closed, compact, orientable manifold of class $C^{1,\alpha}$, $0 < \alpha \leq 1$, and co-dimension 1 in $\mathbb{R}^{n+1}$. It can be represented parametrically by an atlas $\{(\mathcal{V}_i, \mathcal{U}_i, \boldsymbol{\chi}_i)\}_{i\in I}$, where the individual charts $\boldsymbol{\chi}_i : \mathcal{V}_i \to \mathcal{U}_i \cap \gamma \subset \mathbb{R}^{n+1}$ are isomorphisms of class $C^{1,\alpha}$ compatible with the orientation of γ; the open connected sets $\mathcal{V}_i \subset \mathbb{R}^n$ are the parametric domains. Unless stated otherwise, it will be often sufficient to consider a single chart and resort to a partition of the unity. We thus drop the index i for convenience. For $\mathbf{x} \in \mathcal{U} \cap \gamma$, we set $\mathbf{y} := \boldsymbol{\chi}^{-1}(\mathbf{x}) \in \mathcal{V}$.

Let $\partial_j \boldsymbol{\chi}(\mathbf{y})$ be the column vector of jth partial derivatives of $\boldsymbol{\chi}(\mathbf{y})$ for $1 \leq j \leq n$ at $\mathbf{y} \in \mathcal{V}$. By definition, the rank of $D\boldsymbol{\chi}(\mathbf{y}) = (\partial_j \boldsymbol{\chi}(\mathbf{y}))_{j=1}^n \in \mathbb{R}^{(n+1)\times n}$ is n (full rank). This implies that $\{\partial_j \boldsymbol{\chi}(\mathbf{y})\}_{j=1}^n$ are linearly independent and span the tangent hyperplane to γ at $\mathbf{x}$.

The *first fundamental form* is the symmetric and positive definite matrix $\mathbf{g} \in \mathbb{R}^{n\times n}$ defined by

$$\mathbf{g}(\mathbf{y}) := D\boldsymbol{\chi}(\mathbf{y})^t D\boldsymbol{\chi}(\mathbf{y}) \quad \forall \mathbf{y} \in \mathcal{V}. \tag{1}$$

If $\mathbf{g} = (g_{ij})_{i,j=1}^n$, then the components g_{ij} read

$$g_{ij} = \partial_i \boldsymbol{\chi}^t \partial_j \boldsymbol{\chi} = \partial_i \boldsymbol{\chi} \cdot \partial_j \boldsymbol{\chi},$$

which depends on the choice of parametrization. A normal vector $\mathbf{N}(\mathbf{y})$ to γ at $\mathbf{x}$ can be written as $\mathbf{N}(\mathbf{y})=\sum_{j=1}^{n+1}A_j(\mathbf{y})\mathbf{e}_j$, where $A_j := \det(\mathbf{e}_j, D\chi)$ and $\{\mathbf{e}_j\}_{j=1}^{n+1}$ is the canonical basis of $\mathbb{R}^{n+1}$. In fact, since

$$\mathbf{N}\cdot\partial_i\chi=\sum_{j=1}^{n+1}\mathbf{e}_j\cdot\partial_i\chi\det(\mathbf{e}_j, D\chi)=\det\left(\sum_{j=1}^{n+1}(\mathbf{e}_j\cdot\partial_i\chi)\mathbf{e}_j, D\chi\right)=\det(\partial_i\chi, D\chi)=0,$$

and $A_j\neq 0$ for at least one j because $D\chi$ has rank n, we deduce that

$$\boldsymbol{\nu}(\mathbf{y}):=\frac{\mathbf{N}(\mathbf{y})}{|\mathbf{N}(\mathbf{y})|}\quad\forall\mathbf{y}\in\mathcal{V} \tag{2}$$

is a well-defined unit normal vector to γ. Therefore, the matrix

$$\mathbf{T}(\mathbf{y}):=[D\chi(\mathbf{y}),\boldsymbol{\nu}(\mathbf{y})]\in\mathbb{R}^{(n+1)\times(n+1)}\quad\forall\mathbf{y}\in\mathcal{V}$$

has rank $n+1$ and so is invertible. We write its inverse as

$$\mathbf{T}^{-1}=\begin{bmatrix}\mathbf{B}\\ \mathbf{v}^t\end{bmatrix},\quad\mathbf{B}\in\mathbb{R}^{n\times(n+1)},\mathbf{v}\in\mathbb{R}^n,$$

and note that

$$\mathbf{I}_{(n+1)\times(n+1)}=\mathbf{T}^{-1}\mathbf{T}=\begin{bmatrix}\mathbf{B}\,D\chi & \mathbf{B}\mathbf{v}\\ \mathbf{v}^t\;D\chi & \mathbf{v}^t\boldsymbol{\nu},\end{bmatrix}$$

whence

$$\mathbf{B}\,D\chi=\mathbf{I}_{n\times n},\quad\mathbf{v}^tD\chi=0,\quad\mathbf{v}^t\boldsymbol{\nu}=1.$$

The last two equalities imply $\mathbf{v}=\boldsymbol{\nu}$. Reversing the order of multiplication yields

$$\mathbf{I}_{(n+1)\times(n+1)}=\mathbf{T}\,\mathbf{T}^{-1}=D\chi\,\mathbf{B}+\boldsymbol{\nu}\boldsymbol{\nu}^t,$$

whence the *projection matrix* $\Pi\in\mathbb{R}^{(n+1)\times(n+1)}$ on the tangent hyperplane to γ has the form

$$\Pi:=\mathbf{I}-\boldsymbol{\nu}\otimes\boldsymbol{\nu}=D\chi\,\mathbf{B}. \tag{3}$$

To obtain an explicit expression for $\mathbf{B}$ note that

$$D\chi=(\mathbf{I}-\boldsymbol{\nu}\otimes\boldsymbol{\nu})^tD\chi=\mathbf{B}^t\,D\chi^t\,D\chi=\mathbf{B}^t\mathbf{g}\quad\Rightarrow\quad\mathbf{B}=\mathbf{g}^{-1}D\chi^t.$$

This leads to the following useful expression of Π defined in (3):

$$\Pi=D\chi\,\mathbf{g}^{-1}D\chi^t. \tag{4}$$

The *area element* $q(\mathbf{y})$ is the ratio of the infinitesimal volume at $\mathbf{y}\in\mathcal{V}$ and area of γ at $\mathbf{x}=\chi(\mathbf{y})$, namely the volume of the parallelotope in the tangent plane to γ spanned by the vectors $\{\chi_j\}_{j=1}^n$:

$$q(\mathbf{y}):=\det([\boldsymbol{\nu}(\mathbf{y}),D\chi(\mathbf{y})])\quad\forall\mathbf{y}\in\mathcal{V}. \tag{5}$$

To obtain a more familiar form of q we argue as follows:

$$q=\frac{1}{|N|}\det([\mathbf{N},D\chi])=\frac{1}{|N|}\det\left([\mathbf{N},D\chi]^t[\mathbf{N},D\chi]\right)^{\frac{1}{2}}=\sqrt{\det\mathbf{g}}, \tag{6}$$

because $\det([\mathbf{N},D\chi]^t[\mathbf{N},D\chi])=|\mathbf{N}|^2\det(D\chi^tD\chi)=|\mathbf{N}|^2\det\mathbf{g}$. Moreover, exploiting that $|\mathbf{N}|^2=\sum_{j=1}^{n+1}A_j\det([\mathbf{e}_j,D\chi])=\det([\mathbf{N},D\chi])$, we deduce

$$q=|\mathbf{N}|. \tag{7}$$

An integrable function $v:\mathcal{V}\to\mathbb{R}$ induces an integrable function $\widetilde{v}:\gamma\to\mathbb{R}$ by composition $v=\widetilde{v}\circ\chi$, or equivalently $\widetilde{v}(\mathbf{x})=v(\mathbf{y})$ for all $\mathbf{y}\in\mathcal{V}$. The area element allows for integration over γ via the formula

$$\int_\gamma\widetilde{v}=\int_{\mathcal{V}}vq\quad\forall v\in L_1(\mathcal{V}). \tag{8}$$

This definition does not depend on the parametrization: if χ_1,χ_2 are parametrizations of γ, then $\chi_1=\chi_2\circ(\chi_2^{-1}\circ\chi_1)$ and $D\chi_1=D\chi_2D(\chi_2^{-1}\circ\chi_1)$ whence

$$q_1=\left|\det\left(D(\chi_2^{-1}\circ\chi_1)\right)\right|q_2\quad\Rightarrow\quad\int_{\mathcal{V}_1}vq_1=\int_{\mathcal{V}_2}vq_2.$$

2.2 Differential operators

If a function $v:\mathcal{V}\to\mathbb{R}$ is of class C^1, we can define the *tangential (or surface) gradient* of the corresponding function $\widetilde{v}:\gamma\to\mathbb{R}$ as a vector tangent to γ that satisfies the chain rule

$$\nabla v(\mathbf{y})=D\chi(\mathbf{y})^t\nabla_\gamma\widetilde{v}(\mathbf{x})\quad\forall\mathbf{y}\in\mathcal{V}. \tag{9}$$

Since $\nabla_\gamma\widetilde{v}$ is spanned by $\{\partial_j\chi\}_{j=1}^n$, we get $\nabla_\gamma\widetilde{v}=D\chi w$ for some $w\in\mathbb{R}^n$ whence $w=\mathbf{g}^{-1}\nabla v$ and

$$\nabla_\gamma\widetilde{v}(\mathbf{x})=D\chi(\mathbf{y})\mathbf{g}(\mathbf{y})^{-1}\nabla v(\mathbf{y})\quad\forall\mathbf{y}\in\mathcal{V}. \tag{10}$$

If $\widetilde{\mathbf{v}}=(\widetilde{v}_i)_{i=1}^{n+1}:\gamma\to\mathbb{R}^{n+1}$ is a vector field of class C^1, we define its tangential differential $D_\gamma\widetilde{\mathbf{v}}\in\mathbb{R}^{(n+1)\times(n+1)}$ as a matrix whose ith row is $(\nabla_\gamma\widetilde{v}_i)^t$. If γ is of class C^2, then the unit normal vector $\boldsymbol{\nu}$ is of class C^1 and its differential

$$\mathbf{W}(\mathbf{x})=D_\gamma\boldsymbol{\nu}(\mathbf{x})\quad\forall\mathbf{x}\in\gamma \tag{11}$$

is called the *Weingarten* (or shape) map of γ. In addition, the *tangential divergence* of $\widetilde{\mathbf{v}}$ is the trace of $D_\gamma\widetilde{\mathbf{v}}$

$$\mathrm{div}_\gamma\widetilde{\mathbf{v}}(\mathbf{x})=\mathrm{trace}\left(D_\gamma\widetilde{\mathbf{v}}(\mathbf{x})\right)=\sum_{i,j=1}^n g^{ij}(\mathbf{y})\,\partial_i\chi(\mathbf{y})\cdot\partial_j\mathbf{v}(\mathbf{y})\quad\forall\mathbf{y}\in\mathcal{V}, \tag{12}$$

provided $\mathbf{g}^{-1} = (g^{ij})_{i,j=1}^{n}$. If both γ and $v:\gamma \to \mathbb{R}$ are of class C^2, then the *Laplace–Beltrami (or surface Laplace) operator* is now defined to be

$$\Delta_\gamma \widetilde{v} = \frac{1}{q(\mathbf{y})} \operatorname{div} \left(q(\mathbf{y}) \mathbf{g}(\mathbf{y})^{-1} \nabla v(\mathbf{y}) \right) \quad \forall \mathbf{y} \in \mathcal{V}. \tag{13}$$

The following lemma shows that (13) is designed to allow integration by parts on γ, exactly as it happens in flat domains with the Laplace operator Δ.

Lemma 1 (weak form of the Laplace–Beltrami operator). *If $\widetilde{\varphi}$ is of class C^1 with compact support in γ, then*

$$\int_\gamma \widetilde{\varphi}\, \Delta_\gamma \widetilde{v} = -\int_\gamma \nabla_\gamma \widetilde{\varphi} \cdot \nabla_\gamma \widetilde{v}. \tag{14}$$

Proof. In view of (8), which allows us to switch from γ to $\mathcal{V}$ back and forth, we can write

$$\begin{aligned}
\int_\gamma \widetilde{\varphi} \Delta_\gamma \widetilde{v} &= \int_{\mathcal{V}} \varphi \operatorname{div} \left(q \mathbf{g}^{-1} \nabla v \right) \\
&= -\int_{\mathcal{V}} \nabla \varphi \cdot \mathbf{g}^{-1} \nabla v \, q \\
&= -\int_{\mathcal{V}} D\boldsymbol{\chi} \mathbf{g}^{-1} \nabla \varphi \cdot D\boldsymbol{\chi} \mathbf{g}^{-1} \nabla v \, q \\
&= -\int_\gamma \nabla_\gamma \widetilde{\varphi} \cdot \nabla_\gamma \widetilde{v}.
\end{aligned}$$

This proves (14) as desired. □

In view of (14), we are now in the position to introduce the weak formulation for the Laplace–Beltrami operator. We first define the space of square integrable functions on γ with vanishing mean value by

$$L_{2,\#}(\gamma) := \left\{ \widetilde{v} \in L_2(\gamma) \mid \int_\gamma \widetilde{v} = 0 \right\}$$

and its subspace $H^1_\#(\gamma)$ containing square integrable weak derivatives defined as for example in Section 3 of Jerison and Kenig (1995) by

$$H^1_\#(\gamma) := H^1(\gamma) \cap L_{2,\#}(\gamma), \quad H^1(\gamma) := \{ \widetilde{v} \in L_2(\gamma) \mid \nabla(\widetilde{v} \circ \boldsymbol{\chi}) \in [L_2(\mathcal{V})]^n \}.$$

Our next result shows that the natural norm $\|\nabla_\gamma \widetilde{v}\|_{L_2(\gamma)} + \|\widetilde{v}\|_{L_2(\gamma)}$ in $H^1_\#(\gamma)$ is equivalent to the seminorm $\|\nabla_\gamma \widetilde{v}\|_{L_2(\gamma)}$. The proof essentially hinges on the

Peetre–Tartar Lemma (Peetre, 1966; Tartar, 1978), but we proceed with a slightly more direct proof as in Evans (1998, Section 5.8.1).

Lemma 2 (Poincaré–Friedrichs inequality). *Let γ be a compact Lipschitz surface. There exists a constant C only depending on γ such that*

$$\|\widetilde{v}\|_{L_2(\gamma)} \leq C\,\|\nabla_\gamma \widetilde{v}\|_{L_2(\gamma)} \quad \forall\, \widetilde{v} \in H^1_\#(\gamma). \tag{15}$$

Proof. We prove by contradiction the more general estimate

$$\|\widetilde{v}\|_{L_2(\gamma)} \leq C\left(\|\nabla_\gamma \widetilde{v}\|_{L_2(\gamma)} + \left| \int_\gamma \widetilde{v} \right| \right) \quad \forall\, \widetilde{v} \in H^1(\gamma). \tag{16}$$

Suppose that there is a sequence $\widetilde{v}_k \in H^1(\gamma)$ such that

$$\|\widetilde{v}_k\|_{L_2(\gamma)} = 1, \quad \|\nabla_\gamma \widetilde{v}_k\|_{L_2(\gamma)} + \left| \int_\gamma \widetilde{v}_k \right| \to 0$$

as $k \to \infty$. We deduce that $\{\widetilde{v}_k\}_k$ is uniformly bounded in $H^1(\gamma)$. Since the embedding $H^1(\gamma) \subset L_2(\gamma)$ is compact (because $H^1(\mathcal{V}) \subset L_2(\mathcal{V})$ is compact, see the proof of Aubin (1982, Theorem 2.34)), there is a Cauchy subsequence (with abuse of notation not relabelled) of $\{\widetilde{v}_k\}_k$ in $L_2(\gamma)$. This, together with $\|\nabla_\gamma \widetilde{v}_k\|_{L_2(\gamma)} \to 0$, implies that $\{\widetilde{v}_k\}_k$ is a Cauchy sequence in $H^1(\gamma)$. Let $\widetilde{v} \in H^1(\gamma)$ be the limit of $\widetilde{v}_k$ in $H^1(\gamma)$, which yields $\nabla_\gamma \widetilde{v} = \lim_{k\to\infty} \nabla_\gamma \widetilde{v}_k = 0$ whence $\widetilde{v}$ is constant on γ. Moreover, $\int_\gamma \widetilde{v} = \lim_{k\to\infty} \int_\gamma \widetilde{v}_k = 0$ whence $v = 0$. This gives rise to the contradiction $0 = \|\widetilde{v}\|_{L_2(\gamma)} = \lim_{k\to\infty} \|\widetilde{v}_k\|_{L_2(\gamma)} = 1$, and finishes the proof. □

We emphasize that the Poincaré–Friedrichs constant depends on the surface γ. Later we shall consider perturbations Γ of γ and derive Poincaré–Friedrichs type estimates on Γ where the constant depends on γ provided the geometry of γ is minimally approximated by Γ. This is proved in Lemma 18 for Lipschitz surfaces and only requires that the L_2 and H^1 norms on γ and Γ are equivalent.

We will not deal explicitly with functionals in the dual space $H^{-1}_\#(\gamma)$ of $H^1_\#(\gamma)$, but occasionally need its norm for $\widetilde{f} \in L_{2,\#}(\gamma)$

$$\|\widetilde{f}\|_{H^{-1}_\#(\gamma)} = \sup_{\widetilde{v} \in H^1_\#(\gamma)} \frac{\int_\gamma \widetilde{f}\,\widetilde{v}}{\|\nabla_\gamma \widetilde{v}\|_{L_2(\gamma)}}. \tag{17}$$

Lemma 2 (Poincaré–Friedrichs inequality) implies that $\|\widetilde{f}\|_{H^{-1}_\#(\gamma)} \leq C\,\|\widetilde{f}\|_{L_{2,\#}(\gamma)}$. The weak formulation of $-\Delta_\gamma \widetilde{u} = \widetilde{f}$ reads: for $\widetilde{f} \in L_{2,\#}(\gamma)$, seek $\widetilde{u} \in H^1_\#(\gamma)$ so that

$$\int_\gamma \nabla_\gamma \widetilde{u} \cdot \nabla_\gamma \widetilde{v} = \int_\gamma \widetilde{f}\,\widetilde{v} \quad \forall \widetilde{v} \in H^1_\#(\gamma). \tag{18}$$

Since the Dirichlet bilinear form in (18) is coercive, according to Lemma 2, existence and uniqueness of a solution $\widetilde{u} \in H^1_\#(\gamma)$ is a consequence of the Lax–Milgram theorem. We observe that thanks to the property $\widetilde{f} \in L_{2,\#}(\gamma)$, the solution $\widetilde{u} \in H^1_\#(\gamma)$ satisfies

$$\int_\gamma \nabla_\gamma \widetilde{u} \cdot \nabla_\gamma \widetilde{v} = \int_\gamma \widetilde{f}\,\widetilde{v} \quad \forall \widetilde{v} \in H^1(\gamma). \tag{19}$$

It turns out that $\widetilde{u}$ exhibits the usual regularity pick-up provided γ is of class C^2.

Lemma 3 (regularity). *If γ is of class C^2, then there is a constant C only depending on γ such that*

$$\|\widetilde{u}\|_{H^2(\gamma)} \leq C\,\|\widetilde{f}\|_{L_2(\gamma)}. \tag{20}$$

Proof. We use a localization argument to the parametric domain. We assume, without loss of generality, that the atlas $\{(\mathcal{V}_i, \mathcal{U}_i, \chi_i)\}_{i=1}^I$ satisfies the following property: there exist domains $\mathcal{W}_i$ such that $\overline{\mathcal{W}}_i \subset \mathcal{U}_i$ and $\{\mathcal{W}_i\}_{i=1}^I$ is still a covering of γ. Let now $\{\widetilde{\psi}_i\}_{i=1}^I$ be a C^2 partition of unity associated with the covering $\{\mathcal{W}_i\}_{i=1}^I$. The functions $u_i = u\psi_i$ satisfy

$$\Delta_\gamma \widetilde{u}_i = \widetilde{\psi}_i \widetilde{f} + 2\nabla_\gamma \widetilde{u} \cdot \nabla_\gamma \widetilde{\psi}_i + \widetilde{u}\Delta_\gamma \widetilde{\psi}_i =: \widetilde{g}_i.$$

In light of the estimate $\|\nabla_\gamma \widetilde{u}\|_{L_2(\gamma)} \leq \|\widetilde{f}\|_{H_\#^{-1}(\gamma)}$ and (15) we deduce that $\|\widetilde{g}_i\|_{L_2(\gamma)} \leq C\,\|\widetilde{f}\|_{L^2(\gamma \cap \mathcal{U}_i)}$. Recalling (13) we can rewrite $\Delta_\gamma u_i$ in the parametric domain $\mathcal{V}_i$ as

$$\operatorname{div}\left(q_i(\mathbf{y})\mathbf{g}_i(\mathbf{y})^{-1}\nabla u(\mathbf{y})\right) = q_i(y)\widetilde{g}_i(\chi(\mathbf{y})) \quad \forall \mathbf{y} \in \mathcal{V}_i,$$

and observe that this is a uniformly elliptic problem with C^1 coefficients. Applying interior regularity theory (Evans, 1998), we deduce

$$\|u_i\|_{H^2(\chi^{-1}(\mathcal{W}_i))} \leq C\,\|g_i\|_{L_2(\mathcal{U}_i)}.$$

Therefore, adding over i and using the finite overlap property of the sets $\mathcal{U}_i$, we end up with

$$\|\widetilde{u}\|_{H^2(\gamma)} \leq \sum_{i=1}^I \|\widetilde{u}_i\|_{H^2(\mathcal{W}_i)} \leq C\sum_{i=1}^I \|\widetilde{g}_i\|_{L_2(\mathcal{U}_i)} \leq C\,\|\widetilde{f}\|_{L_2(\gamma)},$$

as asserted. □

In view of our discussion below of surfaces of class $C^{1,\alpha}$ with $0 < \alpha \leq 1$, it is natural to ask whether the regularity estimate (20) is still valid in this more general context. We now show that this is indeed the case provided the surface γ is of class W_p^2 with $p > n$, or equivalently the parametrizations $\{\chi_i\}_{i=1}^I$ and partitions of unity $\{\widetilde{\psi}_i\}_{i=1}^I$ subordinate to the covering $\{\mathcal{W}_i\}_{i=1}^I$ of γ are of class W_p^2. In this case a Sobolev embedding implies γ is of class $C^{1,\alpha}$ with $0 < \alpha = 1 - \frac{n}{p} \leq 1$.

Lemma 4 (regularity for W_p^2 surfaces)**.** *If γ is of class W_p^2 with $n < p \leq \infty$, then there is a constant $C > 0$ depending on γ, p and n such that*

$$\|\widetilde{u}\|_{H^2(\gamma)} \leq C\,\|\widetilde{f}\|_{L_2(\gamma)}. \tag{21}$$

Proof. We argue with one chart $(\mathcal{V}, \mathcal{U}, \chi)$ and thus suppress the index i in g, χ, etc. Since $\widetilde{f} \in L_2(\gamma)$ and $\widetilde{u} \in H^1(\gamma)$, the right-hand side $g = \widetilde{g} \circ \chi$ in the proof of Lemma 3 (regularity) satisfies

$$g \in L_{r_0}(\mathcal{V}) \qquad \frac{1}{r_0} = \frac{1}{2} + \frac{1}{p}.$$

On the other hand, the definitions (1) and (5) of the first fundamental form $\mathbf{g}$ and area element q imply that they are bounded in $L_\infty(\mathcal{V})$ as well as

$$\mathbf{g}, q \in W_p^1(\mathcal{V}) \quad \Rightarrow \quad \mathbf{A} := q\mathbf{g}^{-1} \in W_p^1(\mathcal{V}).$$

Therefore, the Laplace–Beltrami equation in the parametric domain $\mathcal{V}$ can be written in nondivergence form as follows:

$$\mathbf{A} : D^2 u = qg - \operatorname{div}(\mathbf{A}) \cdot \nabla u = \ell \in L_{r_0}(\mathcal{V}). \tag{22}$$

Since $\mathbf{A}$ is uniformly continuous, the Calderón–Zygmund regularity theory applies (cf. Gilbarg and Trudinger, 1998, Theorem 9.15 and Lemma 9.17) and gives the interior regularity $u \in W_{r_0}^2(\mathcal{Z})$ with

$$\|u\|_{W_{r_0}^2(\mathcal{Z})} \lesssim \|\ell\|_{L_{r_0}(\mathcal{V})}$$

where $\mathcal{Z} := \chi^{-1}(\mathcal{W})$ and $\overline{\mathcal{W}} \subset \mathcal{U}$ as in the proof Lemma 3 (regularity). Invoking Sobolev embedding again, we deduce

$$u \in W_{t_1}^1(\mathcal{Z}), \qquad \frac{1}{t_1} = \frac{1}{r_0} - \frac{1}{n},$$

and $\widetilde{u} \in W_{t_1}^1(\gamma)$ upon pasting these local estimates together over γ; hence $u \in W_{t_1}^1(\mathcal{V})$. We now iterate this argument and prove a recurrence relation by induction. Suppose that a sequence of real numbers $\{r_k,\, t_k\}$ is governed by the relations $t_0 = 2$ and

$$\frac{1}{r_k} = \frac{1}{p} + \frac{1}{t_k}, \qquad \frac{1}{t_{k+1}} = \frac{1}{r_k} - \frac{1}{n},$$

and the right-hand side of (22) satisfies $\ell \in L_{r_k}(\mathcal{V})$; note that this is the case for $k = 0$. Calderón–Zygmund theory thus implies $u \in W^2_{r_k}(\mathcal{Z})$ with

$$\|u\|_{W^2_{r_k}(\mathcal{Z})} \lesssim \|\ell\|_{L_{r_k}(\mathcal{Z})}.$$

Sobolev embedding in turn yields $u \in W^1_{t_{k+1}}(\mathcal{V})$ whence $\ell \in L_{r_{k+1}}(\mathcal{V})$, which proves the recurrence relation. Iterating these relations we see that for $k \geq 0$

$$\frac{1}{r_k} = \frac{1}{r_{k-1}} + \frac{1}{p} - \frac{1}{n} = \frac{1}{2} + \frac{1}{p} + k\left(\frac{1}{p} - \frac{1}{n}\right),$$

and that every step increases the value of r_k, because $\frac{1}{p} - \frac{1}{n} < 0$. Since $\widetilde{f} \in L_2(\gamma)$, the iteration stops once $r_k \geq 2$ or equivalently

$$k = \left\lceil \frac{n}{p-n} \right\rceil.$$

This concludes the proof. □

2.3 Signed distance function

We now take advantage of the ambient space $\mathbb{R}^{n+1}$ and use standard calculus in a suitable tubular neighbourhood $\mathcal{N}$ of γ to derive useful expressions of geometric quantities; we postpone momentarily the precise definition of $\mathcal{N}$. The surface γ splits $\mathbb{R}^{n+1}$ into two disjoint sets, the interior and exterior of γ. The *signed distance functiond* $: \mathcal{N} \to \gamma$ is defined for every $\mathbf{x} \in \mathcal{N}$ to be the distance of $\mathbf{x}$ to γ, dist($\mathbf{x}$, γ), if $\mathbf{x}$ belongs to the exterior of γ and $-$dist($\mathbf{x}$, γ) if $\mathbf{x}$ belongs to the interior of γ, whence

$$|d(\mathbf{x})| = \text{dist}(\mathbf{x}, \gamma) \quad \forall \mathbf{x} \in \mathcal{N}.$$

It turns out that d belongs to the same regularity class as γ so long as γ is at least C^2, which we henceforth assume in our discussion of d. While the distance function exists for surfaces of regularity less than $C^{1,1}$, as we explain in Section 2.5 its properties are drastically different and it is not immediately useful for purposes of defining and analyzing surface FEM. Returning to the setting of C^2 surfaces, $\nabla d(\mathbf{x})$ is well defined for all $\mathbf{x} \in \mathcal{N}$ and computed on γ gives the unit normal $\boldsymbol{\nu}(\mathbf{x})$ pointing outwards:

$$\boldsymbol{\nu}(\mathbf{x}) = \nabla d(\mathbf{x}) \quad \forall \mathbf{x} \in \gamma.$$

Since ∇d is defined in $\mathcal{N}$ it provides a natural extension of $\boldsymbol{\nu}$ to $\mathcal{N}$. This neighbourhood $\mathcal{N}$ is sufficiently small that for every $\mathbf{x} \in \mathcal{N}$ there is a unique *closest point projection* $\mathbf{P}_d(\mathbf{x}) \in \gamma$ defined by

$$\mathbf{P}_d(\mathbf{x}) = \mathbf{x} - d(\mathbf{x})\nabla d(\mathbf{x}) \quad \forall \mathbf{x} \in \mathcal{N}. \tag{23}$$

An important property is that ∇d coincides at $\mathbf{x} \in \mathcal{N}$ and $\mathbf{P}_d(\mathbf{x}) \in \gamma$:

$$\nabla d(\mathbf{x}) = \nabla d(\mathbf{P}_d(\mathbf{x})) = \nabla d(\mathbf{x} - d(\mathbf{x})\nabla d(\mathbf{x})) \quad \forall \mathbf{x} \in \mathcal{N}. \tag{24}$$

Since $|\nabla d(\mathbf{x})|^2 = 1$, we deduce that the Hessian $D^2 d(\mathbf{x})$ satisfies

$$D^2 d(\mathbf{x})\, \nabla d(\mathbf{x}) = 0 \quad \forall \mathbf{x} \in \mathcal{N}. \tag{25}$$

This implies that $D^2 d(\mathbf{x})$ can be regarded as an operator acting on the tangent hyperplane to γ at $\mathbf{x} \in \gamma$ and thus gives an alternative representation to the Weingarten map (11):

$$\mathbf{W}(\mathbf{x}) = D^2 d(\mathbf{x}) \quad \forall \mathbf{x} \in \gamma.$$

This has two important consequences. First it provides a natural extension of $\mathbf{W}$ to $\mathcal{N}$ and second shows that $\mathbf{W}$ is symmetric, which is not apparent from (11).

Given a generic function $\widetilde{v} : \gamma \to \mathbb{R}$, the distance function d provides a natural way to extend it to the neighbourhood $\mathcal{N}$ upon writing

$$v(\mathbf{x}) = \widetilde{v}(\mathbf{P}_d(\mathbf{x})) = \widetilde{v}(\mathbf{x} - d(\mathbf{x})\nabla d(\mathbf{x})) \quad \forall \mathbf{x} \in \mathcal{N}. \tag{26}$$

Differentiating and using the definition (3) of orthogonal projection, we obtain

$$\begin{aligned}
\nabla v(\mathbf{x}) &= \big(\mathbf{I} - \nabla d(\mathbf{x}) \otimes \nabla d(\mathbf{x}) - d(\mathbf{x}) D^2 d(\mathbf{x})\big) \nabla_\gamma \widetilde{v}(\mathbf{P}_d(\mathbf{x})) \\
&= \big(\Pi(\mathbf{x}) - d(\mathbf{x}) D^2 d(\mathbf{x})\big) \nabla_\gamma \widetilde{v}(\mathbf{P}_d(\mathbf{x})) \\
&= \big(\mathbf{I} - d(\mathbf{x}) D^2 d(\mathbf{x})\big) \Pi(\mathbf{x}) \nabla_\gamma \widetilde{v}(\mathbf{P}_d(\mathbf{x}))
\end{aligned} \tag{27}$$

where the last equality hinges on (24), which implies $\Pi(\mathbf{x}) = \Pi(\mathbf{P}_d(\mathbf{x}))$ and $D^2 d(\mathbf{x}) = D^2 d(\mathbf{x}) \Pi(\mathbf{x})$. In particular, $\nabla v(\mathbf{x}) = \nabla_\gamma \widetilde{v}(\mathbf{P}_d(\mathbf{x}))$ for $\mathbf{x} \in \gamma$ because (26) provides a normal extension of $\widetilde{v}$.

Suppose now that v is an extension of $\widetilde{v}$ to $\mathcal{N}$, but not necessarily in the normal direction. An intrinsic definition of tangential gradient of $\widetilde{v}$ is the orthogonal projection of ∇v to the tangent hyperplane of γ:

$$\nabla_\gamma \widetilde{v}(\mathbf{x}) = (\mathbf{I} - \boldsymbol{\nu}(\mathbf{x}) \otimes \boldsymbol{\nu}(\mathbf{x})) \nabla v(\mathbf{x}) = \Pi(\mathbf{x}) \nabla v(\mathbf{x}) \quad \forall \mathbf{x} \in \gamma. \tag{28}$$

This definition is consistent with (10): $\nabla_\gamma \widetilde{v}(\mathbf{x})$ is orthogonal to $\boldsymbol{\nu}(\mathbf{x})$ and

$$\nabla_\gamma \widetilde{v}(\mathbf{x}) \cdot \partial_i \boldsymbol{\chi}(\mathbf{y}) = \nabla v(\mathbf{x}) \cdot \partial_i \boldsymbol{\chi}(\mathbf{y}) = \partial_i \widetilde{v}(\boldsymbol{\chi}(\mathbf{y}))$$

obeys the chain rule, whence it must coincide with our previous definition based on these two properties. An important consequence of this property follows.

Remark 5 (parametric independence). The definition (28) is independent of the extension: if v_1, v_2 are two extensions of $\widetilde{v}$ then $v_1 - v_2 = 0$ on γ and the only nonvanishing component of $\nabla(v_1 - v_2)$ is in the normal direction $\boldsymbol{\nu}$. Since definitions (28) and (10) agree, we deduce that the tangential gradient

$\nabla_\gamma \widetilde{v}$ is independent of the parametrization χ chosen to described γ. The same happens with the tangential divergence (12) as well as the Laplace–Beltrami operator (13), the latter because of (14) and the fact that (8) is independent of χ.

Given a vector field $\widetilde{\mathbf{v}}:\gamma\to\mathbb{R}^{n+1}$ and corresponding extension to $\mathcal{N}$, the tangential divergence can be written as

$$\operatorname{div}_\gamma(\widetilde{\mathbf{v}}(\mathbf{x}))=\operatorname{trace}\left(\nabla_\gamma\widetilde{\mathbf{v}}(\mathbf{x})\right)=\operatorname{div}(\mathbf{v}(\mathbf{x}))-\boldsymbol{\nu}(\mathbf{x})^t\,\nabla\mathbf{v}(\mathbf{x})\boldsymbol{\nu}(\mathbf{x})\quad\forall\,\mathbf{x}\in\gamma,$$

and gives an alternative expression to (12). Likewise, the Laplace–Beltrami operator $\Delta_\gamma\widetilde{v}=\operatorname{div}_\gamma\left(\nabla_\gamma\widetilde{v}\right)$, written parametrically in (13), can be equivalently written in terms of the extension v as follows

$$\Delta_\gamma\widetilde{v}=\operatorname{trace}((\mathbf{I}-\boldsymbol{\nu}\otimes\boldsymbol{\nu})D^2v)-(\nabla v\cdot\boldsymbol{\nu})\operatorname{div}_\gamma(\boldsymbol{\nu}),$$

because $\nabla_\gamma(\nabla\boldsymbol{\nu}\cdot\boldsymbol{\nu})\cdot\boldsymbol{\nu}=0$. This implies the expression

$$\Delta_\gamma\widetilde{v}(\mathbf{x})=\Delta v(\mathbf{x})-\boldsymbol{\nu}(\mathbf{x})^tD^2v(\mathbf{x})\boldsymbol{\nu}(\mathbf{x})-(\nabla v\cdot\boldsymbol{\nu})(\mathbf{x})\operatorname{div}_\gamma(\boldsymbol{\nu}(\mathbf{x}))\quad\forall\,\mathbf{x}\in\gamma.$$

2.4 Curvatures

We again assume that γ is of class C^2. In view of (11), the Weingarten map is symmetric and its $n+1$ eigenvalues are real. Except for the zero eigenvalue corresponding to the eigenvector $\boldsymbol{\nu}(\mathbf{x})$, according to (25), they are called the *principal curvatures* of γ at $\mathbf{x}$ and are denoted by $\kappa_i(\mathbf{x})$ for $1\le i\le n$. The eigenvectors of $\mathbf{W}$ corresponding to the principal curvatures are called the *principal directions*.

We stress that $\kappa_i(\mathbf{x})$ is well defined for all $\mathbf{x}\in\mathcal{N}$ because so is $\mathbf{W}(\mathbf{x})$. This allows us to make the definition of $\mathcal{N}$ precise. Given $\delta>0$, first let

$$\mathcal{N}(\delta):=\{\mathbf{x}\in\mathbb{R}^{n+1}\;:\;|d(\mathbf{x})|<\delta\}. \tag{29}$$

Let also

$$K(\mathbf{x}):=\max_{1\le i\le n}|\kappa_i(\mathbf{x})|\quad\forall\,\mathbf{x}\in\gamma;\qquad K_\infty:=\|K\|_{L_\infty(\gamma)}. \tag{30}$$

We may now set

$$\mathcal{N}:=\left\{\mathbf{x}\in\mathbb{R}^{n+1}:\operatorname{dist}(\mathbf{x},\gamma)<\frac{1}{2K_\infty}\right\}=\mathcal{N}(K_\infty/2). \tag{31}$$

Note that the distance function, closest point projection, and related properties are defined and hold on the larger set $\mathcal{N}(1/K_\infty)$. We adopt the more limited definition of $\mathcal{N}$ in order to avoid degeneration of some quantities such as curvature of parallel surfaces (see below) that occurs near the boundary of the larger set.

Given ε small so that $|\varepsilon| \leq \frac{1}{2K_\infty}$, we define the parallel surface γ_ε to be

$$\gamma_\varepsilon := \{\mathbf{x} \in \mathcal{N} : d(\mathbf{x}) = \varepsilon\}.$$

The following statement relates the principal curvatures of γ_ε with those of γ.

Lemma 6 (curvatures of parallel surface). *If γ is of class C^2 so are all parallel surfaces γ_ε and their principal curvatures satisfy*

$$\kappa_i(\boldsymbol{x}) = \frac{\kappa_i(\boldsymbol{P}_d(\boldsymbol{x}))}{1 + \varepsilon\,\kappa_i(\boldsymbol{P}_d(\boldsymbol{x}))} \quad \forall\, \mathbf{x} \in \gamma_\varepsilon, \tag{32}$$

whereas the principal directions at $\boldsymbol{x}$ and $\boldsymbol{P}_d(\boldsymbol{x})$ coincide.

Proof. Differentiate (24) to get

$$D^2 d(\mathbf{x}) = D^2 d(\mathbf{P}_d(\mathbf{x}))(\mathbf{I} - \nabla d(\mathbf{x}) \otimes \nabla d(\mathbf{x}) - d(\mathbf{x}) D^2 d(\mathbf{x})),$$

whence, since $\nabla d(\mathbf{x}) = \nabla d(\mathbf{P}_d(\mathbf{x}))$ again from (24),

$$\big(\mathbf{I} + d(\mathbf{x}) D^2 d(\mathbf{P}_d(\mathbf{x}))\big) D^2 d(\mathbf{x}) = D^2 d(\mathbf{P}_d(\mathbf{x}))(\mathbf{I} - \nabla d(\mathbf{x}) \otimes \nabla d(\mathbf{x})) = D^2 d(\mathbf{P}_d(\mathbf{x})).$$

Therefore, for $\mathbf{x} \in \gamma_\varepsilon$ we see that the eigenvalues of $(\mathbf{I} + \varepsilon D^2 d(\mathbf{P}_d(\mathbf{x})))$ are

$$\kappa_i(\mathbf{I} + \varepsilon D^2 d(\mathbf{P}_d(\mathbf{x}))) = 1 + \varepsilon \kappa_i(\mathbf{P}_d(\mathbf{x})) \geq \frac{1}{2},$$

according to (31). This implies that $\mathbf{I} + \varepsilon D^2 d(\mathbf{P}_d(\mathbf{x}))$ is nonsingular and the previous relation reads as follows in terms of the Weingarten map:

$$\mathbf{W}(\mathbf{x}) = (\mathbf{I} + \varepsilon \mathbf{W}(\mathbf{P}_d(\mathbf{x})))^{-1} \mathbf{W}(\mathbf{P}_d(\mathbf{x})).$$

This shows that the eigenvectors of $\mathbf{W}(\mathbf{x})$ and $\mathbf{W}(\mathbf{P}_d(\mathbf{x}))$ coincide and the eigenvalues are related via (32). □

The *second fundamental form* $\mathbf{h} = (h_{ij})_{i,j=1}^n$ of γ is defined by

$$h_{ij}(\mathbf{y}) := -\partial_i \boldsymbol{\nu}(\mathbf{y}) \cdot \partial_i \boldsymbol{\chi}(\mathbf{y}) = \boldsymbol{\nu}(\mathbf{y}) \cdot \partial_{ij} \boldsymbol{\chi}(\mathbf{y}) \quad \forall\, \mathbf{y} \in \mathcal{V},$$

where the last equality relies on the fact that $\boldsymbol{\nu}$ and $\partial_j \boldsymbol{\chi}$ are orthogonal for $1 \leq j \leq n$. The next result connects $\mathbf{h}$ with the Weingarten map (11).

Lemma 7 (second fundamental form). *The symmetric matrix $\mathbf{W} = D_\gamma \boldsymbol{\nu}$ defines a self-adjoint operator on the tangent hyperplane to γ that can be represented in the basis $\{\partial_j \boldsymbol{\chi}\}_{j=1}^n$ by the generally nonsymmetric matrix*

$$\mathbf{s} = -\boldsymbol{h} \boldsymbol{g}^{-1}.$$

The eigenvalues of $\mathbf{s}$ are the principal curvatures of γ.

Proof. Since $D_\gamma \boldsymbol{\nu}\, \boldsymbol{\nu} = 0$, we can regard $D_\gamma \boldsymbol{\nu}$ as an operator acting on the tangent plane to γ and represent its image in terms of $\{\partial_k \boldsymbol{\chi}\}_{k=1}^n$ as follows

$$\partial_i \boldsymbol{\nu}(\mathbf{y}) = D_\gamma \boldsymbol{\nu}(\mathbf{x})\, \partial_i \boldsymbol{\chi}(\mathbf{y}) = \sum_{k=1}^{n} s_{ik}(\mathbf{y})\, \partial_k \boldsymbol{\chi}(\mathbf{y}) \quad \forall\, \mathbf{y} \in \mathcal{V}.$$

Let $\mathbf{s}(\mathbf{y}) := (s_{ij}(\mathbf{y}))_{ij,=1}^n$ and multiply both sides by $\partial_j \boldsymbol{\chi}(\mathbf{y})$ to see that

$$h_{ij}(\mathbf{y}) = -\,\partial_i \boldsymbol{\nu}(\mathbf{y}) \cdot \partial_j \boldsymbol{\chi}(\mathbf{y}) = \sum_{k=1}^{n} s_{ik} \partial_k \boldsymbol{\chi}(\mathbf{y}) \cdot \partial_j \boldsymbol{\chi}(\mathbf{y}) = \sum_{k=1}^{n} s_{ik} g_{kj}.$$

This implies $\mathbf{h} = -\mathbf{sg}$ and thus the assertion. □

2.5 Surface regularity and properties of the distance function

In the previous two sections we have seen that when γ is of class C^2, the closest point projection is uniquely defined in a tubular neighbourhood of γ whose width is related to the principal curvatures of the surface. We shall see below that the closest point projection plays a pivotal role in analyzing finite element methods on C^2 surfaces. On the other hand, some applications may require solving PDE on surfaces that are less regular than C^2. Thus it is natural to ask which properties of the distance function and closest point projection carry over to less regular surfaces. It turns out that the properties of these maps change drastically and fundamentally when crossing the threshold from C^2 to less regular ($C^{1,\alpha}$ with $\alpha < 1$) surfaces.

In order to make this statement precise, we begin by restating for comparison from Gilbarg and Trudinger (1998, Lemma 14.16) some fundamental properties of the distance function for C^k surfaces ($k \geq 2$).

Lemma 8 (properties of distance functions for C^k surfaces). *Let γ be a C^k surface, $k \geq 2$. Then there exists a positive constant δ depending on γ such that $d \in C^k(\mathcal{N}(\delta))$. In addition, the closest point projection $\mathbf{P}_d(\mathbf{x}) = \mathbf{x} - d(\mathbf{x})\nabla d(\mathbf{x})$ is defined and of class C^{k-1} on $\mathcal{N}(\delta)$ with $\delta < \frac{1}{K_\infty}$.*

We now ask whether a similar statement holds for $k < 2$, and in particular for $k = 1$. Note first that the distance function d to *any* closed set $\gamma \subset \mathbb{R}^{n+1}$ is defined and Lipschitz continuous (Federer, 1959, Theorem 4.8.1), so the first question at hand is whether distance functions for $C^{1,\alpha}$ surfaces ($0 \leq \alpha < 1$) are more than Lipshitz continuous.

In order to understand the relationship between surface regularity and the distance function map, we first define the *reach* of a surface γ:

$$\text{reach}(\gamma) := \sup\{\delta \geq 0 : \text{all } \mathbf{x} \in \mathcal{N}(\delta) \text{ have a unique closest point } \mathbf{P}_d(\mathbf{x}) \in \gamma\}.$$

For a C^2 surface γ, we have already seen that reach$(\gamma) = 1/K_\infty$. We now explore the connection between the reach and properties of the distance function for less regular surfaces. We first define

$$U(\gamma) := \{\mathbf{x} \in \mathbb{R}^{n+1} : \ \mathbf{x} \text{ has a unique closest point in } \gamma\}.$$

The following result may be found in Federer (1959, Theorem 4.8.3).

Lemma 9 (properties of differentiable distance functions). *If γ is a C^1 surface, $x \in \mathbb{R}^{n+1} \backslash \gamma$, and d is differentiable at x, then $x \in U(\gamma)$. In particular, if d is differentiable in a neighbourhood of γ, then reach $(\gamma) > 0$.*

Next we define several constants from the technical report (Lucas, 1957). Given $\mathbf{x} \in \gamma$ and $\rho \geq 0$, we first define the closed normal segment

$$S(\mathbf{x}, \rho) := [\mathbf{x} - \rho\boldsymbol{\nu}(\mathbf{x}), \mathbf{x} + \rho\boldsymbol{\nu}(\mathbf{x})].$$

Let $B_\rho(\mathbf{y})$ denotes the ball in $\mathbb{R}^{n+1}$ of center $\mathbf{y}$ and radius $\rho > 0$, and

$$\begin{aligned}
\frac{1}{r_0} &:= \sup_{\mathbf{x},\mathbf{y} \in \gamma, \mathbf{x} \neq \mathbf{y}} \frac{|\boldsymbol{\nu}(\mathbf{x}) - \boldsymbol{\nu}(\mathbf{y})|}{|\mathbf{x} - \mathbf{y}|}, \\
\frac{1}{r_0'} &:= \sup\{\rho \geq 0 : \ S(\mathbf{x}, \rho) \cap S(\mathbf{y}, \rho) = \emptyset \ \ \forall \mathbf{x}, \mathbf{y} \in \gamma, \ \ \mathbf{x} \neq \mathbf{y}\}, \\
\frac{1}{r_0''} &:= \sup\{\rho \geq 0 : \ \overline{B_\rho(\mathbf{x} \pm \rho\boldsymbol{\nu}(\mathbf{x}))} \text{ contain respectively no points} \\
&\qquad \text{interior or exterior to } \gamma \quad \text{for all } \mathbf{x} \in \gamma\}, \\
\frac{1}{r_0'''} &:= \sup_{\mathbf{x},\mathbf{y} \in \gamma, \mathbf{x} \neq \mathbf{y}} \frac{|2(\mathbf{y} - \mathbf{x}) \cdot \boldsymbol{\nu}(\mathbf{x})|}{|\mathbf{y} - \mathbf{x}|^2}.
\end{aligned}$$

Combining Lucas (1957, Theorem 1) and noting that r_0 bounds the Lipschitz constant of γ (cf. the comment on p. 15 of Lucas (1957)), we have the following.

Lemma 10 (further properties of C^1 surfaces). *If the surface γ is of class C^1, then the constants r_0, r_0', r_0'', and r_0''' are all equal. In addition, if $r_0 > 0$ then γ is of class $C^{1,1}$.*

Combining the previous lemmas with the statement in Federer (1959, Theorem 4.18) that $r_0''' = \text{reach}(\gamma)$ yields the following result.

Theorem 11 (C^1 distance function implies $C^{1,1}$ surface). *If the distance function d associated to a C^1 surface γ is continuously differentiable in a tubular neighbourhood $\mathcal{N}(\delta)$ of γ for some $\delta > 0$, then γ is of class $C^{1,1}$. In addition, any C^1 surface with positive reach is of class $C^{1,1}$.*

The preceding results establish that the properties of the distance function and the associated closest point projection for C^2 surfaces that we previously discussed are inherently connected with surfaces of bounded curvature. This can be seen both in Theorem 11 (since the curvatures are defined and bounded almost everywhere on a $C^{1,1}$ surface) and in the definition of the constant r_0'' (since for $\mathbf{x} \in \gamma$, the supremum over the radii ρ for which $\overline{B_\rho(\mathbf{x} \pm \rho \boldsymbol{\nu}(\mathbf{x}))} \cap \gamma = \emptyset$ is the inverse of the maximum principal curvature at $\mathbf{x}$).

For our purposes, Theorem 11 is essentially a negative result in that it establishes that the distance function and closest point projection are of limited immediate use for surfaces that are less regular than $C^{1,1}$. In particular, in this case the closest point projection is not uniquely defined on any tubular neighbourhood of γ. In addition, the regularity of the distance function does *not* vary continuously with that of γ, since for a $C^{1,\alpha}$ surface with $\alpha < 1$ Theorem 11 establishes that d is only Lipschitz. Thus we must use different tools when considering surface finite element methods on less regular surfaces than C^2.

2.6 Divergence theorem on surfaces

We conclude this section with an application of calculus in $\mathbb{R}^{n+1}$ to derive an integration by parts formula on not necessarily closed surfaces.

Proposition 12 (divergence theorem). *Let γ be a compact, oriented surface of class C^2 with Lipschitz boundary $\partial\gamma$. Let $H = \sum_{i=1}^n \kappa_i$ be the total curvature of γ and $\boldsymbol{\mu}$ be the unit outward normal to $\partial\gamma$ lying in the tangent hyperplane to γ. If $\widetilde{v} : \gamma \to \mathbb{R} \in H^1(\gamma)$, then*

$$\int_\gamma \nabla_\gamma \widetilde{v} = \int_\gamma \widetilde{v} H \boldsymbol{\nu} + \int_{\partial\gamma} \widetilde{v} \boldsymbol{\mu}.$$

Proof. Given $\varepsilon < \frac{1}{2K_\infty}$ we define the tubular set

$$\Omega_\varepsilon := \{\mathbf{z} = \mathbf{x} + \rho\boldsymbol{\nu}(\mathbf{x}) : \ \mathbf{x} \in \gamma, |\rho| < \varepsilon\};$$

note that $\mathbf{P}_d(\mathbf{z}) = \mathbf{x}$ for all $\mathbf{z} \in \Omega_\varepsilon$. We decompose the boundary $\partial\Omega_\varepsilon$ of Ω_ε into

$$\gamma_{\pm\varepsilon} := \{\mathbf{x} \pm \varepsilon\boldsymbol{\nu}(\mathbf{x}) : \ \mathbf{x} \in \gamma\}, \ \ \lambda_\varepsilon := \partial\Omega_\varepsilon \backslash (\gamma_\varepsilon \cup \gamma_{-\varepsilon}).$$

The sets $\gamma_{\pm\varepsilon}$ are parallel surfaces to γ whereas λ_ε is the lateral boundary of size 2ε. We first assume that $\widetilde{v}$ is of class C^1, let v be an extension of $\widetilde{v}$ to Ω_ε of class $C^1(\overline{\Omega_\varepsilon})$, and apply the divergence theorem in Ω_ε to obtain

$$\int_{\Omega_\varepsilon} \nabla v = \int_{\partial\Omega_\varepsilon} v\boldsymbol{\nu}_\varepsilon = \int_{\gamma_\varepsilon} v\boldsymbol{\nu} \circ \mathbf{P}_d - \int_{\gamma_{-\varepsilon}} v\boldsymbol{\nu} \circ \mathbf{P}_d + \int_{\lambda_\varepsilon} v\boldsymbol{\mu} \circ \mathbf{P}_d,$$

where $\boldsymbol{\nu}_\varepsilon$ is the unit outward normal of $\partial\Omega_\varepsilon$. We divide both sides of this equality by 2ε, the thickness of Ω_ε, and compute the limits as $\varepsilon \to 0$. According to (27) we first see that

$$\frac{1}{2\varepsilon}\int_{\Omega_\varepsilon}\nabla v=\frac{1}{2\varepsilon}\int_{\Omega_\varepsilon}\big(\mathbf{I}-d(\mathbf{x})D^2d(\mathbf{x})\big)\nabla_\gamma\widetilde{v}(\mathbf{P}_d(\mathbf{x}))d\mathbf{x}\underset{\varepsilon\to 0}{\longrightarrow}\int_\gamma\nabla_\gamma\widetilde{v}.$$

Likewise

$$\frac{1}{2\varepsilon}\int_{\lambda_\varepsilon}v\boldsymbol{\mu}\circ\mathbf{P}_d\underset{\varepsilon\to 0}{\longrightarrow}\int_{\partial\gamma}\widetilde{v}\boldsymbol{\mu}.$$

Moreover, since $\boldsymbol{\nu}\circ\mathbf{P}_d=\nabla d$, we infer that

$$\begin{aligned}\lim_{\varepsilon\to 0}\frac{1}{2\varepsilon}\left(\int_{\gamma_\varepsilon}v\boldsymbol{\nu}\circ\mathbf{P}_d-\int_{\gamma_{-\varepsilon}}v\boldsymbol{\nu}\circ\mathbf{P}_d\right)&=\frac{d}{d\rho}\int_{\gamma_\rho}v\nabla d\ \Big|_{\rho=0}\\&=\frac{d}{d\rho}\int_{\mathcal{V}}v(\mathbf{x})\,\nabla d(\mathbf{x}+\rho\nabla d(\mathbf{x}))\,q_\rho(\mathbf{y})d\mathbf{y}\ \Big|_{\rho=0}\end{aligned}$$

with $\mathbf{x}=\boldsymbol{\chi}(\mathbf{y})\in\gamma$ and $q_\rho(\mathbf{y})$ denotes the infinitesimal area associated with the surface $\gamma_\rho:=\{\mathbf{z}=\mathbf{x}+\rho\boldsymbol{\nu}(\mathbf{x}):\mathbf{x}\in\gamma\}$. Since $\frac{d}{d\rho}\nabla d(\mathbf{x}+\rho\nabla d(\mathbf{x}))=D^2d(\mathbf{x}+\rho\nabla d(\mathbf{x}))\nabla d(\mathbf{x})=0$, it remains to evaluate $\frac{d}{d\rho}q_\rho$. We resort to (53) (shown below) with $\Gamma=\gamma_\rho$ and use that $\boldsymbol{\nu}_\rho\cdot\boldsymbol{\nu}=1$ as well as (32) to write

$$\frac{q_\rho(\mathbf{y})}{q(\mathbf{y})}=\frac{1}{\det\left(\mathbf{I}-\rho D^2d(\mathbf{x})\right)}=\frac{1}{\prod_{i=1}^n(1-\rho\kappa_i(\mathbf{x}))}=\prod_{i=1}^n(1+\rho\kappa_i(\mathbf{P}_d(\mathbf{x}))).$$

We finally observe that

$$\frac{d}{d\rho}q_\rho(\mathbf{y})\Big|_{\rho=0}=q(\mathbf{y})\sum_{i=1}^n\kappa_i(\mathbf{P}_d(\mathbf{x}))=q(\mathbf{y})H(\mathbf{P}_d(\mathbf{x}))$$

to conclude the proof for $\widetilde{v}$ of class C^1. The assertion for $\widetilde{v}\in H^1(\gamma)$ follows by density of $C^1(\overline{\gamma})$ in $\widetilde{v}\in H^1(\gamma)$. □

Applying Proposition 12 (divergence theorem) to a vector field $\widetilde{\mathbf{v}}:\gamma\to\mathbb{R}^{n+1}$ and computing the trace yields the more familiar expression

$$\int_\gamma \operatorname{div}_\gamma\widetilde{\mathbf{v}}=\int_\gamma H\widetilde{\mathbf{v}}\cdot\boldsymbol{\nu}+\int_{\partial\gamma}\widetilde{\mathbf{v}}\cdot\boldsymbol{\mu}.$$

Corollary 13 (integration by parts). *Let γ be a surface of class C^2 with Lipschitz boundary $\partial\gamma$. If $\widetilde{v},\widetilde{w}:\gamma\to\mathbb{R}$ satisfy $\widetilde{v}\in H^2(\gamma)$ and $\widetilde{w}\in H^1(\gamma)$, then*

$$\int_\gamma \widetilde{w}\,\Delta_\gamma\widetilde{v}+\nabla_\gamma\widetilde{w}\cdot\nabla_\gamma\widetilde{w}=\int_{\partial\gamma}\widetilde{w}\,\nabla_\gamma\widetilde{v}\cdot\boldsymbol{\mu}.$$

Proof. Apply the previous equality to $\widetilde{\mathbf{v}}=\widetilde{w}\,\nabla_\gamma\widetilde{v}$. □

3 Perturbation theory

In most surface finite element methods, the approximate problem is not posed on the continuous surface γ. This may occur either for convenience, or because γ is not known precisely. Examples of only incomplete information being present in simulations include free boundary problems such as two-phase flow and cases where γ is reconstructed from some sort of imaging data.

The purpose of this section is to investigate how geometric quantities change under perturbation of the surface γ. To this end, suppose that Γ is a closed Lipschitz surface (not necessarily C^2). We use a subscript Γ to denote geometric quantities associated with Γ: χ_Γ (parametrization), $\mathbf{g}_\Gamma$ (first fundamental form), q_Γ (area element), $\boldsymbol{\nu}_\Gamma$ (unit normal), ∇_Γ (tangential gradient), and Π_Γ (orthogonal projection onto Γ).

Let $\widetilde{u} \in H^1_\#(\gamma)$ solve (19) and $u_\Gamma \in H^1_\#(\Gamma)$ solve

$$\int_\Gamma \nabla_\Gamma u_\Gamma \cdot \nabla_\Gamma v = \int_\Gamma f_\Gamma v \quad \forall v \in H^1_\#(\Gamma), \tag{33}$$

for a given forcing $f_\Gamma \in L_{2,\#}(\Gamma)$. To examine the error between u and u_Γ, we first have to study how the bilinear forms in (19) and (33) change when changing γ. This amounts to deriving expressions for the *error matrices* $\mathbf{E}, \mathbf{E}_\Gamma \in \mathbb{R}^{(n+1)\times(n+1)}$ in the error equations

$$\int_\Gamma \nabla_\Gamma v \cdot \nabla_\Gamma w - \int_\gamma \nabla_\gamma \widetilde{v} \cdot \nabla_\gamma \widetilde{w} = \int_\gamma \nabla_\gamma \widetilde{v} \cdot \mathbf{E}\, \nabla_\gamma \widetilde{w} = \int_\Gamma \nabla_\Gamma v \cdot \mathbf{E}_\Gamma\, \nabla_\Gamma w, \tag{34}$$

valid for all $v, w \in H^1(\Gamma)$ and $\widetilde{v}, \widetilde{w} \in H^1(\gamma)$ the corresponding lifts. We carry out this programme below within two scenarios depending on the regularity of γ. We alert the reader about the following abuse of notation: the matrix $\mathbf{E}$ (resp. $\mathbf{E}_\Gamma$) is defined in γ (resp. Γ), but we will often write them in the parametric domain $\mathcal{V}$ thereby identifying $\mathbf{E}$ (resp. $\mathbf{E}_\Gamma$) with $\mathbf{E} \circ \chi$ (resp. $\mathbf{E}_\Gamma \circ \chi_\Gamma$).

3.1 Perturbation theory for $C^{1,\alpha}$ surfaces

Let γ be of class $C^{1,\alpha}$ and χ and χ_Γ be the parametrizations of γ and Γ. They dictate the relation between $\widetilde{v}$ and v, the former defined on γ and the latter on Γ,

$$v = \widetilde{v} \circ \chi \circ \chi_\Gamma^{-1}.$$

In the sequel, we first establish a relation between $\nabla_\gamma \widetilde{v}$ and $\nabla_\Gamma v$ and next use it to characterize $\mathbf{E}$ and $\mathbf{E}_\Gamma$.

Lemma 14 (relation between tangential gradients). *If $\widetilde{v} : \gamma \to \mathbb{R}$ is of class H^1, then the tangential gradients $\nabla_\gamma \widetilde{v}$ and $\nabla_\Gamma v$ satisfy*

$$\nabla_\Gamma v = D\chi_\Gamma\, \mathbf{g}_\Gamma^{-1}\, D\chi^t\, \nabla_\gamma \widetilde{v}, \qquad \nabla_\gamma \widetilde{v} = D\chi\, \mathbf{g}^{-1}\, D\chi_\Gamma^t\, \nabla_\Gamma v. \tag{35}$$

Proof. We concatenate (10) and (9) to write

$$\nabla_\Gamma v = D\chi_\Gamma\, \mathbf{g}_\Gamma^{-1}\, \nabla(v\circ\chi_\Gamma) = D\chi_\Gamma\, \mathbf{g}_\Gamma^{-1}\, \nabla(\widetilde{v}\circ\chi) = D\chi_\Gamma\, \mathbf{g}_\Gamma^{-1}\, D\chi^t\, \nabla_\gamma \widetilde{v},$$

which is the first asserted expression provided $\widetilde{u}$ is of class C^1. Using the density of $C^1(\gamma)$ in $H^1(\gamma)$ for a surface γ of class $C^{1,\alpha}$, the first assertion follows. The second one follows similarly. □

Lemma 15 (geometric consistency). *The error matrices* $\mathbf{E}$ *and* $\mathbf{E}_\Gamma$ *read on* $\mathcal{V}$

$$\mathbf{E} = D\chi\left(\frac{q_\Gamma}{q}\mathbf{g}_\Gamma^{-1} - \mathbf{g}^{-1}\right)D\chi^t, \tag{36}$$

$$\mathbf{E}_\Gamma = D\chi_\Gamma\left(\mathbf{g}_\Gamma^{-1} - \frac{q}{q_\Gamma}\mathbf{g}^{-1}\right)D\chi_\Gamma^t. \tag{37}$$

Proof. Using (35), together with the definition (1) of $\mathbf{g}_\Gamma = D\chi_\Gamma^t D\chi_\Gamma$, yields

$$\int_\Gamma \nabla_\Gamma v\cdot\nabla_\Gamma w = \int_\gamma \nabla_\gamma \widetilde{v}\cdot\frac{q_\Gamma}{q}(D\chi\mathbf{g}_\Gamma^{-1}D\chi^t)\nabla_\gamma\widetilde{w}.$$

Since $\nabla_\gamma\widetilde{v} = \Pi\nabla_\gamma\widetilde{v} = D\chi\,\mathbf{g}^{-1}\,D\chi^t\nabla_\gamma\widetilde{v}$, according to (4), the first equality in (34) follows immediately. The proof of the second equality is similar. □

Our task now is to relate $\mathbf{g}-\mathbf{g}_\Gamma$ and $q-q_\Gamma$ with $D(\chi-\chi_\Gamma)$. We accomplish this next but first we introduce some additional concepts. For any $\mathbf{y}\in\mathcal{V}$, we denote by $|D\chi(\mathbf{y})|$ (resp. $|D^-\chi(\mathbf{y})|$) the largest (resp. smaller) singular value of $D\chi(\mathbf{y})$. Given the relation $\mathbf{g} = D\chi^t\, D\chi$, these quantities are the square roots of the largest and smallest eigenvalues of $\mathbf{g}$. We define the stability constant

$$S_\chi := \sup_{\mathbf{y}\in\mathcal{V}} \frac{\max\{|D\chi(y)|, |D\chi_\Gamma(y)|\}}{\min\{|D^-\chi(y)|, |D^-\chi_\Gamma(y)|\}} \tag{38}$$

and point out that it is a measure of nondegeneracy of $D\chi$ and $D\chi_\Gamma$. We further define the following relative measure of geometric accuracy

$$\lambda_\infty := \sup_{\mathbf{y}\in\mathcal{V}} \frac{|D(\chi-\chi_\Gamma)(\mathbf{y})|}{\min\{|D^-\chi(\mathbf{y})|, |D^-\chi_\Gamma(\mathbf{y})|\}}. \tag{39}$$

Lemma 16 (error estimates for $\mathbf{g}$ and q). *The following error estimates are valid*

$$\|\boldsymbol{I} - \boldsymbol{g}_\Gamma\boldsymbol{g}^{-1}\|_{L_\infty(\mathcal{V})},\ \|\boldsymbol{I} - \boldsymbol{g}_\Gamma^{-1}\boldsymbol{g}\|_{L_\infty(\mathcal{V})} \lesssim S_\chi\,\lambda_\infty, \tag{40}$$

$$\|1 - q^{-1}q_\Gamma\|_{L_\infty(\mathcal{V})},\ \|1 - q_\Gamma^{-1}q\|_{L_\infty(\mathcal{V})} \lesssim S_\chi^n\,\lambda_\infty. \tag{41}$$

Proof. Since $|D\chi| = |D\chi^t|$, $|\mathbf{g}^{-1}| \leq |D^-\chi|^{-2}$ and

$$(\mathbf{g}-\mathbf{g}_\Gamma)(\mathbf{y}) = D\chi(\mathbf{y})^t D(\chi-\chi_\Gamma)(\mathbf{y}) + D(\chi-\chi_\Gamma)(\mathbf{y})^t D\chi_\Gamma(\mathbf{y}) \quad \forall \mathbf{y} \in \mathcal{V},$$

the first assertion in (40) follows; the second one is similar. To prove (41), we write

$$q(\mathbf{y}) - q_\Gamma(\mathbf{y}) = \frac{\det \mathbf{g}(\mathbf{y}) - \det \mathbf{g}_\Gamma(\mathbf{y})}{q(\mathbf{y}) + q_\Gamma(\mathbf{y})} \quad \forall \mathbf{y} \in \mathcal{V},$$

and note that $q = \sqrt{\det \mathbf{g}} = \sqrt{\prod_{i=1}^n \lambda_i(\mathbf{g})}$ where $\{\lambda_i(\mathbf{g})\}_{i=1}^n$ are the eigenvalues of $\mathbf{g}$. Utilizing the definitions of $|D\chi|$ and $|D^-\chi|$ we end up with

$$|D^-\chi(\mathbf{y})|^n \leq q(\mathbf{y}) \leq |D\chi(\mathbf{y})|^n \quad \forall \mathbf{y} \in \mathcal{V}. \tag{42}$$

Since $\det \mathbf{g} - \det \mathbf{g}_\Gamma$ is the sum of terms of the form $\partial_i\chi \cdot \partial_j\chi - \partial_i\chi_\Gamma \cdot \partial_j\chi_\Gamma$ multiplied by $n-1$ factors bounded by $|D\chi|$, we deduce

$$|q(\mathbf{y})^{-1}(q-q_\Gamma)(\mathbf{y})| \lesssim |D^-\chi(\mathbf{y})|^{-n} \, |D(\chi-\chi_\Gamma)(\mathbf{y})| \, |D\chi(\mathbf{y})|^{n-1} \quad \forall \mathbf{y} \in \mathcal{V}.$$

This is the first assertion in (41) in disguise. The second one is similar. □

Lemma 17 (norm equivalence). *Let γ and Γ be Lipschitz surfaces which are related via a bi-Lipschitz map $\mathbf{P} = \chi \circ \chi_\Gamma^{-1} : \Gamma \to \gamma$. Then there is a constant $C \geq 1$, depending on the stability constant S_χ in (38), such that*

$$C^{-1} \|v\|_{L_2(\Gamma)} \leq \|\widetilde{v}\|_{L_2(\gamma)} \leq C \|v\|_{L_2(\Gamma)} \quad \forall \widetilde{v} \in L_2(\gamma), \tag{43}$$

$$C^{-1} \|\nabla_\Gamma v\|_{L_2(\Gamma)} \leq \|\nabla_\gamma \widetilde{v}\|_{L_2(\gamma)} \leq C \|\nabla_\Gamma v\|_{L_2(\Gamma)} \quad \forall \widetilde{v} \in H^1(\gamma). \tag{44}$$

Proof. Use (9) and (10) in conjunction with (8). □

Lemma 2 (Poincaré–Friedrichs inequality) holds on the perturbed surface Γ but with a constant depending on Γ. In order to avoid this dependence, and thus obtain a uniform constant in Γ, it is only necessary that Lemma 17 (norm equivalence) be valid. Before stating our result, we first define a class of surfaces. Given a Lipschitz surface γ, we let $\mathcal{S}_{eq}$ be the class of Lipschitz surfaces Γ such that Lemma 17 (norm equivalence) holds with uniform equivalence constant C_{eq}. Note that implicit in this definition is the existence of a bi-Lipschitz bijection $\mathbf{P} : \Gamma \to \gamma$ for each $\Gamma \in \mathcal{S}_{eq}$, for instance $\mathbf{P} = \chi \circ \chi_\Gamma^{-1}$.

Lemma 18 (uniform Poincaré–Friedrichs constant). *Given a Lipschitz surface γ, for every $v \in H^1_\#(\Gamma)$ with $\Gamma \in \mathcal{S}_{eq}$ there holds that*

$$\|v\|_{L_2(\Gamma)} \lesssim \|\nabla_\Gamma u\|_{L_2(\Gamma)} \tag{45}$$

with the constant hidden in $\lesssim$ depending only on γ and C_{eq}.

Proof. We argue by contradiction the validity of

$$\|v\|_{L_2(\Gamma)} \leq C \|\nabla_\Gamma v\|_{L_2(\Gamma)} \quad \forall\, v \in H^1(\Gamma)$$

and all $\Gamma \in \mathcal{S}_{eq}$ with uniform constant C. We thus assume the existence of a sequence of surfaces $\Gamma_k \in \mathcal{S}_{eq}$ and functions $v_k \in H^1_\#(\Gamma_k)$ such that

$$\|v_k\|_{L_2(\Gamma_k)} = 1 \qquad \|\nabla_{\Gamma_k} v_k\|_{L_2(\Gamma_k)} \to 0$$

as $k \to \infty$. We denote by $\mathbf{P}_k : \Gamma_k \to \gamma$ the associated bi-Lipschitz bijections and by $\widetilde{v}_k = v_k \circ \mathbf{P}_k^{-1}$ the lifts of the functions v_k to γ. Since $\Gamma_k \in \mathcal{S}_{eq}$, the estimates of Lemma 17 (norm equivalence) hold with uniform constant C_{eq} for each Γ_k, whence $\widetilde{v}_k \in H^1(\gamma)$ and

$$\|\widetilde{v}_k\|_{L_2(\gamma)} \simeq 1, \qquad \|\nabla_\gamma \widetilde{v}_k\|_{L_2(\gamma)} \to 0$$

as $k \to \infty$. Proceeding as in Lemma 2 (Poincaré–Friedrichs inequality), we deduce that a subsequence of $\{\widetilde{v}_k\}_k$, still denoted $\{\widetilde{v}_k\}_k$, converges in $H^1(\gamma)$ to a function $\widetilde{v} \in H^1(\gamma)$ with $\nabla_\gamma \widetilde{v} = 0$; this implies that $\widetilde{v}$ is constant. To show that $\widetilde{v} = 0$, let $\epsilon > 0$ be arbitrary and k sufficiently large so that $\|\widetilde{v}_k - \widetilde{v}\|_{L_2(\gamma)} \leq \epsilon$. Exploiting that $\widetilde{v}$ is constant and $\int_{\Gamma_k} v_k = 0$, we use Lemma 17 to compute

$$\begin{aligned}
|\widetilde{v}| &= |\Gamma_k|^{-1} \left| \int_{\Gamma_k} \widetilde{v} \right| = |\Gamma_k|^{-1} \left| \int_{\Gamma_k} \widetilde{v} - v_k \right| \\
&\leq |\Gamma_k|^{-1/2} \|\widetilde{v} - v_k\|_{L_2(\Gamma_k)} \leq C_{eq} |\Gamma_k|^{-1/2} \|\widetilde{v} - \widetilde{v}_k\|_{L_2(\gamma)} \leq C_{eq} |\Gamma_k|^{-1/2} \epsilon.
\end{aligned}$$

Applying again Lemma 17, now to the function 1, yields $|\Gamma_k| \simeq |\Gamma|$ with constant depending only on C_{eq}, so that $|\widetilde{v}| \lesssim \epsilon$. Since ϵ is arbitrary, we must thus have $\widetilde{v} = 0$. This contradicts $\|\widetilde{v}_k\|_{L_2(\gamma)} \simeq 1$ because $\|\widetilde{v}_k\|_{L_2(\gamma)} \to \|\widetilde{v}\|_{L_2(\gamma)} = 0$. Consequently, the desired statement is proved. □

Lemma 19 (perturbation error estimate for $C^{1,\alpha}$ surfaces). *Let $\widetilde{u} \in H^1_\#(\gamma)$ solve (19) and $u_\Gamma \in H^1_\#(\Gamma)$ solve (33). Then, the following error estimate for $u - u_\Gamma$ holds*

$$\|\nabla_\gamma(\widetilde{u} - \widetilde{u}_\Gamma)\|_{L_2(\gamma)} \lesssim \lambda_\infty \|f_\Gamma\|_{H^{-1}_\#(\Gamma)} + \|fq q_\Gamma^{-1} - f_\Gamma\|_{H^{-1}_\#(\Gamma)}, \tag{46}$$

where the hidden constant depends on S_χ defined in (38).

Proof. We proceed in several steps.

Step 1: error representation. Let $\widetilde{v} = \widetilde{u} - \widetilde{u}_\Gamma$ and make use of (34) to write

$$\|\nabla_\gamma(\widetilde{u} - \widetilde{u}_\Gamma)\|^2_{L_2(\gamma)} = \int_\gamma \nabla_\gamma \widetilde{u} \cdot \nabla_\gamma \widetilde{v} - \int_\Gamma \nabla_\Gamma u_\Gamma \cdot \nabla_\Gamma v + \int_\gamma \nabla_\gamma \widetilde{u}_\Gamma \cdot \mathbf{E}\, \nabla_\gamma \widetilde{v}.$$

We next employ the Eqs. (19) and (33) satisfied by $\widetilde{u}$ and u_Γ to obtain

$$\|\nabla_\gamma(\widetilde{u}-\widetilde{u}_\Gamma)\|^2_{L_2(\gamma)} = \int_\Gamma \left(f\frac{q}{q_\Gamma} - f_\Gamma\right)v + \int_\gamma \nabla_\gamma \widetilde{u}_\Gamma \cdot \mathbf{E}\,\nabla_\gamma \widetilde{v},$$

where we have also employed (8) to switch the domain of integration of f.

Step 2: geometric error matrix. To derive a bound for $\|\mathbf{E}\|_{L_\infty(\gamma)}$, we rewrite $\mathbf{E}$

$$\mathbf{E} = D\chi\left((q^{-1}q_\Gamma - 1)\mathbf{g}_\Gamma^{-1} - \mathbf{g}^{-1}(\mathbf{I} - \mathbf{g}\mathbf{g}_\Gamma^{-1})\right)D\chi^t.$$

Since $|\mathbf{g}^{-1}| = |D^-\chi|^{-2}, |\mathbf{g}_\Gamma^{-1}| = |D^-\chi_\Gamma|^{-2}$, applying (40) and (41) leads to the error estimate

$$\|\mathbf{E}\|_{L_\infty(\gamma)} \lesssim \lambda_\infty. \tag{47}$$

Step 3: final estimates. The Cauchy-Schwarz inequality yields

$$\int_\gamma \nabla \widetilde{u}_\Gamma \cdot \mathbf{E}\,\nabla_\gamma \widetilde{v} \leq \|\nabla_\gamma \widetilde{v}\|_{L_2(\gamma)}\,\|\nabla_\gamma \widetilde{u}_\Gamma\|_{L_2(\gamma)}\,\|\mathbf{E}\|_{L_\infty(\gamma)}.$$

To derive a bound for $\|\nabla_\gamma \widetilde{u}_\Gamma\|_{L_2(\gamma)}$, we first combine (17) with (33) to obtain $\|\nabla_\Gamma u_\Gamma\|_{L_2(\Gamma)} \leq \|f_\Gamma\|_{H_\#^{-1}(\Gamma)}$, and next appeal to Lemma 17 (norm equivalence). On the other hand, we recall that $f\frac{q}{q_\Gamma} - f_\Gamma$ has vanishing mean value on Γ, let $\overline{v} = |\Gamma|^{-1}\int_\Gamma v$ be the mean value of v, and use (17) to arrive at

$$\int_\Gamma \left(f\frac{q}{q_\Gamma} - f_\Gamma\right)v = \int_\Gamma \left(f\frac{q}{q_\Gamma} - f_\Gamma\right)(v - \overline{v}) \leq \|fqq_\Gamma^{-1} - f_\Gamma\|_{H_\#^{-1}(\Gamma)}\|\nabla_\Gamma v\|_{L_2(\Gamma)}.$$

Finally, applying Lemma 17 ends the proof. □

3.2 Perturbation theory for C^2 surfaces

Let γ be of class C^2 and the tubular neighbourhood $\mathcal{N}$ satisfy (31), namely

$$\mathcal{N} = \left\{\mathbf{x} \in \mathbb{R}^{n+1} : \ |d(\mathbf{x})| < \frac{1}{2K_\infty}\right\}, \tag{48}$$

so that parallel surfaces to γ within $\mathcal{N}$ are also C^2. We further assume that $\Gamma \subset \mathcal{N}$ and the distance function projection $\mathbf{P}_d = \mathbf{I} - d\nabla d : \Gamma \to \gamma$ is a bijection. The parametrizations of γ and Γ are given by $\chi := \mathbf{P}_d \circ \chi_\Gamma$ so that

$$v = \widetilde{v} \circ \mathbf{P}_d.$$

Lemma 20 (relation between tangential gradients). *If $\widetilde{v} : \gamma \to \mathbb{R}$ is of class H^1, then the tangential gradients $\nabla_\gamma \widetilde{v}$ and $\nabla_\Gamma v$ satisfy for all $\mathbf{x} \in \Gamma$*

$$\nabla_\Gamma v(\boldsymbol{x}) = \Pi_\Gamma(\boldsymbol{x})\,(\boldsymbol{I} - d\boldsymbol{W})(\boldsymbol{x})\,\Pi(\boldsymbol{x})\nabla_\gamma \widetilde{v}(\boldsymbol{P}_d(\boldsymbol{x})), \tag{49}$$

and

$$\nabla_\gamma \widetilde{v}(\boldsymbol{P}_d(\boldsymbol{x})) = (\boldsymbol{I} - d\boldsymbol{W})^{-1}(\boldsymbol{x})\left(\boldsymbol{I} - \frac{\boldsymbol{\nu}_\Gamma(\boldsymbol{x}) \otimes \boldsymbol{\nu}(\boldsymbol{x})}{\boldsymbol{\nu}_\Gamma(\boldsymbol{x}) \cdot \boldsymbol{\nu}(\boldsymbol{x})}\right)\nabla_\Gamma v(\boldsymbol{x}). \tag{50}$$

Proof. Let us assume that $\widetilde{v} \in C^1(\gamma)$. Recalling (27) and (28), we readily get

$$\nabla_\Gamma v(\mathbf{x}) = \Pi_\Gamma(\mathbf{x})\nabla v(\mathbf{x}) = \Pi_\Gamma(\mathbf{x})\,(\mathbf{I} - d\mathbf{W})(\mathbf{x})\,\Pi(\mathbf{x})\nabla_\gamma \widetilde{v}(\mathbf{P}_d(\mathbf{x})),$$

hence (49). Since $\mathbf{I} - d(\mathbf{x})\mathbf{W}(\mathbf{x})$ is invertible for all $\mathbf{x} \in \mathcal{N}$, according to the definition (31) of $\mathcal{N}$ and shown in Lemma 6 (curvature of parallel surfaces), (27) can be rewritten as

$$\nabla_\gamma \widetilde{v}(\mathbf{P}_d(\mathbf{x})) = (I - d\mathbf{W})(\mathbf{x}))^{-1}\nabla v(\mathbf{x}) \quad \forall \mathbf{x} \in \mathcal{N}.$$

To prove (50) we must relate ∇v and $\nabla_\Gamma v$. First note that for $\mathbf{x} \in \Gamma$

$$\nabla v = (\mathbf{I} - \boldsymbol{\nu}_\Gamma \otimes \boldsymbol{\nu}_\Gamma)\nabla v + \boldsymbol{\nu}_\Gamma \otimes \boldsymbol{\nu}_\Gamma \nabla v = \nabla_\Gamma v + (\nabla v \cdot \boldsymbol{\nu}_\Gamma)\boldsymbol{\nu}_\Gamma.$$

Exploiting next that $\nabla v(\mathbf{x}) \cdot \boldsymbol{\nu}(\mathbf{x}) = 0$, because $v(\mathbf{x})$ is constant in the normal direction to $\mathbf{P}_d(\mathbf{x})$, yields

$$\nabla_\Gamma v \cdot \boldsymbol{\nu} + (\boldsymbol{\nu}_\Gamma \cdot \boldsymbol{\nu})\nabla v \cdot \boldsymbol{\nu}_\Gamma = 0 \;\Rightarrow\; \nabla v \cdot \boldsymbol{\nu}_\Gamma = -\frac{1}{\boldsymbol{\nu}_\Gamma \cdot \boldsymbol{\nu}}\nabla_\Gamma v \cdot \boldsymbol{\nu}.$$

Since $\nabla v = \nabla_\Gamma v + (\nabla v \cdot \boldsymbol{\nu}_\Gamma)\boldsymbol{\nu}_\Gamma$, we deduce

$$\nabla v(\mathbf{x}) = \left(\mathbf{I} - \frac{\boldsymbol{\nu}_\Gamma(\mathbf{x}) \otimes \boldsymbol{\nu}(\mathbf{x})}{\boldsymbol{\nu}_\Gamma(\mathbf{x}) \cdot \boldsymbol{\nu}(\mathbf{x})}\right)\nabla_\Gamma v(\mathbf{x}) \quad \forall \mathbf{x} \in \Gamma.$$

Inserting this into the previous expression for $\nabla_\gamma v(\mathbf{P}_d(\mathbf{x}))$ leads to (50). Finally, a density argument of $C^1(\gamma)$ in $H^1(\gamma)$ for γ of class C^2 concludes the proof. □

The following result mimics Lemma 15 (geometric consistency) except that now it quantifies the effect of perturbing the surface γ on the bilinear forms written in (34) in terms of $\mathbf{P}_d$.

Lemma 21 (geometric consistency). *The error matrices $\boldsymbol{E}, \boldsymbol{E}_\Gamma \in \mathbb{R}^{(n+1)\times(n+1)}$ in (34) are given on Γ by*

$$\boldsymbol{E} \circ \boldsymbol{P}_d := \frac{q_\Gamma}{q}\Pi(\boldsymbol{I} - d\boldsymbol{W})\Pi_\Gamma(\boldsymbol{I} - d\boldsymbol{W})\Pi - \Pi, \tag{51}$$

$$\boldsymbol{E}_\Gamma := \frac{q}{q_\Gamma}\left(I - \frac{\boldsymbol{\nu} \otimes \boldsymbol{\nu}_\Gamma}{\boldsymbol{\nu} \cdot \boldsymbol{\nu}_\Gamma}\right)(\boldsymbol{I} - d\boldsymbol{W})^{-2}\left(I - \frac{\boldsymbol{\nu}_\Gamma \otimes \boldsymbol{\nu}}{\boldsymbol{\nu} \cdot \boldsymbol{\nu}_\Gamma}\right) - \Pi_\Gamma. \tag{52}$$

Proof. In view of (8), (49), and the fact that all matrices involved are symmetric and $\Pi_\Gamma^2 = \Pi_\Gamma$, we can write

$$\int_\Gamma \nabla_\Gamma w \cdot \nabla_\Gamma v = \int_\gamma \nabla_\gamma \widetilde{w} \cdot \left(\frac{q_\Gamma}{q} \Pi (\mathbf{I} - d\mathbf{W}) \Pi_\Gamma (\mathbf{I} - d\mathbf{W}) \Pi \right) \nabla_\gamma \widetilde{v}.$$

Noticing that $\nabla_\gamma \widetilde{w} = \Pi \nabla_\gamma \widetilde{w}$ the first equality on (34) follows immediately. The second equality proceeds along the same lines but using (50) instead. □

It is clear from Lemma 21 that the ratio of area elements q/q_Γ matters. We next derive a representation for q/q_Γ for any dimension n, proved originally for $n = 2, 3$ in Demlow and Dziuk (2007) and Demlow (2009). We stress that, in view of Remark 5 (parametric independence), the solution u of the Laplace–Beltrami equation (18) is independent of the parametrization of γ. This allows us to consider a convenient parametrization χ for theory because it does not change the geometric objects under consideration. We exploit this flexibility next.

Lemma 22 (relation between q *and* q_Γ)**.** *Given any parametrization* χ_Γ *of* Γ, *let* $\chi := \mathbf{P}_d \circ \chi_\Gamma$ *be the parametrization of* γ. *If* $\boldsymbol{\nu}(\mathbf{x}) \cdot \boldsymbol{\nu}_\Gamma(\mathbf{x}) \geq 0$ *for all* $\mathbf{x} \in \Gamma$, *then the ratio of area elements* $q(\mathbf{y})/q_\Gamma(\mathbf{y})$ *with* $\mathbf{y} = \chi_\Gamma^{-1}(\mathbf{x})$ *satisfies*

$$\frac{q(\mathbf{y})}{q_\Gamma(\mathbf{y})} = \det(\boldsymbol{I} - d(\boldsymbol{x})\boldsymbol{W}(\boldsymbol{x}))(\boldsymbol{\nu}(\boldsymbol{x}) \cdot \boldsymbol{\nu}_\Gamma(\boldsymbol{x})) \quad \forall \boldsymbol{x} \in \Gamma. \tag{53}$$

Proof. We start with the formula (5) for the area elements q and q_Γ to get

$$\frac{q}{q_\Gamma} = \det\left([\boldsymbol{\nu}, D\chi] \, [\boldsymbol{\nu}_\Gamma, D\chi_\Gamma]^{-1} \right).$$

We write $[\boldsymbol{\nu}_\Gamma, D\chi_\Gamma]^{-1} = [\mathbf{v}, \mathbf{A}]^t$ for some $\mathbf{v} \in \mathbb{R}^{n+1}$ and $\mathbf{A} \in \mathbb{R}^{(n+1)\times n}$ to be found. The identity $[\mathbf{v}, \mathbf{A}]^t [\boldsymbol{\nu}_\Gamma, D\chi_\Gamma] = \mathbf{I}$ yields $\mathbf{v} = \boldsymbol{\nu}_\Gamma$ while $[\boldsymbol{\nu}_\Gamma, D\chi_\Gamma][\mathbf{v}, \mathbf{A}]^t = \mathbf{I}$ gives $D\chi_\Gamma \mathbf{A}^t = \mathbf{I} - \boldsymbol{\nu}_\Gamma \otimes \boldsymbol{\nu}_\Gamma = \Pi_\Gamma$ and

$$[\boldsymbol{\nu}, D\chi] \, [\boldsymbol{\nu}_\Gamma, D\chi_\Gamma]^{-1} = \boldsymbol{\nu} \otimes \boldsymbol{\nu}_\Gamma + D\chi \, \mathbf{A}^t.$$

To obtain an expression for $D\chi$, let $\mathbf{x} = \chi_\Gamma(\mathbf{y}) \in \Gamma$ and $\chi(\mathbf{y}) = \mathbf{P}_d(\mathbf{x}) = \mathbf{x} - d(\mathbf{x})\nabla d(\mathbf{x}) \in \gamma$, and utilize the chain rule

$$D\chi(\mathbf{y}) = (\mathbf{I} - d(\mathbf{x})\mathbf{W}(\mathbf{x})) \, \Pi(\mathbf{x}) \, D\chi_\Gamma(\mathbf{y}) \quad \forall \mathbf{y} \in \mathcal{V},$$

where we have argued as in (27). Compute now $D\chi \mathbf{A}^t$ and use that $D\chi_\Gamma \mathbf{A}^t = \Pi_\Gamma$ together with $\mathbf{W}\boldsymbol{\nu} = 0$ to arrive at

$$\begin{aligned} \frac{q}{q_\Gamma} &= \det(\boldsymbol{\nu} \otimes \boldsymbol{\nu}_\Gamma + (\mathbf{I} - d\mathbf{W}) \, \Pi \, \Pi_\Gamma) \\ &= \det((\mathbf{I} - d\mathbf{W})(\boldsymbol{\nu} \otimes \boldsymbol{\nu}_\Gamma + \Pi \, \Pi_\Gamma)) = \det((\mathbf{I} - d\mathbf{W})) \det \mathbf{B}. \end{aligned}$$

where $\mathbf{B} := \boldsymbol{\nu} \otimes \boldsymbol{\nu}_\Gamma + \Pi \, \Pi_\Gamma$. It thus remains to show that $\det \mathbf{B} = \boldsymbol{\nu} \cdot \boldsymbol{\nu}_\Gamma$.

We now embark on a spectral analysis of $\mathbf{B}$. We first note that the statement is trivial if $\boldsymbol{\nu} = \boldsymbol{\nu}_\Gamma$. We thus assume that $\{\boldsymbol{\nu}, \boldsymbol{\nu}_\Gamma\}$ are linearly independent and that the space $\mathbb{X} = \text{span}\{\boldsymbol{\nu}, \boldsymbol{\nu}_\Gamma\}$ is generated by two orthonormal vectors $\boldsymbol{\nu}$ and $\mathbf{e}$. We consider the orthogonal decomposition $\mathbb{R}^{n+1} = \mathbb{X} \oplus \mathbb{X}^\perp$ and a rotation $\mathbf{R} \in \mathbb{R}^{(n+1)\times(n+1)}$ on $\mathbb{X}$ that maps $\boldsymbol{\nu}$ into $\boldsymbol{\nu}_\Gamma$, namely

$$\mathbf{R}\boldsymbol{\nu} = \boldsymbol{\nu}_\Gamma = \cos\theta\, \boldsymbol{\nu} + \sin\theta\, \mathbf{e}, \quad \mathbf{R}\mathbf{e} = -\sin\theta\, \boldsymbol{\nu} + \cos\theta\, \mathbf{e};$$

thus the rotation angle θ satisfies $\cos\theta = \boldsymbol{\nu} \cdot \boldsymbol{\nu}_\Gamma$ and $\det \mathbf{R} = 1$. Consequently,

$$\mathbf{B} = (\boldsymbol{\nu} \otimes \boldsymbol{\nu} + \Pi\, \mathbf{R}\, \Pi)\mathbf{R}^t \quad \Rightarrow \quad \det \mathbf{B} = \det(\boldsymbol{\nu} \otimes \boldsymbol{\nu} + \Pi\, \mathbf{R}\, \Pi).$$

The proof concludes upon realizing that $\boldsymbol{\nu}$ and $\mathbf{e}$ are eigenvectors of $\boldsymbol{\nu} \otimes \boldsymbol{\nu} + \Pi\, \mathbf{R}\, \Pi$ with eigenvalues 1 and $\cos\theta$, and the remaining eigenvalues are 1 with eigenspace $\mathbb{X}^\perp$. □

We are now ready to compare solutions u and u_Γ of (19) on two nearby surfaces γ and Γ. In essence, weak solutions u and u_Γ are close in H^1 provided γ and Γ are close in a Lipschitz sense. Therefore, to make this statement quantitative we introduce the following geometric quantities

$$d_\infty := \|d\|_{L_\infty(\Gamma)}, \quad \nu_\infty := \|\boldsymbol{\nu} - \boldsymbol{\nu}_\Gamma\|_{L_\infty(\Gamma)}, \quad K_\infty := \|K\|_{L_\infty(\gamma)}, \tag{54}$$

where $\Gamma \subset \mathcal{N}$ is a Lipschitz surface. Our goal is to bound $\|u - u_\Gamma\|_{H^1_\#(\Gamma)}$ in terms of the forcing functions f, f_Γ, and $d_\infty, \nu_\infty, K_\infty$ in (54).

Lemma 23 (perturbation error estimate for C^2 surfaces). *Let u solve (19) and u_Γ solve (33) with $\Gamma \subset \mathcal{N}$. Let χ_Γ and $\chi := \mathbf{P}_d \circ \chi_\Gamma$ be the parametrizations of Γ and γ that give rise to the area elements q_Γ and q. If the normal vectors satisfy $\boldsymbol{\nu} \cdot \boldsymbol{\nu}_\Gamma \geq c > 0$, then*

$$\|\nabla_\gamma(u - u_\Gamma)\|_{L_2(\gamma)} \lesssim \left(d_\infty K_\infty + \nu_\infty^2\right) \|f_\Gamma\|_{H^{-1}_\#(\Gamma)} + \|f q q_\Gamma^{-1} - f_\Gamma\|_{H^{-1}_\#(\Gamma)}. \tag{55}$$

Proof. We proceed along the lines of Lemma 19 (perturbation error estimate for $C^{1,\alpha}$ surfaces) and realize that Steps 1 and 3 are exactly the same. Therefore, we only deal with the estimate of the geometric error matrix $\mathbf{E}$. If we prove

$$\|\mathbf{E}\|_{L_\infty(\gamma)} \lesssim \nu_\infty^2 + d_\infty K_\infty, \tag{56}$$

then the assertion will readily follow. We first write $\mathbf{E} \circ \mathbf{P}_d = \mathbf{I}_1 + \mathbf{I}_2 + \mathbf{I}_3$ with

$$\begin{aligned}
\mathbf{I}_1 &:= \left(\frac{q_\Gamma}{q} - 1\right)\Pi\,(\mathbf{I} - d\mathbf{W})\,\Pi_\Gamma\,(\mathbf{I} - d\mathbf{W})\,\Pi, \\
\mathbf{I}_2 &:= (\Pi\,(\mathbf{I} - d\mathbf{W})\,\Pi_\Gamma\,(\mathbf{I} - d\mathbf{W})\,\Pi - \Pi\,\Pi_\Gamma\,\Pi), \\
\mathbf{I}_3 &:= (\Pi\,\Pi_\Gamma\,\Pi - \Pi).
\end{aligned}$$

We now estimate these three terms separately. In view of (53) we deduce

$$\frac{q(\mathbf{y})}{q_\Gamma(\mathbf{y})}-1=\left((\boldsymbol{\nu}(\mathbf{x})\cdot\boldsymbol{\nu}_\Gamma(\mathbf{x})-1)\prod_{i=1}^{n}(1-d(\mathbf{x})\kappa_i(\mathbf{x}))\right)+\left(\prod_{i=1}^{n}(1-d(\mathbf{x})\kappa_i(\mathbf{x}))-1\right),$$

where $\mathbf{x}=\chi_\Gamma(\mathbf{y})\in\Gamma$. Since $1-\boldsymbol{\nu}\cdot\boldsymbol{\nu}_\Gamma=\frac{1}{2}|\boldsymbol{\nu}-\boldsymbol{\nu}_\Gamma|^2\le\frac{1}{2}\nu_\infty^2$ and $\Gamma\subset\mathcal{N}$, we readily obtain

$$\left|\frac{q(\mathbf{y})}{q_\Gamma(\mathbf{y})}-1\right|\lesssim\nu_\infty^2+d_\infty K_\infty\quad\forall\,\mathbf{y}\in\mathcal{V},\tag{57}$$

and a similar bound for $\frac{q_\Gamma}{q}$ because $\frac{q_\Gamma}{q}$ is bounded in $\mathcal{V}$ thanks to the assumption $\boldsymbol{\nu}\cdot\boldsymbol{\nu}_\Gamma\ge c>0$. The desired estimate for $\|\mathbf{I}_1\|_{L_\infty(\gamma)}$ follows from the fact that Π, Π_Γ and $\mathbf{W}$ are bounded. This property again, now combined with

$$\mathbf{I}_2=-\Pi\,\Pi_\Gamma\,d\mathbf{W}\,\Pi-\Pi\,d\mathbf{W}\,\Pi_\Gamma\,\Pi+\Pi\,d\mathbf{W}\,\Pi_\Gamma\,d\mathbf{W}\,\Pi,$$

yields $\|\mathbf{I}_2\|_{L_\infty(\gamma)}\lesssim d_\infty K_\infty$. Finally, term $\mathbf{I}_3$ reads

$$\mathbf{I}_3=-\Pi\boldsymbol{\nu}_\Gamma\otimes\Pi\boldsymbol{\nu}_\Gamma=-(\boldsymbol{\nu}_\Gamma-(\boldsymbol{\nu}\cdot\boldsymbol{\nu}_\Gamma)\boldsymbol{\nu})\otimes(\boldsymbol{\nu}_\Gamma-(\boldsymbol{\nu}\cdot\boldsymbol{\nu}_\Gamma)\boldsymbol{\nu})$$

Since $\boldsymbol{\nu}_\Gamma-(\boldsymbol{\nu}\cdot\boldsymbol{\nu}_\Gamma)\boldsymbol{\nu}=(\boldsymbol{\nu}_\Gamma-\boldsymbol{\nu})+(1-\boldsymbol{\nu}\cdot\boldsymbol{\nu}_\Gamma)\boldsymbol{\nu}$ we infer that $\|\mathbf{I}_3\|_{L_\infty(\gamma)}\lesssim\nu_\infty^2$. This ends the proof. □

It is worth comparing Lemmas 19 and 23 (perturbation error estimates). To do so, we next give an estimate for ν_∞ in terms of λ_∞.

Lemma 24 (error estimate for normals). *The errors ν_∞ and λ_∞ defined in (54) and (39) satisfy*

$$\nu_\infty\lesssim\lambda_\infty,\tag{58}$$

where the hidden constant depends on S_χ defined in (38).

Proof. In view of the definition (2) of $\boldsymbol{\nu}$, we realize that

$$\boldsymbol{\nu}-\boldsymbol{\nu}_\Gamma=\frac{\mathbf{N}-\mathbf{N}_\Gamma}{|\mathbf{N}|}+\frac{|\mathbf{N}_\Gamma|-|\mathbf{N}|}{|\mathbf{N}|}\frac{\mathbf{N}_\Gamma}{|\mathbf{N}_\Gamma|}\quad\Rightarrow\quad|\boldsymbol{\nu}-\boldsymbol{\nu}_\Gamma|\le2\frac{|\mathbf{N}-\mathbf{N}_\Gamma|}{|\mathbf{N}|}.$$

Since $\mathbf{N}=\sum_{i=1}^{n+1}\det([\mathbf{e}_i,D\chi])\mathbf{e}_i$ and $\det([\mathbf{e}_i,D\chi])-\det([\mathbf{e}_i,D\chi_\Gamma])$ is a sum of products of $\partial_j(\chi-\chi_\Gamma)\cdot\mathbf{e}_k$ with $k\ne i$ times $n-1$ factors $\partial_\ell\chi_m$, we have

$$|\det([\mathbf{e}_i,D\chi])-\det([\mathbf{e}_i,D\chi_\Gamma])|\lesssim|D(\chi-\chi_\Gamma)|\,|D\chi|^{n-1}.$$

We finally resort to $|\mathbf{N}|=q$, proved in (7), as well as $q\approx|D\chi|^n$, showed in the proof of Lemma 16, to conclude (58). □

We now stress the advantage of using the distance function lift $\mathbf{P}_d$ to represent the error $u - u_\gamma$ whenever the surface γ is of class C^2. Comparing (46) and (55) we see that the geometric error becomes of order $\|d\|_{L_\infty(\Gamma)}$ plus a quadratic term in λ_∞ rather than linear. In the context of finite element methods, Γ is often a polyhedral approximation to γ having faces of diameter h. In this case $\|d\|_{L_\infty(\Gamma)}$ essentially becomes a Lagrange interpolation error measured in L_∞ and λ_∞ a Lagrange interpolation error measured in W^1_∞. The former error is of order h^2 and the latter of order h. Consequently, the perturbation error for C^2 surfaces is of order h^2, whereas for $C^{1,\alpha}$ surfaces with $\alpha < 1$ it is of order h^α from the analysis of the previous subsection. The increased approximation order for C^2 surfaces is a *superconvergence* effect. We also recall from Theorem 11 (C^1 distance function implies $C^{1,1}$ surfaces) that the elegant properties of the distance function and closest point projection that lead to this superconvergence effect are not available when γ is not of class C^2, thus the necessity of developing a separate perturbation theory for less regular surfaces as in the previous subsection. It is not clear whether the order of the perturbation error actually jumps in this manner when crossing from $C^{1,\alpha}$ to C^2 surfaces, or if this jump is an artifact of proof.

3.3 H^2 extensions from C^2 surfaces

The analysis of the trace and narrow band methods that we carry out in later sections requires us to extend the solution $\widetilde{u} \in H^2(\gamma)$ of (19) to tubular neighbourhoods $\mathcal{N}(\delta)$ with the property

$$\|u\|_{H^2(\mathcal{N}(\delta))} \lesssim \delta^{\frac{1}{2}} \|\widetilde{u}\|_{H^2(\gamma)}; \tag{59}$$

we recall that $\mathcal{N}(\delta)$ is defined in (29). The distance function lift $\mathbf{P}_d$ provides a natural way to achieve this upon setting $u = \widetilde{u} \circ \mathbf{P}_d$, namely

$$u(\mathbf{x}) = \widetilde{u}(\mathbf{x} - d(\mathbf{x})\nabla d(\mathbf{x})) \quad \forall \mathbf{x} \in \mathcal{N}(\delta).$$

However, this is problematic because it requires $\mathbf{P}_d$ to be of class C^2, and thus γ of class C^3, for (59) to hold. We now construct an extension that satisfies (59) for γ of class C^2. Our approach employs a regularization $\overline{d}_\varepsilon$ of the signed distance function d and construction of a regularized surface γ_ε close to γ, with the regularization parameter ε appropriately related to the desired value of δ above. We begin by describing properties of this regularization.

3.3.1 Regularization

Recall that given γ of class C^2 there exists a sufficiently thin tubular neighbourhood $\mathcal{N}$ so that the signed distance function d to γ satisfies $d \in C^2(\mathcal{N})$. Let $\delta > 0$ and $\varepsilon = c\delta \leq \frac{\delta}{2}$ be sufficiently small so that the tubular neighbourhood $\mathcal{N}(\delta)$ of width δ satisfies the property

$$\mathcal{N}(\delta+2\varepsilon) \subset \mathcal{N}.$$

Let $B_\varepsilon := B(0, \varepsilon)$ be the ball of center 0 and radius ε, ρ_ε be a smooth and radially symmetric mollifier with support in B_ε and

$$\overline{d}_\varepsilon(\mathbf{x}) := d \star \rho_\varepsilon(\mathbf{x}) = \int_{B_\varepsilon} d(\mathbf{x}-\mathbf{y})\rho_\varepsilon(\mathbf{y})d\mathbf{y} \quad \forall \mathbf{x} \in \mathcal{N}(\delta)$$

be a regularized distance function. This function $\overline{d}_\varepsilon$ induces the smooth surface

$$\gamma_\varepsilon := \{\mathbf{x} \in \mathcal{N} : \ \overline{d}_\varepsilon(\mathbf{x}) = 0\},$$

but is not the signed distance function to γ_ε, which we denote d_ε. The following properties are immediate from the previous definitions.

Lemma 25 (properties of $\overline{d}_\varepsilon$). *If $d \in C^2(\overline{\mathcal{N}})$, then $\overline{d}_\varepsilon$ satisfies*

$$\|d - \overline{d}_\varepsilon\|_{L_\infty(\mathcal{N}(\delta))} + \varepsilon \|\nabla(d - \overline{d}_\varepsilon)\|_{L_\infty(\mathcal{N}(\delta))} \lesssim \varepsilon^2 |d|_{W^2_\infty(\overline{\mathcal{N}})}$$

and $\|D^2\overline{d}_\varepsilon\|_{L_\infty(\mathcal{N}(\delta))} \lesssim |d|_{W^2_\infty(\overline{\mathcal{N}})}$. Moreover, the surface γ_ε is smooth and the Hausdorff distance $\mathrm{dist}_H(\gamma, \gamma_\varepsilon)$ between γ and γ_ε satisfies

$$\mathrm{dist}_H(\gamma, \gamma_\varepsilon) \lesssim \varepsilon^2 |d|_{W^2_\infty(\overline{\mathcal{N}})}$$

provided ε is small enough so that $C\varepsilon|d|_{W^2_\infty(\overline{\mathcal{N}})} \leq \frac{1}{2}$ for a suitable constant C.

Proof. Since ρ_ε is radially symmetric, we have that

$$(d - \overline{d}_\varepsilon)(\mathbf{x}) = \int_{B_\varepsilon} (d(\mathbf{x}) - \nabla d(\mathbf{x}) \cdot \mathbf{y} - d(\mathbf{x}-\mathbf{y}))\rho_\varepsilon(\mathbf{y})d\mathbf{y}$$

and

$$\nabla(d - \overline{d}_\varepsilon)(\mathbf{x}) = \int_{B_\varepsilon} (\nabla d(\mathbf{x}) - \nabla d(\mathbf{x}-\mathbf{y}))\rho_\varepsilon(\mathbf{y})d\mathbf{y}.$$

These relationships imply the first assertion upon employing a Taylor expansion of d and the Lipschitz nature of ∇d, respectively. We also note that

$$D^2\overline{d}_\varepsilon(\mathbf{x}) = \int_{B_\varepsilon} \nabla d(\mathbf{x}-\mathbf{y}) \otimes \nabla\rho_\varepsilon(\mathbf{y})d\mathbf{y} = \int_{B_\varepsilon} \nabla(d(\mathbf{x}-\mathbf{y}) - d(\mathbf{x})) \otimes \nabla\rho_\varepsilon(\mathbf{y})d\mathbf{y}$$

because $\int_{B_\varepsilon} \nabla\rho_\varepsilon(\mathbf{y})d\mathbf{y} = 0$ in view of the radial symmetry of ρ_ε. The second relationship bounding $D^2\overline{d}_\varepsilon$ then follows from the Lipschitz nature of ∇d (i.e., $|\nabla(d(\mathbf{x}-\mathbf{y}) - d(\mathbf{x}))| \lesssim \varepsilon$, $\mathbf{y} \in B_\varepsilon$) and the standard property $\|\nabla\rho_\varepsilon\|_{L_1(B_\varepsilon)} \lesssim \varepsilon^{-1}$ of the mollifier.

To establish the smoothness of γ_ε, note that the closeness of ∇d and $\nabla \overline{d}_\varepsilon$ implies that $\nabla \overline{d}_\varepsilon$ is nondegenerate for $C\varepsilon |d|_{W^2_\infty(\overline{\mathcal{N}})} \leq 1/2$. The smoothness of γ_ε then follows from the implicit function theorem.

The last assertion is a consequence of the nondegeneracy of the distance function: Given $\mathbf{y} \in \gamma_\varepsilon$ let $\mathbf{x} = \mathbf{P}_d(\mathbf{y}) = \mathbf{y} - d(\mathbf{y})\nabla d(\mathbf{y}) \in \gamma$ be the closest point to $\mathbf{y}$ and note that

$$|\mathbf{y}-\mathbf{x}| = |d(\mathbf{y})| = |d(\mathbf{y}) - \overline{d}_\varepsilon(\mathbf{y})| \lesssim \varepsilon^2 |d|_{W^2_\infty(\overline{\mathcal{N}})}.$$

Likewise, given $\mathbf{x} \in \gamma$ let $\mathbf{y}(s) = \mathbf{x} + s\nabla d(\mathbf{x})$. There is $s \in (-\varepsilon, \varepsilon)$ such that $\overline{d}_\varepsilon(\mathbf{y}(s)) = 0$. To see this, note that $d(\mathbf{y}(s)) = \pm\varepsilon$ for $s = \pm\varepsilon$ and

$$\overline{d}_\varepsilon(\mathbf{y}(\varepsilon)) \geq d(\mathbf{y}(\varepsilon)) - C\varepsilon^2 |d|_{W^2_\infty(\overline{\mathcal{N}})} = \varepsilon\Big(1 - \varepsilon|d|_{W^2_\infty(\overline{\mathcal{N}})}\Big) > 0$$

provided $C\varepsilon|d|_{W^2_\infty(\overline{\mathcal{N}})} \leq \frac{1}{2}$; similarly $\overline{d}_\varepsilon(\mathbf{y}(-\varepsilon)) < 0$. Letting $\mathbf{y} = \mathbf{y}(s)$ be such that $\overline{d}_\varepsilon(\mathbf{y}(s)) = 0$, we note that $\mathbf{x} = \mathbf{P}_d(\mathbf{y})$, and so arguing as before we have that

$$|\mathbf{y}-\mathbf{x}| = |d(\mathbf{y})| = |d(\mathbf{y}) - \overline{d}_\varepsilon(\mathbf{y})| \lesssim \varepsilon^2 |d|_{W^2_\infty(\overline{\mathcal{N}})},$$

which concludes the proof. □

We recall that d_ε is the signed distance function to the zero level set γ_ε of $\overline{d}_\varepsilon$. Consider the C^∞ lift

$$\mathbf{P}_\varepsilon(\mathbf{x}) := \mathbf{x} - d_\varepsilon(\mathbf{x})\nabla d_\varepsilon(\mathbf{x}) \quad \forall\, \mathbf{x} \in \mathcal{N}(\delta). \tag{60}$$

It is natural and useful for later considerations to compare tubular neighbourhoods dictated by d and d_ε. Let

$$\mathcal{N}_\varepsilon(\delta_\varepsilon) := \{\mathbf{x} \in \mathbb{R}^{n+1} : |d_\varepsilon(\mathbf{x})| \leq \delta_\varepsilon\},$$

where we choose δ_ε as follows depending on δ and ε. Given $\mathbf{x} \in \mathcal{N}(\delta)$ let $\widetilde{\mathbf{x}} \in \gamma$ be the point at shortest distance, whence $|\mathbf{x} - \widetilde{\mathbf{x}}| \leq \delta$, and let $\mathbf{x}_\varepsilon \in \gamma_\varepsilon$ be a point such that $|\widetilde{\mathbf{x}} - \mathbf{x}_\varepsilon| \leq C|d|_{W^2_\infty(\mathcal{N})}\varepsilon^2$ which is guaranteed to exist because $\mathrm{dist}_H(\gamma, \gamma_\varepsilon) \leq C\varepsilon^2|d| \leq \varepsilon$ in view of Lemma 25 (properties of $\overline{d}_\varepsilon$). Therefore

$$|d_\varepsilon(\mathbf{x})| = \mathrm{dist}(\mathbf{x}, \gamma_\varepsilon) \leq |\mathbf{x} - \mathbf{x}_\varepsilon| \leq |\mathbf{x} - \widetilde{\mathbf{x}}| + |\widetilde{\mathbf{x}} - \mathbf{x}_\varepsilon| \leq \delta + C|d|_{W^2_\infty(\mathcal{N})}\varepsilon^2 \leq \delta + \varepsilon =: \delta_\varepsilon,$$

provided $C\varepsilon|d|_{W^2_\infty(\mathcal{N})} \leq 1$; note that $\delta_\varepsilon \leq \frac{3}{2}\delta$. This implies

$$\mathcal{N}(\delta) \subset \mathcal{N}_\varepsilon(\delta_\varepsilon). \tag{61}$$

Similarly, using again $C\varepsilon|d|_{W^2_\infty(\overline{\mathcal{N}})} \leq 1$ in conjunction with Lemma 25 yields

$$\mathcal{N}_\varepsilon(\delta_\varepsilon) \subset \mathcal{N}(\delta_\varepsilon + \varepsilon) = \mathcal{N}(\delta + 2\varepsilon) \subset \mathcal{N}.$$

The next lemma and corollary study important properties of d_ε and $\mathbf{P}_\varepsilon$, in particular how derivatives degenerate with ε.

Lemma 26 (properties of d_ε). *The function $d_\varepsilon \in C^\infty(\mathcal{N}(\delta))$ and satisfies*

$$\|d\varepsilon\|_{W^2_\infty(\mathcal{N}(\delta))} + \varepsilon\,\|d_\varepsilon\|_{W^3_\infty(\mathcal{N}(\delta))} \lesssim |d|_{W^2_\infty(\mathcal{N})}.$$

Moreover, the following error estimates hold

$$\|\nabla(d - d_\varepsilon)\|_{L_\infty(\mathcal{N}(\delta))} \lesssim \delta|d|_{W^2_\infty(\mathcal{N})}, \quad \|1 - \nabla d\cdot\nabla d_\varepsilon\|_{L_\infty(\mathcal{N}(\delta))} \lesssim \delta^2|d|^2_{W^2_\infty(\mathcal{N})}.$$

Proof. Since $\overline{d}_\varepsilon(\mathbf{x}) = 0$ and $|\nabla\overline{d}_\varepsilon(\mathbf{x})| \geq \frac{1}{2}$ for all $\mathbf{x} \in \gamma_\varepsilon$, fix $\mathbf{x}_0 \in \gamma_\varepsilon$ and a system of coordinates such that $\mathbf{x} = (\mathbf{x}', x_{n+1})$ is a generic point and $\nabla\overline{d}_\varepsilon(\mathbf{x}_0)$ points in the $(n + 1)$th coordinate direction. The Implicit Function Theorem guarantees the existence of a ball B in $\mathbb{R}^n$ centred at $\mathbf{x}'_0$ and a C^∞ function $\psi : B \to \mathbb{R}$ such that

$$\overline{d}_\varepsilon(\mathbf{x}', \psi(\mathbf{x}')) = 0 \quad \forall\,\mathbf{x}' \in B.$$

In other words, γ_ε is locally described in B as a graph $x_{n+1} = \psi(\mathbf{x}')$ for $\mathbf{x}' \in B$. It is not difficult but tedious to see that

$$\|\psi\|_{W^2_\infty(B)} \lesssim \|\overline{d}_\varepsilon\|_{W^2_\infty(\mathcal{N}(\delta))} \lesssim \|d\|_{W^2_\infty(\mathcal{N})},$$
$$\|\psi\|_{W^3_\infty(B)} \lesssim \|\overline{d}_\varepsilon\|_{W^3_\infty(\mathcal{N}(\delta))} \lesssim \frac{1}{\varepsilon}\|d\|_{W^2_\infty(\mathcal{N})},$$

which translates into the first estimates for d_ε

$$\|d_\varepsilon\|_{W^2_\infty(\mathcal{N}(\delta))} \lesssim \|d\|_{W^2_\infty(\mathcal{N})}, \qquad \|d_\varepsilon\|_{W^3_\infty(\mathcal{N}(\delta))} \lesssim \frac{1}{\varepsilon}\|d\|_{W^2_\infty(\mathcal{N})}.$$

To prove the error estimates, let $\mathbf{x} \in \mathcal{N}(\delta) \subset \mathcal{N}$ and note that

$$\nabla d_\varepsilon(\mathbf{x}) = \nabla d_\varepsilon(\mathbf{y}) = \frac{\nabla\overline{d}_\varepsilon(\mathbf{y})}{|\nabla\overline{d}_\varepsilon(\mathbf{y})|}, \qquad \nabla d(\mathbf{x}) = \nabla d(\mathbf{w})$$

with $\mathbf{y} = \mathbf{x} - d_\varepsilon(\mathbf{x})\nabla d_\varepsilon(\mathbf{x}) \in \gamma_\varepsilon$ and $\mathbf{w} = \mathbf{x} - d(\mathbf{x})\nabla d(\mathbf{x}) \in \gamma$. Hence,

$$|\mathbf{w} - \mathbf{y}| \leq |\mathbf{w} - \mathbf{x}| + |\mathbf{y} - \mathbf{x}| \leq \delta + \delta_\varepsilon \leq \frac{5}{2}\delta$$

because of (61). Since $|\nabla d(\mathbf{y})| = 1$, we now write

$$\nabla d(\mathbf{x}) - \nabla d_\varepsilon(\mathbf{x}) = \nabla d(\mathbf{w}) - \nabla d(\mathbf{y}) + \nabla d(\mathbf{y}) - \nabla\overline{d}_\varepsilon(\mathbf{y}) + \frac{\nabla\overline{d}_\varepsilon(\mathbf{y})}{|\nabla\overline{d}_\varepsilon(\mathbf{y})|}\left(|\nabla\overline{d}_\varepsilon(\mathbf{y})| - |\nabla d(\mathbf{y})|\right)$$

and estimate pairs of terms on the right-hand side separately. Since $d \in W^2_\infty(\mathcal{N})$, we get

$$|\nabla d(\mathbf{w}) - \nabla d(\mathbf{y})| \leq |\mathbf{w}-\mathbf{y}|\, |d|_{W^2_\infty(\mathcal{N})} \lesssim \delta |d|_{W^2_\infty(\mathcal{N})},$$

and using Lemma 25 (properties of $\overline{d}_\varepsilon$) we also obtain

$$\big||\nabla d(\mathbf{y})| - |\nabla \overline{d}_\varepsilon(\mathbf{y})|\big| \leq \big|\nabla d(\mathbf{y}) - \nabla \overline{d}_\varepsilon(\mathbf{y})\big| \leq \varepsilon |d|_{W^2_\infty(\mathcal{N})} < \delta |d|_{W^2_\infty(\mathcal{N})},$$

whence the first error estimate follows

$$|\nabla d(\mathbf{x}) - \nabla d_\varepsilon(\mathbf{x})| \lesssim \delta |d|_{W^2_\infty(\mathcal{N})} \quad \forall \mathbf{x} \in \mathcal{N}(\delta).$$

To show the desired second error estimate we observe that $|1 - \nabla d(\mathbf{x}) \cdot \nabla d_\varepsilon(\mathbf{x})| = \frac{1}{2}|\nabla d(\mathbf{x}) - \nabla d_\varepsilon(\mathbf{x})|^2$. This concludes the proof. □

Corollary 27 (property of $\mathbf{P}_\varepsilon$). *The lift $\boldsymbol{P}_\varepsilon$ belongs to $C^\infty(\mathcal{N}(\delta))$ and satisfies*

$$|\boldsymbol{P}_\varepsilon|_{W^2_\infty(\mathcal{N}(\delta))} \lesssim |d|_{W^2_\infty(\mathcal{N})}$$

for suitable constants C_1, C_2 so that $C_1\delta \leq \varepsilon \leq \frac{\delta}{2}$ and $C_2 \varepsilon |d|_{W^2_\infty(\mathcal{N})} \leq 1$.

Proof. Differentiate the kth component of $\mathbf{P}_\varepsilon$ with respect to x_i and x_j to obtain

$$\partial_{ij}\mathbf{P}_{\varepsilon,k} = -\partial^2_{ij} d_\varepsilon\, \partial_k d_\varepsilon - \partial_i d_\varepsilon\, \partial^2_{jk} d_\varepsilon - \partial_j d_\varepsilon\, \partial^2_{ik} d_\varepsilon - d_\varepsilon\, \partial^3_{ijk} d_\varepsilon,$$

whence invoking Lemma 26 (properties of d_ε) yields

$$\|D^2 \mathbf{P}_\varepsilon\|_{L_\infty(\mathcal{N}(\delta))} \lesssim |d|_{W^2_\infty(\mathcal{N})} + \frac{\delta}{\varepsilon} |d|_{W^2_\infty(\mathcal{N})} \lesssim |d|_{W^2_\infty(\mathcal{N})}$$

because of $|\nabla d_\varepsilon| = 1$ and (61). This completes the proof. □

Given a function $\widetilde{u} \in H^2(\gamma)$ we are now ready to introduce an H^2 extension to $\mathcal{N}(\delta)$. For this, we assume that δ is sufficiently small so that (61) is valid. We first define the auxiliary function $u_\varepsilon = \widetilde{u} \circ \mathbf{Q}_\varepsilon : \gamma_\varepsilon \to \mathbb{R}$, where $\mathbf{Q}_\varepsilon = \mathbf{P}_\varepsilon^{-1} : \gamma_\varepsilon \to \gamma$, and then the extension $u = u_\varepsilon \circ \mathbf{P}_\varepsilon : \mathcal{N}(\delta) \to \mathbb{R}$, namely

$$u(\mathbf{x}) := u_\varepsilon(\mathbf{x} - d_\varepsilon(\mathbf{x}) \nabla d_\varepsilon(\mathbf{x})) \quad \forall \mathbf{x} \in \mathcal{N}(\delta). \tag{62}$$

Consequently, we realize that $u = \widetilde{u} \circ \mathbf{Q}_\varepsilon \circ \mathbf{P}_\varepsilon$. We introduce the notation $\mathbf{Q}_\varepsilon$ to avoid confusion between $\mathbf{Q}_\varepsilon \circ \mathbf{P}_\varepsilon : \mathcal{N}(\delta) \to \gamma$ and the identity. We recall that the *coarea formula*

$$\int_{\mathcal{N}(\delta)} g = \int_{\mathcal{N}(\delta)} g |\nabla d| = \int_{-\delta}^{\delta} \int_{\{d^{-1}(s)\}} g d\sigma_s, \tag{63}$$

is valid for any integrable function $g:\mathcal{N}(\delta)\to\mathbb{R}$ [Theorem 3.14, Evans and Gariepy, 2015]. We will use this formula next and later in this chapter.

Proposition 28 (H^2 extension). *Let ε and δ be as in Corollary 27 (property of $\mathbf{P}_\varepsilon$), and assume that $\varepsilon|d|_{W^2_\infty(\mathcal{N})} \le c$ for a sufficiently small constant c. If $\widetilde{u}\in H^2(\gamma)$, then $u\in H^2(\mathcal{N}(\delta))$ and*

$$\|u\|_{H^2(\mathcal{N}(\delta))} \lesssim \delta^{\frac{1}{2}}|d|_{W^2_\infty(\mathcal{N})}\,\|\widetilde{u}\|_{H^2(\gamma)}.$$

Moreover, the trace of u on γ coincides with $\widetilde{u}$, that is u an H^2 extension of $\widetilde{u}$.

Proof. In view of (27), the i-partial derivative of u reads

$$\partial_i u = \sum_{j=1}^{n+1}\left(\delta_{ij} - d_\varepsilon\,\partial^2_{ij}d_\varepsilon\right)\overline{\partial}_j u_\varepsilon \circ \mathbf{P}_\varepsilon$$

where $\overline{\partial}_j u_\varepsilon$ stands for the j-component of $\nabla_{\gamma_\varepsilon}u_\varepsilon$. We use again (27) to obtain

$$\begin{aligned}\nabla\partial_j u = &-\sum_{j=1}^{n+1}\left(\nabla d_\varepsilon\,\partial^2_{ij}d_\varepsilon + d_\varepsilon\,\partial^2_{ij}\nabla d_\varepsilon\right)\overline{\partial}_j u_\varepsilon\circ\mathbf{P}_\varepsilon\\ &+\sum_{j=1}^{n+1}\left(\delta_{ij} - d_\varepsilon\,\partial^2_{ij}d_\varepsilon\right)(\mathbf{I} - d_\varepsilon D^2 d_\varepsilon)\nabla_{\gamma_\varepsilon}\overline{\partial}_j u_\varepsilon \circ \mathbf{P}_\varepsilon.\end{aligned}$$

Setting $\Lambda := 1 + |d|_{W^2_\infty(\mathcal{N})}$ and applying Lemma 26 (properties of d_ε) yields

$$|D^2 u| \lesssim \Lambda\left(|\nabla_{\gamma_\varepsilon}u_\varepsilon\circ\mathbf{P}_\varepsilon| + |\nabla^2_{\gamma_\varepsilon}u_\varepsilon\circ\mathbf{P}_\varepsilon|\right).$$

We reduce the computation of integrals in the bulk $\mathcal{N}(\delta)$ to integrals on parallel surfaces $\gamma_\varepsilon(s) := \{\mathbf{x}\in\mathbb{R}^{n+1} : d_\varepsilon(\mathbf{x}) = s\}$ via the coarea formula (63). Since $|\nabla d_\varepsilon| = 1$ in view of (61) the coarea formula implies

$$\begin{aligned}\int_{\mathcal{N}(\delta)}|D^2u(\mathbf{x})|^2 d\mathbf{x} &\lesssim \Lambda^2\int_{\mathcal{N}(\delta)}\sum_{k=1}^{2}|\nabla^k_{\gamma_\varepsilon}u_\varepsilon(\mathbf{P}_\varepsilon(\mathbf{x}))|^2\,|\nabla d_\varepsilon(\mathbf{x})|\,d\mathbf{x}\\ &\le \Lambda^2\int_{-\delta_\varepsilon}^{\delta_\varepsilon}\int_{\gamma_\varepsilon(s)}\sum_{k=1}^{2}|\nabla^k_{\gamma_\varepsilon}u_\varepsilon(\mathbf{P}_\varepsilon(\mathbf{x}))|^2\,d\sigma_{\varepsilon,s}(\mathbf{x})\,ds\\ &\lesssim \delta\Lambda^2\int_{\gamma_\varepsilon}\sum_{k=1}^{2}|\nabla^k_{\gamma_\varepsilon}u_\varepsilon(\mathbf{x})|^2\,d\sigma_\varepsilon(\mathbf{x}),\end{aligned}$$

Lemma 17 (norm equivalence) immediately yields

$$\int_{\gamma_\varepsilon}|\nabla_{\gamma_\varepsilon}u_\varepsilon(\mathbf{P}_\varepsilon(\mathbf{x}))|^2\,d\sigma_\varepsilon(\mathbf{x}) \lesssim \int_\gamma|\nabla_\gamma\widetilde{u}(\mathbf{x})|^2 d\sigma(\mathbf{x}).$$

In order to relate second derivatives of u_ε on γ_ε to those of $\widetilde{u}$ on γ, we apply (50) with γ_ε playing the role of γ and $\Gamma = \gamma$. Then

$$\nabla_{\gamma_\varepsilon} u_\varepsilon(\mathbf{P}_\varepsilon(\mathbf{x})) = (\mathbf{I} - d_\varepsilon \mathbf{W}_\varepsilon)^{-1}(\mathbf{x}) \left(\mathbf{I} - \frac{\boldsymbol{\nu}_\gamma(\mathbf{x}) \otimes \boldsymbol{\nu}_\varepsilon(\mathbf{x})}{\boldsymbol{\nu}_\gamma(\mathbf{x}) \cdot \boldsymbol{\nu}_\varepsilon(\mathbf{x})} \right) \nabla_\gamma \widetilde{u}(\mathbf{x}) \quad \mathbf{x} \in \gamma,$$

and after applying this formula again to $\nabla_{\gamma_\varepsilon} u_\varepsilon(\mathbf{P}_\varepsilon(\mathbf{x}))$ we obtain

$$|D^2_{\gamma_\varepsilon} u_\varepsilon(\mathbf{P}_\varepsilon(\mathbf{x}))| \leq |D_\gamma \mathbf{M}(\mathbf{x})|\, |\nabla_\gamma \widetilde{u}(\mathbf{x})| + |\mathbf{M}(\mathbf{x})|\, |D^2_\gamma \widetilde{u}(\mathbf{x})|,$$

where $\mathbf{M}(\mathbf{x}) = (\mathbf{I} - d_\varepsilon \mathbf{W}_\varepsilon)^{-1}(\mathbf{x}) \left(\mathbf{I} - \frac{\boldsymbol{\nu}_\gamma(\mathbf{x}) \otimes \boldsymbol{\nu}_\varepsilon(\mathbf{x})}{\boldsymbol{\nu}_\gamma(\mathbf{x}) \cdot \boldsymbol{\nu}_\varepsilon(\mathbf{x})} \right)$. We thus wish to bound $\|M\|_{W^1_\infty(\gamma)}$. First we note that combining the bound on the Hausdorff distance between γ and γ_ε from Lemma 25 (properties of $\overline{d}_\varepsilon$) with $\|d_\varepsilon\|_{W^2_\infty(\mathcal{N}(\delta))} \lesssim |d|_{W^2_\infty(\mathcal{N})}$ from Lemma 26 (properties of d_ε) yields for $\mathbf{x} \in \gamma$ that the eigenvalues of $d_\varepsilon(\mathbf{x})\mathbf{W}_\varepsilon(\mathbf{x})$ are bounded by $C\varepsilon^2 |d|^2_{W^2_\infty(\overline{\mathcal{N}})}$, which is less than $\frac{1}{2}$ under the assumption that $\varepsilon |d|_{W^2_\infty(\mathcal{N})}$ is sufficiently small; thus $\| (\mathbf{I} - d_\varepsilon \mathbf{W}_\varepsilon)^{-1} \|_{L_\infty(\mathcal{N}(\delta))} \leq 2$. In addition, combining the same assumption with $\varepsilon \simeq \delta$ and Lemma 26 yields

$$\|1 - \boldsymbol{\nu}_\gamma \cdot \boldsymbol{\nu}_\varepsilon\|_{L_\infty(\mathcal{N}(\delta))} \lesssim \delta^2 |d|^2_{W^2_\infty(\mathcal{N})} \lesssim \varepsilon^2 |d|^2_{W^2_\infty(\mathcal{N})} \leq \frac{1}{2},$$

so that $\boldsymbol{\nu}_\gamma \cdot \boldsymbol{\nu}_\varepsilon \geq 1/2$ and

$$\left\| \mathbf{I} - \frac{\boldsymbol{\nu}_\gamma \otimes \boldsymbol{\nu}_\varepsilon}{\boldsymbol{\nu}_\gamma \cdot \boldsymbol{\nu}_\varepsilon} \right\|_{L_\infty(\mathcal{N}(\delta))} \lesssim 1;$$

thus $\| \mathbf{M} \|_{L_\infty(\mathcal{N}(\delta))} \lesssim 1$. In order to bound the derivatives of $\mathbf{M}$, we note that for a matrix $\mathbf{A}$ there holds $\partial_i \mathbf{A}^{-1} = -\mathbf{A}^{-1}(\partial_i \mathbf{A})\mathbf{A}^{-1}$. For $\mathbf{A} = \mathbf{I} - d_\varepsilon \mathbf{W}_\varepsilon$, we use Lemmas 25 and 26, $|\nabla d_\varepsilon| = 1$, and the assumption $C_1 \delta \leq \varepsilon$ to deduce in $\mathcal{N}(\delta)$

$$\begin{aligned} |\partial_i \mathbf{A}| &= |(\partial_i d_\varepsilon)\mathbf{W}_\varepsilon + d_\varepsilon\, \partial_i \mathbf{W}_\varepsilon| \\ &\lesssim \|d_\varepsilon\|_{W^2_\infty(\mathcal{N}(\delta)} + \delta\, \|d_\varepsilon\|_{W^3_\infty(\mathcal{N})} \lesssim |d|_{W^2_\infty(\mathcal{N})}. \end{aligned}$$

Since we have already established that $\| \mathbf{A}^{-1} \|_{\mathbf{L}_\infty(\mathcal{N}(\delta))} \lesssim \mathbf{1}$, we infer that $|(\mathbf{I} - d_\varepsilon \mathbf{W}_\varepsilon)^{-1}|_{W^1_\infty(\mathcal{N}(\delta)} \lesssim |d|_{W^2_\infty(\mathcal{N})}$. A similar calculation for $\mathbf{I} - \frac{\boldsymbol{\nu}_\gamma \otimes \boldsymbol{\nu}_\varepsilon}{\boldsymbol{\nu}_\gamma \cdot \boldsymbol{\nu}_\varepsilon}$, while recalling that $\boldsymbol{\nu}_\gamma \cdot \boldsymbol{\nu}_\varepsilon \geq 1/2$, yields $|\mathbf{M}|_{W^1_\infty(\gamma)} \lesssim |d|_{W^2_\infty(\mathcal{N})}$ and, after applying Lemma 17 (norm equivalence), gives

$$\|D^2_{\gamma_\varepsilon} u_\varepsilon\|_{L_2(\gamma_\varepsilon)} \lesssim |d|_{W^2_\infty(\mathcal{N})} \left(\|\nabla_\gamma \widetilde{u}\|_{L_2(\gamma)} + \|D^2_\gamma \widetilde{u}\|_{L_2(\gamma)} \right).$$

The asserted estimate follows from applying again the coarea formula (63), which leads to

$$\int_{\mathcal{N}(\delta)} |u|^2 + |\nabla u|^2 + |D^2 u|^2 \lesssim \delta \Lambda^2 \int_\gamma |\widetilde{u}|^2 + |\nabla_\gamma \widetilde{u}|^2 + |D^2 \widetilde{u}|^2.$$

Finally, we take $\mathbf{x} \in \gamma$, note that $\mathbf{Q}_\varepsilon(\mathbf{P}_\varepsilon(\mathbf{x})) = \mathbf{x}$, and compute

$$u(\mathbf{x}) = \widetilde{u} \circ \mathbf{Q}_\varepsilon \circ \mathbf{P}_\varepsilon(\mathbf{x}) = \widetilde{u}(\mathbf{x})$$

to realize that u is indeed an extension of $\widetilde{u}$ to $\mathcal{N}(\delta)$. □

We now derive the elliptic PDE's satisfied by u_ε on γ_ε and u in $\mathcal{N}(\delta)$. For $\widetilde{u} \in H^2(\gamma)$, let $\widetilde{f} = -\Delta_\gamma \widetilde{u} \in L_{2,\#}(\gamma)$ and consider the extension $\widetilde{f}_\varepsilon$ to γ_ε

$$\widetilde{f}_\varepsilon := \widetilde{f} \circ \mathbf{Q}_\varepsilon.$$

Lemma 29 (PDE satisfied by u_ε). *If γ is closed and of class C^2, then γ_ε is also closed and of class C^∞, and the extension $u_\varepsilon = \widetilde{u} \circ \mathbf{Q}_\varepsilon$ satisfies on γ_ε*

$$-\widetilde{\mu}_\varepsilon \operatorname{div}_{\gamma_\varepsilon}\left(\frac{1}{\widetilde{\mu}_\varepsilon} \widetilde{\mathbf{A}}_\varepsilon \nabla_{\gamma_\varepsilon} u_\varepsilon\right) = \widetilde{f}_\varepsilon,$$

where $\widetilde{\mathbf{A}}_\varepsilon := (\boldsymbol{I} - d_\varepsilon D^2 d_\varepsilon)\Pi(\boldsymbol{I} - d_\varepsilon D^2 d_\varepsilon) \circ \mathbf{Q}_\varepsilon$, Π stands for the orthogonal projection $\Pi = (\mathbf{I} - \nabla d \otimes \nabla d)$ on γ and $\widetilde{\mu}_\varepsilon := \frac{q_\varepsilon}{q \circ \mathbf{Q}_\varepsilon}$ reads

$$\widetilde{\mu}_\varepsilon = \det(\boldsymbol{I} - d_\varepsilon D^2 d_\varepsilon)(\nabla d \cdot \nabla d_\varepsilon) \circ \boldsymbol{Q}_\varepsilon.$$

Proof. Given $\widetilde{v} \in H^1(\gamma)$, let $v = \widetilde{v} \circ \mathbf{Q}_\varepsilon \in H^1(\gamma_\varepsilon)$. We resort to (49) to write

$$\nabla_\gamma \widetilde{u} = \Pi(\mathbf{I} - d_\varepsilon D^2 d_\varepsilon) \nabla_{\gamma_\varepsilon} u_\varepsilon \circ \mathbf{P}_\varepsilon \quad \text{on } \gamma,$$

because $\nabla_{\gamma_\varepsilon} u_\varepsilon \circ \mathbf{P}_\varepsilon = \Pi_\varepsilon \nabla_{\gamma_\varepsilon} u_\varepsilon \circ \mathbf{P}_\varepsilon$. This combined with (8) and Corollary 13 (integration by parts) on the closed surface γ_ε yields

$$\int_\gamma \nabla_\gamma \widetilde{u} \cdot \nabla_\gamma \widetilde{v} = \int_{\gamma_\varepsilon} \frac{1}{\widetilde{\mu}_\varepsilon} \widetilde{\mathbf{A}}_\varepsilon \nabla_{\gamma_\varepsilon} u_\varepsilon \cdot \nabla_{\gamma_\varepsilon} v = -\int_{\gamma_\varepsilon} \operatorname{div}_{\gamma_\varepsilon}\left(\frac{1}{\widetilde{\mu}_\varepsilon} \widetilde{\mathbf{A}}_\varepsilon \nabla_{\gamma_\varepsilon} u_\varepsilon\right) v$$

with $\widetilde{\mu}_\varepsilon = \frac{q_\varepsilon}{q \circ \mathbf{Q}_\varepsilon}$ given by (53). Likewise,

$$\int_\gamma \widetilde{f}\, \widetilde{v} = \int_{\gamma_\varepsilon} \frac{1}{\widetilde{\mu}_\varepsilon} \widetilde{f}_\varepsilon v.$$

Since the last two equalities hold for all $v \in H^1(\gamma_\varepsilon)$, the assertion follows. □

We extend the function $\widetilde{f}_\varepsilon$ to $\mathcal{N}_\varepsilon(\delta_\varepsilon)$ as follows:

$$f_\varepsilon := \widetilde{f}_\varepsilon \circ \mathbf{P}_\varepsilon = \widetilde{f} \circ \mathbf{Q}_\varepsilon \circ \mathbf{P}_\varepsilon.$$

Equivalently, given $\mathbf{x} \in \mathcal{N}_\varepsilon(\delta_\varepsilon)$ let $\widetilde{\mathbf{x}} \in \gamma$ be the unique point such that for some s

$$\widetilde{\mathbf{x}} = \mathbf{x} + s\nabla d_\varepsilon(\mathbf{x}) \;\Rightarrow\; f_\varepsilon(\mathbf{x}) = \widetilde{f}(\widetilde{\mathbf{x}}).$$

Proposition 30 (PDE satisfied by u). *Let ε and δ be as in Corollary 27 (property of $\mathbf{P}_\varepsilon$). The extension $u \in H^2(\mathcal{N}(\delta))$ of $\widetilde{u}$ of Proposition 28 satisfies the PDE*

$$-\frac{1}{\mu_\varepsilon}\operatorname{div}(\mu_\varepsilon \boldsymbol{B}_\varepsilon \nabla u) = f_\varepsilon \quad \text{in} \quad \mathcal{N}(\delta),$$

where

$$\boldsymbol{B}_\varepsilon := (\boldsymbol{I} - d_\varepsilon D^2 d_\varepsilon)^{-1}\Pi_\varepsilon \boldsymbol{A}_\varepsilon \Pi_\varepsilon (\boldsymbol{I} - d_\varepsilon D^2 d_\varepsilon)^{-1},$$

$\boldsymbol{A}_\varepsilon := \widetilde{\boldsymbol{A}}_\varepsilon \circ \boldsymbol{P}_\varepsilon$ with $\widetilde{\mathbf{A}}_\varepsilon$ given in Lemma 29, $\Pi_\varepsilon = \boldsymbol{I} - \nabla d_\varepsilon \otimes \nabla d_\varepsilon$, μ_ε is given by

$$\mu_\varepsilon := \frac{1}{\widetilde{\mu}_\varepsilon \circ \boldsymbol{P}_\varepsilon} \det\left(\boldsymbol{I} - d_\varepsilon D^2 d_\varepsilon\right),$$

and $\widetilde{\mu}_\varepsilon$ is defined in Lemma 29.

Proof. We proceed as in Proposition 28 (H^2 extension). Let $\gamma_\varepsilon(s)$ be a parallel surface to γ_ε at distance s, and let $|s| \leq \delta_\varepsilon$ with $\delta_\varepsilon = \frac{3}{2}\delta$ so that (61) holds. We first employ (50) to obtain the bilinear form for u on $\gamma_\varepsilon(s)$. For δ sufficiently small Lemma 26 (properties of d_ε) guarantees that $(\mathbf{I} - d_\varepsilon D^2 d_\varepsilon)$ is invertible in $\mathcal{N}_\varepsilon(\delta_\varepsilon)$. Hence, if $\mathbf{D}_\varepsilon = (\mathbf{I} - d_\varepsilon D^2 d_\varepsilon)^{-1}\Pi_\varepsilon$ and $v \in C_0^\infty(\mathcal{N}(\delta))$, we restrict v to $\gamma_\varepsilon(s)$, define the auxiliary function $\widetilde{v} := v \circ \mathbf{P}_\varepsilon^{-1} \in C^\infty(\gamma_\varepsilon)$ and observe that (50) reads on $\gamma_\varepsilon(s)$

$$\nabla_{\gamma_\varepsilon} \widetilde{v} \circ \mathbf{P}_\varepsilon = \mathbf{D}_\varepsilon \nabla v,$$

where ∇v is the full gradient of v; this is because of the presence of the projection matrix Π_ε on the tangent hyperplane to $\gamma_\varepsilon(s)$ in the definition of $\mathbf{D}_\varepsilon$. We get

$$\int_{\gamma_\varepsilon} \frac{1}{\widetilde{\mu}_\varepsilon} \widetilde{\mathbf{A}}_\varepsilon \nabla_{\gamma_\varepsilon} u_\varepsilon \cdot \nabla_{\gamma_\varepsilon} \widetilde{v} = \int_{\gamma_\varepsilon(s)} \mu_\varepsilon \mathbf{A}_\varepsilon \mathbf{D}_\varepsilon \nabla u \cdot \mathbf{D}_\varepsilon \nabla v$$

where $\widetilde{\mu}_\varepsilon$ is given in Lemma 29 (PDE satisfied by u_ε) and μ_ε is the surface measure density on $\gamma_\varepsilon(s)$ due to the change of variables, namely

$$\mu_\varepsilon = \frac{1}{\widetilde{\mu}_\varepsilon \circ \mathbf{P}_\varepsilon} \frac{q_\varepsilon}{q_{\varepsilon,s}} = \frac{1}{\widetilde{\mu}_\varepsilon \circ \mathbf{P}_\varepsilon} \det\left(\mathbf{I} - d_\varepsilon D^2 d_\varepsilon\right)$$

according to (53). Similarly, the linear form for the forcing reads

$$\int_{\gamma_\varepsilon} \frac{1}{\widetilde{\mu}_\varepsilon} \widetilde{f}_\varepsilon \widetilde{v} = \int_{\gamma_\varepsilon(s)} \mu_\varepsilon f_\varepsilon v.$$

Since the left-hand sides of the previous integral expressions coincide, in view of Lemma 29, we now integrate over $s \in (-\delta_\varepsilon, \delta_\varepsilon)$ and use the coarea formula (63) to convert the resulting integrals into bulk integrals

$$\begin{aligned}
\int_{\mathcal{N}_\varepsilon(\delta_\varepsilon)} \mu_\varepsilon \mathbf{A}_\varepsilon \mathbf{D}_\varepsilon \nabla u \cdot \mathbf{D}_\varepsilon \nabla v &= \int_{\mathcal{N}_\varepsilon(\delta_\varepsilon)} \mu_\varepsilon \mathbf{A}_\varepsilon \mathbf{D}_\varepsilon \nabla u \cdot \mathbf{D}_\varepsilon \nabla v \, |\nabla d_\varepsilon| \\
&= \int_{-\delta_\varepsilon}^{\delta_\varepsilon} \int_{\gamma_\varepsilon(s)} \mu_\varepsilon \mathbf{A}_\varepsilon \mathbf{D}_\varepsilon \nabla u \cdot \mathbf{D}_\varepsilon \nabla v \, d\sigma_{\varepsilon,s} \, ds \\
&= \int_{-\delta_\varepsilon}^{\delta_\varepsilon} \int_{\gamma_\varepsilon(s)} f_\varepsilon v \, \mu_\varepsilon \, d\sigma_{\varepsilon,s} \, ds \\
&= \int_{\mathcal{N}_\varepsilon(\delta_\varepsilon)} f_\varepsilon v \, \mu_\varepsilon \, |\nabla d_\varepsilon| = \int_{\mathcal{N}_\varepsilon(\delta_\varepsilon)} f_\varepsilon v \, \mu_\varepsilon,
\end{aligned}$$

because $|\nabla d_\varepsilon| = 1$ in $\mathcal{N}_\varepsilon(\delta_\varepsilon)$. Since $\mathcal{N}(\delta) \subset \mathcal{N}_\varepsilon(\delta_\varepsilon)$ according to (61), integration by parts gives

$$-\int_{\mathcal{N}(\delta)} \operatorname{div}(\mu_\varepsilon \mathbf{D}_\varepsilon \mathbf{A}_\varepsilon \mathbf{D}_\varepsilon \nabla u) v = \int_{\mathcal{N}(\delta)} f_\varepsilon v \mu_\varepsilon \quad \forall v \in C_0^\infty(\mathcal{N}(\delta)),$$

whence the desired PDE follows after noticing that $(\mathbf{I} - d_\varepsilon D^2 d_\varepsilon)^{-1}$ and Π_ε commute. This completes the proof. □

4 Parametric finite element method

The parametric method hinges on a surface approximation Γ "interpolating" the exact surface γ. Recall that the latter is assumed to be a closed, compact, orientable hypersurface in $\mathbb{R}^{n+1}$. In the lowest order case of piecewise linear polynomials, this corresponds to a polyhedral surface Γ whose vertices lie on γ or, more generally, sufficiently close to γ. The finite element space is then obtained in a classical way by mapping a finite element triplet defined on a reference element in $\mathbb{R}^n$ to a facet of Γ in $\mathbb{R}^{n+1}$. The FEM requires a bi-Lipschitz map $\mathbf{P} : \Gamma \to \gamma$ which is not necessarily the distance function lift $\mathbf{P}_d$. The latter is used for numerical analysis purposes only even for smooth surfaces.

There are two sources of error: the approximation of the exact surface γ by the polyhedral surface Γ, the so-called *geometric consistency error*, and the *Galerkin error* arising from the actual finite element approximation on Γ. In this section we quantify these two errors depending on the regularity of γ. For the former we rely on the discussion of Section 3 that addresses the effect of perturbing γ. For γ of class $C^{1,\alpha}$ we deal with a generic lift $\mathbf{P} : \Gamma \to \gamma$ and obtain a suboptimal geometric consistency error. For C^2 surfaces, instead, we resort to $\mathbf{P}_d$ for error analysis and restore geometric optimality.

4.1 FEM on Lipschitz parametric surfaces

4.1.1 Lipschitz parametric surfaces

We adopt the viewpoint that the surface γ is described as the deformation of an n-dimensional polyhedral surface Γ by a globally bi-Lipschitz *homeomorphism* $\mathbf{P}:\Gamma\to\gamma\subset\mathbb{R}^{n+1}$. Thus there exists $L > 0$ such that for all $\mathbf{x}_1$, $\mathbf{x}_2 \in \Gamma$

$$L^{-1}|\mathbf{x}_1-\mathbf{x}_2| \leq |\widetilde{\mathbf{x}}_1-\widetilde{\mathbf{x}}_2| \leq L|\mathbf{x}_1-\mathbf{x}_2|, \quad \widetilde{\mathbf{x}}_i=\mathbf{P}(\mathbf{x}_i), \quad i=1,2. \tag{64}$$

If γ is C^2, we may take $\mathbf{P} = \mathbf{P}_d$, but our current definition allows for much more flexibility in the choice of $\mathbf{P}$. For example, if γ has nonempty boundary and is given as the graph of a function $\psi:\Omega\to\mathbb{R}^{n+1}$ with $\Omega\subset\mathbb{R}^n$, then the map between $\mathbf{x} = (x, z) \in \Gamma$ with $x \in \Omega$ and $z \in \mathbb{R}$ could be given by $\mathbf{P}(x, z) = (x, \psi(x)) \in \gamma$, i.e., the "vertical" graph map.

The (closed) facets of Γ are denoted by T, and form the collection $\mathcal{T}=\{T\}$. We assume that these facets are all simplices and denote by $S_{\mathcal{T}}$ the set of interior faces of $\mathcal{T}$. Extension to other element shapes such as n-quadrilaterals and to nonconforming discretizations is possible under reasonable assumptions with minor modifications, but we do not elaborate them further. We let $\mathbf{P}_T : T \to \mathbb{R}^{n+1}$ be the restriction of $\mathbf{P}$ to T. The partition $\mathcal{T}$ of Γ induces the partition $\widetilde{\mathcal{T}} = \{T\}_{T\in\mathcal{T}}$ of γ upon setting

$$\widetilde{T} := \mathbf{P}_T(T) \;\; \forall T \in \mathcal{T}.$$

Note that this *nonoverlapping* parametrization of γ allows for Lipschitz surfaces rougher than globally C^2. We additionally define *macropatches*

$$\omega_T = \cup\{T' : T' \cap T \neq \emptyset\}, \quad \widetilde{\omega}_T = \mathbf{P}(\omega_T). \tag{65}$$

Let $h_T := |T|^{\frac{1}{n}}$ and $\sigma < \infty$ be the triangulation shape-regularity constant, i.e.

$$\sigma := \sup_{\mathcal{T}} \max_{T\in\mathcal{T}} \frac{\operatorname{diam}(T)}{h_T}. \tag{66}$$

We further assume that the number of elements in each patch ω_T is uniformly bounded. This assumption automatically follows from shape regularity for

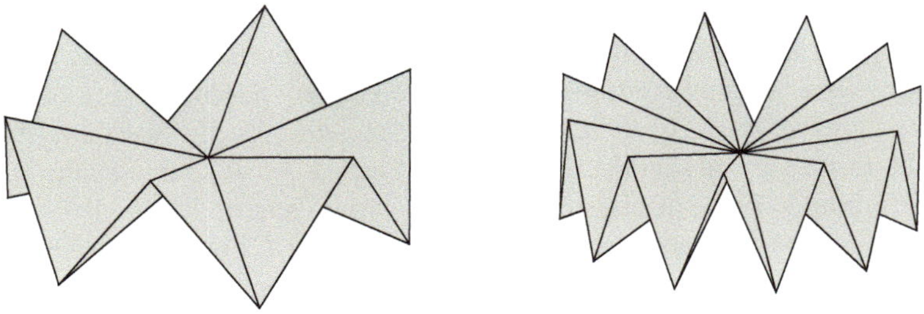

FIG. 1 Two different configurations when $n = 2$ illustrating that the number of elements sharing the same vertex could be arbitrarily large even when using triangles satisfying (67).

triangulations of Euclidean domains, but the situation is more subtle for surface triangulations as illustrated in Fig. 1. Such a bound does for example hold if Γ is obtained by systematic refinement of an initial surface mesh with a uniform bound on the number of elements in a patch (Demlow and Dziuk, 2007), or more generally using adaptive refinement strategies (Bonito et al., 2013, 2016). In addition, this implies that all elements in ω_T have uniformly equivalent diameters, as it happens for shape-regular triangulations on Euclidean domains.

To provide a parametric description, let $\widehat{T}$ be the unit reference simplex, sometimes called the *universal parametric domain*. We denote by $\mathbf{X}_T : \mathbb{R}^n \to \mathbb{R}^{n+1}$ the affine map such that $T = \mathbf{X}_T(\widehat{T})$ and note that (66) implies

$$h_T|\mathbf{w}| \lesssim |D\mathbf{X}_T\mathbf{w}| \lesssim h_T|\mathbf{w}|, \quad \forall \mathbf{w} \in \mathbb{R}^n. \tag{67}$$

Hereafter we omit to specify when the constants (possibly hidden in $\lesssim$ signs) depend on σ. As pointed out in Bonito and Pasciak (2012), even if the initial surface approximation satisfies (67), this property might not hold for refinements unless the initial polyhedral surface approximates the exact surface well. We refer to Bonito et al. (2013, 2016) for a discussion on how to circumvent this in an adaptive strategy. However, since this work focusses on a priori and a posteriori error estimation rather that adaptivity, we assume (67) directly.

We are now ready to introduce the local *nonoverlapping* parametrization χ of γ. Let $\chi_T := \mathbf{P} \circ \mathbf{X}_T : \widehat{T} \to \widetilde{T}$ be the corresponding local parametrization of $\widetilde{T}$ and $\chi := \{\chi_T\}_{T \in \mathcal{T}}$; see Fig. 2. We record for latter use that thanks to the Lipschitz properties (64) and (67), χ_T also satisfies

$$h_T|\mathbf{w}| \lesssim |D\chi_T(\mathbf{y})\mathbf{w}| \lesssim h_T|\mathbf{w}| \quad \forall \mathbf{w} \in \mathbb{R}^n, \ \mathbf{y} \in \widehat{T}. \tag{68}$$

It turns out that it will be convenient to consider χ_T to be defined on a larger domain than T, say $\widehat{\omega}_T \subset \mathbb{R}^n$, so that $\chi_T = \mathbf{P} \circ \mathbf{X}_T : \widehat{\omega}_T \to \widetilde{\omega}_T$ is a bi-Lipschitz local parametrization of γ: there exists a universal constant $L \geq 1$ such that for each fixed $T \in \mathcal{T}$ and for all $\widetilde{\mathbf{x}}_1 = \chi_T(\mathbf{y}_1), \widetilde{\mathbf{x}}_2 = \chi_T(\mathbf{y}_2) \in \widetilde{\omega}_T$,

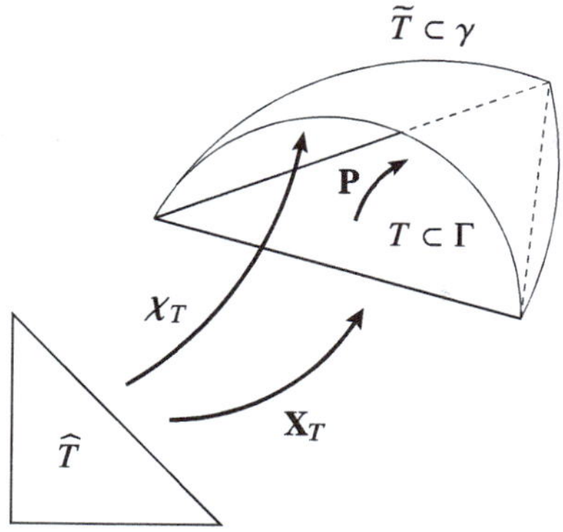

FIG. 2 Nonoverlapping parametrizations $\mathbf{X}_T : \widehat{T} \to T$ of Γ and $\chi_T : \widehat{T} \to \widetilde{T}$ of γ.

$$L^{-1} h_T |\mathbf{y}_1 - \mathbf{y}_2| \leq |\widetilde{\mathbf{x}}_1 - \widetilde{\mathbf{x}}_2| \leq L h_T |\mathbf{y}_1 - \mathbf{y}_2|; \tag{69}$$

in this case $\chi := \{\chi_T\}_{T \in \mathcal{T}}$ is an *overlapping* parametrization of γ. We further assume that $\mathbf{P}(\mathbf{v}) = \mathbf{v}$ for all vertices $\mathbf{v}$ of Γ, so that $\mathbf{X}_T$ is the nodal Lagrange interpolant of χ_T into linears.

We finally note that a function $\widetilde{v}_T : \widetilde{T} \to \mathbb{R}$ defines uniquely two functions $\widehat{v}_T : \widehat{T} \to \mathbb{R}$ and $v_T : T \to \mathbb{R}$ via the maps χ_T and $\mathbf{P}$, namely

$$\widehat{v}_T(\mathbf{y}) := \widetilde{v}_T(\chi_T(\mathbf{y})) \;\; \forall\, \widehat{\mathbf{x}} \in \widehat{T} \quad \text{and} \quad v_T(\mathbf{x}) := \widetilde{v}_T(\mathbf{P}(\mathbf{x})) \;\; \forall\, \mathbf{x} \in T. \tag{70}$$

Moreover, each one of these functions induces the other two uniquely. Accordingly, we will use the symbol v for all three functions if no confusion arises.

4.1.2 Differential geometry on polyhedral surfaces

We use the atlas $\{\widehat{T}, \widetilde{T}, \chi_T\}_{T \in \mathcal{T}}$, induced by the nonoverlapping parametrization $\chi := \{\chi_T\}_{T \in \mathcal{T}}$, to describe γ in the spirit of Section 2. Likewise, we employ the atlas $\{\widehat{T}, T, \mathbf{X}_T\}_{T \in \mathcal{T}}$ to describe the polyhedral surface Γ. In view of (68), the discrete first fundamental form $\mathbf{g}_T$ and area element q_T of Γ are given elementwise by

$$\mathbf{g}_T := (D\mathbf{X}_T)^t D\mathbf{X}_T, \quad q_T := \sqrt{\det \mathbf{g}_T}, \quad \forall\, T \in \mathcal{T}. \tag{71}$$

and satisfy

$$\text{eigen}(\mathbf{g}_T) \approx h_T^2, \quad q_T \approx h_T^n, \quad \forall\, T \in \mathcal{T}. \tag{72}$$

They give rise to the piecewise constant functions $\mathbf{g}_\Gamma := \{\mathbf{g}_T\}_{T \in \mathcal{T}}$ and $q_\Gamma := \{q_T\}_{T \in \mathcal{T}}$. Similar properties are enjoyed by χ, which imply that the stability constant S_χ defined in (38) is purely geometric and independent of mesh size:

$$S_\chi \approx 1. \tag{73}$$

In addition, notice that (68) and (69) imply that

$$C_1 \leq \frac{q}{q_\Gamma} \leq C_2 \tag{74}$$

for constants C_1, C_2 independent of discretization parameters. Moreover, the vector $\mathbf{N}_T := \sum_{i=1}^{n+1} \det([\mathbf{e}_j, D\mathbf{X}_T)])\mathbf{e}_j$ is perpendicular to $T \in \mathcal{T}$ provided $\{\mathbf{e}_j\}_{j=1}^{n+1}$ are the canonical unit vectors of $\mathbb{R}^{n+1}$. This vector satisfies $|\mathbf{N}_T| = q_T$ and yields the unit normal to T

$$\boldsymbol{\nu}_T := \frac{\mathbf{N}_T}{|\mathbf{N}_T|} \quad \forall T \in \mathcal{T},$$

and corresponding piecewise constant unit normal vector $\boldsymbol{\nu}_\Gamma := \{\boldsymbol{\nu}_T\}_{T\in\mathcal{T}}$ to Γ.

Given a function $v:\Gamma \to \mathbb{R}$, its tangential gradient $\nabla_\Gamma v$ and Laplace–Beltrami operator $\Delta_\Gamma v$ over Γ obey the formulas

$$\nabla\widehat{v} = (D\mathbf{X})^t\, \nabla_\Gamma v, \quad \nabla_\Gamma v = (D\mathbf{X})\, \mathbf{g}_\Gamma^{-1}\, \nabla\widehat{v}, \tag{75}$$

and

$$\Delta_\Gamma v = \frac{1}{q_\Gamma} \mathrm{div}\left(q_\Gamma\, \mathbf{g}_\Gamma^{-1}\, \nabla\widehat{v}\right), \tag{76}$$

where $\widehat{v}:\widehat{\mathcal{T}} \to \mathbb{R}$ is defined in (70). The strong form of $\Delta_\Gamma v$ is well defined only elementwise because $q_\Gamma\, \mathbf{g}_\Gamma^{-1}$ is piecewise constant and so discontinuous over $\mathcal{T}$. To find the correct strong form we start from the weak form (19), split the integral over elements and use Corollary 18 (integration by parts) to obtain

$$\begin{aligned} \int_\Gamma \nabla_\Gamma v \cdot \nabla_\Gamma w &= \sum_{T\in\mathcal{T}} -\int_T w\Delta_\Gamma v + \int_{\partial T} w\nabla_\Gamma v \cdot \boldsymbol{\mu}_T \\ &= \sum_{T\in\mathcal{T}} -\int_T w\Delta_\Gamma v + \sum_{S\in\mathcal{S}} \int_S w[\nabla_\Gamma v], \end{aligned}$$

where the jump residual is computed over each face $S \in \mathcal{S}$ of elements of $\mathcal{T}$ via

$$[\nabla_\Gamma v] := \nabla_\Gamma v_+ \cdot \boldsymbol{\mu}_+ + \nabla_\Gamma v_- \cdot \boldsymbol{\mu}_- \tag{77}$$

and $T_\pm \in \mathcal{T}$ are such that $S = T_+ \cap T_-$ and $v_\pm$, $\boldsymbol{\mu}_\pm$ are the restrictions of v and the outer unit normal to $\partial T_\pm$ which is parallel to $T_\pm$. We then see that $\Delta_\Gamma v$ consists of the absolutely continuous part (76) with respect to surface measure defined elementwise and the singular part (77) supported on the skeleton of $\mathcal{T}$. This formula makes sense for functions which are piecewise H^2 and globally H^1, such as continuous piecewise polynomials.

4.1.3 Parametric finite element method

In this work, we focus on continuous piecewise linear finite elements and polyhedral surface approximations. Let $\mathcal{P}$ be the space of linear polynomials and let $\mathbb{V}(\mathcal{T})$ be the space of continuous piecewise linear polynomial functions over Γ, namely

$$\mathbb{V}(\mathcal{T}) := \left\{ V \in C^0(\Gamma) \mid V|_T = \widehat{V} \circ \mathbf{X}^{-1} \text{ for some } \widehat{V} \in \mathcal{P}, \ \ T \in \mathcal{T} \right\}.$$

The finite element space associated with the Laplace–Beltrami equation over Γ is the restriction of $\mathbb{V}(\mathcal{T})$ to functions with vanishing mean

$$\mathbb{V}_{\#}(\mathcal{T}) := \mathbb{V}(\mathcal{T}) \cap L_{2,\#}(\Gamma).$$

We define $\mathcal{I}_{\mathcal{T}} : C^0(\Gamma) \to \mathbb{V}(\mathcal{T})$ to be the Lagrange interpolation operator and use the same notation for vector-valued functions.

We are now ready to introduce the parametric FEM: seek $U := U_{\mathcal{T}} \in \mathbb{V}_{\#}(\mathcal{T})$ that solves

$$\int_\Gamma \nabla_\Gamma U \cdot \nabla_\Gamma V = \int_\Gamma FV \quad \forall V \in \mathbb{V}_{\#}(\mathcal{T}), \tag{78}$$

where $F \in L_{2,\#}(\Gamma)$ is an approximation of $f \in L_{2,\ \#}(\gamma)$ to be specified later. Lax–Milgram theory guarantees that $U \in \mathbb{V}_{\#}(\mathcal{T})$ is well defined. Observe that because $F \in L_{2,\#}(\Gamma)$, we also have

$$\int_\Gamma \nabla_\Gamma U \cdot \nabla_\Gamma V = \int_\Gamma FV \quad \forall V \in \mathbb{V}(\mathcal{T}). \tag{79}$$

Since the exact problem (19) and discrete problem (79) are defined on different domains γ and Γ, the error $u - U$ does not satisfy Galerkin orthogonality in either one. The next statement accounts for geometric inconsistency and uses the convention (70) for the generic lift $\mathbf{P}$.

Lemma 31 (Galerkin quasi-orthogonality). *Let $\boldsymbol{E}$ and $\boldsymbol{E}_\Gamma$ be defined in (36) and (37) via the parametrizations $\chi = \boldsymbol{P} \circ \boldsymbol{X}$ and $\chi_\Gamma = \boldsymbol{X}$. Then, for all $V \in \mathbb{V}(\mathcal{T})$, there holds*

$$\int_\Gamma \nabla_\Gamma (u - U) \cdot \nabla_\Gamma V = \int_\Gamma \left(f \frac{q}{q_\Gamma} - F \right) V + \int_\Gamma \nabla_\Gamma u \cdot \boldsymbol{E}_\Gamma \nabla_\Gamma V, \tag{80}$$

and

$$\int_\gamma \nabla_\gamma (\widetilde{u} - \widetilde{U}) \cdot \nabla_\gamma \widetilde{V} = \int_\gamma \left(\widetilde{f} - \widetilde{F} \frac{q_\Gamma}{q} \right) \widetilde{V} + \int_\gamma \nabla_\gamma \widetilde{U} \cdot \boldsymbol{E} \nabla_\gamma \widetilde{V}. \tag{81}$$

Proof. We only prove (80) as (81) follows similarly. Using the Eq. (79) satisfied by U and the consistency relation (34), we obtain

$$\int_\Gamma \nabla_\Gamma (u - U) \cdot \nabla_\Gamma V = \int_\gamma \nabla_\gamma \widetilde{u} \cdot \nabla_\gamma \widetilde{V} + \int_\Gamma \nabla_\Gamma u \cdot \mathbf{E}_\Gamma \nabla_\Gamma V - \int_\Gamma FV.$$

The first term on the right-hand side equals $\int_\gamma \widetilde{f} \widetilde{V} = \int_\Gamma f \frac{q}{q_\Gamma} V$, in view of (19) and (8), and thus yields (80). □

4.2 Geometric consistency

In this section we study the error inherent to approximating γ with Γ. The polyhedral surface Γ is always represented by a lift $\mathbf{P}$ whose regularity depends on that of γ. We present two scenarios depending on such regularity. We first assume that γ is piecewise $C^{1,\alpha}$ and globally Lipschitz, and later assume that γ is C^2 and exploit the distance function lift $\mathbf{P}_d$ to improve the error estimates.

4.2.1 *Uniform Poincaré–Friedrichs estimate on Γ*

The analysis below takes advantage of the *uniform Poincaré–Friedrichs estimate on Γ*

$$\|v\|_{L_2(\Gamma)} \lesssim \|\nabla v\|_{L_2(\Gamma)} \quad \forall v \in H^1_{\#}(\Gamma), \tag{82}$$

where the constant hidden in the above inequality is independent of Γ. Note that when γ is of class $C^{1,\alpha}$, $0 < \alpha \leq 1$, Lemma 18 (uniform Poincaré–Friedrichs constant) implies that (82) follows from (73) and (74), which in turn are consequences of assumption (64). Furthermore, when γ is of class C^2 and $\mathbf{P} = \mathbf{P}_d$, the discussion in Section 3.2 yields conditions which are also easy to verify: $\Gamma \subset \mathcal{N}(1/2K_\infty)$ and $\boldsymbol{\nu} \cdot \boldsymbol{\nu}_\Gamma \geq c > 0$ on Γ.

4.2.2 *Geometric estimators*

Since γ is described by χ and Γ by $\mathbf{X}$ it is natural to consider the difference $D\chi - D\mathbf{X}$ as a measure of geometric mismatch (Bonito et al., 2016). We thus start with the *geometric element indicator*

$$\lambda_T := \|D(\mathbf{P} - \mathcal{I}_T\mathbf{P})\|_{L_\infty(T)} = \|D\mathbf{P} - \mathbf{I}\|_{L_\infty(T)} \quad \forall T \in \mathcal{T} \tag{83}$$

and the corresponding *geometric estimator*

$$\lambda_{\mathcal{T}}(\Gamma) := \max_{T\in\mathcal{T}} \lambda_T. \tag{84}$$

We have seen that the relative measure of accuracy (39) controls the geometric error. In this vein, we observe that $D\chi_T = D\mathbf{P}\, D\mathbf{X}_T$ because $\chi_T = \mathbf{P} \circ \mathbf{X}_T$, whence such measure satisfies

$$\max_{\mathbf{y}\in\widehat{T}} \frac{|D(\chi_T - \mathbf{X}_T)(\mathbf{y})|}{\min\{|D^-\chi_T(\mathbf{y})|, |D^-\mathbf{X}_T(\mathbf{y})|\}} \leq S_\chi \lambda_T \quad \forall T \in \mathcal{T}, \tag{85}$$

with a stability constant $S_\chi \approx 1$ according to (38). Therefore $\lambda_{\mathcal{T}}(\Gamma)$ is expected to dominate the geometric error for surfaces of class $C^{1,\alpha}$ with $0 < \alpha < 1$. This is consistent with Lemma 19 (perturbation error estimate for $C^{1,\alpha}$ surfaces).

For C^2 surfaces, however, $\lambda_{\mathcal{T}}(\Gamma)$ is suboptimal in that it overestimates the influence of geometry (Bonito and Demlow, 2019). According to Lemma 23 (perturbation error estimate for C^2 surfaces) and Lemma 24 (error estimates for normals), the following quantities should play a crucial role in dealing with geometry via the auxiliary lift $\mathbf{P}_d$

$$\beta_T := \|\mathbf{P} - \mathcal{I}_T \mathbf{P}\|_{L_\infty(T)}, \quad \beta_{\mathcal{T}}(\Gamma) := \max_{T\in\mathcal{T}} \beta_T, \tag{86}$$

and

$$\mu_T := \beta_T + \lambda_T^2, \quad \mu_{\mathcal{T}}(\Gamma) := \max_{T\in\mathcal{T}} \mu_T; \tag{87}$$

we stress that $\mu_{\mathcal{T}}(\Gamma)$ is formally of higher order than $\lambda_{\mathcal{T}}(\Gamma)$. We will show below that $\mu_{\mathcal{T}}(\Gamma)$ indeed controls the geometric error and accounts for the "superconvergence" property associated with the projection $\mathbf{P}_d$ along the normal direction to γ alluded to at the end of Section 3.2.

4.2.3 Geometric consistency error for $C^{1,\alpha}$ surfaces

We now quantify the geometric error incurred when replacing γ by its polygonal approximation Γ.

Corollary 32 (*geometric consistency errors for $C^{1,\alpha}$ surfaces*). *If $\boldsymbol{X}$ and $\boldsymbol{\chi}$ satisfy (67) and (68), then for all $T\in\mathcal{T}$ we have*

$$\|1 - q^{-1} q_\Gamma\|_{L_\infty(\widehat{T})}, \quad \|\boldsymbol{I} - \boldsymbol{g}_\Gamma \boldsymbol{g}^{-1}\|_{L_\infty(\widehat{T})}, \quad \|\boldsymbol{\nu}_\Gamma - \boldsymbol{\nu}\|_{L_\infty(T)} \lesssim \lambda_T, \tag{88}$$

where the hidden constants depend on $S_\chi \approx 1$ defined in (38). Moreover,

$$\|\boldsymbol{E}\|_{L_\infty(\widehat{T})} + \|\boldsymbol{E}_\Gamma\|_{L_\infty(\widehat{T})} \lesssim \lambda_T \quad \forall T\in\mathcal{T}. \tag{89}$$

Proof. We first point out that (67) and (68) yield $S_\chi \approx 1$ according to (73). The asserted estimates follow from Lemma 16 (error estimates for $\mathbf{g}$ and q) and Lemma 24 (error estimate for normals) in conjunction with (47) and (85). □

4.2.4 Geometric consistency error for C^2 surfaces

We now take advantage of the lift $\mathbf{P}_d$ for error representation. We recall that, as in Section 3.2, the parametrizations of γ and Γ are given by $\boldsymbol{\chi} = \mathbf{P}_d \circ \mathbf{X}$ and $\mathbf{X}$. In particular, the infinitesimal area element q of γ is defined using $\mathbf{P}_d$ and so are the consistency matrices $\mathbf{E}$, $\mathbf{E}_\Gamma$; see (51) and (52). To improve upon Corollary 32 (geometric consistency errors for $C^{1,\alpha}$ surfaces) we need more stringent geometric assumptions than simply $\Gamma \subset \mathcal{N}$. These assumptions are somewhat technical but are checkable computationally with information extracted from $\mathbf{P}$ but without access to $\mathbf{P}_d$ (Bonito and Demlow, 2019). We list them now.

- **Distance between γ and Γ.** Invoking the closest point property of the distance function projection $\mathbf{P}_d$ and the definition (86) of $\beta_{\mathcal{T}}(\Gamma)$, we see that $|\mathbf{x}-\mathbf{P}_d(\mathbf{x})| \leq |\mathbf{x}-\mathbf{P}(\mathbf{x})| \leq \beta_{\mathcal{T}}(\Gamma)$ for all $\mathbf{x} \in \Gamma$. We thus assume that Γ is sufficiently close to γ in the sense that

$$\beta_{\mathcal{T}}(\Gamma) < \frac{1}{2K_\infty} \quad \Rightarrow \quad \Gamma \subset \mathcal{N}, \tag{90}$$

according to (48). Therefore, the estimates of Section 3.2 are valid. Moreover, the discrepancy between the two lifts satisfies for all $T \in \mathcal{T}$

$$|\mathbf{P}(\mathbf{x})-\mathbf{P}_d(\mathbf{x})| \leq |\mathbf{P}(\mathbf{x})-\mathbf{x}|+|\mathbf{x}-\mathbf{P}_d(\mathbf{x})| \leq 2|\mathbf{x}-\mathbf{P}(\mathbf{x})| \leq 2\beta_T \quad \forall \mathbf{x} \in T.$$

- **Mismatch between $\mathbf{P}$ and $\mathbf{P}_d$.** We assume that

$$\mathbf{P}_d \circ \mathbf{P}^{-1}(\widetilde{T}) \subset \widetilde{\omega}_T \quad \forall T \in \mathcal{T}, \tag{91}$$

where $\widetilde{\omega}_T$ is the patch around $\widetilde{T}$ within γ. If $\widetilde{\mathbf{x}} = \mathbf{P}(\mathbf{x}) \in \gamma$ for $\mathbf{x} \in \Gamma$, then

$$|\widetilde{\mathbf{x}} - \mathbf{P}_d \circ \mathbf{P}^{-1}(\widetilde{\mathbf{x}})| = |\mathbf{P}(\mathbf{x})-\mathbf{P}_d(\mathbf{x})| \leq 2\beta_T \quad \forall \mathbf{x} \in T. \tag{92}$$

and all $T \in \mathcal{T}$. Since γ is of class C^2, we expect $\frac{\beta_T}{h_T} \to 0$ as $h_T \to 0$ and realize that (91) is always valid asymptotically. We emphasize that it is possible to check (91) computationally without accessing $\mathbf{P}_d$ (Bonito and Demlow, 2019).

Corollary 33 (geometric consistency errors for C^2 surfaces)**.** *If (90) and (74) hold, then so do the following estimates for all $T \in \mathcal{T}$*

$$\|d\|_{L_\infty(T)} \lesssim \beta_T, \quad \|\boldsymbol{\nu}-\boldsymbol{\nu}_\Gamma\|_{L_\infty(T)} \lesssim \lambda_T, \quad \|1-q^{-1}q_\Gamma\|_{L_\infty(T)} \lesssim \mu_T, \tag{93}$$

where all the geometric quantities are defined using the parametrizations $\chi = \mathbf{P}_d \circ \mathbf{X}$ and $\mathbf{X}$. Moreover,

$$\|\mathbf{E}\|_{L_\infty(T)}, \quad \|\mathbf{E}_\Gamma\|_{L_\infty(T)} \lesssim \mu_T \quad \forall T \in \mathcal{T}. \tag{94}$$

Proof. The first estimate in (93) is trivial from the definition (86) of β_T, whereas the second estimate in (93) is a consequence of (58). The third estimate in (93) results from (57) and (58). With these estimates at hand, the estimate for $\mathbf{E}$ in (94) comes from (56) and that for $\mathbf{E}_\Gamma$ is similar. □

We conclude with a technical result assessing the mismatch between $\mathbf{P}$ and $\mathbf{P}_d$. We motivate it with the following simpler L_∞-estimate valid for all $T \in \mathcal{T}$

$$\|\widetilde{w} - \widetilde{w} \circ \mathbf{P}_d \circ \mathbf{P}^{-1}\|_{L_\infty(\widetilde{T})} \lesssim \|\nabla_\gamma \widetilde{w}\|_{L_\infty(\widetilde{\omega}_T)} \beta_T \quad \forall \mathbf{x} \in T.$$

This is a trivial consequence of the property (92) for $\widetilde{\mathbf{x}} \in \widetilde{T}$

$$|\widetilde{w}(\widetilde{\mathbf{x}}) - \widetilde{w}(\mathbf{P}_d \circ \mathbf{P}^{-1}(\widetilde{\mathbf{x}}))| \leq \|\nabla_\gamma \widetilde{w}\|_{L_\infty(\widetilde{\omega}_T)} |\widetilde{\mathbf{x}} - \mathbf{P}_d \circ \mathbf{P}^{-1}(\widetilde{\mathbf{x}})| \leq 2\, \|\nabla_\gamma \widetilde{w}\|_{L_\infty(\widetilde{\omega}_T)} \beta_T.$$

The estimate below is L_2-based and its proof entails regularization by convolution (Bonito and Demlow, 2019, Lemma 3.4).

Proposition 34 (mismatch between **P** and $\mathbf{P}_d$). *Assume that (67) as well as the assumptions (74), (90) and (91) hold. Then there exists $\lambda_* > 0$ such for $\widetilde{w} \in H^1(\gamma)$ and $T \in \mathcal{T}$ we have*

$$\|\widetilde{w} - \widetilde{w} \circ \boldsymbol{P}_d \circ \boldsymbol{P}^{-1}\|_{L_2(\widetilde{T})} \lesssim \beta_T \, \|\widetilde{w}\|_{H^1(\widetilde{\omega}_T)}$$

provided $\lambda_T \leq \lambda_*$ and $\widetilde{\omega}_T$ is a patch in γ around $\widetilde{T}$.

Proof. We proceed in several steps.
Step 1: Reduction to $\mathbb{R}^n$. Fix $T \in \mathcal{T}$ and recall that $\boldsymbol{\chi}_T = \mathbf{P} \circ \mathbf{X}_T$ satisfies (68) and maps the reference patch $\widehat{\omega}_T$ into $\widetilde{\omega}_T$. For notational ease, let

$$\psi = \mathbf{P}_d \circ \mathbf{P}^{-1} : \gamma \to \gamma, \quad \widehat{\psi} = \chi_T^{-1} \circ \psi \circ \chi_T : \widehat{\omega}_T \to \widehat{\omega}_T.$$

Given $\widetilde{w} \in H^1(\gamma)$, let $\widehat{w} = \widetilde{w} \circ \chi_T : \widehat{\omega}_T \to \mathbb{R}$, and note that $\widehat{w} \in H^1(\widehat{\omega}_T)$ because χ_T is Lipschitz. We change variables via χ_T to $\widehat{T}$ and invoke the nondegeneracy property (42) to obtain

$$\|\widetilde{w} - \widetilde{w} \circ \psi\|_{L_2(\widetilde{T})} \lesssim h_T^{n/2} \, \|\widehat{w} - \widehat{w} \circ \widehat{\psi}\|_{L_2(\widehat{T})}.$$

The assumption $\mathbf{P}_d \circ \mathbf{P}^{-1}(\widetilde{T}) \subset \widetilde{\omega}_T$ given in (91) is equivalent to $\widehat{\psi}(\widehat{T}) \subset \widehat{\omega}_T$ and is sufficient to ensure that the quantity on the right-hand side is well defined.

Since $\widehat{w}$ is defined on $\widehat{\omega}_T \subset \mathbb{R}^n$, and its boundary is Lipschitz, there is a universal extension operator $E : H^1(\widehat{\omega}_T) \to H^1(\mathbb{R}^n)$ which is bounded both in L_2 and in the H^1-seminorm (Stein, 1970); this is the so-called Calderón operator. We relabel $E\widehat{w}$ to be $\widehat{w}$, and thus assume it is bounded in $H^1(\mathbb{R}^n)$ while satisfying

$$|\widehat{w}|_{H^1(\mathbb{R}^n)} \lesssim |\widehat{w}|_{H^1(\widehat{\omega}_T)}.$$

Step 2: Mollification. We now regularize $\widehat{w}$ by convolution with a standard smooth mollifier supported in the ball $B(0, \varepsilon)$ centred at 0 with radius $\varepsilon > 0$ to be determined. If $\Omega \subset \mathbb{R}^n$ is an arbitrary domain, it is well known that

$$\|\widehat{w} - \widehat{w}_\varepsilon\|_{L_2(\Omega)} \lesssim \varepsilon |\widehat{w}|_{H^1(\Omega + B(0,\varepsilon))},$$
$$|\widehat{w}_\varepsilon|_{W^1_\infty(\Omega)} \lesssim \varepsilon^{-n/2} |\widehat{w}|_{H^1(\Omega + B(0,\varepsilon))}.$$

We may now write, without restriction on ε, that

$$\|\widehat{w} - \widehat{w} \circ \widehat{\psi}\|_{L_2(\widehat{T})} \lesssim \|\widehat{w} - \widehat{w}_\varepsilon\|_{L_2(\widehat{T})} + \|\widehat{w}_\varepsilon - \widehat{w}_\varepsilon \circ \widehat{\psi}\|_{L_2(\widehat{T})} + \|\widehat{w}_\varepsilon \circ \widehat{\psi} - \widehat{w} \circ \widehat{\psi}\|_{L_2(\widehat{T})}.$$

We estimate the first term using the first formula above for the mollifier

$$\|\widehat{w} - \widehat{w}_\varepsilon\|_{L_2(\widehat{T})} \lesssim \varepsilon |\widehat{w}|_{H^1(\mathbb{R}^n)} \lesssim \varepsilon |\widehat{w}|_{H^1(\widehat{\omega}_T)}.$$

Similarly, changing variables via the map $\widehat{\psi}$, which turns out to be Lipschitz in view of (42) and (68), and applying the restriction $\widehat{\psi}(\widehat{T}) \subset \widehat{\omega}_T$ stated in (91), we find that

$$\|(\widehat{w}_\varepsilon - \widehat{w}) \circ \widehat{\psi}\|_{L_2(\widehat{T})} \lesssim \|\widehat{w}_\varepsilon - \widehat{w}\|_{L_2(\widehat{\omega}_T)} \lesssim \varepsilon |\widehat{w}|_{H^1(\widehat{\omega}_T)}.$$

Step 3: Estimate for $\|\widehat{w}_\varepsilon - \widehat{w}_\varepsilon \circ \widehat{\psi}\|_{L_2(\widehat{T})}$. Let $\{\mathbf{y}_i\}$ be a lattice on $\mathbb{R}^n$ with minimum distance between $\mathbf{y}_i$ and $\mathbf{y}_j$ $(i \neq j)$ proportional to ε and such that $\{B(\mathbf{y}_i, \varepsilon)\}$ covers $\mathbb{R}^n$. The set $\{B(\mathbf{y}_i, M\varepsilon)\}$ then has finite overlap for any $M \geq 1$, with the maximum cardinality of the overlap depending on M. We choose

$$\varepsilon = \sup_{\mathbf{y} \in \widehat{T}} |\mathbf{y} - \widehat{\psi}(\mathbf{y})| \quad \Rightarrow \quad \|\widehat{w}_\varepsilon - \widehat{w}_\varepsilon \circ \widehat{\psi}\|_{L_\infty(B(\mathbf{y}_i, \varepsilon) \cap \widehat{T})} \lesssim \varepsilon |\widehat{w}_\varepsilon|_{W^1_\infty(B(\mathbf{y}_i, 2\varepsilon))}$$

Applying the second property of mollifiers given above yields

$$|\widehat{w}_\varepsilon|_{W^1_\infty(B(\mathbf{y}_i, 2\varepsilon))} \lesssim \varepsilon^{-n/2} |\widehat{w}|_{H^1(B(\mathbf{y}_i, 3\varepsilon))},$$

whence

$$\begin{aligned} \|\widehat{w}_\varepsilon - \widehat{w}_\varepsilon \circ \widehat{\psi}\|^2_{L_2(\widehat{T})} &\lesssim \varepsilon^n \sum_i \|\widehat{w}_\varepsilon - \widehat{w}_\varepsilon \circ \widehat{\psi}\|^2_{L_\infty(B(\mathbf{y}_i, \varepsilon) \cap \widehat{T})} \\ &\lesssim \varepsilon^2 \sum_i |\widehat{w}|^2_{H^1(B(\mathbf{y}_i, 3\varepsilon))} \lesssim \varepsilon^2 |\widehat{w}|^2_{H^1(\mathbb{R}^n)} \lesssim \varepsilon^2 |\widehat{w}|^2_{H^1(\widehat{\omega}_T)}. \end{aligned}$$

Step 4: Bound on ε. Making use of the bi-Lipschitz character (68), we get

$$\begin{aligned} |\mathbf{y} - \widehat{\psi}(\mathbf{y})| &= |\chi_T^{-1}(\chi_T(\mathbf{y})) - \chi_T^{-1}(\psi(\chi(\mathbf{y})))| \\ &\leq L h_T^{-1} |\chi_T(\mathbf{y}) - \psi(\chi(\mathbf{y}))| = L h_T^{-1} |\widetilde{\mathbf{x}} - \mathbf{P}_d \circ \mathbf{P}^{-1}(\widetilde{\mathbf{x}})|, \end{aligned}$$

where $\widetilde{\mathbf{x}} = \chi_T(\mathbf{y})$. Recalling (92) and the definition of ε, we thus obtain

$$\varepsilon \leq 2 L h_T^{-1} \beta_T.$$

We now gather the estimates of Steps 2 and 3. Mapping from $\widehat{T}$ to $\widetilde{T}$ and back via χ_T, and utilizing (42) and (68), yields

$$\begin{aligned} \|\widetilde{w} - \widetilde{w} \circ \psi\|^2_{L_2(\widetilde{T})} &\lesssim h_T^n \|\widehat{w} - \widehat{w} \circ \widehat{\psi}\|^2_{L_2(\widehat{T})} \lesssim h_T^n \varepsilon^2 |\widehat{w}|^2_{H^1(\widehat{\omega}_T)} \\ &\lesssim h_T^n h_T^{-2} \beta_T^2 h_T^{2-n} |\widetilde{w}|^2_{H^1(\widetilde{\omega}_T)} = \beta_T^2 |\widetilde{w}|^2_{H^1(\widetilde{\omega}_T)}. \end{aligned}$$

This completes the proof. □

We conclude this section with a variant of Proposition 34 (mismatch between $\mathbf{P}$ and $\mathbf{P}_d$) which turns out to be instrumental for the study of the Narrow Band method discussed later in Section 6.

Proposition 35 (Lipschitz perturbation). *Let* $\Omega_1, \Omega_2 \subset\subset \Omega \subset \mathbb{R}^{n+1}$ *be Lipschitz bounded domains and* $\mathbf{L} : \Omega_1 \to \Omega_2$ *be a bi-Lipschitz isomorphism. If*

$$r := \max_{x \in \Omega_1} |\boldsymbol{L}(\boldsymbol{x}) - \boldsymbol{x}|$$

is sufficiently small so that $(\Omega_1 \cup \Omega_2) + B(0, r) \subset \Omega$ *then for all* $g \in H^1(\Omega)$ *we have*

$$\| g - g \circ \mathbf{L} \|_{L^2(\Omega_1)} \lesssim r \, \| g \|_{H^1(\Omega)}.$$

Proof. We now proceed as in Proposition 34: let $\varepsilon = r > 0$ and g_ε be a regularization of g by convolution with a standard smooth mollifier supported in the ball $B(0, \varepsilon)$. We write

$$\| g - g \circ \mathbf{L} \|_{L^2(\Omega_1)} \leq \| g - g_\varepsilon \|_{L^2(\Omega_1)} + \| g_\varepsilon - g_\varepsilon \circ \mathbf{L} \|_{L^2(\Omega_1)} + \| g_\varepsilon \circ \mathbf{L} - g \circ \mathbf{L} \|_{L^2(\Omega_1)}$$

and note that

$$\| g - g_\varepsilon \|_{L_2(\Omega_1)} \lesssim \varepsilon \, \| g \|_{H^1(\Omega)}, \quad \| g_\varepsilon \circ \mathbf{L} - g \circ \mathbf{L} \|_{L^2(\Omega_1)} \lesssim \varepsilon \, \| g \|_{H^1(\Omega)}$$

because $\mathbf{L}^{-1}$ is Lipschitz. To estimate $\| g_\varepsilon - g_\varepsilon \circ \mathbf{L} \|_{L^2(\Omega_1)}$, we argue as in Step 3 of Proposition 34 (mismatch between $\mathbf{P}$ and $\mathbf{P}_d$). This completes the proof. □

4.3 A priori error analysis

In this section we derive a priori error estimates in H^1 and L_2, namely estimates expressed in terms of regularity of the exact solution $\tilde{u}$ of (18). Compared to the existing literature these estimates involve two lifts: $\mathbf{P}_d$ and $\mathbf{P}$. The former, based on the distance function d, is only used theoretically or to define a notion of error when comparing U with $\tilde{u}$. The latter is generic and used in practice to define the finite element method, i.e., by setting $F = \tilde{f} \circ \mathbf{P} \frac{q}{q_\Gamma}$ and the discrete parametrization $\mathbf{X}$ to be the interpolant of the continuous one $\chi = \mathbf{P} \circ \mathbf{X}$. Optimal orders of convergence are derived without the need to access the distance function.

We also address a gap in the literature. Existing proofs of optimal a priori estimates for surface FEMs employ the distance function lift $\mathbf{P}_d = \mathbf{x} - d(\mathbf{x}) \nabla d(\mathbf{x})$. However, when γ is C^2, this map is only C^1 because of the presence of ∇d in its definition. Thus given $\tilde{v} \in H^2(\gamma)$, its extension $v = \tilde{v} \circ \mathbf{P}_d$ to Γ is only in H^1 and not piecewise in H^2 as is needed to prove optimal approximation order. Thus existing proofs that only employ the distance function lift require the assumption that γ be of class C^3 in order to obtain optimal-order error estimates in the standard way; cf. the work of Dziuk (1988) in which such error estimates were originally obtained.

As pointed out already in Theorem 11 (C^1 distance function implies $C^{1,1}$ surface), the distance function d to a $C^{1,\alpha}$ surface is no better than Lipschitz in general. Therefore, the aforementioned strategy does not extend to $C^{1,\alpha}$ surfaces. However, the best approximation property of the Galerkin method together with the geometric consistency estimates of Section 4.2 yields a priori error estimates in H^1. We present this discussion after that for C^2 surfaces.

4.3.1 A priori error estimates for C^2 surfaces

The following lemma will be instrumental to prove optimal a priori error estimates for γ of class C^2. It states that a function $\nabla_\Gamma(\widetilde{u}\circ\mathbf{P}_d)$ can be approximated in $H^1(\Gamma)$ to first order for a function $\widetilde{u}\in H^2(\gamma)$. The difficulty is that the composite function $\widetilde{u}\circ\mathbf{P}_d\notin H^2(\Gamma)$ whereas $\nabla_\gamma\widetilde{u}\circ\mathbf{P}_d\in H^1(\Gamma)$. The proof exploits this property to restore optimal approximability of $\nabla_\Gamma(\widetilde{u}\circ\mathbf{P}_d)$ in $H^1(\Gamma)$.

Lemma 36 (approximability in $H^1(\Gamma)$). *Let γ be a surface of class C^2 and $\widetilde{u}\in H^2(\gamma)$. Let K_∞ be defined in (30) and $\beta_\mathcal{T}(\Gamma)$ be given in (86). Then we have*

$$\inf_{V\in\mathbb{V}(\mathcal{T})}\|\nabla_\Gamma(\widetilde{u}\circ\mathbf{P}_d-V)\|_{L_2(\Gamma)}\lesssim h_\mathcal{T}|\widetilde{u}|_{H^2(\gamma)}+\beta_\mathcal{T}(\Gamma)K_\infty\|\nabla_\gamma\widetilde{u}\|_{L_2(\gamma)}. \tag{95}$$

Proof. We know from Veeser (2016) that continuous and discontinuous piecewise polynomial approximations in H^1 are equivalent. Even though this crucial result was originally proved for Euclidean domains, it proofs carries over with essentially no changes to the case of surface meshes

$$\inf_{V\in\mathbb{V}(\mathcal{T})}\|\nabla_\Gamma(\widetilde{u}\circ\mathbf{P}_d-V)\|^2_{L_2(\Gamma)}\lesssim\sum_{T\in\mathcal{T}}\inf_{V_T\in\mathbb{V}(T)}\|\nabla_\Gamma(\widetilde{u}\circ\mathbf{P}_d-V_T)\|^2_{L_2(T)}. \tag{96}$$

We refer to Camacho and Demlow (2015) for related results on surfaces. We thus fix $T\in\mathcal{T}$ and argue over this element hereafter; recall that $\widetilde{T}=\mathbf{P}_d(T)$.

Applying the triangle inequality yields

$$|\nabla_\Gamma(\widetilde{u}\circ\mathbf{P}_d-V_T)|\le\left|\nabla_\Gamma(\widetilde{u}\circ\mathbf{P}_d)-\Pi_\Gamma(\nabla_\gamma\widetilde{u}\circ\mathbf{P}_d)\right|+\left|\Pi_\Gamma(\nabla_\gamma\widetilde{u}\circ\mathbf{P}_d)-\nabla_\Gamma V_T\right|.$$

Using (49), we next find that

$$|\nabla_\Gamma(\widetilde{u}\circ\mathbf{P}_d)-\Pi_\Gamma(\nabla_\gamma\widetilde{u}\circ\mathbf{P}_d)|=\left|\Pi_\Gamma[d\mathbf{W}(\nabla_\gamma\widetilde{u}\circ\mathbf{P}_d)]\right|\le K_\infty|d|\,\left|(\nabla_\gamma\widetilde{u})\circ\mathbf{P}_d\right|,$$

which along with (44) yields

$$\|\nabla_\Gamma(\widetilde{u}\circ\mathbf{P}_d)-\Pi_\Gamma(\nabla_\gamma\widetilde{u}\circ\mathbf{P}_d)\|_{L_2(T)}\lesssim\beta_T K_\infty\|\nabla_\gamma\widetilde{u}\|_{L_2(\widetilde{T})}.$$

Next note that $\Pi_\Gamma=\mathbf{I}-\boldsymbol{\nu}_\Gamma\otimes\boldsymbol{\nu}_\Gamma$ is constant over T. Therefore, $\Pi_\Gamma(\nabla_\gamma\widetilde{u}\circ\mathbf{P}_d)\in[H^1(T)]^{n+1}$ in T because $\widetilde{u}\in H^2(\gamma)$ implies $\nabla_\gamma\widetilde{u}\in[H^1(\gamma)]^{n+1}$ and $\mathbf{P}_d$ is C^1. In addition, $\Pi_\Gamma(\nabla_\gamma\widetilde{u}\circ\mathbf{P}_d)$ is a tangent vector field on Γ.

On the other hand, ∇_Γ maps the affine functions $\mathbb{P}^1$ onto the subspace of $[\mathbb{P}^0]^{n+1}$ tangent to Γ, so standard approximation theory leads to

$$\inf_{V_T\in\mathbb{V}(T)} \|\mathbf{w}-\nabla_\Gamma V_T\|_{L_2(T)} \lesssim h_T|\mathbf{w}|_{H^1(T)}$$

for any tangent vector field $\mathbf{w}\in[H^1(T)]^{n+1}$ to Γ. Using that $\nabla\mathbf{P}_d=\Pi-d\mathbf{W}$ and $\mathbf{W}$ is bounded because γ is of class C^2, together with the fact that Π_Γ is constant in T, we deduce

$$\begin{aligned}\inf_{V_T\in\mathbb{V}(T)} \|\Pi_\Gamma(\nabla_\gamma\widetilde{u}\circ\mathbf{P}_d)-\nabla_\Gamma V_T\|_{L_2(T)} &\lesssim h_T|\Pi_\Gamma(\nabla_\gamma\widetilde{u}\circ\mathbf{P}_d)|_{H^1(T)}\\ &\lesssim h_T\,\|\Pi-d\mathbf{W}\|_{L_\infty(T)}\,\|D^2_\gamma\widetilde{u}\circ\mathbf{P}_d\|_{L_2(T)}\lesssim h_T|\widetilde{u}|_{H_2(\widetilde{T})},\end{aligned}$$

where we used the notation $D^2_\gamma\widetilde{u}:=\nabla_\gamma\nabla_\gamma\widetilde{u}$. This completes the proof. □

This proof reveals that (95) can in fact be written locally:

$$\begin{aligned}\inf_{V\in\mathbb{V}(T)} \|\nabla_\Gamma(\widetilde{u}\circ\mathbf{P}_d-V)\|^2_{L_2(T)} &\lesssim \beta_T^2K_\infty^2\,\|\nabla_\gamma\widetilde{u}\|^2_{L_2(\widetilde{T})}\\ &+\inf_{\mathbf{V}\in\mathbb{P}^0(T)}\|\Pi_\Gamma(\nabla_\gamma\widetilde{u}\circ\mathbf{P}_d)-\mathbf{V}\|^2_{L_2(T)}\,.\end{aligned}$$

We now apply Lemma 36 (approximability in $H^1(\Gamma)$) to derive an a priori error estimate. We present two proofs. The first one is very compact and relies on Lemmas 19 and 23 (perturbation error estimate). The second proof is self-contained and paves the way to the L_2 error estimate that follows. In both cases we rely on Lemma 3 (regularity) for γ of class C^2 and $\widetilde{f}\in L_{2,\#}(\gamma)$:

$$\|\widetilde{u}\|_{H^2(\gamma)}\lesssim\|\widetilde{f}\|_{L_2(\gamma)}.$$

Theorem 37 (H^1 a priori error estimate for C^2 surfaces)**.** *Let γ be of class C^2, $\widetilde{f}\in L_{2,\#}(\gamma)$ and $\widetilde{u}\in H^2(\gamma)$ be the solution of (18). Let $U\in\mathbb{V}_\#(\mathcal{T})$ be the solution to (78) with $F=\widetilde{f}\circ\mathbf{P}\frac{q}{q_\Gamma}$ defined via the lift $\mathbf{P}$. If the geometric assumptions (69), (90), and (91) are valid, then*

$$\|\nabla_\Gamma(\widetilde{u}\circ\mathbf{P}-U)\|_{L_2(\Gamma)}\lesssim(h_\mathcal{T}+\lambda_\mathcal{T}(\Gamma))\,\|\widetilde{f}\|_{L_2(\gamma)}\lesssim h_\mathcal{T}\,\|\widetilde{f}\|_{L_2(\gamma)}$$

as well as

$$\|\nabla_\Gamma(\widetilde{u}\circ\mathbf{P}_d-U)\|_{L_2(\Gamma)}\lesssim(h_\mathcal{T}+\mu_\mathcal{T}(\Gamma))\,\|\widetilde{f}\|_{L_2(\gamma)}\lesssim h_\mathcal{T}\,\|\widetilde{f}\|_{L_2(\gamma)}.$$

Proof 1. We prove the second estimate. Let $f_\Gamma=F$ and $u_\Gamma\in H^1_\#(\Gamma)$ solve (33)

$$\int_\Gamma\nabla_\Gamma u_\Gamma\cdot\nabla_\Gamma v=\int_\Gamma f_\Gamma v\quad\forall\,v\in H^1_\#(\Gamma).$$

Since $U \in \mathbb{V}_{\#}(\mathcal{T})$ is the Galerkin approximation to u_Γ on Γ, we infer that

$$\|\nabla_\Gamma(u_\Gamma - U)\| = \inf_{V\in\mathbb{V}(\mathcal{T})} \|\nabla_\Gamma(u_\Gamma - V)\|.$$

This combined with the triangle inequality yields

$$\|\nabla_\Gamma(\widetilde{u}\circ\mathbf{P}_d - U)\|_{L_2(\Gamma)} \le 2\|\nabla_\Gamma(\widetilde{u}\circ\mathbf{P}_d - u_\Gamma)\|_{L_2(\Gamma)} + \inf_{V\in\mathbb{V}(\mathcal{T})}\|\nabla_\Gamma(\widetilde{u}\circ\mathbf{P}_d - V)\|_{L_2(\Gamma)}.$$

Applying Lemma 36 (approximability of $H^1(\Gamma)$), together with $\|\widetilde{u}\|_{H^2(\gamma)} \lesssim \|\widetilde{f}\|_{L_2(\gamma)}$, readily gives

$$\inf_{V\in\mathbb{V}(\mathcal{T})}\|\nabla_\Gamma(\widetilde{u}\circ\mathbf{P}_d - V)\|_{L_2(\Gamma)} \lesssim (h_\mathcal{T} + \beta_\mathcal{T}(\Gamma))\,\|\widetilde{f}\|_{L_2(\gamma)}.$$

To estimate the remaining term, we resort to Lemma 17 (norm equivalence), Lemma 23 (perturbation error estimate) along with Corollary 33 (geometric consistency errors for C^2 surfaces) to obtain

$$\|\nabla_\Gamma(\widetilde{u}\circ\mathbf{P}_d - u_\Gamma)\|_{L_2(\Gamma)} \lesssim \mu_\mathcal{T}(\Gamma)\,\|F\|_{H^{-1}_{\#}(\Gamma)} + \|f q_d q_\Gamma^{-1} - F\|_{H^{-1}_{\#}(\Gamma)},$$

where q_d denotes the area element induced by the parametrization $\chi = \mathbf{P}_d \circ \mathbf{X}$ of γ. We denote by $\mathbf{P}_d^{-1}$ the inverse of $\mathbf{P}_d$ restricted to Γ, and use Proposition 34 (mismatch between $\mathbf{P}$ and $\mathbf{P}_d$), with $\widetilde{w} = v\circ\mathbf{P}_d^{-1}$ and $v \in H^1_{\#}(\Gamma)$, to get

$$\begin{aligned}\|f q_d\, q_\Gamma^{-1} - F\|_{H^{-1}(\Gamma)} &= \sup_{\|\nabla_\Gamma v\|_{L_2(\Gamma)}=1} \int_\Gamma \left(\widetilde{f}\circ\mathbf{P}_d\frac{q_d}{q_\Gamma} - \widetilde{f}\circ\mathbf{P}\frac{q}{q_\Gamma}\right) v \\ &= \sup_{\|\nabla_\Gamma v\|_{L_2(\Gamma)}=1} \int_\gamma \widetilde{f}\left(v\circ\mathbf{P}_d^{-1} - v\circ\mathbf{P}^{-1}\right) \lesssim \beta_\mathcal{T}(\Gamma)\,\|\widetilde{f}\|_{L_2(\gamma)}.\end{aligned}$$

Combining the previous inequalities with $\|F\|_{H^{-1}(\Gamma)} \lesssim \|\widetilde{f}\|_{L_2(\gamma)}$ completes the proof of the second assertion. The proof of the first one proceeds along the same lines but using Lemma 19 (perturbation error estimate for $C^{1,\alpha}$ surfaces) and Corollary 32 (geometric consistency for $C^{1,\alpha}$ surfaces) instead. □

Proof 2. We closely mimic the proof of Lemmas 19 and 23 (perturbation error estimate) for the solution to the Laplace–Beltrami problem on nearby surfaces, with an additional step needed due to the Galerkin approximation. In addition, the fact that $F = \widetilde{f}\circ\mathbf{P}\frac{q}{q_\Gamma}$ is defined using the map $\mathbf{P}$ while all other quantities are lifted using the closest point projection $\mathbf{P}_d$ adds a twist to our proof as compared with standard proofs of such error estimates. We let $u = \widetilde{u}\circ\mathbf{P}_d(\mathbf{x})$ for all $\mathbf{x}\in\Gamma$ for notational convenience, and focus on the second assertion.

Step 1: Error representation. For $V \in \mathbb{V}(\mathcal{T})$ arbitrary, we let $W = V - U$ to arrive at

$$\| \nabla_\Gamma (V-U) \|_{L_2(\Gamma)}^2 = \int_\Gamma \nabla_\Gamma (u-U) \cdot \nabla_\Gamma W + \int_\Gamma \nabla_\Gamma (V-u) \cdot \nabla_\Gamma W.$$

We now invoke Lemma 31 (Galerkin quasi-orthogonality) to rewrite the first term as follows:

$$\int_\Gamma \nabla_\Gamma (u-U) \cdot \nabla_\Gamma W = \int_\Gamma \left(\widetilde{f} \circ \mathbf{P}_d \frac{q_d}{q_\Gamma} - F \right) W + \int_\Gamma \nabla_\Gamma u \cdot \mathbf{E}_\Gamma \nabla_\Gamma W,$$

where the area element q_d over γ is induced by the parametrization $\chi = \mathbf{P}_d \circ \mathbf{X}$. We thus have the error representation formula

$$\begin{aligned} \| \nabla_\Gamma (V-U) \|_{L_2(\Gamma)}^2 &= \int_\Gamma \nabla_\Gamma u \cdot \mathbf{E}_\Gamma \nabla_\Gamma W + \int_\Gamma \nabla_\Gamma (V-u) \cdot \nabla_\Gamma W \\ &+ \int_\Gamma \left(\widetilde{f} \circ \mathbf{P}_d \frac{q_d}{q_\Gamma} - F \right) W := I + II + III, \end{aligned}$$

and estimate the three terms on the right-hand side separately.

Step 2: Geometric and interpolation errors. According to Corollary 33 (geometric consistency errors for C^2 surfaces), the error matrix satisfies $\| \mathbf{E}_\Gamma \|_{L_\infty(\Gamma)} \lesssim \mu_\mathcal{T}(\Gamma)$. This, together with Lemma 17 (norm equivalence) and the a priori bound $\|\nabla_\gamma \widetilde{u}\|_{L_2(\gamma)} \le \|\widetilde{u}\|_{H^2(\gamma)} \lesssim \|\widetilde{f}\|_{L_2(\gamma)}$, yields

$$I \lesssim \mu_\mathcal{T}(\Gamma) \, \|\widetilde{f}\|_{L_2(\gamma)} \, \|\nabla_\Gamma W\|_{L_2(\Gamma)}.$$

On the other hand, we can choose $V \in \mathbb{V}(\mathcal{T})$ so that Lemma 36 (approximability in $H^1(\Gamma)$) holds, whence

$$II \lesssim (h_\mathcal{T} + \beta_\mathcal{T}(\Gamma)) \, \|\widetilde{f}\|_{L_2(\gamma)} \, \|\nabla_\Gamma W\|_{L_2(\Gamma)}.$$

Step 3: Final estimates. We recall that the discrete forcing is given by $F = \widetilde{f} \circ \mathbf{P} \frac{q}{q_\Gamma}$, where q is the area element in γ induced by the parametrization $\chi = \mathbf{P} \circ \mathbf{X}$. Changing variables to γ via the lifts $\mathbf{P}_d$ and $\mathbf{P}$ for each integral in III gives

$$III = \int_\Gamma \left(\widetilde{f} \circ \mathbf{P}_d \frac{q_d}{q_\Gamma} - \widetilde{f} \circ \mathbf{P} \frac{q}{q_\Gamma} \right) W = \int_\gamma \widetilde{f} \left(W \circ \mathbf{P}_d^{-1} - W \circ \mathbf{P}^{-1} \right),$$

where again $\mathbf{P}_d^{-1}$ denotes the inverse of $\mathbf{P}_d$ restricted to Γ. Since $\widetilde{f}$ has vanishing mean over γ, we can assume that so does W over Γ. This allows us to invoke (82) (uniform Poincaré–Friedrichs constant) to deduce $\|W\|_{H^1(\Gamma)} \lesssim \|\nabla_\Gamma W\|_{L_2(\Gamma)}$ and thus apply Proposition 34 (mismatch between $\mathbf{P}$ and $\mathbf{P}_d$) to obtain

$$III \lesssim \beta_{\mathcal{T}}(\Gamma)\, \|f\|_{L_2(\Gamma)}\, \|\nabla_\Gamma W\|_{L_2(\Gamma)}.$$

Collecting the previous estimates, and using that $\beta_{\mathcal{T}}(\Gamma) \le \mu_{\mathcal{T}}(\Gamma)$, leads to

$$\|\nabla_\Gamma(U-V)\|_{L_2(\Gamma)} \lesssim (h_{\mathcal{T}} + \mu_{\mathcal{T}}(\Gamma))\, \|f\|_{L_2(\Gamma)} \lesssim h_{\mathcal{T}}\, \|\widetilde{f}\|_{L_2(\gamma)}$$

because $\mu_{\mathcal{T}}(\Gamma) \lesssim h_{\mathcal{T}}^2 |d|_{W^2_\infty(\Gamma)}$ according to the definition (87) of $\mu_{\mathcal{T}}(\Gamma)$ and Corollary 33 (geometric consistency for C^2 surfaces). Invoking again Lemma 36 (approximability in $H^1(\Gamma)$) yields the second assertion.

The first statement follows similarly upon replacing $\widetilde{u} \circ \mathbf{P}_d$ by $\widetilde{u} \circ \mathbf{P}$, $\mathbf{P}_d$ by $\mathbf{P}$ and invoking Corollary 32 (geometric consistency errors for $C^{1,\alpha}$ surfaces)

$$\|\mathbf{E}_\Gamma\|_{L_\infty(\Gamma)} \lesssim \lambda_{\mathcal{T}}(\Gamma) \lesssim h_{\mathcal{T}} |\mathbf{P}|_{W^2_\infty(\Gamma)}.$$

This concludes the proof. □

Comparing Corollary 32 (geometric consistency errors for $C^{1,\alpha}$ surfaces) with Corollary 33 (geometric consistency errors for C^2 surfaces) ones sees that using the distance function lift $\mathbf{P}_d$ for error representation gives rise to a quadratic geometric error estimator for surfaces γ of class C^2

$$\mu_{\mathcal{T}}(\Gamma) \lesssim h_{\mathcal{T}}^2 |d|_{W^2_\infty(\Gamma)},$$

even though the FEM is designed in terms of a generic lift $\mathbf{P}$ also of class C^2. Meanwhile the geometric estimator $\lambda_{\mathcal{T}}(\Gamma) \lesssim h_{\mathcal{T}} |\mathbf{P}|_{W^2_\infty(\Gamma)}$ is linear for this regularity class. This does not affect the H^1 a priori error analysis for piecewise linear approximations of γ and u, which is first order, but it is crucial to derive optimal second-order L_2 error estimates by a duality argument. We present next such estimates for surfaces of class C^2 and a FEM based on a generic lift $\mathbf{P}$ also of class C^2, a result that seems to be new in the literature.

Theorem 38 (L_2 a priori error estimate for C^2 surfaces). *Let γ be of class C^2 and be described by a generic lift $\mathbf{P}$ of class C^2. Let the geometric conditions (69), (90), and (91) be satisfied. Let $\widetilde{u} \in H^1_\#(\gamma)$ solve (19) and $U \in \mathbb{V}_\#(\mathcal{T})$ solve (78) with $F = \widetilde{f} \circ \mathbf{P}\frac{q}{q_\Gamma}$. Then*

$$\|\widetilde{u} \circ \mathbf{P} - U\|_{L_2(\Gamma)} \lesssim h_{\mathcal{T}}^2\, \|\widetilde{f}\|_{L_2(\gamma)}, \tag{97}$$

provided $\lambda \le \lambda_*$, where λ_* is as in Proposition 34.

Proof. We employ a standard duality argument, but enforcing compatibility (mean-value-zero) conditions. We use the lift $\mathbf{P}_d$ and its inverse $\mathbf{P}_d^{-1}$ when restricted to Γ to switch from γ to Γ back and forth. To this end we use the notation $\widetilde{w} = w \circ \mathbf{P}_d^{-1} : \gamma \to \mathbb{R}$ and $v = \widetilde{v} \circ \mathbf{P}_d : \Gamma \to \mathbb{R}$ for functions $w : \Gamma \to \mathbb{R}$ and $\widetilde{v} : \gamma \to \mathbb{R}$. We denote q_d the area element induced by $\mathbf{P}_d$. We finally observe that if $\mathbf{P}$ is of class C^2 then

$$\beta_{\mathcal{T}}(\Gamma) \leq \mu_{\mathcal{T}}(\Gamma) \lesssim h_{\mathcal{T}}^2 |\mathbf{P}|_{W_\infty^2(\Gamma)},$$

where $\beta_{\mathcal{T}}(\Gamma)$ and $\mu_{\mathcal{T}}(\Gamma)$ are defined in (86) and (87). We split the proof into several steps.

Step 1: Duality argument. We associate with $U \in \mathbb{V}_\#(\mathcal{T})$ the function $\widetilde{U}_\# = \frac{q_\Gamma}{q}\widetilde{U} \in L_{2,\#}(\gamma)$ with vanishing mean over γ and let $\widetilde{z} \in H_\#^1(\gamma)$ satisfy

$$\int_\gamma \nabla_\gamma \widetilde{z} \cdot \nabla_\gamma \widetilde{w} = \int_\gamma \left(\widetilde{u} - \widetilde{U}_\#\right)\widetilde{w} \quad \forall \widetilde{w} \in H_\#^1(\gamma).$$

Observe that the Lax–Milgram lemma and Lemma 2 (Poincaré–Friedrichs inequality) guarantee existence and uniqueness of $\widetilde{z} \in H_\#^1(\gamma)$. Let also $Z \in \mathbb{V}_\#(\mathcal{T})$ be the Galerkin approximation to $\widetilde{z}$ over Γ, that is

$$\int_\Gamma \nabla_\Gamma Z \cdot \nabla_\Gamma W = \int_\Gamma (u_\# - U)W \quad \forall W \in \mathbb{V}(\mathcal{T}),$$

where $u_\# := \frac{q}{q_\Gamma}u$ has vanishing mean over Γ. Note also that $u_\# - U = (\widetilde{u} - \widetilde{U}_\#) \circ \mathbf{P}_d \frac{q}{q_\Gamma}$ is a compatible right-hand side for Theorem 37 (H^1 a priori error estimate). We thus have

$$\|\widetilde{u} - \widetilde{U}_\#\|_{L_2(\gamma)}^2 = \int_\gamma \nabla_\gamma\left(\widetilde{u} - \widetilde{U}\right) \cdot \nabla_\gamma(\widetilde{z} - \widetilde{Z}) + \int_\gamma \nabla_\gamma\left(\widetilde{u} - \widetilde{U}\right) \cdot \nabla_\gamma \widetilde{Z}.$$

Applying Lemma 31 (Galerkin orthogonality) the second integral becomes

$$\int_\gamma \nabla_\gamma\left(\widetilde{u} - \widetilde{U}\right) \cdot \nabla_\gamma \widetilde{Z} = \int_\gamma \left(\widetilde{f} - \widetilde{F}\frac{q_\Gamma}{q_d}\right)\widetilde{Z} + \int_\gamma \nabla_\gamma \widetilde{U} \cdot \mathbf{E}\,\nabla_\gamma \widetilde{Z},$$

with $\widetilde{F} = F \circ \mathbf{P}_d$. Changing variables first via the lift $\mathbf{P}_d$ and next via $\mathbf{P}$, we get

$$\int_\gamma \widetilde{F}\widetilde{Z}\frac{q_\Gamma}{q_d} = \int_\Gamma FZ = \int_\gamma F \circ \mathbf{P}^{-1} \ Z \circ \mathbf{P}^{-1} \ \frac{q_\Gamma}{q} = \int_\gamma \widetilde{f}\, Z \circ \mathbf{P}^{-1}.$$

Consequently, we have derived the following error representation:

$$\begin{aligned}\|\widetilde{u} - \widetilde{U}_\#\|_{L_2(\gamma)}^2 &= \int_\gamma \nabla_\gamma\left(\widetilde{u} - \widetilde{U}\right) \cdot \nabla_\gamma(\widetilde{z} - \widetilde{Z}) \\ &+ \int_\gamma \widetilde{f}\left(Z \circ \mathbf{P}_d^{-1} - Z \circ \mathbf{P}^{-1}\right) \\ &+ \int_\gamma \nabla_\gamma \widetilde{U} \cdot \mathbf{E}\,\nabla_\gamma \widetilde{Z}.\end{aligned} \tag{98}$$

The first term is standard and the next two account for the mismatch between $\mathbf{P}$ and $\mathbf{P}_d$ and geometric consistency. We examine them separately now.

Step 2: Bounds. Since γ is of class C^2, Lemma 3 (regularity) gives for z

$$\|\widetilde{z}\|_{H^2(\gamma)} \lesssim \|\widetilde{u}-\widetilde{U}_{\#}\|_{L_2(\gamma)}.$$

Combining Theorem 37 (H^1 a priori error estimate) for $\widetilde{z}$ with Lemma 17 (norm equivalence) yields the following estimate in $L_2(\gamma)$ instead of $L_2(\Gamma)$

$$\|\nabla_\gamma(\widetilde{z}-\widetilde{Z})\|_{L_2(\gamma)} \lesssim h_{\mathcal{T}}\, \|\widetilde{u}-\widetilde{U}_{\#}\|_{L_2(\gamma)}.$$

Applying Theorem 37 again, this time for u, implies

$$\int_\gamma \nabla_\gamma\left(\widetilde{u}-\widetilde{U}\right)\cdot\nabla_\gamma(\widetilde{z}-\widetilde{Z}) \lesssim h_{\mathcal{T}}^2\, \|\widetilde{f}\|_{L_2(\gamma)}\, \|\widetilde{u}-\widetilde{U}_{\#}\|_{L_2(\gamma)}.$$

On the other hand, Proposition 34 (mismatch between $\mathbf{P}$ and $\mathbf{P}_d$) with $\widetilde{w}=Z\circ\mathbf{P}_d^{-1}$ leads to

$$\int_\gamma \widetilde{f}\left(Z\circ\mathbf{P}_d^{-1}-Z\circ\mathbf{P}^{-1}\right) \lesssim \beta_{\mathcal{T}}(\Gamma)\, \|\widetilde{f}\|_{L_2(\gamma)}\, \|Z\circ\mathbf{P}^{-1}\|_{H^1(\gamma)}$$
$$\lesssim \beta_{\mathcal{T}}(\Gamma)\, \|\widetilde{f}\|_{L_2(\gamma)}\, \|\nabla_\Gamma Z\|_{L_2(\Gamma)},$$

because Z has a zero mean on Γ. Since $\mathbf{P}$ is of class C^2, one sees that $\beta_{\mathcal{T}}(\Gamma) \lesssim h_{\mathcal{T}}^2|\mathbf{P}|_{W^2_\infty(\Gamma)}$. Hence the a priori bound $\|\nabla_\Gamma Z\|_{L_2(\Gamma)} \lesssim \|u-U\|_{L_2(\Gamma)}$ implies

$$\int_\gamma \widetilde{f}\left(Z\circ\mathbf{P}_d^{-1}-Z\circ\mathbf{P}^{-1}\right) \lesssim h_{\mathcal{T}}^2|\mathbf{P}|_{W^2_\infty(\Gamma)}\, \|\widetilde{f}\|_{L_2(\gamma)}\, \|u_{\#}-U\|_{L_2(\Gamma)}.$$

Finally, Corollary 33 (geometric consistency error for C^2 surfaces), in conjunction with Lemma 17 (norm equivalence), allows us to tackle the geometric error

$$\int_\gamma \nabla_\gamma\widetilde{U}\cdot\mathbf{E}\,\nabla_\gamma\widetilde{Z} \lesssim \|\nabla_\Gamma U\|_{L_2(\Gamma)}\, \|\nabla_\Gamma Z\|_{L_2(\Gamma)}\, \|\mathbf{E}\|_{L_\infty(\gamma)} \lesssim h_{\mathcal{T}}^2\, \|\widetilde{f}\|_{L_2(\gamma)}\, \|u_{\#}-U\|_{L_2(\Gamma)},$$

where again we have used a priori bounds for $\nabla_\Gamma U$ and $\nabla_\Gamma Z$. Lemma 17 (norm equivalence) and the nondegeneracy property (74) of $\frac{q}{q_\Gamma}$ imply that $\|u_{\#}-U\|_{L_2(\Gamma)} \lesssim \|\widetilde{u}-\widetilde{U}_{\#}\|_{L_2(\gamma)}$. Collecting the previous estimates and dividing through by $\|\widetilde{u}-\widetilde{U}_{\#}\|_{L_2(\gamma)}$, we thus arrive at

$$\|\widetilde{u}-\widetilde{U}_{\#}\|_{L_2(\gamma)} \lesssim h_{\mathcal{T}}^2\, \|\widetilde{f}\|_{L_2(\gamma)}.$$

Step 3: Discrepancy between $\widetilde{U}$ and $\widetilde{U}_{\#}$ and final estimates. We still need to deal with the discrepancy between $\widetilde{U}$ and $\widetilde{U}_{\#}=\frac{q_\Gamma}{q}\widetilde{U}$. Using Lemma 4 (geometric consistency errors for C^2 surfaces) and Lemma 17 again, we find that

$$\| \widetilde{U} - \widetilde{U}_{\#} \|_{L_2(\gamma)} \leq \| 1 - q_\Gamma q^{-1} \|_{L_\infty(\gamma)} \| \widetilde{U} \|_{L_2(\gamma)} \lesssim h_{\mathcal{T}}^2 \, \| \widetilde{f} \|_{L_2(\gamma)}.$$

Applying the triangle inequality followed by Lemma 17 gives the intermediate estimate

$$\| \widetilde{u} \circ \mathbf{P}_d - U \|_{L_2(\Gamma)} \lesssim \| \widetilde{u} - \widetilde{U} \|_{L_2(\gamma)} \lesssim h_{\mathcal{T}}^2 \, \| \widetilde{f} \|_{L_2(\gamma)}.$$

To conclude the proof, we simply note that

$$\| \widetilde{u} \circ \mathbf{P}_d - \widetilde{u} \circ \mathbf{P} \|_{L_2(\Gamma)} \approx \| \widetilde{u} - \widetilde{u} \circ \mathbf{P}_d \circ \mathbf{P}^{-1} \|_{L_2(\gamma)} \lesssim \beta_{\mathcal{T}}(\Gamma) \, \| \widetilde{u} \|_{H^1(\gamma)} \lesssim h_{\mathcal{T}}^2 \, \| \widetilde{f} \|_{L_2(\gamma)},$$

according to Proposition 34 (mismatch between $\mathbf{P}$ and $\mathbf{P}_d$) and the estimate $\beta_{\mathcal{T}}(\Gamma) \lesssim h_{\mathcal{T}}^2 \, \| \mathbf{P} \|_{W_\infty^2(\Gamma)}$ for $\mathbf{P}$ of class C^2 (see definition (86) of $\beta_{\mathcal{T}}(\Gamma)$). Finally, the triangle inequality leads to the asserted estimate. □

The estimate (97) is known for surfaces γ of class C^3 and the distance function lift $\mathbf{P}_d$ (Dziuk, 1988). We insist that (97) appears to be new even for $\mathbf{P} = \mathbf{P}_d$ for surfaces of class C^2 and is optimal both in terms of regularity of u and γ as well as order.

The C^2 regularity of γ enters in three distinct places in Step 2 of the proof to tackle the right-hand side of (98) as well as in Step 3. The first instance is via Lemma 3 (regularity) to handle the H^2 regularity of both u and z in terms of the L_2 norm of the forcing terms: it turns out that (20) becomes

$$|\widetilde{u}|_{H^2(\gamma)} \lesssim |d|_{W_\infty^2(\mathcal{N})} \, \| \widetilde{f} \|_{L_2(\gamma)},$$

whence the factor $|d|_{W_\infty^2(\mathcal{N})}^2$ appears. The same happens with the term involving $\| \mathbf{E} \|_{L_\infty(\gamma)}$ in view of (94), whereas a factor $|\mathbf{P}|_{W_\infty^2(\Gamma)}$ shows up for the middle term in (98) and the end of the proof due to Proposition 34 (mismatch between $\mathbf{P}$ and $\mathbf{P}_d$). The complete estimate thus reads

$$\| \widetilde{u} \circ \mathbf{P} - U \|_{L_2(\Gamma)} \lesssim h_{\mathcal{T}}^2 |d|_{W_\infty^2(\mathcal{N})}^2 \, \| \widetilde{f} \|_{L_2(\gamma)}. \tag{99}$$

4.3.2 *A priori error estimates for $C^{1,\alpha}$ surfaces*

We end this section proving H^1 error estimates for surfaces γ of class $C^{1,\alpha}$ and solutions $\widetilde{u}$ of class $H^{1+s}(\gamma)$ for $0 < s \leq 1$. We recall Lemma 4 (regularity for W_p^2 surfaces) that establishes this regularity for $s = 1$, provided $n < p \leq \infty$, along with

$$\| \widetilde{u} \|_{H^2(\gamma)} \lesssim \| \widetilde{f} \|_{L_2(\gamma)}.$$

In general, however, the relation between α and s is not well understood; we refer to Bonito et al. (2019) where it is proved the existence of $s = s(\alpha) > 0$ such that $\widetilde{u} \in H^{1+s}(\gamma)$. We start with a variant of Lemma 36 (approximability in $H^1(\Gamma)$).

Lemma 39 (approximability in $H^1(\Gamma)$). *Let γ be a surface of class $C^{1,\alpha}$ and $\widetilde{u} \in H^{1+s}(\gamma)$, where $0 < s < \alpha < 1$ or $0 < s \leq \alpha = 1$. Then we have*

$$\inf_{V \in \mathbb{V}(\mathcal{T})} \| \nabla_\Gamma(\widetilde{u} \circ \mathbf{P} - V) \|_{L_2(\Gamma)} \lesssim h_{\mathcal{T}}^s |\widetilde{u}|_{H^{1+s}(\gamma)}. \tag{100}$$

Proof. We recall that $u = \widetilde{u} \circ \mathbf{P}$ and $\nabla_\Gamma u \circ \chi_\Gamma = D\chi_\Gamma \mathbf{g}_\Gamma^{-1} D\chi^t \nabla_\gamma \widetilde{u} \circ \chi$, according to (35), and that $D\chi_\Gamma$, $\mathbf{g}_\Gamma^{-1}$ and $D\chi$ are uniformly of class $C^{0,\alpha}$; here $\chi_\Gamma = \mathbf{X}$. Given $T \in \mathcal{T}$, a direct calculation using the definition of the seminorm $|\cdot|_{H^s(T)}$ shows that the composition of a Lipschitz map with a H^s function as well as the product of a $C^{0,\alpha}$ function with a H^s function belong to H^s provided $s < \alpha$ or $s \leq \alpha = 1$. Consequently, we infer that $\nabla_\Gamma u \in H^{1+s}(T)$ for all $T \in \mathcal{T}$ along with

$$|u|_{H^{1+s}(T)} \lesssim |\widetilde{u}|_{H^{1+s}(\widetilde{T})}.$$

A scaling argument guarantees that the constant hidden in this inequality is independent of $T \in \mathcal{T}$. We next apply the localized interpolation estimate of Veeser (2016) to deduce

$$\inf_{V \in \mathbb{V}(\mathcal{T})} \| \nabla_\Gamma(u - V) \|_{L_2(\Gamma)}^2 \lesssim \sum_{T \in \mathcal{T}} \inf_{V \in \mathbb{V}(\mathcal{T})} \| \nabla_\Gamma(u - V) \|_{L_2(T)}^2 \lesssim h_{\mathcal{T}}^{2s} |\widetilde{u}|_{H^{1+s}(\gamma)}^2,$$

which is the asserted estimate. □

We now compare Lemma 39 with Lemma 36 (approximability in $H^1(\Gamma)$). We stress that the lift $\mathbf{P} = \chi \circ \mathbf{X}^{-1}$ is of class $C^{1,\alpha}$ for surfaces of class $C^{1,\alpha}$, whereas the distance lift $\mathbf{P}_d$ is just of class C^1 for surfaces of class C^2. This is why the proof of Lemma 39 is considerably simpler than that of Lemma 36. The virtue of $\mathbf{P}_d$ is reflected in a higher order geometric error $\mu_{\mathcal{T}}(\Gamma)$ in Theorem 37 (H^1 a priori error estimate for C^2 surfaces) relative to the next H^1 error estimate. This is also responsible for the optimal Theorem 38 (L_2 a priori error estimate for C^2 surfaces) which does not have a counterpart in this context.

Theorem 40 (H^1 a priori error estimate for $C^{1,\alpha}$ surfaces). *Let γ be of class $C^{1,\alpha}$, $0 < \alpha \leq 1$, and assume that the geometric assumptions (69), (90), and (91) are valid. Let $\widetilde{f} \in L_{2,\#}(\gamma)$ and $\widetilde{u} \in H^{1+s}(\gamma)$ be the solution of (18) and satisfy*

$$\|\widetilde{u}\|_{H^{1+s}(\gamma)} \lesssim \|\widetilde{f}\|_{L_2(\gamma)},$$

provided $0 < s < \alpha < 1$ or $0 < s \leq \alpha = 1$. If $U \in \mathbb{V}_\#(\mathcal{T})$ is the solution to (78) with $F = \widetilde{f} \circ \mathbf{P} \frac{q}{q_\Gamma}$ defined via the lift $\mathbf{P}$, then

$$\| \nabla_\Gamma(\widetilde{u} \circ \mathbf{P} - U) \|_{L_2(\Gamma)} \lesssim h_{\mathcal{T}}^s \, \|\widetilde{u}\|_{H^{1+s}(\gamma)} + \lambda_{\mathcal{T}}(\Gamma) \, \|\widetilde{f}\|_{L_2(\gamma)} \lesssim h_{\mathcal{T}}^s \, \|\widetilde{f}\|_{L_2(\gamma)}.$$

Proof. We proceed along the lines of Proof of Theorem 37 (H^1 a priori error estimate for C^2 surfaces), which splits the error into an approximation and a perturbation term. For the former we simply resort to Lemma 39 instead of Lemma 36 (approximability in $H^1(\Gamma)$). For the latter we argue exactly as in Theorem 37 and thus employ (82) (uniform Poincaré–Friedrichs constant), Lemma 19 (perturbation error estimate for $C^{1,\alpha}$ surfaces) and Corollary 32 (geometric consistency for $C^{1,\alpha}$ surfaces). This shows the first asserted estimate. The second bound follows from the standard interpolation estimate

$$\lambda_{\mathcal{T}}(\Gamma) \lesssim h_{\mathcal{T}}^{\alpha} \, \|\boldsymbol{\chi}\|_{C^{1,\alpha}(\mathcal{V})}$$

and the condition $\alpha \geq s$. This ends the proof. □

4.4 A posteriori error analysis

In contrast to the previous section, we now derive error estimates in H^1 which rely on information extracted from the computed solution U of (78) and data, but do not make use of the exact solution $\widetilde{u}$ of (18). They are *a posteriori* estimates of residual type, are fully computable, and are instrumental to drive adaptive procedures. In this vein, we mention Bonito et al. (2013, 2016) but we do not elaborate on this issue any longer.

The a posteriori analysis requires a quasi-interpolation operator acting on $H^1(\Gamma)$ functions, i.e., functions without point values. We use the Scott–Zhang operator $\mathcal{I}_{\mathcal{T}}^{sz}: H^1(\Gamma) \to \mathbb{V}(\mathcal{T})$ and recall its local approximability and stability properties for all $T \in \mathcal{T}$

$$\|v - \mathcal{I}_{\mathcal{T}}^{sz} v\|_{L^2(T)} \lesssim h_T \, \|\nabla_\Gamma v\|_{L^2(\omega_T)}, \quad \|\nabla_\Gamma \mathcal{I}_{\mathcal{T}}^{sz} v\|_{L^2(T)} \lesssim \|\nabla_\Gamma v\|_{L^2(\omega_T)}, \qquad (101)$$

where ω_T is a macropatch defined in (65) associated with T. We do not require that $\mathcal{I}_{\mathcal{T}}^{sz} v \in \mathbb{V}_{\#}(\mathcal{T})$ even if $v \in H^1_{\#}(\Gamma)$, as it happened earlier in the a priori error analysis of Section 4.3.

In order to derive a posteriori error estimates, we first introduce the *interior* and *jump residuals* for any $V \in \mathbb{V}(\mathcal{T})$:

$$\begin{aligned} R_T(V) &:= F|_T + \Delta_\Gamma V|_T \quad \forall T \in \mathcal{T} \\ J_S(V) &:= \nabla_\Gamma V^+|_S \cdot \boldsymbol{\mu}_S^+ + \nabla_\Gamma V^-|_S \cdot \boldsymbol{\mu}_S^- \quad \forall S \in S_{\mathcal{T}} \end{aligned}$$

where for $S = \overline{T}^+ \cap \overline{T}^-$ is the face shared by $T^\pm \in \mathcal{T}$ and $\boldsymbol{\mu}_S^\pm := \boldsymbol{\mu}_{T^\pm}$ are pointing outward conormals to the elements $T^\pm$ (see Section 2.6). We point that when using piecewise affine functions $V = \widehat{V} \circ \mathbf{X}^{-1}$ on polyhedral surfaces Γ, the Laplace–Beltrami operator (13) vanishes within elements

$$\Delta_\Gamma V = \frac{1}{q_\Gamma} \operatorname{div}\left(q_\Gamma \mathbf{g}_\Gamma^{-1} \nabla \widehat{V}\right) = 0 \quad \forall T \in \mathcal{T},$$

and that, in contrast to the flat case, $\boldsymbol{\mu}_S^+ \neq -\boldsymbol{\mu}_S^-$ in general. If $J_{\partial T}(V)$ denotes the jump residual on ∂T, then we define the *element indicator* to be

$$\eta_{\mathcal{T}}(V,T)^2 := h_T^2 \, \| R_T(V) \|_{L^2(T)}^2 + h_T \, \| J_{\partial T}(V) \|_{L^2(\partial T)}^2 \quad \forall T \in \mathcal{T},$$

and the *error estimator* to be

$$\eta_{\mathcal{T}}(V)^2 := \sum_{T \in \mathcal{T}} \eta_{\mathcal{T}}(V,T)^2.$$

Theorem 41 (a posteriori upper bound for $C^{1,\alpha}$ surfaces). *Let γ be of class $C^{1,\alpha}$, be parametrized by $\chi = \boldsymbol{P} \circ \boldsymbol{X}$ and satisfy the geometric assumption (69). Let $\widetilde{u} \in H^1_\#(\gamma)$ be the solution to (18) and $U \in \mathbb{V}_\#(\mathcal{T})$ be the solution to (78) with $F = \widetilde{f} \circ \boldsymbol{P} \frac{q}{q_\Gamma} \in L_{2,\#}(\Gamma)$. Then, for $\widetilde{U} := U \circ \boldsymbol{P}^{-1} : \gamma \to \mathbb{R}$ we have*

$$\| \nabla_\gamma (\widetilde{u} - \widetilde{U}) \|_{L^2(\gamma)}^2 \lesssim \eta_{\mathcal{T}}(U)^2 + \lambda_{\mathcal{T}}^2(\Gamma) \, \| \widetilde{f} \|_{L_2(\gamma)}^2 .$$

Proof. Using definitions (19) and (79), along with the consistency relation (34), enables us to write for any $\widetilde{v} \in H^1(\gamma), v = \widetilde{v} \circ \mathbf{P} \in H^1(\Gamma)$ and $V \in \mathbb{V}(\mathcal{T})$

$$\int_\gamma \nabla_\gamma (\widetilde{u} - \widetilde{U}) \cdot \nabla_\gamma \widetilde{v} = I_1 + I_2 + I_3 \tag{102}$$

with

$$I_1 = -\int_\Gamma \nabla_\Gamma U \cdot \nabla_\Gamma (v - V) + \int_\Gamma F(v - V),$$

$$I_2 = \int_\gamma \nabla_\gamma \widetilde{U} \cdot \mathbf{E} \, \nabla_\gamma \widetilde{v},$$

$$I_3 = \int_\gamma \widetilde{f} \, \widetilde{v} - \int_\Gamma F v.$$

Employing the definition $F = \widetilde{f} \circ \mathbf{P} \frac{q}{q_\gamma}$ and changing variables we deduce $I_3 = 0$.

On the one hand, decomposing I_1 over elements $T \in \mathcal{T}$, and resorting to Corollary 13 (integration by parts) on T, leads to

$$I_1 = \sum_{T \in \mathcal{T}} \int_T R_T(U)(v - V) + \sum_{S \in \mathcal{S}} \int_S J_S(U)(v - V) \tag{103}$$

and so

$$I_1 \lesssim \sum_{T \in \mathcal{T}} \eta_{\mathcal{T}}(U,T) \big(h_T^{-1} \, \| v - V \|_{L^2(T)} + \| \nabla_\Gamma (v - V) \|_{L^2(T)} \big),$$

because of the scaled trace inequality

$$\|w\|_{L_2(\partial T)} \lesssim h_T^{-\frac{1}{2}} \|w\|_{L_2(\partial T)} + h_T^{\frac{1}{2}} \|\nabla_\Gamma w\|_{L_2(\partial T)} \quad \forall w \in H^1(T).$$

We now choose $V = \mathcal{I}_{\mathcal{T}}^{sz} v$ to be the Scott–Zhang quasi-interpolant of v. The local approximability and stability properties (101) imply

$$I_1 \lesssim \eta_{\mathcal{T}}(U) \|\nabla_\Gamma v\|_{L^2(\Gamma)} \lesssim \eta_{\mathcal{T}}(U) \|\nabla_\gamma \widetilde{v}\|_{L^2(\gamma)}, \tag{104}$$

where we have used the finite ovelapping properties of the patches $\{\omega_T\}_{T\in\mathcal{T}}$ and Lemma 17 (norm equivalence). Regarding term I_2 we apply Corollary 32 (geometric consistency errors for $C^{1,\alpha}$ surfaces) to arrive at

$$I_2 \lesssim \lambda_{\mathcal{T}}(\Gamma) \; \|\nabla_\gamma \widetilde{U}\|_{L^2(\gamma)} \; \|\nabla_\gamma \widetilde{v}\|_{L^2(\gamma)} \lesssim \lambda_{\mathcal{T}}(\Gamma) \; \|\widetilde{f}\|_{L^2(\gamma)} \; \|\nabla_\gamma \widetilde{v}\|_{L^2(\gamma)},$$

because of the estimates

$$\|\nabla_\gamma \widetilde{U}\|_{L^2(\gamma)} \lesssim \|\nabla_\Gamma U\|_{L^2(\Gamma)} \lesssim \|F\|_{H_\#^{-1}(\Gamma)} \lesssim \|\widetilde{f}\|_{L^2(\gamma)}$$

which are a consequence of Lemma 17 (norm equivalence), $F = \widetilde{f} \circ \mathbf{P} \frac{q}{q_\Gamma}$, Lemma 2 (Poincaré–Friedrich inequality) and $\int_\gamma \widetilde{f} = 0$. Combining the above estimates, we end up with the assertion. □

To assess the tightness of the upper bound in Theorem 41 it is customary to show a lower bound. To this end, we introduce the so-called *data oscillation*

$$\operatorname{osc}_{\mathcal{T}}(F,T)^2 := h_T^2 \, \|F - \overline{F}\|_{L^2(T)}^2, \quad \operatorname{osc}_{\mathcal{T}}(F)^2 := \sum_{T\in\mathcal{T}} \operatorname{osc}_{\mathcal{T}}(F,T)^2,$$

where $\overline{F}$ is the piecewise average of F. This quantity accounts for the fact that the residual is evaluated in a weighted L_2-norm rather than the natural H^{-1}-norm. This in turn makes the estimator $\eta_{\mathcal{T}}(U)$ computable but perhaps at the expense of overestimation. This is the subject of our next estimate, proved in Bonito et al. (2013). We recall that for $T \in \mathcal{T}$, ω_T denotes the union of elements in $\mathcal{T}$ that intersect T and $\widetilde{\omega}_T$ stands for the lift of ω_T to γ via $\mathbf{P}$. Moreover, we set

$$\operatorname{osc}_{\mathcal{T}}(F,\omega_T)^2 := \sum_{T'\subset\omega_T} \operatorname{osc}_{\mathcal{T}}(F,T')^2, \quad \lambda_{\mathcal{T}}^2(\omega_T) := \max_{T'\subset\omega_T} \lambda_{T'}^2.$$

Theorem 42 (a posteriori lower bound for $C^{1,\alpha}$ surfaces). *Under the same conditions of Theorem 41 (a posteriori upper bound for $C^{1,\alpha}$ surfaces), we have*

$$\eta_{\mathcal{T}}(U,T)^2 \lesssim \|\nabla_\gamma(\widetilde{u} - \widetilde{U})\|_{L^2(\widetilde{\omega}_T)}^2 + \operatorname{osc}_{\mathcal{T}}(F,\omega_T)^2 + \lambda_{\mathcal{T}}^2(\omega_T).$$

Proof. The proof of the lower bound is standard and is only sketched here. It relies on an argument due to Verfürth (2013). The starting point is the error relation (102) localized to $T \in \mathcal{T}$ via the test function $v = \overline{F} b_T$, where $b_T \in H_0^1(T)$ is the cubic bubble taking value 1 at the element barycentre. Employing the norm equivalence (44) (valid elementwise), we realize that

$$\|\nabla_\gamma \widetilde{v}\|_{L_2(\widetilde{T})} \lesssim \|\nabla_\Gamma v\|_{L_2(T)} \lesssim h_T^{-1} \|\overline{F}\|_{L_2(T)},$$

whence taking $V = 0$ in (102) yields

$$\|\overline{F}\|_{L_2(T)}^2 \lesssim \int_T \overline{F} v \lesssim h_T^{-1} \left(\|\nabla_\gamma(\widetilde{u} - \widetilde{U})\|_{L_2(\widetilde{T})} + \operatorname{osc}_{\mathcal{T}}(F, T) + \|\mathbf{E}\|_{L_\infty(\widetilde{T})} \right) \|\overline{F}\|_{L_2(T)}$$

upon recalling that $I_3 = 0$ with our choice of F and the expression (103) for I_1. Corollary 32 (geometric consistency errors for $C^{1,\alpha}$ surfaces), combined with a triangle inequality, then leads to the desired estimate for the bulk term

$$h_T^2 \|F\|_{L^2(T)}^2 \lesssim \|\nabla_\gamma(\widetilde{u} - \widetilde{U})\|_{L_2(\widetilde{T})}^2 + \operatorname{osc}_{\mathcal{T}}(F, T)^2 + \lambda_T^2.$$

As for the jump term, we define for a side $S \in \mathcal{S}$ with adjacent elements $T^\pm$, $b_S \in H_0^1(\omega_S)$ as the quadratic bubble taking value 1 at the barycentre of S and 0 at all other quadratic nodes in $\omega_S := T^+ \cup T^-$. We also let $\widetilde{\omega}_S := \mathbf{P}(\omega_S)$ be the lift of ω_S to γ by the map $\mathbf{P}$. Taking $v = J_S(U) b_S$ and $V = 0$ in (102), and recalling the expression (103) for I_1 and that $I_3 = 0$, yields

$$\begin{aligned} \|J_S(U)\|_{L_2(S)}^2 &\lesssim \int_S J_S(U) v \\ &\lesssim \left(\|\nabla_\gamma(\widetilde{u} - \widetilde{U})\|_{L_2(\widetilde{\omega}_S))} + h_T \|F\|_{L_2(\omega_S)} + \max(\lambda_{T^+}, \lambda_{T^-}) \right) \|\nabla_\gamma \widetilde{v}\|_{L_2(\widetilde{\omega}_S))}. \end{aligned}$$

Finally, it suffices to use the preceding estimate for $h_T \|F\|_{L_2(T)}$, together with

$$\|\nabla_\gamma \widetilde{v}\|_{L_2(\widetilde{\omega}_S))} \lesssim \|\nabla_\Gamma v\|_{L_2(\omega_S)} \lesssim h_T^{-1/2} \|J_S(U)\|_{L_2(S)},$$

to conclude the proof. □

One important observation to make is that $\operatorname{osc}_{\mathcal{T}}(F)$ is generically of higher order than $\eta_{\mathcal{T}}(U)$ for $\widetilde{f} \in L_2(\gamma)$, whence this term can be ignored relative to $\eta_{\mathcal{T}}(U)$ asymptotically. However, the geometric estimator $\lambda_{\mathcal{T}}(\Gamma)$ is linear and thus of the same order as $\eta_{\mathcal{T}}(U)$, thereby making the lower bound of Theorem 42 questionable. This estimator comes from the estimate (89) of Corollary 32 (geometric consistency errors for $C^{1,\alpha}$ surfaces), which cannot obviously be improved for surfaces of class $C^{1,\alpha}$. However,

Corollary 33 (geometric consistency errors for C^2 surfaces) shows that this effect becomes of second order for surfaces of class C^2. Practically, the estimator $\lambda_{\mathcal{T}}(\Gamma)$ is pessimistic and leads to unnecessary and thus suboptimal refinements for C^2 surfaces (Bonito and Demlow, 2019). We discuss the impact of this superconvergence estimate next following (Bonito and Demlow, 2019).

Theorem 43 (a posteriori upper bound for C^2 surfaces). *Let γ be of class C^2 and (67), (74), (90), and (91) hold. Let $\widetilde{u}$ be the solution of (18) with $\widetilde{f} \in L_{2,\#}(\gamma)$ and $U \in \mathbb{V}(\mathcal{T})$ be the solution to (78) with $F = \widetilde{f} \circ \boldsymbol{P}\frac{q}{q_\Gamma}$, where q corresponds to the parametrization $\chi = \boldsymbol{P} \circ \boldsymbol{X}$ of γ. Then*

$$\|\nabla_\gamma(\widetilde{u} - U \circ \boldsymbol{P}_d^{-1})\|^2_{L_2(\gamma)} \lesssim \eta_{\mathcal{T}}(U)^2 + \mu^2_{\mathcal{T}}(\Gamma)\, \|\widetilde{f}\,\|^2_{L_2(\gamma)} \,.$$

Proof. We proceed as in the proof of Theorem 41 (a posteriori upper bound for $C^{1,\alpha}$ surfaces) but using the distance function lift to represent the errors. We denote $\widetilde{U} = U \circ \mathbf{P}_d^{-1}$, $v = \widetilde{v} \circ \mathbf{P}_d$ for a generic $\widetilde{v} \in H^1(\gamma)$ and get for any $V \in \mathbb{V}(\mathcal{T})$

$$\int_\gamma \nabla_\gamma\left(\widetilde{u} - \widetilde{U}\right) \cdot \nabla_\gamma \widetilde{v} = I_1 + I_2 + I_3 \tag{105}$$

with

$$\begin{aligned} I_1 &= -\int_\Gamma \nabla_\Gamma U \cdot \nabla_\Gamma (v - V) + \int_\Gamma F(v - V), \\ I_2 &= \int_\gamma \nabla_\gamma \widetilde{U} \cdot \mathbf{E}\, \nabla_\gamma \widetilde{v}, \\ I_3 &= \int_\gamma \widetilde{f}\, \widetilde{v} - \int_\Gamma F v, \end{aligned}$$

where we have used again (34) but with the error matrix $\mathbf{E}$ now defined with respect to $\mathbf{P}_d$ and given by (51) of Lemma 21 (geometric consistency). We tackle I_1 and I_2 exactly as in Theorem 41, thereby obtaining

$$I_1 \lesssim \eta_{\mathcal{T}}(U)\, \|\nabla_\gamma \widetilde{v}\|_{L_2(\gamma)}, \quad I_2 \lesssim \mu_{\mathcal{T}}(\Gamma)\, \|\widetilde{f}\,\|_{L^2(\gamma)}\, \|\nabla_\gamma \widetilde{v}\|_{L_2(\gamma)},$$

except that we resort to (94) of Corollary 33 (geometric consistency errors for C^2 surfaces) to estimate $\mathbf{E}$.

On the other hand, I_3 no longer vanishes because $F = \widetilde{f} \circ \mathbf{P}\frac{q}{q_\Gamma}$ is defined via $\mathbf{P}$ and the function v via $\mathbf{P}_d$. Using $\mathbf{P}$ to change variables back to γ we obtain

$$\int_\Gamma Fv = \int_\Gamma (\widetilde{f}\circ\mathbf{P}) \; (\widetilde{v}\circ\mathbf{P}_d)\frac{q}{q_\Gamma} = \int_\gamma \widetilde{f}(\widetilde{v}\circ\mathbf{P}_d\circ\mathbf{P}^{-1}),$$

whence I_3 becomes

$$I_3 = \int_\gamma \widetilde{f}\left(\widetilde{v} - \widetilde{v}\circ\mathbf{P}_d\circ\mathbf{P}^{-1}\right). \tag{106}$$

Invoking Proposition 34 (mismatch between $\mathbf{P}$ and $\mathbf{P}_d$) yields

$$I_3 \lesssim \beta_{\mathcal{T}}(\Gamma)\; \|\widetilde{f}\|_{L_2(\gamma)}\, \|\nabla_\gamma \widetilde{v}\|_{L_2(\gamma)}$$

and concludes the proof because $\beta_{\mathcal{T}}(\Gamma) \le \mu_{\mathcal{T}}(\Gamma)$. □

We conclude with a lower bound for C^2 surfaces. We point out that, compared with the existing results in the literature, see, e.g., Demlow and Dziuk (2007), we account for the mismatch between the two lifts $\mathbf{P}$ and $\mathbf{P}_d$.

Theorem 44 (a posteriori lower bound for C^2 surfaces). *Under the same conditions as Theorem* 43 (*a posteriori upper bound for C^2 surfaces*), *we have*

$$\eta_{\mathcal{T}}(U,T)^2 \lesssim \|\nabla_\gamma(\widetilde{u} - \widetilde{U})\|^2_{L^2(\widetilde{\omega}_T)} + \mathrm{osc}_{\mathcal{T}}(F,\omega_T)^2 + \mu_{\mathcal{T}}(\omega_T)^2,$$

where $\mu_{\mathcal{T}}(\omega_T) = \max_{T' \subset \omega_T} \mu_{T'}$.

Proof. The proof follows along the lines of Theorem 42 (a posteriori lower bound for $C^{1,\alpha}$ surfaces) with the following variants. We use Corollary 33 instead of Corollary 32 in the error representation (105) to tackle I_2 and account for the fact that $I_3 \neq 0$ via (106) for a generic lift $\mathbf{P}$. □

5 Trace method

In this section we present a class of methods which are known as *trace finite element methods* or *cut finite element methods* (Burman et al., 2015; Olshanskii et al., 2009; Reusken, 2015). The setting for these methods is situations in which a PDE posed on an n-dimensional hypersurface γ embedded in $\mathbb{R}^{n+1}$ must be solved numerically, and a bulk or volume background mesh of some domain $\Omega \subset \mathbb{R}^{n+1}$ is present with $\gamma \subset \Omega$. It is often more convenient to describe γ and solve associated PDE employing the background mesh instead of independently meshing γ. A paradigm physical example is a two-phase flow problem. There Ω is subdivided into subdomains Ω_1 and Ω_2 (one for each phase) and γ is the interface between Ω_1 and Ω_2. In simulations Ω is typically meshed in order to solve equations of fluid dynamics (e.g., Stokes or Navier–Stokes), while accounting for interfacial effects such as surface tension also requires solving a surface PDE on γ. It can be particularly inconvenient to independently mesh Ω and γ in dynamic simulations in which γ evolves as either a specified or free boundary. In addition to the overhead associated with transferring information between unrelated bulk and surface meshes, remeshing is generally necessary from time to time when parametric methods are used to describe dynamic interfaces because mesh degeneracies may occur as the surface deforms.

Trace and cut FEMs were introduced by Olshanskii et al. (2009) and have been further developed over the past decade as one option for circumventing these difficulties. In order to describe them more precisely, first let $\mathcal{T} := \mathcal{T}_\Omega$ be a simplicial decomposition of $\Omega \subset \mathbb{R}^{n+1}$, $n \geq 1$. We let $h_T := |T|^{\frac{1}{n+1}}$ for any $T \in \mathcal{T}$ and set $h := \max_{T \in \mathcal{T}} h_T$ for the mesh size of $\mathcal{T}$. We will omit to mention the explicit dependence on the shape-regularity constant of $\mathcal{T}$

$$\sigma := \max_{T \in \mathcal{T}} \frac{\operatorname{diam}(T)}{h_T}$$

in most estimates below. Assume that $\gamma \subset \Omega$ is a closed, C^2 n-dimensional surface. As outlined in Section 2.3, γ is then the zero level set of a C^2 distance function d defined on a tubular neighbourhood $\mathcal{N}$ of γ. Let $\mathbb{V}(\mathcal{T}) \subset H^1(\Omega)$ consist of the continuous piecewise linear functions over $\mathcal{T}$. In order to fix thoughts, let $d_h \in \mathbb{V}(\mathcal{T})$ be the Lagrange interpolant $\mathcal{I}_\mathcal{T} d$ of d satisfying

$$\|d - d_h\|_{L_\infty(\mathcal{N})} + h\,\|d - d_h\|_{W^1_\infty(\mathcal{N})} \lesssim h^2 |d|_{W^2_\infty(\mathcal{N})}.$$

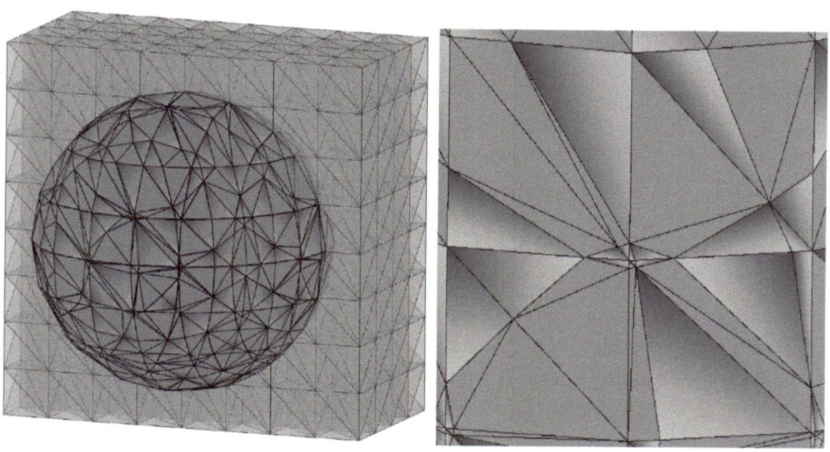

FIG. 3 Bulk mesh cutaway with associated trace mesh (*left*); blowup of a trace mesh showing small and narrow elements (*right*).

The discrete computational surface Γ is then defined by

$$\Gamma := \{\mathbf{x} \in \Omega : d_h(\mathbf{x}) = 0\}.$$

Below we also discuss how to derive Γ from more general implicit representations of γ. Because d_h is piecewise linear, Γ consists of intersections of hyperplanes with simplices and is thus a polyhedron having triangular and quadrilateral faces for $n = 2$ (see Fig. 3). We denote by $\mathcal{F}$ the collection of faces of Γ. In addition, the conditions placed on d_h ensure that $\|d\|_{L_\infty(\Gamma)} + h\,\|\boldsymbol{\nu} - \boldsymbol{\nu}_\Gamma\|_{L_\infty(\Gamma)} \lesssim h^2$, so the perturbation results for C^2 surfaces outlined in Section 3.2 hold on Γ with order h^2 geometric perturbation error.

The surface finite element space $\mathbb{V}(\mathcal{F})$ is simply the restriction of $\mathbb{V}(\mathcal{T})$ to Γ:

$$\mathbb{V}(\mathcal{F}) := \{V|_\Gamma : V \in \mathbb{V}(\mathcal{T})\}.$$

By its definition $\mathbb{V}(\mathcal{F}) \subset H^1(\Gamma)$ consists of the continuous functions which are affine over each face $F \in \mathcal{F}$. We also denote by $\mathbb{V}_\#(\mathcal{F}) := \mathbb{V}(\mathcal{F}) \cap L_{2,\#}(\Gamma)$ its subspace consisting of functions with vanishing mean values. In order to approximate the solution $\widetilde{u}$ to the Laplace–Beltrami problem $-\Delta_\gamma \widetilde{u} = \widetilde{f}$ on γ, we first define a suitable approximation F_Γ to f and then seek $U \in \mathbb{V}_\#(\mathcal{F})$ such that

$$\int_\Gamma \nabla_\Gamma U \cdot \nabla_\Gamma V = \int_\Gamma F_\Gamma V \quad \forall V \in \mathbb{V}_\#(\mathcal{F}). \tag{107}$$

This is the trace method and has two notable advantages:

- **Only single mesh**: The main advantage is that both bulk and interfacial effects can be computed using the same mesh.
- **Error estimates**: Optimal-order and regularity error estimates hold in the H^1 and L_2 norms.

On a practical and theoretical levels the method exhibits three main challenges:

- **Implicit surface representation**: The simplest option of taking the distance function d to define γ and its Lagrange interpolant of d_h to define Γ is not generally practical as d is rarely available in applications. It is generally more practical to assume that the discrete surface Γ is derived from a more general level set representation ϕ of γ. We provide a brief discussion of general level set representations below.
- **Surface integration**: Computing the finite element system is more cumbersome than in standard parametric surface FEMs since both the mesh $\mathcal{F}$ and the finite element space $\mathbb{V}(\mathcal{F})$ are derived from their corresponding bulk counterparts. These difficulties are manageable in the case of the piecewise linear method presented here, but become significantly more cumbersome when a higher order surface approximation is used.
- **Linear algebra and stabilization**: In contrast to parametric surface FEMs there is no obvious practical basis for $\mathbb{V}(\mathcal{F})$, only spanning sets derived from subsets of the bulk space $\mathbb{V}(\mathcal{T})$. In practice such a spanning set is

derived from the degrees of freedom for $\mathbb{V}(\mathcal{T})$ corresponding to elements touching Γ. Degenerate modes arise from this procedure. These are either handled at the linear algebra level or by various stabilization procedures.

Theoretical study of trace FEMs is also more involved than for parametric surface FEMs. One prominent issue is that the surface mesh $\mathcal{F}$ does not consist of shape-regular elements, as is documented in Fig. 3. This is because the faces in $\mathcal{F}$ consist of *arbitrary* intersections of hyperplanes with simplices (planes and tetrahedra for $n = 2$). Thus elements may be arbitrarily small with respect to the bulk mesh size h or fail to satisfy a minimum angle condition, and it is not possible to directly employ standard error estimation techniques. Properties of the "high-quality" bulk mesh $\mathcal{T}$ and finite element space $\mathbb{V}(\mathcal{T})$ must be invoked instead, which in turn requires careful use of extensions and restrictions of functions to and from γ and Γ. For purposes of intuition, it is however useful to note that the surface mesh $\mathcal{F}$ does inherit some structure from the regularity of the bulk mesh $\mathcal{T}$. Elements in $\mathcal{F}$ for example satisfy a maximum-angle condition (Olshanskii et al., 2012), and each element in $\mathcal{F}$ also shares a vertex with a shape-regular element of diameter equivalent to h (Demlow and Olshanskii, 2012).

Below we prove a priori and a posteriori error estimates for a piecewise linear trace FEM. In keeping with the previous section, we concentrate on surface representations and regularity in our discussion. In particular, we only assume that γ is C^2, whereas previous approaches require that γ be C^3. The recent article (Olshanskii and Reusken, 2017) provides a broader survey of trace FEMs, including discussion of topics such as higher order versions, stabilization procedures, and space-time trace FEMs that we omit here.

5.1 Preliminaries

5.1.1 Bulk and surface meshes

Below we need to carefully distinguish between mesh structures defined relative to the surface mesh $\mathcal{F}$ and those defined relative to the volume mesh $\mathcal{T}$. First note that we shall consistently denote by F (n-dimensional) surface elements lying in $\mathcal{F}$, which as we have noted above may not be shape-regular. In addition, T will be used to denote $(n + 1)$-simplices lying in $\mathcal{T}$. Given a face $F \in \mathcal{F}$, we denote by T_F the simplex in which F lies (or one of them if F is a face shared by two bulk elements). In addition, given $T \in \mathcal{T}$ we denote by $\omega^1_{\mathcal{T}}(T)$ the patch of elements of $\mathcal{T}$ surrounding T (first ring)

$$\omega^1_{\mathcal{T}}(T) := \bigcup\{T' \in \mathcal{T} : \quad T' \cap T \neq \emptyset\},$$

and by $\omega^2_{\mathcal{T}}(T)$ the patch of elements of $\mathcal{T}$ surrounding $\omega^1_{\mathcal{T}}(T)$ (second ring). We also define

$$h_F = \text{diam}(T_F) \quad F \in \mathcal{F},$$

hence the local mesh size of the face element F is taken to be the diameter of the corresponding bulk element. Note that it is possible that diam(F) $\ll h_F$. We will also denote by h_T the diameter of elements $T \in \mathcal{T}$. We finally let

$$\mathcal{T}_\Gamma := \{T \in \mathcal{T} : \quad \Gamma \cap T \neq \emptyset \ \text{ or } \ \gamma \cap T \neq \emptyset\}$$

be the set of elements of $\mathcal{T}$ touching either Γ or γ.

5.1.2 Geometric assumptions

Above we described Γ as the zero level set of an approximate distance function d_h. In this section we first place abstract requirements on Γ that are sufficient to obtain optimal-order and regularity a priori error estimates and then prove that these requirements are satisfied on sufficiently fine bulk meshes $\mathcal{T}$ when Γ is built from a suitably general level set description of γ. We now list three main geometric assumptions.

- **Description of Γ.** We assume that Γ is a polyhedral surface whose faces $F \in \mathcal{F}$ consist of the intersection of hyperplanes with simplices $T \in \mathcal{T}$. We further assume that $\Gamma \subset \mathcal{N}$ with $\mathcal{N}$ the tubular neighbourhood defined in (31).
- **Geometric resolution of γ.** Let d be the distance function to γ, $\boldsymbol{\nu} = \nabla d$ and $\boldsymbol{\nu}_\Gamma$ be the outward unit normal on Γ. We assume that

$$\|d\|_{L_\infty(F)} + h_F \, \|\boldsymbol{\nu} - \boldsymbol{\nu}_\Gamma\|_{L_\infty(F)} \lesssim h_F^2 |d|_{W^2_\infty(\mathcal{N})} \quad F \in \mathcal{F}. \tag{108}$$

  This assumption is sufficient to ensure optimal decay of the geometric consistency error in a priori error estimates.
- **Local flattening.** We assume that for each $T \in \mathcal{T}$ with $T \cap \gamma \neq \emptyset$, there is a ball B_R of radius R with $R \simeq 1$ (independent of h_T) such that $T \subset B_{R/2}$ and there is a uniformly bi-C^2 map

$$\Phi : B_R \to \mathbb{R}^3, \quad \Phi(\gamma \cap B_R) \text{ lies in a hyperplane.} \tag{109}$$

The flattening assumption (109) follows from the C^2 nature of the surface γ provided elements $T \in \mathcal{T}$ intersecting γ are sufficiently fine with respect to the inverse of the maximum principal curvature. The flattening map Φ may be constructed by expressing γ as a C^2 graph over tangent hyperplanes of γ, with the radius of the domain of these graphs bounded by the inverse of the maximum principal curvature of γ (cf. Evans (1998, Appendix C) for the construction of Φ; the bound for R follows from the definition of curvature).

5.1.3 Level set representations

While we prove our results below under the abstract geometric resolution assumption (108) involving the distance function, in practice trace methods often build the discrete surface Γ from a more general implicit representation of γ.

Such a representation may be obtained by assuming that γ is the zero level set of a *level set function* $\phi : \mathcal{N} \to \mathbb{R}$

$$\gamma = \{\mathbf{x} \in \mathcal{N} : \quad \phi(\mathbf{x}) = 0\}.$$

Broadening our assumptions concerning implicit representation of γ is important in many practical applications. Because the distance function d has a closed form expression only if γ is a sphere or a torus, there are many settings where γ may easily be represented as a level set even if d is not available. A simple example is the ellipsoid given by $\gamma = \{\mathbf{x} \in \mathbb{R}^3 : \frac{x^2}{a^2} + \frac{y^2}{b^2} + \frac{z^2}{c^2} - 1 = 0\}$. *Level set methods* in which an evolving free boundary is computationally approximated by the level set of a discrete function are also popular in many applications. In this case it is also natural to define γ via a generic level set function ϕ rather than restrict attention to the distance function d.

Our essential assumptions concerning ϕ are that $\phi \in C^2(\mathcal{N})$ and

$$\nabla\phi(\mathbf{x}) \cdot \boldsymbol{\nu}(\mathbf{x}) \geq c_\phi > 0 \quad \forall \mathbf{x} \in \gamma. \tag{110}$$

Because γ is a level set of ϕ, $|\nabla\phi| = |\nabla\phi \cdot \nu|$ on γ, so the assumption (110) is equivalent to assuming that $\nabla\phi$ is nondegenerate on γ and points in the same direction as $\boldsymbol{\nu} = \nabla d$. Let $\phi_h \in \mathbb{V}(\mathcal{T})$ be an approximation to ϕ satisfying

$$\| \phi - \phi_h \|_{L_\infty(T)} + h_T \| \phi - \phi_h \|_{W^1_\infty(T)} \lesssim h_T^2 \| \phi \|_{W^2_\infty(\mathcal{N})} \quad T \in \mathcal{T}, \tag{111}$$

and define the discrete surface Γ by

$$\Gamma := \{\mathbf{x} \in \mathcal{N} : \quad \phi_h(\mathbf{x}) = 0\}.$$

Lemma 45 (geometric resolution). *Let γ be C^2. Under the above assumptions, the inequality* (108) *holds for $h := \max_{T \in \mathcal{T}} h_T$ sufficiently small, namely*

$$\|d\|_{L_\infty(F)} + h_F \, \|\boldsymbol{\nu} - \boldsymbol{\nu}_\Gamma\|_{L_\infty(F)} \lesssim h_F^2 \, \|\phi\|_{W^2_\infty(\mathcal{N})} \quad F \in \mathcal{F}. \tag{112}$$

Proof. First let $\mathbf{x} \in \mathcal{N}$, for which the projection $\mathbf{P}_d(\mathbf{x})$ on γ is uniquely defined. Let $\zeta(s) := \nabla\phi(s\mathbf{x} + (1-s)\mathbf{P}_d(\mathbf{x})) \cdot \boldsymbol{\nu}(\mathbf{x})$ and compute

$$\begin{aligned}
|(\nabla\phi \cdot \boldsymbol{\nu})(\mathbf{x}) - (\nabla\phi \cdot \boldsymbol{\nu})(\mathbf{P}_d(\mathbf{x}))| &= |\zeta(1) - \zeta(0)| = \left| \int_0^1 \zeta'(s) ds \right| \\
&= \left| \int_0^1 \nabla(\nabla\phi(s\mathbf{x} + (1-s)\mathbf{P}_d(\mathbf{x})) \cdot \boldsymbol{\nu}(\mathbf{x})) \cdot (\mathbf{x} - \mathbf{P}_p(\mathbf{x})) \right| \\
&\leq |\mathbf{x} - \mathbf{P}_d(\mathbf{x})| \; \|\nabla(\nabla\phi \cdot \nu)\|_{L_\infty([\mathbf{P}_d(x), \mathbf{x}])}.
\end{aligned}$$

Since $(\nabla\phi \cdot \boldsymbol{\nu})(\mathbf{P}_d(\mathbf{x})) \geq c_\phi > 0$, $\phi \in C^2(\mathcal{N})$, and $\boldsymbol{\nu} \in C^1(\mathcal{N})$, there thus exists a constant $C_\phi \leq \frac{1}{2K_\infty}$ (depending on $\| \phi \|_{W^2_\infty(\mathcal{N})}$ and $|d|_{W^2_\infty(\mathcal{N})}$) such that

$$(\nabla\phi \cdot \boldsymbol{\nu})(\mathbf{x}) \geq \frac{c_\phi}{2} \quad \forall \mathbf{x} \in \mathcal{N}_\phi := \{\mathbf{y} \in \Omega : \; |d(\mathbf{y})| \leq C_\phi\} \subset \mathcal{N},$$

according to (31). Therefore, for any $\mathbf{x} \in \mathcal{N}_\phi$ we have $\phi(\mathbf{P}_d(\mathbf{x})) = 0$ and

$$|\phi(\mathbf{x})| = \left| \int_0^1 \nabla\phi(s\mathbf{x} + (1-s)\mathbf{P}_d(\mathbf{x})) \cdot (\mathbf{x} - \mathbf{P}_d(\mathbf{x})) \right| \simeq |\mathbf{x} - \mathbf{P}_d(\mathbf{x})| = |d(\mathbf{x})|,$$

because $\mathbf{x} - \mathbf{P}_d(\mathbf{x}) = |\mathbf{x} - \mathbf{P}_d(\mathbf{x})| \nu(\mathbf{x})$. Given any face $F \in \mathcal{F}$ of Γ, we realize that $\phi_h(\mathbf{x}) = 0$ for all $\mathbf{x} \in F$ and

$$|\phi(\mathbf{x})| = |\phi(\mathbf{x}) - \phi_h(\mathbf{x})| \lesssim h_F^2 \, \|\phi\|_{W^2_\infty(\mathcal{N})}.$$

If $h \geq h_F$ is sufficiently small, then $\mathbf{x} \in \mathcal{N}_\phi$ and $|d(\mathbf{x})| \simeq |\phi(\mathbf{x})| \lesssim h_F^2 |\phi|_{W^2_\infty(\mathcal{N})}$. This is the desired bound for the first term on the left-hand side of (112).

To prove the remaining bound in (112), we note that for $\mathbf{x} \in F \in \mathcal{F}$, we have $\boldsymbol{\nu}_\Gamma(\mathbf{x}) = \frac{\nabla\phi_h(\mathbf{x})}{|\nabla\phi_h(\mathbf{x})|}$ and $\boldsymbol{\nu}(\mathbf{x}) = \nu(\mathbf{P}_d(\mathbf{x})) = \frac{\nabla\phi(\mathbf{P}_d(\mathbf{x}))}{|\nabla\phi(\mathbf{P}_d(\mathbf{x}))|}$. Consequently, for such $\mathbf{x} \in \Gamma$, we use (111), the bound $|d(\mathbf{x})| \lesssim h_F^2$ already proved, and the C^2 nature of ϕ to obtain

$$\begin{aligned}
|(\boldsymbol{\nu}_\Gamma - \boldsymbol{\nu})(\mathbf{x})| &= \left| \frac{\nabla\phi_h(\mathbf{x})}{|\nabla\phi_h(\mathbf{x})|} - \frac{\nabla\phi(\mathbf{P}_d(\mathbf{x}))}{|\nabla\phi(\mathbf{P}_d(\mathbf{x}))|} \right| \\
&\leq \left| \frac{\nabla\phi_h(\mathbf{x})}{|\nabla\phi_h(\mathbf{x})|} - \frac{\nabla\phi(\mathbf{x})}{|\nabla\phi(\mathbf{x})|} \right| + \left| \frac{\nabla\phi(\mathbf{x})}{|\nabla\phi(\mathbf{x})|} - \frac{\nabla\phi(\mathbf{P}_d(\mathbf{x}))}{|\nabla\phi(\mathbf{P}_d(\mathbf{x}))|} \right| \\
&\lesssim \left(h_F + h_F^2\right) \|\phi\|_{W^2_\infty(\mathcal{N})} \lesssim h_F \, \|\phi\|_{W^2_\infty(\mathcal{N})}.
\end{aligned}$$

This completes the proof. □

Thus we have shown that it is possible to define the discrete surface Γ using a generic level set representation of γ in such a way that Γ has the same geometric approximation properties as if it were derived more directly from the distance function d. Below we assume practical access to the distance function d and associated geometric properties (curvatures and normal vectors) in two further places: the first one is the definition of the right-hand side F_Γ in formulating the trace FEM and the second one is the definition of geometric a posteriori error estimators. As outlined in Demlow and Dziuk (2007), it is computationally feasible to accurately approximate $d(\mathbf{x})$, $\mathbf{P}_d(x)$, and $\nu(\mathbf{x})$ for $\mathbf{x} \in \Gamma$ under the assumption that we have access to a level set function ϕ with the properties assumed above. In outline, the foundational building block of this procedure is a numerical approximation to $\mathbf{P}_d(\mathbf{x})$. Two such algorithms are proposed in Demlow and Dziuk (2007), one being Newton's method and the other an ad hoc first-order method; cf. Grande (2017) for generalizations and analysis of these methods. Once $\mathbf{P}_d(\mathbf{x})$ is computed, we then have

$$|d(\mathbf{x})| = |\mathbf{x} - \mathbf{P}_d(\mathbf{x})|, \quad \nu(\mathbf{x}) = \frac{\nabla\phi(\mathbf{P}_d(\mathbf{x}))}{|\nabla\phi(\mathbf{P}_d(\mathbf{x}))|}, \quad \mathbf{W}(\mathbf{P}_d(\mathbf{x})) = \nabla\frac{\nabla\phi(\mathbf{P}_d(\mathbf{x}))}{|\nabla\phi(\mathbf{P}_d(\mathbf{x}))|}.$$

These relationships allow for the computation of all geometric information required to bound geometric errors in the trace method a posteriori. In addition, because we may reasonably assume access to $\mathbf{P}_d$ it is in turn reasonable to assume a consistent definition of the right-hand side F_Γ, that is, $F_\Gamma = \frac{q}{q_\Gamma} f \circ \mathbf{P}_d$. A different definition of F_Γ would lead to an additional consistency term in the results below.

5.1.4 Harmonic extension and traces

Here we collect instrumental results for our proofs of a priori and a posteriori error estimates. For the latter we use the fractional-order space $H^{3/2}(\Omega)$, so we first define the seminorm of $H^{1+s}(\Omega)$

$$|v|^2_{H^{1+s}(\Omega)} := \sum_{|\alpha|=1} \iint_{\Omega\times\Omega} \frac{|D^\alpha v(x) - D^\alpha v(y)|^2}{|x-y|^{n+2s}} \, \mathrm{d}x\mathrm{d}y$$

for a Lipschitz domain $\Omega \subset \mathbb{R}^n$ and $0 < s < 1$, and corresponding norm

$$\| v \|^2_{H^{1+s}(\Omega)} = \| v \|^2_{H^1(\Omega)} + |v|^2_{H^{1+s}(\Omega)}.$$

Our first lemma is a standard extension result which may for example be found in Grisvard (1985, Theorem 1.4.3.1).

Lemma 46 (H^{1+s} extension). *Let D be a bounded Lipschitz domain in $\mathbb{R}^n$, $n \geq 2$. Then there is an extension operator $E: H^{1+s}(D) \to H^{1+s}(\mathbb{R}^n)$ such that*

$$\|Ev\|_{H^{1+s}(\mathbb{R}^n)} \lesssim \|v\|_{H^{1+s}(D)} \quad \forall s \in [0,1), \quad \forall v \in H^{1+s}(D). \tag{113}$$

We also state a trace result relating $H^1(\mathbb{R}^2)$ and $H^{3/2}(\mathbb{R}^3)$; this is a special case of Adams (1975, Theorem 7.58).

Lemma 47 (trace). *If $v \in H^{3/2}(\mathbb{R}^n)$, $n \geq 2$, and $\mathbb{P}$ is any $(n-1)$-dimensional hyperplane in $\mathbb{R}^n$, then*

$$\|v\|_{H^1(\mathbb{P})} \lesssim \|v\|_{H^{3/2}(\mathbb{R}^n)}. \tag{114}$$

The following is an important technical lemma which expresses traces relationships between norms on surface elements (flat or curved) and corresponding norms on bulk elements. An essential component of these estimates is that they allow for surfaces to cut through bulk elements in an arbitrary fashion. Such estimates were essential in the proof of the first a posteriori estimates for trace methods in Demlow and Olshanskii (2012). In the context of a priori error estimates for trace methods, these provide a substantially simplified proof of error bounds when compared with the original proofs given in Olshanskii et al. (2009) (cf. Burman et al., 2015; Hansbo and Hansbo, 2002, 2004; Reusken, 2015).

Lemma 48 (trace estimates for cut elements). *Let $D \subset \mathbb{R}^n$ $(n \geq 2)$ be a (not necessarily bounded) Lipschitz domain, and let D_{n-1} be the intersection of D with an arbitrary hyperplane of dimension $n-1$. Then*

$$\|v\|_{L_2(D_{n-1})} \lesssim \|v\|_{H^1(D)} \quad \forall v \in H^1(D), \tag{115}$$

where the hidden constant depends on the Lipschitz nature of D but not on the orientation or size of D_{n-1}. In particular, let $F \in \mathcal{F}$ with $F \subset T \in \mathcal{T}$. Then

$$\|v\|_{L_2(F)} \lesssim h_T^{-1/2} \|v\|_{L_2(T)} + h_T^{1/2} \|\nabla v\|_{L_2(T)} \quad \forall v \in H^1(T). \tag{116}$$

In addition, given $T \in \mathcal{T}$ there hold

$$\|v\|_{L_2(T\cap\gamma)} \lesssim h_T^{-1/2} \|v\|_{L_2(T)} + h_T^{1/2} \|\nabla v\|_{L_2(T)} \quad \forall v \in H^1(T), \tag{117}$$

and

$$\begin{aligned} &h_T^{-1} \|v\|_{L_2(T\cap\gamma)} + \|\nabla_\gamma v\|_{L_2(T\cap\gamma)} \\ &\lesssim h_T^{-3/2} \|v\|_{L_2(T)} + h_T^{-1/2} \|\nabla v\|_{L_2(T)} + |v|_{H^{3/2}(T)} \quad \forall v \in H^{3/2}(T). \end{aligned} \tag{118}$$

Proof. The estimate (115) is a special case of Adams (1975, Lemma 5.19). The scaled result (116) follows by a standard scaling argument.

To prove (117) and (118) we employ a flattening argument. First let $\widehat{K}$ be the unit reference simplex in $\mathbb{R}^n$ with standard affine reference mapping $\varphi : \widehat{K} \to T$ satisfying $\|\nabla\varphi\|_{L_\infty(\widehat{K})} \lesssim h_T$ and $\|(\nabla\varphi)^{-1}\|_{L_\infty(T)} \lesssim h_T^{-1}$. Let now Φ be the flattening map in assumption (109). It is possible to extend Φ to all of $\mathbb{R}^n$ so that the resulting extension is also C^2, still flattens $T \cap \gamma$, and has derivative bounded above and below away from 0. To see this, take a smoothly weighted average of Φ and the identity with weight 1 for Φ on $B_{R/2}$ and weight 0 outside of B_R. Having thus extended Φ, we define $\widetilde{\Phi} := \varphi^{-1} \circ \Phi \circ \varphi$. It is easy to check that $\widetilde{\Phi}$ and $\widetilde{\Phi}^{-1}$ are uniformly bounded in C^2 and that $\widetilde{\Phi}(\varphi^{-1}(T\cap\gamma))$ lies in some $(n-1)$-dimensional hyperplane $\mathbb{P}$.

For $v \in H^1(T)$ with $T \in \mathcal{T}$ satisfying $|T \cap \gamma| > 0$, let now $\widehat{v} = v \circ \varphi$. We first prove (118) upon transforming to the reference element back and forth. We start with a simple scaling argument

$$\frac{|\varphi^{-1}(T\cap\gamma)|}{|T\cap\gamma|} \approx h_T^{1-n},$$

regardless of the actual size and orientation of $T \cap \gamma$ relative to $T \in \mathcal{T}$. Hence, applying a standard change of variables involving φ yields

$$h_T^{1-n}\left(\|v\|_{L_2(T\cap\gamma)}^2 + h_T^2\|\nabla_\gamma v\|_{L_2(T\cap\gamma)}^2\right) \approx \|\widehat{v}\|_{H^1(\varphi^{-1}(T\cap\gamma))}^2 .$$

We next resort to the extension operator $E : H^{3/2}(\widehat{K}) \to H^{3/2}(\mathbb{R}^n)$ in Lemma 46 (H^{1+s} extension), the smoothness of $\widetilde{\Phi}^{-1}$, the fact that $\widetilde{\Phi}(\varphi^{-1}(T\cap\gamma)) \subset \mathbb{P}$, the trace inequality (114), the smoothness of $\widetilde{\Phi}^{-1}$ again, and the boundedness (113) of E in $H^{3/2}(\widehat{K})$, in this order, to arrive at

$$\begin{aligned}
\|\widehat{v}\|_{H^1(\varphi^{-1}(T\cap\gamma))} &= \|E\widehat{v}\|_{H^1(\varphi^{-1}(T\cap\gamma))}\\
&\lesssim \|E\widehat{v}\circ\widetilde{\Phi}^{-1}\|_{H^1\left(\widetilde{\Phi}(\varphi^{-1}(T\cap\gamma))\right)}\\
&\lesssim \|E\widehat{v}\circ\widetilde{\Phi}^{-1}\|_{H^1(\mathbb{P})}\\
&\lesssim \|E\widehat{v}\circ\widetilde{\Phi}^{-1}\|_{H^{3/2}(\mathbb{R}^n)}\\
&\lesssim \|E\widehat{v}\|_{H^{3/2}(\mathbb{R}^n)}\\
&\lesssim \|\widehat{v}\|_{H^{3/2}(\widehat{K})}.
\end{aligned}$$

The desired estimate (118) finally follows from a scaling argument from $\widehat{K}$ to T employing again the map φ:

$$\|\widehat{v}\|_{H^{3/2}(\widehat{K})}^2 \lesssim h_T^{-n}\left(\|v\|_{L_2(T)}^2 + h_T^2\|\nabla v\|_{L_2(T)}^2 + h_T^3|v|_{H^{3/2}(T)}^2\right).$$

To prove (117), we argue similarly to above except that we now employ $E : H^1(\widehat{K}) \to H^1(\mathbb{R}^n)$ and (115) instead of (114). Doing so yields

$$\begin{aligned}
h_T^{(1-n)/2}\|v\|_{L_2(T\cap\gamma)} &\lesssim \|\widehat{v}\|_{L_2(\varphi^{-1}(T\cap\gamma))}\\
&= \|E\widehat{v}\|_{L_2(\varphi^{-1}(T\cap\gamma))}\\
&\lesssim \|E\widehat{v}\circ\widetilde{\Phi}^{-1}\|_{L_2(\tilde{\Phi}(\varphi^{-1}(T\cap\gamma)))}\\
&\lesssim \|E\widehat{v}\circ\widetilde{\Phi}^{-1}\|_{L_2(\mathbb{P})}\\
&\lesssim \|E\widehat{v}\circ\widetilde{\Phi}^{-1}\|_{H^1(\mathbb{R}^n)}\\
&\lesssim \|E\widehat{v}\|_{H^1(\mathbb{R}^n)}\\
&\lesssim \|\widehat{v}\|_{H^1(\hat{K})}\\
&\lesssim h_T^{-n/2}\|v\|_{L_2(T)} + h_T^{(2-n)/2}\|\nabla v\|_{L_2(T)}.
\end{aligned}$$

Multiplying both sides by $h_T^{(n-1)/2}$ gives the desired bound (117). □

5.2 A priori error estimates

We recall that we use the notation $h := \max_{T\in\mathcal{T}} h_T$ and that we omit to mention the explicit dependence on the shape-regularity constant of $\mathcal{T}$ in most estimates.

5.2.1 Geometric resolution and extensions

Given a surface γ of class C^2 and $\widetilde{u} \in H^2(\gamma)$, Proposition 28 ($H^2$ extension) yields the existence of an extension u of $\widetilde{u}$ to a tubular neighbourhood $\mathcal{N}(\delta)$ with δ sufficiently small with respect to $\frac{1}{2K_\infty}$ lying in $H^2(\mathcal{N}(\delta))$ and satisfying

$$\|u\|_{H^2(\mathcal{N}(\delta))} \lesssim \delta^{1/2}|d|_{W^2_\infty(\mathcal{N})} \|\widetilde{u}\|_{H^2(\gamma)}. \tag{119}$$

- **First assumption on geometric resolution by the bulk mesh**. We assume

$$\bigcup\{\omega_T^1(T):\ T\in\mathcal{T}_\Gamma\}\subset\mathcal{N}(\delta) \tag{120}$$

  with $\delta \simeq h$ sufficiently small so that (119) holds.
- **Second assumption on geometric resolution by the bulk mesh**. We assume that the layer $D_{\Gamma,\gamma} := \{s\mathbf{x} + (1 - s)\mathbf{P}_d(\mathbf{x}) : \mathbf{x} \in \Gamma \text{ and } 0 \le s \le 1\}$ satisfies

$$D_{\Gamma,\gamma}\subset\bigcup\{T:\ T\in\mathcal{T}_\Gamma\}. \tag{121}$$

  This clearly holds for h sufficiently small because the Hausdorff distance between γ and Γ satisfies $\mathrm{dist}_H(\gamma,\Gamma) \lesssim h^2$ according to (108).
- **Uniform Poincaré–Friedrichs estimate on Γ**. We assume that

$$\|v\|_{L_2(\Gamma)} \lesssim \|\nabla v\|_{L_2(\Gamma)} \quad \forall v\in H^1_\#(\Gamma) \tag{122}$$

  holds with uniform constant. According to the discussion below (82) (uniform Poincaré–Friedrichs constant), this only requires that $\Gamma\subset\mathcal{N}(1/2K_\infty)$ and that $\boldsymbol{\nu}\cdot\boldsymbol{\nu}_\Gamma \ge c > 0$ on Γ. These conditions are easily checkable and valid asymptotically.

5.2.2 Approximation properties of trace finite element space

We next state a fundamental approximation bound for the trace FEM, which we prove under the regularity assumption that γ is of class C^2. We emphasize that this assumption is less restrictive than the hypotheses of previous approximation bounds for trace estimates, which assume that γ is of class C^3.

Lemma 49 (trace approximation). *Let γ be of class C^2 and the geometric resolution assumptions* (108), (109), (120), *and* (121) *hold. Then*

$$\inf_{V\in\mathbb{V}(\mathcal{F})} \|\nabla_\Gamma(\widetilde{u}\circ\boldsymbol{P}_d - V)\|_{L_2(\Gamma)} \lesssim h\,\|\widetilde{u}\|_{H^2(\gamma)}. \tag{123}$$

Proof. Let $\delta \simeq h$ be sufficiently small so that (120) is valid. Let $\mathcal{I}_{\mathcal{T}}^{\mathrm{sz}}$ be the standard Scott–Zhang interpolation operator on $\mathcal{T}$, and take $V = \mathcal{I}_{\mathcal{T}}^{\mathrm{sz}} u$ with $u \in H^2(\mathcal{N}(\delta))$ given by Proposition 28 (H^2 extension) and satisfying (119). We then denote $u_d = \widetilde{u} \circ \mathbf{P}_d, V_d = V \circ \mathbf{P}_d$, add and subtract multiple terms, and apply the triangle inequality to find that

$$\| \nabla_\Gamma (u_d - V) \|_{L_2(\Gamma)} \lesssim \sum_{i=1}^{7} I_i$$

where

$$\begin{aligned}
I_1 &:= \| \nabla_\Gamma (u_d - V_d) \|_{L_2(\Gamma)}, \\
I_2 &:= \| \Pi_\Gamma [\nabla V_d - (\nabla V) \circ \mathbf{P}_d] \|_{L_2(\Gamma)}, \\
I_3 &:= \| \Pi_\Gamma [\nabla V \circ \mathbf{P}_d - \nabla u \circ \mathbf{P}_d] \|_{L_2(\Gamma)}, \\
I_4 &:= \| \Pi_\Gamma [\nabla u \circ \mathbf{P}_d - (\mathcal{I}_{\mathcal{T}}^{\mathrm{sz}} \nabla u) \circ \mathbf{P}_d] \|_{L_2(\Gamma)}, \\
I_5 &:= \| \Pi_\Gamma [(\mathcal{I}_{\mathcal{T}}^{\mathrm{sz}} \nabla u) \circ \mathbf{P}_d - \mathcal{I}_{\mathcal{T}}^{\mathrm{sz}} \nabla u] \|_{L_2(\Gamma)}, \\
I_6 &:= \| \Pi_\Gamma [\mathcal{I}_{\mathcal{T}}^{\mathrm{sz}} \nabla u - \nabla u] \|_{L_2(\Gamma)}, \\
I_7 &:= \| \Pi_\Gamma [\nabla u - \nabla V] \|_{L_2(\Gamma)}.
\end{aligned}$$

Here we have applied the interpolation operator $\mathcal{I}_{\mathcal{T}}^{\mathrm{sz}}$ componentwise to the $(n + 1)$-vector ∇u and used that $\nabla_\Gamma = \Pi_\Gamma \nabla$. We next estimate each term separately.

In order to bound terms I_1 and I_3, we employ Lemma 17 (norm equivalence) between γ and Γ and recall that $|\nabla_\gamma v| \leq |\nabla v|$ pointwise to find that

$$I_1 + I_3 \lesssim \left(\sum_{T \in \mathcal{T}_\Gamma} \| \nabla (u - V) \|_{L_2(T \cap \gamma)}^2 \right)^{1/2}.$$

We next apply the trace estimate (117), utilize standard approximation properties of $\mathcal{I}_{\mathcal{T}}^{\mathrm{sz}}$, and finally use the bound (119) to obtain

$$\begin{aligned}
I_1 + I_3 &\lesssim \left(\sum_{T \in \mathcal{T}_\Gamma} h_T^{-1} \| \nabla (u - V) \|_{L_2(T)}^2 + h_T \| D^2 u \|_{L_2(T)}^2 \right)^{1/2} \\
&\lesssim h^{1/2} \| u \|_{H^2(\mathcal{N}(\delta))} \lesssim h \| \widetilde{u} \|_{H^2(\gamma)}.
\end{aligned}$$

Here we have used that $\nabla \nabla V = 0$ elementwise since V is piecewise linear. Similar arguments lead to the following estimate for I_4

$$I_4 \lesssim \left(\sum_{T \in \mathcal{T}_\Gamma} h_T^{-1} \| \nabla u - \mathcal{I}_{\mathcal{T}}^{\mathrm{sz}} \nabla u \|_{L_2(T)}^2 + h_T \| \nabla (\nabla u - \mathcal{I}_{\mathcal{T}}^{\mathrm{sz}} \nabla u) \|_{L_2(T)}^2 \right)^{1/2},$$

as well as $I_4 \lesssim h\,\|\widetilde{u}\|_{H^2(\gamma)}$ provided $\|\nabla\mathcal{I}_T^{\rm sz}\nabla u\|_{L_2(T)} \lesssim \|D^2u\|_{L_2(\omega_T^1(T))}$. To show this estimate we let $\overline{\nabla u}_T := |\omega_T^1(T)|^{-1}\int_{\omega_T^1(T)}\nabla u$ be the mean value of ∇u in $\omega_T^1(T)$ and exploit the stability of $\mathcal{I}_T^{\rm sz}$ in $H^1(T)$

$$\begin{aligned}\|\nabla\mathcal{I}_T^{\rm sz}\nabla u\|_{L_2(T)} &= \|\nabla\mathcal{I}_T^{\rm sz}[\nabla u-\overline{\nabla u}_T]\|_{L_2(T)} \lesssim h_T^{-1}\,\|\mathcal{I}_T^{\rm sz}[\nabla u-\overline{\nabla u}_T]\|_{L_2(T)}\\ &\lesssim h_T^{-1}\,\|\nabla u-\overline{\nabla u}_T\|_{L_2(\omega_T^1(T))} \lesssim \|D^2u\|_{L_2(\omega_T^1(T))}.\end{aligned}$$

Moreover, applying the trace estimate (116) directly to the terms I_6 and I_7 yields

$$\begin{aligned}I_6 &\lesssim \left(\sum_{F\in\mathcal{F}}\|\mathcal{I}_T^{\rm sz}\nabla u-\nabla u\,\|_{L_2(F)}^2\right)^{1/2}\\ &\lesssim \left(\sum_{T\in\mathcal{T}_\Gamma} h_T^{-1}\,\|\mathcal{I}_T^{\rm sz}\nabla u-\nabla u\,\|_{L_2(T)}^2 + h_T\,\|\nabla[\mathcal{I}_T^{\rm sz}\nabla u-\nabla u]\,\|_{L_2(T)}^2\right)^{1/2} \lesssim h\,\|\widetilde{u}\|_{H^2(\gamma)},\end{aligned}$$

and

$$\begin{aligned}I_7 &\lesssim \left(\sum_{F\in\mathcal{F}}\|\nabla(u-V)\|_{L_2(F)}\right)^{1/2}\\ &\lesssim \left(\sum_{T\in\mathcal{T}_\Gamma} h_T^{-1}\,\|\nabla(u-V)\,\|_{L_2(T)}^2 + h_T\,\|D^2u\,\|_{L_2(T)}^2\right)^{1/2} \lesssim h\,\|\widetilde{u}\|_{H^2(\gamma)}.\end{aligned}$$

In order to bound term I_2, we first note that

$$\Pi_\Gamma[\nabla V_d-(\nabla V)\circ\mathbf{P}_d] = \Pi_\Gamma(\Pi-dD^2d-\mathbf{I})(\nabla V)\circ\mathbf{P}_d.$$

An easy computation using the assumption (108) yields

$$|\Pi_\Gamma(\Pi-dD^2d-\mathbf{I})| \lesssim |\Pi_\Gamma\Pi-\Pi_\Gamma|+|d| = |(\boldsymbol{\nu}\cdot\boldsymbol{\nu}_\Gamma)\boldsymbol{\nu}_\Gamma\otimes\boldsymbol{\nu}-\boldsymbol{\nu}\otimes\boldsymbol{\nu}|+h^2 \lesssim h.$$

Thus employing the equivalence of norms on γ and Γ, the trace estimate (117), the H^1 boundedness of $\mathcal{I}_T^{\rm sz}$, and the boundedness (119) of the extension yields

$$\begin{aligned}I_2 &\lesssim h\,\|\nabla V\circ\mathbf{P}_d\|_{L_2(\Gamma)} \lesssim h\,\|\nabla V\|_{L_2(\gamma)}\\ &\lesssim h^{1/2}\,\|\nabla V\|_{L_2(\mathcal{T}_\Gamma)} \lesssim h^{1/2}\,\|u\|_{H^1(\mathcal{N}(\delta))} \lesssim h\,\|\widetilde{u}\|_{H^2(\gamma)}.\end{aligned}$$

We finally bound term I_5. Given $\mathbf{x} = \mathbf{P}_d(\mathbf{x}) + d(\mathbf{x})\nabla d(\mathbf{P}_d(\mathbf{x})) \in \Gamma$, we infer that

$$|\mathcal{I}_T^{\mathrm{sz}}\nabla u(\mathbf{x}) - \mathcal{I}_T^{\mathrm{sz}}\nabla u(\mathbf{P}_d(\mathbf{x}))| \le \int_0^{d(\mathbf{x})} \left|\nabla\left[\mathcal{I}_T^{\mathrm{sz}}\nabla u(\mathbf{P}_d(\mathbf{x}) + s\nabla d(\mathbf{P}_d(\mathbf{x})))\right]\right| ds$$

and $|d(\mathbf{x})| \lesssim h^2$ according to (108), whence

$$I_5^2 \lesssim h^2 \int_\Gamma \int_0^{d(\mathbf{x})} \left|\nabla\left[\mathcal{I}_T^{\mathrm{sz}}\nabla u(\mathbf{P}_d(\mathbf{x}) + s\nabla d(\mathbf{P}_d(\mathbf{x})))\right]\right|^2 ds d\sigma(\mathbf{x}) \lesssim h^2 \int_{D_{\Gamma,\gamma}} |\nabla \mathcal{I}_T^{\mathrm{sz}}\nabla u|^2.$$

In view of assumptions (121) and (120), and the bound $\|\nabla\mathcal{I}_T^{\mathrm{sz}}\nabla u\|_{L_2(T)} \lesssim \|D^2 u\|_{L_2(\omega_T^1(T))}$, we deduce

$$I_5^2 \lesssim h^2 \, \|D^2 u\,\|^2_{L_2(\mathcal{N}(\delta))} \lesssim h^3 \, \|D^2 \widetilde{u}\,\|^2_{H^2(\gamma)},$$

and conclude the proof. □

Theorem 50 (a priori error estimates)**.** *Let γ be of class C^2 and let Γ be so that the geometric assumptions* (108), (109), (120), (121), *and* (122) *are satisfied. Let $\widetilde{f} \in L_{2,\#}(\gamma)$ and $\widetilde{u} \in H^2(\gamma)$ solve (19). If $U \in \mathbb{V}_\#(\mathcal{F})$ is the finite element solution of (107) with $F_\Gamma = \frac{q}{q_\Gamma}\widetilde{f} \circ \mathbf{P}_d$, then*

$$\|\widetilde{u}\circ\mathbf{P}_d - U\|_{L_2(\Gamma)} + h\,\|\nabla_\Gamma(\widetilde{u}\circ\mathbf{P}_d - U)\|_{L_2(\Gamma)} \lesssim h^2\,\|\widetilde{f}\|_{L_2(\gamma)}.$$

Proof. With the geometric resolution estimate (108) and Lemma 49 (trace approximation) in hand, the proof is nearly identical to those of Theorem 37 (H^1 a priori error estimate) and Theorem 38 (L^2 a priori error estimate) for parametric surface FEM. We thus sketch the proof without details.

Step 1: H^1 error estimate. Let $V \in \mathbb{V}(\mathcal{F})$ achieve the infimum in (123), $W := V - U$, $u = \widetilde{u}\circ\mathbf{P}_d$, and write the error representation formula as

$$\|\nabla_\Gamma(V-U)\,\|^2_{L_2(\Gamma)} = \int_\Gamma \nabla_\Gamma u\cdot\mathbf{E}_\Gamma\nabla_\Gamma W + \int_\Gamma \nabla_\Gamma(V-u)\cdot\nabla_\Gamma W,$$

because $F_\Gamma = \frac{q}{q_\Gamma}\widetilde{f}\circ\mathbf{P}_d$. In view of Lemma 21 (geometric consistency) and the geometric resolution estimate (108) we deduce $|\mathbf{E}_\Gamma| \lesssim h^2 |d|_{W^2_\infty(\mathcal{N})}$ and

$$\left|\int_\Gamma \nabla_\Gamma u\cdot\mathbf{E}_\Gamma\nabla_\Gamma W\right| \lesssim h^2 |d|_{W^2_\infty(\mathcal{N})}\,\|\widetilde{u}\|_{H^1(\gamma)}\,\|\nabla_\Gamma W\|_{L_2(\Gamma)} \lesssim h\,\|\widetilde{f}\|_{L_2(\gamma)}\,\|\nabla_\Gamma W\|_{L_2(\Gamma)}.$$

On the other hand, Lemma 49 (trace approximation) yields

$$\left|\int_\Gamma \nabla_\Gamma(V-u)\cdot\nabla_\Gamma W\right| \lesssim h\,\|\widetilde{u}\|_{H^2(\gamma)}\,\|\nabla_\Gamma W\|_{L_2(\Gamma)}.$$

The desired estimate follows from Lemma 3 (regularity).

Step 2: L_2 error estimate. Let $\mathbf{P}_d^{-1}$ denote the inverse of $\mathbf{P}_d$ restricted to Γ. Let $\widetilde{U} := U \circ \mathbf{P}_d^{-1} : \gamma \to \mathbb{R}$ and $\widetilde{U}_\# := \frac{q_\Gamma}{q} \widetilde{U} \in H^1_\#(\gamma)$; likewise, let $u_\# := \frac{q}{q_\Gamma} u \in H^1_\#(\Gamma)$. We now solve dual problems on γ

$$\widetilde{z} \in H^1_\#(\gamma): \quad \int_\gamma \nabla_\gamma \widetilde{z} \cdot \nabla_\gamma w = \int_\gamma (\widetilde{u} - \widetilde{U}_\#) w \quad \forall w \in H^1_\#(\gamma)$$

and on Γ

$$Z \in \mathbb{V}_\#(\mathcal{F}): \quad \int_\Gamma \nabla_\Gamma Z \cdot \nabla_\Gamma W = \int_\Gamma (u_\# - U) W \quad \forall W \in \mathbb{V}_\#(\mathcal{F}).$$

Note that the right-hand sides $u_\# - U = \frac{q}{q_\Gamma} (\widetilde{u} - \widetilde{U}_\#) \circ \mathbf{P}_d$ are compatible and Step 1 applies. We set $\widetilde{Z} = Z \circ \mathbf{P}_d$ and proceed as in Theorem 38 (L_2 a priori error estimate) to deduce the error representation

$$\|\widetilde{u} - \widetilde{U}_\#\|^2_{L_2(\gamma)} = \int_\gamma \nabla_\gamma (\widetilde{u} - \widetilde{U}) \cdot \nabla_\gamma (\widetilde{z} - \widetilde{Z}) + \int_\gamma \nabla_\gamma \widetilde{U} \cdot \mathbf{E} \nabla_\gamma \widetilde{Z},$$

because $F_\Gamma = \frac{q}{q_\Gamma} \widetilde{f} \circ \mathbf{P}_d$. Applying Lemma 3 (regularity) to $\widetilde{z}$ yields $\|\widetilde{z}\|_{H^2(\gamma)} \lesssim \|\widetilde{u} - \widetilde{U}_\#\|_{L_2(\gamma)}$. This together with Step 1 implies

$$\left| \int_\gamma \nabla_\gamma (\widetilde{u} - \widetilde{U}) \cdot \nabla_\gamma (\widetilde{z} - \widetilde{Z}) \right| \lesssim h^2 \|\widetilde{f}\|_{L_2(\gamma)} \|\widetilde{u} - \widetilde{U}_\#\|_{L_2(\gamma)}.$$

Making use again of Lemma 21 (geometric consistency) and the geometric resolution estimate (108) we deduce $|\mathbf{E}| \lesssim h^2 |d|_{W^2_\infty(\mathcal{N})}$, whence

$$\left| \int_\gamma \nabla_\gamma \widetilde{U} \cdot \mathbf{E} \nabla_\gamma \widetilde{Z} \right| \lesssim h^2 \|\widetilde{f}\|_{L_2(\gamma)} \|u_\# - U\|_{L_2(\Gamma)}.$$

Consequently,

$$\|\widetilde{u} - \widetilde{U}_\#\|^2_{L_2(\gamma)} \lesssim h^2 \|\widetilde{f}\|_{L_2(\gamma)} \Big(\|\widetilde{u} - \widetilde{U}_\#\|_{L_2(\gamma)} + \|u_\# - U\|_{L_2(\Gamma)} \Big)$$

and the asserted bound follows from Lemma 17 (norm equivalence) and the auxiliary estimate

$$\|\widetilde{U} - \widetilde{U}_\#\|_{L_2(\gamma)} \lesssim h^2 \|\widetilde{f}\|_{L_2(\gamma)}.$$

The latter hinges on Corollary 33 (geometric consistency errors for C^2 surfaces) and the geometric resolution estimate (108), as in the proof of Theorem 38. This completes the proof. □

5.3 A posteriori error estimates

A posteriori error estimates for the trace FEM were first proved in Demlow and Olshanskii (2012), while a posteriori estimates for a trace FEM based on octree meshes were proved in Chernyshenko and Olshanskii (2015). The proof of the estimates given in Demlow and Olshanskii (2012) is significantly different than that of the a priori estimates given above. A main reason for the difference is that, in contrast to the framework above that deals with quasi-uniform meshes, we assume that the bulk mesh $\mathcal{T}$ is merely shape-regular. This is necessary to allow for meaningful mesh grading in adaptive algorithms. Moreover, the extension used in Proposition 28 (H^2 extension) is not immediately useful here because the parameter δ specifying the width of the tubular neighbourhood about γ is taken to be proportional to h when $\mathcal{T}$ is quasi-uniform; such a global mesh size parameter is no longer meaningful on graded meshes. A local counterpart of Proposition 28 on graded meshes, that uses the normal extension instead of the regularized normal extension, is employed in Chernyshenko and Olshanskii (2015) to prove a posteriori bounds, but with the drawback that the constants in the estimates depend on the difference in refinement depth between the largest and smallest elements in the bulk mesh. We thus present here the framework of Demlow and Olshanskii (2012), which relies on the harmonic extension of $v \in H^1(\gamma)$ into $H^{3/2}(\mathbb{R}^3)$ instead of either the normal extension v_d or the extension of Proposition 28.

5.3.1 *Assumptions on surface representation*

Because the trace method derives the mesh from a level set representation, it is reasonable to assume that we have computational access to a C^2 level set function ϕ satisfying (110) such that $\gamma = \{\mathbf{x} \in \mathcal{N} : \phi(\mathbf{x}) = 0\}$. As outlined in Demlow and Dziuk (2007), it is then computationally feasible to accurately approximate $d(\mathbf{x})$, $\mathbf{P}_d(x)$, and $\nu(\mathbf{x})$ for $\mathbf{x} \in \Gamma$. In outline, the foundational building block of this procedure is a numerical approximation to $\mathbf{P}_d(\mathbf{x})$. Two such algorithms are proposed in Demlow and Dziuk (2007), one being Newton's method and the other an ad hoc first-order method; cf. Grande (2017) for generalizations and analysis of these methods. Once $\mathbf{P}_d(\mathbf{x})$ is computed, we then have

$$|d(\mathbf{x})| = |\mathbf{x} - \mathbf{P}_d(\mathbf{x})|, \quad \nu(\mathbf{x}) = \frac{\nabla\phi(\mathbf{P}_d(\mathbf{x}))}{|\nabla\phi(\mathbf{P}_d(\mathbf{x}))|},$$

$$\mathbf{W}(\mathbf{P}_d(\mathbf{x})) = \nabla\nu(\mathbf{P}_d(\mathbf{x})) = \nabla\frac{\nabla\phi(\mathbf{P}_d(\mathbf{x}))}{|\nabla\phi(\mathbf{P}_d(\mathbf{x}))|}.$$

These relationships allow for the computation of all geometric information required to bound geometric errors in the trace method a posteriori. In addition, because we may assume access to $\mathbf{P}_d$ it is in turn reasonable to assume a consistent definition of the right-hand side F_Γ, that is, $F_\Gamma = \frac{q}{q_\Gamma} f \circ \mathbf{P}_d$. A different definition of F_Γ would lead to an additional consistency term in the results below.

5.3.2 Notation and surface resolution assumptions

We make the following two assumptions concerning resolution of γ by the bulk mesh $\mathcal{T}$:

- **Resolution of skin layer between γ and Γ.** Given a discrete surface element $F \in \mathcal{F}$, let

  $D_F = \{\mathbf{y} \in \Omega: \ \mathbf{y} = t\mathbf{x} + (1-t)\mathbf{P}_d(\mathbf{x}) \text{ for some } 0 \leq t \leq 1 \text{ and some } \mathbf{x} \in F\}.$

  The set D_F is the collection of all points lying on line segments connecting points in $\mathbf{x} \in F$ and their images $\mathbf{P}_d(\mathbf{x}) \in \gamma$. We assume that

  $$D_F \subset \omega_{\mathcal{T}}^1(T_F), \tag{124}$$

  that is, D_F lies in the volume element patch $\omega_{\mathcal{T}}^1(T_F)$ (first ring) corresponding to the face element F, which is defined in Section 5.1.
- **Normal projections of elements have finite overlap.** We assume that

  $$\mathbf{P}_d\left(\omega_{\mathcal{T}}^1(T_F)\right) \subset \omega_{\mathcal{T}}^2(T_F) \ \ \forall F \in \mathcal{F}, \tag{125}$$

  where the second ring $\omega_{\mathcal{T}}^2(T_F)$ is also defined in Section 5.1.

The above assumptions hold if γ is sufficiently resolved by the bulk mesh $\mathcal{T}$. To see this, note first that $\|d\|_{L_\infty(D_F)} \lesssim h_F^2$ by (108), so that $\text{dist}(\mathbf{y}, F) \lesssim h_F^2$ for all $\mathbf{y} \in D_F$. On the other hand, $\text{dist}(F, \partial\omega_{\mathcal{T}}^1(T_F)) \gtrsim h_F$. Thus there is a constant C such that the assumption (124) is satisfied when $h_F \leq C$; C here depends on geometric properties of γ, the shape-regularity constant of $\mathcal{T}$ and properties of the Lagrange interpolant. In principle an upper bound for C could be computed and this condition checked, but this has not been attempted in the literature and we do not do so here. A similar but more involved argument holds for the assumption (125).

5.3.3 Extension for a posteriori error estimates

The next essential result states that a given a function $\widetilde{v} \in H^1(\gamma)$ can be boundedly extended to $v \in H^{3/2}(\mathbb{R}^{n+1})$.

Lemma 51 (harmonic extension). *Let γ be a closed surface of class C^2 and dimension n embedded in $\mathbb{R}^{n+1}$ for $n \geq 1$. Given $\widetilde{v} \in H^1(\gamma)$, there is $v \in H^{3/2}(\mathbb{R}^n)$ such that* $\text{trace}(v) = \widetilde{v}$ *and*

$$\|v\|_{H^{3/2}(\mathbb{R}^{n+1})} \lesssim \|\widetilde{v}\|_{H^1(\gamma)}. \tag{126}$$

Proof. First let $v \in H^1(D)$ solve $\Delta v = 0$ on the bulk domain D comprising the interior of γ, with $v = \widetilde{v}$ on γ. By Jerison and Kenig (1995, Theorem 5.15), we have that $v \in H^{3/2}(D)$, $\text{trace}(v) = \widetilde{v}$, and $\|v\|_{H^{3/2}(D)} \lesssim \|\widetilde{v}\|_{H^1(\gamma)}$. Boundedly extending v to $H^{3/2}(\mathbb{R}^{n+1})$ via the extension operator E defined in Lemma 46 (H^{1+s} extension) completes the proof. □

5.3.4 Preliminary results

We now give a technical lemma that quantifies the evaluation mismatch between Γ and γ for a discrete function.

Lemma 52 (evaluation mismatch between γ and Γ). *Let $V \in \mathbb{V}(\mathcal{T})$, and let the conditions (108), (109), and (124) hold. For all $F \in \mathcal{F}$, we have*

$$\| V - V \circ \boldsymbol{P}_d \|_{L_\infty(F)} \lesssim h_F^2 |d|_{W^2_\infty(\mathcal{N})} \, \| \nabla V \|_{L_\infty(\omega^1_{\mathcal{T}}(T_F))}. \tag{127}$$

Proof. Fix $\mathbf{x} \in F$, and let $g(t) = V(t\mathbf{x} + (1 - t)\mathbf{P}_d(\mathbf{x}))$, $0 \le t \le 1$. Then $g(0) = V(\mathbf{P}_d(\mathbf{x}))$ and $g(1) = V(\mathbf{x})$. Since $\mathbf{P}_d(\mathbf{x}) = \mathbf{x} - d(\mathbf{x})\nu(\mathbf{x})$, we see that

$$g'(t) = \nabla V(t\mathbf{x} + (1-t)\mathbf{P}_d(\mathbf{x})) \cdot (\mathbf{x} - \mathbf{P}_d(\mathbf{x})) = d(\mathbf{x}) \nabla V(t\mathbf{x} + (1-t)\mathbf{P}_d(\mathbf{x})) \cdot \nu(\mathbf{x}),$$

whence $V(\mathbf{x}) - V(\mathbf{P}_d(\mathbf{x})) = g(0) - g(1) = \int_0^1 g'(t)dt$ and

$$|V(\mathbf{x}) - V(\mathbf{P}_d(\mathbf{x}))| \lesssim |d(\mathbf{x})| \, \| \nabla V \|_{L_\infty(D_F)}.$$

The assertion follows from assumptions (108) and (124). □

5.3.5 A posteriori upper bound

First we define a residual error indicator

$$\eta_{\mathcal{F}}(U,F) := h_F \, \| F_\Gamma + \Delta_\Gamma U \|_{L_2(F)} + h_F^{1/2} \, \| [\![\nabla_\Gamma U]\!] \|_{L_2(\partial F)} \quad \forall F \in \mathcal{F},$$

and corresponding estimator

$$\eta_{\mathcal{F}}(U) := \left(\sum_{F \in \mathcal{F}} \eta_{\mathcal{F}}(U,F)^2 \right)^{1/2}.$$

Here $[\![\cdot]\!]$ denotes the jump in the normal component of the argument over ∂F. Because we have assumed access to the closest point projection $\mathbf{P}_d$, we also employ a geometric indicator that directly accesses information from $\mathbf{P}_d$

$$\xi_F := \| d \|_{L_\infty(F)} \, \| K \|_{L_\infty(\mathbf{P}_d(F))} + \| \boldsymbol{\nu} - \boldsymbol{\nu}_\Gamma \|^2_{L_\infty(F)} \quad \forall F \in \mathcal{F},$$

and corresponding geometric estimator

$$\xi_{\mathcal{F}}(\Gamma) := \max_{F \in \mathcal{F}} \xi_F.$$

Theorem 53 (a posteriori upper estimate). *Let γ be of class C^2 and let Γ be defined so that the geometric assumptions* (108), (109), (124), *and* (125) *hold. Let $\widetilde{f} \in L_{2,\#}(\gamma)$ and $\widetilde{u} \in H^1_\#(\gamma)$ solve (19). If $U \in \mathbb{V}(\mathcal{F})$ is the finite element solution of* (107) *with $F_\Gamma = \frac{q}{q_\Gamma} \widetilde{f} \circ \boldsymbol{P}_d$, and $U_d = U \circ \boldsymbol{P}_d^{-1}$ where $\boldsymbol{P}_d^{-1}$ is the inverse of $\boldsymbol{P}_d$ restricted to Γ, then*

$$\| \nabla_\gamma (\widetilde{u} - U_d) \|_{L_2(\gamma)} \lesssim \eta_{\mathcal{F}}(U) + \xi_{\mathcal{F}}(\Gamma) \, \| \nabla_\Gamma U \|_{L_2(\Gamma)}. \tag{128}$$

Proof. We proceed in several steps.
Step 1: Error representation via the residual equation. First we note that

$$\|\nabla_\gamma(\widetilde{u}-U_d)\|_{L_2(\gamma)}=\sup_{\widetilde{v}\in H^1(\gamma),\,\|\nabla_\gamma\widetilde{v}\|_{L_2(\gamma)}=1}\int_\gamma \nabla_\gamma(\widetilde{u}-U_d)\cdot\nabla_\gamma\widetilde{v}$$

and then write as in (102) that

$$\int_\gamma \nabla_\gamma(\widetilde{u}-U_d)\cdot\nabla_\gamma\widetilde{v}=I_1+I_2+I_3$$

with

$$\begin{aligned}
I_1 &:= -\int_\Gamma \nabla_\Gamma U\cdot\nabla_\Gamma(v_d-V)+\int_\Gamma F_\Gamma(v_d-V),\\
I_2 &:= \int_\gamma \nabla_\gamma U_d\cdot\mathbf{E}\,\nabla_\gamma\widetilde{v},\\
I_3 &:= \int_\gamma \widetilde{f}\,\widetilde{v}-\int_\Gamma F_\Gamma v_d.
\end{aligned}$$

Here $\mathbf{E}$ is as in (51), $v_d=\widetilde{v}\circ\mathbf{P}_d$, and $V\in\mathbb{V}(\mathcal{T})$ is a suitable approximation of the $H^{3/2}$ extension v of $\widetilde{v}$ given by Lemma 51 (harmonic extension). Note that $I_3=0$ because of the definition $F_\Gamma=\frac{q}{q_\Gamma}\widetilde{f}\circ\mathbf{P}_d$.
Step 2: Bounding the geometric error terms. Using (56) (or more accurately the corresponding pointwise bound from which it is derived) directly yields

$$\|\mathbf{E}\|_{L_\infty(F)}\lesssim\xi_F\quad\forall F\in\mathcal{F}.$$

Thus making use of Lemma 17 (norm equivalence) implies

$$|I_2|\lesssim\xi_{\mathcal{F}}(\Gamma)\,\|\nabla_\gamma U_d\|_{L_2(\gamma)}\,\|\nabla_\gamma\widetilde{v}\|_{L_2(\gamma)}\leq\xi_{\mathcal{F}}(\Gamma)\,\|\widetilde{f}\|_{L_2(\gamma)}\,\|\nabla_\gamma\widetilde{v}\|_{L_2(\gamma)}.$$

Step 3: Bounding the residual term. In order to bound I_1, we first decompose the integrals over faces $F\in\mathcal{F}$ and then integrate by parts to arrive at

$$|I_1|\lesssim\sum_{F\in\mathcal{F}}\eta_{\mathcal{F}}(U,F)\Big(h_F^{-1}\,\|v_d-V\|_{L_2(F)}+h_F^{-1/2}\,\|v_d-V\|_{L_2(\partial F)}\Big).$$

We may thus complete the proof upon showing that

$$\left(\sum_{F\in\mathcal{F}}h_F^{-2}\,\|v_d-V\|_{L_2(F)}^2+h_F^{-1}\,\|v_d-V\|_{L_2(\partial F)}^2\right)^{1/2}\lesssim\|\widetilde{v}\|_{H^1(\gamma)}.$$

Given $F \in \mathcal{F}$, we begin by considering the quantity $h_F^{-1} \| v_d - V\|_{L_2(e)}$ for any edge $e \subset \partial F$. We first use the triangle inequality to obtain

$$h_F^{-1/2} \| v_d - V\|_{L_2(e)} \lesssim h_F^{-1/2} \| v_d - V_d\|_{L_2(e)} + h_F^{-1/2} \| V_d - V\|_{L_2(e)}$$

with $V_d = V \circ \mathbf{P}_d$, and examine the last term first. Since e is an $(n-1)$-dimensional edge with $\operatorname{diam}(e) \leq h_F$, combining Hölder inequality and Lemma 52 (evaluation mismatch between γ and Γ) with an inverse estimate over the $(n+1)$-dimensional patch $\omega_{\mathcal{T}}^1(T_F)$ yields

$$\begin{aligned} h_F^{-1/2} \| V_d - V\|_{L_2(e)} &\lesssim h_F^{(n-2)/2} \| V_d - V\|_{L_\infty(F)} \\ &\lesssim h_F^{(n+2)/2} |d|_{W_\infty^2(\mathcal{N})} \|\nabla V\|_{L_\infty(\omega_{\mathcal{T}}^1(T_F))} \\ &\lesssim h_F^{1/2} |d|_{W_\infty^2(\mathcal{N})} \|\nabla V\|_{L_2(\omega_{\mathcal{T}}^1(T_F))}. \end{aligned}$$

For the first term $h_F^{-1/2} \| v_d - V_d\|_{L_2(e)}$ we argue as follows. Let $\mathbb{P}$ be the n-dimensional hyperplane containing F. While it may be that $\operatorname{diam}(e) \ll h_F$, the shape regularity of $\mathcal{T}$ implies that

$$\operatorname{dist}(e, \partial\omega_{\mathcal{T}}^1(T_F)) \geq \operatorname{dist}(T_F, \partial\omega_{\mathcal{T}}^1(T_F) \simeq h_F.$$

Thus there exists an n-dimensional ball $B \subset \mathbb{P} \cap \omega_{\mathcal{T}}^1(T_F) \subset \mathbb{P}$ so that $e \subset B$ and $\operatorname{diam}(B) \simeq h_F$. This ball B is the candidate for applying the h_F-scaled version of (115) of Lemma 48 (trace estimates for cut elements), namely

$$h_F^{-1/2} \| v_d - V_d\|_{L_2(e)} \lesssim h_F^{-1} \| v_d - V_d\|_{L_2(B)} + \| \nabla_{\mathbb{P}}(v_d - V_d)\|_{L_2(B)}.$$

Since $v_d - V_d = (\widetilde{v} - V) \circ \mathbf{P}_d$, we change variables from B to γ while employing Lemma 17 (norm equivalence) to get

$$h_F^{-1/2} \| v_d - V_d\|_{L_2(e)} \lesssim h_F^{-1} \|\widetilde{v} - V\|_{L_2(\mathbf{P}_d(B))} + \|\nabla_\gamma(\widetilde{v} - V)\|_{L_2(\mathbf{P}_d(B))}.$$

We observe that $\mathbf{P}_d(B) \subset \mathbf{P}_d(\omega_{\mathcal{T}}^1(T_F)) \subset \omega_{\mathcal{T}}^2(\mathcal{T}_F)$ in light of (125), whence

$$h_F^{-1/2} \| v_d - V_d\|_{L_2(e)} \lesssim h_F^{-1} \|v - V\|_{L_2(\omega_{\mathcal{T}}^2(T_F) \cap \gamma)} + \|\nabla_\gamma(v - V)\|_{L_2(\omega_{\mathcal{T}}^2(T_F) \cap \gamma)}.$$

We now carry out a similar but more direct computation for the term $h_F^{-1} \| v_d - V\|_{L_2(F)}$ appearing in I_1. Again using Lemma 52 we obtain

$$\begin{aligned} &h_F^{-1} \| v_d - V_d\|_{L_2(F)} \lesssim h_F^{-1} \|\widetilde{v} - V\|_{L_2(T \cap \gamma)} \\ &h_F^{-1} \| V_d - V\|_{L_2(F)} \lesssim h_F^{(n+2)/2} \|\nabla V\|_{L_\infty(\omega_{\mathcal{T}}^1(T_F))} \lesssim \|\nabla V\|_{L_2(\omega_{\mathcal{T}}^1(T_F))}. \end{aligned}$$

Combining the previous estimates we end up with

$$h_F^{-1}\,\|v_d - V\|_{L_2(F)} + h_F^{-1/2}\,\|v_d - V\|_{L_2(\partial F)}$$
$$\lesssim h_F^{-1}\,\|\widetilde{v} - V\|_{L_2(\omega_T^2(T_F))\cap\gamma)} + \|\nabla_\gamma(\widetilde{v} - V)\|_{L_2(\omega_T^2(T_F))\cap\gamma)} + \|\nabla V\|_{L_2(\omega_T^1(T_F))}.$$

Summing over $F\in\mathcal{F}$ while using finite overlap of the patches $\omega_T^2(T_F)$ yields

$$\left(\sum_{F\in\mathcal{F}} h_F^{-2}\,\|v_d - V\,\|_{L_2(F)}^2 + h_F^{-1}\,\|\,v_d - V\,\|_{L_2(\partial F)}^2\right)^{1/2}$$
$$\lesssim \left(\sum_{T\in\mathcal{T}_\Gamma} h_T^{-2}\,\|\widetilde{v} - V\,\|_{L_2(T\cap\gamma)}^2 + \|\nabla_\gamma(\widetilde{v} - V)\,\|_{L_2(T\cap\gamma)}^2\right)^{1/2} + \|\nabla V\|_{L_2(\Omega)}.$$

Step 4: Interpolation. We next apply (118) to the function $v - V$ while realizing that $|V|_{H^{3/2}(T)} = 0$. Doing so yields $|v - V|_{H^{3/2}(T)} = |v|_{H^{3/2}(T)}$ and

$$h_T^{-1}\,\|\widetilde{v} - V\|_{L_2(T\cap\gamma)} + \|\nabla_\gamma(\widetilde{v} - V)\|_{L_2(T\cap\gamma)}$$
$$\lesssim h_T^{-3/2}\,\|v - V\|_{L_2(T)} + h_T^{-1/2}\,\|\nabla(v - V)\|_{L_2(T)} + |v|_{H^{3/2}(T)}.$$

Next let $V = I_T^{sz}v$, where I_T^{sz} is the Scott–Zhang interpolation operator on the bulk space $\mathbb{V}(\mathcal{T})$. Standard approximation theory in $\mathbb{V}(\mathcal{T})$ then yields

$$h_T^{-3/2}\,\|v - V\|_{L_2(T)} + h_T^{-1/2}\,\|\nabla(v - V)\|_{L_2(T)} + |v|_{H^{3/2}(T)} \lesssim \|v\|_{H^{3/2}(\omega_T^1(T))}$$

and

$$\|\nabla V\|_{L_2(T)} \lesssim \|\nabla v\|_{L_2(\omega_T^1(T))}$$

for every $T\in\mathcal{T}_\Gamma$. Using the finite overlap of the patches $\omega_T^1(T)$ and the bound $\|v\|_{H^{3/2}(\mathbb{R}^{n+1})} \lesssim \|\widetilde{v}\|_{H^1(\gamma)}$ of Lemma 51 (harmonic extension), we finally obtain

$$\left(\sum_{F\in\mathcal{F}} h_F^{-2}\,\|\,v_d - V\,\|_{L_2(F)}^2 + h_F^{-1}\,\|\,v_d - V\,\|_{L_2(\partial F)}^2\right)^{1/2} \lesssim \|v\|_{H^{3/2}(\mathbb{R}^3)} \lesssim \|\widetilde{v}\|_{H^1(\gamma)}.$$

This completes the proof. □

Remark 54 (efficiency). In a posteriori error analysis it is standard to prove lower (efficiency) bounds such as those in Theorems 42 and 44. For trace methods such estimates would ideally take the form

$$\eta_{\mathcal{F}}(U,F) \lesssim \|u_d - U\|_{H^1(\omega^1_{\mathcal{F}}(F))} + \text{osc}_{\mathcal{F}}(F_\Gamma, \omega^1_{\mathcal{F}}(F)) + \xi_{\mathcal{F}}(\omega^1_{\mathcal{F}}(F))\, \|\nabla_\Gamma U\|_{L_2(\omega^1_{\mathcal{F}}(F))}.$$

where $\omega^1_{\mathcal{F}}(F)$ is the patch of elements about $F \in \mathcal{F}$ and $\text{osc}_{\mathcal{F}}(F_\Gamma, \omega^1_{\mathcal{F}}(F))$ is a heuristically higher order term measuring the deviation of F_Γ from the piecewise constants. However, the standard proof of this result does not work for trace methods due to the irregular structure of the surface mesh $\mathcal{F}$. The paper (Demlow and Olshanskii, 2012) contains partial efficiency results for the volume residual but none for the jump residual term. Numerical experiments suggest that a local efficiency result may hold, but also show a slight degeneration of the constant as the mesh is refined. Thus it is not clear whether the estimators we have studied for the trace method are efficient, and if so what form an efficiency estimate would take.

6 Narrow band method

In the narrow band approach, the partial differential Eq. (19) on γ

$$-\Delta_\gamma \widetilde{u} = \widetilde{f}$$

is extended to the tubular neighbourhood $\mathcal{N}(\delta)$ of γ defined in (29)

$$\mathcal{N}(\delta) := \{\mathbf{x} \in \mathbb{R}^{n+1}: \ |d(x)| < \delta\} \subset \mathbb{R}^{n+1};$$

we refer to the original papers (Bertalmío et al., 2001; Burger, 2009). The finite element method is then posed over a discrete approximation to $\mathcal{N}(\delta)$. We assume that γ is of class C^2 and $0 < \delta < \frac{1}{2K_\infty}$ so that (61) holds, namely $\mathcal{N}(\delta) \subset \mathcal{N}_\varepsilon(\delta_\varepsilon)$, and all the properties of the distance function detailed in Section 2 are valid in $\mathcal{N}(\delta)$.

A natural / standard way to extend $\widetilde{u}$ and $\widetilde{f}$ to $\mathcal{N}(\delta)$ is to use the constant extensions along the normal direction

$$u = \widetilde{u} \circ \mathbf{P}_d, \quad f = \widetilde{f} \circ \mathbf{P}_d.$$

We use the latter to design the FEM. However, we need $u \in H^2(\mathcal{N}(\delta))$ to derive optimal a priori H^1 error estimates for the FEM, which entails $\gamma \in C^3$ when using the closest point projection $\mathbf{P}_d$. We circumvent this extra regularity on γ via Proposition 28 (H^2 extension), which defines u as a normal extension relative to a perturbation γ_ε of γ constructed as a zero level set of a regularized distance function $\overline{d}_\varepsilon$. We will show below in Lemma 57 (narrow band PDE consistency) that such a function u satisfies

$$\left| \int_{\mathcal{N}(\delta)} \nabla u \cdot \nabla v - \int_{\mathcal{N}(\delta)} f v \right| \lesssim \delta^{3/2} \, \|\widetilde{f}\|_{L_2(\gamma)} \, \|v\|_{H^1(\mathcal{N}(\delta))}. \tag{129}$$

The specific choice of u adds several technicalities to the proof of (129) but reduces the regularity of γ to C^2. This seems to be a new result in the literature consistent with the underlying regularity $\widetilde{u} \in H^2(\gamma)$. This also motivates the narrow band FEM as a straightforward (bulk) finite element approximation of (129) upon replacing $\mathcal{N}(\delta)$ by a polygonal approximation $\mathcal{N}_h(\delta)$ dictated by d_h, the Lagrange interpolant of d in the bulk. We discuss this next. We refer to Olshanskii and Safin (2016) for higher order FEMs and Deckelnick et al. (2010, 2014) for an algorithm based on a level set function, rather that the less practical distance function. The essential ideas, however, are similar to those below but are more technical.

6.1 The narrow band FEM

We assume that $\mathcal{N}$ is enclosed in a $n + 1$ dimensional polyhedral domain D and denote by $\mathcal{T}$ a partition of D made of simplices. We omit to mention the explicit dependence on the shape-regularity constant of $\mathcal{T}$

$$\sigma := \max_{T \in \mathcal{T}} \frac{\operatorname{diam}(T)}{h_T}$$

in most estimates below; we use the notation $h_T = |T|^{\frac{1}{n+1}}$ and $h = \max_{T \in \mathcal{T}} h_T$. Let d_h stand for the Lagrange interpolant of the distance function d by continuous piecewise linear functions over $\mathcal{T}$. The discrete distance function d_h induces the discrete narrow band

$$\mathcal{N}_h(\delta) := \{\mathbf{x} \in D \ : \ |d_h(\mathbf{x})| < \delta\}.$$

Notice that standard interpolation estimates imply

$$\|d - d_h\|_{L_\infty(\mathcal{N})} + h \, \|\nabla(d - d_h)\|_{L_\infty(\mathcal{N})} \le c_I h^2 |d|_{W^2_\infty(\mathcal{N})}, \tag{130}$$

where c_I is a constant only depending on σ. This implies the *nondegeneracy* property

$$|\nabla d_h| \ge ||\nabla d| - |\nabla(d - d_h)|| \ge |1 - |\nabla(d - d_h)|| \ge \frac{1}{2}, \tag{131}$$

provided h is sufficiently small so that $c_I h |d|_{W^2_\infty(\mathcal{N})} \le \frac{1}{2}$. Combining estimates (130) and (131) we deduce that the Hausdorff distance between $\mathcal{N}(\delta)$ and $\mathcal{N}_h(\delta)$ satisfies

$$\operatorname{dist}_H(\mathcal{N}(\delta), \mathcal{N}_h(\delta)) \le 2 c_I h^2 |d|_{W^2_\infty(\mathcal{N})}. \tag{132}$$

Moreover, to guarantee that $\mathcal{N}_h(\delta) \subset \mathcal{N}$, we observe

$$|d(\mathbf{x})| \leq |d_h(\mathbf{x})| + |(d - d_h)(\mathbf{x})| \leq \delta + c_I |d|_{W^2_\infty(\mathcal{N})} h^2 \quad \forall \mathbf{x} \in \mathcal{N}_h(\delta).$$

In view of (31), it thus suffices to restrict δ and h so that

$$\delta + c_I |d|_{W^2_\infty(\mathcal{N})} h^2 \leq \frac{1}{2K_\infty}. \tag{133}$$

Hereafter we make the structural assumption

$$C_1 h \leq \delta \leq C_2 h \tag{134}$$

with $c_I \leq C_1 \leq C_2$ so that (133) holds for h sufficiently small.

We denote by $\mathcal{T}_\delta$ the restriction of $\mathcal{T}$ to $\mathcal{N}_h(\delta)$ in the sense that

$$\mathcal{T}_\delta := \{T \in \mathcal{T} \ : \ T \cap \mathcal{N}_h(\delta) \neq \emptyset\}.$$

The finite element space associated with $\mathcal{T}_\delta$ is then constructed in the usual way

$$\mathbb{V}(\mathcal{T}_\delta) := \left\{ V \in C^0(\overline{\mathcal{N}_h(\delta)}) \ : \ V|_T \in \mathcal{P}, \ T \in \mathcal{T}_\delta \right\},$$

where we recall that $\mathcal{P}$ stands for the space of polynomials of degree 1. The subspace of functions with vanishing mean value is denoted $\mathbb{V}_\#(\mathcal{T}_\delta)$.

With this notation at hand and inspired by (129), we define the narrow band finite element solution $U \in \mathbb{V}_\#(\mathcal{T}_\delta)$ to satisfy

$$\int_{\mathcal{N}_h(\delta)} \nabla U \cdot \nabla V = \int_{\mathcal{N}_h(\delta)} FV, \quad \forall V \in \mathbb{V}_\#(\mathcal{T}_\delta), \tag{135}$$

where F is an approximation to $f = \widetilde{f} \circ \mathbf{P}_d$ satisfying $\int_{\mathcal{N}_h(\delta)} F = 0$. In order to make a convenient choice of F, we first define $\mathbf{M}_h : \mathcal{N}_h(\delta) \to \mathcal{N}(\delta)$ by

$$\mathbf{M}_h(\mathbf{x}) = \mathbf{P}_d(\mathbf{x}) + d_h(\mathbf{x}) \nabla d(\mathbf{x});$$

the properties of $\mathbf{M}_h$ are explored thoroughly later in this section. With this definition in hand, we let

$$F = f \circ \mathbf{M}_h - \frac{1}{|\mathcal{N}_h(\delta)|} \int_{\mathcal{N}_h(\delta)} f \circ \mathbf{M}_h. \tag{136}$$

This requires having access to d, d_h and $\mathbf{P}_d$, which we assume hereafter. Since F has vanishing mean value, (135) is also valid for all $V \in \mathbb{V}(\mathcal{T}_\delta)$. The existence and uniqueness of $U \in \mathbb{V}_\#(\mathcal{T}_\delta)$ follows directly from the Lax–Milgram lemma.

6.2 PDE geometric consistency

We intend to prove (129) for the extension $u \in H^2(\mathcal{N}(\delta))$ in Proposition 28 (H^2 extension) of $\widetilde{u} \in H^2(\gamma)$. We recall Proposition 30 (PDE satisfied by u)

$$-\mathrm{div}\,(\mu_\varepsilon \mathbf{B}_\varepsilon \nabla u) = f_\varepsilon \mu_\varepsilon,$$

multiply by a test function $v \in H^1(\mathcal{N}(\delta))$ and integrate by parts in $\mathcal{N}(\delta)$ to obtain

$$\int_{\mathcal{N}(\delta)} \mathbf{B}_\varepsilon \nabla u \cdot \nabla v \;\; \mu_\varepsilon = \int_{\mathcal{N}(\delta)} f_\varepsilon \;\; v \;\; \mu_\varepsilon + \int_{\partial\mathcal{N}(\delta)} \mathbf{B}_\varepsilon \nabla u \cdot \nabla d \;\; v \;\; \mu_\varepsilon. \tag{137}$$

Notice that we have used that $\boldsymbol{\nu} = \nabla d$ is the outward pointing normal to $\partial\mathcal{N}(\delta)$. We start by estimating geometric quantities appearing in (137).

Lemma 55 (properties of μ_ε and $\mathbf{B}_\varepsilon$). *Let γ be of class C^2 and $C\delta \leq \varepsilon \leq \frac{\delta}{2}$ be sufficiently small. Then for all $\mathbf{x} \in \mathcal{N}(\delta)$ we have*

$$\|1 - \mu_\varepsilon\|_{L_\infty(\mathcal{N}(\delta))} \lesssim \delta |d|_{W^2_\infty(\mathcal{N})} \tag{138}$$

and

$$\| \Pi_\varepsilon - \mathbf{B}_\varepsilon \mu_\varepsilon \|_{L_\infty(\mathcal{N}(\delta))} \lesssim \delta |d|_{W^2_\infty(\mathcal{N})}. \tag{139}$$

Proof. We recall the definitions of μ_ε from Proposition 30 (PDE satisfied by u) and $\widetilde{\mu}_\varepsilon$ from Lemma 29 (PDE satisfies by u_ε)

$$\mu_\varepsilon = \frac{1}{\widetilde{\mu}_\varepsilon \circ \mathbf{P}_\varepsilon} \det\left(\mathbf{I} - d_\varepsilon D^2 d_\varepsilon\right), \;\; \widetilde{\mu}_\varepsilon = \det\left(\mathbf{I} - d_\varepsilon D^2 d_\varepsilon\right)(\nabla d \cdot \nabla d_\varepsilon) \circ \mathbf{Q}_\varepsilon,$$

where $\mathbf{P}_\varepsilon$ is the projection from $\mathcal{N}(\delta)$ onto $\gamma_\varepsilon = \{d_\varepsilon(\mathbf{x}) = 0\}$ and $\mathbf{Q}_\varepsilon$ is its inverse when restricted to γ. Note that in $\mathcal{N}(\delta)$

$$1 - \mu_\varepsilon = \left(1 - \frac{1}{\widetilde{\mu}_\varepsilon \circ \mathbf{P}_\varepsilon}\right) + \frac{1}{\widetilde{\mu}_\varepsilon \circ \mathbf{P}_\varepsilon}\left(1 - \det\left(\mathbf{I} - d_\varepsilon D^2 d_\varepsilon\right)\right).$$

We thus need to examine the eigenvalues $(\zeta_i(\mathbf{x}))_{i=0}^n$ of

$$\mathbf{I} - d_\varepsilon(\mathbf{x}) D^2 d_\varepsilon(\mathbf{x}) \;\; \forall \mathbf{x} \in \mathcal{N}(\delta),$$

with $\zeta_0(\mathbf{x}) = 1$ corresponding to the eigenvector ∇d_ε. We infer that

$$\zeta_i(\mathbf{x}) = 1 - \eta_i(\mathbf{x})$$

where

$$|\eta_i(\mathbf{x})| \lesssim |d_\varepsilon(\mathbf{x})| \;\; |d_\varepsilon|_{W^2_\infty(\mathcal{N}(\delta))} \lesssim \delta |d|_{W^2_\infty(\mathcal{N})}$$

according to Lemma 26 (properties of d_ε) and (61) with $\delta_\varepsilon \leq \frac{3}{2}\delta$. Hence

$$\left|1 - \det\left(\mathbf{I} - d_\varepsilon(\mathbf{x}) D^2 d_\varepsilon(\mathbf{x})\right)\right| = \left|1 - \prod_{i=1}^{n} \zeta_i(\mathbf{x})\right| \lesssim \delta |d|_{W^2_\infty(\mathcal{N})}$$

for all $\mathbf{x} \in \mathcal{N}(\delta)$. This takes care of the second term in the equation for $1 - \mu_\varepsilon$. It remains to estimate $1 - \widetilde{\mu}_\varepsilon \circ \mathbf{P}_\varepsilon$. Since $1 - \widetilde{\mu}_\varepsilon \circ \mathbf{P}_\varepsilon$ reads as follows on γ

$$1 - \widetilde{\mu}_\varepsilon = \left(1 - \det\left(\mathbf{I} - d_\varepsilon D^2 d_\varepsilon\right)\right) + \det\left(\mathbf{I} - d_\varepsilon D^2 d_\varepsilon\right)\left(1 - \nabla d \cdot \nabla d_\varepsilon\right),$$

combining the previous estimate with Lemma 26 (properties of d_ε) yields

$$\left|1 - \widetilde{\mu}_\varepsilon(\mathbf{P}_\varepsilon(\mathbf{x}))\right| \lesssim \delta |d|_{W^2_\infty(\mathcal{N})} \quad \forall \mathbf{x} \in \mathcal{N}(\delta).$$

This implies $|\widetilde{\mu}_\varepsilon(\mathbf{P}_\varepsilon(\mathbf{x}))| \geq \frac{1}{2}$ for δ sufficiently small and thus leads to (138).

We now prove (139) which, in light of (138), reduces to the estimate $\| \Pi_\varepsilon - \mathbf{B}_\varepsilon \|_{L_\infty(\mathcal{N}(\delta))} \lesssim \delta |d|_{W^2_\infty(\mathcal{N})}$. We recall from Proposition 30 that in $\mathcal{N}(\delta)$

$$\mathbf{B}_\varepsilon = \left(\mathbf{I} - d_\varepsilon D^2 d_\varepsilon\right)^{-1} \Pi_\varepsilon \widetilde{\mathbf{A}}_\varepsilon \circ \mathbf{P}_\varepsilon \Pi_\varepsilon \left(\mathbf{I} - d_\varepsilon D^2 d_\varepsilon\right)^{-1}.$$

Since $d_\varepsilon(\mathbf{x}) \leq \widetilde{\delta} \leq \frac{3}{2}\delta$ for $\mathbf{x} \in \mathcal{N}(\delta)$ and $\|D^2 d_\varepsilon\|_{L_\infty(\mathcal{N}(\delta))} \lesssim |d|_{W^2_\infty(\mathcal{N})}$ thanks to Lemma 26 (properties of d_ε), the Taylor expansion of $(\mathbf{I} - t d_\varepsilon D^2 d_\varepsilon)^{-1}$ centred at $t = 0$ and computed at $t = 1$ converges for δ sufficiently small. It reads

$$\left(\mathbf{I} - d_\varepsilon D^2 d_\varepsilon\right)^{-1} = \mathbf{I} + d_\varepsilon \left(\mathbf{I} - \xi d_\varepsilon D^2 d_\varepsilon\right)^{-2} D^2 d_\varepsilon$$

for some $0 < \xi < 1$. The definition of $\widetilde{\mathbf{A}}_\varepsilon$ given in Lemma 29 yields

$$\mathbf{B}_\varepsilon = \Pi_\varepsilon (\Pi \circ \mathbf{Q}_\varepsilon \circ \mathbf{P}_\varepsilon) \Pi_\varepsilon + d_\varepsilon \mathbf{G},$$

where $\mathbf{G} : \mathcal{N}(\delta) \to \mathbb{R}^{(n+1)\times(n+1)}$ satisfies $\| \mathbf{G} \|_{L_\infty(\mathcal{N}(\delta))} \lesssim 1$. Moreover,

$$\Pi_\varepsilon - \Pi_\varepsilon (\Pi \circ \mathbf{Q}_\varepsilon \circ \mathbf{P}_\varepsilon) \Pi_\varepsilon = \Pi_\varepsilon \nabla d \circ (\mathbf{Q}_\varepsilon \circ \mathbf{P}_\varepsilon) \otimes \Pi_\varepsilon \nabla d \circ (\mathbf{Q}_\varepsilon \circ \mathbf{P}_\varepsilon)$$

whence for all $\mathbf{x} \in \mathcal{N}(\delta)$ we see that

$$\Pi_\varepsilon(\mathbf{x}) \nabla d(\mathbf{Q}_\varepsilon(\mathbf{P}_\varepsilon(\mathbf{x}))) = \nabla d(\mathbf{Q}_\varepsilon(\mathbf{P}_\varepsilon(\mathbf{x}))) - \nabla d_\varepsilon(\mathbf{x}) \left(\nabla d(\mathbf{Q}_\varepsilon(\mathbf{P}_\varepsilon(\mathbf{x}))) \cdot \nabla d_\varepsilon(\mathbf{x})\right).$$

Since

$$\begin{aligned} |\nabla d_\varepsilon(\mathbf{x}) - \nabla d(\mathbf{Q}_\varepsilon(\mathbf{P}_\varepsilon(\mathbf{x})))| &\leq |\nabla(d_\varepsilon(\mathbf{x}) - d(\mathbf{x}))| + |\nabla d(\mathbf{x}) - \nabla d(\mathbf{Q}_\varepsilon(\mathbf{P}_\varepsilon(\mathbf{x})))| \\ &\lesssim (\delta + |\mathbf{x} - \mathbf{Q}_\varepsilon(\mathbf{P}_\varepsilon(\mathbf{x}))|) |d|_{W^2_\infty(\mathcal{N})} \lesssim \delta |d|_{W^2_\infty(\mathcal{N})} \end{aligned}$$

thanks to Lemma 26 (properties of d_ε), we get

$$\|\Pi_\varepsilon - \mathbf{B}_\varepsilon\|_{L_\infty(\mathcal{N}(\delta))} \lesssim \delta |d|_{W^2_\infty(\mathcal{N})}$$

as asserted. This concludes the proof. □

Remark 56 (estimate of μ). Lemma 22 (relation between q and q_Γ) gives the expression $\mu(\mathbf{x}) := \det(\mathbf{I} - d(\mathbf{x})D^2 d(\mathbf{x}))$ for the change of infinitesimal area between $\gamma_s := \{d^{-1}(s)\}$ and $\gamma := \{d^{-1}(0)\}$. Proceeding as in the proof of the above lemma, we get

$$\|1-\mu\|_{L_\infty(\mathcal{N}(\delta))} \lesssim \delta\, |d|_{W^2_\infty(\mathcal{N})} \tag{140}$$

provided δ is sufficiently small so that $\mathcal{N}(\delta) \subset \mathcal{N}$.

We are now in position to prove a consistency estimate measuring the discrepancy between f and Δu in $\mathcal{N}(\delta)$.

Lemma 57 (narrow band PDE consistency). *Let γ be of class C^2 and u be the extension of Proposition 28 (H^2 extension) with $C\delta \le \varepsilon \le \frac{\delta}{2}$ sufficiently small. If $\widetilde{f} \in L_2(\gamma)$, then for all $v \in H^1(\mathcal{N}(\delta))$, we have*

$$\left| \int_{\mathcal{N}(\delta)} \nabla u \cdot \nabla v - \int_{\mathcal{N}(\delta)} fv \right| \lesssim \delta^{3/2} |d|^2_{W^2_\infty(\mathcal{N})} \|\widetilde{f}\|_{L_2(\gamma)} \|v\|_{H^1(\mathcal{N}(\delta))}. \tag{141}$$

Proof. In view of (137), we deduce

$$I(v) := \int_{\mathcal{N}(\delta)} \nabla u \cdot \nabla v - \int_{\mathcal{N}(\delta)} fv = I_1(v) + I_2(v) + I_3(v) \quad \forall v \in H^1(\mathcal{N}(\delta)),$$

where

$$I_1(v) := \int_{\mathcal{N}(\delta)} (\mathbf{I} - \mathbf{B}_\varepsilon \mu_\varepsilon) \nabla u \cdot \nabla v,$$

$$I_2(v) := \int_{\mathcal{N}(\delta)} (f_\varepsilon \mu_\varepsilon - f) v,$$

$$I_3(v) := \int_{\partial\mathcal{N}(\delta)} \mathbf{B}_\varepsilon \nabla u \cdot \nabla d \ \ v \ \ \mu_\varepsilon$$

with $f_\varepsilon = \widetilde{f} \circ \mathbf{Q}_\varepsilon \circ \mathbf{P}_\varepsilon$. We now examine these three terms separately.
Step 1: Term $I_1(v)$. Since u is constant along the direction ∇d_ε, we realize that $\nabla u = \Pi_\varepsilon \nabla u$ and Lemma 55 (properties of μ_ε and $\mathbf{B}_\varepsilon$) directly yields

$$|I_1(v)| \lesssim \delta |d|_{W^2_\infty(\mathcal{N})} \|\nabla u\|_{L_2(\mathcal{N}(\delta))} \|\nabla v\|_{L_2(\mathcal{N}(\delta))}.$$

Step 2: Term $I_2(v)$. Let $-\delta < s < \delta$ and consider the isomorphisms

$$\mathbf{R}_s := \mathbf{Q}_\varepsilon \circ \mathbf{P}_\varepsilon \circ \mathbf{Q}_s : \gamma \to \gamma, \quad \mathbf{R}_s^{-1} = \mathbf{P}_d \circ \mathbf{Q}_{\varepsilon,s} \circ \mathbf{P}_\varepsilon : \gamma \to \gamma,$$

where $\mathbf{Q}_s : \gamma \to \gamma_s$ is the inverse of $\mathbf{P}_d$ on γ_s and $\mathbf{Q}_{\varepsilon,s} : \gamma_\varepsilon \to \gamma_s$ is the inverse of $\mathbf{P}_\varepsilon$ on γ_s. Using the coarea formula (63) together with $|\nabla d| = 1$ we write

$$I_2(v) = \int_{-\delta}^{\delta}\int_{\gamma_s} (f_\varepsilon \mu_\varepsilon - f)v,$$

and combining with Lemma 22 (relation between q and q_Γ), we obtain

$$\begin{aligned} I_2(v) &= \int_{-\delta}^{\delta}\int_{\gamma} (\widetilde{f} \circ \mathbf{R}_s)(v \circ \mathbf{Q}_s)(\mu_\varepsilon \circ \mathbf{Q}_s)(\mu^{-1} \circ \mathbf{Q}_s) \\ &\quad - \int_{-\delta}^{\delta}\int_{\gamma} \widetilde{f}(v \circ \mathbf{Q}_s)(\mu^{-1} \circ \mathbf{Q}_s) = II_1(v) + II_2(v) + II_3(v), \end{aligned}$$

where

$$\begin{aligned} II_1(v) &:= \int_{-\delta}^{\delta}\int_{\gamma} (\widetilde{f} \circ \mathbf{R}_s)(v \circ \mathbf{Q}_s) - \widetilde{f}(v \circ \mathbf{Q}_s), \\ II_2(v) &:= \int_{-\delta}^{\delta}\int_{\gamma} \widetilde{f}(v \circ \mathbf{Q}_s)\big(1 - \mu^{-1} \circ \mathbf{Q}_s\big) \\ II_3(v) &:= \int_{-\delta}^{\delta}\int_{\gamma} (\widetilde{f} \circ \mathbf{R}_s)(v \circ \mathbf{Q}_s)\big((\mu_\varepsilon \circ \mathbf{Q}_s)(\mu^{-1} \circ \mathbf{Q}_s) - 1\big). \end{aligned}$$

We proceed to estimate each term separately. To manipulate $II_1(v)$ we first observe that changing variables from γ to γ_s, γ_s to γ_ε, and γ_ε to γ and invoking Lemma 22 (relation between q and q_Γ) yields

$$\begin{aligned} \int_{\gamma} (\widetilde{f} \circ \mathbf{R}_s)(v \circ \mathbf{Q}_s) &= \int_{\gamma} (\widetilde{f} \circ \mathbf{Q}_\varepsilon \circ \mathbf{P}_\varepsilon \circ \mathbf{Q}_s)(v \circ \mathbf{Q}_s) \\ &= \int_{\gamma_s} (\widetilde{f} \circ \mathbf{Q}_\varepsilon \circ \mathbf{P}_\varepsilon) v \mu \\ &= \int_{\gamma_\varepsilon} (\widetilde{f} \circ \mathbf{Q}_\varepsilon)(v\,\mu\,{\mu_\varepsilon}^{-1}) \circ \mathbf{Q}_{\varepsilon,s} \\ &= \int_{\gamma} \widetilde{f}\,(v\,\mu\,{\mu_\varepsilon}^{-1}) \circ \mathbf{Q}_{\varepsilon,s} \circ \mathbf{P}_\varepsilon\,\mu_\varepsilon, \end{aligned} \tag{142}$$

where

$$\mu = \det(\mathbf{I} - dD^2 d), \qquad \mu_\varepsilon = \det(\mathbf{I} - d_\varepsilon D^2 d_\varepsilon)\,(\nabla d \cdot \nabla d_\varepsilon).$$

Therefore, denoting by $\mu_{\mathbf{R}}$ the infinitesimal change in area induced by $\mathbf{R}_s^{-1}$ on γ

$$\mu_{\mathbf{R}} := (\mu\,{\mu_\varepsilon}^{-1}) \circ \mathbf{Q}_{\varepsilon,s} \circ \mathbf{P}_\varepsilon\,\mu_\varepsilon,$$

we infer again from the coarea formula (63) that

$$\begin{aligned}
II_1(v) &= \int_{-\delta}^{\delta}\int_{\gamma} \widetilde{f}\,\big((v\circ \mathbf{Q}_{\varepsilon,s}\circ \mathbf{P}_\varepsilon)\mu_{\mathbf{R}} - v\circ \mathbf{Q}_s\big) \\
&= \int_{-\delta}^{\delta}\int_{\gamma_s} \big(f\,(v\circ \mathbf{Q}_{\varepsilon,s}\circ \mathbf{P}_\varepsilon\circ \mathbf{P}_d)\mu_{\mathbf{R}}\,\mu - fv\mu\big)|\nabla d| \\
&= \int_{\mathcal{N}(\delta)} f\,(v\circ \mathbf{L} - v)\mu_{\mathbf{R}}\mu + \int_{\mathcal{N}(\delta)} f\,v(\mu_{\mathbf{R}}-1)\mu,
\end{aligned}$$

where $\mathbf{L}$ is defined on each γ_s by $\mathbf{L}|_{\gamma_s} := \mathbf{Q}_{\varepsilon,s}\circ \mathbf{P}_\varepsilon\circ \mathbf{P}_d : \gamma_s \to \gamma_s$. Notice that the map $\mathbf{L}:\mathcal{N}(\delta)\to\mathcal{N}(\delta)$ is a bi-Lipschitz perturbation of the identity with perturbation constant

$$r = \|\mathbf{I}-\mathbf{L}\|_{L_\infty(\mathcal{N}(\delta))} \lesssim \delta |d|_{W^2_\infty(\mathcal{N}(\delta))},$$

because

$$\begin{aligned}
&\|\mathbf{I}-\mathbf{P}_d\|_{L_\infty(\mathcal{N}(\delta))} + \|\mathbf{I}-\mathbf{Q}_s\|_{L_\infty(\mathcal{N}(\delta))} \\
&\qquad + \|\mathbf{I}-\mathbf{P}_\varepsilon\|_{L_\infty(\mathcal{N}(\delta))} + \|\mathbf{I}-\mathbf{Q}_{\varepsilon,s}\|_{L_\infty(\mathcal{N}(\delta))} \lesssim \delta |d|_{W^2_\infty(\mathcal{N}(\delta))}.
\end{aligned}$$

Moreover, since $\mu_{\mathbf{R}} - 1 = (\mu\mu_\varepsilon^{-1} - 1)\circ \mathbf{Q}_{\varepsilon,s}\circ \mathbf{P}_\varepsilon\mu_\varepsilon + (\mu_\varepsilon - 1)$, (138) and (140) imply

$$\|\mu_{\mathbf{R}} - 1\|_{L_\infty(\gamma)} \lesssim \delta |d|_{W^2_\infty(\mathcal{N}(\delta))}.$$

These estimates in conjunction in Proposition 35 (Lipschitz perturbation) give

$$|II_1(v)| \lesssim \delta |d|_{W^2_\infty(\mathcal{N}(\delta))}\,\|f\|_{L_2(\mathcal{N}(\delta))}\,\|v\|_{H^1(\mathcal{N}(\delta))};$$

we observe that to apply Proposition 35 we take $\Omega_1=\Omega_2=\mathcal{N}(\delta)$, which are Lipschitz domains, and extend v to $\Omega=\mathcal{N}$ so that $\|v\|_{H^1(\mathcal{N})} \lesssim \|v\|_{H^1(\mathcal{N}(\delta))}$.

Upon utilizing the coarea formula (63) once more, we obtain for $II_2(v)$

$$II_2(v) = \int_{\mathcal{N}(\delta)} fv(1-\mu^{-1})\mu,$$

so that (140) yields

$$|II_2(v)| \lesssim \delta |d|_{W^2_\infty(\mathcal{N}(\delta))}\,\|f\|_{L_2(\mathcal{N}(\delta))}\,\|v\|_{H^1(\mathcal{N}(\delta))}.$$

We proceed similarly for $II_3(v)$ but using in addition that

$$\int_{-\delta}^{\delta} \|\widetilde{f}\circ \mathbf{R}_s\|^2_{L_2(\gamma)}\,ds \lesssim \int_{-\delta}^{\delta} \|\widetilde{f}\|_{L_2(\gamma)} ds \lesssim \|f\|^2_{L_2(\mathcal{N}(\delta))},$$

and

$$\|(\mu_\varepsilon\circ \mathbf{Q}_s)(\mu^{-1}\circ \mathbf{Q}_s) - 1\|_{L_\infty(\gamma)} \lesssim \delta |d|_{W^2_\infty(\mathcal{N}(\delta))}$$

thanks to (138) and (140) again. We thus obtain for $II_3(v)$ an estimate similar to those for $II_1(v)$ and $II_2(v)$, whence

$$|I_2(v)| \lesssim \delta |d|_{W^2_\infty(\mathcal{N}(\delta))}\,\|f\|_{L_2(\mathcal{N}(\delta))}\,\|v\|_{H^1(\mathcal{N}(\delta))}.$$

Step 3: Term $I_3(v)$. In view of $\Pi_\varepsilon = \mathbf{I} - \nabla d_\varepsilon \otimes \nabla d_\varepsilon$, we first note that

$$\nabla d^T \mathbf{B}_\varepsilon \mu_\varepsilon = \nabla d^T(\mathbf{B}_\varepsilon \mu_\varepsilon - \Pi_\varepsilon) + \nabla(d - d_\varepsilon)^T + \nabla d_\varepsilon^T(1 - \nabla d \cdot \nabla d_\varepsilon).$$

Invoking Lemma 55 (properties of μ_ε and $\mathbf{B}_\varepsilon$) and then Lemma 26 (properties of d_ε) yields

$$\|\nabla d^T \mathbf{B}_\varepsilon \mu_\varepsilon\|_{L_\infty(\mathcal{N}(\delta))} \lesssim \delta |d|_{W^2_\infty(\mathcal{N}(\delta))}.$$

It remains to use trace inequalities to obtain

$$\begin{aligned} I_3(v) &\lesssim \delta |d|_{W^2_\infty(\mathcal{N}(\delta))} \|\nabla u\|_{L_2(\partial N(\delta))} \|v\|_{L_2(\partial N(\delta))} \\ &\lesssim \delta |d|_{W^2_\infty(\mathcal{N}(\delta))} \|u\|_{H^2(\mathcal{N}(\delta))} \|v\|_{H^1(\mathcal{N}(\delta))}. \end{aligned}$$

Step 4: Normal extension. Gathering the above estimates we find that

$$I(v) \lesssim \delta |d|_{W^2_\infty(\mathcal{N}(\delta))} \left(\|u\|_{H^2(\mathcal{N}(\delta))} + \|f\|_{L_2(\mathcal{N}(\delta))} \right) \|v\|_{H^1(\mathcal{N}(\delta))}.$$

We finally deduce $\|f\|_{L_2(\mathcal{N}(\delta))} \lesssim \delta^{\frac{1}{2}} |d|_{W^2_\infty(\mathcal{N}(\delta))} \|\tilde{f}\|_{L_2(\gamma)}$ because f is the normal extension of $\tilde{f}$ to γ, and

$$\|\nabla u\|_{L_2(\mathcal{N}(\delta))} \lesssim \delta^{\frac{1}{2}} |d|_{W^2_\infty(\mathcal{N})} \|\tilde{f}\|_{L_2(\gamma)},$$

upon combining Proposition 28 (H^2 extension) with Lemma 3 (regularity). This leads to the desired estimate. □

6.3 Properties of the narrow band FEM

To begin with, we recall the definition of $\mathbf{M}_h : \mathcal{N}_h(\delta) \to \mathcal{N}(\delta)$, that accounts for the mismatch between $\mathcal{N}_h(\delta)$ and $\mathcal{N}(\delta)$:

$$\mathbf{M}_h(\mathbf{x}) = \mathbf{P}_d(\mathbf{x}) + d_h(\mathbf{x})\nabla d(\mathbf{x}) = \mathbf{x} + (d_h(\mathbf{x}) - d(\mathbf{x}))\nabla d(\mathbf{x}). \tag{143}$$

Note that if $\mathbf{x} \in \mathcal{N}_h(\delta) \subset \mathcal{N}$ then $\mathbf{P}_d(\mathbf{M}_h(\mathbf{x})) = \mathbf{P}_d(\mathbf{x})$, because this is what happens with all points in the line $s \mapsto \mathbf{x} + s\nabla d(\mathbf{x})$ within $\mathcal{N}$. Since $|d_h(\mathbf{x})| < \delta$,

$$|d(\mathbf{M}_h(\mathbf{x}))| = |\mathbf{M}_h(\mathbf{x}) - \mathbf{P}_d(\mathbf{M}_h(\mathbf{x}))| = |d_h(\mathbf{x})| \, |\nabla d(\mathbf{x})| < \delta$$

implies that $\mathbf{M}_h(\mathbf{x}) \in \mathcal{N}(\delta)$ and the map M_h is well defined. Further properties of $\mathbf{M}_h$ are discussed next. Before doing so, we mention that the results provided below are not optimal (to avoid technicalities) but are sufficient for our analysis. We refer to Deckelnick et al. (2014) and Olshanskii and Safin (2016) for higher order estimates.

Lemma 58 (properties of $\mathbf{M}_h$). *Let γ be of class C^2 and h be sufficiently small. Then, the map $\boldsymbol{M}_h : \mathcal{N}_h(\delta) \to \mathcal{N}(\delta)$ is bi-Lipschitz with*

$$\|D\boldsymbol{M}_h\|_{L_\infty(\mathcal{N}_h(\delta))} + \|D\boldsymbol{M}_h^{-1}\|_{L_\infty(\mathcal{N}(\delta))} \le L \tag{144}$$

for some constant L independent of h and δ. Moreover, there holds

$$\|\boldsymbol{I}-\boldsymbol{M}_h\|_{L_\infty(\mathcal{N}_h(\delta))}+h\,\|\boldsymbol{I}-D\boldsymbol{M}_h\|_{L_\infty(\mathcal{N}_h(\delta))}\lesssim h^2|d|_{W^2_\infty(\mathcal{N})} \tag{145}$$

and

$$\|\det(D\boldsymbol{M}_h)-1\|_{L_\infty(\mathcal{N}_h(\delta))}\lesssim h|d|_{W^2_\infty(\mathcal{N})}. \tag{146}$$

Proof. From the definition (143) of $\mathbf{M}_h$ and the interpolation estimate (130), we find that

$$|\mathbf{x}-\mathbf{M}_h(\mathbf{x})|\le|d(\mathbf{x})-d_h(\mathbf{x})|\le c_I h^2|d|_{W^2_\infty(\mathcal{N})}.$$

Furthermore, we compute

$$D\mathbf{M}_h(\mathbf{x})=\mathbf{I}+\nabla(d_h(\mathbf{x})-d(\mathbf{x}))\otimes\nabla d(\mathbf{x})+(d_h(\mathbf{x})-d(\mathbf{x}))D^2d(\mathbf{x})$$

to deduce

$$\|\mathbf{I}-D\mathbf{M}_h\|_{L_\infty(\mathcal{N}_h(\delta))}\le c_I h|d|_{W^2_\infty(\mathcal{N})}+c_I h^2|d|^2_{W^2_\infty(\mathcal{N})}.$$

The above two estimates yield (145) because $c_I h|d|_{W^2_\infty(\mathcal{N})}\le\frac{1}{2}$ for h sufficiently small. Exploiting (145), we also deduce that $\mathbf{M}_h$ is invertible, bi-Lipschitz and that (144) holds for h sufficiently small.

We are left to show (146). This follows from $D\det\mathbf{A}=(\det\mathbf{A})\mathbf{A}^{-1}$ for any invertible matrix $\mathbf{A}$ and the first-order Taylor expansion of

$$\psi(t):=\det(\mathbf{I}-t(\nabla(d(\mathbf{x})-d_h(\mathbf{x}))\otimes\nabla d(\mathbf{x})-(d(\mathbf{x})-d_h(\mathbf{x}))D^2d(\mathbf{x})))$$

about $t=0$ and evaluated at $t=1$, along with (145) and the fact that $\psi(1)=\det(D\mathbf{M}_h(\mathbf{x}))$. This concludes the proof. □

The previous lemma is instrumental to estimate the consistency error

$$E_h(V):=\int_{\mathcal{N}_h(\delta)}\nabla u\cdot\nabla V-\int_{\mathcal{N}_h(\delta)}FV\quad\forall V\in\mathbb{V}(\mathcal{T}_\delta), \tag{147}$$

due to the approximation of the narrow band $\mathcal{N}(\delta)$ by $\mathcal{N}_h(\delta)$ and to the use of F in the discrete formulation (135). Since $\mathcal{N}(\delta)\subset\mathcal{N}$ is of class C^2, we assume without loss of generality that the function $u:\mathcal{N}(\delta)\to\mathbb{R}$ constructed in Proposition 28 (H^2 extension) extends to $\mathcal{N}$ and satisfies $\|u\|_{H^2(\mathcal{N})}\lesssim\|u\|_{H^2(\mathcal{N}(\delta))}$. In light of $\mathcal{N}_h(\delta)\subset\mathcal{N}$, the consistency error (147) is well defined.

Lemma 59 (narrow band geometric consistency). *Let γ be of class C^2 and δ and h satisfy the structural condition (134) and be sufficiently small.*

Let $\widetilde{f} \in L_{2,\#}(\gamma)$, $\widetilde{u} \in H^2(\gamma)$ solve (18), and $u \in H^2(\mathcal{N}(\delta))$ be the H^2 extension of $\widetilde{u}$ given by (62) with $C\delta \leq \varepsilon \leq \frac{\delta}{2}$. Let also F be given by (136). Then the consistency error (147) satisfies for all $V \in \mathbb{V}(\mathcal{T}_\delta)$

$$\left| \int_{\mathcal{N}_h(\delta)} \nabla u \cdot \nabla V - \int_{\mathcal{N}_h(\delta)} FV \right| \lesssim \delta^{3/2} |d|_{W^2_\infty(\mathcal{N})} \|\widetilde{f}\|_{L_2(\gamma)} \|V\|_{H^1(\mathcal{N}_h(\delta))}.$$

Proof. We compare the consistency errors (147) and (141) term by term.
Step 1: Dirichlet integrals. Utilizing the change of variables induced by the map $\mathbf{M}_h : \mathcal{N}_h(\delta) \to \mathcal{N}(\delta)$ we end up with

$$\int_{\mathcal{N}_h(\delta)} \nabla u \cdot \nabla V - \int_{\mathcal{N}(\delta)} \nabla u \cdot \nabla (V \circ \mathbf{M}_h^{-1}) = I_1(V) + I_2(V) + I_3(V),$$

where

$$I_1(V) := \int_{\mathcal{N}_h(\delta)} (\nabla u - \nabla u \circ \mathbf{M}_h) \cdot \nabla V \det(D\mathbf{M}_h)$$

$$I_2(V) := \int_{\mathcal{N}_h(\delta)} \nabla u \cdot \nabla V \, (1 - \det(D\mathbf{M}_h))$$

$$I_3(V) := \int_{\mathcal{N}(\delta)} \nabla u \cdot \left(\nabla V \circ \mathbf{M}_h^{-1} - \nabla (V \circ \mathbf{M}_h^{-1}) \right).$$

In view of Proposition 35 (Lipschitz perturbation) and Lemma 58 (properties of $\mathbf{M}_h$), we infer that

$$|I_1(V)|, |I_2(V)| \lesssim h |d|_{W^2_\infty(\mathcal{N})} \|u\|_{H^2(\mathcal{N}(\delta))} \|V\|_{H^1(\mathcal{N}_h(\delta))}.$$

Similarly for $I_3(V)$, we observe that

$$\nabla V \circ \mathbf{M}_h^{-1} - \nabla \left(V \circ \mathbf{M}_h^{-1} \right) = \left(\mathbf{I} - D\mathbf{M}_h^{-1} \right) \nabla V \circ \mathbf{M}_h^{-1},$$

so that employing Lemma 58 (properties of $\mathbf{M}_h$) yields

$$|I_3(V)| \lesssim h |d|_{W^2_\infty(\mathcal{N})} \|\nabla u\|_{L_2(\mathcal{N}(\delta))} \|\nabla V\|_{L_2(\mathcal{N}_h(\delta))}.$$

Recalling the structural assumption $C_1 h \leq \delta$, Lemma 3 (regularity) as well as $\|f\|_{L_2(\mathcal{N}(\delta))} \lesssim \delta^{\frac{1}{2}} \|\widetilde{f}\|_{L_2(\gamma)}$, the estimates for $I_1(V)$, $I_2(V)$ and $I_3(V)$ guarantee

$$\left| \int_{\mathcal{N}_h(\delta)} \nabla u \cdot \nabla V - \int_{\mathcal{N}(\delta)} \nabla u \cdot \nabla (V \circ \mathbf{M}_h^{-1}) \right| \lesssim \delta^{3/2} |d|_{W^2_\infty(\mathcal{N})} \|\widetilde{f}\|_{L_2(\gamma)} \|V\|_{H^1(\mathcal{N}_h(\delta))}.$$

Step 2: Forcing. Recalling (136), we rewrite the forcing term in (147) as

$$\int_{\mathcal{N}_h(\delta)} FV - \int_{\mathcal{N}(\delta)} f V \circ \mathbf{M}_h^{-1} = II_1(V) + II_2(V),$$

where

$$II_1(V) := \int_{\mathcal{N}(\delta)} f V \circ \mathbf{M}_h^{-1} \left(\det(D\mathbf{M}_h)^{-1} - 1 \right),$$

$$II_2(V) := -\frac{1}{|\mathcal{N}_h(\delta)|} \int_{\mathcal{N}_h(\delta)} f \circ \mathbf{M}_h \int_{\mathcal{N}_h(\delta)} V.$$

We make use of (144) and (146), along with a change of variables, to compute

$$\begin{aligned} II_1(V) &\lesssim h|d|_{W^2_\infty(\mathcal{N})} \|f\|_{L_2(\mathcal{N}(\delta))} \|V \circ \mathbf{M}_h^{-1}\|_{L_2(\mathcal{N}(\delta))} \\ &\lesssim h|d|_{W^2_\infty(\mathcal{N})} \|f\|_{L_2(\mathcal{N}(\delta))} \|V\|_{H^1(\mathcal{N}(\delta))}. \end{aligned}$$

Since $|\mathcal{N}_h(\delta)| \simeq |\mathcal{N}(\delta)| \simeq \delta$, the first equivalence resulting from (144) and the second from the coarea formula, using (146) again we obtain

$$\begin{aligned} II_2(V) &\lesssim \delta^{-1/2} \|V\|_{L_2(\mathcal{N}_h(\delta))} \left| \int_{\mathcal{N}_h(\delta)} f \circ \mathbf{M}_h (\det(D\mathbf{M}_h) - 1) - \int_{\mathcal{N}(\delta)} f \right| \\ &\lesssim \|V\|_{L_2(\mathcal{N}_h(\delta))} \left(h|d|_{W^2_\infty(\mathcal{N})} \|f\|_{L_2(\mathcal{N}(\delta))} + \delta^{-1/2} \left| \int_{\mathcal{N}(\delta)} f \right| \right). \end{aligned}$$

To estimate the rightmost term we exploit the fact that $\tilde{f}$ has a vanishing mean on γ. Using the coarea formula (63), we see that

$$\begin{aligned} \int_{\mathcal{N}(\delta)} f = \int_{-\delta}^{\delta} \int_{\gamma_s} f = \int_{-\delta}^{\delta} \int_{\gamma} \tilde{f} \mu_s = \int_{-\delta}^{\delta} \int_{\gamma} \tilde{f} (\mu_s - 1) \\ \leq 2\delta \|\mu_s - 1\|_{L_\infty(\gamma \times [-\delta, \delta])} \|\tilde{f}\|_{L_2(\gamma)}, \end{aligned}$$

where $\mu_s = \det(\mathbf{I} - dD^2 d)^{-1} \circ \mathbf{Q}_s$ according to Lemma 22 (relation between q and q_Γ). Remark 56 (estimate of μ) in turn leads to

$$\left| \int_{\mathcal{N}(\delta)} f \right| \lesssim \delta^2 |d|_{W^2_\infty(\mathcal{N})} \|\tilde{f}\|_{L_2(\gamma)}.$$

Consequently, collecting the previous estimates and using the structural assumption $C_1 h \leq \delta$ again readily gives

$$|II_2(V)| \lesssim \delta^{\frac{3}{2}} |d|_{W^2_\infty(\mathcal{N})} \|\tilde{f}\|_{L_2(\gamma)} \|V\|_{L_2(\mathcal{N}_h(\delta))}.$$

Gathering the bounds for $II_i(V)$ for $i = 1, 2$, we discover

$$\left| \int_{\mathcal{N}_h(\delta)} F V - \int_{\mathcal{N}(\delta)} f V \circ \mathbf{M}_h^{-1} \right| \lesssim \delta^{\frac{3}{2}} |d|_{W^2_\infty(\mathcal{N})} \|\widetilde{f}\|_{L_2(\gamma)} \| V\|_{H^1(\mathcal{N}_h(\delta))}.$$

Step 3: Conclusion. The assertion follows from the bounds derived in Steps 1 and 2 together with the estimate (141) of Lemma 57 (narrow band PDE consistency) with $v = V \circ \mathbf{M}_h^{-1} \in H^1(\mathcal{N}(\delta))$. The proof is complete. □

6.4 A priori error estimates

All of the ingredients for a priori error analysis in the narrow band norm are now in place. We recall that the extension $u : \mathcal{N}(\delta) \to \mathbb{R}$ constructed in Proposition 28 (H^2 extension) is further extended to $\mathcal{N}$ and satisfies

$$\| u\|_{H^2(\mathcal{N})} \lesssim \| u\|_{H^2(\mathcal{N}(\delta))} \lesssim \delta^{\frac{1}{2}} |d|_{W^2_\infty(\mathcal{N})} \|\widetilde{u}\|_{H^2(\gamma)}. \tag{148}$$

Theorem 60 (a priori error estimate). *Let γ be of class C^2 and δ and h satisfy the structural condition (134) and be sufficiently small. Let $\widetilde{u} \in H^1_\#(\gamma)$ be defined by (18) with $\widetilde{f} \in L_{2,\#}(\gamma)$ and u be its extension given by (62) with $C\delta \le \varepsilon \le \frac{\delta}{2}$. Let $U \in \mathbb{V}_\#(\mathcal{T}_\delta)$ be the finite element solution to (135) with F given in (136). Then, the following error estimate is valid*

$$\| \nabla(u - U)\|_{L_2(\mathcal{N}_h(\delta))} \lesssim \inf_{V \in \mathbb{V}(\mathcal{T}_\delta)} \| \nabla(u - V)\|_{L_2(\mathcal{N}_h(\delta))} + h^{\frac{3}{2}} |d|_{W^2_\infty(\mathcal{N})} \|\widetilde{f}\|_{L_2(\gamma)},$$

with hidden constant independent of h and δ.

Proof. The proof consists of a Strang-type argument. For any $V \in \mathbb{V}(\mathcal{T}_\delta)$ the Eq. (135) satisfied by U and the definition (147) of $E_h(.)$ give

$$\| \nabla(V - U) \|^2_{L_2(\mathcal{N}_h(\delta))} = \int_{\mathcal{N}_h(\delta)} \nabla(V - u) \cdot \nabla(V - U) + E_h(V - U).$$

Invoking Lemma 59 (narrow band geometric consistency), together with the structural assumption (134), yields

$$\| \nabla(V - U)\|_{L_2(\mathcal{N}_h(\delta))} \le \| \nabla(V - u)\|_{L_2(\mathcal{N}_h(\delta))} + ch^{\frac{3}{2}} |d|_{W^2_\infty(\mathcal{N})} \|\widetilde{f}\|_{L_2(\gamma)}.$$

The desired error estimate follows from a triangle inequality. □

Corollary 61 (rate of convergence in $\mathcal{N}_h(\delta)$). *Under the assumptions of Theorem 60 we have*

$$\|\nabla(u-U)\|_{L_2(\mathcal{N}_h(\delta))} \lesssim h^{\frac{3}{2}}|d|_{W^2_\infty(\mathcal{N})}\|\widetilde{f}\|_{L_2(\gamma)}.$$

Proof. In view of (148) standard polynomial interpolation theory gives

$$\|\nabla(u-I_h^{\mathrm{sz}}u)\|_{L_2(\mathcal{N}_h(\delta))} \lesssim h\|u\|_{H^2(\mathcal{N})} \lesssim h\|u\|_{H^2(\mathcal{N}(\delta))},$$

where $I_h^{\mathrm{sz}}u$ is the Scott–Zhang interpolant of u. It remains to use Proposition 28 (H^2 extension) and Lemma 3 (regularity) to arrive at

$$\|\nabla(u-I_h^{\mathrm{sz}}u)\|_{L_2(\mathcal{N}_h(\delta))} \lesssim h^{\frac{3}{2}}|d|_{W^2_\infty(\mathcal{N})}\|\widetilde{f}\|_{L_2(\gamma)}.$$

The asserted estimate follows from Theorem 60 (a priori estimate). □

In addition, we follow Olshanskii and Safin (2016) to deduce a rate of convergence for $\|\nabla_\gamma(\widetilde{u}-U)\|_{L_2(\gamma)}$.

Corollary 62 (rate of convergence on γ). *Under the assumptions of Theorem 60 we have*

$$\|\nabla_\gamma(\widetilde{u}-U)\|_{L_2(\gamma)} \lesssim h\|\widetilde{f}\|_{L_2(\gamma)}.$$

Proof. We recall the scaled trace inequality (117): for a bulk triangulation $\mathcal{T}$ there exists a constant C only depending on the mesh shape-regularity constant of $\mathcal{T}$ such that for $T\in\mathcal{T}_\delta$ and $v\in H^1(T)$, one has

$$\|v\|^2_{L_2(T\cap\gamma)} \le C\left(h_T^{-1}\|v\|^2_{L_2(T)} + h_T\|\nabla v\|^2_{L_2(T)}\right),$$

where $h_T=\mathrm{diam}(T)$. We apply this inequality with $v=\nabla(u-U)$, and $h_T\approx h$ sufficiently small, to obtain

$$\|\nabla(u-U)\|^2_{L_2(T\cap\gamma)} \lesssim \left(h^{-1}\|\nabla(u-U)\|^2_{L_2(T)} + h|u|^2_{H^2(T)}\right).$$

Summing up over all $T\in\mathcal{T}_\delta$ with nonempty intersection with γ, Proposition 28 (H^2 extension) and Corollary 61 (rate of convergence in $\mathcal{N}_h(\delta)$) give

$$\|\nabla_\gamma(\widetilde{u}-U)\|_{L_2(\gamma)} \le \|\nabla(u-U)\|_{L_2(\gamma)} \lesssim h|d|_{W^2_\infty(\mathcal{N})}\left(\|\widetilde{f}\|_{L_2(\gamma)} + |\widetilde{u}|_{H^2(\gamma)}\right).$$

The desired estimate follows from Lemma 3 (regularity). □

Acknowledgements

A.B. is partially supported by NSF grant DMS-1817691. A.D. is partially supported by NSF grant DMS-1720369. R.H.N. is partially supported by NSF grant DMS-1411808.

References

Adams, R.A., 1975. Sobolev Spaces, vol. 65. Academic Press [A subsidiary of Harcourt Brace Jovanovich, Publishers], New York–London, p. xviii+268.

Antonietti, P.F., Dedner, A., Madhavan, P., Stangalino, S., Stinner, B., Verani, M., 2015. High order discontinuous Galerkin methods for elliptic problems on surfaces. SIAM J. Numer. Anal. 53 (2), 1145–1171. ISSN 0036-1429. https://doi.org/10.1137/140957172.

Aubin, T., 1982. Nonlinear analysis on manifolds. In: Monge-Amp‘ere equations, Grundlehren der Mathematischen Wissenschaften [Fundamental Principles of Mathematical Sciences], vol. 252. Springer-Verlag, New York. ISBN: 0-387-90704-1, p. xii+204. https://doi.org/10.1007/978-1-4612-5734-9.

Bendali, A., 1984. Numerical analysis of the exterior boundary value problem for the time-harmonic Maxwell equations by a boundary finite element method. II. The discrete problem. Math. Comp. 43 (167), 47–68. ISSN 0025-5718. https://doi.org/10.2307/2007399.

Bertalmío, M., Cheng, L.-T., Osher, S., Sapiro, G., 2001. Variational problems and partial differential equations on implicit surfaces. J. Comput. Phys. 174 (2), 759–780. ISSN 0021-9991. https:/doi.org/10.1006/jcph.2001.6937.

Bonito, A., Demlow, A., 2019. A posteriori error estimates for the Laplace–Beltrami operator on parametric C^2 surfaces. SIAM J. Numer. Anal. 57 (3), 973–996.

Bonito, A., Pasciak, J., 2012. Convergence analysis of variational and non-variational multigrid algorithm for the Laplace-Beltrami operator. Math. Comp. 81, 1263–1288.

Bonito, A., Cascón, J.M., Morin, P., Nochetto, R.H., 2013. AFEM for geometric PDE: the Laplace-Beltrami operator. In: Gianazza, U., Brezzi, F., Franzone, P.C., Gilardi, G. (Eds.), Analysis and Numerics of Partial Differential Equations, vol. 4. Springer INdAM Ser., Springer, Milan, pp. 257–306.

Bonito, A., Cascón, J.M., Mekchay, K., Morin, P., Nochetto, R.H., 2016. High-order AFEM for the Laplace-Beltrami operator: convergence rates. Found. Comput. Math. 16 (6), 1473–1539. ISSN 1615-3375. https://doi.org/10.1007/s10208-016-9335-7.

Bonito, A., Demlow, A., Owen, J., 2019. Convergence and optimality for higher order AFEM for eigenvalue clusters of the Laplace-Beltrami operator (in preparation).

Burger, M., 2009. Finite element approximation of elliptic partial differential equations on implicit surfaces. Comput. Vis. Sci. 12 (3), 87–100. ISSN 1432-9360. https://doi.org/10.1007/s00791-007-0081-x.

Burman, E., Hansbo, P., Larson, M.G., 2015. A stabilized cut finite element method for partial differential equations on surfaces: the Laplace-Beltrami operator. Comput. Methods Appl. Mech. Eng. 285, 188–207. ISSN 0045-7825. https://doi.org/10.1016/j.cma.2014.10.044.

Burman, E., Hansbo, P., Larson, M.G., Massing, A., Zahedi, S., 2016. Full gradient stabilized cut finite element methods for surface partial differential equations. Comput. Methods Appl. Mech. Eng. 310, 278–296. ISSN 0045-7825. https://doi.org/10.1016/j.cma.2016.06.033.

Burman, E., Hansbo, P., Larson, M.G., Massing, A., 2017. A cut discontinuous Galerkin method for the Laplace-Beltrami operator. IMA J. Numer. Anal. 37 (1), 138–169. ISSN 0272-4979. https://doi.org/10.1093/imanum/drv068.

Camacho, F., Demlow, A., 2015. L_2 and pointwise a posteriori error estimates for FEM for elliptic PDEs on surfaces. IMA J. Numer. Anal. 35 (3), 1199–1227. ISSN 0272-4979. https://doi.org/10.1093/imanum/dru036.

Chernyshenko, A.Y., Olshanskii, M.A., 2015. An adaptive octree finite element method for PDEs posed on surfaces. Comput. Methods Appl. Mech. Eng. 291, 146–172. ISSN 0045-7825. https://doi.org/10.1016/j.cma.2015.03.025.

Cockburn, B., Demlow, A., 2016. Hybridizable discontinuous Galerkin and mixed finite element methods for elliptic problems on surfaces. Math. Comp. 85 (302), 2609–2638. ISSN 0025-5718. https://doi.org/10.1090/mcom/3093.

Deckelnick, K., Dziuk, G., Elliott, C.M., Heine, C.-J., 2010. An h-narrow band finite-element method for elliptic equations on implicit surfaces. IMA J. Numer. Anal. 30 (2), 351–376. ISSN 0272-4979. https://doi.org/10.1093/imanum/drn049.

Deckelnick, K., Elliott, C.M., Ranner, T., 2014. Unfitted finite element methods using bulk meshes for surface partial differential equations. SIAM J. Numer. Anal. 52 (4), 2137–2162. ISSN 0036-1429. https://doi.org/10.1137/130948641.

Dedner, A., Madhavan, P., 2016. Adaptive discontinuous Galerkin methods on surfaces. Numer. Math. 132 (2), 369–398. ISSN 0029-599X. https://doi.org/10.1007/s00211-015-0719-4.

Dedner, A., Madhavan, P., Stinner, B., 2013. Analysis of the discontinuous Galerkin method for elliptic problems on surfaces. IMA J. Numer. Anal. 33 (3), 952–973. ISSN 0272-4979. https://doi.org/10.1093/imanum/drs033.

Demlow, A., 2009. Higher-order finite element methods and pointwise error estimates for elliptic problems on surfaces. SIAM J. Numer. Anal. 47 (2), 805–827. ISSN 0036-1429. https://doi.org/10.1137/070708135.

Demlow, A., Dziuk, G., 2007. An adaptive finite element method for the Laplace-Beltrami operator on implicitly defined surfaces. SIAM J. Numer. Anal. 45 (1), 421–442. ISSN 0036-1429 (electronic).

Demlow, A., Olshanskii, M., 2012. An adaptive surface finite element method based on volume meshes. SIAM J. Numer. Anal. 50, 1624–1647.

Dziuk, G., 1988. Finite elements for the Beltrami operator on arbitrary surfaces. In: Hildebrandt, S., Leis, R. (Eds.), Partial Differential Equations and Calculus of Variations, Lecture Notes in Math., vol. 1357. Springer, Berlin, pp. 142–155.

Dziuk, G., Elliott, C.M., 2013. Finite element methods for surface PDEs. Acta Numer. 22, 289–396. ISSN 0962-4929.

Evans, L.C., 1998. Partial Differential Equations, Graduate Studies in Mathematics, vol. 19. American Mathematical Society, Providence, RI. ISBN: 0-8218-0772-2, p. xviii+662.

Evans, L.C., Gariepy, R.F., 2015. Measure Theory and Fine Properties of Functions. Chapman and Hall/CRC.

Federer, H., 1959. Curvature measures. Trans. Am. Math. Soc. 93, 418–491. ISSN 0002-9947. https://doi.org/10.2307/1993504.

Ferroni, A., Formaggia, L., Fumagalli, A., 2016. Numerical analysis of Darcy problem on surfaces. ESAIM Math. Model. Numer. Anal. 50 (6), 1615–1630. ISSN 0764-583X. https://doi.org/10.1051/m2an/2015095.

Gilbarg, D., Trudinger, N.S., 1998. Elliptic Partial Differential Equations of Second Order, second ed. Springer-Verlag, Berlin.

Grande, J., 2017. Analysis of highly accurate finite element based algorithms for computing distances to level sets. SIAM J. Numer. Anal. 55 (1), 376–399. ISSN 0036-1429. https://doi.org/10.1137/15M1011524.

Grande, J., Reusken, A., 2016. A higher order finite element method for partial differential equations on surfaces. SIAM J. Numer. Anal. 54 (1), 388–414. ISSN 0036-1429. https://doi.org/10.1137/14097820X.

Grande, J., Lehrenfeld, C., Reusken, A., 2018. Analysis of a high-order trace finite element method for PDEs on level set surfaces. SIAM J. Numer. Anal. 56 (1), 228–255. ISSN 0036-1429. https://doi.org/10.1137/16M1102203.

Grisvard, P., 1985. Elliptic Problems in Nonsmooth Domains. Pitman Publishing Inc., Marshfield, MA.

Hansbo, A., Hansbo, P., 2002. An unfitted finite element method, based on Nitsche's method, for elliptic interface problems. Comput. Methods Appl. Mech. Eng. 191 (47–48), 5537–5552. ISSN 0045-7825. https://doi.org/10.1016/S0045-7825(02)00524-8.

Hansbo, A., Hansbo, P., 2004. A finite element method for the simulation of strong and weak discontinuities in solid mechanics. Comput. Methods Appl. Mech. Eng. 193 (33–35), 3523–3540. ISSN 0045-7825. https://doi.org/10.1016/j.cma.2003.12.041.

Holst, M., Stern, A., 2012. Geometric variational crimes: Hilbert complexes, finite element exterior calculus, and problems on hypersurfaces. Found. Comput. Math. 12 (3), 263–293. ISSN 1615-3375. https://doi.org/10.1007/s10208-012-9119-7.

Jerison, D., Kenig, C.E., 1995. The inhomogeneous Dirichlet problem in Lipschitz domains. J. Funct. Anal. 130 (1), 161–219. ISSN 0022-1236. https://doi.org/10.1006/jfan.1995.1067.

Lucas, K.R., 1957. Submanifolds of Dimension $n-1$ in $\mathbb{R}^n$ With Normals Satisfying a Lipschitz Condition, Studies in Eigenvalue Problems. University of Kansas.

Nédélec, J.C., 1976. Curved finite element methods for the solution of singular integral equations on surfaces in R^3. Comput. Methods Appl. Mech. Eng. 8 (1), 61–80. ISSN 0045-7825.

Olshanskii, M.A., Reusken, A., 2010. A finite element method for surface PDEs: matrix properties. Numer. Math. 114 (3), 491–520. ISSN 0029-599X. https://doi.org/10.1007/s00211-009-0260-4.

Olshanskii, M.A., Reusken, A., 2017. Trace finite element methods for PDEs on surfaces. In: Bordas, S.P., Burman, E., Larson, M.G., Olshanskii, M.A. (Eds.), Geometrically Unfitted Finite Element Methods and Applications, Lect. Notes Comput. Sci. Eng., vol. 121. Springer, Cham, pp. 211–258.

Olshanskii, M.A., Safin, D., 2016. A narrow-band unfitted finite element method for elliptic PDEs posed on surfaces. Math. Comp. 85 (300), 1549–1570. ISSN 0025-5718. https://doi.org/10.1090/mcom/3030.

Olshanskii, M.A., Reusken, A., Grande, J., 2009. A finite element method for elliptic equations on surfaces. SIAM J. Numer. Anal. 47 (5), 3339–3358.

Olshanskii, M.A., Reusken, A., Xu, X., 2012. On surface meshes induced by level set functions. Comput. Vis. Sci. 15 (2), 53–60. ISSN 1432-9360. https://doi.org/10.1007/s00791-013-0200-9.

Peetre, J., 1966. Espaces d'interpolation et théorème de Soboleff. Ann. Inst. Fourier (Grenoble) 16 (fasc. 1), 279–317. ISSN 0373-0956. http:/www.numdam.org/item?id=AIF_1966__16_1_279_0.

Reusken, A., 2015. Analysis of trace finite element methods for surface partial differential equations. IMA J. Numer. Anal. 35 (4), 1568–1590. ISSN 0272-4979. https://doi.org/10.1093/imanum/dru047.

Ruuth, S.J., Merriman, B., 2008. A simple embedding method for solving partial differential equations on surfaces. J. Comput. Phys. 227 (3), 1943–1961. ISSN 0021-9991. https://doi.org/10.1016/j.jcp.2007.10.009.

Stein, E.M., 1970. Singular Integrals and Differentiability Properties of Functions. Princeton Mathematical Series, No. 30Princeton University Press, Princeton, NJ, p. xiv+290.

Tartar, L., 1978. Topics in Nonlinear Analysis. Publications Mathématiques d'Orsay 78, vol. 13. Université de Paris-Sud, Département de Mathématique, Orsay. ii+271 pp (not consecutively paged).

Veeser, A., 2016. Approximating gradients with continuous piecewise polynomial functions. Found. Comput. Math. 16 (3), 723–750. ISSN 1615-3375. https://doi.org/10.1007/s10208-015-9262-z.

Verfürth, R., 2013. A Posteriori Error Estimation Techniques for Finite Element Methods, Numerical Mathematics and Scientific Computation. Oxford University Press, Oxford. ISBN: 978-0-19-967942-3, p. xx+393. https:/doi.org/10.1093/acprof:oso/9780199679423.001.0001.

Wei, H., Chen, L., Huang, Y., 2010. Superconvergence and gradient recovery of linear finite elements for the Laplace-Beltrami operator on general surfaces. SIAM J. Numer. Anal. 48 (5), 1920–1943. ISSN 0036-1429. https://doi.org/10.1137/100785016.

Chapter 2

The Monge–Ampère equation

Michael Neilan[a,*], Abner J. Salgado[b] and Wujun Zhang[c]
[a]*Department of Mathematics, University of Pittsburgh, Pittsburgh, PA, United States*
[b]*Department of Mathematics, University of Tennessee, Knoxville, TN, United States*
[c]*Department of Mathematics, Rutgers University, Piscataway, NJ, United States*
[*]*Corresponding author: e-mail: neilan@pitt.edu*

Chapter Outline

Abstract

We review recent advances in the numerical analysis of the Monge–Ampère equation. Various computational techniques are discussed including wide stencil finite difference schemes, two-scaled methods, finite element methods, and methods based on geometric

Handbook of Numerical Analysis, Vol. 21. https://doi.org/10.1016/bs.hna.2019.05.003

considerations. Particular focus is the development of appropriate stability and consistency estimates which lead to rates of convergence of the discrete approximations. Finally we present numerical experiments which highlight each method for a variety of test problem with different levels of regularity.

Keywords: Monge–Ampère, Convergence analysis, Error estimates, Comparison principle, Fully nonlinear equations

AMS Classification Codes: 65N12, 65N15, 35B51, 35D40, 35J96

1 Introduction

> All exact science is dominated by the idea of approximation. When a man tells you that he knows the exact truth about anything, you are safe in inferring that he is an inexact man.
>
> Russell (1931)

In this chapter we review recent progress in the numerical treatment of Monge–Ampère type equations. In its simplest form, and assuming Dirichlet boundary conditions, the problem we consider is to seek a scalar function u satisfying the partial differential equation (PDE)

$$\det D^2u(x) = f(x) \quad x \in \Omega, \tag{1a}$$

$$u(x) = g(x) \quad x \in \partial\Omega. \tag{1b}$$

Here, D^2u denotes the Hessian matrix of $u, f \geq 0$, and g are given functions, and $\Omega \subset \mathbb{R}^d$ is a bounded, convex domain. Problem (1) is a prototypical second order, fully nonlinear PDE, and it arises in several broad applications in differential geometry, meteorology, cosmology, economics, and optimal mass transportation problems. Some of these applications are briefly described below.

Despite its growing list of applications, and in contrast to its extensive and mature PDE theory, the construction and analysis of computational methods for (1) is still a relatively new and emerging field in numerical analysis. Numerical algorithms, based on geometric considerations, for the two-dimensional problem ($d = 2$) first appeared in 1988 in Oliker and Prussner (1988), and the extension to practical three-dimensional schemes were not introduced until some 20 years later (Brenner and Neilan, 2012; Feng and Neilan, 2009; Froese and Oberman, 2011a,b). Other early attempts that deserve mention are the least squares and augmented Lagrangian approaches of Dean and Glowinski (2003, 2004, 2005, 2006a,b), and we refer the reader to Feng et al. (2013) for more details on these schemes.

The reasons for this delayed development in numerical methods are plentiful. The most evident obstacle is the full nonlinearity of the problem. However, this is arguably a secondary difficulty, as black-box nonlinear solvers can, at least heuristically, be applied to algebraic systems resulting from discretizations of (1). A rather fundamental difficulty to construct, and especially to analyze, computational methods for Monge–Ampère type equations is the variety of solution concepts and, correspondingly, the low regularity solutions generically possess. As we explain below, weak solutions are not based on variational principles, but rather on either geometric considerations or by monotonicity conditions of test functions that touch the graph of the solution from above or below. These solution concepts are difficult to mimic at the discrete level, and as a result, the construction of convergent schemes is an arduous task. Finally, as if these complications were not enough, the Monge–Ampère equation (1) is usually supplemented by the constraint that the solution u is convex. This is not only because of geometric applications, but in many cases a necessary condition for uniqueness, and for the existence of a well-developed PDE theory. As convexity is a global constraint, it is very difficult to enforce it in a discrete setting.

Nonetheless, an explosion of results and new techniques to develop them in computational methods for (1) have occurred during the last 10 years. These include the construction of monotone, wide stencil finite difference schemes, semi-Lagrangian methods, and finite element methods. Within only the past few years, significant progress has been made in the convergence analysis with an emphasis on the rates of convergence for various discretization schemes.

The main goal of this chapter is to highlight these recent advances in the numerical analysis of the Monge–Ampère problem (1). To this end, we organize the chapter as follows. After stating some geometric applications and a brief PDE theory of the Monge–Ampère problem in this section, we discuss wide stencil finite difference schemes in Section 2. There we introduce the monotone finite difference schemes (Froese and Oberman, 2011a,b; Oberman, 2008b) and the corresponding filtered schemes (Froese and Oberman, 2013), lattice reduction schemes (Benamou et al., 2016), methods based on power diagrams (Mirebeau, 2015), and the so-called two scale methods (Nochetto and Ntogkas, 2018; Nochetto et al., 2019a,b). Of particular focus will be the rates of convergence of these schemes if available. Next, in Section 3, we review the original method of Oliker and Prussner (1988), which in honour of its proponents henceforth we shall call the Oliker–Prussner scheme. This method is based on geometric interpretations of the Monge–Ampère operator and the notion of Alexandrov solutions. Again, the emphasis of the discussion is on consistency error and pointwise rates of convergence recently established in Nochetto and Zhang (2019). Section 4 discusses finite element methods for both smooth and singular solutions. Finally in Section 5 we perform some numerical experiments using some of the methods we discuss in this review for a variety of test problems with different levels of regularity.

We remark that, by design, this review has several major omissions. We intend to minimize the overlap between two other existing, and rather recent, reviews on fully nonlinear problems in general and the Monge–Ampère equation in particular. Namely, the overview of Feng et al. (2013), which is dedicated to the Monge–Ampère equation exclusively, and Neilan et al. (2017) which contains a chapter on the Monge–Ampère equation, and where the reader can find much more details, for instance, on the semi-Lagrangian schemes described in Section 2.8.1.

1.1 Geometric applications

To draw connections with the theme of the current volume in the *Handbook of Numerical Analysis*, and to further emphasize the prevalence of the Monge–Ampère problem, in this section we briefly summarize some applications with a geometric flavour where the Monge–Ampère problem plays an essential role.

1.1.1 Gauss curvature problem

The classic Gauss curvature problem (cf. Bakelman, 1994; Guan and Spruck, 1993; Oliker, 1984) seeks a manifold $\mathcal{M} \subset \mathbb{R}^{n+1}$ with prescribed boundary and Gauss curvature $\mathcal{K}$. We recall that Gauss curvature is the product of the principal curvatures, which themselves are the eigenvalues of the shape operator (or Weingarten map). One may reduce this problem to a PDE problem of Monge–Ampère type if one assumes that the manifold is the graph of a function, i.e.,

$$\mathcal{M} = \{(x, u(x)) : \quad u : \Omega \to \mathbb{R}\}.$$

The shape operator is given by $s = \mathrm{I}^{-1}\,\mathrm{II}$, where I and II denote, respectively, the first and second fundamental forms. In the case that $\mathcal{M}$ is the graph of the function u, we have $\mathrm{I} = I + \nabla u \otimes \nabla u$ and $\mathrm{II} = \frac{D^2 u}{\sqrt{1 + |\nabla u|^2}}$, where I denotes the $d \times d$ identity matrix. Therefore the Gauss curvature is given by

$$\mathcal{K} = \det(s) = \frac{\det(\mathrm{II})}{\det(\mathrm{I})} = \frac{\det D^2 u}{(1 + |\nabla u|^2)^{(d+2)/2}}.$$

Thus, the problem is to find a scalar function $u : \bar{\Omega} \to \mathbb{R}$ satisfying

$$\det D^2 u(x) = \mathcal{K}(x)(1 + |\nabla u(x)|^2)^{(d+2)/2} \quad \text{in } \Omega, \tag{2a}$$

$$u(x) = g(x) \quad \text{on } \partial\Omega. \tag{2b}$$

In conclusion the Gauss curvature problem, in this setting, seeks solutions of a Monge–Ampère type problem with lower-order terms.

1.1.2 Reflector design problem

The reflector design problem (Norris and Westcott, 1976; Oliker, 1987; Oliker and Waltman, 1987; Wang, 1996) can be posed as follows: Let S^2 be the unit sphere in $\mathbb{R}^3$ centred at the origin, and let $\Omega, \mathcal{O}$ be two disjoint domains on S^2. Let f be a positive function defined on $\mathcal{O}$, and suppose that rays originate from the origin with density ρ. We then seek a surface, called Γ, whose radial projection onto S^2 equals Ω, such that the directions of the reflected rays cover $\mathcal{O}$ with distributed density equal to f.

To formulate a PDE model for this problem, we set $\Gamma = \{xm(x) : x \in \Omega\}$, so that if a ray radiates from the origin with direction x, then it is reflected at the point $xm(x)$. This will create a reflected ray in the direction $T(x) \in \mathcal{O}$. Now if we denote by $\boldsymbol{n}$ the unit normal of Γ at $xm\,(x)$, then we have $T(x) - x = -2(x \cdot \boldsymbol{n})\boldsymbol{n}$, and calculations show that $\boldsymbol{n} = (\nabla m - mx)/\sqrt{m^2 + |\nabla m|^2}$. Here, $\nabla = e^{ij}\partial_i x \partial_j$, where x is a smooth parametrization of S^2, $e = e_{ij}\mathrm{d}t^i\mathrm{d}t^j$ is the first fundamental form of S^2, $e^{ij} = (e_{ij})^{-1}$, and $\partial_j = \partial/\partial t^j$. Combining these two identities we find that the direction T is related to m via

$$T(x) = \frac{2m\nabla m + (|\nabla m|^2 - m^2)x}{m^2 + |\nabla m|^2}. \tag{3}$$

Next, if the directions of the reflected light do not overlap and if no loss of energy occurs in the reflection, then we have the energy conservation property

$$\int_E \rho(x)\mathrm{d}x = \int_{T(E)} f(y)\mathrm{d}y = \int_E f(T(x))\frac{|\partial_1 T(x) \times \partial_2 T(x)|}{\det(e_{ij})}\mathrm{d}x$$

for all Borel sets $E \subset \Omega$. Thus we have, at least formally,

$$\frac{|\partial_1 T(x) \times \partial_2 T(x)|}{\det(e_{ij})} = \frac{\rho(x)}{f(T(x))}.$$

Finally, we set $u(x) = 1/m(x)$, and substitute (3) into this last equation to get the following problem of Monge–Ampère type (see Oliker and Newman, 1993; Wang, 1996 for details)

$$\frac{\det(D^2 u + (u - \eta)e_{ij})}{\eta^2 \det(e_{ij})} = \frac{\rho(x)}{f(T(x))} \qquad x \in \Omega,$$

where T is given by (3) and $\eta = (|\nabla u|^2 + u^2)/(2u)$.

1.1.3 Affine plateau problem

Following Trudinger and Wang (2005, 2008) and Calabi (1990), we consider the following problem. Let $\mathcal{M}_0 \subset \mathbb{R}^{d+1}$ be a bounded and connected hypersurface with smooth boundary that is locally uniformly convex We denote by $S[\mathcal{M}_0]$ the set of locally uniformly convex hypersurfaces that can be smoothly deformed from $\mathcal{M}_0$ within the family of locally uniformly convex hypersurfaces and whose Gauss map images lie in that of $\mathcal{M}_0$. As in Section 1.1.1, for a manifold $\mathcal{M}$ we denote by II its second fundamental form and by $\mathcal{K}$ its Gauss curvature. Associated with $\mathcal{M}$ is the *Berwald–Blaschke metric*

$$g = \mathcal{K}^{-1/(d+2)}\mathrm{II},$$

which is an affine invariant Riemannian metric on the surface. The *affine Plateau problem* is then to determine the maximizer of the *affine area functional*

$$A(\mathcal{M}) = \int_{\mathcal{M}} \mathcal{K}^{1/(d+2)} \mathrm{d}\mathcal{M}$$

over $S[\mathcal{M}_0]$.

Recall that if $\mathcal{M} = \mathcal{M}_u$ is the graph of a function $u:\Omega \to \mathbb{R}$, with $\Omega \subset \mathbb{R}^n$, then the Gauss curvature is $\mathcal{K} = \det(D^2u)/(1+|\nabla u|^2)^{(d+2)/2}$, and so, we have by a change of variables,

$$A(\mathcal{M}_u) = \int_{\Omega} (\det D^2 u(x))^{1/(d+2)} \mathrm{d}x.$$

Thus if $\mathcal{M}_0$ is the graph of a locally uniformly convex g, then in the graph case, $S[\mathcal{M}_0]$ consists of the graphs of locally uniformly convex functions $v \in C^2(\Omega) \cap C^0(\bar{\Omega})$ satisfying $v = g$ on $\partial\Omega$ and $\nabla v(\Omega) \subset \nabla g(\Omega)$. In this setting the affine Plateau problem seeks u such that

$$A(\mathcal{M}_u) = \sup\{A(\mathcal{M}_v): \ \mathcal{M}_v \in S[\mathcal{M}_0]\}.$$

Formally taking the Euler–Lagrange equation yields the *affine maximal surface equation*

$$\operatorname{cof} D^2u : D^2 w = 0, \quad w = (\det D^2 u)^{-(d+1)/(d+2)}.$$

1.1.4 Optimal mass transport problem

> *This problem appeared as a generalization of an earlier considered practical problem of assigning production locations on a railway network to consumption locations with minimum total transportation expenses.*
>
> Kantorovich (2004)

The optimal mass transport problem was originally proposed by G. Monge in the 18th century to find the optimal way to move oil to an excavation with minimal transportation cost. In general, the mass transport problem seeks, for two given sets and densities, the optimal mass-preserving mapping between them.

In further detail, given bounded $\Omega, \mathcal{O} \subset \mathbb{R}^d$ and measures $\rho_\Omega : \Omega \to \mathbb{R}$, $\rho_{\mathcal{O}} : \mathcal{O} \to \mathbb{R}$, the optimal transport problem with quadratic cost seeks a map $T: \Omega \to \mathcal{O}$ such that $T_{\#}\rho_\Omega = \rho_{\mathcal{O}}$ that minimizes the functional

$$\frac{1}{2}\int_{\Omega} |x - T(x)|^2 \mathrm{d}\rho_\Omega(x) \tag{4}$$

over all mass-preserving maps. Here, we assume that the measures are absolutely continuous with respect to Lebesgue measure, with $d\rho_\Omega = f_\Omega dx$ and $d\rho_{\mathcal{O}} = f_{\mathcal{O}} dx$, and that the measures satisfy the mass balance condition

$$\int_\Omega f_\Omega(x)dx = \int_{\mathcal{O}} f_{\mathcal{O}}(x)dx.$$

Above, we denoted by $T_{\#}\rho_\Omega$ the pushforward of the measure ρ_Ω under the mapping T, i.e., under the given assumptions, we have

$$\int_E f_{\mathcal{O}}(x)dx = \int_{T^{-1}(E)} f_\Omega(x)dx.$$

Thus, by a change of variables, we have, at least formally,

$$\det(\nabla T(x)) f_{\mathcal{O}}(T(x)) = f_\Omega(x) \quad x \in \Omega, \tag{5}$$

with $T(\Omega) \subset \mathcal{O}$. Thus in summary, we seek a mapping T that minimizes (4) with the constraint (5). One of the fundamental results in the theory of optimal transport (Brenier, 1991; Cuesta and Matrán, 1989; Rüschendorf and Rachev, 1990a,b) is that there exists a unique solution to this problem and that this optimal mapping is characterized as the gradient of some convex function u:

$$T(x) = \nabla u(x).$$

Hence, by substituting this relation into (5), we see that the problem reduces to a Monge–Ampère type PDE

$$f_{\mathcal{O}}(\nabla u(x)) \det D^2 u(x) = f_\Omega(x) \quad x \in \Omega \tag{6}$$

with the constraint $\nabla u(\bar{\Omega}) \subset \bar{\mathcal{O}}$. Thus we find that, with quadratic cost, the optimal mass transport problem reduces to a Monge–Ampère equation with transport boundary conditions.

1.2 Solution concepts for the Monge–Ampère equation

It is impossible to understand an unmotivated definition [...]

Arnol'd (1998)

In order to properly analyze the numerical schemes that we present below, it is important to understand in which sense a function $u : \bar{\Omega} \to \mathbb{R}$ must satisfy the equation and boundary conditions in (1) to be a solution. It is not our intention here to give a survey of the PDE theory regarding the Monge–Ampère equation, and we refer the reader to Gutiérrez (2001), Figalli (2017), and Bakelman (1994) for an in-depth presentation.

1.2.1 Classical solutions

The first definition of a solution to (1) is that of a classical solution. Essentially we require that (1) holds at every point of $\bar{\Omega}$.

Definition 1 (classical solution).
A function $u \in C^2(\Omega) \cap C(\bar{\Omega})$ is called a classical solution of (1) if these identities hold for every $x \in \bar{\Omega}$.

Notice that this necessarily implies that the right-hand side $f : \Omega \to \mathbb{R}$ is continuous. Regarding the existence of classical solutions we have the following result; see Figalli (2017, Section 3.1) for a detailed presentation.
Theorem 1 (existence of classical solutions).
Let $\alpha \in (0, 1)$. *Assume that* Ω *is a bounded and uniformly convex domain, whose boundary is of class* C^3, $f \in C^\alpha(\bar{\Omega})$ *with* $f \geq f_0 > 0$, *and* $g \in C^3(\partial\Omega)$. *Then problem* (1) *has a unique solution* $u \in C^{2,\alpha}(\bar{\Omega})$.

It is important to notice that classical solutions may not always exist, see for instance the counterexample given in Figalli (2017, Section 3.2). This motivates us to introduce weaker notions of solutions.

1.2.2 Viscosity solutions

The Monge–Ampère operator $w \mapsto \det D^2 w$ is a fully nonlinear second order operator, that is it depends nonlinearly on the highest (in this case second) order derivatives that appear in the expression. For this reason, the theory regarding fully nonlinear operators can guide us to develop a notion of solution (viscosity solution) that is weaker than classical. We refer the reader to Gilbarg and Trudinger (2001, Chapter 17), Caffarelli and Cabré (1995), Crandall et al. (1992), and Neilan et al. (2017, Section 2) for additional details.

We begin with a definition that encodes the type of admissible nonlinearities that will allow for the development of the theory of viscosity solutions. Here and in what follows we denote by $\mathbb{S}^d$ the collection of symmetric $d \times d$ matrices. The set $\mathbb{S}^d$ is endowed with a partial order: if $M, N \in \mathbb{S}^d$ then we say that $M \leq N$ if $\boldsymbol{v} \cdot M\boldsymbol{v} \leq \boldsymbol{v} \cdot N\boldsymbol{v}$ for every $\boldsymbol{v} \in \mathbb{R}^d$.
Definition 2 (elliptic operator).
Let $F : \bar{\Omega} \times \mathbb{R} \times \mathbb{S}^d \to \mathbb{R}$ be locally bounded. We say that F is *elliptic* if it satisfies the following *monotonicity condition*: For all $x \in \bar{\Omega}$, $r, s \in \mathbb{R}$ and $M, N \in \mathbb{S}^d$ with $r \geq s$ and $M \leq N$ then

$$F(x, r, M) \leq F(x, s, N).$$

Moreover, we say F is *uniformly elliptic* if for all $r, s \in \mathbb{R}$ and $M \in \mathbb{S}^d$ with $r \geq s$ we have

$$F(x, r, M) \leq F(x, s, M),$$

and, in addition, there are constants $0 < \lambda \leq \Lambda$ such that for all $M \in \mathbb{S}^d$ we have

$$\lambda \|N\|_2 \leq F(x, r, M + N) - F(x, s, M) \leq \Lambda \|N\|_2, \quad \forall N \geq 0.$$

Letting $F:\bar{\Omega}\times\mathbb{R}\times\mathbb{S}^d\to\mathbb{R}$ be an elliptic operator as defined above, we consider the fully nonlinear elliptic problem

$$F(x,u(x),D^2u(x))=0 \ \text{ in } \bar{\Omega}. \tag{7}$$

To be able to properly describe the notion of viscosity solutions we need to recall the following.

Definition 3 (upper and lower semicontinuous envelopes).

Let $w:\bar{\Omega}\to\mathbb{R}$. By $w^\star\in USC(\bar{\Omega})$ and $w_\star\in LSC(\bar{\Omega})$, we denote the upper and lower semicontinuous envelopes, respectively, of the function w. In other words

$$w^\star(x)=\limsup_{y\to x} w(x), \qquad w_\star(x)=\liminf_{y\to x} w(x).$$

Finally, by $USC(\bar{\Omega})$ and $LSC(\bar{\Omega})$ we denote, respectively, the sets of upper and lower semicontinuous functions.

We are now ready to introduce the notion of viscosity solution.

Definition 4 (viscosity solution).

Let F be elliptic in the sense of Definition 2. We say that the locally bounded function $u:\bar{\Omega}\to\mathbb{R}$ is:

1. A *viscosity subsolution* of (7) if whenever $x_0\in\bar{\Omega}$, $\varphi\in C^2(\bar{\Omega})$ and $u^\star-\varphi$ has a local maximum at x_0 we have that

$$F_\star(x_0,\varphi(x_0),D^2\varphi(x_0))\ge 0.$$

2. A *viscosity supersolution* of (7) if whenever $x_0\in\bar{\Omega}$, $\varphi\in C^2(\bar{\Omega})$ and $u_\star-\varphi$ has a local minimum at x_0 we have that

$$F^\star(x_0,\varphi(x_0),D^2\varphi(x_0))\le 0.$$

3. A *viscosity solution* if it is a sub- and supersolution.

The condition "$u^\star-\varphi$ has a local maximum at x_0" is usually phrased as "*φ touches the graph of u from above at x_0*". The reader is encouraged to draw a picture to see why these two have the same meaning. Similarly, "$u_\star-\varphi$ has a local minimum at x_0" is: "*φ touches the graph of u from below at x_0*".

One of the main technical tools in asserting existence and uniqueness of viscosity solutions is a comparison principle.

Definition 5 (comparison principle).

We say that problem (7) satisfies a comparison principle if whenever $\overline{w}\in USC(\bar{\Omega})$ and $\underline{w}\in LSC(\bar{\Omega})$ are sub- and supersolutions, respectively, then we must have

$$\overline{w}\le\underline{w}.$$

Notice now that if we define

$$F_{MA}(x,r,M)=\begin{cases}\det M-f(x), & x\in\Omega,\\ g(x)-r, & x\in\partial\Omega,\end{cases} \tag{8}$$

this operator satisfies the monotonicity conditions given in Definition 2 *only* if we restrict the third argument to the set of positive semidefinite matrices which we denote by $\mathbb{S}^d_+$. Consequently, we need to restrict the class of admissible functions, that define a viscosity solution to (1) to the set of convex functions.

Definition 6 (viscosity solution).
Let $u\in C(\bar{\Omega})$ be a convex function. We say that u is:

1. A *viscosity subsolution* of (1) on the set of convex functions if $u\leq g$ on $\partial\Omega$ and, whenever $x_0\in\Omega$, $\varphi\in C^2(\Omega)$, and $u-\varphi$ has a local maximum at x_0 we have that
$$\det D^2\varphi(x_0)\geq f(x_0).$$
2. A *viscosity supersolution* of (1) on the set of convex functions if $u\geq g$ on $\partial\Omega$ and, whenever $x_0\in\Omega$, $\varphi\in C^2(\Omega)$ is convex, and $u-\varphi$ has a local minimum at x_0 we have that
$$\det D^2\varphi(x_0)\leq f(x_0).$$
3. A *viscosity solution* if it is a sub- and supersolution on the set of convex functions.

The reader may wonder why these definitions are asymmetric. The concept of supersolution requires convexity of the test functions, whereas subsolutions do not. This is due to the fact that, as noted in Gutiérrez (2001, Remark 1.3.2), if u is convex and $u-\varphi$ has a local maximum at x_0, then φ is (locally) convex.

The existence and uniqueness of viscosity solutions will be a consequence of Theorems 2 and 3. Here we mention a remarkable property of viscosity solutions, namely their stability. The following result can be found, for instance, in Nochetto et al. (2019a, Lemma 5.3).

Proposition 1 (continuous dependence).
Let $f_1,f_2\in C(\bar{\Omega})$ with f_1, $f_2\geq 0$ and g_1, $g_2\in C(\partial\Omega)$ and denote by $u_1,u_2\in C(\bar{\Omega})$ the corresponding convex viscosity solutions to (1). Then we have

$$\|u_1-u_2\|_{L^\infty(\Omega)}\leq C\,\|f_1-f_2\|^{1/d}_{L^\infty(\Omega)}+\|g_1-g_2\|_{L^\infty(\partial\Omega)}.$$

In addition, if $f_1\geq f_2\geq 0$ and $g_1\leq g_2$ we have that $u_1\leq u_2$.

Finally we comment that viscosity solutions can be approximated by classical ones over larger, but smooth, domains; see Nochetto et al. (2019a, Lemma 5.4).

Proposition 2 (smooth approximation).
Let Ω be uniformly convex, $f, g \in C(\bar{\Omega})$ with $f \geq 0$, and u the convex viscosity solution to (1). *There exists:*

1. *A decreasing (in the sense of inclusion) sequence of uniformly convex smooth domains Ω_n such that*

$$\mathrm{dist}_H(\Omega_n, \Omega) \to 0, \quad n \to \infty,$$

   where by $\mathrm{dist}_H(A, B)$ we mean the d-dimensional Hausdorff distance between the sets A and B.

2. *A decreasing sequence of smooth functions $f_n : \bar{\Omega}_n \to \mathbb{R}$ with $f_n > 0$ such that*

$$\|f_n - f\|_{L^\infty(\Omega)} \to 0, \quad n \to \infty.$$

3. *A sequence of smooth functions $g_n : \bar{\Omega}_n \to \mathbb{R}$ such that*

$$\|g_n - g\|_{L^\infty(\Omega)} \to 0, \quad n \to \infty.$$

Moreover, if $u_n \in C(\bar{\Omega}_n)$ denotes the convex viscosity solution to (1) *over the domain Ω_n and with data f_n and g_n, then*

$$\|u_n - u\|_{L^\infty(\Omega)} \to 0, \quad n \to \infty.$$

1.2.3 Alexandrov solutions

Besides the concept of solution in the viscosity sense, another type of weak solution to the Monge–Ampère equation is the Alexandrov solution, which is based on a geometric interpretation. To motivate it, let $w \in C^2(\Omega)$ be convex so that the gradient map $\nabla w : \Omega \to \mathbb{R}^d$ is well defined and monotone. In this case, an interesting observation is that $\det D^2 w$ is actually the determinant of the Jacobian of the gradient map. Therefore, for any open (or Borel) subset $E \subset \Omega$, we have

$$\int_E \det D^2 w(x) \mathrm{d}x = \int_{\nabla w(E)} \mathrm{d}y = |\nabla w(E)|,$$

where $|\cdot|$ denotes the d-dimensional Lebesgue measure.

What is remarkable about this simple observation is that to make sense of $\det D^2 u$, we only require $\nabla w(E)$ to be well defined for any Borel set E. This enables us to make sense of the previous identity even if $w \notin C^2(\Omega)$. To define the weak (Alexandrov) solution, we first introduce the subdifferential of a convex function.

Definition 7 (subdifferential).
Let Ω be convex and $w : \Omega \to \mathbb{R}$ be a convex function. The subdifferential of w at point $x \in \Omega$ is the set

$$\partial w(x) := \{\boldsymbol{p} \in \mathbb{R}^d, w(x) + \boldsymbol{p} \cdot (y - x) \leq w(y) \quad \forall y \in \Omega\}.$$

For any Borel set $E \subset \Omega$, we define

$$\partial w(E) = \cup_{x \in E} \partial w(x).$$

In other words, the subdifferential is the collection of slopes of all affine functions that touch the graph of w at $(x, w(x))$ and bound the graph from below. From this observation, it is easy to see that if w is strictly convex and smooth, then $\partial w(x) = \{\nabla w(x)\}$. Here we give an example of subdifferential of a convex (but not strictly convex) function.

Example 1 (subdifferential).

Let $\Omega = B_1(0) \subset \mathbb{R}^2$ and

$$w(x) = |x|.$$

Then at the origin $x = 0$, we note that

$$w(0) + \boldsymbol{p} \cdot y \leq w(y) \quad \forall y \in \Omega$$

provided that the norm of the vector $|\boldsymbol{p}| \leq 1$. Hence, by definition, the subdifferential of w at $x = 0$ is the closed unit ball centred at 0, i.e.

$$\partial w(0) = \overline{B_1(0)}.$$

At any other point $x \in \Omega$, since the function w is differentiable, we note that the inequality

$$w(x) + \boldsymbol{p} \cdot (y - x) \leq w(y) \quad \forall y \in \Omega$$

holds if and only if $\boldsymbol{p} = \nabla w(x)$. Hence, for all $x \in \Omega \backslash \{0\}$,

$$\partial w(x) = \{\nabla w(x)\}.$$

With this motivation at hand we can introduce the so-called Monge–Ampère measure, which will be essential in defining Alexandrov solutions.

Definition 8 (Monge–Ampère measure).

Let $\Omega \subset \mathbb{R}^d$ be convex and $w: \Omega \to \mathbb{R}$ be a convex function. The *Monge–Ampère measure* associated to w is

$$\mu_w(E) = |\partial w(E)|.$$

It can be shown, see Figalli (2017, Theorem 2.3) that this is indeed a locally finite Borel measure on Ω. With this, we are ready to define Alexandrov solutions.

Definition 9 (Alexandrov solution).

Let f be a Borel measure defined in Ω. A convex function $u \in C(\bar{\Omega})$ is an *Alexandrov solution* to the Monge–Ampère equation (1) if $u = g$ on $\partial\Omega$ and $\mu_u = f$, that is,

$$|\partial u(E)| = f(E). \tag{9}$$

for all Borel sets $E \subset \Omega$.

To illustrate the definition of the Alexandrov solution, we consider Example 1. Let $E \subset \Omega$ be Borel, if the set contains the origin, we have the subdifferential

$$\partial u(E) = \cup_{x \in E} \partial u(x) = \overline{B_1(0)},$$

which yields

$$|\partial u(E)| = |B_1(0)| = \pi \quad \text{if } x \in E.$$

On the other hand, if the set does not contain the origin, then the subdifferential

$$\partial u(E) = \cup_{x \in E} \{\nabla u(x)\} \subset \partial B_1(0)$$

Hence, we get $|\partial u(E)| = 0$ if $0 \notin E$. Finally, we conclude that u is an Alexandrov solution of Monge–Ampère equation

$$\det D^2 u(x) = \pi \delta_{\{x=0\}},$$

where $\delta_{\{x=0\}}$ is the Dirac measure at the origin. It is worth mentioning that u is not a viscosity solution because the right-hand side is not a (continuous) function. Also note that the continuity of the source term f is no longer required for (9) to be well defined.

The existence and uniqueness of Alexandrov solutions is summarized in the next theorem, see Gutiérrez (2001, Theorem 1.6.2) and Figalli (2017, Theorem 2.14).

Theorem 2 (existence and uniqueness).
Let $\Omega \subset \mathbb{R}^d$ be a strictly convex domain, let $g \in C(\partial\Omega)$ and f be a nonnegative Borel measure on Ω with $f(\Omega) < \infty$. Then there exists a unique convex function $u \in C(\bar{\Omega})$ that is a solution of (1) *in the sense of Definition* 9.

An important property of Alexandrov solutions is their stability with respect to weak convergence. We refer the reader to Gutiérrez (2001, Lemma 1.2.3) for a proof of the following result.

Lemma 1 (weak convergence).
Let $\{w_k\}_{k=1}^{\infty}, w$ be convex functions on Ω and assume that, as $k \to \infty$, we have $w_k \to w$ uniformly over compact subsets of Ω. Then, the associated Monge–Ampère measures μ_{w_k} tend to μ_w weakly, that is,

$$\int_\Omega \phi(x) \mathrm{d}\mu_{w_k}(x) \to \int_\Omega \phi(x) \mathrm{d}\mu_w(x),$$

for every ϕ continuous with compact support in Ω.

The relation between viscosity and Alexandrov solutions is given in the following result (Gutiérrez, 2001, Propositions 1.3.4 and 1.7.1). Notice that this result not only shows, as we have already pointed out, that the notion of Alexandrov solution is strictly weaker than that of viscosity solutions but, on the basis of Theorem 2, shows existence and uniqueness of viscosity solutions.

Theorem 3 (equivalence).
Let $u \in C(\bar{\Omega})$ be an Alexandrov solution of (1). *If $f \in C(\Omega)$, then u is also a viscosity solution in the sense of Definition* 6. *Conversely, if u is a viscosity solution of* (1) *and $f \in C(\bar{\Omega})$ with $f > 0$, then u is an Alexandrov solution.*

Since it will be useful in the sequel, we introduce here the convex envelope of a function, which is the largest convex function that is bounded above by the given one.

Definition 10 (convex envelope).
Let $\Omega \subset \mathbb{R}^d$ be convex and $w : \bar{\Omega} \to \mathbb{R}$. The *convex envelope* of w, denoted by Γw, is the largest convex function whose graph lies below the graph of w. It can be computed by

$$\Gamma w(x) = \sup\{L(x) : L \text{ affine function and } L(y) \le w(y) \quad \forall y \in \bar{\Omega}\}.$$

To conclude our preliminary discussion we recall the Brunn–Minkowski inequality, a celebrated result in convex geometry. Given two compact sets A, B of $\mathbb{R}^d$, we define their *Minkowski sum*

$$A + B := \{v + w \in \mathbb{R}^d : v \in A \text{ and } w \in B\}. \tag{10}$$

The Brunn–Minkowski inequality relates the Lebesgue measures of compact subsets A, B of Euclidean space $\mathbb{R}^d$ with that of their Minkowski sum $A + B$.

Lemma 2 (Brunn–Minkowski inequality).
Let A and B be two nonempty compact subsets of $\mathbb{R}^d$ for $d \ge 1$. Then the following inequality holds:

$$|A + B|^{1/d} \ge |A|^{1/d} + |B|^{1/d}.$$

2 Wide stencil finite differences

Problems involving the classical linear partial differential equations of mathematical physics can be reduced to algebraic ones of a very much simpler structure by replacing the differentials by difference quotients on some (say rectilinear) mesh.

Courant et al. (1967)

In this section we will study finite difference schemes that aim to approximate the viscosity solution, in the sense of Definition 6, of (1).

2.1 A general framework for approximation schemes

Let us describe a general framework under which convergence of approximation schemes can be shown. Let $F : \bar{\Omega} \times \mathbb{R} \times \mathbb{S}^d \to \mathbb{R}$ be elliptic in the sense of Definition 2 and assume we wish to approximate the viscosity solution to (7). To do so, we introduce a family of approximation schemes, which are described by the collection of maps $\{F_\varepsilon\}_{\varepsilon>0}$, where $F_\varepsilon : \bar{\Omega} \times \mathbb{R} \times B(\bar{\Omega}) \to \mathbb{R}$, and $B(\bar{\Omega})$ denotes the space of bounded functions on $\bar{\Omega}$. The parameter ε can be understood as a discretization parameter. With this family at hand, we seek for $u_\varepsilon \in B(\bar{\Omega})$ such that

$$F_\varepsilon(x, u_\varepsilon(x), u_\varepsilon) = 0, \quad \text{in } \bar{\Omega}. \tag{11}$$

We assume that the approximation schemes satisfy the following assumptions:

1. *Monotonicity*: For all $\varepsilon > 0$, $x \in \bar{\Omega}$, $t \in \mathbb{R}$, and $u, v \in B(\bar{\Omega})$ such that $u \geq v$ we have that

$$F_\varepsilon(x, t, u) \geq F_\varepsilon(x, t, v). \tag{12}$$

2. *Stability*: There is $\varepsilon_0 > 0$ such that if $\varepsilon < \varepsilon_0$, the scheme (11) has a unique solution and there is a constant, independent of ε, such that

$$\| u_\varepsilon \|_{L^\infty(\Omega)} \leq C. \tag{13}$$

3. *Consistency*: For all $x_0 \in \bar{\Omega}$ and $\varphi \in C^2(\bar{\Omega})$ we have

$$\limsup_{\varepsilon \downarrow 0, y \to x_0, \xi \to 0} F_\varepsilon(y, \varphi(y) + \xi, \varphi + \xi) \leq F_\star(x_0, \varphi(x_0), D^2\varphi(x_0)) \tag{14a}$$

$$\liminf_{\varepsilon \downarrow 0, y \to x_0, \xi \to 0} F_\varepsilon(y, \varphi(y) + \xi, \varphi + \xi) \geq F^\star(x_0, \varphi(x_0), D^2\varphi(x_0)). \tag{14b}$$

The main convergence result in this framework is the following; see Barles and Souganidis (1991, Theorem 2.1).

Theorem 4 (Barles–Souganidis).
Assume that the family of approximation schemes (11) *is monotone, stable and consistent, in the sense of* (12), (13), *and* (14), *respectively. Assume, in addition, that problem* (7) *has a comparison principle in the sense of Definition* 5. *Then, as* $\varepsilon \downarrow 0$, *the functions* u_ε, *solution of* (11) *converge locally uniformly to* u, *solution of* (7).

Proof. Define $\overline{u}, \underline{u} \in B(\bar{\Omega})$ by

$$\overline{u}(x) = \limsup_{y \to x, \varepsilon \downarrow 0} u_\varepsilon(y), \qquad \underline{u}(x) = \liminf_{y \to x, \varepsilon \downarrow 0} u_\varepsilon(y).$$

Notice that, by stability, we obtain that these functions are well defined and bounded. In addition, we have that $\overline{u}, \underline{u}$ are upper and lower semicontinuous, respectively.

The idea now is to show that $\overline{u}$ is a subsolution and $\underline{u}$ is a supersolution of (7), for if that is the case we can invoke the comparison principle to see that $\overline{u} \leq \underline{u}$, and so that these must coincide with the viscosity solution of (7). This, in turn, implies the local uniform convergence of u_ε to u.

Let us then show that $\overline{u}$ is a subsolution. Let $\varphi \in C^2(\bar{\Omega})$ and assume that $\overline{u} - \varphi$ has a local maximum at $x_0 \in \bar{\Omega}$ with $\overline{u}(x_0) = \varphi(x_0)$. It can be shown then that there are sequences $\{\varepsilon_n\}_{n=1}^{\infty} \subset \mathbb{R}^+$ and $\{y_n\}_{n=1}^{\infty} \subset \bar{\Omega}$ such that $\varepsilon_n \downarrow 0$, $y_n \to x_0$, $u_{\varepsilon_n}(y_n) \to \overline{u}(x_0)$ and the sequence of functions $u_{\varepsilon_n} - \varphi$ attains its maximum at y_n.

Notice now that, upon denoting $\xi_n = u_{\varepsilon_n}(y_n) - \varphi(y_n)$, we get that $\xi_n \to 0$ and $u_{\varepsilon_n}(x) - \varphi(x) \leq \xi_n$ locally. Monotonicity then implies that

$$
\begin{aligned}
0 = F_{\varepsilon_n}(y_n, u_{\varepsilon_n}(y_n), u_{\varepsilon_n}) &= F_{\varepsilon_n}(y_n, \varphi(y_n) + \xi_n, \varphi + (u_{\varepsilon_n} - \varphi)) \\
&\leq F_{\varepsilon_n}(y_n, \varphi(y_n) + \xi_n, \varphi + \xi_n),
\end{aligned}
$$

which by the consistency condition (14a) yields

$$
F_\star(x_0, \varphi(x_0), D^2\varphi(x_0)) \geq 0,
$$

so that $\overline{u}$ is a subsolution. □

Remark 1 (limitations).
We must remark that, although Theorem 4 seems sufficiently general:

1. It only provides sufficient conditions for convergence. There is no guideline towards the construction of monotone, consistent and stable finite difference schemes.
2. This result, as is, *cannot* be applied to approximate viscosity solutions of the Monge–Ampère equation (1) directly. This is because, as pointed out in Section 1.2.2, the Monge–Ampère operator is only elliptic over $\bar{\Omega} \times \mathbb{R} \times \mathbb{S}_+^d$.
3. The existence of a comparison principle in the sense of Definition 5 is assumed. Notice that, in Jensen and Smears (2018, Proposition 2.1) it is shown that, for a reformulation of the Monge–Ampère problem as a Hamilton Jacobi Bellman equation (which will be discussed in Section 2.8.1), if $f \equiv 0$, there cannot be a comparison principle for this problem. In other words, this is a highly nontrivial assumption.

■

Although not applicable to the Monge–Ampère equation (1), one of the messages of Theorem 4 is that monotonicity of a numerical scheme is a highly desirable property. Thus, it is necessary to explore how to construct monotone approximation schemes. In the context of finite difference schemes it was realized as early as in Motzkin and Wasow (1953) that, even for linear problems, monotonicity of a numerical scheme requires the use of wide stencils, which is rather problematic at points near the boundary. We refer the reader to Neilan et al. (2017, Section 3.2) for more details, and to Mirebeau (2016) for the construction of minimal stencils in two dimensions. For this reason, in the remaining of this section, we will consider wide stencil finite difference schemes to approximate the viscosity solution of (1).

2.2 A variational characterization of the determinant

Let us provide a variational characterization of the determinant that will motivate most of the constructions which will come below. This was originally shown in Froese and Oberman (2011a, Lemma 2).

Lemma 3 (characterization of the determinant).

Let A be a symmetric positive definite $d \times d$ matrix and let

$$\mathcal{V} = \left\{ \{\boldsymbol{w}_i\}_{i=1}^d \subset \mathbb{R}^d : \boldsymbol{w}_i \cdot \boldsymbol{w}_j = \delta_{i,j} \right\},$$

be the set of all orthonormal bases of $\mathbb{R}^d$. Then we have that

$$\det A = \min_{\{\boldsymbol{w}_i\}_{i=1}^d \in \mathcal{V}} \prod_{i=1}^d \boldsymbol{w}_i \cdot A\boldsymbol{w}_i.$$

Proof. To shorten notation, let $M = \min_{\{\boldsymbol{w}_i\}_{i=1}^d \in \mathcal{V}} \prod_{i=1}^d \boldsymbol{w}_i \cdot A\boldsymbol{w}_i$. Then let $\{\boldsymbol{v}_j\}_{j=1}^d$ be an orthonormal set of eigenvectors of A so that

$$\det A = \prod_{i=1}^d \boldsymbol{v}_i \cdot A\boldsymbol{v}_i \geq M.$$

On the other hand, for $\{\boldsymbol{w}_i\}_{i=1}^d \in \mathcal{V}$, we can represent them in the basis of eigenvectors $\boldsymbol{w}_i = \sum_{k=1}^d (\boldsymbol{w}_i \cdot \boldsymbol{v}_k)\boldsymbol{v}_k$. We have

$$\begin{aligned}
-\log \prod_{i=1}^d \boldsymbol{w}_i \cdot A\boldsymbol{w}_i &= -\sum_{i=1}^d \log (\boldsymbol{w}_i \cdot A\boldsymbol{w}_i) \\
&= -\sum_{i=1}^d \log \left(\sum_{m=1}^d (\boldsymbol{w}_i \cdot \boldsymbol{v}_m)\boldsymbol{v}_m \cdot \sum_{k=1}^d (\boldsymbol{w}_i \cdot \boldsymbol{v}_k) A\boldsymbol{v}_k \right) \\
&= -\sum_{i=1}^d \log \left(\sum_{k=1}^d \lambda_k (\boldsymbol{w}_i \cdot \boldsymbol{v}_k)^2 \right),
\end{aligned}$$

where $\sigma(A) = \{\lambda_k\}_{k=1}^d$ is the spectrum of A. Since $|\boldsymbol{w}_i| = 1$ the term $\sum_{k=1}^d \lambda_k (\boldsymbol{w}_i \cdot \boldsymbol{v}_k)^2$ is a convex combination of the elements of $\sigma(A)$. Owing to the convexity of $t \mapsto -\log t$ we can apply Jensen's inequality to obtain that

$$-\log \prod_{i=1}^d \boldsymbol{w}_i \cdot A\boldsymbol{w}_i \leq -\sum_{k=1}^d \log \lambda_k \sum_{i=1}^d (\boldsymbol{w}_i \cdot \boldsymbol{v}_k)^2 = -\sum_{k=1}^d \log \lambda_k = -\log \prod_{i=1}^d \lambda_i.$$

As the function $t \mapsto -\log t$ is decreasing, we conclude that

$$\det A \leq \prod_{i=1}^d \boldsymbol{w}_i \cdot A\boldsymbol{w}_i,$$

which since $\{\boldsymbol{w}_i\}_{i=1}^d \in \mathcal{V}$ was arbitrary implies $\det A \leq M$ and this concludes the proof. □

The previous result allows us to conclude that, if $\varphi \in C^2(\Omega)$ is convex, we can express the determinant of its Hessian at a point in terms of second directional derivatives, that is, if $x_0 \in \Omega$ we have

$$\det D^2\varphi(x_0) = \min_{\{w_i\}_{i=1}^d \in \mathcal{V}} \prod_{i=1}^d w_i \cdot D^2\varphi(x_0) w_i = \min_{\{w_i\}_{i=1}^d \in \mathcal{V}} \prod_{i=1}^d \frac{\partial^2 \varphi}{\partial w_i^2}(x_0).$$

Recall, in addition, that a solution to (1) must be convex. To enforce convexity we then introduce the following operator

$$\mathrm{MA}[\varphi](x_0) = \min_{\{w_i\}_{i=1}^d \in \mathcal{V}} \left[\prod_{i=1}^d \left(\frac{\partial^2 \varphi}{\partial w_i^2}(x_0) \right)^+ - \sum_{i=1}^d \left(\frac{\partial^2 \varphi}{\partial w_i^2}(x_0) \right)^- \right], \tag{15}$$

where $x^+ = \max\{x, 0\}$ and $x^- = (-x)^+$ denote the positive and negative parts of x, respectively. Notice that, if $\varphi \in C^2(\bar{\Omega})$ is convex, $\mathrm{MA}[\varphi] = \det D^2\varphi$. The idea behind (15) is that, if $D^2\varphi(x_0)$ has a negative eigenvalue, then there is $V \in \mathcal{V}$ and $w \in V$ for which $w \cdot D^2\varphi(x_0) w < 0$. Thus,

$$\mathrm{MA}[\varphi](x_0) \leq 0 - (w \cdot D^2\varphi(x_0) w)^- < 0.$$

Consequently, φ cannot be a solution to (1) since, at x_0 we have

$$\det D^2\varphi(x_0) = f(x_0) \geq 0.$$

These ideas are made rigorous in Nochetto et al. (2019a, Lemma 5.6).

Proposition 3 (equivalence of operators).
Let $f \in C(\Omega)$ with $f \geq 0$. The function $u \in C(\bar{\Omega})$ is a convex viscosity solution of (1) *in the sense of Definition* 6 *if and only if it is a viscosity solution, in the sense of Definition* 4, *of the following problem*

$$F_{vMA}(x, u(x), D^2u(x)) = 0 \tag{16}$$

with

$$F_{vMA}(x, u(x), D^2u(x)) = \begin{cases} \mathrm{MA}[u](x) - f(x), & x \in \Omega, \\ g(x) - u(x), & x \in \partial\Omega. \end{cases}$$

One of the advantages of formulation (16) is that it has a comparison principle.

Proposition 4 (comparison principle for the F_{vMA} operator).
The operator F_{vMA}, defined in (16) *has a comparison principle in the sense of Definition* 5.

Proof. It follows from the fact that the operator F_{vMA} satisfies the structural assumptions given, for instance, in Crandall et al. (1992, Theorem 3.3). □

The characterization of the determinant given in Lemma 3 will be the basis of many of the wide stencil schemes we will describe below.

2.3 Wide stencil finite difference schemes

Let us describe the first class of methods that exploit the characterization described in Lemma 3 via the operator introduced in (15) as originally proposed in Froese and Oberman (2011a). Let $h > 0$ be a (spatial) discretization parameter and assume that, up to a linear change of variables, our domain Ω is discretized on a Cartesian grid. In other words, we let

$$\bar{\Omega}_h = \bar{\Omega} \cap \mathbb{Z}_h^d, \quad \mathbb{Z}_h^d = \{h\boldsymbol{e} : \boldsymbol{e} \in \mathbb{Z}^d\}, \quad \partial\Omega_h = \partial\Omega \cap \mathbb{Z}_h^d, \quad \Omega_h = \bar{\Omega}_h \backslash \partial\Omega_h.$$

We set X_h as the space of *grid functions*, that is the collection of functions $w_h : \bar{\Omega}_h \to \mathbb{R}$.

Given $\boldsymbol{e} \in \mathbb{Z}^d$ we call the point $x_h \in \Omega_h$ *interior* with respect to $\boldsymbol{e}$ if $x_h \pm h\boldsymbol{e} \in \bar{\Omega}_h$. We will also say that a point is interior with respect to a subset of $S \subset \mathbb{Z}^d$ if it is interior with respect to all elements of S.

Given $\boldsymbol{e} \in \mathbb{Z}^d$ and an interior point x_h, we define the *second difference* in the direction $\boldsymbol{e}$ to be the operator

$$\Delta_{\boldsymbol{e}} w_h(x_h) = \frac{1}{|e|^2 h^2}(w_h(x_h + h\boldsymbol{e}) - 2w_h(x_h) + w_h(x_h - h\boldsymbol{e})). \tag{17}$$

When x_h is not interior with respect to $\boldsymbol{e}$, it essentially means that x_h is close to $\partial\Omega$. Owing to the convexity of Ω, there are unique $\rho_\pm \in (0, 1]$ such that $x_h \pm \rho_\pm h\boldsymbol{e} \in \partial\Omega$. Thus, we can use the boundary condition (1b) to extend this definition as

$$\Delta_{\boldsymbol{e}} w_h(x_h) = \frac{2}{(\rho_+ + \rho_-)|\boldsymbol{e}|^2 h^2}\left(\frac{\tilde{g}(x_h + \rho_+ h\boldsymbol{e}) - w_h(x_h)}{\rho_+} - \frac{w_h(x_h) - \tilde{g}(x_h + \rho_- h\boldsymbol{e})}{\rho_-}\right), \tag{18}$$

where $\tilde{g}$ is either the boundary condition, or an interpolant of w_h based on neighbouring nodes. With these notions at hand, we would like to define the discretization of the operator MA $[\cdot]$, introduced in (15), as

$$\mathrm{MA}_h^{\mathrm{WS}}[w_h](x_h) = \min_{\{\boldsymbol{w}_i\}_{i=1}^d \in \mathcal{V}} \prod_{i=1}^d (\Delta_{\boldsymbol{w}_i} w_h(x_h))^+.$$

Notice, however, that the given expressions may not be defined for all $\mathcal{V}$, as the points $x_h \pm h\boldsymbol{w}_i$ may not belong to $\bar{\Omega}_h$. Even if they did, it may be very computationally expensive to compute these directional differences at all the nodes. For these reasons, we also need to introduce a discretization of $\mathcal{V}$. To this end we introduce a finite subset $\mathcal{G}_\theta \subset (\mathbb{Z}^d)^d$ such that, if $\{\boldsymbol{\nu}_i\}_{i=1}^d \in \mathcal{G}_\theta$ then the vectors $\boldsymbol{\nu}_i$ are pairwise orthogonal. We call this the *directional* discretization of the Monge–Ampère operator and parametrize it by $\theta > 0$. Thus we define the operator

$$\mathrm{MA}_{h,\theta}^{\mathrm{WS}}[w_h](x_h) = \min_{\{\nu_i\}_{i=1}^d \in \mathcal{G}_\theta} \prod_{i=1}^{d} (\Delta_{\nu_i} w_h(x_h))^+. \tag{19}$$

With this notation at hand, we define the *wide stencil* finite difference approximation scheme of (1) as: Find $u_h \in X_h$ such that

$$\mathrm{MA}_{h,\theta}^{\mathrm{WS}}[u_h](x_h) = f(x_h), \quad \forall x_h \in \Omega_h, \tag{20a}$$

$$u_h(x_h) = g(x_h), \quad \forall x_h \in \partial\Omega_h. \tag{20b}$$

Remark 2 (variant).
We could have also introduced another wide stencil operator via

$$\mathrm{MA}_{h,\theta}^{\mathrm{WS}}[w_h](x_h) = \min_{\{\nu_i\}_{i=1}^d \in \mathcal{G}_\theta} \left[\prod_{i=1}^{d} (\Delta_{\nu_i} w_h(x_h))^+ - \sum_{i=1}^{d} (\Delta_{\nu_i} w_h(x_h))^- \right],$$

see (15). ■

Remark 3 (a regularized version).
Notice that, owing to the presence of the min and max operator in the definition of (19), this operator is not differentiable. This may make it difficult to efficiently solve the ensuing nonlinear systems, since Newton methods are not directly applicable. One could, instead, use semismooth Newton methods (Hintermüller et al., 2002) since these operators are slant differentiable; see Hintermüller et al. (2002, Lemma 3.1). However, if we insist in dealing with smooth operators, Froese and Oberman (2011a, Section 3.5) introduces a regularized version of $\mathrm{MA}_{h,\theta}^{\mathrm{WS}}[\cdot]$ given by

$$\mathrm{MA}_{h,\theta,\delta}^{\mathrm{WS}}[w_h](x_h) = \min_{\{\nu_i\}_{i=1}^d \in \mathcal{G}_\theta}{}^{\delta} \prod_{i=1}^{d} (\Delta_{\nu_i} w_h(x_h))^{+,\delta},$$

where

$$\max{}^\delta\{x,y\} = \frac{1}{2}\left(x + y + \sqrt{(x-y)^2 + \delta^2}\right),$$
$$\min{}^\delta\{x,y\} = \frac{1}{2}\left(x + y - \sqrt{(x-y)^2 + \delta^2}\right),$$
$$\min{}^\delta\{x_1,\ldots,x_n\} = \min{}^\delta\{\min{}^\delta\{x_1,\ldots,x_{n-1}\}, x_n\},$$

and $x^{+,\delta} = \max^\delta\{x,0\}$. The properties of this operator are similar to those of $\mathrm{MA}_{h,\theta}^{\mathrm{WS}}[\cdot]$. ■

Remark 4 (two dimensions).
Given $A \in \mathbb{S}^d$ we have the classical Rayleigh–Ritz relations

$$\lambda_m(A) = \min_{w\in\mathbb{R}^d} \frac{w\cdot Aw}{|w|^2} = \min\sigma(A), \quad \lambda_M(A) = \max_{w\in\mathbb{R}^d} \frac{w\cdot Aw}{|w|^2} = \max\sigma(A),$$

so that, if $d = 2$, we have that

$$\det A = \min_{w\in\mathbb{R}^2} \frac{w\cdot Aw}{|w|^2} \max_{w\in\mathbb{R}^2} \frac{w\cdot Aw}{|w|^2}.$$

This relation was used in Oberman (2008b) to introduce a two-dimensional scheme via

$$\mathrm{MA}_{h,\theta}^{\mathrm{WS,2d}}[w_h](x_h) = \min_{\nu_i\in\{\nu_j\}_{j=1}^d\in\mathcal{G}_\theta} (\Delta_{\nu_i} w_h(x_h))^+ \max_{\nu_i\in\{\nu_j\}_{j=1}^d\in\mathcal{G}_\theta} (\Delta_{\nu_i} w_h(x_h))^+.$$

Note that, although similar to (20), these operators are different. This was illustrated in Froese and Oberman (2011a, Section 3.4) with the following example: Let

$$w(x_1,x_2) = x_1^2 + x_2^2 + x_1^2 x_2^2,$$

which is convex in a neighbourhood of the origin, and

$$\mathcal{G}_\theta = \left\{\left\{\begin{pmatrix}1\\0\end{pmatrix},\begin{pmatrix}0\\1\end{pmatrix}\right\},\left\{\begin{pmatrix}1\\1\end{pmatrix},\begin{pmatrix}-1\\1\end{pmatrix}\right\}\right\}.$$

Computing each of the operators over these directions yields

$$\mathrm{MA}_{h,\theta}^{\mathrm{WS,2d}}[w](0,0) = 4 + 2h^2, \quad \mathrm{MA}_{h,\theta}^{\mathrm{WS}}[w](0,0) = 4.$$

Notice however, that since both operators are consistent with order $\mathcal{O}(h^2)$ we have that, for a convex function v,

$$\left|\mathrm{MA}_{h,\theta}^{\mathrm{WS,2d}}[v](x_h) - \mathrm{MA}_{h,\theta}^{\mathrm{WS}}[v](x_h)\right| = \mathcal{O}(h^2), \quad \forall x_h.$$

■

The analysis of method (20) will be a particular case of the methods and analyses presented in Section 2.7. We just comment that, even for smooth solutions, wide stencils are required in this scheme to assert consistency. Let us illustrate this in a simple case where there is no boundary conditions and in two dimensions ($d = 2$). In other words, given $x_0 \in \Omega$ we assume that it is an interior point for any $e \in \mathbb{Z}^2$. Let now $\varphi(x) = \frac{1}{2}x\cdot Mx$ be a convex quadratic, so that

$$\Delta_e \varphi(x_0) = \frac{1}{|e|^2} e\cdot Me,$$

and therefore

$$\mathrm{MA}_{h,\theta}^{\mathrm{WS}}[\varphi](x_0) = \min_{\{\nu_1,\nu_2\}\in\mathcal{G}_\theta} \frac{1}{|\nu_1|^2|\nu_2|^2}(\nu_1\cdot M\nu_1)(\nu_1\cdot M\nu_2),$$

independently of x_0 and the mesh size. At this point we need to recall that there is $\{w_1, w_2\} \in \mathcal{V}$, namely the normalized eigenvectors of M, for which

$$\det D^2\varphi = \det M = (w_1 \cdot Mw_1)(w_2 \cdot Mw_2).$$

Notice finally, that once w_1 is determined, $w_2 = w_1^\perp$ is obtained by a rotation. In conclusion, to assert consistency, given a $w \in \mathbb{R}^2$ in the unit sphere, for every $\delta > 0$ we must be able to find $e \subset \mathbb{Z}^2$ such that

$$\left| w - \frac{1}{|e|} e \right| < \delta. \tag{21}$$

Indeed, if we denote by e_1 the vector that satisfies this property with respect to w_1, then $e_2 = e_1^\perp$ does so for w_2. Let now $\nu_i = \frac{1}{|e_i|} e_i$ for $i = 1, 2$. Then we have that

$$|\det M - (\nu_1 \cdot M\nu_1)(\nu_2 \cdot M\nu_2)| \leq C(\Lambda)\delta,$$

where $C(\Lambda)$ is a constant that depends polynomially on Λ, the maximal eigenvalue of M.

Notice that, since $e_i \in \mathbb{Z}^2$, then $\nu_i \in \mathbb{Q}^2$, so finding points that satisfy (21) is the problem of *rational approximation in the sphere*. While how to actually find such points is beyond our discussion here, what we are interested in is the size of $|e|$, which would serve as an estimate of the stencil size that guarantees convergence. The following result is a specialization of Schmutz (2008, Lemma 2.1) to the two-dimensional case; we refer the reader to this reference a proof, its generalization to $d > 2$, and to the case of rational approximation orthogonal matrices which is of interest when finding elements of $\mathcal{G}_\theta$.

Proposition 5 (rational approximation).
Let $w \in \mathbb{R}^2$ be such that $|w| = 1$. Then, for every $\delta > 0$, there exists $\nu \in \mathbb{Q}^2$ such that $|\nu| = 1$ and

$$|w - \nu| < \delta.$$

Moreover, if $\nu = (p_1/q_1, p_2/q_2)^\mathsf{T}$ with $p_1, p_2 \in \mathbb{Z}$ and $q_1, q_2 \in \mathbb{N}$ then we have that

$$0 < q_i \leq \frac{64}{\delta^2}.$$

Now, for a given $w \in \mathbb{R}^2$, let ν be as in Proposition 5. This means that $e = \text{hcf}(q_1, q_2)\nu \in \mathbb{Z}^2$ is the smallest vector parallel to ν that satisfies (21) (here, $\text{hcf}(q_1, q_2)$ denotes the highest common factor of q_1 and q_2). Consequently, we have that, generically

$$|e| \leq C \ \mathrm{hcf}(q_1, q_2) \leq C \max\{q_1, q_2\} \leq \frac{C}{\delta^2}.$$

In conclusion, the size of the stencil must grow unboundedly if we restrict ourselves to Cartesian meshes.

2.4 Filtered schemes

The estimates on the stencil size of the previous section are rather pessimistic. This is because they are not assuming anything but convexity of the solution. On the other hand, say in the two-dimensional case ($d = 2$), a standard nine point stencil finite difference approximation can be proposed

$$\mathrm{MA}_h^{\mathrm{FD}}[w_h](x_h) = \Delta_{(1,0)} w_h(x_h) \Delta_{(0,1)} w_h(x_h) - \left(\mathring{\Delta}_{(1,1)} w_h(x_h) \right)^2, \tag{22}$$

where, if $z_h = (x_1, x_2)^{\mathsf{T}}$, then

$$\mathring{\Delta}_{(1,1)} w_h(z_h) = \frac{1}{2h} \left(\frac{w_h(x_1 + h, x_2 + h) - w_h(x_1 - h, x_2 + h)}{2h} - \frac{w_h(x_1 + h, x_2 - h) - w_h(x_1 - h, x_2 - h)}{2h} \right).$$

This formula easily extends to higher dimensions.

It is not difficult to see that $\mathrm{MA}_h^{\mathrm{FD}}[\cdot]$ has second-order consistency, even for nonconvex functions. However, it is not monotone, even if one forgets about boundary conditions. Thus, it does not perform well when used to discretize problems that have singular solutions.

Froese and Oberman (2011b) takes advantage of the simplicity of (22) and the robustness of a wide stencil scheme by proposing a *hybrid* scheme. Locally, it is a convex combination of each one of these schemes, where the weighting is chosen depending on the expected behaviour of the solution. At points where the solution should be smooth the simple scheme (22) is used, whereas if the solution is expected to be singular the robustness of (19) is better suited to capture this behaviour. Summing up, the following discretization is used

$$\begin{aligned} \mathrm{MA}_h^{\mathrm{H}}[w_h](x_h) &= \omega(x_h) \mathrm{MA}_h^{\mathrm{FD}}[w_h](x_h) \\ &\quad + (1 - \omega(x_h)) \mathrm{MA}_{h,\theta}^{\mathrm{WS}}[w_h](x_h). \end{aligned} \tag{23}$$

Here $\omega \in C(\bar{\Omega}, [0,1])$ is a weighting function defined a priori from the data as follows: For $\epsilon > 0$ we let Ω_ϵ be a neighbourhood of the set where the solution u may be singular, that is,

$$\Omega_\epsilon = \{x \in \Omega : 0 \leq f(x) < \epsilon\} \cup \{x \in \partial\Omega : g \notin C^{2,\alpha}(U_x), \text{ or } U_x \cap \partial\Omega \text{ is flat}\},$$

where U_x is a neighbourhood of the point x. We then set $\omega \equiv 0$ in Ω_ϵ and one away from it. This scheme was tested in Froese and Oberman (2011b) for a series of cases, ranging from smooth to singular solutions, and computational experiments suggested that this method is robust and accurate.

This method, however, has a major drawback. The tunable function ω must be described by the user, and its values depend on the behaviour of the problem data. For this reason in Froese and Oberman (2013) it was proposed that instead the difference

$$\left|\mathrm{MA}_{h,\theta}^{\mathrm{WS}}[w_h](x_h) - \mathrm{MA}_h^{\mathrm{FD}}[w_h](x_h)\right|,$$

be used as an a posteriori indicator of accuracy. In regions where this difference is small, it is expected that the solution is smooth, whereas when this is large one expects singularities. On the basis of this, we can choose which scheme to apply. The way to measure this difference is by introducing a filter.

Definition 11 (filter).
A *filter* is a function $S \in C_0(\mathbb{R})$ such that $S(t) = t$ in a neighbourhood of the origin.

For instance, the function

$$S(t) = \begin{cases} x, & |x| \le 1 \\ 0, & |x| \ge 2, \\ 2 - x, & 1 < x < 2, \\ -x - 2, & -2 < x < -1 \end{cases} \tag{24}$$

depicted in Fig. 1 is a possible filter, see Froese and Oberman (2013, Fig. 1.1 and (1.3)). With this at hand, a *filtered* operator can be defined via

$$\begin{aligned} \mathrm{MA}_h^{\mathrm{F}}[w_h](x_h) = {} & \mathrm{MA}_{h,\theta}^{\mathrm{WS}}[w_h](x_h) \\ & + h^\alpha S\left(\frac{\mathrm{MA}_h^{\mathrm{FD}}[w_h](x_h) - \mathrm{MA}_{h,\theta}^{\mathrm{WS}}[w_h](x_h)}{h^\alpha}\right), \end{aligned} \tag{25}$$

where $\alpha \in (0, 2]$ is to be chosen by the user. A filtered scheme seeks $u_h \in X_h$ such that

$$\mathrm{MA}_h^{\mathrm{F}}[u_h](x_h) = f(x_h), \quad \forall x_h \in \Omega_h, \tag{26a}$$

$$u_h(x_h) = g(x_h), \quad \forall x_h \in \partial\Omega_h. \tag{26b}$$

Remark 5 (consistency).
Recall that (Kossaczký et al., 2016; Oberman, 2006) a monotone scheme cannot be more than second-order accurate. Notice, in addition, that by construction we have

$$\left|\mathrm{MA}_h^{\mathrm{F}}[w_h](x_h) - \mathrm{MA}_{h,\theta}^{\mathrm{WS}}[w_h](x_h)\right| \le h^\alpha,$$

so that a filtered scheme is also consistent, up to second order. Moreover, if the parameter α is chosen smaller than the consistency order of both the

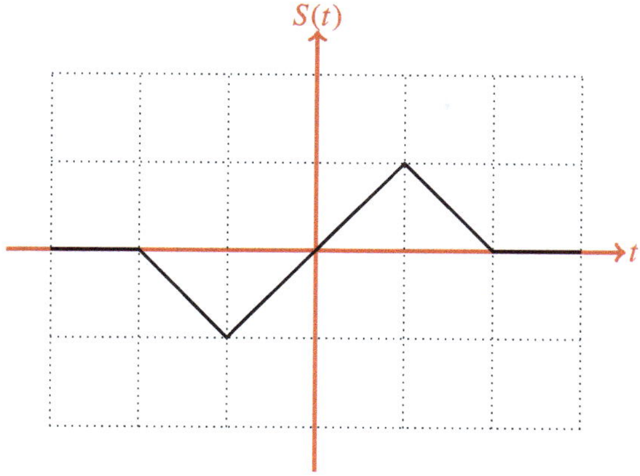

FIG. 1 The function S defined in (24) is a filter.

wide stencil, and the finite difference scheme, and the mesh size h is sufficiently small, it can be shown that

$$\mathrm{MA}_h^{\mathrm{F}}[\varphi](x) = \mathrm{MA}_h^{\mathrm{FD}}[\varphi](x),$$

whenever φ is sufficiently smooth. These two observations serve as a guideline for the choice of α. ■

Remark 6 (motivation).
The construction of a filtered scheme seems to be motivated by similar constructions for conservation laws and first order Hamilton Jacobi equations. For instance, Lions and Souganidis (1995) shows the convergence of filtered finite difference schemes (constructed in a similar way), for Hamilton Jacobi equations. In the realm of hyperbolic conservation laws, several types of limiters or artificial viscosity methods (Bonito et al., 2014; Guermond and Pasquetti, 2011; Guermond et al., 2011, 2018) have been derived from these ideas. ■

As a step towards the analysis of schemes like (26), Froese and Oberman (2013) introduced a class of schemes called *nearly monotone*, and showed that the theory of Section 2.1 also applies to them. To show this, we begin with a definition.

Definition 12 (nearly monotone).
The family of approximation schemes $\{F_\varepsilon\}_{\varepsilon>0}$ where $F_\varepsilon : \bar{\Omega} \times \mathbb{R} \times B(\bar{\Omega})$ is called *nearly monotone*, if every F_ε can be written as

$$F_\varepsilon = F_\varepsilon^M + F_\varepsilon^P,$$

where F_ε^M is monotone in the sense of (12), and the function F_ε^P, called a perturbation, satisfies

$$\lim_{\varepsilon\downarrow 0} |F^P_\varepsilon(x,t,v)| = 0,$$

uniformly on bounded subsets of $\bar{\Omega}\times\mathbb{R}\times B(\bar{\Omega})$.

The convergence of nearly monotone schemes closely follows that of monotone schemes.

Corollary 1 (convergence).
Let $\{F_\varepsilon\}_\varepsilon$ be a family of approximation schemes, that is nearly monotone, in the sense of Definition 12*; consistent, in the sense of* (14)*; and stable in the sense of* (11)*. Assume, in addition, that problem* (7) *has a strong comparison principle. In this setting we have that, as $\varepsilon \downarrow 0$, the functions u_ε, solutions of $F_\varepsilon(x, u_\varepsilon(x), u_\varepsilon) = 0$ converge locally uniformly to u, solution of* (7).

Proof. The proof is a small variation on the proof of Theorem 4. Indeed, with the notation of this proof, we have

$$\begin{aligned} 0 &= F_{\varepsilon_n}(y_n, u_{\varepsilon_n}(y_n), u_{\varepsilon_n}) \\ &= F^M_{\varepsilon_n}(y_n, \varphi(y_n) + \xi_n, \varphi + (u_{\varepsilon_n} - \varphi)) + F^P_{\varepsilon_n}(y_n, u_{\varepsilon_n}(y_n), u_{\varepsilon_n}) \\ &\le F^M_{\varepsilon_n}(y_n, \varphi(y_n) + \xi_n, \varphi + \xi_n) - F^P_{\varepsilon_n}(y_n, u_{\varepsilon_n}(y_n), u_{\varepsilon_n}). \end{aligned}$$

The stability of the scheme allows us to invoke the fact that the perturbation vanishes in the limit. Consequently, we still have that $\overline{u}$ is a subsolution. □

Notice that the same considerations made in Remark 1 apply in this setting.

2.5 Lattice basis reduction scheme

Let us now discuss a two-dimensional method, which was introduced in Benamou et al. (2016) and is termed the *lattice basis reduction* scheme. The aim of this scheme is, for a given stencil, to obtain a different way to compute the determinant, so that the scheme is more accurate. We begin with a definition.

Definition 13 (superbasis).
We will say that a *basis* of $\mathbb{Z}^2$ is a pair of vectors $(\boldsymbol{e}_1, \boldsymbol{e}_2) \in (\mathbb{Z}^2)^2$ that satisfy $|\det(\boldsymbol{e}_1, \boldsymbol{e}_2)| = 1$. A *superbasis* of $\mathbb{Z}^2$ is a triple $(\boldsymbol{e}_0, \boldsymbol{e}_1, \boldsymbol{e}_2) \in (\mathbb{Z}^2)^3$ such that $(\boldsymbol{e}_1, \boldsymbol{e}_2)$ is a basis and $\boldsymbol{e}_0 + \boldsymbol{e}_1 + \boldsymbol{e}_2 = 0$.

We will call a *stencil* a finite subset of $\mathbb{Z}^2\backslash\{0\}$ that is symmetric around the origin. To a stencil S we associate the set of superbases

$$Y(S) = \left\{(\boldsymbol{e}_0, \boldsymbol{e}_1, \boldsymbol{e}_2) \in S^3 : |\det(\boldsymbol{e}_1, \boldsymbol{e}_2)| = 1, \;\; \boldsymbol{e}_0 + \boldsymbol{e}_1 + \boldsymbol{e}_2 = 0\right\}.$$

With these notations at hand, we define the *lattice basis reduction* Monge–Ampère operator

$$\begin{aligned}&\mathrm{MA}_{h,S}^{\mathrm{LBR}}[w_h](x_h)\\&\quad=\min_{(e_0,e_1,e_2)\in Y(S)}\gamma\left((\Delta_{e_0}w_h(x_h))^+,(\Delta_{e_1}w_h(x_h))^+,(\Delta_{e_2}w_h(x_h))^+\right),\end{aligned}\tag{27}$$

where

$$\gamma(\delta_0,\delta_1,\delta_2)=\begin{cases}\delta_{i+1}\delta_{i+2}, & \delta_i\geq\delta_{i+1}+\delta_{i+2},\\ \dfrac{1}{2}(\delta_0\delta_1+\delta_1\delta_2+\delta_0\delta_2)-\dfrac{1}{4}(\delta_0^2+\delta_1^2+\delta_2^2), & \text{otherwise.}\end{cases}$$

This allows us to introduce the following scheme: Find $u_h \in X_h$ such that

$$\mathrm{MA}_{h,S}^{\mathrm{LBR}}[u_h](x_h)=f(x_h),\quad \forall x_h\in\Omega_h,\tag{28a}$$

$$u_h(x_h)=g(x_h),\quad \forall x_h\in\partial\Omega_h.\tag{28b}$$

The motivation for this, at first glance obscure, definition of the operator $\mathrm{MA}_{h,S}^{\mathrm{LBR}}[\cdot]$ is given in Benamou et al. (2016, Remark 1.10). Let $Y=(\boldsymbol{e}_0, \boldsymbol{e}_1, \boldsymbol{e}_2) \in Y(S)$ and notice that for any point x_h that is interior with respect to Y, we have that the convex hull of $\{x_h\pm h\boldsymbol{e}_i\}_{i=0}^2$ is a hexagon. Given a function $w_h\in X_h$ we can associate to it its *local convex envelope*, that is the maximal convex function $\Gamma_{x_h,Y}w_h$ that is bounded from above by w_h at the points $\{x_h\pm h\boldsymbol{e}_i\}_{i=0}^2$. It is then possible to show that $\Gamma_{x_h,Y}w_h$ is a piecewise linear function over a particular triangulation of the aforementioned hexagon. Then we have that

$$\gamma\left((\Delta_{e_0}w_h(x_h))^+,(\Delta_{e_1}w_h(x_h))^+,(\Delta_{e_2}w_h(x_h))^+\right)=|\partial\Gamma_{x_h,Y}w_h(x_h)|,\tag{29}$$

which is consistent with the definition of the Monge–Ampère operator in the sense of Alexandrov given in Definition 9 and hints at the consistency of this scheme.

The consistency analysis of the operator (27) hinges on the following definition.

Definition 14 (*M*–obtuseness).
Let $M\in\mathbb{S}_+^2$. We say that the superbasis $(\boldsymbol{e}_0, \boldsymbol{e}_1, \boldsymbol{e}_2)$ of $\mathbb{Z}^2$ is *M*–obtuse if and only if

$$\boldsymbol{e}_j\cdot M\boldsymbol{e}_i\leq 0,\quad \forall 0\leq i<j\leq 2.$$

From this definition, a necessary and sufficient condition for consistency follows (Benamou et al., 2016, Theorem 1.9).

Theorem 5 (consistency).
Let $\varphi=\frac{1}{2}x\cdot Mx$ be a convex quadratic polynomial. We have that

$$\mathrm{MA}_{h,S}^{\mathrm{LBR}}[\varphi](x)=\det M,\quad \forall x$$

if and only if Y (S) contains an M-obtuse superbais.

Proof. We will follow Benamou et al. (2016, Section 2.1). To simplify the discussion, we set

$$\mathcal{D}=\left\{(a_0,a_1,a_2)\in\mathbb{R}^3 : a_i\le a_{i+1}+a_{i+2},\quad i=0,1,2,\ \mathrm{mod}\ 3\right\},$$

$$\gamma_1(a_0,a_1,a_2)=\frac{1}{2}(a_0a_1+a_1a_2+a_0a_2)-\frac{1}{4}(a_0^2+a_1^2+a_2^2).$$

Notice that $\gamma(a_0, a_1, a_2) = \gamma_1(a_0, a_1, a_2)$ if and only if $(a_0,a_1,a_2)\in\mathcal{D}$, and that if that is not the case, then $\gamma(a_0,a_1,a_2)-\gamma_1(a_0,a_1,a_2)=\frac{1}{4}(a_0-a_1-a_2)^2>0$. In conclusion, we have that

$$\begin{cases}\gamma(a_0,a_1,a_2)\ge\gamma_1(a_0,a_1,a_2),\\ \gamma(a_0,a_1,a_2)=\gamma_1(a_0,a_1,a_2)\Leftrightarrow(a_0,a_1,a_2)\in\mathcal{D}.\end{cases}\tag{30}$$

Given a superbasis $(\boldsymbol{e}_0, \boldsymbol{e}_1, \boldsymbol{e}_2)$ define $\delta_i=\boldsymbol{e}_i\cdot M\boldsymbol{e}_i=(\Delta_{\boldsymbol{e}_i}\varphi(x_h))^+$. For a permutation (i, j, k) of $(0, 1, 2)$ we have

$$\delta_i-\delta_j-\delta_k=(\boldsymbol{e}_j+\boldsymbol{e}_k)\cdot M(\boldsymbol{e}_j+\boldsymbol{e}_k)-\boldsymbol{e}_j\cdot M\boldsymbol{e}_j-\boldsymbol{e}_kM\boldsymbol{e}_k=2\boldsymbol{e}_j\cdot M\boldsymbol{e}_k.$$

Consequently, $(\delta_0,\delta_1,\delta_2)\in\mathcal{D}$ if and only if the superbasis $(\boldsymbol{e}_0, \boldsymbol{e}_1, \boldsymbol{e}_2)$ is M–obtuse.

Let A be the linear transformation that maps $\boldsymbol{e}_1$ and $\boldsymbol{e}_2$ to $\boldsymbol{f}_1=(1,0)^\intercal$ and $\boldsymbol{f}_2=(0,1)^\intercal$, respectively. Then we must have that $\boldsymbol{f}_0=A\boldsymbol{e}_0=(-1,-1)^\intercal$. Thus, $\delta_i=\boldsymbol{e}_i\cdot M\boldsymbol{e}_i=A^{-1}\boldsymbol{f}_i\cdot MA^{-1}\boldsymbol{f}_i$, and so

$$\gamma_1(\mu_0,\mu_1,\mu_2)=\det(A^{-\intercal}MA^{-1}).$$

However, $\det A=|\det(\boldsymbol{e}_1,\boldsymbol{e}_2)/\det(\boldsymbol{f}_1,\boldsymbol{f}_2)|=1$. Combining this with (30) we obtain the claim. □

Essentially, the previous result shows that the operator $\mathrm{MA}_{h,S}^{\mathrm{LBR}}[\cdot]$ systematically overestimates the determinant of the Hessian for quadratic functions, and that we have equality if and only if the stencil S contains a M-obtuse superbasis. For this reason, it is of interest to obtain conditions on the size of the stencil that guarantee that such a superbasis can be found. The following result is a restatement of Benamou et al. (2016, Proposition 1.12).

Proposition 6 (stencil size estimate).
The stencil

$$S=\left\{\boldsymbol{e}\in\mathbb{Z}^2 : \gcd(\boldsymbol{e})=1, |\boldsymbol{e}|\le 2\kappa\right\},$$

contains a M–obtuse superbasis for every matrix $M\in\mathbb{S}_+^2$ *that satisfies*

$$\|M\|_2\,\|M^{-1}\|_2\le\kappa^2.$$

Notice that the cardinality of the stencil stated in Proposition 6 is quite large, approximately κ^2, and that if the solution degenerates, that is

$\det D^2u(x_0) = 0$ at some point, then the stencil size must again grow unboundedly to maintain consistency.

Remark 7. The recent paper (Benamou and Duval, 2018) shows convergence of the lattice basis reduction scheme (28) applied to the optimal transport problem.

2.6 Discretization based on power diagrams

In Mirebeau (2015) the following discretization of the Monge–Ampère operator is proposed and analyzed. Let S be a stencil such that $\operatorname{span} S = \mathbb{R}^d$ and such that its elements have coprime coordinates, that is, if $\boldsymbol{e} = (e_1, \ldots, e_d)^{\mathsf{T}} \in S$, then $\gcd(\boldsymbol{e}) = \gcd(e_1, \ldots, e_d) = 1$. We define

$$\mathrm{MA}_{h,S}^{\mathrm{PD}}[w_h](x_h) = \left|\left\{\boldsymbol{g} \in \mathbb{R}^d : \forall \boldsymbol{e} \in S : 2\boldsymbol{g} \cdot \boldsymbol{e} \leq |\boldsymbol{e}|^2 \Delta_{\boldsymbol{e}} w_h(x_h)\right\}\right|. \tag{31}$$

Here, we denote the Lebesgue measure by $|\cdot|$. With this operator at hand, we define the problem: find $u_h \in X_h$ such that

$$\mathrm{MA}_{h,S}^{\mathrm{PD}}[u_h](x_h) = f(x_h), \quad \forall x_h \in \Omega_h, \tag{32a}$$

$$u_h(x_h) = g(x_h), \quad \forall x_h \in \partial\Omega_h. \tag{32b}$$

Notice that the set entering the definition (31) is a polytope. Efficient ways to compute the volume of a polytope are available. For instance, if the dimension is not too high (and recall that we are mostly interested in the cases $d = 2$ or $d = 3$), one can first triangulate this polytope to then easily compute its volume.

Let us study the consistency of this scheme. To do so, we must introduce a definition.

Definition 15 (Voronoi cells and facets).
Let $M \in \mathbb{S}_+^d$. The Voronoi cell and facet are

$$\mathrm{Vor}(M) = \{\boldsymbol{g} \in \mathbb{R}^d : \forall \boldsymbol{e} \in \mathbb{Z}^d, \ \boldsymbol{g} \cdot M\boldsymbol{g} \leq (\boldsymbol{g} - \boldsymbol{e}) \cdot M(\boldsymbol{g} - \boldsymbol{e})\},$$
$$\mathrm{Vor}(M, \boldsymbol{e}) = \{\boldsymbol{g} \in \mathrm{Vor}(M) : \boldsymbol{g} \cdot M\boldsymbol{g} = (\boldsymbol{g} - \boldsymbol{e}) \cdot M(\boldsymbol{g} - \boldsymbol{e})\}.$$

A M-Voronoi vector is an element $\boldsymbol{e} \in \mathbb{Z}^d \setminus \{0\}$ such that $\mathrm{Vor}(M, \boldsymbol{e}) \neq \emptyset$. It is a strict M-Voronoi vector if the facet $\mathrm{Vor}(M, \boldsymbol{e})$ is $(d - 1)$-dimensional.

Now, the consistency of the operator defined in (31) is as follows.

Proposition 7 (consistency).
Let $\varphi(x) = \frac{1}{2} x \cdot Mx$ be a convex quadratic. Then we have that

$$\mathrm{MA}_{h,S}^{\mathrm{PD}}[\varphi](x) = \det M, \quad \forall x$$

if and only if the stencil S contains all the strict M-Voronoi vectors.

Proof. Let $\kappa = \sqrt{\| M\|_2 \, \| M^{-1}\|_2}$. We divide the proof in several steps.

1. Any point $\boldsymbol{g} \in \mathrm{Vor}(M)$ must satisfy $|\boldsymbol{g}| \le \frac{1}{2}\kappa\sqrt{d}$. Any M-Voronoi vector $\boldsymbol{e}$ satisfies $|\boldsymbol{e}| \le \kappa\sqrt{d}$ and has coprime coordinates:

   Indeed, if $\boldsymbol{g} \in \mathrm{Vor}(M)$, then let $\boldsymbol{e_g} \in \mathbb{Z}^d$ be obtained by rounding its coordinates to the nearest integer, so that $|\boldsymbol{g} - \boldsymbol{e_g}| \le \frac{1}{2}\sqrt{d}$. The estimate

$$\frac{1}{\| M^{-1}\|_2} |\boldsymbol{g}|^2 \le \boldsymbol{g} \cdot M\boldsymbol{g} \le (\boldsymbol{g} - \boldsymbol{e_g}) \cdot M(\boldsymbol{g} - \boldsymbol{e_g}) \le \| M\|_2 |\boldsymbol{g} - \boldsymbol{e_g}|^2 \le \frac{d}{4} \| M\|_2$$

   yields the desired estimate. In addition, if $\boldsymbol{e}$ is a M-Voronoi vector, there is $\boldsymbol{g} \in \mathrm{Vor}(M)$ for which $|\boldsymbol{g}| = |\boldsymbol{e} - \boldsymbol{g}|$ so that

$$|\boldsymbol{e}| \le 2|\boldsymbol{g}| \le \kappa\sqrt{d}.$$

   Finally, to show that the coordinates must be coprime consider $k\boldsymbol{e} \in \mathbb{Z}^d$ with $k \ge 2$ and notice that, for every $\boldsymbol{g} \in \mathbb{R}^d$ we have

$$(k\boldsymbol{e} - \boldsymbol{g}) \cdot M(k\boldsymbol{e} - \boldsymbol{g}) + (k-1)\boldsymbol{g} \cdot M\boldsymbol{g} = k(\boldsymbol{e} - \boldsymbol{g}) \cdot M(\boldsymbol{e} - \boldsymbol{g}) + (k^2 - k)\boldsymbol{e} \cdot M\boldsymbol{e}.$$

   Consequently,

$$(\boldsymbol{e} - \boldsymbol{g}) \cdot M(\boldsymbol{e} - \boldsymbol{g}) < \max\{(k\boldsymbol{e} - \boldsymbol{g}) \cdot M(k\boldsymbol{e} - \boldsymbol{g}), \boldsymbol{g} \cdot M\boldsymbol{g}\},$$

   and $k\boldsymbol{e}$ cannot be a M-Voronoi vector.

2. Let E be the set of strict M-Voronoi vectors, then

$$\mathrm{Vor}(M) \subset \{\boldsymbol{g} \in \mathbb{R}^d : \forall \boldsymbol{e} \in S : 2\boldsymbol{g} \cdot M\boldsymbol{e} \le \boldsymbol{e} \cdot M\boldsymbol{e}\},$$

   with equality if and only if $E \subset S$:

   Notice that $\boldsymbol{g} \cdot M\boldsymbol{g} \le (\boldsymbol{g} - \boldsymbol{e}) \cdot M(\boldsymbol{g} - \boldsymbol{e})$ is equivalent to saying that $2\boldsymbol{g} \cdot M\boldsymbol{e} \le \boldsymbol{e} \cdot M\boldsymbol{e}$. This shows that $\mathrm{Vor}(M)$ is a convex polytope, defined by inequalities of this type where $\boldsymbol{e}$ runs over the set of strict M-Voronoi vectors. The bound established in the previous step shows that there can only be a finite number of them.

3. $|\mathrm{Vor}(M)| = 1$:

   It follows from the observation that $\mathrm{Vor}(M)$ collects all elements $\boldsymbol{g} \in \mathbb{R}^d$ that are closer to the origin (in the metric induced by the matrix M) than to any other point $\boldsymbol{e} \in \mathbb{Z}^d \setminus \{0\}$.

4. Consistency:

   Recall that, for any $\boldsymbol{e} \in S$ we have that $|\boldsymbol{e}|^2 \Delta_{\boldsymbol{e}} \varphi(x) = \boldsymbol{e} \cdot M\boldsymbol{e}$. Consequently,

$$\begin{aligned} \mathrm{MA}^{\mathrm{PD}}_{h,S}[\varphi](x) &= \left|\{\boldsymbol{g} \in \mathbb{R}^d : \forall \boldsymbol{e} \in S : 2\boldsymbol{g} \cdot \boldsymbol{e} \le \boldsymbol{e} \cdot M\boldsymbol{e}\}\right| \\ &= \left|M\{\boldsymbol{g} \in \mathbb{R}^d : \forall \boldsymbol{e} \in S : 2\boldsymbol{g} \cdot M\boldsymbol{e} \le \boldsymbol{e} \cdot M\boldsymbol{e}\}\right|. \end{aligned}$$

   A combination of the second and third steps then yields

$$\mathrm{MA}^{\mathrm{PD}}_{h,S}[\varphi](x) \geq \det M|\mathrm{Vor}(M)| = \det M,$$

with equality if $\mathrm{Vor}(M) = \{\boldsymbol{g} \in \mathbb{R}^d : \forall \boldsymbol{e} \in S : 2\boldsymbol{g} \cdot \boldsymbol{e} \leq \boldsymbol{e} \cdot M\boldsymbol{e}\}$ with equality if $\mathrm{Vor}(M)$ contains all strict M-Voronoi vectors.

This conclude s the proof. □

Since the consistency of the operator $\mathrm{MA}^{\mathrm{PD}}_{h,S}[\cdot]$ requires the stencil to contain all strict Voronoi vectors, it is necessary to provide sufficient conditions for this to happen.

Corollary 2 (stencil size estimate).
Let $\kappa > 0$ and define

$$S = \left\{\boldsymbol{e} \in \mathbb{Z}^d : |\boldsymbol{e}| \leq \sqrt{d}\kappa, \quad \gcd(\boldsymbol{e}) = 1\right\}.$$

Let $\varphi(x) = \frac{1}{2} x \cdot M \cdot x$, then we have that

$$\mathrm{MA}^{\mathrm{PD}}_{h,S}[\varphi](x) = \det M, \quad \forall x$$

provided $\|M\|_2 \|M^{-1}\|_2 \leq \kappa^2$.

Proof. It immediately follows from the norm estimates given in Step 1 in the proof of Proposition 7. □

Let us now provide a convergence analysis of scheme (32), which will follow from the framework provided in Section 2.1. To do so, we introduce the operator $F_{h,S} : \bar{\Omega}_h \times \mathbb{R} \times X_h \to \mathbb{R}$ via

$$F_{h,S}(x_h, t, w) = \begin{cases} \mathrm{MA}^{\mathrm{PD}}_{h,S}[w](x_h) - f(x_h), & x_h \in \Omega_h, \\ g(x_h) - t, & x_h \in \partial\Omega_h, \end{cases} \tag{33}$$

and notice that (32) can be compactly written as

$$F_{h,S}(x_h, u_h(x_h), u_h) = 0, \quad \forall x_h \in \bar{\Omega}_h.$$

Let us also define the operator $F_S : \bar{\Omega} \times \mathbb{R} \times \mathbb{S}^d_+ \to \mathbb{R}$

$$F_S(x, t, M) = \begin{cases} |K(M)| - f(x), & x \in \Omega, \\ g(x) - t, & x \in \partial\Omega, \end{cases} \tag{34}$$

where

$$K(M) = \{\boldsymbol{v} \in \mathbb{R}^d : \forall \boldsymbol{e} \in S, \quad 2\boldsymbol{v} \cdot \boldsymbol{e} \leq \boldsymbol{e} \cdot M\boldsymbol{e}\}.$$

Notice that, if $D^2u(x_0)$ exists for all $x_0 \in \Omega$ and its eigenvalues are properly bounded, see Corollary 2, we have that

$$\det D^2u(x_0) - f(x_0) = F_S(x_0, u(x_0), D^2u(x_0)).$$

For this reason, we will consider the problem: find u that is a viscosity solution of

$$F_S(x,u(x),D^2u(x))=0, \quad x\in\bar{\Omega}. \tag{35}$$

Following Mirebeau (2015, Section 2.3) we will now show the convergence of scheme (32) via Theorem 4. To do so, we must show that scheme (32) is monotone, consistent, and stable in the sense of (12), (14), and (13), respectively. We have shown consistency in Proposition 7. For stability, we refer the reader to Mirebeau (2015, Section 2.2), where stability is shown by proving global convergence of a damped Newton algorithm. We will focus then on the monotonicity of the scheme.

Proposition 8 (monotonicity).
The operator $F_{h,\,S}$, defined in (33) *is monotone in the sense of* (12).

Proof. Notice that, if $x_h \in \partial\Omega_h$, then there is nothing to show. On the other hand, if $x_h \in \Omega_h$, then $\mathrm{MA}^{\mathrm{PD}}_{h,S}[w](x_h)$ is an increasing function of the second differences $\Delta_e w_h(x_h)$. Indeed, increasing this difference makes the polytope larger. Notice also that $\Delta_e w_h(x_h)$ is a linear combination, with positive coefficients, of $w_h(x_h + \boldsymbol{e}h) - w_h(x_h)$ and $w_h(x_h + \boldsymbol{e}h) - w_h(x_h)$, with the obvious modification for points that are not interior with respect to $\boldsymbol{e}$. Thus, we can invoke (Neilan et al., 2017, Lemma 3.11) to conclude the monotonicity. □

Next to be able to apply Theorem 4 we must make sure that the operator F_S satisfies a comparison principle. To establish this we begin with an auxiliary result.

Lemma 4 (polytope comparison).
Let $M_1,M_2\in\mathbb{S}^d_+$ and $x \in \Omega$. If $M_1 \le M_2$ then, for every $t\in\mathbb{R}$ we have that $F_S(x, t, M_1) \le F_S(x, t, M_2)$. In addition,

$$\begin{aligned}(F_S(x,t,M_1+M_2)+f(x))^{1/d} \ge\, &(F_S(x,t,M_1)+f(x))^{1/d}\\ &+(F_S(x,t,M_2)+f(x))^{1/d}.\end{aligned}$$

Proof. Notice that, since $x \in \Omega$ we have, independently of t,

$$F(x,t,M)+f(x)=|K(M)|, \quad K(M)=\{\boldsymbol{v}\in\mathbb{R}^d:\forall\boldsymbol{e}\in S,\ 2\boldsymbol{v}\cdot\boldsymbol{e}\le\boldsymbol{e}\cdot M\boldsymbol{e}\}.$$

Notice, in addition, that $M_1 \le M_2$ implies that $\boldsymbol{e}\cdot M_1\boldsymbol{e} \le \boldsymbol{e}\cdot M_2\boldsymbol{e}$ for every $\boldsymbol{e}\in\mathbb{R}^d$. Consequently, $M_1 \le M_2$ implies $K(M_1)\subset K(M_2)$ from which the first statement follows.

Now, since $\boldsymbol{e}\cdot(M_1 + M_2)\boldsymbol{e} = \boldsymbol{e}\cdot M_1\boldsymbol{e} + \boldsymbol{e}\cdot M_2\boldsymbol{e}$ we have that $K(M_1 + M_2)$ contains $K(M_1) + K(M_2)$. The Brunn–Minkowski inequality given in Lemma 2 allows us to conclude. □

Now we can establish a comparison principle for F_S.

Proposition 9 (comparison).
Let $\overline{u}\in USC(\bar{\Omega})$ and $\underline{u}\in LSC(\bar{\Omega})$ be a sub- and supersolution, respectively, of (35). *Then we have that $\overline{u}\le\underline{u}$.*

Proof. We begin by noticing that, since $F_{S,\star}(x, t, M) \leq F_S(x, t, M)$ for all $x \in \bar{\Omega}$ we obtain that, if $x_0 \in \partial\Omega$ we must have that

$$0 \leq F_{S,\star}(x_0, \varphi(x_0), D^2\varphi(x_0)) \leq F_S(x_0, \varphi(x_0), D^2\varphi(x_0)) = g(x_0) - \varphi(x_0),$$

for every φ sufficiently smooth that satisfies the conditions given in Definition 4. As a consequence, In this case, the condition defining a viscosity subsolution at boundary points reduces to $\overline{u} \leq g$ on $\partial\Omega$. Similarly we can show that for a supersolution we must have $g \leq \underline{u}$ on $\partial\Omega$. In conclusion, at the boundary $\partial\Omega$ we have $\overline{u} \leq \underline{u}$.

By the semicontinuity assumption we can also define $\delta = \sup_{\bar{\Omega}}(\overline{u} - \underline{u}) \in \mathbb{R}$. Additionally, since Ω is bounded, there is $R > 0$ such that $\Omega \subset B_R$. Assume now, for the sake of contradiction, that $\delta > 0$.

Let us define, for $\varepsilon > 0$, the operator $F_{S,\varepsilon} : \bar{\Omega} \times \mathbb{R} \times \mathbb{S}^d \to \mathbb{R}$ by

$$F_{S,\varepsilon}(x,t,M) = \begin{cases} F_S(x,t,M) - \varepsilon(t - \underline{u}(x)), & x \in \Omega, \\ g(x) - t, & x \in \partial\Omega, \end{cases}$$

and notice that this operator satisfies all the conditions of the comparison principle given in Crandall et al. (1992, Theorem 3.3). Moreover, since for all $x \in \bar{\Omega}$ we have that $F_{S,\varepsilon}(x, \underline{u}(x), D^2\underline{u}(x)) = F_S(x, \underline{u}(x), D^2\underline{u}(x))$ we conclude that $\underline{u}$ is a supersolution for the operator $F_{S,\varepsilon}$.

We now construct a subsolution. Define

$$v(x) = \frac{(\varepsilon\delta)^{1/d}}{2}\left(|x|^2 - R^2\right), \quad u_\varepsilon(x) = \overline{u}(x) + v(x)$$

and notice that $u_\varepsilon \in USC(\bar{\Omega})$ and, moreover, $u_\varepsilon \leq \overline{u} \leq g \leq \underline{u}$ on $\partial\Omega$. In addition, we have that, for $x \in \Omega$

$$D^2v(x) = (\varepsilon\delta)^{1/d} I, \quad F_S(x, t, D^2v(x)) + f(x) = \varepsilon\delta,$$

see the proof of Proposition 7. Let now $x \in \Omega$ and, to shorten notation, denote

$$F_S[w] = F_S(x, w(x), D^2w(x)) + f(x).$$

If this is the case we have that, in the viscosity sense

$$\begin{aligned} F_{S,\varepsilon}(x, u_\varepsilon(x), D^2u_\varepsilon(x)) &= F_S[\overline{u} + v] - f(x) - \varepsilon(\overline{u}(x) - \underline{u}(x)) - \varepsilon v(x) \\ &\geq \left(F_S[\overline{u}]^{1/d} + F_S[v]^{1/d}\right)^d - f(x) - \varepsilon(\overline{u}(x) - \underline{u}(x)) \\ &\geq \left(f(x)^{1/d} + F_S[v]^{1/d}\right)^d - f(x) - \varepsilon(\overline{u}(x) - \underline{u}(x)) \\ &\geq F_S[v] - \varepsilon(\overline{u}(x) - \underline{u}(x)) \\ &= \varepsilon(\delta - (\overline{u}(x) - \underline{u}(x))) \geq 0. \end{aligned}$$

where we used Lemma 4, the fact that $v(x) \leq 0$ for all $x \in \bar{\Omega}$, that $\overline{u}$ is a subsolution for the operator F_S, the elementary identity

$$(x+y)^\theta \le x^\theta + y^\theta, \quad \forall x, y \in \mathbb{R}_+, \quad \forall \theta \in (0,1],$$

and the definition of δ. In conclusion, u_ε is a subsolution for the operator $F_{S,\varepsilon}$. The comparison principle of Crandall et al. (1992, Theorem 3.3) then yields that

$$u_\varepsilon(x) = \overline{u}(x) + v(x) \le \underline{u}(x), \quad \forall x \in \bar{\Omega}$$

or that

$$\underline{u}(x) - \overline{u}(x) \ge -\frac{(\varepsilon\delta)^{1/d}}{2} R^2, \quad \forall x \in \bar{\Omega}.$$

Letting $\varepsilon \downarrow 0$ we obtain $\overline{u}(x) \le \underline{u}(x)$, contradicting that $\delta > 0$. □

As a consequence, we have convergence.

Corollary 3 (convergence).

Let $\{u_h\}_{h>0} \subset X_h$ be the solutions to (32). *Then, as $h \downarrow 0$, we have that $u_h \to u$ locally uniformly, where u is the* (*unique*) *viscosity solution of* (35).

Proof. Apply Theorem 4. It is only relevant to mention that owing to the comparison principle showed in Proposition 9, u must necessarily be unique. □

2.7 Two scale methods

We will now present and analyze the so-called *two scale method*, which can be understood as a generalization of the wide stencil schemes presented in Section 2.3 to unstructured meshes (see also Froese (2018)). Here and in what follows we will implicitly assume that Ω is uniformly convex. Additional assumptions will be explicitly stated. Next, for $h > 0$, we introduce a quasiuniform (in the sense of Ciarlet (2002)) simplicial triangulation $\mathcal{T}_h$ of our domain Ω. We denote by Ω_h^i and Ω_h^b the set of interior and boundary nodes, respectively, of $\mathcal{T}_h$. We define X_h to be the set of piecewise linear and continuous functions subject to this triangulation. The mesh size h will constitute the fine scale of discretization. The large scale, denoted by δ, will be the one at which second-order differences will be evaluated. Notice that, since now we are dealing with continuous functions, these can be evaluated at any point. Indeed, given $x_h \in \Omega_h^i$ and $\boldsymbol{w} \in \mathbb{R}^d$ with $|\boldsymbol{w}| = 1$ we define, for $w_h \in X_h$

$$\nabla^2_{\delta\boldsymbol{w}} w_h(x_h) = \frac{w_h(x_h + \rho\delta\boldsymbol{w}) - 2w_h(x_h) + w_h(x_h - \rho\delta\boldsymbol{w})}{\rho^2\delta^2}, \tag{36}$$

where $\rho \in (0, 1]$ is the largest number so that $x_h \pm \rho\delta\boldsymbol{w} \in \bar{\Omega}$; compare with (17) and (18). As a final discretization ingredient, as in the case of the wide stencil schemes of Section 2.3, we need a directional discretization. That is if, as before, $\mathcal{V}$ denotes the set of all orthonormal bases of $\mathbb{R}^d$ we must construct, for $\theta > 0$, a set $\mathcal{V}_\theta$ of collections of d unit vectors such that if $\{\boldsymbol{w}_i\}_{i=1}^d \in \mathcal{V}$, then there is $\{\boldsymbol{w}_i^\theta\}_{i=1}^d \in \mathcal{V}_\theta$ such that

$$\max_{i=1,\ldots,d} |\boldsymbol{w}_i - \boldsymbol{w}_i^\theta| \leq \theta. \tag{37}$$

It is important to notice that the elements of $\mathcal{V}_\theta$ are not required to be orthonormal collections of vectors.

Having defined all the discretization ingredients, which are parametrized by the triple $\varepsilon = (h, \delta, \theta)$, following Nochetto et al. (2019a) we introduce the two scale discrete Monge–Ampère operator by defining, for $w_h \in X_h$, and $x_h \in \Omega_h^i$,

$$\mathrm{MA}_{h,\delta,\theta}^{2\mathrm{S}}[w_h](x_h) = \min_{\{\boldsymbol{w}_i\}_{i=1}^d \in \mathcal{V}_\theta} \left[\prod_{i=1}^d \left(\nabla_{\delta \boldsymbol{w}_i}^2 w_h(x_h) \right)^+ - \sum_{i=1}^d \left(\nabla_{\delta \boldsymbol{w}_i}^2 w_h(x_h) \right)^- \right], \tag{38}$$

compare with the scheme discussed in Remark 2. With these ingredients at hand, the two scale method seeks a function $u_h^\varepsilon \in X_h$ such that

$$\mathrm{MA}_{h,\delta,\theta}^{2\mathrm{S}}[u_h^\varepsilon](x_h) = f(x_h), \quad \forall x_h \in \Omega_h^i, \tag{39a}$$

$$u_h^\varepsilon(x_h) = g(x_h), \quad \forall x_h \in \Omega_h^b. \tag{39b}$$

Remark 8 (generalization).
Starting from the Cartesian mesh Ω_h used to define the wide stencil schemes (20) it is possible to construct a simplicial triangulation of Ω without introducing new vertices: in two dimensions this is accomplished by subdividing each square by its diagonal, and a similar construction is possible in three dimensions. Once this is done, it can be seen that scheme (39) is, after little modifications, a generalization of the wide stencil scheme (20). ■

Remark 9 (domain approximation).
Notice that, since the domain Ω is assumed to be uniformly convex, it is not possible to triangulate it exactly. If we denote $\bar{\Omega}_{\mathcal{T}_h} = \cup_{T \in \mathcal{T}_h} \overline{T}$, then we have $\bar{\Omega}_{\mathcal{T}_h} \subsetneq \bar{\Omega}$. In our discussion we will ignore this fact. This is because we can either replace Ω by $\Omega_{\mathcal{T}_h}$ in all the statements that we shall make, or we can consider all functions in X_h as defined in Ω by extending them to $\Omega \setminus \Omega_{\mathcal{T}_h}$ by a constant in the normal direction to faces. This is a standard construction and we shall not delve into it further. ■

Let us now provide, following Li and Nochetto (2018a); Nochetto et al. (2019a,b), an analysis of (39). We will first introduce a discrete notion of convexity, based on the positivity of the second differences defined in (36).

The operator (38) turns out to have a comparison principle, and acts in a particular way on discretely convex functions. This will allow us to establish existence, uniqueness, and stability of solutions to (39). In addition, since the size large scale δ is reduced near the boundary, the consistency can only hold sufficiently far away from it. For this reason, appropriate barrier functions need to be constructed. All these ingredients will allow us to assert convergence of the method. Finally, using the comparison principle and suitable barriers, we will establish rates of convergence for classical solutions.

2.7.1 Discrete convexity

The second-order differences defined in (36) and the set of directions $\mathcal{V}_\theta$ give a discrete notion of convexity.

Definition 16 (discrete convexity).
We say that the function $w_h \in X_h$ is discretely convex if

$$\nabla^2_{\delta \boldsymbol{w}_j} w_h(x_h) \geq 0, \quad \forall x_h \in \Omega_h^i, \quad \forall \boldsymbol{w}_j \in \{\boldsymbol{w}_i\}_{i=1}^d \in \mathcal{V}_\theta.$$

It is well known that if a function is convex, then its second-order differences are nonnegative. On the other hand, discrete convexity does not imply convexity. This is due, for instance, to the fact that convexity and interpolation are not easily compatible. In other words, if $w \in C(\bar{\Omega})$ is convex, then its Lagrange interpolant $\mathcal{I}_h w \in X_h$ satisfies $\mathcal{I}_h w \geq w$ so that it is discretely convex, but $\mathcal{I}_h w$ is not necessarily convex.

On the other hand, discrete convexity implies nonnegativity of the two scale discrete Monge–Ampère operator; see Nochetto et al. (2019a, Lemma 2.2).

Lemma 5 (discrete convexity).
A function $w_h \in X_h$ is discretely convex if and only if

$$\mathrm{MA}^{2\mathrm{S}}_{h,\delta,\theta}[w_h](x_h) \geq 0, \quad \forall x_h \in \Omega_h^i.$$

Moreover, for a discretely convex function we have that

$$\mathrm{MA}^{2\mathrm{S}}_{h,\delta,\theta}[w_h](x_h) = \min_{\{\boldsymbol{w}_i\}_{i=1}^d \in \mathcal{V}_\theta} \prod_{i=1}^d \nabla^2_{\delta \boldsymbol{w}_i} w_h(x_h).$$

2.7.2 A comparison principle

Let us now show that the operator defined in (38) is monotone and has a comparison principle. From this we will obtain uniqueness of solutions to (39).

Lemma 6 (monotonicity).
Let $v_h, w_h \in X_h$ be such that $v_h - w_h$ attains its maximum at the interior node $x_h \in \Omega_h^i$. Then we have

$$\mathrm{MA}^{2S}_{h,\delta,\theta}[w_h](x_h) \geq \mathrm{MA}^{2S}_{h,\delta,\theta}[v_h](x_h).$$

Proof. Since x_h is the maximum, for suitable $\rho > 0$ and any unit vector $\boldsymbol{w}$ we have

$$v_h(x_h) - w_h(x_h) \geq v_h(x_h \pm \rho\delta\boldsymbol{w}) - w_h(x_h \pm \rho\delta\boldsymbol{w}),$$

which implies that

$$\nabla^2_{\delta w} v_h(x_h) \leq \nabla^2_{\delta w} w_h(x_h).$$

multiplying this inequality as $\boldsymbol{w}$ runs over all elements of $\mathcal{V}_\theta$ allows us to conclude. □

The previous result gives us a comparison principle for the operator (38).
Proposition 10 (comparison).
Let v_h, $w_h \in X_h$ be such that $v_h \leq w_h$ on $\partial\Omega$, and

$$\mathrm{MA}^{2S}_{h,\delta,\theta}[v_h](x_h) \geq \mathrm{MA}^{2S}_{h,\delta,\theta}[w_h](x_h), \quad \forall x_h \in \Omega^i_h,$$

then we must have that $v_h \leq w_h$ in $\bar{\Omega}$.

Proof. We consider two cases for the inequality between the operators:

1. *The inequality is strict.* Let us assume, for the sake of contradiction, $v_h - w_h$ attains a maximum at an interior node. Lemma 6 then gives a contradiction.
2. *The inequality is not strict.* Since Ω is bounded, there is $R > 0$ such that the convex quadratic $q(x) = \frac{1}{2}(|x|^2 - R)$ is nonpositive on $\bar{\Omega}$. Let $q_h = \mathcal{I}_h q \in X_h$. This function is strictly convex and satisfies

$$\nabla^2_{\delta w} q_h(x_h) \geq \nabla^2_{\delta w} q(x_h) = \frac{\partial^2 q(x_h)}{\partial \boldsymbol{w}^2} = 1.$$

We claim now that, for all $\alpha > 0$ and $x_h \in \Omega^i_h$, we have that

$$\mathrm{MA}^{2S}_{h,\delta,\theta}[v_h + \alpha q_h](x_h) \geq \mathrm{MA}^{2S}_{h,\delta,\theta}[v_h](x_h) + \min\left\{\frac{\alpha^d}{2^d} + \frac{\alpha}{2}\right\}. \tag{40}$$

Indeed, fix $\{\boldsymbol{w}_i\} \in \mathcal{V}_\theta$ and assume first that $\nabla^2_{\delta w_i}(v_h(x_h) + \frac{\alpha}{2} q_h(x_h)) \geq 0$ for all i. In this case

$$\begin{aligned}
\prod_{i=1}^d \left(\nabla^2_{\delta w_i} v_h(x_h) + \alpha q_h(x_h)\right) &\geq \prod_{i=1}^d \left(\nabla^2_{\delta w_i}\left(v_h(x_h) + \frac{\alpha}{2} q_h(x_h)\right) + \frac{\alpha}{2}\right) \\
&\geq \min_{\{w_i\} \in \mathcal{V}_\theta} \prod_{i=1}^d \left(\nabla^2_{\delta w_i}\left(v_h(x_h) + \frac{\alpha}{2} q_h(x_h)\right)\right) + \frac{\alpha^d}{2^d} \\
&\geq \left(\prod_{i=1}^d \left(\nabla^2_{\delta w_i} v_h(x_h)\right)^+ - \sum_{i=1}^d \left(\nabla^2_{\delta w_i} v_h(x_h)\right)^-\right) + \frac{\alpha^d}{2^d}.
\end{aligned}$$

On the other hand, if there is $i \in \{1, \ldots, d\}$ for which $\nabla^2_{\delta w_i}(v_h(x_h) + \alpha q_h(x_h)) < 0$, then this implies that $\nabla^2_{\delta w_i} v_h(x_h) < 0$. Thus,

$$\prod_{i=1}^{d} \left(\nabla^2_{\delta w_i} v_h(x_h)\right)^+ = 0,$$

and

$$-\sum_{i=1}^{d} \left(\nabla^2_{\delta w_i}(v_h(x_h) + \alpha q_h(x_h))\right)^- \geq -\sum_{i=1}^{d}\left(\nabla^2_{\delta w_i} v_h(x_h)\right)^- + \frac{\alpha}{2}$$
$$= \left(\prod_{i=1}^{d}\left(\nabla^2_{\delta w_i} v_h(x_h)\right)^+ - \sum_{i=1}^{d}\left(\nabla^2_{\delta w_i} v_h(x_h)\right)^-\right) + \frac{\alpha}{2}.$$

A combination of these two cases, since $\{w_i\}_{i=1}^d \in \mathcal{V}_\theta$ was arbitrary, implies (40).

Finally, since, $v_h + \alpha q_h \leq v_h \leq w_h$ on $\partial\Omega$ and, on the basis of (40), we have

$$\mathrm{MA}^{2S}_{h,\delta,\theta}[v_h + \alpha q_h](x_h) > \mathrm{MA}^{2S}_{h,\delta,\theta}[v_h](x_h) \geq \mathrm{MA}^{2S}_{h,\delta,\theta}[w_h](x_h), \quad \forall x_h \in \Omega_h^i,$$

the previous step then implies that $v_h + \alpha q_h \leq w_h$. Letting $\alpha \downarrow 0$ we can conclude. □

Remark 10 (discrete interior barrier).
Notice, that, in the course of the second case of the proof of this result we effectively constructed a discrete interior barrier. If $q(x) = \frac{1}{2}(|x|^2 - R)$ with $R > 0$ sufficiently large, then we have that

$$\mathcal{I}_h q \leq 0, \quad \text{on } \partial\Omega, \quad \mathrm{MA}^{2S}_{h,\delta,\theta}[\mathcal{I}_h q_h](x_h) \geq 1, \quad \forall x_h \in \Omega_h^i.$$

■

As an immediate consequence, we also have uniqueness of solutions to (39).

Corollary 4 (uniqueness).
Scheme (39) *cannot have more than one solution.*

As a final application of the comparison principle, let us now show existence and uniform bounds on the solution to (39).

Theorem 6 (existence and stability).
For all $\varepsilon = (h, \delta, \theta) > 0$ *scheme* (39) *has a solution* $u_h^\varepsilon \in X_h$. *Moreover, this solution is stable in the sense that* $\| u_h^\varepsilon \|_{L^\infty(\Omega)}$ *is bounded independently of* ε.

Proof. The existence proceeds via Perron's method. For this reason, we will only indicate how to construct a discrete subsolution, that is a function $u_h^0 \in X_h$ such that $u_h^0 = \mathcal{I}_h g$ on $\partial\Omega$ and

$$\mathrm{MA}^{2S}_{h,\delta,\theta}[u_h^0](x_h) \geq f(x_h), \quad \forall x_h \in \Omega_h^i.$$

To construct this function, we define

$$s(x) = \sum_{x_h \in \Omega_h^i} s_{x_h}(x), \qquad s_{x_h}(x) = \frac{\delta \rho_{x_h}}{2} f(x_h)^{1/d} |x - x_h|,$$

where $\rho_{x_h} \in (0,1]$ is the largest number such that, for all $\boldsymbol{w} \in \mathbb{R}^d$ with $|\boldsymbol{w}| = 1$ we have $x_h \pm \rho_{x_h} \boldsymbol{w} \in \bar{\Omega}$. Notice that $\nabla^2_{\delta \boldsymbol{w}} s_{x_h}(y_h) \geq 0$ for all $y_h \in \Omega_h^i$, and that

$$\nabla^2_{\delta \boldsymbol{w}} s_{x_h}(x_h) = f(x_h)^{1/d}, \quad \forall \boldsymbol{w} \in \mathbb{R}^d, \quad |\boldsymbol{w}| = 1.$$

Consequently, for $y_h \in \Omega_h^i$

$$\nabla^2_{\delta \boldsymbol{w}} \mathcal{I}_h s(y_h) \geq \nabla^2_{\delta \boldsymbol{w}} s(y_h) \geq f(y_h)^{1/d} \geq 0,$$

which, by Lemma 5 implies

$$\mathrm{MA}^{2S}_{h,\delta,\theta}[\mathcal{I}_h s](x_h) = \min_{\{\boldsymbol{w}_i\}_{i=1}^d \in \mathcal{V}_\theta} \prod_{i=1}^d \nabla^2_{\delta \boldsymbol{w}_i} \mathcal{I}_h s(x_h) \geq f(x_h), \quad \forall x_h \in \Omega_h^i.$$

Let now $w \in C(\bar{\Omega})$ be the convex envelope of $(\mathcal{I}_h(g - s))|_{\partial\Omega}$, and set $w_h = \mathcal{I}_h w$. By convexity of w we have that

$$\mathrm{MA}^{2S}_{h,\delta,\theta}[w_h](x_h) \geq 0, \quad \forall x_h \in \Omega_h^i.$$

Thus, we define

$$u_h^0 = w_h + \mathcal{I}_h s.$$

This function, by construction, is discretely convex and $u_h^0 = \mathcal{I}_h g$ on $\partial\Omega$. Since the second differences of w_h are nonnegative, then we have that

$$\begin{aligned}
\mathrm{MA}^{2S}_{h,\delta,\theta}[u_h^0](x_h) &= \min_{\{\boldsymbol{w}_i\}_{i=1}^d \in \mathcal{V}_\theta} \prod_{i=1}^d \left[\nabla^2_{\delta \boldsymbol{w}_i} w_h(x_h) + \nabla^2_{\delta \boldsymbol{w}_i} \mathcal{I}_h s(x_h) \right] \\
&\geq \min_{\{\boldsymbol{w}_i\}_{i=1}^d \in \mathcal{V}_\theta} \prod_{i=1}^d \nabla^2_{\delta \boldsymbol{w}_i} \mathcal{I}_h s(x_h) \geq f(x_h),
\end{aligned}$$

and so u_h^0 is a discrete subsolution.

It remains to show the uniform boundedness. To achieve this we will show that every discrete subsolution is uniformly bounded. Let then $w_h \in X_h$ be a discrete subsolution and $b_h = \max_{x \in \partial\Omega} g(x) \in X_h$. We have then that

$$\mathrm{MA}^{2S}_{h,\delta,\theta}[b_h](x_h) = 0 \leq f(x_h) \leq \mathrm{MA}^{2S}_{h,\delta,\theta}[w_h](x_h), \quad \forall x_h \in \Omega_h^i.$$

Since, in addition, we have that $b_h \geq w_h$ on $\partial\Omega$, the comparison principle of Proposition 10 implies that

$$w_h \leq b_h.$$

This is enough since Perron's method shows existence of a solution by constructing an increasing sequence of subsolutions. Thus, u_h^0 is a lower bound for the solution and, evidently, $\|u_h^0\|_{L^\infty(\Omega)}$ is independent of ε. □

2.7.3 Consistency and discrete barriers

Let us now examine the consistency of the operator (38). As we have stated above, the operator can only be consistent at points sufficiently far away from the boundary. For this reason, we define the δ-interior and δ-boundary layer of Ω via

$$\Omega_\delta = \bigcup_{T \in \mathcal{T}_h : \operatorname{dist}(T, \partial\Omega) > \delta} T, \quad (\partial\Omega)_\delta = \bar{\Omega} \setminus \Omega_\delta.$$

For an interior node $x_h \in \Omega_h^i$ its interior patch is

$$\omega_{x_h} = \bigcup_{T \in \mathcal{T}_h : \operatorname{dist}(x_h, T) < \rho\delta} \overline{T},$$

where, as before, $\rho \in (0, 1]$ is the largest number such that, for any $\boldsymbol{w} \in \mathbb{R}^d$ with $|\boldsymbol{w}| = 1$ we have $x_h \pm \rho\delta \boldsymbol{w} \in \bar{\Omega}$.

The following result follows, essentially, by a Taylor expansion argument.

Lemma 7 (consistency of second differences).

Let $x_h \in \Omega_h^i$ and assume that $\varphi \in C^{1,1}(\omega_{x_h})$, then for all $\boldsymbol{w} \in \mathbb{R}^d$ with $|\boldsymbol{w}| = 1$ we have

$$|\nabla_{\delta \boldsymbol{w}}^2 \mathcal{I}_h \varphi(x_h)| \leq C |\varphi|_{C^{1,1}(\omega_{x_h})}.$$

If, in addition, we have that $x_h \in \Omega_\delta$ and $\varphi \in C^{k+2,\alpha}(\omega_{x_h})$ for $k = 0, 1$, and $\alpha \in (0, 1]$ then we also have that

$$\left| \nabla_{\delta \boldsymbol{w}}^2 \mathcal{I}_h \varphi(x_h) - \frac{\partial^2 \varphi(x_h)}{\partial \boldsymbol{w}^2} \right| \leq C \left(|\varphi|_{C^{k+2,\alpha}(\omega_{x_h})} \delta^{k+\alpha} + \frac{h^2}{\delta^2} |\varphi|_{C^{1,1}(\omega_{x_h})} \right).$$

Finally, if φ is, in addition, convex then we have

$$\frac{\partial^2 \varphi(x_h)}{\partial \boldsymbol{w}^2} - \nabla_{\delta \boldsymbol{w}}^2 \mathcal{I}_h \varphi(x_h) \leq C |\varphi|_{C^{k+2,\alpha}(\omega_{x_h})} \delta^{k+\alpha}.$$

The previous result can be applied to obtain interior consistency of (38). The following result was first obtained in Nochetto et al. (2019a, Lemma 4.2) under the assumption that $\mathcal{V}_\theta \subset \mathcal{V}$. This assumption was later removed in Li and Nochetto (2018a, Lemma 2.4).

Theorem 7 (interior consistency).
Let $x_h \in \Omega_h^i$ and $\varphi \in C^{k+2,\alpha}(\omega_{x_h})$ with $k=0,1$ and $\alpha \in (0,1]$ be convex. In this setting the following estimates are valid:

1. *If $\theta \le \frac{1}{4d}$ then, for any $\{\boldsymbol{w}_i\}_{i=1}^d \in \mathcal{V}_\theta$, we have*

$$\det D^2\varphi(x_h) \le \prod_{i=1}^d \frac{\partial^2\varphi(x_h)}{\partial \boldsymbol{w}_i^2}\left(1+16\theta^2(d-1)^2\right).$$

2. *If $\{\boldsymbol{v}_i\}_{i=1}^d \in \mathcal{V}$ realizes the minimum in the variational characterization of the determinant given in Lemma 3, then for any $\{\boldsymbol{v}_i^\theta\}_{i=1}^d \in \mathcal{V}_\theta$ that satisfies (37) we have*

$$\left|\frac{\partial^2\varphi(x_h)}{\partial \boldsymbol{v}_i^2} - \frac{\partial^2\varphi(x_h)}{\partial \boldsymbol{v}_i^{\theta 2}}\right| \le C|\varphi|_{C^{1,1}(\omega_{x_h})}\theta^2.$$

3. *Finally, if $x_h \in \Omega_h^i \cap \Omega_\delta$, then*

$$\left|\det D^2\varphi(x_h) - \mathrm{MA}_{h,\delta,\theta}^{2\mathrm{S}}[\mathcal{I}_h\varphi](x_h)\right| \le C_1\delta^{k+\alpha} + C_2\left(\frac{h^2}{\delta^2}+\theta^2\right),$$

where the constants C_1 and C_2 depend only on the smoothness of φ, the domain Ω, and the dimension d.

Proof. We prove each statement separately.

1. Let $W_\theta = (\boldsymbol{w}_1, \ldots, \boldsymbol{w}_d)$. We have

$$\det(W_\theta^\mathsf{T} W_\theta)\det D^2\varphi(x_h) = \det(W_\theta^\mathsf{T} D^2\varphi(x_h) W_\theta) \le \prod_{i=1}^d \frac{\partial^2\varphi(x_h)}{\partial \boldsymbol{w}_i^2},$$

where, in the last step, we used that $W_\theta^\mathsf{T} D^2\varphi(x_h) W_\theta$ is positive semidefinite and Hadamard's inequality. We now need to estimate the determinant of $W = W_\theta^\mathsf{T} W_\theta$ from below. Write

$$W = \begin{pmatrix} W_0 & \boldsymbol{w} \\ \boldsymbol{w}^\mathsf{T} & 1 \end{pmatrix} = \begin{pmatrix} I & 0 \\ \boldsymbol{w}^\mathsf{T} W_0^{-1} & 1 \end{pmatrix}\begin{pmatrix} W_0 & \boldsymbol{w} \\ 0 & 1-\boldsymbol{w}\cdot W_0\boldsymbol{w} \end{pmatrix}$$

implying that $\det W = (1-\boldsymbol{w}\cdot W_0\boldsymbol{w})\det W_0$, which holds if the submatrix W_0 is nonsingular. Notice, however, that $W_{i,\,i} = 1$ and $|W_{i,\,j}| \le 2\theta$ as the columns of W_θ form an element of $\mathcal{V}_\theta$. This implies, for $\theta \le \frac{1}{4d}$, that $W_0 \ge \frac{1}{2}I$ and $|\boldsymbol{w}| \le 2\theta\sqrt{d-1}$. Thus, $W_0^{-1} \ge 2I$ and

$$|\boldsymbol{w}\cdot W_0\boldsymbol{w}| \le 8\theta^2(d-1) \qquad \det W \ge (1-8\theta^2(d-1))\det W_0,$$

which by repeating this process yields

$$\det W \ge (1-8\theta^2(d-1))^{d-1} \ge 1-8\theta^2(d-1)^2,$$

and using, again the bound on θ

$$\frac{1}{\det W} \le 1+16\theta^2(d-1)^2.$$

2. We begin by noticing that, given the minimization assumption, $\{v_i\}_{i=1}^d$ must be the normalized eigenvectors of $D^2\varphi(x_h)$. Set $v_i^\theta = v_i + w_i$ and write

$$\frac{\partial^2\varphi(x_h)}{\partial v_i^{\theta 2}} = v_i^\theta \cdot D^2\varphi(x_h)v_i^\theta = \frac{\partial^2\varphi(x_h)}{\partial v_i^2} + 2w_i \cdot D^2\varphi(x_h)v_i + w_i \cdot D^2\varphi(x_h)w_i.$$

Since $\{v_i\}_{i=1}^d$ are eigenvectors, $|w_i| \le \theta$, and $|v_i \cdot w_i| \le \frac{1}{2}\theta^2$ we then have

$$\left|\frac{\partial^2\varphi(x_h)}{\partial v_i^{\theta 2}} - \frac{\partial^2\varphi(x_h)}{\partial v_i^2}\right| \le C\theta^2.$$

3. By Lemma 5, since $\mathcal{I}_h\varphi$ is discretely convex, we have that

$$\mathrm{MA}_{h,\delta,\theta}^{2S}[\mathcal{I}_h\varphi](x_h) = \min_{\{w_i\}_{i=1}^d \in \mathcal{V}_\theta} \prod_{i=1}^d \nabla_{\delta w_i}^2 \mathcal{I}_h\varphi(x_h).$$

Let $\{w_i\}_{i=1}^d \in \mathcal{V}_\theta$ be the set that realizes the minimum in this expression. Using Lemma 3 we can write that

$$\begin{aligned}\det D^2\varphi(x_h) - \mathrm{MA}_{h,\delta,\theta}^{2S}[\mathcal{I}_h\varphi](x_h) &\le \prod_{i=1}^d \frac{\partial^2\varphi(x_h)}{\partial w_i^2} - \prod_{i=1}^d \nabla_{\delta w_i}^2 \mathcal{I}_h\varphi(x_h)\\ &\le C\delta^{k+\alpha},\end{aligned}$$

where, in the last step, we used repeatedly Lemma 7.

Let now $\{v_i\}_{i=1}^d \in \mathcal{V}$ be the normalized eigenpairs of $D^2\varphi(x_h)$, and $\{v_i^\theta\}_{i=1}^d \in \mathcal{V}_\theta$ the collection that realizes (37). Then we have

$$\begin{aligned}&\mathrm{MA}_{h,\delta,\theta}^{2S}[\mathcal{I}_h\varphi](x_h) - \det D^2\varphi(x_h)\\ &\le \left(\prod_{i=1}^d \nabla_{\delta v_i^\theta}^2 \mathcal{I}_h\varphi(x_h) - \prod_{i=1}^d \frac{\partial^2\varphi(x_h)}{\partial v_i^{\theta 2}}\right) + \left(\prod_{i=1}^d \frac{\partial^2\varphi(x_h)}{\partial v_i^{\theta 2}} - \prod_{i=1}^d \frac{\partial^2\varphi(x_h)}{\partial v_i{}^2}\right).\end{aligned}$$

The first term can be handled by repeatedly applying Lemma 7, while the second by applying the previous step.

All the estimates have been proved and the interior consistency is thus obtained. □

As mentioned before, the operator is not consistent near the boundary. For this reason we will, instead, construct discrete barriers which will allow us to control the behaviour of the solution near the boundary.

Proposition 11 (discrete barrier I).
Let $E > 0$ be arbitrary and $x_h \in \Omega_h^i$ be such that $dist(x_h, \partial\Omega) \le \delta$. Then, there is $p_h \in X_h$ such that

$$p_h \le 0, \text{ on } \partial\Omega, \quad \mathrm{MA}_{h,\delta,\theta}^{2S}[p_h](y_h) \ge E, \quad \forall y_h \in \Omega_h^i, \quad |p_h(x_h)| \le CE^{1/d}\delta,$$

where the constant C depends only on the domain Ω.

Proof. Without loss of generality, we can assume that $x_h = (0,\ldots,0,z)^{\mathsf{T}}$ with $z > 0$ so that $0 \in \partial\Omega$ and $z = \operatorname{dist}(x_h, \partial\Omega)$. The uniform convexity of Ω shows that there is $R > 0$ such that, in this system of coordinates,

$$\Omega \subset \left\{ x \in \mathbb{R}^d : \sum_{i=1}^{d-1} x_i^2 - (x_d - R)^2 \le R^2 \right\}.$$

Let

$$p(x) = \frac{E^{1/d}}{2} \left(\sum_{i=1}^{d-1} x_i^2 - (x_d - R)^2 - R^2 \right).$$

We claim that $p_h = \mathcal{I}_h p$ is the desired barrier. Indeed, by construction $p_h \le 0$ on the boundary $\partial\Omega$ and, since $z \le \delta$ we have that $|p_h(x_h)| \le CE^{1/d}\delta$. Finally, since p_h is discretely convex, for any interior node y_h we have

$$\mathrm{MA}^{2\mathrm{S}}_{h,\delta,\theta}[p_h](y_h) \ge \mathrm{MA}^{2\mathrm{S}}_{h,\delta,\theta}[p](y_h) = \prod_{i=1}^{d} E^{1/d} = E,$$

as claimed. □

To obtain rates of convergence we shall also require another discrete barrier that was originally introduced in Nochetto and Zhang (2018, Section 6.2). We define

$$\zeta : [0,\infty) \to (-\infty, 0], \quad \zeta(t) = \begin{cases} (t - 2\delta)^2 - (2\delta)^2, & t \in [0, 2\delta], \\ -(2\delta)^2, & t \in (2\delta, \infty). \end{cases}$$

The graph of this function is illustrated in Fig. 2. With this function at hand, we define

$$b(x) = \zeta(\operatorname{dist}(x, \partial\Omega)),$$

and $b_h = \mathcal{I}_h b$. The properties of this barrier are as follows.

Proposition 12 (discrete barrier II).

For $\theta \le \frac{1}{2\sqrt{d}}$ the barrier function b_h satisfies:

1. *For all $x_h \in \Omega_h^i$ and any $\mathbf{w} \in \mathbb{R}^d$ with $|\mathbf{w}| = 1$,*

$$\nabla^2_{\delta\mathbf{w}} b_h(x_h) \ge 0.$$

2. *For all $x_h \in \Omega_h^i \setminus \Omega_\delta$ and $\{\mathbf{w}_i^\theta\}_{i=1}^d \in \mathcal{V}_\theta$,*

$$\max_{i=1,\ldots,d} \nabla^2_{\delta\mathbf{w}_i^\theta} b_h(x_h) \ge \frac{1}{2d}.$$

3. *For all $x \in \bar{\Omega}$*

$$-4\delta^2 \le b_h(x) \le 0.$$

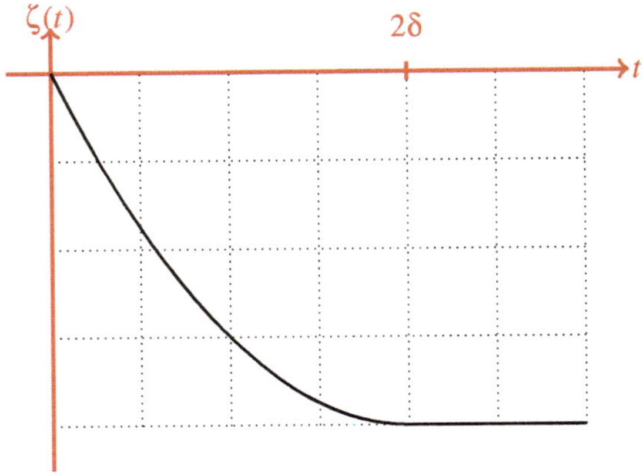

FIG. 2 The function ζ used to define the discrete barrier of Proposition 12.

Proof. We consider each property separately.

1. Let x_+, $x_- \in \Omega$ with $x_+ \neq x_-$. The convexity of Ω ensures that $x_0 = \frac{1}{2}(x_+ + x_-) \in \Omega$. Denote by $y \in \partial\Omega$ the closest point to x_0. Since Ω is convex, there is a supporting hyperplane P at y, whose normal is $\boldsymbol{n} = \frac{1}{|x_0 - y|}(x_0 - y)$. Let now $\boldsymbol{v} = \pm(x_+ - x_-)$, where the sign is chosen so that $\boldsymbol{n} \cdot \boldsymbol{v} \geq 0$. Consequently, see Fig. 3,

$$\operatorname{dist}(x_\pm, \partial\Omega) \leq \operatorname{dist}(x_\pm, P) = \operatorname{dist}(x_0, \partial\Omega) \pm \boldsymbol{n} \cdot \boldsymbol{v}.$$

With this estimate, and using that ζ is nonincreasing, we can compute

$$\begin{aligned} b(x_+) + b(x_-) &\geq \zeta(\operatorname{dist}(x_0, \partial\Omega) + \boldsymbol{v} \cdot \boldsymbol{n}) + \zeta(\operatorname{dist}(x_0, \partial\Omega) - \boldsymbol{v} \cdot \boldsymbol{n}) \\ &\geq 2\zeta(\operatorname{dist}(x_0, \partial\Omega)) = 2b(x_0), \end{aligned}$$

where the second inequality follows directly from the definition of ζ. We then conclude (cf. Krasnosel'skiĭ and Rutickiĭ, 1961, Pages 1–2) that the function b is convex and the stated property of b_h follows.

2. With the notation of the previous step, if we take a node $x_h \in \Omega_h^i \backslash \Omega_\delta$, and $\boldsymbol{v} \in \mathbb{R}^d$ with $|\boldsymbol{v}| = 1$, then $\operatorname{dist}(x_h, \partial\Omega) \pm \rho\delta \boldsymbol{w} \cdot \boldsymbol{n} \in [0, 2\delta]$. Since ζ is nonincreasing and quadratic on that interval

$$\nabla^2_{\delta\boldsymbol{v}} b_h(x_h) \geq \nabla^2_{\delta\boldsymbol{v}} b(x_h) \geq 2\frac{\rho^2\delta^2|\boldsymbol{v} \cdot \boldsymbol{n}|^2}{\rho^2\delta^2} = 2|\boldsymbol{v} \cdot \boldsymbol{n}|^2.$$

Now, if we let $\boldsymbol{v}$ run over $\{\boldsymbol{w}_i^\theta\}_{i=1}^d \in \mathcal{V}_\theta$ we have obtained that

$$\max_{i=1,\ldots,d} \nabla^2_{\delta\boldsymbol{w}_i^\theta} b_h(x_h) \geq 2 \max_{i=1,\ldots,d} |\boldsymbol{w}_i^\theta \cdot \boldsymbol{n}|^2.$$

Let now $\{\boldsymbol{w}_i\}_{i=1}^d \in \mathcal{V}$ be such that it satisfies (37). Since $|\boldsymbol{n}| = 1$ we must have that

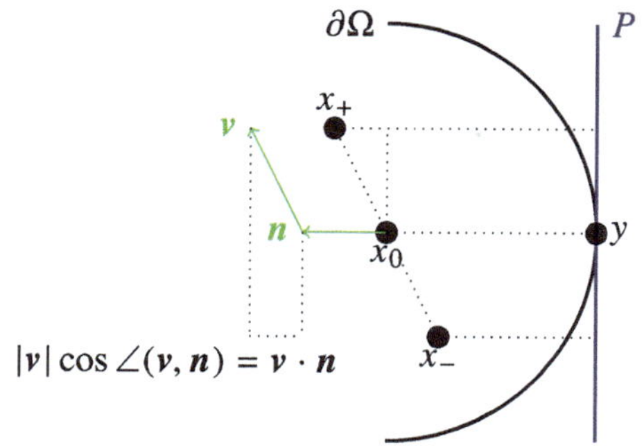

FIG. 3 The construction Proposition 12 that shows that the function b is convex. The distance between x_+ and the supporting hyperplane P equals the sum of the distance from x_0 to the boundary $\partial\Omega$ and the inner product between $\boldsymbol{n}$ and $\boldsymbol{v}$.

$$\sum_{i=1}^{d} |\boldsymbol{n}\cdot\boldsymbol{w}_i|^2 = 1, \quad \Longrightarrow \max_{i=1,\dots,d} |\boldsymbol{n}\cdot\boldsymbol{w}_i| \geq \frac{1}{\sqrt{d}}.$$

Therefore,

$$|\boldsymbol{w}_i^\theta\cdot\boldsymbol{n}| \geq |\boldsymbol{w}_i\cdot\boldsymbol{n}| - |(\boldsymbol{w}_i - \boldsymbol{w}_i^\theta)\cdot\boldsymbol{n}| \geq |\boldsymbol{w}_i\cdot\boldsymbol{n}| - \theta \geq \frac{1}{2\sqrt{d}},$$

where we used that $\theta \leq \frac{1}{2\sqrt{d}}$. This implies the estimate.

3. The last property follows directly from the definition of the function ζ. □

2.7.4 Convergence

Let us now show convergence. We will do so by adapting the arguments developed in Section 2.1 to take into account that test functions must be convex. We will rely on Proposition 3.

Theorem 8 (convergence).
Let Ω be uniformly convex, $f \in C(\bar{\Omega})$ such that $f \geq 0$, and $g \in C(\partial\Omega)$. As $\varepsilon = (h, \delta, \theta) \to 0$ with $h\delta^{-1} \to 0$ we have that the family $\{u_h^\varepsilon\}_\varepsilon$ of solutions of (39) *converges uniformly to $u \in C(\bar{\Omega})$, the solution of* (1).

Proof. In a similar way to Theorem 7 we have that, for all $x_0 \in \Omega$, $x_h \in \Omega_h^i \cap \Omega_\delta$ and all $\varphi \in C^{2,\alpha}(\omega_{x_h})$, it holds that

$$|\mathrm{MA}[\varphi](x_0) - \mathrm{MA}_{h,\delta,\theta}^{2S}[\mathcal{I}_h\varphi](x_h)| \leq C_1(\delta^\alpha + |x_0 - x_h|^\alpha) + C_2\left(\frac{h^2}{\delta^2} + \theta^2\right). \tag{41}$$

Indeed, we only need to use that the operations $t \mapsto t^\pm$ are Lipschitz and with Lipschitz constant equal one.

We now extend the ideas of Theorem 4. As there we define

$$\overline{u}(x) = \limsup_{\varepsilon\to 0, \frac{h}{\delta}\to 0, y\to x} u_h^\varepsilon(y), \quad \underline{u}(x) = \liminf_{\varepsilon\to 0, \frac{h}{\delta}\to 0, y\to x} u_h^\varepsilon(y)$$

and we will show that $\overline{u}$ is a subsolution. For that we assume that $\overline{u} - \varphi$, with $\varphi \in C^{2,\alpha}(\bar{\Omega})$ attains a maximum at $x_0 \in \Omega$. Let $\{x_h\}$ be the sequence of nodes such that $x_h \to x_0$ and $u_h^\varepsilon - \mathcal{I}_h\varphi$ attains a maximum at x_h. By the monotonicity result of Lemma 6 we obtain then that

$$\mathrm{MA}^{2S}_{h,\delta,\theta}[\mathcal{I}_h\varphi](x_h) \geq \mathrm{MA}^{2S}_{h,\delta,\theta}[u_h^\varepsilon](x_h) \geq f(x_h),$$

the consistency, as expressed in (41), implies by passing to the limit that

$$\mathrm{MA}[\varphi](x_0) \geq f(x_0).$$

It remains to understand the boundary behaviour of $\overline{u}$. We will show that the boundary condition is attained in a classical sense, that is $\overline{u} = g$. Let $x \in \partial\Omega$ and p_k be the continuous quadratic constructed during the proof of existence of the boundary barrier function in Proposition 11 with constant $E = k$. As k can be taken arbitrarily large, the sequence of points where $g \pm p_k$ attains a maximum (minimum) over $\partial\Omega$, converges to x.

We now observe that the monotonicity of Lemma 6 implies that if $v_h \in X_h$ is such that $\mathrm{MA}^{2S}_{h,\delta,\theta}[v_h](x_h) > 0$ for all $x_h \in \Omega_h^i$, then v_h attains its maximum on $\partial\Omega$. Since

$$\mathrm{MA}^{2S}_{h,\delta,\theta}[u_h^\varepsilon + \mathcal{I}_h p_k](x_h) > 0, \quad \forall x_h \in \Omega_h^i,$$

we can apply this observation to $u_h^\varepsilon + \mathcal{I}_h p_k$ to obtain that, for $x \in \partial\Omega$,

$$\begin{aligned}
\overline{u}(x) &\leq \limsup_{\varepsilon\to 0, \frac{h}{\delta}\to 0, y\to x} \left(u_h^\varepsilon(y) + \mathcal{I}_h p_k(y)\right) - \liminf_{\varepsilon\to 0, \frac{h}{\delta}\to 0, y\to x} \mathcal{I}_h p_k(y) \\
&\leq \limsup_{\varepsilon\to 0, \frac{h}{\delta}\to 0, y\to x} \max_{z\in\partial\Omega} \mathcal{I}_h(g(z) + p_k(z)) - p_k(x) \leq g(x_k) + p_k(x_k) - p_k(x),
\end{aligned}$$

where x_k is the point where $g + p_k$ attains its maximum over $\partial\Omega$. Letting $k \to \infty$ we conclude $\overline{u} \leq g$. Similarly $\underline{u} \geq g$.

Finally we invoke the comparison principle of Proposition 4 to conclude. □

Remark 11 (convergence by regularization).
It is interesting to note that by invoking the continuous dependence result given in Proposition 1, and the approximation result of Proposition 2, another proof of convergence can be developed. See Nochetto et al. (2019a, Section 5.3) for details. ■

2.7.5 Rates of convergence

The ingredients used to assert the convergence of the two scale method (39) were employed in Nochetto et al. (2019b) to obtain rates of convergence.

The techniques used in this reference were very similar to those that we will describe in Section 3 and so, to avoid repetition, we shall not elaborate on them here. This is further justified by that fact that, although Nochetto et al. (2019b) was the first work to provide rates of convergence for wide stencil-type methods, the rates of convergence obtained in this work were suboptimal.

Let us here, instead, present the results obtained in Li and Nochetto (2018a), where optimal rates of convergence have been obtained. The main tools in this are the comparison principle of Proposition 10 and the discrete barriers constructed in Section 2.7.3.

We begin by noticing that we shall only assume

$$f \geq 0,$$

so that the Monge–Ampère equation (1) may be degenerate. The main result about rates of convergence for classical solutions is the following.

Theorem 9 (error estimate).
Let $u \in C^{2,\alpha}(\bar{\Omega})$, *with* $\alpha \in (0,1]$, *solve* (1) *and* $u_h^\varepsilon \in X_h$ *solve* (39). *If* $\theta \leq \frac{1}{4d}$ *then we have*

$$\|u - u_h^\varepsilon\|_{L^\infty(\Omega)} \leq C\left[h^2\left(1+\delta^{-2}\right)|u|_{C^{1,1}(\bar{\Omega})} + \delta^\alpha |u|_{C^{2,\alpha}(\bar{\Omega})}\right],$$

where the constant C *depends on the domain* Ω, *the dimension* d, *and the shape regularity of the mesh* $\mathcal{T}_h$, *but is independent of* h, *and the solution* u.

Proof. Recall that a standard interpolation estimate yields

$$\| u - \mathcal{I}_h u\|_{L^\infty(\Omega)} \leq Ch^2 |u|_{C^{1,1}(\bar{\Omega})},$$

so that we only need to bound the difference $u_h - \mathcal{I}_h u$. To do so, we will construct a suitable discrete subsolution u_h^- and supersolution u_h^+ and use the comparison principle of Proposition 10.

Let $u_h^- = \mathcal{I}_h u + K_1 q_h \in X_h$, where q_h is the interior barrier of Remark 10 and $K_1 > 0$ is to be chosen later. Notice that, by construction

$$u_h^- \leq \mathcal{I}_h u = \mathcal{I}_h g, \quad \text{on } \partial\Omega.$$

Thus, to guarantee that this is a subsolution we must show that

$$\mathrm{MA}^{2S}_{h,\delta,\theta}[u_h^-](x_h) \geq f(x_h) = \det D^2 u(x_h), \quad \forall x_h \in \Omega_h^i.$$

However, since u_h^- is discretely convex, showing this inequality reduces to showing that, for all $\{w_i\}_{i=1}^d \in \mathcal{V}_\theta$ we have

$$\prod_{i=1}^d \nabla^2_{\delta w_i} u_h^-(x_h) \geq \det D^2 u(x_h), \quad \forall x_h \in \Omega_h^i,$$

see Lemma 5. Using the convexity of u, we have, according to Lemma 7, that

$$\nabla^2_{\delta w_i}\mathcal{I}_h u(x_h) \geq \frac{\partial^2 u(x_h)}{\partial w_i^2} - C|u|_{C^{2,\alpha}(\bar{\Omega})}\delta^\alpha,$$

so that, upon choosing

$$K_1 = C\left[\delta^\alpha |u|_{C^{2,\alpha}(\bar{\Omega})} + \left(\frac{h^2}{\delta^2}+\theta^2\right)|u|_{C^{1,1}(\bar{\Omega})}\right],$$

where C is sufficiently large, we have

$$\begin{aligned}\nabla^2_{\delta w_i}u_h^-(x_h) &\geq \frac{\partial^2 u(x_h)}{\partial w_i^2} - C|u|_{C^{2,\alpha}(\bar{\Omega})}\delta^\alpha + K_1 \geq \frac{\partial^2 u(x_h)}{\partial w_i^2} + C\theta^2|u|_{C^{1,1}(\bar{\Omega})} \\ &\geq \left(1+16\theta^2(d-1)^2\right)\frac{\partial^2 u(x_h)}{\partial w_i^2} \geq \left(1+16\theta^2(d-1)^2\right)^{1/d}\frac{\partial^2 u(x_h)}{\partial w_i^2}.\end{aligned}$$

Finally, since $\theta \leq \frac{1}{4d}$, we multiply this inequality over $i = 1, \ldots, d$ and invoke Theorem 7 item 1 to conclude that u_h^- is a subsolution. The comparison principle of Proposition 10 then yields that

$$\begin{aligned}u_h^\varepsilon \geq u_h^- &= \mathcal{I}_h u + C\left(\delta^\alpha |u|_{C^{1,1}(\bar{\Omega})} + \left(\frac{h^2}{\delta^2}+\theta^2\right)|u|_{C^{1,1}(\bar{\Omega})}\right)q_h \\ &\geq \mathcal{I}_h u - C\left(\delta^\alpha |u|_{C^{1,1}(\bar{\Omega})} + \theta^2|u|_{C^{1,1}(\bar{\Omega})}\right).\end{aligned}$$

We now define

$$u_h^+ = \mathcal{I}_h u - K_1 q_h - K_2 b_h,$$

where q_h and K_1 are as before, b_h is the barrier of Proposition 12 and $K_2 > 0$ is to be chosen. We show that u_h^+ is a supersolution. First of all, because of the choice of signs

$$u_h^+ \geq \mathcal{I}_h u = \mathcal{I}_h g, \quad \text{on } \partial\Omega.$$

Now, to show the inequality between operators we must consider in Ω_δ and outside of it separately. Let $x_h \in \Omega_h^i \cap \Omega_\delta$ and $\{v_i\}_{i=1}^d \in \mathcal{V}$ such that

$$f(x_h) = \det D^2 u(x_h) = \prod_{i=1}^d \frac{\partial^2 u(x_h)}{\partial v_i^2}.$$

Let now $\{v_i^\theta\}_{i=1}^d \in \mathcal{V}_\theta$ satisfy (37). The interior consistency of second differences of Lemma 7, together with the estimate of Theorem 7 item 2 gives us that

$$\left|\nabla^2_{\delta v_i^\theta}\mathcal{I}_h u(x_h) - \frac{\partial^2 u(x_h)}{\partial v_i^2}\right| \leq C\left[\delta^\alpha |u|_{C^{1,1}(\bar{\Omega})} + \left(\frac{h^2}{\delta^2}+\theta^2\right)|u|_{C^{1,1}(\bar{\Omega})}\right],$$

which, using that $\nabla^2_{\delta v_i^\theta} q_h(x_h) \geq 1$, $\nabla^2_{\delta v_i^\theta} b_h(x_h) \geq 0$, and the definition of K_1 immediately implies that

$$\nabla^2_{\delta v_i^\theta} u_h^+(x_h) \leq \frac{\partial^2 u(x_h)}{\partial v_i^2}.$$

Notice now that u_h^+ might not be discretely convex, so that $\nabla^2_{\delta v_i^\theta} u_h^+(x_h)$ might be negative. To deal with this we define the function

$$G : \mathbb{R}^d \to \mathbb{R}, \quad G(z) = \prod_{i=1}^{d} (z \cdot e_i)^+ - \sum_{i=1}^{d} (z \cdot e_i)^-,$$

where $\{e_i\}_{i=1}^d$ is the canonical basis of $\mathbb{R}^d$. Notice that this function is monotone in each coordinate of z. Moreover if, for $\{w_i\}_{i=1}^d \in \mathcal{V}_\theta$ and $w_h \in X_h$, we define the vectors

$$\chi(w_h, \{w_i\}) = \left(\nabla^2_{\delta w_1} w_h(x_h), \ldots, \nabla^2_{\delta w_d} w_h(x_h) \right)^{\mathsf{T}},$$

$$\gamma = \left(\frac{\partial^2 u(x_h)}{\partial v_1^2}, \ldots, \frac{\partial^2 u(x_h)}{\partial v_d^2} \right)^{\mathsf{T}}.$$

Then we have that

$$\mathrm{MA}^{2S}_{h,\delta,\theta}[w_h](x_h) = \min_{\{w_i\}_{i=1}^d \in \mathcal{V}_\theta} G(\chi(w_h, \{w_i\})).$$

Therefore

$$\mathrm{MA}^{2S}_{h,\delta,\theta}[u_h^+](x_h) \leq G(\chi(u_h^+, \{v_i^\theta\})) \leq G(\gamma) = \prod_{i=1}^{d} \frac{\partial^2 u(x_h)}{\partial v_i^2} = f(x_h).$$

Consider now a node close to the boundary, that is $x_h \in \Omega_h^i \setminus \Omega_\delta$, and let $\{w_i^\theta\}_{i=1}^d \in \mathcal{V}_\theta$. Using Proposition 12 item 2 we have that

$$\max_{i=1,\ldots,d} \nabla^2_{\delta w_i^\theta} b_h(x_h) \geq \frac{1}{2d}.$$

Assume that this maximum is attained for index k. Using Lemma 7 we can conclude that

$$\begin{aligned}\nabla^2_{\delta w_k^\theta} u_h^+(x_h) &\leq \nabla^2_{\delta w_k^\theta} \mathcal{I}_h u(x_h) - K_2 \nabla^2_{\delta w_k^\theta} b_h(x_h) \\ &\leq \nabla^2_{\delta w_k^\theta} \mathcal{I}_h u(x_h) - \frac{1}{2d} K_2 \leq C |u|_{C^{1,1}(\bar{\Omega})} - \frac{1}{2d} K_2 \leq 0,\end{aligned}$$

where the last step holds upon choosing K_2 sufficiently large. This shows that

$$\min_{i=1,\ldots,d} \nabla^2_{\delta w_i^\theta} u_h^+(x_h) = 0 \implies \mathrm{MA}^{2S}_{h,\delta,\theta}[u_h^+](x_h) = 0 \leq f(x_h).$$

We have shown that, for all $x_h \in \Omega_h^i$, we have $\mathrm{MA}^{2S}_{h,\delta,\theta}[u_h^+](x_h) \leq f(x_h)$, so that u_h^+ is a supersolution. The discrete comparison principle of Proposition 10 then allows us to conclude that

$$u_h \leq u_h^+ = \mathcal{I}_h u - K_1 q_h - K_2 b_h \leq \mathcal{I}_h u + C_1 K_1 + C_2 \delta^2 K_2,$$

where we used the lower bounds on q_h and b_h. Recalling the choices of K_1 and K_2 allows us to conclude. □

Choosing relations between the discretization parameters h, δ, and θ we can obtain explicit rates of convergence.

Corollary 5 (rates of convergence).

In the setting of Theorem 9, if $\delta = C_1 h^{\frac{2}{2+\alpha}}$ and $\theta = C_2 h^{\frac{2}{2+\alpha}}$, we have

$$\|u - u_h^\varepsilon\|_{L^\infty(\Omega)} \leq C h^{\frac{2\alpha}{2+\alpha}}.$$

On the other hand, choosing $\delta = h^{2/3}$ and $\theta = h^{1/3}$, then we have

$$\|u - u_h^\varepsilon\|_{L^\infty(\Omega)} \leq C h^{\frac{2\alpha}{3}}.$$

In both estimates the hidden constant is independent of h.

Notice that both choices of relations between the coarse parameters and the mesh size h in Corollary 5 have its benefits and drawbacks. While the first choice yields a faster rate of convergence, it requires knowledge of the regularity of u. On the other hand, the second choice yields a slower convergence rate, but does not require a priori knowledge of the smoothness of u.

Remark 12 (error estimates under different assumptions).

The results of Theorem 9 have been extended in Li and Nochetto (2018a) in several directions:

1. *Smoother solutions*: If $u \in C^{3,\alpha}(\bar{\Omega})$*mutatis mutandis* the proof of Theorem 9 it follows a rate of convergence. The discretization parameters can be related to each other in such a way that the error is $\mathcal{O}(h)$, and numerical experiments indicate that this is sharp.
2. *Estimates for solutions with Sobolev regularity*: Assuming that $u \in W^{s,p}(\Omega)$ with $s \leq 3$ and $s - d/p > 2$, and that $D^2u(x) \geq \lambda I$, it has been shown (Li and Nochetto, 2018a, Theorem 5.7) that we have

$$\|u-u_h^\varepsilon\|_{L^\infty(\Omega)} \le C\left(\frac{h^2}{\delta^2}+\theta^2+\delta^2+\frac{\delta^{s-2}}{\lambda}\right),$$

where the constant depends on the smoothness of u. Once again, the discretization parameters can be optimized to obtain a rate $\mathcal{O}(h^{2-4/s})$.

■

2.8 Extensions, generalizations, and applications

We conclude the discussion on finite difference schemes and its variants by briefly describing some connections, extensions, generalizations, and applications of the schemes discussed here.

2.8.1 *Hamilton Jacobi Bellman formulation and semi-Lagrangian schemes*

Let

$$\Lambda = \left\{ \boldsymbol{\lambda} \in \mathbb{R}^d : \boldsymbol{\lambda}_i \ge 0, \quad i=1,\dots,d, \quad \sum_{i=1}^d \boldsymbol{\lambda}_i = 1 \right\}.$$

Define the function $h:\mathbb{S}^d \times \mathbb{R}_+ \to \mathbb{R}$ by

$$h(M,t) = \sup_{\substack{\{w_i\}_{i=1}^d \in \nu \\ \lambda \in \Lambda}} \left[-\frac{1}{d}\sum_{i=1}^d \boldsymbol{\lambda}_i \boldsymbol{w}_i \cdot M\boldsymbol{w}_i + t^{1/d}\left(\prod_{i=1}^d \boldsymbol{\lambda}_i\right)^{1/d} \right].$$

The following result is from Krylov (1987), see also Neilan et al. (2017, Proposition 6.13).

Proposition 13 (determinant).
For $M \in \mathbb{S}^d$ and $\delta \in \mathbb{R}_+$ we have that

$$h(M,\delta) = 0,$$

if and only if $M \in \mathbb{S}_+^d$ and $\det M = \delta$.

This motivates to define the function $F_{HJB} : \bar{\Omega} \times \mathbb{R} \times \mathbb{S}^d \to \mathbb{R}$ by

$$F_{HJB}(x,r,M) = \begin{cases} h(M,f(x)), & x \in \Omega, \\ g(x) - r, & x \in \partial\Omega, \end{cases}$$

and consider the problem: find $u \in C(\bar{\Omega})$ that is a viscosity solution of

$$F_{HJB}(x,u(x),D^2u(x)) = 0, \quad x \in \bar{\Omega}. \tag{42}$$

It turns out that this problem has an intimate connection with (1), as shown in Feng and Jensen (2017, Theorems 3.3 and 3.5).

Theorem 10 (equivalence).

Let $f \in C(\Omega)$ be nonnegative. The function $u \in C(\Omega) \cap B(\bar{\Omega})$ is a viscosity solution of (42)*, in the sense of Definition* 4*, if and only if it is a viscosity solution on the set of convex functions of* (1)*, in the sense of Definition* 6.

It is remarkable that the convexity assumption on the solution is not enforced in (42), it is rather a consequence of the formulation. This motivated Feng and Jensen (2017) to use (42) for numerical purposes. They proposed a so-called semi-Lagrangian scheme which we now describe. Over a triangulation $\mathcal{T}_h$ we introduce X_h as the space of piecewise linear and continuous functions. On the basis of (42) we introduce over X_h the operator

$$\mathrm{MA}_{h,k}^{\mathrm{SL}}[w_h](x_h) = \sup_{\substack{\{w_i\}_{i=1}^d \in \mathcal{V} \\ \lambda \in \Lambda}} \left[-\frac{1}{d}\sum_{i=1}^d \lambda_i \nabla_{kw_i}^2 w_h(x_h) + f(x_h)^{1/d} \left(\prod_{i=1}^d \lambda_i \right)^{1/d} \right],$$

where $x_h \in \Omega_h^i$ and $k > 0$ is a discretization parameter. The semi-Lagrangian scheme then seeks for $u_h \in X_h$ such that

$$\mathrm{MA}_{h,k}^{\mathrm{SL}}[u_h](x_h) = 0, \quad \forall x_h \in \Omega_h^i, \tag{43a}$$

$$u_h(x_h) = g(x_h), \quad \forall x_h \in \Omega_h^b. \tag{43b}$$

Feng and Jensen (2017) showed existence and uniqueness of solutions to (43) as well as, provided $(h, k) \to 0$ with $\frac{h}{k} \to 0$, convergence to the viscosity solution of (42) and, as a consequence of Theorem 10, to the viscosity solution of (1) over the set of convex functions. Rates of convergence, however, were not provided.

Although rates of convergence for general semi-Lagrangian schemes were given in Debrabant and Jakobsen (2013, Corollary 7.3) let us here explore a connection between the solutions of the scheme (43) and the two scale method of Section 2.7 as described in Li and Nochetto (2018a, Section 6). For that one needs to notice, first, that the scheme given in (43) is not fully practical. This is because in the operator $\mathrm{MA}_{h,k}^{\mathrm{SL}}[\cdot]$ the supremum runs over all of $\mathcal{V}$. We need to introduce a directional discretization by, as before, using $\mathcal{V}_\theta$ whose elements satisfy (37). With this we introduce the new operator

$$\mathrm{MA}_{h,k,\theta}^{\mathrm{SL}}[w_h](x_h) = \sup_{\substack{\{w_i\}_{i=1}^d \in \mathcal{V}_\theta \\ \lambda \in \Lambda}} \left[-\frac{1}{d}\sum_{i=1}^d \lambda_i \nabla_{kw_i}^2 w_h(x_h) + f(x_h)^{1/d} \left(\prod_{i=1}^d \lambda_i \right)^{1/d} \right],$$

and denote by $u_h^{(k,\theta)} \in X_h$ the solution to (43) but with this new operator. The following is a rather surprising fact. For a proof see Li and Nochetto (2018a, Proposition 6.2).

Proposition 14 (equivalence).

Let $u_h^\varepsilon \in X_h$ *denote the solution to the two scale method* (39) *and* $u_h^{(k,\theta)} \in X_h$ *the solution to the modified semi-Lagrangian scheme with the operator* $\mathrm{MA}^{\mathrm{SL}}_{h,k,\theta}[\cdot]$. *In this case, we have* $u_h^\varepsilon = u_h^{(k,\theta)}$.

From Proposition 14 and the results of Section 2.7.5, rates of convergence for (43) can be deduced.

Remark 13 (nonconvex domains).

Notice that convexity of the solution is not a constraint in (42) but rather a consequence of it. This has motivated (Jensen, 2018) to explore the possibility of using (42) as an extension of the Monge–Ampère equation to nonconvex domains, or cases with nonconvex data. ■

2.8.2 *Filtered two scale schemes*

In Nochetto and Ntogkas (2018) the ideas of two scale methods of Section 2.7 and those of filtered schemes of Section 2.4 were extended to construct a *filtered two scale scheme*. Let $\mathcal{T}^2_{2h}$ be a quasiuniform triangulation of $\bar{\Omega}$ of size $2h > 0$. The superscript in this triangulation indicates that we are doing a quadratic approximation of the boundary. This can be accomplished, for instance, by the use of isoparametric approximation of the boundary; see Brenner and Scott (2008, Section 10.4) and Ciarlet (2002, Section 4.3). Over this mesh we construct X^2_{2h}, the space of piecewise quadratic and continuous functions. For $w_{2h} \in X^2_{2h}$ and $x_{2h} \in \Omega^i_{2h}$ we define

$$\mathrm{MA}^{2\mathrm{Sq}}_{2h,\tilde{\delta},\tilde{\theta}}[w_{2h}](x_{2h}) = \min_{\{\boldsymbol{w}_i\}_{i=1}^d \in \mathcal{V}_{\tilde{\theta}}} \Bigg[\prod_{i=1}^d \left(\tilde{\nabla}^2_{\tilde{\delta}} \boldsymbol{w}_i \; w_{2h}(x_{2h}) \right)^+ - \sum_{i=1}^d \left(\tilde{\nabla}^2_{\tilde{\delta}} \boldsymbol{w}_i \; w_{2h}(x_{2h}) \right)^- \Bigg], \tag{44}$$

where Ω^i_{2h} denotes the set of internal degrees of freedom of X^2_{2h}, which includes now the vertices and edge midpoints of $\mathcal{T}^2_{2h}$, and $\tilde{\nabla}^2_{\tilde{\delta}\boldsymbol{w}}$ is a more accurate, say using five points, discretization of the second derivative in direction $\boldsymbol{w}$ at scale $\tilde{\delta}$.

Following the ideas presented in Theorem 7 we can show that operator (44) is consistent with order $\mathcal{O}(\tilde{\delta}^{k+\alpha} + \frac{h^3}{\tilde{\delta}^2} + \tilde{\theta}^2)$; see (Nochetto and Ntogkas, 2018, Lemma 5.8). However, this scheme is *not* monotone. It will, instead serve as the two scale analogue of the accurate scheme (22).

By refining in a conforming way once $\mathcal{T}_{2h}^2$ we obtain $\mathcal{T}_h$, over which we can apply the two scale scheme of Section 2.7. Notice that there is a bijection between Ω_{2h}^i and Ω_h^i so that the elements of X_{2h}^2 and X_h can be compared by looking at their nodal values. In light of this observation we alleviate the notation and carry out the rest of the discussion using the scale h.

We combine (44) and (38) into a *filtered* two scale operator: for $w_h \in X_h$ and $x_h \in \Omega_h^i$

$$\mathrm{MA}_{h,\delta,\theta,\tilde{\delta},\tilde{\theta}}^{\mathrm{F}}[w_h](x_h) = \mathrm{MA}_{h,\delta,\theta}^{2\mathrm{S}}[w_h](x_h) + \tau\tilde{S}\left(\frac{\mathrm{MA}_{2h,\tilde{\delta},\tilde{\theta}}^{2\mathrm{Sq}}[w_h](x_h) - \mathrm{MA}_{h,\delta,\theta}^{2\mathrm{S}}[w_h](x_h)}{\tau}\right),$$

where $\tilde{S}(t) = \min\{S(t), 0\}$ and the function S is defined in (24). As explained in Nochetto and Ntogkas (2018, Section 2) the choice of filter function ensures discrete convexity in the case that the right-hand side degenerates, that is if $f(x_h) = 0$, for some $x_h \in \Omega_h^i$.

With these ingredients the filtered two scale scheme seeks for $u_h^F \in X_h$ such that

$$\mathrm{MA}_{h,\delta,\theta,\tilde{\delta},\tilde{\theta}}^{\mathrm{F}}[u_h^F](x_h) = f(x_h), \quad \forall x_h \in \Omega_h^i, \tag{45a}$$

$$u_h^F(x_h) = g(x_h), \quad \forall x_h \in \Omega_h^b. \tag{45b}$$

The theory of almost monotone schemes of Corollary 1 was combined with the convergence results of Section 2.7.4 in Nochetto and Ntogkas (2018, Section 6) to assert the convergence of any solution to (45).

2.8.3 Approximation of convex envelopes

Let us describe the results obtained in Li and Nochetto (2018b) regarding the approximation of the convex envelope of a function, which was introduced in Definition 10. Let $f \in C(\bar{\Omega})$. As shown in Oberman and Ruan (2017) the convex envelope $u = \Gamma f$ of f can be characterized as the viscosity solution of the problem

$$\mathrm{CE}[u](x) = 0, \quad x \in \Omega, \tag{46a}$$

$$u(x) = f(x), \quad x \in \partial\Omega, \tag{46b}$$

where the operator $\mathrm{CE}[\cdot]$ is given by

$$\mathrm{CE}[w](x) = \min\{f(x) - u(x), \min\sigma(D^2 w(x))\}. \tag{47}$$

The intuition behind (46) is clear. First, we have that $u(x) \le f(x)$ for every $x \in \bar{\Omega}$. In addition, if we define the *contact set*

$$\mathcal{C}(f) = \{x \in \bar{\Omega} : u(x) = f(x)\},$$

we obtain, upon denoting $\lambda_1(w) = \min \sigma(D^2 w(x))$, that for $x \in \mathcal{C}(f)$ we must have $\lambda_1(u) \geq 0$. On the other hand, if $x \notin \mathcal{C}(f)$, then we must have $\lambda_1(u) = 0$. In conclusion, u must be convex.

We remark, however, that problem (46) is very degenerate. Indeed, it can be shown, see for instance Li and Nochetto (2018b, Lemma 3.1), that if $\text{dist}(x, \mathcal{C}(f)) \geq d\delta$ and $\boldsymbol{p} \in \partial u(x)$ there is $\boldsymbol{v} \in \mathbb{R}^d$ with $|\boldsymbol{v}| = 1$ such that

$$x_{\pm} = x \pm \delta \boldsymbol{v}, \quad u(x_{\pm}) = u(x) + \delta \boldsymbol{p} \cdot \boldsymbol{v}, \quad \nabla^2_{\delta \boldsymbol{v}} u(x) = 0, \quad \boldsymbol{p} \in \partial u(x_{\pm}).$$

In other words, if we are sufficiently far away from the contact set $\mathcal{C}(f)$, then the graph of u is flat in at least one direction. As a consequence, in general, the convex envelope cannot be arbitrarily smooth, regardless of the smoothness of the domain Ω and data f. Indeed, De Philippis and Figalli (2015) shows that if Ω is strictly convex with $\partial\Omega \in C^{3,1}$, and $f \in C^{3,1}(\bar{\Omega})$, then $u \in C^{1,1}(\bar{\Omega})$, and that this is optimal. This very low regularity is one of the main obstacles in the analysis of numerical schemes for (46).

Formulation (46) was already used for numerical purposes in Oberman (2008a) via wide stencil schemes like those presented in Section 2.3. Let us present here, instead, the two scale methods of Li and Nochetto (2018b). We will follow the notation of Section 2.7. In addition, if $\mathcal{S}$ denotes the unit ball in $\mathbb{R}^d$ we introduce, in full analogy to (37), a discretization $\mathcal{S}_\theta$ of $\mathcal{S}$ such that, for every $\boldsymbol{w} \in \mathcal{S}$ there is $\boldsymbol{w}^\theta \in \mathcal{S}_\theta$ that satisfies

$$|\boldsymbol{w} - \boldsymbol{w}^\theta| \leq \theta.$$

Over the space of piecewise linear functions X_h subordinated to the triangulation $\mathcal{T}_h$ we define

$$\text{CE}_{h,\delta,\theta}[w_h](x_h) = \min \left\{ f(x_h) - w_h(x_h), \min_{\boldsymbol{w} \in \mathcal{S}_\theta} \nabla^2_{\delta \boldsymbol{w}} w_h(x_h) \right\} \tag{48}$$

where $w_h \in X_h$ and $x_h \in \Omega_h^i$. With the aid of this operator we define the discrete convex envelope of a function f as the function $u_h^\varepsilon \in X_h$ that solves

$$\text{CE}_{h,\delta,\theta}[u_h^\varepsilon](x_h) = 0, \quad x_h \in \Omega_h^i, \tag{49a}$$

$$u_h^\varepsilon(x_h) = f(x_h), \quad x_h \in \Omega_h^b. \tag{49b}$$

The analysis of scheme (49) to a large extent follows that of two scale methods presented in Section 2.7. Namely, owing to discrete convexity we can show that the scheme has a comparison principle, from which uniqueness of solutions follows. The existence of solutions is obtained via a discrete Perron method, and the stability by noticing that $u_h^- = \mathcal{I}_h u$ and $u_h^+ = \mathcal{I}_h f$ are discrete sub- and supersolutions, respectively.

The considerations given above show that scheme (49) is monotone and stable. In addition, assuming smoothness of the arguments, one can show its consistency with similar arguments to those of Section 2.7.3. Upon realizing that the operator (47) has a comparison principle in the sense of Definition 5, this is enough to appeal to the theory of Section 2.1 and conclude that the scheme is convergent as $\varepsilon = (h, \delta, \theta) \to 0$, provided $\frac{h}{\delta} \to 0$.

The derivation of rates of convergence, however, requires special attention. This is due to the fact that, as stated above, the best regularity we can expect is $u \in C^{1,1}(\bar{\Omega})$, and this is not enough to exploit the consistency estimates that were used for convergence (which are applied to smooth test functions). To overcome this, one must take advantage of the flatness of the solution outside the contact set. To describe these results we must introduce some notation. Set, for $x_h \in \Omega_h^i$,

$$\delta_{x_h} = \min\{\delta, \operatorname{dist}(x_h, \partial\Omega)\}, \quad B_{x_h} = \bigcup_{T \in \mathcal{T}_h : \operatorname{dist}(x_h, T) < \delta_{x_h}} T,$$

and

$$W_{x_h} = \{x \in \bar{\Omega} : |x - x_h| \leq d\delta\}.$$

The following is Li and Nochetto (2018b, Proposition 3.3).

Proposition 15 (consistency).

Let Ω be strictly convex and u, the solution of (46) *satisfy $u \in C^{k,\alpha}(\bar{\Omega})$ with $k = 0, 1$ and $\alpha \in (0,1]$. For $x_h \in \Omega_h^i$ we have:*

1. *If $dist(x_h, \mathcal{C}(f)) \geq d\delta$, then*

$$\min_{w \in \mathcal{S}_\theta} \nabla^2_{\delta w} \mathcal{I}_h u(x_h) \leq C\left(\frac{(\delta\theta)^{k+\alpha} + h^{k+\alpha}}{\delta^2}\right)|u|_{C^{k,\alpha}(B_{x_h})}.$$

2. *If $dist(x_h, \mathcal{C}(f)) < d\delta$ but $dist(x_h, \partial\Omega) \geq d\delta$, then we have*

$$f(x_h) - u(x_h) \leq C_k \delta^{k+\alpha},$$

   where C_k depends on $|u|_{C^{0,\alpha}(W_{x_h})} + |f|_{C^{0,\alpha}(W_{x_h})}$ for $k = 0$ and on $|f|_{C^{1,\alpha}(W_{x_h})}$ for $k = 1$.

3. *If $0 < dist(x_h, \partial\Omega) < d\delta$, then for all $\mathbf{w} \in \mathcal{S}$ we have*

$$\nabla^2_{\delta w} \mathcal{I}_h u(x_h) \leq C\delta_i^{k+\alpha-2}|u|_{C^{k,\alpha}(B_{x_h})},$$

and the previous item also holds provided k = 0.

To take advantage of this result two new discrete barriers were constructed. One handles the first case, i.e., points sufficiently far away from the contact set. The other barrier handles points near the boundary, that is the third case in the previous result. Without going into details, we present here the main error estimate, and refer the reader to Li and Nochetto (2018b, Theorem 3.7).

Theorem 11 (convergence rate).
Let Ω be strictly convex, u be the viscosity solution of (46), *and u_h^ε the solution of* (49). *If $u \in C^{k,\alpha}(\bar{\Omega})$, with $k = 0, 1$ and $\alpha \in (0, 1]$, then*

$$\|u - u_h^\varepsilon\|_{L^\infty(\Omega)} = \mathcal{O}\left(h^{\frac{(k+\alpha)^2}{k+\alpha+2}}\right),$$

provided, $\delta = \mathcal{O}\left(h^{\frac{k+\alpha}{k+\alpha+2}}\right)$ *and* $\theta = \mathcal{O}\left(h^{\frac{2}{k+\alpha+2}}\right)$. *In particular, if $k = \alpha = 1$, i.e., $u \in C^{1,1}(\bar{\Omega})$, we obtain*

$$\delta = \mathcal{O}(h^{1/2}),\quad \theta = \mathcal{O}(h^{1/2}) \quad \Longrightarrow \quad \|u - u_h^\varepsilon\|_{L^\infty(\Omega)} = \mathcal{O}(h).$$

Similarly, if $k = 0$ and $\alpha = 1$, i.e., $u \in C^{0,1}(\bar{\Omega})$, we get

$$\delta = \mathcal{O}(h^{1/3}),\quad \theta = \mathcal{O}(h^{2/3}) \quad \Longrightarrow \quad \|u - u_h^\varepsilon\|_{L^\infty(\Omega)} = \mathcal{O}(h^{1/3}).$$

2.8.4 The Gauss curvature problem

As an application of the wide stencil finite difference schemes that were presented in Section 2.3 let us here, following Hamfeldt (2018), describe a discretization of the prescribed Gaussian curvature problem (2). To do so, we must begin by defining what is a solution of this problem. In a similar manner to the notion of Alexandrov solutions to Monge–Ampère problem, introduced in Definition 9, we have

Definition 17 (generalized solution).
A convex function $u : \bar{\Omega} \to \mathbb{R}$ is a generalized solution to (2) if the following two conditions hold:

1. It is a generalized solution of (2a). This means that, for all Borel sets $D \subset \Omega$, we have

$$\int_{\partial u(D)} \frac{1}{\left(1 + |\boldsymbol{p}|^2\right)^{(d+2)/2}} \mathrm{d}\boldsymbol{p} = \int_D \mathcal{K}(x) \mathrm{d}x.$$

2. It satisfies

$$\limsup_{y \to x} u(y) \le g(x), \quad \forall x \in \partial\Omega$$

and, if v is any other generalized solution of (2a), then $v \le u$ in Ω.

Under the assumptions of uniform convexity of Ω; continuity of g; continuity, boundedness, and nonnegativity of $\mathcal{K}$; and the compatibility condition

$$\int_{\mathbb{R}^d} \frac{1}{\left(1 + |\boldsymbol{p}|^2\right)^{(d+2)/2}} \mathrm{d}\boldsymbol{p} > \int_\Omega \mathcal{K}(x) \mathrm{d}x;$$

it can be shown that problem (2) has a unique generalized solution; see Bakelman (1994).

It is also possible to extend the notion of viscosity solution presented in Definition 4, by allowing the operators in Definition 2 to also depend on a variable $\boldsymbol{p} \in \mathbb{R}^d$. In doing that, we note that the operator $F_{G\mathcal{K},c} : \bar{\Omega} \times \mathbb{R} \times \mathbb{R}^d \times \mathbb{S}^d \to \mathbb{R}$ defined by

$$F_{G\mathcal{K},c}(x,r,\boldsymbol{p},M) = \begin{cases} \det M - \mathcal{K}(x)\left(1 + |\boldsymbol{p}|^2\right)^{(d+2)/2}, & x \in \Omega, \\ g(x) - r, & x \in \partial\Omega, \end{cases}$$

is, as the Monge–Ampère operator F_{MA} defined in (8), only elliptic when $M \in \mathbb{S}^d_+$, which implies that to have a reasonable notion of viscosity solution, we must require sub- and supersolutions to be convex, and restrict the test functions to be convex, as in Definition 6. As we have seen throughout our discussion, the convexity constraint is rather difficult to impose explicitly during discretization.

Hamfeldt (2018) proposed to consider the following formulation of (2). If for $M \in \mathbb{S}^d$ we set $\sigma(M) = \{\lambda_1(M), \ldots, \lambda_d(M)\}$, where the eigenvalues are counted with multiplicity and arranged in nondecreasing order, then the operator

$$F_{G\mathcal{K}}(x,r,\boldsymbol{p},M) = \begin{cases} F^{\text{in}}_{G\mathcal{K}}(x,\boldsymbol{p},M), & x \in \Omega, \\ g(x) - r, & x \in \partial\Omega, \end{cases}$$

with

$$F^{\text{in}}_{G\mathcal{K}}(x,\boldsymbol{p},M) = \min\left\{\lambda_1(M),\ \prod_{i=1}^{d}\lambda_i(M)^+ - \mathcal{K}(x)\left(1 + |\boldsymbol{p}|^2\right)^{(d+2)/2}\right\}$$

is elliptic in the sense of Definition 2 and, at least formally, it is clear that if

$$F_{G\mathcal{K}}(x,u(x),\nabla u(x),D^2u(x)) = 0, \quad x \in \Omega,$$

then we must have, that either, $\lambda_1(D^2u(x)) > 0$, so that u is convex, and

$$\det D^2u(x) = \mathcal{K}(x)\left(1 + |\nabla u(x)|^2\right)^{(d+2)/2},$$

or $\lambda_1(D^2u(x)) = 0$ and, thus

$$0 = \det D^2u(x) \geq \mathcal{K}(x)\left(1 + |\nabla u(x)|^2\right)^{(d+2)/2} \geq 0.$$

In either case, the convexity of the solution is recovered.

With these constructions we have two options to define viscosity solutions to (2). The first one is, like in Definition 6, to require that it is a viscosity solution, in the set of convex functions, of the problem

$$F_{G\mathcal{K},c}(x,u(x),\nabla u(x),D^2u(x))=0,\quad \forall x\in\bar{\Omega}. \tag{50}$$

The second, as in Definition 4, to require that it is a viscosity solution of

$$F_{G\mathcal{K}}(x,u(x),\nabla u(x),D^2u(x))=0,\quad \forall x\in\bar{\Omega}. \tag{51}$$

In full analogy to Proposition 3 it is shown in Hamfeldt (2018, Section 3) that viscosity subsolutions to problem (51) are convex and that a function is a viscosity solution of (50) over the set of convex function if and only if it is a viscosity solution of (51). In addition it is shown that, under certain assumptions on $\mathcal{K}$, this notion of solution, at least in the interior of the domain Ω, coincides with that of Definition 17.

It is important to note that incorporating the boundary conditions into the definition of the operator is *essential* in this problem, as they may not be realized in a classical sense. The following is Hamfeldt (2018, Example 1).

Example 2 (nonclassical boundary conditions).

Let $d=1$, $\Omega=(0,1)$, and $\mathcal{K}\equiv 1$. We set the boundary conditions $u(0)=-1$ and $u(1)=1$. Then it is possible to show that

$$u(x)=-\sqrt{1-x^2}$$

is a viscosity solution of (51). It is a classical solution over [0, 1) so it remains to understand what happens at $x=1$.

Note that $u'(x)$ grows unboundedly as $x\uparrow 1$ so that it is not possible to find a smooth φ such that $u_\star-\varphi$ has a local minimum at x_0, in other words, the graph of u cannot be touched from below at $x=1$. This makes u automatically a supersolution.

To show that u is also a subsolution we note that $u(1)=0<1$ so that, if φ touches the graph of u from above at $x=1$, we must have $\varphi(1)=u(1)=0$, and

$$(F_{G\mathcal{K}})_\star(1,u(1),\varphi'(1),\varphi''(1))\geq 1-u(1)=1>0.$$

The behaviour of Example 2 was characterized in Hamfeldt (2018, Corollary 24). Namely, if u is a viscosity solution of (50) then at every $x\in\partial\Omega$ we either have that $u_\star(x)=u^\star(x)=g(x)$, or $u_\star(x)\leq u^\star(x)\leq g(x)$ and $\partial u_\star(x)=\varnothing$. The second option here corresponds to the right endpoint in Example 2.

Existence of solutions to (51) was shown using a variant of Perron's method. The usual argument to show uniqueness is obtained via a comparison principle of Definition 5. This problem, however, does not have a comparison principle, as Hamfeldt (2018, Example 3) shows.

Example 3 (lack of comparison).

In the setting of Example 2 we have that $u(x)=-\sqrt{1-x^2}$ is a viscosity solution, so that necessarily it is a supersolution. Let

$$v(x) = \begin{cases} u(x), & x \in [0,1), \\ 1, & x = 1, \end{cases}$$

we see that $v \in USC([0, 1])$ and, as in Example 2, if φ touches from above the graph of u at $x = 1$, then $\varphi(1) = v(1) = 1$ and

$$(F_{GK})_\star(1, v(1), \varphi'(1), \varphi''(1)) \geq 1 - v(1) = 0,$$

showing that v is a subsolution. Note, however, that $u(1) \leq v(1)$ and this problem does not have a comparison principle.

The previous result, combined with the behaviour of solutions at the boundary shows that, in fact, a comparison principle takes place, but only in the interior of the domain; see Hamfeldt (2018, Theorem 7).

Theorem 12 (interior comparison).

Let $\underline{u} \in USC(\bar{\Omega})$ be a subsolution of (51) and $\overline{u} \in LSC(\bar{\Omega})$ a supersolution. Then we have $\underline{u} \leq \overline{u}$ in Ω.

This weakened comparison principle is sufficient to guarantee uniqueness.

Having shown the existence and uniqueness of solutions to (51), it is possible now to construct numerical schemes. This is carried using variants of the wide stencil finite difference schemes of Section 2.3. With the notation introduced there we define, for $w_h \in X_h$,

$$\mathrm{GK}_{h,\theta}[w_h](x_h) = \min\left\{ \min_{\{\nu_i\}_{i=1}^d \in \mathcal{G}_\theta} \Delta_{\nu_i} w_h(x_h),\ \mathrm{MA}_{h,\theta}^{\mathrm{WS}}[w_h](x_h) - \mathcal{K}(x_h)\left(1 + |\nabla_h w_h(x_h)|^2\right)^{(d+2)/2} \right\},$$

where $\mathrm{MA}_{h,\theta}^{\mathrm{WS}}[\cdot]$ was defined in (19) and the vector $\nabla_h w_h(x_h)$ is such that

$$\nabla_h w_h(x_h) \cdot \boldsymbol{e}_i = \max\left\{ \frac{w_h(x_h) - w_h(x_h - h\boldsymbol{e}_i)}{h}, \frac{w_h(x_h) - w_h(x_h + h\boldsymbol{e}_i)}{h}, 0 \right\}, \tag{52}$$

and $\{\boldsymbol{e}_i\}_{i=1}^d$ is the canonical basis of $\mathbb{R}^d$. With this operator, the finite difference approximation of (51) is to find $u_h \in X_h$ such that

$$\mathrm{GK}_{h,\theta}[u_h](x_h) = 0, \quad x_h \in \Omega_h^i, \tag{53a}$$

$$u(x_h) = g(x_h), \quad x_h \in \Omega_h^b. \tag{53b}$$

In Hamfeldt (2018, Section 6) it is shown that scheme (53) is monotone, in the sense of (12), stable, in the sense of (13), and consistent, in the sense of (14).

Notice, however, that as Example 3 shows, problem (51) does not have a comparison principle. As a consequence, Theorem 4 cannot be applied. For this reason, the framework of Section 2.1 was extended in Hamfeldt (2018, Theorem 9) to cases where problem (7) only has an interior comparison principle like that of Theorem 12 and there exist classical sub- and supersolutions. The conclusion is the locally uniform convergence of u_h to u.

2.8.5 Transport boundary conditions

Let us conclude the discussion of wide stencil finite difference schemes by describing how these methods can be used to tackle the optimal transportation problem. Since this will be one of the main topics of chapter "Optimal transport" by Merigot in this volume, we shall be brief.

We recall that, given $\Omega, \mathcal{O} \subset \mathbb{R}^d$, which we assume bounded, with $\mathcal{O}$ convex, and measures $\rho_\Omega : \Omega \to \mathbb{R}$ and $\rho_\mathcal{O} : \mathcal{O} \to \mathbb{R}$, the optimal transportation problem (with quadratic cost) seeks for a map $T : \Omega \to \mathcal{O}$ with $T_\sharp \rho_\Omega = \rho_\mathcal{O}$ that minimizes

$$\frac{1}{2}\int_\Omega |x - T(x)|^2 \mathrm{d}\rho_\Omega(x).$$

We recall that $T_\sharp \mu$ denotes the pushforward of the measure μ under the mapping T. Assuming that the measures are absolutely continuous with respect to Lebesgue measure, with densities $f_\Omega, f_\mathcal{O}$, this condition can be written as

$$\int_E f_\mathcal{O}(x)\mathrm{d}x = \int_{T^{-1}(E)} f_\Omega(x)\mathrm{d}x,$$

and so by a change of variables, $\det(\nabla T(x)) f_\mathcal{O}(T(x)) = f_\Omega(x)$. Finally, we recall that since the cost is quadratic, it can be shown that T is given by the gradient map of a convex potential $u : \Omega \to \mathbb{R}$. This allows us to, at least at the formal level, rewrite the optimal transportation problem as a Monge–Ampère problem: find $u : \bar{\Omega} \to \mathbb{R}$ convex, such that

$$\det D^2 u(x) = F(x, \nabla u(x)), \quad x \in \Omega. \tag{54}$$

where we set $F(x, \boldsymbol{p}) = \rho_\Omega(x)/\rho_\mathcal{O}(\boldsymbol{p})$. This problem is supplemented by the so-called *transport* or *second* boundary condition

$$\nabla u(\bar{\Omega}) = \bar{\mathcal{O}}.$$

Notice that this, more than a boundary condition, is a set of constraints. It can be shown also that this condition can be replaced by

$$\nabla u(\partial\Omega) = \partial\mathcal{O}. \tag{55}$$

Thus, we want to construct numerical schemes to approximate the solution of (54) and (55).

It is clear that the main issue is the discretization of the boundary condition (55). If the boundary of the domain $\mathcal{O}$ is given as the zero level set of some function $\Phi : \mathbb{R}^d \to \mathbb{R}$, then it is clear that (55) can be equivalently written as

$$\Phi(\nabla u(x)) = 0, \quad \forall x \in \partial\Omega.$$

While we would be tempted to discretize this condition directly, the function Φ can be highly nonlinear and nonsmooth, which will make the design of monotone and consistent numerical schemes a daunting task. However, this can be achieved very easily if the domains are rectangles, say $\Omega = (0,1)^2 = \mathcal{O}$. In this case, it is shown in Froese (2012, Section 3.2) that each side must be mapped to itself. If we consider the left side of the square, that is,

$$\{(x_1,x_2) \in \mathbb{R}^2 : x_1 = 0, x_2 \in [0,1]\},$$

then the function that describes this is given by $\Phi(y_1, y_2) = y_1$. Thus, on this side we can write

$$\frac{\partial u(0,x_2)}{\partial x_1} = 0.$$

Similarly, in the right, bottom and top sides, respectively, we can write

$$\frac{\partial u(1,x_2)}{\partial x_1} = 1, \quad \frac{\partial u(x_1,0)}{\partial x_2} = 0, \quad \frac{\partial u(x_1,1)}{\partial x_2} = 1.$$

It is remarkable that on all sides the derivative that appears is actually the normal derivative. This motivated Froese (2012) to replace the boundary condition (55) by a Neumann-type boundary condition

$$\frac{\partial u(x)}{\partial \boldsymbol{n}} = \phi(x)$$

for some unknown function ϕ.

Obviously, the correct choice of function ϕ is $\phi(x) = \nabla u(x) \cdot \boldsymbol{n}(x)$, which motivates the introduction of the following iterative scheme: Given u_0, an initial guess, then

- For $k \geq 0$
  - Define, for $x \in \partial\Omega$,

$$\boldsymbol{p}_k(x) = \mathrm{Proj}_{\partial\mathcal{O}}(\nabla u_k(x)), \tag{56}$$

  where by $\mathrm{Proj}_S(\boldsymbol{w})$ we denoted a projection of the vector $\boldsymbol{w}$ onto the set S.

– Find $u_{k+1} : \bar{\Omega} \to \mathbb{R}$ convex, and $c_{k+1} \in \mathbb{R}$ that satisfy

$$\int_\Omega u_{k+1}(x)\mathrm{d}x = 0, \tag{57a}$$

$$\det D^2 u_{k+1}(x) = c_{k+1} F(x, \nabla u_{k+1}(x)), \quad x \in \Omega, \tag{57b}$$

$$\frac{\partial u_{k+1}(x)}{\partial \boldsymbol{n}} = \boldsymbol{p}_k(x), \quad x \in \partial\Omega. \tag{57c}$$

– `Set` $k \leftarrow k+1$

- `EndFor`

Remark 14 (iterative scheme).
The iterative scheme (56) and (57) deserves several observations.

1. The introduction of the projection $\boldsymbol{p}_k$ in step (56) is due to the fact that there is no reason to expect that $\nabla u_k(\partial\Omega) \subset \partial\mathcal{O}$. Thus, we settle for the closest point on the target boundary.
2. Problem (57) is a Neumann problem for an elliptic equation so that the solution, if it exists, is unique only up to a constant. Condition (57a) forces uniqueness, while the introduction of the number c_{k+1} in (57b) relaxes the equation so that the necessary conditions for existence are fulfilled.
3. The initialization of this scheme can done by choosing $\boldsymbol{p}_0 = Mx \cdot \boldsymbol{n}$, where $\boldsymbol{n}$ is the unit outer normal to $\partial\Omega$ and $M > 0$ is so large that the image of the mapping $\bar{\Omega} \ni x \mapsto Mx \in \mathbb{R}^d$ contains $\bar{\mathcal{O}}$. ■

We are then going to discretize (56) and (57). Notice that now the boundary conditions (57c) are rather standard and can be approximated by, for instance, introducing a layer of ghost nodes near the boundary and computing centred differences.

It remains to discretize (57b). Setting $v_h = u_{h,\,k+1}$, the first alternative, proposed in Froese (2012), is to use

$$\mathrm{MA}_{h,\theta}^{\mathrm{WS}}[v_h](x_h) = F(x_h, \nabla_h v_h), \quad x_h \in \Omega_h^i, \tag{58}$$

where $\mathrm{MA}_{h,\theta}^{\mathrm{WS}}[\cdot](x_h)$ is the operator defined in (19) or Remark 2, and $\nabla_h v_h$ is defined as in (52). Another option, also from Froese (2012), is to take advantage of the directional difference that are already being computed to approximate the Monge–Ampère operator. Notice that if $\{\boldsymbol{\nu}_i\}_{i=1}^d \in \mathcal{V}$, we have that

$$\nabla w = \left(\frac{\partial w}{\partial x_1}, \dots, \frac{\partial w}{\partial x_d}\right)^{\mathsf{T}} = \left(\sum_{i=1}^d \frac{\partial w}{\partial \boldsymbol{\nu}_i}\boldsymbol{\nu}_i \cdot \boldsymbol{e}_1, \dots, \sum_{i=1}^d \frac{\partial w}{\partial \boldsymbol{\nu}_i}\boldsymbol{\nu}_i \cdot \boldsymbol{e}_d\right)^{\mathsf{T}},$$

where, as usual, $\{\boldsymbol{e}_i\}_{i=1}^d$ is the canonical basis of $\mathbb{R}^d$. This allows us to write that

$$
\begin{aligned}
\det D^2 w(x) - F(x, \nabla w(x)) &= \mathrm{MA}[w](x) - F(x, \nabla w(x)) \\
&= \min_{\{w_i\}_{i=1}^d \in \mathcal{V}} \left[\prod_{i=1}^d \left(\frac{\partial^2 w(x)}{\partial w_i^2} \right)^+ - \sum_{i=1}^d \left(\frac{\partial^2 w(x)}{\partial w_i^2} \right)^- \right] \\
&= \min_{\{w_i\}_{i=1}^d \in \mathcal{V}} \left[\prod_{i=1}^d \left(\frac{\partial^2 w(x)}{\partial w_i^2} \right)^+ - \sum_{i=1}^d \left(\frac{\partial^2 w(x)}{\partial w_i^2} \right)^- - F(x, \nabla w(x)) \right] \\
&= \min_{\{w_i\}_{i=1}^d \in \mathcal{V}} \left[\prod_{i=1}^d \left(\frac{\partial^2 w(x)}{\partial w_i^2} \right)^+ - \sum_{i=1}^d \left(\frac{\partial^2 w(x)}{\partial w_i^2} \right)^- \right. \\
&\qquad \left. - F\left(x, \left(\sum_{i=1}^d \frac{\partial w(x)}{\partial \nu_i} \frac{\nu_i \cdot e_1}{|\nu_i|}, \ldots, \sum_{i=1}^d \frac{\partial w(x)}{\partial \nu_i} \frac{\nu_i \cdot e_d}{|\nu_i|} \right)^{\mathsf{T}} \right) \right] \\
&= \min_{\{w_i\}_{i=1}^d \in \mathcal{V}} \mathrm{OT}_{\{w_i\}_{i=1}^d}[w](x).
\end{aligned}
$$

In conclusion, an approximation of (57b) is obtained by setting

$$
\mathrm{OT}_{h,\theta}[v_h](x_h) = 0, \quad \forall x_h \in \Omega_h^i,
$$

where

$$
\mathrm{OT}_{h,\theta}[w_h](x_h) = \min_{\{\nu_i\}_{i=1}^d \in \mathcal{G}\theta} \mathrm{OT}_{\{\nu_i\}_{i=1}^d}[w_h](x_h).
$$

Benamou et al. (2014) considered a different treatment of the boundary condition (55). Since for all $x \in \partial\Omega$ we must have that $\nabla u(x) \in \partial\mathcal{O}$, then we must have

$$
H(\nabla u(x)) = 0, \quad H(y) = \begin{cases} \mathrm{dist}(y, \partial\mathcal{O}), & y \in \mathcal{O}, \\ -\mathrm{dist}(y, \partial\mathcal{O}), & y \notin \mathcal{O}, \end{cases} \tag{59}
$$

where H is nothing but the signed distance function to $\partial\mathcal{O}$. Notice that (59) is a sort of Hamilton Jacobi equation posed on $\partial\Omega$. Exploiting the convexity of $\mathcal{O}$, the authors of Benamou et al. (2014) were able to rewrite the function H as the supremum over linear expressions on y (the supporting hyperplanes of $\mathcal{O}$ at y)

$$
H(y) = \sup_{\boldsymbol{n} \in \mathbb{R}^d : |\boldsymbol{n}| = 1} \{ y \cdot \boldsymbol{n} - H^\star(\boldsymbol{n}) : \boldsymbol{n} \cdot \boldsymbol{n}_x > 0 \},
$$

where $\boldsymbol{n}_x$ is the normal to $\partial\Omega$ at x and $H^\star$ is the support function of $\mathcal{O}$, that is,

$$
H^\star(\boldsymbol{n}) = \sup_{z \in \partial\mathcal{O}} z \cdot \boldsymbol{n}.
$$

This function can be precomputed or evaluated rather cheaply in the discrete setting. The reformulation of the function H can be approximated by replacing the supremum by one over a finite set of directions, and, finally, the gradient appearing in (59) can be discretized as in (52). This gives a discretization of (55). Finally, the discretization of (54) is proposed to be carried similarly to (58).

3 Discretizations based on geometric considerations

In fact, geometrical representations, graphs and diagrams of all sorts, are used in all sciences, not only in physics, chemistry, and the natural sciences, but also in economics, and even in psychology. Using some suitable geometrical representation, we try to express everything in the language of figures, to reduce all sorts of problems to problems of geometry.

Pólya (2014)

In this section we will describe the so-called Oliker–Prussner method, which is a discrete analogue of the notion of solution in the Alexandrov sense. We recall that Alexandrov solutions to the Monge–Ampère equation were introduced in Definition 9. They make a connection between the Monge–Ampère equation and the measure of the subdifferential of its solution. This, very geometric, notion enables us to define solutions that are not smooth, say not $C^2(\Omega)$. The Oliker–Prussner method, in turn, will allow us to approximate these solutions.

3.1 Description of the scheme

To be able to present the Oliker–Prussner method, we must begin by introducing some useful notions.

3.1.1 Nodal set and domain partition

To discretize the domain Ω and its boundary $\partial\Omega$, we introduce a translation invariant nodal set and an open, disjoint partition of the domain. For a parameter $h > 0$, we define the interior nodal set as

$$\Omega_h = \left\{ x_h = h\sum_{j=1}^{d} z^j \tilde{e}_j : z^j \in \mathbb{Z} \right\} \cap \Omega, \tag{60}$$

where $\{\tilde{e}_j\}_{j=1}^d$ is a basis of $\mathbb{R}^d$ with $|e_j| \leq 1$ for all $1 \leq j \leq d$. To discretize the boundary $\partial\Omega$, we set the boundary nodal set $\partial\Omega_h$ as a collection of points on the boundary and require that their spacing is of order h, namely, $\partial\Omega \subset \cup_{x_h \in \partial\Omega_h} B_{h/2}(x_h)$. We set the nodal set $\bar{\Omega}_h = \Omega_h \cup \partial\Omega_h$. We remark that this is a generalization of the finite difference discretizations introduced in Section 2.3. Indeed, in that case the vectors $\{\tilde{e}_j\}_{j=1}^d$ were the canonical basis of $\mathbb{R}^d$.

We define an open, disjoint partition $\{\omega_{x_h}\}_{x_h \in \bar{\Omega}_h}$ of the domain where, for $x_h \in \bar{\Omega}_h$,

$$\omega_{x_h} = \left\{ x_h + \sum_{j=1}^{d} h^j \tilde{e}_j : \; h^j \in \mathbb{R}, \; |h^j| \leq \frac{h}{2} \right\} \cap \Omega. \tag{61}$$

Note that, by construction, the partition is translation invariant, that is, $\omega_{y_h} = y_h - x_h + \omega_{x_h}$ for all x_h, $y_h \in \Omega_h$ with $\omega_{y_h}, \omega_{x_h} \subset \Omega$.

3.1.2 Nodal functions, their subdifferentials, and convex envelopes

On the nodal set Ω_h constructed above, we define a nodal function u_h to approximate the solution of the Monge–Ampère PDE. First, to mimic the convexity constraint for the PDE, we require the notion of convexity for nodal functions (compare to Definition 16).

Definition 18 (nodal convexity).

Let w_h be a (nodal) function that maps the set of nodes $\bar{\Omega}_h$ to $\mathbb{R}$. We say that w_h is *convex* if, for any node $x_h \in \bar{\Omega}_h$, there exist an affine function L, that is, $L(x) = \boldsymbol{p} \cdot (x - x_h) + c$ for some $\boldsymbol{p} \in \mathbb{R}^d$ and $c \in \mathbb{R}$, such that

$$L(y_h) \leq w_h(y_h) \quad \forall y_h \in \bar{\Omega}_h \text{ and } L(x_h) = w_h(x_h). \tag{62}$$

We define the subdifferential of a convex nodal function w_h at a fixed node $x_h \in \bar{\Omega}_h$ as the set

$$\partial w_h(x_h) := \{\boldsymbol{p} \in \mathbb{R}^d : \ \boldsymbol{p} \cdot (y_h - x_h) + w_h(x_h) \leq w_h(y_h) \quad \forall y_h \in \bar{\Omega}_h\}. \tag{63}$$

In other words, this is the collection of slopes of affine functions that satisfy the condition that defines convexity for a nodal function. Note that nodal functions are only defined on $\bar{\Omega}_h$. To extend a nodal function to the domain Ω, we introduce its convex envelope.

Definition 19 (convex envelope of a nodal function).

Let w_h be a nodal function defined on $\bar{\Omega}_h$. The *convex envelope* of w_h is the piecewise linear function

$$\Gamma(w_h)(x) = \sup_L \{L(x) : \ L \text{affine function and} \, L(x_h) \leq w_h(x_h) \quad \forall x_h \in \bar{\Omega}_h\}$$

for any $x \in \Omega$.

We note that, by definition, $\Gamma(w_h)(x_h) \leq w_h(x_h)$ for any node $x_h \in \bar{\Omega}_h$, and equality holds for all interior nodes if w_h is convex. Indeed, if w_h is convex, by (62), for any node $x_h \in \bar{\Omega}_h$, there exists an affine function $L(x)$ satisfying

$$L(y_h) \leq w_h(y_h) \quad \forall y_h \in \bar{\Omega}_h \text{ and } L(x_h) = w_h(x_h).$$

Since $L(x) \leq \Gamma(w_h)(x)$ for any $x \in \Omega$ by Definition 19, we deduce that $w_h(x_h) = L(x_h) \leq \Gamma(w_h)(x_h)$. Combining this inequality with the inequality in the other direction, we have $w_h(x_h) = \Gamma(w_h)(x_h)$ for all interior nodes. Thus, $\Gamma(w_h)$ is a natural extension to $\bar{\Omega}$ of the convex nodal function w_h. With an abuse of notation, we still use w_h to denote the convex envelope of this nodal function.

The convex envelope of a nodal function w_h induces a triangulation of the domain Ω and a piecewise linear function over this triangulation. However, this triangulation is not known a priori. Here we give an example to illustrate this property.

Example 4 (convex envelope and triangulation).
Define the nodal set $\bar{\Omega}_h = \{z_1, \ldots, z_5\}$ with $z_1 = (1, 0)$, $z_2 = (0, 1)$, $z_3 = (-1, 0)$, $z_4 = (0, -1)$, and $z_5 = (0, 0)$. Consider the nodal functions satisfying

$$w_1(z_1) = w_1(z_3) = 1, \quad w_2(z_2) = w_2(z_4) = 1,$$
$$w_3(z_1) = w_3(z_2) = w_3(z_3) = w_3(z_4) = 1,$$

and $w_j(z_i) = 0$ otherwise. The convex envelopes are $\Gamma(w_1) = |x_1|$, $\Gamma(w_2) = |x_2|$, and $\Gamma(w_3) = |x_1| + |x_2|$. The convex envelopes are subordinate to the meshes depicted in Fig. 4. ■

The above example shows that $\Gamma(w_h)$ is a piecewise linear function that induces a mesh $\mathcal{T}_h$ that depends on the values of w_h. The example depicted in Fig. 5 shows that, if w_h is the nodal interpolant of a function w, and if the Hessian $D^2 w$ is degenerate (or nearly degenerate), the induced mesh may be anisotropic.

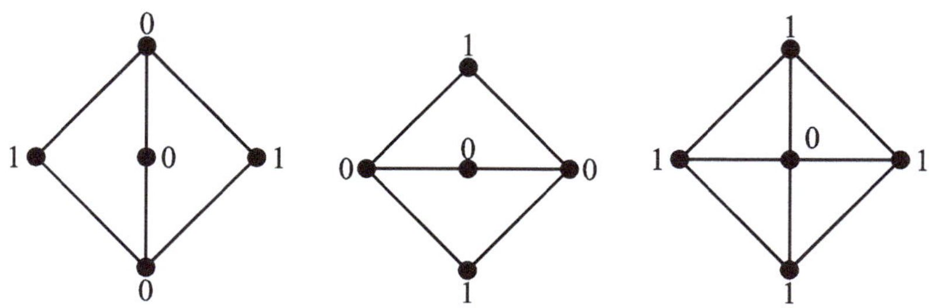

FIG. 4 Meshes corresponding to convex envelopes $\Gamma(w_1) = |x_1|$ (*left*), $\Gamma(w_2) = |x_2|$ (*middle*), and $\Gamma(w_3) = |x_1| + |x_2|$ (*right*).

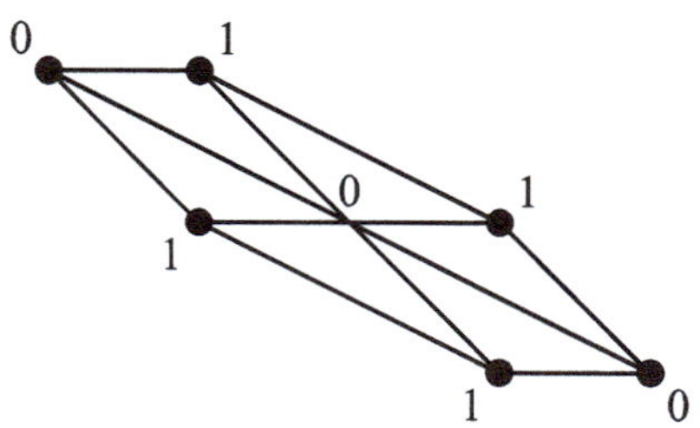

FIG. 5 Mesh induced by the nodal interpolant of $w(x) = (x \cdot e)^2$ where $e = (1,2)^{\mathsf{T}}$. Its convex envelope equals $|x \cdot e|$ in the star of $(0, 0)$.

3.1.3 The Oliker–Prussner method

Now we are ready to introduce the Oliker–Prussner method (Nochetto and Zhang, 2019; Oliker and Prussner, 1988). We seek a convex nodal function u_h satisfying the boundary condition $u_h(x_h) = g(x_h)$ for all $x_h \in \partial\Omega_h$ and

$$|\partial u_h(x_h)| = \int_{\omega_{x_h}} f(x)\mathrm{d}x, \quad \forall x_h \in \Omega_h, \tag{64}$$

Note that, since the partition $\{\omega_{x_h}\}_{x_h \in \Omega_h}$ is nonoverlapping, for all Borel sets $D \subset \Omega$, we have

$$|\partial u_h(D)| = \sum_{x_h \in D} f_{x_h}, \quad \text{where } f_{x_h} = \int_{\omega_{x_h}} f(x)\mathrm{d}x.$$

Thus, the scheme is obtained by replacing f in (9) by a family of Dirac measures supported at the nodes in Ω_h, and by replacing g by its nodal interpolant on the boundary. To implement the method, we need to derive a formula to compute the subdifferential of a nodal function u_h. This is a nontrivial task because it is non local. In fact, it involves computing the convex envelope of u_h. The following observation is useful in the characterization of the subdifferential. For a proof, see Nochetto and Zhang (2018).

Lemma 8 (characterization of subdifferential).
Let w_h be a convex nodal function and $\mathcal{T}_h$ be the mesh induced by its convex envelope $\Gamma(w_h)$. Then the subdifferential of w_h at $x_h \in \Omega_h$ is the convex hull of the constant gradients $\nabla\Gamma(w_h)|_T$ for all $T \in \mathcal{T}_h$ which contain x_h.

Fig. 6 depicts the subdifferential $\partial w_h(x_h)$ of a convex nodal function w_h at node x_h for $d = 2$.

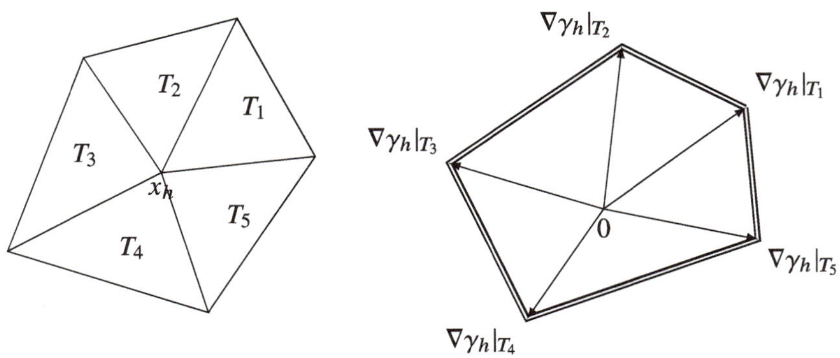

FIG. 6 Star centred at node x_h corresponding to the mesh $\mathcal{T}_h$ induced by the convex envelope $\gamma_h = \Gamma(w_h)$ and subdifferential $\partial w_h(x_h)$ of the convex nodal function w_h at node x_h. The latter is the convex hull of the constant element gradients $\nabla\gamma_h|_{T_j}$ for $1 \le j \le 5$.

3.2 Stability, continuous dependence on data, and discrete maximum principle

The Alexandrov estimate, which establishes the stability and continuous dependence of the Monge–Ampère equation, is a cornerstone in the nonlinear PDE theory. In this subsection, we introduce a discrete version of the Alexandrov estimate suitable for nodal functions. We refer the reader to Nochetto and Zhang (2018) for a complete proof.

Lemma 9 (discrete Alexandrov estimate).
Let w_h be a nodal function with $w_h(x_h) \geq 0$ at all $x_h \in \partial\Omega_h$. Then

$$\sup_{\Omega_h} w_h^- \leq C \left(\sum_{x_h \in \mathcal{C}_h^-(w_h)} |\partial w_h(x_h)| \right)^{1/d}, \tag{65}$$

where $C = C(d, \Omega)$ is proportional to the diameter of Ω and $\mathcal{C}_h^-(w_h)$ is the (lower) contact set:

$$\mathcal{C}_h^-(w_h) := \{x_h \in \Omega_h, \ \ \Gamma(w_h)(x_h) = w_h(x_h)\}. \tag{66}$$

The Alexandrov estimate establishes a lower bound for a nodal function in terms of the measure of the subdifferential at the (lower) contact set. Similarly, one can obtain an upper bound for a nodal function by the measure of the superdifferential at the (upper) contact set.

Applying the discrete Alexandrov estimate, we are ready to compare two arbitrary nodal functions in terms of their subdifferentials. This is instrumental for the error analysis.

Proposition 16 (stability of numerical solution).
Let v_h and w_h be two nodal functions with $v_h \geq w_h$ on $\partial\Omega_h$. Then

$$\sup_{\Omega_h} (v_h - w_h)^- \leq C \left(\sum_{x_h \in \mathcal{C}_h^-(v_h - w_h)} \left(|\partial v_h(x_h)|^{1/d} - |\partial w_h(x_h)|^{1/d} \right)^d \right)^{1/d},$$

where $C = C(d, \Omega)$ is proportional to the diameter of Ω.

Proof. Let v_h, w_h be two nodal functions. We introduce the convex envelope $\Gamma(v_h - w_h)$ as in Definition 19, and the nodal contact set $\mathcal{C}_h^-(v_h - w_h)$ defined in (66). The discrete Alexandrov estimate of Lemma 9 yields

$$\sup_{\Omega_h} (v_h - w_h)^- \leq C \left(\sum_{x_h \in \mathcal{C}_h^-(v_h - w_h)} |\partial \Gamma(v_h - w_h)(x_h)| \right)^{1/d}, \tag{67}$$

whence we only need to estimate $|\partial \Gamma(v_h - w_h)(x_h)|$ for all $x_h \in \mathcal{C}_h^-(v_h - w_h)$. For these nodes, we easily see that

$$\partial\Gamma(v_h - w_h)(x_h) \subset \partial(v_h - w_h)(x_h).$$

We claim that

$$\partial w_h(x_h) + \partial\Gamma(v_h - w_h)(x_h) \subset \partial v_h(x_h) \quad \forall\, x_h \in \mathcal{C}_h^-(v_h - w_h). \tag{68}$$

Fix $x_h \in \mathcal{C}_h^-(v_h - w_h)$, and let $\boldsymbol{p} \in \partial w_h(x_h)$ and $\boldsymbol{q} \in \partial\Gamma(v_h - w_h)(x_h)$, respectively, that is, by definition of the subdifferential (63),

$$\boldsymbol{p} \cdot (y_h - x_h) \leq w_h(y_h) - w_h(x_h)$$

and

$$\boldsymbol{q} \cdot (y_h - x_h) \leq \Gamma(v_h - w_h)(y_h) - \Gamma(v_h - w_h)(x_h)$$

for all nodes $y_h \in \Omega_h$. Adding both inequalities, we get

$$(\boldsymbol{p} + \boldsymbol{q}) \cdot (y_h - x_h) \leq w_h(y_h) + \Gamma(v_h - w_h)(y_h) - (w_h(x_h) + \Gamma(v_h - w_h)(x_h))$$

Since x_h is in the contact set $\mathcal{C}_h^-(v_h - w_h)$, we have $\Gamma(v_h - w_h)(x_h) = (v_h - w_h)(x_h)$. For all other nodes $y_h \in \Omega_h$, we have $\Gamma(v_h - w_h)(y_h) \leq (v_h - w_h)(y_h)$. Hence, we deduce

$$\begin{aligned}(\boldsymbol{p} + \boldsymbol{q}) \cdot (y_h - x_h) &\leq w_h(y_h) + (v_h - w_h)(y_h) - (w_h(x_h) + (v_h - w_h)(x_h)) \\ &= v_h(y_h) - v_h(x_h).\end{aligned}$$

This inequality implies $(\boldsymbol{p} + \boldsymbol{q}) \in \partial v_h(x_h)$ and proves the claim.

The Brunn–Minkowski inequality of Lemma 2 applied to (68) yields

$$\begin{aligned}&|\partial w_h(x_h)|^{1/d} + |\partial\Gamma(v_h - w_h)(x_h)|^{1/d} \\ &\qquad \leq |\partial w_h(x_h) + \partial\Gamma(v_h - w_h)(x_h)|^{1/d} \leq |\partial v_h(x_h)|^{1/d},\end{aligned}$$

whence

$$|\partial\Gamma(v_h - w_h)(x_h)| \leq \left(|\partial v_h(x_h)|^{1/d} - |\partial w_h(x_h)|^{1/d}\right)^d \quad \forall x_h \in \mathcal{C}_h^-(v_h - w_h).$$

This inequality gives us the desired estimate for $|\partial\Gamma(v_h - w_h)(x_h)|$. In view of (67), adding over all $x_h \in \mathcal{C}_h^-(v_h - w_h)$ concludes the proof. □

A direct consequence of this stability result is the maximum principle for nodal functions.

Corollary 6 (discrete maximum principle).

Let v_h and w_h be two nodal functions over the nodal set $\bar{\Omega}_h$. If $v_h(x_h) \geq w_h(x_h)$ at all $x_h \in \partial\Omega_h$ and $|\partial v_h(x_h)| \leq |\partial w_h(x_h)|$ at all $x_h \in \Omega_h$, then

$$w_h(x_h) \leq v_h(x_h) \quad \forall x_h \in \Omega_h.$$

Proof. Since $v_h(y_h) \geq w_h(y_h)$ for all $y_h \in \partial\Omega_h$, then for any node $x_h \in \mathcal{C}_h^-(v_h - w_h)$, we have

$$\partial w_h(x_h) \subset \partial v_h(x_h).$$

Combining this with the assumption that $|\partial v_h(x_h)| \leq |\partial w_h(x_h)|$ for all $x_h \in \Omega_h$, we deduce $|\partial v_h(x_h)| = |\partial w_h(x_h)|$ for all $x_h \in \mathcal{C}_h^-(v_h - w_h)$. Consequently, the stability of Proposition 16 implies

$$\sup_{\Omega_h} (v_h - w_h)^- = 0,$$

whence $v_h - w_h \geq 0$. This completes the proof. □

Proposition 16 yields a lower bound on the difference between two nodal functions in terms of the difference of the measure of their subdifferentials. Similarly, to derive an upper bound, one may consider the functions $-w_h$ and $-v_h$ and derive

$$\sup_{\Omega_h} (w_h - v_h)^- \leq C \left(\sum_{x_h \in \mathcal{C}_h^-(w_h - v_h)} \left(|\partial w_h(x_h)|^{1/d} - |\partial v_h(x_h)|^{1/d} \right)^d \right)^{1/d}.$$

Combining both bounds, we can derive a bound on $\| v_h - w_h \|_{L^\infty(\Omega_h)}$ in terms of $|\partial v_h(x_i)|$ and $|\partial w_h(x_i)|$. In particular, the uniqueness of the solution of the Oliker–Prussner method follows immediately from Proposition 16.

Finally, we notice that Proposition 16 is instrumental to derive error estimates. Define the nodal interpolation of a function w as the nodal function $N_h w$ such that

$$N_h w(x_h) = w(x_h) \quad \forall x_h \in \bar{\Omega}_h. \tag{69}$$

Setting $w_h = u_h$ and $v_h = N_h u$ In Proposition 16, where u_h and u solve (64) and (1), respectively, we can derive an estimate for $\| u_h - N_h u \|_{L^\infty(\Omega)}$. It remains to estimate the discrepancy of the subdifferentials of the two nodal functions. While $|\partial u_h(x_h)| = f_{x_h}$ is known by definition of the scheme (64), the measure of the subdifferential $|\partial N_h u(x_h)|$ remains unknown. Therefore, the goal of our next step is to estimate the quantity $|\partial N_h u(x_h)|^{1/d} - f_{x_h}^{1/d}$ which will then be applied in Proposition 16 to derive a pointwise estimate.

3.3 Consistency

In general, this method (64) is consistent in the sense that the right-hand side of the (64) can be written equivalently as $\sum_{x_h \in \Omega_h} f_{x_h} \delta_{x_h}$ and this converges to f in measure. However, such a concept of convergence is too weak to derive rates of convergence. Fortunately, we realize that if internal nodes are translation invariant, then a reasonable notion of operator consistency holds for any convex quadratic polynomial; see Lemma 12. Such property is shown in Benamou et al. (2016), Mirebeau (2015), and Nochetto and Zhang (2019) for Cartesian nodes, see also Section 2.5. In contrast, we give here an alternative proof of consistency based on the geometric interpretation of subdifferentials of convex quadratic polynomials in the interior of the domain, extend the results to $C^{2,\alpha}$ functions, and further investigate the consistency error in the region close to the boundary. To achieve this we, First, we require a definition.

Definition 20 (adjacent set).
Given a convex nodal function w_h and a node $x_h \in \Omega_h$, the *adjacent set* of x_h, denoted by $A_{x_h}(w_h)$, is the collection of nodes $y_h \in \bar{\Omega}_h$ closest to x_h such that there exists a supporting hyperplane L of w_h and $L(y_h) = w_h(y_h)$. Thus, the set $A_{x_h}(w_h)$ is the collection of nodes in the star associated with x_h in the mesh $\mathcal{T}_h$ induced by $\Gamma(w_h)$.

Lemma 10 (size of adjacent sets).
Let the nodal set Ω_h be translation invariant, and let p be a C^2 convex function defined in $\bar{\Omega}$. If $\lambda I \leq D^2p \leq \Lambda I$ in Ω for some constants $\lambda, \Lambda > 0$ and $p_h := N_h p$ is the nodal function associated with p defined in (69)*, then the adjacent set of nodes $A_{x_h}(p_h)$ satisfies*

$$A_{x_h}(p_h) \subset B_{Rh}(x_h)$$

where $R = \frac{\Lambda}{\lambda} d$, and $B_{Rh}(x_h)$ is the ball centred at x_h with radius Rh.

Proof. Let $z_h \in A_{x_h}(p_h)$ be such that

$$|z_h - x_h| = \max\{|y_h - x_h| : y_h \in A_{x_h}(p_h)\}.$$

Without loss of generality, we may assume that $p(x_h) = 0$ and $\nabla p(x_h) = 0$. Let ω be the convex hull of the nodal set $\{x_{\pm j} := x_h \pm h\tilde{e}_j, j = 1, \ldots, d\}$ where $\{\tilde{e}_j\}_{j=1}^d$ is the basis defined in (60). If $z_h \in \omega$, then the assertion is trivial because $R \geq 1$.

If $z_h \notin \omega$, then there is a constant $\tilde{R} \geq 1$ such that $\tilde{R}^{-1} z_h \in \omega$, which implies that $|z_h| \leq \tilde{R}h$ and $|z_h| \geq \tilde{R} d^{-1/2} h$. Because ω is convex, we may write

$$\tilde{R}^{-1} z_h = \sum_{\substack{j=1 \\ \sigma \in \{+,-\}}}^{d} \alpha_{\sigma j} x_{\sigma j}, \quad \alpha_{\sigma j} \geq 0, \quad \sum_{\substack{=1 \\ \sigma \in \{+,-\}}}^{d} \alpha_{\sigma j} = 1.$$

We next note that $p(x_{\pm j}) \leq \frac{1}{2}\Lambda h^2$ for all $j = 1, \ldots, d$ because $D^2 p \leq \Lambda I$, $|x_{\pm j} - x_h| \leq h$, $p(x_h) = 0$ and $\nabla p(x_h) = 0$. Since $z_h \in A_{x_h}(p_h)$, there exists a supporting hyperplane L at x_h such that

$$L(z_h) = p_h(z_h), \quad L(x_{\pm j}) \leq p_h(x_{\pm j}) \leq \frac{1}{2}\Lambda h^2.$$

Exploiting that L is linear and $p_h(x_h) = 0$ yields

$$p_h(z_h) = L(z_h) = L\left(\tilde{R}\sum_{\substack{j=1\\ \sigma\in\{+,-\}}}^{d}\alpha_{\sigma j}x_{\sigma j}\right) = \tilde{R}\sum_{\substack{j=1\\ \sigma\in\{+,-\}}}^{d}\alpha_{\sigma j}L(x_{\sigma j}) \le \frac{1}{2}\Lambda h^2\tilde{R}.$$

On the other hand, since $D^2p \geq \lambda I$ and $|z_h| \geq \tilde{R}hd^{-1/2}$, we have

$$p_h(z_h) = p(z_h) \geq \frac{\lambda}{2}|z_h|^2 \geq \frac{\lambda}{2}\tilde{R}^2d^{-1}h^2.$$

Combining the last two inequalities implies

$$\tilde{R} \leq R = \frac{\Lambda}{\lambda}d.$$

This completes the proof. □

The previous result shows that for any node x_h with $\mathrm{dist}(x_h, \partial\Omega) > Rh$, all nodes in its adjacent set are contained in Ω_h. We apply this observation to establish the following consistency result.

Lemma 11 (properties of convex interpolation).
Let p be a convex quadratic polynomial such that $\lambda I \leq D^2p \leq \Lambda I$, and let $p_h = N_hp$ be the nodal function defined by (69). *Then the following properties hold:*

1. *For all $x_h \in \Omega_h$ we have $\partial p_h(x_h) \neq \emptyset$.*
2. *If the nodal set Ω_h is translation invariant and dist$(x_h, \partial\Omega_h) \geq Rh$, with $R = \frac{\Lambda}{\lambda}d$, under a uniform refinement from Ω_h to $\Omega_{h/2}$, we have*

$$|\partial p_h(x_h)| = 2^d|\partial p_{h/2}(x_h)|.$$

3. *If the nodal set Ω_h is translation invariant, dist$(x_h, \partial\Omega_h h) \geq Rh$, and dist$(y_h, \partial\Omega_h) \geq Rh$, then $|\partial p_h(x_h)| = |\partial p_h(y_h)|$.*

Proof. To prove the first claim, we only need to note that if ℓ is the tangent plane of p at x_h, then ℓ is a supporting plane of p_h at x_h. Thus $\nabla\ell \in \partial p_h(x_h)$.

To prove the second claim, we may assume that $p(x_h) = 0$, and $\nabla p(x_h) = 0$. Note that for homogeneous quadratic polynomials, we have

$$p(x) = 4p\left(\frac{x}{2}\right).$$

A simple calculation yields

$$\partial p_h(x_h) = 2\partial p_{h/2}(x_h)$$

and therefore $|\partial p_h(x_h)| = 2^d|\partial p_{h/2}(x_h)|$.
To prove the third claim. We consider the function

$$p^{x_h}(x) = p(x) - \nabla p(x_h) \cdot (x - x_h) - p(x_h),$$

obtained by subtracting the tangent plane of p at x_h. Since adding an affine function does not change the measure of the subdifferential, we have $|\partial p_h(x_h)| = |\partial p_h^{x_h}(x_h)|$. Further note that by subtracting the tangent plane at a node y_h, we obtain the same function up to a parallel translation, that is,

$$p^{x_h}(x - x_h) = p^{y_h}(x - y_h).$$

Since the mesh is translation invariant, we have that if L is a supporting plane of $p_h^{x_h}$ at x_h, then by a parallel translation it is also a supporting plane of $p_h^{y_h}$ at y_h. Hence, we have $|\partial p_h^{x_h}(x_h)| = |\partial p_h^{y_h}(y_h)|$. Since $|\partial p_h(x_h)| = |\partial p_h^{x_h}(x_h)|$ for all nodes x_h, we conclude that $|\partial p_h(x_h)| = |\partial p_h(y_h)|$. □

Now we are ready to prove the consistency, for a proof see Nochetto and Zhang (2019, Lemma 5.3).

Lemma 12 (consistency I).

Let p be a convex quadratic polynomial such that $\lambda I \leq D^2 p \leq \Lambda I$, and let $p_h := N_h p$ be the corresponding convex nodal function defined in (69). Let Ω_h be translation invariant. Then

$$|\partial p_h(x_h)| = \int_{\omega_{x_h}} \det D^2 p(x) \mathrm{d}x$$

for any node $x_h \in \Omega_h$ such that dist$(x_h, \partial\Omega) \geq Rh$ with $R = \frac{\Lambda}{\lambda} d$.

Proof. Let ϕ be any continuous function with compact support in Ω. We consider a sequence of nested refinements Ω_{h_n} with $h_n = 2^{-n}H$, for a fixed $H > 0$.

By Lemma 1 we immediately obtain, as $n \to \infty$, that

$$\sum_{y_{h_n} \in \Omega_{h_n}} \phi(y_{h_n}) |\partial p_{h_n}(y_{h_n})| \to \int_\Omega \phi \det D^2 p(x) \mathrm{d}x = \det D^2 p(x) \int_\Omega \phi(x) \mathrm{d}x.$$

Thus, we only need to prove that as $n \to \infty$

$$\sum_{y_{h_n} \in \Omega} \phi(y_{h_n}) |\partial p_{h_n}(y_{h_n})| \to \frac{|\partial p_H(x_H)|}{|\omega_{x_H}|} \int_\Omega \phi(x) \mathrm{d}x.$$

In view of second and third result in Lemma 11, we have

$$|\partial p_{h_n}(y_{h_n})| = |\partial p_{h_n}(x_H)| = 2^{-nd} |\partial p_H(x_H)|.$$

The refinement strategy implies that $|\omega_{y_{h_n}}| = 2^{-nd}|\omega_{x_H}|$. Thus, we infer that

$$\begin{aligned} \sum_{y_{h_n} \in \Omega_{h_n}} \phi(y_{h_n}) |\partial p_{h_n}(y_{h_n})| &= \frac{|\partial p_H(x_H)|}{|\omega_{x_H}|} \sum_{y_{h_n} \in \Omega_{h_n}} \phi(y_{h_n}) |\omega_{y_{h_n}}| \\ &\to \frac{|\partial p_H(x_H)|}{|\omega_{x_H}|} \int_\Omega \phi(x) \mathrm{d}x. \end{aligned}$$

This completes the proof. □

Moreover, for convex cubic polynomials, we have the following consistency error estimate. This result, to our knowledge, has not appeared elsewhere.

Lemma 13 (consistency II).
Let $x_h \in \Omega_h$ and q be a convex cubic polynomial such that $\lambda I \leq D^2 q \leq \Lambda I$ in the ball $B_{Rh} := \overline{B_{Rh}(x_h)} \subset \Omega$, with $R = \frac{\Lambda}{\lambda} d$. Then

$$\left| |\partial N_h q(x_h)| - \int_{\omega_{x_h}} \det D^2 q(x)\, dx \right| \leq C h^{d+2} |q|_{C^3(B_{Rh})}..$$

Proof. Without loss of generality, we may assume that $x_h = 0$ and $q(0) = 0$ and $\nabla q(0) = 0$. We decompose the cubic polynomial $q(x)$ as

$$q(x) = p(x) + hr(x),$$

where $p(x)$ is a quadratic polynomial such that $D^2 p = D^2 q(0)$ and $r(x)$ is a homogeneous cubic polynomial. Since, by Lemma 10, the adjacent set $A_{x_h}(q)$ of the node $x_h = 0$ is contained in a ball of radius Rh we deduce that

$$|p(z_h)| \leq C_q R^2 h^2, \quad |r(z_h)| \leq C_r R^2 h^2 \;\; \forall z_h \in A_{x_h}(q),$$

where C_q and C_r depends on $D^2 p$ and $D^3 r$, respectively. We set

$$q_t(x) = p(x) + tr(x) \quad t \in [-h, h],$$

and note that $\lambda I \leq D^2 q_t(0) \leq \Lambda I$ for all t. Therefore, the adjacent set of q_t at 0 remains in the ball B_{Rh}.

We set the measure of its subdifferential of q_t at x_h as a function of t

$$m(t) = |\partial N_h q_t(x_h)| = |\partial N_h q_t(0)|,$$

and note that we aim to show that

$$\left| m(h) - \int_{\omega_{x_h}} \det D^2 q(x) \mathrm{d}x \right| \leq C h^{d+2} |q|_{C^3(B_{Rh})}.$$

Now we proceed to prove the lemma in the following steps.

1. We aim to show that $m(t)$ is a polynomial of degree d

$$m(t) = \sum_{k=0}^{d} C_k t^k. \tag{70}$$

and the coefficients C_k satisfy $|C_k| \leq C h^d$ where C depends on $|D^2 p|$, $|D^3 r|$, and the dimension d. By the characterization of the subdifferential, given in Lemma 8, the subdifferential of $N_h q_t$ at 0 is the convex hull of the piecewise gradient of its convex envelope $\nabla \Gamma(N_h q_t)|_T$ for all $T \in \mathcal{T}_h$ that have x_h as a vertex; see Fig. 6. We label these simplices as $T_1, \cdots, T_N$ and, to simplify notation, we set the piecewise gradient of $\Gamma(N_h p)$ and $\Gamma(N_h r)$ at T_i as

$$\boldsymbol{v}_i = \nabla\Gamma(N_h p)|_{T_i}, \quad \boldsymbol{w}_i = \nabla\Gamma(N_h r)|_{T_i}, \quad i = 1,\ldots,N.$$

Hence, we have

$$\boldsymbol{u}_i := \nabla\Gamma(N_h q_t)|_{K_i} = \boldsymbol{v}_i + t\boldsymbol{w}_i.$$

To compute the measure of the convex hull of $\{\boldsymbol{u}_i\}$, we may divide the convex hull into a set of disjoint simplices $\{S_i,\ i = 1, \cdots, N\}$ and label the vertices of S_i as $\{\boldsymbol{0}, \boldsymbol{u}_{i_1}, \cdots, \boldsymbol{u}_{i_d}\}$. Thus, we obtain

$$m(t) = \sum_{i=1}^{N} |S_i| \text{ where } |S_i| \text{ denotes the signed volume of } S_i.$$

and so, by the volume formula of simplices, we get

$$m(t) = \frac{1}{d!}\sum_{i=1}^{N} \det \begin{pmatrix} 1 & \boldsymbol{0}^{\mathsf{T}} \\ 1 & \boldsymbol{u}_{i_1}^{\mathsf{T}} \\ \vdots & \vdots \\ 1 & \boldsymbol{u}_{i_d}^{\mathsf{T}} \end{pmatrix}. \qquad \boldsymbol{u}_{i_j} = \boldsymbol{v}_{i_j} + t\boldsymbol{w}_{i_j}. \tag{71}$$

Now, it is clear that

$$|S_i| = \sum_{k=0}^{d} C_k^i t^k$$

is a polynomial of t with degree at most d. Thus, $m(t)$ must be a polynomial with degree at most d as well. Furthermore, by the volume formula of simplices (71), the coefficients $|C_k^i| \le Ch^d$ because both $|\boldsymbol{v}_{i_j}| \le Ch$ and $|\boldsymbol{w}_{i_j}| \le Ch$. Finally, the number N of simplices S_i is finite and bounded by the number of vertices in the adjacent set A.

2. We show that $m'(0) = 0$. To do so, it suffices to show that the function m is even, that is $m(t) = m(-t)$ for all $-h \le t \le h$. Note that if $\boldsymbol{v} \in \partial N_h(p + tr)(0)$, then $-\boldsymbol{v} \in \partial N_h(p - tr)(0)$ for any $t \in (0, h]$. Indeed, since the subdifferential set is determined by the function values on the adjacent set which is contained in the ball $B_{Rh}(0)$, if $\boldsymbol{v} \cdot y_h \le (p + tr)(y_h)$ for all $y_h \in B_{Rh}(0)$, then

$$\boldsymbol{v} \cdot (-y_h) \le (p + tr)(-y_h) \quad \forall y_h \in B_{Rh}(0).$$

Hence, $-\boldsymbol{v} \in \partial N_h(p - tr)(0)$ because $p(y_h) = p(-y_h)$. Thanks to this symmetry property, we deduce that $|\partial N_h(p - tr)(0)| = |\partial N_h(p + tr)(0)|$, i.e., $m(t) = m(-t)$.

3. We show that

$$|m(h) - m(0)| \le Ch^{d+2}.$$

Combining the previous two steps we get that

$$m(t) = m(0) + C_2 t^2 + \cdots + C_d t^d$$

because $C_1 = m'(0) = 0$. Since $|C_j| \le Ch^d$ for $j = 2, \ldots, d$, we deduce that

$$|m(t) - m(0)| \le Ch^{d+2} \quad \forall t \in [0, h].$$

4. It remains to show that

$$\left| \int_{\omega_{x_h}} \det D^2 q(x) \mathrm{d}x - m(0) \right| \le Ch^{d+2}.$$

By the consistency for quadratics given in Lemma 12, we have

$$m(0) = \int_{\omega_{x_h}} \det D^2 p(x) \mathrm{d}x.$$

Therefore, it is sufficient to show that

$$\left| \int_{\omega_{x_h}} (\det D^2 q(x) - \det D^2 p(x)) \mathrm{d}x \right| \le Ch^{d+2}.$$

A Taylor expansion of $\det D^2 q = \det D^2(p + hr)$ reveals that

$$\left| \det D^2 q(x) - \det D^2 p(x) - h \operatorname{cof} D^2 p(x) : D^2 r(x) \right| \le Ch^2.$$

where the constant C depends on $D^2 p$ and $D^3 r$. This implies that

$$\left| \int_{\omega_{x_h}} (\det D^2 q(x) - \det D^2 p(x)) \mathrm{d}x \right| \le h \left| \int_{\omega_{x_h}} \operatorname{cof} D^2 p(x) : D^2 r(x) \mathrm{d}x \right| + Ch^2 |\omega_{x_h}|.$$

Noting that $\operatorname{cof} D^2 p : D^2 r$ is an odd function and ω_{x_h} is symmetric respect to the origin, we obtain

$$\int_{\omega_{x_h}} \operatorname{cof} D^2 p(x) : D^2 r(x) \mathrm{d}x = 0$$

and

$$\left| \int_{\omega_{x_h}} (\det D^2 q(x) - \det D^2 p(x)) \mathrm{d}x \right| \le Ch^{d+2}.$$

This completes the proof. □

Now for any function w that can be approximated locally by a quadratic polynomial such that $w(x) = p(x) + \mathcal{O}(h^{2+\alpha})$ in $B_{Rh}(x_h)$ or by a cubic polynomial such that $w(x) = q(x) + O(h^{3+\alpha})$ in $B_{Rh}(x_h)$, we show that the consistency error of the Oliker–Prussner method is of order $\mathcal{O}(h^{\alpha})$ and $\mathcal{O}(h^{1+\alpha})$, respectively.

Proposition 17 (interior consistency).
Let Ω_h be a translation invariant set of nodes, and $x_h \in \Omega_h$ be such that dist $(x_h, \partial\Omega_h) \geq Rh$ with $R = \frac{\Lambda}{\lambda} d$. If $w \in C^{2+k,\alpha}(\overline{B_{Rh}})$, with $k \in \{0, 1\}$, and $\alpha \in (0, 1]$) is a convex function with $\lambda I \leq D^2 w \leq \Lambda$I, then we have

$$\left| |\partial N_h w(x_h)| - \int_{\omega_{x_h}} \det D^2 w(x) \mathrm{d}x \right| \leq C h^{k+\alpha} |w|_{C^{2+k,\alpha}(\overline{B_{Rh}})} |\omega_{x_h}|,$$

where $C = C(d, \lambda, \Lambda)$.

Proof. We divide the proof into two cases $k = 0$ and $k = 1$.

1. Case $k = 0$: We only need to show the inequality

$$|\partial N_h w(x_h)| \leq \int_{\omega_{x_h}} \det D^2 w(x) \mathrm{d}x + C h^{\alpha} |w|_{C^{2,\alpha}(\overline{B_{Rh}})} |\omega_{x_h}|,$$

because the reverse inequality can be derived similarly. Since $w \in C^{2,\alpha}(\overline{B_{Rh}})$, we estimate w by a quadratic polynomial p so that

$$w(x) \leq p(x) \quad \forall x \in B_{Rh}(x_h),$$

where $p(x_h) = w(x_h)$, $\nabla p(x_h) = \nabla w(x_h)$ and

$$D^2 p = D^2 w(x_h) + C h^{\alpha} |w|_{C^{2,\alpha}(\overline{B_{Rh}})} I$$

for a fixed, and sufficiently large, constant C. Let $p_h = N_h p$, and note that

$$|\partial N_h w(x_h)| \leq |\partial p_h(x_h)|.$$

It remains to show that

$$|\partial p_h(x_h)| \leq \int_{\omega_{x_h}} \det D^2 w(x) \mathrm{d}x + C h^{\alpha} |w|_{C^{2,\alpha}(\overline{B_{Rh}})} |\omega_{x_h}|.$$

Since $(\lambda + Ch^{\alpha})I \leq D^2 p \leq (\Lambda + Ch^{\alpha})I$ and

$$\frac{\Lambda + Ch^{\alpha}}{\lambda + Ch^{\alpha}} \leq \frac{\Lambda}{\lambda} \quad \text{because } \Lambda \geq \lambda,$$

invoking the consistency of Lemma 12 we obtain

$$|\partial p_h(x_h)| = \int_{\omega_{x_h}} \det D^2 p(x) \mathrm{d}x$$

provided that $\mathrm{dist}(x_h, \partial\Omega_h) \geq Rh$. Recalling that $w \in C^{2,\alpha}(\overline{B_{Rh}})$, we can write $D^2 p = D^2 w(x) + E(x)$ for all $x \in \overline{B_{Rh}}$, where $|E(x)| \leq Ch^\alpha |w|_{C^{2,\alpha}(\overline{B_{Rh}})}$. A Taylor expansion yields

$$|\partial p_h(x_h)| \leq \int_{\omega_{x_h}} \det D^2 w(x) \mathrm{d}x + Ch^\alpha |w|_{C^{2,\alpha}(\overline{B_{Rh}})} |\omega_{x_h}|.$$

2. **Case** $k = 1$: If $w \in C^{3,\alpha}(\overline{B}_{Rh})$, we approximate w by a cubic polynomial q so that

$$w(x) \leq q(x) \quad \forall x \in B_{Rh}(x_h),$$

where $q(x_h) = w(x_h)$, $\nabla q(x_h) = \nabla w(x_h)$,

$$D^2 q(x_h) = D^2 w(x_h) + Ch^{1+\alpha} |w|_{C^{3,\alpha}(\overline{B_{Rh}})},$$

and $D^3 q = D^3 w(x_h)$ with universal constant C. The rest of the proof is similar to the previous case.

Combing both cases, we conclude the proof of the estimate. □

3.4 Pointwise error estimate

We are now ready to show a pointwise error estimate for the method (64) under suitable regularity assumptions on the solution u. We aim to apply the stability of the numerical scheme shown in Proposition 16 to derive a lower bound of the difference $v_h - u_h$, for a suitable convex piecewise linear function v_h.

Assume that the convex solution u of the Monge–Ampère equation (1) is $C^{k,\alpha}$ near the boundary of the domain Ω where $k \in \{2, 3\}$ and $\alpha \in (0, 1]$. We first extend the solution to a larger convex domain

$$\Omega_{4Rh} = \{x \in \mathbb{R}^d, \mathrm{dist}(x, \Omega) \leq 4Rh\}$$

such that, for sufficiently small h, the extended function, which we still denote as u, remains $C^{k,\alpha}$-continuous in the extended region and satisfies

$$\frac{\lambda}{2} I \leq D^2 u(x) \leq 2\Lambda I \quad \text{for any } x \in \Omega_{4Rh}. \tag{72}$$

Next, we extend the translation invariant interior nodal set Ω_h to the extended domain Ω_{4Rh} and, by an abuse of notation, we still denote the set as Ω_h, that is,

$$\Omega_h = \left\{ x_h = \sum_{j=1}^{d} z^j \tilde{\boldsymbol{e}}_j : \ z^j \in \mathbb{Z} \right\} \cap \Omega_{4Rh}.$$

We construct the piecewise linear function $v_h = \Gamma(N_h u)$ by taking the convex envelope of the nodal interpolation of the solution u on Ω_h in the extended domain and then restrict the piecewise linear function v_h to the domain Ω. Thus, this procedure yields a piecewise linear function v_h defined on the domain Ω.

We claim that the piecewise linear function v_h satisfies the following two conditions which are useful in the error estimate. First, the adjacent set size estimate of Lemma 10 and the bound of D^2u given in (72) imply that for any interior node $x_h \in \Omega_h \cap \Omega$, its adjacent set $A_{x_h}(v_h)$ is contained in the extended domain Ω_{4Rh}. Second, we notice that $|v_h(x) - u(x)| \leq Ch^2$ on the boundary $\partial\Omega$ where the constant C depends on $\|u\|_{C^2(\Omega)}$. This is simply due to the fact that the diameter of any patch of a node $z \in \Omega_h \cap \Omega$ is bounded by $4Rh$ and interpolation theory of piecewise linear function.

Now we are ready to derive the main error estimate.

Theorem 13 (error estimate).
Let u be the solution of the Monge–Ampère equation (1), $0 \leq \lambda I \leq D^2 u \leq \Lambda I$ *and $u \in C^{2+k,\alpha}(\bar{\Omega})$ with $k \in \{0, 1\}$ and $\alpha \in (0, 1]$. Let Ω_h be a translation invariant nodal set satisfying* (60), *and let u_h be the solution of discrete Monge–Ampère equation* (64) *defined on Ω_h. Then we have*

$$\|u - u_h\|_{L^\infty(\Omega)} \leq Ch^{k+\alpha},$$

where the constant C depends only on $\|u\|_{C^{2+k,\alpha}(\Omega)}$, λ, Λ, diam(Ω), and space dimension d.

Proof. Let v_h be the interpolation of the extension of the solution u defined above. Since $|v_h - u_h| \leq Ch^2$ on the boundary $\partial\Omega$, we have $v_h + Ch^2 \geq u_h$. By the stability of the numerical solution, Proposition 16, we obtain

$$\sup_{\Omega_h}(v_h + Ch^2 - u_h)^- \leq C\left(\sum_{x_i \in C_h^-(v_h - u_h)} \left(|\partial v_h(x_i)|^{1/d} - |\partial u_h(x_i)|^{1/d}\right)^d\right)^{1/d}.$$

Invoking the consistency error estimate, Proposition 17, we immediately obtain

$$\sup_{\Omega_h}(v_h + Ch^2 - u_h)^- \leq Ch^{k+\alpha}.$$

By a simple algebraic manipulation, the estimate yields a lower bound for the error $v_h - u_h \geq -Ch^2 - Ch^{k+\alpha}$. Similarly, an estimate for the upper bound follows by considering the function $u_h + Ch^2 - v_h$. Combining both estimates, we get the desired result. □

3.5 $W^{2,p}$ error estimate

The results and arguments of the previous section have recently been extended to the derivation of $W^{2,p}$ error estimates of the Oliker–Prussner scheme (Neilan and Zhang, 2018). Here, the discrete $W^{2,p}$ norm is taken to be the sum of weighted second-order differences:

$$\|v\|_{W_f^{2,p}} = \left(\sum_{x_h \in \Omega_h} f_{x_h} |\Delta_e v(x_h)|^p \right)^{1/p}.$$

The starting point is a simple observation that the contact set of a nodal function contains information of its second-order difference. In particular, if u_h is the solution to (64) and v_h is some approximation to u, then we can define the perturbed error

$$w_h^\epsilon = v_h - (1-\epsilon)u_h \tag{73}$$

with parameter $\epsilon \in (0, 1)$. Now, by using the identity $\Delta_e w_h^\epsilon(x_h) \geq \Delta_e \Gamma w_h^\epsilon(x_h) \geq 0$ for $x_h \in \mathcal{C}_h^-(w_h^\epsilon)$, we have, after some algebraic manipulations,

$$\Delta_e(u_h - v_h)(x_h) \leq \frac{\epsilon}{1-\epsilon} \Delta_e v_h(x_h) \quad \forall x_h \in \mathcal{C}_h^-(w_h^\epsilon).$$

The right-hand side of this expression is uniformly bounded for appropriate v_h if u is sufficiently smooth, and therefore we find that the error $\Delta_e(u_h - v_h)(x_h)$ is controlled on the contact set $\mathcal{C}_h^-(w_h^\epsilon)$. However, noting that w_h^ϵ is not necessary convex, we must estimate $\Delta_e(u_h - v_h)(x_h)$ on the complement set

$$E^\epsilon := \Omega_h \backslash \mathcal{C}_h^-(w_h^\epsilon). \tag{74}$$

This is done by estimating its cardinality in terms of the consistency of the method.

Lemma 14 (size of complement set).
Let u_h and v_h be convex nodal functions with $u_h = v_h$ on $\partial\Omega_h$ and $u_h \leq v_h$ on Ω_h. Set

$$|\partial u_h(x_h)| = f_{x_h} \quad \text{and} \quad |\partial v_h(x_h)| = g_{x_h} \quad x_h \in \Omega_h.$$

Then there exists a constant $C > 0$ depending only on f such that

$$\sum_{x_h \in E^\epsilon} f_{x_h} \leq C \frac{(1-\epsilon)}{\epsilon} \|f^{1/d} - g^{1/d}\|_{\ell^d(\mathcal{C}_h^-(w_h^\epsilon))},$$

where w_h^ϵ and E^ϵ are defined by (73) and (74), respectively.

The last ingredient to develop $W^{2,p}$ estimates is a simple result of the discrete L^1 norm of a nodal function in terms of its level sets. Roughly speaking this result gives a relation between Riemann and Lebesgue sums; see Neilan and Zhang (2018, Lemma 5.1)

Lemma 15. *Let s_h be a nodal function with $|s_h(x_h)| \le M$ for some $M > 0$. Then, for any $\sigma > 0$,*

$$\sum_{x_h \in \Omega_h} f_{x_h} |s_h(x_h)| \le \sigma \sum_{k=0}^{M} \sum_{x_h \in A_k} f_{x_h},$$

where

$$A_k = \{x_h \in \Omega_h : \quad |s_h(x_h)| \ge k\sigma\}.$$

Theorem 14 ($W^{2,p}$ error estimate).

Suppose that the conditions of Theorem 14 are satisfied with $k + \alpha = 2$. Then there holds

$$\|u - u_h\|_{W_f^{2,p}} \le \begin{cases} Ch^{1/p} & p \in (d, \infty) \\ C|\log h|^{1/d} h^{1/d} & p \in (1, d]. \end{cases}$$

We now give a sketch of the main ideas to prove Theorem 14 and refer the reader to Neilan and Zhang (2018) for details. To communicate the main ideas, we make the simplifying assumption that the consistency estimate in Proposition 17 holds up to the boundary. We also assume homogeneous boundary conditions, i.e., $g = 0$ in (1b). These assumptions, which do not hold in general, allow us to derive better rates of convergence than those stated in Theorem 14.

As a first step we set $v_h = (1 - Ch^2)^{1/d} N_h u$, where $C > 0$ is sufficiently large such that (cf. Proposition 17)

$$g_{x_h} = |\partial v_h(x_h)| = (1 - Ch^2) |\partial N_h u(x_h)| \le f_{x_h}.$$

Therefore by the comparison principle in Corollary 6, we have $v_h \ge u_h$ on Ω_h. We also have $|f_{x_h} - g_{x_h}| \le Ch^{2+d}$.

To deduce the estimate, it suffices bound

$$\sum_{x_h \in \Omega_h} f_{x_h} (\Delta_e (u_h - v_h)(x_h))^+.$$

Bounding the negative part of the error can be obtained by similar arguments.

For parameter ϵ_k with $\epsilon_k/(1 - \epsilon_k) = Ck^{1/p} h^2$, we define

$$A_k = \left\{ x_h \in \Omega_h : \quad \Delta_e (u_h - v_h)(x_h) \ge \frac{\epsilon_k}{1 - \epsilon_k} \Delta_e v_h(x_h) \right\},$$

and note that $A_k \subset E^{\epsilon_k}$. Let $s_h(x_h) = |(\Delta_e(u_h - v_h))^+|^p$, and note that $|s_h(x_h)| \le Ch^{-2p}$ because u_h and v_h are bounded. Applying Lemma 15, with $\sigma = Ch^{2p}$, we have

$$\sum_{x_h \in \Omega_h} f_{x_h} |(\Delta_e(u_h - v_h)(x_h))^+|^p \le Ch^{2p} \left(1 + \sum_{k=1}^{Ch^{-2p}} \sum_{x_h \in A_k} f_{x_h} \right).$$

On the other had, using Lemma 14 and the consistency of the scheme yields, for h sufficiently small,

$$\begin{aligned} \sum_{x_h \in A_k} f_{x_h} \le \sum_{x_h \in E^{\epsilon_k}} f_{x_h} &\le C \frac{1 - \epsilon_k}{\epsilon_k} \| f^{1/d} - g^{1/d} \|_{\ell^d(C_h^-(w_h^{\epsilon_k}))} \\ &\le Ch^2 \frac{1 - \epsilon_k}{\epsilon_k} = Ck^{-1/p}. \end{aligned}$$

Thus, we find that

$$\begin{aligned} \sum_{x_h \in \Omega_h} f_{x_h} |(\Delta_e(u_h - v_h)(x_h))^+|^p &\le Ch^{2p} \left(1 + \sum_{k=1}^{Ch^{-2p}} \frac{1}{k^{1/p}} \right) \\ &\le C \begin{cases} h^2 |\log h| & \text{if } p = 1, \\ h^2 & \text{if } p > 1. \end{cases} \end{aligned}$$

In certain settings, Theorems 13 and 14 immediately give us $W^{1,p}$ error estimates as well. To make this precise, we assume that the basis $\{\tilde{e}_j\}_{j=1}^d = \{e_j\}_{j=1}^d$ defined in (60) is the canonical one. We then define the backward difference operator

$$D_e^- v(x_h) = \frac{v(x_h) - v(x_h - eh)}{h},$$

and the discrete norms/semi-norms, for $p \in (1, \infty)$,

$$\|v\|_{L_h^p(\Omega_h)} = \left(h^d \sum_{x_h \in \Omega_h} |v(x_h)|^p \right)^{1/p},$$

$$\|v\|_{W_h^{1,p}(\Omega_h)} = \left(\|v\|_{L_h^p(\Omega_h)}^p + h^d \sum_{j=1}^d \| D_{e_j}^- v \|_{L_h^p(\Omega_h)}^p \right)^{1/p},$$

$$\|v\|_{W_h^{2,p}(\Omega_h)} = \left(\|v\|_{W_h^{1,p}(\Omega_h)}^p + h^d \sum_{j=1}^d \left(\| \Delta_{e_j} v \|_{L_h^p(\Omega_h)}^p + \sum_{\substack{i=1 \\ j \ne i}}^d \| D_{e_i}^- D_{e_j}^- v \|_{L_h^p(\Omega_h)}^p \right) \right)^{1/p}.$$

We then have (Jovanović and Süli, 2014, Lemmas 2.60–2.61)

$$\|v\|_{W_h^{1,p}(\Omega_h)} \le C\,\|v\|_{L_h^p(\Omega_h)}^{1/2}\|v\|_{W_h^{2,p}(\Omega_h)}^{1/2}.$$

Therefore noting that $\|v\|_{L_h^p(\Omega_h)} \le C\,\|v\|_{L^\infty(\Omega)}$, and,

$$D_{e_i}^- D_{e_j}^- v(x) = \frac{1}{2}\big(\Delta_{e_i} v(x-he_i) + \Delta_{e_j} v(x-he_j) - \Delta_{\tilde{e}_{i,j}} v(x-h(e_i+e_j))\big)$$

with $\tilde{e}_{i,j} = e_i - e_j$, we have the following, by Theorems 13 and 14.

Corollary 7. *Suppose that the conditions in Theorem 13 are satisfied with $k + \alpha = 2$, and assume that $f \ge f_0 > 0$ in Ω. Then there holds*

$$\|u-u_h\|_{W_h^{1,p}(\Omega_h)} \le \begin{cases} Ch^{1+\frac{1}{2p}} & p \in (d,\infty), \\ C|\log h|^{\frac{1}{2d}} h^{1+\frac{1}{2d}} & p \in (1,d]. \end{cases}$$

Remark 15 (extensions).
In this section we showed that the stability estimate given in Proposition 16 provides a powerful tool to develop error estimates for the Monge–Ampère equation, as it allows us to derive L^∞ and W_p^2 error estimates when the solution enjoys regularity $u \in C^{2+k,\alpha}(\bar{\Omega})$. Thanks to this stability estimate, it also possible to extend these estimates if the solution is of lower regularity and/or degenerate. The key observation is that the stability estimate measures the consistency error in the ℓ^d-norm. If the solution is rough in a region of small measure and smooth elsewhere, so that the consistency error is small in ℓ^d-norm, then by the stability estimate, we may still derive a rate of convergence for the low regularity case. This is explored in Nochetto and Zhang (2019, Theorem 6.3) to prove a rate of convergence for solutions in $C^{1,1}(\Omega)$, but not in $C^2(\Omega)$. ■

4 Finite Element Methods

It will be found that most classical mathematical approximation procedures as well as the various direct approximations used in engineering fall into this category. It is thus difficult to determine the origins of the finite element method and the precise moment of its invention.

Zienkiewicz and Taylor (2000)

In this section, we summarize recent developments of finite element methods for the Monge–Ampère problem with Dirichlet boundary conditions (1). For simplicity, throughout this section, we assume that boundary conditions in (1) are homogeneous, i.e., $g = 0$. The extension to nonhomogenous boundary conditions is straightforward.

The main difficulty to construct (and analyze) finite element schemes for fully nonlinear problems is that the PDEs are nonvariational. Recall that a finite element method is typically derived by

(i) multiplying the PDE by a test function;
(ii) integrating the resulting product over the domain;
(iii) performing integration by parts to arrive at a variational formulation;
(iv) posing the variational formulation on a finite dimensional space, usually consisting of piecewise polynomials.

Note that the third step usually requires some structure conditions of the PDE, e.g., that the PDE is in divergence-form, which is not present for fully nonlinear problems. Another obvious difficult to construct convergent finite element schemes is that the notion of viscosity solutions, given in Definition 4, and Alexandrov solutions, as in Definition 9 for the Monge–Ampère equation are nonvariational, and it is unclear how this solution concept can be adopted within a finite element framework.

We must remark, however, that the Monge–Ampère operator (1a) does possess a divergence-form. Using well-known algebraic identities and the divergence-free property of cofactor matrices, there holds $\det D^2u = \frac{1}{d}\nabla\cdot(\operatorname{cof} D^2u\nabla u)$. Note however that variational formulations based on this identity would still involve second-order derivatives, and therefore, at this time, it is unclear whether numerical methods based on this approach are advantageous.

Nonetheless, assuming some regularity of the solution, well-defined finite element methods can be formulated and analyzed for fully nonlinear PDEs. One approach is to omit the third step of the four-step process described above. For example, multiplying the Monge–Ampère equation (1a) by a function v and integrating over Ω yields the identity

$$\int_\Omega (f - \det D^2u)v\,\mathrm{d}x = 0. \tag{75}$$

A simple calculation involving Hölder's inequality and Sobolev embeddings show that expression (75) is well-defined provided $u,\ v \in W^{2,d}(\Omega)$. Finite element methods can then be constructed based on the identity (75). Namely, an obvious finite element method based on the identity (75) seeks $u_h \in V_h$ satisfying

$$\int_\Omega (f - \det D^2u_h)v_h\,\mathrm{d}x = 0 \quad \forall v_h \in X_h, \tag{76}$$

where X_h is a finite dimensional space consisting of piecewise polynomials with respect to a partition of Ω that vanish on the boundary. While this method may be convergent (cf. Awanou, 2014, 2015c, 2017b; Böhmer, 2008; Davydov and Saeed, 2013; Neilan, 2014b), the appearance of global second-order derivatives in the method necessitates the use of C^1 finite element spaces which can be arduous to implement and are not found in most finite element software packages. In addition, C^1 finite element generally require high-degree polynomial bases, resulting in a relatively large algebraic system.

Because of the many disadvantages of the finite element method (76) several finite element methods with simpler spaces have been developed. These include C^0 penalty methods, discontinuous Galerkin (DG) methods, mixed finite element methods, and methods based on high order regularizations. We now discuss these methods in the subsequent sections.

4.1 Continuous finite element methods

Here we summarize finite element methods presented in Brenner et al. (2011), Brenner and Neilan (2012), and Neilan (2013) for the Monge–Ampère equation which employ spaces consisting of continuous, piecewise polynomials, i.e., the Lagrange finite element space. These are arguably the simplest finite element spaces and are available on virtually all finite element software programs and libraries. In addition, we provide a slightly new and improved convergence analysis based on recent results for finite element methods for linear nondivergence form PDEs (Feng et al., 2017). To describe these methods and their accompanying analysis, we require some notation.

As before, we assume that $\Omega \subset \mathbb{R}^d$ $(d=2,3)$ is a bounded, convex domain. Let $\mathcal{T}_h$ denote a shape-regular and simplicial triangulation of Ω. We denote the sets of interior and boundary $(d-1)$–dimensional faces of $\mathcal{T}_h$ by $\mathcal{F}_h^I$ and $\mathcal{F}_h^B$, restrictively. The jump of a vector valued function $\boldsymbol{v}$ across an interior face $F = \partial T_+ \cap \partial T_- \in \mathcal{F}_h^I$ is given by

$$[\![\boldsymbol{v}]\!] = \frac{1}{2}(\boldsymbol{v}_+ \otimes \boldsymbol{n}_+ + \boldsymbol{n}_+ \otimes \boldsymbol{v}_+ + \boldsymbol{v}_- \otimes \boldsymbol{n}_- + \boldsymbol{n}_- \otimes \boldsymbol{v}_-), \tag{77}$$

where $\boldsymbol{n}_\pm$ is the outward unit normal of $\partial T_\pm$, and $\boldsymbol{v}_\pm = \boldsymbol{v}|_{T_\pm}$. We also define the average of B (a scalar, vector, or matrix-valued function) across F as

$$\{\!\{B\}\!\} = \frac{1}{2}(B_+ + B_-). \tag{78}$$

If $F = \partial T_+ \cap \partial\Omega \in \mathcal{F}_h^B$, then we define

$$[\![\boldsymbol{v}]\!] = \frac{1}{2}(\boldsymbol{v}_+ \otimes \boldsymbol{n}_+ + \boldsymbol{n}_+ \otimes \boldsymbol{v}_+), \quad \{\!\{B\}\!\} = B_+. \tag{79}$$

For an integer $r \geq 2$, the Lagrange finite element space with homogeneous boundary conditions is given by

$$V_h=\{v_h\in W_0^{1,\infty}(\Omega):\ \ v_h|_T\in\mathbb{P}_r(T)\ \ \forall T\in\mathcal{T}_h\},$$

where $\mathbb{P}_r(T)$ is the space of polynomials with degree less than or equal to r with domain T. In addition, for a number $p\in(1,\infty)$ and integer m, we define

$$W^{m,p}(\mathcal{T}_h)=\prod_{T\in\mathcal{T}_h}W^{m,p}(T),\qquad V_p=W_0^{1,p}(\Omega)\cap W^{2,p}(\mathcal{T}_h),$$

and note that $V_h\subset V_p$ for all $p\in(1,\infty)$. We also set $H^m(\mathcal{T}_h)=W^{m,2}(\mathcal{T}_h)$.

Because of the noninclusion $V_h\not\subset W^{2,d}(\Omega)$, the finite element formulation (76) is not well defined if X_h is taken to be the Lagrange finite element space. A naïve approach to bypass this issue is to redefine this formulation so that integration is done piecewise over the mesh, i.e., to consider

$$\sum_{T\in\mathcal{T}_h}\int_T(f-\det D^2u_h)v_h\mathrm{d}x\qquad\forall v_h\in V_h.\tag{80}$$

While this method is well defined (i.e., all quantities are defined and bounded), it is easy to see that the scheme is ill-posed. For example, if $w_h\in V_h$ is strictly piecewise linear, then $\det D^2w_h=0$ on each $T\in\mathcal{T}_h$, and consequently, uniqueness (and stability) is dramatically lost.

The arguments given in Brenner et al. (2011) offer an alternative explanation on why the formulation (80) leads to an ill-posed problem. Namely, the main point in Brenner et al. (2011) is that the linearization of the discrete problem (80) is not consistent with respect to the linearization of the continuous problem (1a). Instead, to ensure consistency and stability, finite element methods for the Monge–Ampère problem should be designed such that the discrete linearization at the solution u is a coercive operator over the finite element space. We now explain how to construct methods with stable linearizations. To do so, we first assume that the exact solution to the Monge–Ampère equation satisfies $u\in C^{k,\alpha}(\bar{\Omega})$ with $k+\alpha>2$ and is strictly convex.

Define

$$\mathbb{F}[u]=f-\det D^2u$$

to be the Monge–Ampère operator, and let L be the linearization of F at the solution u, i.e.,

$$Lw=\lim_{t\to0}\frac{\mathbb{F}[u+tw]-\mathbb{F}[u]}{t}=-\mathrm{cof}\,D^2u:D^2w,\tag{81}$$

where cof D^2u denotes the cofactor matrix of D^2u, and ":" denotes the Frobenius inner product. The assumptions on u imply that matrix cof D^2u is positive definite on $\bar{\Omega}$ and uniformly continuous.

A consistent discretization of linear operators in nondivergence form (such as L) was introduced in Feng et al. (2017). In the case that the linear problem is given by (81), the discretization is given by $L_h : V_p \to V_h'$ with

$$
\begin{aligned}
\langle L_h v, w_h \rangle = & -\sum_{T \in \mathcal{T}_h} \int_T \left(\operatorname{cof} D^2 u : D^2 v\right) w_h \mathrm{d}x \\
& + \sum_{F \in \mathcal{F}_h^I} \int_F \{\!\{\operatorname{cof} D^2 u\}\!\} : [\![\nabla v]\!] w_h \mathrm{d}s,
\end{aligned} \tag{82}
$$

where $\langle \cdot, \cdot \rangle$ denotes the dual pairing between some Banach space and its dual. The operator L_h is clearly consistent with L: If $v \in W^{2,p}(\Omega) \cap W_0^{1,p}(\Omega)$, then $\langle L_h v, w_h \rangle = \langle Lv, w_h \rangle$ for all $w_h \in V_h$. In addition, the discrete operator is stable as the next lemma shows. We refer the reader to Feng et al. (2017) for a proof.

Lemma 16 (stability).
Define the discrete $W^{2,\,p}$-norm

$$
\begin{aligned}
\|v\|^p_{W_h^{2,p}(\Omega)} &:= \| D_h^2 v \|^p_{L^p(\Omega)} + \sum_{F \in \mathcal{F}_h^I} h_F^{1-p} \| [\![\nabla v]\!] \|^p_{L^p(F)} \qquad 1 < p < \infty, \\
\|v\|_{W_h^{2,\infty}(\Omega)} &:= \| D_h^2 v\|_{L^\infty(\Omega)} + \max_{F \in \mathcal{F}_h^I} h_F^{-1} \| [\![\nabla v]\!] \|_{L^\infty(F)},
\end{aligned}
$$

where $D_h^2 v$ is the piecewise Hessian of v. Assume that $u \in C^2(\bar{\Omega})$ and is strictly convex over $\bar{\Omega}$. Then there exists $h_0 > 0$ depending on the modulus of continuity of $D^2 u$, such that for $h \in (0, h_0]$, there holds the following inf-sup condition $(2 \le p < \infty)$

$$
\|w_h\|_{W_h^{2,p}(\Omega)} \le C \| L_h w_h \|_{L_h^p(\Omega)} := \sup_{v_h \in V_h \setminus \{0\}} \frac{\langle L_h w_h, v_h \rangle}{\| v_h \|_{L^{p'}(\Omega)}} \qquad \forall w_h \in V_h,
$$

where $1/p + 1/p' = 1$

Based on the definition of L_h and the stability results stated in Lemma 16 we can develop a consistent discretization for the Monge–Ampère problem as well as a convergence theory. Essentially, its construction is based on the observations that the expressions $\int_T (\operatorname{cof} D^2 u : D^2 v) w_h \mathrm{d}x$ and $\int_F \{\!\{\operatorname{cof} D^2 u\}\!\} : [\![\nabla v]\!] w_h \, ds$ are the linearizations of $\int_T (f - \det D^2 v) w_h$ and $\int_F \{\!\{\operatorname{cof} D^2 v\}\!\} : [\![\nabla v]\!] w_h \mathrm{d}s$, respectively, about the solution u. With this in mind, we define the discrete operator $\mathbb{F}_h : V \to V_h'$ via

$$
\langle \mathbb{F}_h[v], w \rangle = \sum_{T \in \mathcal{T}_h} \int_T (f - \det D^2 v) w_h \mathrm{d}x + \sum_{F \in \mathcal{F}_h^I} \int_F \{\!\{\operatorname{cof} D^2 v\}\!\} : [\![\nabla v]\!] w_h \mathrm{d}s,
$$

and consider the finite element method: Find $u_h \in V_h$ such that

$$\begin{array}{ccc} \mathbb{F}(u)=0 & \xrightarrow{\text{Discretize}} & \mathbb{F}_h(u_h)=0 \\ \big\downarrow \text{Linearize} & & \big\downarrow \text{Linearize} \\ L(w)=0 & \xrightarrow{\text{Discretize}} & L_h(w_h)=0 \end{array}$$

FIG. 7 A commuting diagram connecting the nonlinear problems and their discretizations.

$$\langle \mathbb{F}_h[u_h], v_h\rangle = 0 \quad \forall v_h \in V_h. \tag{83}$$

We immediately see that method (83) is consistent: There holds $[\![\nabla u]\!]|_F = 0$ over all interior faces F, and therefore $\langle \mathbb{F}_h[u], v_h\rangle = 0$ for all $v_h \in V_h$. Furthermore, the proceeding discussion implies that L_h is the linearization of $\mathbb{F}_h$:

$$L_h w = \lim_{t\to 0} \frac{\mathbb{F}_h[u+tw] - \mathbb{F}_h[u]}{t} \quad \text{in } V_h'.$$

In summary the diagram given in Fig. 7 commutes. We now show that this property (along with the regularity and convexity assumptions of u) implies that there exists a locally unique solution to (83) with optimal rates of convergence.

As a first step, we first point out that Lemma 16 implies that $L_h|_{V_h}$ is bijective. Therefore, the mapping $M_h : V_p \to V_h$ given by

$$M_h = \left(L_h|_{V_h}\right)^{-1} (L_h - \mathbb{F}_h) \tag{84}$$

is well defined. The existence of a solution to the finite element method (83) is proven by showing that M_h has a fixed point in a ball centred at $u_{c,h}$, where $u_{c,h}$ is the elliptic projection of u given by

$$u_{c,h} := \left(L_h|_{V_h}\right)^{-1} L_h u. \tag{85}$$

The basis of this argument is provided in the next lemma.

Lemma 17 (M_h is Lipschitz).

Assume that the convex solution of the Monge–Ampère equation satisfies $u \in C^{k,\alpha}(\bar{\Omega})$ *with* $k + \alpha > 2$. *Then there holds, for all* $p \in [2, \infty)$ *and all* $v_1, v_2 \in V_p$,

$$\|M_h v_1 - M_h v_2\|_{W_h^{2,p}(\Omega)} \le C_1 \left\| u - \frac{1}{2}(v_1 + v_2)\right\|_{W_h^{2,\infty}(\Omega)} \|v_1 - v_2\|_{W_h^{2,p}(\Omega)},$$

where $C_1 > 0$ *depends on* p *and* u, *but is independent of* h.

Proof. We give the proof of the two-dimensional case $d = 2$; the arguments in three dimensions are similar and can be found in Brenner and Neilan (2012).

We first use Taylor's Theorem and the fact that F_h is quadratic in two dimensions, to get

$$\mathbb{F}_h[v] = \mathbb{F}_h[u] + L_h(v-u) + R_h[v-u] = L_h(v-u) + R_h[v-u],$$

where $R_h : V \to V_h'$ is quadratic in its arguments and independent of u. Using this expansion into the mapping M_h yields

$$\begin{aligned} M_h[v_1] - M_h[v_2] &= \left(L_h|_{V_h}\right)^{-1} (L_h v_1 - L_h v_2 - (\mathbb{F}_h[v_1] - \mathbb{F}_h[v_2])) \\ &= \left(L_h|_{V_h}\right)^{-1} (R_h[v_2 - u] - R_h[v_1 - u]). \end{aligned} \tag{86}$$

Since R_h is quadratic there holds

$$\begin{aligned} R_h[v_2 - u] - R_h[v_1 - u] &= \int_0^1 DR_h[t(v_2 - u) + (1-t)(v_1 - u)](v_2 - v_1)\mathrm{d}t \\ &= DR_h(\frac{1}{2}(v_2 + v_1) - u)(v_2 - v_1), \end{aligned}$$

where by DR_h we denoted the derivative of R_h. Therefore, by (86) and Lemma 16 we have

$$\|M_h v_1 - M_h v_2\|_{W_h^{2,p}(\Omega)} \leq C \left\| DR_h\left(\frac{1}{2}(v_2 + v_1) - u\right)(v_2 - v_1)\right\|_{L_h^p(\Omega)}.$$

Several applications of Hölder's inequality yields (cf. Neilan, 2013, Lemma 4.2)

$$\|DR_h(w)(q)\|_{L_h^p(\Omega)} \leq C\,\|w\|_{W_h^{2,\infty}(\Omega)}\,\|q\|_{W_h^{2,p}(\Omega)},$$

and therefore

$$\|M_h v_1 - M_h v_2\|_{W_h^{2,p}(\Omega)} \leq C \left\|\frac{1}{2}(v_1 + v_2) - u\right\|_{W_h^{2,\infty}(\Omega)} \|v_1 - v_2\|_{W_h^{2,p}(\Omega)}.$$

□

Lemma 18 (contraction).
Assume that the hypotheses of Lemma 17 *are satisfied. For fixed* $\rho > 0$ *and* $p \in [2, \infty)$*, define the closed ball*

$$B_{\rho,p} = \left\{v_h \in V_h : \quad \|u_{c,h} - v_h\|_{W_h^{2,p}(\Omega)} \leq \rho\right\},$$

where $u_{c,h} \in V_h$ *is defined by* (85). *Then, for all* $v_1, v_2 \in B_{\rho,p}$*, there holds*

$$\|M_h v_1 - M_h v_2\|_{W_h^{2,p}(\Omega)} \leq C_2 h^{-d/p}\left(h^{\ell+\alpha} + \rho\right)\|v_1 - v_2\|_{W_h^{2,p}(\Omega)},$$

where $\ell = \min\{r-2, k-2\}$.

Proof. First, the smoothness assumptions on u allows us to conclude that the elliptic projection $u_{c,h}$ satisfies (Feng et al., 2017, Theorem 3.2)

$$\|u - u_{c,h}\|_{W_h^{2,p}(\Omega)} \le C_3 h^{\ell+\alpha} \quad p \in [2, \infty), \tag{87}$$

where $C_3 > 0$ depends on p and $\|u\|_{C^{k,\alpha}(\bar{\Omega})}$. Consequently, there holds by an inverse estimate, for any $w_h \in V_h$,

$$\begin{aligned}\|u - u_{c,h}\|_{W_h^{2,\infty}(\Omega)} &\le \|u - w_h\|_{W_h^{2,\infty}(\Omega)} + Ch^{-d/p}\, \|u_{c,h} - w_h\|_{W_h^{2,p}(\Omega)} \\ &\le \|u - w_h\|_{W_h^{2,\infty}(\Omega)} + Ch^{-d/p}\left(\|u - u_{c,h}\|_{W_h^{2,p}(\Omega)} + \|u - w_h\|_{W_h^{2,p}(\Omega)}\right).\end{aligned}$$

Taking w_h to be the nodal interpolant of u yields

$$\|u - u_{c,h}\|_{W_h^{2,\infty}(\Omega)} \le C_4 h^{\ell+\alpha-d/p}. \tag{88}$$

Applying this result to Lemma 17 and using an inverse estimate, we obtain

$$\begin{aligned}&\|M_h v_1 - M_h v_2\|_{W_h^{2,p}(\Omega)} \\ &\quad\le C\left(\|u - u_{c,h}\|_{W_h^{2,\infty}(\Omega)} + h^{-d/p}\, \|u_{c,h} - \frac{1}{2}(v_1 + v_2)\|_{W_h^{2,p}(\Omega)}\right) \|v_1 - v_2\|_{W_h^{2,p}(\Omega)} \\ &\quad\le Ch^{-d/p}\left(h^{\ell+\alpha} + \rho\right) \|v_1 - v_2\|_{W_h^{2,p}(\Omega)}\end{aligned}$$

for all $v_1,\ v_2 \in B_{\rho,p}$. □

Theorem 15 (error estimate).
Assume that $u \in C^{k,\alpha}(\bar{\Omega})$ with $k + \alpha > 2$ and is strictly convex. Set $\ell = \min\{r-2, k-2\}$. There exists $h_1 > 0$ such that for $h \le h_1$, there exists a solution to (83) *satisfying*

$$\|u - u_h\|_{W_h^{2,p}(\Omega)} \le Ch^{\ell+\alpha}. \tag{89}$$

Moreover, if $\tilde{u}_h$ is another solution to (83) *then there holds $\|u - \tilde{u}_h\|_{W_h^{2,\infty}(\Omega)} \ge C$, with the constant $C > 0$ independent of h.*

Proof. Fix $p \in [2, \infty)$ such that $\ell + \alpha - d/p > 0$, and let

$$h_1 = \min\{1/(4C_2), 1/(2C_1C_2C_3C_4)\}^{1/(\alpha+\ell-d/p)}.$$

Then, for $h \le \min\{h_0, h_1\}$, where h_0 was defined in Lemma 16, set $\rho_1 = h^{\ell+\alpha}/(4C_2)$. Lemma 18 then shows that, for $v_1, v_2 \in B_{\rho_1,p}$,

$$\begin{aligned}\|M_h v_1 - M_h v_2\|_{W_h^{2,p}(\Omega)} &\le C_2\left(h^{\ell+\alpha-d/p} + h^{-d/p}\rho_1\right) \|v_1 - v_2\|_{W_h^{2,p}(\Omega)} \\ &\le 2C_2 h_1^{\alpha+\ell-d/p}\, \|v_1 - v_2\|_{W_h^{2,p}(\Omega)} \le \frac{1}{2}\|v_1 - v_2\|_{W_h^{2,p}(\Omega)},\end{aligned}$$

and therefore $M_h|_{V_h}$ is a contraction mapping on $B_{\rho_1,p}$. Likewise, we can use Lemma 17 and the fact that $u_{c,h} = M_h u$ to get (cf. (87) and (88))

$$\begin{aligned}\|u_{c,h} - M_h v\|_{W_h^{2,p}(\Omega)} &= \|M_h u - M_h v\|_{W_h^{2,p}(\Omega)} \\ &\le \frac{C_1}{2} \|u - u_{c,h}\|_{W_h^{2,\infty}(\Omega)} \|u - u_{c,h}\|_{W_h^{2,p}(\Omega)} \\ &\le \frac{C_1 C_3 C_4 h^{2\alpha+2\ell-d/p}}{2} \le \frac{h^{\ell+\alpha}}{4C_2} = \rho_1.\end{aligned}$$

Therefore M_h maps $B_{\rho_1,p}$ to itself. By Banach's fixed point theorem, we conclude that M_h has a fixed point in $B_{\rho_1,p}$, and this fixed point is a solution to (83). The error estimate for $\alpha - d/p > 0$ (89) follows from the inclusion $u_h \in B_{\rho_1,p}$ and the definition of ρ_1. The other cases $\ell + \alpha - d/p \le 0$ then follow from Hölder's inequality.

Finally, if $\tilde{u}_h \in V_h$ is another solution to (83), then there holds $M_h \tilde{u}_h = \tilde{u}_h$. Therefore, by Lemma 17 we conclude that

$$\begin{aligned}\|\tilde{u}_h - u_h\|_{W_h^{2,p}(\Omega)} &= \|M_h \tilde{u}_h - M_h u_h\|_{W_h^{2,p}(\Omega)} \\ &\le \frac{C_1}{2}\left(\|u - u_h\|_{W_h^{2,\infty}(\Omega)} + \|u - \tilde{u}_h\|_{W_h^{2,\infty}(\Omega)}\right) \|u_h - \tilde{u}_h\|_{W_h^{2,p}(\Omega)}.\end{aligned}$$

Now applying similar arguments as those found in Lemma 17, we conclude that $\|u - u_h\|_{W_h^{2,\infty}(\Omega)} \le C h^{\alpha - d/p} \to 0$. Therefore, by dividing by $\|u_h - \tilde{u}_h\|_{W_h^{2,p}(\Omega)}$, we get $C \le \|u - \tilde{u}_h\|_{W_h^{2,\infty}(\Omega)}$ for h sufficiently small. □

Remark 16 (extensions).
The proposed method and the conclusion of Theorem 15 deserve the following comments:

- As mentioned earlier, the analysis given here slightly improves the results given in Brenner et al. (2011) and Neilan (2013). Namely, the paper (Brenner et al., 2011) requires $d = 2$, $r \ge 3$, and $u \in H^s(\Omega)$ for $s > 3$ (implying that $u \in C^{2,\alpha}(\bar{\Omega})$ by a Sobolev embedding). The paper (Neilan, 2013) requires $r \ge 2$ and regularity $u \in W^{3,\infty}(\Omega)$ to carry out the analysis.
- Discontinuous Galerkin methods have also been developed under this methodology in Neilan (2013). The analysis carried out in this section can be applied to these methods using the recent results for nondivergence PDEs given in Feng et al. (2018).
- A two-grid method to solve the nonlinear method has recently been proposed in Awanou et al. (2018). ■

4.2 Mixed formulations

In this section we describe mixed finite element formulations for the Monge–Ampère equation proposed in Lakkis and Pryer (2011), Neilan (2014a), Awanou (2015a); Awanou and Li (2014), Awanou (2017a), and Kawecki et al. (2018). Essentially, the main idea in these approaches is to introduce the Hessian matrix of u as an additional auxiliary unknown in the formulation of the Monge–Ampère problem, that is, we write the PDE (1a) as

$$\sigma = D^2 u, \qquad \det\sigma = f \quad \text{in}\, \Omega. \tag{90}$$

As before, assuming regularity $u \in W^{2,d}(\Omega)$ so that $\sigma \in L^d(\Omega)$, we can multiply the second equation by a smooth test function and integrate over the domain:

$$\int_\Omega (f - \det\sigma) v\mathrm{d}x = 0 \tag{91}$$

for all $v \in L^\infty(\Omega)$.

The direct analogue of this formulation in the discrete setting requires C^1 finite element spaces by the same reasons that the method described in Section 4.1 does. In other words, to ensure that the discrete version of (91) is well–defined, we require that the Hessian of the discrete approximation u_h has (global) second-order derivatives in $L^d(\Omega)$; if u_h is a piecewise polynomial, then this restriction implies that $u \in C^1(\Omega)$. To relax this restriction on the finite element spaces, one can instead develop finite element methods that only employ continuous (or discontinuous) bases based on this formulation by introducing the notion of a discrete Hessian (also known as a finite element Hessian (Lakkis and Pryer, 2011)). The discrete Hessian is defined globally via an integration by parts procedure rather than a piecewise fashion. This idea has been carried out for (linear) Kirchhoff plates in Huang et al. (2010), and its formulation is reminiscent of the construction of local discontinuous Galerkin methods for second-order problems (Arnold et al., 2002; Cockburn and Shu, 1998).

To motivate the definition of the discrete Hessian, we introduce the auxiliary space

$$\Sigma_h = \{\tau_h \in L^\infty(\Omega;\mathbb{R}^{d\times d}) : \ \tau_h|_T \in \mathbb{P}_r(T;\mathbb{R}^{d\times d}) \ \ \forall T \in \mathcal{T}_h\},$$

and note the following integration by parts identity

$$\begin{aligned}\sum_{T\in\mathcal{T}_h} \int_T D^2 w : \tau_h \mathrm{d}x = &- \sum_{T\in\mathcal{T}_h} \int_T (\nabla\cdot\tau_h)\cdot\nabla w \mathrm{d}x \\ &+ \sum_{T\in\mathcal{T}_h} \int_{\partial T} (\tau_h \boldsymbol{n}_T)\cdot\nabla w \mathrm{d}s,\end{aligned} \tag{92}$$

for all $w \in H^2(\Omega)$ and $\tau_h \in \Sigma_h$. Here, $\boldsymbol{n}_T$ is the outward unit normal of ∂T, and the divergence acting on a matrix is performed row-wise. We may then write the integral boundary terms in (92) using the jump and average operators. In addition to (77)–(79), we define the jump of a matrix-valued function τ across $F = \partial T_+ \cap T_- \in \mathcal{F}_h^I$ as

$$[\![\tau]\!] = \tau_+ \boldsymbol{n}_+ + \tau_- \boldsymbol{n}_-,$$

and define $[\![\tau]\!] = \tau_+ \boldsymbol{n}_+$ if $F = \partial T_+ \cap \partial\Omega \in \mathcal{F}_h^B$. We then have

$$\begin{aligned}\sum_{T\in\mathcal{T}_h}\int_{\partial T}(\tau_h \boldsymbol{n}_T)\cdot\nabla w \mathrm{d}s &= \sum_{F\in\mathcal{F}_h^I}\int_F \{\!\{\tau_h\}\!\} : [\![\nabla w]\!]\mathrm{d}s + \sum_{F\in\mathcal{F}_h}\int_F [\![\tau_h]\!]\cdot\{\!\{\nabla w\}\!\}\mathrm{d}s \\ &= \sum_{F\in\mathcal{F}_h}\int_F [\![\tau_h]\!]\cdot\{\!\{\nabla w\}\!\}\mathrm{d}s,\end{aligned}$$

where we used that $[\![\nabla w]\!]|_F = 0$ for all $F \in \mathcal{F}_h^I$ due to the regularity $w \in H^2(\Omega)$. Combining this identity with (92), we arrive at

$$\sum_{T\in\mathcal{T}_h}\int_T D^2 w : \tau_h \mathrm{d}x = -\sum_{T\in\mathcal{T}_h}\int_T (\nabla\cdot\tau_h)\cdot\nabla w \mathrm{d}x + \sum_{F\in\mathcal{F}_h}\int_F [\![\tau_h]\!]\cdot\{\!\{\nabla w\}\!\}\mathrm{d}s.$$

This identity leads to the following definitions of the discrete Hessian.

Definition 21 (discontinuous discrete Hessian).

The *discontinuous discrete Hessian* is the operator $\mathbb{H}_h : H^1(\Omega)\cap H^2(\mathcal{T}_h) \to \Sigma_h$ uniquely defined by the conditions

$$\int_\Omega \mathbb{H}_h(w) : \tau_h \mathrm{d}x = -\sum_{T\in\mathcal{T}_h}\int_T (\nabla\cdot\tau_h)\cdot\nabla w \mathrm{d}x + \sum_{F\in\mathcal{F}_h}\int_F [\![\tau_h]\!]\cdot\{\!\{\nabla w\}\!\}\mathrm{d}s$$

for all $\tau_h \in \Sigma_h$.

Remark 17 (characterization through liftings).

Define the lifting operator

$$\Theta : L^2(\mathcal{F}_h^I;\mathbb{R}^d) \to \Sigma_h$$

via

$$\int_\Omega \Theta(v) : \tau_h \mathrm{d}x = -\sum_{F\in\mathcal{F}_h^I}\int_F \{\!\{\tau_h\}\!\} : [\![v]\!]\mathrm{d}s \quad \forall \tau_h \in \Sigma_h.$$

Integrating by parts we obtain

$$\begin{aligned}\sum_{T\in\mathcal{T}_h}\int_T \mathbb{H}_h(w) : \tau_h \mathrm{d}x &= \sum_{T\in\mathcal{T}_h}\int_T D^2 w : \tau_h \mathrm{d}x - \sum_{F\in\mathcal{F}_h^I}\int_F \{\!\{\tau_h\}\!\} : [\![\nabla w]\!]\mathrm{d}s \\ &= \sum_{T\in\mathcal{T}_h}\int_T \left(D^2 w + \Theta(\nabla w)\right) : \tau_h \mathrm{d}x.\end{aligned}$$

Recalling that $D_h^2 w$ denotes the piecewise Hessian of w, and that V_h is the (scalar) Lagrange space of degree r, we then have $D_h^2 V_h \subset \Sigma_h$, and therefore

$$\mathbb{H}_h(w_h) = D_h^2 w_h + \Theta(\nabla w_h) \quad \forall w_h \in V_h.$$

■

The notion of the discrete Hessian and the formal identities (90) and (91) lead to the following scheme introduced in Neilan (2014a): Find $u_h \in V_h$ such that

$$\int_\Omega (f - \det \mathbb{H}_h(u_h)) v_h \mathrm{d}x \quad \forall v_h \in V_h. \tag{93}$$

Remark 18 (mixed formulation).
While (93) is written in primal form, the problem is in fact a mixed finite element method. Introducing $\sigma_h = \mathbb{H}_h(u_h) \in \Sigma_h$, we see from the definition of the discrete Hessian that (93) is equivalent to the system

$$\int_\Omega \sigma_h : \tau_h \mathrm{d}x + \int_\Omega (\nabla \cdot \tau_h) \cdot u_h \mathrm{d}x - \sum_{F \in \mathcal{F}_h} \int_F [\![\tau_h]\!] \cdot \{\!\{\nabla u_h\}\!\} \mathrm{d}s = 0, \tag{94a}$$

$$\int_\Omega (f - \det \sigma_h) v_h \mathrm{d}x = 0, \tag{94b}$$

for all $(\tau_h, v_h) \in \Sigma_h \times V_h$. Note that the matrix representation of the form $(\sigma_h, \tau_h) \to \int_\Omega \sigma_h : \tau_h \mathrm{d}x$ is symmetric positive definite, and more importantly, block-diagonal because Σ_h does not have any continuity constraints. As a result, the Schur complement (i.e., the primal method (93)) represents a sparse algebraic system of equations. ■

Theorem 16 (error estimate).
Assume that $d = 2$, and that (1) *has a unique strictly convex solution $u \in C^{r+3,\,\alpha}(\Omega)$ with $r \geq 3$ and $\alpha > 0$. Then for h sufficiently small, there exists a locally unique solution to the finite element method* (93). *Moreover, there holds*

$$\|u - u_h\|_{H^1(\Omega)} + h\,\|\sigma - \sigma_h\|_{L^2(\Omega)} \leq Ch^r. \tag{95}$$

Proof. See Neilan (2014a, Theorem 4.2). □

Remark 19 (regularity).
The regularity assumptions on u in Theorem 16 can be relaxed using the stability analysis for linear nondivergence form PDES found in Neilan (2017). There it is shown that, assuming $u \in C^2(\bar{\Omega})$,

$$\|w_h\|_{W_h^{2,2}(\Omega)} \leq C\,\|L_h w_h\|_{L_h^2(\Omega)} \quad \forall w_h \in V_h,$$

with

$$\langle L_h w_h, v_h \rangle = -\int_\Omega \operatorname{cof} D^2 u : \mathbb{H}_h(w_h) v_h \mathrm{d}x.$$

By applying the same techniques found in the previous section, it is simple to show that the solution to (93) satisfies $\|u-u_h\|_{W_h^{2,2}(\Omega)} \le Ch^{\ell+\alpha}$ with $\ell = \min\{r-2, k-2\}$ provided that $u \in C^{k,\alpha}(\bar{\Omega})$ with $k + \alpha > 3$, $r \ge 3$, and h is sufficiently small.

To reduce the number of unknowns in the mixed system (94), continuity constraints can be added in the matrix–valued space Σ_h. This is the idea of the method proposed in Lakkis and Pryer (2013). There, the auxiliary space is defined as the matrix-valued Lagrange space, i.e.,

$$\Sigma_h^c := \Sigma_h \cap H^1(\Omega;\mathbb{R}^{d\times d}) = \{\tau_h \in H^1(\Omega):\ \ \tau_h \in \mathbb{P}_r(T;\mathbb{R}^{d\times d})\ \ \forall T \in \mathcal{T}_h\}.$$

Restricting Definition 21 to Σ_h^c leads to the following notation of the discrete Hessian.

Definition 22 (continuous discrete Hessian).
The *continuous discrete Hessian* is the operator $\mathbb{H}_h^c : H^1(\Omega) \cap H^2(\mathcal{T}_h) \to \Sigma_h^c$ uniquely defined by the conditions

$$\int_\Omega \mathbb{H}_h^c(w) : \tau_h \mathrm{d}x \ = -\int_\Omega (\nabla \cdot \tau_h) \cdot \nabla w \mathrm{d}x + \int_{\partial\Omega} (\tau_h \boldsymbol{n}) \cdot \nabla w \mathrm{d}s$$

for all $\tau_h \in \Sigma_h^c$.

This definition leads to a finite element method proposed in Lakkis and Pryer (2013) which similar to (93), but with the continuous version of the discrete Hessian.

$$\int_\Omega (f - \det \mathbb{H}_h^c(u_h)) v_h \mathrm{d}x = 0 \quad \forall v_h \in V_h. \tag{96}$$

As before, we may set $\sigma_h = \mathbb{H}_h^c(u_h)$ as an auxiliary variable, and deduce from Definition 22 that (96) is equivalent to the mixed method

$$\int_\Omega \sigma_h : \tau_h \mathrm{d}x + \int_\Omega (\nabla \cdot \tau_h) \cdot \nabla u_h \mathrm{d}x - \int_{\partial\Omega} (\tau_h \boldsymbol{n}) \cdot \nabla u_h \mathrm{d}s \ = 0, \tag{97a}$$

$$\int_\Omega (f - \det \sigma_h) v_h \mathrm{d}x \ = 0, \tag{97b}$$

for all $(\tau_h, v_h) \in \Sigma_h^c \times V_h$ Compared with the formulation using the discontinuous discrete Hessian, the mixed problem (97) has significantly less unknowns than (94) due to the continuity restrictions of Σ_h^c. On the other hand, the (mass) matrix associated with the form $(\sigma_h, \tau_h) \to \int_\Omega \sigma_h : \tau_h$ is not block-diagonal, and therefore the Schur complement of (97) (i.e., the algebraic system representing the primal problem (96)) is dense.

Existence, (local) uniqueness, and error estimates for method (97) are similar to the statements given in Theorem 16.

Theorem 17 (error estimates).
Assume that $d \in \{2, 3\}$, and that (1) *has a unique strictly convex solution $u \in H^{r+3}(\Omega)$ with $r \geq d$. Then for h sufficiently small, there exists a locally unique solution to the finite element method* (97). *Moreover, there holds*

$$\|u - u_h\|_{H^1(\Omega)} + h\,\|\sigma - \sigma_h\|_{L^2(\Omega)} \leq Ch^r. \tag{98}$$

Proof. See Awanou and Li (2014, Theorem 3.13) and Awanou (2015a, 2017a, Theorem 1). □

Remark 20 (extension to optimal transport).
The mixed finite element method (97) has recently been extended to the optimal transport problem in Kawecki et al. (2018). ■

Remark 21 (historical remark).
Our presentation follows a reverse chronological order. The first Galerkin-type method based on the concept of discrete Hessians was that of Lakkis and Pryer (2013), where they used the continuous Hessian of Definition 22. The DG version was introduced later. ■

4.3 Galerkin methods for singular solutions

The analysis of the Galerkin methods discussed thus far require relatively stringent regularity conditions to carry out the analysis (e.g., $u \in C^{2,\alpha}(\Omega)$). While numerical experiments indicate that regularity assumptions can be relaxed somewhat, they also indicate that some regularity of the solution is required for the methods to converge. For example, the numerical experiments in Brenner et al. (2011) indicate that the C^0 penalty method (83) does not converge if $u \notin H^2(\Omega)$ in two dimensions. In this section, we discuss various ways to modify the Galerkin methods and the analysis such that the resulting numerical scheme is robust with respect to the solution's regularity.

The first approach, introduced in Feng and Neilan (2009), regularizes the problem at the PDE level by adding a higher order perturbation, resulting in a fourth-order, quasi-linear problem. The motivation of this approach is that solutions of the regularized problem are defined via variational principles, so that weak formulations can be obtained via integration by parts, and therefore the resulting PDE framework is amenable to Galerkin methods. Applying this methodology to the Monge–Ampère problem results in

$$-\epsilon\Delta^2 u^\epsilon + \det D^2 u^\epsilon = f \qquad \text{in}\,\Omega, \tag{99a}$$

$$u = 0 \qquad \text{on}\,\partial\Omega, \tag{99b}$$

where $\epsilon > 0$ and $\Delta^2 = \Delta\Delta$ denotes the biharmonic operator. Note that, due to the higher order of the PDE, the Dirichlet boundary condition is no longer sufficient to close the system. In Feng and Neilan (2009), the following additional boundary conditions are proposed:

$$\Delta u^\epsilon = 0, \quad \text{or} \quad \frac{\partial \Delta u^\epsilon}{\partial \boldsymbol{n}} = 0 \quad \text{on}\,\partial\Omega. \tag{99c}$$

These conditions are chosen so that the resulting boundary layer is minimized; see Feng and Neilan (2009) for details. For the sake of illustration, we take the first boundary condition in (99c) in the discussion below.

Since the problem (99) is quasi–linear and in divergence–form, the notion of weak solutions is easily defined.

Definition 23 (weak solution).
A function $u \in W^{2,d}(\Omega) \cap W_0^{1,d}(\Omega)$ is a weak solution to (99) provided that

$$-\epsilon \int_\Omega \Delta u^\epsilon \Delta v \mathrm{d}x + \int_\Omega v \det D^2 u^\epsilon \mathrm{d}x = \int_\Omega f v \mathrm{d}x \quad \forall v \in W^{2,d}(\Omega) \cap W_0^{1,d}(\Omega). \tag{100}$$

The function $u = \lim_{\epsilon \downarrow 0} u^\epsilon$, if it exists, is called a weak (resp., strong) moment solution to the Monge–Ampère problem if convergence holds in a $W^{1,d}$-weak (resp., $W^{2,d}$-weak) topology.

Remark 22 (relation to other solution concepts).
Except in very simple settings (e.g., radially symmetric solutions (Feng and Neilan, 2014)), the existence of moment solutions and their relation with viscosity and Alexandrov solutions is an open problem. Nonetheless, numerical experiments indicate that this methodology leads to robust numerical methods with respect to regularity of the solution of the Monge–Ampère equation. For example, numerical methods applied to problem (99) are able to capture viscosity/Alexandrov solutions that are merely Lipschitz continuous. ■

Constructing methods for the regularized problem (100) can be done by applying any of the above Galerkin methods described above; one only needs to tack on a consistent and stable discretization of the biharmonic operator to the discrete formulation. For example, the simplest method, at least in theory, is to restrict the variational formulation (100) onto a finite dimensional subspace of $W^{2,d}(\Omega) \cap W_0^{1,d}(\Omega)$. This results in the method to find $u_h^\epsilon \in X_h$ satisfying

$$\epsilon \int_\Omega \Delta u_h^\epsilon \Delta v_h \mathrm{d}x + \int_\Omega (f - \det D^2 u_h^\epsilon) v_h \, dx = 0 \quad \forall v_h \in X_h, \tag{101}$$

with $X_h \subset C^1(\Omega) \cap W_0^{1,d}(\Omega)$. A convergence analysis of this discrete problem has been done in Feng and Neilan (2011). There it is shown that, if there exists a moment solution with sufficient regularity, then there exists a locally unique solution to the discrete problem (101).

Analogously, combining the C^0 finite element method (83) with the symmetric C^0 interior penalty method for the biharmonic problem introduced in Engel et al. (2002) and Brenner and Sung (2005) results in the method: Find $u_h^\epsilon \in V_h$ satisfying

$$\begin{aligned} &\epsilon \sum_{T \in \mathcal{T}_h} \int_T \Delta u_h^\epsilon \Delta v_h \mathrm{d}x \\ &- \epsilon \sum_{F \in \mathcal{F}_h^I} \int_F \Big(\{\!\{\Delta u_h^\epsilon\}\!\} (I : [\![\nabla v_h]\!]) + \{\!\{\Delta v_h\}\!\} (I : [\![\nabla u_h^\epsilon]\!]) \\ &- \frac{\sigma}{h_F} [\![\nabla u_h^\epsilon]\!] : [\![\nabla v_h]\!] \Big) \mathrm{d}s + \sum_{T \in \mathcal{T}_h} \int_T (f - \det D^2 u_h^\epsilon) v_h \mathrm{d}x \\ &+ \sum_{F \in \mathcal{F}_h^I} \int_F \{\!\{\operatorname{cof} D^2 u_h^\epsilon\}\!\} : [\![\nabla u_h^\epsilon]\!] v_h \mathrm{d}s = 0 \end{aligned} \tag{102}$$

for all $v_h \in V_h$. Here, $\sigma > 0$ is a penalty parameter, and we recall that I denotes the $d \times d$ identity matrix and V_h is the Lagrange finite element space of degree $r \geq 2$ with homogeneous Dirichlet boundary conditions. The method (102) can be written succinctly as

$$\epsilon \langle A_h u_h^\epsilon, v_h \rangle + \langle \mathbb{F}_h[u_h^\epsilon], v_h \rangle = 0 \quad \forall v_h \in V_h,$$

where the operator $\mathbb{F}_h$ is defined by (83), and A_h is a consistent discretization of the biharmonic operator given by

$$\begin{aligned} \langle A_h w, v_h \rangle = &\sum_{T \in \mathcal{T}_h} \int_T \Delta w \Delta v_h \mathrm{d}x - \sum_{F \in \mathcal{F}_h^I} \int_F \Big(\{\!\{\Delta w\}\!\} (I_d : [\![\nabla v_h]\!]) \\ &+ \{\!\{\Delta v_h\}\!\} (I_d : [\![\nabla w_h]\!]) - \frac{\sigma}{h_F} [\![\nabla w]\!] : [\![\nabla v_h]\!] \Big) \mathrm{d}s. \end{aligned}$$

Arguments given in Brenner and Sung (2005); Engel et al. (2002) show that there exists $\sigma_0 > 0$, independent of h, such that $\langle A_h v_h, v_h \rangle \geq C \| v_h \|^2_{W_h^{2,2}(\Omega)}$ for all $v_h \in V_h$ provided that $\sigma \geq \sigma_0$. Moreover, there holds $\epsilon \langle A_h u^\epsilon, v_h \rangle + \langle \mathbb{F}_h[u^\epsilon], v_h \rangle = 0$ for all $v_h \in V_h$ provided that $u^\epsilon \in H^s(\Omega)$ for some $s > 5/2$. Thus, the method (102) is consistent.

While a convergence analysis of the regularized PDE (99) and the discretization (102) is an open problem, we show, via numerical experiments in the next section, that the method is able to capture nonsmooth solutions for the

Monge–Ampère problem in a variety of settings. In addition, as shown in Brenner et al. (2011), Newton's method is robust for the regularized solution, which allows a natural way to construct initial guesses for the (unregularized) problem (83).

4.3.1 Convergence of interior discretizations

Recent results given in Awanou (2015b, 2016, 2017b); Awanou and Awi (2016) argue that, in certain settings, standard discretizations (both finite element and finite difference) for the Monge–Ampère equation converge to the Alexandrov solution as the discretization parameter tends to zero. Here, in this section, we summarize these results and the techniques to obtain them.

As always, we assume that Ω is convex. More importantly, we assume also that the Dirichlet boundary conditions can be extended to a function $\tilde{g}$ that is convex on Ω. Note that the existence of $\tilde{g}$ is guaranteed if the domain is strictly convex. However, due to our assumption that $u|_{\partial\Omega} = 0$, we may simply take $\tilde{g} \equiv 0$ in our setting. We further assume that $f \in C(\bar{\Omega})$ with $f \geq C > 0$ on Ω. Let $\{f_m\}_{m=0}^{\infty} \subset C^{\infty}(\bar{\Omega})$ be a sequence of approximations of f with $f_m \to f$ uniformly on $\bar{\Omega}$ and $f_m \geq C > 0$ for all m. We then consider the PDE problem

$$\det D^2 u_m = f_m \quad \text{in } \Omega, \tag{103a}$$

$$u_m = 0 \quad \text{on } \partial\Omega. \tag{103b}$$

Even though the source data of this problem is smooth, in general there does not exist smooth solutions to (103) because Ω is not necessarily strictly convex nor smooth, see Theorem 1. Nonetheless, there exists a unique (convex) Alexandrov solution $u_m \in C(\bar{\Omega})$.

Let $\tilde{\Omega} \subset \Omega$ be a strict subdomain of Ω that is polyhedral and convex, and let $\tilde{\mathcal{T}}_h$ be a simplicial triangulation of $\tilde{\Omega}$. Finally, we denote by $\tilde{X}_h$ a $C^1(\bar{\tilde{\Omega}})$-conforming finite element space consisting of piecewise polynomials with respect to $\tilde{\mathcal{T}}_h$. We then consider the finite element method: Find $\tilde{u}_h \in \tilde{X}_h$ satisfying $\tilde{u}_{m,h} = u_m$ on $\partial\tilde{\Omega}$ and

$$\int_{\tilde{\Omega}} (f_m - \det D^2 \tilde{u}_{m,h}) v_h \, dx = 0 \quad \forall v_h \in \tilde{X}_h \cap W_0^{1,d}(\tilde{\Omega}). \tag{104}$$

This method is similar to (76), the differences being

(i) the problem is posed on $\tilde{\Omega}$ instead of Ω;
(ii) the source function has been regularized;
(iii) the homogeneous Dirichlet boundary conditions have been replaced by

$$\tilde{u}_{m,h}|_{\partial\tilde{\Omega}} = u_m|_{\partial\tilde{\Omega}}.$$

It is clear that method (104) is a discretization of the PDE problem

$$\det D^2\tilde{u}_m = f_m \qquad \text{in } \tilde{\Omega}, \tag{105a}$$

$$\tilde{u}_m = u_m \qquad \text{on } \partial\tilde{\Omega}, \tag{105b}$$

which, similar to (103), has a unique Alexandrov solution and is generally nonsmooth. In fact, it is simple to see that, due to the inclusion $\tilde{\Omega} \subset \Omega$ and the uniqueness of Alexandrov solutions, that $\tilde{u}_m = u_m$ on $\tilde{\Omega}$.

Theorem 18 (interior convergence).
There exists $h_0 > 0$, which depends on $\text{dist}\{\partial\Omega, \partial\tilde{\Omega}\}$, such that for $h \leq h_0$, there exists a locally unique solution to (104). *In addition, as $h \to 0$, $\tilde{u}_{m,h}$ converges uniformly to $\tilde{u}_m$ (the solution to* (105)) *on compact subsets of $\tilde{\Omega}$.*

Proof. The proof relies on a series of smooth approximations to problem (105). Let $\{\Omega_s\}_{s=0}^{\infty}$ be a sequence of strictly convex and smooth domains such that $\Omega_s \subset \Omega_{s+1} \subset \Omega$ for all s, and $\Omega_s \to \Omega$ as $s \to \infty$; see Fig. 8. Consider the problem

$$\begin{aligned} \det D^2 u_{ms} &= f_m \quad \text{in } \Omega_s, \\ u_{ms} &= 0 \quad \text{on } \partial\Omega_s. \end{aligned}$$

Note that, because the data is regular, and since Ω_s is uniformly convex with smooth boundary, the solution to this problem is smooth. In particular,

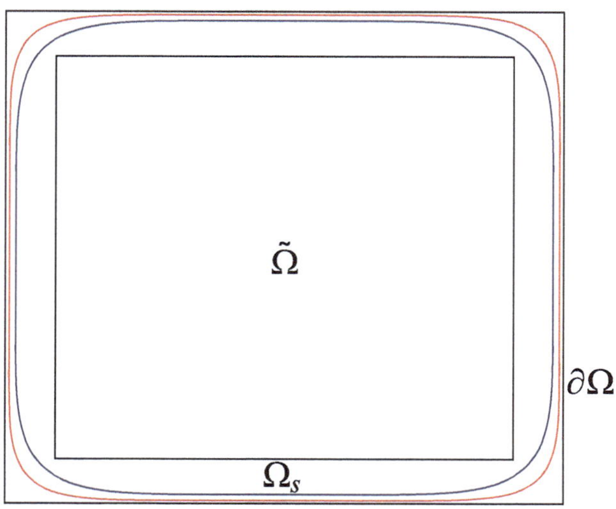

FIG. 8 Pictorial description of the proof of Theorem 18. Here, $\tilde{\Omega} \subset \Omega_s \subset \Omega_{s+1} \subset \Omega$, where Ω is the physical domain, $\tilde{\Omega}$ is the computational domain, and $\{\Omega_s\}$ are smooth and uniformly convex approximations to Ω.

interior Schauder estimates (Gilbarg and Trudinger, 2001, Section 6.1) show that, for any $D \subset\subset \Omega_s$,

$$\|u_{ms}\|_{C^{r+1}(D)} \le C_m,$$

where $C_m > 0$ depends on m, f_m, D, and $\text{dist}\{D, \partial\Omega_s\} \le \text{dist}\{D, \partial\Omega\}$. Moreover, results in Savin (2013) show that u_{ms} (up to subsequence) converges uniformly on compact subsets of Ω as $s \to \infty$. Now, because u_{ms} is smooth, and because the derivatives of u_{ms} are uniformly bounded on $\tilde{\Omega}$ (with respect to s), arguments similar those given in the previous section (see Awanou, 2015d; Böhmer, 2008) show that, for $h \le h_0$ with h_0 sufficiently small, there exists a locally unique and convex solution to the following discrete problem: Find $\tilde{u}_{ms,h} \in \tilde{X}_h$ satisfying $\tilde{u}_{ms,h}|_{\partial\tilde{\Omega}} = u_{ms}|_{\partial\tilde{\Omega}}$ and

$$\int_{\tilde{\Omega}} (f_m - \det D^2 \tilde{u}_{ms,h}) v_h \mathrm{d}x = 0 \quad \forall v_h \in \tilde{X}_h \cap W_0^{1,d}(\tilde{\Omega}).$$

Furthermore, there holds $\|u_{ms} - \tilde{u}_{ms,h}\|_{W^{2,2}(\tilde{\Omega})} \le Ch^{r-1}$ where $C > 0$ depends on $\|u_{ms}\|_{C^{r+1}(\tilde{\Omega})}$ but is independent of s. Because $\|u_{ms}\|_{C^{r+1}(\tilde{\Omega})}$ is uniformly bounded with respect to s, it follows from a Sobolev embedding theorem that $\tilde{u}_{ms,h}$ is uniformly bounded. Thus, since $u_{ms,h}$ is convex and uniformly bounded, the sequence $\{\tilde{u}_{ms,h}\}_s$ is locally uniformly equicontinuous, and thus has a pointwise convergent subsequence. Standard arguments, along with $u_{ms} \to u_m$ on $\partial\tilde{\Omega}$, then show that this limit is a solution to the discrete problem (104). □

Remark 23 (interior convergence).
Regarding Theorem 18 note that:

1. The ideas and techniques given in this section has been applied to standard finite difference discretizations of the Monge–Ampère problem in Awanou (2016).
2. While the results and techniques of Theorem 18 are interesting, it is not immediately clear how to obtain the Dirichlet boundary condition $\tilde{u}_{m,h} = u_m$ on $\partial\tilde{\Omega}$ since u_m is not given data. One can alternatively use $\tilde{u}_{m,h}|_{\partial\tilde{\Omega}} = 0$, but this condition is not consistent with problem (103). We also point out that h_0 depends on $\text{dist}\{\partial\tilde{\Omega}, \partial\Omega\}$, and therefore Theorem 18 suggests we cannot take $\tilde{\Omega}$ to be arbitrarily close to Ω. ■

5 Numerical examples

The high point of this classical algorithmic age was perhaps reached in the work of Leonhard Euler [...] Innumerable numerical examples are dispersed in the (so far) seventy volumes of his collected works, showing that Euler always kept foremost in his mind the immediate numerical use of his formulas and algorithms.

Henrici (1964)

In this section we perform some simple numerical examples to show the efficiency and accuracy of some of the numerical schemes discussed in the previous sections. We consider three different test problems, each reflecting different scenarios of regularity. These are computed using the wide stencil finite difference scheme (20), the analogous filtered scheme (26), Oliker–Prussner method (64), the C^0 finite element method (83), and its regularized version using the vanishing moment methodology (102). We emphasize that these tests are not meant to form comparisons, but rather to highlight their advantages in different situations.

5.1 Example 1: Smooth solution

In the first set of experiments, we take the data such that the Monge–Ampère equation has a $C^\infty(\Omega)$ solution: $\Omega = (-1,\ 1)^2$,

$$f(x_1,x_2) = (1+x_1^2+x_2^2)e^{x_1^2+x_2^2},\quad u(x_1,x_2) = e^{\frac{x_1^2+x_2^2}{2}}. \tag{106}$$

In this setting, the Galerkin methods discussed in Sections 4.1 and 4.2 are advantageous due to their relative high order. We implement the C^0 finite element method (83) and the Oliker–Prussner method (64) on a sequence of mesh refinements and report the resulting errors in Fig. 9. In agreement with Theorem 89 (with $\ell = r - 2$ and $\alpha = 1$), the plots show optimal order convergence in $W_h^{2,p}$-norm with respect to the discretization parameter h for the Galerkin methods. In terms of the degrees of freedom (DOFs), the errors scale like

$$\|u - u_h\|_{W_h^{2,p}(\Omega)} = \mathcal{O}(DOFs^{(1-r)/2}).$$

The errors in L^∞ converge with optimal order provided that the polynomial degree is sufficiently high. Fig. 9 shows that

$$\begin{aligned}
\|u - u_h\|_{L^\infty(\Omega)} &= \mathcal{O}(DOFs^{(-1-r)/2})\quad r = 3,4,\\
\|u - u_h\|_{L^\infty(\Omega)} &= \mathcal{O}(DOFs^{-1})\quad r = 2.
\end{aligned}$$

These rates are proven in Neilan (2013). For the Oliker–Prussner method and finite difference methods defined on translation invariant meshes, we define its $W^{2,p}$ error on the nodal set as

$$\|u - u_h\|_{W_h^{2,p}(\Omega_h)} = \left(h^d \sum_{x_h\in\Omega_h^i,\, e_j\in S} |\Delta_{e_j}(u-u_h)(x_h)|^p\right)^{1/p}$$

where S is the 9-points stencil in two space dimensions and $\Delta_{e_j}v(x_h)$ denotes the centred second difference, defined in (17), of the function v at node x_h in the direction $\boldsymbol{e}_j$. We observe in Fig. 9 that, for the Oliker–Prussner method,

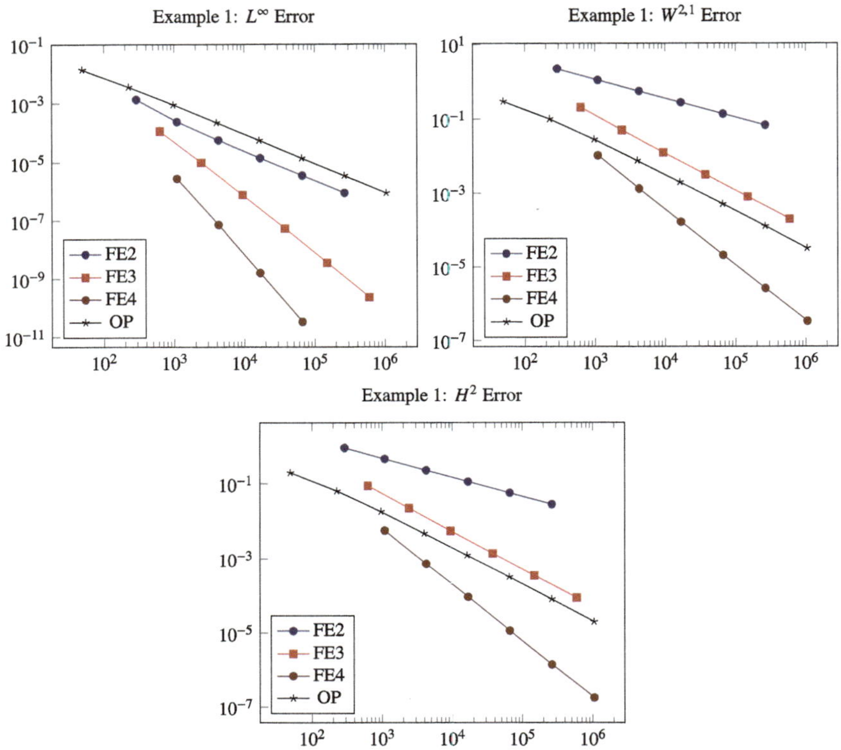

FIG. 9 Example 1: Errors versus degrees of freedom for the C^0 finite element method (83) with polynomial degrees $r = 2, 3, 4$, and the Oliker–Prussner method (64) applied to the smooth test problem (106).

$$\|u - u_h\|_{W_h^{2,p}(\Omega_h)} = \mathcal{O}(DOFs^{-1}) \quad \text{and} \quad \|u - u_h\|_{L^\infty(\Omega)} = \mathcal{O}(DOFs^{-1})$$

These results on $W^{2,\,p}$ error are consistent with the theorems proven in Neilan and Zhang (2018) and Theorem 13.

5.2 Example 2: Nonclassical solution

In this set of experiments, we again take $\Omega = (-1,\ 1)^2$, but choose the data such that the resulting solution is not a classical one:

$$f(x_1,x_2) = \begin{cases} 16, & |x| \le 1/2, \\ 64 - 16|x|^{-1}, & |x| > 1/2. \end{cases}$$

$$u(x_1,x_2) = \begin{cases} 2|x|^2, & |x| \le 1/2, \\ 2(|x| - 1/2)^2 + 2|x|^2, & |x| > 1/2. \end{cases}$$

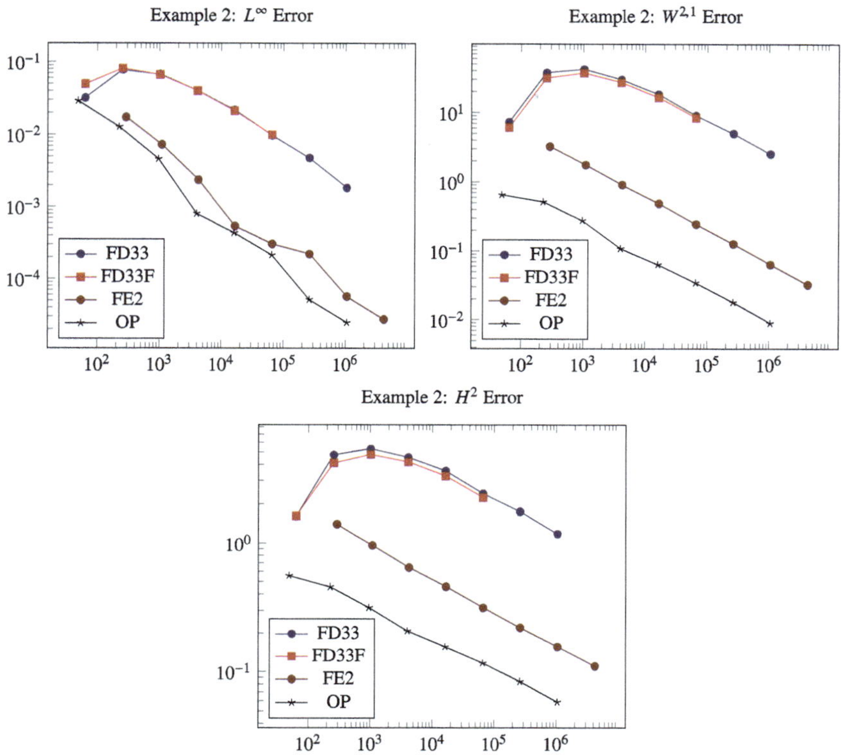

FIG. 10 Example 2: Errors versus degrees of freedom for the 33-point wide stencil scheme, 33-point wide stencil filtered scheme, the quadratic C^0 finite element method and Oliker–Prussner method.

One easily finds that $u \notin C^{1,1}(\bar{\Omega}) \backslash C^2(\Omega)$. We implement the C^0 finite element method (83), the wide stencil finite difference scheme (20) with a stencil size that consists of 33 grid points, and the Oliker–Prussner method (64). We also compare the results with the filtered scheme (26) The errors, depicted in Fig. 10, show that all methods converge with similar rates, although the finite element scheme and Oliker–Prussner method have smaller errors with similar DOFs. While the rate of convergence in the L^∞ norm is not obvious from the tests, Fig. 10 clearly shows that all three methods converge in the $W^{2,p}$-norms with rates

$$\|u - u_h\|_{H_h^2(\Omega)} = \mathcal{O}(DOFs^{-1/4}), \quad \|u - u_h\|_{W_h^{2,1}(\Omega)} = \mathcal{O}(DOFs^{-1/2}). \tag{107}$$

We note that, for the finite element, these rates seem to be the same rates of interpolation errors. Indeed, let $\mathcal{T}_h^\Gamma$ denote the set of triangles in $\mathcal{T}_h$ intersect the circle $|x| = 1/2$. Likewise, we let $\mathcal{F}_h^\Gamma$ denote the set of edges in $\mathcal{F}_h^I$ that intersect Γ. Finally, we denote by $\mathcal{I}_h u$ the nodal interpolant of u.

Because u is smooth on both $\Omega\cap\{x\in\Omega:\ |x|<1/2\}$ and $\Omega\cap\{x\in\Omega:\ x>1/2\}$, we have by standard interpolation estimates,

$$\begin{aligned}\|u-\mathcal{I}_h u\|^p_{W_h^{2,p}(\Omega)} &\le Ch^{p(r-1)} + \sum_{T\in\mathcal{T}_h^\Gamma} \|D^2(u-\mathcal{I}_h u)\|^p_{L^p(T)} \\ &\quad + \sum_{F\in\mathcal{F}_h^\Gamma} h_F^{1-p} \|[\![\nabla(u-\mathcal{I}_h u)]\!]\|^p_{L^p(F)} \\ &\le C\left(h^{p(r-1)} + \sum_{T\in\mathcal{T}_h^\Gamma}\left(\|D^2(u-\mathcal{I}_h u)\|^p_{L^p(T)} + h_T^{-p}\|\nabla(u-\mathcal{I}_h u)\|^p_{L^p(T)}\right)\right),\end{aligned}$$

where we used a standard trace inequality. Applying interpolation estimates and Hölder's inequality, noting that $u\in W^{2,\infty}(\Omega)$, yields

$$\begin{aligned}\|u-\mathcal{I}_h u\|^p_{W_h^{2,p}(\Omega)} &\le C\left(h^{p(r-1)} + \sum_{T\in\mathcal{T}_h^\Gamma}\|D^2 u\|^p_{L^p(T)}\right) \\ &\le C\left(h^{p(r-1)} + \sum_{T\in\mathcal{T}_h^\Gamma} h_T^2 \|D^2 u\|^p_{L^\infty(T)}\right) \\ &\le Ch^{p(r-1)} + Ch,\end{aligned}$$

where we used that the cardinality of $\mathcal{T}_h^\Gamma$ is $\mathcal{O}(h^{-1})$. We then take the pth root of this inequality to deduce that $\|u-\mathcal{I}_h u_h\|_{W_h^{2,p}(\Omega)} = \mathcal{O}(h^{1/p}) = \mathcal{O}(DOFs^{-1/(2p)})$, which is the same rates as (107).

5.3 Example 3: Lipschitz and degenerate solution

In our last set of experiments, we take the domain to be $\Omega=(-1,1)^2$ with data

$$f(x_1,x_2)=\begin{cases}36-9x_2^2x_1^{-6}, & |x_2|\le|x_1|^3,\\ \dfrac{8}{9}-\dfrac{5}{9}x_1^2x_2^{-\frac{2}{3}}, & |x_2|>|x_1|^3,\end{cases}$$

$$u(x_1,x_2)=\begin{cases}|x_1|^4+\dfrac{3x_2^2}{2x_1^2}, & |x_2|\le|x_1|^3,\\ \dfrac{1}{2}x_1^2x_2^{\frac{2}{3}}+2x_2^{\frac{4}{3}}, & |x_2|>|x_1|^3.\end{cases}$$

Similar to the previous example, u is not a classical solution to (1) as it only has regularity $u\in C^{0,1}(\Omega)$ and $u\notin W^{2,p}(\Omega)$ for any $p>2$ (Wang, 1995). Moreover, a simple calculation shows that $|D^2u(x)|\to\infty$ as $x\to 0$. Since the determinant in two dimensions is the product of two eigenvalues of the Hessian and $\det D^2u(x)=f(x)$ is bounded in the domain, the largest eigenvalue blows up while the other eigenvalue of $D^2u(x)$ approaches zero as $x\to 0$. Hence, the Hessian of the solution degenerates as $x\to 0$.

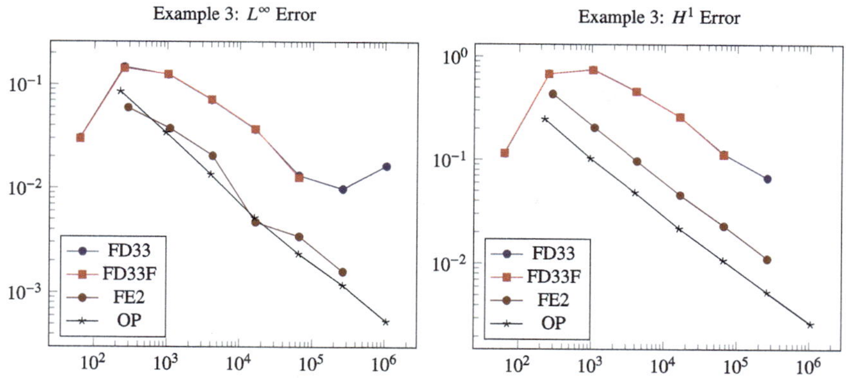

FIG. 11 Example 3: Errors versus degrees of freedom for the 33-point wide stencil scheme, the 33-point wide stencil filtered scheme, and the quadratic C^0 finite element method with regularization and Oliker–Prussner method.

While the monotone finite difference schemes presented in Section 2 are robust for problems with low regularity, Galerkin methods generally fail to capture solutions whose second derivatives are not square integrable; our numerical tests show that Newton's method applied to (83) does not converge for this example even when using very generous initial guesses. In fact, even for the monotone finite difference schemes and the Oliker–Prussner method, Newton's method is very sensitive with respect to the initial guess and the convexity of the iterates for this problem. In our implementation, we found that at each iteration, we require the solution to remain convex. As Newton's method may not give a convex solution in general, we applied, if necessary, the algorithm proposed in Oberman (2008a) to preserve convexity.

In addition to the 33-point finite difference scheme and Oliker–Prussner method, we implement the fourth-order regularization of the C^0 finite element method (83) with parameters $\sigma = 100$ and $\epsilon = 0.1h^2$. The resulting errors measured in the L^∞ and H^1 norms are plotted in Fig. 11. Similar to the previous series of experiments, the plots show that both methods have similar behaviour rates. While the rate in the L^∞ is not clear, the second plot in Fig. 11 shows that

$$\|u - u_h\|_{H^1(\Omega)} = \mathcal{O}(DOFs^{-1/2}).$$

6 Concluding remarks

"And if anyone knows anything about anything" said Bear to himself, "it's Owl who knows something about something," he said, "or my name is not Winnie-the-Pooh," he said. "Which it is," he added. "So there you are."

Hoff (1982)

In this work we have reviewed the progress that has been made concerning the approximation and numerical analysis of the Monge–Ampère problem. In doing so we highlighted how to develop a convergence analysis of wide stencil finite difference schemes as well as their generalizations, schemes based on geometric considerations, and finite element methods. A focus that we have taken, and one of recent development, is the derivation of rates of convergence for these discretizations.

Despite fundamental advances in only the past decade, there still remain several open problems in the analysis of computational methods for Monge–Ampère problems. One of these is the derivation of rates of convergence for the Oliker–Prussner scheme on unstructured grids. Another basic problem is rates of convergence of any of the schemes presented in this work assuming that the solution is not a classical one, i.e., without the assumption $u \in C^{2,\alpha}(\bar{\Omega})$. In most of the error analyses we have presented, it is assumed that $0 < \lambda I \leq D^2 u(x) \leq \Lambda I$ for all $x \in \Omega$. However, if the function $f(x)$ is discontinuous, the Hessian of the solution may be degenerate as the third example in the numerics section illustrates. The design and analysis of robust and high order numerical schemes to capture degenerate solutions remains a challenging problem. A posteriori error estimation, and adaptive methods based on the existing schemes are nonexistent. Finally let us mention that, as far as we are aware, except for the recent work (Berman, 2018), rates of convergence are restricted to the Dirichlet problem (1); extensions to, e.g., the applications discussed in Section 1.1 is still unchartered territory.

In conclusion, we know something about the numerical analysis of the Monge–Ampère problem, but there is much more that needs to be developed. It is our hope that this overview will encourage the numerical analysis community to work on the interesting, and challenging, problems found in geometry in general, and those that the Monge–Ampère equation in particular present to us.

Acknowledgements

The work of M.J.N. was supported by NSF Grant DMS–1719829. The work of A.J.S. was supported by NSF Grant DMS–1720213. The work of W.Z. was supported by NSF Grant DMS–1818861.

References

Arnol'd, V., 1998. On the teaching of mathematics. Uspekhi Mat. Nauk 53 (1), 229–234. ISSN 0042-1316. https://doi.org/10.1070/rm1998v053n01ABEH000005.

Arnold, D., Brezzi, F., Cockburn, B., Marini, L., 2002. Unified analysis of discontinuous Galerkin methods for elliptic problems. SIAM J. Numer. Anal. 39 (5), 1749–1779. ISSN 0036-1429. https://doi.org/10.1137/S0036142901384162.

Awanou, G., 2014. Isogeometric method for the elliptic Monge-Ampère equation. In: Approximation Theory XIV: San Antonio 2013, Springer Proc. Math. Stat, vol. 83. Springer, Cham, pp. 1–13. https://doi.org/10.1007/978-3-319-06404-8_1.

Awanou, G., 2015. Quadratic mixed finite element approximations of the Monge-Ampère equation in 2D. Calcolo 52 (4), 503–518. ISSN 0008-0624. https://doi.org/10.1007/s10092-014-0127-7.

Awanou, G., 2015. Smooth approximations of the Aleksandrov solution of the Monge-Ampère equation. Commun. Math. Sci. 13 (2), 427–441. ISSN 1539-6746. https://doi.org/10.4310/CMS.2015.v13.n2.a8.

Awanou, G., 2015. Spline element method for Monge-Ampère equations. BIT 55 (3), 625–646. ISSN 0006-3835. https://doi.org/10.1007/s10543-014-0524-y.

Awanou, G., 2015. Standard finite elements for the numerical resolution of the elliptic Monge-Ampère equations: classical solutions. IMA J. Numer. Anal. 35 (3), 1150–1166. ISSN 0272-4979. https://doi.org/10.1093/imanum/dru028.

Awanou, G., 2016. On standard finite difference discretizations of the elliptic Monge-Ampère equation. J. Sci. Comput. 69 (2), 892–904. ISSN 0885-7474. https://doi.org/10.1007/s10915-016-0220-y.

Awanou, G., 2017. Erratum to: Quadratic mixed finite element approximations of the Monge-Ampère equation in 2D [MR3421667]. Calcolo 54 (1), 281–297. ISSN 0008-0624. https://doi.org/10.1007/s10092-016-0187-y.

Awanou, G., 2017. Standard finite elements for the numerical resolution of the elliptic Monge-Ampère equation: Aleksandrov solutions. ESAIM Math. Model. Numer. Anal. 51 (2), 707–725. ISSN 0764-583X. https://doi.org/10.1051/m2an/2016037.

Awanou, G., Awi, R., 2016. Convergence of finite difference schemes to the Aleksandrov solution of the Monge-Ampère equation. Acta Appl. Math. 144, 87–98. ISSN 0167-8019. https://doi.org/10.1007/s10440-016-0041-x.

Awanou, G., Li, H., 2014. Error analysis of a mixed finite element method for the Monge-Ampère equation. Int. J. Numer. Anal. Model. 11 (4), 745–761. ISSN 1705-5105.

Awanou, G., Li, H., Malitz, E., 2018. A two-grid method for the C0 interior penalty discretization of the Monge-Ampère equation. Preprint.

Bakelman, I., 1994. Convex Analysis and Nonlinear Geometric Elliptic Equations. Springer-Verlag, Berlin, ISBN: 3-540-13620-7, pp. xxii–510. https://doi.org/10.1007/978-3-642-69881-1.

Barles, G., Souganidis, P.E., 1991. Convergence of approximation schemes for fully nonlinear second order equations. Asymptotic Anal. 4 (3), 271–283. ISSN 0921-7134.

Benamou, J., Duval, V., 2018. Minimal convex extensions and finite difference discretisation of the quadratic Monge–Kantorovich problem. ArXiv:1710.05594 [math.NA].

Benamou, J.-D., Froese, B., Oberman, A., 2014. Numerical solution of the optimal transportation problem using the Monge-Ampère equation. J. Comput. Phys. 260, 107–126. ISSN 0021-9991. https://doi.org/10.1016/j.jcp.2013.12.015.

Benamou, J.-D., Collino, F., Mirebeau, J.-M., 2016. Monotone and consistent discretization of the Monge-Ampère operator. Math. Comp. 85 (302), 2743–2775. ISSN 0025-5718. https://doi.org/10.1090/mcom/3080.

Berman, R., 2018. Convergence rates for discretized Monge-Ampère equations and quantitative stability of optimal transport. ArXiv:1803.00785 [math.NA].

Böhmer, K., 2008. On finite element methods for fully nonlinear elliptic equations of second order. SIAM J. Numer. Anal. 46 (3), 1212–1249. ISSN 0036-1429. https://doi.org/10.1137/040621740.

Bonito, A., Guermond, J.-L., Popov, B., 2014. Stability analysis of explicit entropy viscosity methods for non-linear scalar conservation equations. Math. Comp. 83 (287), 1039–1062. ISSN 0025-5718. https://doi.org/10.1090/S0025-5718-2013-02771-8.

Brenier, Y., 1991. Polar factorization and monotone rearrangement of vector-valued functions. Comm. Pure Appl. Math. 44 (4), 375–417. ISSN 0010-3640. https://doi.org/10.1002/cpa.3160440402.

Brenner, S., Neilan, M., 2012. Finite element approximations of the three dimensional Monge-Ampère equation. ESAIM Math. Model. Numer. Anal. 46 (5), 979–1001. ISSN 0764-583X. https://doi.org/10.1051/m2an/2011067.

Brenner, S., Scott, L., 2008. The Mathematical Theory of Finite Element Methods. In: Texts in Applied Mathematics, third ed 15. Springer, New York. ISBN: 978-0-387-75933-3, pp. xviii–397. https://doi.org/10.1007/978-0-387-75934-0.

Brenner, S., Sung, L.-Y., 2005. C^0 interior penalty methods for fourth order elliptic boundary value problems on polygonal domains. J. Sci. Comput. 22/23, 83–118. ISSN 0885-7474. https://doi.org/10.1007/s10915-004-4135-7.

Brenner, S., Gudi, T., Neilan, M., Sung, L.-Y., 2011. C^0 penalty methods for the fully nonlinear Monge-Ampère equation. Math. Comp. 80 (276), 1979–1995. ISSN 0025-5718. https://doi.org/10.1090/S0025-5718-2011-02487-7.

Caffarelli, L., Cabré, X., 1995. Fully Nonlinear Elliptic Equations. American Mathematical Society Colloquium Publications, vol. 43American Mathematical Society, Providence, RI. ISBN: 0-8218-0437-5, p. vi+104.

Calabi, E., 1990. Affine differential geometry and holomorphic curves. In: Villani, V. (Ed.), Complex Geometry and Analysis (Pisa, 1988), Lecture Notes in Math., vol. 1422. Springer, Berlin, pp. 15–21. 10.1007/BFb0089401.

Ciarlet, P., 2002. The Finite Element Method for Elliptic Problems. Classics in Applied Mathematics, vol. 40. Society for Industrial and Applied Mathematics (SIAM), Philadelphia, PA. ISBN: 0-89871-514-8, p. xxviii+530. https://doi.org/10.1137/1.9780898719208.

Cockburn, B., Shu, C.-W., 1998. The local discontinuous Galerkin method for time-dependent convection-diffusion systems. SIAM J. Numer. Anal. 35 (6), 2440–2463. ISSN 0036-1429. https://doi.org/10.1137/S0036142997316712.

Courant, R., Friedrichs, K., Lewy, H., 1967. On the partial difference equations of mathematical physics. IBM J. Res. Develop. 11, 215–234. ISSN 0018-8646. https://doi.org/10.1147/rd.112.0215.

Crandall, M., Ishii, H., Lions, P.-L., 1992. User's guide to viscosity solutions of second order partial differential equations. Bull. Amer. Math. Soc. (N.S.) 27 (1), 1–67. ISSN 0273-0979. https://doi.org/10.1090/S0273-0979-1992-00266-5.

Cuesta, J., Matrán, C., 1989. Notes on the Wasserstein metric in Hilbert spaces. Ann. Probab. 17 (3), 1264–1276. ISSN 0091-1798. http:/links.jstor.org/sici?sici=0091-1798(198907)17:3%3C1264:NOTWMI%3E2.0.CO;2-J&origin=MSN.

Davydov, O., Saeed, A., 2013. Numerical solution of fully nonlinear elliptic equations by Böhmer's method. J. Comput. Appl. Math. 254, 43–54. ISSN 0377-0427. https://doi.org/10.1016/j.cam.2013.03.009.

De Philippis, G., Figalli, A., 2015. Optimal regularity of the convex envelope. Trans. Amer. Math. Soc. 367 (6), 4407–4422. ISSN 0002-9947. https://doi.org/10.1090/S0002-9947-2014-06306-X.

Dean, E., Glowinski, R., 2003. Numerical solution of the two-dimensional elliptic Monge-Ampère equation with Dirichlet boundary conditions: an augmented Lagrangian approach. C. R. Math. Acad. Sci. Paris 336 (9), 779–784. ISSN 1631-073X. https://doi.org/10.1016/S1631-073X(03)00149-3.

Dean, E., Glowinski, R., 2004. Numerical solution of the two-dimensional elliptic Monge-Ampère equation with Dirichlet boundary conditions: a least-squares approach. C. R. Math. Acad. Sci. Paris 339 (12), 887–892. ISSN 1631-073X. https://doi.org/10.1016/j.crma.2004.09.018.

Dean, E., Glowinski, R., 2005. On the numerical solution of a two-dimensional Pucci's equation with Dirichlet boundary conditions: a least-squares approach. C. R. Math. Acad. Sci. Paris 341 (6), 375–380. ISSN 1631-073X. https://doi.org/10.1016/j.crma.2005.08.002.

Dean, E., Glowinski, R., 2006a. An augmented Lagrangian approach to the numerical solution of the Dirichlet problem for the elliptic Monge-Amp'ere equation in two dimensions. Electron. Trans. Numer. Anal. 22, 71–96. ISSN 1068-9613 (electronic).

Dean, E., Glowinski, R., 2006. Numerical methods for fully nonlinear elliptic equations of the Monge-Ampère type. Comput. Methods Appl. Mech. Engrg. 195 (13–16), 1344–1386. ISSN 0045-7825. https://doi.org/10.1016/j.cma.2005.05.023.

Debrabant, K., Jakobsen, E., 2013. Semi-Lagrangian schemes for linear and fully non-linear diffusion equations. Math. Comp. 82 (283), 1433–1462. ISSN 0025-5718. https://doi.org/10.1090/S0025-5718-2012-02632-9.

Engel, G., Garikipati, K., Hughes, T., Larson, M., Mazzei, L., Taylor, R., 2002. Continuous/discontinuous finite element approximations of fourth-order elliptic problems in structural and continuum mechanics with applications to thin beams and plates, and strain gradient elasticity. Comput. Methods Appl. Mech. Engrg. 191 (34), 3669–3750. ISSN 0045-7825. https://doi.org/10.1016/S0045-7825(02)00286-4.

Feng, X., Jensen, M., 2017. Convergent semi-Lagrangian methods for the Monge-Ampère equation on unstructured grids. SIAM J. Numer. Anal. 55 (2), 691–712. ISSN 0036-1429. https://doi.org/10.1137/16M1061709.

Feng, X., Neilan, M., 2009. Vanishing moment method and moment solutions for fully nonlinear second order partial differential equations. J. Sci. Comput. 38 (1), 74–98. ISSN 0885-7474. https://doi.org/10.1007/s10915-008-9221-9.

Feng, X., Neilan, M., 2011. Analysis of Galerkin methods for the fully nonlinear Monge-Ampère equation. J. Sci. Comput. 47 (3), 303–327. ISSN 0885-7474. https://doi.org/10.1007/s10915-010-9439-1.

Feng, X., Neilan, M., 2014. Convergence of a fourth-order singular perturbation of the n-dimensional radially symmetric Monge-Ampère equation. Appl. Anal. 93 (8), 1626–1646. ISSN 0003-6811. https://doi.org/10.1080/00036811.2013.842228.

Feng, X., Glowinski, R., Neilan, M., 2013. Recent developments in numerical methods for fully nonlinear second order partial differential equations. SIAM Rev. 55 (2), 205–267. ISSN 0036-1445. https://doi.org/10.1137/110825960.

Feng, X., Hennings, L., Neilan, M., 2017. Finite element methods for second order linear elliptic partial differential equations in non-divergence form. Math. Comp. 86 (307), 2025–2051. ISSN 0025-5718. https://doi.org/10.1090/mcom/3168.

Feng, X., Neilan, M., Schnake, S., 2018. Interior penalty discontinuous Galerkin methods for second order linear non-divergence form elliptic PDEs. J. Sci. Comput. 74 (3), 1651–1676. ISSN 0885-7474. https://doi.org/10.1007/s10915-017-0519-3.

Figalli, A., 2017. The Monge-Ampère Equation and Its Applications. Zurich Lectures in Advanced MathematicsEuropean Mathematical Society (EMS), Zürich. ISBN: 978-3-03719-170-5, p. x+200. https://doi.org/10.4171/170.

Froese, B., 2012. A numerical method for the elliptic Monge-Ampère equation with transport boundary conditions. SIAM J. Sci. Comput. 34 (3), A1432–A1459. ISSN 1064-8275. https://doi.org/10.1137/110822372.

Froese, B., 2018. Meshfree finite difference approximations for functions of the eigenvalues of the Hessian. Numer. Math. 138 (1), 75–99. ISSN 0029-599X. https://doi.org/10.1007/s00211-017-0898-2.

Froese, B., Oberman, A., 2011. Convergent finite difference solvers for viscosity solutions of the elliptic Monge-Ampère equation in dimensions two and higher. SIAM J. Numer. Anal. 49 (4), 1692–1714. ISSN 0036-1429. https://doi.org/10.1137/100803092.

Froese, B., Oberman, A., 2011. Fast finite difference solvers for singular solutions of the elliptic Monge-Ampère equation. J. Comput. Phys. 230 (3), 818–834. ISSN 0021-9991. https://doi.org/10.1016/j.jcp.2010.10.020.

Froese, B., Oberman, A., 2013. Convergent filtered schemes for the Monge-Ampère partial differential equation. SIAM J. Numer. Anal. 51 (1), 423–444. ISSN 0036-1429. https://doi.org/10.1137/120875065.

Gilbarg, D., Trudinger, N., 2001. Elliptic Partial Differential Equations of Second Order. Classics in MathematicsSpringer-Verlag, Berlin. ISBN: 3-540-41160-7, p. xiv+517.

Guan, B., Spruck, J., 1993. Boundary-value problems on S^n for surfaces of constant Gauss curvature. Ann. of Math. (2) 138 (3), 601–624. ISSN 0003-486X. https://doi.org/10.2307/2946558.

Guermond, J.-L., Pasquetti, R., 2011. Entropy viscosity method for high-order approximations of conservation laws. In: Bittencourt, M., Dumont, N., Hesthaven, J.S. (Eds.), Spectral and High Order Methods for Partial Differential Equations, Lect. Notes Comput. Sci. Eng., vol. 76. Springer, Heidelberg, pp. 411–418.

Guermond, J.-L., Pasquetti, R., Popov, B., 2011. Entropy viscosity method for nonlinear conservation laws. J. Comput. Phys. 230 (11), 4248–4267. ISSN 0021-9991. https://doi.org/10.1016/j.jcp.2010.11.043.

Guermond, J.-L., Nazarov, M., Popov, B., Tomas, I., 2018. Second-order invariant domain preserving approximation of the Euler equations using convex limiting. SIAM J. Sci. Comput. 40 (5), A3211–A3239. ISSN 1064-8275. https://doi.org/10.1137/17M1149961.

Gutiérrez, C., 2001. The Monge-Ampère Equation. Progress in Nonlinear Differential Equations and their Applications, 44Birkhäuser Boston, Inc., Boston, MA, . ISBN: 0-8176-4177-7xii+127. https://doi.org/10.1007/978-1-4612-0195-3.

Hamfeldt, B., 2018. Convergent approximation of non-continuous surfaces of prescribed Gaussian curvature. Commun. Pure Appl. Anal. 17 (2), 671–707. ISSN 1534-0392. https://doi.org/10.3934/cpaa.2018036.

Henrici, P., 1964. Elements of Numerical Analysis. John Wiley & Sons, Inc., New York, London, Sydney, p. xv+328.

Hintermüller, M., Ito, K., Kunisch, K., 2002. The primal-dual active set strategy as a semismooth Newton method. SIAM J. Optim 13 (3), 865–888. ISSN 1052-6234. (2003)10.1137/S1052623401383558.

Hoff, B., 1982. The Tao of Pooh. Penguin Books, p. xii+158 0 14 0 0.6747 7.

Huang, J., Huang, X., Han, W., 2010. A new C^0 discontinuous Galerkin method for Kirchhoff plates. Comput. Methods Appl. Mech. Engrg. 199 (23-24), 1446–1454. ISSN 0045-7825. https://doi.org/10.1016/j.cma.2009.12.012.

Jensen, M., 2018. Numerical solution of the simple Monge-Ampère equation with nonconvex Dirichlet data on nonconvex domains. In: Kalise, D., Kunisch, K., Rao, Z. (Eds.), Hamilton-Jacobi-Bellman Equations: Numerical Methods and Applications in Optimal Control, Radon Series on Computational and Applied Mathematics, vol. 21. De Gryuter, Boston, Berlin, pp. 129–142.

Jensen, M., Smears, I., 2018. On the notion of boundary conditions in comparison principles. In: Kalise, D., Kunisch, K., Rao, Z. (Eds.), Hamilton-Jacobi-Bellman Equations: Numerical

Methods and Applications in Optimal Control, Radon Series on Computational and Applied Mathematics, vol. 21. De Gryuter, Boston, Berlin, pp. 143–154.

Jovanović, B.S., Süli, E., 2014. Analysis of Finite Difference Schemes, Springer Series in Computational Mathematics. 46, Springer, London, pp. xiv–408. 978-1-4471-5459-4; 978-1-4471-5460-010.1007/978-1-4471-5460-0.

Kantorovich, L., 2004. On a problem of Monge. Zap. Nauchn. Sem. S.-Peterburg. Otdel. Mat. Inst. Steklov. (POMI) 312, 15–16. ISSN 0373-2703. https://doi.org/10.1007/s10958-006-0050-9 (Teor. Predst. Din. Sist. Komb. i Algoritm. Metody. 11).

Kawecki, E., Lakkis, O., Pryer, T., 2018. A finite element method for the Monge-Ampère equation with transport boundary conditions. ArXiv:1807.03535.

Kossaczký, I., Ehrhardt, M., Günther, M., 2016. On the non-existence of higher order monotone approximation schemes for HJB equations. Appl. Math. Lett. 52, 53–57. ISSN 0893-9659. https://doi.org/10.1016/j.aml.2015.08.005.

Krasnosel'skiı˘, M., Rutickiı˘, J., 1961. Convex Functions and Orlicz Spaces. Translated from the First Russian edition by Leo F. BoronP. Noordhoff Ltd., Groningen, p. xi+249

Krylov, N., 1987. Nonlinear Elliptic and Parabolic Equations of the Second Order. Mathematics and its Applications (Soviet Series), vol. 7D. Reidel Publishing Co., Dordrecht. ISBN: 90-277-2289-7, p. xiv+462. https://doi.org/10.1007/978-94-010-9557-0

Lakkis, O., Pryer, T., 2011. A finite element method for second order nonvariational elliptic problems. SIAM J. Sci. Comput. 33 (2), 786–801. ISSN 1064-8275. https://doi.org/10.1137/100787672.

Lakkis, O., Pryer, T., 2013. A finite element method for nonlinear elliptic problems. SIAM J. Sci. Comput. 35 (4), A2025–A2045. ISSN 1064-8275. https://doi.org/10.1137/120887655.

Li, W., Nochetto, R., 2018. Optimal pointwise error estimates for two-scale methods for the Monge-Ampère equation. SIAM J. Numer. Anal. 56 (3), 1915–1941. ISSN 0036-1429. https://doi.org/10.1137/18M1165670.

Li, W., Nochetto, R., 2018b. Two-scale methods for convex envelopes. ArXiv:1812.11519 [math.NA].

Lions, P.-L., Souganidis, P., 1995. Convergence of MUSCL and filtered schemes for scalar conservation laws and Hamilton-Jacobi equations. Numer. Math. 69 (4), 441–470. ISSN 0029-599X. https://doi.org/10.1007/s002110050102.

Mirebeau, J.-M., 2015. Discretization of the 3D Monge-Ampere operator, between wide stencils and power diagrams. ESAIM Math. Model. Numer. Anal. 49 (5), 1511–1523. ISSN 0764-583X.

Mirebeau, J.-M., 2016. Minimal stencils for discretizations of anisotropic PDEs preserving causality or the maximum principle. SIAM J. Numer. Anal. 54 (3), 1582–1611. ISSN 0036-1429. https://doi.org/10.1137/16M1064854.

Motzkin, T., Wasow, W., 1953. On the approximation of linear elliptic differential equations by difference equations with positive coefficients. J. Math. Physics 31, 253–259.

Neilan, M., 2013. Quadratic finite element approximations of the Monge-Ampère equation. J. Sci. Comput. 54 (1), 200–226. ISSN 0885-7474. https://doi.org/10.1007/s10915-012-9617-4.

Neilan, M., 2014. Finite element methods for fully nonlinear second order PDEs based on a discrete Hessian with applications to the Monge-Ampère equation. J. Comput. Appl. Math. 263, 351–369. ISSN 0377-0427. https://doi.org/10.1016/j.cam.2013.12.027.

Neilan, M., 2014. A unified analysis of three finite element methods for the Monge-Ampère equation. Electron. Trans. Numer. Anal. 41, 262–288. ISSN 1068-9613.

Neilan, M., 2017. Convergence analysis of a finite element method for second order non-variational elliptic problems. J. Numer. Math. 25 (3), 169–184. ISSN 1570-2820. https://doi.org/10.1515/jnma-2016-1017.

Neilan, M., Zhang, W., 2018. Rates of convergence in W_p^2-norm for the Monge-Ampère equation. SIAM J. Numer. Anal. 56 (5), 3099–3120. ISSN 0036-1429. https://doi.org/10.1137/17M1160409.

Neilan, M., Salgado, A., Zhang, W., 2017. Numerical analysis of strongly nonlinear PDEs. Acta Numer. 26, 137–303. ISSN 0962-4929. https://doi.org/10.1017/S0962492917000071.

Nochetto, R., Ntogkas, D., 2018. Convergent two-scale filtered scheme for the Monge-Ampère equation. arXiv:1807.04866.

Nochetto, R., Zhang, W., 2018. Discrete ABP estimate and convergence rates for linear elliptic equations in non-divergence form. Found. Comput. Math. 18 (3), 537–593. ISSN 1615-3375. https://doi.org/10.1007/s10208-017-9347-y.

Nochetto, R., Zhang, W., 2019. Pointwise rates of convergence for the Oliker-Prussner method for the Monge-Ampère equation. Numer. Math. https://doi.org/10.1007/s0021. To appear.

Nochetto, R., Ntogkas, D., Zhang, W., 2019. Two-scale method for the Monge-Ampère equation: convergence to the viscosity solution. Math. Comp. 88 (316), 637–664. ISSN 0025-5718. https://doi.org/10.1090/mcom/3353.

Nochetto, R., Ntogkas, D., Zhang, W., 2019. Two-scale method for the Monge-Ampère equation: pointwise error estimates. IMA J. Numer. Anal. https://doi.org/10.1093/imanum/dry026. To appear.

Norris, A., Westcott, B., 1976. Computation of reflector surfaces for bivariate beamshaping in the elliptic case. J. Phys. A: Math. Gen. 9 (12), 2159. http://stacks.iop.org/0305-4470/9/i=12/a=020.

Oberman, A., 2006. Convergent difference schemes for degenerate elliptic and parabolic equations: Hamilton-Jacobi equations and free boundary problems. SIAM J. Numer. Anal. 44 (2), 879–895. ISSN 0036-1429. (electronic). 10.1137/S0036142903435235.

Oberman, A., 2008. Computing the convex envelope using a nonlinear partial differential equation. Math. Models Methods Appl. Sci. 18 (5), 759–780. ISSN 0218-2025. https://doi.org/10.1142/S0218202508002851.

Oberman, A., 2008. Wide stencil finite difference schemes for the elliptic Monge-Ampère equation and functions of the eigenvalues of the Hessian. Discrete Contin. Dyn. Syst. Ser. B 10 (1), 221–238. ISSN 1531-3492. https://doi.org/10.3934/dcdsb.2008.10.221.

Oberman, A., Ruan, Y., 2017. A partial differential equation for the rank one convex envelope. Arch. Ration. Mech. Anal. 224 (3), 955–984. ISSN 0003-9527. https://doi.org/10.1007/s00205-017-1092-5.

Oliker, V., 1984. Hypersurfaces in $\mathbf{R}^{n+1}$ with prescribed Gaussian curvature and related equations of Monge-Ampère type. Comm. Partial Differential Equations 9 (8), 807–838. ISSN 0360-5302. https://doi.org/10.1080/03605308408820348.

Oliker, V., 1987. Near radially symmetric solutions of an inverse problem in geometric optics. Inverse Problems 3 (4), 743–756. ISSN 0266-5611. http:/stacks.iop.org/0266-5611/3/743.

Oliker, V., Newman, E., 1993. The energy conservation equation in the reflector mapping problem. Appl. Math. Lett. 6 (1), 91–95. ISSN 0893-9659. https://doi.org/10.1016/0893-9659(93)90156-H.

Oliker, V., Prussner, L., 1988. On the numerical solution of the equation $(\partial^2 z/\partial x^2)(\partial^2 z/\partial y^2) - ((\partial^2 z/\partial x \partial y))^2 = f$ and its discretizations. I. Numer. Math. 54 (3), 271–293. ISSN 0029-599X. https://doi.org/10.1007/BF01396762.

Oliker, V., Waltman, P., 1987. Radially symmetric solutions of a Monge-Ampère equation arising in a reflector mapping problem. In: Knowles, I.W., Saitō, Y. (Eds.), Differential Equations and Mathematical Physics, Lecture Notes in Math, 1285, Springer, Berlin, pp. 361–374. https:/doi.org/10.1007/BFb0080616.

Pólya, G., 2014. How to Solve It. Princeton Science Library. Princeton University Press, Princeton, NJ, ISBN: 978-0-691-16407-6, p. xxviii+253.

Rüschendorf, L., Rachev, S., 1990. A characterization of random variables with minimum L^2-distance. J. Multivariate Anal. 32 (1), 48–54. ISSN 0047-259X. https://doi.org/10.1016/0047-259X(90)90070-X.

Rüschendorf, L., Rachev, S., 1990. Corrigendum: "A characterization of random variables with minimum L^2-distance" J. Multivariate Anal. 34 (1), 156. ISSN 0047-259X. https://doi.org/10.1016/0047-259X(90)90066-Q.

Russell, B., 1931. The Scientific Outlook. Routledge.

Savin, O., 2013. Pointwise $C^{2,\alpha}$ estimates at the boundary for the Monge-Ampère equation. J. Amer. Math. Soc. 26 (1), 63–99. ISSN 0894-0347. https://doi.org/10.1090/S0894-0347-2012-00747-4.

Schmutz, E., 2008. Rational points on the unit sphere. Cent. Eur. J. Math. 6 (3), 482–487. ISSN 1895-1074. https://doi.org/10.2478/s11533-008-0038-4.

Trudinger, N., Wang, X.-J., 2005. The affine Plateau problem. J. Am. Math. Soc. 18 (2), 253–289. ISSN 0894-0347. https://doi.org/10.1090/S0894-0347-05-00475-3.

Trudinger, N., Wang, X.-J., 2008. Boundary regularity for the Monge-Ampère and affine maximal surface equations. Ann. Math. 167 (3), 993–1028. ISSN 0003-486X. https://doi.org/10.4007/annals.2008.167.993.

Wang, X.-J., 1995. Some counterexamples to the regularity of Monge-Ampére equations. Proc. Amer. Math. Soc. 123 (3), 841–845.

Wang, X.-J., 1996. On the design of a reflector antenna. Inverse Problems 12 (3), 351–375. ISSN 0266-5611. https://doi.org/10.1088/0266-5611/12/3/013.

Zienkiewicz, O., Taylor, R., 2000. The Finite Element Method, fifth ed. Vol. 1. Butterworth-Heinemann, Oxford. ISBN: 0-7506-5049-4, p. xvi+689.

Chapter 3

Finite element simulation of nonlinear bending models for thin elastic rods and plates

Sören Bartels*
Abteilung für Angewandte Mathematik, Albert-Ludwigs-Universität Freiburg, Freiburg im Breisgau, Germany
Corresponding author: e-mail: bartels@mathematik.uni-freiburg.de

Chapter Outline

Abstract

Nonlinear bending phenomena of thin elastic structures arise in various modern and classical applications. Characterizing low energy states of elastic rods has been investigated by Bernoulli in 1738 and related models are used to determine configurations of DNA strands. The bending of a piece of paper has been described mathematically by Kirchhoff in 1850 and extensions of his model arise in nanotechnological applications such as the development of externally operated microtools. A rigorous mathematical framework that identifies these models as dimensionally reduced limits from three-dimensional hyperelasticity has only recently been established. It provides

Handbook of Numerical Analysis, Vol. 21. https://doi.org/10.1016/bs.hna.2019.06.003

a solid basis for developing and analyzing numerical approximation schemes. The fourth-order character of bending problems and a pointwise isometry constraint for large deformations require appropriate discretization techniques which are discussed in this article. Methods developed for the approximation of harmonic maps are adapted to discretize the isometry constraint and gradient flows are used to decrease the bending energy. For the case of elastic rods, torsion effects and a self-avoidance potential that guarantees injectivity of deformations are incorporated. The devised and rigorously analyzed numerical methods are illustrated by means of experiments related to the relaxation of elastic knots, the formation of singularities in a Möbius strip, and the simulation of actuated bilayer plates.

Keywords: Nonlinear bending, Elasticity, Finite element methods, Convergence, Iterative solution

AMS Classification Codes: 65N12, 65N15, 65N30

1 Introduction

Thin elastic structures occur in various practical applications and in fact truly three-dimensional objects are hardly ever used. Important reasons for this are the reduction of weight and cost but also the special mechanical features of rods and plates. Correspondingly, their numerical treatment is expected to be more efficient when such structures can be described as lower-dimensional objects. Because of the different mechanical behaviour they cannot be treated like three-dimensional objects and new discretization techniques are needed. Typical large bending deformations of rods and plates are distinct from those of three-dimensional objects and are illustrated in Fig. 1.

In this article we address the numerical approximation of dimensionally reduced models for describing large deformations of thin elastic rods and plates. These models result from rigorous limiting processes of general three-dimensional hyperelastic material descriptions when the diameter of a circular rod or the thickness of a plate is small compared to the length or diameter and when the acting forces lead to deformations with energies

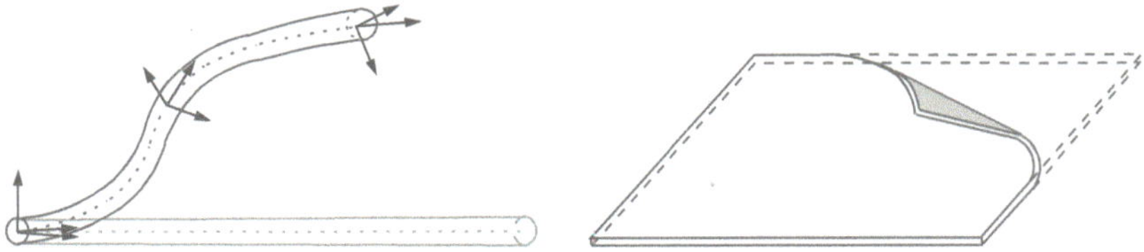

FIG. 1 The mathematical description of large bending deformations of thin objects requires the use of appropriate geometric quantities: deformed rod with circular cross section together with an orthonormal frame that allows to measure bending and torsion effects (*left*); deformation of a flat plate that preserves angle and length relations (*right*).

comparable to the third power of the diameter or thickness. Examples of such situations are the bending of a springy wire or sheet of paper.

Characteristic for large nonlinear bending phenomena is that nearly no shearing or stretching effects of the object occur and that curvature quantities define the amount of energy required for particular deformations. These aspects become explicitly apparent in the dimensionally reduced models: the energy functionals depend on curvature quantities and an isometry condition arises in the vanishing thickness or diameter limit. In particular, this condition implies that length and angle relations remain unchanged by a deformation.

The employed models for elastic rods and plates result from dimension reductions of general descriptions for hyperelastic material behaviour. We thus consider an energy density $W : \mathbb{R}^{3\times 3} \to \mathbb{R}$ and a corresponding energy minimization of

$$I_{3d}^{\delta}[y] = \int_{\Omega_\delta} W(\nabla y)\, dx$$

in a set $\mathcal{A} \subset W^{1,p}(\Omega_\delta; \mathbb{R}^3)$ of admissible deformations $y : \Omega_\delta \to \mathbb{R}^3$ that includes boundary conditions. The parameter $\delta > 0$ indicates a small diameter or thickness of the reference configuration $\Omega_\delta \subset \mathbb{R}^3$, e.g., $\Omega_\delta = (0, L) \times \delta S$ for a thin rod with cross section $\delta S \subset \mathbb{R}^2$ containing zero or $\Omega_\delta = \omega \times (-\delta/2, \delta/2)$ for a thin plate with midplane $\omega \subset \mathbb{R}^2$. Assuming that the minimal energies are comparable to δ^3, i.e.,

$$\min_{y \in \mathcal{A}} I_{3d}^{\delta}[y] = \mathcal{O}(\delta^3)$$

as $\delta \to 0$, and following the contributions (Friesecke et al., 2002c,b; Pantz, 2003; Mora and Müller, 2003), it is possible to identify limiting, dimensionally reduced theories that determine the corresponding limits of solutions as $\delta \to 0$. The particular cubic scaling characterizes bending phenomena of the elastic body and excludes membrane effects, we refer the reader to Friesecke et al. (2006), Conti and Maggi (2008), and Freddi et al. (2016) for discussions of models corresponding to other scaling regimes. We outline the numerical methods that have been developed for simulating nonlinear bending behaviour of rods and plates after a discussion of the related literature.

Throughout this article we use energy minimization principles to determine deformations subject to boundary conditions and external forces. For other approaches to the modelling of rods and plates via equilibria of forces or conservation of momentum we refer the reader to Antman (2005), Ciarlet (1997), and Audoly and Pomeau (2010). Only a few numerical methods have been discussed mathematically for the numerical solution of nonlinear bending models with inextensibility or isometry constraint. The articles (Wardetzky et al., 2007; Bergou et al., 2008) devise various methods to compute discrete curvature quantities. The focus of this article is on the reliability of methods, i.e., the accuracy of finite element discretizations and the convergence of iterative solution methods for the discrete problems. The methods discussed here use techniques developed for the approximation of harmonic maps into surfaces in the articles (Alouges, 1997; Bartels, 2005, 2016). We review the numerical treatment of rods following Bartels (2013c) and Bartels and Reiter (2019) and plates as proposed in Bartels (2013b) and Bartels (2015). For the efficient iterative solution we adopt ideas from Nash and Sofer (1996) and Kraus et al. (2018). We discuss the treatment of bilayer plates following Bartels et al. (2017) and Bartels et al. (2018a), illustrate a method that enforces injectivity of deformations in the case of rods following Bartels et al. (2018b) and Bartels and Reiter (2018), and propose methods for the numerical solution of bending deformations with shearing effects following ideas from Bartels (2017). The problems considered in this article have similarities with problems related to the length-preserving elastic flow of curves and the surface area and volume preserving Willmore–Helfrich flow of closed surfaces but require different numerical methods. For contributions related to those problems we refer the reader to Dziuk et al. (2002), Deckelnick et al. (2005), Barrett et al. (2007), Barrett et al. (2012), Sander et al. (2016), Pozzi and Stinner (2017), and Bonito et al. (2019); for examples of modern applications of nonlinear bending phenomena including the construction of micromachining fingers, the fabrication of nanotubes, the occurrence of wrinkling in plastic sheets, and the description of certain properties of DNA molecules, we refer the reader to Smela et al. (1993), Schmidt and Eberl (2001), Sharon et al. (2007), and Thompson et al. (2004).

1.1 Bending of elastic rods

We consider an elastic rod, e.g., a springy wire, which in its reference configuration occupies the region $(0, L) \times \{0\} \subset \mathbb{R}^3$. A low energy deformation

$$y : (0, L) \to \mathbb{R}^3$$

leaves distances of pairs of points on the rod unchanged. This is described by the inextensibility (and incompressibility) condition

$$|y'(x_1)| = 1$$

for almost every $x_1 \in (0, L)$. For appropriate boundary conditions a deformation then minimizes the bending energy

$$I_{\text{rod}}[y] = \frac{1}{2} \int_0^L |y''(x_1)|^2 \, dx_1.$$

The inextensibility condition implies that y defines an arclength parametrization of the deformed rod and hence its curvature is given by the second derivative of y. The simple energy functional I_{rod}, which has been proposed by Bernoulli in 1738, ignores torsion effects and arises as a special case of the dimension reduction from three-dimensional hyperelasticity.

It is interesting to see that the dimension reduction leads to significant changes in the nature of the energy functionals. The three-dimensional model depends on strains, is not constrained, and often provides existence of unique solutions. The dimensionally reduced functional I_{rod} depends on curvature, is constrained, and is singular in the sense that the set of admissible deformations may be empty, e.g., for extensive boundary conditions, and that solutions may be nonunique, e.g., for simple compressive boundary conditions. These aspects are related to the presence of a critical nonlinearity via a Lagrange multiplier for the inextensibility constraint in the Euler–Lagrange equations for critical points of I_{rod}, i.e.,

$$(y'', w'') = (\lambda y', w') \iff y^{(4)} = (\lambda y')',$$

where the scalar function λ depends nonlinearly on y. The explicit presence of a Lagrange multiplier can be avoided if only test functions are considered that satisfy the linearized inextensibility condition $y' \cdot w' = 0$. This corresponds to normal, i.e., nontangential perturbations of a curve in the energy minimization.

The inextensibility condition requires a suitable numerical treatment to avoid locking phenomena or other artefacts. For a partitioning of the interval $(0, L)$ with nodes

$$0 = z_0 < z_1 < \cdots < z_N = L$$

and a subordinated conforming finite element space $\mathcal{A}_h \subset H^2(0, L; \mathbb{R}^3)$ we impose the inextensibility condition only at these nodes, i.e.,

$$|y_h'(z_i)| = 1$$

for $i = 0, 1, \ldots, N$. Since $y_h \in H^2(0, L; \mathbb{R}^3)$ we obtain linear convergence with respect to the mesh-size h of the constraint violation error away from the nodes. The discrete minimization problem then seeks a minimizer $y_h \in \mathcal{A}_h$ for the functional

$$y_h \mapsto I_{\text{rod}}^h[y_h] = \frac{1}{2}\int_0^L |y_h''(x_1)|^2 \, dx_1,$$

subject to the nodal constraints $|y_h'(z_i)| = 1$ for $i = 0, 1, \ldots, N$. A possible choice of a finite element space uses piecewise cubic, continuously differentiable functions. This space has the advantage that its degrees of freedom are the positions and tangent vectors at the nodes, i.e.,

$$y_h \equiv \left(y_h(z_i), y_h'(z_i)\right)_{i=0,\ldots,N}.$$

The discretized inextensibility condition can thus be explicitly imposed on certain degrees of freedom, cf. Fig. 2.

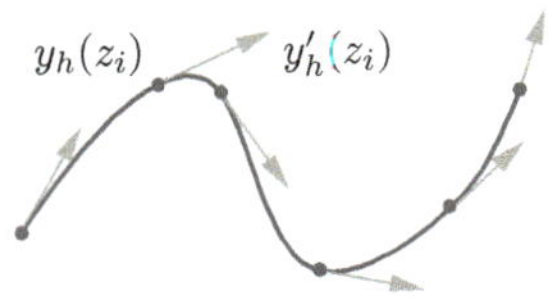

FIG. 2 Continuously differentiable, piecewise cubic curves are defined by positions and tangent vectors at nodes $z_0 < z_1 < \cdots < z_N$.

We iteratively solve the discrete minimization problem by using a gradient flow, i.e., on the continuous level we consider a family $y:[0,T]\times(0,L)\to\mathbb{R}^3$ of deformations that solve the evolution equation

$$(\partial_t y, w)_\star = -(y'', w'')_{L^2}, \quad y(0) = y_0,$$

subject to the linearized inextensibility conditions

$$\partial_t y'(t, x_1)\cdot y'(t, x_1) = 0, \quad w'(x_1)\cdot y'(t, x_1) = 0.$$

Provided that we have $|y_0{}'(x_1)|^2 = 1$ it follows that

$$|y'(s, x_1)|^2 - 1 = \int_0^s \frac{d}{dt}|y'(t, x_1)|^2\,\mathrm{d}t = 2\int_0^s \partial_t y'(t, x_1)\cdot y'(t, x_1)\,\mathrm{d}t = 0,$$

i.e., the inextensibility condition is satisfied. We use an implicit discretization of the evolution equation and a semiimplicit treatment of the linearized constraint, i.e., with the backward difference quotient operator d_t we consider the time-stepping scheme

$$(d_t y_h^k, w_h)_\star = -([y_h^k]'', [w_h]'')_{L^2}, \quad y_h^0 = y_{0,h},$$

subject to the linearized constraints evaluated at the nodes, i.e.,

$$[d_t y_h^k]'(x_i)\cdot[y_h^{k-1}]'(x_i) = 0, \quad [w_h]'(x_i)\cdot[y_h^{k-1}]'(x_i) = 0$$

for $i = 0, 1, \ldots, N$. The scheme is unconditionally energy-decreasing and convergent to a stationary configuration, i.e., choosing the admissible test function $w_h = d_t y_h^k$ directly shows

$$I_{\mathrm{rod}}^h[y_h^k] + \tau \| d_t y_h^k \|_\star^2 \le I_{\mathrm{rod}}^h[y_h^{k-1}].$$

The inextensibility constraint will not be satisfied exactly at the nodes but its violation is controlled by the step size τ and the initial energy. A proof in the discrete setting imitates the continuous argument given above. Using the orthogonality $[d_t y_h^k]'(z_i)\cdot[y_h^{k-1}]'(z_i) = 0$ and the property $|[y_h^0]'(z_i)|^2 = 1$ we have

$$\begin{aligned} |[y_h^k]'(z_i)|^2 - 1 &= |[y_h^{k-1}](z_i)'|^2 + \tau^2|[d_t y_h^k]'(z_i)|^2 - 1 \\ &= \cdots = \tau^2\sum_{\ell=1}^k |[d_t y_h^\ell]'(z_i)|^2. \end{aligned}$$

Because of the unconditional energy stability the term on the right-hand side is of order $\mathcal{O}(\tau)$. We will show below that these properties are also valid if torsion effects are taken into account.

1.2 Elastic plates

The mathematical description and numerical treatment of elastic plates generalizes that of elastic rods. In the dimensionally reduced model we consider deformations of a two-dimensional midplane

$$y:\omega \to \mathbb{R}^3$$

that leave angle and area relations unchanged, i.e., they satisfy the isometry condition

$$(\nabla y)^{\mathsf{T}} \nabla y = I_2$$

almost everywhere in $\omega \subset \mathbb{R}^2$ with the identity matrix $I_2 \in \mathbb{R}^{2\times 2}$. This is equivalent to saying that the tangent vectors $\partial_1 y$ and $\partial_2 y$ of the deformed plate and the normal vector $b = \partial_1 y \times \partial_2 y$ define an orthonormal basis for $\mathbb{R}^3$ in almost every point $x' \in \omega$. The actual deformation for appropriate boundary conditions minimizes the bending energy proposed by Kirchhoff in 1850,

$$I_{\text{plate}}[y] = \frac{1}{2}\int_\omega |D^2 y|^2 \, \mathrm{d}x'.$$

Because of the isometry condition, the integrand coincides with the mean curvature of the deformed plate while its Gaussian curvature vanishes. For a minimizing or critical isometry y we have that

$$(D^2 y, D^2 w) = 0$$

for all test fields w satisfying appropriate homogeneous boundary conditions and the linearized isometry condition

$$(\nabla y)^{\mathsf{T}} \nabla w + (\nabla w)^{\mathsf{T}} \nabla y = 0.$$

A finite element discretization uses a possibly nonconforming finite element space such as so-called discrete Kirchhoff triangles and imposes the isometry condition in the set of nodes $\mathcal{N}_h$, i.e., for all $z \in \mathcal{N}_h$ we have

$$(\nabla y_h(z))^{\mathsf{T}} \nabla y_h(z) = I_2.$$

With a discrete Hessian D_h^2 the numerical minimization is then realized for the functional

$$I_{\text{plate}}^h[y_h] = \frac{1}{2}\int_\omega |D_h^2 y_h|^2 \, \mathrm{d}x'.$$

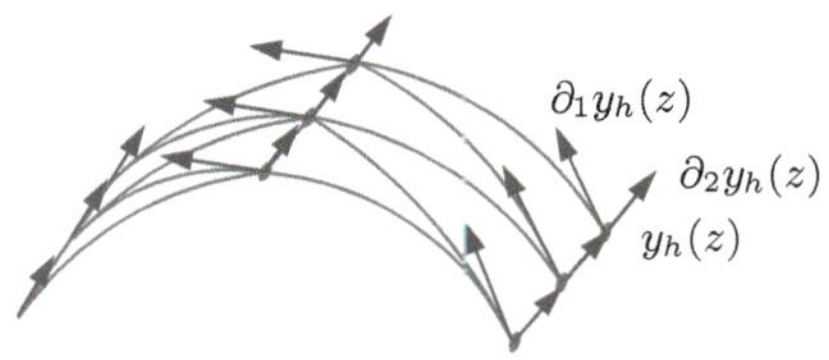

FIG. 3 Discrete deformations defined by discrete Kirchhoff triangles are defined by positions of nodes and tangent vectors at the displaced nodes.

In the case of the discrete Kirchhoff triangle, which may be seen as a natural generalization of the space of one-dimensional cubic C^1 functions, the degrees of freedom are the deformations and the deformation gradients in the nodes, i.e., the quantities

$$(y_h(z), \nabla y_h(z))_{z\in\mathcal{N}_h}.$$

An image of a discrete Kirchhoff deformation is depicted in Fig. 3.

The isometry constraint is thus imposed directly on certain degrees of freedom. The iterative numerical minimization follows closely the approach used in the one-dimensional situation. For an initial isometry y_0, we consider the continuous evolution problem

$$(\partial_t y, w)_\star = -(D^2 y, D^2 w), \quad y(0) = y_0,$$

for appropriate test functions $w \in H^2(\omega; \mathbb{R}^3)$ subject to the linearized isometry condition

$$L^{\text{iso}}_{\nabla y}[\partial_t \nabla y] = 0, \quad L^{\text{iso}}_{\nabla y}[\nabla w] = 0,$$

with the linearized isometry operator

$$L^{\text{iso}}_{A}[B] = A^{\mathsf{T}} B + B^{\mathsf{T}} A.$$

A semiimplicit discretization of this constrained evolution problem leads to a sequence of linearly constrained problems: given an admissible $y_h^0 \in \mathcal{A}_h$ compute the sequence $(y_h^k)_{k=0,1,\ldots}$ via $y_h^k = y_h^{k-1} + \tau d_t y_h^k$, where $d_t y_h^k$ solves

$$(d_t y_h^k, w_h)_\star = -(D_h^2 y_h^k, D_h^2 w_h)$$

subject to the conditions

$$L^{\text{iso}}_{\nabla y_h^{k-1}}[d_t \nabla y_h^k] = 0, \quad L^{\text{iso}}_{\nabla y_h^{k-1}}[\nabla w_h] = 0.$$

Again, straightforward calculations show that the iteration is energy decreasing and convergent, and that the constraint violation is of order $\mathcal{O}(\tau)$.

1.3 Outline of the article

The article is organized as follows. In Section 2 we discuss the arguments that lead to dimensionally reduced models for elastic rods and plates in the case of small energies. We review partial Γ-convergence results and explain the occurrence of the inextensibility and isometry constraints. Section 3 is devoted to the convergent and practical finite element discretization of the one- and two-dimensional minimization problems describing the elastic deformation of rods and plates. The rigorous justification of the finite element methods will be established via showing Γ-convergence of the discretized functionals to the continuous one as discretization parameters tend to zero. Difficulties arise in the appropriate treatment of nonlinear constraints and higher order derivatives. The practical minimization of the discretized energy functionals is addressed in Section 4. We use appropriately discretized gradient flows that lead to sequences of linear systems of equations together with guaranteed energy decrease. Moreover, we verify that they converge to stationary configurations. In view of the nonuniqueness and limited additional regularity properties of solutions this appears to be the best attainable result if no further assumptions are made. The special saddle-point structure of the linear systems of equations that arise in the time steps is investigated in Section 5. It turns out that the nodal constraints can be incorporated in the solution space leading to reduced linear systems with symmetric and positive definite system matrices. Section 6 is concerned with extensions and modifications of the models and solution methods. In particular, we discuss the numerical treatment of bilayer bending problems, the inclusion of a self-avoidance potential, and a bending problem allowing for the formation of wrinkles. The final section provides a summary and conclusions of our considerations. We end the introduction with an overview of employed notation.

1.4 Notation

Throughout this article we use standard notation for derivatives and integrals, matrices and inner products, Lebesgue, Sobolev, and finite element spaces. The list in Table 1 provides an overview of the most important symbols.

2 Formal dimension reductions

Following the articles Friesecke et al. (2002c,b) and Mora and Müller (2003) we illustrate in this section how the dimensionally reduced minimization problems can be obtained from a general three-dimensional hyperelastic energy minimization problem:

$$\begin{cases} \text{Minimize } I_{3\mathrm{d}}[y] = \displaystyle\int_\Omega W(\nabla y)\,\mathrm{d}x \\ \text{in the set } \mathcal{A} \subset W^{1,p}(\Omega;\mathbb{R}^3). \end{cases} \tag{1}$$

TABLE 1 Frequently used notation.

$(0, L)$, ω, Ω	One-, two-, and three-dimensional domains
$L^p(A;\mathbb{R}^\ell)$	Lebesgue functions with values in $\mathbb{R}^\ell$
$W^{k,p}(A,\mathbb{R}^\ell)$	Sobolev functions with values in $\mathbb{R}^\ell$
$H^k(A,\mathbb{R}^\ell)$	Sobolev space with $p = 2$
$x = (x_1, x_2, x_3)$	Spatial variable
$x' = (x_1, x_2)$	Planar component of spatial variable
$\lvert \cdot \rvert$	Euclidean or Frobenius norm of a vector or matrix
$(\cdot,\cdot)$, $\lVert \cdot \rVert$	Scalar product and norm in L^2
$x \cdot y$, $A : B$	Scalar products of vectors and matrices
I_ℓ	Identity matrix in $\mathbb{R}^{\ell\times\ell}$
$\mathrm{sym}(A)$, $\mathrm{tr}(A)$	Symmetric part and trace of a matrix
$SO(3)$	Orthogonal matrices with positive determinant
y'	One-dimensional derivative
$\nabla y = [\partial_1 y, \partial_2 y, \partial_3 y]$	Gradient of a vector field
$\nabla' y$	Planar component of gradient
G', I'	Total or Fréchet derivative
$P_k(A)$	Polynomials of degree at most k on a set A
h, $h_{\min}$	Maximal and minimal mesh-sizes
$\mathcal{T}_h$	Triangulation with intervals or triangles
$\mathcal{N}_h$, $\mathcal{S}_h$	Nodes and sides in a triangulation
z, z_T, z_S	Vertices and midpoints of elements, midpoints of sides
$\mathcal{S}^{k,\ell}(\mathcal{T}_h)$	Elementwise degree k polynomials in C^ℓ
$\mathcal{I}_h^{1,0}$, $\widehat{\mathcal{I}}_h^{1,0}$	Global and elementwise nodal $P1$ interpolants
Q_h	Elementwise averaging operator
$\lVert \cdot \rVert_{L_h^p}$, $\lVert \cdot \rVert_h$, $(\cdot,\cdot)_h$	Discrete L^p norms, case $p = 2$, discrete L^2 product
τ	Step size
$d_t a^k = (a^k - a^{k-1})/\tau$	Backward difference quotient for step size $\tau > 0$
δ	Small thickness parameter
$I[y]$	Energy functional
$L[y]$	Linear operator

TABLE 1 Frequently used notation.—Cont'd

W	Energy density
$Q_3, Q_{\text{rod}}, Q_{\text{plate}}$	Quadratic forms
λ, μ	Lamé parameters
c_b, c_t	Bending and torsion rigidity
$\mathcal{A}, \mathcal{A}_h$	Sets of admissible deformations
$\mathcal{F}[y], \mathcal{F}_h[y_h]$	(Shifted) tangent spaces
$(\cdot, \cdot)_\star, (\cdot, \cdot)_+$	Metrics used to define gradient flows
$c, c', c'', \ldots$	Generic constants

The set of admissible deformations $\mathcal{A}$ is assumed to be a weakly closed subset of a Sobolev space $W^{1,p}(\Omega;\mathbb{R}^3)$ and required to include appropriate boundary conditions which imply a coercivity property. In an abstract way these are defined by a bounded linear operator

$$L_{\text{bc}} : W^{1,p}(\Omega;\mathbb{R}^3) \to Y$$

and given data $\ell_{\text{bc}} \in Y$ for a suitable linear space Y, e.g., traces of functions in $W^{1,p}(\Omega;\mathbb{R}^3)$ restricted to a subset Γ_{D} of $\partial\Omega$. We assume that the energy density $W \in C^2(\mathbb{R}^{3\times 3})$ satisfies the following standard requirements:

- W is frame-indifferent, i.e., for all $F \in \mathbb{R}^{3\times 3}$ and $Q \in SO(3)$ we have

$$W(QF) = W(F),$$

- W vanishes at the identity $I_3 \in \mathbb{R}^{3\times 3}$ and grows at least quadratically away from $SO(3)$, i.e., for all $F \in \mathbb{R}^{3\times 3}$ we have

$$W(I_3) = 0, \quad W(F) \geq c\,\text{dist}^2(F, SO(3)),$$

- W is isotropic, i.e., for all $F \in \mathbb{R}^{3\times 3}$ and $R \in SO(3)$ we have

$$W(FR) = W(F).$$

From the first two conditions we have that $W'(I_3) = 0$ and a Taylor expansion yields

$$W(I_3 + G) = \frac{1}{2} Q_3(G) + o(|G|^2),$$

where $Q_3(G) = W''(I_3)[G, G]$ is the quadratic form defined by the second variation of W at the identity matrix. Incorporating the implicitly assumed homogeneity of the underlying material we have

$$Q_3(G) = \mathbb{C}G : G,$$

where the linear operator $\mathbb{C} : \mathbb{R}^{3\times3} \to \mathbb{R}^{3\times3}$ is given by

$$\mathbb{C}A = 2\mu \,\mathrm{sym}(A) + \lambda \,\mathrm{tr}(A) I_3,$$

with the Lamé parameters λ, $\mu > 0$, the symmetric part $\mathrm{sym}(A) = (A + A^{\mathsf{T}})/2$, and the trace $\mathrm{tr}(A) = A : I_3$.

2.1 Elastic rods

We assume that the deformation $y : (0, L) \to \mathbb{R}^3$ of an elastic rod of vanishing thickness and length L preserves distances, i.e., satisfies $|y'(x_1)| = 1$, and complement the vector field y' by normal vector fields $b, d : (0, L) \to \mathbb{R}^3$ to an orthonormal frame

$$[y', b, d] : (0, L) \to SO(3).$$

We then consider a three-dimensional deformation y_δ obtained from extending the deformation of the centreline $(0, L)$ to the three-dimensional body $\Omega_\delta = (0, L) \times \delta S$ with scaled cross section $S \subset \mathbb{R}^2$ containing zero, i.e.,

$$y_\delta(x_1, x_2, x_3) = y(x_1) + x_2 b(x_1) + x_3 d(x_1) + \delta^2 \beta(x),$$

with a correction function $\beta : \Omega_\delta \to \mathbb{R}^3$. Inserting this deformation into the three-dimensional energy functional and considering the limit as $\delta \to 0$, we expect to identify a dimensionally reduced functional minimized by y and the normal fields b and d. We note that $x_2, x_3 = \mathcal{O}(\delta)$ and that we expect $\partial_2\beta, \partial_3\beta = \mathcal{O}(\delta^{-1})$. We therefore set

$$\begin{aligned}
\nabla y_\delta &= [y', b, d] + \left[x_2 b' + x_3 d', \delta^2 \partial_2\beta, \delta^2 \partial_3\beta\right] + \delta^2 [\partial_1\beta, 0, 0] \\
&= R + \delta B + \delta^2 C.
\end{aligned}$$

with the matrix $R = [y', b, d] \in SO(3)$. The matrix $R^{\mathsf{T}} \nabla y_\delta$ is thus a perturbation of the identity matrix I_3 and a Taylor expansion of the energy density yields with

$$R^{\mathsf{T}} \nabla y_\delta = I_3 + \delta R^{\mathsf{T}} B + \delta^2 R^{\mathsf{T}} C$$

that we have

$$\begin{aligned}
W(\nabla y_\delta) &= W(R^{\mathsf{T}} \nabla y_\delta) \\
&= \frac{1}{2} Q_3\left(R^{\mathsf{T}} \left[x_2 b' + x_3 d', \delta^2 \partial_2\beta, \delta^2 \partial_3\beta\right]\right) + o(\delta^2).
\end{aligned}$$

Letting $\alpha = \delta^2 R^{\mathsf{T}} \beta$ and noting that R does not depend on x_2 and x_3 we have

$$R^{\mathsf{T}}\left[x_2 b' + x_3 d', \delta^2 \partial_2 \beta, \delta^2 \partial_3 \beta\right] = R^{\mathsf{T}} R' \begin{bmatrix} 0 \\ x_2 \\ x_3 \end{bmatrix} + [0, \partial_2 \alpha, \partial_3 \alpha].$$

We note that since $R^{\mathsf{T}} R = I_3$ we have $(R^{\mathsf{T}})' R = -R^{\mathsf{T}} R'$ so that $R^{\mathsf{T}} R'$ is skew-symmetric. A minimization of the integral of the energy density over the cross section δS motivates defining for skew-symmetric matrices $A = [a_1, a_2, a_3] \in \mathbb{R}^{3\times 3}$ the reduced quadratic form Q_{rod} via

$$Q_{\mathrm{rod}}(A) = \min_{\alpha \in H^1(S;\mathbb{R}^3)} \int_{\delta S} Q_3([x_2 a_2 + x_3 a_3, \partial_2 \alpha, \partial_3 \alpha])\, \mathrm{d}x_2\, \mathrm{d}x_3.$$

With the particular representation of Q_3 by the Lamé parameters one finds with the entries $a_{ij} = -a_{ji}$ of A for a circular cross section $S = B_{1/\pi}(0)$ that

$$Q_{\mathrm{rod}}(A) = \frac{1}{2\pi} \frac{\mu(3\lambda + 2\mu)}{\lambda + \mu}(a_{12}^2 + a_{13}^2) + \frac{\mu}{2\pi} a_{23}^2.$$

The constant factors on the right-hand side define the bending and torsion rigidities and are abbreviated by

$$c_{\mathrm{b}} = \frac{1}{2\pi} \frac{\mu(3\lambda + 2\mu)}{\lambda + \mu}, \quad c_{\mathrm{t}} = \frac{\mu}{2\pi}.$$

We always assume $\lambda, \mu > 0$ so that $c_{\mathrm{b}} \geq 2c_{\mathrm{t}}$. For the particular matrix $A = R^{\mathsf{T}} R'$ and $R = [y', b, d] \in SO(3)$ we have that

$$a_{12} = y'' \cdot b, \quad a_{13} = y'' \cdot d, \quad a_{23} = b' \cdot d.$$

Noting that $y'' \cdot y' = 0$ we have that

$$a_{12}^2 + a_{13}^2 = |y''|^2$$

is the squared curvature of the deformed rod and that

$$a_{23}^2 = (b' \cdot d)^2 = (d' \cdot b)^2$$

is its squared torsion. We eliminate the variable d via the identity $d = y' \times b$ in what follows. We thus expect that the deformation $y : (0, L) \to \mathbb{R}^3$ of the centreline of a thin rod and the unit normal vector field $b:(0,L)\to\mathbb{R}^3$ solve the following dimensionally reduced problem:

$$\begin{cases} \text{Minimize} \\ I_{\text{rod}}[y,b]=\dfrac{c_{\text{b}}}{2}\displaystyle\int_0^L |y''|^2\,\mathrm{d}x_1+\dfrac{c_{\text{t}}}{2}\displaystyle\int_0^L (b'\cdot(y'\times b))^2\,\mathrm{d}x_1 \\ \text{in the set} \\ \mathcal{A}=\left\{(y,b)\in V_{\text{rod}}\,:\,L_{\text{bc}}^{\text{rod}}[y,b]=\ell_{\text{bc}}^{\text{rod}},\ |y'|=|b|=1, y'\cdot b=0\right\}. \end{cases} \tag{2}$$

Here, we abbreviate

$$V_{\text{rod}}=H^2(0,L;\mathbb{R}^3)\times H^1(0,L;\mathbb{R}^3).$$

The second part for justifying the dimensionally reduced model consists in showing that for any sequence $(y_\delta)_{\delta>0}$ of three-dimensional deformations with $I_{3\text{d}}[y_\delta]\le c\delta^3$ there exists an appropriate limit $(y,b)\in\mathcal{A}$ such that

$$\lim_{\delta\to 0}\inf I_{3\text{d}}[y\delta]\ge I_{\text{rod}}[y,b].$$

This so-called compactness property is proved in Mora and Müller (2003) which provides the complete rigorous dimension reduction in a more general setting. We refer the reader to Langer and Singer (1996) for further aspects of the description of elastic rods.

Remark 1. To illustrate that the inextensibility or isometry condition $|y'|=1$ arises naturally in the dimension reduction we consider the planar deformation of a two-dimensional thin beam $\Omega=(0,L)\times(-\delta/2,\delta/2)$ with the simple energy density

$$W(F)=\text{dist}^2(F,SO(2))\approx(1/4)|F^{\mathsf{T}}F-I|^2.$$

We assume that the deformation is given by

$$y_\delta(x_1,x_2)=y(x_1)+x_2b(x_1)$$

for a deformation $y:(0,L)\to\mathbb{R}^2$ of the centreline and a corresponding normal field $b:(0,L)\to\mathbb{R}^2$, i.e., we have $y'(x_1)\cdot b(x_1)=0$. Noting that

$$\nabla y_\delta=[y'+x_2b',b]$$

we find that

$$\begin{aligned}(\nabla y_\delta)^{\mathsf{T}}\nabla y_\delta-I_2&=\begin{bmatrix}|y'|^2-1 & 0\\ 0 & |b|^2-1\end{bmatrix}+x_2\begin{bmatrix}2y'\cdot b' & b\cdot b'\\ b\cdot b' & 0\end{bmatrix}+x_2^2\begin{bmatrix}|b'|^2 & 0\\ 0 & 0\end{bmatrix}\\ &=A+x_2B+x_2^2C.\end{aligned}$$

We insert this expression into the energy functional and carry out the integration in vertical direction, i.e.

$$I_\delta[y_\delta] \approx \frac{1}{4}\int_0^L \int_{-\delta/2}^{\delta/2} |A + x_2 B + x_2^2 C|^2 \, dx_2 \, dx_1$$
$$= \frac{1}{4}\int_0^L \delta|A|^2 + \frac{\delta^3}{12}|B|^2 + \frac{\delta^5}{80}|C|^2 + \frac{\delta^3}{12} 2A : C \, dx_1.$$

For a cubic scaling of the elastic energy we need $A = 0$, i.e., $|y'|^2 = 1$ and $|b|^2 = 1$. This implies $b' \cdot b = 0$, hence $|B|^2 = 4|y' \cdot b'|^2$, and shows that up to terms of order δ^5 we have

$$I_\delta[y_\delta] \approx \frac{\delta^3}{12}\int_0^L |y''|^2 \, dx_1,$$

where we used the identities $y' \cdot b = -y'' \cdot b$ and $y'' \cdot y' = 0$ in combination with the fact that (y', b) is an orthonormal basis in $\mathbb{R}^2$ so that $(y' \cdot b)' = 0$.

2.2 Elastic plates

To explain the derivation of the bending model for elastic plates we consider an isometry $y : \omega \to \mathbb{R}^3$ and a corresponding unit normal field $b : \omega \to \mathbb{R}^3$, i.e., we have

$$\partial_i y(x') \cdot \partial_j y(x') = \delta_{ij},$$

for $1 \le i, j \le 2$ and

$$|b(x')|^2 = 1, \quad \partial_j y(x') \cdot b(x') = 0$$

for almost every $x' \in \omega$ and $j = 1, 2$. We define a deformation y_δ of the three-dimensional body $\Omega_\delta = \omega \times (-\delta/2, \delta/2)$ by extending y in normal direction, i.e.,

$$y_\delta(x', x_3) = y(x') + x_3 b(x') + (x_3^2/2)\beta(x'),$$

with a quadratic correction term $\beta : \omega \to \mathbb{R}^3$. Using the planar gradient $\nabla' = [\partial_1, \partial_2]$ we have that

$$R = [\nabla' y, b] \in SO(3)$$

and

$$\nabla y_\delta = [\nabla' y_\delta, \partial_3 y_\delta] = \left[\nabla' y + x_3 \nabla' b + (x_3^2/2)\nabla'\beta, b + x_3\beta\right]$$
$$= R + x_3[\nabla' b, \beta] + (x_3^2/2)[\nabla'\beta, 0].$$

This implies that

$$R^\mathsf{T}\nabla y_\delta - I = x_3 R^\mathsf{T}[\nabla' b, \beta] + (x_3^2/2)[\nabla'\beta, 0].$$

We insert the deformation y_δ into the hyperelastic energy functional, use the Taylor expansion $W(I+x_3G)=Q_3(x_3G)+o(x_3^2|G|^2)$, with $G = R^\mathsf{T}\nabla y_\delta$, and carry out the integration in vertical direction. This leads to

$$\begin{aligned}\int_\omega\int_{-\delta/2}^{\delta/2} W(\nabla y_\delta)\,\mathrm{d}x_3\,\mathrm{d}x' &= \int_\omega\int_{-\delta/2}^{\delta/2} W(R^\mathsf{T}\nabla y_\delta)\,\mathrm{d}x_3\,\mathrm{d}x'\\ &= \frac{1}{2}\int_\omega\int_{-\delta/2}^{\delta/2} Q_3\left(x_3R^\mathsf{T}[\nabla' b,\beta]+(x_3^2/2)[\nabla'\beta,0]\right)\mathrm{d}x_3\,\mathrm{d}x' + o(\delta^3)\\ &= \frac{\delta^3}{24}\int_\omega Q_3(R^\mathsf{T}[\nabla' b,\beta])\,\mathrm{d}x' + o(\delta^3).\end{aligned}$$

The correction field $\beta:\omega\to\mathbb{R}^3$ is eliminated via a pointwise minimization, i.e., for $M\in\mathbb{R}^{2\times 2}$ extended by a vanishing third row to a matrix $\hat{M}\in\mathbb{R}^{3\times 2}$, we define

$$Q_{\text{plate}}(M) = \min_{c\in\mathbb{R}^3} Q_3([\hat{M},c]).$$

Since we assume a homogeneous and isotropic material one obtains for a symmetric matrix $M\in\mathbb{R}^{2\times 2}$ that

$$Q_{\text{plate}}(M) = 2\mu|M|^2 + \frac{\lambda\mu}{\mu+\lambda/2}\operatorname{tr}(M)^2.$$

For the matrix $\widehat{M}=R^\mathsf{T}\nabla' b\in\mathbb{R}^{3\times 2}$ the third row vanishes and its upper 2×2 submatrix coincides with the second fundamental form II of the surface parametrized by y, i.e.,

$$II_{ij}(x') = \partial_i y(x')\cdot\partial_j b(x') = -\partial_i\partial_j y(x')\cdot b(x'),$$

for $i, j = 1, 2$, where in fact $b = \pm\partial_1 y\times\partial_2 y$. Using that y is an isometry we have that the squared mean curvature is up to a fixed factor given by the identical expressions

$$|II|^2 = \operatorname{tr}(II)^2 = |D^2 y|^2 = |\Delta y|^2.$$

Hence, the dimensionally reduced problem seeks an isometric deformation that minimizes the integral of the squared Hessian:

$$\begin{cases}\text{Minimize } I_{\text{plate}}[y] = \dfrac{c_\mathrm{b}}{2}\displaystyle\int_\omega |D^2 y|^2\,\mathrm{d}x' \text{ in the set}\\ \mathcal{A}=\left\{y\in H^2(\omega;\mathbb{R}^3):(\nabla y)^\mathsf{T}\nabla y = I_2,\ L_{\mathrm{bc}}^{\text{plate}}[y]=\ell_{\mathrm{bc}}^{\text{plate}}\right\}.\end{cases} \tag{3}$$

The bending rigidity is defined by $c_\mathrm{b} = 2\mu + \lambda 2\mu/(2\mu+\lambda)$. As in the case of rods, a rigorous derivation additionally requires showing that the functional defines a general lower bound, i.e., establishing a lim-inf inequality, and we refer the reader to Friesecke et al. (2002b,c) for details. Analogously to the functional, also the boundary conditions change their nature in the dimension reduction. A fixed part of the lateral boundary leads to a clamped boundary condition in the reduced model which imposes a condition on the deformation and its gradient.

3 Convergent finite element discretizations

We discuss in this section the discretization of the dimensionally reduced nonlinear bending models using appropriate finite element methods. Challenges are the treatment of higher order derivatives and a nonlinear pointwise constraint. We establish the correctness of the discretizations by showing that the discrete functionals I^h converge in the sense of Γ-convergence with respect to weak convergence on a space X, cf., e.g., Dal Maso (1993), to the continuous, dimensionally reduced functional I. We use the terminology *almost-minimizing* for a sequence of objects that are minimizers of a sequence of functions up to tolerances that converge to zero with h. This follows from verifying the following three conditions:

(a) *Well posedness or equicoercivity:* The discrete functionals are uniformly coercive, i.e., if $I^h[y_h] \leq c$ then it follows that $\|y_h\|_X \leq c'$ with h-independent constants $c, c' \geq 0$, and admit discrete minimizers.
(b) *Stability or lim-inf inequality:* If $(y_h)_{h>0} \subset X$ is a bounded sequence of discrete almost-minimizers then every weak accumulation point y belongs to the set of admissible deformations $\mathcal{A}$ and we have

$$I[y] \leq \liminf_{h\to 0} I^h[y_h].$$

(c) *Consistency or lim-sup inequality:* For every $y \in \mathcal{A}$ there exists a sequence $(y_h)_{h>0}$ of admissible discrete deformations such that $y_h \rightharpoonup y$ in X and

$$I[y] \geq \limsup_{h\to 0} I^h[y_h].$$

It is an immediate consequence of (a)–(c) that sequences of discrete almost-minimizers accumulate at minimizers of the continuous problem. Well posedness typically follows from coercivity properties of the functional I while the stability is established with the help of lower semicontinuity properties of I. If the union of discrete sets of admissible deformations is dense in the set of admissible deformations then consistency is obtained via continuity properties of I. We specify these concepts for the finite element approximation of elastic deformations of rods and plates in what follows. We always use a regular triangulation $\mathcal{T}_h$ of the domain $A = (0, L)$ or $A = \omega$ into intervals or triangles, respectively, with a set of nodes (vertices of elements) denoted $\mathcal{N}_h$, i.e.,

$$\mathcal{N}_h = \{z_1, z_2, \ldots, z_N\}, \quad \mathcal{T}_h = \{T_1, T_2, \ldots, T_M\},$$

We often use numerical integration or quadrature, defined with the elementwise applied nodal interpolation via

$$(v,w)_h = \int_A \hat{\mathcal{I}}_h[(v \cdot w)]\,\mathrm{d}x$$

for elementwise continuous functions $v,w : A \to \mathbb{R}^\ell$ with $A \subset \mathbb{R}^d$ and

$$\|v\|_{L^p_h}^p = \sum_{T \in \mathcal{T}_h} \frac{|T|}{d+1} \sum_{z \in \mathcal{N}_h \cap T} |v(z)|^p.$$

If $p = 2$ we write $\|v\|_h$ instead of $\|v\|_{L^2_h}$. We note that these expressions define equivalent scalar products and norms on function spaces containing elementwise polynomials of bounded degree, cf., e.g., Bartels (2015).

3.1 Elastic rods

For a discretization of the bending-torsion model (2) for elastic rods we first derive a suitable reformulation of the minimization problem. Recalling that for an admissible pair $(y,b) \in \mathcal{A}$ and the vector $d = y' \times b$ we have that $[y', b, d] \in SO(3)$ almost everywhere in the interval $(0, L)$, we deduce that

$$|b'|^2 = (b' \cdot y')^2 + (b' \cdot d)^2 + (b' \cdot b)^2.$$

Since $|b|^2 = 1$ the last term on the right-hand side vanishes while the orthogonality $b \cdot y' = 0$ implies that $b' \cdot y' = -b \cdot y''$. We thus have that

$$(b' \cdot d)^2 = |b'|^2 - (b \cdot y'')^2.$$

This identity leads to the equivalent representation

$$I_{\mathrm{rod}}[y,b] = \frac{c_{\mathrm{b}}}{2} \int_0^L |y''|^2\,\mathrm{d}x_1 + \frac{c_{\mathrm{t}}}{2} \int_0^L |b'|^2\,\mathrm{d}x_1 - \frac{c_{\mathrm{t}}}{2} \int_0^L (b \cdot y'')^2\,\mathrm{d}x_1.$$

The dimension reduction of Section 2.1 shows that we have $c_{\mathrm{b}} \geq 2c_{\mathrm{t}}$ so that the last term is controlled by the first one and the coercivity of I^h_{rod} becomes explicit. Another advantage of this representation is that the last term is separately concave which allows for an effective iterative treatment. To define the discrete functional I^h_{rod} we consider a partitioning of the reference interval $(0, L)$ defined by sets of nodes $\mathcal{N}_h$ and elements $\mathcal{T}_h$. For this partitioning we define the linear and cubic finite element spaces with different differentiability requirements via

$$\mathcal{S}^{1,0}(\mathcal{T}_h) = \{\phi_h \in C^0([0,L]) : \phi_h|_T \in P_1(T) \text{ for all } T \in \mathcal{T}_h\},$$
$$\mathcal{S}^{3,1}(\mathcal{T}_h) = \{v_h \in C^1([0,L]) : v_h|_T \in P_3(T) \text{ for all } T \in \mathcal{T}_h\},$$

with sets of polynomials of degree at most k on T given by $P_k(T)$. The degrees of freedom of the finite element spaces are depicted in Fig. 4.

FIG. 4 Degrees of freedom of piecewise linear, continuous and piecewise cubic, continuously differentiable finite element functions. *Filled dots* indicate function values and circle evaluations of derivatives.

It is straightforward to verify that there exist nodal bases $(\varphi_z)_{z\in\mathcal{N}_h}$ and $(\psi_{z,j})_{z\in\mathcal{N}_h,j=1,2}$ such that for all $\phi_h \in \mathcal{S}^{1,0}(\mathcal{T}_h)$ and $v_h \in \mathcal{S}^{3,1}(\mathcal{T}_h)$ we have

$$\phi_h = \sum_{z\in\mathcal{N}_h} \phi_h(z)\varphi_z,$$
$$v_h = \sum_{z\in\mathcal{N}_h} v_h(z)\psi_{z,0} + \sum_{z\in\mathcal{N}_h} v_h'(z)\psi_{z,1}.$$

The right-hand sides define nodal interpolation operators $\mathcal{I}_h^{1,0}$ and $\mathcal{I}_h^{3,1}$ on $C^0([0, L])$ and $C^1([0, L])$, respectively. For ease of notation we use the abbreviation

$$V_{\mathrm{rod}}^h = \mathcal{S}^{3,1}(\mathcal{T}_h)^3 \times \mathcal{S}^{1,0}(\mathcal{T}_h)^3.$$

For efficient numerical quadrature we introduce the elementwise averaging operator Q_h defined for a vector field $v \in L^1(0, L; \mathbb{R}^3)$ and every element $T \in \mathcal{T}_h$ via

$$Q_h v|_T = |T|^{-1} \int_T v \, \mathrm{d}x_1.$$

With the product finite element space V_{rod}^h and the operator Q_h we consider the following discretization of the minimization problem (2) in which the pointwise orthogonality relation $y' \cdot b = 0$ is approximated via a penalty term:

$$\begin{cases} \text{Minimize } I_{\mathrm{rod}}^{h,\varepsilon}[y_h, b_h] = \dfrac{c_{\mathrm{b}}}{2}\displaystyle\int_0^L |y_h''|^2 \, \mathrm{d}x_1 + \dfrac{c_{\mathrm{t}}}{2}\displaystyle\int_0^L |b_h'|^2 \, \mathrm{d}x_1 \\ \quad -\dfrac{c_{\mathrm{t}}}{2}\displaystyle\int_0^L (Q_h b_h \cdot y_h'')^2 \, \mathrm{d}x_1 + \dfrac{1}{2\varepsilon}\displaystyle\int_0^L \mathcal{I}_h^{1,0}[(y_h' \cdot b_h)^2] \, \mathrm{d}x_1 \\ \text{in the set } \mathcal{A}_h = \big\{(y_h, b_h) \in V_{\mathrm{rod}}^h : L_{\mathrm{bc}}^{\mathrm{rod}}[y_h, b_h] = \ell_{\mathrm{bc}}^{\mathrm{rod}}, \\ \qquad |y_h'(z)| = |b_h(z)| = 1 \text{ f.a. } z \in \mathcal{N}_h\big\}. \end{cases} \quad (4)$$

Note that the constraints are imposed on particular degrees of freedom which makes the method practical. We have the following existence and convergence result.

Proposition 1 (Convergent approximation).
For every pair $(h, \varepsilon) > 0$ there exists a minimizer $(y_h, b_h) \in \mathcal{A}_h$ for $I_{rod}^{h,\varepsilon}$ satisfying

$$\|y_h\|_{H^2} + \|b_h\|_{H^1} \le c.$$

As $(h, \varepsilon) \to 0$ we have that every accumulation point of a sequence of discrete almost-minimizers is a minimizer for I_{rod} in $\mathcal{A}$.

Proof (sketched). We outline the main arguments of the proof and refer the reader to Bartels (2013c) and Bartels and Reiter (2019) for details.

(a) Let $(y_h, b_h) \in \mathcal{A}_h$ with $I^{h,\varepsilon}_{\mathrm{rod}}[y_h, b_h] \leq c$. The coercivity of the discretized functional follows from the fact that $c_\mathrm{b} \geq 2c_\mathrm{t}$ and the identity

$$I^{h,\varepsilon}_{\mathrm{rod}}[y_h, b_h] = \frac{c_\mathrm{b} - c_\mathrm{t}}{2} \int_0^L |y_h''|^2 \, \mathrm{d}x_1 + \frac{c_\mathrm{t}}{2} \int_0^L |b_h'|^2 \, \mathrm{d}x_1$$
$$+ \frac{c_\mathrm{t}}{2} \int_0^L [y_h'']^\mathsf{T} P_{b_h} [y_h''] \, \mathrm{d}x_1 + \frac{1}{2\varepsilon} \int_0^L \mathcal{I}_h[(y_h' \cdot b_h)^2] \, \mathrm{d}x_1,$$

with the positive semidefinite matrix

$$P_{b_h} = I_3 - (Q_h b_h) \otimes (Q_h b_h).$$

The discrete coercivity and the continuity properties of the functional $I^{h,\varepsilon}_{\mathrm{rod}}$ imply the existence of discrete minimizers.

(b) Given a sequence of discrete almost-minimizers $(y_h, b_h)_{h,\varepsilon>0}$ one first checks that accumulation points (y, b) as $(h, \varepsilon) \to 0$ belong to the continuous admissible set $\mathcal{A}$. Noting that $b_h \to b$ strongly in L^∞ we find that

$$I_{\mathrm{rod}}[y, b] \leq \liminf_{(h,\varepsilon)\to 0} I^{h,\varepsilon}_{\mathrm{rod}}[y_h, b_h].$$

(c) It remains to show that $I_{\mathrm{rod}}[y, b]$ is minimal. For this, we choose a smooth almost-minimizing pair $(\tilde{y}, \tilde{b}) \in \mathcal{A}$ obtained from an appropriate regularization of a minimizing pair and verify that the sequence of interpolants $(\tilde{y}_h, \tilde{b}_h)$ satisfies $\lim_{(h,\varepsilon)\to 0} I^{h,\varepsilon}_{\mathrm{rod}}[y_h, b_h] = I_{\mathrm{rod}}[\tilde{y}, \tilde{b}]$.

□

Remark 2. We note that the result of the proposition can be also be established if the orthogonality relation $y_h' \cdot b_h = 0$ is imposed exactly in the nodes of the triangulation. For an efficient numerical solution of the minimization problem the approximation via a separately convex term is advantageous as this allows for a decoupled treatment of the variables.

3.2 Elastic plates

Constructing finite element spaces that provide convergent second-order derivatives and which are efficiently implementable is significantly more challenging in two space dimensions. Among the various possibilities is the *discrete Kirchhoff triangle*, cf., e.g., Braess (2007), which defines a nonconforming finite element method in the sense that its elements do not belong to H^2. The space can be seen as a natural generalization of the space of one-dimensional cubic C^1 functions since the degrees of freedom are the deformations and deformation gradients at the nodes of a triangulation which are appropriately interpolated on the individual elements. To define this finite element space we choose a triangulation $\mathcal{T}_h$ of ω into triangles and set

$$\mathcal{S}^{\text{dkt}}(\mathcal{T}_h) = \{w_h \in C(\overline{\omega}) : w_h|_T \in P_{3-}(T) \text{ for all } T \in \mathcal{T}_h,$$
$$\nabla w_h \text{ continuous at all } z \in \mathcal{N}_h\},$$
$$\mathcal{S}^{2,0}(\mathcal{T}_h) = \{q_h \in C(\overline{\omega}) : q_h|_T \in P_2(T) \text{ for all } T \in \mathcal{T}_h\}.$$

Here, P_{3-} denotes the subset of cubic polynomials on T obtained by eliminating the degree of freedom associated with the midpoint z_T of T, i.e., we have

$$P_{3-}(T) = \left\{ p \in P_3(T) : p(z_T) = \frac{1}{3} \sum_{z \in \mathcal{N}_h \cap T} [p(z) + \nabla p(z) \cdot (z_T - z)] \right\}.$$

The degrees of freedom in $\mathcal{S}^{\text{dkt}}(\mathcal{T}_h)$ are the function values and the derivatives at the vertices of the elements. It is interesting to note that a particular basis for $\mathcal{S}^{\text{dkt}}(\mathcal{T}_h)$ will not be needed. A canonical interpolation operator $\mathcal{I}_h^{\text{dkt}} : C^1(\overline{\omega}) \to \mathcal{S}^{\text{dkt}}(\mathcal{T}_h)$ is defined by requiring that the identities

$$\mathcal{I}_h^{\text{dkt}} w(z) = w(z), \quad \nabla \mathcal{I}_h^{\text{dkt}} w(z) = \nabla w(z)$$

hold at all nodes $z \in \mathcal{N}_h$. The employed finite element spaces are depicted in Fig. 5.

Crucial for the finite element discretization of the bending problem is the definition of a discrete gradient operator

$$\nabla_h : \mathcal{S}^{\text{dkt}}(\mathcal{T}_h) \to \mathcal{S}^{2,0}(\mathcal{T}_h)^2$$

which allows us to define discrete second-order derivatives of functions $w_h \in \mathcal{S}^{\text{dkt}}$ via

$$D_h^2 w_h = \nabla \nabla_h w_h.$$

Here we make use of the fact that $\nabla_h w_h \in H^1(\Omega; \mathbb{R}^2)$. The discrete gradient operator ∇_h is for given $w_h \in \mathcal{S}^{\text{dkt}}(\mathcal{T}_h)$ defined as the unique piecewise quadratic, continuous vector field $q_h \in \mathcal{S}^{2,0}(\mathcal{T}_h)^2$ that satisfies the condition

$$q_h(z) = \nabla w_h(z)$$

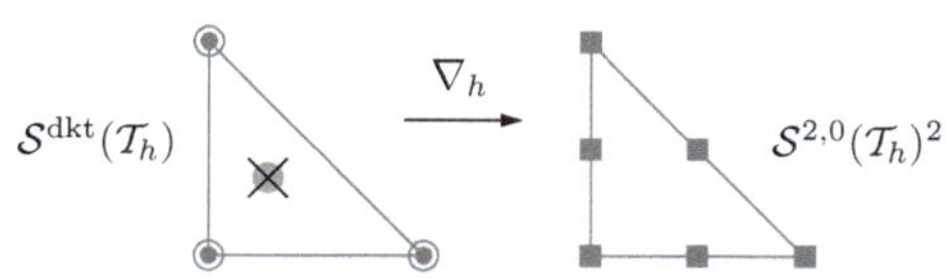

FIG. 5 Degrees of freedom of finite element spaces using reduced cubic polynomials and quadratic vector fields. *Filled dots* indicate function values, circle evaluations of derivatives, and squares vectorial function values. One degree of freedom is eliminated from the set of cubic polynomials.

for all $z \in \mathcal{N}_h$ while the degrees of freedom associated with the sides of elements are defined by the two conditions

$$q_h(z_S) \cdot n_S = \frac{1}{2}\left(\nabla w_h(z_S^1) + \nabla w_h(z_S^2)\right) \cdot n_S,$$
$$q_h(z_S) \cdot t_S = \nabla w_h(z_S) \cdot t_S,$$

for all sides $S = [z_S^1, z_S^2] \in \mathcal{S}_h$ with normals n_S, tangent vectors t_S, and midpoints $z_S = (z_S^1 + z_S^2)/2$. For $w \in C^1(\overline{\omega})$, we set

$$\nabla_h w = \nabla_h \mathcal{I}_h^{\mathrm{dkt}} w.$$

With the discrete second derivatives we are in a position to state the finite element discretization of the plate bending model:

$$\begin{cases} \text{Minimize } I_{\text{plate}}^h[y_h] = \frac{c_b}{2}\int_\omega |D_h^2 y_h|^2 \, \mathrm{d}x' \text{ in the set} \\ \mathcal{A}_h = \Big\{ y_h \in \mathcal{S}^{\mathrm{dkt}}(\mathcal{T}_h)^3 : L_{\mathrm{bc}}^{\mathrm{plate}}[y_h] = \ell_{\mathrm{bc}}^{\mathrm{plate}}, \\ \qquad [\nabla y_h(z)]^{\mathsf{T}} \nabla y_h(z) = I_2 \text{ f.a. } z \in \mathcal{N}_h \Big\}. \end{cases} \tag{5}$$

To prove the correctness of this discretization we show that existing finite element minimizers accumulate at admissible isometries, incorporate that the bending energy is weakly lower semicontinuous, and use that isometries can be approximated by smooth isometries which is a result proved in Pakzad (2004) and Hornung (2011).

Theorem 1 (Convergent approximation).
For every $h > 0$ there exists a minimizer $y_h \in \mathcal{A}_h$. If $(y_h)_{h>0}$ is a sequence of almost-minimizers, then $\|\nabla y_h\| \leq c$, for all $h > 0$, and every accumulation point $y \in H^1(\omega; \mathbb{R}^3)$ of the sequence is a strong accumulation point, belongs to the continuous admissible set $\mathcal{A}$ and is minimal for I_{plate}.

Proof (sketched). We follow the typical steps for establishing a Γ-convergence result provided in Bartels (2013a, 2015).

(a) Using the boundary conditions included in the set $\mathcal{A}_h$ it follows that the mapping $z_h \mapsto \| D_h^2 z_h \|$ is a norm on the subset of $\mathcal{S}^{\mathrm{dkt}}(\mathcal{T}_h)^3$ with functions satisfying corresponding homogeneous boundary conditions. This leads to a coercivity property and the existence of discrete solutions.
(b) The uniform discrete coercivity property implies that the sequences $(D_h^2 y_h)_{h>0}$ and $(\nabla y_h)_{h>0}$ have weak accumulation points ξ and ∇y in L^2 which are compatible in the sense that $\xi = D^2 y$. Moreover, we have that $y \in \mathcal{A}$ and weak lower semicontinuity of the L^2 norm shows that

$$I_{\text{plate}}[y] \leq \liminf_{h \to 0} I_{\text{plate}}^h[y_h].$$

(c) Let $y \in \mathcal{A}$ be a minimizer for I_{plate}. The continuity of the functional I_{plate} with respect to the strong topology in H^2 in combination with the density results for smooth isometries established in Hornung (2011) allow us to assume that y is smooth. We may thus define an approximating sequence of finite element functions by setting $\tilde{y}_h = \mathcal{I}_h^{\text{dkt}}[y]$. Approximation properties of the interpolation operator lead to the inequality

$$I_{\text{plate}}[y] \geq \limsup_{h \to 0} I_{\text{plate}}^h[\tilde{y}_h],$$

which proves the statement.

□

4 Iterative solution via constrained gradient flows

The practical solution of the finite element discretizations of the nonlinear bending problems is nontrivial due to the presence of nonlinear pointwise constraints and the corresponding lack of higher regularity properties. To provide a reliable strategy that decreases the energy we adopt gradient flow strategies. Our estimates show that these converge to stationary, low energy configurations. We will always use a linearized treatment of the constraints which is then discretized semiimplicitly. This makes the iterative scheme practical. To illustrate the main idea, consider the following abstract minimization problem in a Hilbert space $X \subset L^2(\Omega; \mathbb{R}^\ell)$:

$$\begin{cases} \text{Minimize } I[y] \\ \text{in } X \text{ subject to } G[y] = 0. \end{cases} \tag{6}$$

Here, we assume that the constraint is understood pointwise with a function $G: \mathbb{R}^\ell \to \mathbb{R}$. The Euler–Lagrange equations for the problem are then formally given by the identity

$$I'[y; w] + (\lambda, G'[y; w]) = 0$$

for all $w \in X$ with a Lagrange multiplier $\lambda \in L^1(\Omega)$. Note that the term involving λ disappears if w satisfies $G'[y; w] = 0$ and that this is sufficient to characterize a stationary point subject to the constraint. The corresponding gradient flow is formally defined via

$$\partial_t y = -\nabla_X I[y] - \lambda G'[y, \cdot] \quad \text{subject to} \quad G'[y; \partial_t y] = 0.$$

Our corresponding time-stepping scheme uses the backward difference quotient operator $d_t a^k = (a^k - a^{k-1})/\tau$ for a step size $\tau > 0$ and determines iterates via the linearly constrained problems

$$d_t y^k = -\nabla_X I[y^k] - \lambda^k G'[y^{k-1}, \cdot] \quad \text{subject to} \quad G'[y^{k-1}; d_t y^k] = 0.$$

We specify the meaning of the iterative scheme in the following algorithm.

Algorithm 1 (Abstract constrained gradient descent).
Let $y^0 \in X$ be such that $G[y^0] = 0$ and $I[y^0] < \infty$ and choose $\tau > 0$, set $k = 1$.
(1) Compute $y^k \in X$ such that

$$(d_t y^k, w)_X + I'[y^k; w] = 0$$

for all $w \in X$ under the constraints $d_t y^k$, $w \in \ker G'[y^{k-1}]$, i.e.,

$$G'[y^{k-1}; d_t y^k] = 0, \quad G'[y^{k-1}; w] = 0.$$

(2) Stop the iteration if $\|d_t y^k\|_X \le \varepsilon_{\text{stop}}$; otherwise increase $k \to k + 1$ and continue with (1).

We remark that it is useful to regard $d_t y^k$ rather than y^k as the unknown in the iteration steps. In particular, we may eliminate y^k via the identity $y^k = y^{k-1} + \tau d_t y^k$ with the known function y^{k-1}. A geometric interpretation of the iteration is that given an iterate y^{k-1} the correction $d_t y^k$ is computed in the tangent space of the level set $\mathcal{M}^{k-1}$ of G defined by the value $G[y^{k-1}]$. Note that we do not use a projection step onto the zero level set of G since this will in general not be energy stable. Fig. 6 illustrates the conceptual idea of Algorithm 1.

The following theorem states the main features of Algorithm 1, i.e., its unconditional energy stability with the resulting convergence to a stationary configuration of lower energy and the control of the constraint violation by the step size.

Theorem 2 (Convergent iteration).
Assume that I is convex, coercive, continuous, and Fréchet differentiable on X and $G : \mathbb{R}^\ell \to \mathbb{R}$ is twice differentiable with uniformly bounded second derivative, i.e., we have $G''[r; s, s] \le 2c_G|s|^2$ for all $r, s \in \mathbb{R}^\ell$. Then, the iterates of Algorithm 1 are uniquely defined and satisfy for $L = 0, 1, 2, \ldots$ the energy estimate

$$I[y^L] + \tau \sum_{k=1}^{L} \|d_t y^k\|_X^2 \le I[y^0].$$

Moreover, if $\|\cdot\|_H$ is a norm on X with the property $\| |z|^2 \|_H \le c_H \|z\|_X^2$ for all $z \in X$ then we have the constraint violation bound

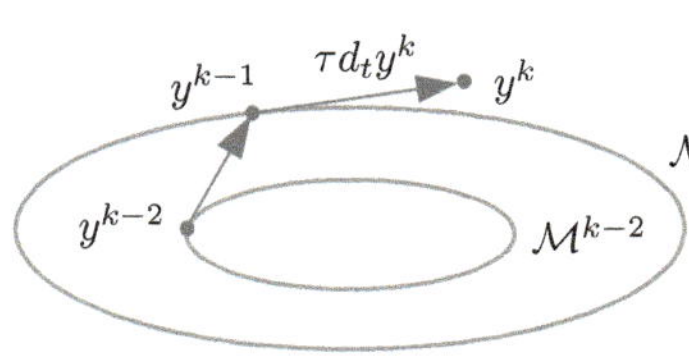

FIG. 6 Illustration of the iteration of Algorithm 1: corrections are computed in tangent spaces of level sets $\mathcal{M}^\ell$ of G defined by the values $G[y^\ell]$.

$$\max_{k=1,\ldots,L} \|G(y^k)\|_H \le \tau c_G c_H e_0,$$

where $e_0 = I[y^0]$. *In particular, we have that* $\|d_t y^k\|_X \to 0$ *and Algorithm* 1 *terminates within a finite number of iterations. The output* $y^L \in X$ *satisfies the residual estimate*

$$\sup_{\substack{w\in X\setminus\{0\} \\ G'[y^{L-1};w]=0}} \frac{|I'[y^L;w]|}{\|w\|_X} \le \varepsilon_{\text{stop}}.$$

Proof.

(a) The existence of the iterates follows by applying the direct method of the calculus of variations to the minimization problems:

$$\begin{aligned} &\text{Minimize } z \mapsto \frac{1}{2\tau}\|z-y^{k-1}\|_X^2 + I[z] \\ &\text{subject to } G'[y^{k-1};z]=0. \end{aligned}$$

The solution is unique and the corresponding Euler–Lagrange equation coincides with the equation that defines the iterates in Algorithm 1.

(b) Choosing the admissible test function $w = d_t y^k$ and using the convexity of I leads to

$$\|d_t y^k\|_X^2 + \frac{1}{\tau}\left(I[y^k] - I[y^{k-1}]\right) \le 0.$$

A multiplication by τ and summation over $k = 1, 2, \ldots, L$ prove the energy stability.

(c) With a Taylor expansion of G about the iterate y^{k-1} and the imposed identity $\delta G[y^{k-1}; d_t y^k] = 0$ we find that

$$G[y^k] = G[y^{k-1}] + \frac{1}{2}\tau^2 G''[\xi; d_t y^k, d_t y^k].$$

Repeating this argument and noting that $G[y^0] = 0$ leads to the estimate

$$|G[y^\ell]| \le c_G \tau^2 \sum_{k=1}^{\ell} |d_t y^k|^2.$$

Applying the norm $\|\cdot\|_H$ to the estimate, using the triangle inequality, and incorporating the assumed bound $\|\,|z|^2\|_H \le c_H \|z\|_X^2$ as well as the energy bound proves the estimate for the constraint violation error.

(d) The estimate for $I'[y^L; w]$ is an immediate consequence of the bound $\|d_t y^L\|_X \le \varepsilon_{\text{stop}}$.

Examples of pairs of norms $\|\cdot\|_H$ and $\|\cdot\|_X$ that satisfy the assumed estimate are the L^1 norm in combination with the L^2 norm or the L^∞ norm together with a Sobolev norm in H^s with s sufficiently large. It is remarkable that the violation of the constraint is independent of the number of iterations. An explanation for this is that the updates $d_t y^k$ converge quickly to zero in the gradient flow iteration.

4.1 Elastic rods

We apply the abstract framework for constrained minimization problems to the energy functional describing the bending-torsion behaviour of elastic rods. For a vector field $y_h \in \mathcal{S}^{3,1}(\mathcal{T}_h)^3$ we set

$$\mathcal{F}_h[y_h] = \{w_h \in \mathcal{S}^{3,1}(\mathcal{T}_h)^3 : L^{\mathrm{rod}}_{\mathrm{bc},y}[w_h] = 0,\ y_h'(z) \cdot w_h'(z) = 0 \text{ f.a. } z \in \mathcal{N}_h\}$$

while for a vector field $b_h \in \mathcal{S}^{1,0}(\mathcal{T}_h)^3$ we define

$$\mathcal{E}_h[b_h] = \{v_h \in \mathcal{S}^{1,0}\ (\mathcal{T}_h)^3 : L^{\mathrm{rod}}_{\mathrm{bc},b}[v_h] = 0,\ v_h(z) \cdot b_h(z) = 0 \text{ f.a. } z \in \mathcal{N}_h\}.$$

The functionals $L^{\mathrm{rod}}_{\mathrm{bc},y}$ and $L^{\mathrm{rod}}_{\mathrm{bc},b}$ are the components of $L^{\mathrm{rod}}_{\mathrm{bc}}$ corresponding to the variables y and b, respectively. We generate a sequence $(y_h^k, b_h^k)_{k=0,1,\ldots}$ that approximates a stationary configuration for $I^{h,\varepsilon}_{\mathrm{rod}}$ with the following algorithm.

Algorithm 2 (Gradient descent for elastic rods).

Choose an initial pair $(y_h^0, b_h^0) \in \mathcal{A}_h$ and a step size $\tau > 0$, set $k = 1$.

(1) Compute $d_t y_h^k \in \mathcal{F}_h[y_h^{k-1}]$ such that for all $w_h \in \mathcal{F}_h[y_h^{k-1}]$ we have

$$(d_t y_h^k, w_h)_\star + c_{\mathrm{b}}([y_h^k]'', w_h'') + \varepsilon^{-1}([y_h^k]' \cdot b_h^{k-1}, w_h' \cdot b_h^{k-1})_h$$
$$= c_{\mathrm{t}}\big([Q_h b_h^{k-1}] \cdot [y_h^{k-1}]'', [Q_h b_h^{k-1}] \cdot [w_h]''\big).$$

(2) Compute $d_t b_h^k \in \mathcal{E}_h[b_h^{k-1}]$ such that for all $r_h \in \mathcal{E}_h[b_h^{k-1}]$ we have

$$(d_t b_h^k, r_h)_\dagger + c_{\mathrm{t}}([b_h^k]', r_h') + \varepsilon^{-1}([y_h^k]' \cdot b_h^k, [y_h^k]' \cdot r_h)_h$$
$$= c_{\mathrm{t}}\big([Q_h b_h^{k-1}] \cdot [y_h^k]'', Q_h r_h \cdot [y_h^k]''\big).$$

(3) Stop the iteration if

$$\| d_t y_h^k \|_\star + \| d_t b_h^k \|_\dagger \le \varepsilon_{\mathrm{stop}};$$

otherwise, increase $k \to k+1$ and continue with (1).

Again, it is useful to view $d_t y_h^k$ and $d_t b_h^k$ as the unknowns in Steps (1) and (2) instead of $y_h^k = y_h^{k-1} + \tau d_t y_h^k$ and $b_h^k = b_h^{k-1} + \tau d_t b_h^k$. The algorithm exploits the fact that the penalty term is separately convex while the nonquadratic contribution to the torsion term is separately concave. Therefore, the decoupled semiimplicit treatment of these terms is natural and unconditionally energy stable.

Proposition 2 (Convergent iteration).
Assume that we have

$$\|w_h'\|_h \le c_\star \|w_h\|_\star, \qquad \|r_h\|_h \le c_\dagger \|r_h\|_\dagger$$

for all $(w_h, r_h) \in V_{\rm rod}^h$ *with* $L_{bc}^{rod}[w_h, r_h] = 0$. *Algorithm 2 is well defined and produces a sequence* $(y_h^k, b_h^k)_{k=0,1,\dots}$ *such that for all* $L \ge 0$ *we have*

$$I_{\rm rod}^{h,\varepsilon}[y_h^L, b_h^L] + \tau \sum_{k=1}^{L} \left(\|d_t y_h^k\|_\star^2 + \|d_t b_h^k\|_\dagger^2 \right) \le I_{\rm rod}^{h,\varepsilon}[y_h^0, b_h^0].$$

The iteration controls the unit-length violation via

$$\max_{k=0,\dots,L} \| \, |[y_h^k]'|^2 - 1\|_{L_h^1} + \| \, |b_h^k|^2 - 1\|_{L_h^1} \le \tau c_{\star,\dagger} e_{0,h},$$

where $e_{0,h} = I_{\rm rod}^{h,\varepsilon}[y_h^0, b_h^0]$. *In particular, the algorithm terminates within a finite number of iterations.*

Proof.

(a) To prove the stability estimate we note that the functional

$$G_h[y_h, b_h] = \frac{c_t}{2} \int_0^L (Q_h b_h \cdot y_h'')^2 \, dx_1$$

is separately convex, i.e., convex in y_h and in b_h. Therefore, we have that

$$\partial_y G_h[y_h^{k-1}, b_h^{k-1}; y_h^k - y_h^{k-1}] + G_h[y_h^{k-1}, b_h^{k-1}] \le G_h[y_h^k, b_h^{k-1}],$$
$$\partial_b G_h[y_h^k, b_h^{k-1}; b_h^k - b_h^{k-1}] + G_h[y_h^k, b_h^{k-1}] \le G_h[y_h^k, b_h^k],$$

which by summation leads to the inequality

$$\partial_y G_h[y_h^{k-1}, b_h^{k-1}; d_t y_h^k] + \partial_b G_h[y_h^k, b_h^{k-1}; d_t b_h^k] \le d_t G_h[y_h^k, b_h^k].$$

Similarly, the functional

$$P_{h,\varepsilon}[y_h, b_h] = \frac{1}{2\varepsilon} \int_0^L \mathcal{I}_h^{1,0}[(y_h' \cdot b_h)^2] \, dx_1$$

is separately convex and we have

$$\partial_y P_{h,\varepsilon}[y_h^k, b_h^{k-1}; d_t y_h^k] + \partial_b P_{h,\varepsilon}[y_h^k, b_h^k; d_t b_h^k] \ge d_t P_{h,\varepsilon}[y_h^k, b_h^k].$$

By choosing $w_h = d_t y_h^k$ and $r_h = d_t b_h^k$ in the equations of Steps (2) and (3) of Algorithm 2 we thus find that

$$\begin{aligned} \|d_t y_h^k\|_\star^2 + \|d_t b_h^k\|_\dagger^2 + d_t \left(\frac{c_b}{2} \| [y_h^k]'' \|^2 + \frac{c_t}{2} \| [b_h^k]' \|^2 \right) + d_t P_{h,\varepsilon}[y_h^k, b_h^k] \\ + \tau \left(\frac{c_b}{2} \| [d_t y_h^k]'' \|^2 + \frac{c_t}{2} \| [d_t b_h^k]' \|^2 \right) \\ \le \partial_y G_h[y_h^{k-1}, b_h^{k-1}; d_t y_h^k] + \partial_b G_h[y_h^k, b_h^{k-1}; d_t b_h^k] \le d_t G_h[y_h^k, b_h^k]. \end{aligned}$$

Since

$$I_{\text{rod}}^{h,\varepsilon}[y_h^k, b_h^k] = \frac{c_b}{2} \| [y_h^k]'' \|^2 + \frac{c_t}{2} \| [b_h^k]' \|^2 - G_h[y_h^k, b_h^k] + P_{h,\varepsilon}[y_h^k, b_h^k]$$

we deduce the asserted estimate.

(b) The nodal orthogonality conditions encoded in the spaces $\mathcal{F}_h[y_h^{k-1}]$ and $\mathcal{E}_h[b_h^{k-1}]$ lead to the relations

$$|[y_h^k]'(z)|^2 = |[y_h^{k-1}]'(z)|^2 + \tau^2 |[d_t y_h^k]'(z)|^2,$$
$$|b_h^k(z)|^2 = |b_h^{k-1}(z)|^2 + \tau^2 |d_t b_h^k(z)|^2.$$

By induction and incorporation of the stability estimate we deduce the asserted estimates for the constraint violation. □

We illustrate the performance of Algorithm 2 via a numerical experiment showing the relaxation of a twisted initially flat curve which is clamped at both ends. We plotted in the bottom of Fig. 7 the total energy and the torsion contribution defined by

$$T_{\text{tor}}^h[y_h, b_h] = \frac{c_t}{2} \int_0^L |b_h'|^2 \, dx_1 - \frac{c_t}{2} \int_0^L (Q_h b_h \cdot y_h'')^2 \, dx_1.$$

We observe that the curve quickly releases its large energy and becomes a spatial curve attaining a stationary configuration with equilibrated curvature after approximately 2000 iterations. The model parameters used in the simulation were set to $c_b = 2$ and $c_t = 1$. We used a partition into 1006 subintervals corresponding to a mesh-size $h = 1/80$. The step size τ and the penalty parameter ε were chosen proportional to h.

4.2 Elastic plates

We next apply the conceptual approach for solving nonlinearly constrained minimization problems outlined above to the case of approximating bending isometries. For this we recall that the discrete bending energy is given by

$$I_{\text{plate}}^h[y_h] = \frac{c_b}{2} \int_\omega |D_h^2 y_h|^2 \, dx'$$

with the set of admissible discrete deformations defined as

$$\mathcal{A}_h = \Big\{ y_h \in \mathcal{S}^{\text{dkt}}(\mathcal{T}_h)^3 : L_{\text{bc}}^{\text{plate}}[y_h] = \ell_{\text{bc}}^{\text{plate}}, \\ [\nabla y_h(z)]^{\mathsf{T}} \nabla y_h(z) = I_2 \text{ f.a. } z \in \mathcal{N}_h \Big\}.$$

Using the linearized isometry operator

$$L_A^{\text{iso}}[B] = A^{\mathsf{T}} B + B^{\mathsf{T}} A$$

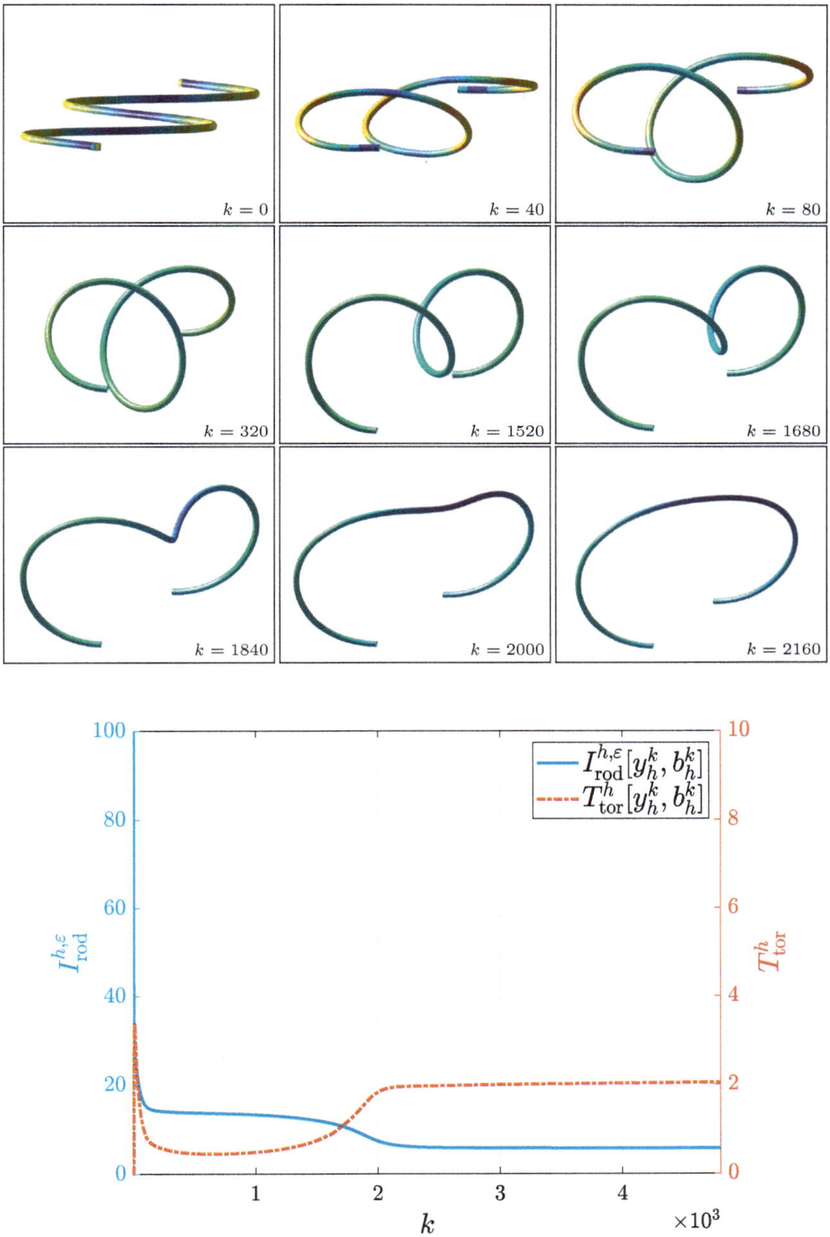

FIG. 7 Snapshots of an evolution from an initially flat but twisted curve colored by curvature after different numbers of iterations (*top*): the curve equilibrates its curvature to relax the initially dominant bending energy, afterwards deforms into a spatial helix, and finally attains a large stationary configuration. The total energy decreases monotonically while the contribution due to torsion increases (*bottom*).

we define the (shifted) tangent space of the set of discrete isometric deformations $\mathcal{A}_h$ at a deformation y_h via

$$\mathcal{F}_h[y_h] = \Big\{ w_h \in \mathcal{S}^{\mathrm{dkt}}(\mathcal{T}_h)^3 : L_{\mathrm{bc}}^{\mathrm{plate}}[w_h] = 0, \\ L_{\nabla y_h}^{\mathrm{iso}}[\nabla w_h](z) = 0 \text{ f.a. } z \in \mathcal{N}_h \Big\}.$$

We then decrease the bending energy for given boundary conditions by iterating the steps of the following algorithm.

Algorithm 3 (Gradient descent for elastic plates).

Choose an initial $y_h^0 \in \mathcal{A}_h$ and a step size $\tau > 0$, set $k = 1$.

(1) Compute $d_t y_h^k \in \mathcal{F}_h[y_h^{k-1}]$ such that for all $w_h \in \mathcal{F}_h[y_h^{k-1}]$ we have

$$(d_t y_h^k, w_h)_\star + (D_h^2 y_h^k, D_h^2 w_h) = 0.$$

(2) Stop if $\| d_t y_h^k \|_\star \le \varepsilon_{\mathrm{stop}}$; otherwise, increase $k \to k+1$ and continue with (1).

The iterates $(y_h^k)_{k=0,1,\dots}$ will in general not satisfy the nodal isometry constraint exactly, but the violation is again independent of the number of iterations and controlled by the step size τ.

Proposition 3 (Convergent iteration).

The iterates $(y_h^k)_{k=0,1,\dots}$ of Algorithm 3 are well defined and satisfy for every $L \ge 0$ the energy estimate

$$I_{\mathrm{plate}}^h[y_h^L] + \tau \sum_{k=1}^{L} \| d_t y_h^k \|_\star^2 \le I_{\mathrm{plate}}^h[y_h^0].$$

Moreover, if $\|\nabla w_h\|_h \le c_\star \|w_h\|_\star$ for all $w_h \in \mathcal{S}^{\mathrm{dkt}}(\mathcal{T}_h)^3$ with $L_{bc}^{plate}[w_h] = 0$, then we have the constraint violation bound

$$\max_{k=0,\dots,L} \| (\nabla y_h^k)^{\mathsf{T}} \nabla y_h^k - I_2 \|_{L_h^1} \le c \tau e_{0,h},$$

where $e_{0,h} = I_{plate}^h[y_h^0]$.

Proof (sketched).

(a) Since for any $y_h^{k-1} \in \mathcal{S}^{\mathrm{dkt}}(\mathcal{T}_h)^3$ we have that $\mathcal{F}_h[y_h^{k-1}]$ is a nonempty linear space the Lax–Milgram lemma implies that there exists a unique solution $y_h^k \in \mathcal{F}_h[y_h^{k-1}]$.

(b) The energy decay property is an immediate consequence of choosing $w_h = d_t y_h^k$ and using the binomial formula

$$2(D_h^2 y_h^k, D_h^2 d_t y_h^k) = d_t \| D_h^2 y_h^k \|^2 + \tau \| D_h^2 d_t y_h^k \|^2.$$

(c) The error bound for the isometry violation follows from the orthogonality defined by the linearized isometry condition and the energy decay property as in the proof of Proposition 2.

□

Fig. 8 illustrates the discrete evolution defined by Algorithm 3 via snapshots of different iterates. The clamped boundary conditions imposed at the ends $\gamma_D = \{0, L\}\times [0, w]$ of the strip $\omega = (0, L) \times (0, w)$ with length $L = 10$ and width $w = 1$ are defined via the operator

$$L_{bc}^{plate}[y] = \left[y|_{\gamma_D}, \nabla y|_{\gamma_D}\right]$$

and functions $y_D \in L^2(\gamma_D; \mathbb{R}^3)$ and $\hat{y}_D \in L^2(\gamma_D; \mathbb{R}^{3\times 2})$. These functions are constructed in such a way that the segment $\{L\}\times [0, w]$ is rotated and mapped onto the fixed opposite segment $\{0\}\times [0, w]$. In this way the formation of a Möbius strip is enforced, we included a small linear forcing term to avoid certain nonuniqueness effects. We observe that the nonsmooth choice of the starting value with large bending energy does not influence the robustness of the iteration and that within less than 10.000 iterations a stationary configuration for the stopping parameter $\varepsilon_{stop} = 5 \cdot 10^{-3}$ and evolution metric with $\|\cdot\|_\star = \| D_h^2 \cdot \|$ is attained. We also observe that we obtain a satisfactory stationary shape for coarse triangulations and that the energy decreases monotonically as predicted, with a small violation of the isometry constraint indicated by the quantity

$$\delta_{iso}^{\infty}[y_h^k] = \| (\nabla y_h^k)^{\mathsf{T}} \nabla y_h^k - I_2 \|_{L_h^\infty}$$

which appears to be nearly independent of the iteration. We finally remark that we observe concentrations of curvature at boundary points corresponding to certain singularities discussed in Bartels and Hornung (2015).

5 Linear finite element systems with nodal constraints

The discrete gradient flows devised in the previous sections lead to linear systems of equations in the time steps that have a special saddle point structure. In particular, they are given in the form

$$\begin{bmatrix} A & B^{\mathsf{T}} \\ B & 0 \end{bmatrix} \begin{bmatrix} x \\ \lambda \end{bmatrix} = \begin{bmatrix} f \\ 0 \end{bmatrix} \tag{7}$$

with a fixed positive definite symmetric matrix $A \in \mathbb{R}^{n\times n}$ and a matrix $B \in \mathbb{R}^{p\times n}$ that changes in the iteration steps. The matrix B is of full rank and block diagonal, i.e.,

$$B = \begin{bmatrix} b_1^{\mathsf{T}} & & & \\ & b_2^{\mathsf{T}} & & \\ & & \ddots & \\ & & & b_p^{\mathsf{T}} \end{bmatrix}$$

with vectors $b_i \in \mathbb{R}^\ell \setminus \{0\}$ for $i = 1, 2, \ldots, p$ and $n = p\ell$. We note that the solution $x \in \mathbb{R}^n$ of the linear system of equations above is equivalently characterized by the system

$$z^{\mathsf{T}} A x = z^{\mathsf{T}} f$$

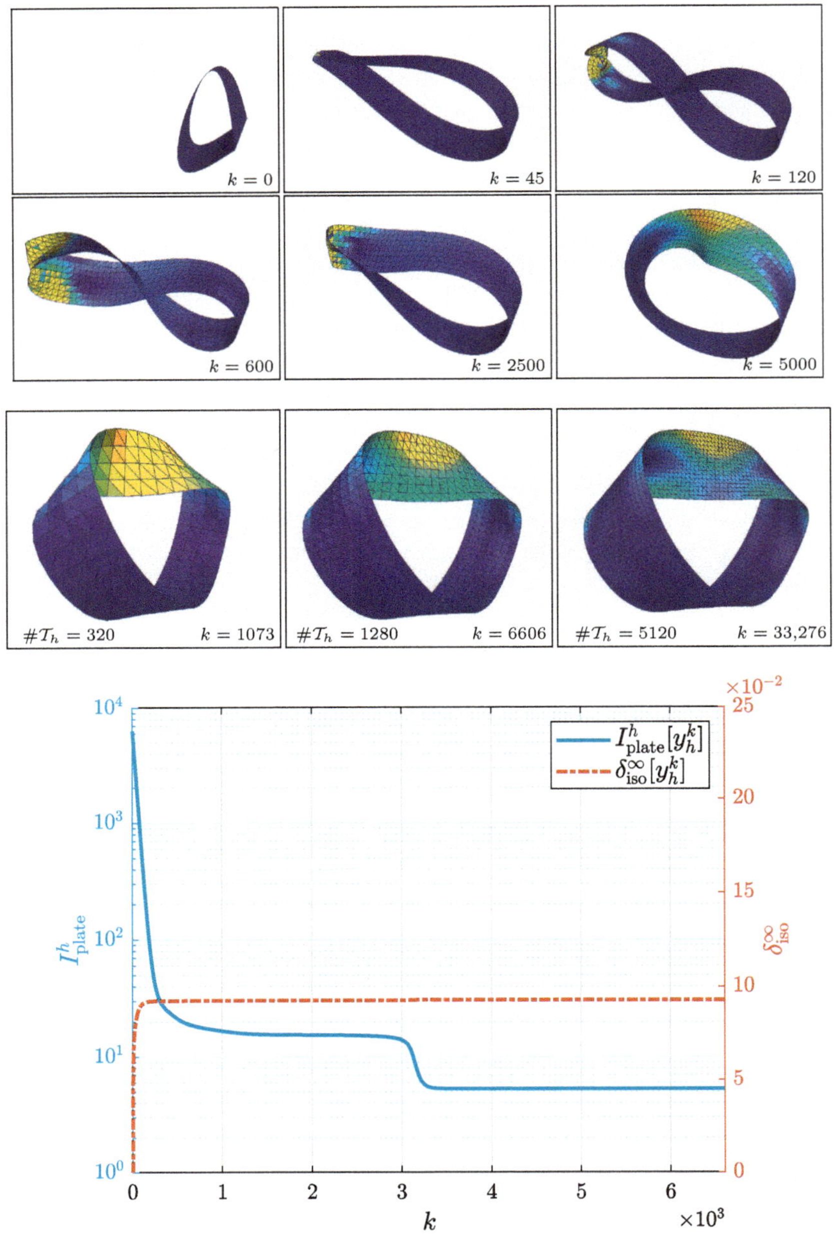

FIG. 8 Snapshots of the iteration to minimize bending energy of an elastic strip with boundary conditions leading to the formation of a Möbius strip coloured by its mean curvature (*top*); stationary configurations for different triangulations (*middle*); energy decay and constraint violation throughout the iteration (*bottom*).

for all $z \in \mathbb{R}^n$ belonging to the kernel of B, i.e., subject to the conditions

$$x, z \in \ker B.$$

To obtain a simpler, unconstrained system of equations we choose for each $i = 1, 2, \ldots, p$ a set of orthonormal vectors $(c_2^i, c_3^i, \ldots, c_\ell^i) \subset \mathbb{R}^\ell$ that are orthogonal to b_i, i.e., we have

$$\{b_i\}^\perp = \mathrm{span}\{c_2^i, c_3^i, \ldots, c_\ell^i\}.$$

We may then represent the kernel of B by the image of the matrix $C \in \mathbb{R}^{p\ell \times p(\ell-1)}$ defined by

$$C = \begin{bmatrix} c_2^1 \ \ldots \ c_\ell^1 & & & \\ & c_2^2 \ \ldots \ c_\ell^2 & & \\ & & \ddots \quad \ddots \quad \ddots & \\ & & & c_2^p \ \ldots \ c_\ell^p \end{bmatrix}.$$

The matrix C defines an isomorphism $C : \mathbb{R}^{p(\ell-1)} \to \ker B$ and we may thus reformulate the linear system of equations as

$$C^\mathsf{T} A C \hat{x} = C^\mathsf{T} f. \tag{8}$$

With the solution $\hat{x}$ we obtain the solution x of the original system (7) via $x = C\hat{x}$. Since the columns of C are linearly independent the matrix C has full rank. Hence, the symmetric matrix $\hat{A} = C^\mathsf{T} A C$ is positive definite and the linear system of equations $\hat{A}\hat{x} = \hat{f}$ can be solved efficiently with a preconditioned conjugate gradient scheme. The construction of suitable preconditioners has been discussed in Nash and Sofer (1996) and the main issue in their justification is to understand how accurate the approximation

$$(C^\mathsf{T} A C)^{-1} \approx C^\mathsf{T} A^{-1} C$$

is. Straightforward manipulations show that the solution x of the saddle-point system (7) satisfies

$$x = A^{-1} f - A^{-1} B^\mathsf{T} (B A^{-1} B^\mathsf{T})^{-1} B A^{-1} f.$$

With the equivalent formulation $(C^\mathsf{T} A C)\hat{x} = C^\mathsf{T} f$ and $x = C\hat{x}$ we find that

$$x = C (C^\mathsf{T} A C)^{-1} C^\mathsf{T} f.$$

Since C is orthogonal in the sense that $C^\mathsf{T} C = I_{p(\ell-1)}$ it follows that

$$(C^\mathsf{T} A C)^{-1} = C^\mathsf{T} A^{-1} C - C^\mathsf{T} (A^{-1} B^\mathsf{T} B A^{-1} B^\mathsf{T})^{-1} B A^{-1} C.$$

This justifies the approximation $(C^\mathsf{T} A C)^{-1} \approx C^\mathsf{T} A^{-1} C$ if the second term on the right-hand side is small. Another relation is obtained by formally applying the Neumann series

$$T^{-1}=\sum_{j=0}^{\infty}(I-T)^j$$

to the product $T = \varrho(C^{\mathsf{T}}AC)(C^{\mathsf{T}}A^{-1}C)$ with a suitable parameter $0 < \varrho \leq 1$ to deduce that

$$(C^{\mathsf{T}}AC)^{-1}=\varrho(C^{\mathsf{T}}A^{-1}C)\left(I+(I-T)+(I-T)^2+\cdots\right).$$

Accepting the approximation $(C^{\mathsf{T}}AC)^{-1} \approx C^{\mathsf{T}}A^{-1}C$ the next step is to choose a preconditioner P for A, i.e., P approximates A^{-1} and the multiplication by P is inexpensive, and use the matrix $\hat{P}=C^{\mathsf{T}}PC$ as a preconditioner for $\hat{A}$. Noting that C is orthogonal this is justified by the spectral norm estimate from Nash and Sofer (1996),

$$\|C^{\mathsf{T}}A^{-1}C-C^{\mathsf{T}}PC\|\leq\|A^{-1}-P\|.$$

A further discussion of related preconditioners in the context of micromagnetics can be found in Kraus et al. (2018). We illustrate the construction of the bases of the null space and the performance of different solution strategies in the context of harmonic maps into spheres.

5.1 Application to harmonic maps

Harmonic maps are vector fields with values in a given manifold which are stationary for the Dirichlet energy. In the case of the unit sphere, the vector field satisfies a pointwise unit-length constraint. Already this simple case illustrates analytical difficulties when dealing with constrained partial differential equations. In particular, harmonic maps are nonunique even for fixed boundary values and may be discontinuous everywhere (cf. Rivière, 1995). Partial regularity results can be proved if a harmonic map is energy minimizing (cf. Schoen and Uhlenbeck, 1982). These observations underline the importance of computing harmonic maps with low energy. We aim at applying the concepts of the previous sections to the approximation of harmonic maps and consider the following model problem defined with a function $u_{\mathrm{D}} \in C(\partial\Omega;\mathbb{R}^d)$ with $|u_{\mathrm{D}}(x)| = 1$ for all $x \in \partial\Omega$ which is the trace of a function $\tilde{u}_{\mathrm{D}} \in H^1(\Omega;\mathbb{R}^d)$:

$$\begin{cases} \text{Minimize } \dfrac{1}{2}\displaystyle\int_\Omega |\nabla u|^2\,\mathrm{d}x \text{ in the set} \\ \mathcal{A}=\left\{u\in H^1(\Omega;\mathbb{R}^d): |u(x)|^2=1 \text{ f.a.e. } x\in\Omega,\ u|_{\partial\Omega}=u_{\mathrm{D}}\right\}. \end{cases} \tag{9}$$

Following Alouges (1997) and Bartels (2005) we discretize the admissible set $\mathcal{A}$ by piecewise affine vector fields and impose the unit-length constraint in the nodes of the underlying triangulation. This leads to the following discrete minimization problem:

$$\left\{\begin{array}{l}\text{Minimize } \dfrac{1}{2}\displaystyle\int_\Omega |\nabla u_h|^2\,\mathrm{d}x \text{ in the set}\\ \mathcal{A}_h = \left\{u_h \in \mathcal{S}^{1,0}(\mathcal{T}_h)^d : |u_h(z)|^2 = 1 \text{ f.a. } z \in \mathcal{N}_h, u_h|_{\partial\Omega} = u_{\mathrm{D},h}\right\}.\end{array}\right. \tag{10}$$

For a function $u_h \in \mathcal{S}^{1,0}(\mathcal{T}_h)^d$ with nonvanishing nodal values we define

$$\mathcal{F}_h[u_h] = \left\{w_h \in \mathcal{S}^{1,0}(\mathcal{T}_h)^d : w_h(z)\cdot u_h(z) = 0 \text{ f.a. } z \in \mathcal{N}_h,\ \ w_h|_{\partial\Omega} = 0\right\}.$$

With a gradient flow approach and a linearized treatment of the nodal constraint we are led to the following algorithm.
Algorithm 4 (Gradient descent for harmonic maps).
Choose $u_h^0 \in \mathcal{A}_h$ and $\tau > 0$, set $k = 1$.
(1) Compute $d_t u_h^k \in \mathcal{F}_h[u_h^{k-1}]$ such that for all $w_h \in \mathcal{F}_h[u_h^{k-1}]$ we have

$$(d_t u_h^k, w_h)_\star + (\nabla u_h^k, \nabla w_h) = 0.$$

(2) Stop if $\|d_t u_h^k\|_\star \le \varepsilon_{\mathrm{stop}}$; otherwise, increase $k \to k+1$ and continue with (1).

The Lax–Milgram lemma shows that the iteration is well defined. Because of the orthogonality condition we have that

$$|u_h^k(z)|^2 = |u_h^{k-1}(z) + \tau d_t u_h^k(z)|^2 \ge |u_h^{k-1}(z)|^2$$

which by induction gives $|u_h^k(z)| \ge \cdots \ge |u_h^0(z)| = 1$ for $k = 0, 1, \ldots, K$. If the restriction of the inner product $(\cdot,\cdot)_\star$ to $\mathcal{S}^{1,0}(\mathcal{T}_h)^d$ is represented by the matrix M then the linear systems in the iteration can be written as

$$\begin{bmatrix} M+\tau S & B^\top \\ B & 0 \end{bmatrix}\begin{bmatrix} V^k \\ \Lambda \end{bmatrix} = \begin{bmatrix} -SU^{k-1} \\ 0 \end{bmatrix}$$

with the vectorial finite element stiffness matrix S and the constraint matrix

$$B = \begin{bmatrix} u_h^{k-1}(z_1)^\top & & & \\ & u_h^{k-1}(z_2)^\top & & \\ & & \ddots & \\ & & & u_h^{k-1}(z_p)^\top \end{bmatrix}.$$

To define the matrix C that provides an isomorphism from $\mathbb{R}^{p(d-1)}$ onto the kernel of B we need to compute for a given vector $b \in \mathbb{R}^d \setminus \{0\}$ an orthonormal basis $(c_2, \ldots, c_d)$ of its orthogonal complement. We proceed as follows: if b is parallel to e_1 then we choose the remaining canonical basis vectors $(e_2, \ldots, e_d)$; otherwise, we set

$$(\tilde{c}_2, \ldots, \tilde{c}_d) = \begin{cases} b^\perp & \text{if } d = 2, \\ (b\times e_1, b\times(b\times e_1)) & \text{if } d = 3, \end{cases}$$

where $b^{\perp}$ denotes the rotation of b by $\pi/2$, followed by the normalization $c_j = \tilde{c}_j/|\tilde{c}_j|$ for $j = 2, \ldots, d$.

Example 1. Let $\Omega = (-1/2, 1/2)^3$ and define u_D on $\Gamma_D = \partial\Omega$ for $x \in \partial\Omega$ via

$$u_D(x) = \frac{x}{|x|}.$$

The employed triangulations $\mathcal{T}_\ell$ result from ℓ uniform refinements of a reference triangulation of Ω into six tetrahedra. The starting value $u_h^0 \in \mathcal{S}^{1,0}(\mathcal{T}_\ell)^3$ is defined by generating random nodal values of unit length.

Table 2 compares the following general strategies to solve the linear problems in the iterative solution of harmonic maps:

(1) Solve the original indefinite saddle-point system (7) with a direct solution method for sparse systems.

(2) Solve the reduced positive definite system (8) with a direct solution method for sparse linear systems.

(3) Solve the reduced positive definite system (8) with the preconditioned conjugate gradient method using a diagonal preconditioning of the full system matrix $C^{\mathsf{T}}AC$.

(4) Solve the reduced system (8) with the preconditioned conjugate gradient method using the preconditioner $C^{\mathsf{T}}(L_{ic}L_{ic}^{\mathsf{T}})^{-1}C$ with the incomplete Cholesky factorization $A \approx L_{ic}L_{ic}^{\mathsf{T}}$ computed once.

TABLE 2 Average runtimes in seconds and average iteration numbers of the preconditioned conjugate gradient scheme (pcg) per iteration in parentheses for solution strategies (1)–(4) for the constrained linear systems of equations arising in the iterative approximation of harmonic maps on triangulations $\mathcal{T}_\ell$ of the unit cube with mesh-sizes $h_\ell \approx 2^{-\ell}$, $\ell = 2, 3, \ldots, 6$.

$\#\mathcal{N}_\ell$	Saddle, direct (backslash)	Reduced, direct (backslash)	Reduced, pcg (diagonal $C^{\mathsf{T}}AC$)	Reduced, pcg (incompl. Chol.)
125	7.4×10^{-4}	3.2×10^{-4}	8.4×10^{-5} (11.9)	3.0×10^{-5} (6.0)
729	1.5×10^{-2}	6.5×10^{-3}	1.9×10^{-4} (25.4)	7.4×10^{-5} (10.2)
4913	6.0×10^{-1}	2.4×10^{-1}	9.2×10^{-4} (52.5)	3.9×10^{-4} (19.9)
35,937	3.4×10^{1}	8.3	8.1×10^{-3} (105.9)	3.2×10^{-3} (38.8)
274,625	—	—	1.1×10^{-1} (212.1)	5.9×10^{-2} (76.1)

We measured the average time needed for different discretizations to solve the linear systems of equations using strategies (1)–(4). We always chose the step size $\tau = h$, the stopping parameter $\varepsilon_{pcg} = 10^{-8}$ for the relative residuals in the preconditioned conjugate gradient scheme, and the stopping criterion $\varepsilon_{stop} = h_\ell/10$ in Algorithm 4. We employed the H^1 inner product to define $(\cdot, \cdot)_\star$ and used MATLAB's backslash operator as a model direct solver for sparse linear systems. From the obtained numbers we see that the preconditioner particularly designed for the structure of the constrained problems with changing constraints clearly outperforms all other approaches. However, it does not lead to mesh-size independent iteration numbers for this example with a nonsmooth solution. We also remark that we did not observe a further improvement when additional terms in the Neumann series were used with a damping factor $\varrho = 1$. In a two-dimensional setting with a smooth solution the complete Cholesky factorization led to mesh-independent iteration numbers.

6 Applications, modifications, and extensions

We address in this section the application of the developed methods to the simulation of extended problems. In particular, we devise a numerical method for simulating bilayer plates, discuss how injectivity can be enforced in the case of elastic rods, and investigate a model that involves the thickness parameter and thereby allows for describing situations in which low energy membrane and bending effects occur simultaneously.

6.1 Bilayer plates

An important class of applications of nonlinear plate bending is related to composite materials, i.e., structures such as bilayer plates which are made of two sheets with slightly different mechanical features that are glued on top of each other. The difference in the material properties, e.g., concerning the response to environmental changes such as temperature, allows for externally controlled large bending effects. In bimetal strips one of the metals contracts while the other one expands leading in combination to the formation of rolls with prescribed curvature (Fig. 9).

The preferred curvature is defined by the difference in elastic properties of the involved materials and is explicitly visible in the dimensionally reduced

FIG. 9 Schematical description of the bilayer bending effect: the upper sheet contracts while the lower one expands leading in combination to a large isometric deformation.

model identified in Schmidt (2007a). Following that work we consider a thin plate $\Omega_\delta = \omega \times (-\delta/2, \delta/2)$ and the inhomogeneous energy density

$$W_{\text{bil}}(x_3, F) = \begin{cases} W\left((1+\delta\alpha)^{-1}F\right) & \text{for } x_3 > 0, \\ W\left((1-\delta\alpha)^{-1}F\right) & \text{for } x_3 < 0, \end{cases}$$

with a material parameter $\alpha \in \mathbb{R}$. The energy density W is assumed to satisfy the standard requirements stated in Section 2 so that W_{bil} is minimal for deformation gradients F which are multiples of rotations with $\det F = (1+\delta\alpha)^3 > 1$ if $x_3 > 0$ and $\det F = (1-\delta\alpha)^3 < 1$ for $x_3 < 0$ corresponding to expansive and contractive behaviour in the upper and lower sheets, respectively. It has been shown in Schmidt (2007a) that for deformations with energy proportional to δ^3 a limiting, dimensionally reduced problem is given by the following minimization problem:

$$\begin{cases} \text{Minimize } I_{\text{bil}}[y] = \displaystyle\int_\omega Q_{\text{plate}}(II - \alpha I_2)\,\mathrm{d}x' \text{ in the set} \\ \mathcal{A} = \left\{ y \in H^2(\omega; \mathbb{R}^3) : (\nabla y)^\mathsf{T} \nabla y = I_2,\, L_{\text{bc}}^{\text{bil}}[y] = \ell_{\text{bc}}^{\text{bil}} \right\}. \end{cases} \tag{11}$$

Here, Q_{plate} is the reduced quadratic form introduced in Section 2.2 which is applied to the difference of the second fundamental form II and a multiple of the identity matrix with a factor given by half of the jump in the material difference divided by δ. This material mismatch introduces a term that is similar to so-called spontaneous curvature terms in the context of biomembranes described via the Helfrich-Willmore model. Note that in the definition of W_{bil} the difference is $2\delta\alpha$ and its proportionality to the thickness δ of the plates is important to avoid delamination effects. The reduced model prefers deformations y that define surfaces with mean curvature α and vanishing Gaussian curvature due to the isometry constraint. Hence, rolling effects in one direction occur. In Schmidt (2007b) global minimizers for I_{bil} have been shown to be given by cylinders with radius $r = 1/\alpha$. Recalling that the quadratic form Q_{plate} is given by the scalar product

$$\langle M, N \rangle_{\text{plate}} = c_1 M : N + c_2 \operatorname{tr}(M) \operatorname{tr}(N),$$

we do not have the property that the integrand is proportional to $|D^2 y|^2$ as in the case of a single layer or two identical materials corresponding to the case $\alpha = 0$. Instead we infer

$$Q_{\text{plate}}(II - \alpha I_2) = Q_{\text{plate}}(II) - 2\alpha \langle II, I_2 \rangle_{\text{plate}} + \alpha^2 Q_{\text{plate}}(I_2).$$

Because of the isometry condition the first term is proportional to $|D^2 y|^2$ while the third term is a constant c_α which is irrelevant for the minimization process. For the second term on the right-hand side we have

$$-2\alpha \langle II, I_2 \rangle_{\text{plate}} = -2\alpha(c_1 + 2c_2) \operatorname{tr}(II).$$

Therefore, using $II_{ij} = -\partial_i \partial_j y \cdot (\partial_1 y \times \partial_2 y)$ we have

$$Q_{\text{plate}}(II - \alpha I_2) = c_{\text{b}} |D^2 y|^2 + 2\alpha c_{\text{sc}} \Delta y \cdot (\partial_1 y \times \partial_2 y) + c_\alpha.$$

Convergent finite element discretizations and iterative strategies for the numerical solution of the bilayer plate bending problem have been devised in the articles (Bartels et al., 2017, 2018a) extending the ideas described in the Sections 3.2 and 4.2. The iterative scheme of Bartels et al. (2017) is unconditionally stable but requires a subiteration, i.e., the solution of a nonlinear system of equations in every time step, which limits its practical applicability, while the scheme used in Bartels et al. (2018a) is explicit and efficient but can only be expected to be conditionally stable. We devise here a new semiimplicit scheme with improved stability properties. The set of admissible deformations $\mathcal{A}_h$ and the tangent spaces $\mathcal{F}_h[y_h]$ are defined as in Section 3.2 and we set

$$\Delta_h v_h = \operatorname{tr} D_h^2 v_h$$

for a function $v_h \in \mathcal{S}^{\text{dkt}}(\mathcal{T}_h)$ with a componentwise application in the case of a vector field. The discretized energy functional describing actuated bilayer plates is therefore given by

$$I_{\text{bil}}^h[y_h] = \frac{c_{\text{b}}}{2} \int_\omega |D_h^2 y_h|^2 \, \mathrm{d}x' + \alpha c_{\text{sc}} \int_\omega \widehat{\mathcal{I}}_h^{1,0} [\Delta_h y_h \cdot (\partial_1 y_h \times \partial_2 y_h)] \, \mathrm{d}x'.$$

A uniform discrete coercivity property of this discretization has been established in Bartels et al. (2017). The new iterative minimization from Bartels and Palus (2019) uses an explicit treatment of the spontaneous curvature term.
Algorithm 5 (Gradient descent for bilayer plates).
Choose $y_h^0 \in \mathcal{A}_h$ and $\tau > 0$, set $k = 1$.

(1) Compute $d_t y_h^k \in \mathcal{F}_h[y_h^{k-1}]$ such that for all $w_h \in \mathcal{F}_h[y_h^{k-1}]$ we have

$$\begin{aligned}\left(d_t y_h^k, w_h\right)_\star + c_{\text{b}}\left(D_h^2 y_h^k, D_h^2 w_h\right) &= -\alpha c_{\text{sc}} \int_\omega \widehat{\mathcal{I}}_h^{1,0}\left[\Delta_h w_h \cdot \left(\partial_1 y_h^{k-1} \times \partial_2 y_h^{k-1}\right)\right] \mathrm{d}x' \\ &\quad - \alpha c_{\text{sc}} \int_\omega \widehat{\mathcal{I}}_h^{1,0}\left[\Delta_h y_h^{k-1} \cdot \left(\partial_1 y_h^{k-1} \times \partial_2 w_h + \partial_1 w_h \times \partial_2 y_h^{k-1}\right)\right] \mathrm{d}x'.\end{aligned}$$

(2) Stop if $\|d_t y_h^k\|_\star \le \varepsilon_{\text{stop}}$; otherwise increase $k \to k + 1$ and continue with (1).

We show that the iteration of Algorithm 5 is energy stable on finite time intervals under a moderate condition on the step size resulting from the use of an inverse estimate. To formulate it, we first note that we assume that the boundary conditions imply a Poincaré type estimate

$$\|\nabla w_h\|_h \le c_{\text{P}} \|D_h^2 w_h\|$$

for all $w_h \in \mathcal{S}^{\mathrm{dkt}}(\mathcal{T}_h)^3$ with $L_{\mathrm{bc}}^{\mathrm{bil}}[w_h] = 0$. With this estimate one verifies that the mesh-dependent estimate

$$\|\nabla w_h\|_{L_h^\infty} \leq c_{\mathrm{inv}} |\log h_{\min}| \, \|D_h^2 w_h\|_h,$$

holds for all $w_h \in \mathcal{S}^{\mathrm{dkt}}(\mathcal{T}_h)^3$ with $L_{\mathrm{bc}}^{\mathrm{bil}}[w_h] = 0$ and with the minimal mesh-size $h_{\min}$.

Proposition 4 (Convergent iteration).

The iterates of Algorithm 5 are well defined. Assume that we have

$$\|D_h^2 w_h\|_h \leq c_\star \, \|w_h\|_\star$$

for all $w_h \in \mathcal{S}^{\mathrm{dkt}}(\mathcal{T}_h)^3$ with $L_{bc}^{bil}[w_h] = 0$. Then there exist constants c_{bil}, $c'_{bil} > 0$ such that if $\tau|\log h_{\min}|^2 \leq c'_{\mathrm{bil}}$ for $L = 0, 1, \ldots, K \leq T/\tau$ we have

$$I_{bil}^h[y_h^L] + (1 - c_{bil}\tau|\log h_{min}|)\tau \sum_{k=1}^{L} \|d_t y_h^k\|_\star^2 \leq I_{bil}^h[y_h^0],$$

and

$$\max_{k=0,\ldots,L} \|[\nabla y_h^k]^{\mathsf{T}} \nabla y_h^k - I_2\|_{L_h^\infty} \leq 2\tau c_{inv}^2 |\log h_{min}|^2 e_{0,h},$$

where $e_{0,h} = I_{bil}^h[y_h^0]$.

Proof. We argue by induction and assume that the estimate has been established for $L - 1$. Choosing $w_h = d_t y_h^k$ and using Hölder's inequality shows that for $k \leq L$ we have

$$\begin{aligned}
&\|d_t y_h^k\|_\star^2 + \frac{c_{\mathrm{b}}}{2} d_t \|D_h^2 y_h^k\|^2 \\
&\leq \alpha c_{\mathrm{sc}} \|\Delta_h d_t y_h^k\|_h \|\partial_1 y_h^{k-1}\|_h \|\partial_2 y_h^{k-1}\|_{L_h^\infty} \\
&\quad + \alpha c_{\mathrm{sc}} \|\Delta_h y_h^{k-1}\|_h \Big(\|\partial_1 y_h^{k-1}\|_{L_h^\infty} \|\partial_2 d_t y_h^k\|_h + \|\partial_2 y_h^{k-1}\|_{L_h^\infty} \|\partial_1 d_t y_h^k\|_h \Big).
\end{aligned}$$

Absorbing terms involving $d_t y_h^k$ on the left-hand side and noting that terms involving y_h^{k-1} are bounded because of the bounds for $k \leq L - 1$, we find that

$$\frac{\tau}{2} \|d_t y_h^k\|_\star^2 + \frac{c_{\mathrm{b}}}{2} \|D_h^2 y_h^k\|_h^2 \leq \frac{c_{\mathrm{b}}}{2} \|D_h^2 y_h^{k-1}\|_h^2 + \tau c \leq c'.$$

Here, we also used that the discrete energy is coercive. With this intermediate estimate and the orthogonality condition encoded in the definition of $\mathcal{F}_h[y_h^{k-1}]$ we infer with $y_h^k = y_h^{k-1} + \tau d_t y_h^k$ that

$$[\nabla y_h^k(z)]^{\mathsf{T}} \nabla y_h^k(z) = [\nabla y_h^{k-1}(z)]^{\mathsf{T}} \nabla y_h^{k-1}(z) + \tau^2 [\nabla d_t y_h^k(z)]^{\mathsf{T}} \nabla d_t y_h^k(z)$$

and hence with the inverse estimate we obtain the suboptimal estimate that

$$\|[\nabla y_h^k]^{\mathsf{T}}\nabla y_h^k - I_2\|_{L_h^\infty} \le c\tau|\log h_{\min}|^2 e_{0,h} \le c''.$$

To obtain the energy bound we note that a discrete product rule shows that the discrete time derivative of the spontaneous curvature term is given by

$$\begin{aligned} d_t \int_\omega \widehat{\mathcal{I}}_h^{1,0}\left[\Delta_h y_h^k \cdot (\partial_1 y_h^k \times \partial_2 y_h^k)\right] \mathrm{d}x' &= \int_\omega \widehat{\mathcal{I}}_h^{1,0}\left[\Delta_h d_t y_h^k \cdot (\partial_1 y_h^k \times \partial_2 y_h^k)\right] \mathrm{d}x' \\ &+ \int_\omega \widehat{\mathcal{I}}_h^{1,0}\left[\Delta_h y_h^{k-1} \cdot (\partial_1 y_h^{k-1} \times \partial_2 d_t y_h^k + \partial_1 d_t y_h^k \times \partial_2 y_h^k\right] \mathrm{d}x'. \end{aligned}$$

Comparing this expression with the right-hand side of the equation in Step (1) of Algorithm 5 with $w_h = d_t y_h^k$ leads to

$$\begin{aligned} \|d_t y_h^k\|_\star^2 &+ d_t I_{\mathrm{bil}}^h[y_h^k] + \tau\frac{c_{\mathrm{b}}}{2}\|d_t D_h^2 y_h^k\|^2 \\ &= \alpha c_{\mathrm{sc}} \int_\omega \widehat{\mathcal{I}}_h^{1,0}\left[\Delta_h d_t y_h^k \cdot \left(\partial_1 y_h^k \times \partial_2 y_h^k - \partial_1 y_h^{k-1} \times \partial_2 y_h^{k-1}\right)\right] \mathrm{d}x' \\ &+ \alpha c_{\mathrm{sc}} \int_\omega \widehat{\mathcal{I}}_h^{1,0}\left[\Delta_h y_h^{k-1} \cdot \left(\partial_1 d_t y_h^k \times \partial_2 \left(y_h^k - y_h^{k-1}\right)\right)\right] \mathrm{d}x' \\ &\le c\,\|\Delta_h d_t y_h^k\|_h \Big(\|\partial_1 y_h^k\|_{L_h^\infty}\tau\,\|\partial_2 d_t y_h^k\|_h + \|\partial_1 y_h^{k-1}\|_{L_h^\infty}\tau\,\|\partial_2 d_t y_h^k\|_h\Big) \\ &+ c\,\|\Delta_h y_h^{k-1}\|_h \tau\,\|\partial_1 d_t y_h^k\|_h\,\|\partial_2 d_t y_h^k\|_{L_h^\infty}. \end{aligned}$$

We use the estimates $\|\partial_j d_t y_h^k\|_{L_h^\infty} \le c|\log h_{\min}|\,\|d_t y_h^k\|_\star$ and $\|\partial_j y_h^{k-\ell}\|_{L_h^\infty} \le c$ for $j = 1, 2$ and $\ell = 0, 1$, to bound the right-hand side by $\tau|\log h_{\min}|c'''\,\|d_t y_h^k\|_\star^2$. This leads to

$$(1 - c'''\tau|\log h_{\min}|)\,\|d_t y_h^k\|_\star^2 + d_t I_{\mathrm{bil}}^h[y_h^k] \le 0$$

and proves the asserted energy estimate. With this bound we also obtain the improved bound for the constraint violation error. □

The good stability properties of the newly proposed numerical scheme are confirmed by a numerical experiment whose outcome is visualized in Fig. 10. The setup uses a rectangular strip of length $L = 10$ and width $w = 4$ that is clamped at one end. The spontaneous curvature parameter is $\alpha = -1$ and we have $c_{\mathrm{b}} = 1$ and $c_{\mathrm{sc}} = 1$. The figure shows snapshots of the evolution on a grid with medium mesh-size and stationary states for different triangulations with mesh-sizes h proportional to $2^{-\ell}$, $\ell = 1, 2, 3$. The step size was always set to $\tau = h/20$. The stopping criterion was chosen as $\varepsilon_{\mathrm{stop}} = 10^{-3}$. Because of the extreme geometry and the large deformation the asymmetry of the underlying triangulations is reflected in the numerical solutions but this effect disappears for smaller mesh-sizes. The theoretical results about the energy monotonicity and controlled constraint violation are confirmed by the bottom plot of Fig. 10.

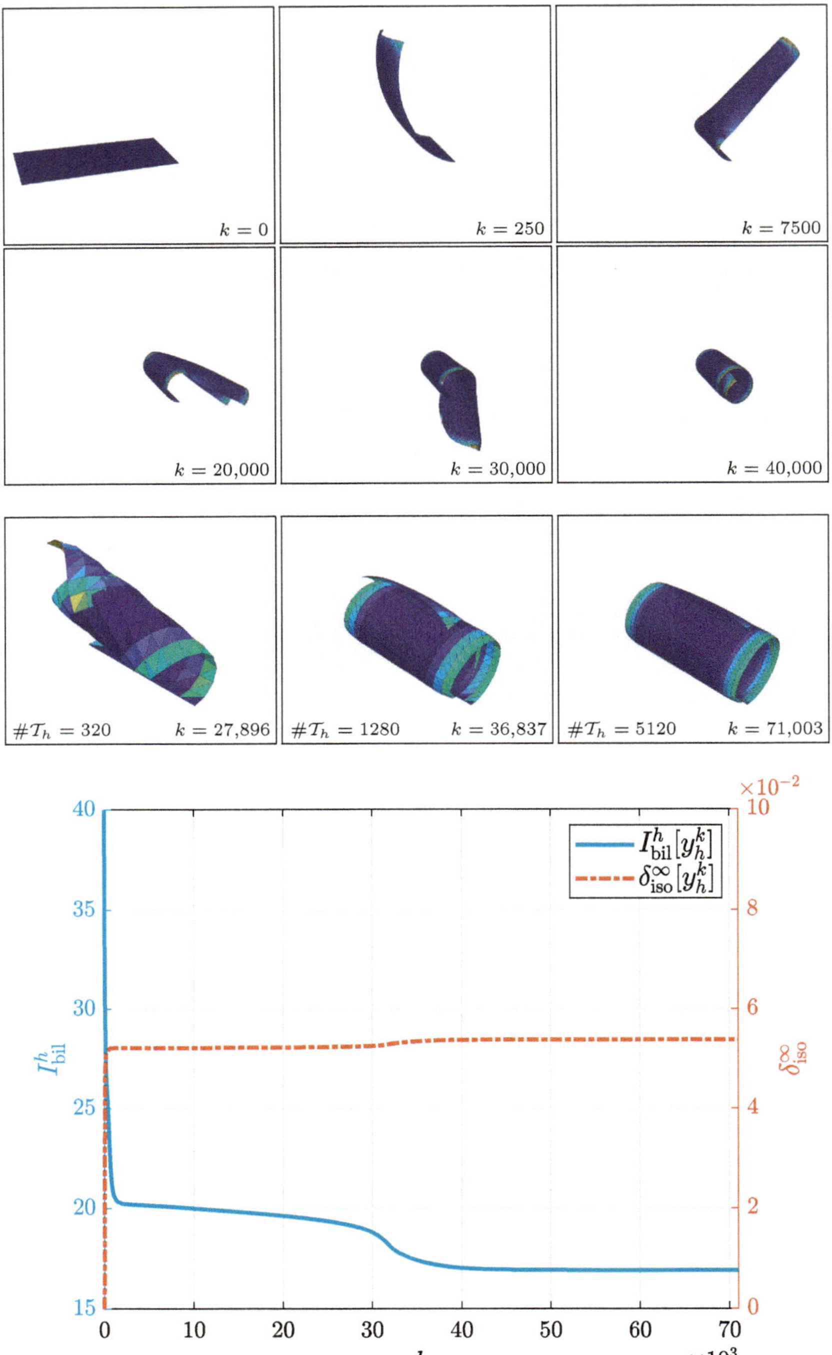

FIG. 10 Iterates in a bilayer bending problem with clamped boundary condition on one end of the strip. The plate immediately bends everywhere and evolves into a multiply covered tube (*top*); discretization effects like asymmetries disappear for finer triangulations (*middle*); the energy decreases monotonically and becomes stationary (*bottom*).

6.2 Self-avoiding curves and elastic knots

A strategy for finding useful representatives of knot classes, i.e., closed curves within a given isotopy class with particular features, is to minimize or decrease the bending energy within the given class via continuous evolutions defined by a gradient flow. To ensure that the flow does not change the topological properties of the curve, appropriate terms have to be included in the mathematical formulation. To this end we add a self-avoidance potential to the bending energy that prevents the curve from self-intersecting or pulling tight, i.e., we consider flows determined by functionals

$$I^{\mathrm{tot}}[y] = \frac{c_{\mathrm{b}}}{2}\int_0^L |y''|^2 \, \mathrm{d}x_1 + \varrho \mathrm{TP}[y]$$

in sets of curves satisfying periodic or, e.g., clamped boundary conditions. We refer the reader to O'Hara (2003) for a general discussion of appropriate functionals. A functional that turned out to have several advantageous features is the tangent-point functional proposed in Gonzalez and Maddocks (1999). It is for a C^1 curve $y:[0,L]\to\mathbb{R}^3$ defined via the tangent-point radius $r_y(x, z)$ which is the radius of the circle that is tangent to y in $y(x)$ and intersects the curve in the point $y(z)$. For curves that are parametrized by arclength we have

$$r_y(x,z) = \frac{1}{2}\frac{|y(x)-y(z)|}{|y'(x)\times(y(x)-y(z))|}$$

which follows from investigating the geometrical configuration sketched in Fig. 11. If $z \to x$ then we have that $r_y(x, z)$ converges to the inverse of the curvature of the curve y at x. If otherwise $y(x) \to y(z)$ for fixed $x \neq z$ then the radius converges to zero as depicted in Fig. 12. An alternative choice is the use of the Menger curvature which is discussed in Strzelecki et al. (2013).

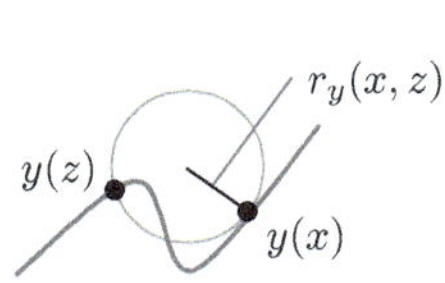

FIG. 11 The tangent-point radius $r_y(x, z)$ of a curve y is the radius of the circle that is tangent to y in $y(x)$ and intersects the curve in $y(z)$.

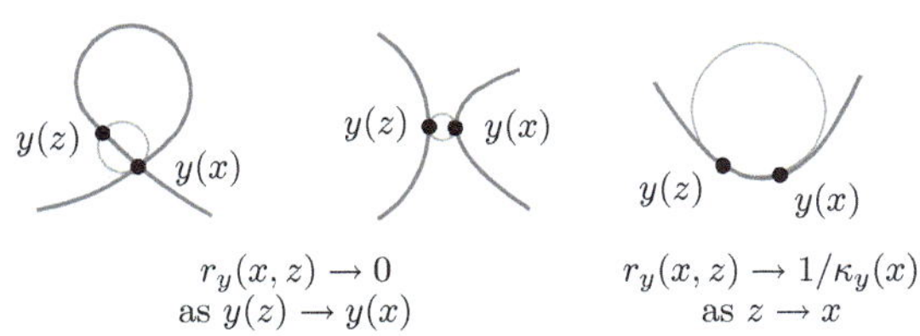

FIG. 12 Situations in which the tangent-point radius approaches zero leading to a singularity in the tangent-point functional TP (*left and middle*); as $z \to x$ the tangent-point radius approximates the inverse of the curvature of y at x (*right*).

The tangent-point functional TP is for an exponent $q \geq 1$ defined via

$$\mathrm{TP}[y] = \frac{2^q}{q} \int_0^L \int_0^L \frac{1}{r_y(x,z)^q} \,\mathrm{d}x\,\mathrm{d}z.$$

The exponent has to be suitably chosen so that singularities corresponding to a vanishing tangent-point radius are sufficiently strong to lead to an infinite value of the functional. Thereby, an energy barrier is defined that separates different isotopy classes. The functional TP has important and remarkable features.

Remark 3..

(i) If the arclength-parametrized curve $y \in C^1([0,L];\mathbb{R}^3)$ is injective then $\mathrm{TP}[y]$ is finite if and only if $y \in W^{2-1/q,q}(0, L; \mathbb{R}^3)$ (cf. Blatt, 2013).

(ii) For all $M > 0$ and $q > 2$ there exists a constant $c_{M,q} > 0$ such that for all arclength-parametrized curves $y \in C^1([0, L]; \mathbb{R}^3)$ with $\mathrm{TP}[y] \leq M$ we have the bi-Lipschitz estimate $|x - z| \leq c_{M,q}|y(x) - y(z)|$ for all $x, z \in [0, L]$ (cf. Blatt and Reiter, 2015). In particular, TP is a knot energy in the sense that for a family of curves $(y_k)_{k\in\mathbb{N}}$ converging pointwise to a curve with self-intersection, the values $\mathrm{TP}[y_k]$ blowup (cf. von der Mosel, 1998; O'Hara, 2003; Strzelecki and von der Mosel, 2012; Blatt and Reiter, 2015).

(iii) The functional TP is twice continuously differentiable with bounded variations and L^1 integrands (cf. Blatt and Reiter, 2015; Bartels and Reiter, 2018).

To decrease the total energy I^{tot} of a given initial curve y_0 within its isotopy class we use a discretization of the evolution

$$(\partial_t y, w)_\star + c_\mathrm{b}(y'', w'') + \varrho \mathrm{TP}'[y; w] = 0$$

subject to the linearized arclength conditions $\partial_t y' \cdot y' = 0$ and $w' \cdot y' = 0$. The discretization uses the ideas outlined in Sections 3.1 and 4.1 with an explicit treatment of the potential.

Algorithm 6 (Gradient descent for self-avoiding curves).
Choose an initial $y_h^0 \in \mathcal{A}_h$ and a step size $\tau > 0$, set $k = 1$.

(1) Compute $d_t y_h^k \in \mathcal{F}_h[y_h^{k-1}]$ such that for all $w_h \in \mathcal{F}_h[y_h^{k-1}]$ we have

$$(d_t y_h^k, w_h)_\star + c_\mathrm{b}([y_h^k]'', w_h'') = -\varrho \mathrm{TP}_h'[y_h^{k-1}; w_h].$$

(2) Stop the iteration if $\|d_t y_h^k\|_\star \leq \varepsilon_{\mathrm{stop}}$; otherwise, increase $k \to k+1$ and continue with (1).

The explicit treatment of the potential has several advantages. First, we obtain linear problems in the time steps. Second, we avoid the inversion of fully populated matrices. Third, the assembly of the vector on the right-hand side can be fully parallelized without communication costs. Finally, we do not change the stability of the iteration compared to a fully implicit time-stepping scheme since the potential does not have obvious convexity properties. Using the differentiability properties of the functional TP and assuming that the flow metric is the H^2 scalar product, it has been shown in Bartels and Reiter (2018) that the conditional energy decay property

$$I_h^{\text{tot}}[y_h^L] + (1 - c_{\text{TP}}\tau)\tau \sum_{k=1}^{L} \| d_t y_h^k \|_\star^2 \le I_h^{\text{tot}}[y_h^0]$$

holds for all $0 \le L \le K \le T/\tau$ with a finite time horizon $T > 0$. This inequality implies the convergence of the iteration to a stationary configuration. Since the total energy is bounded uniformly, the values of the potential remain controlled so that in case of a sufficient resolution no self-contact or topology changes can occur.

The good stability properties and corresponding selfavoidance behaviour of the numerical scheme are illustrated by the snapshots of configurations and the corresponding energy decay shown for a generic setting in Fig. 13. We observe that the curve relaxes to a curve with equilibrated curvature and which is nearly flat. While the total energy decreases the (unscaled) tangent-point functional increases. In the experiments we used the model parameters $c_b = 10$, $\varrho = 10^{-3}$, $q = 3.9$. The initial curve belongs to the knot class 8_{10} (cf. Rolfsen, 1990), and has a total length $L \approx 45.45$ which is accurately preserved during the evolution. To compute the evolution we used a partitioning of the reference interval into 551 subintervals. The step size τ was chosen proportional to the mesh-size h.

6.3 Föppl–von Kármán model

The nonlinear bending model discussed in the previous sections is not suitable to describe certain phenomena such as the formation of wrinkles which occur for very small energies. In order to describe these effects in a dimensionally reduced model it is essential to involve the thickness parameter $\delta > 0$ as this number determines the oscillating or nonsmooth behaviour of deformations. A model that captures such effects is the Föppl–von Kármán model which can be identified via different scalings of the in-plane and out-of-plane components of a three-dimensional deformation when δ is small. It determines a planar displacement $u : \omega \to \mathbb{R}^2$ and a deflection $w : \omega \to \mathbb{R}$ as a minimizing pair for the energy functional

$$I_{\text{fvk}}[u,w] = \frac{\delta^2}{2} \int_\omega |D^2 w|^2 \, dx + \frac{1}{2} \int_\omega |\varepsilon(u) + \nabla w \otimes \nabla w|^2 \, dx$$

in a set $\mathcal{A}$ of admissible pairs contained in the product space $H^1(\omega; \mathbb{R}^2) \times H^2(\omega)$. Here, we use the symmetric gradient

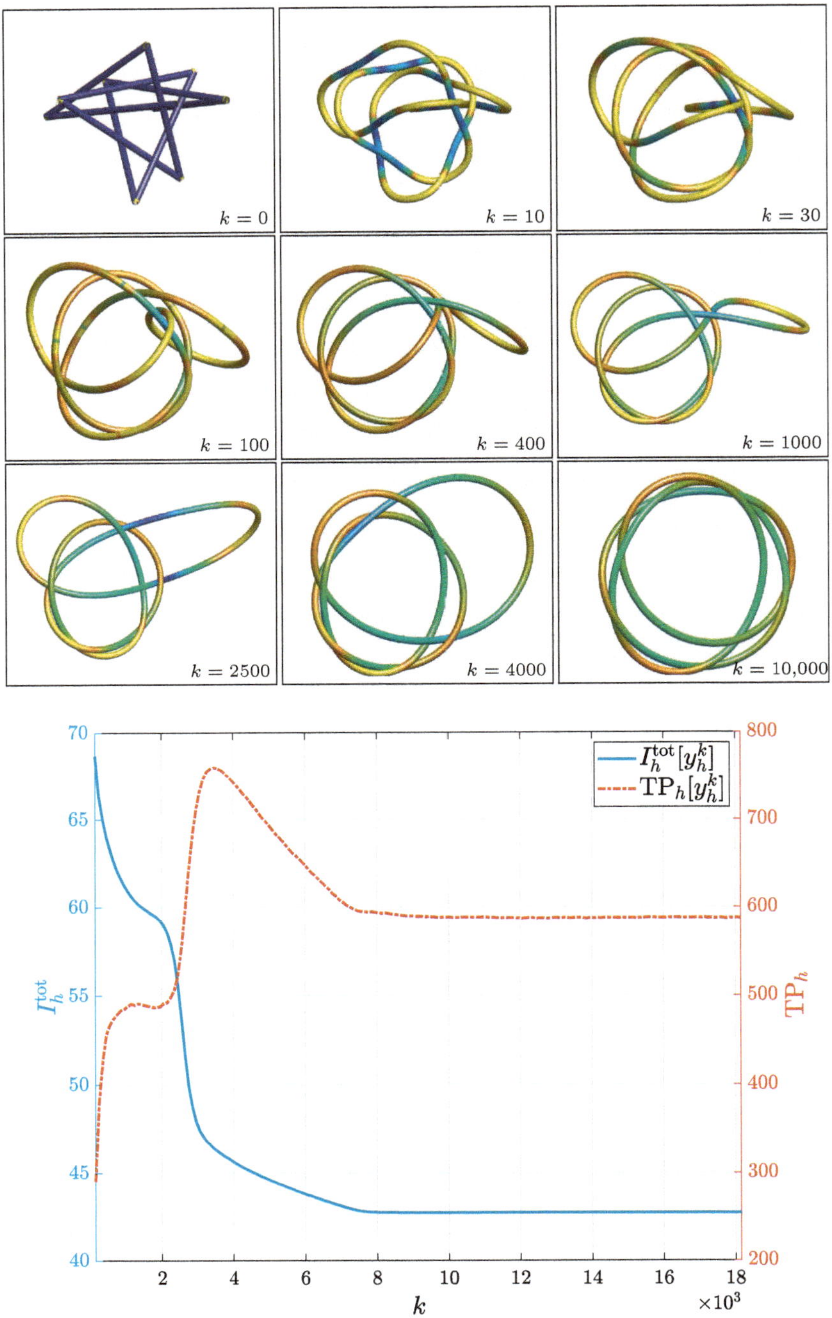

FIG. 13 Snapshots of an evolution from a polygonal initial curve after different numbers of iterations. The curve relaxes to a configuration that is close to a multiple covering of a flat circle (*top*). The total energy decreases monotonically while the contribution of the tangent-point functional remains controlled (*bottom*) so that the curve preserves the isotopy class defined by its initial configuration.

$$\tilde{\varepsilon}(u) = 2\,\mathrm{sym}(\nabla u) = (\nabla u)^{\mathsf{T}} + \nabla u$$

and the dyadic product

$$\nabla w \otimes \nabla w = (\nabla w)(\nabla w)^{\mathsf{T}},$$

where ∇w is understood as a column vector. We refer the reader to Ciarlet (1980), Ciarlet (1997), Friesecke et al. (2002a), and Friesecke et al. (2006) for justifications of this minimization problem as a simplification of three-dimensional hyperelastic material models. It has been shown rigorously in Müller and Olbermann (2014) and Venkataramani (2004) that the presence of the parameter δ leads to minimizers with wrinkling patterns whose geometry is determined by the parameter δ. This is done by identifying optimal scaling laws of the energy in terms of δ, cf. Bella and Kohn (2014) and Conti et al. (2017) for further examples. Only a few numerical methods for minimizing the Föppl–von Kármán energy are available, cf. (Ciarlet et al., 2005) for an abstract investigation.

To minimize the energy for given boundary conditions we follow Bartels (2017) and again adopt a gradient flow approach defined by the system

$$\begin{aligned}(\partial_t w, v)_\star &= -\gamma^2 (D^2 w, D^2 v) - 2\Big(|\nabla w|^2 \nabla w + \tilde{\varepsilon}(u)\nabla w, \nabla v\Big),\\ (\partial_t u, z)_\dagger &= -(\tilde{\varepsilon}(u), \tilde{\varepsilon}(z)) - (\nabla w \otimes \nabla w, \tilde{\varepsilon}(z)).\end{aligned}$$

Here, $(\cdot,\cdot)_\star$ and $(\cdot,\cdot)_\dagger$ are inner products on $H^2(\omega)$ and $H^1(\omega;\mathbb{R}^2)$, respectively, and we used the identities $|a \otimes a|^2 = |a|^4$ and $M : (a \otimes b) = (Ma)\cdot b = (Mb)\cdot a$ for a symmetric matrix $M \in \mathbb{R}^{2\times 2}$ and $a, b \in \mathbb{R}^2$. The temporal discretization of the system decouples the equations via a semiimplicit evaluation of different terms. In particular, given (u^{k-1}, w^{k-1}) we compute (u^k, w^k) such that

$$\begin{aligned}\left(d_t w^k, v\right)_\star &= -\gamma^2 (D^2 w^k, D^2 v) - 2\Big(|\nabla w^k|^2 \nabla w^k + \tilde{\varepsilon}(u^{k-1})\nabla w^{k-1/2}, \nabla v\Big),\\ \left(d_t u^k, z\right)_\dagger &= -\left(\tilde{\varepsilon}(u^k), \tilde{\varepsilon}(z)\right) - \left(\nabla w^k \otimes \nabla w^k, \tilde{\varepsilon}(z)\right),\end{aligned}$$

for all (v, z) satisfying appropriate homogeneous boundary conditions. The average $w^{k-1/2}$ is defined for subsequent approximations w^k and w^{k-1} via

$$w^{k-1/2} = \frac{1}{2}\left(w^k + w^{k-1}\right).$$

The unconditional stability and energy decay of the iteration follows from choosing $v = d_t w^k$ and $z = d_t u^k$ and exploiting a discrete product rule, cf. Bartels (2017) for details. The problems in the time steps determine minimizers of certain functionals and if $0 < \tau \le \tau_0$ then these functionals are strongly convex and minimizers are uniquely defined. Correspondingly, we expect the Newton scheme to converge provided that the step size τ is sufficiently small.

For a full discretization we use the set of admissible pairs

$$\mathcal{A}_h = \left\{ (u_h, w_h) \in \mathcal{S}^1(\mathcal{T}_h)^2 \times \mathcal{S}^{\mathrm{dkt}}(\mathcal{T}_h) : L_{\mathrm{bc}}[u_h, w_h] = \ell_{\mathrm{bc}} \right\}$$

and the corresponding homogeneous space

$$\mathcal{F}^0_{h,w} = \left\{ v_h \in \mathcal{S}^{\mathrm{dkt}}(\mathcal{T}_h) : L_{\mathrm{bc},w}[v_h] = 0 \right\},$$
$$\mathcal{F}^0_{h,u} = \left\{ z_h \in \mathcal{S}^1(\mathcal{T}_h)^2 : L_{\mathrm{bc},u}[z_h] = 0 \right\}.$$

To avoid an unnecessary small uniform step size τ we instead use variable step sizes $(\tau_k)_{k=0,1,\ \ldots}$ that are adjusted according to the performance of the Newton scheme using the following rules:

(i) decrease τ_k until Newton scheme terminates within $N_{\max}$ iterations,
(ii) set $\tau_{k+1} = \min\{2\tau_k,\ 10^r\}$ for next time step.

The parameter $r \geq 0$ defines an upper bound for the step sizes and avoids a numerical overflow. The ideas lead to the following algorithm.

Algorithm 7 (Gradient descent for Föppl–von Kármán functional).
Choose $(u_h^0, w_h^0) \in \mathcal{A}_h$, an integer $N_{\max} > 0$, stopping tolerances $\varepsilon_{\mathrm{stop}}$, $\varepsilon_{\mathrm{N}} > 0$, and an initial step size $\tau_1 > 0$, set $k = 1$.

(1a) Repeatedly decrease τ_k until the Newton scheme terminates within $N_{\max}$ steps and tolerance ε_{N} to determine $d_t w_h^k \in \mathcal{F}^0_{h,w}$ such that

$$\begin{aligned}(d_t w_h^k, v_h)_\star = &-\gamma^2 (D_h^2 w_h^k, D_h^2 v_h) \\ &- 2\left(|\nabla w_h^k|^2 \nabla w_h^k + \tilde{\varepsilon}(u_h^{k-1}) \nabla w_h^{k-1/2}, \nabla v_h \right)_h.\end{aligned}$$

for all $v_h \in \mathcal{F}^0_{h,w}$.

(1b) Compute $d_t u_h^k \in \mathcal{F}^0_{h,u}$ such that

$$\left(d_t \tilde{\varepsilon}(u_h^k), \tilde{\varepsilon}(z_h) \right)_\dagger = -\left(\tilde{\varepsilon}(u_h^k), \tilde{\varepsilon}(z_h) \right) - \left(\nabla w_h^k \otimes \nabla w_h^k, \tilde{\varepsilon}(z_h) \right)_h$$

for all $z_h \in \mathcal{F}^0_{h,u}$.

(2) Stop if $\|d_t w_h^k\|_\star + \|d_t \tilde{\varepsilon}(u_h^k)\|_\dagger \leq \varepsilon_{\mathrm{stop}} \min\{1, \tau_k\}$; otherwise, define

$$\tau_{k+1} = \min\{2\tau_k, 10^r\},$$

increase $k \to k + 1$, and continue with (1).

Note that in the algorithm a stopping criterion is used that is proportional to the step size which is important in the case of small step sizes. To illustrate the performance of the algorithm and features of the mathematical model we follow Bartels (2017) and consider the compression of a plate along one of its sides. Particularly, we let $\omega = (-1/2,\ 1/2) \times (0,\ 1)$ and compress the side $\{0\} \times\ [-1/2,\ 1/2]$ by 10 percent which defines the boundary data for the in-plane displacement u. We use homogeneous clamped boundary conditions along the same side for the deflection w. The results shown in Fig. 14 confirm

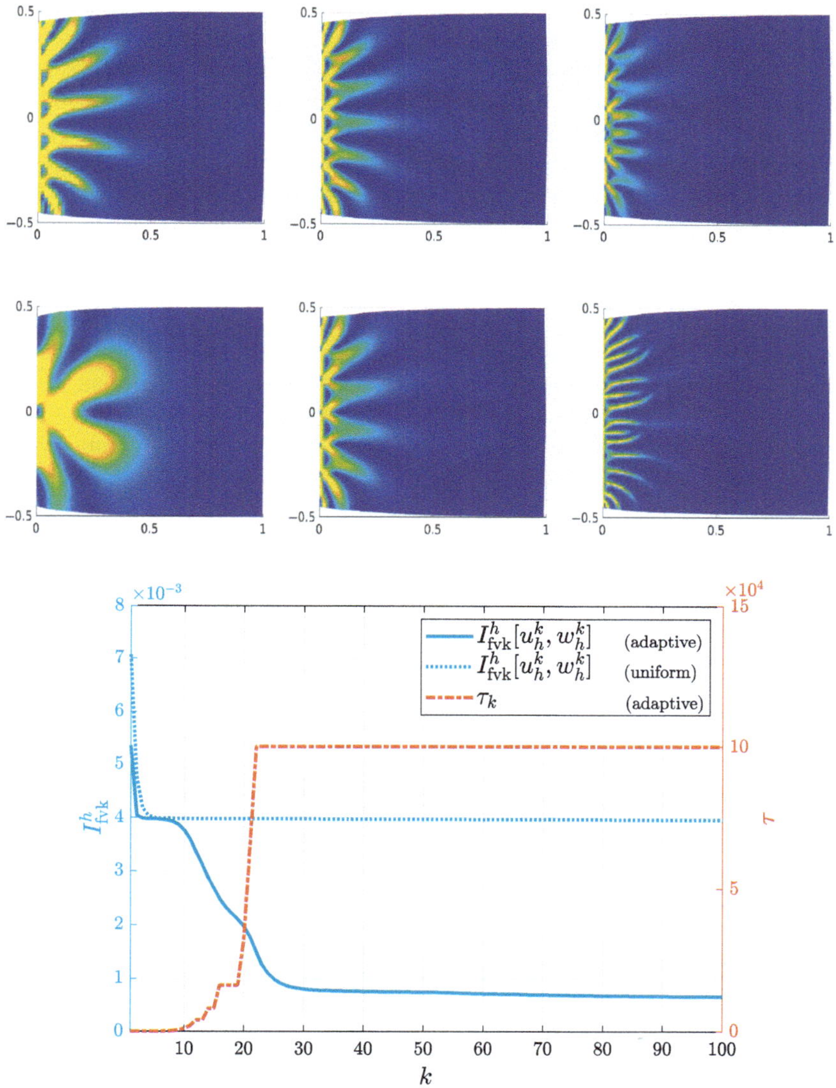

FIG. 14 In-plane deformation and bending energy density $|D^2 w_h|$ of the numerical approximations for the Föppl–von Kármán model with compressive boundary conditions for different triangulations with mesh-sizes $h = 2^{-\ell}$, $\ell = 6, 7, 8$, and fixed $\delta = 1/200$ (*top*); for $\delta = 1/40$, $1/160$, and $1/640$ on a fixed triangulation with $h = 2^{-7}$ (*middle*); energies and step sizes for uniform and adaptive time-stepping (*bottom*).

the formation of wrinkling structures. These depend on both the numerical resolution and the thickness parameter. Moreover, the energy decay shown in the bottom plot of Fig. 14 indicates that these are only obtained within a reasonable number of iterations if an adaptive step size strategy is used.

7 Conclusions

We have addressed in this article the accurate finite element discretization and reliable iterative solution of nonlinear models for describing large deformations of elastic rods and plates. To avoid unjustified regularity assumptions we have adopted the concept of Γ-convergence and thereby proved convergence of approximations. To compute stationary configurations of low and possibly minimal energy we have used gradient flow discretizations that are guaranteed to decrease the elastic energy and preserve inextensibility and isometry constraints appropriately. These two concepts leave a theoretical gap in the sense that the framework of Γ-convergence is concerned with global energy minimizers while gradient flows can only be expected to determine stationary configurations. However, Γ-convergence goes beyond global energy minimization and gradient flows typically avoid unstable critical configurations. These properties are convincingly confirmed by our numerical experiments which did not employ any a priori knowledge about expected solutions. The state of the art of reliable methods for computing nonlinear bending phenomena leaves open several aspects such as the development of optimal preconditioners, adaptive refinement of stress concentrations, determination of convergence rates, effective combination with membrane effects, or simulation of realistic dynamic models open. We believe that the methods presented in this article can be useful in investigating them in future research.

Acknowledgements

The author wishes to thank his coworkers in various collaborations that led to the results presented in this article.

References

Alouges, F., 1997. A new algorithm for computing liquid crystal stable configurations: the harmonic mapping case. SIAM J. Numer. Anal. 34 (5), 1708–1726. ISSN 0036-1429. https://doi.org/10.1137/S0036142994264249.

Antman, S.S., 2005. Nonlinear Problems of Elasticity. Applied Mathematical Sciences, vol. 107. Springer, New York. ISBN: 0-387-20880-1, p. xviii+831 second.

Audoly, B., Pomeau, Y., 2010. Elasticity and Geometry. Oxford University Press, Oxford. ISBN: 978-0-19-850625-6, p. x+586.

Barrett, J.W., Garcke, H., Nürnberg, R., 2007. A parametric finite element method for fourth order geometric evolution equations. J. Comput. Phys. 222 (1), 441–462. ISSN 0021-9991. https://doi.org/10.1016/j.jcp.2006.07.026.

Barrett, J.W., Garcke, H., Nürnberg, R., 2012. Parametric approximation of isotropic and anisotropic elastic flow for closed and open curves. Numer. Math. 120 (3), 489–542. ISSN 0029-599X 10.1007/s00211-011-0416-x.

Bartels, S., 2005. Stability and convergence of finite-element approximation schemes for harmonic maps. SIAM J. Numer. Anal. 43 (1), 220–238. ISSN 0036-1429. https://doi.org/10.1137/040606594.

Bartels, S., 2013. Approximation of large bending isometries with discrete Kirchhoff triangles. SIAM J. Numer. Anal. 51 (1), 516–525. ISSN 0036-1429. https://doi.org/10.1137/110855405.

Bartels, S., 2013. Finite element approximation of large bending isometries. Numer. Math. 124 (3), 415–440. ISSN 0029-599X. https://doi.org/10.1007/s00211-013-0519-7.

Bartels, S., 2013. A simple scheme for the approximation of the elastic flow of inextensible curves. IMA J. Numer. Anal. 33 (4), 1115–1125. ISSN 0272-4979. 10.1093/imanum/drs041.

Bartels, S., 2015. Numerical Methods for Nonlinear Partial Differential Equations. Springer Series in Computational Mathematics, vol. 47. Springer, Cham, p. x+393. https://doi.org/10.1007/978-3-319-13797-1. 978-3-319-13796-4; 978-3-319-13797-1.

Bartels, S., 2016. Projection-free approximation of geometrically constrained partial differential equations. Math. Comp. 85 (299), 1033–1049. ISSN 0025-5718. https://doi.org/10.1090/mcom/3008.

Bartels, S., 2017. Numerical solution of a Föppl-von Kármán model. SIAM J. Numer. Anal. 55 (3), 1505–1524. ISSN 0036-1429. https://doi.org/10.1137/16M1069791.

Bartels, S., Hornung, P., 2015. Bending paper and the Möbius strip. J. Elasticity 119 (1-2), 113–136. ISSN 0374-3535. https://doi.org/10.1007/s10659-014-9501-6.

Bartels, S., Palus, C., 2019. Stable iterative solution of bilayer bending problems. (in preparation).

Bartels, S., Reiter, P., 2018. Stability of a simple scheme for the approximation of elastic knots and self-avoiding inextensible curves. arXiv (arXiv:1804.02206).

Bartels, S., Reiter, P., 2019. Numerical solution of a bending-torsion model for elastic rods. (in preparation).

Bartels, S., Bonito, A., Nochetto, R.H., 2017. Bilayer plates: model reduction, Γ-convergent finite element approximation, and discrete gradient flow. Comm. Pure Appl. Math. 70 (3), 547–589. ISSN 0010-3640. https://doi.org/10.1002/cpa.21626.

Bartels, S., Bonito, A., Muliana, A.H., Nochetto, R.H., 2018. Modeling and simulation of thermally actuated bilayer plates. J. Comput. Phys. 354, 512–528. ISSN 0021-9991. https://doi.org/10.1016/j.jcp.2017.10.044.

Bartels, S., Reiter, P., Riege, J., 2018. A simple scheme for the approximation of self-avoiding inextensible curves. IMA J. Numer. Anal. 38 (2), 543–565. ISSN 0272-4979. https://doi.org/10.1093/imanum/drx021.

Bella, P., Kohn, R.V., 2014. Metric-induced wrinkling of a thin elastic sheet. J. Nonlinear Sci. 24 (6), 1147–1176. ISSN 0938-8974. https://doi.org/10.1007/s00332-014-9214-9.

Bergou, M., Wardetzky, M., Robinson, S., Audoly, B., Grinspun, E., 2008. Discrete elastic rods. ACM Trans. Graph. (SIGGRAPH) 27 (3), 63:1–63:12.

Blatt, S., 2013. The energy spaces of the tangent point energies. J. Topol. Anal. 5 (3), 261–270. ISSN 1793-5253. https://doi.org/10.1142/S1793525313500131.

Blatt, S., Reiter, P., 2015. Regularity theory for tangent-point energies: the non-degenerate subcritical case. Adv. Calc. Var. 8 (2), 93–116. ISSN 1864-8258. https://doi.org/10.1515/acv-2013-0020.

Bonito, A., Nochetto, R. H., Ntogkas, D., 2019. DG approach to large bending plate deformations with isometry constraint. (in preparation).

Braess, D., 2007. Finite Elements, third ed. Cambridge University Press, Cambridge, xviii+365. https://doi.org/10.1017/CBO9780511618635. 978-0-521-70518-9; 0-521-70518-5.

Ciarlet, P.G., 1980. A justification of the von Kármán equations. Arch. Rational Mech. Anal. 73 (4), 349–389. ISSN 0003-9527. https://doi.org/10.1007/BF00247674.

Ciarlet, P.G., 1997. Mathematical Elasticity. Vol. II. Studies in Mathematics and its Applications, vol. 27. North-Holland Publishing Co., Amsterdam. ISBN: 0-444-82570-3, p. lxiv+497.

Ciarlet, P.G., Gratie, L., Kesavan, S., 2005. Numerical analysis of the generalized von Kármán equations. C. R. Math. Acad. Sci. Paris 341 (11), 695–699. ISSN 1631-073X. https://doi.org/10.1016/j.crma.2005.09.031.

Conti, S., Maggi, F., 2008. Confining thin elastic sheets and folding paper. Arch. Ration. Mech. Anal. 187 (1), 1–48. ISSN 0003-9527. https://doi.org/10.1007/s00205-007-0076-2.

Conti, S., Olbermann, H., Tobasco, I., 2017. Symmetry breaking in indented elastic cones. Math. Models Methods Appl. Sci. 27 (2), 291–321. ISSN 0218-2025. https://doi.org/10.1142/S0218202517500026.

Dal Maso, G., 1993. An introduction to Γ-convergence. Progress in Nonlinear Differential Equations and their Applications, vol. 8. Birkhäuser Boston, Inc., Boston, MA. ISBN: 0-8176-3679-X, p. xiv+340. https://doi.org/10.1007/978-1-4612-0327-8

Deckelnick, K., Dziuk, G., Elliott, C.M., 2005. Computation of geometric partial differential equations and mean curvature flow. Acta Numer. 14, 139–232. ISSN 0962-4929. https://doi.org/10.1017/S0962492904000224.

Dziuk, G., Kuwert, E., Schätzle, R., 2002. Evolution of elastic curves in *mathbbR*^{n}: existence and computation. SIAM J. Math. Anal. 33 (5), 1228–1245. ISSN 0036-1410. https://doi.org/10.1137/S0036141001383709.

Freddi, L., Hornung, P., Mora, M.G., Paroni, R., 2016. A corrected Sadowsky functional for inextensible elastic ribbons. J. Elasticity 123 (2), 125–136. ISSN 0374-3535. https://doi.org/10.1007/s10659-015-9551-4.

Friesecke, G., James, R.D., Müller, S., 2002. The Föppl-von Kármán plate theory as a low energy Γ-limit of nonlinear elasticity. C. R. Math. Acad. Sci. Paris 335 (2), 201–206. ISSN 1631-073X. https://doi.org/10.1016/S1631-073X(02)02388-9.

Friesecke, G., James, R.D., Müller, S., 2002. A theorem on geometric rigidity and the derivation of nonlinear plate theory from three-dimensional elasticity. Comm. Pure Appl. Math. 55 (11), 1461–1506. ISSN 0010-3640. https://doi.org/10.1002/cpa.10048.

Friesecke, G., Müller, S., James, R.D., 2002. Rigorous derivation of nonlinear plate theory and geometric rigidity. C. R. Math. Acad. Sci. Paris 334 (2), 173–178. ISSN 1631-073X. https://doi.org/10.1016/S1631-073X(02)02133-7.

Friesecke, G., James, R.D., Müller, S., 2006. A hierarchy of plate models derived from nonlinear elasticity by gamma-convergence. Arch. Ration. Mech. Anal. 180 (2), 183–236. ISSN 0003-9527. https://doi.org/10.1007/s00205-005-0400-7.

Gonzalez, O., Maddocks, J.H., 1999. Global curvature, thickness, and the ideal shapes of knots. Proc. Natl. Acad. Sci. USA 96 (9), 4769–4773. ISSN 1091-6490. https://doi.org/10.1073/pnas.96.9.4769.

Hornung, P., 2011. Approximation of flat $W^{2,\,2}$ isometric immersions by smooth ones. Arch. Ration. Mech. Anal. 199 (3), 1015–1067. ISSN 0003-9527. https://doi.org/10.1007/s00205-010-0374-y.

Kraus, J., Pfeiler, C.-M., Praetorius, D., Ruggeri, M., Stiftner, B., 2018. Iterative solution and preconditioning for the tangent plane scheme in computational micromagnetics. arXiv (1808.10281).

Langer, J., Singer, D.A., 1996. Lagrangian aspects of the Kirchhoff elastic rod. SIAM Rev. 38 (4), 605–618. ISSN 0036-1445. https://doi.org/10.1137/S0036144593253290.

Mora, M.G., Müller, S., 2003. Derivation of the nonlinear bending-torsion theory for inextensible rods by Γ-convergence. Calc. Var. Partial Differential Equations 18 (3), 287–305. ISSN 0944-2669. https://doi.org/10.1007/s00526-003-0204-2.

Müller, S., Olbermann, H., 2014. Almost conical deformations of thin sheets with rotational symmetry. SIAM J. Math. Anal. 46 (1), 25–44. ISSN 0036-1410. https://doi.org/10.1137/120901362.

Nash, S.G., Sofer, A., 1996. Preconditioning reduced matrices. SIAM J. Matrix Anal. Appl. 17 (1), 47–68. ISSN 0895-4798. https://doi.org/10.1137/S0895479893245371.

O'Hara, J., 2003. Energy of knots and conformal geometry. Series on Knots and Everything, vol. 33. World Scientific Publishing Co., Inc., River Edge, NJ. ISBN: 981-238-316-6, p. xiv+288. https://doi.org/10.1142/9789812795304

Pakzad, M.R., 2004. On the Sobolev space of isometric immersions. J. Differential Geom. 66 (1), 47–69. ISSN 0022-040X. http:/projecteuclid.org/euclid.jdg/1090415029.

Pantz, O., 2003. On the justification of the nonlinear inextensional plate model. Arch. Ration. Mech. Anal. 167 (3), 179–209. ISSN 0003-9527. https://doi.org/10.1007/s00205-002-0238-1.

Pozzi, P., Stinner, B., 2017. Curve shortening flow coupled to lateral diffusion. Numer. Math. 135 (4), 1171–1205. ISSN 0029-599X. https://doi.org/10.1007/s00211-016-0828-8.

Rivière, T., 1995. Everywhere discontinuous harmonic maps into spheres. Acta Math. 175 (2), 197–226. ISSN 0001-5962. https://doi.org/10.1007/BF02393305.

Rolfsen, D., 1990. Knots and Links. Mathematics Lecture Series, vol. 7. Publish or Perish, Inc., Houston, TX. ISBN: 0-914098-16-0, p. xiv+439

Sander, O., Neff, P., Bîrsan, M., 2016. Numerical treatment of a geometrically nonlinear planar Cosserat shell model. Comput. Mech. 57 (5), 817–841. ISSN 0178-7675. https://doi.org/10.1007/s00466-016-1263-5.

Schmidt, B., 2007. Minimal energy configurations of strained multi-layers. Calc. Var. Partial Differential Equations 30 (4), 477–497. ISSN 0944-2669. https://doi.org/10.1007/s00526-007-0099-4.

Schmidt, B., 2007. Plate theory for stressed heterogeneous multilayers of finite bending energy. J. Math. Pures Appl. (9) 88 (1), 107–122. ISSN 0021-7824. https://doi.org/10.1016/j.matpur.2007.04.011.

Schmidt, O.G., Eberl, K., 2001. Thin solid films roll up into nanotubes. Nature 410, 168.

Schoen, R., Uhlenbeck, K., 1982. A regularity theory for harmonic maps. J. Differential Geom. 17 (2), 307–335. ISSN 0022-040X. http:/projecteuclid.org/euclid.jdg/1214436923.

Sharon, E., Roman, B., Swinney, H., 2007. Geometrically driven wrinkling observed in free plastic sheets and leaves. Phys. Rev. E Stat. Nonlin. Soft Matter Phys. 75, 046211. https://doi.org/10.1103/PhysRevE.75.046211.

Smela, E., Inganös, O., Pei, Q., Lundström, I., 1993. Electrochemical muscles: micromachining fingers and corkscrews. Adv. Mater. 5 (9), 630–632. ISSN 1521-4095. https://doi.org/10.1002/adma.19930050905.

Strzelecki, P., von der Mosel, H., 2012. Tangent-point self-avoidance energies for curves. J. Knot Theory Ramifications 21(5). 1250044, 28. https://doi.org/10.1142/S0218216511009960.

Strzelecki, P., Szumańska, M., von der Mosel, H., 2013. On some knot energies involving Menger curvature. Topology Appl. 160 (13), 1507–1529. ISSN 0166-8641. https://doi.org/10.1016/j.topol.2013.05.022.

Thompson, J.M.T., Coleman, B.D., Swigon, D., 2004. Theory of self-contact in Kirchhoff rods with applications to supercoiling of knotted and unknotted DNA plasmids. Phil. Trans. R. Soc. London. A: Math. Phys. Eng. Sci. 362 (1820), 1281–1299. https://doi.org/10.1098/rsta.2004.1393.

Venkataramani, S.C., 2004. Lower bounds for the energy in a crumpled elastic sheet–a minimal ridge. Nonlinearity 17 (1), 301–312. ISSN 0951-7715. https://doi.org/10.1088/0951-7715/17/1/017.

von der Mosel, H., 1998. Minimizing the elastic energy of knots. Asymptot. Anal. 18 (1–2), 49–65. ISSN 0921-7134.

Wardetzky, M., Bergou, M., Harmon, D., Zorin, D., Grinspun, E., 2007. Discrete quadratic curvature energies. Comput. Aided Geom. Design 24 (8–9), 499–518. ISSN 0167-8396. https://doi.org/10.1016/j.cagd.2007.07.006.

Chapter 4

Parametric finite element approximations of curvature-driven interface evolutions

John W. Barrett[a], **Harald Garcke**[b,*] **and Robert Nürnberg**[a]
[a]*Department of Mathematics, Imperial College London, London, United Kingdom*
[b]*Fakultät für Mathematik, Universität Regensburg, Regensburg, Germany*
[*]*Corresponding author: e-mail: harald.garcke@mathematik.uni-regensburg.de*

Chapter Outline

Handbook of Numerical Analysis, Vol. 21. https://doi.org/10.1016/bs.hna.2019.05.002

Abstract

Parametric finite elements lead to very efficient numerical methods for surface evolution equations. We introduce several computational techniques for curvature driven evolution equations based on a weak formulation for the mean curvature. The approaches discussed, in contrast to many other methods, have good mesh properties that avoid mesh coalescence and very nonuniform meshes. Mean curvature flow, surface diffusion, anisotropic geometric flows, solidification, two-phase flow, Willmore and Helfrich flow as well as biomembranes are treated. We show stability results as well as results explaining the good mesh properties.

Keywords: Parametric finite elements, Geometric flows, Tangential motion, Free boundary problems, Mean curvature flow, Surface diffusion, Anisotropy, Crystal growth, Two-phase flow, Willmore flow, Biomembranes

AMS Classification Codes: 65M60, 53C44, 35K55, 65M12, 74E10, 74N05, 76D05, 92C05

1 Introduction

Interfaces separating different regions in space frequently appear in the natural and imaging sciences as well as in mathematics. In many situations these interfaces evolve in time and the evolution laws contain curvature quantities. Often an interface carries a surface energy and in the simplest case the surface energy is given by the total surface area of the interface. Frequently the surface energy decreases in time and as the negative mean curvature is the first variation of the surface area, the mean curvature naturally appears in the context of evolving interfaces. In the simplest such situation the evolution of the interface is the L^2-gradient flow of the surface area. This leads to mean curvature flow, which will play a prominent role in this chapter. Many recent applications involve surface energies, which contain surface integrals of curvature quantities. An example is the Willmore functional, which is the integrated squared mean curvature. Evolution problems for surfaces involving the first variation of the Willmore functional are particularly challenging and will also be discussed in this work.

Often, however, the evolution of the interface is given by laws which couple quantities on the interface to quantities which are solutions of partial differential equations in the bulk, i.e., in the surrounding regions. This happens, for example, in solidification phenomena in which the temperature, which solves a diffusion equation in the liquid and solid phases, influences the evolution of the interface. Other examples are interfaces driven by a flow field, which is, for example, the solution of a Stokes or Navier–Stokes system. These situations are much more involved and lead to challenging problems in analysis and computation.

This work deals with parametric finite element methods for evolving interfaces. We first present the main ideas for simple geometric evolution equations like the mean curvature flow. We will then couple equations on the interface to bulk equations using an unfitted finite element method, which uses a surface mesh that is independent from the bulk mesh. As examples, solidification phenomena and problems arising in two-phase fluid flow will be discussed. Upon introducing ideas on the approximation of the Willmore energy and the associated Willmore flow, we end this chapter with a discussion of the numerical approximation of evolution problems for biomembranes. For other numerical approaches for geometric partial differential equations, like level set methods, phase field methods and other front tracking methods, we refer to the review article Deckelnick et al. (2005a), as well as to the following chapters in this handbook: "Free boundary problems in fluids and materials" by Bänsch and Schmidt; "Finite element simulation of nonlinear bending models for thin elastic rods and plates" by Bartels; "Finite element methods for the Laplace-Beltrami operator" by Bonito et al.; "The phase field method for geometric moving interfaces and their numerical approximations" by Du and Feng; "A review of level set methods to model interfaces moving under complex physics: Recent challenges and advances" by Saye and Sethian; "Numerical simulation and benchmarking of drops and bubbles" by Turek and Mierka.

2 Geometry of surfaces

In this section we review basic facts about surfaces, which will be necessary later on, when we develop numerical algorithms for curvature-driven flows of hypersurfaces. We will define surfaces, differential operators on surfaces, important curvature quantities as well as a divergence theorem on hypersurfaces. For a more detailed discussion of these themes we refer to do Carmo (1976), Deckelnick et al. (2005a), Kühnel (2015) and Walker (2015). Readers familiar with these concepts may skip this section.

2.1 Surfaces in $\mathbb{R}^d$

We will first define the term surface. Here, and throughout, let $d \geq 2$.

Definition 1.

(i) A subset $\Gamma \subset \mathbb{R}^d$ is called an n-dimensional C^k-surface, for $0 \leq n \leq d$ and $k \geq 1$, if for every point $\vec{p} \in \Gamma$ there exists an open neighbourhood $V \subset \mathbb{R}^d$ of $\vec{p}$ and a bijective C^k-map $\vec{\mathfrak{x}} : U \to V \cap \Gamma$, with $U \subset \mathbb{R}^n$ open and connected, such that the Jacobian $\nabla \vec{\mathfrak{x}} = (\partial_1 \vec{\mathfrak{x}}, \ldots, \partial_n \vec{\mathfrak{x}}) : U \to \mathbb{R}^{d \times n}$ of $\vec{\mathfrak{x}}$ has full rank in U.

(ii) The map $\vec{\mathfrak{x}}$ is called a local parameterization of Γ.

(iii) If $n = d - 1$, then we call Γ a hypersurface.

We now define what we mean by a mapping defined on a hypersurface being differentiable.

Definition 2. Let $\Gamma \subset \mathbb{R}^d$ be an n-dimensional C^k-surface. A function $f : \Gamma \to \mathbb{R}$ is a C^k-function, denoted by $f \in C^k(\Gamma)$, if $f \circ \vec{\mathfrak{x}} : U \to \mathbb{R}$ is of class C^k for all C^k-parameterizations $\vec{\mathfrak{x}} : U \to \Gamma$.

With the help of a local parameterization we can define a basis of the tangent space.

Definition 3. Let Γ be an n-dimensional C^1-surface with a local parameterization $\vec{\mathfrak{x}} : U \to \mathbb{R}^d$. For $\vec{p} \in \Gamma$ let $\vec{u} \in U$ be such that $\vec{p} = \vec{\mathfrak{x}}(\vec{u})$. The vectors

$$\vec{\partial}_1 = (\partial_1 \vec{\mathfrak{x}})(\vec{u}), \ldots, \vec{\partial}_n = (\partial_n \vec{\mathfrak{x}})(\vec{u})$$

are linearly independent and span an n-dimensional space, which is called the tangent space and will be denoted by

$$\mathrm{T}_{\vec{p}}\,\Gamma = \mathrm{span}\{\vec{\partial}_1, \ldots, \vec{\partial}_n\}.$$

Its orthogonal complement $\mathrm{N}_{\vec{p}}\,\Gamma = (\mathrm{T}_{\vec{p}}\,\Gamma)^{\perp}$ in $\mathbb{R}^d$ is called the normal space of Γ at $\vec{p}$.

Mainly oriented hypersurfaces will be of interest, which we define next.

Definition 4. A hypersurface Γ is called orientable, if a continuous normal vector field $\vec{\nu} : \Gamma \to \mathbb{S}^{d-1}$, with $\vec{\nu}(\vec{p}) \in \mathrm{N}_{\vec{p}}\,\Gamma$ for all $\vec{p} \in \Gamma$, exists, where $\mathbb{S}^{d-1}$ is the $(d-1)$-dimensional sphere in $\mathbb{R}^d$.

We will frequently use the following differential operators on a surface Γ.

Definition 5. Let $\Gamma \subset \mathbb{R}^d$ be an n-dimensional C^1-surface, and let $f:\Gamma \to \mathbb{R}$, $\vec{f}:\Gamma \to \mathbb{R}^d$ be C^1-functions.

(i) For $\vec{p} \in \Gamma$ and $\vec{\tau} \in \mathrm{T}_{\vec{p}}\,\Gamma$ we define the directional derivative as

$$(\partial_{\vec{\tau}}\, f)(\vec{p}) = \lim_{\varepsilon \to 0} \frac{f(\vec{y}(\varepsilon)) - f(\vec{y}(0))}{\varepsilon} = (f \circ \vec{y})'(0),$$

where $\vec{y}:(-1,1) \to \Gamma$ parameterizes a curve on Γ with $\vec{y}(0) = \vec{p}$ and $\vec{y}'(0) = \vec{\tau}$.

(ii) The surface gradient of f on Γ at the point $\vec{p} \in \Gamma$ is defined as

$$(\nabla_s\, f)(\vec{p}) = \sum_{i=1}^{n} (\partial_{\vec{\tau}_i}\, f)(\vec{p})\, \vec{\tau}_i,$$

where $\{\vec{\tau}_1, \ldots, \vec{\tau}_n\}$ is an orthonormal basis of $\mathrm{T}_{\vec{p}}\,\Gamma$.

(iii) The surface divergence of $\vec{f}$ at the point $\vec{p} \in \Gamma$ is defined as

$$(\nabla_s \cdot \vec{f})(\vec{p}) = \sum_{i=1}^{n} \vec{\tau}_i \cdot (\partial_{\vec{\tau}_i}\, \vec{f})(\vec{p}),$$

where $\{\vec{\tau}_1, \ldots, \vec{\tau}_n\}$ is again an orthonormal basis of $\mathrm{T}_{\vec{p}}\,\Gamma$.

(iv) The surface Jacobian of $\vec{f}$ at the point $\vec{p} \in \Gamma$ is defined as

$$(\nabla_s\, \vec{f})(\vec{p}) = \sum_{i=1}^{n} (\partial_{\vec{\tau}_i} \vec{f})(\vec{p}) \otimes \vec{\tau}_i,$$

where $\{\vec{\tau}_1, \ldots, \vec{\tau}_n\}$ is again an orthonormal basis of $\mathrm{T}_{\vec{p}}\,\Gamma$.

(v) If Γ is a C^2-surface, and $f \in C^2(\Gamma)$, then we define the Laplace–Beltrami operator and the surface Hessian of f on Γ as

$$\Delta_s\, f = \nabla_s \cdot (\nabla_s\, f) \quad \text{and} \quad \nabla_s^2\, f = \nabla_s\, (\nabla_s\, f),$$

respectively.

(vi) The Laplace–Beltrami operator of $\vec{f} \in [C^2(\Gamma)]^d$, for a C^2-surface Γ, is defined component-wise, i.e., $(\Delta_s\, \vec{f}) \cdot \vec{e}_i = \Delta_s\, (\vec{f} \cdot \vec{e}_i)$, $i = 1, \ldots, d$. Here, and throughout, we let $\vec{e}_i$ denote the i-th standard basis vector in $\mathbb{R}^d$.

(vii) We define the generalized symmetrized surface Jacobian

$$\underline{\underline{D}}_s(\vec{f}) = \frac{1}{2} \underline{\underline{P}}_\Gamma\, (\nabla_s\, \vec{f} + (\nabla_s\, \vec{f})^{\mathsf{T}})\, \underline{\underline{P}}_\Gamma \quad \text{on}\, \Gamma,$$

where $\underline{\underline{P}}_\Gamma$ is the orthogonal projection onto the tangent space of Γ. That is,

$$\underline{\underline{P}}_\Gamma(\vec{p}) = \sum_{i=1}^{n} \vec{\tau}_i \otimes \vec{\tau}_i \tag{1}$$

for an orthonormal basis $\{\vec{\tau}_1, \ldots, \vec{\tau}_n\}$ of $\mathrm{T}_{\vec{p}}\,\Gamma$.

(viii) The surface divergence of a tensor $\underline{\underline{f}} \in [C^1(\Gamma)]^{d \times d}$ is defined via $(\nabla_s \cdot \underline{\underline{f}}) \cdot \vec{e}_i = \nabla_s \cdot (\underline{\underline{f}}^{\mathsf{T}}\, \vec{e}_i)$ on Γ, for $i = 1, \ldots, d$.

Remark 6.

(i) It is easy to check that Definition 5(i) does not depend on the choice of the parameterization $\vec{y}$. Parameterizations $\vec{y}$ of curves, as 1-dimensional surfaces, are always assumed to be regular in the sense that $\vec{y}$ is continuously differentiable and such that $\vec{y}' \neq \vec{0}$ holds everywhere, recall Definition 1. A curve is an equivalence class of regular parameterizations, see, e.g., Kühnel (2015, p. 8), where the equivalence relation is given by parameter transformations. For simplicity, from now on, we will often also call a parameterization $\vec{y}$ a curve.

(ii) It is easy to check that Definitions 5(ii), 5(iii) and 5(iv) do not depend on the choice of the orthonormal basis of $\mathrm{T}_{\vec{p}}\,\Gamma$.

(iii) For $\vec{p} \in \Gamma$ it holds that

$$(\nabla_s \vec{f})(\vec{p})\,\vec{\tau} = (\partial_{\vec{\tau}}\,\vec{f})(\vec{p}) \quad \forall \vec{\tau} \in \mathrm{T}_{\vec{p}}\,\Gamma$$
$$\text{and} \quad (\nabla_s \vec{f})(\vec{p})\,\vec{\mathrm{v}} = \vec{0} \quad \forall \vec{\mathrm{v}} \in \mathrm{N}_{\vec{p}}\,\Gamma.$$

Moreover, it holds that the ith row of $\nabla_s \vec{f}$ is the surface gradient of the ith component of $\vec{f}$, i.e., $(\nabla_s \vec{f})^{\top}\,\vec{e}_i = \nabla_s (\vec{f} \cdot \vec{e}_i)$, $i = 1, \ldots, d$.

(iv) We remark that the second projection $\underline{\underline{P}}_{\Gamma}$ in Definition 5(vii) ensures that

$$(\underline{\underline{D}}_s(\vec{f}))(\vec{p})\,\vec{\mathrm{v}} = \vec{0} \quad \forall \vec{\mathrm{v}} \in \mathrm{N}_{\vec{p}}\,\Gamma, \;\; \vec{p} \in \Gamma,$$

similarly to (iii). The first projection then ensures that $(\underline{\underline{D}}_s(\vec{f}))(\vec{p})$ is symmetric.

(v) It follows directly from Definitions 5(ii), 5(iv) and (1) that

$$\nabla_s\, f = \underline{\underline{P}}_{\Gamma}\, \nabla_s\, f \quad \text{and} \quad \nabla_s\, \vec{f} = (\nabla_s\, \vec{f})\, \underline{\underline{P}}_{\Gamma} \quad \text{on}\, \Gamma.$$

(vi) We observe that if Γ is a hypersurface, then

$$\underline{\underline{P}}_{\Gamma} = \underline{\underline{\mathrm{Id}}} - \vec{\nu} \otimes \vec{\nu} \quad \text{on}\, \Gamma,$$

with $\vec{\nu}(p)$ denoting a unit vector in $\mathrm{N}_{\vec{p}}\,\Gamma$ for $\vec{p} \in \Gamma$.

The following product rules and identities hold.

Lemma 7. *Let $\Gamma \subset \mathbb{R}^d$ be an n-dimensional C^1-surface, and let $f : \Gamma \to \mathbb{R}$, $\vec{f}, \vec{g} : \Gamma \to \mathbb{R}^d$ and $\underline{\underline{f}} : \Gamma \to \mathbb{R}^{d\times d}$ be C^1-functions. Then it holds:*

(i)
$$\nabla_s \cdot (f\,\vec{g}) = \vec{g} \cdot \nabla_s\, f + f\, \nabla_s \cdot \vec{g} \quad \mathit{on}\, \Gamma,$$

(ii)
$$\nabla_s\, (\vec{f} \cdot \vec{g}) = (\nabla_s\, \vec{f})^{\top}\, \vec{g} + (\nabla_s\, \vec{g})^{\top}\, \vec{f} \quad \mathit{on}\, \Gamma,$$

(iii)
$$\nabla_s\, (f\,\vec{g}) = \vec{g} \otimes \nabla_s\, f + f\, \nabla_s\, \vec{g} \quad \mathit{on}\, \Gamma,$$

(iv)
$$\nabla_s \cdot \left(\underline{f}^{\mathsf{T}} \vec{g}\right) = \vec{g} \cdot \left(\nabla_s \cdot \underline{f}\right) + \underline{f} : \nabla_s \vec{g} \quad on\,\Gamma,$$

(v) *and*
$$\operatorname{tr}(\nabla_s \vec{f}\,) = \nabla_s \cdot \vec{f} \quad on\,\Gamma.$$

Proof.
(i)–(iv) are direct analogues of their flat counterparts, and follow easily from Definition 5.
(v) Using the fact that the trace is invariant under basis changes, we obtain from Remark 6(iii) and Definition 5(iii) that for $\vec{p} \in \Gamma$

$$\operatorname{tr}\left((\nabla_s \vec{f}\,)(\vec{p})\right) = \sum_{i=1}^{n} \vec{\tau}_i \cdot (\nabla_s \vec{f}\,)(\vec{p})\, \vec{\tau}_i = \sum_{i=1}^{n} \vec{\tau}_i \cdot (\partial_{\vec{\tau}_i} \vec{f}\,)(\vec{p}) = (\nabla_s \cdot \vec{f}\,)(\vec{p}),$$

where $\{\vec{\tau}_1, \ldots, \vec{\tau}_n\}$ is an orthonormal basis of $\mathrm{T}_{\vec{p}}\,\Gamma$. □

It is often convenient to consider alternative representations of the differential operators introduced in Definition 5.

Remark 8.

(i) For a local parameterization $\vec{\mathfrak{x}} : U \to \Gamma$ one can define the first fundamental form, or metric tensor, $(g_{ij})_{i,\, j=1,\ldots,n}$ as
$$g_{ij} = \vec{\partial}_i \cdot \vec{\partial}_j = \partial_i\, \vec{\mathfrak{x}} \cdot \partial_j\, \vec{\mathfrak{x}},$$
and a little linear algebra shows that
$$(\nabla_s f) \circ \vec{\mathfrak{x}} = \sum_{i,\, j=1}^{n} g^{ij}\, \partial_i\, (f \circ \vec{\mathfrak{x}})\, \vec{\partial}_j \quad \text{in}\, U,$$
$$(\nabla_s \cdot \vec{f}\,) \circ \vec{\mathfrak{x}} = \sum_{i,\, j=1}^{n} g^{ij}\, \partial_i\, (\vec{f} \circ \vec{\mathfrak{x}}) \cdot \vec{\partial}_j \quad \text{in}\, U,$$
where $(g^{ij})_{i,\, j=1,\ldots,n}$ is the inverse of $(g_{ij})_{i,\, j=1,\ldots,n}$. Moreover, it holds that
$$(\Delta_s f) \circ \vec{\mathfrak{x}} = \frac{1}{\sqrt{g}} \sum_{i,\, j=1}^{n} \partial_i\, (\sqrt{g}\, g^{ij}\, \partial_j\, (f \circ \vec{\mathfrak{x}})) \quad \text{in}\, U,$$
where
$$g = \det\left((g_{ij})_{i,\, j=1,\, \ldots,\, n}\right) \tag{2}$$
is the square of the local area element, see Amann and Escher (2009, Section XI.6).

(ii) Denoting by $\nabla_s = (\partial_{s_1}, \ldots, \partial_{s_d})^{\mathsf{T}}$ the surface gradient on Γ, it holds that $\nabla_s \cdot \vec{f} = \sum_{i=1}^{d} \partial_{s_i} (\vec{f} \cdot \vec{e}_i)$, $\nabla_s \vec{f} = (\partial_{s_j} \vec{f} \cdot \vec{e}_i)_{i,\, j=1}^{d}$ and $\nabla_s^2 f = (\partial_{s_j}\, \partial_{s_i} f)_{i,\, j=1}^{d}$. Moreover, it follows from Lemma 7(v) that the Laplace–Beltrami operator satisfies $\Delta_s = \sum_{i=1}^{d} \partial_{s_i}^2$.

(iii) If Γ is an orientable hypersurface with normal vector field $\vec{\nu}$, we obtain for an extension of the functions f and $\vec{f}$ to an open neighbourhood of Γ the formulas

$$\nabla_s f = \nabla f - (\nabla f \cdot \vec{\nu})\,\vec{\nu} \quad \text{and} \quad \nabla_s \cdot \vec{f} = \nabla \cdot \vec{f} - ((\nabla \vec{f})\,\vec{\nu}) \cdot \vec{\nu} \quad \text{on}\,\Gamma,$$

where ∇ and $\nabla\,\cdot$ are the gradient and the divergence in $\mathbb{R}^d$. These identities follow directly from Definitions 5(ii), 5(iii) and the representation of ∇f and $\nabla \cdot \vec{f}$ with the help of an orthonormal basis.

Lemma 9. *Let Γ be a C^1-hypersurface. Then the identity map $\vec{\mathrm{id}}$ satisfies the following.*

(i)

$$\nabla_s \vec{\mathrm{id}} = \underline{\underline{P}}_\Gamma \quad \text{on}\,\Gamma.$$

(ii) *For $\vec{f} \in [C^1(\Gamma)]^d$ it holds that*

$$\nabla_s \vec{\mathrm{id}} : \nabla_s \vec{f} = \nabla_s \cdot \vec{f} \quad \text{on}\,\Gamma,$$

where $\underline{\underline{A}} : \underline{\underline{B}} = \mathrm{tr}(\underline{\underline{A}}^{\mathrm{T}}\,\underline{\underline{B}})$ denotes the Hilbert–Schmidt inner product for matrices $\underline{\underline{A}}$ and $\underline{\underline{B}}$.

(iii) *If Γ is an orientable C^2-surface with normal vector field $\vec{\nu}$, then*

$$\Delta_s \vec{\mathrm{id}} = -\,(\nabla_s \cdot \vec{\nu})\,\vec{\nu} \quad \text{on}\,\Gamma.$$

Proof.
(i) The result follows from Definitions 5(iv), 5(i) and (1).
(ii) It follows from (i), Remark 6(iii) and Lemma 7(v) that $\nabla_s \vec{\mathrm{id}} : \nabla_s \vec{f} = \mathrm{tr}(\nabla_s \vec{f}) = \nabla_s \cdot \vec{f}$.
(iii) Using Remark 8(iii), and on recalling Definitions 5(vi) and 5(v), we compute for $i = 1, \ldots, d$

$$\begin{aligned}(\Delta_s \vec{\mathrm{id}}) \cdot \vec{e}_i &= \Delta_s\,(\vec{\mathrm{id}} \cdot \vec{e}_i) = \nabla_s \cdot \left[\nabla_s\,(\vec{\mathrm{id}} \cdot \vec{e}_i)\right] = \nabla_s \cdot [\vec{e}_i - (\vec{e}_i \cdot \vec{\nu})\,\vec{\nu}] \\ &= -\nabla_s \cdot [(\vec{\nu} \cdot \vec{e}_i)\,\vec{\nu}] = -(\vec{\nu} \cdot \vec{e}_i)\,\nabla_s \cdot \vec{\nu} \quad \text{on}\,\Gamma,\end{aligned}$$

where we have noted that $\nabla_s\,(\vec{\nu} \cdot \vec{e}_i)$ is orthogonal to $\vec{\nu}$. □

2.2 Curvature

We now define the fundamental curvature quantities needed for the material presented in the remainder of this chapter.

Definition 10. Let Γ be an orientable C^2-hypersurface with normal vector field $\vec{\nu}$.

(i) The Weingarten map at $\vec{p} \in \Gamma$ is defined through

$$\mathrm{W}_{\vec{p}} : \mathrm{T}_{\vec{p}}\,\Gamma \to \mathrm{T}_{\vec{p}}\,\Gamma, \quad \mathrm{W}_{\vec{p}}(\vec{\tau}) = -(\partial_{\vec{\tau}}\,\vec{\nu})(\vec{p}).$$

(ii) The corresponding bilinear form is called the second fundamental form and is given by

$$\mathrm{II}_{\vec{p}}(\vec{\tau}_1,\vec{\tau}_2)=\mathrm{W}_{\vec{p}}(\vec{\tau}_1)\cdot\vec{\tau}_2=-(\partial_{\vec{\tau}_1}\vec{\nu})(\vec{p})\cdot\vec{\tau}_2$$

for all $\vec{\tau}_1, \vec{\tau}_2 \in \mathrm{T}_{\vec{p}}\,\Gamma$, $\vec{p}\in\Gamma$. It can be shown that $\mathrm{W}_{\vec{p}}$ is self-adjoint, which implies that $\mathrm{II}_{\vec{p}}$ is symmetric, see, e.g., Kühnel (2015, Section 3.9). Hence there exists an orthonormal basis $\{\vec{\mathrm{t}}_1, \ldots, \vec{\mathrm{t}}_{d-1}\}$ of $\mathrm{T}_{\vec{p}}\,\Gamma$, consisting of eigenvectors of $\mathrm{W}_{\vec{p}}$ with corresponding eigenvalues $\varkappa_1, \ldots, \varkappa_{d-1}$.

Definition 11.

(i) The eigenvalues $\varkappa_1, \ldots, \varkappa_{d-1}$ of $\mathrm{W}_{\vec{p}}$ are called the principal curvatures of Γ at $\vec{p}$.

(ii) The mean curvature $\varkappa$ of Γ at $\vec{p}$ is defined to be the trace of $\mathrm{W}_{\vec{p}}$, i.e.,

$$\varkappa(\vec{p})=\mathrm{tr}\mathrm{W}_{\vec{p}}=\varkappa_1+\cdots+\varkappa_{d-1}.$$

(iii) The mean curvature vector $\vec{\varkappa}$ of Γ at $\vec{p}$ is defined as

$$\vec{\varkappa}(\vec{p})=\varkappa(\vec{p})\,\vec{\nu}(\vec{p}).$$

(iv) For $d=3$ the Gaussian curvature $\mathcal{K}$ at $\vec{p}$ is the determinant of $\mathrm{W}_{\vec{p}}$, i.e., the product of the principal curvatures. This means we set

$$\mathcal{K}(\vec{p})=\det\mathrm{W}_{\vec{p}}=\varkappa_1\,\varkappa_2.$$

Lemma 12. *Let Γ be an orientable C^2-hypersurface with normal vector field $\vec{\nu}$. The Jacobian $(\nabla_s\vec{\nu})(\vec{p})$, for $\vec{p}\in\Gamma$, induces a self-adjoint linear map from $\mathbb{R}^d$ to $\mathbb{R}^d$ that collapses to $-\mathrm{W}_{\vec{p}}$ on $T_{\vec{p}}\,\Gamma$ and that maps $N_{\vec{p}}\,\Gamma$ to zero. As a consequence, the following hold:*

(i)

$$(\nabla_s\vec{\nu})^{\mathsf{T}}\,\vec{\nu}=(\nabla_s\vec{\nu})\,\vec{\nu}=\vec{0}\quad\textit{on }\Gamma,$$

(ii)

$$(\nabla_s\vec{\nu})^{\mathsf{T}}=\nabla_s\vec{\nu}\quad\textit{on }\Gamma,$$

(iii)

$$\varkappa=-\,\mathrm{tr}(\nabla_s\vec{\nu})\quad\textit{on }\Gamma,$$

(iv) *and*

$$|\nabla_s\vec{\nu}|^2=\nabla_s\vec{\nu}:\nabla_s\vec{\nu}=\varkappa_1^2+\cdots+\varkappa_{d-1}^2\quad\textit{on }\Gamma.$$

Proof. Let $\vec{p}\in\Gamma$. It follows from Remark 6(iii) and Definition 10 that $\mathrm{W}_{\vec{p}}(\vec{\tau})=-(\nabla_s\vec{\nu})(\vec{p})\,\vec{\tau}=-((\nabla_s\vec{\nu})(\vec{p}))^{\mathsf{T}}\vec{\tau}$ for all $\vec{\tau}\in\mathrm{T}_{\vec{p}}\,\Gamma$. Moreover, on denoting $\vec{\nu}=(\nu_1, \ldots, \nu_d)^{\mathsf{T}}$, and on noting from Definition 5(iv) that the ith column of $(\nabla_s\vec{\nu})^{\mathsf{T}}$ is $\nabla_s\nu_i$, we have that

$$(\nabla_s\vec{\nu})^{\mathsf{T}}\,\vec{\nu}=\sum_{i=1}^{d}\nu_i\,\nabla_s\,\nu_i=\frac{1}{2}\,\nabla_s\,|\vec{\nu}|^2=0.$$

Combining this with Remark 6(iii) yields that $(\nabla_s \vec{\nu})(\vec{p})\,\vec{\nu} = ((\nabla_s \vec{\nu})(\vec{p}))^{\top}\,\vec{\nu} = \vec{0}$, and so the linear map induced by $(\nabla_s \vec{\nu})(\vec{p})$ is self-adjoint. This proves (i) and (ii). Furthermore, if $\{\vec{t}_1, \ldots, \vec{t}_{d-1}\}$ is an orthonormal basis of $\mathrm{T}_{\vec{p}}\,\Gamma$, consisting of eigenvectors of $\mathrm{W}_{\vec{p}}$ with corresponding eigenvalues $\varkappa_1, \ldots, \varkappa_{d-1}$, then $\{\vec{t}_1, \ldots, \vec{t}_{d-1}, \vec{\nu}(\vec{p})\}$ is an orthonormal basis of $\mathbb{R}^d$, consisting of eigenvectors of $(\nabla_s \vec{\nu})(\vec{p})$ with corresponding eigenvalues $\varkappa_1, \ldots, \varkappa_{d-1}, 0$. This implies (iii), on recalling Definition 11(ii), and (iv). □

The next lemma shows that while the sign of the mean curvature $\varkappa$ depends on the choice of the unit normal $\vec{\nu}$, the mean curvature vector $\vec{\varkappa}$ is an invariant under the change of the sign of the normal.

Lemma 13. *Let Γ be an orientable C^2-hypersurface with normal vector field $\vec{\nu}$. The following formulas for the mean curvature and the mean curvature vector hold true.*

(i) *For the mean curvature it holds that*

$$\varkappa = -\nabla_s \cdot \vec{\nu} \quad on\,\Gamma.$$

(ii) *For the mean curvature vector it holds that*

$$\vec{\varkappa} = \varkappa\,\vec{\nu} = \Delta_s\,\vec{\mathrm{id}} \quad on\,\Gamma.$$

Proof.
(i) This follows from Lemma 12(iii) and Lemma 7(v).
(ii) The result follows directly from (i), Definition 11(iii) and Lemma 9(iii). □

Lemma 14. *Let Γ be an orientable C^2-hypersurface with normal vector field $\vec{\nu}$. Let $\vec{f} \in [C^1(\Gamma)]^d$ and set $\vec{f}_{\mathrm{T}} = \underline{\underline{P}}_{\Gamma}\vec{f}$ on Γ. Then it holds that*

(i)
$$\nabla_s \cdot \vec{f} = -(\vec{f}\cdot\vec{\nu})\,\varkappa + \nabla_s \cdot \vec{f}_{\mathrm{T}} \quad on\,\Gamma,$$

(ii)
$$\nabla_s \vec{f} = (\vec{f}\cdot\vec{\nu})\,\nabla_s \vec{\nu} + \vec{\nu} \otimes \nabla_s (\vec{f}\cdot\vec{\nu}) + \nabla_s \vec{f}_{\mathrm{T}} \quad on\,\Gamma,$$

(iii)
$$\underline{\underline{D}}_s(\vec{f}) = (\vec{f}\cdot\vec{\nu})\,\nabla_s \vec{\nu} + \frac{1}{2}\,(\underline{\underline{P}}_{\Gamma}\,\nabla_s \vec{f}_{\mathrm{T}} + (\nabla_s \vec{f}_{\mathrm{T}})^{\top}\,\underline{\underline{P}}_{\Gamma}) \quad on\,\Gamma.$$

Proof.
(i) Using Lemma 7(i), Definition 5(ii) and Lemma 13(i), we compute that

$$\begin{aligned}\nabla_s \cdot \vec{f} &= \nabla_s \cdot \left((\vec{f}\cdot\vec{\nu})\,\vec{\nu}\right) + \nabla_s \cdot \vec{f}_{\mathrm{T}} = (\vec{f}\cdot\vec{\nu})\,\nabla_s \cdot \vec{\nu} + \nabla_s \cdot \vec{f}_{\mathrm{T}} \\ &= -(\vec{f}\cdot\vec{\nu})\,\varkappa + \nabla_s \cdot \vec{f}_{\mathrm{T}}.\end{aligned}$$

(ii) Similarly, it follows from Lemma 7(iii) that

$$\nabla_s \vec{f} = \nabla_s \left((\vec{f} \cdot \vec{\nu})\, \vec{\nu} \right) + \nabla_s \vec{f}_{\rm T} = (\vec{f} \cdot \vec{\nu})\, \nabla_s \vec{\nu} + \vec{\nu} \otimes \nabla_s\, (\vec{f} \cdot \vec{\nu}) + \nabla_s \vec{f}_{\rm T}.$$

(iii) The desired result follows immediately from (ii), Definitions 5(iv) and 5(vii) and the fact that $\nabla_s \vec{\nu}$ is a symmetric mapping that maps tangent vectors to tangent vectors, recall Lemma 12. □

In the derivation of some relevant formulas it is sometimes helpful to be able to extend functions defined on a hypersurface Γ to a neighbourhood of Γ. This then frequently allows us to use the calculus in $\mathbb{R}^d$ to compute for quantities on the surface, recall Remark 8(iii). Let $\Gamma \subset \mathbb{R}^d$ be a compact orientable C^2-hypersurface without boundary and with normal vector field $\vec{\nu}$. For $\delta > 0$ we define a tubular neighbourhood

$$\mathcal{N}_\delta = \{\vec{z} \in \mathbb{R}^d : \vec{z} = \vec{p} + \eta\, \vec{\nu}(\vec{p}), \quad \vec{p} \in \Gamma, \quad |\eta| < \delta\}$$

of Γ. In Gilbarg and Trudinger (1983, Appendix 14.6) it is shown that there is a bijective relation between $\vec{z} \in \mathcal{N}_\delta$ and $(\vec{p}, \eta) \in \Gamma \times (-\delta, \delta)$, provided that δ is small enough. We can hence define the functions $\vec{\Pi}_\Gamma(\vec{z}) = \vec{p}$ and $d_\Gamma(\vec{z}) = \eta$, and it turns out that $|d_\Gamma(\vec{z})|$ is the distance of $\vec{z}$ to Γ, where $d_\Gamma(\vec{z})$ is positive if $\vec{z}$ lies on the side towards which $\vec{\nu}$ is pointing, and negative otherwise. We call d_Γ the signed distance function to Γ, and $\vec{\Pi}_\Gamma(\vec{z})$ is the projection of $\vec{z}$ onto Γ, i.e.,

$$\vec{\Pi}_\Gamma(\vec{z}) = \arg\min_{\vec{y} \in \Gamma} |\vec{y} - \vec{z}|.$$

For later use, we recall from Gilbarg and Trudinger (1983, Appendix 14.6) that, for δ sufficiently small,

$$d_\Gamma \in C^2(\mathcal{N}_\delta). \tag{3a}$$

It also holds that

$$\nabla\, d_\Gamma = \vec{\nu} \circ \vec{\Pi}_\Gamma \quad \Rightarrow \quad |\nabla\, d_\Gamma| = 1 \quad \text{in}\, \mathcal{N}_\delta, \tag{3b}$$

which can be shown as follows. It clearly holds that

$$\vec{\rm id} = \vec{\Pi}_\Gamma + d_\Gamma\, \vec{\nu} \circ \vec{\Pi}_\Gamma \quad \text{and} \quad d_\Gamma = \left(\vec{\rm id} - \vec{\Pi}_\Gamma \right) \cdot (\vec{\nu} \circ \vec{\Pi}_\Gamma) \quad \text{in}\, \mathcal{N}_\delta.$$

Differentiating the second identity, and observing the first, yields for $k = 1, \ldots, d$

$$\begin{aligned}
\partial_k\, d_\Gamma &= (\vec{e}_k - \partial_k\, \vec{\Pi}_\Gamma) \cdot (\vec{\nu} \circ \vec{\Pi}_\Gamma) - \left(\vec{\rm id} - \vec{\Pi}_\Gamma \right) \cdot \partial_k (\vec{\nu} \circ \vec{\Pi}_\Gamma) \\
&= \vec{e}_k \cdot \vec{\nu} \circ \vec{\Pi}_\Gamma - d_\Gamma (\vec{\nu} \circ \vec{\Pi}_\Gamma) \cdot \partial_k (\vec{\nu} \circ \vec{\Pi}_\Gamma) \\
&= \vec{e}_k \cdot \vec{\nu} \circ \vec{\Pi}_\Gamma - d_\Gamma\, \frac{1}{2}\, \partial_k |\vec{\nu} \circ \vec{\Pi}_\Gamma|^2 = \vec{e}_k \cdot \vec{\nu} \circ \vec{\Pi}_\Gamma \quad \text{in}\, \mathcal{N}_\delta,
\end{aligned}$$

where we have noted that $\partial_k\, \vec{\Pi}_\Gamma$ is tangential in $\mathcal{N}_\delta$. This proves (3b).

We now extend a function f defined on Γ to $\mathcal{N}_\delta$, for δ sufficiently small, via

$$f = f \circ \vec{\Pi}_\Gamma \quad \text{in}\, \mathcal{N}_\delta. \tag{4}$$

This means that we extend f constantly in the normal direction, and so we obtain from Remark 8(iii) that

$$\nabla f \cdot \vec{\nu} = 0, \quad \text{and hence} \quad \nabla f = \nabla_s f \quad \text{on } \Gamma. \tag{5}$$

In the flat case the Schwarz theorem yields that the Hessian is symmetric. In the curved case, however, this is no longer the case. In fact, the following result holds.

Lemma 15. *Let Γ be an orientable C^2-hypersurface with normal vector field $\vec{\nu}$, and let $f \in C^2(\Gamma)$. For the surface Hessian it holds that*

$$\nabla_s^2 f - (\nabla_s^2 f)^{\top} = [(\nabla_s \vec{\nu})\, \nabla_s f] \otimes \vec{\nu} - \vec{\nu} \otimes [(\nabla_s \vec{\nu})\, \nabla_s f] \quad \textit{on } \Gamma.$$

Proof. Recalling from Remark 8(ii) the notation $\nabla_s = (\partial_{s_1}, \ldots, \partial_{s_d})^{\top}$, and denoting $\vec{\nu} = (\nu_1, \ldots, \nu_d)^{\top}$, the claim can be equivalently written as

$$\partial_{s_j} \partial_{s_i} f - \partial_{s_i} \partial_{s_j} f = [(\nabla_s \vec{\nu})\, \nabla_s f]_i\, \nu_j - [(\nabla_s \vec{\nu})\, \nabla_s f]_j\, \nu_i \quad \text{on}\, \Gamma, \tag{6}$$

for all $i, j \in \{1, \ldots, d\}$. In order to prove this, we extend f to a neighbourhood of Γ as in (4). It follows from (5) that $\partial_{s_i}(\nabla f \cdot \vec{\nu}) = 0$ on Γ, and so we compute, using Remark 8(iii), for all $i, j = 1, \ldots, d$ that

$$\partial_{s_i} \partial_{s_j} f = \partial_{s_i} (\partial_j f - (\nabla f \cdot \vec{\nu})\, \nu_j) = \partial_{s_i} \partial_j f = \partial_i \partial_j f - (\nabla \partial_j f \cdot \vec{\nu})\, \nu_i.$$

In addition we have, on using the Schwarz theorem and on extending $\vec{\nu}$ to the neighbourhood of Γ similarly to (4), and so (5) yields $\partial_j \vec{\nu} = \partial_{s_j} \vec{\nu}$, that

$$\begin{aligned} 0 = \partial_{s_j} (\nabla f \cdot \vec{\nu}) &= \partial_j (\nabla f \cdot \vec{\nu}) - (\nabla (\nabla f \cdot \vec{\nu}) \cdot \vec{\nu}) \nu_j \\ &= \nabla \partial_j f \cdot \vec{\nu} + \nabla f \cdot \partial_j \vec{\nu} - (\nabla (\nabla f \cdot \vec{\nu}) \cdot \vec{\nu}) \nu_j = \nabla \partial_j f \cdot \vec{\nu} + \nabla_s f \cdot \partial_{s_j} \vec{\nu}. \end{aligned}$$

Combining the above with the Schwarz theorem and Lemma 12(ii) yields that

$$\begin{aligned} \partial_{s_j} \partial_{s_i} f - \partial_{s_i} \partial_{s_j} f &= (\nabla_s f \cdot \partial_{s_i} \vec{\nu})\, \nu_j - (\nabla_s f \cdot \partial_{s_j} \vec{\nu})\, \nu_i \\ &= (\nabla_s f \cdot \nabla_s \nu_i)\, \nu_j - (\nabla_s f \cdot \nabla_s \nu_j)\, \nu_i \\ &= [(\nabla_s \vec{\nu})\, \nabla_s f]_i\, \nu_j - [(\nabla_s \vec{\nu})\, \nabla_s f]_j\, \nu_i. \end{aligned}$$

This proves (6). □

Lemma 16. *Let Γ be an orientable C^3-hypersurface with normal vector field $\vec{\nu}$. It holds that*

$$\nabla_s \varkappa = -\Delta_s \vec{\nu} - |\nabla_s \vec{\nu}|^2\, \vec{\nu} \quad \textit{on}\, \Gamma.$$

Proof. Using the same notation as in the proof of Lemma 15, it follows from Lemmas 12(ii), 15, 13(i) and 12(i) that, for $j = 1, \ldots, d$,

$$
\begin{aligned}
\Delta_s \nu_j &= \sum_{i=1}^{d} \partial_{s_i} \partial_{s_i} \nu_j = \sum_{i=1}^{d} \partial_{s_i} \partial_{s_j} \nu_i \\
&= \sum_{i=1}^{d} \left(\partial_{s_j} \partial_{s_i} \nu_i - [(\nabla_s \vec{\nu}) \nabla_s \nu_i]_i \nu_j + [(\nabla_s \vec{\nu}) \nabla_s \nu_i]_j \nu_i \right) \\
&= \partial_{s_j} (\nabla_s \cdot \vec{\nu}) - \nu_j \sum_{i=1}^{d} |\nabla_s \nu_i|^2 + (\nabla_s \nu_j) \cdot \sum_{i=1}^{d} (\nabla_s \nu_i) \nu_i \\
&= \partial_{s_j} (\nabla_s \cdot \vec{\nu}) - \nu_j |\nabla_s \vec{\nu}|^2 + (\nabla_s \nu_j) \cdot (\nabla_s \vec{\nu})^{\mathsf{T}} \vec{\nu} = -\partial_{s_j} \varkappa - \nu_j |\nabla_s \vec{\nu}|^2.
\end{aligned}
$$

This yields the claim. □

2.3 The divergence theorem

Definition 17.

(i) A subset $\Gamma \subset \mathbb{R}^d$ is called an n-dimensional C^k-surface with boundary, for $1 \leq n \leq d$ and $k \geq 1$, if for each point $\vec{p} \in \Gamma$ one of the following conditions is satisfied.

  (a) There exist an open neighbourhood $V \subset \mathbb{R}^d$ of $\vec{p}$ and a bijective C^k-map $\vec{\mathfrak{x}} : U \to V \cap \Gamma$, with $U \subset \mathbb{R}^n$ open and connected, such that the Jacobian $\nabla \vec{\mathfrak{x}} : U \to \mathbb{R}^{d \times n}$ of $\vec{\mathfrak{x}}$ has full rank in U.

  (b) There exist open and connected sets $U \subset \mathbb{R}^n$, $V \subset \mathbb{R}^d$, with $\vec{0} \in U$ and $\vec{p} \in V$, and an injective C^k-map $\vec{\mathfrak{x}} : U \to \mathbb{R}^d$, such that $\vec{\mathfrak{x}}(\vec{0}) = \vec{p}$ and

$$\vec{\mathfrak{x}}(U \cap (\mathbb{R}^{n-1} \times \mathbb{R}_{\geq 0})) = V \cap \Gamma.$$

(ii) The maps $\vec{\mathfrak{x}} : U \to \mathbb{R}^d$ in (ia) and (ib) are called local parameterizations of Γ. In the latter case, we call $\vec{\mathfrak{x}}$ a local boundary parameterization of Γ.

(iii) If $n = d - 1$, then we call Γ a hypersurface with boundary.

Remark 18.

(i) A boundary point of Γ is a point on Γ for which the second condition in the definition is fulfilled. The set of all boundary points is called the boundary of Γ and is denoted by $\partial\Gamma$.

(ii) If Γ is a C^k-hypersurface with boundary, its boundary $\partial\Gamma$ is either empty or a $(d-2)$-dimensional surface without boundary, see Agricola and Friedrich (2002, Section 3.1) for details.

(iii) If $\partial\Gamma$ is empty, we say that Γ is a surface without boundary. If, in addition, Γ is compact, then we call Γ a closed surface. Here we note that Definition 17 implies that any bounded hypersurface is compact.

(iv) All the definitions and results in Sections 2.1 and 2.2 have been stated for $\vec{p} \in \Gamma \backslash \partial\Gamma$. However, they easily generalize to $\vec{p} \in \partial\Gamma$ for a surface with boundary. The only required changes are as follows. In Definition 2, U is replaced by $U \cap (\mathbb{R}^{n-1} \times \mathbb{R}_{\geq 0})$ for local boundary parameterizations $\vec{\mathfrak{x}}$, and similarly in Remark 8(i). In addition, in Definition 5(i), we choose a curve $\vec{y} : (-1,0] \to \Gamma$, or $\vec{y} : [0,1) \to \Gamma$, and take the natural one-sided limit in the definition of $(\partial_{\vec{\tau}} f)(\vec{p})$.

Definition 19. Let Γ be an n-dimensional C^1-surface in $\mathbb{R}^d$, $f : \Gamma \to \mathbb{R}$ a function and let $\vec{\mathfrak{x}} : U \to \Gamma$ be a local parameterization of $\Gamma \setminus \partial\Gamma$. Then, on recalling (2), we define

$$\int_{\vec{\mathfrak{x}}(U)} f \, \mathrm{d}\mathcal{H}^n = \int_U f \circ \vec{\mathfrak{x}} \, \sqrt{g} \, \mathrm{d}\mathcal{L}^n, \tag{7}$$

for all functions f such that $f \circ \vec{\mathfrak{x}}$ is integrable, where $\mathcal{L}^n$ is the n-dimensional Lebesgue measure. For a local boundary parameterization $\vec{\mathfrak{x}}$ of Γ, we replace U with $U \cap (\mathbb{R}^{n-1} \times \mathbb{R}_{\geq 0})$ in (7). The integral $\int_\Gamma f \, \mathrm{d}\mathcal{H}^n$ is defined using a partition of unity and the definition (7), see Amann and Escher (2009, Section XII.1) for details. The induced measure, defined by $\mathcal{H}^n(\Gamma) = \int_\Gamma 1 \, \mathrm{d}\mathcal{H}^n$, is called the n-dimensional Hausdorff measure in $\mathbb{R}^d$.

Definition 20. For a C^1-hypersurface Γ with boundary it holds that for $\vec{p} \in \partial\Gamma$ the tangent space $\mathrm{T}_{\vec{p}}\, \partial\Gamma$ is $(d-2)$-dimensional, $\mathrm{T}_{\vec{p}}\, \Gamma$ is $(d-1)$-dimensional and $\mathrm{T}_{\vec{p}}\, \partial\Gamma \subset \mathrm{T}_{\vec{p}}\, \Gamma$. We can hence choose a unique vector $\vec{\mu}(\vec{p}) \in \mathrm{T}_{\vec{p}}\, \Gamma$, which we call the outer unit conormal, such that

(i) $|\vec{\mu}(\vec{p})| = 1$,
(ii) $\vec{\mu}(\vec{p}) \in \mathrm{N}_{\vec{p}}\, \partial\Gamma$,
(iii) there exists a curve $\vec{y} : (-1,0] \to \Gamma$ on Γ with $\vec{y}(0) = \vec{p}$ and $\vec{y}\,'(0) = \vec{\mu}(\vec{p})$.

We will frequently use the following generalization of the divergence theorem on hypersurfaces.

Theorem 21. *Let Γ be a compact orientable C^2-hypersurface with normal vector field $\vec{\nu}$, and let $\vec{f} \in [C^1(\Gamma)]^d$. Then it holds that*

$$\int_\Gamma \nabla_s \cdot \vec{f} + \varkappa \vec{f} \cdot \vec{\nu} \, \mathrm{d}\mathcal{H}^{d-1} = \int_{\partial\Gamma} \vec{f} \cdot \vec{\mu} \, \mathrm{d}\mathcal{H}^{d-2},$$

where $\vec{\mu}$ is the outer unit conormal to $\partial\Gamma$.

Proof. The result is well known for tangential vector fields $\vec{f}$, for which $\varkappa \vec{f} \cdot \vec{\nu}$ is not present. This special case can be shown on surfaces similarly to the case of flat domains in $\mathbb{R}^d$, see, e.g., Amann and Escher (2009, Section XII.3) or Agricola and Friedrich (2002, Section 3.8). In the case of a nontangential vector field $\vec{f}$, we define the tangential vector field

$$\vec{h} = \vec{f} - (\vec{f} \cdot \vec{\nu})\, \vec{\nu}$$

and compute, on using Lemma 7(i), $\nabla_s (\vec{f} \cdot \vec{\nu})$ being orthogonal to $\vec{\nu}$ and Lemma 13(i), that

$$\nabla_s \cdot \vec{h} = \nabla_s \cdot \vec{f} - (\vec{f} \cdot \vec{\nu})\, \nabla_s \cdot \vec{\nu} = \nabla_s \cdot \vec{f} + \varkappa \vec{f} \cdot \vec{\nu}.$$

Using the divergence theorem for $\vec{h}$, taking into account that $\vec{h} \cdot \vec{\mu} = \vec{f} \cdot \vec{\mu}$, we obtain the assertion of the theorem. □

Remark 22.

(i) The following integration by parts rule is a direct consequence of Theorem 21, Lemma 7(i) and Definition 5(ii). Let $f \in C^2(\Gamma)$ and $\eta \in C^1(\Gamma)$. Then it holds that

$$\int_\Gamma \eta\, \Delta_s f + \nabla_s f \cdot \nabla_s \eta \; d\mathcal{H}^{d-1} = \int_{\partial\Gamma} \eta\, \nabla_s f \cdot \vec{\mu} \; d\mathcal{H}^{d-2}.$$

(ii) Moreover, for $f \in C^1(\Gamma)$ we obtain the Gauss–Green formula

$$\int_\Gamma \nabla_s f + f\, \varkappa\, \vec{\nu} \; d\mathcal{H}^{d-1} = \int_{\partial\Gamma} f\, \vec{\mu} \; d\mathcal{H}^{d-2},$$

by choosing $\vec{f} = f\, \vec{e}_i$, $i = 1, \ldots, d$, in Theorem 21, and applying Lemma 7(i).

(iii) Let $\vec{f} \in [C^2(\Gamma)]^d$ and $\vec{\eta} \in [C^1(\Gamma)]^d$. Then it follows from (i) and Remark 6(iii) that

$$\int_\Gamma \vec{\eta} \cdot \Delta_s \vec{f} + \nabla_s \vec{f} : \nabla_s \vec{\eta} \; d\mathcal{H}^{d-1} = \int_{\partial\Gamma} \vec{\eta} \cdot (\nabla_s \vec{f})\, \vec{\mu} \; d\mathcal{H}^{d-2}.$$

On noting that for symmetric matrices $\underline{\underline{A}} \in \mathbb{R}^{d\times d}$ it holds that $\underline{\underline{P}}_\Gamma \underline{\underline{A}}\, \underline{\underline{P}}_\Gamma : \underline{\underline{B}} = \underline{\underline{P}}_\Gamma \underline{\underline{A}}\, \underline{\underline{P}}_\Gamma : \frac{1}{2} \underline{\underline{P}}_\Gamma (\underline{\underline{B}} + \underline{\underline{B}}^\top)\, \underline{\underline{P}}_\Gamma$ on Γ for all $\underline{\underline{B}} \in \mathbb{R}^{d\times d}$, we obtain furthermore that

$$\int_\Gamma \vec{\eta} \cdot \left(\nabla_s \cdot \underline{\underline{D}}_s(\vec{f}) \right) + \underline{\underline{D}}_s(\vec{f}) : \underline{\underline{D}}_s(\vec{\eta}) \; d\mathcal{H}^{d-1} = \int_{\partial\Gamma} \vec{\eta} \cdot \underline{\underline{D}}_s(\vec{f})\, \vec{\mu} \; d\mathcal{H}^{d-2},$$

where we have recalled Definition 5(vii) and used Theorem 21 with Lemma 7(iv) and Remark 6(iv).

(iv) Of fundamental importance in the development of numerical approximations for curvature-driven evolution equations is the following identity. Let $\vec{\eta} \in [C^1(\Gamma)]^d$. Then it follows from (iii), Lemmas 13(ii), 9(i) and Definition 20 that

$$\int_\Gamma \varkappa\, \vec{\eta} \cdot \vec{\nu} + \nabla_s \, \vec{\mathrm{id}} : \nabla_s \vec{\eta} \; d\mathcal{H}^{d-1} = \int_{\partial\Gamma} \vec{\eta} \cdot \vec{\mu} \; d\mathcal{H}^{d-2}.$$

(v) For numerical approximations of the Weingarten map, recall Definition 10(i), the following identity is of use. Let $\underline{\eta} \in [C^1(\Gamma)]^{d\times d}$. Then it follows from Theorem 21, with $\vec{f} = \underline{\eta}^{\top}\,\vec{\nu}$, on recalling Lemma 7(iv), that

$$\int_{\Gamma} \nabla_s \vec{\nu} : \underline{\eta} + \vec{\nu} \cdot [\varkappa\, \underline{\eta}\, \vec{\nu} + \nabla_s \cdot \underline{\eta}] \, \mathrm{d}\mathcal{H}^{d-1} = \int_{\partial\Gamma} \vec{\nu} \cdot \underline{\eta}\, \vec{\mu} \, \mathrm{d}\mathcal{H}^{d-2}.$$

2.4 Evolving surfaces and transport theorems

We are mainly interested in evolving hypersurfaces. Hence we now consider hypersurfaces which evolve in a time interval $[0, T]$ and define the term evolving hypersurface.

Definition 23.

(i) Let $(\Gamma(t))_{t\in[0,T]}$ be a family of C^k-hypersurfaces (with or without boundary), for $k \geq 1$. The set

$$\mathcal{G}_T = \bigcup_{t\in[0,T]} (\Gamma(t) \times \{t\})$$

is called a C^k-evolving hypersurface if it is a C^k-hypersurface with boundary in $\mathbb{R}^{d+1}$, such that $\mathrm{T}_{(\vec{p},t)}\, \mathcal{G}_T \neq \mathbb{R}^d \times \{0\}$ for all $(\vec{p}, t) \in \mathcal{G}_T$. We will often identify $\mathcal{G}_T$ with $(\Gamma(t))_{t\in[0,T]}$, and call the latter also a C^k-evolving hypersurface.

(ii) Let $\mathcal{G}_T$ be a C^1-evolving hypersurface and $(\vec{p}_0, t_0) \in \mathcal{G}_T$. We assume that $\mathcal{G}_T$ allows for a continuous vector field $\vec{\nu} : \mathcal{G}_T \to \mathbb{R}^d$, such that $\vec{\nu}(\cdot, t)$ is a unit normal to $\Gamma(t)$. Furthermore, let $\vec{y} : (t_0 - \delta, t_0 + \delta) \to \mathbb{R}^d$, for some $\delta > 0$, with $\vec{y}(t) \in \Gamma(t)$ and $\vec{y}(t_0) = \vec{p}_0$ be a smooth curve in $\mathbb{R}^d$. Then the normal velocity of $\Gamma(t_0)$ at $\vec{p}_0$ is defined as

$$\mathcal{V}(\vec{p}_0, t_0) = \vec{\nu}(\vec{p}_0, t_0) \cdot \vec{y}\,'(t_0).$$

(iii) Let $\mathcal{G}_T$ be a C^k-evolving hypersurface satisfying the assumptions in (ii). Then we call $\mathcal{G}_T$ a C^k-evolving orientable hypersurface.

Remark 24. Let $\mathcal{G}_T$ be a C^1-evolving orientable hypersurface.

(i) The condition $\mathrm{T}_{(\vec{p},t)}\, \mathcal{G}_T \neq \mathbb{R}^d \times \{0\}$, for all $(\vec{p}, t) \in \mathcal{G}_T$, guarantees the existence of a curve $\vec{y}$ in Definition 23(ii), recall Remark 6(i).

(ii) It is easy to show that $\mathcal{V}$ does not depend on the curve $\vec{y}$. To see this, we note that $(\vec{y}\,'(t_0), 1) \in \mathrm{T}_{(\vec{p}_0,t_0)}\, \mathcal{G}_T$. Hence, on letting $\{\vec{\tau}_1, \ldots, \vec{\tau}_{d-1}\}$ denote a basis of $\mathrm{T}_{\vec{p}_0}\, \Gamma(t_0)$, we have that $\{(\vec{\tau}_1, 0), \ldots, (\vec{\tau}_{d-1}, 0), (\vec{y}\,'(t_0), 1)\}$ is a basis of $\mathrm{T}_{(\vec{p}_0,t_0)}\, \mathcal{G}_T$. Then it follows from Definition 23(ii) that $\vec{y}\,'(t_0) = \mathcal{V}(\vec{p}_0, t_0)\, \vec{\nu}(\vec{p}_0, t_0) + \sum_{i=1}^{d-1} \alpha_i\, \vec{\tau}_i$ for some $\alpha_i \in \mathbb{R}$, $i = 1, \ldots, d-1$, and so

$(\mathcal{V}(\vec{p}_0,t_0)\,\vec{\nu}(\vec{p}_0,t_0),1)\in \mathrm{T}_{(\vec{p}_0,t_0)}\,\mathcal{G}_T$. In addition, we observe that there exists a unique vector $(\vec{\omega},1)\in \mathrm{T}_{(\vec{p}_0,t_0)}\,\mathcal{G}_T$ with $\vec{\omega}$ parallel to $\vec{\nu}(\vec{p}_0,t_0)$. Therefore we have that $\vec{\omega} = \mathcal{V}(\vec{p}_0,t_0)\,\vec{\nu}(\vec{p}_0,t_0)$, and so $\mathcal{V}(\vec{p}_0,t_0)$ does not depend on $\vec{y}$.

(iii) Moreover, we observe that $(1+\mathcal{V}^2)^{-\frac{1}{2}}(\vec{\nu},-\mathcal{V})$ is a continuous normal vector field on $\mathcal{G}_T$, and so $\mathcal{G}_T$ is an orientable hypersurface in $\mathbb{R}^{d+1}$.

(iv) If $\mathcal{G}_T$ is a C^2-evolving hypersurface, then it is even easier to show that $\mathcal{V}$ does not depend on the curve $\vec{y}$. In order to do so, we choose a C^1-function f defined in a small neighbourhood of $\mathcal{G}_T$ such that $f=0$ on $\mathcal{G}_T$, and such that $(\nabla\, f)(\vec{z},t)\neq\vec{0}$ for all $(\vec{z},t)\in\mathcal{G}_T$. A possible choice for $f(\cdot,t)$ is the signed distance function $d_\Gamma(\cdot,t)$ to $\Gamma(t)$, recall (3). We then compute for a curve $\vec{y}$ as in Definition 23(ii) that

$$0=\frac{\mathrm{d}}{\mathrm{d}t}\,f(\vec{y}(t),t)=(\nabla f)(\vec{y}(t),t)\cdot\vec{y}^{\,\prime}(t)+(\partial_t\, f)(\vec{y}(t),t).$$

As $\vec{\nu}=\nabla f/|\nabla f|$, we obtain

$$\mathcal{V}(\vec{p}_0,t_0)=\vec{\nu}(\vec{p}_0,t_0)\cdot\vec{y}^{\,\prime}(t_0)=-\left(\frac{\partial_t\, f}{|\nabla f|}\right)(\vec{p}_0,t_0),$$

where the right-hand side does not depend on $\vec{y}$.

Typically we will consider evolving hypersurfaces that are given by a global parameterization as follows.

Definition 25. Let $\mathcal{G}_T$ be a C^k-evolving hypersurface, and let Υ be a C^k-hypersurface in $\mathbb{R}^d$, with $k\geq 1$.

(i) A C^k-map $\vec{x}:\Upsilon\times[0,T]\to\mathbb{R}^d$ such that $\vec{x}(\,\cdot\,,t)$ is a diffeomorphism from Υ to $\Gamma(t)$ for all $t\in[0,T]$ is called a global parameterization of $\mathcal{G}_T$.

(ii) The full velocity of $\Gamma(t)$ on $\mathcal{G}_T$, induced by the parameterization $\vec{x}$, is defined by

$$\vec{\mathcal{V}}(\vec{x}(\vec{q},t),t)=(\partial_t\,\vec{x})(\vec{q},t)\quad\forall\,(\vec{q},t)\in\Upsilon\times[0,T].$$

(iii) The tangential velocity of $\Gamma(t)$ on $\mathcal{G}_T$, induced by the parameterization $\vec{x}$, is defined by

$$\vec{\mathcal{V}}_{\mathrm{T}}=\underline{\underline{P}}_\Gamma\,\vec{\mathcal{V}}\quad\text{on}\,\Gamma(t),$$

recall (1).

(iv) We define the rate of deformation tensor of $\Gamma(t)$ by

$$\underline{\underline{D}}_s(\vec{\mathcal{V}})\quad\text{on}\,\Gamma(t),$$

recall Definition 5(vii).

Remark 26.

(i) For the normal velocity of an evolving orientable hypersurface $\mathcal{G}_T$ defined in Definition 23(ii) it holds that

$$\mathcal{V} = \vec{\mathcal{V}} \cdot \vec{\nu} \quad \text{on } \mathcal{G}_T.$$

Hence we have that

$$\vec{\mathcal{V}} = \mathcal{V}\,\vec{\nu} + \vec{\mathcal{V}}_{\mathrm{T}} \quad \text{on } \mathcal{G}_T.$$

Here we stress that $\mathcal{V}$ does not depend on the parameterization $\vec{x}$, while $\vec{\mathcal{V}}_{\mathrm{T}}$ clearly does.

(ii) The expression "rate of deformation tensor" for $\underline{\underline{D}}_s(\vec{\mathcal{V}})$ is justified, because it encodes how $\Gamma(t)$ is locally deformed due to the motion induced by $\vec{x}$. This will be made rigorous in Lemma 30.

For the velocity field $\vec{\mathcal{V}}$ on $\mathcal{G}_T$ we have the following properties. Here and throughout, for notational convenience, we often identify $\Gamma(t) \times \{t\}$ with $\Gamma(t)$.

Lemma 27. *Let $\mathcal{G}_T$ be a C^2-evolving orientable hypersurface with a global parameterization leading to the velocity field $\vec{\mathcal{V}}$.*

(i) *It holds that*

$$\nabla_s \cdot \vec{\mathcal{V}} = -\mathcal{V}\,\varkappa + \nabla_s \cdot \vec{\mathcal{V}}_{\mathrm{T}} \quad \text{on } \Gamma(t).$$

(ii) *Moreover, it holds that*

$$\nabla_s \vec{\mathcal{V}} = \mathcal{V}\,\nabla_s \vec{\nu} + \vec{\nu} \otimes \nabla_s \mathcal{V} + \nabla_s \vec{\mathcal{V}}_{\mathrm{T}} \quad \text{on } \Gamma(t).$$

(iii) *Finally, for the rate of deformation tensor it holds that*

$$\underline{\underline{D}}_s(\vec{\mathcal{V}}) = \mathcal{V}\,\nabla_s \vec{\nu} + \frac{1}{2}\left(\underline{\underline{P}}_\Gamma \nabla_s \vec{\mathcal{V}}_{\mathrm{T}} + (\nabla_s \vec{\mathcal{V}}_{\mathrm{T}})^{\mathsf{T}} \underline{\underline{P}}_\Gamma\right) \quad \text{on } \Gamma(t).$$

Proof. The desired results follow immediately from Lemma 14, on noting $\vec{\mathcal{V}}(\cdot, t) \in [C^1(\Gamma(t))]^d$ and Remark 26(i). □

It is often convenient to also consider local parameterizations of $\Gamma(t)$, as defined in Definition 17. To this end, let $\vec{\varphi} : U \to \mathbb{R}^d$, $U \subset \mathbb{R}^{d-1}$ open and connected, be a local parameterization of Υ. Then

$$\vec{\mathfrak{x}}(t) = \vec{x}(\cdot, t) \circ \vec{\varphi} \quad \text{in } U, \;\; t \in [0, T], \tag{8}$$

defines a local parameterization of $\Gamma(t)$.

We now define the time derivative of a function $f : \mathcal{G}_T \to \mathbb{R}$. We cannot differentiate $f(\vec{p}, t)$ directly with respect to t due to the fact that $\vec{p}$ might not lie on $\Gamma(t)$ for different times t. When differentiating with respect to t, we need to move the point $\vec{p}$. There is hence some ambiguity in defining the time derivative.

Definition 28. Let $\mathcal{G}_T$ be a C^1-evolving hypersurface with a global parameterization $\vec{x} : \Upsilon \times [0, T] \to \mathbb{R}^d$, and let $f \in C^1(\mathcal{G}_T)$.

(i) The expression

$$(\partial_t^\circ f)(\vec{x}(\vec{q}, t), t) = \frac{\mathrm{d}}{\mathrm{d}t} f(\vec{x}(\vec{q}, t), t) \quad \forall (\vec{q}, t) \in \Upsilon \times [0, T]$$

is the time derivative following the parameterizations $\vec{x}(\cdot, t)$ of f on $\Gamma(t)$. It is also called the material time derivative induced by $\vec{x}$.

(ii) The normal time derivative of f on $\Gamma(t)$ is defined as

$$\partial_t^\square f = \partial_t^\circ f - \vec{\mathcal{V}}_\mathrm{T} \cdot \nabla_s f.$$

Remark 29.

(i) The quantity $\partial_t^\circ f$ depends on the parameterization $\vec{x}$. Moreover, it holds that $\vec{\mathcal{V}} = \partial_t^\circ \,\vec{\mathrm{id}}$ on $\Gamma(t)$ and $\vec{\mathcal{V}} \circ \vec{\mathfrak{X}} = \partial_t \vec{\mathfrak{X}}$ in U.

For the following observations, we assume that f is extended to a neighbourhood of $\mathcal{G}_T$.

(ii) It holds that

$$\partial_t^\circ f = \partial_t f + \vec{\mathcal{V}} \cdot \nabla f \quad \text{on}\, \Gamma(t),$$

where ∇f denotes the gradient in $\mathbb{R}^d$ of the extension f.

(iii) Moreover, we have that

$$\partial_t^\square f = \partial_t f + \mathcal{V} \vec{\nu} \cdot \nabla f \quad \text{on}\, \Gamma(t).$$

Taking Remark 24(ii) into account, we observe, in particular, that $\partial_t^\square f$ does not depend on $\vec{x}$. In fact, $\partial_t^\square f$ is the derivative of f in the direction $(\mathcal{V}\vec{\nu}, 1)$, where the vector $(\mathcal{V}\vec{\nu}, 1)$ is a space time tangential vector of the evolving surface $\mathcal{G}_T$. I.e.,

$$\partial_t^\square f = \partial_{(\mathcal{V}\vec{\nu}, 1)} f \quad \text{on}\, \mathcal{G}_T,$$

recall Definition 5(i).

(iv) Taking a curve $t \mapsto \vec{y}(t) \in \mathbb{R}^d$ such that $\vec{y}(t) \in \Gamma(t)$ and $\vec{y}\,'(t) = (\mathcal{V}\vec{\nu}) \circ \vec{y}$, we obtain from (iii) that

$$(\partial_t^\square f)(\vec{y}(t), t) = \frac{\mathrm{d}}{\mathrm{d}t} f(\vec{y}(t), t).$$

For the time-dependent metric tensor $(g_{ij}(t))_{i,j=1,\ldots,d-1}$, which, similarly to Remark 8(i), is defined via

$$g_{ij}(t) = \partial_i \vec{\mathfrak{X}}(t) \cdot \partial_j \vec{\mathfrak{X}}(t) \quad \text{in}\, U, \tag{9}$$

for $\vec{\mathfrak{X}}(t) = \vec{x}(\cdot, t) \circ \vec{\varphi}$, recall (8), we obtain the following lemma. Here and throughout, for notational convenience, we often omit the dependence on t.

Lemma 30. *Let $\mathcal{G}_T$ be a C^2-evolving hypersurface with a global parameterization $\vec{x}$ leading to the velocity field $\vec{\mathcal{V}}$, and let the metric tensor be defined by (9). Then it holds that*

$$\partial_t g_{ij} = 2\,((\underline{\underline{D}}_s(\vec{\mathcal{V}}) \circ \vec{\mathfrak{x}})\,\partial_i \vec{\mathfrak{x}}) \cdot \partial_j \vec{\mathfrak{x}} \quad \text{in } U,$$

where $\underline{\underline{D}}_s(\vec{\mathcal{V}})$ is the rate of deformation tensor, recall Definition 25(iv).

Proof. We introduce the shorthand notation $\vec{\partial}_i = \partial_i \vec{\mathfrak{x}}$, $i = 1, \ldots, d-1$, and recall from Remark 29(i) that $\vec{\mathcal{V}} \circ \vec{\mathfrak{x}} = \partial_t \vec{\mathfrak{x}}$ in U. Then we compute, using Remark 6(iii),

$$\begin{aligned}
\partial_t g_{ij} &= \partial_t (\partial_i \vec{\mathfrak{x}} \cdot \partial_j \vec{\mathfrak{x}}) = \partial_i (\vec{\mathcal{V}} \circ \vec{\mathfrak{x}}) \cdot \vec{\partial}_j + \vec{\partial}_i \cdot \partial_j (\vec{\mathcal{V}} \circ \vec{\mathfrak{x}}) \\
&= ((\nabla_s \vec{\mathcal{V}}) \circ \vec{\mathfrak{x}}\, \vec{\partial}_i) \cdot \vec{\partial}_j + \vec{\partial}_i \cdot (\nabla_s \vec{\mathcal{V}}) \circ \vec{\mathfrak{x}}\, \vec{\partial}_j \\
&= (\nabla_s \vec{\mathcal{V}} + (\nabla_s \vec{\mathcal{V}})^{\top}) \circ \vec{\mathfrak{x}}\, \vec{\partial}_i \cdot \vec{\partial}_j.
\end{aligned}$$

As $\vec{\partial}_i$ and $\vec{\partial}_j$ are tangential, the claim follows. $\square$

In order to compute the first variation of area, it is crucial to know how the area element $\sqrt{g}$ evolves in time, recall Definition 19. This is studied in the next lemma.

Lemma 31. *Let $\mathcal{G}_T$ be a C^2-evolving orientable hypersurface with a global parameterization $\vec{x}$ leading to the velocity field $\vec{\mathcal{V}}$, and let the metric tensor be defined by (9). It holds that*

$$\partial_t \sqrt{g} = (\nabla_s \cdot \vec{\mathcal{V}}) \circ \vec{\mathfrak{x}}\, \sqrt{g} = (-\mathcal{V}\varkappa + \nabla_s \cdot \vec{\mathcal{V}}_{\mathrm{T}}) \circ \vec{\mathfrak{x}}\, \sqrt{g} \quad \text{in } U,$$

where

$$g(t) = \det\left((g_{ij}(t))_{i,j=1,\ldots,d-1} \right).$$

Proof. Jacobi's formula for the derivative of the determinant, see, e.g., Eck et al. (2017, Lemma 5.3), for $\underline{\underline{G}}(t) = (g_{ij}(t))_{i,j=1,\ldots,d-1}$, gives

$$\partial_t \det \underline{\underline{G}}(t) = \det \underline{\underline{G}}(t)\, \mathrm{tr}(\underline{\underline{G}}^{-1}(t)\, \partial_t \underline{\underline{G}}(t)).$$

As $g(t) = \det \underline{\underline{G}}(t)$, and since $\underline{\underline{G}}(t)$ is symmetric, we obtain from the proof of Lemma 30 that

$$\begin{aligned}
\partial_t \sqrt{g} = \partial_t \sqrt{\det \underline{\underline{G}}} &= \frac{1}{2} \sqrt{g}\, \mathrm{tr}(\underline{\underline{G}}^{-1}\, \partial_t \underline{\underline{G}}) = \frac{1}{2} \sqrt{g}\, \underline{\underline{G}}^{-1} : \partial_t \underline{\underline{G}} \\
&= \frac{1}{2} \sqrt{g} \sum_{i,\, j=1}^{d-1} g^{ij}\, \partial_t g_{ij} = \sqrt{g} \sum_{i,\, j=1}^{d-1} g^{ij}\, \partial_i (\vec{\mathcal{V}} \circ \vec{\mathfrak{x}}) \cdot \partial_j \vec{\mathfrak{x}} \\
&= \sqrt{g}\, (\nabla_s \cdot \vec{\mathcal{V}}) \circ \vec{\mathfrak{x}},
\end{aligned}$$

where in the last step we have recalled Remark 8(i). The second identity is then a direct consequence of Lemma 27(i). $\square$

We can now prove a transport theorem for evolving hypersurfaces, which will be crucial for many arguments that follow.

Theorem 32. *Let $\mathcal{G}_T$ be a compact C^2-evolving orientable hypersurface with a global parameterization leading to the velocity field $\vec{\mathcal{V}}$, and let $f \in C^1(\mathcal{G}_T)$. Then it holds that*

$$\begin{aligned}
\frac{\mathrm{d}}{\mathrm{d}t}\int_{\Gamma(t)} f \, \mathrm{d}\mathcal{H}^{d-1} &= \int_{\Gamma(t)} (\partial_t^\circ f + f\,\nabla_s \cdot \vec{\mathcal{V}}) \, \mathrm{d}\mathcal{H}^{d-1} \\
&= \int_{\Gamma(t)} (\partial_t^\circ f + f\,\nabla_s \cdot \vec{\mathcal{V}}_T - f\,\mathcal{V}\,\varkappa) \, \mathrm{d}\mathcal{H}^{d-1} \\
&= \int_{\Gamma(t)} (\partial_t^\square f - f\,\mathcal{V}\,\varkappa) \, \mathrm{d}\mathcal{H}^{d-1} + \int_{\partial\Gamma(t)} f\,\vec{\mathcal{V}} \cdot \vec{\mu} \, \mathrm{d}\mathcal{H}^{d-2},
\end{aligned}$$

where $\vec{\mu}$ is the outer unit conormal to $\partial\Gamma$.

Proof. We first consider an f with support in the image of a single time-dependent local parameterization $\vec{\mathfrak{x}}(t) = \vec{x}(\cdot, t) \circ \vec{\varphi}$, recall (8). In this case we can compute from Definitions 19, 28(i) and Lemma 31 that

$$\begin{aligned}
\frac{\mathrm{d}}{\mathrm{d}t}\int_{\Gamma(t)} f \, \mathrm{d}\mathcal{H}^{d-1} &= \frac{\mathrm{d}}{\mathrm{d}t}\int_{\vec{x}(\vec{\varphi}(U),t)} f \, \mathrm{d}\mathcal{H}^{d-1} \\
&= \frac{\mathrm{d}}{\mathrm{d}t}\int_{U} f(\vec{x}(\vec{\varphi}(\vec{u}),t),t) \sqrt{g} \, \mathrm{d}\mathcal{L}^{d-1} \\
&= \int_{U} (\partial_t^\circ f + f\,\nabla_s \cdot \vec{\mathcal{V}}) \circ \vec{\mathfrak{x}} \sqrt{g} \, \mathrm{d}\mathcal{L}^{d-1} \\
&= \int_{\Gamma(t)} (\partial_t^\circ f + f\,\nabla_s \cdot \vec{\mathcal{V}}) \, \mathrm{d}\mathcal{H}^{d-1}.
\end{aligned}$$

Using a partition of unity argument now proves the first identity in the claim. Lemma 27(i) yields the second identity, and Definition 28(ii) and Lemma 7(i) then give the last identity, on noting from Theorem 21 and Definition 20 that

$$\begin{aligned}
\int_{\Gamma(t)} \nabla_s \cdot (f\,\vec{\mathcal{V}}_{\mathrm{T}}) \, \mathrm{d}\mathcal{H}^{d-1} &= -\int_{\Gamma(t)} \varkappa f\,\vec{\mathcal{V}}_{\mathrm{T}} \cdot \vec{\nu} \, \mathrm{d}\mathcal{H}^{d-1} + \int_{\partial\Gamma(t)} f\,\vec{\mathcal{V}}_{\mathrm{T}} \cdot \vec{\mu} \, \mathrm{d}\mathcal{H}^{d-2} \\
&= \int_{\partial\Gamma(t)} f\,\vec{\mathcal{V}} \cdot \vec{\mu} \, \mathrm{d}\mathcal{H}^{d-2}.
\end{aligned}$$

□

We also have the following transport theorem for moving domains.

Theorem 33. *Let $\mathcal{G}_T$ be a compact C^2-evolving orientable hypersurface, such that $\Gamma(t)$ is bounding a domain $\Omega(t) \subset \mathbb{R}^d$, for $t \in [0,T]$. We assume that $\vec{\nu}(t)$ is the outer unit normal to $\Omega(t)$ on $\Gamma(t)$, and that $f \in C^1(\overline{\mathfrak{O}_T})$, where*

$$\mathfrak{O}_T = \bigcup_{t\in[0,T]} (\Omega(t) \times \{t\}).$$

Then it holds that

$$\frac{\mathrm{d}}{\mathrm{d}t}\int_{\Omega(t)} f \,\mathrm{d}\mathcal{L}^d = \int_{\Omega(t)} \partial_t f \,\mathrm{d}\mathcal{L}^d + \int_{\Gamma(t)} f\,\mathcal{V}\,\mathrm{d}\mathcal{H}^{d-1}. \tag{10}$$

Proof. A proof can be found in Prüss and Simonett (2016, Chapter 2). If the moving domain is transported with a velocity field $\vec{\mathcal{V}}_{\mathfrak{O}} : \mathfrak{O}_T \to \mathbb{R}^d$, with $\vec{\mathcal{V}}_{\mathfrak{O}} \cdot \vec{\nu} = \mathcal{V}$ on $\mathcal{G}_T$, then the theorem can be shown similarly to Theorem 32. An alternative proof would integrate (10), and use the divergence theorem in space-time, see, e.g., Eck et al. (2017, Section 7.3). □

Remark 34. If the moving domain $\mathfrak{O}_T$ is transported with a velocity field $\vec{\mathcal{V}}_{\mathfrak{O}} : \mathfrak{O}_T \to \mathbb{R}^d$, we obtain the Reynolds transport theorem

$$\begin{aligned}\frac{\mathrm{d}}{\mathrm{d}t}\int_{\Omega(t)} f \,\mathrm{d}\mathcal{L}^d &= \int_{\Omega(t)} \partial_t f \,\mathrm{d}\mathcal{L}^d + \int_{\Gamma(t)} f\,\vec{\mathcal{V}}_{\mathfrak{O}} \cdot \vec{\nu}\,\mathrm{d}\mathcal{H}^{d-1} \\ &= \int_{\Omega(t)} \partial_t f + \nabla\cdot(f\,\vec{\mathcal{V}}_{\mathfrak{O}})\,\mathrm{d}\mathcal{L}^d,\end{aligned}$$

where we have used the divergence theorem in $\mathbb{R}^d$, and where we once again assumed that $\vec{\nu}(t)$ is the outer unit normal to $\Omega(t)$ on $\Gamma(t)$.

Using Theorem 33, one can also show a transport theorem for two-phase moving domains. To this end, let $\Omega \subset \mathbb{R}^d$ be a fixed, bounded domain. Suppose that $\mathcal{G}_T$ is a compact C^2-evolving orientable hypersurface with $\Gamma(t) \subset \Omega$, such that $\Gamma(t)$ encloses a region $\Omega_-(t) \subset \Omega$, with $\Gamma(t) = \partial\Omega_-(t)$, for all $t \in [0,T]$. Let $\Omega_+(t) = \Omega \setminus \overline{\Omega_-(t)}$, and let $\vec{\nu}(t)$ denote the outer normal to $\Omega_-(t)$ on $\Gamma(t)$, see Fig. 1. We have

$$\overline{\Omega} = \overline{\Omega_-(t)} \cup \overline{\Omega_+(t)}, \quad \overline{\Omega_-(t)} \cap \overline{\Omega_+(t)} = \Gamma(t), \quad \partial\Omega_+(t) = \Gamma(t) \cup \partial\Omega.$$

Now let

$$\mathfrak{O}_{\pm,T} = \bigcup_{t\in[0,T]} (\Omega_\pm(t) \times \{t\}),$$

and let $f_\pm : \mathfrak{O}_{\pm,T} \to \mathbb{R}$ be given such that each $f_\pm$ has a continuous extension to $\overline{\mathfrak{O}_{\pm,T}}$. Defining $f : \mathfrak{O}_{-,T} \cup \mathfrak{O}_{+,T} \to \mathbb{R}$ as

$$f(\cdot, t) = f_-(\cdot, t)\,\mathcal{X}_{\Omega_-(t)} + f_+(\cdot, t)\,\mathcal{X}_{\Omega_+(t)} \quad \forall t \in [0, T], \tag{11}$$

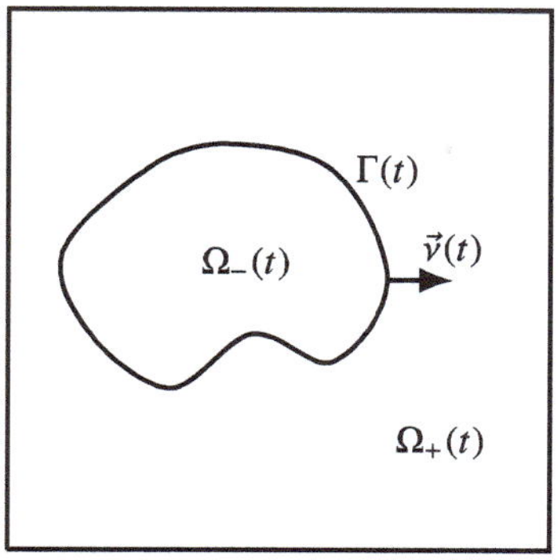

FIG. 1 A sketch of the domain Ω in the case $d = 2$.

where, here and throughout, $\mathcal{X}_{\mathfrak{A}}$ defines the characteristic function for a set $\mathfrak{A}$, we let

$$[f]_{-}^{+}(\vec{z}, t) = \lim_{\substack{\vec{y} \to \vec{z} \\ \vec{y} \in \Omega_{+}(t)}} f(\vec{y}, t) - \lim_{\substack{\vec{y} \to \vec{z} \\ \vec{y} \in \Omega_{-}(t)}} f(\vec{y}, t) \quad \forall (\vec{z}, t) \in \mathcal{G}_T. \tag{12}$$

Then we obtain that following result.

Theorem 35. *Let $\mathcal{G}_T \subset \Omega \times [0, T]$ be a compact C^2-evolving orientable hypersurface, such that $\Gamma(t)$ is bounding a domain $\Omega_{-}(t) \subset \mathbb{R}^d$, for $t \in [0, T]$. We assume that $\vec{\nu}(t)$ is the outer unit normal to $\Omega_{-}(t)$ on $\Gamma(t)$, and that $f_{\pm} \in C^1(\overline{\mathfrak{O}_{\pm,T}})$. Then, for f as defined in (11), it holds that*

$$\frac{\mathrm{d}}{\mathrm{d}t} \int_{\Omega} f \, \mathrm{d}\mathcal{L}^d = \int_{\Omega_{-}(t)} \partial_t f \, \mathrm{d}\mathcal{L}^d + \int_{\Omega_{+}(t)} \partial_t f \, \mathrm{d}\mathcal{L}^d - \int_{\Gamma(t)} [f]_{-}^{+} \, \mathcal{V} \, \mathrm{d}\mathcal{H}^{d-1}.$$

Proof. The claim follows directly from Theorem 33. □

Remark 36. We naturally extend (12) to vector-valued quantities. For example, if f, as defined in (11), has a continuous extension to $\overline{\Omega} \times [0, T]$, and such that each $\nabla f_{\pm}$ has a continuous extension to $\overline{\mathfrak{O}_{\pm,T}}$, then we define

$$[\partial_{\vec{\nu}} f]_{-}^{+}(\vec{z}, t) = \lim_{\substack{\vec{y} \to \vec{z} \\ \vec{y} \in \Omega_{+}(t)}} (\nabla f_{+})(\vec{y}, t) \cdot \vec{\nu}(\vec{z}, t) - \lim_{\substack{\vec{y} \to \vec{z} \\ \vec{y} \in \Omega_{-}(t)}} (\nabla f_{-})(\vec{y}, t) \cdot \vec{\nu}(\vec{z}, t) \quad \forall (\vec{z}, t) \in \mathcal{G}_T. \tag{13}$$

For such an f, integration by parts in $\mathbb{R}^d$ immediately yields that

$$\int_{\Omega_-(t)\cup\Omega_+(t)} \eta\,\Delta\,f\,\mathrm{d}\mathcal{L}^d$$
$$= \int_{\partial\Omega} \eta\,\partial_{\vec{\nu}_\Omega} f\,\mathrm{d}\mathcal{H}^{d-1} - \int_{\Gamma(t)} \eta\,[\partial_{\vec{\nu}} f]_-^+\,\mathrm{d}\mathcal{H}^{d-1} - \int_\Omega \nabla f\cdot\nabla\eta\,\mathrm{d}\mathcal{L}^d$$

for all $\eta\in C^1(\overline{\Omega})$, where $\vec{\nu}_\Omega$ denotes the outer unit normal to Ω on $\partial\Omega$.

2.5 Time derivatives of the normal

We also frequently need time derivatives of the normal. The relevant results are stated in the following lemma.

Lemma 37. *Let $\mathcal{G}_T$ be a C^2-evolving orientable hypersurface.*
(i) *Let $\vec{\mathcal{V}}$ be the velocity field induced by a a global parameterization of $\mathcal{G}_T$. Then it holds that*

$$\partial_t^\circ\,\vec{\nu} = -(\nabla_s\,\vec{\mathcal{V}})^\top\,\vec{\nu}\quad \textit{on}\,\Gamma(t).$$

(ii) *The normal time derivative of $\vec{\nu}$ satisfies*

$$\partial_t^\square\,\vec{\nu} = -\nabla_s\,\mathcal{V}\quad \textit{on}\,\Gamma(t).$$

Proof.
(i) For $\vec{p}\in\Gamma(t)$, with $\vec{p}=\vec{x}(\vec{\varphi}(\vec{u},t),t)=\vec{\mathfrak{X}}(\vec{u},t)$, recall (8), we define a basis $\{\vec{\tau}_1,\ldots,\vec{\tau}_{d-1}\}$ of the tangent space $\mathrm{T}_{\vec{p}}\,\Gamma(t)$ via $\vec{\tau}_i(\vec{p})=\partial_i\,\vec{\mathfrak{X}}(\vec{u},t)$, $i=1\,\ldots,\,d-1$. We also recall from Remark 29(i) that $\vec{\mathcal{V}}\circ\vec{\mathfrak{X}}=\partial_t\,\vec{\mathfrak{X}}$ and hence note that

$$\partial_t^\circ\,\vec{\tau}_i=\partial_t\,[(\partial_i\,\vec{\mathfrak{X}})\circ\vec{\mathfrak{X}}^{-1}]=[\partial_i\,(\partial_t\,\vec{\mathfrak{X}})]\circ\vec{\mathfrak{X}}^{-1}=[\partial_i\,(\vec{\mathcal{V}}\circ\vec{\mathfrak{X}})]\circ\vec{\mathfrak{X}}^{-1}=\partial_{\vec{\tau}_i}\,\vec{\mathcal{V}}$$
$$=(\nabla_s\,\vec{\mathcal{V}})\,\vec{\tau}_i\ \ \text{on}\,\Gamma(t),$$

where for the last step we have recalled Remark 6(iii). As $\vec{\nu}\cdot\vec{\tau}_i=0$, it follows for $i=1,\ldots,d-1$ that

$$(\partial_t^\circ\,\vec{\nu})\cdot\vec{\tau}_i=-\vec{\nu}\cdot\partial_t^\circ\,\vec{\tau}_i=-\vec{\nu}\cdot((\nabla_s\,\vec{\mathcal{V}})\,\vec{\tau}_i)=-((\nabla_s\,\vec{\mathcal{V}})^\top\,\vec{\nu})\cdot\vec{\tau}_i\ \ \text{on}\,\Gamma(t).$$

Since $\{\vec{\tau}_1,\ldots,\vec{\tau}_{d-1}\}$ is a basis of the tangent space $\mathrm{T}_{\vec{p}}\,\Gamma(t)$, we obtain, on using $(\partial_t^\circ\,\vec{\nu})\cdot\vec{\nu}=\frac{1}{2}\,\partial_t^\circ\,|\vec{\nu}|^2=0$ and the fact that $\vec{\nu}\cdot(\nabla_s\,\vec{\mathcal{V}})^\top\,\vec{\nu}=(\nabla_s\,\vec{\mathcal{V}})\,\vec{\nu}\cdot\vec{\nu}=0$, recall Remark 6(iii), that

$$\partial_t^\circ\,\vec{\nu}=-(\nabla_s\,\vec{\mathcal{V}})^\top\,\vec{\nu}\quad \text{on}\,\Gamma(t).$$

(ii) Let $\vec{x}$ be an arbitrary global parameterization of $\mathcal{G}_T$, with induced velocity field $\vec{\mathcal{V}}$. Using Definition 28(ii), Lemma 12, the result from (i) and Lemma 7(ii), we compute

$$\begin{aligned}\partial_t^{\square}\,\vec{\nu} = \partial_t^{\circ}\,\vec{\nu} - (\nabla_s\vec{\nu})\,\vec{\mathcal{V}}_{\rm T} &= \partial_t^{\circ}\,\vec{\nu} - (\nabla_s\vec{\nu})^{\rm T}\,\vec{\mathcal{V}}_{\rm T} = -(\nabla_s\vec{\mathcal{V}})^{\rm T}\vec{\nu} - (\nabla_s\vec{\nu})^{\rm T}\vec{\mathcal{V}}_{\rm T}\\ &= -(\nabla_s\vec{\mathcal{V}})^{\rm T}\vec{\nu} - (\nabla_s\vec{\nu})^{\rm T}\vec{\mathcal{V}} = -\nabla_s\,(\vec{\mathcal{V}}\cdot\vec{\nu})\\ &= -\nabla_s\,\mathcal{V} \quad \text{on}\,\Gamma(t).\end{aligned}$$

□

2.6 Time derivatives of the mean curvature

In order to be able to compute the first variation of energies that depend on the mean curvature, for example, with the help of Theorem 32, we need expressions for the time derivatives of curvature.

We begin with the following commutator rule for time derivatives and surface differential operators.

Lemma 38. *Let $\mathcal{G}_T$ be a C^2-evolving orientable hypersurface with a global parameterization leading to the velocity field $\vec{\mathcal{V}}$, and let $f:\mathcal{G}_T\to\mathbb{R}$, $\vec{f}:\mathcal{G}_T\to\mathbb{R}^d$ be C^1-functions. Then we have the following results:*

(i)
$$\partial_t^{\circ}\,\nabla_s f - \nabla_s\,\partial_t^{\circ} f = [\nabla_s\vec{\mathcal{V}} - 2\,\underline{\underline{D}}_s(\vec{\mathcal{V}})]\,\nabla_s f \quad \text{on}\,\Gamma(t).$$

(ii)
$$\partial_t^{\circ}\,\nabla_s \vec{f} - \nabla_s\,\partial_t^{\circ}\vec{f} = (\nabla_s\vec{f})\,[\nabla_s\vec{\mathcal{V}} - 2\,\underline{\underline{D}}_s(\vec{\mathcal{V}})]^{\rm T} \quad \text{on}\,\Gamma(t).$$

(iii)
$$\partial_t^{\circ}\,(\nabla_s\cdot\vec{f}) - \nabla_s\cdot(\partial_t^{\circ}\vec{f}) = [\nabla_s\vec{\mathcal{V}} - 2\,\underline{\underline{D}}_s(\vec{\mathcal{V}})] : \nabla_s\vec{f} \quad \text{on}\,\Gamma(t).$$

Proof.
(i) We extend f to a neighbourhood of $\mathcal{G}_T$, such that the extension from $\Gamma(t)$ is constant in the $\vec{\nu}$-direction, recall (4). On noting (5) and Lemma 37(i), it holds on $\Gamma(t)$ that

$$\begin{aligned}0 = \partial_t^{\circ}\,(\nabla f\cdot\vec{\nu}) &= \partial_t^{\circ}\,(\nabla_s f\cdot\vec{\nu}) = (\partial_t^{\circ}\,\nabla_s f)\cdot\vec{\nu} + \nabla_s f\cdot\partial_t^{\circ}\,\vec{\nu}\\ &= (\partial_t^{\circ}\,\nabla_s f)\cdot\vec{\nu} - (\nabla_s f)\cdot(\nabla_s\vec{\mathcal{V}})^{\rm T}\,\vec{\nu} = (\partial_t^{\circ}\,\nabla_s f)\cdot\vec{\nu} - ((\nabla_s\vec{\mathcal{V}})\,\nabla_s f)\cdot\vec{\nu}.\end{aligned}$$

As $\nabla_s\,\partial_t^{\circ} f$ is tangential, we obtain

$$(\partial_t^{\circ}\,\nabla_s f - \nabla_s\,\partial_t^{\circ}\,f)\cdot\vec{\nu} = ((\nabla_s\vec{\mathcal{V}})\,\nabla_s f)\cdot\vec{\nu}.$$

We now identify the tangential part of $\partial_t^{\circ}\,\nabla_s f - \nabla_s\,\partial_t^{\circ} f$, recall (1). We compute, on noting (5), Remark 6(v), Remark 29(ii), Remark 8(iii), Lemma 7(ii) and Remark 6(iii) that

$$\begin{aligned}\underline{\underline{P}}_\Gamma\,(\partial_t^\circ \nabla_s f - \nabla_s \partial_t^\circ f) &= \underline{\underline{P}}_\Gamma\,\partial_t^\circ \nabla f - \nabla_s \partial_t^\circ f\\ &= \underline{\underline{P}}_\Gamma\,(\partial_t \nabla f + (\nabla^2 f)\,\vec{\mathcal{V}}) - \nabla_s\,(\partial_t f + \vec{\mathcal{V}} \cdot \nabla f)\\ &= \underline{\underline{P}}_\Gamma\,(\partial_t \nabla f + (\nabla^2 f)\,\vec{\mathcal{V}}) - \underline{\underline{P}}_\Gamma\,\nabla\,\partial_t f - \nabla_s\,(\vec{\mathcal{V}} \cdot \nabla f)\\ &= \underline{\underline{P}}_\Gamma\,(\nabla^2 f)\,\vec{\mathcal{V}} - (\nabla_s \vec{\mathcal{V}})^\top\,\nabla f - \underline{\underline{P}}_\Gamma \sum_{i=1}^{d} \vec{\mathcal{V}} \cdot \vec{e}_i\,\nabla\,\partial_i f\\ &= -(\nabla_s \vec{\mathcal{V}})^\top\,\nabla_s f.\end{aligned}$$

Combining the above, on recalling Definition 5(vii) and Remark 6(v), yields that

$$\begin{aligned}&\partial_t^\circ \nabla_s f - \nabla_s \partial_t^\circ f = \underline{\underline{P}}_\Gamma\,(\partial_t^\circ \nabla_s f - \nabla_s \partial_t^\circ f) + (\underline{\underline{\mathrm{Id}}} - \underline{\underline{P}}_\Gamma)\,(\partial_t^\circ \nabla_s f - \nabla_s \partial_t^\circ f)\\ &= -(\nabla_s \vec{\mathcal{V}})^\top\,\nabla_s f + (\underline{\underline{\mathrm{Id}}} - \underline{\underline{P}}_\Gamma)\,(\nabla_s \vec{\mathcal{V}})\,\nabla_s f = [\nabla_s \vec{\mathcal{V}} - 2\,\underline{\underline{D}}_s(\vec{\mathcal{V}})]\,\nabla_s f,\end{aligned}$$

which shows the claim.
(ii) It follows from (i) and Remark 6(iii) that

$$\begin{aligned}(\partial_t^\circ \nabla_s \vec{f} - \nabla_s \partial_t^\circ \vec{f})^\top\,\vec{e}_i &= (\partial_t^\circ \nabla_s - \nabla_s \partial_t^\circ)\,(\vec{f} \cdot \vec{e}_i)\\ &= [\nabla_s \vec{\mathcal{V}} - 2\,\underline{\underline{D}}_s(\vec{\mathcal{V}})]\,(\nabla_s \vec{f})^\top\,\vec{e}_i,\end{aligned}$$

for $i = 1, \ldots, d$, and this proves the desired result.
(iii) Using (ii) and Lemma 7(v), we compute

$$\begin{aligned}&\partial_t^\circ\,(\nabla_s \cdot \vec{f}) - \nabla_s \cdot (\partial_t^\circ \vec{f}) = \mathrm{tr}\Big(\partial_t^\circ\,(\nabla_s \vec{f}) - \nabla_s\,(\partial_t^\circ \vec{f})\Big)\\ &= \mathrm{tr}\Big((\nabla_s \vec{f})\,[\nabla_s \vec{\mathcal{V}} - 2\,\underline{\underline{D}}_s(\vec{\mathcal{V}})]^\top\Big) = [\nabla_s \vec{\mathcal{V}} - 2\,\underline{\underline{D}}_s(\vec{\mathcal{V}})] : \nabla_s \vec{f},\end{aligned}$$

which yields the desired result. □

We now obtain formulas for the time derivatives of the mean curvature.

Lemma 39. *Let $\mathcal{G}_T$ be a C^3-evolving orientable hypersurface.*
(i) *Let $\vec{\mathcal{V}}_\mathrm{T}$ be the tangential velocity field induced by a a global parameterization of $\mathcal{G}_T$. Then it holds that*

$$\partial_t^\circ\,\varkappa = \Delta_s\,\mathcal{V} + \mathcal{V}\,|\nabla_s \vec{\nu}|^2 + \vec{\mathcal{V}}_T \cdot \nabla_s \varkappa \quad \text{on}\,\Gamma(t).$$

(ii) *The normal time derivative of the mean curvature satisfies*

$$\partial_t^\square\,\varkappa = \Delta_s\,\mathcal{V} + \mathcal{V}\,|\nabla_s \vec{\nu}|^2 \quad \text{on}\,\Gamma(t).$$

Proof.
(i) It follows from Lemmas 13(i), 38(iii), 37(i), 27(ii), 27(iii) and 12 that

$$\begin{aligned}\partial_t^\circ\,\varkappa &= -\partial_t^\circ\,(\nabla_s \cdot \vec{\nu}) = -\nabla_s \cdot (\partial_t^\circ\,\vec{\nu}) - [\nabla_s \vec{\mathcal{V}} - 2\,\underline{\underline{D}}_s(\vec{\mathcal{V}})] : \nabla_s \vec{\nu}\\ &= \nabla_s \cdot ((\nabla_s \vec{\mathcal{V}})^\top\,\vec{\nu}) + \mathcal{V}\,|\nabla_s \vec{\nu}|^2 + \nabla_s \vec{\mathcal{V}}_\mathrm{T} : \nabla_s \vec{\nu}\\ &= \nabla_s \cdot (\nabla_s \mathcal{V} + (\nabla_s \vec{\mathcal{V}})^\top\,\vec{\nu}) + \mathcal{V}\,|\nabla_s \vec{\nu}|^2 + \nabla_s \vec{\mathcal{V}}_\mathrm{T} : \nabla_s \vec{\nu}.\end{aligned}$$

As $\vec{\mathcal{V}}_{\rm T} \cdot \vec{\nu} = 0$, we obtain from Lemmas 7(ii) and 12(ii) that $(\nabla_s \vec{\mathcal{V}}_{\rm T})^{\rm T}\, \vec{\nu} = -(\nabla_s \vec{\nu})\, \vec{\mathcal{V}}_{\rm T}$, and hence Lemma 7(i) implies that

$$
\begin{aligned}
\partial_t^{\circ}\, \varkappa &= \Delta_s\, \mathcal{V} - \nabla_s \cdot ((\nabla_s \vec{\nu})\, \vec{\mathcal{V}}_{\rm T}) + \mathcal{V}\, |\nabla_s \vec{\nu}|^2 + \nabla_s \vec{\mathcal{V}}_{\rm T} : \nabla_s \vec{\nu} \\
&= \Delta_s\, \mathcal{V} - (\Delta_s \vec{\nu}) \cdot \vec{\mathcal{V}}_{\rm T} - \nabla_s \vec{\nu} : \nabla_s \vec{\mathcal{V}}_{\rm T} + \mathcal{V}\, |\nabla_s \vec{\nu}|^2 + \nabla_s \vec{\mathcal{V}}_{\rm T} : \nabla_s \vec{\nu} \\
&= \Delta_s\, \mathcal{V} - (\Delta_s \vec{\nu}) \cdot \vec{\mathcal{V}}_{\rm T} + \mathcal{V}\, |\nabla_s \vec{\nu}|^2.
\end{aligned}
$$

Combining this with Lemma 16 yields the desired result.
(ii) On choosing an arbitrary global parameterization of $\mathcal{G}_T$, with induced tangential velocity field $\vec{\mathcal{V}}_{\rm T}$, the claim follows from (i) and Definition 28(ii). □

2.7 Gauss–Bonnet theorem

In the case $d = 3$, we can consider curves $\mathfrak{g}$ on a hypersurface Γ. For any curve in $\mathbb{R}^3$, we can define the curvature vector

$$
\vec{\varkappa}_{\mathfrak{g}} = \vec{\rm id}_{ss} \quad \text{on } \mathfrak{g}, \tag{14}
$$

where ∂_s^2 denotes the second derivative with respect to arclength on γ. We note that (14) is invariant under a change of parameterization of the curve.

Later we will need the Gauss–Bonnet theorem, which uses (14) for the special case $\mathfrak{g} = \partial\Gamma \subset \Gamma$.

Theorem 40 (Gauss–Bonnet).
Let Γ be a compact orientable C^2-hypersurface in $\mathbb{R}^3$. Then it holds that

$$
\int_\Gamma \mathcal{K}\, \mathrm{d}\mathcal{H}^2 = 2\,\pi\, m(\Gamma) + \int_{\partial\Gamma} \vec{\varkappa}_{\partial\Gamma} \cdot \vec{\mu}\, \mathrm{d}\mathcal{H}^1,
$$

where $\vec{\mu}$ is the outer unit conormal to $\partial\Gamma$, and where where $m(\Gamma) \in \mathbb{Z}$ is the Euler characteristic of Γ.

For a definition of the Euler characteristic and a proof of the Gauss–Bonnet formula we refer to Kühnel (2015, Section 4F).

3 Parametric finite elements

In this section we discuss the main concepts that are necessary for the numerical approximation of curvature-driven evolution equations with the help of parametric finite elements. Readers familiar with these concepts, and readers more interested in the actual finite element approximations discussed in this chapter, may skip this section and go directly to the appropriate sections of interest, below.

3.1 Polyhedral surfaces

In order to approximate a smooth surface we use polyhedral surfaces as follows. The idea of using polyhedral surfaces to approximate the curvature-driven evolution of hypersurfaces goes back to Dziuk (1991).

Definition 41.

(i) A subset $\Gamma^h \subset \mathbb{R}^d$ is called an n-dimensional polyhedral surface for $1 \leq n \leq d$, with or without boundary, if it is the finite union of closed, nondegenerate n-simplices, where the intersection of any two simplices is either empty or a common k-simplex, $0 \leq k < n$.

(ii) If $n = d - 1$, then we call Γ^h a polyhedral hypersurface.

(iii) If $n = 1$, then we call Γ^h a polygonal curve.

Remark 42.

(i) The boundary of a polyhedral surface Γ^h, as a C^0-surface, is defined as in Definition 17, and is denoted by $\partial\Gamma^h$. Of course, $\partial\Gamma^h$ is given as the union of all $(n-1)$-simplices, that form part of the boundary of exactly one of the n-simplices that make up Γ^h. If $\partial\Gamma^h$ is empty, then we call Γ^h a closed polyhedral surface.

(ii) For a polyhedral hypersurface Γ^h with boundary, the outer conormal to Γ^h is well-defined almost everywhere on $\partial\Gamma^h$, where locally we use the definition Definition 20.

(iii) In order to define geometric quantities for, and finite element spaces on polyhedral surfaces, it is often convenient to define Γ^h in terms of a triangulation. To this end, from now on, we let an n-dimensional polyhedral surface be given by

$$\Gamma^h = \bigcup_{j=1}^{J} \overline{\sigma_j},$$

where $\{\sigma_j\}_{j=1}^J$ is a family of disjoint, (relatively) open n-simplices, such that $\overline{\sigma_i} \cap \overline{\sigma_j}$ for $i \neq j$ is either empty or a common k-simplex of $\overline{\sigma_i}$ and $\overline{\sigma_j}$, $0 \leq k < n$. For later use, we denote the vertices of Γ^h by $\{\vec{q}_k\}_{k=1}^K$, and assume that the vertices of σ_j are given by $\{\vec{q}_{j,k}\}_{k=1}^{n+1}$, $j = 1, \ldots, J$.

Definition 43. Let $\Gamma^h = \bigcup_{j=1}^J \overline{\sigma_j}$ be an n-dimensional polyhedral surface, as described in Remark 42(iii), with vertices $\{\vec{q}_k\}_{k=1}^K$.

(i) We define the finite element spaces of continuous piecewise linear functions on Γ^h via

$$V(\Gamma^h) = \{\chi \in C(\Gamma^h) : \chi_{|_{\sigma_j}} \text{ is affine for } j = 1, \ldots, J\},$$

$$\underline{V}(\Gamma^h) = [V(\Gamma^h)]^d, \quad \underline{\underline{V}}(\Gamma^h) = [V(\Gamma^h)]^{d\times d}.$$

We let $\{\phi_k^{\Gamma^h}\}_{k=1}^K$ denote the standard basis of $V(\Gamma^h)$, i.e.,

$$\phi_i^{\Gamma^h}(\vec{q}_j) = \delta_{ij}, \quad i,j = 1, \ldots, K.$$

Moreover, we let $\pi_{\Gamma^h} : C(\Gamma^h) \to V(\Gamma^h)$ be the standard interpolation operator, i.e.,

$$\pi_{\Gamma^h}\, \eta = \sum_{k=1}^{K} \eta(\vec{q}_k)\, \phi_k^{\Gamma^h} \quad \forall \eta \in C(\Gamma^h),$$

and similarly $\vec{\pi}_{\Gamma^h} : [C(\Gamma^h)]^d \to \underline{V}(\Gamma^h)$.

(ii) We define the spaces of piecewise constant functions on Γ^h via

$$V_c(\Gamma^h) = \{\chi \in L^\infty(\Gamma^h) : \chi_{|_{\sigma_j}} \text{ is constant for } j = 1, \ldots, J\},$$
$$\underline{V}_c(\Gamma^h) = [V_c(\Gamma^h)]^d, \quad \underline{\underline{V}}_c(\Gamma^h) = [V_c(\Gamma^h)]^{d \times d}.$$

(iii) We let $\langle \cdot, \cdot \rangle_{\Gamma^h}$ denote the L^2-inner product on Γ^h, with $|\cdot|_{\Gamma^h}$ the associated L^2-norm, and we extend these definitions to any n-dimensional piecewise C^1-surface Γ. For piecewise continuous functions, u, $\mathrm{v} \in L^\infty(\Gamma^h)$, with possible jumps across the edges of $\{\sigma_j\}_{j=1}^J$, we introduce the mass-lumped inner product $\langle \cdot, \cdot \rangle_{\Gamma^h}^h$ as

$$\langle u, \mathrm{v} \rangle_{\Gamma^h}^h = \frac{1}{n+1} \sum_{j=1}^{J} \mathcal{H}^n(\sigma_j) \sum_{k=1}^{n+1} (u\,\mathrm{v})((\vec{q}_{j,k})^-), \tag{15}$$

where $u((\vec{q})^-) = \lim_{\sigma_j \ni \vec{p} \to \vec{q}} u(\vec{p})$. The definition (15) is naturally extended to vector- and tensor-valued functions. We also let

$$|u|_{\Gamma^h}^h = \left(\langle u, u \rangle_{\Gamma^h}^h\right)^{\frac{1}{2}}, \tag{16}$$

which on $V(\Gamma^h)$ defines the norm induced by (15). We extend (16) to vector-valued functions to obtain a norm on $\underline{V}(\Gamma^h)$.

Remark 44. It follows from Definition 43 that

$$\langle \eta, 1 \rangle_{\Gamma^h}^h = \langle \pi_{\Gamma^h}\, \eta, 1 \rangle_{\Gamma^h} \quad \forall \eta \in C(\Gamma^h).$$

3.1.1 Orientation

In order to discuss the orientation of a polyhedral hypersurface, we begin with the definition of the wedge product in $\mathbb{R}^d$.

Definition 45. Let $\vec{\mathrm{v}}_1, \ldots, \vec{\mathrm{v}}_{d-1} \in \mathbb{R}^d$. Then the wedge product

$$\vec{z} = \vec{\mathrm{v}}_1 \wedge \cdots \wedge \vec{\mathrm{v}}_{d-1}$$

is the unique vector $\vec{z} \in \mathbb{R}^d$ such that $\vec{b} \cdot \vec{z} = \det(\vec{\mathrm{v}}_1, \ldots, \vec{\mathrm{v}}_{d-1}, \vec{b})$ for all $\vec{b} \in \mathbb{R}^d$.

Remark 46.

(i) The wedge product is the usual cross product of two vectors in $\mathbb{R}^3$, and the anticlockwise rotation through $\frac{\pi}{2}$ of a vector in $\mathbb{R}^2$.

(ii) The wedge product $\vec{v}_1 \wedge \cdots \wedge \vec{v}_{d-1}$ is perpendicular to each of the $(d-1)$ vectors $\vec{v}_1, \ldots, \vec{v}_{d-1}$, and has length equal to the volume of the parallelotope spanned by them.

(iii) The measure of a $(d-1)$-simplex σ, with vertices $\{\vec{q}_k\}_{k=1}^d$, can be computed via

$$\mathcal{H}^{d-1}(\sigma) = \frac{1}{d-1} |(\vec{q}_2 - \vec{q}_1) \wedge \cdots \wedge (\vec{q}_d - \vec{q}_1)|.$$

We recall the following definition of orientable polyhedral hypersurface from Matveev (2006, p. 20).

Definition 47. We say that the polyhedral hypersurface $\Gamma^h = \bigcup_{j=1}^J \overline{\sigma_j}$ is orientable, if it is possible to consistently orientate the simplices $\{\sigma_j\}_{j=1}^J$, e.g., by choosing the order $\{\vec{q}_{j,k}\}_{k=1}^d$ for the vertices of σ_j, $j = 1, \ldots, J$, in such a way, that on nonempty intersections $\overline{\sigma_i} \cap \overline{\sigma_j}$ that form a $(d-2)$-simplex, the two orientations induced by σ_i and σ_j are opposite to each other.

Remark 48. For a polyhedral hypersurface in $\mathbb{R}^3$, each triangle is oriented by choosing a direction around the boundary of the triangle. On each triangle, this gives a direction to every edge of the triangle. If this can be done in such a way, that two neighbouring edges are always pointing in the opposite direction, then the surface is orientable. An example for a nonorientable polyhedral hypersurface in $\mathbb{R}^3$ is a triangulation of the Möbius strip. Of course, for $d = 2$ we are dealing with polygonal curves, and they are always orientable.

Definition 49. Let $\Gamma^h = \bigcup_{j=1}^J \overline{\sigma_j}$ be an orientable polyhedral hypersurface Γ^h, with a consistent ordering of the vertices $\{\vec{q}_{j,k}\}_{k=1}^d$ for each σ_j, $j = 1, \ldots, J$. Then we define the consistent, piecewise constant unit normal $\vec{\nu}^h \in \underline{V}_c(\Gamma^h)$ via

$$\vec{\nu}^h = \frac{(\vec{q}_{j,2} - \vec{q}_{j,1}) \wedge \cdots \wedge (\vec{q}_{j,d} - \vec{q}_{j,1})}{|(\vec{q}_{j,2} - \vec{q}_{j,1}) \wedge \cdots \wedge (\vec{q}_{j,d} - \vec{q}_{j,1})|} \quad \text{on } \sigma_j,$$

$j = 1, \ldots, J$, recall Definition 45.

Remark 50. Of course, changing the orientation of Γ^h will change the sign of $\vec{\nu}^h$. For the majority of the approximations introduced in this chapter, the choice of normal is not important. We will clearly state the choice of the sign of $\vec{\nu}^h$ in situations where it is critical.

Definition 51. Let Γ^h be an orientable polyhedral hypersurface with unit normal $\vec{\nu}^h$. Then we define the vertex normal vector $\vec{\omega}^h \in \underline{V}(\Gamma^h)$ to be the mass-lumped L^2-projection of $\vec{\nu}^h$ onto $\underline{V}(\Gamma^h)$, i.e.,

$$\left\langle \vec{\omega}^h, \vec{\varphi} \right\rangle_{\Gamma^h}^h = \left\langle \vec{\nu}^h, \vec{\varphi} \right\rangle_{\Gamma^h} \quad \forall \vec{\varphi} \in \underline{V}(\Gamma^h). \tag{17}$$

Remark 52. It is easy to see that, for $k = 1, \ldots, K$,

$$\vec{\omega}^h(\vec{q}_k) = \frac{1}{\mathcal{H}^{d-1}(\Lambda_k)} \sum_{\sigma_j \in \mathcal{T}_k} \mathcal{H}^{d-1}(\sigma_j)\, \vec{\nu}^h_{|_{\sigma_j}}, \tag{18}$$

where

$$\Lambda_k = \bigcup_{\sigma_j \in \mathcal{T}_k} \overline{\sigma_j} \quad \text{and} \quad \mathcal{T}_k = \{\sigma_j : \vec{q}_k \in \overline{\sigma_j}\}.$$

In particular, we note that one can interpret $\vec{\omega}^h(\vec{q}_k)$ as a weighted normal at the vertex $\vec{q}_k$ of Γ^h. It follows from (15) and (17) that

$$\left\langle \chi\, \vec{\omega}^h, \vec{\varphi} \right\rangle_{\Gamma^h}^h = \left\langle \chi\, \vec{\nu}^h, \vec{\varphi} \right\rangle_{\Gamma^h}^h \quad \forall \chi \in V(\Gamma^h), \ \ \vec{\varphi} \in \underline{V}(\Gamma^h). \tag{19}$$

Combining (19) and (17) yields that

$$\left\langle \vec{\omega}^h, \vec{\varphi} \right\rangle_{\Gamma^h}^h = \left\langle \vec{\nu}^h, \vec{\varphi} \right\rangle_{\Gamma^h}^h = \left\langle \vec{\nu}^h, \vec{\varphi} \right\rangle_{\Gamma^h} \quad \forall \vec{\varphi} \in \underline{V}(\Gamma^h). \tag{20}$$

3.1.2 Polygonal curves

Most of the above definitions simplify dramatically when Γ^h is a polygonal curve. Given a closed polygonal curve $\Gamma^h = \bigcup_{j=1}^J \overline{\sigma_j}$, we can parameterize Γ^h with the help of a finite element function defined on the periodic unit interval $\mathbb{I} = \mathbb{R}/\mathbb{Z}$.

Definition 53. Let $\mathbb{I} = \bigcup_{j=1}^J I_j$ be decomposed into the intervals $I_j = [q_{j-1}, q_j]$, given by the nodes $q_j = j\, h$, $h = J^{-1}$, for $j = 0, \ldots, J$. We make use of the periodicity of $\mathbb{R}/\mathbb{Z}$, i.e., $q_J = q_0$, $q_{J+1} = q_1$ and so on.

(i) We define the finite element spaces of periodic, continuous piecewise linear functions in $\mathbb{I}$ via

$$V^h(\mathbb{I}) = \{\chi \in C(\mathbb{I}) : \chi_{|_{I_j}} \text{ is affine for } j = 1, \ldots, J\},$$

$$\underline{V}^h(\mathbb{I}) = [V^h(\mathbb{I})]^d.$$

(ii) We let $\langle \cdot, \cdot \rangle_{\mathbb{I}}$ denote the L^2-inner product on $\mathbb{I}$. For piecewise continuous functions, $u, v \in L^\infty(\mathbb{I})$, with possible jumps at the nodes $\{q_j\}_{j=1}^J$, we introduce the mass lumped inner product on $\mathbb{I}$ as

$$\langle u, v \rangle_{\mathbb{I}}^h = \frac{1}{2} h \sum_{j=1}^J \left[(u\, v)(q_j^-) + (u\, v)(q_{j-1}^+) \right].$$

Remark 54. Given a closed polygonal curve $\Gamma^h = \bigcup_{j=1}^J \overline{\sigma_j}$, we can now find a function $\vec{X}^h \in \underline{V}^h(\mathbb{I})$ such that $\Gamma^h = \vec{X}^h(\mathbb{I})$. Then the following hold.

(i) It follows from Remark 8(i) that

$$(\nabla_s f) \circ \vec{X}^h = \partial_s (f \circ \vec{X}^h) \vec{X}_s^h \qquad \text{in } I_j,$$
$$(\nabla_s \cdot \vec{f}) \circ \vec{X}^h = \partial_s (\vec{f} \circ \vec{X}^h) \cdot \vec{X}_s^h \qquad \text{in } I_j,$$

for $j = 1, \ldots, J$, where $\vec{X}_s^h = \partial_s \vec{X}^h$ and $\partial_s = |\partial_1 \vec{X}^h|^{-1} \partial_1$ denotes differentiation with respect to the arclength of Γ^h. From now on we define the shorthand notation $\vec{X}_\rho^h = \partial_1 \vec{X}^h$, i.e., $\rho \in \mathbb{I}$ plays the role of the parameterization variable.

(ii) We have that

$$\langle f, 1 \rangle_{\Gamma^h} = \langle f \circ \vec{X}^h, |\vec{X}_\rho^h| \rangle_{\mathbb{I}} \quad \text{and} \quad \langle f, 1 \rangle_{\Gamma^h}^h = \langle f \circ \vec{X}^h, |\vec{X}_\rho^h| \rangle_{\mathbb{I}}^h.$$

(iii) For the normal $\vec{\nu}^h$ on Γ^h defined as in Definition 49 and Remark 46(i), it holds that

$$\vec{\nu}^h \circ \vec{X}^h = -(\vec{X}_s^h)^\perp \quad \text{in } I_j, \tag{21}$$

if $\overline{\sigma_j} = [\vec{q}_{j,1}^h, \vec{q}_{j,2}^h] = [\vec{X}^h(q_{j-1}), \vec{X}^h(q_j)]$, for $j = 1, \ldots, J$. Here $\cdot^\perp$, acting on $\mathbb{R}^2$, denotes clockwise rotation by $\frac{\pi}{2}$.

(iv) In order to find a simple expression for the vertex normal $\vec{\omega}^h$ on Γ^h defined as in Definition 51, we let $\vec{h}_j = \vec{X}^h(q_j) - \vec{X}^h(q_{j-1})$, $j = 1, \ldots, J+1$, which according to Definition 41(i) are nonzero. Then (21) reduces to

$$(\vec{\nu}^h \circ \vec{X}^h)_{|_{I_j}} = \vec{\nu}_j^h = -\frac{\vec{h}_j^\perp}{|\vec{h}_j|} \qquad j = 1, \ldots, J+1.$$

Hence, on recalling (18), we obtain the weighted vertex normals

$$\begin{aligned} \vec{\omega}^h(\vec{X}^h(q_j)) &= \frac{|\vec{h}_j|\, \vec{\nu}_j^h + |\vec{h}_{j+1}|\, \vec{\nu}_{j+1}^h}{|\vec{h}_j| + |\vec{h}_{j+1}|} = -\frac{\left(\vec{h}_j + \vec{h}_{j+1}\right)^\perp}{|\vec{h}_j| + |\vec{h}_{j+1}|} \\ &= -\frac{\left(\vec{X}^h(q_{j+1}) - \vec{X}^h(q_{j-1})\right)^\perp}{|\vec{h}_j| + |\vec{h}_{j+1}|} \qquad j = 1, \ldots, J. \end{aligned} \tag{22}$$

3.2 Stability estimates

Lemma 55. *Let $\Gamma^h = \bigcup_{j=1}^J \overline{\sigma_j}$ be a two-dimensional polyhedral surface. Then we have for $j = 1, \ldots, J$ that*

$$\frac{1}{2} \int_{\sigma_j} |\nabla_s \vec{X}|^2 \, \mathrm{d}\mathcal{H}^2 \geq \mathcal{H}^2(\vec{X}(\sigma_j)) \quad \forall \vec{X} \in \underline{V}(\Gamma^h) \tag{23}$$

with equality for $\vec{X} = \vec{\mathrm{id}}_{|_{\Gamma^h}} \in \underline{V}(\Gamma^h)$.

Proof. We note that the integrands in (23) are constant. In particular, we recall from Definition 5(iv) that, for $\vec{X} \in \underline{V}(\Gamma^h)$,

$$\nabla_s \vec{X} = \sum_{i=1}^{2} (\partial_{\vec{\tau}_i} \vec{X}) \otimes \vec{\tau}_i \quad \text{and} \quad |\nabla_s \vec{X}|^2 = \sum_{i=1}^{2} |\partial_{\vec{\tau}_i} \vec{X}|^2 \quad \text{on } \sigma_j, \tag{24a}$$

and so

$$\nabla_s \vec{\mathrm{id}} = \sum_{i=1}^{2} \vec{\tau}_i \otimes \vec{\tau}_i \quad \text{and} \quad |\nabla_s \vec{\mathrm{id}}|^2 = 2 \quad \text{on } \sigma_j, \tag{24b}$$

where $\{\vec{\tau}_1, \vec{\tau}_2\}$ is an orthonormal basis for the tangent plane of σ_j. Moreover, it holds that

$$\mathcal{H}^2(\vec{X}(\sigma_j)) = \int_{\sigma_j} \sqrt{g} \, \mathrm{d}\mathcal{H}^2, \tag{25}$$

where, similarly to (7) and (2),

$$g = \det\left(\partial_{\vec{\tau}_i} \vec{X} \cdot \partial_{\vec{\tau}_j} \vec{X}\right)_{i,j=1,2} = |\partial_{\vec{\tau}_1} \vec{X}|^2 \, |\partial_{\vec{\tau}_2} \vec{X}|^2 - \left(\partial_{\vec{\tau}_1} \vec{X} \cdot \partial_{\vec{\tau}_2} \vec{X}\right)^2.$$

Next, we note that

$$\sqrt{g} \le |\partial_{\vec{\tau}_1} \vec{X}| \, |\partial_{\vec{\tau}_2} \vec{X}| \le \frac{1}{2}\left(|\partial_{\vec{\tau}_1} \vec{X}|^2 + |\partial_{\vec{\tau}_2} \vec{X}|^2\right), \tag{26}$$

with equality if and only if $\partial_{\vec{\tau}_1} \vec{X} \cdot \partial_{\vec{\tau}_2} \vec{X} = 0$ and $|\partial_{\vec{\tau}_1} \vec{X}| = |\partial_{\vec{\tau}_2} \vec{X}|$. The desired results (23) then follow immediately on combining (24), (25) and (26). □

Remark 56. A result like that in Lemma 55 is not true for an n-dimensional polyhedral surface with $n \neq 2$. In this case

$$\mathcal{H}^n(\vec{X}(\sigma_j)) = \int_{\sigma_j} \sqrt{g} \, \mathrm{d}\mathcal{H}^n,$$

where $g = \det\left(\partial_{\vec{\tau}_i} \vec{X} \cdot \partial_{\vec{\tau}_j} \vec{X}\right)_{i,j=1,\ldots,n}$, with $\{\vec{\tau}_1, \ldots, \vec{\tau}_n\}$ being an orthonormal basis for the tangent space of σ_j, scales with respect to $\vec{X}$ with the power n. Whereas

$$|\nabla_s \vec{X}|^2 = \sum_{i=1}^{n} |\partial_{\vec{\tau}_i} \vec{X}|^2$$

scales to the power two. Hence a simple scaling argument shows that there can be no constant c_0 such that

$$c_0 \int_{\sigma_j} |\nabla_s \vec{X}|^2 \, \mathrm{d}\mathcal{H}^n \ge \mathcal{H}^n(\vec{X}(\sigma_j)) \quad \forall \vec{X} \in \underline{V}(\Gamma^h).$$

This shows that the estimate in Lemma 55 can only be used for 2-dimensional polyhedral surfaces.

Despite the above remark, we are able to prove the following crucial stability bound, which will be extensively used in later sections, for $n = 1$ as well as for $n = 2$.

Lemma 57. *Let* $\Gamma^h = \bigcup_{j=1}^J \overline{\sigma_j}$ *be an n-dimensional polyhedral surface, and let* $\vec{X} \in \underline{V}(\Gamma^h)$*. Then it holds, in the case* $n = 1$*, that*

$$\left\langle \nabla_s \vec{X}, \nabla_s (\vec{X} - \vec{\mathrm{id}}) \right\rangle_{\Gamma^h} \geq \mathcal{H}^1(\vec{X}(\Gamma^h)) - \mathcal{H}^1(\Gamma^h) + \left| |\nabla_s \vec{X}| - 1 \right|^2_{\Gamma^h}.$$

Moreover, in the case $n = 2$*, we have that*

$$\left\langle \nabla_s \vec{X}, \nabla_s (\vec{X} - \vec{\mathrm{id}}) \right\rangle_{\Gamma^h} \geq \mathcal{H}^2(\vec{X}(\Gamma^h)) - \mathcal{H}^2(\Gamma^h) + \frac{1}{2}\left|\nabla_s (\vec{X} - \vec{\mathrm{id}})\right|^2_{\Gamma^h}.$$

Proof. For $n = 2$ it follows from Lemma 55 that

$$\begin{aligned}
&\left\langle \nabla_s \vec{X}, \nabla_s (\vec{X} - \vec{\mathrm{id}}) \right\rangle_{\Gamma^h} \\
&\quad = \frac{1}{2}\left[\left|\nabla_s \vec{X}\right|^2_{\Gamma^h} - \left|\nabla_s \vec{\mathrm{id}}\right|^2_{\Gamma^h} + \left|\nabla_s (\vec{X} - \vec{\mathrm{id}})\right|^2_{\Gamma^h} \right] \\
&\quad \geq \mathcal{H}^2(\vec{X}(\Gamma^h)) - \mathcal{H}^2(\Gamma^h) + \frac{1}{2}\left|\nabla_s (\vec{X} - \vec{\mathrm{id}})\right|^2_{\Gamma^h}.
\end{aligned}$$

For $n = 1$, we let $\vec{h}_j = \vec{q}_{j,2} - \vec{q}_{j,1}$, and similarly $\vec{h}_j^{\vec{X}} = \vec{X}(\vec{q}_{j,2}) - \vec{X}(\vec{q}_{j,1})$, for $j = 1, \ldots, J$. Then, on using ideas from Dziuk (1999b, Theorem 2), it follows from the Cauchy–Schwarz inequality that

$$\begin{aligned}
\left\langle \nabla_s \vec{X}, \nabla_s (\vec{X} - \vec{\mathrm{id}}) \right\rangle_{\Gamma^h} &= \sum_{j=1}^J \left[\frac{|\vec{h}_j^{\vec{X}}|^2 - \vec{h}_j^{\vec{X}} \cdot \vec{h}_j}{|\vec{h}_j|} \right] \\
&= \sum_{j=1}^J \left[\frac{(|\vec{h}_j^{\vec{X}}| - |\vec{h}_j|)^2 + |\vec{h}_j^{\vec{X}}|\,|\vec{h}_j| - \vec{h}_j^{\vec{X}} \cdot \vec{h}_j}{|\vec{h}_j|} + |\vec{h}_j^{\vec{X}}| - |\vec{h}_j| \right] \\
&\geq \sum_{j=1}^J \left[|\vec{h}_j^{\vec{X}}| - |\vec{h}_j| + \left(\frac{|\vec{h}_j^{\vec{X}}|}{|\vec{h}_j|} - \frac{|\vec{h}_j|}{|\vec{h}_j|} \right)^2 |\vec{h}_j| \right] \\
&= \mathcal{H}^1(\vec{X}(\Gamma^h)) - \mathcal{H}^1(\Gamma^h) + \left| |\nabla_s \vec{X}| - 1 \right|^2_{\Gamma^h}.
\end{aligned} \tag{27}$$

□

The result in Lemma 57 is relevant for semiimplicit time discretizations. For fully implicit discretizations we need the following result.

Lemma 58. *Let* $\Gamma^h = \bigcup_{j=1}^J \overline{\sigma_j}$ *be a polygonal curve, and let* $\vec{X} \in \underline{V}(\Gamma^h)$*. Then it holds that*

$$\left\langle \nabla_s \vec{\mathrm{id}}, \nabla_s (\vec{\mathrm{id}} - \vec{X}) \right\rangle_{\Gamma^h} \geq \mathcal{H}^1(\Gamma^h) - \mathcal{H}^1(\vec{X}(\Gamma^h)).$$

Proof. Using the same notation as in the proof of Lemma 57, we have that

$$\begin{aligned}\left\langle \nabla_s \vec{\mathrm{id}}, \nabla_s (\vec{\mathrm{id}} - \vec{X}) \right\rangle_{\Gamma^h} &= \sum_{j=1}^J \left[\frac{|\vec{h}_j|^2 - \vec{h}_j \cdot \vec{h}_j^{\vec{X}}}{|\vec{h}_j|} \right] \\ &\geq \sum_{j=1}^J \left[|\vec{h}_j| - |\vec{h}_j^{\vec{X}}| \right] = \mathcal{H}^1(\Gamma^h) - \mathcal{H}^1(\vec{X}(\Gamma^h)). \qquad \square\end{aligned}$$

Remark 59. A result analogous to Lemma 58 for n-dimensional polyhedral surfaces with $n > 1$ is not true. To see this, we construct the following counterexample. Let Γ^h be given by a single n-simplex, and let $\{\vec{\tau}_1, \ldots, \vec{\tau}_n\}$ be an orthonormal basis for the tangent space of Γ^h. Now choose $\vec{X} \in \underline{V}(\Gamma^h)$ such that $\partial_{\vec{\tau}_1} \vec{X} = \alpha \vec{\tau}_1$ and $\partial_{\vec{\tau}_i} \vec{X} = \varepsilon \vec{\tau}_i$, $i = 2, \ldots, n$, for $\alpha, \varepsilon \in \mathbb{R}_{>0}$. Then it holds that $\mathcal{H}^n(\vec{X}(\Gamma^h)) = \int_{\Gamma^h} \sqrt{\det\left(\partial_{\vec{\tau}_i} \vec{X} \cdot \partial_{\vec{\tau}_j} \vec{X}\right)_{i,j=1,\ldots,n}} \, \mathrm{d}\mathcal{H}^n = \alpha\, \varepsilon^{n-1}\, \mathcal{H}^n(\Gamma^h)$. Moreover, $|\nabla_s \vec{\mathrm{id}}|^2 = n$ and $\nabla_s \vec{\mathrm{id}} : \nabla_s \vec{X} = \alpha + (n-1)\, \varepsilon$ on Γ^h, and so

$$\left\langle \nabla_s \vec{\mathrm{id}}, \nabla_s (\vec{\mathrm{id}} - \vec{X}) \right\rangle_{\Gamma^h} \geq \mathcal{H}^n(\Gamma^h) - \mathcal{H}^n(\vec{X}(\Gamma^h))$$

is equivalent to $n - (\alpha + (n-1)\, \varepsilon) \geq 1 - \alpha\, \varepsilon^{n-1}$, and hence to $(n-1)(1 - \varepsilon) \geq \alpha\, (1 - \varepsilon^{n-1})$. Choosing $\varepsilon \in (0, 1)$ and $\alpha > (n-1)\, \frac{1-\varepsilon}{1-\varepsilon^{n-1}}$ yields a contradiction.

3.3 Curvature approximations

Given a polyhedral hypersurface $\Gamma^h = \bigcup_{j=1}^J \overline{\sigma_j}$, it is clear from Definition 5 that first order differential operators are well-defined almost everywhere on Γ^h, for example, for functions in $V(\Gamma^h)$ or $\underline{V}(\Gamma^h)$. However, second order operators are not.

That means that discrete curvature approximations need to be defined in a suitable way. For everything that follows we assume that Γ^h is a closed hypersurface.

One way is to define the discrete Laplace–Beltrami operator $\Delta_s^{\Gamma^h} : V(\Gamma^h) \to V(\Gamma^h)$ via

$$\left\langle \Delta_s^{\Gamma^h} \chi, \zeta \right\rangle_{\Gamma^h}^h = -\langle \nabla_s \chi, \nabla_s \zeta \rangle_{\Gamma^h} \quad \forall\, \zeta \in V(\Gamma^h), \tag{28}$$

which is a discrete analogue of Remark 22(i). As usual, for $\vec{\chi} \in \underline{V}(\Gamma^h)$, we define $\Delta_s^{\Gamma^h} \vec{\chi}$ component-wise. Then a possible approximation to the curvature vector, recall Lemma 13(ii), is $\vec{\kappa}^h = \Delta_s^{\Gamma^h} \vec{\mathrm{id}}$, i.e., $\vec{\kappa}^h \in \underline{V}(\Gamma^h)$ is the unique solution to

$$\langle \vec{\kappa}^h, \vec{\eta} \rangle_{\Gamma^h}^h = -\left\langle \nabla_s \vec{\mathrm{id}}, \nabla_s \vec{\eta} \right\rangle_{\Gamma^h} \quad \forall \vec{\eta} \in \underline{V}(\Gamma^h). \tag{29}$$

Of course, $\vec{\kappa}^h$ gives both an approximation to the mean curvature, as well as a notion of a vertex normal direction, which in general will be different to the direction defined by Definition 51.

Some special polyhedral hypersurfaces allow an alternative definition of mean curvature.

Definition 60. A closed orientable polyhedral hypersurface Γ^h, with unit normal $\vec{\nu}^h$, is called a conformal polyhedral hypersurface, if there exists a $\kappa^h \in V(\Gamma^h)$ such that

$$\langle \kappa^h \vec{\nu}^h, \vec{\eta} \rangle_{\Gamma^h}^h = -\left\langle \nabla_s \vec{\mathrm{id}}, \nabla_s \vec{\eta} \right\rangle_{\Gamma^h} \quad \forall \vec{\eta} \in \underline{V}(\Gamma^h). \tag{30}$$

Remark 61.

(i) For a conformal polyhedral hypersurface Γ^h, the two vertex normal directions defined by $\vec{\kappa}^h$ in (29) and $\vec{\omega}^h$ in Definition 51 agree, i.e., the two vectors are parallel at each vertex of Γ^h. In particular, on recalling (19), it holds that

$$\vec{\pi}_{\Gamma^h} \left[\kappa^h \vec{\omega}^h \right] = \vec{\kappa}^h.$$

(ii) It is discussed in Barrett et al. (2008a, Section 4.1) that for $d = 3$ the geometric property from (i) means that the triangulation of Γ^h is characterized by a good mesh quality. In the case $d = 2$ it holds that any conformal polygonal curve is weakly equidistributed, see the following theorem.

Theorem 62. *Let Γ^h be a closed conformal polygonal curve in $\mathbb{R}^2$, as defined in Definition 60. Then any two neighbouring elements on Γ^h either have equal length, or they are parallel.*

Proof. We choose a $\vec{X}^h \in \underline{V}^h(\mathbb{I})$ with $\Gamma^h = \vec{X}^h(\mathbb{I})$. Then it follows from Remark 54 and Definition 60 that there exists a $\kappa^h \in V^h(\mathbb{I})$ such that

$$\left\langle \kappa^h \vec{\omega}^h \circ \vec{X}^h, \vec{\eta}\, |\vec{X}_\rho^h| \right\rangle_{\mathbb{I}}^h = -\left\langle \vec{X}_s^h, \vec{\eta}_s\, |\vec{X}_\rho^h| \right\rangle_{\mathbb{I}} \quad \forall \vec{\eta} \in \underline{V}^h(\mathbb{I}). \tag{31}$$

On using the notation from Remark 54(iv), we fix a $j \in \{1, \ldots, J\}$, and then need to show that

$$|\vec{h}_j| = |\vec{h}_{j+1}| \text{ if } \vec{h}_j \nparallel \vec{h}_{j+1}. \tag{32}$$

We recall from Definition 41(i) that $\vec{h}_j$ and $\vec{h}_{j+1}$ are nonzero. If $\vec{h}_j + \vec{h}_{j+1} = \vec{0}$, then (32) directly follows. Otherwise, we observe from (22) that

$$\vec{\omega}^h(\vec{X}^h(q_j)) = -\frac{(\vec{X}^h(q_{j+1}) - \vec{X}^h(q_{j-1}))^\perp}{|\vec{h}_j| + |\vec{h}_{j+1}|}.$$

Hence choosing an $\vec{\eta} \in \underline{V}^h(\mathbb{I})$ in (31) with

$$\vec{\eta}(q_i) = \delta_{ij}(\vec{X}^h(q_{j+1}) - \vec{X}^h(q_{j-1})) = \delta_{ij}\left(\vec{h}_j + \vec{h}_{j+1}\right),$$

for $i = 1, \ldots, J$, we obtain

$$0 = \left(\frac{\vec{h}_{j+1}}{|\vec{h}_{j+1}|} - \frac{\vec{h}_j}{|\vec{h}_j|}\right) \cdot \left(\vec{h}_j + \vec{h}_{j+1}\right) = \frac{|\vec{h}_{j+1}| - |\vec{h}_j|}{|\vec{h}_j|\,|\vec{h}_{j+1}|}\left(|\vec{h}_j|\,|\vec{h}_{j+1}| - \vec{h}_j \cdot \vec{h}_{j+1}\right).$$

The Cauchy–Schwarz inequality now implies that $|\vec{h}_j| = |\vec{h}_{j+1}|$ if $\vec{h}_j$ and $\vec{h}_{j+1}$ are not parallel. □

For polyhedral hypersurfaces that do not satisfy the special property in Definition 60, we can introduce a discrete mean curvature as follows. Find $(\vec{X}, \kappa^h) \in \underline{V}(\Gamma^h) \times V(\Gamma^h)$ such that

$$\left\langle \vec{X} - \vec{\mathrm{id}}, \chi \vec{\nu}^h \right\rangle_{\Gamma^h}^h = 0 \quad \forall \chi \in V(\Gamma^h), \tag{33a}$$

$$\left\langle \kappa^h \vec{\nu}^h, \vec{\eta} \right\rangle_{\Gamma^h}^h + \left\langle \nabla_s \vec{X}, \nabla_s \vec{\eta} \right\rangle_{\Gamma^h} = 0 \quad \forall \vec{\eta} \in \underline{V}(\Gamma^h). \tag{33b}$$

Remark 63.

(i) The system (33) can be viewed as a linearization of (30), where in some sense we allow Γ^h to deform slightly, by moving vertices tangentially.

(ii) If $(\vec{\mathrm{id}}_{|_{\Gamma^h}}, \kappa^h) \in \underline{V}(\Gamma^h) \times V(\Gamma^h)$ solves (33), then κ^h solves (30), and so Γ^h is a conformal polyhedral hypersurface.

Under a mild assumption, there exists a unique solution to the system (33).

Assumption 64. Let Γ^h be an orientable polyhedral hypersurface with unit normal $\vec{\nu}^h$ and vertex normal vector $\vec{\omega}^h \in \underline{V}(\Gamma^h)$.

(i) Let $\dim \operatorname{span}\{\vec{\omega}^h(\vec{q}_k)\}_{k=1}^K = d$.

(ii) Let $\vec{\omega}^h(\vec{q}_k) \neq \vec{0}$, $k = 1, \ldots, K$.

Remark 65. Assumption 64(i) means that the discrete vertex normals of Γ^h span the whole space $\mathbb{R}^d$. On recalling (17) we observe that Assumption 64(i) is equivalent to $\dim\{\int_{\Gamma^h}\chi\,\vec{\nu}^h\,\mathrm{d}\mathcal{H}^{d-1}:\chi\in V(\Gamma^h)\}=d$. Clearly, Assumption 64 is only violated in very rare occasions. For example, it always holds for surfaces Γ^h without self-intersections.

Lemma 66. *Let Γ^h satisfy Assumption 64. Then there exists a unique solution $(\vec{X},\kappa^h)\in\underline{V}(\Gamma^h)\times V(\Gamma^h)$ to (33).*

Proof. As (33) is a linear system, where the number of unknowns equals the number of equations, it is enough to show uniqueness. We hence consider the homogeneous system and assume that $(\vec{X}_0,\kappa_0)\in\underline{V}(\Gamma^h)\times V(\Gamma^h)$ is such that

$$\left\langle\vec{X}_0,\chi\,\vec{\nu}^h\right\rangle_{\Gamma^h}^h=0\quad\forall\chi\in V(\Gamma^h),\tag{34a}$$

$$\left\langle\kappa_0\,\vec{\nu}^h,\vec{\eta}\right\rangle_{\Gamma^h}^h+\left\langle\nabla_s\vec{X}_0,\nabla_s\vec{\eta}\right\rangle_{\Gamma^h}=0\quad\forall\vec{\eta}\in\underline{V}(\Gamma^h).\tag{34b}$$

Choosing $\chi=\kappa_0\in V(\Gamma^h)$ in (34a) and $\vec{\eta}=\vec{X}_0\in\underline{V}(\Gamma^h)$ in (34b) yields that $|\nabla_s\vec{X}_0|_{\Gamma^h}^2=0$, and so $\vec{X}_0$ is constant, i.e., $\vec{X}_0=\vec{X}^c$ on Γ^h for $\vec{X}^c\in\mathbb{R}^d$. In particular, choosing $\vec{\eta}=\vec{\pi}_{\Gamma^h}[\kappa_0\,\vec{\omega}^h]$ in (34b) yields, on recalling (19) and (16), that

$$0=\left\langle\kappa_0\,\vec{\omega}^h,\kappa_0\,\vec{\omega}^h\right\rangle_{\Gamma^h}^h=\left(\left|\kappa_0\,\vec{\omega}^h\right|_{\Gamma^h}^h\right)^2,$$

and so $\vec{\pi}_{\Gamma^h}[\kappa_0\,\vec{\omega}^h]=\vec{0}$. Now Assumption 64(ii) yields that $\kappa_0=0$. Moreover, it follows from (34a), on recalling (20), that

$$0=\left\langle\vec{X}^c,\chi\,\vec{\nu}^h\right\rangle_{\Gamma^h}^h=\left\langle\vec{X}^c,\chi\,\vec{\nu}^h\right\rangle_{\Gamma^h}=\vec{X}^c\cdot\int_{\Gamma^h}\chi\,\vec{\nu}^h\,\mathrm{d}\mathcal{H}^{d-1}\quad\forall\chi\in V(\Gamma^h),$$

and so Assumption 64(i), recall Remark 65, implies that $\vec{X}^c=\vec{0}$. Hence we have shown that there exists a unique solution $(\vec{X},\kappa^h)\in\underline{V}(\Gamma^h)\times V(\Gamma^h)$ to (33). □

Further discrete curvature approximations can be obtained with the help of approximations to the Weingarten map, recall Definition 10 and Lemma 12. In particular, using Remark 22(v) leads to the following discretization of $\nabla_s\vec{\nu}$, which goes back to Heine (2004, (3.2)). Given a closed polyhedral hypersurface Γ^h and a curvature vector approximation $\vec{\kappa}^h\in\underline{V}(\Gamma^h)$, find $\underline{\underline{W}}^h\in\underline{\underline{V}}(\Gamma^h)$ such that

$$\left\langle\underline{\underline{W}}^h,\underline{\underline{\chi}}\right\rangle_{\Gamma^h}=-\left\langle\vec{\kappa}^h,\underline{\underline{\chi}}\,\vec{\nu}^h\right\rangle_{\Gamma^h}-\left\langle\vec{\nu}^h,\nabla_s\cdot\underline{\underline{\chi}}\right\rangle_{\Gamma^h}\quad\forall\underline{\underline{\chi}}\in\underline{\underline{V}}(\Gamma^h).\tag{35a}$$

For example, $\vec{\kappa}^h$ can be defined via (29), or via (29) without mass lumping, which corresponds to the choice in Heine (2004, (3.1)). We note that $\underline{\underline{W}}^h$ is not necessarily symmetric, whereas $\nabla_s \vec{\nu}$ is, recall Lemma 12(ii). An alternative approximation of $\nabla_s \vec{\nu}$ replaces (35a) with

$$\left\langle \underline{\underline{W}}^h, \underline{\underline{\chi}} \right\rangle_{\Gamma^h}^h = -\frac{1}{2}\left\langle \vec{\nu}^h, (\underline{\underline{\chi}} + \underline{\underline{\chi}}^\top)\,\vec{\kappa}^h + \nabla_s \cdot (\underline{\underline{\chi}} + \underline{\underline{\chi}}^\top) \right\rangle_{\Gamma^h}^h \quad \forall \underline{\underline{\chi}} \in \underline{\underline{V}}(\Gamma^h), \tag{35b}$$

which yields $(\underline{\underline{W}}^h)^\top = \underline{\underline{W}}^h$, and which was considered, for example, in Barrett et al. (2017d, (4.12b)).

A slightly modified version of (35a) has been utilized in Barrett et al. (2008d), and is given as follows. Given Γ^h and a mean curvature approximation $\kappa^h \in V(\Gamma^h)$, find $\underline{\underline{W}}^h \in \underline{\underline{V}}(\Gamma^h)$ such that

$$\left\langle \underline{\underline{W}}^h, \underline{\underline{\chi}} \right\rangle_{\Gamma^h}^h = -\left\langle \kappa^h\, \vec{\nu}^h, \underline{\underline{\chi}}\, \vec{\nu}^h \right\rangle_{\Gamma^h}^h - \left\langle \vec{\nu}^h, \nabla_s \cdot \underline{\underline{\chi}} \right\rangle_{\Gamma^h} \quad \forall \underline{\underline{\chi}} \in \underline{\underline{V}}(\Gamma^h). \tag{35c}$$

Finally, piecewise constant approximations to $\nabla_s \vec{\nu}$ can be defined by

$$\nabla_s \vec{\omega}^h \in \underline{\underline{V}}_c(\Gamma^h) \quad \text{and} \quad \nabla_s \left(\vec{\pi}_{\Gamma^h} \frac{\vec{\omega}^h}{|\vec{\omega}^h|} \right) \in \underline{\underline{V}}_c(\Gamma^h), \tag{35d}$$

the latter of which clearly needs Assumption 64(ii) to hold, and has been employed in, e.g., Barrett et al. (2008d). We note that the two approximations in (35d) are in general not symmetric.

Remark 67. For the case of curves, $d = 2$, and adopting the notation of Section 3.1.2, we set $\Gamma^h = \vec{X}^h(\mathbb{I})$, where $\vec{X}^h \in \underline{V}^h(\mathbb{I})$ interpolates $\vec{x}$ with $\Gamma = \vec{x}(\mathbb{I})$. Then it is shown in Deckelnick and Dziuk (2009, Lemma 2.2) that the approximation $\vec{\kappa}^h$ from (29) approximates the true curvature vector, $\vec{\varkappa}$, of Γ with order $O(h)$ in $[L^2(\mathbb{I})]^2$ for smooth Γ. Unfortunately, for the case $d = 3$ it is shown in Heine (2004) that (29) and (35a) are not convergent on general meshes, see also Hildebrandt et al. (2006). Similar conclusions can be drawn from the numerical experiments in Barrett et al. (2008d, Section 4.2.1), and also apply to (35b) and (35c). However, we note that the approximations (35d) and (33) behave better in practice, see Tables 2 and 5 and Tables 3 and 6 in Barrett et al. (2008d), respectively, for closely related approximations. Moreover, one can prove convergence for higher order piecewise polynomial approximations of Γ and $\vec{\varkappa}$, see Heine (2004). Even though (29) may not be convergent for continuous piecewise linears, it turns out that the use of such an approximation does lead to convergence in approximating geometric flows; see, e.g., Section 4.7.1.

3.4 Evolving polyhedral surfaces and transport theorems

We now define discrete analogues to evolving hypersurfaces, their velocity fields and material time derivatives. For more details we refer to Dziuk and Elliott (2013, Section 5.4).

Definition 68.

(i) Let $(\Gamma^h(t))_{t\in[0,T]}$ be a family of polyhedral hypersurfaces, such that each $\Gamma^h(t)$ admits a triangulation of the form Remark 42(iii) for fixed J and K, and such that the position of each vertex $\vec{q}_k$, $k = 1, \ldots, K$, is a C^1-function in time. Then the set

$$\mathcal{G}_T^h = \bigcup_{t\in[0,T]} (\Gamma^h(t) \times \{t\})$$

is called an evolving polyhedral hypersurface. We will often identify $\mathcal{G}_T^h$ with $(\Gamma^h(t))_{t\in[0,T]}$, and call the latter also an evolving polyhedral hypersurface.

(ii) The velocity of $\Gamma^h(t)$ on $\mathcal{G}_T^h$ is defined by

$$\vec{\mathcal{V}}^h(\vec{z}, t) = \sum_{k=1}^{K} \left[\frac{\mathrm{d}}{\mathrm{d}t}\vec{q}_k(t)\right] \phi_k^{\Gamma^h(t)}(\vec{z}) \quad \forall (\vec{z}, t) \in \mathcal{G}_T^h,$$

where we have recalled the notation from Definition 43(i).

(iii) We define the finite element spaces

$$V(\mathcal{G}_T^h) = \{\chi \in C(\mathcal{G}_T^h) : \chi(\cdot, t) \in V(\Gamma^h(t)) \quad \forall t \in [0, T]\}$$

and $\underline{V}(\mathcal{G}_T^h) = [V(\mathcal{G}_T^h)]^d$.

(iv) Let $f \in L^\infty(\mathcal{G}_T^h)$, with $f \in C^1(\mathcal{S}_{j,T}^h)$ for $j = 1, \ldots, J$, where

$$\mathcal{S}_{j,T}^h = \bigcup_{t\in[0,T]} (\overline{\sigma_j(t)} \times \{t\})$$

is a C^1-evolving hypersurface. For $j \in \{1, \ldots, J\}$, let $\vec{x} : \Upsilon \times [0, T] \to \mathbb{R}^d$ be a global parameterization of $\mathcal{S}_{j,T}^h$ such that $\vec{x}(\cdot, t) : \Upsilon \to \overline{\sigma_j(t)}$ is an affine function. Then we define the discrete time derivative of f by

$$\partial_t^{\circ,h} f = \partial_t^\circ f \quad on\ \mathcal{S}_{j,T}^h,$$

recall Definition 28(i).

(v) We define the finite element spaces

$$V_T(\mathcal{G}_T^h) = \{\chi \in V(\mathcal{G}_T^h) : \partial_t^{\circ,h} \chi \in C(\mathcal{G}_T^h)\}$$

and $\underline{V}_T(\mathcal{G}_T^h) = [V_T(\mathcal{G}_T^h)]^d$ of finite element functions on $\mathcal{G}_T^h$ with a continuous material derivative.

Remark 69.

(i) On introducing the short hand notation $\phi_k^h(\cdot, t) = \phi_k^{\Gamma^h(t)}$, it holds that

$$\partial_t^{\circ,h} \phi_k^h = 0 \quad \text{on } \mathcal{G}_T^h, \quad k = 1, \ldots, K.$$

(ii) In general the discrete material derivative $\partial_t^{\circ,h} f$ is only defined piecewise on $\mathcal{G}_T^h$. But a direct consequence of (i) is that for $\chi \in V(\mathcal{G}_T^h)$, with $\chi(\vec{q}_k(\cdot), \cdot) \in C^1([0,T])$, $k = 1, \ldots, K$, it holds that

$$(\partial_t^{\circ,h} \chi)(\vec{z}, t) = \sum_{k=1}^{K} \left[\frac{\mathrm{d}}{\mathrm{d}t} \chi(\vec{q}_k(t), t)\right] \phi_k^{\Gamma^h(t)}(\vec{z}) \quad \forall (\vec{z}, t) \in \mathcal{G}_T^h,$$

i.e., we can choose a continuous representation of $\partial_t^{\circ,h} \chi$, and hence $\chi \in V_T(\mathcal{G}_T^h)$.

(iii) We have that $\vec{\mathcal{V}}^h \in \underline{V}(\mathcal{G}_T^h)$ with $\vec{\mathcal{V}}^h = \partial_t^{\circ,h} \vec{\mathrm{id}}$ on $\mathcal{G}_T^h$.

(iv) On extending f to a neighbourhood of $\mathcal{S}_{j,T}^h$, $j = 1, \ldots, J$, it holds that

$$\partial_t^{\circ,h} f = \partial_t f + \vec{\mathcal{V}}^h \cdot \nabla f \quad \text{on } \mathcal{S}_{j,T}^h,$$

recall Remark 29(ii).

Theorem 70. *Let $\mathcal{G}_T^h$ be an evolving polyhedral hypersurface, and let $\eta, \zeta \in V_T(\mathcal{G}_T^h)$.*

(i) *It holds that*

$$\frac{\mathrm{d}}{\mathrm{d}t} \langle \eta, \zeta \rangle_{\Gamma^h(t)} = \langle \partial_t^{\circ,h} \eta, \zeta \rangle_{\Gamma^h(t)} + \langle \eta, \partial_t^{\circ,h} \zeta \rangle_{\Gamma^h(t)} + \left\langle \eta\, \zeta, \nabla_s \cdot \vec{\mathcal{V}}^h \right\rangle_{\Gamma^h(t)}.$$

(ii) *It holds that*

$$\frac{\mathrm{d}}{\mathrm{d}t} \langle \eta, \zeta \rangle_{\Gamma^h(t)}^h = \langle \partial_t^{\circ,h} \eta, \zeta \rangle_{\Gamma^h(t)}^h + \langle \eta, \partial_t^{\circ,h} \zeta \rangle_{\Gamma^h(t)}^h + \left\langle \eta\, \zeta, \nabla_s \cdot \vec{\mathcal{V}}^h \right\rangle_{\Gamma^h(t)}^h.$$

Proof.

(i) Using the transport theorem, Theorem 32, on each evolving simplex $\sigma_j(t)$ of $\Gamma^h(t)$, and using the assumptions $\eta, \zeta \in V_T(\mathcal{G}_T^h)$, leads to

$$\begin{aligned}
\frac{\mathrm{d}}{\mathrm{d}t} \langle \eta, \zeta \rangle_{\Gamma^h(t)} &= \frac{\mathrm{d}}{\mathrm{d}t} \sum_{j=1}^{J} \int_{\sigma_j(t)} \eta\, \zeta \, \mathrm{d}\mathcal{H}^{d-1} \\
&= \sum_{j=1}^{J} \int_{\sigma_j(t)} \partial_t^{\circ,h} (\eta\, \zeta) + \eta\, \zeta\, \nabla_s \cdot \vec{\mathcal{V}}^h \, \mathrm{d}\mathcal{H}^{d-1} \\
&= \langle \partial_t^{\circ,h} \eta, \zeta \rangle_{\Gamma^h(t)} + \langle \eta, \partial_t^{\circ,h} \zeta \rangle_{\Gamma^h(t)} + \left\langle \eta\, \zeta, \nabla_s \cdot \vec{\mathcal{V}}^h \right\rangle_{\Gamma^h(t)}.
\end{aligned}$$

(ii) This proof is analogous to (i) and can be found in Barrett et al. (2015b, Lemma 3.1). □

Theorem 71. *Let $\mathcal{G}_T^h$ be an evolving polyhedral hypersurface, such that $\Gamma^h(t)$ is bounding a domain $\Omega^h(t) \subset \mathbb{R}^d$, for $t \in [0,T]$. We assume that $\vec{\nu}^h(t)$ is the outer unit normal to $\Omega^h(t)$ on $\Gamma^h(t)$, and that $f \in C^1(\overline{\mathfrak{O}_T^h})$, where*

$$\mathfrak{O}_T^h = \bigcup_{t \in [0,T]} (\Omega^h(t) \times \{t\}).$$

Then it holds that

$$\frac{\mathrm{d}}{\mathrm{d}t} \int_{\Omega^h(t)} f \, \mathrm{d}\mathcal{L}^d = \int_{\Omega^h(t)} \partial_t f \, \mathrm{d}\mathcal{L}^d + \left\langle f, \vec{\mathcal{V}}^h \cdot \vec{\nu}^h \right\rangle_{\Gamma^h(t)}.$$

Proof. This follows as in Eck et al. (2017, Section 7.3) using a variant of the divergence theorem for Lipschitz domains. □

3.5 Further results for evolving polyhedral surfaces

We state discrete analogues of Lemmas 37 and 38.

Lemma 72. *Let $\mathcal{G}_T^h$ be an evolving polyhedral hypersurface. Then it holds that*

$$\partial_t^{\circ,h} \vec{\nu}^h = -(\nabla_s \vec{\mathcal{V}}^h)^\top \vec{\nu}^h \quad a.e. \ on \, \Gamma^h(t).$$

Proof. Similarly to the proof of Theorem 70(i), we appeal to Lemma 37(i) on each evolving simplex $\sigma_j(t)$ of $\Gamma^h(t)$. □

Lemma 73. *Let $\mathcal{G}_T^h$ be an evolving polyhedral hypersurface, and let $\eta \in V_T(\mathcal{G}_T^h)$, $\vec{\eta} \in \underline{V}_T(\mathcal{G}_T^h)$. Then we have the following results, where we recall from Definition 5(vii) that $\underline{\underline{D}}_s(\vec{\mathcal{V}}^h) = \frac{1}{2} \underline{\underline{P}}_{\Gamma^h} (\nabla_s \vec{\mathcal{V}}^h + (\nabla_s \vec{\mathcal{V}}^h)^\top) \underline{\underline{P}}_{\Gamma^h}$ almost everywhere on $\Gamma^h(t)$.*

(i) $$\partial_t^{\circ,h} \nabla_s \eta - \nabla_s \partial_t^{\circ,h} \eta = [\nabla_s \vec{\mathcal{V}}^h - 2 \underline{\underline{D}}_s(\vec{\mathcal{V}}^h)] \nabla_s \eta \quad a.e. \ on \, \Gamma^h(t).$$

(ii) $$\partial_t^{\circ,h} \nabla_s \vec{\eta} - \nabla_s \partial_t^{\circ,h} \vec{\eta} = (\nabla_s \vec{\eta}) \, [\nabla_s \vec{\mathcal{V}}^h - 2 \underline{\underline{D}}_s(\vec{\mathcal{V}}^h)]^\top \quad a.e. \ on \, \Gamma^h(t).$$

(iii) $$\partial_t^{\circ,h} (\nabla_s \cdot \vec{\eta}) - \nabla_s \cdot (\partial_t^{\circ,h} \vec{\eta}) = [\nabla_s \vec{\mathcal{V}}^h - 2 \underline{\underline{D}}_s(\vec{\mathcal{V}}^h)] : \nabla_s \vec{\eta} \quad a.e. \ on \, \Gamma^h(t).$$

Proof. Similarly to the proof of Theorem 70(i), we appeal to Lemma 38 on each evolving simplex $\sigma_j(t)$ of $\Gamma^h(t)$. □

4 Mean curvature flow

The main ideas needed to numerically solve curvature-driven evolution equations are most easily introduced with the help of the mean curvature flow. For a family of closed evolving hypersurfaces $(\Gamma(t))_{t \in [0,T]}$ in $\mathbb{R}^d$, $d \geq 2$, we consider at each time the total surface area $|\Gamma(t)| = \mathcal{H}^{d-1}(\Gamma(t))$. In this section we will only consider closed surfaces $\Gamma(t)$, i.e., surfaces which are compact and without boundary. The transport theorem, Theorem 32, gives

$$\frac{\mathrm{d}}{\mathrm{d}t}|\Gamma(t)| = -\langle \varkappa, \mathcal{V} \rangle_{\Gamma(t)}, \tag{36}$$

where $\varkappa$ is the mean curvature of $\Gamma(t)$, $\mathcal{V}$ is its normal velocity, and where we recall that $\langle \cdot, \cdot \rangle_{\Gamma(t)}$ denotes the L^2-inner product on $\Gamma(t)$. Here we used that $\Gamma(t)$ has no boundary. We hence obtain that the geometric evolution law for an evolving hypersurface

$$\mathcal{V} = \varkappa \quad \text{on}\,\Gamma(t) \tag{37}$$

most efficiently decreases the surface area, and hence it is also called the L^2-gradient flow of $|\Gamma(t)|$, see, e.g., Mantegazza (2011) and Garcke (2013) for details. This law is the most fundamental curvature-driven evolution law and has been studied in detail both analytically and numerically, we refer to Huisken (1984), Gage and Hamilton (1986), Giga (2006), Mantegazza (2011), Garcke (2013), Deckelnick et al. (2005a) and the references therein, for details.

4.1 Weak formulation

In this section we want to introduce, for mean curvature flow, ideas introduced by the present authors to approximate curvature-driven evolution laws for hypersurfaces. These ideas lead to stable approximations, which, in addition, are such that the quality of the mesh approximating the evolving surface in general remains good. In fact, the latter property is crucial, as many parametric approaches suffer from mesh degeneracies during the evolution. These degeneracies may even lead to situations, where the resulting algorithms break down during the evolution. The ideas presented in this section will then be the basis for more complex evolution laws studied later on. We will also compare our approach to other methods in the literature dealing with mean curvature flow.

The basis of our approach is a weak formulation, which we introduce next. The goal is to write the evolution law $\mathcal{V} = \varkappa$ in a weak form. To this end, we firstly note that the evolution law, for a global parameterization $\vec{x} : \Upsilon \times [0, T] \to \mathbb{R}^d$, and corresponding orientable hypersurfaces $\Gamma(t) = \vec{x}(\Upsilon, t)$, recall Definition 25, can be written as

$$\vec{\mathcal{V}} \cdot \vec{\nu} = \varkappa, \quad \varkappa\,\vec{\nu} = \Delta_s\,\vec{\mathrm{id}} \quad \text{on}\,\Gamma(t), \tag{38}$$

where we have noted Remark 26(i) and Lemma 13(ii). On recalling from Remark 22(iv) the weak formulation of the second identity, we propose the following weak formulation for mean curvature flow. Given a closed hypersurface $\Gamma(0)$, find an evolving hypersurface $(\Gamma(t))_{t\in[0,T]}$ with a global parameterization and induced velocity field $\vec{\mathcal{V}}$, and $\varkappa \in L^2(\mathcal{G}_T)$ as follows. For almost all $t \in (0,T)$, find $(\vec{\mathcal{V}}(\cdot, t), \varkappa(\cdot, t)) \in [L^2(\Gamma(t))]^d \times L^2(\Gamma(t))$ such that

$$\left\langle \vec{\mathcal{V}}, \chi \vec{\nu} \right\rangle_{\Gamma(t)} - \langle \varkappa, \chi \rangle_{\Gamma(t)} = 0 \quad \forall \chi \in L^2(\Gamma(t)), \tag{39a}$$

$$\langle \varkappa \vec{\nu}, \vec{\eta} \rangle_{\Gamma(t)} + \left\langle \nabla_s \vec{\mathrm{id}}, \nabla_s \vec{\eta} \right\rangle_{\Gamma(t)} = 0 \quad \forall \vec{\eta} \in [H^1(\Gamma(t))]^d. \tag{39b}$$

We note here that we consider closed surfaces and hence no boundary term appears in (39b). Using the weak formulation (39b) of $\varkappa \vec{\nu} = \Delta_s \vec{\mathrm{id}}$ was the fundamental idea of Dziuk (1991), which made it possible to approximate smooth surfaces and their mean curvature by piecewise smooth surfaces. Finally we remark that $H^1(\Gamma(t))$ denotes the usual Sobolev space of square integrable functions on $\Gamma(t)$ with square integrable surface gradient, and we refer to Wloka (1987, I Section 4) for an introduction to Sobolev spaces on surfaces.

4.2 Finite element approximation

Given an initial polyhedral hypersurface Γ^0 the plan is to construct polyhedral hypersurfaces Γ^m, $m = 1, \ldots, M$, which approximate the true continuous solution $\Gamma(t_m)$ to the mean curvature flow at times $0 = t_0 < t_1 < \ldots < t_M = T$, which form a partition of a time interval $[0, T]$ with time steps

$$\Delta t_m = t_{m+1} - t_m, \qquad m = 0, \ldots, M-1. \tag{40}$$

An idea going back to Dziuk (1991) is to parameterize Γ^{m+1} over Γ^m with the help of parameterizations $\vec{X}^{m+1} : \Gamma^m \to \mathbb{R}^d$. We recall the definitions of the finite element spaces and inner products from Definition 43, and then recall the following finite element approximation of (39) for mean curvature flow from Barrett et al. (2007b, 2008a).

Let the closed polyhedral hypersurface Γ^0 be an approximation of $\Gamma(0)$. Then, for $m = 0, \ldots, M-1$, find $(\vec{X}^{m+1}, \kappa^{m+1}) \in \underline{V}(\Gamma^m) \times V(\Gamma^m)$ such that

$$\left\langle \frac{\vec{X}^{m+1} - \vec{\mathrm{id}}}{\Delta t_m}, \chi \vec{\nu}^m \right\rangle_{\Gamma^m}^h - \left\langle \kappa^{m+1}, \chi \right\rangle_{\Gamma^m}^h = 0 \quad \forall \chi \in V(\Gamma^m), \tag{41a}$$

$$\left\langle \kappa^{m+1} \vec{\nu}^m, \vec{\eta} \right\rangle_{\Gamma^m}^h + \left\langle \nabla_s \vec{X}^{m+1}, \nabla_s \vec{\eta} \right\rangle_{\Gamma^m} = 0 \quad \forall \vec{\eta} \in \underline{V}(\Gamma^m) \tag{41b}$$

and set $\Gamma^{m+1} = \vec{X}^{m+1}(\Gamma^m)$.

Here we recall from Definition 25(ii) that $\frac{\vec{X}^{m+1} - \vec{\mathrm{id}}}{\Delta t_m}$ on Γ^m is a natural approximation of $\vec{\mathcal{V}}$ on $\Gamma(t_m)$. We also remark that although the original problem was highly nonlinear, the system (41) is linear and easy to solve. The main reason for this is that the geometry, which enters the weak formulations via the area element, the normal vector and the surface gradients, is taken explicitly.

Remark 74 (Surfaces with boundary).
We recall that (37) was derived as the L^2-gradient flow for $|\Gamma(t)|$ from (36) for an evolving hypersurface without boundary. If we allow $\partial\Gamma(t)$ to be nonempty, on the other hand, then (36) needs to be adapted to

$$\frac{\mathrm{d}}{\mathrm{d}t}|\Gamma(t)| = -\left\langle \varkappa\,\vec{\nu}, \vec{\mathcal{V}} \right\rangle_{\Gamma(t)} + \left\langle \vec{\mathcal{V}}, \vec{\mu} \right\rangle_{\partial\Gamma(t)}, \tag{42}$$

where $\vec{\mathcal{V}}$ is the velocity field induced by a global parameterization of the evolving hypersurface, and $\vec{\mu}(t)$ denotes the outer unit conormal on $\partial\Gamma(t)$, recall Theorem 32. Now choosing a boundary condition that makes the last term in (42) vanish will ensure that (37) is still the L^2-gradient flow for $|\Gamma(t)|$. The simplest such boundary condition fixes

$$\partial\Gamma(t) = \partial\Gamma(0) \quad \forall\, t \in (0, T]. \tag{43}$$

The approximation (41) can be easily generalized to this situation, by replacing the space $\underline{V}(\Gamma^m)$ in (41b) with

$$\underline{V}_{\mathrm{D}}(\Gamma^m) = \left\{\vec{\eta} \in \underline{V}(\Gamma^m) : \vec{\eta} = \vec{0} \text{ on } \partial\Gamma^m \right\}, \tag{44}$$

and by seeking $\vec{X}^{m+1}$ such that $\vec{X}^{m+1} - \vec{\mathrm{id}}_{|_{\Gamma^m}} \in \underline{V}_{\mathrm{D}}(\Gamma^m)$. More complicated boundary conditions can be handled in a similar way, for example, the case of prescribed contact angles when the boundary $\partial\Gamma(t)$ is allowed to move along the boundary of a fixed given domain. A further related example is the evolution of a cluster of hypersurfaces, where the boundaries of a number of surfaces are required to remain attached to each other. Typically triple junction points in the plane, and triple junction lines in $\mathbb{R}^3$ are of interest, and these situations can be naturally approximated with the methods presented in this chapter both for mean curvature flow, as well as for more general geometric evolution equations. We refer the interested reader to the series of papers Barrett et al. (2007a,b, 2008b, 2010a,d).

Remark 75 (Implementation).
We note that implementing the system (41) is not difficult. For $d = 2$ an equivalent finite difference formulation can be derived, see Section 4.3.1 below. For $d = 3$ the scheme (41) can either be implemented directly in a high level computing environment like MATLAB, or within a finite element toolbox that allows the approximation of PDEs on two-dimensional hypersurfaces in $\mathbb{R}^3$. Examples for such toolboxes are ALBERTA (Schmidt and Siebert, 2005); Dune (Dedner et al., 2010) and FELICITY (Walker, 2018). An advantage of these toolboxes is that they allow a nearly dimension-independent implementation of the scheme, and so the cases $d = 2$ and $d = 3$ can be treated together. For the piecewise linear approximation (41), the only difference

to standard problems on flat, stationary domains in $\mathbb{R}^{d-1}$ is then, that the vertices of the initial mesh, on which the PDE is to be approximated, have d coordinates, rather than $d-1$, and that the vertices of the mesh are moved after each time step.

The initial mesh can either be created with the help of a simple, coarse macro-triangulation, that is then refined and transformed with the help of the capabilities of the chosen finite element toolbox. Or it can be created by using sophisticated 3D volume mesh generators that allow to extract the surface mesh of the generated 3D volume mesh. Examples for such volume mesh generators are Gmsh (Geuzaine and Remacle, 2009); CGAL (Rineau and Yvinec, 2019); TIGER (Walker, 2013); Cleaver (CIBC, 2016) and NETGEN (Schöberl, 1997).

4.3 Discrete linear systems

We now describe the linear systems arising from (41). We introduce the matrices $\vec{N}_{\Gamma^m} \in (\mathbb{R}^d)^{K\times K}$, $M_{\Gamma^m}, A_{\Gamma^m} \in \mathbb{R}^{K\times K}$ and $\underline{\underline{A}}_{\Gamma^m} \in (\mathbb{R}^{d\times d})^{K\times K}$ with entries

$$[M_{\Gamma^m}]_{kl} = \left\langle \phi_k^{\Gamma^m}, \phi_l^{\Gamma^m} \right\rangle_{\Gamma^m}^h, \quad [\vec{N}_{\Gamma^m}]_{kl} = \int_{\Gamma^m} \pi_{\Gamma^m}\left[\phi_k^{\Gamma^m}\, \phi_l^{\Gamma^m}\right] \vec{\nu}^m \, \mathrm{d}\mathcal{H}^{d-1},$$
$$[A_{\Gamma^m}]_{kl} = \left\langle \nabla_s \phi_k^{\Gamma^m}, \nabla_s \phi_l^{\Gamma^m} \right\rangle_{\Gamma^m} \tag{45}$$

and $\left[\underline{\underline{A}}_{\Gamma^m}\right]_{kl} = [A_{\Gamma^m}]_{kl}\, \underline{\underline{\mathrm{Id}}}$, where we have recalled Definition 43(i). Assembling these matrices is similar to the situation of finite element methods for domains in $\mathbb{R}^{d-1}$. In addition to computing the volume of a simplex, one has to compute its normal and the surface gradients of the basis functions. The latter can be computed using the formula

$$\nabla_s \phi_k^{\Gamma^m} = \sum_{i=1}^{d-1} \left(\partial_{\vec{\tau}_i} \phi_k^{\Gamma^m}\right) \vec{\tau}_i,$$

where $\{\vec{\tau}_1, \ldots, \vec{\tau}_{d-1}\}$ is an orthonormal basis of the tangent space to the simplex, recall Definition 5(ii). We can then formulate (41) as: Find $(\delta\vec{X}^{m+1}, \kappa^{m+1}) \in (\mathbb{R}^d)^K \times \mathbb{R}^K$ such that

$$\begin{pmatrix} \Delta t_m M_{\Gamma^m} & -\vec{N}_{\Gamma^m}^{\mathsf{T}} \\ \vec{N}_{\Gamma^m} & \underline{\underline{A}}_{\Gamma^m} \end{pmatrix} \begin{pmatrix} \kappa^{m+1} \\ \delta\vec{X}^{m+1} \end{pmatrix} = \begin{pmatrix} 0 \\ -\underline{\underline{A}}_{\Gamma^m}\vec{X}^m \end{pmatrix}, \tag{46}$$

where, with the obvious abuse of notation, $\delta\vec{X}^{m+1} = (\delta\vec{X}_1^{m+1}, \ldots, \delta\vec{X}_K^{m+1})^{\mathsf{T}}$, $\kappa^{m+1} = (\kappa_1^{m+1}, \ldots, \kappa_K^{m+1})^{\mathsf{T}}$, and $\vec{X}^m = (\vec{X}_1^m, \ldots, \vec{X}_K^m)^{\mathsf{T}}$ are the vectors of coefficients with respect to the standard basis for $\vec{X}^{m+1} - \vec{\mathrm{id}}_{|_{\Gamma^m}}$, κ^{m+1} and $\vec{\mathrm{id}}_{|_{\Gamma^m}}$, respectively. In the following section we will show that the above system, under very mild assumptions, is invertible. Hence it can be solved, for example, with a sparse direct solution method like UMFPACK, see Davis (2004), in an efficient way. It is also possible to solve the system iteratively, by first using a Schur complement approach to (46), in order to eliminate κ^{m+1}, and then use a (precondioned) conjugate gradient solver, see, e.g., Barrett et al. (2008a) for details. In fact, the Schur complement approach essentially boils down to the system (47), which we derive next.

It is also possible to rewrite the system (41) as an equation for $\vec{X}^{m+1}$ as the only unknown. To this end, let $\vec{\omega}^m \in \underline{V}(\Gamma^m)$ be the vertex normal to Γ^m, recall Definition 51. Then it follows from (19) that we can compute κ^{m+1} from (41a) via

$$\kappa^{m+1} = \frac{1}{\Delta t_m} \pi_{\Gamma^m}\left[\left(\vec{X}^{m+1} - \vec{\mathrm{id}}_{|\Gamma^m}\right) \cdot \vec{\omega}^m\right],$$

recall Definition 43(i). Hence we can rewrite (41) as: Find $\vec{X}^{m+1} \in \underline{V}(\Gamma^m)$ such that

$$\left\langle \frac{\vec{X}^{m+1} - \vec{\mathrm{id}}}{\Delta t_m} \cdot \vec{\omega}^m, \vec{\eta} \cdot \vec{\omega}^m \right\rangle_{\Gamma^m}^h + \langle \nabla_s \vec{X}^{m+1}, \nabla_s \vec{\eta} \rangle_{\Gamma^m} = 0 \quad \forall \vec{\eta} \in \underline{V}(\Gamma^m). \tag{47}$$

4.3.1 Curves in the plane

The system (41) is particularly simple in the case of closed curves. On recalling Definition 53, we can reformulate it as follows.

Let $\vec{X}^0 \in \underline{V}^h(\mathbb{I})$ be such that $\Gamma^0 = \vec{X}^0(\mathbb{I})$ is a polygonal approximation of $\Gamma(0)$. Then, for $m = 0, \ldots, M-1$, find $(\vec{X}^{m+1}, \kappa^{m+1}) \in \underline{V}^h(\mathbb{I}) \times V^h(\mathbb{I})$ such that

$$\left\langle \frac{\vec{X}^{m+1} - \vec{X}^m}{\Delta t_m}, \chi \vec{\nu}^m \circ \vec{X}^m \, |\vec{X}^m_\rho| \right\rangle_{\mathbb{I}}^h - \left\langle \kappa^{m+1}, \chi \, |\vec{X}^m_\rho| \right\rangle_{\mathbb{I}}^h = 0 \quad \forall \chi \in V^h(\mathbb{I}), \tag{48a}$$

$$\left\langle \kappa^{m+1} \, \vec{\nu}^m \circ \vec{X}^m, \vec{\eta} \, |\vec{X}^m_\rho| \right\rangle_{\mathbb{I}}^h + \left\langle \vec{X}^{m+1}_\rho, \vec{\eta}_\rho \, |\vec{X}^m_\rho|^{-1} \right\rangle_{\mathbb{I}} = 0 \quad \forall \vec{\eta} \in \underline{V}^h(\mathbb{I}) \tag{48b}$$

and set $\Gamma^{m+1} = \vec{X}^{m+1}(\mathbb{I})$.

It is now straightforward to rewrite (48) as a finite difference scheme. Let κ^{m+1}_j, $\vec{X}^{m+1}_j$, $\vec{X}^m_j$ and $\vec{\omega}^m_j$ be the values of κ^{m+1}, $\vec{X}^{m+1}$, $\vec{X}^m$ and $\vec{\omega}^m \circ \vec{X}^m$ at the node q_j, for $j = 0, \ldots, J+1$. Let $\vec{h}^m_j = \vec{X}^m_j - \vec{X}^m_{j-1}$, $j = 1, \ldots, J+1$. Then, on recalling (22), we obtain the weighted vertex normals

$$\vec{\omega}^m_j = -\frac{\left(\vec{X}^m_{j+1} - \vec{X}^m_{j-1}\right)^\perp}{|\vec{h}^m_j| + |\vec{h}^m_{j+1}|} \quad j = 1, \ldots, J. \tag{49}$$

Eq. (48a) can hence be rewritten in the following finite difference form

$$\frac{1}{\Delta t_m}\left(\vec{X}_j^{m+1} - \vec{X}_j^m\right) \cdot \vec{\omega}_j^m = \kappa_j^{m+1}, \quad j = 1, \ldots, J. \tag{50a}$$

Testing (48b) with the standard basis functions gives the finite difference type equations

$$\kappa_j^{m+1}\,\vec{\omega}_j^m = \frac{2}{|\vec{h}_j^m| + |\vec{h}_{j+1}^m|}\left(\frac{\vec{X}_{j+1}^{m+1} - \vec{X}_j^{m+1}}{|\vec{h}_{j+1}^m|} - \frac{\vec{X}_j^{m+1} - \vec{X}_{j-1}^{m+1}}{|\vec{h}_j^m|}\right) \quad j = 1, \ldots, J. \tag{50b}$$

Of course, it is now also possible to use (50a) in order to reduce (50b) to an equation for the nodal values of $\vec{X}^{m+1}$, in order to obtain a finite difference analogue of (47).

4.4 Existence and uniqueness

In order to show existence and uniqueness of solutions to (41), we recall Assumption 64.

Theorem 76. *Let* Γ^m *satisfy Assumption 64(i). Then there exists a unique solution* $(\vec{X}^{m+1}, \kappa^{m+1}) \in \underline{V}(\Gamma^m) \times V(\Gamma^m)$ *to the system (41).*

Proof. Similarly to the proof of Lemma 66, existence follows from uniqueness, and so we consider the homogeneous system

$$\left\langle \vec{X}, \chi\,\vec{\nu}^m \right\rangle_{\Gamma^m}^h - \Delta t_m \langle \kappa, \chi \rangle_{\Gamma^m}^h = 0 \quad \forall \chi \in V(\Gamma^m), \tag{51a}$$

$$\langle \kappa\,\vec{\nu}^m, \vec{\eta} \rangle_{\Gamma^m}^h + \left\langle \nabla_s \vec{X}, \nabla_s \vec{\eta} \right\rangle_{\Gamma^m} = 0 \quad \forall \vec{\eta} \in \underline{V}(\Gamma^m), \tag{51b}$$

for unknowns $(\vec{X}, \kappa) \in \underline{V}(\Gamma^m) \times V(\Gamma^m)$. Choosing $\chi = \kappa \in V(\Gamma^m)$ in (51a) and $\vec{\eta} = \vec{X} \in \underline{V}(\Gamma^m)$ in (51b) yields that

$$\left|\nabla_s \vec{X}\right|_{\Gamma^m}^2 + \Delta t_m \left(|\kappa|_{\Gamma^m}^h\right)^2 = 0, \tag{52}$$

where we have recalled (16). It follows from (52) that $\kappa = 0$ and $\vec{X} = \vec{X}^c$ on Γ^m, for $\vec{X}^c \in \mathbb{R}^d$. Together with (51a) this implies, on recalling (20), that

$$\left\langle \vec{X}^c, \chi\,\vec{\nu}^m \right\rangle_{\Gamma^m} = 0 \quad \forall \chi \in V(\Gamma^m). \tag{53}$$

As in the proof of Lemma 66 it follows from (53) and Assumption 64(i) that $\vec{X}^c = \vec{0}$. Hence we have shown that there exists a unique solution $(\vec{X}^{m+1}, \kappa^{m+1}) \in \underline{V}(\Gamma^m) \times V(\Gamma^m)$ to (41). □

4.5 Stability

For $d = 2$ and $d = 3$ it can be shown that the scheme (41) is unconditionally stable.

Theorem 77. *Let $d = 2$ or $d = 3$. Let $(\vec{X}^{m+1}, \kappa^{m+1}) \in \underline{V}(\Gamma^m) \times V(\Gamma^m)$ be a solution to (41). Then we have that*

$$\left|\Gamma^{m+1}\right| + \Delta t_m \left(\left|\kappa^{m+1}\right|^h_{\Gamma^m} \right)^2 \leq \left|\Gamma^m\right|.$$

Proof. Choosing $\chi = \kappa^{m+1} \in V(\Gamma^m)$ in (41a) and $\vec{\eta} = \frac{1}{\Delta t_m} (\vec{X}^{m+1} - \vec{\mathrm{id}}_{|_{\Gamma^m}}) \in \underline{V}(\Gamma^m)$ in (41b) yields that

$$\left\langle \nabla_s \vec{X}^{m+1}, \nabla_s \left(\vec{X}^{m+1} - \vec{\mathrm{id}} \right) \right\rangle_{\Gamma^m} + \Delta t_m \left(\left|\kappa^{m+1}\right|^h_{\Gamma^m} \right)^2 = 0. \tag{54}$$

In addition, we have from Lemma 57 that

$$\left\langle \nabla_s \vec{X}^{m+1}, \nabla_s \left(\vec{X}^{m+1} - \vec{\mathrm{id}} \right) \right\rangle_{\Gamma^m} \geq \left|\Gamma^{m+1}\right| - \left|\Gamma^m\right|. \tag{55}$$

Combining (54) and (55) then yields the claim. □

Remark 78. Clearly, if $\{(\vec{X}^{m+1}, \kappa^{m+1})\}_{m=0}^{M-1}$ denote solutions to (41) for $m = 0, \ldots, M-1$, then the above theorem immediately yields the energy bound

$$\left|\Gamma^k\right| + \sum_{m=0}^{k-1} \Delta t_m \left(\left|\kappa^{m+1}\right|^h_{\Gamma^m} \right)^2 \leq \left|\Gamma^0\right|.$$

for $k = 1, \ldots, M$.

4.6 Equipartition property

Many numerical methods that use polyhedral surfaces to approximate an evolving hypersurface suffer from the fact that the meshes will have very inhomogeneous properties during the evolution. For example, mesh points can come very close to each other during the evolution, or some angles in the mesh can become rather small or rather large. This leads to very unstable and inaccurate situations. In addition, the condition number of the linear systems to be solved at each time level will become very large. In some situations, the computations cannot even be continued after some time. It is an important property of the approach discussed in this section so far that the mesh typically behaves very well. We will discuss this below for mean curvature flow, which is the simplest possible situation. However, similar results hold true for the extension of this scheme to more complicated flows discussed in the remaining sections of this chapter.

The behaviour of the mesh is best explained with the help of a semidiscrete version of the scheme (41), where we recall Definition 68. Given the closed polyhedral hypersurface $\Gamma^h(0)$, find an evolving polyhedral hypersurface $\mathcal{G}_T^h$, with induced velocity $\vec{\mathcal{V}}^h \in \underline{V}(\mathcal{G}_T^h)$, and $\kappa^h \in V(\mathcal{G}_T^h)$, i.e., $\vec{\mathcal{V}}^h(\cdot,t) \in \underline{V}(\Gamma^h(t))$ and $\kappa^h(\cdot, t) \in V(\Gamma^h(t))$ for all $t \in [0,T]$, such that, for all $t \in (0, T]$,

$$\left\langle \vec{\mathcal{V}}^h, \chi \vec{\nu}^h \right\rangle_{\Gamma^h(t)}^h - \langle \kappa^h, \chi \rangle_{\Gamma^h(t)} = 0 \quad \forall \chi \in V(\Gamma^h(t)), \tag{56a}$$

$$\langle \kappa^h \vec{\nu}^h, \vec{\eta} \rangle_{\Gamma^h(t)}^h + \left\langle \nabla_s \vec{\mathrm{id}}, \nabla_s \vec{\eta} \right\rangle_{\Gamma^h(t)} = 0 \quad \forall \vec{\eta} \in \underline{V}(\Gamma^h(t)). \tag{56b}$$

We can prove that the scheme (56) is stable, i.e., a semidiscrete analogue of Theorem 77 holds, and that its tangential motion ensures that any solution satisfies the property in Definition 60.

Theorem 79. *Let $(\mathcal{G}_T^h, \kappa^h)$ be a solution of (56).*
(i) *It holds that*

$$\frac{\mathrm{d}}{\mathrm{d}t} |\Gamma^h(t)| + \left(|\kappa^h|_{\Gamma^h(t)}^h \right)^2 = 0.$$

(ii) *For any $t \in (0,T]$, it holds that $\Gamma^h(t)$ is a conformal polyhedral surface. In particular, for $d = 2$, any two neighbouring elements of the curve $\Gamma^h(t)$ either have equal length, or they are parallel.*

Proof.
(i) Choosing $\chi = \kappa^h(\cdot,t) \in V(\Gamma^h(t))$ in (56a) and $\vec{\eta} = \vec{\mathcal{V}}^h(\cdot,t) \in \underline{V}(\Gamma^h(t))$ in (56b) gives, on recalling Theorem 70 and Lemma 9(ii), that

$$\frac{\mathrm{d}}{\mathrm{d}t} |\Gamma^h(t)| = \left\langle 1, \nabla_s \cdot \vec{\mathcal{V}}^h \right\rangle_{\Gamma^h(t)} = \left\langle \nabla_s \vec{\mathrm{id}}, \nabla_s \vec{\mathcal{V}}^h \right\rangle_{\Gamma^h(t)} = -\langle \kappa^h, \kappa^h \rangle_{\Gamma^h(t)}^h,$$

which is the claim.
(ii) This follows directly from Definition 60 and Theorem 62. □

Remark 80. In general it is not clear whether solutions to (56) exist. For example, for $d \geq 3$ the topology of the triangulation fixed by $\Gamma^h(0)$ may be such that no solutions satisfying both (56a) and (56b) exist. In those situations, the fully discrete scheme (41), even for very small time step sizes Δt_m, may exhibit mesh defects that would not be expected for surfaces satisfying Definition 60, recall Remark 61. An example for such defects was recently presented in Elliott and Fritz (2017, Fig. 27).

In the case of curves, i.e., $d = 2$, the scheme (56) can be easily interpreted as a differential-algebraic system of equations. To this end, we adopt the notation from Section 4.3.1 for $\vec{X}^h(t) \in \underline{V}^h(\mathbb{I})$, and let $\vec{h}_j(t) = \vec{X}_j^h(t) - \vec{X}_{j-1}^h(t)$, $j = 1, \ldots, J + 1$. Then we obtain the obvious analogue of (49) for the vertex normals $\vec{\omega}_j^h(t)$ and, similarly to (50), the system (56) can then be written as the following differential-algebraic system of equations (DAEs). Given $\vec{X}^h(0) \in \underline{V}^h(\mathbb{I})$, for $t \in (0,T]$ find $\vec{X}^h(t) \in \underline{V}^h(\mathbb{I})$ and $\kappa^h(t) \in V^h(\mathbb{I})$ such that

$$-\left(\frac{\mathrm{d}}{\mathrm{d}t}\vec{X}_j^h\right) \cdot \left(\frac{\left(\vec{X}_{j+1}^h - \vec{X}_{j-1}^h\right)^\perp}{\left|\vec{h}_j\right| + \left|\vec{h}_{j+1}\right|}\right) = \kappa_j^h, \tag{57a}$$

$$-\kappa_j^h \frac{\left(\vec{X}_{j+1}^h - \vec{X}_{j-1}^h\right)^\perp}{\left|\vec{h}_j\right| + \left|\vec{h}_{j+1}\right|} = \frac{2}{\left|\vec{h}_j\right| + \left|\vec{h}_{j+1}\right|}\left(\frac{\vec{X}_{j+1}^h - \vec{X}_j^h}{\left|\vec{h}_{j+1}\right|} - \frac{\vec{X}_j^h - \vec{X}_{j-1}^h}{\left|\vec{h}_j\right|}\right), \tag{57b}$$

for $j = 1, \ldots, J$.

Remark 81. Eq. (57a) only specifies one direction of $\frac{\mathrm{d}}{\mathrm{d}t}\vec{X}_j^h$, while the evolution of the remaining direction is enforced via the algebraic conditions in (57b). Hence the overall system is a highly nonlinear and degenerate differential-algebraic system of equations. The defining equations become singular where vertices coalesce. However, as is shown in Theorem 79(ii), in regions where the curve is not straight, this does not happen.

So far we have discussed the linear fully discrete scheme (41), which is obtained as a possible time discretization of (56). The fully discrete scheme (41) will not inherit the conformality property of Theorem 79(ii), but solutions to (41) still exhibit a tangential motion that leads to a good distribution of vertices. We now discuss a fully implicit time discretization of (56), where the solutions are conformal polyhedral surfaces at every time step. These ideas were first presented in Barrett et al. (2011), in the case of curves.

Let the closed polyhedral hypersurface Γ^0 be an approximation of $\Gamma(0)$. Then, for $m = 0, \ldots, M-1$, find a polyhedral hypersurface Γ^{m+1}, and $(\vec{X}^m, \kappa^{m+1}) \in \underline{V}(\Gamma^{m+1}) \times V(\Gamma^{m+1})$ with $\Gamma^m = \vec{X}^m(\Gamma^{m+1})$, such that

$$\left\langle \frac{\vec{\mathrm{id}} - \vec{X}^m}{\Delta t_m}, \chi\,\vec{\nu}^{m+1} \right\rangle_{\Gamma^{m+1}}^h - \left\langle \kappa^{m+1}, \chi \right\rangle_{\Gamma^{m+1}}^h = 0 \quad \forall \chi \in V(\Gamma^{m+1}), \tag{58a}$$

$$\left\langle \kappa^{m+1}\,\vec{\nu}^{m+1}, \vec{\eta} \right\rangle_{\Gamma^{m+1}}^h + \left\langle \nabla_s \vec{\mathrm{id}}, \nabla_s \vec{\eta} \right\rangle_{\Gamma^{m+1}} = 0 \quad \forall \vec{\eta} \in \underline{V}(\Gamma^{m+1}). \tag{58b}$$

Theorem 82. *Let Γ^{m+1} and $(\vec{X}^m, \kappa^{m+1}) \in \underline{V}(\Gamma^{m+1}) \times V(\Gamma^{m+1})$ be a solution to (58).*

(i) *Then it holds that Γ^{m+1} is a conformal polyhedral surface. In particular, for $d = 2$, any two neighbouring elements of the curve Γ^{m+1} either have equal length, or they are parallel.*

(ii) *In the case $d = 2$ it holds that*

$$\left|\Gamma^{m+1}\right| + \Delta t_m \left(\left|\kappa^{m+1}\right|_{\Gamma^{m+1}}^h\right)^2 \leq \left|\Gamma^m\right|. \tag{59}$$

Proof.
(i) This follows directly from Definition 60 and Theorem 62.
(ii) Choosing $\chi = \kappa^{m+1} \in V(\Gamma^{m+1})$ in (58a) and $\vec{\eta} = \frac{1}{\Delta t_m}(\vec{\mathrm{id}}_{|_{\Gamma^{m+1}}} - \vec{X}^m) \in \underline{V}(\Gamma^{m+1})$ in (58b) yields that

$$\left\langle \nabla_s \vec{\mathrm{id}}, \nabla_s \left(\vec{\mathrm{id}} - \vec{X}^m \right) \right\rangle_{\Gamma^{m+1}} + \Delta t_m \left(\left| \kappa^{m+1} \right|_{\Gamma^{m+1}}^h \right)^2 = 0. \tag{60}$$

In addition, we have from Lemma 58 that

$$\left\langle \nabla_s \vec{\mathrm{id}}, \nabla_s \left(\vec{\mathrm{id}} - \vec{X}^m \right) \right\rangle_{\Gamma^{m+1}} \geq \left| \Gamma^{m+1} \right| - \left| \Gamma^m \right|. \tag{61}$$

Combining (60) and (61) then yields the claim. □

Remark 83. Similarly to Remark 80, in general it is not clear whether solutions to (58) exist. However, for $d = 2$ solutions in general appear to exist, and a small adaptation of (58) discussed below yields an unconditionally stable scheme that equidistributes.

For the case of curves, $d = 2$, and building on (58), in Barrett et al. (2011) the present authors introduced fully discrete parametric finite element discretizations for $\mathcal{V} = \varkappa$ that are unconditionally stable and that intrinsically equidistribute the vertices at each time level. Using the notation from Section 4.3.1, see also Remark 54, we can reformulate (58) in the case of curves as follows.

Let $\vec{X}^0 \in \underline{V}^h(\mathbb{I})$ be such that $\Gamma^0 = \vec{X}^0(\mathbb{I})$ is a polygonal approximation of $\Gamma(0)$. Then, for $m = 0, \ldots, M-1$, find $\Gamma^{m+1} = \vec{X}^{m+1}(\mathbb{I})$, with $\vec{X}^{m+1} \in \underline{V}^h(\mathbb{I})$, and $\kappa^{m+1} \in V^h(\mathbb{I})$ such that for all $\chi \in V^h(\mathbb{I})$, $\vec{\eta} \in \underline{V}^h(\mathbb{I})$

$$\left\langle \frac{\vec{X}^{m+1} - \vec{X}^m}{\Delta t_m}, \chi \left(\vec{X}_\rho^{m+1} \right)^\perp \right\rangle_{\mathbb{I}}^h + \left\langle \kappa^{m+1}, \chi \left| \vec{X}_\rho^{m+1} \right| \right\rangle_{\mathbb{I}}^h = 0, \tag{62a}$$

$$\left\langle \kappa^{m+1} \left(\vec{X}_\rho^{m+1} \right)^\perp, \vec{\eta} \right\rangle_{\mathbb{I}}^h - \left\langle \vec{X}_\rho^{m+1}, \vec{\eta}_\rho \left| \vec{X}_\rho^{m+1} \right|^{-1} \right\rangle_{\mathbb{I}} = 0. \tag{62b}$$

We recall from Theorem 82 that solutions to (62) are equidistributed, at least where elements of the curve Γ^{m+1} are not locally parallel. This result can be sharpened to full equidistribution by considering the following adapted version of (62). For $m = 0, \ldots, M-1$, find $\Gamma^{m+1} = \vec{X}^{m+1}(\mathbb{I})$, with $\vec{X}^{m+1} \in \underline{V}^h(\mathbb{I})$, and $\kappa^{m+1} \in V^h(\mathbb{I})$ such that for all $\chi \in V^h(\mathbb{I})$, $\vec{\eta} \in \underline{V}^h(\mathbb{I})$

$$\left\langle \frac{\vec{X}^{m+1} - \vec{X}^m}{\Delta t_m}, \chi \left(\vec{X}_\rho^{m+1} \right)^\perp \right\rangle_{\mathbb{I}}^h + \left| \Gamma^{m+1} \right| \left\langle \kappa^{m+1}, \chi \right\rangle_{\mathbb{I}}^h = 0, \tag{63a}$$

$$\left\langle \kappa^{m+1} \left(\vec{X}_\rho^{m+1} \right)^\perp, \vec{\eta} \right\rangle_{\mathbb{I}}^h - \left| \Gamma^{m+1} \right|^{-1} \left\langle \vec{X}_\rho^{m+1}, \vec{\eta}_\rho \right\rangle_{\mathbb{I}} = 0. \tag{63b}$$

Theorem 84. *Let* $(\vec{X}^{m+1}, \kappa^{m+1}) \in \underline{V}^h(\mathbb{I}) \times V^h(\mathbb{I})$ *be a solution of (63). Then it holds that*

$$\left|\vec{X}^{m+1}_\rho\right| = \left|\Gamma^{m+1}\right| \quad in\, I_j, \quad j = 1, \ldots, J. \tag{64}$$

Moreover, $(\vec{X}^{m+1}, \kappa^{m+1})$ *solves (62), and satisfies the stability estimate*

$$\left|\Gamma^{m+1}\right| + \Delta t_m \left|\Gamma^{m+1}\right| \left\langle \kappa^{m+1}, \kappa^{m+1} \right\rangle^h_{\mathbb{I}} \leq \left|\Gamma^m\right|. \tag{65}$$

Proof. If $(\vec{X}^{m+1}, \kappa^{m+1}) \in \underline{V}^h(\mathbb{I}) \times V^h(\mathbb{I})$ is a solution of (63), then on using the notation from Remark 54(iv), and similarly to the proof of Theorem 62, we fix $j \in \{1, \ldots, J\}$ and choose $\vec{\eta} \in \underline{V}^h(\mathbb{I})$ in (58b) with

$$\vec{\eta}(q_i) = \delta_{ij}\left(\vec{X}^{m+1}(q_{j+1}) - \vec{X}^{m+1}(q_{j-1})\right) = \delta_{ij}\left(\vec{h}^{m+1}_j + \vec{h}^{m+1}_{j+1}\right),$$

for $i = 1, \ldots, J$, in order to obtain

$$0 = \left(\vec{h}^{m+1}_{j+1} - \vec{h}^{m+1}_j\right) \cdot \left(\vec{h}^{m+1}_{j+1} + \vec{h}^{m+1}_j\right) = \left|\vec{h}^{m+1}_{j+1}\right|^2 - \left|\vec{h}^{m+1}_j\right|^2. \tag{66}$$

Since (66) holds for $j = 1, \ldots, J$, we obtain (64). It follows from (64) that $(\vec{X}^{m+1}, \kappa^{m+1}) \in \underline{V}^h(\mathbb{I}) \times V^h(\mathbb{I})$ also solves (62), and $(\vec{X}^m \circ (\vec{X}^{m+1})^{-1}, \kappa^{m+1} \circ (\vec{X}^{m+1})^{-1}) \in \underline{V}(\Gamma^{m+1}) \times V(\Gamma^{m+1})$ is a solution to (58), satisfying the stability bound (59). Hence we obtain (65). □

Remark 85.

(i) Clearly, (64) means that any solution to (63) is truly equidistributed.

(ii) The system (63) is nonlinear, and so it can be solved either by a Newton method or with the following iteration. Given $\Gamma^{m+1,0} = \vec{X}^{m+1,0}(\mathbb{I})$, with $\vec{X}^{m+1,0} \in \underline{V}^h(\mathbb{I})$, we seek for $i \geq 0$ solutions $(\vec{X}^{m+1,i+1}, \kappa^{m+1,i+1}) \in \underline{V}^h(\mathbb{I}) \times V^h(\mathbb{I})$ such that for all $\chi \in V^h(\mathbb{I})$, $\vec{\eta} \in \underline{V}^h(\mathbb{I})$

$$\left\langle \frac{\vec{X}^{m+1,i+1} - \vec{X}^m}{\Delta t_m}, \chi\left(\vec{X}^{m+1,i}_\rho\right)^\perp \right\rangle^h_{\mathbb{I}} + \left|\Gamma^{m+i,i}\right| \left\langle \kappa^{m+1,i+1}, \chi \right\rangle^h_{\mathbb{I}} = 0,$$

$$\left\langle \kappa^{m+1,i+1}\left(\vec{X}^{m+1,i}_\rho\right)^\perp, \vec{\eta} \right\rangle^h_{\mathbb{I}} - \left|\Gamma^{m+1,i}\right|^{-1} \left\langle \vec{X}^{m+1,i+1}_\rho, \vec{\eta}_\rho \right\rangle = 0,$$

and set $\Gamma^{m+1,i+1} = \vec{X}^{m+1,i+1}(\mathbb{I})$. This system has a unique solution and can be solved similarly to the discussion in Section 4.3. We refer to Barrett et al. (2011) for details.

As an example, we show an evolution of curve shortening flow, i.e., mean curvature flow in the case $d = 2$, computed with the scheme (41) in Fig. 2.

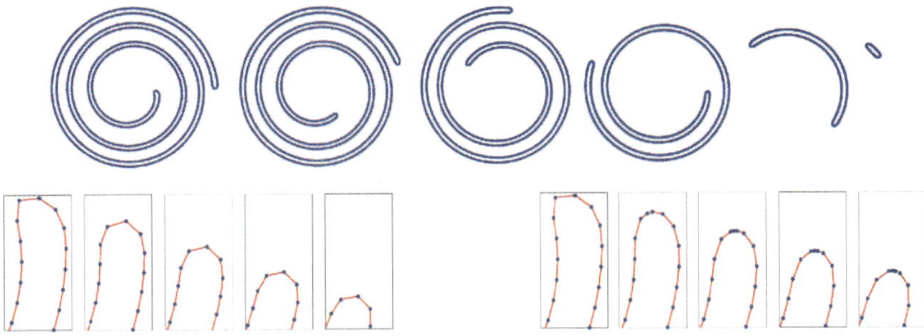

FIG. 2 Mean curvature flow for a spiral. Numerical solution for the scheme (41) with $J = 1024$ and $\Delta t_m = \Delta t = 10^{-7}$, $m = 0, \ldots, M-1$. The discrete solution is shown at times $t = 0, 0.001, 0.005, 0.01, 0.02, 0.024$. Below we show details of the vertex distributions for the scheme (41) close to the inner tip (*left*), and compare that with the classical Dziuk scheme (68) (*right*). Coalescence for the latter scheme means that the simulation breaks down and cannot be continued.

4.7 Alternative parametric methods

The discussion so far was focussed on contributions of the present authors. We will now discuss other numerical methods for the numerical approximation of mean curvature flow, focussing on parametric finite element methods.

4.7.1 The classical Dziuk approach

The approach introduced by Dziuk (1991) is based on a weak formulation of

$$\vec{\mathcal{V}} = \vec{\varkappa}, \quad \vec{\varkappa} = \Delta_s \vec{\mathrm{id}} \quad \text{on}\, \Gamma(t) \quad \Rightarrow \quad \vec{\mathcal{V}} = \Delta_s \vec{\mathrm{id}} \quad \text{on}\, \Gamma(t), \tag{67}$$

rather than (38). Since $\Delta_s \vec{\mathrm{id}} = \vec{\varkappa} = \varkappa \vec{\nu}$ is collinear to the normal, recall Lemma 13(ii), the evolving surface always moves in a direction collinear to the normal. The system to be solved, using the notation of Section 4.2, is then, given the closed polyhedral hypersurface Γ^m, to find $\vec{X}^{m+1} \in \underline{V}(\Gamma^m)$ such that

$$\left\langle \frac{\vec{X}^{m+1} - \vec{\mathrm{id}}}{\Delta t_m}, \vec{\eta} \right\rangle_{\Gamma^m} + \left\langle \nabla_s \vec{X}^{m+1}, \nabla_s \vec{\eta} \right\rangle_{\Gamma^m} = 0 \quad \forall \vec{\eta} \in \underline{V}(\Gamma^m). \tag{68}$$

Existence, uniqueness and stability of this fully discrete linear system can be shown similarly to the results in Sections 4.4 and 4.5. For the case $d = 2$ an error estimate for a semidiscrete continuous-in-time variant of (68) with mass lumping is shown in Dziuk (1994), on assuming that the approximated solution is sufficiently smooth. Even though for the case $d = 3$ no convergence results have been shown for either (68) or (41), in practice both approximations appear to be convergent, see, e.g., Barrett et al. (2008a, Table 1).

4.7.2 *The convergent finite element algorithm of Kovács, Li, Lubich*

Very recently, a first proof of convergence for a numerical method for mean curvature flow for $d = 3$ was given by Kovács et al. (2019). They prove error estimates for semi- as well as full-discretizations of mean curvature flow, again assuming that the approximated solution is sufficiently smooth. They use the system

$$\vec{\mathcal{V}} = \varkappa\,\vec{\nu}, \;\; \partial_t^{\circ}\,\vec{\nu} = \Delta_s\,\vec{\nu} + |\nabla_s\,\vec{\nu}|^2\vec{\nu}, \;\; \partial_t^{\circ}\,\varkappa = \Delta_s\,\varkappa + |\nabla_s\,\vec{\nu}|^2\varkappa \qquad \text{on}\,\Gamma(t).$$

The last two equations can be derived using basic geometric analysis and the law $\vec{\mathcal{V}} = \varkappa\,\vec{\nu}$; see Lemmas 37(i), 16, 7(iii), 12(i) and 39(i). Compared to (41) the approach has the disadvantages that meshes can still deteriorate, as the approximated flow is normal, similarly to the classical Dziuk approach, and that an additional system has to be solved. Of course, a clear advantage is that it is presently the only method, for which convergence can be shown for $d = 3$.

4.7.3 *Alternative numerical methods that equidistribute*

For the mean curvature flow of curves in the plane, the scheme (63) equidistributes the vertices of the polygonal approximation of the curve at every time step. In this section we discuss some alternative numerical methods that achieve the same goal.

As discussed before, it is desirable to control the tangential distribution of mesh points, and so to prevent numerical singularities such as coalescence or swallow tails, i.e., numerically induced self-intersections. In practice, many authors use mesh smoothing methods, see, e.g., Sethian (1985), Dziuk et al. (2002) and Bänsch et al. (2005). However, mesh smoothing has some undesirable features, i.e., they might smoothen the solution excessively and stability results that hold for the original approximation are in general lost. For geometric evolution laws that have some conservation properties, for example, flows where the enclosed area is conserved, great care must be taken to maintain the conservation properties after mesh smoothing.

In the case of an evolving curve $(\Gamma(t))_{t\in[0,T]}$, we consider parameterizations $\vec{x}:\mathbb{I}\times[0,T]\to\mathbb{R}^2$, where $\mathbb{I}=\mathbb{R}/\mathbb{Z}$ is the periodic interval $[0,1]$, such that $\Gamma(t)=\vec{x}(\mathbb{I},t)$, $t\in[0,T]$ are solutions of the curvature flow equation $\mathcal{V}=\varkappa$. With s we denote the arclength of $\Gamma(t)$, so that the unit tangent is given by $\vec{x}_s(\rho,t)=\frac{\vec{x}_\rho(\rho,t)}{|\vec{x}_\rho(\rho,t)|}$, $\rho\in\mathbb{I}$, and a possible normal on $\Gamma(t)$ is $\vec{\nu}\circ\vec{x}=-\vec{x}_s^{\perp}$. The basic idea is to only consider parameterizations that satisfy

$$|\vec{x}_\rho(\rho,\,t)| = |\Gamma(t)| \quad \forall\rho\in\mathbb{I}, \;\; t\in[0,T], \tag{69}$$

which means that ρ, up to a constant, is arclength. First such approaches are due to Kessler et al. (1984), and were analyzed and modified in Strain (1989),

Hou et al. (1994) and Mikula and Ševčovič (2001). It is also possible to discretize an evolution law for the position vector $\vec{x}$ in the form

$$\partial_t \vec{x} = (\mathcal{V}\vec{\nu}) \circ \vec{x} + \alpha \vec{x}_s \quad \text{in } \mathbb{I}, \tag{70}$$

where the tangential velocity is chosen such that the vertices of the polygonal approximation of $\Gamma(t)$ remain close to being equidistributed. This idea has been used in Kimura (1994) and Mikula and Ševčovič (2001) show that choosing α such that

$$\alpha_s = (\mathcal{V}\varkappa) \circ \vec{x} - \frac{1}{|\Gamma(t)|} \langle \mathcal{V}, \varkappa \rangle_{\Gamma(t)},$$

leads to the property (69), if (69) is fulfilled at the initial time. Also Deckelnick and Dziuk (1995) use (70) with

$$\alpha = \frac{\vec{x}_{\rho\rho} \cdot \vec{x}_\rho}{|\vec{x}_\rho|^3} \quad \text{in } \mathbb{I}, \tag{71}$$

and observe well distributed polygonal meshes in practice. In addition, an error analysis can be performed for a semidiscrete continuous-in-time finite element approximation. The resulting fully discrete scheme is particularly simple, and can be formulated as follows. Given $\vec{X}^m \in \underline{V}^h(\mathbb{I})$, find $\vec{X}^{m+1} \in \underline{V}^h(\mathbb{I})$ such that

$$\left\langle \frac{\vec{X}^{m+1} - \vec{X}^m}{\Delta t_m}, \vec{\eta} \left| \vec{X}^m_\rho \right|^2 \right\rangle_{\mathbb{I}} + \left\langle \vec{X}^{m+1}_\rho, \vec{\eta}_\rho \right\rangle_{\mathbb{I}} = 0 \quad \forall \vec{\eta} \in \underline{V}^h(\mathbb{I}), \tag{72}$$

where we observe that the numerical computations presented in Deckelnick and Dziuk (1995) use mass lumping for the first term in (72). The authors in Pan and Wetton (2008) and Pan (2008), on the other hand, solve $\mathcal{V} = \varkappa$ with the side constraint

$$\vec{x}_\rho(\rho, t) \cdot \vec{x}_{\rho\rho}(\rho, t) = 0 \quad \forall \rho \in \mathbb{I}, \ t \in [0, T],$$

which leads to $(\frac{1}{2}|\vec{x}_\rho|^2)_\rho = \vec{x}_\rho \cdot \vec{x}_{\rho\rho} = 0$ and hence to equidistributed parameterizations. However, this approach leads to very nonlinear fully discrete problems.

4.7.4 Using the DeTurck trick to obtain good mesh properties

Mean curvature flow $\mathcal{V} = \varkappa$, as a parabolic equation, has certain degeneracies, which basically stem from the fact that a reparameterization of a solution leads to another solution. It was an idea by DeTurck (1983) to use solutions of the harmonic map heat flow, in order to reparameterize the evolution of curvature flows in such a way that the reparameterized equations are strongly parabolic. Elliott and Fritz (2017) used the DeTurck trick to solve the mean curvature flow numerically. The resulting algorithm has the remarkable property, that provided the initial surface is approximated by a triangulation with a good mesh quality, the meshes will remain good during the evolution. This, roughly speaking, is due to the fact that, under appropriate assumptions, harmonic maps are conformal maps, which hence preserve angles.

The algorithm of Elliott and Fritz (2017), for the case $d = 3$, may at times lead to better meshes than the method discussed in Section 4.2. However, the systems to be solved are more complicated, and the approach also seems to be less flexible, e.g., when it comes to anisotropy, triple junctions and the coupling of the interface evolution to other fields. We refer to Elliott and Fritz (2017) for more details about this approach and for an error analysis in the case $d = 2$ for a semidiscrete continuous-in-time variant. This error analysis is an extension of that in Deckelnick and Dziuk (1995) for the variant with the specific tangential velocity (71). The error analysis in Elliott and Fritz (2017) was extended to the fully discrete case in Barrett et al. (2017a).

4.7.5 Other numerical approaches

In some situations the surface to be computed can be written as a graph. Finite element approximations in this case have been analyzed in Dziuk (1999b) and Deckelnick and Dziuk (1999). For the case of axisymmetric surfaces moving by mean curvature flow, numerical schemes have been presented in Mayer and Simonett (2002) and Barrett et al. (2019a). We also mention Mikula et al. (2014), who proposed methods for a tangential redistribution of points on evolving surfaces. They use volume-oriented and length-oriented tangential redistribution methods, which can be applied to manifolds in any dimension. Completely different approaches are the level set method and the phase field method, which are discussed in other chapters of this handbook.

5 Surface diffusion and other flows

In this section we extend the ideas from Section 4 to more general evolution laws for evolving hypersurfaces in $\mathbb{R}^d$, $d \geq 2$.

5.1 Properties of the surface diffusion flow

Fourth order geometric evolution equations for hypersurfaces have played an important role in the last twenty years. A basic example, which can be formulated in a very simple way, is motion by surface diffusion

$$\mathcal{V} = -\Delta_s \varkappa \quad \text{on}\, \Gamma(t).$$

This flow was introduced by Mullins (1957) to describe diffusion at interfaces in alloys. The flow has the remarkable property that for closed surfaces it is surface area decreasing and the enclosed volume is preserved. These two properties follow from the transport theorems as follows. If we define $\Omega(t)$ to be the domain enclosed by the closed hypersurface $\Gamma(t)$, with $\vec{\nu}(t)$ denoting the outer unit normal to $\Omega(t)$ on $\Gamma(t)$, then we obtain with the help of Theorem 33 that

$$\frac{\mathrm{d}}{\mathrm{d}t}\mathcal{L}^d(\Omega(t)) = \frac{\mathrm{d}}{\mathrm{d}t}\int_{\Omega(t)} 1 \,\mathrm{d}\mathcal{L}^d = \langle 1, \mathcal{V}\rangle_{\Gamma(t)} = -\langle 1, \Delta_s \varkappa\rangle_{\Gamma(t)} = 0, \tag{73}$$

where the last identity follows from the divergence theorem on manifolds, recall Remark 22(i). Furthermore, for the total surface area, $|\Gamma(t)|$, we obtain from Theorem 32 that

$$\frac{\mathrm{d}}{\mathrm{d}t}|\Gamma(t)| = -\langle \varkappa, \mathcal{V}\rangle_{\Gamma(t)} = \langle \varkappa, \Delta_s \varkappa\rangle_{\Gamma(t)} = -|\nabla_s \varkappa|^2_{\Gamma(t)} \leq 0, \tag{74}$$

where for the last identity we recall once more Remark 22(i). In fact, the properties (73) and (74) are closely related to the fact that mathematically surface diffusion can be interpreted as the H^{-1}-gradient flow of $|\Gamma(t)|$, see, e.g., Taylor and Cahn (1994).

Using the formulation

$$\vec{\mathcal{V}} \cdot \vec{\nu} = -\Delta_s \varkappa, \quad \varkappa \vec{\nu} = \Delta_s \vec{\mathrm{id}} \quad \text{on } \Gamma(t), \tag{75}$$

it is now possible to formulate a finite element approximation.

5.2 Finite element approximation for surface diffusion

Using the same notation as in Section 4.2, we recall the following finite element approximation of (75) for surface diffusion from Barrett et al. (2007a, 2008a). Let the closed polyhedral hypersurface Γ^0 be an approximation of $\Gamma(0)$. Then, for $m = 0, \ldots, M-1$, find $(\vec{X}^{m+1}, \kappa^{m+1}) \in \underline{V}(\Gamma^m) \times V(\Gamma^m)$ such that

$$\left\langle \frac{\vec{X}^{m+1} - \vec{\mathrm{id}}}{\Delta t_m}, \chi \vec{\nu}^m \right\rangle^h_{\Gamma^m} - \langle \nabla_s \kappa^{m+1}, \nabla_s \chi\rangle_{\Gamma^m} = 0 \quad \forall \chi \in V(\Gamma^m), \tag{76a}$$

$$\langle \kappa^{m+1}\vec{\nu}^m, \vec{\eta}\rangle^h_{\Gamma^m} + \langle \nabla_s \vec{X}^{m+1}, \nabla_s \vec{\eta}\rangle_{\Gamma^m} = 0 \quad \forall \vec{\eta} \in \underline{V}(\Gamma^m) \tag{76b}$$

and set $\Gamma^{m+1} = \vec{X}^{m+1}(\Gamma^m)$. Existence and uniqueness of discrete solutions for this scheme can be shown as in Section 4.2.

Theorem 86. *Let Γ^m satisfy Assumption 64(i). Then there exists a unique solution $(\vec{X}^{m+1}, \kappa^{m+1}) \in \underline{V}(\Gamma^m) \times V(\Gamma^m)$ to the system (76).*

Proof. The proof is very similar to the proof of Theorem 76. For a solution $(\vec{X}, \kappa) \in \underline{V}(\Gamma^m) \times V(\Gamma^m)$ of the homogeneous system

$$\langle \vec{X}, \chi\,\vec{\nu}^m \rangle_{\Gamma^m}^h - \Delta t_m \langle \nabla_s \kappa, \nabla_s \chi \rangle_{\Gamma^m} = 0 \quad \forall \chi \in V(\Gamma^m), \tag{77a}$$

$$\langle \kappa\,\vec{\nu}^m, \vec{\eta} \rangle_{\Gamma^m}^h + \langle \nabla_s \vec{X}, \nabla_s \vec{\eta} \rangle_{\Gamma^m} = 0 \quad \forall \vec{\eta} \in \underline{V}(\Gamma^m) \tag{77b}$$

we first prove $\kappa = 0$, and then proceed as in the proof of Lemma 66. Choosing $\chi = \kappa \in V(\Gamma^m)$ in (77a) and $\vec{\eta} = \vec{X} \in \underline{V}(\Gamma^m)$ in (77b) yields that

$$\left| \nabla_s \vec{X} \right|_{\Gamma^m}^2 + \Delta t_m \left| \nabla_s \kappa \right|_{\Gamma^m}^2 = 0.$$

We hence obtain that $\kappa = \kappa^c$ and $\vec{X} = \vec{X}^c$ on Γ^m, for $\kappa^c \in \mathbb{R}$ and $\vec{X}^c \in \mathbb{R}^d$, and so it follows from (77b) and (19) that

$$0 = \langle \kappa^c\,\vec{\nu}^m, \vec{\eta} \rangle_{\Gamma^m}^h = \kappa^c \langle \vec{\omega}^m, \vec{\eta} \rangle_{\Gamma^m}^h \quad \forall \vec{\eta} \in \underline{V}(\Gamma^m). \tag{78}$$

If we assume that $\kappa^c \neq 0$, then choosing $\vec{\eta} = \vec{\omega}^m \in \underline{V}(\Gamma^m)$ in (78) yields, on recalling (16), that $|\vec{\omega}^m|_{\Gamma_m}^h = 0$, and so $\vec{\omega}^m = \vec{0}$. However, that clearly contradicts Assumption 64(i) and hence we have that $\kappa^c = 0$. Moreover, as in the proof of Lemma 66, we obtain from (77a) and Assumption 64(i) that $\vec{X}^c = 0$. Hence we have shown that there exists a unique solution $(\vec{X}^{m+1}, \kappa^{m+1}) \in \underline{V}(\Gamma^m) \times V(\Gamma^m)$ to (76). □

Similarly to Section 4.5, it can be shown that the scheme (76) is unconditionally stable for $d = 2$ and $d = 3$.

Theorem 87. *Let $d = 2$ or $d = 3$. Let $(\vec{X}^{m+1}, \kappa^{m+1}) \in \underline{V}(\Gamma^m) \times V(\Gamma^m)$ be a solution to (76). Then we have that*

$$\left| \Gamma^{m+1} \right| + \Delta t_m \left| \nabla_s \kappa^{m+1} \right|_{\Gamma^m}^2 \leq |\Gamma^m|.$$

Proof. Choosing $\chi = \kappa^{m+1} \in V(\Gamma^m)$ in (76a) and $\vec{\eta} = \frac{1}{\Delta t_m}(\vec{X}^{m+1} - \vec{\mathrm{id}}_{|\Gamma^m}) \in \underline{V}(\Gamma^m)$ in (76b) yields that

$$\left\langle \nabla_s \vec{X}^{m+1}, \nabla_s \left(\vec{X}^{m+1} - \vec{\mathrm{id}} \right) \right\rangle_{\Gamma^m} + \Delta t_m \left| \nabla_s \kappa^{m+1} \right|_{\Gamma^m}^2 = 0. \tag{79}$$

Combining (79) and Lemma 57 yields the claim. □

5.3 Volume conservation for the semidiscrete scheme

An important aspect of discretizations for geometric evolution equations is to mimic decay and conservation properties of the evolution laws on the discrete level. In this section we discuss the volume conservation properties of the scheme (76) for surface diffusion, recall (73).

On recalling Definition 68, we consider the following semidiscrete variant of (76). Given the closed polyhedral hypersurface $\Gamma^h(0)$, find an evolving polyhedral hypersurface $\mathcal{G}_T^h$ with induced velocity $\vec{\mathcal{V}}^h \in \underline{V}(\mathcal{G}_T^h)$, and $\kappa^h \in V(\mathcal{G}_T^h)$, i.e., $\vec{\mathcal{V}}^h(\cdot,t) \in \underline{V}(\Gamma^h(t))$ and $\kappa^h(\cdot, t) \in V(\Gamma^h(t))$ for all $t \in [0,T]$, such that, for all $t \in (0,T]$,

$$\left\langle \vec{\mathcal{V}}^h, \chi\, \vec{\nu}^h \right\rangle_{\Gamma^h(t)}^h - \langle \nabla_s \kappa^h, \nabla_s \chi \rangle_{\Gamma^h(t)} = 0 \quad \forall \chi \in V(\Gamma^h(t)), \tag{80a}$$

$$\langle \kappa^h \vec{\nu}^h, \vec{\eta} \rangle_{\Gamma^h(t)}^h + \left\langle \nabla_s \vec{\mathrm{id}}, \nabla_s \vec{\eta} \right\rangle_{\Gamma^h(t)} = 0 \quad \forall \vec{\eta} \in \underline{V}(\Gamma^h(t)). \tag{80b}$$

Theorem 88. *Let $(\mathcal{G}_T^h, \kappa^h)$ be a solution of (80), and let $\Omega^h(t)$ be the domain enclosed by $\Gamma^h(t)$, for $t \in [0,T]$.*

(i) *It holds that*

$$\frac{\mathrm{d}}{\mathrm{d}t} |\Gamma^h(t)| + |\nabla_s \kappa^h|_{\Gamma^h(t)}^2 = 0.$$

(ii) *It holds that*

$$\frac{\mathrm{d}}{\mathrm{d}t} \mathcal{L}^d(\Omega^h(t)) = 0.$$

(iii) *For any $t \in (0,T]$, it holds that $\Gamma^h(t)$ is a conformal polyhedral surface. In particular, for $d = 2$, any two neighbouring elements of the curve $\Gamma^h(t)$ either have equal length, or they are parallel.*

Proof.
(i) Similarly to the proof of Theorem 79(i), choosing $\chi = \kappa^h(\cdot, t) \in V(\Gamma^h(t))$ in (80a) and $\vec{\eta} = \vec{\mathcal{V}}^h(\cdot,t) \in \underline{V}(\Gamma^h(t))$ in (80b) gives

$$\frac{\mathrm{d}}{\mathrm{d}t} |\Gamma^h(t)| = \left\langle \nabla_s \vec{\mathrm{id}}, \nabla_s \vec{\mathcal{V}}^h \right\rangle_{\Gamma^h(t)} = -|\nabla_s \kappa^h|_{\Gamma^h(t)}^2,$$

which is the claim.
(ii) Choosing $\chi = 1$ in (80a) yields, on recalling Theorem 71, $\vec{\mathcal{V}}^h(\cdot,t) \in \underline{V}(\Gamma^h(t))$ and (20), that

$$\frac{\mathrm{d}}{\mathrm{d}t} \mathcal{L}^d(\Omega^h(t)) = \pm \left\langle \vec{\mathcal{V}}^h, \vec{\nu}^h \right\rangle_{\Gamma^h(t)} = \pm \left\langle \vec{\mathcal{V}}^h, \vec{\nu}^h \right\rangle_{\Gamma^h(t)}^h = 0,$$

where the sign $\pm$ depends on whether $\vec{\nu}^h$ here is the outer/inner normal of $\Omega^h(t)$ on $\Gamma^h(t)$.
(iii) This follows directly from Definition 60 and Theorem 62. □

Remark 89.

(i) The result in Theorem 88(i) is the semidiscrete analogue of the stability result Theorem 87 for the fully discrete scheme (76), and it mimics the energy law (74) on the discrete level.

(ii) The result in Theorem 88(ii) mimics the volume conservation property (73) on the discrete level. Moreover, while the fully discrete scheme (76) does not conserve the enclosed volume exactly, it does satisfy

$$\left\langle \vec{X}^{m+1} - \vec{\mathrm{id}}, \vec{\nu}^m \right\rangle_{\Gamma^m} = \left\langle \vec{X}^{m+1} - \vec{\mathrm{id}}, \vec{\nu}^m \right\rangle_{\Gamma^m}^h = 0,$$

recall (20). On noting Theorem 71 this can be interpreted as a fully discrete analogue of Theorem 88(ii). In fact, in practice the scheme (76) exhibits excellent volume conservation properties, with the relative enclosed volume loss tending to zero as the time step size goes to zero.

5.4 Generalizations to other flows

The finite element approximation (76) for surface diffusion can easily be adapted to more general situations. The resultant discretizations will, under certain circumstances, again satisfy a stability result.

Here we consider flows of the form

$$\mathcal{V} = \mathfrak{F}(\varkappa) \quad \text{on}\, \Gamma(t), \tag{81}$$

where $\mathfrak{F}$ maps functions on the closed hypersurface $\Gamma(t)$ to functions on $\Gamma(t)$. For mean curvature flow we have $\mathfrak{F}(\chi) = \chi$, while surface diffusion is obtained with $\mathfrak{F}(\chi) = -\Delta_s \chi$. If the map $\mathfrak{F}$ is such that

$$\langle \mathfrak{F}(\varkappa), \varkappa \rangle_{\Gamma(t)} \geq 0, \tag{82}$$

then we have, on recalling Theorem 32, that

$$\frac{\mathrm{d}}{\mathrm{d}t} |\Gamma(t)| = -\langle \mathcal{V}, \varkappa \rangle_{\Gamma(t)} = -\langle \mathfrak{F}(\varkappa), \varkappa \rangle_{\Gamma(t)} \leq 0. \tag{83}$$

The inequality (82) clearly is true for mean curvature flow, where we obtain

$$\langle \mathfrak{F}(\varkappa), \varkappa \rangle_{\Gamma(t)} = |\varkappa|^2_{\Gamma(t)} \geq 0,$$

and also for surface diffusion, where

$$\langle \mathfrak{F}(\varkappa), \varkappa \rangle_{\Gamma(t)} = |\nabla_s \varkappa|^2_{\Gamma(t)} \geq 0,$$

recall (74). Our goal now is to obtain a stability result like that in Theorem 87 for discretizations of flows of the form (81) that satisfy the energy bound (83). As a consistency condition for a discrete formulation, we require that there is a map $\mathfrak{F}^m : V(\Gamma^m) \to V(\Gamma^m)$ which approximates $\mathfrak{F}$ and is such that

$$\langle \mathfrak{F}^m(\chi), \chi \rangle_{\Gamma^m}^h \geq 0 \quad \forall \chi \in V(\Gamma^m). \tag{84}$$

We consider now the following finite element approximation of (81). Let the closed polyhedral hypersurface Γ^0 be an approximation of $\Gamma(0)$. Then, for $m = 0, \ldots, M-1$, find $(\vec{X}^{m+1}, \kappa^{m+1}) \in \underline{V}(\Gamma^m) \times V(\Gamma^m)$ such that

$$\left\langle \frac{\vec{X}^{m+1} - \vec{\mathrm{id}}}{\Delta t_m}, \chi\,\vec{\nu}^m \right\rangle_{\Gamma^m}^h - \left\langle \mathfrak{F}^m(\kappa^{m+1}), \chi \right\rangle_{\Gamma^m}^h = 0 \quad \forall \chi \in V(\Gamma^m), \tag{85a}$$

$$\left\langle \kappa^{m+1}\,\vec{\nu}^m, \vec{\eta} \right\rangle_{\Gamma^m}^h + \left\langle \nabla_s \vec{X}^{m+1}, \nabla_s \vec{\eta} \right\rangle_{\Gamma^m} = 0 \quad \forall \vec{\eta} \in \underline{V}(\Gamma^m) \tag{85b}$$

and set $\Gamma^{m+1} = \vec{X}^{m+1}(\Gamma^m)$. We obtain the following stability theorem, together with an existence and uniqueness result in the linear case.

Theorem 90.

(i) *Let Γ^m satisfy Assumption 64, and let $\mathfrak{F}^m : V(\Gamma^m) \to V(\Gamma^m)$ be a linear map such that (84) holds. Then there exists a unique solution $(\vec{X}^{m+1}, \kappa^{m+1}) \in \underline{V}(\Gamma^m) \times V(\Gamma^m)$ to the system (85).*

(ii) *Let $d = 2$ or $d = 3$. Let $(\vec{X}^{m+1}, \kappa^{m+1}) \in \underline{V}(\Gamma^m) \times V(\Gamma^m)$ be a solution to (85). Then we have that*

$$|\Gamma^{m+1}| + \Delta t_m \left\langle \mathfrak{F}^m(\kappa^{m+1}), \kappa^{m+1} \right\rangle_{\Gamma^m}^h \le |\Gamma^m|,$$

where both terms on the left-hand side are nonnegative if $\mathfrak{F}^m$ satisfies (84).

Proof.
(i) The proof is analogous to the proof of Lemma 66.
(ii) The proof is analogous to the proof of Theorem 87. □

Remark 91 (Examples).

(i) Choosing $\mathfrak{F}(\varkappa) = -\Delta_s \varkappa$ and $\mathfrak{F}^m(\chi) = -\Delta_s^{\Gamma^m} \chi$, recall (28), we recover the results in Theorems 86 and 87 for surface diffusion. Similarly, for $\mathfrak{F}(\varkappa) = \varkappa$ and $\mathfrak{F}^m(\chi) = \chi$ we obtain Theorems 76 and 77 for mean curvature flow.

(ii) Of interest are also nonlinear flows $\mathcal{V} = f(\varkappa)$, i.e., $\mathfrak{F}(\varkappa) = f(\varkappa)$, where $f : (a,b) \to \mathbb{R}$ with $-\infty \le a < b \le \infty$ is a strictly monotonically increasing continuous function such that $f(0) = 0$. An example is $f(r) = |r|^{\beta-1} r$, with $\beta \in \mathbb{R}_{>0}$, which has applications in image analysis, see Alvarez et al. (1993), Sapiro and Tannenbaum (1994) and Mikula and Ševčovič (2001). Here we define $\mathfrak{F}^m(\chi) = \pi_{\Gamma^m}[f(\chi)]$, recall Definition 43(i). For the resulting nonlinear scheme (85) we obtain the stability result Theorem 90(ii). Existence and uniqueness results for (85) are shown in Barrett et al. (2008a, Theorem 2.1).

(iii) Also the volume conserving flow

$$\mathcal{V} = \mathfrak{F}(\varkappa) = \varkappa - \fint_{\Gamma(t)} \varkappa \quad \text{on } \Gamma(t), \tag{86}$$

where we define the average

$$\fint_{\Gamma(t)} \varkappa = \frac{1}{|\Gamma(t)|} \int_{\Gamma(t)} \varkappa \, \mathrm{d}\mathcal{H}^{d-1}, \tag{87}$$

falls into the class of functions for which stability can be shown. Here, for all $\chi \in V(\Gamma^m)$, we choose $\mathfrak{F}^m(\chi) = \chi - \fint_{\Gamma^m} \chi \in V(\Gamma^m)$ and obtain

$$\langle \mathfrak{F}^m(\chi), \chi \rangle^h_{\Gamma^m} = \left(|\chi - \fint_{\Gamma^m} \chi|^h_{\Gamma^m} \right)^2 \geq 0,$$

and so both results in Theorem 90 hold.

(iv) In materials science (86) is called surface attachment limited kinetics (SALK). A flow interpolating between SALK and surface diffusion is

$$\mathcal{V} = -\Delta_s \left(\frac{1}{\alpha} - \frac{1}{\xi} \Delta_s \right)^{-1} \varkappa \quad \text{on } \Gamma(t),$$

where $\alpha, \xi \in \mathbb{R}_{>0}$. This flow was proposed by Taylor and Cahn (1994) and analyzed by Elliott and Garcke (1997) and Escher et al. (2001) as an evolution law for interfaces in materials science. Introducing the new variable y, this flow can be restated as

$$\vec{\mathcal{V}} \cdot \vec{\nu} = -\Delta_s \, y, \quad \left(\frac{1}{\alpha} - \frac{1}{\xi} \Delta_s \right) y = \varkappa, \quad \varkappa \vec{\nu} = \Delta_s \, \vec{\mathrm{id}} \quad \text{on } \Gamma(t).$$

Hence, for any $\chi \in V(\Gamma^m)$, the function $\mathfrak{F}^m(\chi) \in V(\Gamma^m)$ is defined via

$$\langle \mathfrak{F}^m(\chi), \eta \rangle^h_{\Gamma^m} = \langle \nabla_s Y(\chi), \nabla_s \eta \rangle_{\Gamma^m} \quad \forall \eta \in V(\Gamma^m), \tag{88}$$

where $Y \in V(\Gamma^m)$ solves

$$\frac{1}{\xi} \langle \nabla_s Y, \nabla_s \varphi \rangle_{\Gamma^m} + \frac{1}{\alpha} \langle Y, \varphi \rangle^h_{\Gamma^m} = \langle \chi, \varphi \rangle^h_{\Gamma^m} \quad \forall \varphi \in V(\Gamma^m). \tag{89}$$

Choosing $\eta = \chi$ in (88) and $\varphi = \chi - \frac{1}{\alpha} Y$ in (89) then yields that

$$\langle \mathfrak{F}^m(\chi), \chi \rangle^h_{\Gamma^m} = \langle \nabla_s Y, \nabla_s \chi \rangle_{\Gamma^m} = \frac{1}{\alpha} |\nabla_s Y|^2_{\Gamma^m} + \xi \left(\left| \chi - \frac{1}{\alpha} Y \right|^h_{\Gamma^m} \right)^2 \geq 0,$$

and hence we obtain a stable discretization, satisfying the two results in Theorem 90, also in this case.

Remark 92 (Semidiscrete schemes and tangential motion).

(i) Similarly to Section 5.3, semidiscrete variants of the schemes discussed in Remark 91 can also be considered. In each case they will satisfy Theorem 88(iii), the appropriate analogue of Theorem 88(i), as well as

Theorem 88(ii) when the approximated flow is volume preserving, provided that the semidiscrete analogue of $\mathfrak{F}^m$ satisfies $\langle \mathfrak{F}^h(\chi), 1 \rangle^h_{\Gamma^h(t)} = 0$ for all $\chi \in V(\Gamma^h(t))$.

(ii) In the case of curves, $d = 2$, for all the situations in Remark 91, fully discrete fully implicit schemes along the lines of (63) can be considered. In each case these schemes will inherit the stability properties from Remark 91 and, in addition, they will satisfy the strong equidistribution property (64). For further details we refer to Barrett et al. (2011).

5.5 Approximations with reduced or induced tangential motion

In this section we consider an alternative to (85b) that allows to either reduce the tangential motion or encourage tangential motion in selected directions. To this end, we assume that Γ^m satisfies Assumption 64(ii) and let $\vec{\tau}_i^m \in \underline{V}(\Gamma^m)$, for $i = 1, \ldots, d-1$, be such that $\left\{ \frac{\vec{\omega}^m(\vec{q}_k^m)}{|\vec{\omega}^m(\vec{q}_k^m)|}, \vec{\tau}_1^m(\vec{q}_k^m), \ldots, \vec{\tau}_{d-1}^m(\vec{q}_k^m) \right\}$ is an orthonormal basis of $\mathbb{R}^d$ for $k = 1, \ldots, K$. Moreover, we choose coefficients $0 \leq \alpha_i^m, \delta_i^m \in V(\Gamma^m)$, $i = 1, \ldots, d-1$, and forcing terms $c_i^m \in V(\Gamma^m)$, $i = 1, \ldots, d-1$. Then, in place of (85) we consider the following approximation.

Let the closed polyhedral hypersurface Γ^0 be an approximation of $\Gamma(0)$. Then, for $m = 0, \ldots, M-1$, find $(\vec{X}^{m+1}, \kappa^{m+1}) \in \underline{V}(\Gamma^m) \times V(\Gamma^m)$ and $\beta_i^{m+1} \in V(\Gamma^m)$, $i = 1, \ldots, d-1$, such that

$$\left\langle \frac{\vec{X}^{m+1} - \vec{\mathrm{id}}}{\Delta t_m}, \chi \vec{\nu}^m \right\rangle^h_{\Gamma^m} - \langle \mathfrak{F}^m(\kappa^{m+1}), \chi \rangle^h_{\Gamma^m} = 0 \quad \forall \chi \in V(\Gamma^m), \tag{90a}$$

$$\left\langle \alpha_i^m \frac{\vec{X}^{m+1} - \vec{\mathrm{id}}}{\Delta t_m}, \xi \vec{\tau}_i^m \right\rangle^h_{\Gamma^m} - \langle \alpha_i^m [\delta_i^m \beta_i^{m+1} + c_i^m], \xi \rangle^h_{\Gamma^m} = 0 \quad \forall \xi \in V(\Gamma^m), \quad i = 1, \ldots, d-1, \tag{90b}$$

$$\left\langle \kappa^{m+1} \vec{\nu}^m + \sum_{i=1}^{d-1} \alpha_i^m \beta_i^{m+1} \vec{\tau}_i^m, \vec{\eta} \right\rangle^h_{\Gamma^m} + \langle \nabla_s \vec{X}^{m+1}, \nabla_s \vec{\eta} \rangle_{\Gamma^m} = 0 \quad \forall \vec{\eta} \in \underline{V}(\Gamma^m) \tag{90c}$$

and set $\Gamma^{m+1} = \vec{X}^{m+1}(\Gamma^m)$.

Theorem 93.

(i) *Let Γ^m satisfy Assumption 64 and let $\mathfrak{F}^m : V(\Gamma^m) \to V(\Gamma^m)$ be a linear map such that (84) holds. Then there exists a solution $(\vec{X}^{m+1}, \kappa^{m+1}, \beta_1^{m+1}, \ldots, \beta_{d-1}^{m+1}) \in \underline{V}(\Gamma^m) \times [V(\Gamma^m)]^d$ to the system (90), with $(\vec{X}^{m+1}, \kappa^{m+1}, \pi_{\Gamma^m}[\alpha_1^m \beta_1^{m+1}], \ldots, \pi_{\Gamma^m}[\alpha_{d-1}^m \beta_{d-1}^{m+1}])$ being unique.*

(ii) *Let $d=2$ or $d=3$. Let $(\vec{X}^{m+1}, \kappa^{m+1}, \beta_1^{m+1}, \ldots, \beta_{d-1}^{m+1}) \in \underline{V}(\Gamma^m) \times [V(\Gamma^m)]^d$ be a solution to (90). Then we have that*

$$|\Gamma^{m+1}| + \Delta t_m \langle \mathfrak{F}^m(\kappa^{m+1}), \kappa^{m+1} \rangle_{\Gamma^m}^h + \Delta t_m \langle \alpha_i^m [\delta_i^m \beta_i^{m+1} + c_i^m], \beta_i^{m+1} \rangle_{\Gamma^m}^h \leq |\Gamma^m|. \tag{91}$$

Proof.
(i) Let

$$B_i^m(\Gamma^m) = \{\xi \in V(\Gamma^m) : \xi(\vec{q}_k^m) = 0 \quad \text{if } \alpha_i^m(\vec{q}_k^m) = 0, k = 1, \ldots, K\},$$

for $i = 1, \ldots, d-1$. We now prove the desired results by showing existence of a unique solution to the system (90) with $\beta_i^{m+1} \in V(\Gamma^m)$ replaced by $\beta_i^{m+1} \in B_i^m(\Gamma^m)$, and with $V(\Gamma^m)$ in (90b) replaced by $B_i^m(\Gamma^m)$. As usual, existence to this adapted linear system follows from uniqueness, and so we let $(\vec{X}, \kappa, b_1, \ldots, b_{d-1}) \in \underline{V}(\Gamma^m) \times V(\Gamma^m) \times B_1^m(\Gamma^m) \times \cdots \times B_{d-1}^m(\Gamma^m)$ be such that

$$\langle \vec{X}, \chi \vec{\omega}^m \rangle_{\Gamma^m}^h - \Delta t_m \langle \mathfrak{F}^m(\kappa), \chi \rangle_{\Gamma^m}^h = 0 \quad \forall \chi \in V(\Gamma^m), \tag{92a}$$

$$\langle \alpha_i^m \vec{X}, \xi \vec{\tau}_i^m \rangle_{\Gamma^m}^h - \Delta t_m \langle \alpha_i^m \delta_i^m b_i, \xi \rangle_{\Gamma^m}^h = 0 \quad \forall \xi \in B_i^m(\Gamma^m), \quad i = 1, \ldots, d-1, \tag{92b}$$

$$\left\langle \kappa \vec{\omega}^m + \sum_{i=1}^{d-1} \alpha_i^m b_i \vec{\tau}_i^m, \vec{\eta} \right\rangle_{\Gamma^m}^h + \langle \nabla_s \vec{X}, \nabla_s \vec{\eta} \rangle_{\Gamma^m} = 0 \quad \forall \vec{\eta} \in \underline{V}(\Gamma^m), \tag{92c}$$

where we have observed (19). Choosing $\chi = \kappa \in V(\Gamma^m)$ in (92a), $\xi = b_i \in B_i^m(\Gamma^m)$ in (92b) for $i = 1, \ldots, d-1$, and $\vec{\eta} = \vec{X} \in \underline{V}(\Gamma^m)$ in (92c) yields that

$$\Delta t_m \langle \mathfrak{F}^m(\kappa), \kappa \rangle_{\Gamma^m}^h + \Delta t_m \langle \alpha_i^m \delta_i^m b_i, b_i \rangle_{\Gamma^m}^h + |\nabla_s \vec{X}|_{\Gamma^m}^2 = 0. \tag{93}$$

It immediately follows from (93), (84) and our assumptions on α_i^m and δ_i^m that $\vec{X}$ is constant on Γ^m. Hence choosing $\vec{\eta} = \vec{\tau}_i^m$ in (92c) yields that $|\pi_{\Gamma^m}[\alpha_i^m b_i]|_{\Gamma^m}^h = 0$, and so $\pi_{\Gamma^m}[\alpha_i^m b_i] = 0$, which implies that $b_i = 0 \in B_i^m(\Gamma^m)$, for $i = 1, \ldots, d-1$. The remainder of the proof proceeds as the proof of Lemma 66 to show that $\kappa = 0$ and $\vec{X} = \vec{0}$. This proves the uniqueness of a solution to (92), and hence the existence of a solution $(\vec{X}^{m+1}, \kappa^{m+1}, \beta_1^{m+1}, \ldots, \beta_{d-1}^{m+1}) \in \underline{V}(\Gamma^m) \times V(\Gamma^m) \times B_1^m(\Gamma^m) \times \cdots \times B_{d-1}^m(\Gamma^m)$ to the modified system discussed at the beginning of the proof. Clearly, this solution also solves the original system (90a), with only the values of β_i^{m+1} arbitrary where α_i^m vanishes.
(ii) Choosing $\chi = \kappa^{m+1} \in V(\Gamma^m)$ in (90a), $\xi = \beta_i^{m+1} \in V(\Gamma^m)$ in (90b) for $i = 1, \ldots, d-1$ and $\vec{\eta} = \frac{1}{\Delta t_m}(\vec{X}^{m+1} - \vec{\mathrm{id}}_{|\Gamma^m}) \in \underline{V}(\Gamma^m)$ in (90c) yields the desired result on recalling Lemma 57. □

Remark 94.

(i) Clearly, (90) with $\alpha_i^m = 0$, for $i = 1, \ldots, d-1$, reduces to the original scheme (85).

(ii) In the case $d = 2$ or $d = 3$, if $\mathfrak{F}^m$ satisfies (84), and if $c_i^m = 0$ for $i = 1, \ldots, d-1$, then (91) provides a stability estimate for (90).

(iii) Choosing $\delta_i^m(\vec{q}_k^m) = 1$ and $c_i^m(\vec{q}_k^m) = 0$ for $i = 1, \ldots, d-1$, it follows intuitively from (91) that tangential motion of $\vec{q}_k^m$ in the direction of $\vec{\tau}_i^m(\vec{q}_k^m)$ will be suppressed if $\alpha_i^m(\vec{q}_k^m)$ is large. Conversely, it is also clear from (90b) that choosing $\delta_i^m(\vec{q}_k^m) = 0$ and $\alpha_i^m(\vec{q}_k^m) > 0$ allows us to fix the tangential motion. These observations form an ansatz to control tangential movement in the discrete evolution of geometric flows. In particular, in Barrett et al. (2008d) the following strategies have been proposed and employed.

(S1) $\alpha_i^m = \alpha \in \mathbb{R}_{\geq 0}, \quad \delta_i^m = 1, \quad c_i^m = 0;$

(S2) $\alpha_i^m = \alpha \in \mathbb{R}_{>0}, \quad \delta_i^m = \delta \in \mathbb{R}_{>0},$

$$c_i^m(\vec{q}_k^m) = \tfrac{1}{\Delta t_m}(\vec{z}_k^m - \vec{q}_k^m) \cdot \vec{\tau}_i^m(\vec{q}_k^m), \quad k = 1, \ldots, K;$$

(S3) $\alpha_i^m = 1, \quad \delta_i^m = 0,$

$$c_i^m(\vec{q}_k^m) = \tfrac{1}{\Delta t_m}(\vec{z}_k^m - \vec{q}_k^m) \cdot \vec{\tau}_i^m(\vec{q}_k^m), \quad k = 1, \ldots, K;$$

for $i = 1, \ldots, d-1$, where $\vec{z}_k^m$ is the average of the neighbouring nodes of $\vec{q}_k^m$. The effect of these strategies can be summarized as follows. With increasing $\alpha > 0$, (S1) leads to smaller $|\beta_i^{m+1}|$, $i = 1 \to d-1$, and hence to less tangential motion, see (90c). Strategy (S2), on the other hand, is intended to induce a tangential movement towards the "barycentres" $\vec{z}_k^m$. Lastly, (S3) completely fixes the tangential motion, so that after the time step, each vertex $\vec{X}^{m+1}(\vec{q}_k^m)$ has the same tangential components as $\vec{z}_k^m$.

(iv) Similarly to Section 5.3, a semidiscrete variant of the scheme (90) can also be considered. It will satisfy the appropriate analogue of Theorem 88(i), as well as Theorem 88(ii) when the approximated flow is volume preserving.

Remark 95.

(i) In place of (85) and (90) it is possible to consider two closely related variants, where in (85) and (90) the terms $\vec{\nu}^m$ are replaced by $\frac{\vec{\omega}^m}{|\vec{\omega}^m|}$. On recalling (19) this is a minor change to the original schemes, where at each vertex $\vec{\omega}^m$ is now normalized. For these variants all the results in Theorems 90 and 93 still hold true, as well as most of the results in Remarks 91(i) and 94. The only property in the latter two remarks that no longer holds is the volume preservation. As can be seen from the proof of Theorem 88(ii), volume conservation requires $\vec{\nu}^m$ to be present in (85) and (90), and it is for this reason that in general we prefer these schemes in their original form.

(ii) In the case of mean curvature flow, when $\mathfrak{F}^m(\chi)=\chi$, the solution $\vec{X}^{m+1}\in \underline{V}(\Gamma^m)$ to (90) for the strategy (S1) from Remark 94(iii) satisfies

$$\frac{1}{\Delta t_m}\left\langle \left[\left(\vec{X}^{m+1}-\vec{\mathrm{id}}\right)\cdot\vec{\omega}^m\right]\vec{\omega}^m+\alpha\sum_{i=1}^{d-1}\left[\left(\vec{X}^{m+1}-\vec{\mathrm{id}}\right)\cdot\vec{\tau}_i^m\right]\vec{\tau}_i^m,\vec{\eta}\right\rangle_{\Gamma^m}^h + \left\langle \nabla_s\vec{X}^{m+1},\nabla_s\vec{\eta}\right\rangle_{\Gamma^m}=0 \quad \forall\vec{\eta}\in\underline{V}(\Gamma^m). \tag{94}$$

A closely related approximation is obtained by normalizing $\vec{\omega}^m$ in (94) to yield

$$\frac{1}{\Delta t_m}\left\langle \left[\left(\vec{X}^{m+1}-\vec{\mathrm{id}}\right)\cdot\vec{\omega}^m\right]\frac{\vec{\omega}^m}{|\vec{\omega}^m|^2}+\alpha\sum_{i=1}^{d-1}\left[\left(\vec{X}^{m+1}-\vec{\mathrm{id}}\right)\cdot\vec{\tau}_i^m\right]\vec{\tau}_i^m,\vec{\eta}\right\rangle_{\Gamma^m}^h + \left\langle \nabla_s\vec{X}^{m+1},\nabla_s\vec{\eta}\right\rangle_{\Gamma^m}=0 \quad \forall\vec{\eta}\in\underline{V}(\Gamma^m), \tag{95}$$

and this corresponds to the variant of (90) discussed in (i). On introducing $\underline{\underline{Q}}_\theta^m\in\underline{V}(\Gamma^m)$, for $\theta\in\mathbb{R}_{\geq 0}$, with

$$\underline{\underline{Q}}_\theta^m(\vec{q}_k^m)=\theta\,\underline{\underline{\mathrm{Id}}}+(1-\theta)\,\frac{\vec{\omega}^m(\vec{q}_k^m)\otimes\vec{\omega}^m(\vec{q}_k^m)}{|\vec{\omega}^m(\vec{q}_k^m)|^2},\quad k=1,\ldots,K, \tag{96}$$

it immediately follows from the fact that $\{\frac{\vec{\omega}^m(\vec{q}_k^m)}{|\vec{\omega}^m(\vec{q}_k^m)|},\vec{\tau}_1^m(\vec{q}_k^m),\ldots,\vec{\tau}_{d-1}^m(\vec{q}_k^m)\}$ is an orthonormal basis, that for every $k=1,\ldots,K$ and for every $\vec{z}\in\mathbb{R}^d$

$$\begin{aligned}&\left[\vec{z}\cdot\vec{\omega}^m(\vec{q}_k^m)\right]\frac{\vec{\omega}^m(\vec{q}_k^m)}{|\vec{\omega}^m(\vec{q}_k^m)|^2}+\alpha\sum_{i=1}^{d-1}\left[\vec{z}\cdot\vec{\tau}_i^m(\vec{q}_k^m)\right]\vec{\tau}_i^m(\vec{q}_k^m)\\&=(1-\alpha)\,\vec{z}\cdot\vec{\omega}^m(\vec{q}_k^m)\,\frac{\vec{\omega}^m(\vec{q}_k^m)}{|\vec{\omega}^m(\vec{q}_k^m)|^2}+\alpha\,\vec{z}=\underline{\underline{Q}}_\alpha^m(\vec{q}_k^m)\,\vec{z}.\end{aligned}$$

Hence, for $\alpha=\theta$, (95) is equivalent to

$$\left\langle \underline{\underline{Q}}_\theta^m\frac{\vec{X}^{m+1}-\vec{\mathrm{id}}}{\Delta t_m},\vec{\eta}\right\rangle_{\Gamma^m}^h+\left\langle \nabla_s\vec{X}^{m+1},\nabla_s\vec{\eta}\right\rangle_{\Gamma^m}=0\quad\forall\vec{\eta}\in\underline{V}(\Gamma^m). \tag{97}$$

Clearly, $\theta=1$ collapses to (68) with mass lumping, while $\theta=0$ is (47) with normalized $\vec{\omega}^m$, recall (i). For $\theta\in(0,1)$ we obtain some interpolation between the two, with the tangential motion as described in (S1) from Remark 94(iii) for $\alpha=\theta$. We also observe that the main idea in Elliott and Fritz (2017), for the case $d=2$, is to introduce $\theta\,\underline{\underline{\mathrm{Id}}}+(1-\theta)\,\vec{\nu}^m\otimes\vec{\nu}^m$ into the first term in (72), similarly to how (97) relates to (68).

Remark 96 (Discrete linear systems).
It is a simple matter to adapt the definitions and techniques in Section 4.3 in order to derive the discrete linear systems that need to be solved at each time level for the approximations (76), (85) and (90). The most efficient way is employing a sparse direct solution method such as UMFPACK, see Davis (2004).

5.6 Alternative parametric methods

An alternative parametric finite element approximation of surface diffusion is given in Bänsch et al. (2005). Their scheme is based on a discretization of the formulation

$$\vec{\mathcal{V}} = \mathcal{V}\,\vec{\nu},\ \ \mathcal{V} = -\Delta_s\,\varkappa,\ \ \varkappa = \vec{\varkappa}\cdot\vec{\nu},\ \ \vec{\varkappa} = \Delta_s\,\vec{\mathrm{id}},\qquad \text{on}\,\Gamma(t),$$

in contrast to (75). Compared to (76) there are two more variables, and as the surface always moves in a direction collinear to the normal, the surface meshes will in general deteriorate.

We refer also to Bänsch et al. (2004), where, on assuming a sufficiently smooth solution, an error analysis is presented for a semidiscrete finite element approximation by continuous piecewise polynomials of degree $k \geq 1$ for a graph formulation of surface diffusion. A semidiscrete finite element approximation of axisymmetric surface diffusion by continuous piecewise linears in a graph formulation has been considered in Coleman et al. (1996). The corresponding error analysis, on assuming a sufficiently smooth solution, is presented in Deckelnick et al. (2003). In addition, a parametric finite element approximation of axisymmetric surface diffusion has been considered in Barrett et al. (2019b).

6 Anisotropic flows

6.1 Derivation of the governing equations

In many interface problems the energy density is directionally dependent. This can result, for example, from a material's directional dependence of a physical property. This appears, for example, in a crystal where the energy of an interface depends on how its direction is related to the crystal lattice orientations. A typical anisotropic surface energy has the form

$$|\Gamma|_\gamma = \int_\Gamma \gamma(\vec{\nu})\ \mathrm{d}\mathcal{H}^{d-1}, \tag{98}$$

where Γ is a closed orientable C^1-hypersurface in $\mathbb{R}^d$, $d \geq 2$, with a unit normal field $\vec{\nu}$, and $\gamma : \mathbb{S}^{d-1} \to \mathbb{R}_{>0}$ is a given anisotropic energy density. It is mathematically convenient, to extend γ to a function on $\mathbb{R}^d$. In particular, denoting the extension again with γ, we assume that it is absolutely homogeneous of degree one, i.e.,

$$\gamma(\lambda\,\vec{p}) = |\lambda|\,\gamma(\vec{p})\quad \forall\,\vec{p} \in \mathbb{R}^d,\ \lambda \in \mathbb{R}.$$

Assuming from now on that $\gamma \in C^2(\mathbb{R}^d\backslash\{\vec{0}\})$, we can differentiate this identity with respect to λ to obtain

$$\gamma'(\vec{p})\cdot\vec{p} = \gamma(\vec{p})\quad \forall\,\vec{p} \in \mathbb{R}^d\backslash\{\vec{0}\}, \tag{99}$$

where γ' is the gradient of γ. In the isotropic case, $\gamma(\vec{p}) = |\vec{p}|$, and so $\gamma'(\vec{p}) = \frac{\vec{p}}{|\vec{p}|}$. We refer to Taylor et al. (1992), Bellettini et al. (1999) and Giga (2006) for more details on anisotropic energies in materials science and geometry.

Lemma 97. *Let $(\Gamma(t))_{t\in[0,T]}$ be a C^2-evolving closed orientable hypersurface with normal vector field $\vec{\nu}$. Then we have the anisotropic version of (36)*

$$\frac{\mathrm{d}}{\mathrm{d}t}|\Gamma(t)|_\gamma = -\langle \varkappa_\gamma, \mathcal{V}\rangle_{\Gamma(t)},$$

where $\varkappa_\gamma = -\nabla_s \cdot \vec{\nu}_\gamma$ is the weighted mean curvature and $\vec{\nu}_\gamma = \gamma'(\vec{\nu})$ is the Cahn–Hoffmann vector, the anisotropic version of Lemma 13(i).

Proof. It follows from the transport theorem, Theorem 32, on noting (98), (99), Lemma 37(ii), Theorem 21 and Lemma 7(i), that

$$\begin{aligned}\frac{\mathrm{d}}{\mathrm{d}t}|\Gamma(t)|_\gamma &= \frac{\mathrm{d}}{\mathrm{d}t}\langle 1, \gamma(\vec{\nu})\rangle_{\Gamma(t)} = \left\langle 1, \partial_t^\square \gamma(\vec{\nu}) - \gamma(\vec{\nu})\, \mathcal{V}\varkappa\right\rangle_{\Gamma(t)} \\ &= \left\langle 1, \gamma'(\vec{\nu})\cdot \partial_t^\square \vec{\nu} - \gamma'(\vec{\nu})\cdot\vec{\nu}\,\mathcal{V}\varkappa\right\rangle_{\Gamma(t)} \\ &= \langle 1, -\gamma'(\vec{\nu})\cdot\nabla_s \mathcal{V} + \nabla_s\cdot(\gamma'(\vec{\nu})\,\mathcal{V})\rangle_{\Gamma(t)} \\ &= \langle \nabla_s\cdot\gamma'(\vec{\nu}), \mathcal{V}\rangle_{\Gamma(t)} = -\langle \varkappa_\gamma, \mathcal{V}\rangle_{\Gamma(t)}.\end{aligned}$$

□

We can hence define anisotropic versions of mean curvature flow and surface diffusion as

$$\text{(a)}\ \ \mathcal{V} = \beta(\vec{\nu})\,\varkappa_\gamma \quad \text{and} \quad \text{(b)}\ \ \mathcal{V} = -\nabla_s\cdot\left(\beta(\vec{\nu})\,\nabla_s\varkappa_\gamma\right), \tag{100}$$

where $\beta : \mathbb{S}^{d-1} \to \mathbb{R}_{>0}$ is a smooth kinetic coefficient. We refer to Taylor et al. (1992), Cahn and Taylor (1994), Taylor and Cahn (1994) and Giga (2006) for a derivation and more information about these evolution laws.

For anisotropic mean curvature flow we obtain from Lemma 97 and (100) that

$$\frac{\mathrm{d}}{\mathrm{d}t}|\Gamma(t)|_\gamma = -\langle\varkappa_\gamma, \mathcal{V}\rangle_{\Gamma(t)} = -\left\langle \beta(\vec{\nu}), \varkappa_\gamma^2\right\rangle_{\Gamma(t)} \le 0, \tag{101}$$

and for anisotropic surface diffusion it holds with the help of Lemma 7(i), Definition 5(ii) and the divergence theorem, Theorem 21, that

$$\frac{\mathrm{d}}{\mathrm{d}t}|\Gamma(t)|_\gamma = -\langle\varkappa_\gamma, \mathcal{V}\rangle_{\Gamma(t)} = -\left\langle \beta(\vec{\nu}), |\nabla_s\varkappa_\gamma|^2\right\rangle_{\Gamma(t)} \le 0. \tag{102}$$

Of course, also for the surface energy $|\Gamma|_\gamma$, an isoperimetric problem can be formulated. Here one wants to find the shape, which minimizes $|\Gamma|_\gamma$ under all shapes with a given enclosed volume. In order to do so, one defines the dual function

$$\gamma^*(\vec{q}) = \sup_{\vec{p}\in\mathbb{R}^d\setminus\{\vec{0}\}} \frac{\vec{p}\cdot\vec{q}}{\gamma(\vec{p})} \quad \forall \vec{q}\in\mathbb{R}^d.$$

Then the solution of the isoperimetric problem is, up to a scaling, the Wulff shape

$$\mathcal{W} = \{\vec{q}\in\mathbb{R}^d : \gamma^*(\vec{q}) \le 1\}.$$

This is the 1-ball of γ^* and we also define the 1-ball of γ

$$\mathcal{F} = \{\vec{p}\in\mathbb{R}^d : \gamma(\vec{p}) \le 1\},$$

which is called Frank diagram. We refer to Fig. 3 for examples. Also surface energies for which the Frank diagram or the Wulff shape have flat parts, edges and corners are of interest. These are called crystalline surface energies, and we will be able to approximate these also in a stable manner.

6.2 Suitable weak formulations

Coming up with stable discretizations is a very difficult task as the flows

$$\mathcal{V} = \beta(\vec{\nu})\,\varkappa_\gamma \quad \text{and} \quad \mathcal{V} = -\nabla_s\cdot\left(\beta(\vec{\nu})\,\nabla_s\,\varkappa_\gamma\right) \quad \text{on}\,\Gamma(t)$$

are even more nonlinear than their isotropic counterparts. A major difficulty is that an analogue of the identity

$$\vec{\varkappa} = \varkappa\,\vec{\nu} = \Delta_s\,\vec{\mathrm{id}} \quad \text{on}\,\Gamma(t), \tag{103}$$

recall Lemma 13(ii), is no longer true in general. However, in many practical situations the present authors were able to come up with an anisotropic version of (103).

FIG. 3 Frank diagram and Wulff shape in $\mathbb{R}^2$ for the l^1-norm, $\gamma(\vec{p}) = \sum_{i=1}^2 |p_i|$, *left*, and for the weighted norm $\gamma(\vec{p}) = (\vec{p}\cdot\underline{G}\,\vec{p})^{\frac{1}{2}}$, $\underline{G} = \frac{1}{4}\begin{pmatrix}4 & 0\\ 0 & 1\end{pmatrix}$, *right.*

The main observation is that (103) remains true if we replace the standard Euclidean inner product in $\mathbb{R}^d$ by an inner product

$$(\vec{u},\vec{v})_{\widetilde{G}} = \vec{u}\cdot\widetilde{\underline{\underline{G}}}\,\vec{v} \quad \forall\, \vec{u},\vec{v}\in\mathbb{R}^d, \tag{104}$$

where $\widetilde{\underline{\underline{G}}}\in\mathbb{R}^{d\times d}$ is symmetric and positive definite. One only has to replace the mean curvature vector $\vec{\varkappa}$ and the Laplace–Beltrami operator Δ_s by versions which are appropriate for this new inner product. In fact, one just has to consider the canonical Laplace–Beltrami operator on $\Gamma(t)$ with respect to the Riemannian metric given by the new inner product. However, the surface energy density related to the new inner product needs to be identified. It can be shown, see Barrett et al. (2008c, Lemma 2.1), that

$$\gamma(\vec{\nu}) = \sqrt{\vec{\nu}\cdot\underline{\underline{G}}\,\vec{\nu}} \tag{105}$$

leads to $\widetilde{\underline{\underline{G}}} = [\det\underline{\underline{G}}]^{\frac{1}{d-1}}\underline{\underline{G}}^{-1}$ and a weighted mean curvature $\varkappa_\gamma$ for which we have a relationship similar to that in the isotropic case, recall (103). In fact, it is shown that

$$\varkappa_\gamma\,\vec{\nu} = \gamma(\vec{\nu})\,\widetilde{\underline{\underline{G}}}\,\Delta_s^{\widetilde{G}}\,\vec{\mathrm{id}} \quad \text{on}\,\Gamma(t),$$

see Barrett et al. (2008c, (2.33), (2.37)), where $\Delta_s^{\widetilde{G}} = \nabla_s^{\widetilde{G}}\cdot\nabla_s^{\widetilde{G}}$ is the Laplace–Beltrami operator induced by the inner product (103) with $(\nabla_s^{\widetilde{G}}\cdot)$ and $\nabla_s^{\widetilde{G}}$ the associated tangential divergence and tangential gradient, respectively. A suitable generalized divergence theorem on manifolds then allows one to introduce a weak formulation of $\gamma(\vec{\nu})\,\widetilde{\underline{\underline{G}}}\,\Delta_s^{\widetilde{G}}\,\vec{\mathrm{id}}$. Unfortunately, simple anisotropies of the form (105) only lead to ellipsoidal Wulff shapes as on the right hand side of Fig. 3, and of course we would like to handle more general situations.

We now consider a larger class of surface energy densities, which are given as suitable norms of the ellipsoidal anisotropies. In particular, we choose

$$\gamma(\vec{p}) = \left(\sum_{\ell=1}^{L}[\gamma_\ell(\vec{p})]^r\right)^{\frac{1}{r}}, \quad \gamma_\ell(\vec{p}) = \sqrt{\vec{p}\cdot\underline{\underline{G}}_\ell\,\vec{p}}, \quad \ell = 1,\ldots,L, \tag{106}$$

so that

$$\gamma'(\vec{p}) = [\gamma(\vec{p})]^{1-r}\sum_{\ell=1}^{L}[\gamma_\ell(\vec{p})]^{r-1}\gamma'_\ell(\vec{p}).$$

Here $r\in[1,\infty)$ and $\underline{\underline{G}}_\ell\in\mathbb{R}^{d\times d}$, $\ell = 1,\ldots,L$, are symmetric and positive definite. It turns out that most energies of relevance can be approximated by the above class of energies. In particular, hexagonal and cubic anisotropies can be modelled with appropriate choices of r, L and $\{\underline{\underline{G}}_\ell\}_{\ell=1}^{L}$, see Fig. 4. We remark that in the planar case, $d = 2$, from a modelling point of view,

FIG. 4 Frank diagram and Wulff shape in $\mathbb{R}^3$ for a regularized l^1-norm, $\gamma(\vec{p}) = \sum_{\ell=1}^{3} [\varepsilon^2 |\vec{p}|^2 + p_\ell^2(1-\varepsilon^2)]^{\frac{1}{2}}$, $\varepsilon = 0.01$, *left*, a cubic anisotropy, $\gamma(\vec{p}) = [\sum_{\ell=1}^{3} [\varepsilon^2 |\vec{p}|^2 + p_\ell^2(1-\varepsilon^2)]^{\frac{r}{2}}]^{\frac{1}{r}}$, $\varepsilon = 0.01$, $r = 30$, *middle*, and a hexagonal anisotropy, $\gamma(\vec{p}) = \sum_{\ell=1}^{4} [\vec{p} \cdot \underline{\underline{R}}_\ell^{\mathrm{T}} \operatorname{diag}(1, \varepsilon^2, \varepsilon^2)\, \underline{\underline{R}}_\ell \vec{p}]^{\frac{1}{2}}$, $\varepsilon = 0.01$, *right*. Here $\underline{\underline{R}}_\ell$, $\ell = 1, \ldots, 4$ are suitable rotation matrices.

there is no benefit in choosing $r > 1$. Moreover, the choice $r = 1$ has the advantage that it leads to linear schemes, i.e., a linear system of equations needs to be solved at each time level.

Most of the surface calculus discussed in Section 2 can be repeated in the context of the energies discussed above. We obtain, for example, see Barrett et al. (2008c, Theorem 2.1), that for a C^2-hypersurface Γ it holds that

$$\varkappa_\gamma \vec{\nu} = \sum_{\ell=1}^{L} \gamma_\ell(\vec{\nu})\, \underline{\underline{\widetilde{G}}}_\ell\, \nabla_s^{\widetilde{G}_\ell} \cdot \left[\left[\frac{\gamma_\ell(\vec{\nu})}{\gamma(\vec{\nu})} \right]^{r-1} \nabla_s^{\widetilde{G}_\ell}\, \vec{\mathrm{id}} \right] \quad \text{on } \Gamma, \tag{107}$$

where $\underline{\underline{\widetilde{G}}}_\ell = \left[\det \underline{\underline{G}}_\ell\right]^{\frac{1}{d-1}} \underline{\underline{G}}_\ell^{-1}$, $\ell = 1, \ldots, L$, and $\nabla_s^{\widetilde{G}_\ell}$, $\left(\nabla_s^{\widetilde{G}_\ell} \cdot\right)$ are defined as follows. Let $\Gamma \subset \mathbb{R}^d$ be a C^1-hypersurface and let $\vec{p} \in \Gamma$. Let $\{\vec{\tau}_1, \ldots, \vec{\tau}_{d-1}\}$ be an orthonormal basis of $\mathrm{T}_{\vec{p}}\Gamma$ with respect to the inner product $(\cdot,\cdot)_{\widetilde{G}_\ell}$ on $\mathbb{R}^d$ induced by $\underline{\underline{\widetilde{G}}}_\ell$, recall (104). Let $f : \Gamma \to \mathbb{R}$, $\vec{f} : \Gamma \to \mathbb{R}^d$ be C^1-functions. The anisotropic surface gradient of f, the anisotropic surface divergence of $\vec{f}$ and the anisotropic surface Jacobian of $\vec{f}$ at the point $\vec{p} \in \Gamma$ are defined as

$$(\nabla_s^{\widetilde{G}_\ell} f)(\vec{p}) = \sum_{i=1}^{d-1} (\partial_{\vec{\tau}_i} f)(\vec{p})\, \vec{\tau}_i, \tag{108a}$$

$$(\nabla_s^{\widetilde{G}_\ell} \cdot \vec{f})(\vec{p}) = \sum_{i=1}^{d-1} \underline{\underline{\widetilde{G}}}_\ell\, \vec{\tau}_i \cdot (\partial_{\vec{\tau}_i} \vec{f})(\vec{p}), \tag{108b}$$

$$(\nabla_s^{\widetilde{G}_\ell} \vec{f})(\vec{p}) = \sum_{i=1}^{d-1} (\partial_{\vec{\tau}_i} \vec{f})(\vec{p}) \otimes \underline{\underline{\widetilde{G}}}_\ell\, \vec{\tau}_i. \tag{108c}$$

The definitions (108) are anisotropic versions of Definition 5(ii)–(iv) in the case $n = d - 1$. If, in addition, $\vec{g} \in C^1(\Gamma)$, we also define

$$\left(\nabla_s^{\widetilde{G}_\ell} \vec{f}, \nabla_s^{\widetilde{G}_\ell} \vec{g}\right)_{\widetilde{G}_\ell}(\vec{p}) = \sum_{i=1}^{d-1} \left((\partial_{\vec{\tau}_i} \vec{f})(\vec{p}), (\partial_{\vec{\tau}_i} \vec{g})(\vec{p}) \right)_{\widetilde{G}_\ell}.$$

Using a generalized divergence theorem, see Barrett et al. (2008c, Lemma 2.8), we can obtain the following weak formulations of the anisotropic flows (100) in the case that the anisotropic energy density γ is of the form (106). Given a closed hypersurface $\Gamma(0)$, find an evolving hypersurface $(\Gamma(t))_{t\in[0,T]}$ with a global parameterization and induced velocity field $\vec{\mathcal{V}}$, and $\varkappa_\gamma \in L^2(\mathcal{G}_T)$ as follows. For almost all $t\in(0,T)$, find $\vec{\mathcal{V}}(\cdot,t)\in[L^2(\Gamma(t))]^d$ and $\varkappa_\gamma(\cdot,t)\in L^2(\Gamma(t))$, respectively, $H^1(\Gamma(t))$, such that

$$\left\langle \vec{\mathcal{V}}, \chi\,\vec{\nu}\right\rangle_{\Gamma(t)} = \begin{cases} \langle \beta(\vec{\nu})\,\varkappa_\gamma, \chi\rangle_{\Gamma(t)} & \forall \chi\in L^2(\Gamma(t)) \\ \langle \beta(\vec{\nu})\,\nabla_s \varkappa_\gamma, \nabla_s\chi\rangle_{\Gamma(t)} & \forall \chi \in H^1(\Gamma(t)) \end{cases}, \tag{109a}$$

$$\langle \varkappa_\gamma\,\vec{\nu}, \vec{\eta}\rangle_{\Gamma(t)} + \left\langle \nabla_s^{\widetilde{G}}\,\vec{\mathrm{id}}, \nabla_s^{\widetilde{G}}\,\vec{\eta}\right\rangle_{\Gamma(t),\gamma} = 0 \quad \forall \vec{\eta}\in[H^1(\Gamma(t))]^d, \tag{109b}$$

where

$$\left\langle \nabla_s^{\widetilde{G}}\,\vec{\chi}, \nabla_s^{\widetilde{G}}\,\vec{\eta}\right\rangle_{\Gamma(t),\gamma} = \sum_{\ell=1}^{L}\int_{\Gamma(t)}\left[\frac{\gamma_\ell(\vec{\nu})}{\gamma(\vec{\nu})}\right]^{r-1}(\nabla_s^{\widetilde{G}_\ell}\,\vec{\chi}, \nabla_s^{\widetilde{G}_\ell}\,\vec{\eta})_{\widetilde{G}_\ell}\,\gamma_\ell(\vec{\nu})\,\mathrm{d}\mathcal{H}^{d-1} \tag{110}$$

for all $\vec{\chi},\vec{\eta}\in[H^1(\Gamma(t))]^d$. In the above weak formulations only derivatives up to first order appear, and hence once again the equations can be discretized with the help of continuous piecewise linear finite elements.

6.3 Finite element approximation

The weak formulations (109) can be used to formulate finite element approximations of (100). Discretizing the velocity and the geometry, similarly as in (41) for isotropic mean curvature flow, we recall the following finite element approximations from Barrett et al. (2008b,c).

Let the closed polyhedral hypersurface Γ^0 be an approximation of $\Gamma(0)$. Then, for $m=0,\ldots,M-1$, find $\Gamma^{m+1}=\vec{X}^{m+1}(\Gamma^m)$, for $\vec{X}^{m+1}\in \underline{V}(\Gamma^m)$, with unit normal $\vec{\nu}^{m+1}$, and $\kappa_\gamma^{m+1}\in V(\Gamma^m)$ such that

$$\left\langle \frac{\vec{X}^{m+1}-\vec{\mathrm{id}}}{\Delta t_m}, \chi\,\vec{\nu}^m\right\rangle_{\Gamma^m}^h = \begin{cases} \left\langle \beta(\vec{\nu}^m)\,\kappa_\gamma^{m+1}, \chi\right\rangle_{\Gamma^m}^h \\ \left\langle \beta(\vec{\nu}^m)\,\nabla_s\kappa_\gamma^{m+1}, \nabla_s\chi\right\rangle_{\Gamma^m} \end{cases} \quad \forall \chi\in V(\Gamma^m), \tag{111a}$$

$$\left\langle \kappa_\gamma^{m+1}\,\vec{\nu}^m, \vec{\eta}\right\rangle_{\Gamma^m}^h + \left\langle \nabla_s^{\widetilde{G}_\ell}\,\vec{X}^{m+1}, \nabla_s^{\widetilde{G}_\ell}\,\vec{\eta}\right\rangle_{\Gamma^m,\gamma} = 0 \quad \forall \vec{\eta}\in \underline{V}(\Gamma^m). \tag{111b}$$

In the above, we have used the notation

$$\left\langle \nabla_s^{\widetilde{G}_\ell} \vec{X}^{m+1}, \nabla_s^{\widetilde{G}_\ell} \vec{\eta} \right\rangle_{\Gamma^m,\gamma} = \sum_{\ell=1}^{L} \int_{\Gamma^m} \left[\frac{\gamma_\ell(\vec{\nu}^{m+1} \circ \vec{X}^{m+1})}{\gamma(\vec{\nu}^{m+1} \circ \vec{X}^{m+1})} \right]^{r-1} (\nabla_s^{\widetilde{G}_\ell} \vec{X}^{m+1}, \nabla_s^{\widetilde{G}_\ell} \vec{\eta})_{\widetilde{G}_\ell} \gamma_\ell(\vec{\nu}^m) \, d\mathcal{H}^{d-1},$$

i.e., a discrete analogue of (110).

Remark 98 (Curves in the plane).
On using the notation from Section 3.1.2, and similarly to (48), the systems (111) simplify considerably in the case $d=2$, i.e., for the evolution of curves in the plane, recall also Barrett et al. (2008c, (3.5)). In particular, we can reformulate (111) as follows. Let $\vec{X}^0 \in \underline{V}^h(\mathbb{I})$ be such that $\Gamma^0 = \vec{X}^0(\mathbb{I})$ is a polygonal approximation of $\Gamma(0)$. Then, for $m = 0, \ldots, M-1$, find $(\vec{X}^{m+1}, \kappa_\gamma^{m+1}) \in \underline{V}^h(\mathbb{I}) \times V^h(\mathbb{I})$ such that

$$-\left\langle \frac{\vec{X}^{m+1} - \vec{X}^m}{\Delta t_m}, \chi \, [\vec{X}^m_\rho]^\perp \right\rangle_{\mathbb{I}}^h = \begin{cases} \left\langle \beta\left(-\dfrac{[\vec{X}^m_\rho]^\perp}{|\vec{X}^m_\rho|}\right) \kappa_\gamma^{m+1}, \chi \, |\vec{X}^m_\rho| \right\rangle_{\mathbb{I}}^h \\ \left\langle \beta\left(-\dfrac{[\vec{X}^m_\rho]^\perp}{|\vec{X}^m_\rho|}\right) (\kappa_\gamma^{m+1})_\rho, \chi_\rho \, |\vec{X}^m_\rho|^{-1} \right\rangle_{\mathbb{I}} \end{cases}, \tag{112a}$$

$$-\left\langle \kappa_\gamma^{m+1} \, [\vec{X}^m_\rho]^\perp, \vec{\eta} \right\rangle_{\mathbb{I}}^h + \sum_{\ell=1}^{L} \left\langle \left[\frac{\gamma_\ell([\vec{X}^{m+1}_\rho]^\perp)}{\gamma([\vec{X}^{m+1}_\rho]^\perp)} \right]^{r-1} \frac{\underline{\underline{G}}_\ell \, [\vec{X}^{m+1}_\rho]^\perp}{\gamma_\ell([\vec{X}^m_\rho]^\perp)}, \vec{\eta}_\rho^\perp \right\rangle_{\mathbb{I}} = 0 \tag{112b}$$

for all $\chi \in V^h(\mathbb{I})$ and $\vec{\eta} \in \underline{V}^h(\mathbb{I})$.

For $r=1$ in (106) the systems (111) do not depend on $\vec{\nu}^{m+1}$, and so they reduce to linear systems for the unknowns $\vec{X}^{m+1}$ and κ_γ^{m+1}. For these discrete systems we have the following existence and uniqueness result, which can be shown by generalizing the proofs of the corresponding isotropic cases, see Theorems 76 and 86.

Theorem 99. *Let Γ^m satisfy Assumption 64(i), and let γ be of the form (106) with $r=1$. Then there exists a unique solution $(\vec{X}^{m+1}, \kappa_\gamma^{m+1}) \in \underline{V}(\Gamma^m) \times V(\Gamma^m)$ to the systems (111).*

Proof. The two results can be shown as in the proofs of Theorems 76 and 86, see also Barrett et al. (2008c, Theorem 3.1). □

Remark 100. For the nonlinear discretizations with $r > 1$, neither existence nor uniqueness results are known. However, in practice there are no difficulties in finding solutions to the nonlinear systems (111), and the employed iterative solvers always converged.

The main property of the schemes (111) is that they can be shown to be stable. This enables one to compute solutions for strongly anisotropic, nearly crystalline anisotropies. Something which is difficult with other schemes in the literature. The main insight is that a local inequality for the anisotropic energy is true, which is similar to the isotropic version stated in Lemma 55 for $d = 3$. In fact, we have the following result.

Lemma 101. *Let $d = 3$ and let $\Gamma^h = \bigcup_{j=1}^{J} \overline{\sigma_j}$ be a polyhedral hypersurface, with unit normal $\vec{\nu}^h$, in $\mathbb{R}^d$. Then we have for $j = 1, \ldots, J$ and $\ell = 1, \ldots, L$ that*

$$\frac{1}{2} \int_{\sigma_j} \gamma_\ell(\vec{\nu}^h)\, |\nabla_s^{\widetilde{G}_\ell} \vec{X}|^2_{\widetilde{G}_\ell} \, d\mathcal{H}^2 \geq \int_{\vec{X}(\sigma_j)} \gamma_\ell(\vec{\nu}^h_{\vec{X}}) \, d\mathcal{H}^2 \quad \forall \vec{X} \in \underline{V}(\Gamma^h),$$

with equality for $\vec{X} = \vec{\mathrm{id}}_{|_{\Gamma^h}} \in \underline{V}(\Gamma^h)$. Here $\vec{\nu}^h_{\vec{X}}$ is the unit normal on the polyhedral hypersurface $\vec{X}(\Gamma^h)$ and $|\cdot|^2_{\widetilde{G}_\ell} = (\cdot, \cdot)_{\widetilde{G}_\ell}$.

Proof. See Barrett et al. (2008c, Lemma 3.1). □

Lemma 102. *Let γ be of the form (106) and let $d = 2$ or $d = 3$. Let $\Gamma^h = \bigcup_{j=1}^{J} \overline{\sigma_j}$ be a polyhedral hypersurface, with unit normal $\vec{\nu}^h$, in $\mathbb{R}^d$. Let γ be of the form (106) and let $\vec{X} \in \underline{V}(\Gamma^h)$. Then it holds that*

$$\left\langle \nabla_s^{\widetilde{G}} \vec{X}, \nabla_s^{\widetilde{G}} (\vec{X} - \vec{\mathrm{id}}) \right\rangle_{\Gamma^h,\gamma} \geq |\vec{X}(\Gamma^h)|_\gamma - |\Gamma^h|_\gamma,$$

where $|\Gamma^h|_\gamma = \langle 1, \gamma(\vec{\nu}^h) \rangle_{\Gamma^h}$.

Proof. The proof for $d = 3$ hinges on Lemma 101 and can be found in Barrett et al. (2008c, Theorem 3.2). On recalling Remark 98, the result for $d = 2$, and $r = 1$, can be shown by using ideas similar to (27) in the proof of Lemma 57, see Barrett et al. (2008b, Theorem 2.5). It can easily be extended to the case $r \geq 1$, on using the techniques in Barrett et al. (2008c, Theorem 3.2). □

Theorem 103. *Let γ be of the form (106) and let $d = 2$ or $d = 3$. Let $(\vec{X}^{m+1}, \kappa_\gamma^{m+1}) \in \underline{V}(\Gamma^m) \times V(\Gamma^m)$ be a solution of (111). Then it holds that*

$$|\Gamma^{m+1}|_\gamma + \Delta t_m \begin{cases} \left\langle \beta(\vec{\nu}^m)\, \kappa_\gamma^{m+1}, \kappa_\gamma^{m+1} \right\rangle^h_{\Gamma^m} \\ \left\langle \beta(\vec{\nu}^m)\, \nabla_s \kappa_\gamma^{m+1}, \nabla_s \kappa_\gamma^{m+1} \right\rangle_{\Gamma^m} \end{cases} \leq |\Gamma^m|_\gamma, \tag{113}$$

where we recall (98).

Proof. Similarly to the proof of Theorem 77, we can choose $\chi = \kappa_\gamma^{m+1} \in V(\Gamma^m)$ in (111a) and $\vec{\eta} = \frac{1}{\Delta t_m} (\vec{X}^{m+1} - \vec{\mathrm{id}}_{|\Gamma^m}) \in \underline{V}(\Gamma^m)$ in (111b), and then combine with Lemma 102, in order to obtain the desired results. □

The inequalities (113) are the discrete analogues of (101) and (102), respectively, and anisotropic analogues of Theorems 77 and 87, respectively.

6.4 Solution method and discrete systems

In comparison to the isotropic case the assembly of the matrices is only slightly more complicated. In addition to the matrices in Section 4.3, we define $M_{\Gamma^m,\beta}$, $A_{\Gamma^m,\beta} \in \mathbb{R}^{K\times K}$ with entries

$$[M_{\Gamma^m,\beta}]_{kl} = \langle \beta(\vec{\nu}^m)\, \phi_k^{\Gamma^m}, \phi_l^{\Gamma^m} \rangle_{\Gamma^m}^h, \quad [A_{\Gamma^m,\beta}]_{kl} = \langle \beta(\vec{\nu}^m)\, \nabla_s \phi_k^{\Gamma^m}, \nabla_s \phi_l^{\Gamma^m} \rangle_{\Gamma^m}.$$

Moreover, given a $\vec{\mathrm{v}} \in \underline{V}_c(\Gamma^m)$, we introduce $\underline{\underline{A}}_{\Gamma^m,\gamma}(\vec{\mathrm{v}}) \in (\mathbb{R}^{d\times d})^{K\times K}$ with entries

$$[\underline{\underline{A}}_{\Gamma^m,\gamma}(\vec{\mathrm{v}})]_{kl} = \sum_{\ell=1}^{L} \left\langle \left[\frac{\gamma_\ell(\vec{\mathrm{v}})}{\gamma(\vec{\mathrm{v}})}\right]^{r-1} \gamma_\ell(\vec{\nu}^m)\, \nabla_s^{\widetilde{G}_\ell} \phi_k^{\Gamma^m}, \nabla_s^{\widetilde{G}_\ell} \phi_l^{\Gamma^m} \right\rangle_{\Gamma^m} \underline{\underline{\widetilde{G}}}_\ell,$$

where we have noted (110) and (104). Assembling these matrices is straightforward, since assembling, e.g., $\left(\left\langle \nabla_s^{\widetilde{G}_\ell} \phi_k^{\Gamma^m}, \nabla_s^{\widetilde{G}_\ell} \phi_l^{\Gamma^m} \right\rangle_{\Gamma^m}\right)_{k,l=1}^{K}$ is very similar to assembling A_{Γ^m} in (45). Using the notation from (46), we can then formulate (111) as: Find $(\delta\vec{X}^{m+1}, \kappa_\gamma^{m+1}) \in (\mathbb{R}^d)^K \times \mathbb{R}^K$ such that

$$\begin{pmatrix} \Delta t_m \begin{cases} M_{\Gamma^m,\beta} \\ A_{\Gamma^m,\beta} \end{cases} & -\vec{N}_{\Gamma^m}^{\mathsf{T}} \\ \vec{N}_{\Gamma^m} & \underline{\underline{A}}_{\Gamma^m,\gamma}(\vec{\mathrm{v}}^{m+1}) \end{pmatrix} \begin{pmatrix} \kappa_\gamma^{m+1} \\ \delta\vec{X}^{m+1} \end{pmatrix} = \begin{pmatrix} 0 \\ -\underline{\underline{A}}_{\Gamma^m,\gamma}(\vec{\mathrm{v}}^{m+1})\, \vec{X}^m \end{pmatrix}, \tag{114}$$

where we have used the notation $\vec{\mathrm{v}}^{m+1} = \vec{\nu}^{m+1} \circ \vec{X}^{m+1}$, and where we recall that $\delta\vec{X}^{m+1}$ is the vector of coefficients with respect to the standard basis for $\vec{X}^{m+1} - \vec{\mathrm{id}}_{|\Gamma^m} \in \underline{V}(\Gamma^m)$.

If γ is of the form (106) with $r = 1$, then the system (114) does not depend on $\vec{\mathrm{v}}^{m+1}$ and so is linear. Hence it can be solved analogously to (46), i.e., either with a sparse direct solver or with the help of a Schur complement approach and a preconditioned conjugate gradient solver, see Barrett et al. (2008c, Section 4) for details.

If $r > 1$, on the other hand, then the nonlinear system (114) can be solved with a lagged fixed-point type iteration, where each iterate is obtained by solving a linear system of the form (114), with $\vec{\mathrm{v}}^{m+1}$ replaced by a given quantity $\vec{\mathrm{v}}^{m+1,i}$. We note that in practice such an iteration always converges, and refer to Barrett et al. (2008c, Section 4) for further details.

6.5 Volume conservation for semidiscrete schemes

For the semidiscrete continuous-in-time variants of (111) it is straightforward to prove the obvious anisotropic analogues of Theorems 79(i) and 88(i), as well as Theorem 88(ii) in the case of anisotropic surface diffusion.

Remark 104 (Tangential motion).
In the anisotropic situation the tangential motion induced by the semidiscrete analogue of (111b) will no longer lead to surfaces satisfying Definition 60. In particular, Theorem 79(ii) will in general not hold. In the planar case, and if γ is of the form (106) with $L = 1$ and $r = 1$, then equidistribution with respect to γ can be shown, see Barrett et al. (2008b, Remark 2.7). However, the fully discrete schemes (111), for any γ of the form (106) and for $d = 2$ and $d = 3$, exhibit good quality meshes in practice, without coalescence that could lead to a breakdown of the schemes. See, for example, the numerical simulations in Barrett et al. (2008c, Section 5).

6.6 Alternative numerical approaches

For the case $d = 2$ a numerical scheme for anisotropic mean curvature flow has been proposed in Dziuk (1999a), and an error estimate for a semidiscrete continuous-in-time variant is shown, on assuming that the approximated solution is sufficiently smooth.

Moreover, and still for $d = 2$, fully implicit fully discrete schemes, similar to (63), for anisotropic evolution equations are considered in Barrett et al. (2011).

The extension of the methods discussed in this section to anisotropic evolution laws of surface clusters and surfaces with boundary has been discussed in Barrett et al. (2008b, 2010a,d).

For parametric methods for anisotropic curvature-driven flows in higher codimension we refer to Pozzi (2007, 2008) and Barrett et al. (2010b).

We refer also to Deckelnick and Dziuk (1999), where, on assuming a sufficiently smooth solution, an error analysis is presented for a semidiscrete finite element approximation by continuous piecewise linears of a graph formulation of anisotropic mean curvature flow. We refer to Deckelnick et al. (2005b) for the corresponding error analysis for a fully discrete finite element approximation of a graph formulation of anisotropic surface diffusion.

We also refer to Deckelnick et al. (2005a) for the approximation of anisotropic mean curvature flow in the context of level set and phase field methods. For the former we mention Burger et al. (2007) and Clarenz et al. (2005), while for the latter we refer to Caginalp and Lin (1987), Garcke et al. (1999), Beneš (2003), Gräser et al. (2013) and the present authors' work in Barrett et al. (2013b, 2014a) on stable approximations of anisotropic mean curvature flow and anisotropic surface diffusion.

Finally we mention that for strongly anisotropic surface energies, some authors propose a curvature energy regularization, which leads to higher order flows with similarities to Willmore flow, see Burger (2005), Haußer and Voigt (2005) and Torabi et al. (2009).

7 Coupling bulk equations to geometric equations on the surface and applications to crystal growth

We now consider how curvature-driven interface evolutions can involve quantities defined in the bulk regions surrounding the interface.

7.1 The Mullins–Sekerka problem

Let us start with one of the simplest problems that couple geometric equations on the interface to equations in the bulk. Let $\Omega \subset \mathbb{R}^d$, $d \geq 2$, be a domain with a Lipschitz boundary $\partial\Omega$, which is separated by an interface into two different phases. The interface is at each time t assumed to be a closed C^2-hypersurface $\Gamma(t) \subset \Omega$. We adopt the notation in Section 2.4, recall Fig. 1, and assume that two phases occupy regions $\Omega_-(t)$ and $\Omega_+(t) = \Omega \setminus \overline{\Omega_-(t)}$ with $\Gamma(t) = \partial\Omega_-(t)$, and with $\vec{\nu}$ denoting the outer unit normal to $\Omega_-(t)$ on $\Gamma(t)$. Then the Mullins–Sekerka problem is given as follows. Given $\Gamma(0)$, find $u : \Omega \times [0,T] \to \mathbb{R}$ and the evolving interface $(\Gamma(t))_{t\in[0,T]}$ such that for all $t \in (0,T]$ the following conditions hold

$$-\Delta u = 0 \quad \text{in } \Omega_-(t), \qquad -\Delta u = 0 \quad \text{in } \Omega_+(t), \tag{115a}$$

$$u = \varkappa \quad \text{on } \Gamma(t), \qquad [\partial_{\vec{\nu}}\, u]_-^+ = -\mathcal{V} \quad \text{on } \Gamma(t), \tag{115b}$$

$$\partial_{\vec{\nu}_\Omega}\, u = 0 \quad \text{on } \partial\Omega, \tag{115c}$$

where we have recalled the notation (13) and $\vec{\nu}_\Omega$ is the outer unit normal to Ω on $\partial\Omega$.

7.1.1 Weak formulation of the Mullins–Sekerka problem

A finite element approximation of the Mullins–Sekerka problem needs a suitable weak formulation, which is given as follows. Given a closed hypersurface $\Gamma(0) \subset \Omega$, we seek an evolving hypersurface $(\Gamma(t))_{t\in[0,T]}$ that separates Ω into $\Omega_-(t)$ and $\Omega_+(t)$, with a global parameterization and induced velocity field $\vec{\mathcal{V}}$, and $\varkappa \in L^2(\mathcal{G}_T)$ as well as $u : \Omega \times [0,T] \to \mathbb{R}$ as follows. For almost all $t \in (0,T)$, find $(u(\cdot,t), \vec{\mathcal{V}}(\cdot,t), \varkappa(\cdot,t)) \in H^1(\Omega) \times [L^2(\Gamma(t))]^d \times L^2(\Gamma(t))$ such that

$$(\nabla u, \nabla \phi) = \left\langle \vec{\mathcal{V}}, \phi\, \vec{\nu} \right\rangle_{\Gamma(t)} \quad \forall \phi \in H^1(\Omega), \tag{116a}$$

$$\langle u, \chi \rangle_{\Gamma(t)} = \langle \varkappa, \chi \rangle_{\Gamma(t)} \quad \forall \chi \in L^2(\Gamma(t)), \tag{116b}$$

$$\langle \varkappa\, \vec{\nu}, \vec{\eta} \rangle_{\Gamma(t)} + \left\langle \nabla_s\, \vec{\mathrm{id}}, \nabla_s\, \vec{\eta} \right\rangle_{\Gamma(t)} = 0 \quad \forall \vec{\eta} \in [H^1(\Gamma(t))]^d. \tag{116c}$$

Here $(\cdot,\cdot) = \langle\cdot,\cdot\rangle_\Omega$ is the L^2-inner product on Ω. It is easy to show that a sufficiently smooth solution to (116) solves (115). To this end, we observe that it follows from (116a) and Remark 36, on using a density argument, that

$$\begin{aligned}\langle \mathcal{V}, \phi\rangle_{\Gamma(t)} &= \left\langle \vec{\mathcal{V}}, \phi\,\vec{\nu}\right\rangle_{\Gamma(t)} = (\nabla u, \nabla \phi)\\ &= -\int_{\Omega_-(t)\cup\Omega_+(t)} \phi\,\Delta u \,\mathrm{d}\mathcal{L}^d + \langle \partial_{\vec{\nu}_\Omega} u, \phi\rangle_{\partial\Omega} - \left\langle [\partial_{\vec{\nu}}\, u]_-^+, \phi\right\rangle_{\Gamma(t)} \qquad (117)\\ &\qquad \forall\,\phi\in H^1(\Omega).\end{aligned}$$

Now the fundamental lemma of the calculus of variations, together with (117), yields that (115a), (115c) and the second condition in (115b) hold. Similarly, it follows from (116b) and (116c), on recalling Remark 22(iv), that the first condition in (115b) is also satisfied.

Remark 105. For a sufficiently smooth weak solution of (116), we can formally prove the following results.

(i) The Mullins–Sekerka problem decreases the surface area of the interface. This follows from the transport theorem, Theorem 32, (116b) with $\chi = \mathcal{V}$ and (116a) with $\phi = u$, as

$$\frac{\mathrm{d}}{\mathrm{d}t}|\Gamma(t)| = -\langle \varkappa, \mathcal{V}\rangle_{\Gamma(t)} = -\langle u, \mathcal{V}\rangle_{\Gamma(t)} = -\left\langle \vec{\mathcal{V}}, u\,\vec{\nu}\right\rangle_{\Gamma(t)} = -|\nabla u|_\Omega^2 \le 0.$$

(ii) The Mullins–Sekerka problem preserves the volume $\mathcal{L}^d(\Omega_-(t))$, and hence $\mathcal{L}^d(\Omega_+(t))$, as the transport theorem, Theorem 33, yields that

$$\frac{\mathrm{d}}{\mathrm{d}t}\mathcal{L}^d(\Omega_-(t)) = \langle \mathcal{V}, 1\rangle_{\Gamma(t)} = \left\langle \vec{\mathcal{V}}, \vec{\nu}\right\rangle_{\Gamma(t)} = (\nabla u, \nabla 1) = 0,$$

where we have chosen $\phi = 1$ in (116a).

Remark 106. Given the hypersurface $\Gamma(t) \subset \Omega$, and hence its normal $\vec{\nu}(t)$ and curvature $\varkappa(t)$, we can solve (115a), (115c) and the first equation in (115b) to obtain a bulk function $u(t)$ that depends on $\varkappa(t)$. Defining now

$$\mathfrak{F}(\varkappa) = -[\partial_{\vec{\nu}}\, u]_-^+,$$

we obtain that

$$\langle \mathfrak{F}(\varkappa), \varkappa\rangle_{\Gamma(t)} = -\left\langle [\partial_{\vec{\nu}}\, u]_-^+, \varkappa\right\rangle_{\Gamma(t)} = -\left\langle [\partial_{\vec{\nu}}\, u]_-^+, u\right\rangle_{\Gamma(t)} = |\nabla u|_\Omega^2 \ge 0,$$

where we have observed the first equation in (115b), and the choice $\phi = u$ in (117), on noting (115a) and (115c). Therefore, the Mullins–Sekerka problem (115) can be written in the general form (81) satisfying (82).

7.1.2 An unfitted finite element approximation of the Mullins–Sekerka problem

In addition to the discretization of the evolving hypersurface, we need a discretization of the domain Ω. For simplicity we assume that Ω is a polyhedral domain. Although a generalization to a curved domain Ω is possible using suitable boundary approximations, see Ciarlet (1978). For all $m \geq 0$, let $\mathcal{T}^m$ be a regular partitioning of Ω into disjoint open simplices, so that $\overline{\Omega} = \cup_{o \in \mathcal{T}^m} \overline{o}$, see Ciarlet (1978) for a definition of regular partitioning and further details about finite elements. Note that we allow for time-dependent bulk triangulations. Associated with $\mathcal{T}^m$ is the finite element space

$$S^m = \left\{ \chi \in C(\overline{\Omega}) : \chi|_o \text{ is affine } \forall\, o \in \mathcal{T}^m \right\} \subset H^1(\Omega). \tag{118}$$

Let K_{S^m} be the number of nodes of $\mathcal{T}^m$ and let $\{\vec{p}_k^m\}_{k=1}^{K_{S^m}}$ be the coordinates of these nodes. Let $\{\phi_k^{S^m}\}_{k=1}^{K_{S^m}}$ be the standard basis functions for S^m. We introduce $I^m : C(\overline{\Omega}) \to S^m$, the interpolation operator, such that $(I^m \eta)(\vec{p}_k^m) = \eta(\vec{p}_k^m)$ for $k = 1, \ldots, K_{S^m}$. A discrete semiinner product on $C(\overline{\Omega})$ is then defined by

$$(\eta_1, \eta_2)_m^h = (I^m[\eta_1\, \eta_2], 1) \quad \forall\, \eta_1, \eta_2 \in C(\overline{\Omega}),$$

with the induced seminorm given by $|\eta|_{\Omega,m} = [\,(\eta, \eta)_m^h\,]^{\frac{1}{2}}$ for $\eta \in C(\overline{\Omega})$.

We can now introduce the finite element approximation of the Mullins–Sekerka problem from Barrett et al. (2010c), based on the weak formulation (116), as follows. Let the closed polyhedral hypersurface Γ^0 be an approximation of $\Gamma(0)$ and recall the time interval partitioning (40). Then, for $m = 0, \ldots, M-1$, find $(U^{m+1}, \vec{X}^{m+1}, \kappa^{m+1}) \in S^m \times \underline{V}(\Gamma^m) \times V(\Gamma^m)$ such that

$$(\nabla U^{m+1}, \nabla \varphi) = \left\langle \pi_{\Gamma^m}\left[\frac{\vec{X}^{m+1} - \vec{\mathrm{id}}}{\Delta t_m} \cdot \vec{\omega}^m\right], \varphi \right\rangle_{\Gamma^m} \quad \forall\, \varphi \in S^m, \tag{119a}$$

$$\left\langle \kappa^{m+1}, \chi \right\rangle_{\Gamma^m}^h = \left\langle U^{m+1}, \chi \right\rangle_{\Gamma^m} \quad \forall\, \chi \in V(\Gamma^m), \tag{119b}$$

$$\left\langle \kappa^{m+1}\, \vec{\nu}^m, \vec{\eta} \right\rangle_{\Gamma^m}^h + \left\langle \nabla_s \vec{X}^{m+1}, \nabla_s \vec{\eta} \right\rangle_{\Gamma^m} = 0 \quad \forall\, \vec{\eta} \in \underline{V}(\Gamma^m) \tag{119c}$$

and set $\Gamma^{m+1} = \vec{X}^{m+1}(\Gamma^m)$. In the above we have recalled (17).

Remark 107 (Implementation).
The approximation (119) is called unfitted because the surface mesh is totally independent of the bulk mesh, and is not fitted to the bulk mesh in the sense that the surface mesh does not consist of faces of the bulk mesh, see Barrett and Elliott (1982). As a consequence, special quadrature rules need to be employed in order to calculate the right-hand sides in (119a) and (119b). Here the most challenging aspect is to compute intersections $\sigma^m \cap o^m$ between an arbitrary element $\sigma^m \subset \Gamma^m$ and an element $o^m \in \mathcal{T}^m$ of the bulk mesh.

An algorithm that describes how these intersections can be calculated is given in Barrett et al. (2010c, p. 6284), see also Figures 4 and 5 in Barrett et al. (2010c).

Before stating our next result, we need an assumption on the compatibility between bulk and surface mesh.

Assumption 108. Let Γ^m and $\mathcal{T}^m$ be such that

$$\dim\left\{\int_{\Gamma^m} \varphi\, \vec{\omega}^m \,\mathrm{d}\mathcal{H}^{d-1} : \varphi \in S^m\right\} = d.$$

Theorem 109. *Let Γ^m and $\mathcal{T}^m$ satisfy Assumption 108. Then there exists a unique solution $(U^{m+1}, \vec{X}^{m+1}, \kappa^{m+1}) \in S^m \times \underline{V}(\Gamma^m) \times V(\Gamma^m)$ to (119). In addition, if $d = 2$ or $d = 3$, then a solution to (119) satisfies the stability estimate*

$$|\Gamma^{m+1}| + \Delta t_m |\nabla U^{m+1}|_\Omega^2 \leq |\Gamma^m|.$$

Proof. Existence follows from uniqueness and hence we consider the homogeneous linear system. Find $(U, \vec{X}, \kappa) \in S^m \times \underline{V}(\Gamma^m) \times V(\Gamma^m)$ such that

$$\Delta t_m (\nabla U, \nabla \varphi) - \left\langle \pi_{\Gamma^m}[\vec{X} \cdot \vec{\omega}^m], \varphi \right\rangle_{\Gamma^m} = 0 \quad \forall \varphi \in S^m, \tag{120a}$$

$$\langle \kappa, \chi \rangle_{\Gamma^m}^h - \langle U, \chi \rangle_{\Gamma^m} = 0 \quad \forall \chi \in V(\Gamma^m), \tag{120b}$$

$$\langle \kappa\, \vec{\omega}^m, \vec{\eta} \rangle_{\Gamma^m}^h + \left\langle \nabla_s \vec{X}, \nabla_s \vec{\eta} \right\rangle_{\Gamma^m} = 0 \quad \forall \vec{\eta} \in \underline{V}(\Gamma^m), \tag{120c}$$

where we have noted (19) in (120c). Choosing $\varphi = U$ in (120a), $\chi = \pi_{\Gamma^m}[\vec{X} \cdot \vec{\omega}^m]$ in (120b) and $\vec{\eta} = \vec{X}$ in (120c) gives

$$\Delta t_m |\nabla U|_\Omega^2 + \left|\nabla_s \vec{X}\right|_{\Gamma^m}^2 = 0.$$

We hence obtain that U is constant in Ω and that $\vec{X}$ is constant on Γ^m. Now (120a) and Assumption 108 yield that $\vec{X} = \vec{0}$. Moreover, it follows from (120b) that $\kappa = U$ is constant on Γ^m. Hence we can proceed as in the proof of Theorem 86 to show that $\kappa = U = 0$, and so we obtain uniqueness.

It remains to prove the stability bound. Here we choose $\varphi = U^{m+1}$ in (119a), $\chi = \pi_{\Gamma^m}[(\vec{X}^{m+1} - \vec{\mathrm{id}}) \cdot \vec{\omega}^m]$ in (119b) and $\vec{\eta} = \vec{X}^{m+1} - \vec{\mathrm{id}}_{|\Gamma^m}$ in (119c) and obtain

$$\Delta t_m |\nabla U^{m+1}|_\Omega^2 + \left\langle \nabla_s \vec{X}^{m+1}, \nabla_s (\vec{X}^{m+1} - \vec{\mathrm{id}}) \right\rangle_{\Gamma^m} = 0.$$

Now Lemma 57 yields the claim. □

Remark 110 (Semidiscrete scheme).
We note that the above stability bound is a discrete analogue of Remark 105(i). However, the fully discrete scheme (119) will in general not satisfy an exact discrete analogue of the volume conservation property Remark 105(ii). That is because choosing $\varphi = 1$ in (119a) only leads to

$$\begin{aligned} 0 &= \left\langle \pi_{\Gamma^m}\left[\left(\vec{X}^{m+1} - \vec{\mathrm{id}}\right)\cdot\vec{\omega}^m\right], 1\right\rangle_{\Gamma^m} = \left\langle \vec{X}^{m+1} - \vec{\mathrm{id}}, \vec{\omega}^m\right\rangle^h_{\Gamma^m} \\ &= \left\langle \vec{X}^{m+1} - \vec{\mathrm{id}}, \vec{\nu}^m\right\rangle^h_{\Gamma^m} = \left\langle \vec{X}^{m+1} - \vec{\mathrm{id}}, \vec{\nu}^m\right\rangle_{\Gamma^m}, \end{aligned} \tag{121}$$

where we have noted (19) and (20). In general, the terms in (121) do not equal the discrete volume change $\mathcal{L}^d(\Omega_-^{m+1}) - \mathcal{L}^d(\Omega_-^m)$, where Ω_-^m denotes the interior of Γ^m. However, on recalling Remark 89(ii) it is possible to prove an exact discrete analogue of Remark 105(ii) for a semidiscrete version of the scheme (119). Moreover, this semidiscrete scheme will satisfy the mesh property Theorem 88(iii).

Remark 111 (Discrete linear systems).
On recalling the notation from Section 4.3, we can formulate the linear systems of equations to be solved at each time level for (119) as follows. Find $(U^{m+1}, \kappa^{m+1}, \delta\vec{X}^{m+1}) \in \mathbb{R}^{K_{S^m}} \times \mathbb{R}^K \times (\mathbb{R}^d)^K$ such that

$$\begin{pmatrix} A_\Omega & 0 & -\dfrac{1}{\Delta t_m}\vec{N}^{\top}_{\Omega,\Gamma^m} \\ -M_{\Gamma^m,\Omega} & M_{\Gamma^m} & 0 \\ 0 & \vec{N}_{\Gamma^m} & \underline{\underline{A}}_{\Gamma^m} \end{pmatrix} \begin{pmatrix} U^{m+1} \\ \kappa^{m+1} \\ \delta\vec{X}^{m+1} \end{pmatrix} = \begin{pmatrix} 0 \\ 0 \\ -\underline{\underline{A}}_{\Gamma^m}\vec{X}^m \end{pmatrix}, \tag{122}$$

where we use a similar abuse of notation as in (46). The definitions of the matrices in (122) are either given in (45), or they follow directly from (119), see also Barrett et al. (2010c, Section 4.1) for details. In practice, the linear system (122) can either be solved with a sparse direct solution method like UMFPACK, see Davis (2004), or by first using a Schur complement approach and then use a (precondioned) conjugate gradient solver. Once again, we refer to Barrett et al. (2010c, Section 4.1) for more details.

7.2 The Stefan problem with a (kinetic) Gibbs–Thomson law

In general crystal growth phenomena involve more complex models, in comparison to the above Mullins–Sekerka model. The overall model is the Stefan problem with a Gibbs–Thomson law and kinetic undercooling with anisotropic surface energy (98) taken into account. It is given as follows with the same notation and conventions as for (115).

Given $\Gamma(0)$ and, if $\vartheta > 0$, $u_0 : \Omega \to \mathbb{R}$, find $u : \Omega \times [0,T] \to \mathbb{R}$ and the evolving interface $(\Gamma(t))_{t\in[0,T]}$ such that $\vartheta u(\cdot, 0) = \vartheta u_0$ and for all $t \in (0,T]$ the following conditions hold

$$\vartheta\,\partial_t\, u - \mathcal{K}_-\, \Delta u = f \quad \text{in}\, \Omega_-(t), \quad \vartheta\,\partial_t\, u - \mathcal{K}_+\, \Delta u = f \quad \text{in}\, \Omega_+(t), \tag{123a}$$

$$\frac{\rho\,\mathcal{V}}{\beta(\vec{\nu})} = \alpha\,\varkappa_\gamma - a\,u \quad \text{on}\, \Gamma(t), \quad [\mathcal{K}\,\partial_{\vec{\nu}}\, u]_-^+ = -\lambda\,\mathcal{V} \quad \text{on}\, \Gamma(t), \tag{123b}$$

$$\partial_{\vec{\nu}_\Omega} u = 0 \qquad \text{on}\, \partial_{\rm N}\Omega, \quad u = u_{\rm D} \qquad \text{on}\, \partial_{\rm D}\Omega, \tag{123c}$$

where $[\mathcal{K}\,\partial_{\vec{\nu}}\, u]_-^+$ is defined similarly to (13), and $\partial\Omega = \overline{\partial_{\rm N}\Omega} \cup \overline{\partial_{\rm D}\Omega}$ with $\partial_{\rm N}\Omega \cap \partial_{\rm D}\Omega = \emptyset$. In the above system u denotes the deviation from the melting temperature at a planar interface, f describes heat sources, $\vartheta \in \mathbb{R}_{\geq 0}$ is the volumetric heat capacity, and $\mathcal{K}(\cdot, t) = \mathcal{K}_+\, \mathcal{X}_{\Omega_+(t)} + \mathcal{K}_-\, \mathcal{X}_{\Omega_-(t)}$, with $\mathcal{K}_\pm \in \mathbb{R}_{>0}$, is the phase-dependent thermal conductivity, recall (11). Moreover, $\lambda \in \mathbb{R}_{>0}$ is the latent heat per unit volume, $\alpha \in \mathbb{R}_{>0}$ is an interfacial energy density per surface area, $\rho \in \mathbb{R}_{\geq 0}$ is a kinetic coefficient and $a \in \mathbb{R}_{>0}$ is a coefficient having the dimension entropy/volume. In addition, as in Section 6.1, $\beta : \mathbb{S}^{d-1} \to \mathbb{R}_{>0}$ is a dimensionless mobility function, which allows one to describe the dependence of the mobility on the local orientation of the interface. Clearly, on choosing $\vartheta = \rho = 0$, $\mathcal{K} = a = \alpha = \lambda = 1$ and $\partial\Omega = \partial_{\rm N}\Omega$ then (123) in the isotropic case collapses to (115).

In order to state the weak form of (123), we introduce the function spaces

$$S_0 = \{\phi \in H^1(\Omega) : \phi = 0 \quad \text{on}\, \partial_{\rm D}\Omega\} \text{ and } S_{\rm D} = \{\phi \in H^1(\Omega) : \phi = u_{\rm D} \quad \text{on}\, \partial_{\rm D}\Omega\}, \tag{124}$$

where we assume for simplicity from now on that $u_{\rm D} \in \mathbb{R}$, with $u_{\rm D} = 0$ in the case $\partial_{\rm N}\Omega = \partial\Omega$. Now a weak formulation of (123) for an anisotropic surface energy density of the form (106) can be obtained by generalizing the weak formulation of the Mullins–Sekerka problem, (116), as follows.

Given a closed hypersurface $\Gamma(0) \subset \Omega$ and, if $\vartheta > 0$, $u_0 \in L^2(\Omega)$, we seek an evolving hypersurface $(\Gamma(t))_{t\in[0,T]}$ that separates Ω into $\Omega_-(t)$ and $\Omega_+(t)$, with a global parameterization and induced velocity field $\vec{\mathcal{V}}$, and $\varkappa \in L^2(\mathcal{G}_T)$ as well as $u \in H^1(0,T; L^2(\Omega)) \cap L^2(0,T; H^1(\Omega)))$ with $\vartheta u(\cdot, 0) = \vartheta u_0$ as follows. For almost all $t \in (0,T)$, find $(u(\cdot, t), \vec{\mathcal{V}}(\cdot, t), \varkappa(\cdot, t)) \in S_{\rm D} \times [L^2(\Gamma(t))]^d \times L^2(\Gamma(t))$ such that

$$\vartheta\,(\partial_t\, u, \phi) + (\mathcal{K}\,\nabla\, u, \nabla\,\phi) - (f, \phi) = \lambda \left\langle \vec{\mathcal{V}}, \phi\,\vec{\nu} \right\rangle_{\Gamma(t)} \quad \forall\, \phi \in S_0, \tag{125a}$$

$$\rho \left\langle \frac{\vec{\mathcal{V}}}{\beta(\vec{\nu})}, \chi\,\vec{\nu} \right\rangle_{\Gamma(t)} = \langle \alpha\,\varkappa_\gamma - a\,u, \chi \rangle_{\Gamma(t)} \quad \forall\, \chi \in L^2(\Gamma(t)), \tag{125b}$$

$$\langle \varkappa_\gamma\,\vec{\nu}, \vec{\eta} \rangle_{\Gamma(t)} + \left\langle \nabla_s^{\widetilde{G}}\, \vec{\rm id}, \nabla_s^{\widetilde{G}}\, \vec{\eta} \right\rangle_{\Gamma(t),\gamma} = 0 \quad \forall\, \vec{\eta} \in [H^1(\Gamma(t))]^d, \tag{125c}$$

where we have adopted the notation (110).

We can establish the following formal a priori bound for a sufficiently smooth solution of (125). Choosing $\phi = u - u_{\mathrm{D}}$ in (125a), $\chi = \frac{\lambda}{a}\vec{\mathcal{V}}\cdot\vec{\nu}$ in (125b) and $\vec{\eta} = \frac{\alpha\lambda}{a}\vec{\mathcal{V}}$ in (125c) we obtain, on noting the transport theorem, Theorem 33, and Lemma 97 that

$$\begin{aligned}\frac{\mathrm{d}}{\mathrm{d}t}\left(\frac{\vartheta}{2}|u-u_{\mathrm{D}}|_{\Omega}^{2}+\frac{\alpha\lambda}{a}\left|\Gamma(t)\right|_{\gamma}+\lambda\, u_{\mathrm{D}}\,\mathcal{L}^{d}(\Omega_{-}(t))\right)&\\ +(\mathcal{K}\nabla u,\nabla u)+\frac{\lambda\rho}{a}\langle\beta(\vec{\nu}),\mathcal{V}^{2}\rangle_{\Gamma(t)}=(f,u-u_{\mathrm{D}}).&\end{aligned}\tag{126}$$

We now introduce the finite element approximations of S_0 and S_{D}. On recalling (118) and (124), we set

$$S_0^m = S^m \cap S_0 \quad\text{and}\quad S_{\mathrm{D}}^m = S^m \cap S_{\mathrm{D}}.\tag{127}$$

Then a finite element approximation of (125) from Barrett et al. (2010c), combining the ideas introduced for mean curvature flow in Section 4, for anisotropic flows in Section 6 and for the Mullins–Sekerka problem above, is given as follows. Let the closed polyhedral hypersurface Γ^0 be an approximation of $\Gamma(0)$, and, if $\vartheta > 0$, let $U^0 \in S_{\mathrm{D}}^0$ be an approximation of u_0, e.g., $U^0 = I^0 u_0$ if $u_0 \in C(\overline{\Omega})$. Then, for $m = 0, \ldots, M-1$, find $(U^{m+1}, \vec{X}^{m+1}, \kappa_\gamma^{m+1}) \in S_{\mathrm{D}}^m \times \underline{V}(\Gamma^m) \times V(\Gamma^m)$ such that

$$\begin{aligned}\vartheta\left(\frac{U^{m+1}-U^{m}}{\Delta t_{m}},\varphi\right)_{m}^{h}+(\mathcal{K}\nabla U^{m+1},\nabla\varphi)-\lambda\left\langle\pi_{\Gamma^{m}}\left[\frac{\vec{X}^{m+1}-\vec{\mathrm{id}}}{\Delta t_{m}}\cdot\vec{\omega}^{m}\right],\varphi\right\rangle_{\Gamma^{m}}&\\ =(f^{m+1},\varphi)_{m}^{h}\quad\forall\,\varphi\in S_{0}^{m},&\end{aligned}\tag{128a}$$

$$\begin{aligned}\rho\left\langle[\beta(\vec{\nu}^{m})]^{-1}\,\frac{\vec{X}^{m+1}-\vec{\mathrm{id}}}{\Delta t_{m}},\chi\,\vec{\omega}^{m}\right\rangle_{\Gamma^{m}}^{h}-\alpha\left\langle\kappa_{\gamma}^{m+1},\chi\right\rangle_{\Gamma^{m}}^{h}+a\langle U^{m+1},\chi\rangle_{\Gamma^{m}}=0&\\ \forall\,\chi\in V(\Gamma^{m}),&\end{aligned}\tag{128b}$$

$$\left\langle\kappa_{\gamma}^{m+1}\,\vec{\nu}^{m},\vec{\eta}\right\rangle_{\Gamma^{m}}^{h}+\left\langle\nabla_{s}^{\widetilde{G}}\,\vec{X}^{m+1},\nabla_{s}^{\widetilde{G}}\,\vec{\eta}\right\rangle_{\Gamma^{m},\gamma}=0\quad\forall\,\vec{\eta}\in\underline{V}(\Gamma^{m})\tag{128c}$$

and set $\Gamma^{m+1} = \vec{X}^{m+1}(\Gamma^m)$. In the above, we have introduced $f^{m+1} = f(\cdot, t_{m+1})$, where we assume for convenience that $f(\cdot, t) \in C(\overline{\Omega})$ for all $t \in [0, T]$. Moreover, we have recalled (17) and observe that using $\vec{\omega}^m$ in (128b), and not $\vec{\nu}^m$, is necessary in order be able to prove existence, uniqueness and stability results for (128), see the proof of Theorem 113 below.

Remark 112 (Implementation).
Compared to the computational challenges discussed in Remark 107 for (119), the only new difficulty in (128) is the term $(\mathcal{K}\nabla U^{m+1}, \nabla\varphi)$. In order to compute this in the case $\mathcal{K}_+ \neq \mathcal{K}_-$, it is necessary to calculate $\mathcal{L}^d(o \cap \Omega_-^m)$ for every

$o \in \mathcal{T}^m$, where Ω_-^m denotes the interior of Γ^m. This can be done as described in Barrett et al. (2013a, Remark 4.2), but turns out to be computationally expensive due to the unfitted nature of the finite element approximation. Alternatively, suitable numerical approximations of the integral $(\mathcal{K}\nabla U^{m+1}, \nabla \varphi)$ can be introduced, similarly to what we present in Sections 7.3 and 8. The case of phase-dependent forcings $f_\pm$ in (123a) can be dealt with similarly, on introducing a suitable analogue of $(f^{m+1}, \varphi)_m^h$ in (128a).

On defining

$$\mathcal{E}^m(U, \vec{X}) = \frac{\vartheta}{2} |U - u_{\rm D}|^2_{\Omega,m} + \frac{\alpha\,\lambda}{a} |\vec{X}(\Gamma^m)|_\gamma$$

for $U \in C(\overline{\Omega})$ and $\vec{X} \in \underline{V}(\Gamma^m)$, we have the following results.

Theorem 113. *Let Γ^m and $\mathcal{T}^m$ satisfy Assumptions 64(ii) and 108, and let $U^m \in C(\overline{\Omega})$. Let γ be of the form (106) with $r = 1$. Then there exists a unique solution $(U^{m+1}, \vec{X}^{m+1}, \kappa_\gamma^{m+1}) \in S_{\rm D}^m \times \underline{V}(\Gamma^m) \times V(\Gamma^m)$ to (128). Moreover, if $d = 2$ or $d = 3$ and if γ is of the form (106) with $r \in [1, \infty)$, then a solution to (128) satisfies*

$$\begin{aligned}
&\mathcal{E}^m(U^{m+1}, \vec{X}^{m+1}) + \lambda\, u_{\rm D} \left\langle \vec{X}^{m+1} - \vec{\rm id}, \vec{\nu}^m \right\rangle_{\Gamma^m} + \frac{\vartheta}{2} |U^{m+1} - U^m|^2_{\Omega,m} \\
&\quad + \Delta t_m (\mathcal{K}\nabla U^{m+1}, \nabla U^{m+1}) + \Delta t_m \frac{\lambda\,\rho}{a} \left(\left| [\beta(\vec{\nu}^m)]^{-\frac{1}{2}} \frac{\vec{X}^{m+1} - \vec{\rm id}}{\Delta t_m} \cdot \vec{\omega}^m \right|^h_{\Gamma^m} \right)^2 \\
&\quad \le \mathcal{E}^m(U^m, \vec{\rm id}) + \Delta t_m (f^{m+1}, U^{m+1} - u_{\rm D})^h_m.
\end{aligned} \tag{129}$$

Proof. The existence and uniqueness proof for $r = 1$ proceeds analogously to the proof of Theorem 109. In particular, if $(U, \vec{X}, \kappa_\gamma) \in S_0^m \times \underline{V}(\Gamma^m) \times V(\Gamma^m)$ denotes a solution to the homogeneous analogue of (128), similarly to (120), then choosing $\varphi = U$, $\chi = \frac{\lambda}{a}\, \pi_{\Gamma^m} [\vec{X} \cdot \vec{\omega}^m]$ and $\vec{\eta} = \frac{\alpha\,\lambda}{a} \vec{X}$ yields

$$\begin{aligned}
&\vartheta |U|^2_{\Omega,m} + \Delta t_m (\mathcal{K}\nabla U, \nabla U) + \frac{\lambda\,\rho}{\Delta t_m\, a} \left(\left| [\beta(\vec{\nu}^m)]^{-\frac{1}{2}} \vec{X} \cdot \vec{\omega}^m \right|^h_{\Gamma^m} \right)^2 \\
&\quad + \frac{\alpha\,\lambda}{a} \left\langle \nabla_s^{\widetilde{G}} \vec{X}, \nabla_s^{\widetilde{G}} \vec{X} \right\rangle_{\Gamma^m,\gamma} = 0,
\end{aligned}$$

which implies that U is constant in Ω, with $U = 0$ if $\vartheta > 0$ or $S_0^m \ne S^m$, and that $\vec{X}$ is constant on Γ^m, since all the involved physical parameters are non-negative, with $\mathcal{K}$, α, a and λ being positive. As in the proof of Theorem 109 we can then infer from Assumption 108 that $\vec{X} = \vec{0}$. This implies that $\kappa_\gamma = \frac{a}{\alpha} U$ is constant on Γ^m and so we can proceed as in the proof of Theorem 109 to show that $\kappa_\gamma = U = 0$. Overall this proves existence of a unique solution $(U^{m+1}, \vec{X}^{m+1}, \kappa_\gamma^{m+1}) \in S_{\rm D}^m \times \underline{V}(\Gamma^m) \times V(\Gamma^m)$ to (128).

It remains to establish (129). Choosing $\varphi = U^{m+1} - u_{\rm D}$ in (128a), $\chi = \frac{\lambda}{a}\,\pi_{\Gamma^m}[(\vec{X}^{m+1} - \vec{\rm id}_{|\Gamma^m}) \cdot \vec{\omega}^m]$ in (128b) and $\vec{\eta} = \frac{\alpha\,\lambda}{a}\,(\vec{X}^{m+1} - \vec{\rm id}_{|\Gamma^m})$ in (128c) yields that

$$\begin{aligned}
&\vartheta\,(U^{m+1} - U^m, U^{m+1} - u_{\rm D})_m^h + \Delta t_m (\mathcal{K}\,\nabla\,U^{m+1}, \nabla\,U^{m+1}) \\
&\quad + \frac{\alpha\,\lambda}{a}\left\langle \nabla_s^{\widetilde{G}}\,\vec{X}^{m+1}, \nabla_s^{\widetilde{G}}\,(\vec{X}^{m+1} - \vec{\rm id}) \right\rangle_{\Gamma^m,\gamma} \\
&\quad + \Delta t_m\,\frac{\lambda\,\rho}{a}\left(\left| [\beta(\vec{\nu}^m)]^{-\frac{1}{2}}\,\frac{\vec{X}^{m+1} - \vec{\rm id}}{\Delta t_m} \cdot \vec{\omega}^m \right|_{\Gamma^m}^h \right)^2 \\
&= -\lambda\,u_{\rm D} \left\langle \vec{X}^{m+1} - \vec{\rm id}, \vec{\omega}^m \right\rangle_{\Gamma^m}^h + \Delta t_m (f^{m+1}, U^{m+1} - u_{\rm D})_m^h
\end{aligned}$$

and hence (129) follows immediately from Lemma 102 and (20). □

Remark 114 (Semidiscrete scheme).
We note that (129) closely mimics the corresponding continuous energy law (126). The reason why it is not an exact discrete analogue of (126) has been discussed in Remark 110 already. However, on recalling Remark 89(ii) it is possible to prove an exact discrete analogue of (126) for a semidiscrete version of the scheme (128), see also Barrett et al. (2010c, Remark 3.5). In addition, this semidiscrete scheme will feature the tangential motion discussed in Remark 104.

Remark 115 (Discrete systems).
It is a simple matter to combine the techniques in Remark 111 and Section 6.4 in order to derive the discrete systems that need to be solved at each time level for (128). If γ is of the form (106) with $r = 1$, then the systems are linear and can hence it can be solved analogously to Remark 111, i.e., either with a sparse direct solver or with the help of a Schur complement approach and a preconditioned conjugate gradient solver. If $r > 1$, on the other hand, then the systems are nonlinear and can be solved with a lagged fixed-point type iteration. We refer to Barrett et al. (2010c, Section 4) for further details.

7.3 One-sided free boundary problems

In this section we consider the situation where diffusion is restricted to one phase. We hence study one-sided versions of the Mullins–Sekerka or Stefan problems. This is, for example, relevant for snow crystal growth, where diffusion can be restricted to the gas phase. In this case we only find the unknown

u in one-phase, which occupies at time t a domain $\Omega_+(t)$. Once again, for simplicity of the presentation, we always assume that the region $\Omega_-(t) = \Omega \setminus \overline{\Omega_+(t)}$ is compactly contained in Ω. The problem now reads as follows. Given $\Gamma(0)$ and, if $\vartheta > 0$, $u_0 : \Omega_+(0) \to \mathbb{R}$, find the evolving interface $(\Gamma(t))_{t\in[0,T]}$ and $u(\cdot, t) : \Omega_+(t) \to \mathbb{R}$, $t \in [0, T]$, such that $\vartheta u(\cdot, 0) = \vartheta u_0(\cdot)$ and for all $t \in (0, T]$ the following conditions hold

$$\vartheta\, \partial_t\, u - \mathcal{K}_+\, \Delta u = f \quad \text{in}\, \Omega_+(t), \qquad u = u_{\rm D} \qquad \text{on}\, \partial\Omega, \tag{130a}$$

$$\frac{\rho\, \mathcal{V}}{\beta(\vec{\nu})} = \alpha\, \varkappa_\gamma - a\, u \;\; \text{and} \;\; \mathcal{K}_+\, \partial_{\vec{\nu}}\, u = -\lambda\, \mathcal{V} \qquad \text{on}\, \Gamma(t), \tag{130b}$$

where the given data satisfies the same conditions as in (123). For snow crystal growth $-u$ is a suitably scaled concentration with $-u_{\rm D}$ being the scaled supersaturation, see Libbrecht (2005) and Barrett et al. (2012a, 2013a).

As we wish to model snow crystal growth, we now state a relevant hexagonal surface energy. Let $d = 3$ and $l_\varepsilon(\vec{p}) = \left[\varepsilon^2\, |\vec{p}|^2 + p_1^2\, (1 - \varepsilon^2)\right]^{\frac{1}{2}}$ for $\varepsilon > 0$. We then introduce the rotation matrices

$$\underline{\underline{R}}_1(\theta) = \begin{pmatrix} \cos\theta & \sin\theta & 0 \\ -\sin\theta & \cos\theta & 0 \\ 0 & 0 & 1 \end{pmatrix} \;\; \text{and} \;\; \underline{\underline{R}}_2(\theta) = \begin{pmatrix} \cos\theta & 0 & \sin\theta \\ 0 & 1 & 0 \\ -\sin\theta & 0 & \cos\theta \end{pmatrix}.$$

Then setting

$$\gamma(\vec{p}) = l_\varepsilon\left(\underline{\underline{R}}_2\left(\frac{\pi}{2}\right)\vec{p}\right) + \frac{1}{\sqrt{3}} \sum_{\ell=1}^{3} l_\varepsilon\left(\underline{\underline{R}}_1\left(\theta_0 + \frac{\ell\,\pi}{3}\right)\vec{p}\right),$$

where $\theta_0 \in [0, \frac{\pi}{3})$ rotates the anisotropy in the $x_1 - x_2$ plane, defines a density of the form (106), with $r = 1$ and $L = 4$, that approximates a crystalline surface energy density with a regular hexagonal prism as its Wulff shape, where each face of the Wulff shape has the same distance to the origin.

We recall from Barrett et al. (2013a) an unfitted finite element approximation of the above one-sided problem, following ideas of Barrett and Elliott (1982). Given an approximation of the interface Γ^m, we let Ω_+^m denote the exterior of Γ^m and let Ω_-^m denote the interior of Γ^m, so that $\Gamma^m = \partial\Omega_-^m = \overline{\Omega_-^m} \cap \overline{\Omega_+^m}$. We now introduce the appropriate discrete trial and test function spaces. To this end, let $\Omega_+^{m,h}$ be an approximation to Ω_+^m and set $\Omega_-^{m,h} = \Omega \setminus \overline{\Omega_+^{m,h}}$. We stress that $\Omega_+^{m,h}$ need not necessarily be a union of elements from $\mathcal{T}^m$. Then we define, on recalling (118) and (127), the finite element spaces

$$S_+^m = \left\{\chi \in S^m : \chi(\vec{p}_k^m) = 0 \;\text{if supp}\, \phi_k^{S^m} \subset \overline{\Omega_-^{m,h}},\;\; k = 1, \ldots, K_{S^m}\right\},$$
$$S_{0,+}^m = S_0^m \cap S_+^m, \quad S_{{\rm D},+}^m = S_{\rm D}^m \cap S_+^m. \tag{131}$$

Our finite element approximation of (130) is then given as follows. Let the closed polyhedral hypersurface Γ^0 be an approximation of $\Gamma(0)$, and, if $\vartheta > 0$, let $U^0 \in S^0_D$ be an approximation of u_0, e.g., $U^0 = I^0 u_0$ if $u_0 \in C(\overline{\Omega})$ is an extension of the given $u_0 \in C(\overline{\Omega_+(0)})$. Then, for $m = 0, \ldots, M-1$, find $(U^{m+1}, \vec{X}^{m+1}, \kappa_\gamma^{m+1}) \in S^m_{D,+} \times \underline{V}(\Gamma^m) \times V(\Gamma^m)$ such that

$$\vartheta\left(\frac{U^{m+1}-U^m}{\Delta t_m}, \varphi\right)^h_{m,+} + \mathcal{K}_+(\nabla U^{m+1}, \nabla \varphi)_{m,+}$$
$$-\lambda\left\langle \pi_{\Gamma^m}\left[\frac{\vec{X}^{m+1}-\vec{\mathrm{id}}}{\Delta t_m}\cdot \vec{\omega}^m\right], \varphi\right\rangle_{\Gamma^m} = (f^{m+1}, \varphi)^h_{m,+} \quad \forall \varphi \in S^m_{0,+}, \tag{132a}$$

$$\rho\left\langle [\beta(\vec{\nu}^m)]^{-1}\frac{\vec{X}^{m+1}-\vec{\mathrm{id}}}{\Delta t_m}, \chi\, \vec{\omega}^m\right\rangle^h_{\Gamma^m} - \alpha\left\langle \kappa_\gamma^{m+1}, \chi\right\rangle^h_{\Gamma^m} + a\langle U^{m+1}, \chi\rangle_{\Gamma^m} = 0$$
$$\forall \chi \in V(\Gamma^m), \tag{132b}$$

$$\left\langle \kappa_\gamma^{m+1}\, \vec{\nu}^m, \vec{\eta}\right\rangle^h_{\Gamma^m} + \left\langle \nabla_s^{\widetilde{G}} \vec{X}^{m+1}, \nabla_s^{\widetilde{G}} \vec{\eta}\right\rangle_{\Gamma^m,\gamma} = 0 \quad \forall \vec{\eta} \in \underline{V}(\Gamma^m) \tag{132c}$$

and set $\Gamma^{m+1} = \vec{X}^{m+1}(\Gamma^m)$. Here we define for all χ, $\varphi \in S^m$

$$(\nabla \chi, \nabla \varphi)_{m,+} = \int_{\Omega_+^{m,h}} \nabla \chi \cdot \nabla \varphi \, \mathrm{d}\mathcal{L}^d$$
$$= \sum_{o \in \mathcal{T}^m} \frac{\mathcal{L}^d(o \cap \Omega_+^{m,h})}{\mathcal{L}^d(o)} \int_o \nabla \chi \cdot \nabla \varphi \, \mathrm{d}\mathcal{L}^d \tag{133}$$

and, in a similar fashion,

$$(\chi, \varphi)^h_{m,+} = \sum_{o \in \mathcal{T}^m} \frac{\mathcal{L}^d(o \cap \Omega_+^{m,h})}{\mathcal{L}^d(o)} \int_o I^m[\chi\, \varphi] \, \mathrm{d}\mathcal{L}^d.$$

It follows immediately from (131) and (133) that

$$(\nabla \varphi, \nabla \varphi)_{m,+} > 0 \quad \forall \varphi \in S^m_{0,+} \setminus \{0\}.$$

Remark 116.

(i) We note that for $\vartheta > 0$ the approximation (132) is only meaningful when the discrete solid region does not shrink, see Barrett et al. (2013a, Remark 3.1). In practice this technical constraint is not very restrictive, since in most physically relevant applications the solid region grows.

(ii) Existence and uniqueness for $r = 1$, as well as stability for $r \in [1, \infty)$, can be shown for (132) under appropriate assumptions, see Barrett et al. (2013a, Theorems 3.1 and 3.2) for the proofs for the case $r = 1$. We note that the stability proof in Barrett et al. (2013a, Theorem 3.2), for $d = 2$ and $d = 3$, immediately carries over to the case $r > 1$, on recalling Lemma 102.

(iii) The main new computational challenge in implementing the scheme (132), compared to (128) with $\mathcal{K}_+ = \mathcal{K}_-$, is the determination of $S^m_{0,+}$ and the calculation of the terms involving $(\cdot, \cdot)_{m,+}$ and $(\cdot, \cdot)^h_{m,+}$ at each time step. Clearly, the involved difficulty crucially depends on the choice of $\Omega^{m,h}_+$ as an approximation to Ω^m_+. A thorough discussion of possible choices can be found in Barrett et al. (2013a, Section 4.1).

We present two numerical snow crystal growth simulations based on the scheme (132) with a hexagonal surface energy density γ, and with a time-dependent mobility β, in Fig. 5.

7.4 Alternative numerical approaches

Other authors who numerically studied crystal growth with the help of a front tracking type method are Roosen and Taylor (1991), Yokoyama (1993), Almgren (1993), Schmidt (1993, 1996, 1998), Juric and Tryggvason (1996) and Bänsch and Schmidt (2000). A level set method for crystal growth was studied in Sethian and Strain (1992). In the context of phase field approximations, we refer to Kobayashi (1993), Wheeler et al. (1993),

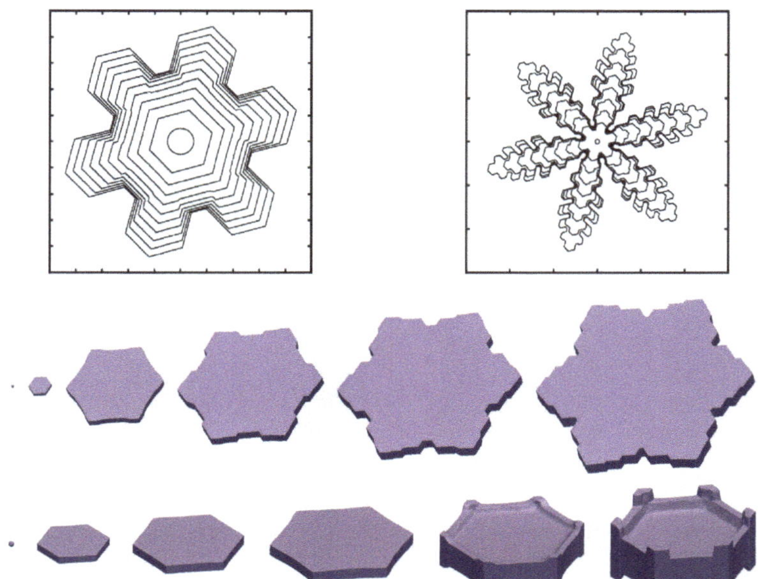

FIG. 5 Visualizations of the numerical experiments from Barrett et al. (2013a, Fig. 12), *above*, Barrett et al. (2013a, Fig. 22), *middle*, and Barrett et al. (2013a, Fig. 27 (left)), *below*. The 2D solutions are shown at times $t = 0, 0.5, \ldots, 5$ (*left*) and $t = 0, 5, \ldots, 40$ (*right*). The snapshots for the 3D simulations are taken at times $t = 0, 5, 50, 100, 150, 200$ (*middle*) and $t = 0, 10, \ldots, 50$ (*below*).

Karma and Rappel (1998), Debierre et al. (2003) and Nestler (2005), as well as to the works by the present authors Barrett et al. (2013b, 2014a,b). Solidification phenomena can also be described with the help of cellular automata and we refer to Reiter (2005), Libbrecht (2008) and Gravner and Griffeath (2009) for details.

8 Two-phase flow

We now study free boundary problems appearing in fluid flow. Here we consider two immiscible fluids that are separated by an interface, which moves in time with a velocity given by that of the fluid. We start with the simplest situation in which the evolution in the two phases is given by Stokes flow. For the discretization of the evolving hypersurface we consider the parametric finite element approximation introduced above. The good mesh properties are particularly helpful in two-phase flow as the mesh, which is moved by the fluid velocity, can change quite drastically. Many other approaches, using a surface mesh, have problems in such situations due the mesh deteriorating. This is not the case, in most situations, for the approach stated below. Moreover, on using a simple XFEM pressure space enrichment, we obtain exact volume conservation for the two phase regions. Furthermore, our fully discrete finite element approximation can be shown to be unconditionally stable. A common feature in two-phase flow is that nonphysical velocities appear in numerical approximations. If surface tension effects are taken into account, a jump discontinuity in the pressure results, and this poses serious challenges for the numerical method. This can lead to nonphysical velocities. These so-called spurious velocities are typically avoided by the XFEM approach discussed in this section.

8.1 Two-phase Stokes flow

In this section we consider the flow of viscous, incompressible, immiscible two fluid systems in a low Reynolds regime, i.e., we can neglect inertia terms. The governing equations for the velocity $\vec{u}$ and the pressure p are given by the momentum equation and conservation of mass, i.e.,

$$-\mu_\pm \, \Delta \, \vec{u} + \nabla \, p = \vec{f}, \quad \nabla \cdot \vec{u} = 0 \quad \text{in} \, \Omega_\pm(t), \tag{134}$$

where $\vec{f}$ is a possible forcing, and $\Omega_+(t)$ and $\Omega_-(t)$ are the time-dependent regions occupied by the two fluid phases as in Fig. 1 in Section 2.4, where $\Omega \subset \mathbb{R}^d$ is again a fixed domain, with $d \geq 2$. As usual, we let $\vec{\nu}$ denote the outer unit normal to $\Omega_-(t)$ on $\Gamma(t) = \partial\Omega_-(t)$. We consider the case in which the viscosity of the two fluids can be different and introduce $\mu(\cdot,t) = \mu_+ \, \mathcal{X}_{\Omega_+(t)} + \mu_- \, \mathcal{X}_{\Omega_-(t)}$, with $\mu_\pm \in \mathbb{R}_{>0}$ denoting the fluid viscosities, recall (11). We now define the conditions that have to hold on the interface. Therefore, we introduce the stress tensor

$$\underline{\underline{\sigma}} = \mu \, (\nabla \, \vec{u} + (\nabla \, \vec{u})^{\mathsf{T}}) - p \, \underline{\underline{\mathrm{Id}}} = 2 \, \mu \, \underline{\underline{D}}(\vec{u}) - p \, \underline{\underline{\mathrm{Id}}}, \tag{135}$$

where $\underline{\underline{D}}(\vec{u}) = \frac{1}{2}\left(\nabla \vec{u} + (\nabla \vec{u})^{\top}\right)$ is the rate of deformation tensor. Using the fact that the velocity is divergence free, we can rewrite (134) as

$$-\nabla \cdot \underline{\underline{\sigma}} = \vec{f}, \quad \nabla \cdot \vec{u} = 0 \quad \text{in } \Omega_{\pm}(t).$$

On the moving interface, we require

$$[\vec{u}]_-^+ = \vec{0}, \quad [\underline{\underline{\sigma}}\,\vec{\nu}]_-^+ = -\gamma_0\,\varkappa\,\vec{\nu}, \quad \mathcal{V} = \vec{u}\cdot\vec{\nu} \quad \text{on } \Gamma(t),$$

where $\gamma_0 \in \mathbb{R}_{>0}$ represents surface tension. To close the system we prescribe initial data $\Gamma(0)$ and the boundary condition $\vec{u} = \vec{0}$ on $\partial\Omega$. Overall, we can rewrite the total system as follows. Given $\Gamma(0)$, find $\vec{u} : \Omega \times [0,T] \to \mathbb{R}^d$, $p : \Omega \times [0,T] \to \mathbb{R}$ and the evolving interface $(\Gamma(t))_{t\in[0,T]}$ such that for all $t \in (0,T]$ the following conditions hold

$$-2\,\mu\,\nabla\cdot\underline{\underline{D}}(\vec{u}) + \nabla\,p = \vec{f}, \quad \nabla\cdot\vec{u} = 0 \;\text{ in } \Omega_{\pm}(t), \quad \vec{u} = \vec{0} \;\text{ on } \partial\Omega, \tag{136a}$$

$$[\vec{u}]_-^+ = \vec{0}, \quad \left[\left(2\,\mu\,\underline{\underline{D}}(\vec{u}) - p\,\underline{\underline{\mathrm{Id}}}\right)\vec{\nu}\right]_-^+ = -\gamma_0\,\varkappa\,\vec{\nu}, \quad \mathcal{V} = \vec{u}\cdot\vec{\nu} \;\text{ on } \Gamma(t). \tag{136b}$$

A finite element approximation of (136) needs a suitable weak formulation. Here a choice has to be made regarding the tangential velocity $\vec{\mathcal{V}}_{\mathrm{T}}$, recall Definition 25(iii), which will have important repercussions on the discrete level. Given our results derived in Sections 4 and 5, it is natural to not fix $\vec{\mathcal{V}}_{\mathrm{T}}$ explicitly and to use the formulations

$$\vec{\mathcal{V}}\cdot\vec{\nu} = \vec{u}\cdot\vec{\nu} \;\text{ and }\; \varkappa\,\vec{\nu} = \Delta_s\,\vec{\mathrm{id}} \quad \text{on } \Gamma(t), \tag{137}$$

compare with (38). An alternative approach would let $\vec{\mathcal{V}}_{\mathrm{T}}$ be given by the fluid flow, and so use the formulations

$$\vec{\mathcal{V}} = \vec{u} \;\text{ and }\; \vec{\varkappa} = \Delta_s\,\vec{\mathrm{id}} \quad \text{on } \Gamma(t),$$

compare with (67). We will consider this alternative approach in Section 8.1.4. However, our main focus is on the formulation (137), which will lead to good mesh properties for the discrete schemes.

We begin with the following simple generalization of Remark 36

$$\begin{aligned}
&-\int_{\Omega_-(t)\cup\Omega_+(t)} (\nabla\cdot\underline{\underline{\sigma}})\cdot\vec{\xi}\,\mathrm{d}\mathcal{L}^d \\
&= 2\left(\mu\,\underline{\underline{D}}(\vec{u}), \underline{\underline{D}}(\vec{\xi})\right) - \left(p, \nabla\cdot\vec{\xi}\right) + \left\langle [\underline{\underline{\sigma}}\,\vec{\nu}]_-^+, \vec{\xi}\right\rangle_{\Gamma(t)} \quad \forall\,\vec{\xi} \in [H_0^1(\Omega)]^d,
\end{aligned} \tag{138}$$

where we have noted for symmetric matrices $\underline{\underline{A}} \in \mathbb{R}^{d\times d}$ that $\underline{\underline{A}} : \underline{\underline{B}} = \underline{\underline{A}} : \frac{1}{2}\left(\underline{\underline{B}} + \underline{\underline{B}}^{\top}\right)$ for all $\underline{\underline{B}} \in \mathbb{R}^{d\times d}$. Now (138) leads to the following weak formulation of (136), using the formulations (137). Given a closed hypersurface $\Gamma(0) \subset \Omega$, we seek an evolving hypersurface $(\Gamma(t))_{t\in[0,T]}$ that separates Ω into $\Omega_-(t)$ and $\Omega_+(t)$, with a global parameterization and induced velocity field $\vec{\mathcal{V}}$, and $\varkappa \in L^2(\mathcal{G}_T)$ as well as $\vec{u} : \Omega \times [0,T] \to \mathbb{R}^d$ and $p : \Omega \times [0,T] \to \mathbb{R}$ as follows.

For almost all $t \in (0,T)$, find $(\vec{u}(\cdot,t), p(\cdot,t), \vec{\mathcal{V}}(\cdot,t), \varkappa(\cdot,t)) \in [H_0^1(\Omega)]^d \times L^2(\Omega) \times [L^2(\Gamma(t))]^d \times L^2(\Gamma(t))$ such that

$$2\left(\mu\,\underline{\underline{D}}(\vec{u}), \underline{\underline{D}}(\vec{\xi})\right) - \left(p, \nabla\cdot\vec{\xi}\right) = \left(\vec{f}, \vec{\xi}\right) + \gamma_0\left\langle \varkappa\,\vec{\nu}, \vec{\xi}\right\rangle_{\Gamma(t)} \quad \forall \vec{\xi} \in [H_0^1(\Omega)]^d, \tag{139a}$$

$$(\nabla\cdot\vec{u}, \varphi) = 0 \quad \forall \varphi \in L^2(\Omega), \tag{139b}$$

$$\left\langle \vec{\mathcal{V}}\cdot\vec{\nu}, \chi\right\rangle_{\Gamma(t)} = \langle \vec{u}\cdot\vec{\nu}, \chi\rangle_{\Gamma(t)} \quad \forall \chi \in L^2(\Gamma(t)), \tag{139c}$$

$$\langle \varkappa\,\vec{\nu}, \vec{\eta}\rangle_{\Gamma(t)} + \left\langle \nabla_s\,\vec{\mathrm{id}}, \nabla_s\,\vec{\eta}\right\rangle_{\Gamma(t)} = 0 \quad \forall \vec{\eta} \in [H^1(\Gamma(t))]^d. \tag{139d}$$

Remark 117. For a sufficiently smooth weak solution of (139), we can formally prove the following results.

(i) The two-phase Stokes flow, in the absence of external forcings, decreases the surface area of the interface. This follows from the transport theorem, Theorem 32, (139c) with $\chi = \varkappa$, (139a) with $\xi = \vec{u}$ and (139b) with $\varphi = p$, as one then obtains the energy identity

$$\begin{aligned}\gamma_0 \frac{\mathrm{d}}{\mathrm{d}t}|\Gamma(t)| &= -\gamma_0\left\langle \varkappa\,\vec{\nu}, \vec{\mathcal{V}}\right\rangle_{\Gamma(t)} = -\gamma_0\langle \varkappa\,\vec{\nu}, \vec{u}\rangle_{\Gamma(t)} \\ &= -2\left(\mu\,\underline{\underline{D}}(\vec{u}), \underline{\underline{D}}(\vec{u})\right) + \left(\vec{f}, \vec{u}\right).\end{aligned} \tag{140}$$

(ii) The two-phase Stokes flow preserves $\mathcal{L}^d(\Omega_\pm(t))$, as the transport theorem, Theorem 33, (139c) with $\chi = 1$, the divergence theorem and (139b) with $\varphi = \mathcal{X}_{\Omega_-(t)}$ yield that

$$\frac{\mathrm{d}}{\mathrm{d}t}\mathcal{L}^d(\Omega_-(t)) = \left\langle \vec{\mathcal{V}}, \vec{\nu}\right\rangle_{\Gamma(t)} = \langle \vec{u}, \vec{\nu}\rangle_{\Gamma(t)} = \left(\nabla\cdot\vec{u}, \mathcal{X}_{\Omega_-(t)}\right) = 0. \tag{141}$$

Remark 118. In the case $\vec{f} = \vec{0}$ the evolution (139) decreases the surface area of the interface $\Gamma(t)$, recall (140), and can be written in the general form (81) satisfying (82). This can be seen as follows. For a given interface $\Gamma(t)$, with normal $\vec{\nu}(t)$ and curvature $\varkappa(t)$, we solve (139a), (139b) for $(\vec{u}(\cdot,t), p(\cdot,t)) \in [H_0^1(\Omega)]^d \times L^2(\Omega)$ and set

$$\mathfrak{F}(\varkappa) = \vec{u}\cdot\vec{\nu}.$$

We then compute, similarly to (140),

$$\langle \mathfrak{F}(\varkappa), \varkappa\rangle_{\Gamma(t)} = \langle \vec{u}\cdot\vec{\nu}, \varkappa\rangle_{\Gamma(t)} = \frac{2}{\gamma_0}\left(\mu\,\underline{\underline{D}}(\vec{u}), \underline{\underline{D}}(\vec{u})\right) \geq 0. \tag{142}$$

8.1.1 Finite element approximation

In order to approximate the velocity and pressure on $\mathcal{T}^m$, recall the assumptions and notation above (118), we introduce finite element spaces $\mathbb{U}^m \subset [H_0^1(\Omega)]^d$ and $\mathbb{P}^m \subset L^2(\Omega)$. We require also the space $\widehat{\mathbb{P}}^m = \mathbb{P}^m \cap \widehat{\mathbb{P}}$, where $\widehat{\mathbb{P}} = \{\varphi \in L^2(\Omega) : (\varphi, 1) = 0\}$. The velocity/pressure finite element spaces $(\mathbb{U}^m, \mathbb{P}^m)$ satisfy the LBB inf-sup condition if there exists a $C \in \mathbb{R}_{>0}$ independent of $h^m = \max_{o \in \mathcal{T}^m}\{\operatorname{diam}(o)\}$, such that

$$\inf_{\varphi \in \hat{\mathbb{P}}^m} \sup_{\vec{\xi} \in \mathbb{U}^m} \frac{\left(\varphi, \nabla \cdot \vec{\xi}\right)}{|\varphi|_\Omega \, \|\vec{\xi}\|_{1,\Omega}} \geq C > 0, \tag{143}$$

where $\|\cdot\|_{1,\Omega}^2 = |\cdot|_\Omega^2 + |\nabla \cdot|_\Omega^2$ defines the H^1-norm on Ω, see, e.g., Girault and Raviart (1986, p. 114).

Next we introduce a generalization of (118) to

$$\mathcal{S}_k^m = \left\{\chi \in C(\overline{\Omega}) : \chi_{|_o} \in \mathcal{P}_k(o) \quad \forall o \in \mathcal{T}^m\right\} \subset H^1(\Omega), \tag{144}$$

where, for $k \in \mathbb{N}$, $\mathcal{P}_k(o)$ denotes the space of polynomials of degree k on o. In addition, we denote by $\mathcal{S}_0^m$ the space of piecewise constant functions on $\mathcal{T}^m$. Then, for example, we may choose the lowest order Taylor–Hood element P2-P1, the P2-P0 element, or the P2-(P1+P0) element on setting $\mathbb{U}^m = [\mathcal{S}_2^m]^d \cap \mathbb{U}$, and $\mathbb{P}^m = \mathcal{S}_1^m, \mathcal{S}_0^m$ or $\mathcal{S}_1^m + \mathcal{S}_0^m$, respectively. It is well known that the P2-P1 element satisfies the LBB condition (143) for $d = 2$ and $d = 3$, where the latter requires the weak constraint that all simplices have a vertex in Ω, see Boffi (1997). While the other two choices, P2-P0 and P2-(P1+P0), satisfy it for $d = 2$.

Moreover, as in earlier sections, we approximate the interface at time t_m by a polyhedral surface Γ^m with the same notation. Similarly to Section 7.3, let Ω_+^m denote the exterior of Γ^m and Ω_-^m the interior of Γ^m, so that $\Gamma^m = \partial\Omega_-^m = \overline{\Omega_-^m} \cap \overline{\Omega_+^m}$. For simplicity we assume that the unit normal $\vec{\nu}^m$ on Γ^m points into Ω_+^m. Due to the phase-dependent viscosity, $\mu_\pm$, we now subdivide the elements of the bulk mesh $\mathcal{T}^m$ into exterior, interior and interfacial elements as follows. Let

$$\mathcal{T}_\pm^m = \{o \in \mathcal{T}^m : o \subset \Omega_\pm^m\}, \quad \mathcal{T}_{\Gamma^m}^m = \{o \in \mathcal{T}^m : o \cap \Gamma^m \neq \emptyset\}. \tag{145}$$

The disjoint partition $\mathcal{T}^m = \mathcal{T}_-^m \cup \mathcal{T}_+^m \cup \mathcal{T}_{\Gamma^m}^m$ can easily be found, e.g., with the Algorithm 4.1 in Barrett et al. (2013a).

Similarly to Section 7.3, we use an unfitted finite element approximation of (139). We define the discrete viscosity $\mu^m \in \mathcal{S}_0^m$, for $m \geq 0$, as

$$\mu^m_{|_o} = \begin{cases} \mu_- & o \in \mathcal{T}_-^m, \\ \mu_+ & o \in \mathcal{T}_+^m, \\ \frac{1}{2}(\mu_- + \mu_+) & o \in \mathcal{T}_{\Gamma^m}^m. \end{cases} \tag{146}$$

Then the unfitted finite element approximation of (139) from Barrett et al. (2013c) is given as follows. Let the closed polyhedral hypersurface Γ^0 be an approximation of $\Gamma(0)$ and recall the time interval partitioning (40). Then, for $m = 0, \ldots, M-1$, find $(\vec{U}^{m+1}, P^{m+1}, \vec{X}^{m+1}, \kappa^{m+1}) \in \mathbb{U}^m \times \widehat{\mathbb{P}}^m \times \underline{V}(\Gamma^m) \times V(\Gamma^m)$ such that

$$2\left(\mu^m \underline{\underline{D}}(\vec{U}^{m+1}), \underline{\underline{D}}(\vec{\xi})\right) - \left(P^{m+1}, \nabla \cdot \vec{\xi}\right) = \left(\vec{f}^{m+1}, \vec{\xi}\right) + \gamma_0 \left\langle \kappa^{m+1} \vec{\nu}^m, \vec{\xi} \right\rangle_{\Gamma^m} \quad \forall \vec{\xi} \in \mathbb{U}^m, \tag{147a}$$

$$\left(\nabla \cdot \vec{U}^{m+1}, \varphi\right) = 0 \quad \forall \varphi \in \widehat{\mathbb{P}}^m, \tag{147b}$$

$$\left\langle \frac{\vec{X}^{m+1} - \vec{\mathrm{id}}}{\Delta t_m} \cdot \vec{\nu}^m, \chi \right\rangle_{\Gamma^m}^h = \left\langle \vec{U}^{m+1} \cdot \vec{\nu}^m, \chi \right\rangle_{\Gamma^m} \quad \forall \chi \in V(\Gamma^m), \tag{147c}$$

$$\left\langle \kappa^{m+1} \vec{\nu}^m, \vec{\eta} \right\rangle_{\Gamma^m}^h + \left\langle \nabla_s \vec{X}^{m+1}, \nabla_s \vec{\eta} \right\rangle_{\Gamma^m} = 0 \quad \forall \vec{\eta} \in \underline{V}(\Gamma^m) \tag{147d}$$

and set $\Gamma^{m+1} = \vec{X}^{m+1}(\Gamma^m)$. Here, with the help of the interpolation operator $\vec{I}_2^m : [C(\overline{\Omega})]^d \to [S_2^m]^d$, the natural generalization of $I^m : C(\overline{\Omega}) \to S^m$, we define $\vec{f}^{m+1} = \vec{I}_2^m \vec{f}(\cdot, t_{m+1})$.

We note that the scheme (147) is linear in the unknowns $(\vec{U}^{m+1}, P^{m+1}, \vec{X}^{m+1}, \kappa^{m+1})$. For the mathematical analysis of (147) it is convenient to introduce a reduced version. Given $\mathbb{U}^m$ and $\mathbb{P}^m$, we define

$$\mathbb{U}_0^m = \{\vec{U} \in \mathbb{U}^m : (\nabla \cdot \vec{U}, \varphi) = 0 \quad \forall \varphi \in \widehat{\mathbb{P}}^m\}. \tag{148}$$

Then the discrete pressure can be eliminated from (147) to yield the following reduced variant. For $m = 0, \ldots, M-1$, find $(\vec{U}^{m+1}, \vec{X}^{m+1}, \kappa^{m+1}) \in \mathbb{U}_0^m \times \underline{V}(\Gamma^m) \times V(\Gamma^m)$ such that

$$2\left(\mu^m \underline{\underline{D}}(\vec{U}^{m+1}), \underline{\underline{D}}(\vec{\xi})\right) = \left(\vec{f}^{m+1}, \vec{\xi}\right) + \gamma_0 \left\langle \kappa^{m+1} \vec{\nu}^m, \vec{\xi} \right\rangle_{\Gamma^m} \quad \forall \vec{\xi} \in \mathbb{U}_0^m, \tag{149a}$$

$$\left\langle \frac{\vec{X}^{m+1} - \vec{\mathrm{id}}}{\Delta t_m} \cdot \vec{\nu}^m, \chi \right\rangle_{\Gamma^m}^h = \left\langle \vec{U}^{m+1} \cdot \vec{\nu}^m, \chi \right\rangle_{\Gamma^m} \quad \forall \chi \in V(\Gamma^m), \tag{149b}$$

$$\left\langle \kappa^{m+1} \vec{\nu}^m, \vec{\eta} \right\rangle_{\Gamma^m}^h + \left\langle \nabla_s \vec{X}^{m+1}, \nabla_s \vec{\eta} \right\rangle_{\Gamma^m} = 0 \quad \forall \vec{\eta} \in \underline{V}(\Gamma^m) \tag{149c}$$

and set $\Gamma^{m+1} = \vec{X}^{m+1}(\Gamma^m)$.

We now show existence, uniqueness and stability results for the two schemes (149) and (147).

Theorem 119.

(i) *Let* Γ^m *satisfy Assumption 64. Then there exists a unique solution* $(\vec{U}^{m+1}, \vec{X}^{m+1}, \kappa^{m+1}) \in \mathbb{U}_0^m \times \underline{V}(\Gamma^m) \times V(\Gamma^m)$ *to (149).*

(ii) *If* $(\vec{U}^{m+1}, P^{m+1}, \vec{X}^{m+1}, \kappa^{m+1}) \in \mathbb{U}^m \times \widehat{\mathbb{P}}^m \times \underline{V}(\Gamma^m) \times V(\Gamma^m)$ *solves (147), then* $(\vec{U}^{m+1}, \vec{X}^{m+1}, \kappa^{m+1})$ *is a solution to (149).*

(iii) *Let* $(\mathbb{U}^m, \widehat{\mathbb{P}}^m)$ *satisfy the LBB condition (143) and let* Γ^m *satisfy Assumption 64. Then there exists a unique solution* $(\vec{U}^{m+1}, P^{m+1}, \vec{X}^{m+1}, \kappa^{m+1}) \in \mathbb{U}^m \times \widehat{\mathbb{P}}^m \times \underline{V}(\Gamma^m) \times V(\Gamma^m)$ *to (147).*

Proof.

(i) Existence follows from uniqueness and hence we consider the homogeneous linear system. Find $(\vec{U}, \vec{X}, \kappa) \in \mathbb{U}_0^m \times \underline{V}(\Gamma^m) \times V(\Gamma^m)$ such that

$$2\left(\mu^m \underline{\underline{D}}(\vec{U}), \underline{\underline{D}}(\vec{\xi})\right) = \gamma_0 \left\langle \kappa \vec{\nu}^m, \vec{\xi} \right\rangle_{\Gamma^m} \quad \forall \vec{\xi} \in \mathbb{U}_0^m, \tag{150a}$$

$$\langle \vec{X} \cdot \vec{\nu}^m, \chi \rangle_{\Gamma^m}^h = \Delta t_m \langle \vec{U} \cdot \vec{\nu}^m, \chi \rangle_{\Gamma^m} \quad \forall \chi \in V(\Gamma^m), \tag{150b}$$

$$\langle \kappa \vec{\nu}^m, \vec{\eta} \rangle_{\Gamma^m}^h + \langle \nabla_s \vec{X}, \nabla_s \vec{\eta} \rangle_{\Gamma^m} = 0 \quad \forall \vec{\eta} \in \underline{V}(\Gamma^m). \tag{150c}$$

We now need to show that the zero solution is the only possible solution. Choosing $\vec{\xi} = \Delta t_m \vec{U}$ in (150a), $\chi = \gamma_0 \kappa$ in (150b) and $\vec{\eta} = \gamma_0 \vec{X}$ in (150c), we obtain

$$2\, \Delta t_m \left(\mu^m \underline{\underline{D}}(\vec{U}), \underline{\underline{D}}(\vec{U})\right) + \gamma_0 \left|\nabla_s \vec{X}\right|_{\Gamma^m}^2 = 0.$$

Now Korn's inequality, see, e.g., Zeidler (1988, Section 62.15), yields $\vec{U} = \vec{0}$. Hence it follows from (150b), (150c) and the proof of Lemma 66 that $\vec{X} = \vec{0}$ and $\kappa = 0$.

(ii) The claim follows trivially from (147b) and the definition of $\mathbb{U}_0^m$.

(iii) Upon considering the homogeneous system, for $(\vec{U}, P, \vec{X}, \kappa) \in \mathbb{U}^m \times \widehat{\mathbb{P}}^m \times \underline{V}(\Gamma^m) \times V(\Gamma^m)$ it follows immediately from (ii) and (i) that $\vec{U} = \vec{0}$, $\vec{X} = \vec{0}$ and $\kappa = 0$. Then the LBB inf-sup condition (143) yields that $P = 0$. □

Theorem 120. *Let* $d = 2$ *or* $d = 3$. *Let* $(\vec{U}^{m+1}, \vec{X}^{m+1}, \kappa^{m+1}) \in \mathbb{U}_0^m \times \underline{V}(\Gamma^m) \times V(\Gamma^m)$ *be a solution to (149), or let* $(\vec{U}^{m+1}, P^{m+1}, \vec{X}^{m+1}, \kappa^{m+1}) \in \mathbb{U}^m \times \widehat{\mathbb{P}}^m \times \underline{V}(\Gamma^m) \times V(\Gamma^m)$ *be a solution to (147). Then it holds that*

$$\gamma_0 \left|\Gamma^{m+1}\right| + 2\, \Delta t_m \left(\mu^m \underline{\underline{D}}(\vec{U}^{m+1}), \underline{\underline{D}}(\vec{U}^{m+1})\right) \leq \gamma_0 |\Gamma^m| + \Delta t_m \left(\vec{f}^{m+1}, \vec{U}^{m+1}\right). \tag{151}$$

Proof. It follows from Theorem 119(ii) that we only need to consider (149). Choosing $\vec{\xi}=\vec{U}^{m+1}\in\mathbb{U}_0^m$ in (149a), $\chi=\gamma_0\,\kappa^{m+1}$ in (149b) and $\vec{\eta}=\gamma_0\,(\vec{X}^{m+1}-\vec{\mathrm{id}}_{|\Gamma^m})$ in (149c) yields that

$$2\,\Delta t_m\,(\mu^m\,\underline{\underline{D}}(\vec{U}^{m+1}),\underline{\underline{D}}(\vec{U}^{m+1}))+\gamma_0\left\langle\nabla_s\,\vec{X}^{m+1},\nabla_s\,(\vec{X}^{m+1}-\vec{\mathrm{id}})\right\rangle_{\Gamma^m}$$
$$=\Delta t_m\left(\vec{f}^{m+1},\vec{U}^{m+1}\right).$$

Hence (151) follows immediately, on recalling Lemma 57. □

Remark 121.

(i) The stability bound (151) is a natural fully discrete analogue of (140).

(ii) In the case $\vec{f}=\vec{0}$, it is possible to derive the stability bound (151) with the help of the general strategy from Section 5.4, on recalling (142). To this end, for a given Γ^m and $\kappa\in V(\Gamma^m)$, we determine $\vec{U}\in\mathbb{U}_0^m$ as the unique solution to

$$2\left(\mu^m\,\underline{\underline{D}}(\vec{U}),\underline{\underline{D}}(\vec{\xi})\right)=\gamma_0\left\langle\kappa\,\vec{\nu}^m,\vec{\xi}\right\rangle_{\Gamma^m}\quad\forall\,\vec{\xi}\in\mathbb{U}_0^m. \tag{152}$$

We then define $\mathfrak{F}^m(\kappa)\in V(\Gamma^m)$ such that $\langle\mathfrak{F}^m(\kappa),\chi\rangle_{\Gamma^m}^h=\left\langle\vec{U},\chi\,\vec{\nu}^m\right\rangle_{\Gamma^m}^h$ for all $\chi\in V(\Gamma^m)$. Choosing $\vec{\xi}=\vec{U}$ in (152) we obtain that

$$\langle\mathfrak{F}^m(\kappa),\kappa\rangle_{\Gamma^m}^h=\frac{2}{\gamma_0}\,(\mu^m\,\underline{\underline{D}}(\vec{U}),\underline{\underline{D}}(\vec{U}))\geq 0\quad\forall\,\kappa\in V(\Gamma^m).$$

Now Theorem 90(ii) implies the desired stability result.

(iii) The scheme (147) leads to well-behaved meshes for Γ^m, $m=1,\ldots,M$, which follows as in Section 4.6 by considering a semidiscrete version, see Barrett et al. (2013c, Remark 3). In particular, we obtain equidistribution in two space dimensions and conformal polyhedral hypersurfaces in three space dimensions.

Remark 122 (Discrete linear systems).
We recall the notations and definitions from Remark 111. Moreover, as is standard practice for the solution of linear systems arising from discretizations of (Navier–)Stokes equations, we avoid the complications of the constrained pressure space $\widehat{\mathbb{P}}^m$ by considering an overdetermined linear system with $\mathbb{P}^m$ instead. Introducing the obvious abuse of notation, the linear system (147), with $\widehat{\mathbb{P}}^m$ replaced by $\mathbb{P}^m$, can be formulated as: Find $(\vec{U}^{m+1},P^{m+1},\kappa^{m+1},\delta\vec{X}^{m+1})\in(\mathbb{R}^d)^{K_{\mathbb{U}^m}}\times\mathbb{R}^{K_{\mathbb{P}^m}}\times\mathbb{R}^K\times(\mathbb{R}^d)^K$ such that

$$\begin{pmatrix} \underline{\underline{B}}_\Omega & \vec{C}_\Omega & -\gamma_0\,\vec{N}_{\Gamma^m,\Omega} & 0 \\ \vec{C}_\Omega^\top & 0 & 0 & 0 \\ \vec{N}_{\Gamma^m,\Omega}^\top & 0 & 0 & -\frac{1}{\Delta t_m}\,\vec{N}_{\Gamma^m}^\top \\ 0 & 0 & \vec{N}_{\Gamma^m} & \underline{\underline{A}}_{\Gamma^m} \end{pmatrix} \begin{pmatrix} \vec{U}^{m+1} \\ P^{m+1} \\ \kappa^{m+1} \\ \delta\vec{X}^{m+1} \end{pmatrix} = \begin{pmatrix} \underline{\underline{M}}_\Omega\,\vec{f}^{m+1} \\ 0 \\ 0 \\ -\underline{\underline{A}}_{\Gamma^m}\,\vec{X}^m \end{pmatrix}, \tag{153}$$

where $K_{\mathbb{U}^m}$ and $K_{\mathbb{P}^m}$ denote the degrees of freedom for the finite element spaces $\mathbb{U}^m$ and $\mathbb{P}^m$, respectively. The definitions of the matrices in (153) are either given in (45), or they follow directly from (147), see also Barrett et al. (2015a, Section 5) for details.

The overdetermined linear system (153) can either be solved directly, with the help of a sparse *QR* factorization method such as SPQR, see Davis (2011). Or it can be solved with the help of a Schur complement approach that eliminates $(\kappa^{m+1}, \delta\vec{X}^{m+1})$ from (153), and then uses an iterative solver for the remaining system in $(\vec{U}^{m+1}, P^{m+1})$. This approach has the advantage that for the reduced system well-known solution methods for finite element discretizations for the standard (Navier–)Stokes equations may be employed. In particular, we let

$$\Xi_{\Gamma^m} = \begin{pmatrix} 0 & -\frac{1}{\Delta t_m}\,\vec{N}_{\Gamma^m}^\top \\ \vec{N}_{\Gamma^m} & \underline{\underline{A}}_{\Gamma^m} \end{pmatrix}$$

and recall from Lemma 66 that if Assumption 64 holds, then the matrix Ξ_{Γ^m} is nonsingular. On defining $\underline{\underline{T}}_\Omega = (\vec{N}_{\Gamma^m,\Omega}\;\; 0)\,\Xi_{\Gamma^m}^{-1}\begin{pmatrix}\vec{N}_{\Gamma^m,\Omega}^\top \\ 0\end{pmatrix}$, we can reduce (153) to

$$\begin{pmatrix} \underline{\underline{B}}_\Omega + \gamma_0\,\underline{\underline{T}}_\Omega & \vec{C}_\Omega \\ \vec{C}_\Omega^\top & 0 \end{pmatrix} \begin{pmatrix} \vec{U}^{m+1} \\ P^{m+1} \end{pmatrix} = \begin{pmatrix} \underline{\underline{M}}_\Omega\,\vec{f}^{m+1} - \gamma_0\,(\vec{N}_{\Gamma^m,\Omega}\;\; 0)\,\Xi_{\Gamma^m}^{-1}\begin{pmatrix} 0 \\ \underline{\underline{A}}_{\Gamma^m}\,\vec{X}^m \end{pmatrix} \\ 0 \end{pmatrix} \tag{154}$$

and $\begin{pmatrix}\kappa^{m+1} \\ \delta\vec{X}^{m+1}\end{pmatrix} = \Xi_{\Gamma^m}^{-1}\begin{pmatrix} -\vec{N}_{\Gamma^m,\Omega}^\top\,\vec{U}^{m+1} \\ -\underline{\underline{A}}_{\Gamma^m}\,\vec{X}^m \end{pmatrix}$. The linear system (154) can be solved, for example, with preconditioned GMRES iterative solvers for standard (Navier–)Stokes discretizations, see, e.g., Elman et al. (2005) for some examples. For particular preconditioners for (154) and further details on possible solution procedures, we refer to Barrett et al. (2015a, Section 5).

8.1.2 Semidiscrete finite element approximation

In this section we introduce a continuous-in-time semidiscrete variant of (147). Similarly to (118) and (144), for a fixed regular partitioning $\mathcal{T}^h$ of Ω, with $\overline{\Omega} = \cup_{o\in\mathcal{T}^h}\overline{o}$, we introduce the finite element spaces

$$\mathcal{S}_k^h = \left\{\chi \in C(\overline{\Omega}) : \chi_{|_o} \in \mathcal{P}_k(o) \ \ \forall o \in \mathcal{T}^h\right\} \subset H^1(\Omega), \quad k \in \mathbb{N}.$$

As before, we let $\mathcal{S}_0^h$ denote the space of piecewise constant functions on $\mathcal{T}^h$. Let $\mathbb{U}^h \subset [H_0^1(\Omega)]^d$ and $\mathbb{P}^h(t) \subset L^2(\Omega)$ be the finite element spaces for the semidiscrete velocity and pressure approximations, and set $\widehat{\mathbb{P}}^h(t) = \mathbb{P}^h(t) \cap \widehat{\mathbb{P}}$. Note that while $\mathbb{U}^h$ is fixed, for later developments we allow a time-dependent discrete pressure space $\mathbb{P}^h(t)$, see Section 8.1.3. In addition, we use the notation of Section 3.4 for evolving polyhedral surfaces, the corresponding finite element spaces, and discrete time derivatives. In addition, we define

$$\widehat{\mathbb{P}}_T^h = \left\{\varphi \in L^2(0,T;L^2(\Omega)) : \varphi(t) \in \widehat{\mathbb{P}}^h(t) \ \ \forall t \in (0,T]\right\}. \tag{155}$$

Given $\Gamma^h(t)$, we denote by $\Omega_+^h(t)$ the exterior of $\Gamma^h(t)$ and by $\Omega_-^h(t)$ the interior of $\Gamma^h(t)$, so that $\Gamma^h(t) = \partial\Omega_-^h(t) = \overline{\Omega_-^h(t)} \cap \overline{\Omega_+^h(t)}$. The elements of the bulk mesh $\mathcal{T}^h$ are partitioned into interior, exterior and interfacial elements precisely as in (145), and the discrete viscosity $\mu^h(t) \in \mathcal{S}_0^h$ is defined as the natural semidiscrete analogue of (146).

Then we can formulate the semidiscrete analogue of (147) as follows. Given the closed polyhedral hypersurface $\Gamma^h(0)$, find an evolving polyhedral hypersurface $\mathcal{G}_T^h$ with induced velocity $\vec{\mathcal{V}}^h \in \underline{V}(\mathcal{G}_T^h)$, $\kappa^h \in V(\mathcal{G}_T^h)$, $\vec{U}^h \in L^2(0,T;\mathbb{U}^h)$ and $P^h \in \widehat{\mathbb{P}}_T^h$ as follows. For all $t \in (0,T]$, find $(\vec{U}^h(\cdot,t), P^h(\cdot,t), \vec{\mathcal{V}}^h(\cdot,t), \kappa^h(\cdot,t)) \in \mathbb{U}^h \times \widehat{\mathbb{P}}^h(t) \times \underline{V}(\Gamma^h(t)) \times V(\Gamma^h(t))$ such that

$$2\left(\mu^h \underline{\underline{D}}(\vec{U}^h), \underline{\underline{D}}(\vec{\xi})\right) - \left(P^h, \nabla\cdot\vec{\xi}\right) = \left(\vec{f}^h, \vec{\xi}\right) + \gamma_0\left\langle \kappa^h\,\vec{\nu}^h, \vec{\xi}\right\rangle_{\Gamma^h(t)} \quad \forall \vec{\xi} \in \mathbb{U}^h, \tag{156a}$$

$$\left(\nabla\cdot\vec{U}^h, \varphi\right) = 0 \quad \forall \varphi \in \widehat{\mathbb{P}}^h(t), \tag{156b}$$

$$\left\langle \vec{\mathcal{V}}^h\cdot\vec{\nu}^h, \chi\right\rangle_{\Gamma^h(t)}^h = \langle \vec{U}^h\cdot\vec{\nu}^h, \chi\rangle_{\Gamma^h(t)} \quad \forall \chi \in V(\Gamma^h(t)), \tag{156c}$$

$$\langle \kappa^h\,\vec{\nu}^h, \vec{\eta}\rangle_{\Gamma^h(t)}^h + \left\langle \nabla_s\,\mathrm{i\vec{d}}, \nabla_s\,\vec{\eta}\right\rangle_{\Gamma^h(t)} = 0 \quad \forall \vec{\eta} \in \underline{V}(\Gamma^h(t)), \tag{156d}$$

where $\vec{f}^h$ is the natural semidiscrete analogue of the fully discrete forcings $\vec{f}^{m+1}$, $m = 0, \ldots, M-1$.

Theorem 123. *Let $(\mathcal{G}_T^h, \kappa^h, \vec{U}^h, P^h)$ be a solution of (156).*

(i) *It holds that*

$$\gamma_0 \frac{\mathrm{d}}{\mathrm{d}t} |\Gamma^h(t)| + 2(\mu^h \underline{\underline{D}}(\vec{U}^h), \underline{\underline{D}}(\vec{U}^h)) = \left(\vec{f}^h, \vec{U}^h\right).$$

(ii) *If $\mathcal{X}_{\Omega_-^h(t)} \in \mathbb{P}^h(t)$, then it holds that*

$$\frac{\mathrm{d}}{\mathrm{d}t} \mathcal{L}^d(\Omega_-^h(t)) = 0.$$

(iii) *For any $t \in (0,T]$, it holds that $\Gamma^h(t)$ is a conformal polyhedral surface. In particular, for $d = 2$, any two neighbouring elements of the curve $\Gamma^h(t)$ either have equal length, or they are parallel.*

Proof.

(i) Similarly to the proof of Theorem 79(i), choosing $\vec{\xi} = \vec{U}^h(\cdot, t) \in \mathbb{U}^h$, $\varphi = P^h(\cdot, t) \in \widehat{\mathbb{P}}^h(t)$, $\chi = \kappa^h(\cdot, t) \in V(\Gamma^h(t))$ $\vec{\eta} = \vec{\mathcal{V}}^h(\cdot, t) \in \underline{V}(\Gamma^h(t))$ in (156) gives

$$\gamma_0 \frac{\mathrm{d}}{\mathrm{d}t} |\Gamma^h(t)| = \gamma_0 \left\langle \nabla_s \,\vec{\mathrm{id}}, \nabla_s \vec{\mathcal{V}}^h \right\rangle_{\Gamma^h(t)} = -\gamma_0 \left\langle \vec{\mathcal{V}}^h, \kappa^h \vec{\nu}^h \right\rangle_{\Gamma^h(t)}^h$$
$$= -\gamma_0 \langle \vec{U}^h, \kappa^h \vec{\nu}^h \rangle_{\Gamma^h(t)} = \left(\vec{f}^h, \vec{U}^h\right) - 2(\mu^h \underline{\underline{D}}(\vec{U}^h), \underline{\underline{D}}(\vec{U}^h)),$$

which is the claim.

(ii) Similarly to the proof of Theorem 88(ii), choosing $\chi = 1$ in (156c) and $\varphi = \mathcal{X}_{\Omega_-^h(t)} - \frac{\mathcal{L}^d(\Omega_-^h(t))}{\mathcal{L}^d(\Omega)} \in \widehat{\mathbb{P}}^h(t)$ in (156b) yields, on using the divergence theorem, that

$$\frac{\mathrm{d}}{\mathrm{d}t} \mathcal{L}^d(\Omega_-^h(t)) = \left\langle \vec{\mathcal{V}}^h, \vec{\nu}^h \right\rangle_{\Gamma^h(t)}^h = \langle \vec{U}^h, \vec{\nu}^h \rangle_{\Gamma^h(t)} = \left(\nabla \cdot \vec{U}^h, \mathcal{X}_{\Omega_-^h(t)}\right)$$
$$= \left(\nabla \cdot \vec{U}^h, \mathcal{X}_{\Omega_-^h(t)} - \frac{\mathcal{L}^d(\Omega_-^h(t))}{\mathcal{L}^d(\Omega)}\right) = 0,$$

in a discrete analogue to (141).

(iii) This follows directly from Definition 60 and Theorem 62. □

8.1.3 XFEM$_\Gamma$ for conservation of the phase volumes

Conservation of the total mass, equivalent to the conservation of $\mathcal{L}^d(\Omega_-(t))$, (141), is clearly a desirable property on the discrete level. We have seen in Theorem 123(ii) that the semidiscrete scheme (156) conserves $\mathcal{L}^d(\Omega_-^h(t))$ only if the time-dependent discrete pressure spaces $\mathbb{P}^h(t)$ contain the characteristic function of the discrete inner phase $\mathcal{X}_{\Omega_-^h(t)}$ for all $t \in [0,T]$. Hence, for the fully discrete approximation (147) it was suggested in Barrett et al. (2013c, Section 3.4) to extend the pressure space $\mathbb{P}^m$ by one single basis function, namely $\mathcal{X}_{\Omega_-^m}$. There, we referred to this as the XFEM$_\Gamma$ approach, because the extra contributions to (147a) and (147b) coming from $\mathcal{X}_{\Omega_-^m} - \frac{\mathcal{L}^d(\Omega_-^m)}{\mathcal{L}^d(\Omega)} \in \widehat{\mathbb{P}}^m$ can be written in terms of integrals over Γ^m, on noting from the divergence theorem that

$$\left(\nabla \cdot \vec{\xi}, \mathcal{X}_{\Omega_-^m} - \frac{\mathcal{L}^d(\Omega_-^m)}{\mathcal{L}^d(\Omega)}\right) = \left(\nabla \cdot \vec{\xi}, \mathcal{X}_{\Omega_-^m}\right) = \left\langle \vec{\nu}^m, \vec{\xi} \right\rangle_{\Gamma^m} \quad \forall \vec{\xi} \in \mathbb{U}^m. \tag{157}$$

For the fully discrete approximation (147), even with the XFEM$_\Gamma$ pressure space extension, it is not possible to show that the total mass is conserved. However, we observe that combining (157), with $\vec{\xi} = \vec{U}^{m+1}$, (147b) and (147c) leads to

$$\left\langle \vec{X}^{m+1} - \vec{\mathrm{id}}, \vec{\nu}^m \right\rangle_{\Gamma^m} = 0,$$

which means that in practice this fully discrete approximation conserves the volume of the two phases well, see also Remark 89(ii).

Moreover, it turns out that the XFEM$_\Gamma$ approach avoids spurious velocities. To make this precise, we state the following theorem.

Theorem 124.

(i) *Let $d = 2$ or $d = 3$. Let $(\vec{U}^{m+1}, \vec{X}^{m+1}, \kappa^{m+1}) \in \mathbb{U}^m \times \underline{V}(\Gamma^m) \times V(\Gamma^m)$ be a solution to (149) with $\vec{f}^{m+1} = \vec{0}$. If $\vec{X}^{m+1} = \vec{\mathrm{id}}_{|\Gamma^m}$, then $\vec{U}^{m+1} = \vec{0}$.*

(ii) *Let $\mathcal{X}_{\Omega_-^m} \in \mathbb{P}^m$, let Γ^m satisfy Assumption 64, and let Γ^m be a polyhedral surface with constant discrete mean curvature, i.e., there exists a constant $\overline{\kappa} \in \mathbb{R}$ such that*

$$\overline{\kappa} \langle \vec{\nu}^m, \vec{\eta} \rangle_{\Gamma^m} + \left\langle \nabla_s \vec{\mathrm{id}}, \nabla_s \vec{\eta} \right\rangle_{\Gamma^m} = 0 \quad \forall \vec{\eta} \in \underline{V}(\Gamma^m).$$

Then Γ^m is a conformal polyhedral surface, recall Definition 60, and $(\vec{U}^{m+1}, \vec{X}^{m+1}, \kappa^{m+1}) = (\vec{0}, \vec{X}^m, \overline{\kappa}) \in \mathbb{U}_0^m \times \underline{V}(\Gamma^m) \times V(\Gamma^m)$ is the unique solution to (149) with $\vec{f}^{m+1} = \vec{0}$.

(iii) *Let the assumptions in (ii) hold and let $(\mathbb{U}^m, \widehat{\mathbb{P}}^m)$ satisfy the LBB condition (143). Then Γ^m is a conformal polyhedral surface, and $(\vec{U}^{m+1}, P^{m+1}, \vec{X}^{m+1}, \kappa^{m+1}) = (\vec{0}, -\gamma_0 \overline{\kappa} [\mathcal{X}_{\Omega_-^m} - \frac{\mathcal{L}^d(\Omega_-^m)}{\mathcal{L}^d(\Omega)}], \vec{\mathrm{id}}_{|\Gamma^m}, \overline{\kappa}) \in \mathbb{U}^m \times \widehat{\mathbb{P}}^m \times \underline{V}(\Gamma^m) \times V(\Gamma^m)$ is the unique solution to (147) with $\vec{f}^{m+1} = \vec{0}$.*

Proof.

(i) It follows from Theorem 120 that the solution fulfils (151) with Γ^{m+1} replaced by Γ^m and $\vec{f}^{m+1} = \vec{0}$. Hence we obtain $(\mu^m \underline{\underline{D}}(\vec{U}^{m+1}), \underline{\underline{D}}(\vec{U}^{m+1})) = 0$, and so Korn's inequality implies $\vec{U}^{m+1} = \vec{0}$.

(ii) It immediately follows from (20) that Γ^m is a conformal polyhedral surface. Theorem 119(i) implies that in order to establish the remaining result, we only need to show that $(\vec{U}^{m+1}, \vec{X}^{m+1}, \kappa^{m+1}) = (\vec{0}, \vec{\mathrm{id}}_{|\Gamma^m}, \overline{\kappa})$ is a solution to (149) with $\vec{f}^{m+1} = \vec{0}$. But this follows immediately from $\overline{\kappa} \langle \vec{\nu}^m, \vec{\eta} \rangle_{\Gamma^m} = \langle \overline{\kappa}\, \vec{\nu}^m, \vec{\eta} \rangle_{\Gamma^m}^h$ for all $\vec{\eta} \in \underline{V}(\Gamma^m)$, and

$$\left\langle \vec{\nu}^m, \vec{\xi} \right\rangle_{\Gamma^m} = \left(\nabla \cdot \vec{\xi}, \mathcal{X}_{\Omega^m_-} - \frac{\mathcal{L}^d(\Omega^m_-)}{\mathcal{L}^d(\Omega)} \right) = 0 \quad \forall \vec{\xi} \in \mathbb{U}^m_0,$$

where we have recalled (157) and (148).
(iii) On recalling Theorem 119(iii), the proof is analogous to the proof of (ii). It holds that

$$\gamma_0 \left\langle \kappa^{m+1} \vec{\nu}^m, \vec{\xi} \right\rangle_{\Gamma^m} + \left(P^{m+1}, \nabla \cdot \vec{\xi} \right)$$
$$= \gamma_0 \,\overline{\kappa} \left\langle \vec{\nu}^m, \vec{\xi} \right\rangle_{\Gamma^m} - \gamma_0 \,\overline{\kappa} \left(\mathcal{X}_{\Omega^m_-} - \frac{\mathcal{L}^d(\Omega^m_-)}{\mathcal{L}^d(\Omega)}, \nabla \cdot \vec{\xi} \right) = 0 \quad \forall \vec{\xi} \in \mathbb{U}^m,$$

and this proves the claim. □

Remark 125. It follows from Theorem 124 that, independently of the choice of $\mu_\pm$, no spurious velocities appear for discrete stationary solutions, $\Gamma^{m+1} = \Gamma^m$. Moreover, for the XFEM$_\Gamma$ approach it holds that polyhedral surfaces with constant discrete mean curvature are discrete stationary solutions. In particular, spherical bubbles can be approximated by such polyhedral surfaces, and so our method admits a stationary solution with zero velocity in these situations. This is not the case for many other discretizations and is one of the reasons for spurious velocities in simple situations like a spherical bubble.

8.1.4 Approximations based on the fluidic tangential velocity

Let us briefly discuss an alternative approximation of two-phase Stokes flow, that is based on a weak formulation of (136a) and

$$[\vec{u}]_-^+ = \vec{0}, \quad [(2\,\mu\,\underline{\underline{D}}(\vec{u}) - p\,\underline{\underline{\mathrm{Id}}})\vec{\nu}]_-^+ = -\gamma_0\,\vec{\varkappa}, \quad \vec{\varkappa} = \Delta_s\,\vec{\mathrm{id}}, \quad \vec{\mathcal{V}} = \vec{u} \quad \text{on}\,\Gamma(t)$$

as opposed to (136) with (137). The semidiscrete finite element approximation, in line with (156), then features the equations

$$2\left(\mu^h\,\underline{\underline{D}}(\vec{U}^h), \underline{\underline{D}}(\vec{\xi})\right) - \left(P^h, \nabla \cdot \vec{\xi}\right) = \left(\vec{f}^h, \vec{\xi}\right) + \gamma_0 \left\langle \vec{\kappa}^h, \vec{\xi} \right\rangle^h_{\Gamma^h(t)} \quad \forall \vec{\xi} \in \mathbb{U}^h, \tag{158a}$$

$$\left(\nabla \cdot \vec{U}^h, \varphi\right) = 0 \quad \forall \varphi \in \widehat{\mathbb{P}}^h(t), \tag{158b}$$

$$\left\langle \vec{\mathcal{V}}^h, \vec{\chi} \right\rangle^h_{\Gamma^h(t)} = \langle \vec{U}^h, \vec{\chi} \rangle^h_{\Gamma^h(t)} \quad \forall \vec{\chi} \in \underline{V}(\Gamma^h(t)), \tag{158c}$$

$$\langle \vec{\kappa}^h, \vec{\eta} \rangle^h_{\Gamma^h(t)} + \left\langle \nabla_s\,\vec{\mathrm{id}}, \nabla_s\,\vec{\eta} \right\rangle_{\Gamma^h(t)} = 0 \quad \forall \vec{\eta} \in \underline{V}(\Gamma^h(t)), \tag{158d}$$

for $(\vec{U}^h(\cdot,t), P^h(\cdot,t), \vec{\mathcal{V}}^h(\cdot,t), \vec{\kappa}^h(\cdot,t)) \in \mathbb{U}^h \times \widehat{\mathbb{P}}^h(t) \times \underline{V}(\Gamma^h(t)) \times \underline{V}(\Gamma^h(t))$. We note that a variant of (158) without numerical integration can also be considered, see Barrett et al. (2013c, Section 3.6) for details on the fully discrete case. It is a simple matter to prove that solutions to (158) satisfy the stability result Theorem 123(i). However, since $\vec{\nu}^h(\cdot,t)$ is not a valid test function in (158c), it is not possible to prove the volume conservation result in Theorem 123(ii) for a solution of (158), even if $\mathcal{X}_{\Omega^h_-(t)} \in \mathbb{P}^h(t)$. However, on choosing $\vec{\chi} = \vec{\omega}^h(\cdot,t) \in \underline{V}(\Gamma^h(t))$ in (158c), it follows from Theorem 71 and (20) that

$$\frac{\mathrm{d}}{\mathrm{d}t}\mathcal{L}^d(\Omega^h_-(t)) = \left\langle \vec{\mathcal{V}}^h, \vec{\nu}^h \right\rangle^h_{\Gamma^h(t)} = \left\langle \vec{\mathcal{V}}^h, \vec{\omega}^h \right\rangle^h_{\Gamma^h(t)} = \langle \vec{U}^h, \vec{\omega}^h \rangle^h_{\Gamma^h(t)}. \tag{159}$$

Hence, by enforcing the needed condition

$$\langle \vec{U}^h, \vec{\omega}^h \rangle^h_{\Gamma^h(t)} = 0$$

directly, together with a suitable Lagrange multiplier $P^h_{\mathrm{sing}}(t) \in \mathbb{R}$, we can introduce the following semidiscrete approximation of two-phase Stokes flow that satisfies Theorem 123(i), (ii). Note that in order for (159) to hold, it is crucial to employ numerical integration on $\Gamma^h(t)$ throughout.

Given the closed polyhedral hypersurface $\Gamma^h(0)$, find an evolving polyhedral hypersurface $\mathcal{G}^h_T$ with induced velocity $\vec{\mathcal{V}}^h \in \underline{V}(\mathcal{G}^h_T)$, $\vec{\kappa}^h \in \underline{V}(\mathcal{G}^h_T)$, $\vec{U}^h \in L^2(0,T;\mathbb{U}^h)$, $P^h \in \widehat{\mathbb{P}}^h_T$ and $P^h_{\mathrm{sing}} \in L^2(0,T;\mathbb{R})$ as follows. For all $t \in (0,T]$, find $(\vec{U}^h(\cdot,t), P^h(\cdot,t), P^h_{\mathrm{sing}}(t), \vec{\mathcal{V}}^h(\cdot,t), \vec{\kappa}^h(\cdot,t)) \in \mathbb{U}^h \times \widehat{\mathbb{P}}^h(t) \times \mathbb{R} \times \underline{V}(\Gamma^h(t)) \times \underline{V}(\Gamma^h(t))$ such that

$$\begin{aligned} 2\left(\mu^h \underline{\underline{D}}(\vec{U}^h), \underline{\underline{D}}(\vec{\xi})\right) - \left(P^h, \nabla\cdot\vec{\xi}\right) - P^h_{\mathrm{sing}}\left\langle \vec{\omega}^h, \vec{\xi}\right\rangle^h_{\Gamma^h(t)} \\ = \left(\vec{f}^h, \vec{\xi}\right) + \gamma_0\left\langle \vec{\kappa}^h, \vec{\xi}\right\rangle^h_{\Gamma^h(t)} \quad \forall \vec{\xi} \in \mathbb{U}^h, \end{aligned} \tag{160a}$$

$$(\nabla\cdot\vec{U}^h, \varphi) = 0 \quad \forall \varphi \in \widehat{\mathbb{P}}^h(t) \quad \text{and} \quad \langle \vec{U}^h, \vec{\omega}^h \rangle^h_{\Gamma^h(t)} = 0, \tag{160b}$$

$$\left\langle \vec{\mathcal{V}}^h, \vec{\chi} \right\rangle^h_{\Gamma^h(t)} = \langle \vec{U}^h, \vec{\chi} \rangle^h_{\Gamma^h(t)} \quad \forall \vec{\chi} \in \underline{V}(\Gamma^h(t)), \tag{160c}$$

$$\langle \vec{\kappa}^h, \vec{\eta} \rangle^h_{\Gamma^h(t)} + \left\langle \nabla_s \vec{\mathrm{id}}, \nabla_s \vec{\eta} \right\rangle_{\Gamma^h(t)} = 0 \quad \forall \vec{\eta} \in \underline{V}(\Gamma^h(t)). \tag{160d}$$

We note that in terms of pressure space enrichment, the above procedure may be viewed as a virtual element method, see, e.g., Beirão da Veiga et al. (2013).

8.2 Two-phase Navier–Stokes flow

In Barrett et al. (2015a) the present authors extended the approximation (147) to two-phase Navier–Stokes flow, which is given by the model (136) with the first equation in (136a) replaced by

$$\rho\,(\partial_t \vec{u} + (\vec{u}\cdot\nabla)\,\vec{u}) - 2\,\mu\,\nabla\cdot\underline{\underline{D}}(\vec{u}) + \nabla\,p = \vec{f} \quad \text{in}\,\Omega_\pm(t),$$

where $\rho(\cdot,t) = \rho_+\,\mathcal{X}_{\Omega_+(t)} + \rho_-\,\mathcal{X}_{\Omega_-(t)}$, with $\rho_\pm \in \mathbb{R}_{\geq 0}$ denoting the two fluid densities, and with the additional initial condition $\rho(\cdot,0)\,\vec{u}(\cdot,0) = \rho(\cdot,0)\,\vec{u}_0$ in Ω. The treatment of the interface evolution, and its coupling to the quantities in the bulk, remains unchanged, and for the approximation of the fluid flow in the bulk standard techniques for the finite element approximation of one-phase Navier–Stokes flow can be employed, see, e.g., Temam (2001). In the following, we recall the fully discrete approximation from Barrett et al. (2015a), which is based on the weak formulation of two-phase Navier–Stokes flow defined by (139) with the additional terms

$$\frac{1}{2}\,\frac{\mathrm{d}}{\mathrm{d}t}\left(\rho\,\vec{u},\vec{\xi}\right) + \frac{1}{2}\left(\rho\,\partial_t\,\vec{u},\vec{\xi}\right) + \frac{1}{2}\left(\rho,[(\vec{u}\cdot\nabla)\,\vec{u}]\cdot\vec{\xi} - [(\vec{u}\cdot\nabla)\,\vec{\xi}]\cdot\vec{u}\right) \tag{161}$$

on the left-hand side of (139a), recall Barrett et al. (2015a, (3.9)).

Let $\rho^m \in S_0^m$ be defined analogously to (146), for $m \geq 0$, and set $\rho^{-1} = \rho^0$. In addition we define the standard projection operator $I_0^m : L^1(\Omega) \to S_0^m$, such that $(I_0^m \eta)|_o = \frac{1}{\mathcal{L}^d(o)}\int_o \eta \,\mathrm{d}\mathcal{L}^d$ for all $o \in \mathcal{T}^m$. In this section, we consider the partitioning $t_m = m\,\Delta t$, $m = 0, \ldots, M$, of $[0,T]$ into uniform time steps $\Delta t = \frac{T}{M}$. Uniform time steps are required in order to be able to introduce a consistent fully discrete approximation of the time derivative terms in (161). Let the closed polyhedral hypersurface Γ^0 be an approximation of $\Gamma(0)$, and let $\vec{U}^0 \in \mathbb{U}^0$ be an approximation to $\vec{u}_0$. Then, for $m = 0, \ldots, M-1$, find $(\vec{U}^{m+1}, P^{m+1}, \vec{X}^{m+1}, \kappa^{m+1}) \in \mathbb{U}^m \times \widehat{\mathbb{P}}^m \times \underline{V}(\Gamma^m) \times V(\Gamma^m)$ such that

$$\begin{aligned}
&\frac{1}{2}\left(\frac{\rho^m\,\vec{U}^{m+1} - (I_0^m\,\rho^{m-1})\,\vec{I}_2^m\,\vec{U}^m}{\Delta t} + (I_0^m\,\rho^{m-1})\,\frac{\vec{U}^{m+1} - \vec{I}_2^m\,\vec{U}^m}{\Delta t},\vec{\xi}\right)\\
&+\frac{1}{2}\left(\rho^m,[(\vec{I}_2^m\,\vec{U}^m\cdot\nabla)\,\vec{U}^{m+1}]\cdot\vec{\xi} - [(\vec{I}_2^m\,\vec{U}^m\cdot\nabla)\,\vec{\xi}]\cdot\vec{U}^{m+1}\right)\\
&+2\left(\mu^m\,\underline{\underline{D}}(\vec{U}^{m+1}),\underline{\underline{D}}(\vec{\xi})\right) - \left(P^{m+1},\nabla\cdot\vec{\xi}\right) = \left(\vec{f}^{m+1},\vec{\xi}\right) + \gamma_0\left\langle\kappa^{m+1}\,\vec{\nu}^m,\vec{\xi}\right\rangle_{\Gamma^m}\\
&\qquad\forall\,\vec{\xi}\in\mathbb{U}^m,
\end{aligned} \tag{162a}$$

$$\left(\nabla\cdot\vec{U}^{m+1},\varphi\right) = 0 \quad \forall\,\varphi\in\widehat{\mathbb{P}}^m, \tag{162b}$$

$$\left\langle\frac{\vec{X}^{m+1} - \vec{\mathrm{id}}}{\Delta t}\cdot\vec{\nu}^m,\chi\right\rangle_{\Gamma^m}^h = \left\langle\vec{U}^{m+1}\cdot\vec{\nu}^m,\chi\right\rangle_{\Gamma^m} \quad \forall\,\chi\in V(\Gamma^m), \tag{162c}$$

$$\left\langle\kappa^{m+1}\,\vec{\nu}^m,\vec{\eta}\right\rangle_{\Gamma^m}^h + \left\langle\nabla_s\,\vec{X}^{m+1},\nabla_s\,\vec{\eta}\right\rangle_{\Gamma^m} = 0 \quad \forall\,\vec{\eta}\in\underline{V}(\Gamma^m), \tag{162d}$$

and set $\Gamma^{m+1} = \vec{X}^{m+1}(\Gamma^m)$. Clearly, in the case $\rho_- = \rho_+ = 0$, the approximation (162) collapses to the scheme (147), with uniform time steps, for two-phase Stokes flow.

Theorem 126.

(i) *Let* $(\mathbb{U}^m, \widehat{\mathbb{P}}^m)$ *satisfy the LBB condition (143), let* Γ^m *satisfy Assumption 64 and let* $\vec{U}^m \in [C(\overline{\Omega})]^d$. *Then there exists a unique solution* $(\vec{U}^{m+1}, P^{m+1}, \vec{X}^{m+1}, \kappa^{m+1}) \in \mathbb{U}^m \times \widehat{\mathbb{P}}^m \times \underline{V}(\Gamma^m) \times V(\Gamma^m)$ *to (162).*

(ii) *Let* $d = 2$ *or* $d = 3$. *Then a solution to (162) satisfies*

$$\begin{aligned}
&\frac{1}{2}\left(\rho^m, |\vec{U}^{m+1}|^2\right) + \gamma_0\,|\Gamma^{m+1}| + \frac{1}{2}\left(I_0^m \rho^{m-1}, |\vec{U}^{m+1} - \vec{I}_2^m\, \vec{U}^m|^2\right)\\
&\qquad + 2\,\Delta t\,(\mu^m\, \underline{\underline{D}}(\vec{U}^{m+1}), \underline{\underline{D}}(\vec{U}^{m+1}))\\
&\quad \leq \frac{1}{2}\left(I_0^m\, \rho^{m-1}, |\vec{I}_2^m\, \vec{U}^m|^2\right) + \gamma_0\,|\Gamma^m| + \Delta t\left(\vec{f}^{m+1}, \vec{U}^{m+1}\right).
\end{aligned} \tag{163}$$

Proof.

(i) The result can be shown as in the proof of Theorem 119.

(ii) We choose $\vec{\xi} = \vec{U}^{m+1}$ in (162a), $\varphi = P^{m+1}$ in (162b), $\chi = \gamma_0\, \kappa^{m+1}$ in (162c) and $\vec{\eta} = \gamma_0\,(\vec{X}^{m+1} - \vec{\mathrm{id}}_{|_{\Gamma^m}})$ in (162d) to obtain

$$\begin{aligned}
&\frac{1}{2}(\rho^m\, \vec{U}^{m+1}, \vec{U}^{m+1}) + \frac{1}{2}((I_0^m\, \rho^{m-1})\,(\vec{U}^{m+1} - \vec{I}_2^m\, \vec{U}^m), \vec{U}^{m+1} - \vec{I}_2^m\, \vec{U}^m)\\
&\quad + 2\,\Delta t(\mu^m\, \underline{\underline{D}}(\vec{U}^{m+1}), \underline{\underline{D}}(\vec{U}^{m+1})) + \gamma_0\left\langle \nabla_s\, \vec{X}^{m+1}, \nabla_s\,(\vec{X}^{m+1} - \vec{\mathrm{id}})\right\rangle_{\Gamma^m}\\
&= \frac{1}{2}((I_0^m\, \rho^{m-1})\, \vec{I}_2^m\, \vec{U}^m, \vec{I}_2^m\, \vec{U}^m) + \Delta t\left(\vec{f}^{m+1}, \vec{U}^{m+1}\right).
\end{aligned}$$

Hence (163) follows immediately, on recalling Lemma 57. □

Remark 127.

(i) If $d = 2$ or $d = 3$ then, on assuming that

$$\left(I_0^m \rho^{m-1}, |\vec{I}_2^m\, \vec{U}^m|^2\right) \leq \left(\rho^{m-1}, |\vec{U}^m|^2\right) \quad \text{for } m = 1, \ldots, M-1,$$

we can prove an unconditional stability bound for the scheme (162), see Barrett et al. (2015a, Theorem 4.2). The condition is always satisfied if no bulk mesh coarsening in time is performed.

(ii) If

$$\mathcal{X}_{\Omega_-^m} \in \mathbb{P}^m \quad \text{for } m = 0, \ldots, M-1,$$

then a semidiscrete continuous-in-time version of (162) conserves the volume of the two phase exactly, which follows from the direct discrete analogue of Remark 117(ii), as discussed previously in Section 8.1.3.

(iii) The scheme (162) leads to well-behaved meshes, which follows as usual by considering a semidiscrete version, see Section 4.6. In particular, we obtain equidistribution in two space dimensions and conformal polyhedral hypersurfaces in three space dimensions.

(iv) It is a simple matter to extend the scheme (162), and hence (147), to more general boundary conditions than $\vec{u} = \vec{0}$ on $\partial\Omega$ for the fluid flow.

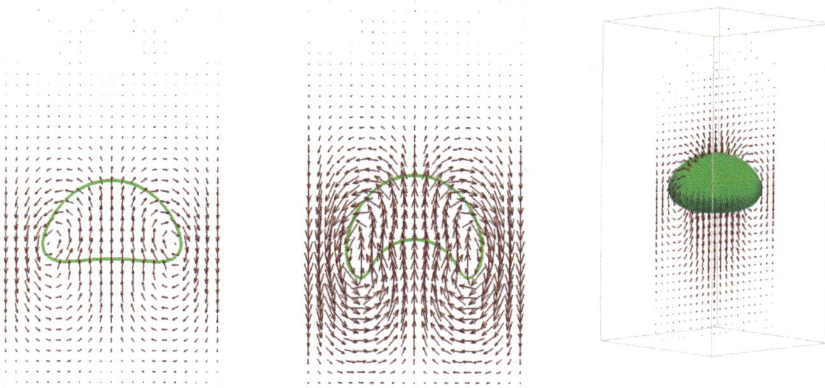

FIG. 6 Visualization of the numerical results for the rising droplet experiments shown in Figures 3, 7 and 11 in Barrett et al. (2015a). Each plot shows the interface Γ^m and the velocity $\vec{U}^m$ at time $t = 1.5$. For the 3D experiment, the fluid velocity is only visualized within a 2D cut through Ω.

Apart from this no-slip condition, also free-slip and stress-free boundary conditions, as well as their inhomogeneous analogues, may be considered. See Barrett et al. (2013c, 2015a, 2016a) for details.

(v) The discrete linear systems arising from (162) can be solved as described in Remark 122.

In Fig. 6 we show some numerical results for a generalization of the scheme (162) to include, for example, free-slip boundary conditions on parts of the boundary $\partial\Omega$ and gravitational forces $\vec{f} = \rho \vec{f}_1$. In Fig. 6 we show the interface of a rising bubble together with a visualization of the fluid flow for three different simulations from Barrett et al. (2015a). The two 2D simulations have density values $10\ \rho_- = \rho_+ = 10^3$ and $10^3\ \rho_- = \rho_+ = 10^3$, respectively, while the 3D simulation has $10\ \rho_- = \rho_+ = 10^3$. As the density of the inner fluid is chosen smaller than the density of the outer fluid in each case, the bubble rises in the presence of gravity.

8.3 Alternative numerical approaches

Numerical methods based on interface tracking methods using an indicator function to describe the interface are also popular methods to numerically solve two-phase flow problems. The volume of fluid (VOF) method uses a characteristic function of one of the phases to evolve the interface and has been used by Hirt and Nichols (1981) and Renardy and Renardy (2002). Another interface tracking method is the level set method, which uses a level set function to track the interface. We refer to Sussman et al. (1994) and to Groß and Reusken (2011) and the references therein for details. Phase field methods, which are also called diffuse interface methods in this context, have been studied numerically by Kim et al. (2004), Kay et al. (2008), Grün and Klingbeil (2014) and Garcke et al. (2016). Other parametric methods, which use a polyhedral mesh to directly represent the interface, are discussed in Unverdi and Tryggvason (1992), Bänsch (2001), Tryggvason et al. (2001), Ganesan et al. (2007) and Agnese and Nürnberg (2016, 2019).

It is possible to generalize the approximation (162) to the case when surfactants are present. Then the surface tension γ_0 depends on the local concentration of surface active agents on the moving interface. The cases of insoluble and soluble surfactants have been considered by the authors in Barrett et al. (2015b) and Barrett et al. (2015c), respectively. Other approaches to two phase flow with surfactants are discussed in James and Lowengrub (2004), Groß and Reusken (2011), Ganesan and Tobiska (2012) and Aland et al. (2017) and the references therein.

9 Willmore flow

9.1 Derivation of the flow

Willmore flow is the L^2-gradient flow of the Willmore energy

$$E(\Gamma) = \frac{1}{2}\int_\Gamma \varkappa^2 \, \mathrm{d}\mathcal{H}^{d-1},$$

for a sufficiently smooth hypersurface Γ in $\mathbb{R}^d$, $d \geq 2$. We remark that in the case $d = 2$ this evolution law is often called elastic flow. In order to derive Willmore flow, we need the first variation of $E(\Gamma)$, which is given in the following lemma. Here we make use of the notations and conventions introduced in Section 2.4.

Lemma 128. *Let $\mathcal{G}_T$ be a closed C^4-evolving orientable hypersurface. Then it holds that*

$$\frac{\mathrm{d}}{\mathrm{d}t} E(\Gamma(t)) = \left\langle \Delta_s \varkappa + \varkappa \, |\nabla_s \vec{\nu}|^2 - \frac{1}{2}\varkappa^3, \mathcal{V} \right\rangle_{\Gamma(t)}.$$

Proof. Using Theorem 32, Lemma 39(ii) and Remark 22(i), we compute

$$\begin{aligned}\frac{\mathrm{d}}{\mathrm{d}t} E(\Gamma(t)) &= \left\langle \varkappa, \partial_t^\square \varkappa - \frac{1}{2}\varkappa^2 \mathcal{V} \right\rangle_{\Gamma(t)} = \left\langle \varkappa, \Delta_s \mathcal{V} + \mathcal{V}\, |\nabla_s \vec{\nu}|^2 - \frac{1}{2}\varkappa^2 \mathcal{V} \right\rangle_{\Gamma(t)} \\ &= \left\langle \Delta_s \varkappa + \varkappa \, |\nabla_s \vec{\nu}|^2 - \frac{1}{2}\varkappa^3, \mathcal{V} \right\rangle_{\Gamma(t)},\end{aligned} \tag{164}$$

where we have used the fact that $\Gamma(t)$ has no boundary. □

Hence we obtain that an evolving hypersurface $(\Gamma(t))_{t\in[0,T]}$, with

$$\mathcal{V} = -\Delta_s \varkappa - \varkappa\,|\nabla_s \vec{\nu}|^2 + \frac{1}{2}\varkappa^3 \quad \text{on}\, \Gamma(t), \tag{165}$$

most efficiently decreases the Willmore energy, and this evolution law is called Willmore flow. Therefore, (165) is the L^2-gradient flow of the Willmore energy.

9.2 A finite element approximation of Willmore flow

We begin with finite element approximations of Willmore flow from Barrett et al. (2007a, 2008d) or in the spirit of those papers. They are based on the following formulation of Willmore flow

$$\vec{\mathcal{V}}\cdot\vec{\nu} = -\Delta_s \varkappa - \varkappa\,|\nabla_s \vec{\nu}|^2 + \frac{1}{2}\varkappa^3, \qquad \varkappa\,\vec{\nu} = \Delta_s\, \vec{\mathrm{id}} \quad \text{on}\, \Gamma(t). \tag{166}$$

Comparing (75) with (166), we note that once a suitable approximation of $|\nabla_s \vec{\nu}|^2$ is given, then it is a simple matter to extend the techniques in Section 5 in order to derive finite element approximations for Willmore flow. For $d=2$, $|\nabla_s \vec{\nu}|^2$ collapses to $\varkappa^2$, recall Lemma 12(iv), and so we can consider the scheme in Barrett et al. (2007a, Section 2.3). In general, we rely on one of the approximations in (35) to obtain a discrete approximation of the Weingarten map $\nabla_s \vec{\nu}$. Hence, we introduce the following finite element approximations for this formulation of Willmore flow. Let the closed polyhedral hypersurface Γ^0 be an approximation to $\Gamma(0)$, and let $\kappa^0_{\Gamma^0} \in V(\Gamma^0)$ be an approximation to its mean curvature. We also recall the time interval partitioning (40). Then, for $m=0,\ldots,M-1$, first find $\underline{\underline{W}}^{m+1} \in \underline{\underline{V}}(\Gamma^m)$, or $\underline{\underline{W}}^{m+1} \in \underline{\underline{V}}_{\mathrm{c}}(\Gamma^m)$, as an approximation of the Weingarten map on Γ^m, and then find $(\vec{X}^{m+1}, \kappa^{m+1}) \in \underline{V}(\Gamma^m) \times V(\Gamma^m)$ such that

$$\begin{aligned}
&\left\langle \frac{\vec{X}^{m+1} - \vec{\mathrm{id}}}{\Delta t_m}, \chi\,\vec{\nu}^m \right\rangle^h_{\Gamma^m} - \left\langle \nabla_s \kappa^{m+1}, \nabla_s \chi \right\rangle_{\Gamma^m} - \frac{1}{2}\left\langle (\kappa^m_{\Gamma^m})^2\,\kappa^{m+1}, \chi \right\rangle^h_{\Gamma^m} \\
&\qquad = -\left\langle \kappa^m_{\Gamma^m}\,|\underline{\underline{W}}^{m+1}|^2, \chi \right\rangle^h_{\Gamma^m} \quad \forall \chi \in V(\Gamma^m),
\end{aligned} \tag{167a}$$

$$\left\langle \kappa^{m+1}\,\vec{\nu}^m, \vec{\eta} \right\rangle^h_{\Gamma^m} + \left\langle \nabla_s \vec{X}^{m+1}, \nabla_s \vec{\eta} \right\rangle_{\Gamma^m} = 0 \quad \forall \vec{\eta} \in \underline{V}(\Gamma^m) \tag{167b}$$

and set $\Gamma^{m+1} = \vec{X}^{m+1}(\Gamma^m)$ and $\kappa^{m+1}_{\Gamma^{m+1}} = \kappa^{m+1} \circ (\vec{X}^{m+1})^{-1} \in V(\Gamma^{m+1})$. For the definition of $\underline{\underline{W}}^{m+1}$ we may, for example, choose one of the formulations (35), based on Γ^m and possibly $\kappa^m_{\Gamma^m}$ or $\vec{\kappa}^m = \vec{\pi}_{\Gamma^m}[\kappa^m_{\Gamma^m}\,\vec{\omega}^m]$.

Theorem 129. *Let* Γ^m *satisfy Assumption 64(i), let* $\kappa^m_{\Gamma^m} \in V(\Gamma^m)$ *and* $\underline{\underline{W}}^{m+1} \in \underline{\underline{V}}(\Gamma^m)$, *or* $\underline{\underline{W}}^{m+1} \in \underline{\underline{V}}_{\mathrm{c}}(\Gamma^m)$, *be given. Then there exists a unique solution* $(\vec{X}^{m+1}, \kappa^{m+1}) \in \underline{V}(\Gamma^m) \times V(\Gamma^m)$ *to (167).*

Proof. The desired result follows similarly to the proof of Theorem 86. □

Remark 130.

(i) Similarly to Section 5.3, a semidiscrete variant of the scheme (167) can also be considered, and it will satisfy Theorem 88(iii).

(ii) Similarly to Section 5.5, one can consider a variant of the scheme (167) with reduced or induced tangential motion.

(iii) The discrete linear systems arising at each time level of (167) are very similar to the ones induced by (76). They can be solved, for example, with the help of a sparse factorization package such as UMFPACK, see Davis (2004).

9.3 A stable approximation of Willmore flow

Unfortunately, it does not seem possible to prove a stability result for the fully discrete approximation (167) of (166). However, the important paper Dziuk (2008) introduced a stable semidiscrete finite element approximation of Willmore flow. The discretization is based on an alternative formulation of the first variation of the Willmore energy, which leads to a weak formulation of Willmore flow. In order to derive the weak formulation, we prove the following result, which is inspired by Dziuk (2008, Lemma 3).

Lemma 131. *Let $\mathcal{G}_T$ be a closed C^3-evolving orientable hypersurface. Let $\vec{\mathcal{V}}$ be the velocity field induced by a global parameterization of $\mathcal{G}_T$. Then it holds that*

$$\frac{\mathrm{d}}{\mathrm{d}t}E(\Gamma(t)) = -\left\langle \frac{1}{2}|\vec{\varkappa}|^2 + \nabla_s\cdot\vec{\varkappa}, \nabla_s\cdot\vec{\mathcal{V}}\right\rangle_{\Gamma(t)} + \left\langle \nabla_s\vec{\varkappa}, 2\underline{\underline{D}}_s(\vec{\mathcal{V}}) - \nabla_s\vec{\mathcal{V}}\right\rangle_{\Gamma(t)},$$

where $\underline{\underline{D}}_s(\vec{\mathcal{V}})$ is the rate of deformation tensor, recall Definition 25(iv).

Proof. It follows from Definition 11(iii) and Theorem 32 that

$$\frac{\mathrm{d}}{\mathrm{d}t}E(\Gamma(t)) = \frac{1}{2}\frac{\mathrm{d}}{\mathrm{d}t}\left\langle |\vec{\varkappa}|^2, 1\right\rangle_{\Gamma(t)} = \langle \vec{\varkappa}, \partial_t^\circ\,\vec{\varkappa}\rangle_{\Gamma(t)} + \frac{1}{2}\left\langle |\vec{\varkappa}|^2, \nabla_s\cdot\vec{\mathcal{V}}\right\rangle_{\Gamma(t)}. \tag{168}$$

We now slightly modify the arguments in the proof of Dziuk (2008, Lemma 3), in order to compute the term $\langle \vec{\varkappa}, \partial_t^\circ\,\vec{\varkappa}\rangle_{\Gamma(t)}$.

Let $\vec{x} : \Upsilon \times [0,T] \to \mathbb{R}^d$ be a global parameterization of $\mathcal{G}_T$ as described in Definition 25(i). For a given $\vec{\eta}_0 \in [C^1(\Upsilon)]^d$, we define $\vec{\eta} : \mathcal{G}_T \to \mathbb{R}^d$ via

$$\vec{\eta}(\vec{x}(\vec{z},t),t) = \vec{\eta}_0(\vec{z}) \quad \text{for all} \quad \forall\,(\vec{z},t) \in \Upsilon \times [0,T], \tag{169}$$

i.e., the values of $\vec{\eta}$ are transported with the map $\vec{x}$. Hence it follows from Definition 28(i) that

$$\partial_t^\circ\,\vec{\eta} = \vec{0} \quad \text{on } \mathcal{G}_T. \tag{170}$$

On recalling Remark 22(iv), Definition 11(iii) and a density argument, we have that

$$\langle \vec{\varkappa}, \vec{\eta} \rangle_{\Gamma(t)} + \left\langle \nabla_s \, \vec{\mathrm{id}}, \, \nabla_s \vec{\eta} \right\rangle_{\Gamma(t)} = 0 \quad \forall \vec{\eta} \in [H^1(\Gamma(t))]^d, \tag{171}$$

since we assume $\Gamma(t)$ to be closed. Differentiating (171) with respect to t, we obtain from Theorem 32, on noting (170), that

$$\langle \partial_t^\circ \, \vec{\varkappa}, \vec{\eta} \rangle_{\Gamma(t)} + \left\langle \vec{\varkappa} \cdot \vec{\eta}, \nabla_s \cdot \vec{\mathcal{V}} \right\rangle_{\Gamma(t)} + \frac{\mathrm{d}}{\mathrm{d}t} \left\langle \nabla_s \, \vec{\mathrm{id}}, \nabla_s \vec{\eta} \right\rangle_{\Gamma(t)} = 0. \tag{172}$$

Using Lemma 9(ii), Theorem 32, Lemma 38(iii) and (170), we now compute

$$\begin{aligned} \frac{\mathrm{d}}{\mathrm{d}t} \left\langle \nabla_s \, \vec{\mathrm{id}}, \nabla_s \vec{\eta} \right\rangle_{\Gamma(t)} &= \frac{\mathrm{d}}{\mathrm{d}t} \langle \nabla_s \cdot \vec{\eta}, 1 \rangle_{\Gamma(t)} \\ &= \left\langle \nabla_s \cdot \vec{\eta}, \nabla_s \cdot \vec{\mathcal{V}} \right\rangle_{\Gamma(t)} + \left\langle \nabla_s \vec{\eta}, \nabla_s \vec{\mathcal{V}} - 2 \, \underline{\underline{D}}_s(\vec{\mathcal{V}}) \right\rangle_{\Gamma(t)}. \end{aligned} \tag{173}$$

On combining (172) and (173), and then choosing $\vec{\eta} = \vec{\varkappa}$, we obtain

$$\langle \partial_t^\circ \, \vec{\varkappa}, \vec{\varkappa} \rangle_{\Gamma(t)} = -\left\langle |\vec{\varkappa}|^2 + \nabla_s \cdot \vec{\varkappa}, \nabla_s \cdot \vec{\mathcal{V}} \right\rangle_{\Gamma(t)} - \left\langle \nabla_s \vec{\varkappa}, \nabla_s \vec{\mathcal{V}} - 2 \, \underline{\underline{D}}_s(\vec{\mathcal{V}}) \right\rangle_{\Gamma(t)}.$$

Together with (168) this yields the desired result. □

A weak formulation of Willmore flow based on Lemma 131 is then the following. Given a closed hypersurface $\Gamma(0)$, we seek an evolving hypersurface $(\Gamma(t))_{t \in [0,T]}$, with a global parameterization and induced velocity field $\vec{\mathcal{V}}$, and $\vec{\varkappa} \in [L^2(\mathcal{G}_T)]^d$ as follows. For almost all $t \in (0,T)$, find $(\vec{\mathcal{V}}(\cdot,t), \vec{\varkappa}(\cdot,t)) \in [L^2(\Gamma(t))]^d \times [H^1(\Gamma(t))]^d$ such that

$$\begin{aligned} \left\langle \vec{\mathcal{V}}, \vec{\chi} \right\rangle_{\Gamma(t)} &= \langle \nabla_s \vec{\varkappa}, \nabla_s \vec{\chi} - 2 \, \underline{\underline{D}}_s(\vec{\chi}) \rangle_{\Gamma(t)} + \langle \nabla_s \cdot \vec{\varkappa}, \nabla_s \cdot \vec{\chi} \rangle_{\Gamma(t)} \\ &\quad + \frac{1}{2} \left\langle |\vec{\varkappa}|^2, \nabla_s \cdot \vec{\chi} \right\rangle_{\Gamma(t)} \quad \forall \vec{\chi} \in [H^1(\Gamma(t))]^d, \end{aligned} \tag{174a}$$

$$\langle \vec{\varkappa}, \vec{\eta} \rangle_{\Gamma(t)} + \left\langle \nabla_s \, \vec{\mathrm{id}}, \nabla_s \vec{\eta} \right\rangle_{\Gamma(t)} = 0 \quad \forall \vec{\eta} \in [H^1(\Gamma(t))]^d, \tag{174b}$$

where we have noted a density argument.

Remark 132 (Comparison to Dziuk (2008)).
We note that our notation is such that $\nabla_s \vec{\chi} = (\nabla_\Gamma \vec{\chi})^\top$, with $\nabla_\Gamma \vec{\chi}$ defined as in Dziuk (2008, (2.4)). In addition, our $\underline{\underline{D}}_s(\vec{\chi}) = \frac{1}{2} \underline{\underline{P}}_\Gamma \, \underline{\underline{D}}_\Gamma(\vec{\chi}) \, \underline{\underline{P}}_\Gamma$, where we recall Definition 5(vii) and $\underline{\underline{D}}_\Gamma(\vec{\chi}) = \nabla_\Gamma \vec{\chi} + (\nabla_\Gamma \vec{\chi})^\top$ as defined in Dziuk (2008, (3.14)). Hence, it is easily deduced from Remark 6(v) and Lemma 9(i) that $2 \, \nabla_s \vec{f} : \underline{\underline{D}}_s(\vec{\chi}) = \nabla_\Gamma \vec{f} : \underline{\underline{D}}_\Gamma(\vec{\chi}) \, \nabla_\Gamma \, \vec{\mathrm{id}}$ for all $\vec{f}, \vec{\chi} \in [H^1(\Gamma(t)]^d$, which implies that (174) agrees with Dziuk (2008, Problem 2).

We now relate the weak formulation (174) to the strong formulation (166).

Lemma 133. *A sufficiently smooth solution of (174) is a solution of the strong formulation (166) of Willmore flow.*

Proof. First of all, (174b) and Remark 22(iv) imply that $\vec{\varkappa} = \varkappa\,\vec{\nu}$ and the second equation in (166) holds.

Next, we have, on noting Definition 5(vii), Remark 6(v), Lemmas 7(iii) and 12(ii), that

$$\nabla_s \vec{\varkappa} : (\nabla_s \vec{\chi} - 2\,\underline{\underline{D}}_s(\vec{\chi})) = (\vec{\nu} \otimes \nabla_s \varkappa) : \nabla_s \vec{\chi} - \varkappa\,\nabla_s \vec{\nu} : \nabla_s \vec{\chi}.$$

Hence, it follows from Remarks 22(i), 22(iii) and Lemmas 7(ii), 7(iii) that the first term on the right-hand side of (174a) can be rewritten as

$$\langle \nabla_s \vec{\varkappa}, \nabla_s \vec{\chi} - 2\,\underline{\underline{D}}_s(\vec{\chi})\rangle_{\Gamma(t)} = \langle \varkappa\,\Delta_s \vec{\nu} - (\Delta_s \varkappa)\,\vec{\nu}, \vec{\chi}\rangle_{\Gamma(t)}. \tag{175}$$

In addition, Definition 5(ii) and Lemma 13(i) yield that $\nabla_s \cdot \vec{\varkappa} = \nabla_s \cdot (\varkappa\,\vec{\nu}) = \varkappa\,\nabla_s \cdot \vec{\nu} = -\varkappa^2$. Hence, we obtain from the second and third terms on the right-hand side of (174a), on noting Lemma 7(i) and Theorem 21, that

$$\begin{aligned}\left\langle \nabla_s \cdot \vec{\varkappa} + \frac{1}{2}\,|\vec{\varkappa}|^2, \nabla_s \cdot \vec{\chi} \right\rangle_{\Gamma(t)} &= -\frac{1}{2}\langle \varkappa^2, \nabla_s \cdot \vec{\chi}\rangle_{\Gamma(t)} \\ &= \langle \varkappa\,\nabla_s \varkappa, \vec{\chi}\rangle_{\Gamma(t)} + \frac{1}{2}\langle \varkappa^3, \vec{\chi} \cdot \vec{\nu}\rangle_{\Gamma(t)}.\end{aligned} \tag{176}$$

Combining (174a), (175) and (176) yields that

$$\vec{\mathcal{V}} = \left(-\Delta_s \varkappa + \frac{1}{2}\,\varkappa^3\right)\vec{\nu} + \varkappa\,(\Delta_s \vec{\nu} + \nabla_s \varkappa) = \left(-\Delta_s \varkappa - \varkappa\,|\nabla_s \vec{\nu}|^2 + \frac{1}{2}\,\varkappa^3\right)\vec{\nu}, \tag{177}$$

where we have recalled Lemma 16. Therefore, we obtain the desired first result in (166). □

We now introduce a natural semidiscrete variant of (174). Given the closed polyhedral hypersurface $\Gamma^h(0)$, find an evolving polyhedral hypersurface $\mathcal{G}_T^h$ with induced velocity $\vec{\mathcal{V}}^h \in \underline{V}(\mathcal{G}_T^h)$, and $\vec{\kappa}^h \in \underline{V}(\mathcal{G}_T^h)$, i.e., $(\vec{\mathcal{V}}^h(\cdot, t), \vec{\kappa}^h(\cdot, t)) \in \underline{V}(\Gamma^h(t)) \times \underline{V}(\Gamma^h(t))$ for all $t \in [0, T]$, such that, for all $t \in (0, T]$,

$$\begin{aligned}\left\langle \vec{\mathcal{V}}^h, \vec{\chi} \right\rangle_{\Gamma^h(t)}^h &= \langle \nabla_s \vec{\kappa}^h, \nabla_s \vec{\chi} - 2\,\underline{\underline{D}}_s(\vec{\chi})\rangle_{\Gamma^h(t)} + \langle \nabla_s \cdot \vec{\kappa}^h, \nabla_s \cdot \vec{\chi}\rangle_{\Gamma^h(t)} \\ &\quad + \frac{1}{2}\left\langle |\vec{\kappa}^h|^2, \nabla_s \cdot \vec{\chi} \right\rangle_{\Gamma^h(t)}^h \quad \forall\,\vec{\chi} \in \underline{V}(\Gamma^h(t)),\end{aligned} \tag{178a}$$

$$\langle \vec{\kappa}^h, \vec{\eta}\rangle_{\Gamma^h(t)}^h + \left\langle \nabla_s \,\mathrm{i\vec{d}}, \nabla_s \vec{\eta} \right\rangle_{\Gamma^h(t)} = 0 \quad \forall\,\vec{\eta} \in \underline{V}(\Gamma^h(t)), \tag{178b}$$

where we recall the definition of $\underline{\underline{D}}_s$ on $\Gamma^h(t)$ from Lemma 73.

We now prove the following stability theorem.

Theorem 134. *Let $(\mathcal{G}_T^h, \vec{\kappa}^h)$ be a solution of (178), and let $\vec{\kappa}^h \in \underline{V}_T(\mathcal{G}_T^h)$. Then it holds that*

$$\frac{\mathrm{d}}{\mathrm{d}t}\frac{1}{2}\left(|\vec{\kappa}^h|_{\Gamma^h(t)}^h\right)^2 = -\left(|\vec{\mathcal{V}}^h|_{\Gamma^h(t)}^h\right)^2 \leq 0.$$

Proof. We argue as in the proof of Lemma 131. Similarly to (170), we extend test functions $\vec{\eta} \in \underline{V}(\Gamma^h(t))$ in (178b) to $\vec{\eta} \in \underline{V}_T(\mathcal{G}_T^h)$ such that $\partial_t^{\circ,h}\vec{\eta} = \vec{0}$ on $\mathcal{G}_T^h$. Then taking the time derivative of (178b), on noting Theorem 70 and Lemma 73(iii), yields

$$\begin{aligned}&\left\langle \partial_t^{\circ,h}\vec{\kappa}^h, \vec{\eta}\right\rangle_{\Gamma^h(t)}^h + \left\langle \vec{\kappa}^h\cdot\vec{\eta}, \nabla_s\cdot\vec{\mathcal{V}}^h\right\rangle_{\Gamma^h(t)}^h + \left\langle \nabla_s\cdot\vec{\mathcal{V}}^h, \nabla_s\cdot\vec{\eta}\right\rangle_{\Gamma^h(t)} \\ &\quad + \left\langle \nabla_s\vec{\mathcal{V}}^h, \nabla_s\vec{\eta}\right\rangle_{\Gamma^h(t)} - 2\left\langle \underline{\underline{D}}_s(\vec{\mathcal{V}}^h), \nabla_s\vec{\eta}\right\rangle_{\Gamma^h(t)} = 0.\end{aligned} \tag{179}$$

We now choose $\vec{\eta} = \vec{\kappa}^h$ in (179) and $\vec{\chi} = \vec{\mathcal{V}}^h$ in (178a), to obtain that

$$\left\langle \partial_t^{\circ,h}\vec{\kappa}^h, \vec{\kappa}^h\right\rangle_{\Gamma^h(t)}^h + \frac{1}{2}\left\langle |\vec{\kappa}^h|^2, \nabla_s\cdot\vec{\mathcal{V}}^h\right\rangle_{\Gamma^h(t)}^h + \left\langle \vec{\mathcal{V}}^h, \vec{\mathcal{V}}^h\right\rangle_{\Gamma^h(t)}^h = 0.$$

The desired result then follows on noting Theorem 70(ii) and (16). □

Remark 135.

(i) For the case $d = 2$ an error analysis of (178) can be found in Deckelnick and Dziuk (2009).

(ii) We note the version of (178) proposed in Dziuk (2008, Problem 3) is without mass lumping. Nevertheless, either version of (178), in contrast to (167), does not have good mesh properties. This is due to the fact that the mesh movements are almost exclusively in the normal direction, which in general leads to bad meshes. This is because (178a) approximates (174a), which is a weak formulation of

$$\vec{\mathcal{V}} = \left[-\Delta_s\varkappa - \varkappa\,|\nabla_s\vec{\nu}|^2 + \frac{1}{2}\varkappa^3\right]\vec{\nu} \quad \text{on } \Gamma(t),$$

where we have recalled (177).

We now wish to derive a weak formulation, which leads to semidiscretizations that are both stable and have good mesh properties. A main ingredient is to ensure that the equation

$$\langle \varkappa\,\vec{\nu}, \vec{\eta}\rangle_{\Gamma(t)} + \left\langle \nabla_s\,\vec{\mathrm{id}}, \nabla_s\vec{\eta}\right\rangle_{\Gamma(t)} = 0 \quad \forall\,\vec{\eta} \in [H^1(\Gamma(t))]^d \tag{180}$$

holds. This then leads to good meshes on the discrete level, recall Section 4.6. We hence want to compute the time derivative of $\frac{1}{2}\langle \varkappa, \varkappa\rangle_{\Gamma(t)}$ by taking the constraint (180) into account. To this end, we use the calculus of PDE constrained optimization, see, e.g., Hinze et al. (2009) and Tröltzsch (2010), and define the Lagrangian

$$L(\Gamma(t), \varkappa^\star, \vec{y}) = \frac{1}{2}\langle \varkappa^\star, \varkappa^\star\rangle_{\Gamma(t)} - \langle \varkappa^\star \vec{\nu}, \vec{y}\rangle_{\Gamma(t)} - \left\langle \nabla_s \vec{\mathrm{id}}, \nabla_s \vec{y}\right\rangle_{\Gamma(t)}, \tag{181}$$

where $\vec{y} \in [H^1(\Gamma(t)]^d$ is the Lagrange multiplier associated with the constraint (180) with $\varkappa$ replaced by $\varkappa^\star$. We note at this stage that $\varkappa^\star(t)$ is an independent variable, and not the mean curvature, $\varkappa(t)$, of $\Gamma(t)$. We now need to take variations of $L(\Gamma(t), \varkappa^\star, \vec{y})$ with respect to $\Gamma(t)$, $\varkappa^\star$ and $\vec{y}$.

To this end, for any $\vec{\chi} \in [H^1(\Gamma(t)]^d$ and for any $\varepsilon \in \mathbb{R}$ let

$$\Gamma_\varepsilon(t) = \vec{\Phi}(\Gamma(t), \varepsilon), \quad \text{where } \vec{\Phi}(\cdot, \varepsilon) = \vec{\mathrm{id}}_{|\Gamma(t)} + \varepsilon\vec{\chi}. \tag{182}$$

For $\varepsilon_0 > 0$, we now consider a smooth function f defined on $\bigcup_{\varepsilon\in[-\varepsilon_0,\varepsilon_0]} (\Gamma_\varepsilon(t) \times \{\varepsilon\})$. Then, similarly to Definition 28(i), we define

$$(\partial_\varepsilon^\circ f)(\vec{\Phi}(\vec{z}, \varepsilon), \varepsilon) = \frac{\mathrm{d}}{\mathrm{d}\varepsilon} f(\vec{\Phi}(\vec{z}, \varepsilon), \varepsilon) \ \ \forall\, (\vec{z}, \varepsilon) \in \Gamma(t) \times [-\varepsilon_0, \varepsilon_0], \tag{183}$$

and note that

$$\partial_\varepsilon^\circ f = \left(\frac{\mathrm{d}}{\mathrm{d}\varepsilon} f(\vec{\Phi}(\cdot, \varepsilon), \varepsilon)\right)_{|\varepsilon=0} = \lim_{\varepsilon\to 0} \frac{f(\vec{\Phi}(\cdot, \varepsilon), \varepsilon) - f}{\varepsilon} \quad \text{on } \Gamma(t). \tag{184}$$

In what follows, and similarly to Section 2.4, we often identify $\Gamma_\varepsilon(t) \times \{\varepsilon\}$ with $\Gamma_\varepsilon(t)$, and hence $(\Gamma_\varepsilon(t) \times \{\varepsilon\})_{|\varepsilon=0}$ with $\Gamma(t)$. With the help of a direct analogue of Theorem 32, we then obtain that the first variation of $\langle f, 1\rangle_{\Gamma(t)}$ is given by

$$\left[\frac{\delta}{\delta\Gamma}\langle f, 1\rangle_{\Gamma(t)}\right](\vec{\chi}) = \left(\frac{\mathrm{d}}{\mathrm{d}\varepsilon}\langle f, 1\rangle_{\Gamma_\varepsilon(t)}\right)_{|\varepsilon=0} = \langle \partial_\varepsilon^\circ f + f\, \nabla_s \cdot \vec{\chi}, 1\rangle_{\Gamma(t)}. \tag{185}$$

In addition, if $\vec{\nu}_\varepsilon$ is the unit normal on $\Gamma_\varepsilon(t)$, corresponding to $\vec{\nu}$ on $\Gamma(t)$, then a direct analogue of Lemma 37(i) yields that

$$\left[\frac{\delta}{\delta\Gamma}\vec{\nu}\right](\vec{\chi}) = (\partial_\varepsilon^\circ \vec{\nu}_\varepsilon)_{|\varepsilon=0} = -(\nabla_s \vec{\chi})^{\mathsf{T}}\, \vec{\nu} \quad \text{on } \Gamma(t). \tag{186}$$

Similarly to (169), for any $\eta \in L^\infty(\Gamma(t))$, one can define $\eta_{(\varepsilon)}$ on $\bigcup_{\varepsilon\in[-\varepsilon_0,\varepsilon_0]} (\Gamma_\varepsilon(t) \times \{\varepsilon\})$ via

$$\eta_{(\varepsilon)}(\vec{\Phi}(\vec{z}, \varepsilon), \varepsilon) = \eta(\vec{z}) \ \ \forall\, (\vec{z}, \varepsilon) \in \Gamma(t) \times [-\varepsilon_0, \varepsilon_0], \tag{187}$$

and analogously for $\vec{\eta} \in [L^\infty(\Gamma(t))]^d$. Similarly to (170), it follows from this extension and (184) that

$$\left[\frac{\delta}{\delta\Gamma}\vec{\eta}\right](\vec{\chi}) = \left(\partial_\varepsilon^\circ \vec{\eta}_{(\varepsilon)}\right)_{|\varepsilon=0} = \vec{0} \quad \text{on } \Gamma(t).$$

In particular, we have the following analogue of (173),

$$\begin{aligned}\left[\frac{\delta}{\delta\Gamma}\left\langle\nabla_s\,\vec{\mathrm{id}},\nabla_s\,\vec{\eta}\right\rangle_{\Gamma(t)}\right](\vec{\chi}) &= \left(\frac{\mathrm{d}}{\mathrm{d}\varepsilon}\left\langle\nabla_s\,\vec{\mathrm{id}},\nabla_s\,\vec{\eta}_{(\varepsilon)}\right\rangle_{\Gamma_\varepsilon(t)}\right)_{|\varepsilon=0} \\ &= \langle\nabla_s\cdot\vec{\eta},\nabla_s\cdot\vec{\chi}\rangle_{\Gamma(t)} + \langle\nabla_s\,\vec{\eta},\nabla_s\,\vec{\chi} - 2\,\underline{\underline{D}}_s(\vec{\chi})\rangle_{\Gamma(t)}.\end{aligned} \tag{188}$$

We now consider the variations of $L(\Gamma(t),\varkappa^\star,\vec{y})$ with respect to $\Gamma(t)$, $\varkappa^\star$ and $\vec{y}$. In particular, for all $\vec{\chi}\in[H^1(\Gamma(t))]^d$, $\xi\in L^2(\Gamma(t))$ and $\vec{\eta}\in[H^1(\Gamma(t))]^d$, we let

$$\left[\frac{\delta}{\delta\Gamma}L\right](\vec{\chi}) = \left(\frac{\mathrm{d}}{\mathrm{d}\varepsilon}L(\Gamma_\varepsilon(t),\varkappa^\star_{(\varepsilon)},\vec{y}_{(\varepsilon)})\right)_{|\varepsilon=0}, \tag{189a}$$

$$\left[\frac{\delta}{\delta\varkappa^\star}L\right](\xi) = \left(\frac{\mathrm{d}}{\mathrm{d}\varepsilon}L(\Gamma(t),\varkappa^\star + \varepsilon\,\xi,\vec{y})\right)_{|\varepsilon=0}, \tag{189b}$$

$$\left[\frac{\delta}{\delta\vec{y}}L\right](\vec{\eta}) = \left(\frac{\mathrm{d}}{\mathrm{d}\varepsilon}L(\Gamma(t),\varkappa^\star,\vec{y} + \varepsilon\,\vec{\eta})\right)_{|\varepsilon=0}, \tag{189c}$$

where $\varkappa^\star_{(\varepsilon)}\in L^2(\Gamma_\varepsilon(t))$ and $\vec{y}_{(\varepsilon)}\in[H^1(\Gamma_\varepsilon(t))]^d$ are defined by transporting the values of $\varkappa^\star\in L^2(\Gamma(t))$ and $\vec{y}\in[H^1(\Gamma(t))]^d$ as defined in (187), recall (182) for the $\vec{\chi}$ at hand. Setting the variation $\left[\frac{\delta}{\delta\vec{y}}L\right](\vec{\eta})=0$, yields (180) with $\varkappa$ replaced by $\varkappa^\star$. Hence, comparing this with the original (180), we obtain that $\varkappa^\star=\varkappa$. Setting $\left[\frac{\delta}{\delta\varkappa^\star}L\right](\xi)=0$, yields, on noting $\varkappa^\star=\varkappa$, that

$$\varkappa = \vec{y}\cdot\vec{\nu} \quad \text{on } \Gamma(t). \tag{190}$$

Finally, setting the variation $\left[\frac{\delta}{\delta\Gamma}L\right](\vec{\chi}) = -\left\langle\vec{\mathcal{V}}\cdot\vec{\nu},\vec{\chi}\cdot\vec{\nu}\right\rangle_{\Gamma(t)}$ and noting (185), (186), (188) and that $\varkappa^\star=\varkappa$, we obtain

$$\begin{aligned}\left\langle\vec{\mathcal{V}}\cdot\vec{\nu},\vec{\chi}\cdot\vec{\nu}\right\rangle_{\Gamma(t)} &= \langle\nabla_s\,\vec{y},\nabla_s\,\vec{\chi} - 2\,\underline{\underline{D}}_s(\vec{\chi})\rangle_{\Gamma(t)} + \langle\nabla_s\cdot\vec{y},\nabla_s\cdot\vec{\chi}\rangle_{\Gamma(t)} \\ &\quad - \left\langle\varkappa\left(\frac{1}{2}\varkappa - \vec{y}\cdot\vec{\nu}\right),\nabla_s\cdot\vec{\chi}\right\rangle_{\Gamma(t)} - \left\langle\varkappa\,\vec{y},(\nabla_s\,\vec{\chi})^{\mathsf{T}}\,\vec{\nu}\right\rangle_{\Gamma(t)} \\ &\qquad \forall\,\vec{\chi}\in[H^1(\Gamma(t))]^d.\end{aligned} \tag{191}$$

Therefore we have the following weak formulation. Given a closed hypersurface $\Gamma(0)$, we seek an evolving hypersurface $(\Gamma(t))_{t\in[0,T]}$, with a global parameterization and induced velocity field $\vec{\mathcal{V}}$, $\varkappa\in L^2(\mathcal{G}_T)$ and $\vec{y}\in[L^2(\mathcal{G}_T)]^d$ as follows. For almost all $t\in(0,T)$, find $(\vec{\mathcal{V}}(\cdot,t),\varkappa(\cdot,t),\vec{y}(\cdot,t))\in[L^2(\Gamma(t))]^d\times L^2(\Gamma(t))\times[H^1(\Gamma(t))]^d$ such that (191), (190) and (180) hold.

Remark 136.

(i) Using the techniques in Barrett et al. (2017c, Appendix A) one can show, similarly to Lemma 133, that a sufficiently smooth solution of this weak formulation is a solution of the strong formulation (166).

(ii) Clearly, using (190) one can eliminate $\varkappa$ from (191) and (180) in this weak formulation.

We now consider a discrete analogue of (191), (190) and (180), by first introducing the discrete analogue of (181)

$$L^h(\Gamma^h(t), \kappa^h, \vec{Y}^h) = \frac{1}{2}\langle \kappa^h, \kappa^h \rangle^h_{\Gamma^h(t)} - \langle \kappa^h \vec{\nu}^h, \vec{Y}^h \rangle^h_{\Gamma^h(t)} - \left\langle \nabla_s \vec{\mathrm{id}}, \nabla_s \vec{Y}^h \right\rangle_{\Gamma^h(t)}, \quad (192)$$

where $\vec{Y}^h(\cdot, t) \in \underline{V}(\Gamma^h(t))$ is the Lagrange multiplier associated with the constraint $\kappa^h(\cdot, t) \in V(\Gamma^h(t))$ satisfying

$$\langle \kappa^h \vec{\nu}^h, \vec{\eta} \rangle^h_{\Gamma^h(t)} + \left\langle \nabla_s \vec{\mathrm{id}}, \nabla_s \vec{\eta} \right\rangle_{\Gamma^h(t)} = 0 \quad \forall \vec{\eta} \in \underline{V}(\Gamma^h(t)). \quad (193)$$

Setting $\left[\frac{\delta}{\delta \vec{Y}^h} L^h\right](\vec{\eta}) = 0$ yields (193). Setting $\left[\frac{\delta}{\delta \kappa^h} L^h\right](\xi) = 0$ yields that

$$\langle \kappa^h - \vec{Y}^h \cdot \vec{\nu}^h, \xi \rangle^h_{\Gamma^h(t)} = 0 \quad \forall \xi \in V(\Gamma^h(t)). \quad (194)$$

Finally, we need to take the variation of $L^h(\Gamma^h(t), \kappa^h, \vec{Y}^h)$ with respect to $\Gamma^h(t)$. To this end, we have the following discrete analogue of (182). For any $\vec{\chi} \in \underline{V}(\Gamma^h(t))$ and for any $\varepsilon \in \mathbb{R}$, let

$$\Gamma^h_\varepsilon(t) = \vec{\Phi}^h(\Gamma^h(t), \varepsilon), \quad \text{where } \vec{\Phi}^h(\cdot, \varepsilon) = \vec{\mathrm{id}}_{|_{\Gamma^h(t)}} + \varepsilon \vec{\chi}.$$

We also define $\partial_\varepsilon^{\circ,h}$ to be the discrete analogue of (183). Similarly to Theorem 70(ii), we then have that

$$\begin{aligned}\left[\frac{\delta}{\delta \Gamma^h} \langle \kappa^h, \vec{Y}^h \cdot \vec{\nu}^h \rangle^h_{\Gamma^h(t)}\right](\vec{\chi}) &= \left(\frac{\mathrm{d}}{\mathrm{d}\varepsilon} \left\langle \kappa^h_{(\varepsilon)}, \vec{Y}^h_{(\varepsilon)} \cdot \vec{\nu}^h_\varepsilon \right\rangle^h_{\Gamma^h_\varepsilon(t)}\right)_{|_{\varepsilon=0}} \\ &= \left(\left\langle \kappa^h_{(\varepsilon)}, \vec{Y}^h_{(\varepsilon)} \cdot \partial_\varepsilon^{\circ,h} \vec{\nu}^h_\varepsilon \right\rangle^h_{\Gamma^h_\varepsilon(t)}\right)_{|_{\varepsilon=0}} + \langle \kappa^h \vec{Y}^h \cdot \vec{\nu}^h, \nabla_s \cdot \vec{\chi} \rangle^h_{\Gamma^h(t)} \\ &= \left\langle \kappa^h, \vec{Y}^h \cdot \left[\frac{\delta}{\delta \Gamma^h} \vec{\nu}^h\right](\vec{\chi}) \right\rangle^h_{\Gamma^h(t)} + \langle \kappa^h \vec{Y}^h \cdot \vec{\nu}^h, \nabla_s \cdot \vec{\chi} \rangle^h_{\Gamma^h(t)},\end{aligned} \quad (195)$$

where $\kappa^h_{(\varepsilon)}$ and $\vec{Y}^h_{(\varepsilon)}$ are defined similarly to (187), with the help of the map $\vec{\Phi}^h$, and $\vec{\nu}^h_\varepsilon$ is the unit normal on $\Gamma^h_\varepsilon(t)$. Hence, setting the variation $\left[\frac{\delta}{\delta \Gamma^h} L^h\right](\vec{\chi}) = -\left\langle \vec{\mathcal{V}}^h \cdot \vec{\omega}^h, \vec{\chi} \cdot \vec{\omega}^h \right\rangle^h_{\Gamma(t)}$, on noting (195), the discrete analogue of (186) for $\vec{\nu}^h$ on $\Gamma^h(t)$, compare Lemma 72, and (188) for $\Gamma^h(t)$, we obtain

$$\left\langle \vec{\mathcal{V}}^h \cdot \vec{\omega}^h, \vec{\chi} \cdot \vec{\omega}^h \right\rangle^h_{\Gamma^h(t)} = \langle \nabla_s \vec{Y}^h, \nabla_s \vec{\chi} - 2\,\underline{\underline{D}}_s(\vec{\chi}) \rangle_{\Gamma^h(t)} + \langle \nabla_s \cdot \vec{Y}^h, \nabla_s \cdot \vec{\chi} \rangle_{\Gamma^h(t)}$$
$$- \left\langle \kappa^h \left(\frac{1}{2}\kappa^h - \vec{Y}^h \cdot \vec{\nu}^h\right), \nabla_s \cdot \vec{\chi} \right\rangle^h_{\Gamma^h(t)} - \left\langle \kappa^h \vec{Y}^h, (\nabla_s \vec{\chi})^{\mathsf{T}}\, \vec{\nu}^h \right\rangle^h_{\Gamma^h(t)}$$
$$\forall \vec{\chi} \in \underline{V}(\Gamma^h(t)). \tag{196}$$

Therefore we have the following semidiscrete finite element approximation of (191), (190) and (180). Given the closed polyhedral hypersurface $\Gamma^h(0)$, find an evolving polyhedral hypersurface $\mathcal{G}^h_T$ with induced velocity $\vec{\mathcal{V}}^h \in \underline{V}(\mathcal{G}^h_T)$, and $\kappa^h \in V(\mathcal{G}^h_T)$, $\vec{Y}^h \in \underline{V}(\mathcal{G}^h_T)$, i.e., $(\vec{\mathcal{V}}^h(\cdot,t), \kappa^h(\cdot,t), \vec{Y}^h(\cdot,t)) \in \underline{V}(\Gamma^h(t)) \times V(\Gamma^h(t)) \times \underline{V}(\Gamma^h(t))$ for all $t \in [0,T]$, such that, for all $t \in (0,T]$, (196), (194) and (193) hold.

Remark 137.

(i) Similarly to Remark 136(ii), using (194), one can eliminate κ^h from (196) and (193) in this semidiscrete finite element approximation.
(ii) The scheme (196), (194) and (193) satisfies Theorem 88(iii).

We have the following stability result.

Theorem 138. *Let $(\mathcal{G}^h_T, \kappa^h, \vec{Y}^h)$ be a solution of (196), (194) and (193), and let $\kappa^h \in V_T(\mathcal{G}^h_T)$. Then it holds that*

$$\frac{\mathrm{d}}{\mathrm{d}t}\frac{1}{2}\left(|\kappa^h|^h_{\Gamma^h(t)}\right)^2 = -\left(\left|\vec{\mathcal{V}}^h \cdot \vec{\omega}^h\right|^h_{\Gamma^h(t)}\right)^2 \le 0.$$

Proof. Similarly to (179), we extend test functions $\vec{\eta} \in \underline{V}(\Gamma^h(t))$ in (193) to $\vec{\eta} \in \underline{V}_T(\mathcal{G}^h_T)$ such that $\partial_t^{\circ,h}\,\vec{\eta} = \vec{0}$ on $\mathcal{G}^h_T$. Then taking the time derivative of (193), on noting Theorem 70 and Lemma 73(iii), yields

$$\left\langle \partial_t^{\circ,h}\,\kappa^h, \vec{\nu}^h \cdot \vec{\eta} \right\rangle^h_{\Gamma^h(t)} + \left\langle \kappa^h\, \partial_t^{\circ,h}\,\vec{\nu}^h, \vec{\eta} \right\rangle^h_{\Gamma^h(t)} + \left\langle \kappa^h\, \vec{\nu}^h \cdot \vec{\eta}, \nabla_s \cdot \vec{\mathcal{V}}^h \right\rangle^h_{\Gamma^h(t)}$$
$$+ \left\langle \nabla_s \cdot \vec{\mathcal{V}}^h, \nabla_s \cdot \vec{\eta} \right\rangle_{\Gamma^h(t)} + \left\langle \nabla_s \vec{\mathcal{V}}^h, \nabla_s \vec{\eta} \right\rangle_{\Gamma^h(t)} - 2\left\langle \underline{\underline{D}}_s(\vec{\mathcal{V}}^h), \nabla_s \vec{\eta} \right\rangle_{\Gamma^h(t)} = 0. \tag{197}$$

We now take $\vec{\eta} = \vec{Y}^h$ in (197) and $\vec{\chi} = \vec{\mathcal{V}}^h$ in (196), to obtain, on noting (194), (19) and Lemma 72, that

$$\left\langle \partial_t^{\circ,h}\,\kappa^h, \kappa^h \right\rangle^h_{\Gamma^h(t)} + \frac{1}{2}\left\langle |\kappa^h|^2, \nabla_s \cdot \vec{\mathcal{V}}^h \right\rangle^h_{\Gamma^h(t)} + \left\langle \vec{\mathcal{V}}^h \cdot \vec{\omega}^h, \vec{\mathcal{V}}^h \cdot \vec{\omega}^h \right\rangle^h_{\Gamma^h(t)} = 0.$$

The desired result then follows on noting Theorem 70(ii) and (16). □

We now state a fully discrete version of (196), (194) and (193), on recalling (19).

Let the closed polyhedral hypersurface Γ^0 be an approximation to $\Gamma(0)$, and let $\kappa^0_{\Gamma^0} \in V(\Gamma^0)$ and $\vec{Y}^0_{\Gamma^0} \in \underline{V}(\Gamma^0)$ be approximations to its mean curvature and mean curvature vector, respectively. We also recall the time interval partitioning (40). Then, for $m = 0, \ldots, M-1$, find $(\vec{X}^{m+1}, \kappa^{m+1}, \vec{Y}^{m+1}) \in \underline{V}(\Gamma^m) \times V(\Gamma^m) \times \underline{V}(\Gamma^m)$ such that

$$\left\langle \frac{\vec{X}^{m+1} - \vec{\mathrm{id}}}{\Delta t_m} \cdot \vec{\omega}^m, \vec{\chi} \cdot \vec{\omega}^m \right\rangle^h_{\Gamma^m} - \langle \nabla_s \vec{Y}^{m+1}, \nabla_s \vec{\chi} \rangle_{\Gamma^m}$$
$$= -\left\langle \kappa^m_{\Gamma^m} \left(\frac{1}{2} \kappa^m_{\Gamma^m} - \vec{Y}^m_{\Gamma^m} \cdot \vec{\nu}^m \right), \nabla_s \cdot \vec{\chi} \right\rangle^h_{\Gamma^m} - \left\langle \kappa^m_{\Gamma^m} \vec{Y}^m_{\Gamma^m}, (\nabla_s \vec{\chi})^{\mathsf{T}} \vec{\nu}^m \right\rangle^h_{\Gamma^m} \tag{198a}$$
$$+ \langle \nabla_s \cdot \vec{Y}^m_{\Gamma^m}, \nabla_s \cdot \vec{\chi} \rangle_{\Gamma^m} - 2 \langle \nabla_s \vec{Y}^m_{\Gamma^m}, \underline{\underline{D}}_s(\vec{\chi}) \rangle_{\Gamma^m} \quad \forall \vec{\chi} \in \underline{V}(\Gamma^m),$$

$$\kappa^{m+1} = \pi_{\Gamma^m} \left[\vec{Y}^{m+1} \cdot \vec{\omega}^m \right], \tag{198b}$$

$$\langle \kappa^{m+1} \vec{\nu}^m, \vec{\eta} \rangle^h_{\Gamma^m} + \langle \nabla_s \vec{X}^{m+1}, \nabla_s \vec{\eta} \rangle_{\Gamma^m} = 0 \quad \forall \vec{\eta} \in \underline{V}(\Gamma^m) \tag{198c}$$

and set $\Gamma^{m+1} = \vec{X}^{m+1}(\Gamma^m)$, $\kappa^{m+1}_{\Gamma^{m+1}} = \kappa^{m+1} \circ (\vec{X}^{m+1})^{-1} \in V(\Gamma^{m+1})$ and $\vec{Y}^{m+1}_{\Gamma^{m+1}} = \vec{Y}^{m+1} \circ (\vec{X}^{m+1})^{-1} \in \underline{V}(\Gamma^{m+1})$. We have the following result.

Theorem 139. *Let Γ^m satisfy Assumption 64(i) and let $\kappa^m_{\Gamma^m} \in V(\Gamma^m)$ and $\vec{Y}^m_{\Gamma^m} \in \underline{V}(\Gamma^m)$. Then there exists a unique solution $(\vec{X}^{m+1}, \kappa^{m+1}, \vec{Y}^{m+1}) \in \underline{V}(\Gamma^m) \times V(\Gamma^m) \times \underline{V}(\Gamma^m)$ to (198).*

Proof. The proof is very similar to the proof of Theorem 86. We consider a solution $(\vec{X}, \kappa, \vec{Y}) \in \underline{V}(\Gamma^m) \times V(\Gamma^m) \times \underline{V}(\Gamma^m)$ of the homogeneous system

$$\langle \vec{X} \cdot \omega^m, \vec{\chi} \cdot \vec{\omega}^m \rangle^h_{\Gamma^m} - \Delta t_m \langle \nabla_s \vec{Y}, \nabla_s \vec{\chi} \rangle_{\Gamma^m} = 0 \quad \forall \vec{\chi} \in \underline{V}(\Gamma^m), \tag{199a}$$

$$\kappa - \pi_{\Gamma^m} \left[\vec{Y} \cdot \vec{\omega}^m \right] = 0, \tag{199b}$$

$$\langle \kappa \vec{\nu}^m, \vec{\eta} \rangle^h_{\Gamma^m} + \langle \nabla_s \vec{X}, \nabla_s \vec{\eta} \rangle_{\Gamma^m} = 0 \quad \forall \vec{\eta} \in \underline{V}(\Gamma^m). \tag{199c}$$

Choosing $\vec{\chi} = \vec{Y} \in \underline{V}(\Gamma^m)$ in (199a) and $\vec{\eta} = \vec{X} \in \underline{V}(\Gamma^m)$ in (199c) yields, on noting (19) and (199b), that

$$|\nabla_s \vec{X}|^2_{\Gamma^m} + \Delta t_m |\nabla_s \vec{Y}|^2_{\Gamma^m} = 0.$$

We deduce from this that $\vec{X} = \vec{X}^c$ and $\vec{Y} = \vec{Y}^c$ on Γ^m, for $\vec{X}^c$, $\vec{Y}^c \in \mathbb{R}^d$. Hence, on substituting (199b) into (199c), it follows from (199) and (19) that

$$\left(|\vec{X}^c \cdot \vec{\omega}^m|^h_{\Gamma^m} \right)^2 + \left(|\vec{Y}^c \cdot \vec{\omega}^m|^h_{\Gamma^m} \right)^2 = 0,$$

i.e., $\vec{X}^c \cdot \vec{\omega}^m = \vec{Y}^c \cdot \vec{\omega}^m = 0$. Therefore Assumption 64(i) yields that $\vec{X}^c = \vec{Y}^c = \vec{0}$, and hence $\kappa = 0$. □

Remark 140 (Stability).
Unfortunately, it does not seem possible to prove a stability bound for the fully discrete scheme (198) or its generalizations. A similar comment applies to a fully discrete approximation of the Dziuk scheme (178).

Remark 141 (Discrete linear systems).
On recalling the notation from Section 4.3, we can formulate the linear systems of equations to be solved at each time level for (198) as follows. Find $(\vec{Y}^{m+1}, \kappa^{m+1}, \delta\vec{X}^{m+1}) \in (\mathbb{R}^d)^K \times \mathbb{R}^K \times (\mathbb{R}^d)^K$ such that

$$\begin{pmatrix} \underline{\underline{A}}_{\Gamma^m} & 0 & -\frac{1}{\Delta t_m}\underline{\underline{\mathcal{M}}}_{\Gamma^m} \\ -\vec{N}^{\mathsf{T}}_{\Gamma^m} & M_{\Gamma^m} & 0 \\ 0 & \vec{N}_{\Gamma^m} & \underline{\underline{A}}_{\Gamma^m} \end{pmatrix} \begin{pmatrix} \vec{Y}^{m+1} \\ \kappa^{m+1} \\ \delta\vec{X}^{m+1} \end{pmatrix} = \begin{pmatrix} \vec{g}_{\Gamma^m} \\ 0 \\ -\underline{\underline{A}}_{\Gamma^m}\vec{X}^m \end{pmatrix}, \tag{200}$$

where we use a similar abuse of notation as in (46). The definitions of the matrices and vectors in (200) are either given in (45), or they follow directly from (198). In practice, the linear system (200) can be efficiently solved with a sparse direct solution method like UMFPACK, see Davis (2004).

9.4 Willmore flow with spontaneous curvature and area difference elasticity effects

Curvature energies also play an important role for vesicles and biomembranes. As in many membrane elastic energies, the curvature of the membrane enters the elastic energy density. However, for biomembranes, additional effects play a role, which we would like to discuss now. In the original curvature energies for biomembranes a possible asymmetry of the membrane in the normal direction was taken into account, which can result, for example, from a different chemical environment on the two sides of the membrane. This top-down asymmetry makes it necessary to generalize the Willmore energy, and one example of such a model is the spontaneous curvature model introduced by Canham (1970) and Helfrich (1973). The simplest such energy, in a nondimensional form, is

$$E_{\overline{\varkappa}}(\Gamma) = \frac{1}{2}\int_\Gamma (\varkappa - \overline{\varkappa})^2 \,\mathrm{d}\mathcal{H}^{d-1},$$

where $\overline{\varkappa} \in \mathbb{R}$ is the given so-called spontaneous curvature. Biomembranes consist of two layers of lipids, and the number of lipid molecules is conserved. In addition, there are osmotic pressure effects, arising from the chemistry around the lipid. These two effects lead to constraints on the possible membrane configurations. Early models for bilayer membranes modelled this by taking hard constraints on the total surface area and the enclosed volume of the membrane into account. The fact that it is difficult to exchange molecules between the two layers imply that the total number of lipids in each layer is conserved, and hence a surface area difference between the two layers will appear. The area difference is to first order given by the total integrated mean curvature. This follows from the first variation formula for surface area, recall (36). Different models incorporate this area difference either by a hard constraint or penalize deviations from an optimal area difference. In the latter case, we obtain the energy

$$E_{\overline{\varkappa},\beta}(\Gamma) = E_{\overline{\varkappa}}(\Gamma) + \frac{1}{2}\beta\big(\langle \varkappa, 1\rangle_\Gamma - M_0\big)^2 \tag{201}$$

with given constants $\beta \in \mathbb{R}_{\geq 0}$, $M_0 \in \mathbb{R}$. A model based on the energy (201) is called an area difference elasticity (ADE) model, see Seifert (1997). We now extend Lemma 128 to $E_{\overline{\varkappa},\beta}(\Gamma(t))$, where for notational convenience we define

$$A(t) = \langle \varkappa, 1\rangle_{\Gamma(t)} - M_0. \tag{202}$$

Lemma 142. *Let $\mathcal{G}_T$ be a closed C^4-evolving orientable hypersurface. Then it holds that*

$$\frac{\mathrm{d}}{\mathrm{d}t} E_{\overline{\varkappa},\beta}(\Gamma(t)) = \left\langle \Delta_s \varkappa + (\varkappa - \overline{\varkappa})\, |\nabla_s \vec{\nu}|^2 - \frac{1}{2}(\varkappa - \overline{\varkappa})^2 \varkappa + \beta A(t)\left(|\nabla_s \vec{\nu}|^2 - \varkappa^2\right), \mathcal{V} \right\rangle_{\Gamma(t)}.$$

Proof. Generalizing (164) to $E_{\overline{\varkappa}}(\Gamma(t)$ yields that

$$\begin{aligned} \frac{\mathrm{d}}{\mathrm{d}t} E_{\overline{\varkappa}}(\Gamma(t)) &= \left\langle \varkappa - \overline{\varkappa}, \partial_t^\square \varkappa - \frac{1}{2}(\varkappa - \overline{\varkappa})\, \varkappa\, \mathcal{V} \right\rangle_{\Gamma(t)} \\ &= \left\langle \varkappa - \overline{\varkappa}, \Delta_s \mathcal{V} + \mathcal{V}\, |\nabla_s \vec{\nu}|^2 - \frac{1}{2}(\varkappa - \overline{\varkappa})\, \varkappa\, \mathcal{V} \right\rangle_{\Gamma(t)} \\ &= \left\langle \Delta_s \varkappa + (\varkappa - \overline{\varkappa})\, |\nabla_s \vec{\nu}|^2 - \frac{1}{2}(\varkappa - \overline{\varkappa})^2 \varkappa, \mathcal{V} \right\rangle_{\Gamma(t)}. \end{aligned} \tag{203}$$

Next, we compute, on recalling Theorem 32 and Lemma 39(ii), that

$$\begin{aligned} \frac{\mathrm{d}}{\mathrm{d}t}\left[\frac{1}{2}\left(\langle \varkappa, 1\rangle_{\Gamma(t)} - M_0\right)^2\right] &= A(t)\,\frac{\mathrm{d}}{\mathrm{d}t}\langle \varkappa, 1\rangle_{\Gamma(t)} = A(t)\langle \partial_t^\square \varkappa - \varkappa^2 \mathcal{V}, 1\rangle_{\Gamma(t)} \\ &= A(t)\left\langle |\nabla_s \vec{\nu}|^2 - \varkappa^2, \mathcal{V}\right\rangle_{\Gamma(t)}. \end{aligned}$$

Combining the above with (203) yields the claim, on noting (201). □

Hence the L^2-gradient flow of $E_{\overline{\varkappa},\beta}(\Gamma(t))$, (201), is given as

$$\mathcal{V} = -\Delta_s \varkappa - (\varkappa - \overline{\varkappa})\, |\nabla_s \vec{\nu}|^2 + \frac{1}{2}(\varkappa - \overline{\varkappa})^2 \varkappa - \beta A(t)\left(|\nabla_s \vec{\nu}|^2 - \varkappa^2\right) \quad \text{on } \Gamma(t). \tag{204}$$

Taking our discussion above into account, the volume preserving and surface area preserving flows are also of interest. In the case $\beta = 0$, the volume and surface area preserving flow is called Helfrich flow. We consider the two side constraints

$$\langle \varkappa, \mathcal{V}\rangle_{\Gamma(t)} = 0 \quad \text{and} \quad \langle 1, \mathcal{V}\rangle_{\Gamma(t)} = 0 \tag{205}$$

for surface area and volume preservation, where we have recalled Theorems 32 and 33, respectively. In particular, we introduce Lagrange multipliers $(\lambda_A(t), \lambda_V(t))^\top \in \mathbb{R}^2$ and then, on writing (204) as $\mathcal{V} = f$, we adapt (204) to

$$\mathcal{V} = f + \lambda_A\, \varkappa + \lambda_V \quad \text{on}\, \Gamma(t). \tag{206}$$

We see that the constraints (205) are satisfied if $(\lambda_A(t), \lambda_V(t))^\top \in \mathbb{R}^2$ solve the symmetric system

$$\begin{pmatrix} \langle \varkappa, \varkappa\rangle_{\Gamma(t)} & \langle \varkappa, 1\rangle_{\Gamma(t)} \\ \langle \varkappa, 1\rangle_{\Gamma(t)} & \langle 1, 1\rangle_{\Gamma(t)} \end{pmatrix} \begin{pmatrix} \lambda_A \\ \lambda_V \end{pmatrix} = -\begin{pmatrix} \langle f, \varkappa\rangle_{\Gamma(t)} \\ \langle f, 1\rangle_{\Gamma(t)} \end{pmatrix}. \tag{207}$$

We observe that the matrix in (207) is symmetric positive semidefinite, and it is singular if and only if $\varkappa$ is constant, i.e., $\Gamma(t)$ is sphere. In the case of just one constraint, the Lagrange multiplier corresponding to the other constraint is set to zero and the corresponding equation is removed from (207). Equivalently to (207), in the two constraint case and on assuming that $\varkappa$ is not constant, $\lambda_A = -\left\langle f, \varkappa - \fint_{\Gamma(t)} \varkappa \right\rangle_{\Gamma(t)} / |\varkappa - \fint_{\Gamma(t)} \varkappa|^2_{\Gamma(t)}$ and λ_V is obtained from the second equation in (207). Here, we have recalled the definition of $\fint_{\Gamma(t)} \varkappa$ from (87).

The corresponding changes to the finite element approximation (167) is to replace (167a) by

$$\begin{aligned} &\left\langle \frac{\vec{X}^{m+1} - \vec{\mathrm{id}}}{\Delta t_m}, \chi\, \vec{\nu}^m \right\rangle^h_{\Gamma^m} - \langle \nabla_s\, \kappa^{m+1}, \nabla_s\, \chi\rangle_{\Gamma^m} \\ &\qquad\qquad -\frac{1}{2}\left\langle (\kappa^m_{\Gamma^m} - \overline{\varkappa})^2\, \kappa^{m+1}, \chi\right\rangle^h_{\Gamma^m} - [\lambda^m_A]_+ \langle \kappa^{m+1}, \chi\rangle^h_{\Gamma^m} \\ &= \langle g^m, \chi\rangle^h_{\Gamma^m} + [\lambda^m_A]_- \left\langle \kappa^m_{\Gamma^m}, \chi\right\rangle^h_{\Gamma^m} + \lambda^m_V \langle 1, \chi\rangle^h_{\Gamma^m} \quad \forall \chi \in V(\Gamma^m), \end{aligned} \tag{208}$$

where $g^m \in V(\Gamma^m)$ is such that

$$\begin{aligned} \langle g^m, \chi\rangle^h_{\Gamma^m} &= -\left\langle (\kappa^m_{\Gamma^m} - \overline{\varkappa})\, |\underline{\underline{W}}^{m+1}|^2, \chi\right\rangle^h_{\Gamma^m} - \beta\, A^m \left\langle |\underline{\underline{W}}^{m+1}|^2 - (\kappa^m_{\Gamma^m})^2, \chi\right\rangle^h_{\Gamma^m} \\ &\qquad\qquad \forall \chi \in V(\Gamma^m), \\ A^m &= \left\langle \kappa^m_{\Gamma^m}, 1\right\rangle^h_{\Gamma^m} - M_0. \end{aligned}$$

In addition, $[\lambda^m_A]_\pm = \pm \max\{\pm \lambda^m_A, 0\}$ and the Lagrange multipliers $(\lambda^m_A, \lambda^m_V)^\top \in \mathbb{R}^2$ satisfy the symmetric system

$$\begin{pmatrix} \left\langle \kappa^m_{\Gamma^m}, \kappa^m_{\Gamma^m} \right\rangle^h_{\Gamma^m} & \left\langle \kappa^m_{\Gamma^m}, 1 \right\rangle_{\Gamma^m} \\ \left\langle \kappa^m_{\Gamma^m}, 1 \right\rangle_{\Gamma^m} & \langle 1,1 \rangle_{\Gamma^m} \end{pmatrix} \begin{pmatrix} \lambda^m_A \\ \lambda^m_V \end{pmatrix} = - \begin{pmatrix} \left\langle g^m + \frac{1}{2}\left(\kappa^m_{\Gamma^m} - \overline{\varkappa}\right)^2 \kappa^m_{\Gamma^m}, \kappa^m_{\Gamma^m} \right\rangle^h_{\Gamma^m} + \left|\nabla_s \kappa^m_{\Gamma^m}\right|^2_{\Gamma^m} \\ \left\langle g^m + \frac{1}{2}\left(\kappa^m_{\Gamma^m} - \overline{\varkappa}\right)^2 \kappa^m_{\Gamma^m}, 1 \right\rangle^h_{\Gamma^m} \end{pmatrix}. \tag{209}$$

Once again, we observe that the matrix in (209) is symmetric positive semidefinite, and it is singular if and only if $\kappa^m_{\Gamma^m}$ is constant. Similarly, in the case of just one constraint, the Lagrange multiplier corresponding to the other constraint is set to zero and the corresponding equation removed from (209). Moreover, this modified scheme satisfies the suitably modified versions of Theorem 129 and Remark 130; see Barrett et al. (2008d) for details.

We now discuss generalizations of the stable approximations of Willmore flow in Section 9.3 to the case of Willmore flow with spontaneous curvature and area difference effects. In Barrett et al. (2016b), the present authors extended the approach of Dziuk (2008), see (178), to the case of nonzero $\overline{\varkappa}$ and β.

First of all, we consider the variation of $E_{\overline{\varkappa},\beta}(\Gamma(t))$ subject to the side constraint (171). Similarly to (181), we introduce the Lagrangian

$$\widehat{L}_{\overline{\varkappa},\beta}(\Gamma(t),\vec{\varkappa}^\star,\vec{y}) = \frac{1}{2}\left\langle |\vec{\varkappa}^\star - \overline{\varkappa}\,\vec{\nu}|^2, 1\right\rangle_{\Gamma(t)} + \frac{1}{2}\beta\left(\langle \vec{\varkappa}^\star,\vec{\nu}\rangle_{\Gamma(t)} - M_0\right)^2 - \langle \vec{\varkappa}^\star,\vec{y}\rangle_{\Gamma(t)} - \left\langle \nabla_s \,\vec{\mathrm{id}}, \nabla_s \vec{y}\right\rangle_{\Gamma(t)},$$

where $\vec{y}(\cdot,t) \in [H^1(\Gamma(t))]^d$ is the Lagrange multiplier for (171) with $\vec{\varkappa}(\cdot,t) \in [L^2(\Gamma(t))]^d$ replaced by $\vec{\varkappa}^\star(\cdot,t) \in [L^2(\Gamma(t))]^d$. As in (189), we consider the variations, for all $\vec{\chi} \in [H^1(\Gamma(t))]^d$, $\vec{\xi} \in [L^2(\Gamma(t))]^d$ and $\vec{\eta} \in [H^1(\Gamma(t))]^d$,

$$\left[\frac{\delta}{\delta \Gamma}\widehat{L}_{\overline{\varkappa},\beta}\right](\vec{\chi}) = \left(\frac{\mathrm{d}}{\mathrm{d}\varepsilon}\widehat{L}_{\overline{\varkappa},\beta}(\Gamma_\varepsilon(t),\varkappa^\star_{(\varepsilon)},\vec{y}_{(\varepsilon)})\right)_{|\varepsilon=0},$$
$$\left[\frac{\delta}{\delta \vec{\varkappa}^\star}\widehat{L}_{\overline{\varkappa},\beta}\right](\vec{\xi}) = \left(\frac{\mathrm{d}}{\mathrm{d}\varepsilon}\widehat{L}_{\overline{\varkappa},\beta}(\Gamma(t),\vec{\varkappa}^\star + \varepsilon\,\vec{\xi},\vec{y})\right)_{|\varepsilon=0},$$
$$\left[\frac{\delta}{\delta \vec{y}}\widehat{L}_{\overline{\varkappa},\beta}\right](\vec{\eta}) = \left(\frac{\mathrm{d}}{\mathrm{d}\varepsilon}\widehat{L}_{\overline{\varkappa},\beta}(\Gamma(t),\vec{\varkappa}^\star,\vec{y} + \varepsilon\,\vec{\eta})\right)_{|\varepsilon=0},$$

where $\vec{\varkappa}^\star_{(\varepsilon)}(\cdot,t), \vec{y}_{(\varepsilon)}(\cdot,t) \in [H^1(\Gamma_\varepsilon(t))]^d$ are defined by transporting the values of $\vec{\varkappa}^\star$ and $\vec{y}$ as in (187). Setting $\left[\frac{\delta}{\delta \vec{y}}\widehat{L}_{\overline{\varkappa},\beta}\right](\vec{\eta}) = 0$ and comparing with (171) leads to $\vec{\varkappa}^\star = \vec{\varkappa}$. Moreover, on setting $\left[\frac{\delta}{\delta \vec{\varkappa}^\star}\widehat{L}_{\overline{\varkappa},\beta}\right](\vec{\xi}) = 0$ and $\langle \vec{\mathcal{V}},\vec{\chi}\rangle_{\Gamma(t)} = -\left[\frac{\delta}{\delta \Gamma}\widehat{L}_{\overline{\varkappa},\beta}\right](\vec{\chi})$, we obtain, on noting (185), (186) and (188), the following weak formulation of $\vec{\mathcal{V}} = f\,\vec{\nu}$ and Lemma 13(ii), where f is the right-hand side of (204); recall Remark 135(ii) in the absence of spontaneous curvature and ADE effects. Given a closed hypersurface $\Gamma(0)$, we seek an evolving hypersurface $(\Gamma(t))_{t\in[0,T]}$, with a global parameterization and induced velocity field $\vec{\mathcal{V}}$, $\vec{\varkappa} \in [L^2(\mathcal{G}_T)]^d$ and $\vec{y} \in [L^2(\mathcal{G}_T)]^d$ as follows. For almost all $t \in (0,T)$, find $(\vec{\mathcal{V}}(\cdot,t), \vec{\varkappa}(\cdot,t), \vec{y}(\cdot,t)) \in [L^2(\Gamma(t))]^d \times [L^2(\Gamma(t))]^d \times [H^1(\Gamma(t))]^d$ such that

$$
\begin{aligned}
\left\langle \vec{\mathcal{V}}, \vec{\chi} \right\rangle_{\Gamma(t)} &= \langle \nabla_s \vec{y}, \nabla_s \vec{\chi} - 2\,\underline{\underline{D}}_s(\vec{\chi}) \rangle_{\Gamma(t)} + \langle \nabla_s \cdot \vec{y}, \nabla_s \cdot \vec{\chi} \rangle_{\Gamma(t)} \\
&\quad - \left\langle \frac{1}{2}\, |\vec{\varkappa} - \overline{\varkappa}\,\vec{\nu}|^2 - \vec{\varkappa} \cdot (\vec{y} - \beta\, A(t)\,\vec{\nu}), \nabla_s \cdot \vec{\chi} \right\rangle_{\Gamma(t)} \\
&\quad + (\beta\, A(t) - \overline{\varkappa}) \left\langle \vec{\varkappa}, (\nabla_s \vec{\chi})^{\mathsf{T}}\, \vec{\nu} \right\rangle_{\Gamma(t)} \quad \forall \vec{\chi} \in [H^1(\Gamma(t))]^d,
\end{aligned} \tag{210a}
$$

$$
\left\langle \vec{\varkappa} + (\beta\, A(t) - \overline{\varkappa})\vec{\nu} - \vec{y}, \vec{\xi} \right\rangle_{\Gamma(t)} = 0 \quad \forall \vec{\xi} \in [L^2(\Gamma(t))]^d, \tag{210b}
$$

$$
\langle \vec{\varkappa}, \vec{\eta} \rangle_{\Gamma(t)} + \left\langle \nabla_s\, \mathrm{id}, \nabla_s\, \vec{\eta} \right\rangle_{\Gamma(t)} = 0 \quad \forall \vec{\eta} \in [H^1(\Gamma(t))]^d \tag{210c}
$$

with

$$
A(t) = \langle \vec{\varkappa}, \vec{\nu} \rangle_{\Gamma(t)} - M_0. \tag{210d}
$$

We note that if $\overline{\varkappa} = \beta = 0$ then $\vec{y} = \vec{\varkappa}$, and so the system (210) reduces to (174).

We now introduce a semidiscrete variant of (210) with the help of the first variation of the discrete energy

$$
E^h_{\overline{\varkappa},\beta}(\Gamma^h(t)) = \frac{1}{2} \left\langle |\vec{\kappa}^h - \overline{\varkappa}\,\vec{\nu}^h|^2, 1 \right\rangle^h_{\Gamma^h(t)} + \frac{1}{2}\,\beta \left(\langle \vec{\kappa}^h, \vec{\nu}^h \rangle_{\Gamma^h(t)} - M_0 \right)^2 \tag{211}
$$

subject to the side constraint (178b). Hence, we define the Lagrangian

$$
\widehat{L}^h_{\overline{\varkappa},\beta}(\Gamma^h(t), \vec{\kappa}^h, \vec{Y}^h) = E^h_{\overline{\varkappa},\beta}(\Gamma^h(t)) - \langle \vec{\kappa}^h, \vec{Y}^h \rangle^h_{\Gamma^h(t)} - \left\langle \nabla_s\, \mathrm{id}, \nabla_s\, \vec{Y}^h \right\rangle_{\Gamma^h(t)},
$$

where $\vec{Y}^h(\cdot,t) \in \underline{V}(\Gamma^h(t))$ is the Lagrange multiplier for (178b). Similarly to (192), we set $\left[\frac{\delta}{\delta\Gamma^h}\, \widehat{L}^h_{\overline{\varkappa},\beta}\right](\vec{\chi}) = -\langle \vec{\mathcal{V}}^h, \vec{\chi} \rangle_{\Gamma^h(t)}$ for $\vec{\chi} \in \underline{V}(\Gamma^h(t))$, $\left[\frac{\delta}{\delta\vec{\kappa}^h}\, \widehat{L}^h_{\overline{\varkappa},\beta}\right](\vec{\xi}) = 0$ for $\vec{\xi} \in \underline{V}(\Gamma^h(t))$ and $\left[\frac{\delta}{\delta\vec{Y}^h}\, \widehat{L}^h_{\overline{\varkappa},\beta}\right](\vec{\eta}) = 0$ for $\vec{\eta} \in \underline{V}(\Gamma^h(t))$. Therefore, on noting the analogue of (195), the discrete analogue of (186) for $\vec{\nu}^h$ on $\Gamma^h(t)$ and (188) for $\Gamma^h(t)$, we have the following semidiscrete finite element approximation of (210). Given the closed polyhedral hypersurface $\Gamma^h(0)$, find an evolving polyhedral hypersurface $\mathcal{G}^h_T$ with induced velocity $\vec{\mathcal{V}}^h \in \underline{V}(\mathcal{G}^h_T)$, $\vec{\kappa}^h \in \underline{V}(\mathcal{G}^h_T)$ and $\vec{Y}^h \in \underline{V}(\mathcal{G}^h_T)$ as follows. For all $t \in (0,T]$, find $(\vec{\mathcal{V}}^h(\cdot,t), \vec{\kappa}^h(\cdot,t), \vec{Y}^h(\cdot,t)) \in \underline{V}(\Gamma^h(t)) \times \underline{V}(\Gamma^h(t)) \times \underline{V}(\Gamma^h(t))$ such that

$$\left\langle \vec{\mathcal{V}}^h, \vec{\chi} \right\rangle_{\Gamma^h(t)}^h = \langle \nabla_s \vec{Y}^h, \nabla_s \vec{\chi} - 2\,\underline{\underline{D}}_s(\vec{\chi}) \rangle_{\Gamma^h(t)} + \langle \nabla_s \cdot \vec{Y}^h, \nabla_s \cdot \vec{\chi} \rangle_{\Gamma^h(t)}$$
$$- \left\langle \frac{1}{2} |\vec{\kappa}^h - \overline{\varkappa}\, \vec{\nu}^h|^2 - \vec{\kappa}^h \cdot (\vec{Y}^h - \beta\, A^h(t)\, \vec{\nu}^h), \nabla_s \cdot \vec{\chi} \right\rangle_{\Gamma^h(t)}^h$$
$$+ (\beta\, A^h(t) - \overline{\varkappa}) \left\langle \vec{\kappa}^h, (\nabla_s \vec{\chi})^\top \vec{\nu}^h \right\rangle_{\Gamma^h(t)}^h \quad \forall \vec{\chi} \in \underline{V}(\Gamma^h(t)), \tag{212a}$$

$$\left\langle \vec{\kappa}^h + (\beta\, A^h(t) - \overline{\varkappa}) \vec{\nu}^h - \vec{Y}^h, \vec{\xi} \right\rangle_{\Gamma^h(t)}^h = 0 \quad \forall \vec{\xi} \in \underline{V}(\Gamma^h(t)), \tag{212b}$$

$$\langle \vec{\kappa}^h, \vec{\eta} \rangle_{\Gamma^h(t)}^h + \left\langle \nabla_s \vec{\mathrm{id}}, \nabla_s \vec{\eta} \right\rangle_{\Gamma^h(t)} = 0 \quad \forall \vec{\eta} \in \underline{V}(\Gamma^h(t)), \tag{212c}$$

where
$$A^h(t) = \langle \vec{\kappa}^h, \vec{\nu}^h \rangle_{\Gamma^h(t)}^h - M_0. \tag{212d}$$

We note that (212b) and (19) imply that $\vec{\kappa}^h + (\beta\, A^h(t) - \overline{\varkappa}) \vec{\omega}^h = \vec{Y}^h$ on $\Gamma^h(t)$. Of course, $\vec{\kappa}^h$ can be eliminated from (212).

We now prove the analogue of Theorem 134 for the semidiscrete scheme (212).

Theorem 143. *Let $(\mathcal{G}_T^h, \vec{\kappa}^h, \vec{Y}^h)$ be a solution of (212), and let $\vec{\kappa}^h \in \underline{V}_T(\mathcal{G}_T^h)$. Then it follows that*
$$\frac{\mathrm{d}}{\mathrm{d}t} E_{\overline{\varkappa},\beta}^h(\Gamma^h(t)) = -\left(\left| \vec{\mathcal{V}}^h \right|_{\Gamma^h(t)}^h \right)^2 \le 0. \tag{213}$$

Proof. As in the proof of Theorem 134, on taking the derivative with respect to t of (212c) we obtain (179). Choosing $\vec{\eta} = \vec{Y}^h$ in (179) leads to

$$\left\langle \partial_t^{\circ,h} \vec{\kappa}^h, \vec{Y}^h \right\rangle_{\Gamma^h(t)}^h + \left\langle \vec{\kappa}^h \cdot \vec{Y}^h, \nabla_s \cdot \vec{\mathcal{V}}^h \right\rangle_{\Gamma^h(t)}^h + \left\langle \nabla_s \cdot \vec{\mathcal{V}}^h, \nabla_s \cdot \vec{Y}^h \right\rangle_{\Gamma^h(t)}$$
$$+ \left\langle \nabla_s \vec{\mathcal{V}}^h, \nabla_s \vec{Y}^h \right\rangle_{\Gamma^h(t)} - 2 \left\langle \underline{\underline{D}}_s(\vec{\mathcal{V}}^h), \nabla_s \vec{Y}^h \right\rangle_{\Gamma^h(t)} = 0. \tag{214}$$

Combining (212a) with $\vec{\chi} = \vec{\mathcal{V}}^h$ and (214), we obtain, on noting Lemma 72 and (212b), that

$$\left\langle \vec{\mathcal{V}}^h, \vec{\mathcal{V}}^h \right\rangle_{\Gamma^h(t)}^h + \left\langle \frac{1}{2} |\vec{\kappa}^h - \overline{\varkappa}\, \vec{\nu}^h|^2 + \beta\, A^h(t)\, \vec{\kappa}^h \cdot \vec{\nu}^h, \nabla_s \cdot \vec{\mathcal{V}}^h \right\rangle_{\Gamma^h(t)}^h$$
$$+ (\beta\, A^h(t) - \overline{\varkappa}) \langle \vec{\kappa}^h, \partial_t^{\circ,h} \vec{\nu}^h \rangle_{\Gamma^h(t)}^h$$
$$= -\langle \partial_t^{\circ,h} \vec{\kappa}^h, \vec{Y}^h \rangle_{\Gamma^h(t)}^h = -\langle \partial_t^{\circ,h} \vec{\kappa}^h, \vec{\kappa}^h + (\beta\, A^h(t) - \overline{\varkappa}) \vec{\nu}^h \rangle_{\Gamma^h(t)}^h. \tag{215}$$

The desired result (213) then follows from (215), (211), (16), Theorem 70 and (212d), where we have observed that

$$\frac{1}{2}\frac{\mathrm{d}}{\mathrm{d}t}\left(\langle\vec{\kappa}^h,\vec{\nu}^h\rangle^h_{\Gamma^h(t)}-M_0\right)^2 = A^h(t)\left[\left\langle\partial_t^{\circ,h}\vec{\kappa}^h,\vec{\nu}^h\right\rangle^h_{\Gamma^h(t)}+\left\langle\vec{\kappa}^h,\partial_t^{\circ,h}\vec{\nu}^h\right\rangle^h_{\Gamma^h(t)}+\left\langle\vec{\kappa}^h\cdot\vec{\nu}^h,\nabla_s\cdot\vec{\mathcal{V}}^h\right\rangle^h_{\Gamma^h(t)}\right]. \square$$

We now state a fully discrete variant of the semidiscrete scheme (212). Let the closed polyhedral hypersurface Γ^0 be an approximation to $\Gamma(0)$, and let $\vec{\kappa}^0_{\Gamma^0},\vec{Y}^0_{\Gamma^0}\in\underline{V}(\Gamma^0)$ and $A^0\in\mathbb{R}$ be given. Then, for $m=0,\ldots,M-1$, find $(\vec{X}^{m+1},\vec{\kappa}^{m+1},\vec{Y}^{m+1})\in\underline{V}(\Gamma^m)\times\underline{V}(\Gamma^m)\times\underline{V}(\Gamma^m)$ such that

$$\begin{aligned}\left\langle\frac{\vec{X}^{m+1}-\vec{\mathrm{id}}}{\Delta t_m},\vec{\chi}\right\rangle^h_{\Gamma^m} &= \langle\nabla_s\vec{Y}^{m+1},\nabla_s\vec{\chi}\rangle_{\Gamma^m}+\langle\nabla_s\cdot\vec{Y}^m_{\Gamma^m},\nabla_s\cdot\vec{\chi}\rangle^h_{\Gamma^m}\\ &\quad -2\langle\nabla_s\vec{Y}^m_{\Gamma^m},\underline{\underline{D}}_s(\vec{\chi})\rangle_{\Gamma^m}+(\beta A^m-\overline{\varkappa})\left\langle\vec{\kappa}^m_{\Gamma^m},(\nabla_s\vec{\chi})^{\mathsf{T}}\,\vec{\nu}^m\right\rangle^h_{\Gamma^m}\\ &\quad -\left\langle\frac{1}{2}|\vec{\kappa}^m_{\Gamma^m}-\overline{\varkappa}\,\vec{\nu}^m|^2-\vec{\kappa}^m_{\Gamma^m}\cdot(\vec{Y}^m_{\Gamma^m}-\beta A^m\,\vec{\nu}^m),\nabla_s\cdot\vec{\chi}\right\rangle^h_{\Gamma^m}\\ &\qquad \forall\,\vec{\chi}\in\underline{V}(\Gamma^m),\end{aligned}\tag{216a}$$

$$\vec{\kappa}^{m+1}=\vec{Y}^{m+1}-(\beta A^m-\overline{\varkappa})\vec{\omega}^m,\tag{216b}$$

$$\langle\vec{\kappa}^{m+1},\vec{\eta}\rangle^h_{\Gamma^m}+\langle\nabla_s\vec{X}^{m+1},\nabla_s\vec{\eta}\rangle_{\Gamma^m}=0\quad\forall\,\vec{\eta}\in\underline{V}(\Gamma^m)\tag{216c}$$

and set $\Gamma^{m+1}=\vec{X}^{m+1}(\Gamma^m)$, $A^{m+1}=\langle\vec{\kappa}^{m+1},\vec{\nu}^m\rangle_{\Gamma^m}-M_0$, $\vec{\kappa}^{m+1}_{\Gamma^{m+1}}=\vec{\kappa}^{m+1}\circ(\vec{X}^{m+1})^{-1}\in\underline{V}(\Gamma^{m+1})$ and $\vec{Y}^{m+1}_{\Gamma^{m+1}}=\vec{Y}^{m+1}\circ(\vec{X}^{m+1})^{-1}\in\underline{V}(\Gamma^{m+1})$. Of course, $\vec{\kappa}^{m+1}$ can be eliminated from (216).

Theorem 144. *Let $\vec{\kappa}^m_{\Gamma^m},\vec{Y}^m_{\Gamma^m}\in\underline{V}(\Gamma^m)$ and $A^m\in\mathbb{R}$. Then there exists a unique solution $(\vec{X}^{m+1},\vec{\kappa}^{m+1},\vec{Y}^{m+1})\in\underline{V}(\Gamma^m)\times\underline{V}(\Gamma^m)\times\underline{V}(\Gamma^m)$ to (216).*

Proof. It is a simple matter to adapt the proof of Theorem 139. □

Remark 145 (Discrete linear systems).
Similarly to Remark 141, the linear systems of equations to be solved at each time level for (216) can be formulated as follows. Find $(\vec{Y}^{m+1},\vec{\kappa}^{m+1},\delta\vec{X}^{m+1})\in(\mathbb{R}^d)^K\times(\mathbb{R}^d)^K\times(\mathbb{R}^d)^K$ such that

$$\begin{pmatrix}\underline{\underline{A}}_{\Gamma^m} & 0 & -\frac{1}{\Delta t_m}\underline{\underline{M}}_{\Gamma^m}\\ -\underline{\underline{M}}_{\Gamma^m} & \underline{\underline{M}}_{\Gamma^m} & 0\\ 0 & \underline{\underline{M}}_{\Gamma^m} & \underline{\underline{A}}_{\Gamma^m}\end{pmatrix}\begin{pmatrix}\vec{Y}^{m+1}\\ \vec{\kappa}^{m+1}\\ \delta\vec{X}^{m+1}\end{pmatrix}=\begin{pmatrix}\vec{\mathfrak{g}}_{\Gamma^m}\\ -(\beta\,A^m-\overline{\varkappa})\,\underline{\underline{M}}_{\Gamma^m}\,\vec{\omega}^m\\ -\underline{\underline{A}}_{\Gamma^m}\,\vec{X}^m\end{pmatrix},$$

and this system can be efficiently solved with a sparse direct solution method like UMFPACK, see Davis (2004).

Similarly to Remark 135(ii), the semidiscrete scheme (212) and its fully discrete version (216) do not have good mesh properties. In order to obtain a scheme with good mesh properties, we extend the semidiscrete approximation (196), (194) and (193) to take into account spontaneous curvature and ADE effects. First, we extend the weak formulation (191), (190) and (180) to include spontaneous curvature and ADE effects. In order to do so, we extend the Lagrangian (181) to

$$\begin{aligned} L_{\overline{\varkappa},\beta}(\Gamma(t),\varkappa^\star,\vec{y}) =& \frac{1}{2}\langle \varkappa^\star - \overline{\varkappa}, \varkappa^\star - \overline{\varkappa}\rangle_{\Gamma(t)} + \frac{1}{2}\beta\Big(\langle \varkappa^\star, 1\rangle_{\Gamma(t)} - M_0\Big)^2 \\ & - \langle \varkappa^\star \vec{\nu}, \vec{y}\rangle_{\Gamma(t)} - \left\langle \nabla_s \vec{\mathrm{id}}, \nabla_s \vec{y}\right\rangle_{\Gamma(t)}. \end{aligned} \tag{217}$$

Once again, setting the variation $\left[\frac{\delta}{\delta \vec{y}} L_{\overline{\varkappa},\beta}\right](\vec{\eta}) = 0$, yields (180) with $\varkappa$ replaced by $\varkappa^\star$, and so $\varkappa^\star = \varkappa$. Setting $\left[\frac{\delta}{\delta \varkappa^\star} L_{\overline{\varkappa},\beta}\right](\xi) = 0$, yields, on noting $\varkappa^\star = \varkappa$ and (202), that

$$\varkappa - \overline{\varkappa} = \vec{y}\cdot\vec{\nu} - \beta A(t) \quad \text{on}\,\Gamma(t). \tag{218}$$

Finally, on setting the variation $\left[\frac{\delta}{\delta \Gamma} L_{\overline{\varkappa},\beta}\right](\vec{\chi}) = -\left\langle \vec{\mathcal{V}}\cdot\vec{\nu}, \vec{\chi}\cdot\vec{\nu}\right\rangle_{\Gamma(t)}$ and noting (185), (186), (188) and that $\varkappa^\star = \varkappa$, we obtain

$$\begin{aligned} \left\langle \vec{\mathcal{V}}\cdot\vec{\nu}, \vec{\chi}\cdot\vec{\nu}\right\rangle_{\Gamma(t)} =& \langle \nabla_s \vec{y}, \nabla_s \vec{\chi} - 2\,\underline{\underline{D}}_s(\vec{\chi})\rangle_{\Gamma(t)} + \langle \nabla_s\cdot\vec{y}, \nabla_s\cdot\vec{\chi}\rangle_{\Gamma(t)} \\ & - \left\langle \frac{1}{2}(\varkappa - \overline{\varkappa})^2 - \varkappa(\vec{y}\cdot\vec{\nu} - \beta A(t)), \nabla_s\cdot\vec{\chi}\right\rangle_{\Gamma(t)} \\ & - \left\langle \varkappa\vec{y}, (\nabla_s\vec{\chi})^{\mathsf{T}}\vec{\nu}\right\rangle_{\Gamma(t)} \quad \forall \vec{\chi} \in [H^1(\Gamma(t))]^d. \end{aligned} \tag{219}$$

Therefore we have the following weak formulation. Given a closed hypersurface $\Gamma(0)$, we seek an evolving hypersurface $(\Gamma(t))_{t\in[0,T]}$, with a global parameterization and induced velocity field $\vec{\mathcal{V}}$, $\varkappa \in L^2(\mathcal{G}_T)$ and $\vec{y} \in [L^2(\mathcal{G}_T)]^d$ as follows. For almost all $t \in (0,T)$, find $(\vec{\mathcal{V}}(\cdot,t), \varkappa(\cdot,t), \vec{y}(\cdot,t)) \in [L^2(\Gamma(t))]^d \times L^2(\Gamma(t)) \times [H^1(\Gamma(t))]^d$ such that (219), (218) and (180) hold.

Once again, a remark equivalent to Remark 136 for this weak formulation holds. In addition, one can extend the semidiscrete finite element approximation (196), (194) and (193) to approximate the weak formulation (219), (218) and (180). Moreover, one can extend Remark 137 and Theorem 138 to this approximation. Furthermore, one can also incorporate volume and surface area constraints and still prove stability of the approximation.

In Barrett et al. (2016b), we considered an extension of the semidiscrete approximation (196), (194) and (193), and its extension to incorporate spontaneous curvature and ADE effects, which possibly reduces the tangential motion. This is based on the Lagrangian

$$\widehat{L}^h_{\overline{\varkappa},\beta,\theta}(\Gamma^h(t),\vec{\kappa}^h,\vec{Y}^h) = E^h_{\overline{\varkappa},\beta}(\Gamma^h(t)) - \left\langle \underline{\underline{Q}}^h_\theta\,\vec{\kappa}^h,\vec{Y}^h\right\rangle^h_{\Gamma^h(t)} - \left\langle \nabla_s\,\vec{\mathrm{id}},\nabla_s\,\vec{Y}^h\right\rangle_{\Gamma^h(t)}, \tag{220}$$

where $\vec{Y}^h(\cdot,t) \in \underline{V}(\Gamma^h(t))$ is the Lagrange multiplier associated with the constraint $\vec{\kappa}^h(\cdot,t) \in \underline{V}(\Gamma^h(t))$ satisfying

$$\left\langle \underline{\underline{Q}}^h_\theta\,\vec{\kappa}^h,\vec{\eta}\right\rangle^h_{\Gamma^h(t)} + \left\langle \nabla_s\,\vec{\mathrm{id}},\nabla_s\,\vec{\eta}\right\rangle_{\Gamma^h(t)} = 0 \quad \forall\,\vec{\eta} \in \underline{V}(\Gamma^h(t)). \tag{221}$$

Here, $\underline{\underline{Q}}^h_\theta$, for a given $\theta \in [0, 1]$, is the semidiscrete version of (96), where $\vec{\omega}^m$ is replaced by $\vec{\omega}^h$, and where we assume that Assumption 64(ii) holds.

Setting $\left[\frac{\delta}{\delta\vec{Y}^h}\widehat{L}^h_{\overline{\varkappa},\beta,\theta}\right](\vec{\eta}) = 0$, for all $\vec{\eta} \in \underline{V}(\Gamma^h(t))$, yields (221). Setting $\left[\frac{\delta}{\delta\vec{\kappa}^h}\widehat{L}^h_{\overline{\varkappa},\beta,\theta}\right](\vec{\xi}) = 0$, for all $\vec{\xi} \in \underline{V}(\Gamma^h(t))$, yields that $\vec{\kappa}^h + (\beta\,A^h(t) - \overline{\varkappa})\,\vec{\omega}^h = \vec{\pi}_{\Gamma^h}\left[\underline{\underline{Q}}^h_\theta\,\vec{Y}^h\right]$, where $A^h(t)$ is given by (212d). Finally, we set $\left[\frac{\delta}{\delta\Gamma^h}\widehat{L}^h_{\overline{\varkappa},\beta,\theta}\right](\vec{\chi}) = -\langle \underline{\underline{Q}}^h_\theta\,\vec{\mathcal{V}}^h,\vec{\chi}\rangle^h_{\Gamma(t)}$, for all $\vec{\chi} \in \underline{V}(\Gamma^h(t))$. If $\theta = 1$, the resulting scheme collapses to the semidiscrete Dziuk scheme (212). If $\theta = 0$, this scheme collapses to a variant of (196), (194) and (193), which takes spontaneous curvature and ADE effects into account, and this scheme still satisfies Theorem 88(iii). For any $\theta \in (0, 1)$, the scheme interpolates between these two extremes. Moreover, this scheme, for any given $\theta \in [0, 1]$, satisfies a stability bound, a generalization of Theorem 143; see Barrett et al. (2016b, Theorem 3.3) for details. In addition, this stability bound holds for a variant involving volume and surface area constraints.

Remark 146 (Surfaces with boundary).
For an evolving surface with a boundary, the result of Lemma 142 can be generalized as follows, where we recall the shorthand notation $\mathcal{V} = f$ for (204). On noting Theorem 32, Lemma 39(ii) and Theorem 21, it holds that

$$\begin{aligned}\frac{\mathrm{d}}{\mathrm{d}t}\,E_{\overline{\varkappa},\beta}(\Gamma(t)) = &-\langle f,\mathcal{V}\rangle_{\Gamma(t)} + \int_{\partial\Gamma(t)} \left(\frac{1}{2}\,(\varkappa - \overline{\varkappa})^2 + \beta\,A(t)\,\varkappa\right)\vec{\mathcal{V}}\cdot\vec{\mu}\;\mathrm{d}\mathcal{H}^{d-2} \\ &- \int_{\partial\Gamma(t)} \mathcal{V}\,\vec{\mu}\cdot\nabla_s\,\varkappa\;\mathrm{d}\mathcal{H}^{d-2} + \int_{\partial\Gamma(t)} (\varkappa - \overline{\varkappa} + \beta\,A(t))\,\vec{\mu}\cdot\nabla_s\,\mathcal{V}\;\mathrm{d}\mathcal{H}^{d-2},\end{aligned} \tag{222}$$

where $\vec{\mathcal{V}}$ is the velocity field induced by a global parameterization of the evolving hypersurface, and $\vec{\mu}(t)$ denotes the outer unit conormal on $\partial\Gamma(t)$. Hence, in order to ensure that (204) is still the L^2-gradient flow of $E_{\overline{\varkappa},\beta}(\Gamma(t))$, conditions need to be prescribed at the boundary so that the boundary terms in

(222) vanish. In particular, the following boundary conditions may be considered. In the simplest situation the boundary is kept fixed, i.e., $\partial\Gamma(t) = \partial\Gamma(0)$ for all $t \in (0,T]$, which is the same as (43). This leads to the first two boundary terms vanishing in (222). The third boundary term vanishes if either $\varkappa - \overline{\varkappa} + \beta A(t) = 0$ or $\vec{\mu} \cdot \nabla_s \mathcal{V} = 0$ on $\partial\Gamma(t)$. Setting $\partial\Gamma(t) = \partial\Gamma(0)$ and $\varkappa - \overline{\varkappa} + \beta A(t) = 0$ on $\partial\Gamma(t)$ for all $t \in (0,T]$ are called Navier boundary conditions. Whereas, for a given $\vec{\zeta} \in C(\partial\Gamma(0), \mathbb{S}^{d-1})$, setting $\partial\Gamma(t) = \partial\Gamma(0)$ and $\vec{\mu}(t) = \vec{\zeta}$ on $\partial\Gamma(t)$ for all $t \in (0,T]$ are called clamped boundary conditions. It is a simple matter to show that clamped boundary conditions lead to $\vec{\mu} \cdot \nabla_s \mathcal{V} = 0$ on $\partial\Gamma(t)$, and hence to all three boundary terms vanishing in (222).

A third type of possible boundary conditions are called free boundary conditions. Here the boundary can move freely and the three boundary terms in (222) vanish on imposing

$$\frac{1}{2}(\varkappa - \overline{\varkappa})^2 + \beta A(t)\, \varkappa = 0, \quad \vec{\mu} \cdot \nabla_s \varkappa = 0, \quad \varkappa - \overline{\varkappa} + \beta A(t) = 0 \quad \text{on } \partial\Gamma(t).$$

We refer to Barrett et al. (2012b) for more details in the case $d = 2$, and to Barrett et al. (2017c) for more details in the case $d = 3$. The discussion in the latter article easily generalizes to $d > 3$. Moreover, in Barrett et al. (2017c) the present authors also included a boundary energy, $\varsigma \mathcal{H}^{d-2}(\partial\Gamma(t))$ for $\varsigma \in \mathbb{R}_{\geq 0}$, which is called line energy in the case $d = 3$, and Gaussian curvature effects. Here the inclusion of the Gaussian curvature effects is achieved via the Gauss–Bonnet theorem, recall Theorem 40.

In the case of Navier boundary conditions, one can easily adapt the finite element approximation (208), (167b) by replacing $\underline{V}(\Gamma^m)$ in (167b) by $\underline{V}_{\rm D}(\Gamma^m)$, recall (44), by replacing $V(\Gamma^m)$ in (208) by $V_{\rm D}(\Gamma^m)$, where $\underline{V}_{\rm D}(\Gamma^m) = [V_{\rm D}(\Gamma^m)]^d$, and by seeking $(\vec{X}^{m+1}, \kappa^{m+1}) \in \underline{V}(\Gamma^m) \times V(\Gamma^m)$ such that $\vec{X}^{m+1} - \vec{\rm id}_{|_{\Gamma^m}} \in \underline{V}_{\rm D}(\Gamma^m)$ and $\kappa^{m+1} - (\overline{\varkappa} - \beta A^m) \in V_{\rm D}(\Gamma^m)$. For clamped and free boundary conditions this finite element approximation cannot be adapted. However, the approaches based on (217) and (220) can be adapted to all three types of boundary conditions. For the latter approach, we once again refer to Barrett et al. (2012b, 2017c) for more details, including stability results for semidiscrete discretizations. Furthermore, in Barrett et al. (2018) we extended this approach to approximate gradient flows for two-phase biomembranes, where instead of boundary conditions certain matching conditions need to hold across interfaces between different phases on the hypersurface. Finally, a related problem is the evolution of curve networks under elastic flow, where two or more curves meet at junction points. This problem has been considered by the authors in Barrett et al. (2012c).

9.5 Alternative numerical approaches

We note that Rusu (2005) first introduced a mixed variational form for Willmore flow, based on position and mean curvature vectors, which allowed for the approximation by continuous piecewise linear finite elements. This approach was extended to surfaces with boundaries and applied to surface restoration problems in Clarenz et al. (2004). We refer also to Deckelnick and Dziuk (2006), Deckelnick and Schieweck (2010) and Deckelnick et al. (2015), where, on assuming a sufficiently smooth solution, an error analysis is presented for semi-discrete finite element approximations for a graph formulation of Willmore flow.

Other parametric numerical methods are discussed in Dziuk et al. (2002), Mayer and Simonett (2002), Bobenko and Schröder (2005), Bonito et al. (2010), Elliott and Stinner (2010, 2013), Bartels et al. (2012) and Bartezzaghi et al. (2019). In addition, a numerical method based on a level set approach has been studied in Droske and Rumpf (2004), while phase field type approaches are discussed in Du et al. (2005), Esedoḡlu et al. (2014) and Bretin et al. (2015).

10 Biomembranes

In this section we will discuss how the first variation of curvature energies discussed in Section 9 appear in evolution laws involving also bulk quantities. This is relevant for the evolution of vesicles and biomembranes, where the curvature energy interacts with a fluid surrounding the membrane. Biomembranes are lipid bilayers and they form bag-like structures containing fluid and they are surrounded by a possibly different fluid. Membranes appear in a multitude of biological systems and a proper understanding of the form and the evolution of biomembranes is of major interest in the life sciences. As an example, we mention that the biconcave shapes of red blood cells appear as stationary states of the curvature energies discussed in Section 9.

10.1 A model for the dynamics of fluidic biomembranes

We will introduce a model, which is a variant of a model introduced by Arroyo and DeSimone (2009), that couples the first variation of a curvature energy to the (Navier–)Stokes equations in the bulk and on the surface. To be more precise a bulk (Navier–)Stokes system is coupled to a tangential (Navier–)Stokes system on the membrane. The tangential (Navier–)Stokes system also includes a surface incompressibility condition, which will lead to good mesh properties on the discrete level even in situations where the fluid velocity leads to a high deformation of the surface.

A generalized elastic energy for a biomembrane is given by

$$\int_\Gamma \frac{1}{2}\,\alpha(\varkappa-\overline{\varkappa})^2+\alpha^G\,\mathcal{K}\,\mathrm{d}\mathcal{H}^{d-1}+\frac{1}{2}\,\alpha\beta\big(\langle\varkappa,1\rangle_\Gamma-M_0\big)^2 \tag{223}$$

for a compact hypersurface $\Gamma\subset\mathbb{R}^d$ without boundary, $d\geq 2$. Here $\alpha\in\mathbb{R}_{>0}$ and $\alpha^G\in\mathbb{R}$ are the bending and Gaussian bending rigidities. As before, $\overline{\varkappa}$ is the spontaneous curvature, which arises, for example, from local inhomogeneities within the membrane and the parameters $\beta\in\mathbb{R}_{\geq 0}$, $M_0\in\mathbb{R}$ relate the area difference elasticity (ADE) model, recall Section 9.4. Moreover, $\mathcal{K}$ is the

Gaussian curvature. It is discussed in Nitsche (1993) that the most general form of a curvature energy of the form $\int_\Gamma q(\varkappa_1, \dots, \varkappa_{d-1})\, \mathrm{d}\mathcal{H}^{d-1}$, with q being at most quadratic in the principal curvatures and invariant under permutations of its arguments, has the form $q(\varkappa_1, \dots, \varkappa_{d-1}) = \frac{1}{2}\alpha\,\varkappa^2 + \alpha^G\,\mathcal{K} + \alpha_1\,\varkappa + \alpha_2$, which leads to the first term in (223) by choosing $\alpha_1 = -\alpha\,\overline{\varkappa}$ and $\alpha_2 = \frac{1}{2}\alpha\,\overline{\varkappa}^2$. If α^G is constant we obtain from the Gauss–Bonnet theorem, recall Theorem 40, that, as the surface has no boundary, the contribution $\int_\Gamma \alpha^G\,\mathcal{K}\,\mathrm{d}\mathcal{H}^{d-1}$ is constant in a fixed topological class, and hence will always disappear in a first variation. For now, we will hence set $\alpha^G = 0$ and only if one considers inhomogeneous membranes, or open membranes with boundary, one has to consider this term, see Elliott and Stinner (2010) and Barrett et al. (2017c,d, 2018). Hence, overall, we consider from now on $\alpha\,E_{\overline{\varkappa},\beta}(\Gamma(t))$ as the elastic energy for an evolving biomembrane, where $E_{\overline{\varkappa},\beta}(\Gamma(t))$ is as defined in (201).

We adopt the notation in Section 2.4, and assume that two fluids occupy regions $\Omega_-(t)$ and $\Omega_+(t) = \Omega\setminus\overline{\Omega_-(t)}$ with $\Gamma(t) = \partial\Omega_-(t)$, and with $\vec{\nu}$ denoting the outer unit normal to $\Omega_-(t)$ on $\Gamma(t)$, recall Fig. 1 in Section 2.4. Here $\Omega \subset \mathbb{R}^d$ is a fixed domain, with $d \geq 2$. In these two fluid regions we require, as in Section 8, the incompressible Stokes equations with a no-slip boundary condition on $\partial\Omega$, i.e.,

$$-\nabla\cdot\underline{\underline{\sigma}} = \vec{0}, \quad \nabla\cdot\vec{u} = 0 \quad \text{in}\,\Omega_\pm(t), \qquad \vec{u} = \vec{0} \quad \text{on}\,\partial\Omega \tag{224}$$

with the stress tensor $\underline{\underline{\sigma}}$ as defined in (135). Once again, we refer to Remark 127(iv) for more general boundary conditions.

We remark that for biomembranes scales are typically such that the Stokes approximation of the Navier–Stokes equations is a very good approximation, see Barrett et al. (2015e). Therefore, we will consider the Stokes system in the bulk in what follows and refer to Barrett et al. (2016a, 2017b) and Section 8 for a discussion on how to generalize the following considerations to Navier–Stokes flow.

We now follow and generalize an approach of Arroyo and DeSimone (2009), who used the theory of interfacial fluid dynamics, which goes back to Scriven (1960), to formulate evolution laws that take the fluidic behaviour of vesicles and biomembranes into account. In fact, these lipid bilayer membranes can be described as a two-dimensional surface, where the lipid molecules have a fluid-like behaviour in the tangential direction. However, elastic forces stemming from the elastic bending energy act in normal direction. For more details on the modelling and the analysis of fluidic interfaces we refer to Slattery et al. (2007) and Bothe and Prüss (2010). For our purposes it suffices to mention that in the absence of mass transfer to/from the interface from/to the bulk, and on assuming a no-slip condition between the outer and inner fluid at the interface, it is natural to assume that the bulk velocity is continuous across the interface and that the interface is moved with the bulk velocity. We refer to Barrett et al. (2015d, p. 1830) for more details.

Overall the following conditions need to hold on the free surface $\Gamma(t)$:

$$[\vec{u}]_-^+ = \vec{0} \qquad \text{on}\,\Gamma(t), \tag{225a}$$

$$\rho_\Gamma\, \partial_t^\bullet\, \vec{u} - \nabla_s \cdot \underline{\underline{\sigma}}_\Gamma = \left[\underline{\underline{\sigma}}\,\vec{\nu}\right]_-^+ + \alpha \vec{f}_\Gamma \qquad \text{on}\,\Gamma(t), \tag{225b}$$

$$\nabla_s \cdot \vec{u} = 0 \qquad \text{on}\,\Gamma(t), \tag{225c}$$

$$\mathcal{V} = \vec{u}\cdot\vec{\nu} \qquad \text{on}\,\Gamma(t), \tag{225d}$$

where $\rho_\Gamma \in \mathbb{R}_{\geq 0}$ denotes the surface material density, $\alpha \in \mathbb{R}_{>0}$ is the bending rigidity and $\vec{f}_\Gamma = f_\Gamma\,\vec{\nu}$, with f_Γ denoting minus the first variation of the bending energy $E_{\overline{\varkappa},\beta}$, see (201), i.e.,

$$f_\Gamma = -\Delta_s\,\varkappa - (\varkappa - \overline{\varkappa})|\nabla_s\,\vec{\nu}|^2 + \frac{1}{2}(\varkappa - \overline{\varkappa})^2\,\varkappa - \beta\,A(t)\left(|\nabla_s\,\vec{\nu}|^2 - \varkappa^2\right), \tag{226}$$

where $\varkappa$ satisfies Lemma 13(ii) and $A(t) = \langle \varkappa,\, 1\rangle_{\Gamma(t)} - M_0$, recall Lemma 142. Moreover,

$$\partial_t^\bullet f = \partial_t f + \vec{u}\cdot\nabla f \tag{227}$$

is the material time derivative with respect to the fluid velocity $\vec{u}$. This definition is related to Remark 29(ii), where the parameterizations $\vec{x}(\,\cdot\,,t)$ here are defined such that

$$\vec{u}(\vec{x}(\vec{q},t),t) = (\partial_t\,\vec{x})(\vec{q},t) \qquad \forall\,(\vec{q},t) \in \Upsilon \times [0,T], \tag{228}$$

as opposed to Definition 25(ii), so that $\vec{x}(\vec{q},t)$ really is the trajectory of a material point. Eq. (225c) is conservation of mass of the interface, similarly to standard incompressible (Navier–)Stokes in the bulk, recall (224). In particular, it implies that the membrane is locally incompressible. This can be seen by considering a surface patch $\sigma(t) \subset \Gamma(t)$ that is transported with the fluid velocity $\vec{u}$, i.e., $\sigma(t) = \vec{x}(\sigma_\Upsilon, t)$ with $\sigma_\Upsilon \subset \Upsilon$ and $\vec{x}$ satisfying (228). Then it follows with the help of the transport theorem, Theorem 32, that

$$\frac{\mathrm{d}}{\mathrm{d}t}\,\mathcal{H}^{d-1}(\sigma(t)) = \int_{\sigma(t)} \nabla_s \cdot \vec{\mathcal{V}}\;\mathrm{d}\mathcal{H}^{d-1} = \int_{\sigma(t)} \nabla_s \cdot \vec{u}\;\mathrm{d}\mathcal{H}^{d-1} = 0,$$

see also Arroyo and DeSimone (2009). The balance of momentum on the surface implies (225b), see Arroyo and DeSimone (2009), Köhne and Lengeler (2018) and Lengeler (2018). Here the surface stress tensor is defined by

$$\underline{\underline{\sigma}}_\Gamma = 2\,\mu_\Gamma\,\underline{\underline{D}}_s(\vec{u}) - p_\Gamma\,\underline{\underline{P}}_\Gamma \qquad \text{on}\,\Gamma(t), \tag{229}$$

where p_Γ is the surface pressure, $\mu_\Gamma \in \mathbb{R}_{>0}$ is the surface shear viscosity, $\underline{\underline{P}}_\Gamma$ is the projection onto the tangent space and $\underline{\underline{D}}_s(\vec{u})$ is the surface rate of deformation tensor, recall Definition 5(vii). The tensor $\underline{\underline{D}}_s(\vec{u})$ is the relevant tensor for measuring the rate of change of lengths and angles through the metric tensor, cf. Lemma 30.

Finally, the term $\left[\underline{\underline{\sigma}}\,\vec{\nu}\right]_-^+$, recall (12) and its generalization to vector-valued quantities in Remark 36, is the force exerted by the bulk on the surface. The total bending energy considered for now is given by $\alpha\, E_{\overline{\varkappa},\beta}(\Gamma)$ with $E_{\overline{\varkappa},\beta}$, the dimensionless energy defined in (201), and α, having the dimension of energy, is the bending rigidity. As in Section 8.1, (225d) states that the interface evolves with the normal component of the fluid velocity.

The system (224), (225) is closed with an initial condition for $\Gamma(0)$ and

$$\rho_\Gamma\, \vec{u}(\cdot, 0) = \rho_\Gamma\, \vec{u}_0 \quad \text{on}\, \Gamma(0),$$

where for simplicity we assume that $\vec{u}_0 : \Omega \to \mathbb{R}^d$ is a given initial velocity.

Lemma 147. *Let $(\mathcal{G}_T, \vec{u}, p)$ be a sufficiently smooth solution of (224), (225) and (226).*

(i) *It holds that*

$$\frac{\mathrm{d}}{\mathrm{d}t} |\Gamma(t)| = 0.$$

(ii) *It holds that*

$$\frac{\mathrm{d}}{\mathrm{d}t} \mathcal{L}^d(\Omega_-(t)) = 0.$$

(iii) *It holds that*

$$\frac{\mathrm{d}}{\mathrm{d}t}\left[\alpha\, E_{\overline{\varkappa}}, \beta(\Gamma(t)) + \frac{1}{2}\rho_\Gamma |\vec{u}|^2_{\Gamma(t)}\right]$$
$$= -2\,\mu_\Gamma \langle \underline{\underline{D}}_s(\vec{u}), \underline{\underline{D}}_s(\vec{u})\rangle_{\Gamma(t)} - 2\langle \mu\, \underline{\underline{D}}(\vec{u}), \underline{\underline{D}}(\vec{u})\rangle \leq 0.$$

Proof. Let $\vec{x}$ be an arbitrary global parameterization of $\mathcal{G}_T$, with induced velocity field $\vec{\mathcal{V}}$.

(i) Combining the transport theorem, Theorem 32, and the divergence theorem, Theorem 21, with the tangential fields $\vec{f} = \vec{\mathcal{V}}_\mathrm{T}$ and $\vec{f} = \underline{\underline{P}}_\Gamma\, \vec{u}$, gives

$$\begin{aligned}\frac{\mathrm{d}}{\mathrm{d}t}\langle 1, 1\rangle_{\Gamma(t)} &= \left\langle 1, \nabla_s \cdot \vec{\mathcal{V}}\right\rangle_{\Gamma(t)} = \left\langle 1, \nabla_s \cdot \left(\vec{\mathcal{V}} - \vec{\mathcal{V}}_\mathrm{T}\right)\right\rangle_{\Gamma(t)} = \langle 1, \nabla_s \cdot (\mathcal{V}\,\vec{\nu})\rangle_{\Gamma(t)} \\ &= \langle 1, \nabla_s \cdot ((\vec{u}\cdot\vec{\nu})\,\vec{\nu})\rangle_{\Gamma(t)} = \langle 1, \nabla_s \cdot (\vec{u} - \underline{\underline{P}}_\Gamma\,\vec{u})\rangle_{\Gamma(t)} \\ &= \langle 1, \nabla_s \cdot \vec{u}\rangle_{\Gamma(t)} = 0,\end{aligned} \tag{230}$$

which is the claim.

(ii) As in Remark 117(ii), the result follows from the transport theorem, Theorem 33, and the facts that $\vec{\mathcal{V}}\cdot\vec{\nu}=\vec{u}\cdot\vec{\nu}$ on $\Gamma(t)$ and $\nabla\cdot\vec{u}=0$ in $\Omega_-(t)$.
(iii) From Lemma 142, (226) and (225d) we obtain

$$\begin{aligned}&\frac{\mathrm{d}}{\mathrm{d}t}E_{\overline{\varkappa},\beta}(\Gamma(t))\\&=\left\langle\Delta_s\varkappa+(\varkappa-\overline{\varkappa})\,|\nabla_s\vec{\nu}|^2-\frac{1}{2}(\varkappa-\overline{\varkappa})^2\varkappa+\beta A(t)\left(|\nabla_s\vec{\nu}|^2-\varkappa^2\right),\mathcal{V}\right\rangle_{\Gamma(t)}\\&=-\langle f_\Gamma,\mathcal{V}\rangle_{\Gamma(t)}=-\langle f_\Gamma\,\vec{\nu},\vec{u}\rangle_{\Gamma(t)}=-\left\langle\vec{f}_\Gamma,\vec{u}\right\rangle_{\Gamma(t)}.\end{aligned}\tag{231}$$

It remains to analyse the right-hand side in (231). To this end, we first of all note that it follows from Lemmas 9(i), 9(ii), Definition 5(viii), Lemma 7(iii), the divergence theorem on hypersurfaces, i.e., Theorem 21 together with a density argument, and Remark 6(vi) that

$$\begin{aligned}\left\langle p_\Gamma,\nabla_s\cdot\vec{\xi}\right\rangle_{\Gamma(t)}&=\left\langle p_\Gamma\,\underline{\underline{P}}_\Gamma,\nabla_s\vec{\xi}\right\rangle_{\Gamma(t)}\\&=\left\langle\nabla_s\cdot\left(p_\Gamma\,\underline{\underline{P}}_\Gamma\,\vec{\xi}\right),1\right\rangle_{\Gamma(t)}-\left\langle\nabla_s\cdot(p_\Gamma\,\underline{\underline{P}}_\Gamma),\vec{\xi}\right\rangle_{\Gamma(t)}\\&=-\left\langle\nabla_s\cdot(p_\Gamma\,\underline{\underline{P}}_\Gamma),\vec{\xi}\right\rangle_{\Gamma(t)}\quad\forall\,\vec{\xi}\in[H^1(\Gamma(t))]^d.\end{aligned}\tag{232}$$

Hence it follows from (232), Remark 22(iii), together with a density argument, (229), (225b), (138) and (224) that

$$\begin{aligned}&\rho_\Gamma\left\langle\partial_t^\bullet\,\vec{u},\vec{\xi}\right\rangle_{\Gamma(t)}+2\,\mu_\Gamma\left\langle\underline{\underline{D}}_s(\vec{u}),\underline{\underline{D}}_s(\vec{\xi})\right\rangle_{\Gamma(t)}-\left\langle p_\Gamma,\nabla_s\cdot\vec{\xi}\right\rangle_{\Gamma(t)}\\&=\left\langle\rho_\Gamma\,\partial_t^\bullet\,\vec{u}-\nabla_s\cdot\underline{\underline{\sigma}}_\Gamma,\vec{\xi}\right\rangle_{\Gamma(t)}=\left\langle\left[\underline{\underline{\sigma}}\,\vec{\nu}\right]_-^+,\vec{\xi}\right\rangle_{\Gamma(t)}+\alpha\left\langle\vec{f}_\Gamma,\vec{\xi}\right\rangle_{\Gamma(t)}\\&=-2\left(\mu\,\underline{\underline{D}}(\vec{u}),\underline{\underline{D}}(\vec{\xi})\right)+\left(p,\nabla\cdot\vec{\xi}\right)+\alpha\left\langle\vec{f}_\Gamma,\vec{\xi}\right\rangle_{\Gamma(t)}\quad\forall\,\vec{\xi}\in\mathbb{V}_{\Gamma(t)},\end{aligned}\tag{233}$$

where

$$\mathbb{V}_{\Gamma(t)}=\left\{\vec{\xi}\in[H_0^1(\Omega)]^d:\underline{\underline{P}}_\Gamma\,\vec{\xi}_{|_{\Gamma(t)}}\in[H^1(\Gamma(t))]^d\right\}.\tag{234}$$

Here we recall from Lemmas 14(i), 14(iii), together with a density argument, and a trace theorem for $\Omega_-(t)$ that all the terms in (233) are well-defined for $\vec{\xi}\in\mathbb{V}_{\Gamma(t)}$. Combining (231) and (233) with $\vec{\xi}=\vec{u}$, on noting (224) and (225c), yields that

$$\begin{aligned}&\alpha\,\frac{\mathrm{d}}{\mathrm{d}t}E_{\overline{\varkappa},\beta}(\Gamma(t))+\rho_\Gamma\langle\partial_t^\bullet\,\vec{u},\vec{u}\rangle_{\Gamma(t)}\\&=-2\,\mu_\Gamma\langle\underline{\underline{D}}_s(\vec{u}),\underline{\underline{D}}_s(\vec{u})\rangle_{\Gamma(t)}-2\left(\mu\,\underline{\underline{D}}(\vec{u}),\underline{\underline{D}}(\vec{u})\right).\end{aligned}\tag{235}$$

Next, similarly to (230), it follows from (227), Remark 29(ii), (225d), Remark 8(iii), Theorem 21, Lemma 7(i), (225c) and Theorem 32 that

$$\begin{aligned}\langle \partial_t^\bullet \vec{u},\vec{u}\rangle_{\Gamma(t)} &= \langle \partial_t^\circ \vec{u},\vec{u}\rangle_{\Gamma(t)} + \left\langle \left[\left(\vec{u}-\vec{\mathcal{V}}\right)\cdot\nabla\right]\vec{u},\vec{u}\right\rangle_{\Gamma(t)} \\ &= \frac{1}{2}\left\langle \partial_t^\circ |\vec{u}|^2,1\right\rangle_{\Gamma(t)} + \left\langle \left[\underline{\underline{P}}_\Gamma\left(\vec{u}-\vec{\mathcal{V}}\right)\cdot\nabla_s\right]\vec{u},\vec{u}\right\rangle_{\Gamma(t)} \\ &= \frac{1}{2}\left\langle \partial_t^\circ |\vec{u}|^2,1\right\rangle_{\Gamma(t)} - \frac{1}{2}\left\langle \nabla_s\cdot\left[\underline{\underline{P}}_\Gamma\left(\vec{u}-\vec{\mathcal{V}}\right)\right],|\vec{u}|^2\right\rangle_{\Gamma(t)} \\ &= \frac{1}{2}\left\langle \partial_t^\circ |\vec{u}|^2,1\right\rangle_{\Gamma(t)} - \frac{1}{2}\left\langle \nabla_s\cdot\left(\vec{u}-\vec{\mathcal{V}}\right),|\vec{u}|^2\right\rangle_{\Gamma(t)} = \frac{1}{2}\frac{\mathrm{d}}{\mathrm{d}t}\left\langle |\vec{u}|^2,1\right\rangle_{\Gamma(t)}.\end{aligned}$$

Combining this with (235) yields the desired result. □

We note, in particular, that the conservation properties Lemma 147(i), (ii) follow from the continuity equations $\nabla_s\cdot\vec{u}=0$ on $\Gamma(t)$ and $\nabla\cdot\vec{u}=0$ in $\Omega_\pm(t)$, respectively. Hence, in contrast to the situation in, e.g., Section 9.4, no additional side constraints for surface area and enclosed volume are required.

10.2 A weak formulation for the dynamics of biomembranes

The most natural weak formulation for the model introduced in Section 10.1 uses the tangential velocity of the fluid for the evolution of $\Gamma(t)$, and hence (225d) is replaced by $\vec{\mathcal{V}}=\vec{u}$ on $\Gamma(t)$. This mimics the procedure in Section 8.1.4 for two-phase Stokes flow. However, contrary to the situation there, the presence of the continuity equation $\nabla_s\cdot\vec{u}=0$ on $\Gamma(t)$ will lead to good mesh properties for discretizations based on this natural weak formulation.

Now (233) leads to the following weak formulation of the system (224), (225) and (226). Given a closed hypersurface $\Gamma(0)$ and, if $\rho_\Gamma>0$, $\vec{u}_0\in[H_0^1(\Omega)]^d$, we seek an evolving hypersurface $(\Gamma(t))_{t\in[0,T]}$, with a global parameterization and induced velocity field $\vec{\mathcal{V}}$, $\vec{\varkappa}\in[L^2(\mathcal{G}_T)]^d$, $p_\Gamma\in L^2(\mathcal{G}_T)$, $\vec{f}_\Gamma\in[L^2(\mathcal{G}_T)]^d$, as well as $\vec{u}:\Omega\times[0,T]\to\mathbb{R}^d$, with $\vec{u}_{|\mathcal{G}_T}\in[H^1(\mathcal{G}_T)]^d$ and $\rho_\Gamma\,\vec{u}(\cdot,0)_{|\Gamma(0)}=\rho_\Gamma\,(\vec{u}_0)_{|\Gamma(0)}$, and $p:\Omega\times[0,T]\to\mathbb{R}$ as follows. For almost all $t\in(0,T)$, find $(\vec{\mathcal{V}}(\cdot,t),\vec{\varkappa}(\cdot,t),p_\Gamma(\cdot,t),\vec{f}_\Gamma(\cdot,t),\vec{u}(\cdot,t),p(\cdot,t))\in[L^2(\Gamma(t))]^d\times[H^1(\Gamma(t))]^d\times L^2(\Gamma(t))\times[L^2(\Gamma(t))]^d\times\mathbb{V}_{\Gamma(t)}\times L^2(\Omega)$ such that

$$\begin{aligned}&2\left(\mu\,\underline{\underline{D}}(\vec{u}),\underline{\underline{D}}(\vec{\xi})\right)-\left(p,\nabla\cdot\vec{\xi}\right)+\rho_\Gamma\left\langle\partial_t^\bullet\vec{u},\vec{\xi}\right\rangle_{\Gamma(t)}+2\,\mu_\Gamma\left\langle\underline{\underline{D}}_s(\vec{u}),\underline{\underline{D}}_s(\vec{\xi})\right\rangle_{\Gamma(t)}\\&\quad-\left\langle p_\Gamma,\nabla_s\cdot\vec{\xi}\right\rangle_{\Gamma(t)}=\alpha\left\langle\vec{f}_\Gamma,\vec{\xi}\right\rangle_{\Gamma(t)}\quad\forall\vec{\xi}\in\mathbb{V}_{\Gamma(t)},\end{aligned}\tag{236a}$$

$$(\nabla\cdot\vec{u},\varphi)=0\quad\forall\varphi\in L^2(\Omega),\tag{236b}$$

$$\langle\nabla_s\cdot\vec{u},\eta\rangle_{\Gamma(t)}=0\quad\forall\eta\in L^2(\Gamma(t)),\tag{236c}$$

$$\left\langle \vec{\mathcal{V}}, \vec{\chi} \right\rangle_{\Gamma(t)} = \langle \vec{u}, \vec{\chi} \rangle_{\Gamma(t)} \quad \forall \vec{\chi} \in [L^2(\Gamma(t))]^d, \tag{236d}$$

$$\langle \vec{\varkappa}, \vec{\eta} \rangle_{\Gamma(t)} + \left\langle \nabla_s \, \vec{\mathrm{id}}, \nabla_s \, \vec{\eta} \right\rangle_{\Gamma(t)} = 0 \quad \forall \vec{\eta} \in [H^1(\Gamma(t))]^d, \tag{236e}$$

together with an equation for $\vec{f}_\Gamma$. In the simplest case, when $\overline{\varkappa} = \beta = 0$, we recall Lemma 131 and (174a), and set

$$\begin{aligned} \left\langle \vec{f}_\Gamma, \vec{\chi} \right\rangle_{\Gamma(t)} &= \langle \nabla_s \, \vec{\varkappa}, \nabla_s \, \vec{\chi} - 2 \, \underline{\underline{D}}_s(\vec{\chi}) \rangle_{\Gamma(t)} + \langle \nabla_s \cdot \vec{\varkappa}, \nabla_s \cdot \vec{\chi} \rangle_{\Gamma(t)} \\ &\quad + \frac{1}{2} \left\langle |\vec{\varkappa}|^2, \nabla_s \cdot \vec{\chi} \right\rangle_{\Gamma(t)} \quad \forall \vec{\chi} \in [H^1(\Gamma(t))]^d. \end{aligned} \tag{236f}$$

In the general case, for nonzero $\overline{\varkappa}$ or β, we recall (210). Hence we require, instead of (236f), that in addition a $\vec{y} \in [L^2(\mathcal{G}_T)]^d$ exists such that for almost all $t \in (0, T)$ it holds that $\vec{y}(\cdot, t) \in [H^1(\Gamma(t))]^d$ and

$$\begin{aligned} \left\langle \vec{f}_\Gamma, \vec{\chi} \right\rangle_{\Gamma(t)} &= \langle \nabla_s \, \vec{y}, \nabla_s \, \vec{\chi} - 2 \, \underline{\underline{D}}_s(\vec{\chi}) \rangle_{\Gamma(t)} + \langle \nabla_s \cdot \vec{y}, \nabla_s \cdot \vec{\chi} \rangle_{\Gamma(t)} \\ &\quad - \left\langle \frac{1}{2} |\vec{\varkappa} - \overline{\varkappa} \, \vec{\nu}|^2 - \vec{\varkappa} \cdot (\vec{y} - \beta A(t) \, \vec{\nu}), \nabla_s \cdot \vec{\chi} \right\rangle_{\Gamma(t)} \\ &\quad + (\beta A(t) - \overline{\varkappa}) \left\langle \vec{\varkappa}, (\nabla_s \, \vec{\chi})^{\mathsf{T}} \, \vec{\nu} \right\rangle_{\Gamma(t)} \quad \forall \vec{\chi} \in [H^1(\Gamma(t))]^d, \end{aligned} \tag{237a}$$

$$\left\langle \vec{\varkappa} + (\beta A(t) - \overline{\varkappa}) \vec{\nu} - \vec{y}, \vec{\xi} \right\rangle_{\Gamma(t)} = 0 \quad \forall \vec{\xi} \in [H^1(\Gamma(t))]^d, \tag{237b}$$

with $A(t)$ defined by (210d).

We now argue that if no connected component of $\Gamma(t)$ is spherical, then the surface pressure is unique and the bulk pressure is unique up to an additive constant. Recall that Fig. 1 shows the special case of $\Gamma(t)$ having just a single connected component. We assume that two solutions to (236) are found that only differ in the pressure. Say one solution features the pressure pair $(p_1, p_{\Gamma,1})$ and the other $(p_2, p_{\Gamma,2})$. Let $\overline{p} = p_1 - p_2$ and $\overline{p}_\Gamma = p_{\Gamma,1} - p_{\Gamma,2}$. Then we obtain

$$\left(\overline{p}, \nabla \cdot \vec{\xi}\right) + \left\langle \overline{p}_\Gamma, \nabla_s \cdot \vec{\xi} \right\rangle_{\Gamma(t)} = 0 \quad \forall \vec{\xi} \in [H_0^1(\Omega)]^d,$$

which first of all implies that $\overline{p}$ is equal to some constants $\overline{p}_\pm$ in each connected component of $\Omega_\pm(t)$. Applying the divergence theorem in the bulk regions $\Omega_\pm(t)$, Lemma 7(i), and the divergence theorem on $\Gamma(t)$, Theorem 21, yields that

$$\nabla_s \, \overline{p}_\Gamma + \varkappa \, \overline{p}_\Gamma \, \vec{\nu} = -[\overline{p} \, \vec{\nu}]_-^+ \quad \text{on } \Gamma(t).$$

Since $\nabla_s \, \overline{p}_\Gamma$ is tangential, we obtain that $\nabla_s \, \overline{p}_\Gamma = \vec{0}$, and hence $\overline{p}_\Gamma$ is constant on connected components of $\Gamma(t)$. In addition, we obtain that $\varkappa \, \overline{p}_\Gamma = -[\overline{p}]_-^+$ on

$\Gamma(t)$. If $\varkappa$ is not constant on a connected component of $\Gamma(t)$, which is the case if it is not a sphere, then we have $\overline{p}_\Gamma = [\overline{p}]_-^+ = 0$ on this component, and so $\overline{p}_+ = \overline{p}_-$ for the associated connected components of $\Omega_\pm(t)$. Repeating this argument for all the connected components of $\Gamma(t)$ yields, if no connected component of $\Gamma(t)$ is a sphere, that p_Γ is unique on $\Gamma(t)$ and that p is unique in Ω up to an additive constant. If $\varkappa$ is constant on a connected component of $\Gamma(t)$, however, i.e., if this component of $\Gamma(t)$ is a sphere, then on this component p_Γ is only unique up to an additive constant.

Taking the above into account we formulate the following LBB-type condition. If $\Gamma(t)$ and $\partial\Omega$ are sufficiently smooth, and provided that $\Gamma(t)$ does not contain a sphere, then there exists a constant $C \in \mathbb{R}_{>0}$ such that

$$\inf_{(\varphi,\eta)\in\widehat{\mathbb{P}}\times L^2(\Gamma(t))} \sup_{\vec{\xi}\in\mathbb{V}_{\Gamma(t)}} \frac{\left(\varphi, \nabla\cdot\vec{\xi}\right) + \left\langle \eta, \nabla_s\cdot\vec{\xi}\right\rangle_{\Gamma(t)}}{\left(|\varphi|_\Omega + |\eta|_{\Gamma(t)}\right)\left(\left\|\vec{\xi}\right\|_{1,\Omega} + \left\|\underline{\underline{P}}_\Gamma\,\vec{\xi}_{|_{\Gamma(t)}}\right\|_{1,\Gamma(t)}\right)} \geq C > 0. \tag{238}$$

Here we have recalled (234) as well as the definitions of the space $\widehat{\mathbb{P}}$ and the H^1-norm $\|\cdot\|_{1,\Omega}$ from (143). In addition, we let $\|\cdot\|_{1,\Gamma(t)}^2 = |\cdot|_{\Gamma(t)}^2 + |\nabla_s\cdot|_{\Gamma(t)}^2$ define the H^1-norm on $\Gamma(t)$. The LBB-type condition (238) can be deduced from the pressure reconstruction result in Lengeler (2015).

10.3 Semidiscrete finite element approximation

We now introduce a finite element version of the weak formulation of (236). We have discussed previously in this chapter the importance of the choice of the discrete tangential velocity of $\Gamma^h(t)$. In general, the mesh quality will deteriorate when using a velocity in the direction of the discrete normal. Similarly, for many two-phase fluid flow problems, using the tangential velocity induced by the surrounding fluid flow also leads to bad meshes. However, here it turns out that due to the approximation of the local area conservation property, $\nabla_s\cdot\vec{u} = 0$ on $\Gamma(t)$, the mesh quality typically will remain very good during the evolution even if we choose $\vec{\mathcal{V}} = \vec{u}$.

We recall the definitions of the semidiscrete finite element spaces in Section 8.1.2, and in particular the space (155). In addition, we define the space

$$\mathbb{U}^h_{\Gamma^h} = \Big\{\vec{\phi} \in H^1(0,T;\mathbb{U}^h) : \\ \exists\,\vec{\chi} \in \underline{V}_T(\mathcal{G}^h_T) \text{ with } \vec{\chi}(\cdot,t) = \vec{\pi}_{\Gamma^h}\left[\vec{\phi}_{|_{\Gamma^h(t)}}\right] \;\forall t \in [0,T]\Big\}.$$

Overall, we then obtain the following semidiscrete continuous-in-time finite element approximation of (236). First, we will state it for the simpler situation when $\overline{\varkappa} = \beta = 0$. To this end, we recall the scheme (178).

Given the closed polyhedral hypersurface $\Gamma^h(0)$, find an evolving polyhedral hypersurface $\mathcal{G}^h_T$ with induced velocity $\vec{\mathcal{V}}^h \in \underline{V}(\mathcal{G}^h_T)$, $\vec{\kappa}^h \in \underline{V}(\mathcal{G}^h_T)$, $\vec{U}^h \in \mathbb{U}^h_{\Gamma^h}$, $P^h \in \widehat{\mathbb{P}}^h_T$, $P^h_\Gamma \in V(\mathcal{G}^h_T)$ and $\vec{F}^h_\Gamma \in \underline{V}(\mathcal{G}^h_T)$ as follows. For all $t \in (0,T]$, find

$(\vec{U}^h(\cdot, t), P^h(\cdot, t), P_\Gamma^h(\cdot, t), \vec{\mathcal{V}}^h(\cdot, t), \vec{\kappa}^h(\cdot, t), \vec{F}_\Gamma^h(\cdot, t)) \in \mathbb{U}^h \times \widehat{\mathbb{P}}^h(t) \times V(\Gamma^h(t)) \times \underline{V}(\Gamma^h(t)) \times \underline{V}(\Gamma^h(t)) \times \underline{V}(\Gamma^h(t))$ such that

$$2\left(\mu^h \underline{\underline{D}}(\vec{U}^h), \underline{\underline{D}}(\vec{\xi})\right) - \left(P^h, \nabla \cdot \vec{\xi}\right) + \rho_\Gamma \left\langle \partial_t^{\circ,h} \vec{\pi}_{\Gamma^h} \vec{U}^h, \vec{\xi} \right\rangle_{\Gamma^h(t)}^h$$
$$+ 2\mu_\Gamma \left\langle \underline{\underline{D}}_s(\vec{\pi}_{\Gamma^h} \vec{U}^h), \underline{\underline{D}}_s(\vec{\pi}_{\Gamma^h} \vec{\xi}) \right\rangle_{\Gamma^h(t)} - \left\langle P_\Gamma^h, \nabla_s \cdot (\vec{\pi}_{\Gamma^h} \vec{\xi}) \right\rangle_{\Gamma^h(t)}$$
$$= \alpha \left\langle \vec{F}_\Gamma^h, \vec{\xi} \right\rangle_{\Gamma^h(t)}^h \quad \forall \vec{\xi} \in \mathbb{U}^h, \tag{239a}$$

$$\left(\nabla \cdot \vec{U}^h, \varphi\right) = 0 \quad \forall \varphi \in \widehat{\mathbb{P}}^h(t), \tag{239b}$$

$$\left\langle \nabla_s \cdot (\vec{\pi}_{\Gamma^h} \vec{U}^h), \eta \right\rangle_{\Gamma^h(t)} = 0 \quad \forall \eta \in V(\Gamma^h(t)), \tag{239c}$$

$$\left\langle \vec{\mathcal{V}}^h, \vec{\chi} \right\rangle_{\Gamma^h(t)}^h = \left\langle \vec{U}^h, \vec{\chi} \right\rangle_{\Gamma^h(t)}^h \quad \forall \vec{\chi} \in \underline{V}(\Gamma^h(t)), \tag{239d}$$

$$\left\langle \vec{\kappa}^h, \vec{\eta} \right\rangle_{\Gamma^h(t)}^h + \left\langle \nabla_s \,\vec{\mathrm{id}}, \nabla_s \vec{\eta} \right\rangle_{\Gamma^h(t)} = 0 \quad \forall \vec{\eta} \in \underline{V}(\Gamma^h(t)), \tag{239e}$$

$$\left\langle \vec{F}_\Gamma^h, \vec{\chi} \right\rangle_{\Gamma^h(t)}^h = \left\langle \nabla_s \vec{\kappa}^h, \nabla_s \vec{\chi} - 2\,\underline{\underline{D}}_s(\vec{\chi}) \right\rangle_{\Gamma^h(t)} + \left\langle \nabla_s \cdot \vec{\kappa}^h, \nabla_s \cdot \vec{\chi} \right\rangle_{\Gamma^h(t)}$$
$$+ \frac{1}{2} \left\langle |\vec{\kappa}^h|^2, \nabla_s \cdot \vec{\chi} \right\rangle_{\Gamma^h(t)}^h \quad \forall \vec{\chi} \in \underline{V}(\Gamma^h(t)). \tag{239f}$$

For this semidiscrete approximation we have the following stability and conservation results.

Theorem 148. *Let $(\mathcal{G}_T^h, \vec{U}^h, P^h, P_\Gamma^h, \vec{\kappa}^h, \vec{F}_\Gamma^h)$ be a solution to (239), and let $\vec{\kappa}^h \in \underline{V}_T(\mathcal{G}_T^h)$. Then it holds that*

$$\frac{1}{2} \frac{\mathrm{d}}{\mathrm{d}t} \left(\rho_\Gamma \left(|\vec{U}^h|_{\Gamma^h(t)}^h \right)^2 + \alpha \left(|\vec{\kappa}^h|_{\Gamma^h(t)}^h \right)^2 \right) + 2\left(\mu^h \underline{\underline{D}}(\vec{U}^h), \underline{\underline{D}}(\vec{U}^h)\right)$$
$$+ 2\mu_\Gamma \left\langle \underline{\underline{D}}_s(\vec{\pi}_{\Gamma^h} \vec{U}^h), \underline{\underline{D}}_s(\vec{\pi}_{\Gamma^h} \vec{U}^h) \right\rangle_{\Gamma^h(t)} = 0. \tag{240}$$

Moreover, it holds that

$$\frac{\mathrm{d}}{\mathrm{d}t} \left\langle \phi_k^{\Gamma^h(t)}, 1 \right\rangle_{\Gamma^h(t)} = 0, \quad k = 1, \ldots, K, \tag{241}$$

and hence that

$$\frac{\mathrm{d}}{\mathrm{d}t} \left|\Gamma^h(t)\right| = 0. \tag{242}$$

Proof. Choosing $\vec{\xi} = \vec{U}^h(\cdot, t)$ in (239a), $\varphi = P^h(\cdot, t)$ in (239b) and $\eta = P_\Gamma^h(\cdot, t)$ in (239c), we obtain

$$2\left(\mu^h\,\underline{\underline{D}}(\vec{U}^h),\,\underline{\underline{D}}(\vec{U}^h)\right)+\rho_\Gamma\left\langle\partial_t^{\circ,h}\,\vec{\pi}_{\Gamma^h}\,\vec{U}^h,\vec{U}^h\right\rangle^h_{\Gamma^h(t)}$$
$$+2\,\mu_\Gamma\langle\underline{\underline{D}}_s(\vec{\pi}_{\Gamma^h}\,\vec{U}^h),\underline{\underline{D}}_s(\vec{\pi}_{\Gamma^h}\,\vec{U}^h)\rangle_{\Gamma^h(t)}=\alpha\left\langle\vec{F}^h_\Gamma,\vec{U}^h\right\rangle^h_{\Gamma^h(t)}. \tag{243}$$

In addition, arguing as in the proof of Theorem 134 on (239e) and (239f), recall (178), yields

$$\frac{1}{2}\,\frac{\mathrm{d}}{\mathrm{d}t}\left(|\vec{\kappa}^h|^h_{\Gamma^h(t)}\right)^2=-\left\langle\vec{F}^h_\Gamma,\vec{\mathcal{V}}^h\right\rangle^h_{\Gamma^h(t)}=-\left\langle\vec{F}^h_\Gamma,\vec{U}^h\right\rangle^h_{\Gamma^h(t)}, \tag{244}$$

where the last equality follows from choosing $\vec{\chi}=\vec{F}^h_\Gamma(\,\cdot\,,t)$ in (239d). Moreover, it follows from (239d) that

$$\vec{\mathcal{V}}^h(\,\cdot\,,t)=\vec{\pi}_{\Gamma^h}\left[\vec{U}^h_{|_{\Gamma^h(t)}}\right]. \tag{245}$$

Noting this, Theorem 70(ii) and (239c) with $\eta=\pi_{\Gamma^h}\left[\left|\vec{U}^h_{|_{\Gamma^h(t)}}\right|^2\right]$ yields

$$\begin{aligned}\frac{1}{2}\,\frac{\mathrm{d}}{\mathrm{d}t}\,\langle\vec{U}^h,\vec{U}^h\rangle^h_{\Gamma^h(t)}&=\frac{1}{2}\left\langle\partial_t^{\circ,h}\,\vec{\pi}_{\Gamma^h}\left[|\vec{U}^h|^2\right],1\right\rangle^h_{\Gamma^h(t)}+\frac{1}{2}\left\langle\nabla_s\cdot\vec{\mathcal{V}}^h,|\vec{U}^h|^2\right\rangle^h_{\Gamma^h(t)}\\&=\langle\partial_t^{\circ,h}\,\vec{\pi}_{\Gamma^h}\,\vec{U}^h,\vec{U}^h\rangle^h_{\Gamma^h(t)}+\frac{1}{2}\left\langle\nabla_s\cdot(\vec{\pi}_{\Gamma^h}\,\vec{U}^h),|\vec{U}^h|^2\right\rangle^h_{\Gamma^h(t)}\\&=\langle\partial_t^{\circ,h}\,\vec{\pi}_{\Gamma^h}\,\vec{U}^h,\vec{U}^h\rangle^h_{\Gamma^h(t)}.\end{aligned} \tag{246}$$

Combining (243), (244) and (246) yields (240).

Similarly to (240), the identity (241) follows directly from Theorem 70(ii), Remark 69(i), (245) and choosing $\eta=\phi_k^{\Gamma^h(t)}$ in (239c). Finally, (242) follows by adding (241) for $k=1,\ldots,K$. □

Remark 149 (Spontaneous curvature and ADE effects).
In the case that $\overline{\varkappa}$ or β are nonzero, we replace (239f) in the semidiscrete approximation (239) with

$$\left\langle\vec{\kappa}^h+(\beta\,A^h(t)-\overline{\varkappa})\vec{\nu}^h-\vec{Y}^h,\vec{\xi}\right\rangle^h_{\Gamma^h(t)}=0\quad\forall\,\vec{\xi}\in\underline{V}(\Gamma^h(t)), \tag{247a}$$

$$\begin{aligned}\left\langle\vec{F}^h_\Gamma,\vec{\chi}\right\rangle^h_{\Gamma^h(t)}&=\langle\nabla_s\vec{Y}^h,\nabla_s\vec{\chi}-2\,\underline{\underline{D}}_s(\vec{\chi})\rangle_{\Gamma^h(t)}+\langle\nabla_s\cdot\vec{Y}^h,\nabla_s\cdot\vec{\chi}\rangle_{\Gamma^h(t)}\\&\quad-\left\langle\frac{1}{2}|\vec{\kappa}^h-\overline{\varkappa}\,\vec{\nu}^h|^2-\vec{\kappa}^h\cdot(\vec{Y}^h-\beta\,A^h(t)\,\vec{\nu}^h),\nabla_s\cdot\vec{\chi}\right\rangle^h_{\Gamma^h(t)}\\&\quad+(\beta\,A^h(t)-\overline{\varkappa})\left\langle\vec{\kappa}^h,(\nabla_s\vec{\chi})^{\mathsf{T}}\,\vec{\nu}^h\right\rangle^h_{\Gamma^h(t)}\quad\forall\,\vec{\chi}\in\underline{V}(\Gamma^h(t)),\end{aligned} \tag{247b}$$

$$A^h(t)=\langle\vec{\kappa}^h,\vec{\nu}^h\rangle^h_{\Gamma^h(t)}-M_0, \tag{247c}$$

and seek in addition a $\vec{Y}^h \in \underline{V}(\mathcal{G}_T^h)$. Here we recall (212). Then a stability result as in Theorem 148, with $\frac{1}{2}\left(\left|\vec{\kappa}^h\right|^h_{\Gamma^h(t)}\right)^2$ replaced by $E^h_{\bar{\varkappa},\beta}(\Gamma^h(t))$ can be shown for the system (239a–e), (247), on combining the proofs of Theorems 148 and 143; see also Barrett et al. (2017b, Theorem 4.2).

Remark 150.

(i) The virtual element approach introduced in Section 8.1.4 can be used for the semidiscrete approximation (239) and its generalization (239a–e), (247). In this case we obtain, in addition to $\frac{\mathrm{d}}{\mathrm{d}t}\left|\Gamma^h(t)\right| = 0$, also

$$\frac{\mathrm{d}}{\mathrm{d}t}\mathcal{L}^d(\Omega^h_-(t)) = 0,$$

which means that we obtain an approximation which conserves both the surface area and the enclosed volume, and also fulfils an energy identity.

(ii) The identity (241) ensures that the measure of the support of each basis function on $\Gamma^h(t)$ is conserved. In the case of two space dimensions, and for the number of elements/vertices $J = K$ being odd, this is equivalent to each element maintaining its length. In particular, if $\Gamma^h(0)$ is equidistributed, then $\Gamma^h(t)$ will remain equidistributed throughout. For a slight modification of (239) one can prove that the measure of each element on $\Gamma^h(t)$ is conserved, i.e., that $\frac{\mathrm{d}}{\mathrm{d}t}\mathcal{H}^{d-1}\left(\sigma_j(t)\right) = 0, j = 1, \ldots, J$, and hence (242), for arbitrary $J \geq 2$ and $d \geq 2$. To achieve this, one needs to replace $V(\Gamma^h(t))$ in (239) by $V_c(\Gamma^h(t))$, recall Definition 43(ii), i.e., piecewise constants are used for the trial space for $P^h_\Gamma(\cdot, t)$ and for the test space in (239c). However, this constraint for $d \geq 3$ can be too severe, see Barrett et al. (2016a, Remark 4.1) for more details in the case $d = 3$.

(iii) The approximation (239) and its generalization (239a–e), (247) are based on continuous piecewise linear approximations of the surface velocity and surface pressure, and so are unlikely to satisfy a discrete version of the LBB condition (238) with a constant C independent of the mesh parameters. In fact, in practice it can lead to oscillatory surface pressure approximations, see Barrett et al. (2016a, Fig. 5). This can be avoided by using a continuous piecewise quadratic interpolation of the bulk velocity on the surface. This leads to better behaved approximations of the surface pressure. Although one can still prove (240) and its generalization for the modified versions of (239) and (239a–e), (247) if $\rho_\Gamma = 0$, one can no longer show (241) and (242). This can lead to poor surface area conservation for $d = 3$ in practice, see Barrett et al. (2016a, Remark 4.1) for more details.

Finally, we state a fully discrete equivalent of (239). Let the closed polyhedral hypersurface Γ^0 be an approximation of $\Gamma(0)$, and let $\vec{\kappa}^0_{\Gamma^0} \in \underline{V}(\Gamma^0)$ be an approximation to its mean curvature vector. If $\rho_\Gamma > 0$, let $\vec{U}^0_{\Gamma^0} \in \underline{V}(\Gamma^0)$ be an approximation to $(\vec{u}_0)_{|_{\Gamma^0}}$. We also recall the time interval partitioning (40).

Then, for $m=0,\ldots,M-1$, find $\vec{U}^{m+1}\in\mathbb{U}^m$, $P^{m+1}\in\widehat{\mathbb{P}}^m$, $P_\Gamma^{m+1}\in V(\Gamma^m)$, $\vec{X}^{m+1}\in\underline{V}(\Gamma^m)$, $\vec{\kappa}^{m+1}\in\underline{V}(\Gamma^m)$ and $\vec{F}_\Gamma^{m+1}\in\underline{V}(\Gamma^m)$ such that

$$
\begin{aligned}
&2\left(\mu^m\,\underline{\underline{D}}(\vec{U}^{m+1}),\underline{\underline{D}}(\vec{\xi})\right)-\left(P^{m+1},\nabla\cdot\vec{\xi}\right)+\rho_\Gamma\left\langle\frac{\vec{U}^{m+1}-\vec{U}^m_{\Gamma^m}}{\Delta t_m},\vec{\xi}\right\rangle^h_{\Gamma^m}\\
&\quad+2\,\mu_\Gamma\left\langle\underline{\underline{D}}_s(\vec{\pi}_{\Gamma^m}\,\vec{U}^{m+1}),\,\underline{\underline{D}}_s(\vec{\pi}_{\Gamma^m}\,\vec{\xi})\right\rangle_{\Gamma^m}-\left\langle P_\Gamma^{m+1},\nabla_s\cdot(\vec{\pi}_{\Gamma^m}\,\vec{\xi})\right\rangle_{\Gamma^m}\\
&\quad=\alpha\left\langle\vec{F}_\Gamma^{m+1},\vec{\xi}\right\rangle^h_{\Gamma^m}\quad\forall\,\vec{\xi}\in\mathbb{U}^m,
\end{aligned}
\tag{248a}
$$

$$
\left(\nabla\cdot\vec{U}^{m+1},\varphi\right)=0\quad\forall\,\varphi\in\widehat{\mathbb{P}}^m,
\tag{248b}
$$

$$
\left\langle\nabla_s\cdot(\vec{\pi}_{\Gamma^m}\,\vec{U}^{m+1}),\eta\right\rangle_{\Gamma^m}=0\quad\forall\,\eta\in V(\Gamma^m),
\tag{248c}
$$

$$
\left\langle\frac{\vec{X}^{m+1}-\vec{\mathrm{id}}}{\Delta t_m},\vec{\chi}\right\rangle^h_{\Gamma^m}=\left\langle\vec{U}^{m+1},\vec{\chi}\right\rangle^h_{\Gamma^m}\quad\forall\,\vec{\chi}\in\underline{V}(\Gamma^m),
\tag{248d}
$$

$$
\left\langle\vec{\kappa}^{m+1},\vec{\eta}\right\rangle^h_{\Gamma^m}+\left\langle\nabla_s\vec{X}^{m+1},\nabla_s\vec{\eta}\right\rangle_{\Gamma^m}=0\quad\forall\,\vec{\eta}\in\underline{V}(\Gamma^m),
\tag{248e}
$$

$$
\begin{aligned}
&\left\langle\vec{F}_\Gamma^{m+1},\vec{\chi}\right\rangle^h_{\Gamma^m}=\left\langle\nabla_s\vec{\kappa}^{m+1},\nabla_s\vec{\chi}\right\rangle_{\Gamma^m}+\left\langle\nabla_s\cdot\vec{\kappa}^m_{\Gamma^m},\nabla_s\cdot\vec{\chi}\right\rangle_{\Gamma^m}\\
&\quad+\frac{1}{2}\left\langle\left|\vec{\kappa}^m_{\Gamma^m}\right|^2,\nabla_s\cdot\vec{\chi}\right\rangle^h_{\Gamma^m}-2\left\langle\nabla_s\vec{\kappa}^m_{\Gamma^m},\underline{\underline{D}}_s(\vec{\chi})\right\rangle_{\Gamma^m}\quad\forall\,\vec{\chi}\in\underline{V}(\Gamma^m),
\end{aligned}
\tag{248f}
$$

and set $\Gamma^{m+1}=\vec{X}^{m+1}(\Gamma^m)$, $\vec{\kappa}^{m+1}_{\Gamma^{m+1}}=\vec{\kappa}^{m+1}\circ(\vec{X}^{m+1})^{-1}\in\underline{V}(\Gamma^{m+1})$ and $\vec{U}^{m+1}_{\Gamma^{m+1}}=\vec{\pi}_{\Gamma^m}\left[\vec{U}^{m+1}_{|\Gamma^m}\right]\circ(\vec{X}^{m+1})^{-1}\in\underline{V}(\Gamma^{m+1})$.

The cases in which one or both of $\overline{\varkappa}$ and β are different from zero can be handled in a similar way. Moreover, on using the techniques from Section 8.2, the approximations introduced above can also be generalized to the case of Navier–Stokes flow in the bulk. In addition, a fully discrete equivalent of the approach mentioned in Remark 150(i) ensures good volume conservation in practice. We refer to Barrett et al. (2016a, 2017b) for more details. Moreover, on assuming a discrete version of the LBB condition (238), one can prove that there exists a unique solution to (248) and to the corresponding fully discrete approximation of (239a–e), (247), as well as to their extensions discussed above. The necessary techniques can be found in the proofs of Barrett et al. (2016a, Theorem 5.1) and Barrett et al. (2017b, Theorem 5.1). Finally, we observe that the linear systems resulting from (248) and its extensions can be solved with the help of a Schur complement approach, on combining the techniques presented in Remarks 122 and 145. We refer to Barrett et al. (2016a, Section 6) and Barrett et al. (2017b, Section 6) for more details.

In Fig. 7 we show a numerical simulation for an extension of the scheme (248) to Navier–Stokes flow in the bulk. The domain Ω is chosen to have a constriction, and the chosen boundary conditions model a Poiseuille-type flow.

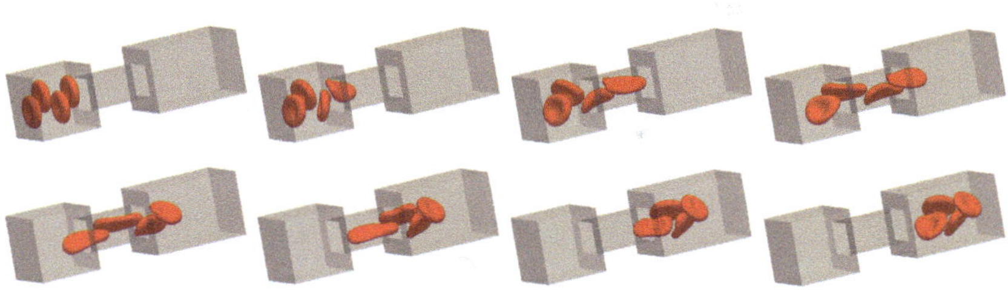

FIG. 7 Visualization of the numerical experiment shown in Barrett et al. (2016a, Fig. 12). The snapshots are taken at times $t = 0, 0.5, 1, 1.5, 2, 2.5, 3, 4$.

10.4 Two-phase biomembranes

It is also possible to consider two-phase biomembranes. In this case we can introduce an order parameter $\mathfrak{c}$, which takes the values ± 1 in the two different phases, and this parameter is related to the composition of a chemical species within the membrane. On the surface we then use a phase field model to approximate the interfacial energy by the Ginzburg–Landau functional

$$\varsigma \int_\Gamma \frac{1}{2}\,\varepsilon\,|\nabla_s\,\mathfrak{c}|^2 + \varepsilon^{-1}\,\Psi(\mathfrak{c})\;\mathrm{d}\mathcal{H}^{d-1}.$$

Here Ψ is a double well potential, $\varsigma > 0$ is related to the tension of the interface between the two phases on the membrane, often called line tension in the special case $d = 3$, and $\varepsilon > 0$ is related to the interfacial thickness of a diffuse layer between the two phases.

In the different phases α, $\overline{\varkappa}$ and α^G, cf. (223), will take different values, and we will interpolate these values obtaining functions $\alpha(\mathfrak{c}) > 0$, $\overline{\varkappa}(\mathfrak{c})$ and $\alpha^G(\mathfrak{c})$. The total energy, in the absence of ADE effects, will hence have the form

$$E(\Gamma,\mathfrak{c}) = \int_\Gamma b(\varkappa,\mathfrak{c}) + \alpha^G(\mathfrak{c})\,\mathcal{K} + \varsigma\,b_{GL}(\mathfrak{c})\;\mathrm{d}\mathcal{H}^{d-1}, \tag{249}$$

where

$$b(\varkappa,\mathfrak{c}) = \frac{1}{2}\,\alpha(\mathfrak{c})\,(\varkappa - \overline{\varkappa}(\mathfrak{c}))^2 \quad \text{and} \quad b_{GL}(\mathfrak{c}) = \frac{1}{2}\,\varepsilon\,|\nabla_s\,\mathfrak{c}|^2 + \varepsilon^{-1}\,\Psi(\mathfrak{c}).$$

In Barrett et al. (2017d) the present authors generalized the model for the dynamics of fluidic biomembranes to the two-phase case in the absence of ADE effects. Let us state here some basic ingredients. One now has to introduce an appropriate evolution law for the species concentration, $\mathfrak{c}$, on the membrane. To this end, we considered the following Cahn–Hilliard dynamics on $\Gamma(t)$

$$\vartheta\,\partial_t^\bullet\,\mathfrak{c} = \Delta_s\,\mathfrak{m}, \quad \mathfrak{m} = -\varsigma\,\varepsilon\,\Delta_s\,\mathfrak{c} + \varsigma\,\varepsilon^{-1}\,\Psi'(\mathfrak{c}) + (\partial_{\mathfrak{c}}\,b)(\varkappa,\mathfrak{c}) + (\alpha^G)'(\mathfrak{c})\,\mathcal{K},$$

where $\partial_t^\bullet$ is the material time derivative as defined in (227), $\mathfrak{m}$ denotes the chemical potential and $\vartheta \in \mathbb{R}_{>0}$ is a kinetic coefficient. We note here that $\mathfrak{m} = \frac{\delta}{\delta\mathfrak{c}}\,E(\Gamma,\mathfrak{c})$ is the first variation of the total energy with respect to $\mathfrak{c}$.

In addition, the force $\vec{f}_\Gamma$ in (225b) needs to be modified to take into account the concentration $\mathfrak{c}$. The generalized force is given by minus the first variation of the energy (249) with respect to Γ, i.e.,

$$\begin{aligned}\vec{f}_\Gamma &= -\frac{\delta}{\delta\Gamma} E(\Gamma,\mathfrak{c}) \\ &= \Big[-\Delta_s\left[\alpha(\mathfrak{c})\,(\varkappa-\overline{\varkappa}(\mathfrak{c}))\right]-\alpha(\mathfrak{c})\,(\varkappa-\overline{\varkappa}(\mathfrak{c}))\,|\nabla_s\vec{\nu}|^2+b(\varkappa,\mathfrak{c})\,\varkappa \\ &\quad -\nabla_s\cdot\left(\left[\varkappa\,\underline{\underline{\mathrm{Id}}}+\nabla_s\vec{\nu}\right]\nabla_s\alpha^G(\mathfrak{c})\right)\Big]\vec{\nu}+\left[(\partial_\mathfrak{c}\,b)(\varkappa,\mathfrak{c})+(\alpha^G)'(\mathfrak{c})\,\mathcal{K}\right]\nabla_s\,\mathfrak{c} \\ &\quad +\varsigma\left[b_{GL}(\mathfrak{c})\,\varkappa\,\vec{\nu}+\nabla_s\,b_{GL}(\mathfrak{c})-\varepsilon\,\nabla_s\cdot((\nabla_s\,\mathfrak{c})\otimes(\nabla_s\,\mathfrak{c}))\right],\end{aligned}$$

where we note that

$$(\partial_\mathfrak{c}\,b)(\varkappa,\mathfrak{c})=\frac{1}{2}\,\alpha'(\mathfrak{c})\,(\varkappa-\overline{\varkappa}(\mathfrak{c}))^2-\alpha(\mathfrak{c})\,(\varkappa-\overline{\varkappa}(\mathfrak{c}))\,\overline{\varkappa}'(\mathfrak{c}).$$

We observe that in contrast to situations where the energy density does not depend on a species concentration, we now have tangential contributions to $\vec{f}_\Gamma$, which gives rise to a Marangoni-type effect. These equations now have to be coupled to (224) and (225). A stable semidiscrete formulation of the resulting total system has been derived in Barrett et al. (2017d), where also several numerical computations showing the influence of the occurrence of the two different phases on the evolution of the membrane are shown.

10.5 Alternative numerical approaches

Let us now mention other contributions that take local incompressibility and/or fluid effects into account in the evolution of vesicles and membranes. In Bonito et al. (2011) a fluid-membrane system, in which forces resulting from the Willmore energy act on an interior flow, has been studied. Local incompressibility conditions on the membrane have been addressed by Salac and Miksis (2011), Laadhari et al. (2014), Gera and Salac (2018) within a level set context, by Jamet and Misbah (2007) and Aland et al. (2014) with the help of a phase field approach and by Hu et al. (2014) and Heintz (2015) using an immersed boundary method. Moreover, Rahimi and Arroyo (2012) and Rodrigues et al. (2015) presented numerical results using a surface Stokes system without taking the bulk fluid flow into account. There the volume conservation is enforced by a global Lagrange multiplier. Arroyo et al. (2010) simultaneously take surface and bulk viscosity effects in the fluidic membrane evolution into account. Their numerical results are in an axisymmetric setting.

Acknowledgement

The authors gratefully acknowledge the support of the Regensburger Universitätsstiftung Hans Vielberth.

References

Agnese, M., Nürnberg, R., 2016. Fitted finite element discretization of two-phase Stokes flow. Int. J. Numer. Methods Fluids 82, 709–729.

Agnese, M., Nürnberg, R., 2019. Fitted front tracking methods for two-phase incompressible Navier-Stokes flow: Eulerian and ALE finite element discretizations. Int. J. Numer. Anal. Model. (to appear).

Agricola, I., Friedrich, T., 2002. Global Analysis: Differential Forms in Analysis, Geometry, and Physics. Graduate Studies in Mathematics, vol. 52. American Mathematical Society, Providence, RI, pp. xiii+343.

Aland, S., Egerer, S., Lowengrub, J., Voigt, A., 2014. Diffuse interface models of locally inextensible vesicles in a viscous fluid. J. Comput. Phys. 277, 32–47.

Aland, S., Hahn, A., Kahle, C., Nürnberg, R., 2017. Comparative simulations of Taylor flow with surfactants based on sharp- and diffuse-interface methods. In: Bothe, D., Reusken, A. (Eds.), Transport Processes at Fluidic Interfaces. Birkhäuser, Berlin, pp. 639–679.

Almgren, R.F., 1993. Variational algorithms and pattern formation in dendritic solidification. J. Comput. Phys. 106, 337–354.

Alvarez, L., Guichard, F., Lions, P.-L., Morel, J.-M., 1993. Axioms and fundamental equations of image processing. Arch. Rational Mech. Anal. 123, 199–257.

Amann, H., Escher, J., 2009. Analysis III. Birkhäuser, Basel, pp. xii+468.

Arroyo, M., DeSimone, A., 2009. Relaxation dynamics of fluid membranes. Phys. Rev. E 79, 031915.

Arroyo, M., DeSimone, A., Heltai, L., 2010. The role of membrane viscosity in the dynamics of fluid membranes. arXiv:1007.4934.

Bänsch, E., 2001. Finite element discretization of the Navier-Stokes equations with a free capillary surface. Numer. Math. 88, 203–235.

Bänsch, E., Schmidt, A., 2000. Simulation of dendritic crystal growth with thermal convection. Interfaces Free Bound. 2, 95–115.

Bänsch, E., Morin, P., Nochetto, R.H., 2004. Surface diffusion of graphs: variational formulation, error analysis, and simulation. SIAM J. Numer. Anal. 42, 773–799.

Bänsch, E., Morin, P., Nochetto, R.H., 2005. A finite element method for surface diffusion: the parametric case. J. Comput. Phys. 203, 321–343.

Barrett, J.W., Elliott, C.M., 1982. A finite element method on a fixed mesh for the Stefan problem with convection in a saturated porous medium. In: Morton, K.W., Baines, M.J. (Eds.), Numerical Methods for Fluid Dynamics, Academic Press (London), pp. 389–409.

Barrett, J.W., Garcke, H., Nürnberg, R., 2007a. A parametric finite element method for fourth order geometric evolution equations. J. Comput. Phys. 222, 441–462.

Barrett, J.W., Garcke, H., Nürnberg, R., 2007b. On the variational approximation of combined second and fourth order geometric evolution equations. SIAM J. Sci. Comput. 29, 1006–1041.

Barrett, J.W., Garcke, H., Nürnberg, R., 2008a. On the parametric finite element approximation of evolving hypersurfaces in $\mathbb{R}^3$. J. Comput. Phys. 227, 4281–4307.

Barrett, J.W., Garcke, H., Nürnberg, R., 2008b. Numerical approximation of anisotropic geometric evolution equations in the plane. IMA J. Numer. Anal. 28, 292–330.

Barrett, J.W., Garcke, H., Nürnberg, R., 2008c. A variational formulation of anisotropic geometric evolution equations in higher dimensions. Numer. Math. 109, 1–44.

Barrett, J.W., Garcke, H., Nürnberg, R., 2008d. Parametric approximation of Willmore flow and related geometric evolution equations. SIAM J. Sci. Comput. 31, 225–253.

Barrett, J.W., Garcke, H., Nürnberg, R., 2010a. Parametric approximation of surface clusters driven by isotropic and anisotropic surface energies. Interfaces Free Bound. 12, 187–234.

Barrett, J.W., Garcke, H., Nürnberg, R., 2010b. Numerical approximation of gradient flows for closed curves in $\mathbb{R}^d$. IMA J. Numer. Anal. 30, 4–60.

Barrett, J.W., Garcke, H., Nürnberg, R., 2010c. On stable parametric finite element methods for the Stefan problem and the Mullins-Sekerka problem with applications to dendritic growth. J. Comput. Phys. 229, 6270–6299.

Barrett, J.W., Garcke, H., Nürnberg, R., 2010d. Finite element approximation of coupled surface and grain boundary motion with applications to thermal grooving and sintering. Eur. J. Appl. Math. 21, 519–556.

Barrett, J.W., Garcke, H., Nürnberg, R., 2011. The approximation of planar curve evolutions by stable fully implicit finite element schemes that equidistribute. Numer. Methods Partial Differ. Equ. 27, 1–30.

Barrett, J.W., Garcke, H., Nürnberg, R., 2012a. Numerical computations of faceted pattern formation in snow crystal growth. Phys. Rev. E 86, 011604.

Barrett, J.W., Garcke, H., Nürnberg, R., 2012b. Parametric approximation of isotropic and anisotropic elastic flow for closed and open curves. Numer. Math. 120, 489–542.

Barrett, J.W., Garcke, H., Nürnberg, R., 2012c. Elastic flow with junctions: variational approximation and applications to nonlinear splines. Math. Models Methods Appl. Sci. 22, 1250037.

Barrett, J.W., Garcke, H., Nürnberg, R., 2013a. Finite element approximation of one-sided Stefan problems with anisotropic, approximately crystalline, Gibbs-Thomson law. Adv. Differ. Equ. 18, 383–432.

Barrett, J.W., Garcke, H., Nürnberg, R., 2013b. On the stable discretization of strongly anisotropic phase field models with applications to crystal growth. ZAMM Z. Angew. Math. Mech. 93, 719–732.

Barrett, J.W., Garcke, H., Nürnberg, R., 2013c. Eliminating spurious velocities with a stable approximation of viscous incompressible two-phase Stokes flow. Comput. Methods Appl. Mech. Eng. 267, 511–530.

Barrett, J.W., Garcke, H., Nürnberg, R., 2014a. Stable phase field approximations of anisotropic solidification. IMA J. Numer. Anal. 34, 1289–1327.

Barrett, J.W., Garcke, H., Nürnberg, R., 2014b. Phase field models versus parametric front tracking methods: Are they accurate and computationally efficient? Commun. Comput. Phys. 15, 506–555.

Barrett, J.W., Garcke, H., Nürnberg, R., 2015a. A stable parametric finite element discretization of two-phase Navier-Stokes flow. J. Sci. Comp. 63, 78–117.

Barrett, J.W., Garcke, H., Nürnberg, R., 2015b. On the stable numerical approximation of two-phase flow with insoluble surfactant. M2AN Math. Model. Numer. Anal. 49, 421–458.

Barrett, J.W., Garcke, H., Nürnberg, R., 2015c. Stable finite element approximations of two-phase flow with soluble surfactant. J. Comput. Phys. 297, 530–564.

Barrett, J.W., Garcke, H., Nürnberg, R., 2015d. Stable numerical approximation of two-phase flow with a Boussinesq-Scriven surface fluid. Commun. Math. Sci. 13, 1829–1874.

Barrett, J.W., Garcke, H., Nürnberg, R., 2015e. Numerical computations of the dynamics of fluidic membranes and vesicles. Phys. Rev. E 92, 052704.

Barrett, J.W., Garcke, H., Nürnberg, R., 2016a. A stable numerical method for the dynamics of fluidic biomembranes. Numer. Math. 134, 783–822.

Barrett, J.W., Garcke, H., Nürnberg, R., 2016b. Computational parametric Willmore flow with spontaneous curvature and area difference elasticity effects. SIAM J. Numer. Anal. 54, 1732–1762.

Barrett, J.W., Deckelnick, K., Styles, V., 2017a. Numerical analysis for a system coupling curve evolution to reaction diffusion on the curve. SIAM J. Numer. Anal. 55, 1080–1100.

Barrett, J.W., Garcke, H., Nürnberg, R., 2017b. Finite element approximation for the dynamics of asymmetric fluidic biomembranes. Math. Comp. 86, 1037–1069.

Barrett, J.W., Garcke, H., Nürnberg, R., 2017c. Stable variational approximations of boundary value problems for Willmore flow with Gaussian curvature. IMA J. Numer. Anal. 37, 1657–1709.

Barrett, J.W., Garcke, H., Nürnberg, R., 2017d. Finite element approximation for the dynamics of fluidic two-phase biomembranes. M2AN Math. Model. Numer. Anal. 51, 2319–2366.

Barrett, J.W., Garcke, H., Nürnberg, R., 2018. Gradient flow dynamics of two-phase biomembranes: sharp interface variational formulation and finite element approximation. SMAI J. Comput. Math. 4, 151–195.

Barrett, J.W., Garcke, H., Nürnberg, R., 2019a. Variational discretization of axisymmetric curvature flows. Numer. Math. 141, 791–837.

Barrett, J.W., Garcke, H., Nürnberg, R., 2019b. Finite element methods for fourth order axisymmetric geometric evolution equations. J. Comput. Phys. 376, 733–766.

Bartels, S., Dolzmann, G., Nochetto, R.H., Raisch, A., 2012. Finite element methods for director fields on flexible surfaces. Interfaces Free Bound. 14, 231–272.

Bartezzaghi, A., Dedè, L., Quarteroni, A., 2019. Biomembrane modeling with isogeometric analysis. Comput. Methods Appl. Mech. Eng. 347, 103–119.

Beirão da Veiga, L., Brezzi, F., Cangiani, A., Manzini, G., Marini, L.D., Russo, A., 2013. Basic principles of virtual element methods. Math. Models Methods Appl. Sci. 23, 199–214.

Bellettini, G., Novaga, M., Paolini, M., 1999. Facet-breaking for three-dimensional crystals evolving by mean curvature. Interfaces Free Bound. 1, 39–55.

Beneš, M., 2003. Diffuse-interface treatment of the anisotropic mean-curvature flow. Appl. Math. 48, 437–453.

Bobenko, A.I., Schröder, P., 2005. Discrete Willmore flow. In: Fujii, J. (Ed.), SIGGRAPH '05. ACM SIGGRAPH 2005 Courses, ACM, New York, NY.

Boffi, D., 1997. Three-dimensional finite element methods for the Stokes problem. SIAM J. Numer. Anal. 34, 664–670.

Bonito, A., Nochetto, R.H., Pauletti, M.S., 2010. Parametric FEM for geometric biomembranes. J. Comput. Phys. 229, 3171–3188.

Bonito, A., Nochetto, R.H., Pauletti, M.S., 2011. Dynamics of biomembranes: effect of the bulk fluid. Math. Model. Nat. Phenom. 6, 25–43.

Bothe, D., Prüss, J., 2010. On the two-phase Navier-Stokes equations with Boussinesq-Scriven surface fluid. J. Math. Fluid Mech. 12, 133–150.

Bretin, E., Masnou, S., Oudet, E., 2015. Phase-field approximations of the Willmore functional and flow. Numer. Math. 131, 115–171.

Burger, M., 2005. Numerical simulation of anisotropic surface diffusion with curvature-dependent energy. J. Comput. Phys. 203, 602–625.

Burger, M., Haußer, F., Stöcker, C., Voigt, A., 2007. A level set approach to anisotropic flows with curvature regularization. J. Comput. Phys. 225, 183–205.

Caginalp, G., Lin, J.-T., 1987. A numerical analysis of an anisotropic phase field model. IMA J. Appl. Math. 39, 51–66.

Cahn, J.W., Taylor, J.E., 1994. Surface motion by surface diffusion. Acta Metall. Mater. 42, 1045–1063.

Canham, P.B., 1970. The minimum energy of bending as a possible explanation of the biconcave shape of the human red blood cell. J. Theor. Biol. 26, 61–81.

Ciarlet, P.G., 1978. The Finite Element Method for Elliptic Problems. North-Holland Publishing Co., Amsterdam, pp. xix+530.

CIBC, 2016. Cleaver: A MultiMaterial Tetrahedral Meshing Library and Application. Scientific Computing and Imaging Institute (SCI).

Clarenz, U., Diewald, U., Dziuk, G., Rumpf, M., Rusu, R., 2004. A finite element method for surface restoration with smooth boundary conditions. Comput. Aided Geom. Design 21, 427–445.

Clarenz, U., Haußer, F., Rumpf, M., Voigt, A., Weikard, U., 2005. On level set formulations for anisotropic mean curvature flow and surface diffusion. In: Multiscale Modeling in Epitaxial Growth, Int. Ser. Numer. Math., vol. 149. Birkhäuser, Basel, pp. 227–237.

Coleman, B.D., Falk, R.S., Moakher, M., 1996. Space-time finite element methods for surface diffusion with applications to the theory of the stability of cylinders. SIAM J. Sci. Comput. 17, 1434–1448.

Davis, T.A., 2004. Algorithm 832: UMFPACK V4.3—an unsymmetric-pattern multifrontal method. ACM Trans. Math. Software 30, 196–199.

Davis, T.A., 2011. Algorithm 915, SuiteSparseQR: Multifrontal multithreaded rank-revealing sparse QR factorization. ACM Trans. Math. Software 38, 1–22.

Debierre, J.-M., Karma, A., Celestini, F., Guérin, R., 2003. Phase-field approach for faceted solidification. Phys. Rev. E 68, 041604.

Deckelnick, K., Dziuk, G., 1995. On the approximation of the curve shortening flow. In: Bandle, C., Bemelmans, J., Chipot, M., Paulin, S.J., Shafrir, I. (Eds.), Calculus of Variations, Applications and Computations (Pont-à-Mousson, 1994), Pitman Res. Notes Math. Ser., vol. 326. Longman Sci. Tech., Harlow, pp. 100–108.

Deckelnick, K., Dziuk, G., 1999. Discrete anisotropic curvature flow of graphs. M2AN Math. Model. Numer. Anal. 33, 1203–1222.

Deckelnick, K., Dziuk, G., 2006. Error analysis of a finite element method for the Willmore flow of graphs. Interfaces Free Bound. 8, 21–46.

Deckelnick, K., Dziuk, G., 2009. Error analysis for the elastic flow of parametrized curves. Math. Comp. 78, 645–671.

Deckelnick, K., Schieweck, F., 2010. Error analysis for the approximation of axisymmetric Willmore flow by C^1-finite elements. Interfaces Free Bound. 12, 551–574.

Deckelnick, K., Dziuk, G., Elliott, C.M., 2003. Error analysis of a semidiscrete numerical scheme for diffusion in axially symmetric surfaces. SIAM J. Numer. Anal. 41, 2161–2179.

Deckelnick, K., Dziuk, G., Elliott, C.M., 2005a. Computation of geometric partial differential equations and mean curvature flow. Acta Numer. 14, 139–232.

Deckelnick, K., Dziuk, G., Elliott, C.M., 2005b. Fully discrete finite element approximation for anisotropic surface diffusion of graphs. SIAM J. Numer. Anal. 43, 1112–1138.

Deckelnick, K., Katz, J., Schieweck, F., 2015. A C^1-finite element method for the Willmore flow of two-dimensional graphs. Math. Comp. 84, 2617–2643.

Dedner, A., Klöfkorn, R., Nolte, M., Ohlberger, M., 2010. A generic interface for parallel and adaptive discretization schemes: abstraction principles and the DUNE-FEM module. Computing 90, 165–196.

DeTurck, D.M., 1983. Deforming metrics in the direction of their Ricci tensors. J. Differential Geom. 18, 157–162.

do Carmo, M. P., 1976. Differential Geometry of Curves and Surfaces. Prentice-Hall Inc., Englewood Cliffs, NJ, pp. viii+503.

Droske, M., Rumpf, M., 2004. A level set formulation for Willmore flow. Interfaces Free Bound. 6, 361–378.

Du, Q., Liu, C., Ryham, R., Wang, X., 2005. A phase field formulation of the Willmore problem. Nonlinearity 18, 1249–1267.

Dziuk, G., 1991. An algorithm for evolutionary surfaces. Numer. Math. 58, 603–611.

Dziuk, G., 1994. Convergence of a semi-discrete scheme for the curve shortening flow. Math. Models Methods Appl. Sci. 4, 589–606.

Dziuk, G., 1999a. Discrete anisotropic curve shortening flow. SIAM J. Numer. Anal. 36, 1808–1830.

Dziuk, G., 1999b. Numerical schemes for the mean curvature flow of graphs. In: Variations of domain and free-boundary problems in solid mechanics (Paris, 1997). Solid Mech. Appl., vol. 66. Kluwer Acad. Publ., Dordrecht, pp. 63–70.

Dziuk, G., 2008. Computational parametric Willmore flow. Numer. Math. 111, 55–80.

Dziuk, G., Elliott, C.M., 2013. Finite element methods for surface PDEs. Acta Numer. 22, 289–396.

Dziuk, G., Kuwert, E., Schätzle, R., 2002. Evolution of elastic curves in $\mathbb{R}^n$: existence and computation. SIAM J. Math. Anal. 33, 1228–1245.

Eck, C., Garcke, H., Knabner, P., 2017. Mathematical Modeling. Springer Undergraduate Mathematics Series. Springer, Cham, pp. xv+509.

Elliott, C.M., Fritz, H., 2017. On approximations of the curve shortening flow and of the mean curvature flow based on the DeTurck trick. IMA J. Numer. Anal. 37, 543–603.

Elliott, C.M., Garcke, H., 1997. Diffusional phase transitions in multicomponent systems with a concentration dependent mobility matrix. Phys. D 109, 242–256.

Elliott, C.M., Stinner, B., 2010. Modeling and computation of two phase geometric biomembranes using surface finite elements. J. Comput. Phys. 229, 6585–6612.

Elliott, C.M., Stinner, B., 2013. Computation of two-phase biomembranes with phase dependent material parameters using surface finite elements. Commun. Comput. Phys. 13, 325–360.

Elman, H.C., Silvester, D.J., Wathen, A.J., 2005. Finite Elements and Fast Iterative Solvers: With Applications in Incompressible Fluid Dynamics. Numerical Mathematics and Scientific Computation. Oxford University Press, New York, pp. xiv+400.

Escher, J., Giga, Y., Ito, K., 2001. On a limiting motion and self-intersections of curves moved by the intermediate surface diffusion flow. Nonlinear Anal. 47, 3717–3728.

Esedoḡlu, S., Rätz, A., Röger, M., 2014. Colliding interfaces in old and new diffuse-interface approximations of Willmore-flow. Commun. Math. Sci. 12, 125–147.

Gage, M., Hamilton, R.S., 1986. The heat equation shrinking convex plane curves. J. Differ. Geom. 23, 69–96.

Ganesan, S., Tobiska, L., 2012. Arbitrary Lagrangian-Eulerian finite-element method for computation of two-phase flows with soluble surfactants. J. Comput. Phys. 231, 3685–3702.

Ganesan, S., Matthies, G., Tobiska, L., 2007. On spurious velocities in incompressible flow problems with interfaces. Comput. Methods Appl. Mech. Eng. 196, 1193–1202.

Garcke, H., 2013. Curvature driven interface evolution. Jahresber. Dtsch. Math.-Ver. 115, 63–100.

Garcke, H., Stoth, B., Nestler, B., 1999. Anisotropy in multi-phase systems: a phase field approach. Interfaces Free Bound. 1, 175–198.

Garcke, H., Hinze, M., Kahle, C., 2016. A stable and linear time discretization for a thermodynamically consistent model for two-phase incompressible flow. Appl. Numer. Math. 99, 151–171.

Gera, P., Salac, D., 2018. Modeling of multicomponent three-dimensional vesicles. Comput. Fluids 172, 362–383.

Geuzaine, C., Remacle, J.-F., 2009. Gmsh: A 3-D finite element mesh generator with built-in pre- and post-processing facilities. Int. J. Numer. Methods Eng. 79, 1309–1331.

Giga, Y., 2006. Surface Evolution Equations. Birkhäuser, Basel, pp. xii+264.

Gilbarg, D., Trudinger, N.S., 1983. Elliptic Partial Differential Equations of Second Order. In: Grundlehren der Mathematischen Wissenschaften [Fundamental Principles of Mathematical Sciences], second ed. vol. 224. Springer-Verlag, Berlin, pp. xiii+513.

Girault, V., Raviart, P.-A., 1986. Finite Element Methods for Navier-Stokes Equations. Springer Series in Computational Mathematics, vol. 5. Springer-Verlag, Berlin, pp. x+374.

Gräser, C., Kornhuber, R., Sack, U., 2013. Time discretizations of anisotropic Allen-Cahn equations. IMA J. Numer. Anal. 33, 1226–1244.

Gravner, J., Griffeath, D., 2009. Modeling snow-crystal growth: a three-dimensional mesoscopic approach. Phys. Rev. E 79, 011601.

Groß, S., Reusken, A., 2011. Numerical Methods for Two-Phase Incompressible Flows. Springer Series in Computational Mathematics. Springer-Verlag, Berlin, pp. xviii+480.

Grün, G., Klingbeil, F., 2014. Two-phase flow with mass density contrast: stable schemes for a thermodynamic consistent and frame-indifferent diffuse-interface model. J. Comput. Phys. 257, 708–725.

Haußer, F., Voigt, A., 2005. A discrete scheme for regularized anisotropic surface diffusion: a 6th order geometric evolution equation. Interfaces Free Bound. 7, 353–369.

Heine, C.-J., 2004. Isoparametric finite element approximation of curvature on hypersurfaces. Preprint, University Freiburg.

Heintz, A., 2015. A numerical method for simulation dynamics of incompressible lipid membranes in viscous fluid. J. Comput. Appl. Math. 289, 87–100.

Helfrich, W., 1973. Elastic properties of lipid bilayers: theory and possible experiments. Z. Naturforsch. C 28, 693–703.

Hildebrandt, K., Polthier, K., Wardetzky, M., 2006. On the convergence of metric and geometric properties of polyhedral surfaces. Geom. Dedicata 123, 89–112.

Hinze, M., Pinnau, R., Ulbrich, M., Ulbrich, S., 2009. Optimization With PDE Constraints. Mathematical Modelling: Theory and Applications, vol. 23. Springer-Verlag, New York, pp. xii+270.

Hirt, C.W., Nichols, B.D., 1981. Volume of fluid (VOF) method for the dynamics of free boundaries. J. Comput. Phys. 39, 201–225.

Hou, T.Y., Lowengrub, J.S., Shelley, M.J., 1994. Removing the stiffness from interfacial flows with surface tension. J. Comput. Phys. 114, 312–338.

Hu, W.-F., Kim, Y., Lai, M.-C., 2014. An immersed boundary method for simulating the dynamics of three-dimensional axisymmetric vesicles in Navier-Stokes flows. J. Comput. Phys. 257, 670–686.

Huisken, G., 1984. Flow by mean curvature of convex surfaces into spheres. J. Differ. Geom. 20, 237–266.

James, A.J., Lowengrub, J., 2004. A surfactant-conserving volume-of-fluid method for interfacial flows with insoluble surfactant. J. Comput. Phys. 201, 685–722.

Jamet, D., Misbah, C., 2007. Towards a thermodynamically consistent picture of the phase-field model of vesicles: local membrane incompressibility. Phys. Rev. E 76, 051907.

Juric, D., Tryggvason, G., 1996. A front-tracking method for dendritic solidification. J. Comput. Phys. 123, 127–148.

Karma, A., Rappel, W.-J., 1998. Quantitative phase-field modeling of dendritic growth in two and three dimensions. Phys. Rev. E 57, 4323–4349.

Kay, D., Styles, V., Welford, R., 2008. Finite element approximation of a Cahn-Hilliard-Navier-Stokes system. Interfaces Free Bound. 10, 15–43.

Kessler, D.A., Koplik, J., Levine, H., 1984. Numerical simulation of two-dimensional snowflake growth. Phys. Rev. A 30, 2820–2823.

Kim, J., Kang, K., Lowengrub, J., 2004. Conservative multigrid methods for Cahn-Hilliard fluids. J. Comput. Phys. 193, 511–543.

Kimura, M., 1994. Accurate numerical scheme for the flow by curvature. Appl. Math. Lett. 7, 69–73.

Kobayashi, R., 1993. Modeling and numerical simulations of dendritic crystal growth. Phys. D 63, 410–423.

Köhne, M., Lengeler, D., 2018. Local well-posedness for relaxational fluid vesicle dynamics. J. Evol. Equ. 18, 1787–1818.

Kovács, B., Li, B., Lubich, C., 2019. A convergent evolving finite element algorithm for mean curvature flow of closed surfaces. Numer. Math. 143, 797–853.

Kühnel, W., 2015. Differential Geometry: Curves - Surfaces - Manifolds. Student Mathematical Library, vol. 77. American Mathematical Society, Providence, RI, pp. xii+402.

Laadhari, A., Saramito, P., Misbah, C., 2014. Computing the dynamics of biomembranes by combining conservative level set and adaptive finite element methods. J. Comput. Phys. 263, 328–352.

Lengeler, D., 2015. On a Stokes-type system arising in fluid vesicle dynamics. arXiv:1506.08991.

Lengeler, D., 2018. Asymptotic stability of local Helfrich minimizers. Interfaces Free Bound. 20, 533–550.

Libbrecht, K.G., 2005. The physics of snow crystals. Rep. Progr. Phys. 68, 855–895.

Libbrecht, K.G., 2008. Physically derived rules for simulating faceted crystal growth using cellular automata. arXiv:0807.2616.

Mantegazza, C., 2011. Lecture Notes on Mean Curvature Flow. Progress in Mathematics, vol. 290. Birkhäuser/Springer Basel AG, Basel, pp. xii+166.

Matveev, S.V., 2006. Lectures on algebraic topology. EMS Series of Lectures in MathematicsEuropean Mathematical Society, Zürich, pp. viii+99.

Mayer, U.F., Simonett, G., 2002. A numerical scheme for axisymmetric solutions of curvature-driven free boundary problems, with applications to the Willmore flow. Interfaces Free Bound. 4, 89–109.

Mikula, K., Ševčovič, D., 2001. Evolution of plane curves driven by a nonlinear function of curvature and anisotropy. SIAM J. Appl. Math. 61, 1473–1501.

Mikula, K., Remešíková, M., Sarkoci, P., Ševčovič, D., 2014. Manifold evolution with tangential redistribution of points. SIAM J. Sci. Comput. 36, 1384–1414.

Mullins, W.W., 1957. Theory of thermal grooving. J. Appl. Phys. 28, 333–339.

Nestler, B., 2005. A 3D parallel simulator for crystal growth and solidification in complex alloy systems. J. Cryst. Growth 275, e273–e278.

Nitsche, J.C.C., 1993. Boundary value problems for variational integrals involving surface curvatures. Quart. Appl. Math. 51, 363–387.

Pan, Z., 2008. Simulation and Analysis of Coupled Surface and Grain Boundary Motion (Ph.D. thesis). University of British Columbia, Vancouver.

Pan, Z., Wetton, B., 2008. A numerical method for coupled surface and grain boundary motion. Eur. J. Appl. Math. 19, 311–327.

Pozzi, P., 2007. Anisotropic curve shortening flow in higher codimension. Math. Methods Appl. Sci. 30, 1243–1281.

Pozzi, P., 2008. Anisotropic mean curvature flow for two dimensional surfaces in higher codimension: a numerical scheme. Interfaces Free Bound. 10, 539–576.

Prüss, J., Simonett, G., 2016. Moving interfaces and quasilinear parabolic evolution equations. Monographs in Mathematics, vol. 105. Birkhäuser/Springer, pp. xix+609.

Rahimi, M., Arroyo, M., 2012. Shape dynamics, lipid hydrodynamics, and the complex viscoelasticity of bilayer membranes. Phys. Rev. E 86, 011932.

Reiter, C.A., 2005. A local cellular model for snow crystal growth. Chaos Soliton. Fract. 23, 1111–1119.

Renardy, Y., Renardy, M., 2002. PROST: a parabolic reconstruction of surface tension for the volume-of-fluid method. J. Comput. Phys. 183, 400–421.

Rineau, L., Yvinec, M., 2019. 3D surface mesh generation. In: CGAL 4.14.1 User and Reference Manual, CGAL Editorial Board.

Rodrigues, D.S., Ausas, R.F., Mut, F., Buscaglia, G.C., 2015. A semi-implicit finite element method for viscous lipid membranes. J. Comput. Phys. 298, 565–584.

Roosen, A.R., Taylor, J.E., 1991. Simulation of crystal growth with facetted interfaces. Mater. Res. Soc. Symp. Proc. 237, 25–36.

Rusu, R.E., 2005. An algorithm for the elastic flow of surfaces. Interfaces Free Bound. 7, 229–239.

Salac, D., Miksis, M., 2011. A level set projection model of lipid vesicles in general flows. J. Comput. Phys. 230, 8192–8215.

Sapiro, G., Tannenbaum, A., 1994. On affine plane curve evolution. J. Funct. Anal. 119, 79–120.

Schmidt, A., 1993. Die Berechnung dreidimensionaler Dendriten mit Finiten Elementen (Ph.D. thesis). University Freiburg, Freiburg.

Schmidt, A., 1996. Computation of three dimensional dendrites with finite elements. J. Comput. Phys. 195, 293–312.

Schmidt, A., 1998. Approximation of crystalline dendrite growth in two space dimensions. In: Kačur, J., Mikula, K. (Eds.), Proceedings of the Algoritmy'97 Conference on Scientific Computing (Zuberec), vol. 67. Bratislava, pp. 57–68.

Schmidt, A., Siebert, K.G., 2005. Design of Adaptive Finite Element Software: The Finite Element Toolbox ALBERTA. Lecture Notes in Computational Science and Engineering, vol. 42. Springer-Verlag, Berlin pp. xii+315.

Schöberl, J., 1997. NETGEN: an advancing front 2D/3D-mesh generator based on abstract rules. Comput. Vis. Sci. 1, 41–52.

Scriven, L.E., 1960. Dynamics of a fluid interface: equation of motion for Newtonian surface fluids. Chem. Eng. Sci. 12, 98–108.

Seifert, U., 1997. Configurations of fluid membranes and vesicles. Adv. Phys. 46, 13–137.

Sethian, J.A., 1985. Curvature and the evolution of fronts. Comm. Math. Phys. 101, 487–499.

Sethian, J.A., Strain, J., 1992. Crystal growth and dendritic solidification. J. Comput. Phys. 98, 231–253.

Slattery, J.C., Sagis, L., Oh, E.-S., 2007. Interfacial Transport Phenomena, second ed. Springer, New York, pp. xviii+827.

Strain, J., 1989. A boundary integral approach to unstable solidification. J. Comput. Phys. 85, 342–389.

Sussman, M., Semereka, P., Osher, S., 1994. A level set approach for computing solutions to incompressible two-phase flow. J. Comput. Phys. 114, 146–159.

Taylor, J.E., Cahn, J.W., 1994. Linking anisotropic sharp and diffuse surface motion laws via gradient flows. J. Statist. Phys. 77, 183–197.

Taylor, J.E., Cahn, J.W., Handwerker, C.A., 1992. Geometric models of crystal growth. Acta Metall. Mater. 40, 1443–1474.

Temam, R., 2001. Navier-Stokes Equations. American Mathematical Society, Providence, RI, pp. xiv+408.

Torabi, S., Lowengrub, J., Voigt, A., Wise, S., 2009. A new phase-field model for strongly anisotropic systems. Proc. R. Soc. Lond. Ser. A Math. Phys. Eng. Sci. 465, 1337–1359.

Tröltzsch, F., 2010. Optimal Control of Partial Differential Equations: Theory, Methods and Applications. Graduate Studies in Mathematics, vol. 112. Am. Math. Soc., Providence, RI, pp. xvi+399.

Tryggvason, G., Bunner, B., Esmaeeli, A., Juric, D., Al-Rawahi, N., Tauber, W., Han, J., Nas, S., Jan, Y.J., 2001. A front-tracking method for the computations of multiphase flow. J. Comput. Phys. 169, 708–759.

Unverdi, S.O., Tryggvason, G., 1992. A front-tracking method for viscous, incompressible multi-fluid flows. J. Comput. Phys. 100, 25–37.

Walker, S.W., 2013. Tetrahedralization of isosurfaces with guaranteed-quality by edge rearrangement (TIGER). SIAM J. Sci. Comput. 35, 294–326.

Walker, S.W., 2015. The Shapes of Things: A Practical Guide to Differential Geometry and the Shape Derivative. Advances in Design and Control, vol. 28. Society for Industrial and Applied Mathematics (SIAM), Philadelphia, PA, pp. ix+154.

Walker, S.W., 2018. FELICITY: A MATLAB/C++ toolbox for developing finite element methods and simulation modeling. SIAM J. Sci. Comput. 40, 234–257.

Wheeler, A.A., Murray, B.T., Schaefer, R.J., 1993. Computation of dendrites using a phase field model. Phys. D 66, 243–262.

Wloka, J., 1987. Partial Differential Equations. Cambridge University Press, Cambridge, pp. xii+518.

Yokoyama, E., 1993. Formation of patterns during growth of snow crystals. J. Cryst. Growth 128, 251–257.

Zeidler, E., 1988. Nonlinear Functional Analysis and its Applications IV, Applications to Mathematical Physics. Springer-Verlag, Berlin, pp. xxiv+975.

Chapter 5

The phase field method for geometric moving interfaces and their numerical approximations

Qiang Du[a,*] **and Xiaobing Feng**[b]
[a]*Department of Applied Physics and Applied Mathematics and the Data Science Institute, Columbia University, New York, NY, United States*
[b]*Department of Mathematics, The University of Tennessee, Knoxville, TN, United States*
[*]*Corresponding author: e-mail: qd2125@columbia.edu*

Chapter Outline

Handbook of Numerical Analysis, Vol. 21. https://doi.org/10.1016/bs.hna.2019.05.001

Abstract
This chapter surveys recent numerical advances in the phase field method for geometric surface evolution and related geometric nonlinear partial differential equations (PDEs). Instead of describing technical details of various numerical methods and their analyses, the chapter presents a holistic overview about the main ideas of phase field modelling, its mathematical foundation, and relationships between the phase field formalism and other mathematical formalisms for geometric moving interface problems, as well as the current state of the art of numerical approximations of various phase field models with an emphasis on discussing the main ideas of numerical analysis techniques. The chapter also reviews recent development on adaptive grid methods and various applications of the phase field modelling and their numerical methods in materials science, fluid mechanics, biology and image science.

Keywords: Phase field method, Geometric law, Curvature-driven flow, Geometric nonlinear PDEs, Finite difference methods, Finite element methods, Spectral methods, Discontinuous Galerkin methods, Adaptivity, Coarse and fine error estimates, Convergence of numerical interfaces, Nonlocal and stochastic phase field models, Microstructure evolution, Biology and image science applications

AMS Classification Codes: 65M06, 65M60, 65M70, 65M20, 65M12, 65M15, 65Z05, 35B25, 35K20, 35K35, 53C44, 74A50, 74N20, 76T99, 80A22, 82C26, 92C17, 92C37, 92C55

1 Introduction

The idea of *the phase field* method could be traced back to Lord Rayleigh, Gibbs and Van der Waals and was used to describe material interfaces during phase transitions. It represents material interfaces as thin layers of finite thickness over which material properties vary smoothly. Such a thin layer of the width $O(\varepsilon)$ is often referred as a diffuse interface, and by the design the exact interface is guaranteed to be within the thin layer. In other words, this is amount to smear the exact sharp interface into a thin diffuse interface layer. For this very reason, the phase field method is also known as *the diffuse interface method* in the literature.

The phase field method was first introduced to model solid–liquid phase transition in which surface tension and nonequilibrium thermodynamics behaviour become important at the interface. Computationally, the phase field method has features in common with the level set method (Chen et al., 1991; Evans and Spruck, 1991; Osher and Fedkiw, 2003; Osher and Sethian, 1988; Sethian, 1999) in that explicit tracking of the interface can be avoided in both mathematical formulation and in numerical computations, which has advantages in mesh generation and in capturing topological changes of the interface. The phase field method uses an auxiliary phase variable/function u^ε, also known as the order parameter, to indicate phases. The phase function assumes distinct values in the bulk phases away from the diffuse interface (or the interfacial region); the after-sought interface can be identified with an intermediate level set (e.g. the zero-level set) of the phase function u^ε. It should be noted that although the level set method and the phase field method are intimately related, they differ fundamentally because the former tracks the exact sharp interface without introducing any diffuse layer.

It is clear that the idea of the phase field method does not restrict to the material phase transition problems, it is applicable to any moving interface (or free boundary) problems. Indeed, in the past forty years the phase field (or the diffuse interface) method has been developed into a major and general methodology for moving interface problems arising from astrophysics, biology, differential geometry, image processing, multiphase fluid mechanics, chemical and petroleum engineering, materials phase transition and solidification. One common theme of these application problems is that interfacial energy plays an important role in each of these moving interface evolutions. Mathematically, interfacial energies such as the surface tension are characterized by the curvature(s) of the interface such as the mean and Gauss curvatures. It turns out that the mean and Gauss curvatures can be conveniently expressed in term of the phase function, which makes the phase field method very effective for modelling the interfacial energetics, particularly the surface tension effect.

Moving interface problems under the influence of the surface tension or other interfacial energy belong to a larger class of so-called *geometric moving interface problems* in which the motion of the interfaces is driven by some curvature-dependent *geometric law*. The phase field formulations of such problems often give rise interesting and difficult geometric partial differential equations (PDEs), which is a main subject of this chapter. Among many geometric moving interface problems, the best known one perhaps is the *mean curvature flow* (MCF) whose governing geometric law is: $V_n(t) = H_{\Gamma(t)}$, where V_n and $H_{\Gamma(t)}$, respectively, stand for the normal velocity and the mean curvature of the moving interface $\Gamma(t)$ at time t. As the MCF is purely a geometric problem, it can be described by various formulations including parametric, level set and phase field formulations. It is well known that the best known phase field formulation for the MCF is the Allen–Cahn equation: $u_t^\varepsilon - \Delta u^\varepsilon + \varepsilon^{-2}((u^\varepsilon)^3 - u^\varepsilon) = 0$, which is the simplest geometric PDE. Here the small positive parameter ε measures the width of the diffuse interface layer. One of the most important theoretical results for the phase field method is to prove that the zero-level set $\Gamma^{\varepsilon}(t) := \{x \in \mathbf{R}^d;\ u^{\varepsilon}(x, t) = 0\}$ evolves according to the geometric law of the MCF and $\Gamma^{\varepsilon}(t)$ converges (as a set) to the exact MCF interface $\Gamma(t)$ as $\varepsilon \to 0^+$ (Evans et al., 1992).

Looking at the PDE, it may not be clear why the Allen–Cahn equation is related to the MCF as its phase field model. This was indeed an open question for more than 10 years. It took a considerable amount of effort by many researchers before the above convergence result was proved (Bronsard and Kohn, 1991; Chen, 1992; de Mottoni and Schatzman, 1995; Evans et al., 1992; Kohn and Sternberg, 1989; Rubinstein et al., 1989). Although rigorous proofs of the convergence of the phase field models/formulations of moving interface problems is generally difficult, there is a formal procedure which has been widely used for constructing/deriving phase field model/formulation for a given moving interface problem. This procedure is based on an energetic approach in which one first construct/postulate a Helmholtz free energy functional associated with the underlying moving interface problem, a desired phase field model/formulation then is obtained as the gradient flow for the energy functional in an appropriately chosen topology. The Helmholtz free energy functional often can be written as the sum of two parts: the first part is called the bulk energy and the second part is the interfacial (or mixing) energy. As its name indicates, the interfacial/mixing energy measures the energy stored in the interfacial layer and can be defined and computed with the help of the phase function. As expected, it would depend on the width of the interfacial layer. On the other hand, the bulk energy is usually problem-dependent and may consist of several other types of energies such as kinetic and potential/gradient energy. It should be emphasized that this energetic approach makes the phase field modelling become systematic and methodical instead of ad hoc. It shifts the main task of modelling to the construction of the Helmholtz free energy functional and the driving forces that provide the mechanism of energy dissipation.

After the phase field model is obtained, several important questions naturally arise and must be addressed. Among them we mention the following three: (1) Is the model mathematically well-posed as a geometric PDE problem? (2) Does the phase field model converge to the original sharp interface model as the width of the interfacial layer goes to zero? If it does, how fast or what is the convergence rate? (3) How to efficiently solve the phase field model (i.e., the geometric PDE problem) numerically? It turns out that these questions are not all easy to answer. For question (2) the main difficulties are to derive required uniform in ε estimates for the phase function and to characterize the solution of the limiting model. Such a rigorous proof of convergence is still missing for many moving interface problems from materials science and multiphase fluid mechanics. For question (3) several important issues must be considered. First, in order to preserve the gradient flow structure at the discrete level, it is desirable to ensure the numerical method to be energy stable besides to be accurate, which is a nontrivial task, especially if one aims to have high order (time) schemes (see Yang (2016), Yang et al. (2017), and Shen et al. (2018) for some recent advances in this direction). Second, the phase field method is used to formulate/approximate a moving interface problem (i.e., the underlying sharp interface problem) and numerical methods are constructed to solve the phase field model. If the goal is to solve the sharp interface problem, a critical question is that whether the numerical solution of the phase field model converges to the solution of the underlying sharp interface model as the width of the diffuse interface tends to zero (i.e., $\varepsilon \to 0$). One may further ask in what sense the above convergence should be interpreted. It turns out that these issues are quite difficult to address and remain open for many phase field models and their numerical approximations. The successful proofs all require to carry out some delicate and nonstandard error analyses (Feng and Li, 2015; Feng and Prohl, 2003a, 2004a,c, 2005; Feng et al., 2016). Finally, in order to resolve the thin interfacial layer, one must use spatial (and temporal) mesh sizes that are much smaller than the layer width ε for computer simulations (note that $\varepsilon \to 0$). For such extremely small mesh sizes, the resulting large linear and nonlinear algebraic problems require huge computational effort to solve. This is especially true in the three-dimensional case with uniform spatial meshes. To overcome the difficulty, one may rely on highly scalable algorithms (Takashi et al., 2011; Zhang et al., 2016a) or use adaptive methods that only use fine meshes in the interfacial layer and much coarser meshes away from the layer. The use of adaptive mesh is a natural choice considering the fact that the phase function has a nice structured profile that assumes distinct (close-to-constant) values in bulk phases away from the diffuse interface and has a large gradient in the diffuse interface.

Although the phase field method was introduced much earlier as a modelling tool for material phase transition (in particular, solidification), it was proposed as a computational technique in the early 1980s by J. S. Langer, G. Fix, G. Caginalp and others. Fix (1983) was perhaps the first to use the phase field method to numerically solve moving interface (or free boundary) problems. Since then the phase field method has garnered a lot of interest and become more and more popular. It has been developed into a general numerical methodology for solving various moving interface (i.e., free boundary) problems from differential geometry, materials science, fluid mechanics and biology, just to name a few. Practically, the methodology is well understood, however, theoretically, many difficult questions remain to be addressed both at the PDE level and at the discrete level. Moreover, computer simulations of the phase field method, especially in high dimensions, remains to be challenging because of shear amount of computations involved. Indeed, phase field simulations on extreme scales have been used as a test of computing power on world's leading high performance computers as exemplified by the appearances on the Gordon Bell prize competition Takashi et al. (2011) (2011 Gordon Bell prize winner) and Zhang et al. (2016a) (2016 Gordon Bell prize finalist, which reported the largest 3D phase field simulations to date that reached 40 Petaflops on the world's fastest supercomputer at the time, employing over 300 billion spatial grid points (6700^3) and utilizing over 10 million cores). With the increasing popularity of phase field modelling in more and more application domains, it is imperative that efficient numerical algorithms and solvers such as adaptive grid methods and preconditioning techniques must be developed and utilized.

The phase field method provides a diffusive interface description to many physical processes and geometric models. Its capability goes beyond codimensional one surfaces. For example, using vector fields or tensor fields, they can be used to describe the characteristics of geometric information with higher codimension such as point defects in two dimensions and curves in three dimensions. Mathematical models of this kind include the celebrated Ginzburg–Landau models of superconductivity, Landau–deGennes model for liquid crystal flow, generalized Ohta-Kawasaki model of diblock and triblock copolymers, among others (Davis and Gartland, 1998; Du et al., 1992; Kim and Lowengrub, 2005; Ohta and Kawasaki, 1986; Priestly, 2012; Tinkham, 2004). While many numerical methods designed for scalar equations can be extended to systems, there are new challenges associated with invariance and symmetries and choices of gauge (Du, 2005) in addition to added computational complexity due to additional quantities of interests.

This chapter consists of additional seven sections. Section 2 devotes to the mathematical foundation of the phase field method. This includes its formulations for specific geometric moving interface problems, its intimate connection to the level set method, convergence of the phase field method as the defuse interface width tends to zero, fine properties of solutions of phase field models and phase function representations of geometric quantities. Section 3 focuses on time-stepping schemes for the phase field models, in addition to ensure their accuracy, the emphasis is on presenting various energy stable schemes of different orders which include the classical energy splitting schemes and stabilized schemes as well as the newly developed IEQ (invariant energy quadratization) and SAV (scalar auxiliary variable) schemes. Section 4 devotes to spatial discretization methods, all the classical methodologies are covered. Among them are finite difference methods, finite element methods, spectral methods, discontinuous Galerkin methods and isogeometric methods. Section 5 focuses on two convergence theories of fully discrete numerical methods for phase field models. The first theory analyzes rates of convergence (or a priori error estimates) for a fixed diffuse interface width ε, while the second theory considers the convergence to the solution of the underlying sharp interface problem when both the mesh parameters and the physical parameter ε all tend zero. Section 6 addresses a posteriori error estimates which are used to design adaptive methods. The emphasis will be on the residual-based a posteriori estimators for finite element discretizations. All three types of popular adaptive methodologies will be discussed, they include *h*-, *hp*- and *r*-adaptive methods. Section 7 lists a host of applications of the phase field method ranging from materials science, fluid mechanics to biology. It also discusses some extensions of the classical phase field method such as nonlocal and stochastic phase field models. The chapter is ended with a brief summary and a few concluding remarks given in Section 8.

2 Mathematical foundation of the phase field method

2.1 Geometric surface evolution

Mathematically, a hypersurface or an interface in the Euclidean space $\mathbf{R}^{d+1}$ refers to codimension one subset Γ of $\mathbf{R}^{d+1}$. A moving interface (or a free boundary) refers to one-parameter family of interfaces $\{\Gamma_t\}_{t\geq 0}$ which evolves in the space $\mathbf{R}^{d+1}$ and are governed by some explicit or implicit mechanism/law, often called a *geometric law*. The parameter t represents the time. Such a geometric law determines how the interface moves. Most commonly seen geometric laws specify the velocity V or normal velocity V_n of the points (thinking them as particles) on the interface at each given time t. It may depend on intrinsic features (such as curvatures) of the interface and/or on external factors (such as flow velocity) of the environment where the interface exists.

Specifically, geometric surface evolution concerns with the following question: given an initial hypersurface $\Gamma_0 \in \mathbf{R}^{d+1}$, find $\Gamma_t \in \mathbf{R}^{d+1}$ such that

$$V_n(t) = F_{\text{int}}(\lambda_1, \ldots, \lambda_d) + F_{\text{ext}} \quad \text{on}\,\Gamma_t, \tag{1}$$

where $V_n = V \cdot n$ denotes the (outward) normal velocity of Γ_t and $n = n_{\Gamma_t}$ stands for the outward normal to Γ_t. $\{\lambda_j(t)\}_{j=1}^d$ are the principle curvatures of Γ_t and F_{int} is a given function of $\{\lambda_j\}_{j\geq 1}$. F_{ext} denotes an external (source) function. The geometric law (1) says that the normal velocity V_n of the interface is determined/driven by the sum of an internal (F_{int}) and an external (Fext) "forces". We note that problem (1) is stated purely as a geometric problem, it may be embedded in a more complicated moving interface problem arisen from a particular application.

2.2 Examples of geometric surface evolution

The following is a list of some best known geometric moving interface problems in the literature. We rephrase them in the framework of problem (1) by explicitly describing the functions F_{int} and Fext for each example.
Example 1 (*Mean curvature flow*).
The mean curvature flow (MCF) is defined by setting

$$F_{\text{int}} = H := \sum_{j=1}^{d} \lambda_j, \quad F_{\text{ext}} \equiv 0$$

in (1). Where $H = H(t)$ denotes the mean curvature of Γ_t. So the MCF seeks a family of hypersurfaces whose normal velocity at every point on the hypersurface is equal to the mean curvature of the hypersurface at that point for all time $t > 0$. The MCF is a curve-shortening (in 2D) and an area minimization flow (in higher dimensions) in the sense that it can be interpreted as a gradient flow for the curve length functional (in 2D) and the surface area functional (in higher dimensions) (Ecker, 2004; Giga, 2006; Zhu, 2002). The MCF finds applications in many fields such as materials science, image processing, and multiphase fluids.

Example 2 (*A generalized mean curvature flow*).
The generalized mean curvature flow (gMCF) to be considered in this chapter refers to the case with

$$F_{\text{int}} = H, \quad F_{\text{ext}} \equiv \mathbf{v} \cdot n_{\Gamma_t} + g$$

in (1). Clearly, the difference between and the MCF and the gMCF is that the latter has a nonzero external force F_{ext}. Here $\mathbf{v}$ represents the background fluid velocity, and g denotes the combined other external forces.

Example 3 (*Inverse mean curvature flow*).
The inverse mean curvature flow (IMCF) is defined by setting

$$F_{\text{int}} = \frac{1}{H}, \quad F_{\text{ext}} = 0$$

in (1), it describes the evolution of a family of hypersurfaces whose normal velocity at every point on the hypersurface is equal to the reciprocal of its mean curvature at that point for all time $t > 0$. The IMCF finds application in general relativity and was used as a main tool to prove the *Penrose Inequality* by Huisken and Ilmanen (2001).

Example 4 (*Surface diffusion flow*).
The surface diffusion flow refers to the case with

$$F_{\text{int}} = \Delta_{\Gamma_t} H, \quad F_{\text{ext}} = 0$$

in (1). Where Δ_{Γ_t} denotes the surface Laplace operator on Γ_t, so the flow requires that the normal velocity of the hypersurface equals the surface Laplace of its mean curvature at every point on the surface for all time $t > 0$. The surface diffusion arises from applications in materials science, image processing and cell biology (Chopp and Sethian, 1999; Elliott and Stinner, 2010a; Escher et al., 1998).

Example 5 (*Willmore flow*).
The Willmore flow is defined by setting

$$F_{\text{int}} = -\Delta_{\Gamma_t} H - 2H(H^2 - K), \quad K := \prod_{j=1}^{d} \lambda_j, \quad F_{\text{ext}} = 0$$

in (1). Note that K denotes the Gauss curvature of the surface. The Willmore flow can be interpreted as a gradient flow for the total quadratic mean curvature functional $\int_{\Gamma_t} H^2$ and find applications in biology and materials science (Bretin et al., 2015a; Du et al., 2004; Elliott and Stinner, 2010a; Marques and Neves, 2014; Simonett, 2001; Wang and Du, 2008; Zhang et al., 2009).

Example 6 (*Hele-Shaw flow*).
To define the Hele-Shaw flow, we set

$$F_{\text{int}} = 0, \quad F_{\text{ext}} = \frac{1}{2}\left[\frac{\partial w}{\partial n}\right]_{\Gamma_t},$$

in (1), where w is defined by

$$\begin{aligned} \Delta w &= 0 \quad \text{in } \Omega \backslash \Gamma_t, \\ w &= \sigma H \quad \text{on } \Gamma_t, \end{aligned}$$

here $\Omega \in \mathbf{R}^d$ is a bounded domain, $\frac{\partial w}{\partial n}$ stands for the normal derivative of w and $[w]_{\Gamma_t}$ denotes the jump of a function w across the interface Γ_t. σ is a positive constant and is called the surface tension in the literature. Note that we omit the outer boundary condition for w. The Hele-Shaw flow says that the normal velocity of the interface equals the jump of the normal derivative of the pressure field w across the interface. Since the pressure depends on the mean curvature H, then the normal velocity implicitly depends on the mean curvature H. The Hele-Shaw flow arises in the study of two-phase fluids. It should be noted that the identical model also arises in materials sciences and is known as the Mullins–Sekerka model, though function w stands for the temperature, not the pressure, in the Mullins–Sekerka model (Hele-Shaw, 1898b; Mullins and Sekerka, 1963).

Example 7 (*Generalized Stefan problem*).
The generalized Stefan problem can be defined by setting

$$F_{\text{int}} = 0, \quad F_{\text{ext}} = \frac{1}{2}\left[\frac{\partial \phi}{\partial n}\right]_{\Gamma_t},$$

in (1), where ϕ is defined by

$$\begin{aligned} \frac{\partial \phi}{\partial t} - \Delta \phi &= 0 \quad \text{in } \Omega \backslash \Gamma_t, \\ \phi &= \sigma(H - \alpha V_n) \quad \text{on } \Gamma_t, \end{aligned}$$

here α is another positive constant. Compared with the Hele-Shaw flow, the temperature ϕ now satisfies the heat equation in $\Omega \backslash \Gamma_t$ and the normal velocity V_n appears in the boundary condition for ϕ (thus, it appears in both sides of (1)). Note that we omit the outer boundary condition and the initial condition for ϕ. The generalized Stefan problem describes the solidification process of fluids with consideration of the surface diffusion (characterized by σ) and the super/under cooling effect (characterized by α) (Caginalp and Chen, 1998; Fix, 1983; Langer, 1986).

Example 8 (*Two-phase immiscible fluids*).
The sharp interface two-phase immiscible flow model is often written as

$$
\begin{aligned}
\rho\left(\frac{\partial \mathbf{u}}{\partial t} + (\mathbf{u}\cdot\nabla)\mathbf{u}\right) - \operatorname{div} T(\mathbf{u}, p) &= \mathbf{f} && \text{in } \Omega_T \backslash \Gamma_t,\\
\operatorname{div} \mathbf{u} &= 0 && \text{in } \Omega_T \backslash \Gamma_t,\\
[T(\mathbf{u}, p)n]_{\Gamma_t} &= \sigma H && \text{on } \Gamma_t,\\
[\mathbf{u}]_{\Gamma_t} &= 0 && \text{on } \Gamma_t,\\
T(\mathbf{u}, p) := \nu(\nabla \mathbf{u} + \nabla \mathbf{u}^T) - pI, &&&
\end{aligned}
$$

where $\Omega_T := \Omega \times (0, T)$, $\mathbf{u}$ and p denote the fluid velocity and pressure. $\nu(> 0)$ is the viscosity coefficient. Again, $[\cdot]_{\Gamma_t}$ denotes the jump function. The first two equations in the PDE system are called the Navier–Stokes equations. $T(\mathbf{u}, p)$ is called the deformation tensor which can be replaced by more complicated nonlinear version in the case when complex fluids are involved (Anderson et al., 1998; Feng et al., 2005a; Jacqmin, 1999; Liu and Shen, 2003a; Lowengrub and Truskinovsky, 1998; Yue et al., 2004). To fit the above model into the form (1), we simply set

$$F_{\text{int}} = 0, \qquad F_{\text{ext}} = [T(\mathbf{u}, p)n]_{\Gamma_t}.$$

It follows from (1) and the first interface condition in the PDE system is $V_n = \sigma H$, hence the fluid interface Γ_t evolves as a scaled mean curvature flow.

Example 9 (*Two-phase immiscible fluids*).
An interesting special case of two-phase flows is so-called the two-phase Hele-Shaw flow which refers to the motion of (one or more) viscous fluids between two flat parallel plates separated by an infinitesimally small gap. Such a physical setup is often called a Hele-Shaw cell and was originally designed by Hele-Shaw to study two dimensional potential flows (Feng and Wise, 2012; Han and Wang, 2018; Hele-Shaw, 1898b; Kim and Lowengrub, 2005; Wang and Zhang, 2013). The sharp interface model of two-phase Hele-Shaw flows is given by

$$
\begin{aligned}
\mathbf{u} &= -\frac{1}{12\eta}(\nabla p - \rho \mathbf{g}) && \text{in } \Omega_T \backslash \Gamma_t,\\
\operatorname{div} \mathbf{u} &= 0 && \text{in } \Omega_T \backslash \Gamma_t,\\
[p] &= \gamma H && \text{on } \Gamma_t,\\
[\mathbf{u}\cdot n] &= 0 && \text{on } \Gamma_t,
\end{aligned}
$$

where the first interface condition is called the Laplace–Young condition in which γ is the dimensionless surface tension coefficient and H stands for the mean curvature of Γ_t. To fit the above model into the form (1), we set

$$F_{\text{int}} = 0, \qquad F_{\text{ext}} = [p]_{\Gamma_t},$$

which says that the fluid interface Γ_t evolves as a scaled mean curvature flow.

2.3 Mathematical formulations and methodologies

As Eq. (1) is a geometric problem, in order to analyze and approximate its solutions, one needs to choose an appropriate formulation and to define the meaning of solutions, which in turn then lead to different mathematical and numerical methods for the problem.

Several mathematical formulations of (1) have been proposed and developed in the past 40 years. Among them the best known one is the parametric formulation (cf. Ecker (2004) and Zhu (2002) and the references therein), in which the coordinates of all points on the hypersurface are written as functions of chosen parameters. All best know geometric quantities can be computed in terms of the coordinate functions. The parametric formulation has been widely used to study smooth geometric surface evolutions. It has also been used to develop numerical methods for computing the solutions (i.e., the coordinate functions). In order to deal with nonsmooth geometric surface evolutions, other formulations and weak solution concepts have been introduced and investigated. One of such formulations is the varifold formulation developed by Brakke (1978), especially for the mean curvature flow. Its measure-valued weak solutions are called varifold solutions. Another weak formulation and weak solution notion, called the theory of minimal barriers, were introduced by De Giorgi (1994) and were further developed by Bellettini and Novaga (1997) and Bellettini (2013) including establishing the connection between barrier solutions and other type weak solutions. The fourth formulation, which has been very popular for both PDE analysis and numerical approximation, is the level set formulation (Osher and Fedkiw, 2003; Osher and Sethian, 1988; Sethian, 1999). The corresponding weak solutions are called level set solutions (Chen et al., 1991; Evans and Spruck, 1991). The level set formulation provides a convenient and effective formalism not only for analyzing curvature-driven flows such as the mean curvature flow but also for approximating their (level set) solutions numerically (see more details below). Finally, the fifth formulation of (1) is the phase field formulation (Anderson et al., 1998; Fife, 1988; Fix, 1983; Langer, 1986; McFadden, 2002), which is the main subject of this chapter. Like the level set formulation, the phase field formulation also provides a convenient and effective formalism for both mathematical analysis and numerical approximation of problem (1), especially, in the case when $F_{\text{ext}} \not\equiv 0$.

2.4 Level set and phase field formulations of the MCF

We now use the mean curvature flow (MCF) as an example to demonstrate the derivations of both the level set and phase field formulations of the MCF.

The level set method was introduced by Osher and Sethian (1988) for general moving interface problems. As its name indicates, the main idea of the level set formulation/method is to represent the hypersurface Γ_t as a zero-level set of a function u in $\mathbf{R}^{d+1}$, that is

$$\Gamma_t := \{x(t) \in \Omega; u(x(t), t) = 0\}, \tag{2}$$

and then to evolve the level set function u, instead of the interface Γ_t. To transfer Eq. (1) for Γ_t to an equation for u, we formally differentiate equation $u(x(t), t) = 0$ with respect to t, while treating $x = x(t)$ as an implicit function, and using the chain rule to get

$$\frac{\partial u}{\partial t} + \nabla u \cdot \frac{dx}{dt} = 0.$$

Since $V = \frac{dx}{dt}$ is the velocity of the surface, then

$$\frac{\partial u}{\partial t} + \nabla u \cdot V = 0. \tag{3}$$

Eq. (3) is often called the level set equation, and it is determined by the velocity field V and the initial condition u_0 such that $\Gamma_0 = \{x \in \mathbb{R}^{d+1}; u_0(x) = 0\}$. To illustrate, we consider the mean curvature flow described in Example 1 whose geometric law is

$$V_n(t, \cdot) = H(t, \cdot).$$

By the differential geometry facts we have

$$n = -\frac{\nabla u}{|\nabla u|} \quad \text{and} \quad H = -\mathrm{div}(n),$$

where n stands for the inward normal to Γ_t. Then the level set equation (3) becomes

$$0 = \frac{\partial u}{\partial t} + \nabla u \cdot V = \frac{\partial u}{\partial t} - |\nabla u| V_n = \frac{\partial u}{\partial t} - |\nabla u| H = \frac{\partial u}{\partial t} - |\nabla u| \,\mathrm{div}\left(\frac{\nabla u}{|\nabla u|}\right),$$

or

$$\frac{\partial u}{\partial t} - |\nabla u| \,\mathrm{div}\left(\frac{\nabla u}{|\nabla u|}\right) = 0. \tag{4}$$

Eq. (4) is the well-known level set formulation of the mean curvature flow (Chen et al., 1991; Evans and Spruck, 1991; Giga, 2006).

It should be pointed out that here we only showed that Eq. (4) holds on Γ_t, but it easy to verify that it actually holds on all level sets, hence, holds in $\mathbf{R}^{d+1}$. It is also important to note that the level set equation (3) remains the same for *all* moving interface problems although the velocity V may be different for different problems.

The phase field method for the moving interface problems can be traced back to Rayleigh (1899), Rowlinson (1979) and Edwards et al. (1991). It was originally developed as a model for solidification, but has been used for many other applications, such as crack propagation, electromigration, crystal and tumour growth. The main idea of phase field method is to introduce thickness to the interface, more precisely, it seeks a phase field function u^ε such that the interface lies in the narrow region (called the diffuse interface)

$$\Gamma_t \subset Q_t^\varepsilon := \{x(t) \in \mathbf{R}^{d+1} : |u^\varepsilon(x(t),t)| \leq 1 - \mathcal{O}(\varepsilon)\}. \tag{5}$$

Here ε is a small positive constant, which controls the width of Q_t^ε. The phase field function takes two distinct value +1 and −1, which represent two distinct phases, with a smooth change between −1 and +1 in Q_t^ε. The zero-level set $\Gamma_t^\varepsilon := \{x(t) \in \mathbf{R}^{d+1};\ u^\varepsilon(x(t), t) = 0\}$ of u^ε, which is contained in the diffuse interface Q_t^ε, is often chosen to represent Γ_t approximately. Like the level set method, this diffuse interface approach provides a convenient mathematical formalism for numerically approximating moving interface problems because explicitly tracking the interface is not needed in the formulation. Unlike the level set method, there is no master phase field equation which is valid for all moving interface problems, instead, the phase field formulation is problem-dependent and are often difficult to derive. The difficulty is due to the fact that the interface lies inside Q_t^ε, but the specified location of the interface is unknown, so the curvature at the interface cannot be calculated exactly as in the level set method. Below we again use the mean curvature flow (MCF) as an example to show a formal derivation of its phase field formulation when the profile of the interface is postulated. To this end, let $d(x)$ denote the signed distance function between point x and the interface Γ_t, and consider the fact that the solution approximates the $\tanh(\cdot)$ function, we heuristically postulate that

$$u^\varepsilon(x, t) := \tanh\left(\frac{d(x)}{\sqrt{2}\varepsilon}\right), \tag{6}$$

because it matches with the desired profile for the phase function u^ε. Then we have

$$\tanh'(s) = 1 - \tanh^2(s), \quad \tanh''(s) = -2\tanh(s)(1 - \tanh^2(s)),$$

and

$$\nabla u^\varepsilon(x) = \frac{\tanh\left(\frac{d(x)}{\sqrt{2}\varepsilon}\right)}{\sqrt{2}\varepsilon} \nabla d(x),$$

$$D^2 d(x) = \frac{\sqrt{2}\varepsilon}{1-(u^\varepsilon(x))^2}\left(D^2 u^\varepsilon(x) + \frac{2u^\varepsilon(x)}{1-(u^\varepsilon(x))^2}\nabla u^\varepsilon(x)\otimes\nabla u^\varepsilon(x)\right).$$

By differential geometry facts we have

$$|\nabla d(x)| = 1, \quad |\nabla u^\varepsilon(x)|^2 = \frac{1}{2\varepsilon^2}\left(1-(u^\varepsilon(x))^2\right)^2,$$

hence,

$$H = \mathrm{tr}(D^2 d(x)) = \frac{\sqrt{2}\varepsilon}{1-(u^\varepsilon(x))^2}\left(\Delta u^\varepsilon(x) + \frac{1}{\varepsilon^2}(u^\varepsilon(x) - (u^\varepsilon(x))^3)\right). \tag{7}$$

Recall that the approximate moving interface Γ_t^ε is represented by the zero-level set of u_t^ε, that is, $u_t^\varepsilon(x, t) = 0$ on Γ_t^ε. As in the derivation of the level set equation, we formally differentiate equation $u^\varepsilon(x, t) = 0$ with respect to t, while treating $x = x(t)$ as an implicit function, and using the chain rule to get

$$0 = \frac{\partial u^\varepsilon}{\partial t} + \nabla u^\varepsilon \cdot V = \frac{\partial u^\varepsilon}{\partial t} - |\nabla u^\varepsilon| V_n = \frac{\partial u^\varepsilon}{\partial t} - |\nabla u^\varepsilon| H, \tag{8}$$

here we have used the facts that $V = \frac{dx(t)}{dt}$ and $n = -\frac{\nabla u^\varepsilon}{|\nabla u^\varepsilon|}$. Combining (7) and (8), we obtain the follow phase field equation for the MCF:

$$\frac{\partial u^\varepsilon}{\partial t} - \Delta u^\varepsilon + \frac{1}{\varepsilon^2}\left((u^\varepsilon)^3 - u^\varepsilon\right) = 0. \tag{9}$$

Eq. (9) is called the Allen–Cahn equation in the literature, it was introduced by Allen and Cahn (1979) as a model to describe the phase separation process of a binary alloy at a fixed temperature. In the original Allen–Cahn equation u^ε denotes the concentration of one of the two species of the alloy. We also remark that Eq. (9) differs from the original Allen–Cahn equation in the time scale, t here represents $\frac{t}{\varepsilon^2}$ in the original Allen–Cahn equation, hence, it is a fast time.

The Allen–Cahn equation can also be derived using the modern *energetic approach*. To this end, we introduce the following Cahn–Hilliard free energy functional

$$\mathcal{J}(u) := \int_\Omega \left(\frac{1}{2}|\nabla u|^2 + \frac{1}{\varepsilon^2}F(u)\right) dx, \quad F(u) := \frac{1}{4}(u^2-1)^2. \tag{10}$$

where the first term in $\mathcal{J}$ is called the bulk energy and the second term is called the interfacial (or potential) energy. Then (9) can be interpreted as the L^2-gradient flow for $\mathcal{J}$, that is,

$$\frac{\partial u}{\partial t} = -\mathcal{J}'(u) \quad \text{in } L^2(\Omega), \tag{11}$$

where $\mathcal{J}'(u)$ denotes the Gâteaux derivative of $\mathcal{J}$ at u in the specified topological space. For the $L^2(\Omega)$ space, it is easy to verify that

$$\mathcal{J}'(u) = -\Delta u + \frac{1}{\varepsilon^2} f(u), \tag{12}$$

where $f(u) := F'(u) = u^3 - u$ for the double well potential given in (10). We thus get a concrete form of (11) given by

$$\frac{\partial u}{\partial t} = \Delta u - \frac{1}{\varepsilon^2} f(u). \tag{13}$$

Clearly, the above energetic approach is quite simple and systematic compared to the traditional derivation which could be involved for complex moving interface problems. The key ingredient/step is to design a "correct" energy functional $\mathcal{J}$ and to choose the "right" topology (or inner product) for a specific problem, the resulting phase field equation is then given by the general gradient flow equation (11). For this reason we may regard Eq. (11) as the (master) phase field equation.

2.5 Phase field formulations of other moving interface problems

In this section we list the known phase field formulations for all moving interface problems introduced in Section 2.2.

For the generalized mean curvature flow defined in Example 2, its phase field formulation is given by the following *convective* Allen–Cahn equation (Caginalp and Chen, 1998; Chen et al., 1998):

$$\frac{\partial u^\varepsilon}{\partial t} - \Delta u^\varepsilon + \mathbf{v} \cdot \nabla u^\varepsilon + \frac{1}{\varepsilon^2} f(u^\varepsilon) = g. \tag{14}$$

For the surface diffusion flow defined in Example 4, its phase field formulation is the following *degenerate* Cahn–Hilliard equation (Cahn et al., 1996):

$$\frac{\partial u^\varepsilon}{\partial t} + \operatorname{div}\left(b(u^\varepsilon)\nabla\left(\Delta u^\varepsilon - \frac{1}{\varepsilon^2} f(u^\varepsilon)\right)\right) = 0, \quad b(z) := 1 - z^2 \tag{15}$$

where $b(z) := 1 - z^2$.

For the Willmore flow introduced in Example 5, its phase field formulation is given by the following fourth-order PDE (Bellettini, 2013; Bellettini and Novaga, 1997; Bretin et al., 2015b; Du et al., 2004):

$$\frac{\partial u^\varepsilon}{\partial t} + \Delta\left(\Delta u^\varepsilon - \frac{1}{\varepsilon^2} f(u^\varepsilon)\right) - \frac{1}{\varepsilon^2} f'(u)\left(\Delta u^\varepsilon - \frac{1}{\varepsilon^2} f(u^\varepsilon)\right) = 0, \tag{16}$$

corresponding to the following phase field relaxation of the Willmore energy functional:

$$\mathcal{W}(u) := \frac{1}{2\varepsilon}\int_{\Omega}\left(\varepsilon\Delta u^{\varepsilon} - \frac{1}{\varepsilon}f(u^{\varepsilon})\right)^2 dx. \tag{17}$$

For the Hele-Shaw flow defined in Example 6, its phase field formulation is given by the following well-known Cahn–Hilliard equation (Alikakos et al., 1994; Cahn and Hilliard, 1958; Pego, 1989):

$$\frac{\partial u^{\varepsilon}}{\partial t} + \Delta\left(\varepsilon\Delta u^{\varepsilon} - \frac{1}{\varepsilon}f(u^{\varepsilon})\right) = 0, \tag{18}$$

which can also be obtained from (15) after setting $b(z) \equiv 1$.

For the generalized Stefan problem introduced in Example 7 of Section 2.2, its phase field formulation is given by the following so-called classical phase field model (Caginalp and Chen, 1998):

$$\alpha(\varepsilon)\frac{\partial u^{\varepsilon}}{\partial t} = \varepsilon\Delta u^{\varepsilon} - \frac{1}{\varepsilon}f(u^{\varepsilon}) + s(\varepsilon)\phi^{\varepsilon} \tag{19}$$

$$c(\varepsilon)\frac{\partial \phi^{\varepsilon}}{\partial t} = \Delta\phi^{\varepsilon} - \frac{\partial u^{\varepsilon}}{\partial t}. \tag{20}$$

Note that u^{ε} denotes the phase function in the model.

Remark 1. There is also a phase field model for generalized anisotropic Stefan problem which is widely used in the materials science community (Boettinger et al., 2002; McFadden, 2002; Plapp and Karma, 2000; Provatas and Elder, 2011).

For the two-phase-fluid diffuse interface model introduced in Example 8 of Section 2.2, when $\rho = 1$ and $\nu = 1$, its phase field formulation is given by the following coupled Navier–Stokes and Cahn–Hilliard system (Anderson et al., 1998; Jacqmin, 1999; Liu and Shen, 2003a):

$$\frac{\partial \mathbf{u}^{\varepsilon}}{\partial t} + (\mathbf{u}^{\varepsilon}\cdot\nabla)\mathbf{u}^{\varepsilon} - \Delta\mathbf{u}^{\varepsilon} + \nabla p^{\varepsilon} = \mu^{\varepsilon}\nabla\varphi^{\varepsilon} + \mathbf{f} \quad \text{in } \Omega_T, \tag{21}$$

$$\operatorname{div}\mathbf{u}^{\varepsilon} = 0 \quad \text{in } \Omega_T, \tag{22}$$

$$\frac{\partial \varphi^{\varepsilon}}{\partial t} - \Delta\mu^{\varepsilon} + \mathbf{u}^{\varepsilon}\cdot\nabla\varphi^{\varepsilon} = 0 \quad \text{in } \Omega_T, \tag{23}$$

$$\mu^{\varepsilon} = -\varepsilon\Delta\varphi^{\varepsilon} + \frac{1}{\varepsilon}f(\varphi^{\varepsilon}) \quad \text{in } \Omega_T. \tag{24}$$

Note that φ^{ε} is used to denote the phase function in the model.

Finally, the phase field formulation for the two-phase Hele-Shaw flow model described in Example 9 of Section 2.2 is given by the following coupled Darcy and Cahn–Hilliard system (Chen et al., 2017; Diegel et al., 2015; Feng and Wise, 2012; Han and Wang, 2018; Wang and Zhang, 2013):

$$\mathbf{u}^\varepsilon = -\nabla p^\varepsilon - \gamma\varphi^\varepsilon\nabla\mu^\varepsilon \qquad \text{in } \Omega_T, \tag{25}$$

$$\operatorname{div}\mathbf{u}^\varepsilon = 0 \qquad \text{in } \Omega_T, \tag{26}$$

$$\frac{\partial\varphi^\varepsilon}{\partial t} - \Delta\mu^\varepsilon + \mathbf{u}^\varepsilon\cdot\nabla\varphi^\varepsilon = 0 \qquad \text{in } \Omega_T, \tag{27}$$

$$\mu^\varepsilon = -\varepsilon\Delta\varphi^\varepsilon + \frac{1}{\varepsilon}f(\varphi^\varepsilon) \qquad \text{in } \Omega_T. \tag{28}$$

Again, φ^ε is used to denote the phase function.

Remark 2. Huisken and Ilmanen (2001) proposed a sublevel set formulation for the inverse mean curvature flow (IMCF), as expected, this formulation does not have the form of the general level set equation (3), which would give the following unusual PDE:

$$\operatorname{div}\left(\frac{\nabla u}{|\nabla u|}\right)\cdot\frac{\partial u}{\partial t} = |\nabla u|. \tag{29}$$

To the best of our knowledge, no phase field formulation for the IMCF was proposed in the literature, however, for a phase function u^ε there holds the approximate mean curvature formula (Bellettini, 2013; Bretin et al., 2014; Du et al., 2005):

$$H^\varepsilon = \frac{1}{2}\left(\Delta u^\varepsilon - \frac{1}{\varepsilon^2}f(u^\varepsilon)\right). \tag{30}$$

Using this formula, the chain rule and the geometric law for the IMCF, we then get the following phase field formulation for the IMCF:

$$\frac{1}{2}\left(\Delta u^\varepsilon - \frac{1}{\varepsilon^2}f(u^\varepsilon)\right)\cdot\frac{\partial u^\varepsilon}{\partial t} = |\nabla u^\varepsilon|, \tag{31}$$

which is also an unusual PDE.

Finally, we like to point out that similar to the formulation of the Allen–Cahn equation as an L^2-gradient flow discussed earlier in (11) and (12), many of the phase field models presented here can also be obtained by the energetic approach. For example it is well known (Alikakos et al., 1994; Chen, 1996; Fife, 1988) that the Cahn–Hilliard equations given in Example 6 is the H^{-1}-gradient flow of the Cahn–Hilliard free energy $\widehat{\mathcal{J}} := \varepsilon\mathcal{J}$. That is,

$$\frac{\partial u}{\partial t} = -\widehat{\mathcal{J}}'(u) \quad \text{in } H^{-1}(\Omega), \tag{32}$$

where the H^{-1} inner product is defined by

$$(u,v)_{H^{-1}} = \left((-\Delta)^{-1}u, v\right)_{L^2},$$

where $(\cdot\,,\cdot)_{L^2}$ denotes the standard L^2-inner product, and $(-\Delta)^{-1}$ denotes the inverse of the negative Laplacian (subject to appropriate boundary and normalization conditions).

The gradient flow structure implies in particular the dynamic energy law

$$\frac{\partial \widehat{\mathcal{J}}(u)}{\partial t} = -\left\|\frac{\partial u}{\partial t}\right\|^2, \tag{33}$$

where the $\|\cdot\|$ denotes the norm induced by the inner product that defines the gradient flow.

In the same fashion, the degenerate Cahn–Hilliard equation given in (33) can also be interpreted as a weighted H^{-1}-gradient flow of the free energy $\widehat{\mathcal{J}}$ with the energy law

$$\frac{\partial \widehat{\mathcal{J}}(u)}{\partial t} = -\,\| \, b(u)\nabla\mu \, \|_{L^2}^2, \tag{34}$$

where $\mu := -\varepsilon\Delta u + \frac{1}{\varepsilon}f(u)$ is often called the *chemical potential*. On the other hand, Eq. (16) in Example 5 is the L^2-gradient flow of the Willmore energy functional (17). The energy law for the coupled Navier–Stokes and Cahn–Hilliard system in Example 8 can also be formulated using the idea of flow map for the total energy that is the sum of the phase field interfacial energy and the fluid kinetic energy.

2.6 Relationships between phase field and other formulations

Among five mathematical formulations/methodologies for moving interface problems, the first four, namely, the parametric formulation, the (Brakke's) varifold formulation, the (De Georgi's) barrier formulation and the level set formulation, all intend to represent and to seek the sharp interface exactly. For this reason we call them *sharp interface formulations/methodologies*. In contrast, the phase field formulation and methodology does not represent the interface exactly but only seeks an approximate interface, it introduces a width to the interface so the interface is diffused. For this reason, it is often called a *diffuse interface* formulation/methodology. So the phase field methodology is fundamentally different from other methodologies. On the other hand, if we use how the interface is represented and sought as a yardstick, we also can divide the five formulations/methodologies differently as *direct* and *indirect* formulations/methodologies. The parametric, the varifold and the barrier's formulations/methodologies belong to the direct approach camp because they represent and seek the interface directly, while the level set and the phase field methodologies belong to the indirect approach camp because they both represent and seek the interface indirectly as a level set of an auxiliary function (i.e., the level set function and phase field function/variable). Comparative studies of phase field methods and other methods are of much interests, see for example, a comparison of phase field and parametric front tracking reported in Barrett et al. (2014a).

Indeed, the level set method and the phase field method are intimately connected. Their connection is revealed in the limiting process as the parameter $\varepsilon \to 0^+$ in the phase field formulations/equations. It can be shown (rigorously or formally) that the diffuse interface converges (in some sense) to the sharp interface defined by the level set formulation (Abels, 2015; Abels and Lengeler, 2014; Abels and Liu, 2018; Alikakos et al., 1994; Chen, 1992, 1996; de Mottoni and Schatzman, 1995; Evans et al., 1992; Pego, 1989). It should be noted that the convergence may hold beyond onset of singularities. This is quite important because it says that theoretically the level set formulation provides a nice mathematical framework and a good solution notion for the underlying moving interface problem which the phase field method aims to approximate and to solve, although numerically both methods provide different formalisms and platforms for developing efficient numerical methods for moving interface problems.

On the other hand, the level set method and the phase field method are fundamentally different methodologies because they are based on different principles/ideas (i.e., sharp interface vs diffuse interface). The difference also reflects on their respective PDE models. The level set PDEs often have stronger nonlinearities than their phase field counterparts. For example, the level set PDE for the mean curvature flow, see (4), is a nondivergence form quasilinear PDE whose solutions are often defined in the viscosity sense (Chen et al., 1991; Evans and Spruck, 1991), while the phase field PDE for the mean curvature flow is a semilinear PDE with constant coefficient whose solutions are defined simply by using integration by parts. Compared to the phase field method, the level set method has advantages of being simple to obtain the level set equation and to provide sharp representation of the interface. On the other hand, it often does not satisfy mass conservation, which may lead to the nonphysical motions of the interface, and strong nonlinearity of the level set PDE makes it harder to solve numerically. Since the phase field function usually has physical meaning, which makes the phase field method more convenient to handle physically or biologically charged interface motion. However, although the phase field equation is easier to discretize, to resolve the thin diffuse interface one is forced to use adaptive mesh techniques (Feng and Wu, 2005, 2008; Kessler et al., 2004; Provatas and Elder, 2011; Provatas et al., 1999) because the computation becomes intractable, especially in three-dimensional cases if uniform meshes are used. Both the level set method and the phase field method share an important advantage over the direct methods, that is, they can handle with ease singularities (or topological changes) of the interfaces.

We note that the phase field formulation also relates to other sharp interface formulations. Ilmanen (1993) proved the convergence of the solution of a phase field equation to the Brakke's varifold solution for the mean curvature flow, and in Bellettini (2013), Bellettini and Novaga (1997), etc., established the convergence of the solution of another phase field equation to the De Georgi's barrier solution for the Willmore flow.

2.7 Phase function representations of geometric quantities

To obtain a phase field formulation for a given geometric law, it is very important to be able to represent well-known geometric quantities such as the mean curvature and Gauss curvature in terms of the phase function, this then requires to develop so-called the phase field calculus (cf. Du, 2011).

First, it is well known that the volume formula of a geometric domain in terms of a phase function u^ε can be derived from the integral of the phase field u^ε, or more generally, the integral of $g^\varepsilon(u^\varepsilon)$ where g^ε could take on an approximation of indicator function such that $g^\varepsilon(u^\varepsilon)$ is approximately one inside the geometric domain and zero outside (as determined by the values of u^ε).

Second, the interface area formula can be expressed, as discussed before by the Cahn–Hilliard energy functional given in (10).

Other frequently used geometric quantities include the interface normal, expressed by $\nabla u^\varepsilon / |\nabla u^\varepsilon|$, and as already mentioned above, the formula for the mean curvature of the interface represented by the zero-level set of the phase field function u^ε as given in (30) (Du et al., 2004, 2006). In addition, the phase field version of Gauss curvature can also be computed (Bellettini, 2013; Bellettini and Mugnai, 2010; Bretin et al., 2014; Du et al., 2005). A form similar to (30) is given in Bellettini and Mugnai (2010) as

$$K^\varepsilon = \frac{1}{2\varepsilon}\left[\left(\varepsilon\Delta u^\varepsilon - \frac{1}{\varepsilon}f(u^\varepsilon)\right)^2 - \left|\varepsilon\nabla^2 u^\varepsilon - \frac{1}{\varepsilon}f(u^\varepsilon)\frac{\nabla u}{|\nabla u|}\otimes\frac{\nabla u}{|\nabla u|}\right|^2\right],$$

where $|\cdot|$ denotes the standard Euclidean norm. Consequently, the phase field Euler–Poincaré index can be computed by integrating the Gauss curvature to retrieve topological information of the implicitly defined zero-level set surface of the phase field function (Du et al., 2005). A simple form in 2D can be found in Du et al. (2005) and is given by

$$\chi^\varepsilon = \frac{C_2}{\varepsilon}\int_\Omega \left(\Delta u^\varepsilon - \frac{1}{\varepsilon^2}f(u^\varepsilon)\right)dx, \tag{35}$$

with the 3D version given in Du et al. (2007) by

$$\chi^\varepsilon = \frac{C_2}{\varepsilon}\int_\Omega \left(\Delta u^\varepsilon - \frac{1}{\varepsilon^2}f(u^\varepsilon)\right)p(u^\varepsilon)dx$$

for a suitably chosen function p such as $p(t) = -2n(1-t^2)^{n-1}t$, $n \geq 1$ and normalization constants C_2 and C_3. Other studies concerning topological constraints in phase field models can be found in Dondl et al. (2017)

2.8 Convergence of the phase field formulation

Since the phase field method is based on a diffuse interface idea/approach, a fundamental mathematical question is whether the diffuse interface converges to a sharp interface when the width of the diffuse interface tends to zero (i.e., $\varepsilon \longrightarrow 0$). It turns out that this is very difficult question to rigorously answer for all moving interface problems. It remains an open problem for many phase field models. To address this question, one first needs to know or to guess what is the limiting sharp interface formulation.

Many people contributed to the convergence proof for the Allen–Cahn problem to the MCF (cf. Bronsard and Kohn, 1991; Chen, 1992; de Mottoni and Schatzman, 1995; Kohn and Sternberg, 1989; Rubinstein et al., 1989), it was finally proved by Evans et al. (1992) that

$$\text{dist}_H\left(\Gamma_t^\varepsilon, \Gamma_t\right) = O(\varepsilon), \quad \text{as } \varepsilon \to 0,$$

where dist_H denotes the Hausdorff distance between two sets.

The connection between the Cahn–Hilliard equation and the Hele-Shaw problem was first formally established by Pego (1989). It was later proved by Alikakos et al. (1994) that if u^ε and μ^ε satisfy the Cahn–Hilliard system

$$\frac{\partial u^\varepsilon}{\partial t} = \Delta \mu^\varepsilon, \quad \mu^\varepsilon = -\varepsilon \Delta u^\varepsilon + \frac{1}{\varepsilon} f(u^\varepsilon),$$

then

$$\mu^\varepsilon \to w, \quad \text{dist}_H\left(\Gamma_t^\varepsilon, \Gamma_t\right) \to 0 \quad \text{as } \varepsilon \to 0$$

before onset of singularities (or topological changes). Where w denotes the solution of the Hele-Shaw problem. Chen also proved in Chen (1996) using the energy method the convergence of the solution of the Cahn–Hilliard problem to a (very) weak solution of the Hele-Shaw problem.

For *the surface diffusion flow* $V_n = -\alpha \Delta_\Gamma H$, the convergence was formally proved by Cahn et al. (1996) if u^ε and μ^ε satisfy the following *degenerate* Cahn–Hilliard equation/system:

$$\frac{\partial u^\varepsilon}{\partial t} = \text{div}\left(b(u^\varepsilon)\nabla \mu^\varepsilon\right),$$

$$\mu^\varepsilon = -\Delta u^\varepsilon + \frac{1}{\varepsilon^2} f(u^\varepsilon), \quad b(z) := 1 - z^2.$$

For *the generalized Stefan problem*, Caginalp and Chen (1998) proved that if φ^ε and u^ε satisfies the following parabolic system (which is often called the classical phase field model)

$$\alpha(\varepsilon)\frac{\partial \varphi^{\varepsilon}}{\partial t} = \varepsilon \Delta \varphi^{\varepsilon} - \frac{1}{\varepsilon} f(\varphi^{\varepsilon}) + s(\varepsilon) u^{\varepsilon},$$
$$c(\varepsilon)\frac{\partial u^{\varepsilon}}{\partial t} = \Delta u^{\varepsilon} - \frac{\partial \varphi^{\varepsilon}}{\partial t},$$

then

$$u^{\varepsilon} \to \phi, \quad \mathrm{dist}_H\left(\Gamma_t^{\varepsilon}, \Gamma_t\right) \to 0 \quad \text{as } \varepsilon \to 0$$

before onset of singularities (or topological changes) for various combinations of $\alpha(\varepsilon)$ and $s(\varepsilon)$. Where ϕ denotes the solution (temperature) of the generalized Stefan problem.

Other works concerning the sharp interface limit of Cahn–Hilliard equations and its variants can be found in Antonopoulou et al. (2018a), Dai and Du (2012, 2014), Chen et al. (2014), Dai et al. (2018), Dai and Promislow (2013), Garcke and Novick-Cohen (2000), Lee et al. (2016), and Novick-Cohen (2000). Recently a convergence proof of the Navier–Stokes–Cahn–Hilliard phase field model to the sharp interface model for two-phase fluids was carried out in Abels and Lengeler (2014), Abels et al. (2017), and Abels and Liu (2018) and for a Cahn–Larché phase field model approximating an elasticity sharp interface problem in Abels (2015).

We conclude this section by summing up the main points of the above convergence results and also make a few comments. First, the level set method is a *sharp interface* method, while phase field method is a *diffuse interface* method. The latter converges to the former as $\varepsilon \to 0$. Second, the level set function of each moving interface problem must satisfy a master equation, which is the following Hamilton–Jacobi equation:

$$\varphi_t + V_n|\nabla \varphi| = 0 \quad \text{on } \Gamma_t.$$

However, there is no such a general master equation for phase field models. Phase field equations are usually not unique and different for different moving interface problems. Third, level set formulations often give rise to quasilinear or fully nonlinear PDEs while phase field formulations result in semilinear or quasilinear PDEs which are of singularly perturbed type and involve a small parameter ε. Lastly, level set functions are purely mathematical objects and may not have physical meaning while phase field functions often represent physical quantities such as densities and concentrations.

3 Time-stepping schemes for phase field models

Simulating the time evolution of phase field models has many important applications such as phase transition, microstructure coarsening and cell motion. Typically the dynamic process governed by phase field models involve a number of stages that exhibit features on multiple time and spatial scales. The main focuses on effective time discretization have largely been devoted to the construction of methods which can offer both stable long time simulations and adequate resolutions of transient phenomena. Preserving, on the discrete level, some mathematical features associated with the continuum phase field dynamics (such as the energy law and uniform pointwise bounds on the phase field variable) has taken a centre stage in much of the numerical analysis literature. It is also desirable that these feature preserving schemes can provide high level resolution (which is not limited to just a formal high order truncation error) and can be efficiently implemented (from effective solver to scalable algorithms). While a number of classical time integrators have been studied, various splitting and stabilization techniques have been proposed to address these concerns. We refer to, for example, Barrett et al. (1999), Chen and Shen (1998), Chen et al. (2000), Du and Nicolaides (1991), Du and Yang (2017), Zhang and Du (2009), Elliott and French (1987, 1989), Eyre (1998), Feng and Prohl (2003a, 2004a,c, 2005), Feng and Li (2015), Feng et al. (2016), Wu and Li (2018), Gomez and van der Zee (2018), Li et al. (2016a), Shen and Yang (2010, 2019), Shen et al. (2018), Wise et al. (2009), Yang et al. (2018) and the references cited therein.

We note in particular that past works have utilized the gradient flow structure of many phase field equations which we focus on here. As an illustration, following (32), we formulate a generic phase field equation as $u_t = -\mathcal{J}'(u)$ with u_t being the time derivative and $\mathcal{J}'$ representing the variation of the phase field energy with respect to a generic inner product (topology) that could be either L^2 (for Allen–Cahn dynamics) or H^{-1} (for Cahn–Hilliard dynamics). We use $< \cdot, \cdot >$ to represent the associated inner product for convenience.

To present time discretization schemes, we let $\{t_n\}_{n=1}^N$ denote the discrete time steps with t_0 be the initial time and $t_N = T$ be the terminal time. We use τ to denote the discrete time step size, generically τ can be dependent on t_n, that is, $\tau = \tau_n$, which gives nonuniform time steps or adaptive time steps. For the sake of simple notation, we drop the explicit dependence on n for the discussion in this section. We use u^n to denote the numerical approximation of $u(t_n)$, the function value of the phase field variable u at time t_n.

3.1 Classical schemes

By viewing phase field equations as abstract dynamic systems, many classical time discretization schemes can be used. With the gradient flow formulation, one may offer various interpretations to these classical schemes.

Example 10 (*Fully explicit/forward Euler Scheme*).

Standard fully explicit Euler schemes for the gradient system $u_t = -\mathcal{J}'(u)$ can be written as

$$u^{n+1} = u^n - \tau \mathcal{J}'(u^n). \tag{36}$$

It can be viewed as a steepest descent iteration of the phase field energy, in the corresponding topology, with a step size τ. While popular among domain scientists due to the simplicity in the numerical implementation, the limitation on the time step size, due to stability considerations, severely affects its performance for long time integration.

Example 11 (*Fully implicit/backward Euler Scheme*).
To deal with the stiff nature of phase field models, implicit schemes that often better stability properties are better choices for long time integration than the conventional explicit Euler scheme.

The standard backward/implicit Euler scheme can be written as

$$u^{n+1} = u^n - \tau \mathcal{J}'(u^{n+1}), \tag{37}$$

which may be reformulated in a variational form as

$$\min\left\{ \mathcal{J}(u) + \frac{1}{2\tau} \| u - u^n \|^2 \right\},$$

where $\|\cdot\|$ is the norm corresponding to the same inner product $< \cdot, \cdot >$ that defines the variational derivative of J.

Thus, by interpreting $1/\tau$ as a penalty constant, one can view the backward/implicit Euler scheme as to look for the minimum of J near u^n. A consequence is that

$$\mathcal{J}(u^{n+1}) + \frac{1}{2\tau} \|u^{n+1} - u^n\|^2 = \min\left\{ \mathcal{J}(u) + \frac{1}{2\tau} \|u - u^n\|^2 \right\} \le \mathcal{J}(u^n).$$

In particular, this leads to the decrease of energy for any step size. Nevertheless, to compute u^{n+1}, a nonconvex problem needs to be solved and condition of τ may need to be imposed to ensure the uniqueness of the minimizer.

Example 12 (*Crank–Nicolson and its variant*).
For second-order Crank–Nicolson type schemes, we may consider

$$u^{n+1} = u^n - \tau \mathcal{J}'\left(\frac{u^{n+1} + u^n}{2} \right),$$

which may be obtained via the extrapolation $2u_*^{n+1} - u^n$ where u_*^{n+1} is the backward/implicit Euler solution. Other variants include

$$u^{n+1} = u^n - \frac{\tau}{2} \mathcal{J}'(u^{n+1}) - \frac{\tau}{2} \mathcal{J}'(u^n),$$

which solves

$$\min\left\{ \mathcal{J}(u) + \frac{1}{\tau} \|u - u^n\|^2 + \langle u - u^n, \mathcal{J}'(u^n) \rangle \right\},$$

instead, and a modified scheme

$$u^{n+1} = u^n - \tau\phi(u^n, u^{n+1}), \tag{38}$$

$$\langle\phi(v, w), v - w\rangle = \mathcal{J}(v) - \mathcal{J}(w) \quad \forall v, w. \tag{39}$$

For the latter, we have

$$\frac{1}{\tau}\left(\mathcal{J}(u^{n+1}) - \mathcal{J}(u^n)\right) + \frac{1}{\tau}\|u^{n+1} - u^n\|^2 = 0.$$

While for other schemes one can often only derive an inequality version the energy law, the above can be seen as an exact time discrete analogue of the energy law associated with the continuum model.

The use of the discrete variation $\phi(v, w)$ has been considered in other works such as Du et al. (2011) and Furihata (2001). Different variants have also been studied based on Taylor expansions (Gomez and Hughes, 2011).

Besides classical schemes considered above, there are also works on numerical discretization of phase field models based on traditional linear multistep methods (Akrivis et al., 1998) and Runge–Kutta (RK) methods (Shin et al., 2017; Song and Shu, 2017) which also can be studied for general gradient system (Humphries and Stuart, 1994).

3.2 Convex splitting and stabilized schemes

It is desirable to have time discretization of the gradient flows satisfying the energy stability which refers to

$$\mathcal{J}(u^{n+1}) \leq \mathcal{J}(u^n).$$

In particular, it is often viewed that the decay of energy not only preserves the physical feature but also provides stability to numerical discretization. For some phase models, preserving the pointwise bound enjoyed by the continuum model is also an important feature of numerical schemes. While a number of classical schemes mentioned in the above share such properties, there are more variants that offer more efficient implementation and better performance.

Example 13 (*Convex splitting*).

The fully implicit schemes enjoy energy stability but at the expense of solving nonlinear nonconvex problems in general. An alternative is the convex splitting for gradient flows (Elliott and Stuart, 1993; Eyre, 1998). Here, based on a decomposition $J = J_1 + J_2$ where J_1 is convex and J_2 is concave, the convex splitting scheme is given by

$$u^{n+1} = u^n - \tau\mathcal{J}_1'(u^{n+1}) - \tau\mathcal{J}_2'(u^n), \tag{40}$$

which also has an equivalent variations formulation

$$\min\left\{\mathcal{J}_1(u)+\frac{1}{2\tau}\|u-u^n\|^2-\langle u-u_n,\mathcal{J}_2'(u^n)\rangle\right\}.$$

The convexity of $\mathcal{J}_1$ assures the unique solution for any $\tau > 0$. The energy stability also follows, that is,

$$\mathcal{J}(u^{n+1})+\frac{1}{2\tau}\|u^{n+1}-u^n\|^2\le\mathcal{J}(u^n).$$

There are a number of higher order extensions, for example, by combining the convex splitting with Runge–Kutta methods.

Example 14 (*Linearly-implicit stabilized schemes*).
To make the computational task simpler for each time step than even that associated with the convex splitting schemes, a popular technique for getting energy stable scheme is to use a semiimplicit or coupled implicit-explicit scheme that lead to only linear systems of equation.

Among the first-order scheme, this generically amounts to a linear splitting of the $\mathcal{J}'(u)$ into the sum $\mathcal{J}'(u)-Au=N(u)$ and Au for some linear operator A where the first term is treated explicitly while the second is treated implicitly. That is,

$$(I+\tau A)u^{n+1}=u^n-\tau N(u^n), \tag{41}$$

Typically Au is taken to be a scalar linear combination of the variation of the quadratic energy involving $|\nabla u|$ and u itself. The scalars in the combination are choosing to make A sufficiently coercive so that (41) becomes dissipative and energy stability can be assured.

The energy stability and the need for only linear solvers per time step have made the first-order stabilized scheme popular in practice. Similar work on second-order BDF scheme can be found in (Yan et al., 2018). For studies on other higher order stabilization, we refer to Li and Qiao (2017) and Song and Shu (2017).

Example 15 (*Exponential integrator*).
Another way of utilizing the linear splitting of the $\mathcal{J}'(u)=Au+N(u)$ is to develop exponential integrators (Cox and Matthews, 2002; Hochbruck and Ostermann, 2010; Kassam and Trefethen, 2005). The central idea is to use an equivalent formulation of the phase field equation

$$u(t+\tau)=e^{\tau A}u(t)+\int_0^\tau e^{(\tau-s)A}N(u(t+s))\,ds, \tag{42}$$

or

$$e^{-\tau A}u(t+\tau)=u(t)+\int_0^\tau e^{sA}N(u(t+s))\,ds. \tag{43}$$

By applying different quadratures to the integral over time, various discrete integration schemes can be derived. For (42), if the quadratures are based polynomial interpolations of $N(u(t + s))$ using the discrete values of u from previous time steps, we end up with the ETD (exponential time difference) linear multistep schemes. Meanwhile, the ETD-RK type scheme is to approximate $N(u(t + s))$ via polynomial interpolations of $\{N(u(t + s_i))\}$ where $\{s_i\}$ are the quadrature nodes in $(0, \tau)$ and the nodal values are constructed stage-wise through the lower order ETD-RK schemes. On the other hand, if we consider (43), a polynomial interpolation can be done for the entire integrand $e^{sA}N(u(t + s))$ instead, which would lead to the IF-RK schemes if a multistage interpolation is used.

As an illustration, the ETD-RK1 scheme is simply

$$u^{n+1}=e^{\tau A}u^n+(e^{\tau A}-I)A^{-1}N(u^n), \tag{44}$$

and the ETD-RK2 scheme is given by

$$u^{n+1}=\tilde{u}^{n+1}+\tau^{-1}(e^{-\tau A}-1+\tau A)A^{-2}\left[N(\tilde{u}^{n+1})-N(u^n)\right], \tag{45}$$

where $\tilde{u}^{n+1}$ is obtained from (44).

For both schemes, stability can be established for various phase field models (Du and Zhu, 2005; Du et al., 2018a; Kassam and Trefethen, 2005). Long time integrations and large scale simulations have been carried out (Ju et al., 2015a; Wang et al., 2016; Zhang et al., 2016a).

We note that if a first-order Taylor expansion of the exponential is used, then we recover the first-order explicit Euler scheme if $e^{\tau A} \sim I + \tau A$, or the first-order semiimplicit Euler $(I - \tau A)u^{n+1} = u^n + \tau N(u^n)$ if $e^{-\tau A} \sim I - \tau A$.

3.3 Schemes using Lagrangian multipliers

Recently an interesting approach was proposed in Guillén-González and Tierra (2013) for constructing unconditional stable linear schemes, it is based on a Lagrange multiplier approach introduced in Badia et al. (2011). This approach has lately been extensively developed and expanded, two new families of time-stepping schemes have been obtained, namely, *invariant energy quadratization* (IEQ) (Yang, 2016; Zhao et al., 2016b) and *scalar auxiliary variable schemes* (SAV) schemes (Shen et al., 2018), which are applicable to a large class of free energies. Below we explain the ideas of each family using the constant mobility Cahn–Hilliard equation (18) as an example.

Example 16 (*Invariant energy quadratization schemes*).
Consider the Cahn–Hilliard equation (18) and suppose that $F(u) \geq -C_0$ for some positive constant C_0 (note that $F'(u) = f(u) = u^3 - u$ and $C_0 = 0$ for the Cahn–Hilliard equation). Let $q := \sqrt{F(u) + C_0}$ and rewrite (18) as the system

$$u_t = \Delta\mu,$$

$$\mu = -\varepsilon\Delta u + \frac{1}{\varepsilon}G(u)q,$$

$$q_t = \frac{1}{2}G(u)u_t, \quad G(u) := \frac{f(u)}{\sqrt{F(u) + C_0}}.$$

Based on the above reformulation, the following linear implicit-explicit (IMEX) scheme, which is referred as an IEQ scheme, can be easily obtained

$$u^{n+1} - u^n = \tau\Delta\mu^{n+1}, \tag{46a}$$

$$\mu^{n+1} = -\varepsilon\Delta u^{n+1} + \frac{1}{\varepsilon}G(u^n)q^{n+1}, \tag{46b}$$

$$q^{n+1} - q^n = \frac{1}{2}G(u^n)\left(u^{n+1} - u^n\right). \tag{46c}$$

It is easy to check that the above IEQ scheme has $O(\tau)$ truncation error and it is unconditional energy stable for the discrete energy functional

$$E^n := \frac{\varepsilon}{2}\|\nabla u^n\|_{L^2}^2 + \varepsilon\|q^n\|_{L^2}^2.$$

It needs to be noted that the above discrete energy functional is a modification of the original energy functional and both are the same at the PDE level. For its detailed analysis and higher order extensions as well as applications to other problems, we refer to Yang and Ju (2017b) and Yang (2019) and the references therein.

Example 17 (*Scalar auxiliary variable schemes*).
Notice that although the above IEQ only solves a linear system (after a spatial discretization is applied) at each time step, its coefficient matrix varies at each time step. To remedy this drawback, the SAV approach starts with the induction of an auxiliary scalar function

$$r := \sqrt{E_1(u) + C_1}, \quad \text{where } E_1(u) := \int_\Omega F(u)\,dx,$$

assuming that $E_1(u) \geq -C_1$. Then the original Cahn–Hilliard equation can be rewritten as the following system:

$$
\begin{aligned}
u_t &= \Delta\mu, \\
\mu &= -\varepsilon\Delta u + \frac{1}{\varepsilon}H(u)r, \\
r_t &= \frac{1}{2}(H(u), u_t), \quad H(u) := \frac{f(u)}{\sqrt{E_1(u)+C_1}}.
\end{aligned}
$$

Based on the above reformulation, the following linear IMEX scheme, which is referred as a SAV scheme, can be easily obtained:

$$u^{n+1} - u^n = \tau\Delta\mu^{n+1}, \tag{47a}$$

$$\mu^{n+1} = -\varepsilon\Delta u^{n+1} + \frac{1}{\varepsilon}H(u^n)r^{n+1}, \tag{47b}$$

$$r^{n+1} - r^n = \frac{1}{2}\left(H(u^n), u^{n+1} - u^n\right), \tag{47c}$$

It should be noted that constant coefficient equations only need to be solved at each time step because auxiliary variable r^{n+1} can be eliminated. It is easy to check that the above SAV scheme has $O(\tau)$ truncation error and it is unconditional energy stable for the discrete energy functional

$$G^n := \frac{\varepsilon}{2} \| \nabla u^n \|_{L^2}^2 + \varepsilon(r^n)^2.$$

It also needs to be pointed out that the above discrete energy functional is a modification of the original energy functional and both are the same at the PDE level. For its detailed analysis and higher order extensions as well as applications to other problems, we refer to Cheng and Shen (2018), Shen and Xu (2018) and Shen et al. (2018) and the references therein.

3.4 Further considerations

To design effective time integrator, there are other important aspects besides the construction of the time discretization. For example, spatial approximations require particular attention due to the appearance of high order spatial derivatives in many phase field models such as the Cahn–Hilliard equation, see discussions in Section 4. A related matter is the design of efficient linear and nonlinear solvers to find the discrete solutions at each time step (from Newton's methods, Quasi-Newton, optimization-based methods, preconditioner techniques, multilevel methods, and domain decomposition and scalable algorithms). While some time discretizations lead to difficult nonlinear systems, others (such as ETD schemes) avoids the linear solver completely. The latter, when combined with domain decomposition techniques, can offer high scalability as demonstrated recently (Ju et al., 2015b; Zhang et al., 2016a).

Concerning long time integration, there is much potential to get more effective simulations via variable time steps. This is discussion in the context of adaptive methods in Section 6. Another aspect is the study of time integrators for coupling phase field and other physical models such as Cahn–Hilliard–Navier–Stokes equations and Cahn–Hilliard-Microelasticity models. This will be discussed later in Section 7. We also note that the time integration of stochastic phase field models such as stochastic Allen–Cahn and Cahn–Hilliard equations are also topics studied in the literature, we refer the reader to Otto et al. (2014), Röger and Weber (2013), Furihata et al. (2018), Kovács et al. (2014, 2015) and the references therein for detailed discussions.

Related to the subject, there are algorithmic issues in search of transition states and critical nuclei of the phase field energy, which can help shed light on processes of phase transformation. Instead of gradient dynamics that lead to equilibria, gentlest ascend dynamics and shrinking dimer dynamics have been developed for converging to saddle points (Zhang and Du, 2012; Zhou et al., 2010). Various optimization algorithms and time-stepping schemes have also been developed (Zhang et al., 2016b).

4 Spatial discretization methods for phase field models

Spatial discretization of phase field models has been widely studied, ranging from conventional finite difference, finite element and spectral methods, to specialized techniques like radial basis functions and isogeometric analysis. In this section, we present the basic formulations of a number of spatial discretization schemes.

4.1 Spatial finite difference discretization

For numerical solutions of PDE models, finite difference methods are often the ones studied first before other discretization techniques. Along with the discussion of phase field modelling and simulations of free boundary problems in the 1980s, there have been works on their spatial approximations by finite difference methods (Caginalp and Lin, 1987; Lin, 1988). Much of the difference approximations are standard, for example, second-order centre differences are used to approximate the second-order spatial derivatives. Higher order spatial derivatives are then approximated by the compositions of differences.

For example, for the Allen–Cahn equation (13) and the Cahn–Hilliard equation (18) with a constant mobility, the semidiscrete in space approximation can be given, respectively, by

$$\frac{\partial u_j^h}{\partial t} = (\Delta_h u^h)_j - \frac{1}{\varepsilon^2} f(u_j^h).$$

and

$$\frac{\partial u_j^h}{\partial t} + \Delta_h\left(\varepsilon\Delta_h u_j^h - \frac{1}{\varepsilon} f(u_j^h)\right) = 0,$$

Here u_j^h denotes the numerical solution at spatial grid point (labelled by the subscript j) with a typical uniform grid spacing h. Δ_h denotes the standard discrete Laplacian based on the second-order centre difference. Coupled with time integration schemes, one can get fully discrete approximation, see for example the work in Du and Nicolaides (1991) on the Cahn–Hilliard equation that can preserve the energy law at the discrete level. Further numerical analysis work on the difference approximations can also be found in Sun (1995) and Furihata (2001). Analysis of difference approximations in space under periodic conditions coupled with some BDF energy stable time discretization was presented in Cheng et al. (2019). For the Allen–Cahn equation, the analysis of difference approximations can be found in Chen et al. (1998).

Starting from the earlier attempts, finite difference approximations have been adopted in many subsequent computational investigations of spinodal decomposition, dendritic growth, solidification and so on, see for example, Rogers et al. (1988). In particular, there are discussions, largely based on numerical experiments, on the choices of time steps and spatial grid size with respect to the diffuse interfacial width parameter ε, see for example, Wheeler et al., 1993. In the application domain, difference approximations remain popular for many practical applications of phase field methods in recent years as exemplified by the Gordon Bell prize winning work (Takashi et al., 2011).

On the algorithm development and numerical analysis side, more recent focus has been on effective linear and nonlinear solvers, including the development of various adaptive and multigrid techniques, see Kim et al. (2004), Rosam et al. (2008) and Wise et al. (2007) and discussion in Sections 6 and 7. High order compact difference spatial approximations have also been studied (Li et al., 2016c) which, when coupled with high order time integrators, have led to record-breaking extreme scale 3D simulations of coarsening dynamics based on the Cahn–Hilliard model.

4.2 Spatial Galerkin discretizations

Unlike the finite difference method that is based on the idea of approximating derivatives in a PDE by discrete finite difference operators, another larger class of numerical methods, called the Galerkin method, for PDEs are based on a very different idea, namely, to employ a variational principle (or a weak formulation) and to approximate an infinite-dimensional Banach (or Hilbert) space by a sequence of finite-dimensional spaces which may or may not be subspaces of the infinite-dimensional Banach space.

Again, we use the Allen–Cahn equation (13) and the Cahn–Hilliard equation (18) as two examples to demonstrate the formulation of their Galerkin approximations and the widely used four types of Galerkin methods, that is, traditional finite element methods, spectral methods, discontinuous Galerkin methods and isogeometric analysis.

A simple integration by parts immediately leads to the following weak formulation for the Allen–Cahn equation (13) complemented by the Neumann boundary condition: Find $u : (0, T) \to V := H^1(\Omega)$ such that

$$(u_t, v) + a(u, v) + \frac{1}{\varepsilon^2}(f(u), v) = 0 \quad \forall v \in V, \tag{48}$$

where and throughout this section $(\cdot, \cdot) = (\cdot, \cdot)_\Omega$ denotes the $L^2(\Omega)$-inner product and

$$a(u, v) := (\nabla u, \nabla v).$$

For the Cahn–Hilliard equation (18) with no-flux boundary conditions, its weak formulation is defined as finding $u : (0, T) \to \widetilde{V} = \left\{v \in H^2(\Omega);\ \frac{\partial v}{\partial n} = 0 \text{ on } \Omega\right\}$ such that

$$(u_t, v) + \varepsilon \widetilde{a}(u, v) + \frac{1}{\varepsilon} c(u, v) = 0 \quad \forall v \in \widetilde{V}, \tag{49}$$

where

$$\widetilde{a}(w, v) := (\Delta w, \Delta v), \quad c(w, v) := a(f(w), v).$$

Let $N(\gg 1)$ be a positive integer and V_N denote an N-dimensional approximation of V. In addition, let $a_N(\cdot, \cdot)$, which is defined on $V \cup V_N \times V \cup V_N$, denote an approximation of $a_N(\cdot, \cdot)$, then the (semidiscrete in space) Galerkin method for (48) is defined as seeking $u_N : (0, T) \to V_N$ such that

$$\left((u_N)_t, v_N\right) + a_N(u_N, v_N) + \frac{1}{\varepsilon^2}(f(u_N), v_N) = 0 \quad \forall v_N \in V_N. \tag{50}$$

Similarly, the (semidiscrete in space) Galerkin method for (49) is defined as seeking $u_N : (0, T) \to \widetilde{V}_N$ such that

$$\left((u_N)_t, v_N\right) + \varepsilon \widetilde{a}_N(u_N, v_N) + \frac{1}{\varepsilon} c_N(u_N, v_N) = 0 \quad \forall v_N \in \widetilde{V}_N. \tag{51}$$

To define specific Galerkin methods, one needs to choose some N-dimensional spaces V_N and $\widetilde{V}_N$, specific bilinear forms $a_N(\cdot\,.\cdot)$, $\widetilde{a}_N(\cdot\,.\cdot)$ and a nonlinear form $c_N(\cdot, \cdot)$. It turns out that the choice of the latter often depends on the choice of the former and different combinations of V_N (resp., $\widetilde{V}_N$) and $a_N(\cdot, \cdot)$ (resp., $\widetilde{V}_N(\cdot, \cdot)$) lead to different Galerkin methods for problem (48) (resp., (49)). Below we briefly discuss four types of Galerkin methods which are widely used in the literature to approximate phase field models.

4.2.1 *Galerkin discretization via finite element methods*

To define the finite element spatial discretization, let $\mathcal{T}_h$ denote a quasi-uniform (conforming) triangular or rectangular mesh for the physical domain Ω and $h(>0)$ denote the mesh size. The finite element method can be viewed as a Galerkin method in which the finite-dimensional space V_N (resp., $\widetilde{V}_N$), denoted by V_h (resp., $\widetilde{V}_h$), consists of piecewise polynomial functions (of a fixed degree) over $\mathcal{T}_h$. Here one may think that $h=O(N^{-\frac{1}{d}})$. There are two scenarios to be considered separately. First, $V_h \subset V$ (resp., $\widetilde{V}_h \subset \widetilde{V}$). In this case, we define $a_N(\cdot,\cdot) := a(\cdot,\cdot)$ (resp., $\widetilde{a}_N(\cdot,\cdot) := \widetilde{a}(\cdot,\cdot)$) and the method is called a *conforming finite element method*. Second, $V_h \not\subset V$ (resp., $\widetilde{V}_h \not\subset \widetilde{V}$), the resulting method is called a *nonconforming finite element method*. In this case, $a_N(\cdot,\cdot) = a(\cdot,\cdot)$ (resp., $\widetilde{a}_N(\cdot,\cdot) := \widetilde{a}(\cdot,\cdot)$) does not work anymore because it may not even be defined on $V_h \times V_h$ (resp., $\widetilde{V}_h \times \widetilde{V}_h$). The remedy for coping with this difficulty depends on the structure of V_h (resp., $\widetilde{V}_h$), however, the choice $a_N(\cdot,\cdot) = a_h(\cdot,\cdot)$ (resp., $\widetilde{a}_N(\cdot,\cdot) = \widetilde{a}_h(\cdot,\cdot)$ and $c_N(\cdot,\cdot) = c_h(\cdot,\cdot)$) works most of the time, where

$$a_h(w,v) := \sum_{K\in\mathcal{T}_h} a_K(w,v) \quad \forall v,w \in V_h \cup V, \tag{52}$$

$$\widetilde{a}_h(w,v) := \sum_{K\in\mathcal{T}_h} \widetilde{a}_K(w,v) \quad \forall v,w \in \widetilde{V}_h \cup \widetilde{V}, \tag{53}$$

$$c_h(w,v) := \sum_{K\in\mathcal{T}_h} c_K(w,v) \quad \forall v,w \in \widetilde{V} \cup \widetilde{V}, \tag{54}$$

and $a_K(\cdot,\cdot) := a(\cdot,\cdot)|_K$, $\widetilde{a}_K(\cdot,\cdot) := \widetilde{a}(\cdot,\cdot)|_K$ and $c_K(\cdot,\cdot) := c(\cdot,\cdot)|_K$, they are, respectively, the restrictions of $a(\cdot,\cdot)$, $\widetilde{a}(\cdot,\cdot)$ and $c(\cdot,\cdot)$ on the element $K \in \mathcal{T}_h$. It should be noted that for the Cahn–Hilliard equation, the construction of both conforming and nonconforming finite element spaces $\widetilde{V}_h$ has been a nontrivial problem, especially in the 3D case (Ciarlet, 1978; Hu and Zhang, 2017).

Conforming and nonconforming finite element methods for the Allen–Cahn equation were studied in Chen et al. (1998), Bates et al. (2009) and Feng and Prohl (2003a). Conforming and nonconforming finite element methods for the Cahn–Hilliard equation were studied in Barrett and Blowey (1999), Barrett et al. (1999), Elliott and French (1987, 1989), Du and Nicolaides (1991), Feng and Prohl (2004c, 2005), Du et al. (2011), Wu and Li (2018) and Zhang and Wang (2010).

4.2.2 *Galerkin discretization via spectral methods*

The spectral element method can be viewed as a Galerkin method in which the finite dimensional space V_N (resp., $\widetilde{V}_N$) is chosen as the space of (global algebraic or trigonometric) polynomials whose degree does not exceed $m(\gg 1)$ in each coordinate direction. Hence, $N = (m+1)^d$. In practice, V_N (resp., $\widetilde{V}_N$) is often written as a tensor product space and is expanded by an orthogonal polynomial basis such as Legendre polynomials and Chebyshev polynomials. Since functions in V_N (resp., $\widetilde{V}_N$) are smooth, there always hold $V_N \subset V$ and $\widetilde{V}_N \subset \widetilde{V}$. Hence, all spectral methods are confirming methods and one naturally chooses $a_N(\cdot,\cdot) = a(\cdot,\cdot)$ and $\widetilde{a}_N(\cdot,\cdot) := \widetilde{a}(\cdot,\cdot)$ in the general Galerkin framework.

For problems defined on regular computational domains but not with periodic boundary conditions, other spectral implementations are possible such as Chebyshev spectral methods for Dirichlet boundary conditions, see discussions in the following books and recent surveys (Bernardi and Maday, 1997; Boyd, 2001; Canuto et al., 2007; Gottlieb and Orszag, 1983; Guo, 1998; Luo et al., 2016; Shen et al., 2011).

Optimal error estimates can be found in Shen and Yang (2010) for the above approximation and a number of other methods involving spectral approximations in space and semiimplicit or fully implicit in time schemes. These estimates show that for a fixed ε, the convergence rate is exponential in terms of the number of the Fourier modes or the dimension of the associated finite dimensional spaces.

Spectral methods can be extended to more complex phase field models such as those involving elastic interactions via microelasticity theory (Hu and Chen, 2001). Adaptive spectral approximations for phase field models have been implemented in the context of moving mesh methods (Feng et al., 2006).

4.2.3 Galerkin discretization via DG methods

Although conceptually discontinuous Galerkin (DG) methods can be viewed as nonconforming finite element methods, it is often presented as a distinct class of Galerkin methods because they use *totally discontinuous* piecewise polynomials, while classical nonconforming finite element methods use piecewise polynomial functions which are not totally discontinuous across element edges/faces. So all DG methods (for second and higher order PDEs) are nonconforming Galerkin methods. In addition, as expected, the convenience of using totally discontinuous piecewise polynomial functions should add some complications for designing bilinear forms $a_N(\cdot,\cdot)$ and $\widetilde{a}_N(\cdot,\cdot)$ as well as the nonlinear form $c_N(\cdot,\cdot)$.

Again, let $\mathcal{T}_h$ denote a (conforming or nonconforming) triangular or rectangular mesh for the physical domain Ω and $h(>0)$ denote the mesh size. Also let $\mathcal{E}^I$ and $\mathcal{E}^B$, respectively, denote the interior and boundary edge/face sets and define $\mathcal{E} := \mathcal{E}^I \cup \mathcal{E}^B$, The finite dimensional space V_N (resp., $\widetilde{V}_N$), denoted by $V_h^{\rm DG}$ (resp., $\widetilde{V}_h^{\rm DG} = V_h^{\rm DG}$) and called it the DG space, consists of piecewise polynomial functions (of a fixed degree) over $\mathcal{T}_h$ which are totally discontinuous across element edges/faces. Here one may think that $h = O(N^{-\frac{1}{d}})$.

The choices for $a_N(\cdot,\cdot)$ and $\widetilde{a}_N(\cdot,\cdot)$ are more complicated, the following are two widely used DG discretizations of $a(\cdot,\cdot)$ and $\widetilde{a}(\cdot,\cdot)$ (Feng and Karakashian, 2007; Feng and Li, 2015):

$$a_h(w,v) := \sum_{K\in\mathcal{T}_h} a_K(w,v) - \sum_{e\in\mathcal{E}^I}\left(\langle\{\partial_n w\},[v]\rangle_e + \lambda\langle[w],\{\partial_n v\}\rangle_e + \frac{\sigma_e}{h_e}\langle[w],[v]\rangle_e\right),$$

$$\begin{aligned}\widetilde{a}(w,v) := &\sum_{K\in\mathcal{T}_h}\widetilde{a}_K(w,v) + \sum_{e\in\mathcal{E}^I}\big(\langle\{\partial_n\Delta w\},[v]\rangle_e + \lambda\langle\{\partial_n\Delta v\},[w]\rangle_e \\ &+\gamma_e h_e^{-3}\langle[w],[v]\rangle_e\big) - \sum_{e\in\mathcal{E}}\big(\langle\{\Delta w\},[\partial_n v]\rangle_e + \lambda\langle\{\Delta v\},[\partial_n w]\rangle_e \\ &-\beta_e h_e^{-1}\langle[\partial_n w],[\partial_n v]\rangle_e\big)\end{aligned}$$

for all $v,\ w\in V_h^{\mathrm{DG}}$. Where $\langle\cdot,\ \cdot\rangle_e$ stands for the L^2-inner product on e, $\partial_n v$ denotes the normal derivative (on e), $[\cdot]$ and $\{\cdot\}$ denote, respectively, the jump and average operators (on e). Moreover, σ_e, β_e and γ_e are under-determined positive constants, and $\lambda = \{-1,\ 0,\ 1\}$ with $\lambda = 1$ giving two symmetric DG bilinear forms.

The choices for $c_N(\cdot,\ \cdot)$ are various. The simplest one is

$$c_h(w,v) := \sum_{K\in\mathcal{T}_h} c_K(w,v) \qquad \forall v, w\in V_h^{\mathrm{DG}}.$$

Another more involved choice (Feng and Karakashian, 2007) is

$$\begin{aligned}c_h(w,v) := -\sum_{K\in\mathcal{T}_h}(f(u),\Delta v)_K &+ \sum_{e\in\mathcal{E}}\langle f(\{u\}),[\partial_n v]\rangle_e \\ &-\sum_{e\in\mathcal{E}^I}\langle f'(\{u\})\{\partial_n u\},[v]\rangle_e.\end{aligned}$$

DG methods for the Allen–Cahn equation were studied in Feng and Li (2015), Guo and Xu (2018), Guo et al. (2016a) and Xia et al. (2009) and for the Cahn–Hilliard equation in Feng and Karakashian (2007), Xia et al. (2007), Kay et al. (2009), Aristotelous et al. (2015), Feng et al. (2016) and Song and Shu (2017). We also note that a weak Galerkin method was recently proposed and analyzed for the Cahn–Hilliard equation in Wang et al. (2019).

4.2.4 Isogeometric analysis

In recent years, the technique of isogeometric analysis has been developed with the aim at integrating traditional numerical discretization like the finite element method with computer-aided geometric design tools. Isogeometric analysis has been studied in many application domains including the discretization of phase field models. The first study in the latter direction is given in Gómez et al. (2008) for Cahn–Hilliard equations. Subsequent studies include applications to phase field models for topology optimization (Dedè et al., 2012), phase field models of brittle fracture (Borden et al., 2014), high order equations on surfaces (Bartezzaghi et al., 2015), Darcy flows (Dedè and Quarteroni, 2018), chemotaxis (Moure and Gomez, 2018) and electrochemical reactions (Zhao et al., 2016c). Convergence studies have also been carried out, see for example (Kästner et al., 2016). While many works on isogeometric analysis are based on the Galerkin approach, there are also works that adopted other formulations like collocation method (Schillinger et al., 2015) and discontinuous Galerkin method (Zhang et al., 2015).

Isogeometric analysis adopts C^1 NURBS-based finite dimensional spaces so that conforming finite element framework can be utilized for spatially high order differential equations. For phase field models with regularized solutions, the high order approximations can be effective using NURBS basis as in the case of spectral method and high order DG methods. For a recent review and additional references, we refer to Gomez and van der Zee (2018).

4.3 Spatial mixed discretization

To overcome the difficulty caused by discretization of the biharmonic operator in the Cahn–Hilliard equation, especially in 3D case, one popular approach is to discretize the Cahn–Hilliard by mixed finite element or mixed DG methods, which are based on rewriting (18) as a system of two second-order PDEs given by

$$u_t + \Delta\mu = 0, \tag{55}$$

$$\mu - \varepsilon\Delta u + \frac{1}{\varepsilon} f(u) = 0, \tag{56}$$

and discretizing the above system either by finite element methods or by DG methods.

A semidiscrete approximation can be formulated as: finding (u^h, μ^h) such that for any (v^h, q^h), it holds

$$(u_t^h, v^h) - a(\mu^h, v^h) = 0, \tag{57}$$

$$(\mu^h, q^h) + \varepsilon a(u^h, q^h) + \frac{1}{\varepsilon}(f(u^h), q^h) = 0. \tag{58}$$

One of the early works on the fully discrete conforming finite element approximations to the mixed weak formulation of (55) and (56) was given in Copetti and Elliott (1992) which has also noted an even earlier work by Du in an unpublished preprint. The latter adopted the same modified Crank–Nicolson in time discretization as in Du and Nicolaides (1991) (also see Du et al. (2011) for similar discussions on solving the Cahn–Hilliard equation on a sphere), while the former took the backward Euler in time discretization. Mixed finite element methods for the Cahn–Hilliard equation were also studied in Feng and Prohl (2004c) and Feng and Prohl (2005) with the aim of deriving refined estimates in the sharp interface limit. Mixed least square finite element was studied in Dean et al. (1996). Mixed DG methods were developed in Kay et al. (2009), Aristotelous et al. (2015) and Feng et al. (2016). For Cahn–Hilliard equations on evolving surfaces, mixed finite element methods were formulated and analyzed in Elliott and Ranner (2015). Local DG methods were also studied for Allen–Cahn and Cahn–Hilliard equations (Guo et al., 2016b; Xia et al., 2007, 2009).

4.4 Implementations and advantages of high order methods

Given the smooth phase field functions, it is expected that high order methods such as high order DG and spectral methods can yield more competitive numerical algorithms. For instance, errors of spectral approximations can be reduced exponentially as the number of spectral basis functions grow, which is known in the literature as spectral accuracy. Based on such observations, Fourier spectral methods, for example, have become particularly effective for phase field simulations in a periodic cell.

Since spatial differential operators subject to periodic conditions are diagonalizable using Fourier modes, the coupling of spectral spatial discretization with semiimplicit in time schemes, as well as ETD and IF exponential type time integration schemes, becomes particularly popular. The treatment of nonlinearity can also be greatly simplified with a collocation implementation or pseudo-spectral approximation (Chen and Shen, 1998; Cheng et al., 2016).

Taking, for example, the Fourier spectral method for the Allen–Cahn model over a periodic domain $[-\pi, \pi]^d$, we let k denote a d-dimensional index with $|k|$ denoting the ℓ^∞ norm and consider a Fourier approximation given by

$$u_k^n(x) = \sum_{|k|\le K} \hat{u}_k^n e^{ikx} \in \widehat{B}_K = \text{span}\{e^{ikx}\}_{|k|\le K}.$$

We can use the first-order stabilized ETD-RK1 scheme (44) to get (Du and Zhu, 2004, 2005)

$$\hat{u}_k^{n+1} = e^{-\tau k\cdot k\hat{u}_k^n - \tau\alpha\varepsilon^{-2}} + \frac{1-e^{-\tau k\cdot k-\tau\alpha\varepsilon^{-2}}}{\tau k\cdot k\varepsilon^2+\tau\alpha}(P_K[\widehat{f(u_k^n)}]_k) - \alpha\hat{u}_k^n),$$

where τ is a step size, $\alpha > 0$ is a stabilizing constant that is often taken to be larger than the one half of the Lipschitz constant of the nonlinear term f, and P_K denotes the projection to $\widehat{B}_K$.

Replacing $e^{\tau k\cdot k+\tau\alpha\varepsilon^{-2}}$ by $1 + \tau k \cdot k + \tau\alpha\varepsilon^{-2}$, we recover

$$(1+\tau k\cdot k+\tau\alpha\varepsilon^{-2})\hat{u}_k^{n+1} = (1+\tau\alpha\varepsilon^{-2})\hat{u}_k^n - \tau\varepsilon^{-2}P_K[\widehat{f(u_k^n)}]_k.$$

Computationally, Fourier spectral methods can be implemented via FFT so that the complexity per step is $O(K\log K)$. Pseudo-spectral or spectral collocation methods are often adopted to treat the nonlinear term instead of the spectral projection, that is, $f(u_k^n)$ is sampled at the discrete spatial grid points, then the FFT is applied to obtained a representation in the Fourier space.

The low computational cost and high spectral accuracy make spectral methods, when applicable, highly attractive for phase field simulations. An interesting question pertains to their performance with $\varepsilon \ll 1$, that is, the sharp interface limit. In Chen and Shen (1998), it has been shown for a typical benchmark setting concerning a single spherical droplet; Fourier approximations can be much more effective than low order finite difference even as ε gets significantly reduced. In other words, high order methods are competitive even in the sharp interface limit. The latter is intriguing as the sharp interface limit of phase field variables are generically discontinuous functions that often do not share high order approximations due to the lack of regularity. A justification was given in Zhang and Du (2009) when mesh and time step refinement are considered as ε is getting smaller. If the sharp interface limit is viewed as the quantity of interests to be sought after, then the typical error expected is of the form

$$O(\varepsilon^{\beta}) + O(\tau^{\gamma}, \varepsilon^{-\gamma}) + O(h^{\delta}, \varepsilon^{-\delta})$$

for a finite difference or finite element methods and

$$O(\varepsilon^{\beta}) + O(\tau^{\gamma}, \varepsilon^{-\gamma}) + O(e^{-cK/\varepsilon})$$

for spectral methods where $O(\varepsilon^{\beta})$ accounts for the diffuse interface approximation error while the other two terms are due to time and spatial discretization. Thus as $\varepsilon \to 0$, to have a balanced total error, asymptotically the need for refinement of spectral approximation is insignificant in comparison with a low order spatial approximation.

5 Convergence theories of fully discrete numerical methods

In this section we shall first present a few exemplary fully discrete numerical schemes for phase field models by combining the time-stepping schemes of Section 3 and the spatial discretization methods of Section 4, we then discuss the convergence results of these fully discrete numerical methods. Below we divide those results into two groups: the first group addresses convergence and error estimates for a fixed diffusive interfacial width ε, the second group concerns the convergence and error estimates for the numerical solutions as well as the numerical interfaces as $\varepsilon, \tau, h \to 0$.

5.1 Construction of fully discrete numerical schemes

As for most evolution equations, a fully discrete numerical method for phase field models can be easily constructed by any combination of a time-stepping scheme of Section 4 and a spatial discretization method of Section 5. Such a construction is also known as the *method of lines* in the literature (cf. Reddy and Trefethen, 1992; Schiesser, 1991 and the references therein). Below we shall use the Allen–Cahn equation (13) and the Cahn–Hilliard equation (18) to demonstrate the procedure.

We start with the spatial semidiscrete finite element method for the Allen–Cahn equation (with the Neumann boundary condition), which is given by (50) with (52), since V_h is finite dimensional, it is easy to see that (50) with (52) is a system of ODEs, which must be complemented by the initial condition

$$(u_h(0), v_h) = (u_0, v_h) \quad \forall v_h \in V_h.$$

Applying the convex splitting scheme to (50) then leads to the following fully discrete convex splitting finite element method: find $u_h^{n+1} \in V_h$ such that

$$\left(d_t u_h^{n+1}, v_h\right) + a_h\left(u_h^{n+1}, v_h\right) + \frac{1}{\varepsilon^2}\left(\left(u_h^{n+1}\right)^3, v_h\right) = \frac{1}{\varepsilon^2}\left(u_h^n, v_h\right) \tag{59}$$

for all $v_h \in V_h$. Where $d_t u_h^{n+1} := (u_h^{n+1} - u_h^n)/\tau$. As expected, it is easy to check that the above method is unconditionally (in h and τ) and uniformly (in ε) energy stable (cf. Feng and Li, 2015; Shen and Yang, 2010).

Similarly, applying the implicit-explicit (IMEX) scheme to (50) gives the following fully discrete IMEX finite element method: find $u_h^{n+1} \in V_h$ such that

$$\left(d_t u_h^{n+1}, v_h\right) + a_h\left(u_h^{n+1}, v_h\right) = \frac{1}{\varepsilon^2}\left(u_h^n - \left(u_h^n\right)^3, v_h\right), \quad \forall v_h \in V_h. \tag{60}$$

As expected, this IMEX finite element method is not unconditional stable. To strengthen its stability, the following popular stabilized method was proposed in the literature (Feng et al., 2015; Li et al., 2016a,b; Shen and Yang, 2010; Xu and Tang, 2006): find $u_h^{n+1} \in V_h$ such that

$$\left(d_t u_h^{n+1}, v_h\right) + a_h\left(u_h^{n+1}, v_h\right) + S\left(u_h^{n+1} - u_h^n, v_h\right) = \frac{1}{\varepsilon^2}\left(u_h^n - \left(u_h^n\right)^3, v_h\right) \tag{61}$$

for all $v_h \in V_h$. Where S is an under-determined constant or linear operator.

Now adopting the finite element method for the spatial discretization in the IEQ scheme (46) then leads to the following fully discrete IEQ finite element method for the Cahn–Hilliard equation (18): find $(u_h^{n+1}, \mu_h^{n+1}, q_h^{n+1}) \in [V_h]^3$ such that

$$\begin{aligned}
\left(u_h^{n+1} - u_h^n, v_h\right) &= -\tau a_h\left(\mu_h^{n+1}, v_h\right) \quad \forall v_h \in V_h, \\
\left(\mu_h^{n+1}, \lambda_h\right) &= \varepsilon a_h\left(u_h^{n+1}, \lambda_h\right) + \frac{1}{\varepsilon}\left(G(u_h^n) q_h^{n+1}, \lambda_h\right) \quad \forall \lambda_h \in V_h, \\
\left(q_h^{n+1} - q_h^n, p_h\right) &= \frac{1}{2}\left(G(u_h^n)\left(u_h^{n+1} - u_h^n\right), p_h\right) \quad \forall p_h \in V_h,
\end{aligned}$$

where $a_h(\cdot, \cdot)$ denotes the discrete bilinear form for $-\Delta$ as defined in (52).

Finally, using the finite element method for the spatial discretization in the SAV scheme (47) one immediately obtains the following fully discrete SAV finite element method for the Cahn–Hilliard equation (18): find $(u_h^{n+1}, \mu_h^{n+1}, r_h^{n+1}) \in [V_h]^2 \times \mathbf{R}$ such that

$$\begin{aligned}
\left(u_h^{n+1} - u_h^n, v_h\right) &= -\tau a_h\left(\mu_h^{n+1}, v_h\right) \quad \forall v_h \in V_h,\\
\left(\mu_h^{n+1}, \lambda_h\right) &= \varepsilon a_h\left(u_h^{n+1}, \lambda_h\right) + \frac{1}{\varepsilon}\left(H(u_h^n), \lambda_h\right) r_h^{n+1} \quad \forall \lambda_h \in V_h,\\
r_h^{n+1} - r_h^n &= \frac{1}{2}\left(H(u_h^n), u_h^{n+1} - u_h^n\right),
\end{aligned}$$

where $a_h(\cdot, \cdot)$ is the same as above.

5.2 Types of convergence and a priori error estimates

As for many numerical methods, three important theoretical issues to be addressed for above fully discrete numerical methods are *stability, convergence* and *rates of convergence*. As the stability issue has already been considered when constructing the time-stepping schemes, and it is not difficult to show that the similar stability properties also hold for the corresponding fully discrete schemes, especially when Galerkin spatial discretization methods are used, in this section we only focus on the issues related to convergence and rates of convergence (i.e., error estimates) of fully discrete numerical methods for phase field models in two cases: (i) $\varepsilon > 0$ is fixed and $h, \tau \to 0$; (ii) $\varepsilon, h, \tau \to 0$. It should be emphasized that the stability is only a necessary criterion for screening "good" numerical methods, however, it does not guarantee the convergence, as a result, it is important to examine the convergence (and rates of convergence) for stable methods, see Xu et al. (2019) for further discussions in this direction.

Let u^ε denote the PDE solution of a underlying phase field model and $u_{h,\tau}^\varepsilon$ denote a fully discrete numerical solution. The goal of error estimates for a fixed $\varepsilon > 0$ is to derive the following type of error bounds:

$$\|u^\varepsilon - u_{h,\tau}^\varepsilon\|_* \le C_1(\varepsilon, u^\varepsilon)\tau^{\ell_1} + C_2(\varepsilon, u^\varepsilon)h^{\ell_2} \tag{62}$$

for some ε- and u^ε-dependent positive constants C_1 and C_2 and positive numbers (mostly integers) ℓ_1 and ℓ_2. Where $\|\cdot\|_*$ denotes a (function space) norm which is problem-dependent.

The precise dependence of C_1 and C_2 on ε are usually complicated and difficult to obtain, but they are expected to grow in $\frac{1}{\varepsilon}$ as $\varepsilon \to 0$, this is because C_1 and C_2 depend on high order ($\ge$2) space and time derivatives of u^ε and those derivatives can be proved to grow *polynomially* in $\frac{1}{\varepsilon}$ as $\varepsilon \to 0$ (cf. Abels, 2015; Abels and Lengeler, 2014; Abels and Liu, 2018; Abels et al., 2017; Feng, 2006; Feng and Prohl, 2003a, 2004a, 2005; Feng et al., 2007a).

Here we distinguish two cases based on how C_1 and C_2 depend on $\frac{1}{\varepsilon}$. If at least one of C_1 and C_2 grow *exponentially* in $\frac{1}{\varepsilon}$, we then call (62) a *coarse error estimate*. On the other hand, if both C_1 and C_2 grow *polynomially* in $\frac{1}{\varepsilon}$, the estimate is called a *fine error estimate*. Obviously, in both cases, for a fixed $\varepsilon > 0$, (62) implies the convergence

$$\lim_{h,\tau\to 0} \|u^\varepsilon - u^\varepsilon_{h,\tau}\|_* = 0,$$

that is, $u^\varepsilon_{h,\tau}$ converges to u^ε in the $\|\cdot\|_*$-norm. Hence, when $\varepsilon > 0$ is fixed, both types of coarse and fine error estimates imply the convergence of the numerical solution as $h, \tau \to 0$.

On the other hand, the situation is very different if the convergence of the numerical interface to the sharp interface of the underlying geometric moving interface problem is a main concern. In that case, one must study the limiting behaviours of the numerical solution $u^\varepsilon_{h,\tau}$ and the numerical interface

$$\Gamma_t^{\varepsilon,h,\tau} := \left\{x \in \Omega;\, u^\varepsilon_{h,\tau}(x,t) = 0\right\}$$

as $\varepsilon \to 0$. First, it is easy to see that the coarse error estimates will become useless because they fail to provide any useful information about the convergence. To see the point, we notice that τ^{ℓ_1} and h^{ℓ_1} only decrease in polynomial orders but C_1 and/or C_2 grow in exponential orders, hence, the limit of the right-hand side of (62) is $+\infty$ as $\varepsilon, h, \tau \to 0$. Second, on contrary the fine error estimates are still valuable, this is because C_1 and C_2 grow in polynomially orders, h and τ can be chosen as powers of ε so that the right-hand side of (62) is guaranteed to converge to 0 as $\varepsilon \to 0$. Let u^0 denote the solution of the underlying moving sharp interface problem (which is defined by its level set formulation), by the triangle inequality we have

$$\|u^0 - u^\varepsilon_{h,\tau}\|_\dagger \le \|u^0 - u^\varepsilon\|_\dagger + \|u^\varepsilon - u^\varepsilon_{h,\tau}\|_\dagger,$$

where $\|\cdot\|_\dagger$ stands for another (function space) norm which is a topology used to measure the convergence of u^ε to u^0 as described in Section 2. We note that $\|\cdot\|_\dagger$ is usually a weaker norm compared to the $\|\cdot\|_*$-norm. Thus, combining the PDE convergence result and the above numerical convergence result we get

$$\lim_{\varepsilon\to 0} \|u^0 - u^\varepsilon_{h,\tau}\|_\dagger = 0,$$

provided that h and τ are chosen as appropriate powers of ε. Moreover, it turns out that (Feng and Li, 2015; Feng and Prohl, 2003a, 2004a, 2005; Feng et al., 2016) such a convergence result is good enough to infer the following convergence of the numerical interface:

$$\lim_{\varepsilon\to 0} \mathrm{dist}_H\left(\Gamma_t,\, \Gamma_t^{\varepsilon,h,\tau}\right) = 0,$$

as well as its rates of convergence in powers of ε if Γ_t is sufficiently smooth. Where dist_H denotes the Hausdorff distance between two sets.

We shall focus on coarse error estimates in the next section and discuss fine error estimates and the convergence of numerical interfaces in the subsequent section.

5.3 Coarse error estimates for a fixed value $\varepsilon > 0$

For a fixed $\varepsilon > 0$, all phase field equations are either semilinear or quasi-linear parabolic PDEs, the desired error estimates for their fully discrete (and spatial semidiscrete) methods can be obtained by following the standard perturbation procedure which consists of three main steps: (i) decomposing the global error with help of an elliptic projection of the exact PDE solution; (ii) using the energy method to obtain an error equation/inequality for the error between the numerical solution and the elliptic projection; (iii) completing the error estimate using a discrete Gronwall's inequality and the triangle inequality. Below we shall demonstrate this perturbation procedure using scheme (59).

Let

$$R^{n+1} := u_t(t_{n+1}) - d_t u^{n+1}$$

denote the truncation error of the backward difference operator d_t which is well known to be of $O(\tau)$ order provided that $u_{tt} \in L^2((0, T); H^{-1}(\Omega))$. Subtracting (59) from the weak formulation of (13) immediately yields the following error equation:

$$\begin{aligned}\left(d_t e_h^{n+1}, v_h\right) + a_h\left(e_h^{n+1}, v_h\right) + \frac{1}{\varepsilon^2}\left(\left(u(t_{n+1})\right)^3 - (u_h^{n+1})^3, v_h\right)\\ = \frac{1}{\varepsilon^2}\left(e_h^{n+1} + \tau d_t u_h^{n+1}, v_h\right) + R^{n+1} \quad \forall v_h \in V_h,\end{aligned} \tag{63}$$

where $e_h^n := u(t_n) - u_h^n$ denotes the global error at t_n. Note that the super-index ε is suppressed on all functions.

Now, let $P_h : H^1(\Omega) \to V_h$ denote the standard elliptic projection operator associated with the discrete bilinear form $a_h(\cdot, \cdot)$ and set

$$e_h^n := \eta_h^n + \xi_h^n; \;\; \eta_h^n := u(t_n) - P_h u(t_n), \;\; \xi_h^n := P_h u(t_n) - u_h^n.$$

Then (63) implies that

$$\begin{aligned}\left(d_t \xi_h^{n+1}, v_h\right) + a_h\left(\xi_h^{n+1}, v_h\right) + \frac{1}{\varepsilon^2}\left(\left(P_h u(t_{n+1})\right)^3 - (u_h^{n+1})^3 - \xi_h^{n+1}, v_h\right)\\ = \frac{1}{\varepsilon^2}\left(\eta_h^{n+1} + \tau d_t u_h^{n+1}, v_h\right) + R^{n+1} - \left(d_t \eta_h^{n+1}, v_h\right)\\ - \frac{1}{\varepsilon^2}\left(\left(u(t_{n+1})\right)^3 - (P_h u_h^{n+1})^3, v_h\right) \quad \forall v_h \in V_h.\end{aligned} \tag{64}$$

Setting $v_h = \xi_h^{n+1}$ and using the monotonicity of cubic power function and the stability estimates of the numerical solution u_h^ε (not shown here) as well as the approximation properties of P_h we get

$$\frac{1}{2} d_t \|\xi_h^{n+1}\|_{L^2}^2 + a_h\left(\xi_h^{n+1}, \xi_h^{n+1}\right) \le \frac{1}{\varepsilon^2} \|\xi_h^{n+1}\|_{L^2}^2 + c_1\tau^2 + c_2 h^{r+1} \tag{65}$$

for some positive constants $c_i = c_i(\varepsilon)(i = 1, 2)$. Where r, a positive integer, denotes the order of the underlying finite element method (i.e., the degree of polynomial shape functions). Here we have also used the fact that

$$\|\eta_h^{n+1}\|_{L^2} \le C h^{r+1}.$$

Finally, applying a discrete Gronwall's inequality to (65) yields

$$\|\xi_h^m\|_{L^2}^2 + \tau \sum_{n=0}^{m-1} a_h\left(\xi_h^{n+1}, \xi_h^{n+1}\right) \le C \exp\left(\varepsilon^{-2}\right)\left(c_1\tau^2 + c_2 h^{r+1}\right) \tag{66}$$

for some positive constant $C = C(T)$, which and the triangle inequality infer

$$\|e_h^m\|_{L^2}^2 \le C \exp\left(\varepsilon^{-2}\right)\left(c_1\tau^2 + c_2 h^{r+1}\right), \tag{67}$$

$$\tau \sum_{n=0}^{m-1} a_h\left(e_h^{n+1}, e_h^{n+1}\right) \le C \exp\left(\varepsilon^{-2}\right)\left(c_1\tau^2 + c_2 h^{r}\right) \tag{68}$$

for all $1 \le m \le M$.

We remark that error estimates (67) and (68) are two coarse error estimates for the fully discrete implicit Euler finite element method for the Allen–Cahn equation. It turns out that they are typical error estimates for all fully discrete numerical methods for various phase field models, which can be abstractly written into the form (62). Although for different models, the norms used to measure errors and the orders of errors (indicated by ℓ_1 and ℓ_2 in (62)) may be different, the nature of the exponential dependence on $\frac{1}{\varepsilon}$ of the constants C_1 and C_2 remains the same for all these methods and models when the above standard perturbation procedure is used to derive the error estimates, see Elliott and French (1987, 1989), Du and Nicolaides (1991), Barrett and Blowey (1999), Barrett et al. (1999), Xu and Tang (2006), Feng et al. (2007a, 2015), Liu and Shen (2003a), Du et al. (2004), Feng (2006), Feng and Karakashian (2007), Wise et al. (2009), Li et al. (2009), Kay et al. (2009), Shen and Yang (2010), Aristotelous et al. (2015), Li et al. (2016a), Diegel et al. (2017a), Yang (2016), Song and Shu (2017) and Shen et al. (2018).

5.4 Fine error estimates and convergence of numerical interfaces as ε, h, $\tau \to 0$

As explained earlier, the coarse error estimates obtained in the previous section are not useful to study the convergence as $\varepsilon \to 0$. A couple natural questions arise: (i) Are the error estimates sharp in terms of ε? (ii) If not, what are sharp error estimates and how to establish such estimates? As all numerical experiments (cf. Provatas et al., 1999; Zhang and Du, 2009 and the references therein) indicate that the above coarse error estimates are not sharp in terms of ε (otherwise, the phase field methodology would be an impractical method to compute moving interface problems!), then the focus is on addressing the second question. The good news is that there are a couple successful numerical analysis techniques which have been developed in the last 30 years, the bad news is that they are the only known techniques for deriving so-called fine error estimates, and the applicability of both techniques is restrictive although one technique fares better than the other. Below we shall again use scheme (59) for the Allen–Cahn equation as an example to explain the ideas of both techniques.

The first technique, which was developed in Nochetto and Verdi (1997) and Nochetto et al. (1994), is to use the discrete maximum principle to derive the L^∞-norm *fine* error estimate for $u^\varepsilon - u^\varepsilon_{h,\tau}$ for the implicit Euler P_1-conforming finite element (i.e., $r = 1$), such an estimate in turn leads to the convergence proof of the numerical interface $\Gamma_t^{\varepsilon,h,\tau}$ to its underlying sharp interface limit Γ_t. However, this maximum principle technique only applies to the Allen–Cahn equation which does possesses a maximum principle and only to P_1-conforming finite element method in order to ensure a discrete maximum principle. The technique does not work for high order phase field model such as the Cahn–Hilliard equation nor for phase field systems which do not have a maximum principle. Moreover, it does not apply to higher order finite element methods either because higher order finite element methods do not have a discrete maximum principle in general even the underlying PDE (such as the Allen–Cahn equation) does.

Before introducing the second technique, we first have a closer look at the derivation of the coarse error estimate given in the previous section to find out the guilty part/step which leads to the exponential dependence of the error constants on $\frac{1}{\varepsilon}$. First, since the error equation (64) is an identity, no error is introduced there. Second, after the inequality (65) is reached, noticing that the coefficient $\frac{1}{\varepsilon^2}$ in the first term on the right-hand side, then it is too late to improve the estimate because (65) and the Gronwall's inequality certainly lead to a coarse error estimate! This simple observation suggests that the guilty step is from the error equation (64) to the error inequality (65), in other words, the transition from equality to inequality is too loose which makes the right-hand side of (65) becomes a too large bound for its left-hand side. To improve the coarse error estimate, we must refine this step and to obtain a tight upper bound for the left-hand side of (65). It turns out that this is exactly what needs to be done to derive a fine error estimate for the Allen–Cahn equation, and a similar situation occurs for several other phase field models.

Moreover, since the right-hand side of (64) only involve the truncation error and the projection error, they contribute $c_1\tau^2$ and c_2h^{r+1} terms in (65), which is good. Hence, the desired improvement (if possible) must comes from bounding the last term on the left-hand side of (64). To obtain (65) we bounded that term as two separate terms that inevitably produces the first term on the right-hand side of (65). A key idea of the second technique of the fine error analysis, which was developed in Feng and Prohl (2003a, 2004a, 2005), Feng and Li (2015) and Feng et al. (2016), is not to separate that term, instead, to estimate it together with the proceeding term. We demonstrate the steps of this techniques below.

Let $f(u) := u^3 - u$, then $f'(u) = 3u^2 - 1$. The last term on the left-hand side of (64) with $v_h = \xi_h^{n+1}$ can be written as

$$\begin{aligned}\frac{1}{\varepsilon^2}\left(f'(P_h u(t_{n+1}))\xi_h^{n+1}, \xi_h^{n+1}\right) + \frac{1}{\varepsilon^2}\|\xi_h^{n+1}\|_{L^4}^4 - \frac{3}{\varepsilon^2}\left(P_h u(t_{n+1}), (\xi_h^{n+1})^3\right) \\ \geq \frac{1}{\varepsilon^2}\left(f'(P_h u(t_{n+1}))\xi_h^{n+1}, \xi_h^{n+1}\right) + \frac{1}{\varepsilon^2}\|\xi_h^{n+1}\|_{L^4}^4 \\ - \frac{3}{\varepsilon^2}\|P_h u(t_{n+1})\|_{L^\infty}\|\xi_h^{n+1}\|_{L^3}^3 .\end{aligned}$$

Substituting the above inequality into (64) yields

$$\begin{aligned}\frac{1}{2}d_t\|\xi_h^{n+1}\|_{L^2}^2 + \varepsilon^2\left[a_h\left(\xi_h^{n+1}, \xi_h^{n+1}\right) + \frac{1}{\varepsilon^2}\left(f'(P_h u(t_{n+1}))\xi_h^{n+1}, \xi_h^{n+1}\right)\right] \\ + (1-\varepsilon^2)\left[a_h\left(\xi_h^{n+1}, \xi_h^{n+1}\right) + \frac{1}{\varepsilon^2}\left(f'(P_h u(t_{n+1}))\xi_h^{n+1}, \xi_h^{n+1}\right)\right] + \frac{1}{\varepsilon^2}\|\xi_h^{n+1}\|_{L^4}^4 \\ \leq \frac{C}{\varepsilon^2}\|\xi_h^{n+1}\|_{L^3}^3 + c_1\tau^2 + c_2h^{r+1},\end{aligned} \tag{69}$$

which replaces (65). One key step of the technique is to show that there exists an ε-independent positive constant $\widehat{C}_0$ such that

$$a_h\left(\xi_h^{n+1}, \xi_h^{n+1}\right) + \frac{1}{\varepsilon^2}\left(f'(P_h u(t_{n+1}))\xi_h^{n+1}, \xi_h^{n+1}\right) \geq -\widehat{C}_0\|\xi_h^{n+1}\|_{L^2}^2, \tag{70}$$

which is a corollary of the following PDE spectral estimate result (Chen, 1994; de Mottoni and Schatzman, 1995):

$$\lambda_{\min} := \inf_{\varphi \in H^1} \frac{(\nabla\varphi, \nabla\varphi) + \frac{1}{\varepsilon^2}\left(f'(u^\varepsilon)\varphi, \varphi\right)}{\|\varphi\|_{L^2}} \geq -C_0$$

for some positive constant C_0 and $0 < \varepsilon \ll 1$. Notice that $\lambda_{\min}$ is the principal eigenvalue of the linearized Allen–Cahn operator

$$L_{\mathrm{AC}}(\varphi) := -\Delta\varphi + \frac{1}{\varepsilon^2} f'(u^\varepsilon)\varphi.$$

Another key step of the technique is to bound the first term on the right-hand side of (69) as follows:

$$\frac{C}{\varepsilon^2}\|\xi_h^{n+1}\|_{L^3}^3 \leq \frac{\varepsilon^2}{2}\|\nabla\xi_h^{n+1}\|_{L^2}^2 + \widehat{C}_0\varepsilon^2\|\xi_h^{n+1}\|_{L^2}^2 + C\varepsilon^{\frac{-2(4+d)}{4-d}}\|\xi_h^{n+1}\|_{L^2}^{\frac{2(6-d)}{4-d}}.$$

Combining the above with (69) and (70) yields

$$\begin{aligned} d_t\|\xi_h^{n+1}\|_{L^2}^2 + \varepsilon^2 a_h(\xi_h^{n+1},\xi_h^{n+1}) + \frac{2}{\varepsilon^2}\|\xi_h^{n+1}\|_{L^4}^4 \\ \leq C\varepsilon^{\frac{-2(4+d)}{4-d}}\|\xi_h^{n+1}\|_{L^2}^{\frac{2(6-d)}{4-d}} + 2\left(1+\widehat{C}_0\right)\|\xi_h^{n+1}\|_{L^2}^2 + c_1\tau^2 + c_2h^{r+1}. \end{aligned} \tag{71}$$

Notice that the exponent $\frac{2(6-d)}{4-d} > 2$ for all $d \geq 1$, the final key step of the technique is to obtain the desired fine error bound for ξ_h^n (and hence for e_h^n) from (71) by using a discrete generalized Gronwall's inequality. See Feng and Prohl (2003a) for finite element approximations and Feng and Li (2015) for discontinuous Galerkin approximations.

From the above derivation one can see that the most important step of the second technique for fine error estimates is to establish a discrete spectral estimate based on the PDE spectral estimate for the linearized operator of the underlying PDE operator. It was proved that such a spectral estimate also holds for the Cahn–Hilliard equation/operator and for the general phase field model for the generalized Stefan problem (Alikakos and Fusco, 1993; Bates and Fife, 1990; Chen, 1994), and similar fine error estimates to above estimate were also established in Feng and Prohl (2004c, 2005) (mixed finite element approximations), Feng et al. (2016) (mixed discontinuous Galerkin approximations), Wu and Li (2018) and Li (2019) (Morley nonconforming finite element approximation) for the Cahn–Hilliard equation and in Feng and Prohl (2004a) for the general phase field model based on those spectral estimates. It should be emphasized that this technique is applicable to any phase field model as long as its linearized operator satisfies a desired spectral estimate. Hence, this technique essentially reduces a numerical problem of deriving fine error estimates for a phase field model into a problem of proving a PDE spectral estimate for the linearized phase field operator. Finally, we also note that a duality argument was recently proposed in Chrysafinos (2019) to derive unconditional stability for a space-time finite element method, it may have a potential to offer an alternative technique for fine error estimates.

6 A posteriori error estimates and adaptive methods

6.1 Spatial and temporal adaptivity

Since phase field models are singularly perturbed equations which involve a small scale parameter ε and the solutions of phase field models have distinctive profiles in the sense that they take a close-to-constant value in each bulk region, which indicates a phase of the material or fluid mixture, and vary smoothly but sharply in a thin layer (called diffuse interface) of width $O(\varepsilon)$ between bulk regions, to resolve this kind of (solution) functions, it is necessary to use spatial mesh size smaller than ε (sometimes much smaller) in the thin layer and it is preferable to use much coarser meshes in the bulk regions. This seems to provide an ideal situation for using adaptive grid methods for efficient numerical simulations. Moreover, should a uniform mesh be used for simulations, the mesh must be very fine over the whole domain, this then results in huge nonlinear algebraic systems to solve, which may not be feasible for large scale and long time simulations, especially in high dimensions. Furthermore, the time scale in most phase field models represents so-called fast time scale, to capture the dynamics of underlying physical or biological phase transition phenomena, it is necessary to use very fine time step size. If the physical or biological processes are slowly varying, it often takes a long time for them to reach equilibrium states, when that happens, long time numerical simulations are required and adaptive time-stepping schemes must be used to make such numerical simulations feasible (Chen and Shen, 2016; Li et al., 2017; Wodo and Ganapathysubramanian, 2011; Zhang and Qiao, 2012).

Since the method of lines is the preferable approach in practice, as explained in Sections 3–5, after a spatial discretization is done, each phase field model then reduces into a system of nonlinear ordinary differential equations (ODEs), then the existing well-developed adaptive time-stepping schemes for ODEs can be utilized for temporal adaptivity. For this reason we shall not further discuss temporal adaptivity in this section and refer the reader to Iserles (2009) and Lambert (1991), instead, below we shall focus on discussing the ideas of various approaches of spatial adaptivity for phase field models.

To design spatial adaptive methods/algorithms for PDE problem, three major approaches have been extensively developed in the past 30 years, they are h-, hp- and r-adaptivity (see Eriksson et al. (1996), Ainsworth and Oden (2000) and Huang and Russell (2011) for detailed expositions). The h-adaptivity starts with a relatively coarse initial mesh and then refines or coarsens the mesh successively based on a chosen error estimator, the sought-after adaptive method, if done correctly, should be able to automatically refine or coarsen the mesh where a refinement or coarsening should be done, so the method successively searches an "optimal" mesh and computes a satisfactory numerical solution on the mesh. The h-adaptivity often goes together with finite element and finite difference methods, and the same finite element or finite difference method is often employed on each mesh successively generated by the adaptive algorithm. That is, the degree of the finite element space does not change in the algorithm. A typical h-adaptive algorithm consists of successive loops of the following sequence:

$$\boxed{\textbf{SOLVE}} \rightarrow \boxed{\textbf{ESTIMATE}} \rightarrow \boxed{\textbf{MARK}} \rightarrow \boxed{\textbf{REFINE/COARSEN}}$$

The hp-adaptivity is often associated with the hp-finite element method (Schwab, 1998), it only differs from the h-adaptivity in one aspect, that is, the refinement or coarsening is not only done for the mesh but also for the degree of finite element spaces. Hence, the finite element spaces may be different on the successive meshes generated by the adaptive method. The r-adaptivity, which is significantly different from the h- and hp-adaptivity, uses a fixed number of mesh points in the adaptive algorithm, however, the distribution of these mesh points is allowed to move, hence, a new mesh is generated by a new distribution of the mesh points, in particular, a locally fine mesh can be obtained by clustering a large (enough) number of the mesh points to where a fine mesh is desired. Clearly, a central issue for r-adaptive methods is how to develop an automatic strategy to move the mesh points. In Feng et al. (2006, 2009), Yu et al. (2008), and Shen and Yang (2009a), r-adaptivity was developed for spectral approximations of several phase field models. For more strategies on r-adaptivity, we refer to Huang and Russell (2011), Tan et al. (2006), Zhang and Tang (2007), Shen et al. (2011), Hu et al. (2009) and Di et al. (2008).

In the rest of this section, we shall only discuss h-adaptive finite element methods for phase field models. Indeed, there has been a lot of interest in spatial adaptive simulations for various phase field models (Braun and Murray, 1997; Ceniceros and Roma, 2007; Ceniceros et al., 2010; Du and Zhang, 2008; Joshi and Jaiman, 2018; Provatas et al., 2005; Rosam et al., 2008; Stogner et al., 2008).

We refer the reader to Ainsworth and Oden (2000), Shen et al. (2011) and Huang and Russell (2011) for detailed expositions for the other two spatial adaptive methodologies.

6.2 Coarse and fine a posteriori error estimates for phase field models

As mentioned above the key step for developing an h-adaptive method is to design a "good" error indicator which can successively predict where to refine and where to coarsen the mesh. As expected, such an indicator should be computable and must relate to the error of numerical solutions, so the mesh is refined where the error is large and coarsened where the error is too small. Error indicators of this kind is known as a posteriori error estimates in the literature.

The general form of a posteriori error estimates is given by

$$\|u^\varepsilon - u^\varepsilon_{h,\tau}\|_\# \leq \mathcal{E}(\varepsilon, h, \tau, u^\varepsilon_h) \tag{72}$$

for some positive functional $\mathcal{E}$. Where $\|\cdot\|_\#$ denotes a (function space) norm which is problem-dependent. Compare to the error estimate (62), there is one important difference, that is, $\mathcal{E}(\varepsilon, h, \tau, u^\varepsilon_h)$ is computable because it depends on the numerical solution u^ε_h, not the PDE solution u^ε; on the other hand, the error bound in (62) is not computable because it depends on the unknown PDE solution u^ε. For this very reason, (62) and (72) are called, respectively, a priori and a posteriori error estimates.

The primary objective of a posteriori error estimates is to derive a tight error bound $\mathcal{E}(\varepsilon, h, \tau, u^\varepsilon_h)$ which is relatively easy to compute. Similar to the classification of a priori error estimates, we also divide a posteriori error estimates into two groups for phase field models. The first group of estimates, called coarse estimates, depends on ε^{-1} exponentially and the second group, called fine estimates, depends on ε^{-1} polynomially. Both groups of a posteriori error estimates have been developed for phase field models although the majority of them belong to the first group. Moreover, two main techniques have been used to derive the a posteriori error bound/functional $\mathcal{E}(\varepsilon, h, \tau, u^\varepsilon_h)$. The first technique gives rise so-called residual-based error bounds (Binev et al., 2004; Eriksson et al., 1996; Morin et al., 2002; Nochetto et al., 2009), and the second one is a duality-based technique, called the *Dual Weighted Residual* method, which leads to goal-oriented error estimates (Bangerth and Rannacher, 2003). Below we shall only focus on discussing the residual-based a posteriori error estimation because of its popularity and superiority for phase field models.

The residual-based a posteriori error estimates give an error bound/functional $\mathcal{E}(\varepsilon, h, \tau, u^\varepsilon_h)$ which typically depends on local residuals of u^ε_h and the jumps of the flux $\nabla u^\varepsilon_h \cdot n$ along each interior element edge/face. It turns out that residual-based error estimators are quite effective for phase field models, this is because the distinctive profiles of all phase field solutions, which are almost constant-valued in each bulk region and vary smoothly but sharply in the thin diffuse interface. We refer to Provatas and Elder (2011), Provatas et al. (1999), Du and Zhang (2008) and Elliott and Stinner (2010a) for detailed discussions about adaptive simulations of phase field models from materials science and biology applications. It should be noted that although coarse a posteriori error estimates were not rigorously derived in those cited works, they could be obtained by following the standard residual-based a posteriori estimate techniques (Ainsworth and Oden, 2000; Eriksson et al., 1996; Nochetto et al., 2009). Moreover, we note that an alternative technique based elliptic reconstruction was also developed for deriving a posteriori error estimates for parabolic PDEs including the Allen–Cahn equation (Georgoulis et al., 2011; Lakkis and Makridakis, 2006; Lakkis et al., 2015).

Like in the case of a priori error estimates for phase field models, the coarse a posteriori error bounds is not useful anymore if one is interested in knowing the limiting behaviour of the solution error and the error for numerical interfaces when $\varepsilon \to 0$. To this end, one must obtain fine a posteriori error estimates. As expected, the only known technique for deriving such error estimates is to utilize the PDE spectral estimate for the linearized operator of the underlying phase field PDE operator. Indeed, the desired fine a posteriori error estimates were obtained in Kessler et al. (2004) and Feng and Wu (2005) for the Allen–Cahn equation and in Feng and Wu (2008) for the Cahn–Hilliard equation. We note that the arguments and the usages of the PDE spectral estimate are quite different in Kessler et al. (2004) and in Feng and Wu (2005) and Feng and Wu (2008). A topological argument was used in Kessler et al. (2004), in which the PDE spectral estimate is embedded, to obtain a fine a posteriori error bound; while a sharp continuous dependence estimate for the PDE solution was used, in which the PDE spectral estimate was crucially utilized, to derive fine a posteriori error estimates in Feng and Wu (2005, 2008) (also see Cockburn, 2003). Because this continuous dependence argument is very simple and may be applicable to other evolution PDEs, we briefly explain it below.

Let V be a Hilbert space and $\mathcal{L}$ be an operator from $D(\mathcal{L})$ ($\subset V$), the domain of $\mathcal{L}$, to V^*, the dual space of V. Consider the abstract evolution problem

$$\frac{\partial u}{\partial t} + \mathcal{L}(u) = g \quad \text{in}\, \Omega_T := \Omega \times (0,T), \tag{73}$$

$$u(0) = u_0 \quad \text{in}\, \Omega. \tag{74}$$

Assume that the above PDE problem satisfies the continuous dependence estimate in the sense that if $u^{(j)}$ is the (unique) solution of (73) and (74) with respect to the data $(g^{(j)}, u_0^{(j)})$ for $j = 1, 2$, then there holds

$$\|u^{(1)} - u^{(2)}\|_{L^p(0,T;V)} \leq \Phi\big(g^{(1)} - g^{(2)}\big) + \Psi\left(u_0^{(1)} - u_0^{(2)}\right) \tag{75}$$

for some (monotone increasing) nonnegative functionals $\Phi(\cdot)$ and $\Psi(\cdot)$. and $1 \leq p \leq \infty$. Let u^A be an approximation of u with the initial value u_0^A, it is easy to show that (Feng and Wu, 2005)

$$\|u - u^A\|_{L^p(0,T;V)} \leq \Phi(R(u^A)) + \Psi\big(u_0 - u_0^A\big), \tag{76}$$

where

$$R(u^A) := g - \frac{\partial u^A}{\partial t} - \mathcal{L}(u^A).$$

Clearly, $R(u^A)$ denotes the residual of the approximation u^A. To utilize the above continuous dependence result to derive a posteriori error estimates for a spatial numerical discretization method, one only needs to set $u^A = u_h$, the space-semidiscrete numerical solution and to get an upper bound for the residual $R(u_h)$ usually in terms of the element-wise local residuals and the jumps of the flux across the interior element edges/faces (cf. Feng and Wu, 2005, 2008). The biggest advantage of this approach for a posteriori error estimates is that it converts a numerical error estimate problem into a PDE continuous dependence (or stability) problem, which may be easier to cope with, especially, for deriving fine a posteriori error estimates for phase field models. In that case, the functionals $\Phi(\cdot)$ and $\Psi(\cdot)$ are expected to grow in $\frac{1}{\varepsilon}$ polynomially as shown in Feng and Wu (2005) and Feng and Wu (2008).

Finally, we point out that the spectral estimate idea has also been successfully used as an a posteriori indicator for detecting singularities (such as topological changes) of the underlying sharp interface limit of a phase field model in Antil and Bartels (2017), Bartels (2015), Bartels and Müller (2010, 2011a,b) and Bartels et al. (2011). The rationale for this approach is that if no singularity occurs, the discrete principal eigenvalue of the linearized operator, which is computable, should have a finite negative low bound. Hence, when such a low bound ceases to exist (i.e., it tends to $-\infty$ faster than $O(|\ln\varepsilon|)$ order as $\varepsilon \to 0$), then it flags a possible singularity of the underlying moving interface problem, and suggests that a fine mesh should be used to resolve the singularity.

7 Applications and extensions

The phase field (diffuse interface) method has many applications ranging from mathematical subjects like differential geometry, to image processing and geometric modelling, and to physical sciences like astrophysics, cell biology, multiphase fluid mechanics and (of course) materials science. We provide a sampler of some of the applications here.

7.1 Materials science applications

Applications to problems in materials science are among the early and most widely recognized successes of phase field models. From the works of Lord Rayleigh, Gibbs and Van de Walls to the seminal contributions of John W. Cahn, ideas of using diffuse interface and phase order parameters have drawn much interests in the materials science community in the mesoscopic modelling of materials structure. Generically, a set of conserved field variables $c_1, c_2, \ldots, c_m$ and nonconserved field variables $\eta_1, \eta_2, \ldots \eta_n$ are often used to describe the compositional/structural domains and the interfaces, and the total free energy of an inhomogeneous microstructure system is formulated as

$$E_{total} = \int [\sum_{i=1}^{m} \alpha_i (\nabla c_i)^2 + \sum_{i=1}^{3}\sum_{j=1}^{3}\sum_{k=1}^{n} \beta_{ij} \nabla_i \eta_k \nabla_j \eta_k + f(c_1, c_2, \ldots, c_m, \eta_1, \eta_2, \ldots \eta_n)] dx + \int\int G(x - x', \vec{c}, \vec{\eta}) dx dx', \tag{77}$$

where the gradient coefficient α_i and β_{ij} can be used to reflect the interfacial energy anisotropy and the function f corresponds to the local free energy density. The last integral in the above equation represents a nonlocal term that includes a general long-range interaction such as elastic interactions in solids. The time evolution is governed by either the nonconserved (Allen–Cahn type for c_j's) or conserved dynamics (Cahn–Hilliard type for η_j's).

Phase field modelling is arguably one of the most popular methods for modelling and simulation of microstructure evolution under different driving forces such as compositional gradients, temperature, stress/strain and electric and magnetic fields (Biner, 2017). Phase field modelling has now been used to study not only binary phases and single components but also materials with multicomponents multiphases. It also can account for mechanical, thermodynamic, electric and magnetic interactions. It can deal with spinodal decomposition, dendritic growth, Ostwald ripening and coalescence, adhesion and dewetting, sintering, directional solidification, nucleation and coarsening, grain boundary motion, pattern formation in thin films, epitaxial growth, electromigration and many other processes. For additional references and more comprehensive reviews, we refer to Barrett et al. (2004, 2006, 2007, 2013, 2014b), Biner (2017), Bañas and Nürnberg (2008, 2009), Boettinger et al. (2002), Chen (2002), Dai and Du (2016), Provatas and Elder (2011), Shu et al. (2007), Garcke et al. (2004), Singer-Loginova and Singer (2008), Steinbach (2009) and Wang and Li (2010).

As an example, we mention one application on the study of nucleation based on the phase field approach (Zhang et al., 2016b). The search for index-1 saddle point of the phase field free energy (Heo et al., 2010; Zhang et al., 2007) can help characterize the transition states that lead to critical nuclei. For solid state phase transformation, anisotropic elastic energy plays a critical role, which leads to formation of nonspherical nuclei. A particular form of the total energy in the phase field set up is given by

$$E_{total} = \int (|\nabla \eta|^2 + f(\eta))dx + \frac{1}{2}\int C_{ijkl}C_{ijkl}\varepsilon_{ij}^{el}\varepsilon_{kl}^{el}dx,$$

where the elastic strain ε^{el} is the difference between the total strain and stress-free strain since stress-free strain does not contribute to the total elastic energy. In Zhang et al. (2007, 2008), different critical nuclei have been computed as the driving forces change and the elastic effects vary. Numerical algorithms for index-1 saddle points are constructed by essentially reformulating the problem into a problem of optimization in an extended configuration space which includes not only the state variable (point on the energy landscape) but also the most unstable direction. In other words, one is effectively doing minimization along the stable directions (descending) while maximization along the (most) unstable direction (ascending) directions. This is the essential behind various saddle-point algorithms such as the Shrinking Dimer Methods (Zhang and Du, 2012) and other methods reviewed in Zhang et al. (2016b). The predicted phase field nuclei can also be utilized in dynamic simulations and extended to investigate complex nucleation phenomena (Heo et al., 2010).

Let us also mention other variations of phase fields model related to materials science applications. For example, phase field modelling of polymeric materials has received much attention in theoretical chemistry and materials science. Phase field models of diblock copolymers, such as the model due to Ohta and Kawasaki (1986), have had a long history and is an active area of research (Ceniceros and Fredrickson, 2004; Cheng et al., 2017; Choksi and Ren, 2003; Li et al., 2013; Ren and Wei, 2007; Shi, 2019; Wanner, 2017).

In addition, as noted previously, there are many other extended phase field models but with vector or tensor valued fields as phase field variables. These models can describe physical phenomena and geometric objects with higher codimension such as point defects in two dimensions and curves in three dimensions.

7.2 Fluid and solid mechanics applications

Applications of phase field models and diffuse interface approaches to mechanical problems is also a major area of research.

Concerning fluid mechanics, a review was given in Anderson et al. (1998) on phase field/diffuse interface models of hydrodynamics and their application to a wide variety of interfacial phenomena involving fluid flows. Phase field models have been developed in various cases of fluid flows with a length scale commensurate with the diffuse interfacial width. Examples include small-scale flows and multiphase flows. The latter subject, typically involving breakup and coalescence such as fluid jets and droplets as well as fluid mixing and other interfacial phenomena, has become particularly important in microfluidics and micro-process engineering, see Prosperetti and Tryggvason (2009) and Wörner (2012) for reviews on the various numerical approaches. Phase field modelling of multiphase flows has been developed by a number of groups along with the development and numerical analysis, see, for instance (Abels et al., 2012; Badalassi et al., 2003; Bañas and Nürnberg, 2017; Diegel et al., 2017a; Feng et al., 2005a; Garcke et al., 2000; Grun, 2013; Liu and Shen, 2003b; Lowengrub and Truskinovsky, 1998) and the references cited therein. Various phase field based simulations of drop formation in microfluidic channels have been carried out, see for example, (Yue et al., 2004). Related optimal control problems and their simulations have been considered in Hintermüller and Wegner (2012) and Hintermüller et al. (2017). Three-dimensional phase field simulations of interfacial dynamics in Newtonian and viscoelastic fluids can be found in Zhou et al. (2010). Algorithm development and numerical analysis of three phase flows can be found in Yang et al. (2017).

Phase field modelling and simulation of multiphase flows with additional complications have also been made, studies include, for example, the wetting phenomena (Jacqmin, 2004), two-phase flow in porous media (Cogswell and Szulczewski, 2017). A problem of considerable interests is the moving contact line problem for the interaction of fluid–fluid interface with a solid wall is a widely studied subject, see Pomeau (2002) for a review. Ideas of diffuse interface and phase field models have also been considered (Jacqmin, 2000; Pismen, 2001). A phase field model with generalized Navier boundary condition (GNBC) is proposed in Qian et al. (2003). Further computational studies can be found in He et al. (2011), Bao et al. (2012), Shi et al. (2013), Gao and Wang (2014) and Zhang and Wang (2016).

Phase field modelling has also received much attention in solid mechanics. Representative works in this direction includes the modelling of brittle fracture (Borden et al., 2012; Conti et al., 2016; Miehe et al., 2010b). For example, in addition to the fracture energy of the type given by the Cahn–Hilliard functional as proposed in Bourdin et al. (2000), one can define the elastic energy density as

$$\psi_e(\varepsilon,u)=[(1-k)u^2+k]\psi_e^+(\varepsilon)k\psi_e^-(\varepsilon),$$

where u is the phase field variable, $\psi_e^\pm$ are the strain energies computed from the positive and negative components of the strain tensor $\varepsilon^\pm$, respectively, defined through a spectral decomposition of strain (Miehe et al., 2010a), that is,

$$\psi_e^+(\varepsilon)=\frac{1}{2}\lambda[(\mathrm{Tr}\varepsilon)_{\cdot}^{+\cdot 2}+\mu\mathrm{Tr}[(\varepsilon^+)^2],$$

$$\psi_e^-(\varepsilon)=\frac{1}{2}\lambda[\mathrm{Tr}\varepsilon(\mathrm{Tr}\varepsilon)_{\cdot}^{+\cdot 2}+\mu\mathrm{Tr}[(\varepsilon-\varepsilon^+)^2].$$

The phase field variable is only applied to the tensile part of the elastic energy density, so that crack propagation under compression can be prevented (Miehe et al., 2010a,b). One may further incorporate other features to model ductile fracture (Ambati et al., 2016; Miehe et al., 2016) and fluid-filled fracture (Wheeler et al., 2014), and utilize various discretization techniques, see recent discussions in Heister et al. (2015) and Zhang et al. (2018).

7.3 Image and data processing applications

Ideas of phase field modelling are used not only for physical sciences but also for other domain sciences. Imaging science is a good example where variational methods have been widely used for imaging and data analysis (Aubert and Kornprobst, 2006; Chan and Shen, 2005). Phase field models and diffuse interface descriptions of geometric features form one major type of variational formulations for various image processing tasks. Taking, for example, the task of image restoration and image deblurring, a diffuse interface relaxation to Mumford–Shah model can be formulated by Brett et al. (2014)

$$\mathcal{J}(u) := \int_{\Omega} \left(\frac{1}{2} |\nabla u|^2 + \frac{1}{\varepsilon^2}(u^2 - 1)^2 \right) dx + \|Su - f\|^2, \tag{78}$$

where S can be a deblurring operator. Other examples include image classification and restoration (Samson et al., 2000), edge detection (Ambrosio and Tortorelli, 1990), motion estimation (Preusser et al., 2007), image segmentation (Burke et al., 2013; Jung et al., 2007; Li and Kim, 2011), curve smoothing (Zhu et al., 2010), active contour (Rochery et al., 2005), clustering of vector fields (Garcke et al., 2001) and so on.

For image segmentation, a phase field relaxation to Mumford–Shah can be formulated as (March and Dozio, 1997)

$$\mathcal{J}(v, u) := E(u) + \lambda \int_{\Omega} u^2 |\nabla v|^2 dx + \|v - f\|^2, \tag{79}$$

where f is the image under consideration, v represents a piecewise smooth approximation to f, and $E(u)$ is a diffuse interfacial energy of the phase field variable u.

While a standard Cahn–Hilliard energy is the most popular choice, a more general form has also been used as well:

$$E(u) := \alpha \int_{\Omega} \left(\frac{1}{2} |\nabla u|^2 + \frac{1}{\varepsilon^2}(u^2 - 1)^2 \right) dx + \beta \int_{\Omega} \left| \Delta u + \frac{1}{\varepsilon^2}(u - u^3) \right|^2 dx. \tag{80}$$

The first term in the above is the standard phase field form of the surface tension, the second term corresponds to a diffuse interface relaxation of the Euler-elastica (integral of the square of curvature along a planar curve) as discussed in De Giorgi (1991).

Discussions on issues related to the application of similar phase field energy to image inpainting can be found in Esedoglu and Shen (2002). More discussions on the three-dimensional analogue of the phase field Euler-elastica, a special form of the Helfrich bending energy, are given later in this chapter for phase field modelling of vesicle membrane.

Phase field approaches have also been used in data analysis and discrete graph modelling, see, for example, Bertozzi and Flenner (2012), Van Gennip et al. (2014) and Li and Kim (2011).

7.4 Biology applications

In more recent years, phase field modelling and simulations have also become very successful tools to study problems arising in life sciences, including problems in biophysics, bioengineering, ecology, biology and so on. Some examples include the development of phase field models and computational methods for the deformation and dynamics of bio-membranes and cell vesicles, cellular activities such as motility and cell division, fluid-structure interactions in blood flows and other biological fluids, and tumour adhesion and growth. These problems span across scales from cellular and molecular levels to macroscopic systems.

On the cellular level, activities such as cell crawling, cell motion on patterned substrate and other migration processes are also important biological questions. They often involve deformable geometry and associated mechanical interactions. These questions have been studied using various phase field approaches in recent years, see, for example, (Camley et al., 2013; Löber et al., 2014; Marth and Voigt, 2014; Moure and Gomez, 2017; Najem and Grant, 2016; Shao et al., 2012; Ziebert and Aranson, 2016). Phase field modelling has also found many applications in biological molecular processes such as conformational change, molecular recognition and molecular assembly. For example, Sun et al. (2015) presented phase field models for the implicit solvation of charged molecules with a coupling to Poisson–Boltzmann electrostatics. Various refined phase field models, numerical methods on both spatial discretization and time integration have also been studied (Dai et al., 2018; Li and Zhao, 2013; Zhao et al., 2018).

Bilayer vesicles are models systems for cell membranes. Their deformation has often been modelled by the Helfrich bending elasticity model for fluid membranes with contributions from the surface integrals of the mean curvature squared and the Gaussian curvature, thus leading to natural connections with the Willmore energy and Willmore flow. The diffuse interface formulation has been introduced in Du et al. (2004, 2006). Other variants have also been studied (Aland et al., 2014; Barrett et al., 2017; Biben et al., 2005; Bretin et al., 2015a; Campelo and Hernandez-Machado, 2007; Du et al., 2009; Esedoglu et al., 2014; Jamet and Misbah, 2007), together with analytical studies (Bellettini and Mugnai, 2010; Bretin et al., 2015b; Mugnai, 2013) and numerical approximations (Chen et al., 2015; Colli and Laurençot, 2011; Du and Zhang, 2008; Du and Zhu, 2006; Yang and Ju, 2017a) for the deformation and dynamics of vesicles. The effective phase field modelling of Gaussian curvature energy also leads to an interesting development of the diffuse interface Euler–Poincaré characteristics (Bellettini and Mugnai, 2010; Du et al., 2005) that can be used to detect topological change of the implicitly defined interface within the phase field framework beyond the biophysical applications. Phase field formulation has also been developed for vesicle–substrate and vesicle–vesicle interactions (Gu et al., 2016; Zhang et al., 2009). For multicomponent membranes, Elliott and Stinner (2010a) and Lowengrub et al. (2009b) studied phase field models defined on an evolving surface. Using two-phase field variables, Wang and Du (2008) developed phase field models for two-component membranes and obtained interesting patterns mimicking to experimental observations. Given the complexities involved in these systems, effective numerical algorithms such as those utilizing high order spectral methods and adaptive discretizations (Du and Zhang, 2008; Du et al., 2004) and energy law preserving schemes (Hua et al., 2011) can be very useful for large scale and long time simulations.

Another example of broad applications of phase field modelling and simulations in biomedical field is on the cancer and tumour growth models. This is a subject that have been studied by a number of authors (Garcke et al., 2018; Lima et al., 2014; Vilanova et al., 2017), see Lowengrub et al. (2009a) for a review on the models and numerical methods.

7.5 Other variants of phase field models

There are many other variants of phase field models. We provide a couple of examples here that focus on the modelling of the nonlocal and stochastic aspects of the underlying physical processes.

7.5.1 Nonlocal and factional order phase field models

Discussion on nonlocal interactions in the form of integral operators may be tracked back to the work of Van der Waals (1894), see discussions made in Pismen (2001). The usual differential equation form of the local phase field energy can be derived from the nonlocal version via the so-called Landau expansion (Landau and Lifshitz, 2013), assuming a smooth and spatially slowly varying field. A number of studies on nonlocal Allen–Cahn and nonlocal Cahn–Hilliard can be found in Bates et al. (2006), Bates (2006), Benešová et al. (2014), Choksi et al. (2009, 2011), Du et al. (2019), Fife (2003) and Jeong and Kim (2015). More studies on nonlocal modelling, analysis and computation can be found in Du (2019). A special class of nonlocal models are fractional phase field models where fractional derivatives are used to replace the integer order derivatives in the conventional local phase field models. Such models have attracted the attention of community in recent years (Akagi et al., 2016; Gui and Zhao, 2015; Milovanov and Rasmussen, 2005; Tarasov and Zaslavsky, 2005; Valdinoci, 2013). Algorithmic development and numerical analysis concerning these nonlocal models (including the fractional ones in either space or time or both) can be found in Ainsworth and Mao (2017), Bates et al. (2009), Du and Yang (2016), Du et al. (2018b), Guan et al. (2014), Hou et al. (2017), Song et al. (2016) and Zhai et al. (2016). Related algorithmic studies with respect to different applications were presented in Antil and Bartels (2017).

For spatially nonlocal phase field models, due to the spread of nonlocal interactions, the phase field variables may no longer be as smooth as their local counterpart. They may also lead to narrower interfacial region and permit singularities across the interface or at the defects (Du and Yang, 2016; Gui and Zhao, 2015; Song et al., 2016).

7.5.2 Stochastic phase field models

Uncertainty may arise from various sources such as thermal fluctuation, impurities of the materials and the intrinsic instabilities of the deterministic evolutions. Therefore, the evolution of interfaces under influence of noise is of great importance in applications such as lattice models, the scaling limit of lattice models and derivations of continuum equations, it is necessary and interesting to consider stochastic effects, and to study the impact of noise on phase field modelling and on their solutions, especially on their long time behaviours. This then leads to considering the stochastic phase field models. However, how to incorporate noises correctly into those models is a nontrivial matter, which turns out to be both a science and an art.

A few approaches have been known in the literature. The first one, which is the simplest, is to add (small) noise terms to the existing deterministic models and to study and simulate those stochastically perturbed models. Some recent works on modelling and PDE analyses in this direction can be found in Weinan and Liu (2002), Weber (2010a,b), Hairer et al. (2012), Otto et al. (2014), Antonopoulou et al. (2018a,c), Debussche and Zambotti (2007), Debussche and Goudenège (2011), Da Prato and Debussche (1996) and Röckner et al. (2018) and the references therein. Finite element approximations have been studied in Kovács et al. (2011), Kovács et al. (2014) and Furihata et al. (2018) for the Cahn–Hilliard–Cook equation (which is a stochastic Cahn–Hilliard equation with additive noise) and in Kovács et al. (2015) and Feng et al. (2018) for stochastic Allen–Cahn models. Relevant time discretizations have been discussed in Printems (2001). Parallel algorithms and numerical simulations of the Cahn–Hilliard–Cook equation have also been reported in Zheng et al. (2015). Stochastic Cahn–Hilliard dynamics have also been used to study nucleation in microstructure evolution in Heo et al. (2010) and Li et al. (2012, 2013). It should be noted that the added noises may not always have physical meaning and those stochastically perturbed models may not be associate with sharp interface models.

An alternative approach is to directly consider stochastic sharp interface problems such as stochastic mean curvature flows and to derive (formally) corresponding phase field models, subsequently, to study and simulate the resulting stochastic models. Recent PDE studies in this direction can be found in Röger and Weber (2013), Dirr et al. (2001), Souganidis and Yip (2004) and Yip (1998, 2002), those stochastic PDEs involve gradient-type multiplicative noise and thus have stronger nonlinearity. Finite element approximations have been carried out in Feng et al. (2017) for a stochastic Allen–Cahn equation with gradient-type multiplicative noise and in Feng et al. (2019) for a related stochastic Cahn–Hilliard equation with gradient-type multiplicative noise. Not surprisingly, their sharp interface models are expected to be a stochastic mean curvature flow (Röger and Weber, 2013) and a stochastic Hele-Shaw model (Antonopoulou et al., 2018b; Feng et al., 2019), respectively. However, their rigorous convergence proofs are still missing although some partial results were reported in Röger and Weber (2013) and Antonopoulou et al. (2018b) and positive numerical results were given in Feng et al. (2014, 2017, 2019).

An additional approach to model stochasticity is to study stochastic variants of coupled phase field equations with other physical models. Stochasticity can enter through other physical laws such as equations for fluid flows in multiphase flows and flow structure interactions (Chaudhri et al., 2014; Du and Li, 2011).

8 Conclusion

In this chapter, we presented a holistic review about some basic elements of phase field modelling, particularly related to curvature-driven geometric interfacial motion. We discussed the relevant mathematical theory, numerical approximations and selected applications, as well as the relationship between the phase field methodology and other methodologies (such as the level set methodology) for geometric moving interface problems. Instead of presenting much involved technical details, we focused on discussing the main elements and ideas of the phase field modelling, analysis and their numerical approximations, with extensive references provided on each of these aspects. They should be helpful for the interested reader to do further reading for specific details.

Over last several decades, the phase field method has been developed into a powerful and versatile general methodology for interface problems arising from various scientific and engineering applications including biology, differential geometry, fluid and solid mechanics, image processing and materials science. It has garnered tremendous attention and popularity among researchers and practitioners, the trend will likely continue and make even broader impact in more and more fields. Yet, there are still many challenging issues in the phase field methodology, from modelling, analysis, approximation and application, that should be further investigated. On the modelling side, addressing different geometric features and connecting to microscopic physics are important future research topics. Effective recovery of geometric and topological features and statistical information from the phase field approach are of both theoretical and practical interests. On the computational side, there are much need and urgent demand for developing effective and robust adaptive algorithms, particularly those involving anisotropic spatial adaptivity and locally adapted time steps. Developing efficient and fast linear and nonlinear solvers, preconditioning techniques, high order and stable time-stepping schemes and scalable algorithms remains a focus of future research. Effective methods for computing saddle points or transition states, rather than equilibrium, are also very limited currently and are in high demand. Methods for characterizing the captured interfacial geometry and quantifying statistical features based on the phase field models are also helpful for many practical applications. On the analysis front, a lot of open questions remain to be answered. For the PDE analysis, convergence studies of many phase field models to their respective sharp interface limits are still missing. The lack of analysis techniques and machineries along with singularities of solutions of the sharp interface limits seem to be one of the primary hurdles to overcome. For the numerical analysis, analyzing the approximation errors and their dependencies on the diffuse interface parameter ε remains an interesting and challenging issue. Finally, while extreme scale simulations have been achieved for some model phase field equations, simulating more realistic and more complex phase field systems still require much future effort.

Acknowledgements

The work of the first author was partially supported by the NSF grant DMS-1719699, and US Army Research Office MURI grant W911NF-15-1-0562. The work of the second author was partially supported by the NSF grants DMS-0410266 and DMS-1620168.

References

Abels, H., 2015. Sharp interface limit for the Cahn-Larché system. Asymptot. Anal. 91 (3–4), 283–340. ISSN 0921-7134.

Abels, H., Lengeler, D., 2014. On sharp interface limits for diffuse interface models for two-phase flows. Interfaces Free Bound. 16 (3), 395–418. ISSN 1463-9963. https://doi.org/10.4171/IFB/324.

Abels, H., Liu, Y., 2018. Sharp interface limit for a Stokes/Allen-Cahn system. Arch. Ration. Mech. Anal. 229 (1), 417–502. ISSN 0003-9527. https://doi.org/10.1007/s00205-018-1220-x.

Abels, H., Garcke, H., Grün, G., 2012. Thermodynamically consistent, frame indifferent diffuse interface models for incompressible two-phase flows with different densities. Math. Models Methods Appl. Sci. 22 (03), 1150013.

Abels, H., Liu, Y., Schöttl, A., 2017. Sharp interface limits for diffuse interface models for two-phase flows of viscous incompressible fluids. In: Transport processes at fluidic interfaces. Adv. Math. Fluid Mech, Birkhäuser/Springer, Cham, pp. 231–253.

Ainsworth, M., Mao, Z., 2017. Analysis and approximation of a fractional Cahn-Hilliard equation. SIAM J. Numer. Anal. 55 (4), 1689–1718.

Ainsworth, M., Oden, J.T., 2000. A Posteriori Error Estimation in Finite Element Analysis, Pure and Applied Mathematics (New York). Wiley-Interscience [John Wiley & Sons], New York, p. xx+240. ISBN 0-471-29411-X. https://doi.org/10.1002/9781118032824.

Akagi, G., Schimperna, G., Segatti, A., 2016. Fractional Cahn-Hilliard, Allen-Cahn and porous medium equations. J. Differ. Equ. 261 (6), 2935–2985.

Akrivis, G., Crouzeix, M., Makridakis, C., 1998. Implicit-explicit multistep finite element methods for nonlinear parabolic problems. Math. Comput. Am. Math. Soc. 67 (222), 457–477.

Aland, S., Egerer, S., Lowengrub, J., Voigt, A., 2014. Diffuse interface models of locally inextensible vesicles in a viscous fluid. J. Comput. Phys. 277, 32–47.

Alikakos, N.D., Fusco, G., 1993. The spectrum of the Cahn-Hilliard operator for generic interface in higher space dimensions. Indiana Univ. Math. J. 42 (2), 637–674. ISSN 0022-2518.

Alikakos, N.D., Bates, P.W., Chen, X., 1994. Convergence of the Cahn-Hilliard equation to the Hele-Shaw model. Arch. Rational Mech. Anal. 128 (2), 165–205. ISSN 0945-8396.

Allen, S., Cahn, J.W., 1979. A microscopic theory for antiphase boundary motion and its application to antiphase domain coarsening. Acta Metall. 27, 1084–1095.

Ambati, M., Kruse, R., De Lorenzis, L., 2016. A phase-field model for ductile fracture at finite strains and its experimental verification. Comput. Mech. 57 (1), 149–167.

Ambrosio, L., Tortorelli, V.M., 1990. Approximation of functionals depending on jumps by elliptic functionals via Γ-convergence. Comm. Pure Appl. Math. 43 (8), 999–1036. ISSN 0010-3640. https://doi.org/10.1002/cpa.3160430805.

Anderson, D.M., McFadden, G.B., Wheeler, A.A., 1998. Diffuse-interface methods in fluid mechanics. In: Annual review of fluid mechanics, Annual Review of Fluid Mechanics, vol. 30. Annual Reviews, Palo Alto, CA, pp. 139–165.

Antil, H., Bartels, S., 2017. Spectral approximation of fractional PDEs in image processing and phase field modeling. Comput. Methods Appl. Math. 17 (4), 661–678.

Antonopoulou, D.C., Blömker, D., Karali, G.D., 2018a. The sharp interface limit for the stochastic Cahn-Hilliard equation. Ann. Inst. Henri Poincaré Probab. Stat. 54 (1), 280–298. ISSN 0246-0203. https://doi.org/10.1214/16-AIHP804.

Antonopoulou, D.C., Blömker, D., Karali, G.D., 2018b. The sharp interface limit for the stochastic Cahn-Hilliard equation. Ann. Inst. Henri Poincaré Probab. Stat. 54 (1), 280–298. ISSN 0246-0203. https://doi.org/10.1214/16-AIHP804.

Antonopoulou, D.C., Farazakis, D., Karali, G., 2018c. Malliavin calculus for the stochastic Cahn-Hilliard/Allen-Cahn equation with unbounded noise diffusion. J. Differ. Equ. 265 (7), 3168–3211. ISSN 0022-0396. https://doi.org/10.1016/j.jde.2018.05.004.

Aristotelous, A.C., Karakashian, O.A., Wise, S.M., 2015. Adaptive, second-order in time, primitive-variable discontinuous Galerkin schemes for a Cahn-Hilliard equation with a mass source. IMA J. Numer. Anal. 35 (3), 1167–1198. ISSN 0272-4979. https://doi.org/10.1093/imanum/dru035.

Aubert, G., Kornprobst, P., 2006. Mathematical Problems in Image processing: Partial Differential Equations and the Calculus of Variations. vol. 147. Springer Science & Business Media.

Badalassi, V.E., Ceniceros, H.D., Banerjee, S., 2003. Computation of multiphase systems with phase field models. J. Comput. Phys. 190 (2), 371–397.

Badia, S., Guillén-González, F., Gutiérrez-Santacreu, J.V., 2011. Finite element approximation of nematic liquid crystal flows using a saddle-point structure. J. Comput. Phys. 230 (4), 1686–1706. ISSN 0021-9991. https://doi.org/10.1016/j.jcp.2010.11.033.

Bañas, L., Nürnberg, R., 2008. Finite element approximation of a three dimensional phase field model for void electromigration. J. Sci. Comp. 37 (2), 202–232. https://doi.org/10.1007/s10915-008-9203-y.

Bañas, L., Nürnberg, R., 2009. Phase field computations for surface diffusion and void electromigration in $\mathbb{R}^3$. Comput. Vis. Sci. 12 (7), 319–327. ISSN 1432-9360. https://doi.org/10.1007/s00791-008-0114-0.

Bañas, L., Nürnberg, R., 2017. Numerical approximation of a non-smooth phase-field model for multicomponent incompressible flow. ESAIM: Math. Model. Numer. Anal. 51 (3), 1089–1117.

Bangerth, W., Rannacher, R., 2003. Adaptive Finite Element Methods for Differential Equations, Lectures in Mathematics ETH Zürich. Birkhäuser Verlag, Basel, Basel. ISBN: 3-7643-7009-2, p. viii+207. https://doi.org/10.1007/978-3-0348-7605-6.

Bao, K., Shi, Y., Sun, S., Wang, X.-P., 2012. A finite element method for the numerical solution of the coupled Cahn-Hilliard and Navier-Stokes system for moving contact line problems. J. Comput. Phys. 231 (24), 8083–8099. ISSN 0021-9991. https://doi.org/10.1016/j.jcp.2012.07.027.

Barrett, J.W., Blowey, J.F., 1999. Finite element approximation of the Cahn-Hilliard equation with concentration dependent mobility. Math. Comp. 68 (226), 487–517. ISSN 0025-5718.

Barrett, J.W., Blowey, J.F., Garcke, H., 1999. Finite element approximation of the Cahn-Hilliard equation with degenerate mobility. SIAM J. Numer. Anal. 37 (1), 286–318. ISSN 1095-7170 (electronic).

Barrett, J.W., Nürnberg, R., Styles, V., 2004. Finite element approximation of a phase field model for void electromigration. SIAM J. Numer. Anal. 42 (2), 738–772. ISSN 0036-1429. https://doi.org/10.1137/S0036142902413421.

Barrett, J.W., Garcke, H., Nürnberg, R., 2006. Finite element approximation of a phase field model for surface diffusion of voids in a stressed solid. Math. Comp. 75 (253), 7–41. ISSN 0025-5718. https://doi.org/10.1090/S0025-5718-05-01802-8.

Barrett, J.W., Garcke, H., Nürnberg, R., 2007. A phase field model for the electromigration of intergranular voids. Interfaces Free Bound. 9 (2), 171–210. ISSN 1463-9963. https://doi.org/10.4171/IFB/161.

Barrett, J.W., Garcke, H., Nürnberg, R., 2013. On the stable discretization of strongly anisotropic phase field models with applications to crystal growth. ZAMM Z. Angew. Math. Mech. 93 (10–11), 719–732. https://doi.org/10.1002/zamm.201200147.

Barrett, J.W., Garcke, H., Nürnberg, R., 2014a. Phase field models versus parametric front tracking methods: are they accurate and computationally efficient? Commun. Comput. Phys. 15 (2), 506–555. https://doi.org/10.4208/cicp.190313.010813a.

Barrett, J.W., Garcke, H., Nürnberg, R., 2014b. Stable phase field approximations of anisotropic solidification. IMA J. Numer. Anal. 34 (4), 1289–1327. https://doi.org/10.1093/imanum/drt044.

Barrett, J.W., Garcke, H., Nürnberg, R., 2017. Finite element approximation for the dynamics of fluidic two-phase biomembranes. M2AN Math. Model. Numer. Anal. 51 (6), 2319–2366. https://doi.org/10.1051/m2an/2017037.

Bartels, S., 2015. Robustness of error estimates for phase field models at a class of topological changes. Comput. Methods Appl. Mech. Eng. 288, 75–82. ISSN 0045-7825. https://doi.org/10.1016/j.cma.2014.11.005.

Bartels, S., Müller, R., 2010. A posteriori error controlled local resolution of evolving interfaces for generalized Cahn-Hilliard equations. Interfaces Free Bound. 12 (1), 45–73. ISSN 1463-9963. https://doi.org/10.4171/IFB/226.

Bartels, S., Müller, R., 2011a. Error control for the approximation of Allen-Cahn and Cahn-Hilliard equations with a logarithmic potential. Numer. Math. 119 (3), 409–435. ISSN 0029-599X. https://doi.org/10.1007/s00211-011-0389-9.

Bartels, S., Müller, R., 2011b. Quasi-optimal and robust a posteriori error estimates in $L^\infty(L^2)$ for the approximation of Allen-Cahn equations past singularities. Math. Comp. 80 (274), 761–780. ISSN 0025-5718. https://doi.org/10.1090/S0025-5718-2010-02444-5.

Bartels, S., Müller, R., Ortner, C., 2011. Robust a priori and a posteriori error analysis for the approximation of Allen-Cahn and Ginzburg-Landau equations past topological changes. SIAM J. Numer. Anal. 49 (1), 110–134. ISSN 0036-1429. https://doi.org/10.1137/090751530.

Bartezzaghi, A., Dedè, L., Quarteroni, A., 2015. Isogeometric analysis of high order partial differential equations on surfaces. Comput. Methods Appl. Mech. Eng. 295, 446–469. ISSN 0045-7825. https://doi.org/10.1016/j.cma.2015.07.018.

Bates, P.W., 2006. On some nonlocal evolution equations arising in materials science. Nonlinear Dyn. Evol. Equ. 48, 13–52.

Bates, P.W., Fife, P.C., 1990. Spectral comparison principles for the Cahn-Hilliard and phase-field equations, and time scales for coarsening. Physica D 43 (2–3), 335–348. ISSN 0167-2789.

Bates, P.W., Han, J., Zhao, G., 2006. On a nonlocal phase-field system. Nonlinear Anal. Theory Methods Appl. 64 (10), 2251–2278.

Bates, P.W., Brown, S., Han, J., 2009. Numerical analysis for a nonlocal Allen-Cahn equation. Int. J. Numer. Anal. Model 6 (1), 33–49.

Bellettini, G., 2013. Lecture Notes on Mean Curvature Flow, Barriers and Singular Perturbations, Appunti. In: Scuola Normale Superiore di Pisa (Nuova Serie) [Lecture Notes. Scuola Normale Superiore di Pisa (New Series)], vol. 12. Edizioni della Normale, Pisa, p. xviii+325. https:/doi.org/10.1007/978-88-7642-429-8. ISBN 978-88-7642-428-1; 978-88-7642-429-8.

Bellettini, G., Mugnai, L., 2010. Approximation of Helfrich's functional via diffuse interfaces. SIAM J. Math. Anal. 42 (6), 2402–2433. ISSN 0036-1410. https://doi.org/10.1137/09077549X.

Bellettini, G., Novaga, M., 1997. Minimal barriers for geometric evolutions. J. Differ. Equ. 139 (1), 76–103. ISSN 0022-0396.

Benesová, B., Melcher, C., Süli, E., 2014. An implicit midpoint spectral approximation of nonlocal Cahn-Hilliard equations. SIAM J. Numer. Anal. 52 (3), 1466–1496.

Bernardi, C., Maday, Y., 1997. Spectral methods. In: Handbook of Numerical Analysis, vol. 5. Elsevier, pp. 209–485.

Bertozzi, A.L., Flenner, A., 2012. Diffuse interface models on graphs for classification of high dimensional data. Multiscale Model. Simul. 10 (3), 1090–1118.

Biben, T., Kassner, K., Misbah, C., 2005. Phase-field approach to three-dimensional vesicle dynamics. Phys. Rev. E 72 (4), 041921.

Biner, S.B., 2017. Programming Phase-Field Modeling. Springer.

Binev, P., Dahmen, W., DeVore, R., 2004. Adaptive finite element methods with convergence rates. Numer. Math. 97 (2), 219–268. ISSN 0029-599X. https://doi.org/10.1007/s00211-003-0492-7.

Boettinger, W.J., Warren, J.A., Beckermann, C., Karma, A., 2002. Phase-field simulation of solidification. Ann. Rev. Mater. Res. 32 (1), 163–194.

Borden, M.J., Verhoosel, C.V., Scott, M.A., Hughes, T.J.R., Landis, C.M., 2012. A phase-field description of dynamic brittle fracture. Comput. Methods Appl. Mech. Eng. 217, 77–95.

Borden, M.J., Hughes, T.J.R., Landis, C.M., Verhoosel, C.V., 2014. A higher-order phase-field model for brittle fracture: formulation and analysis within the isogeometric analysis framework. Comput. Methods Appl. Mech. Eng. 273, 100–118.

Bourdin, B., Francfort, G.A., Marigo, J.-J., 2000. Numerical experiments in revisited brittle fracture. J. Mech. Phys. Solids 48 (4), 797–826.

Boyd, J.P., 2001. Chebyshev and Fourier Spectral Methods. Courier Corporation.

Brakke, K.A., 1978. The Motion of a Surface by its Mean Curvature. Mathematical Notes, vol. 20. Princeton University Press, Princeton, NJ. ISBN: 0-691-08204-9, p. i+252.

Braun, R.J., Murray, B.T., 1997. Adaptive phase-field computations of dendritic crystal growth. J. Cryst. Growth 174 (1–4), 41–53.

Bretin, E., Masnou, S., Oudet, E., 2014. Phase-field models for the approximation of the Willmore functional and flow. In: Congrès SMAI 2013, ESAIM Proc. Surveys, vol. 45. EDP Sci., Les Ulis, pp. 118–127.

Bretin, E., Masnou, S., Oudet, É., 2015a. Phase-field approximations of the Willmore functional and flow. Numer. Math. 131 (1), 115–171.

Bretin, E., Masnou, S., Oudet, E., 2015b. Phase-field approximations of the Willmore functional and flow. Numer. Math. 131 (1), 115–171. ISSN 0029-599X. https://doi.org/10.1007/s00211-014-0683-4.

Brett, C., Elliott, C.M., Dedner, A.S., 2014. Phase field methods for binary recovery. In: Optimization With PDE Constraints, Springer, pp. 25–63.

Bronsard, L., Kohn, R.V., 1991. Motion by mean curvature as the singular limit of Ginzburg-Landau dynamics. J. Differ. Equ. 90 (2), 211–237. ISSN 0022-0396. https://doi.org/10.1016/0022-0396(91)90147-2.

Burke, S., Ortner, C., Süli, E., 2013. An adaptive finite element approximation of a generalized Ambrosio-Tortorelli functional. Math. Models Methods Appl. Sci. 23 (9), 1663–1697. ISSN 0218-2025. https://doi.org/10.1142/S021820251350019X.

Caginalp, G., Chen, X., 1998. Convergence of the phase field model to its sharp interface limits. Eur. J. Appl. Math. 9 (4), 417–445. ISSN 0956-7925.

Caginalp, G., Lin, J.T., 1987. A numerical analysis of an anisotropic phase field model. IMA J. Appl. Math. 39 (1), 51–66.

Cahn, J.W., Hilliard, J.E., 1958. Free energy of a nonuniform system I. Interfacial free energy. J. Chem. Phys. 28, 258–267.

Cahn, J.W., Elliott, C.M., Novick-Cohen, A., 1996. The Cahn-Hilliard equation with a concentration dependent mobility: motion by minus the Laplacian of the mean curvature. Eur. J. Appl. Math. 7 (3), 287–301. ISSN 0956-7925.

Camley, B.A., Zhao, Y., Li, B., Levine, H., Rappel, W.-J., 2013. Periodic migration in a physical model of cells on micropatterns. Phys. Rev. Lett. 111 (15), 158102.

Campelo, F., Hernandez-Machado, A., 2007. Shape instabilities in vesicles: a phase-field model. Eur. Phys. J. Spec. Top. 143 (1), 101–108.

Canuto, C., Hussaini, M.Y., Quarteroni, A., Zang, T.A., 2007. Spectral Methods: Evolution to Complex Geometries and Applications to Fluid Dynamics. Springer Science & Business Media.

Ceniceros, H.D., Fredrickson, G.H., 2004. Numerical solution of polymer self-consistent field theory. Multiscale Model. Simul. 2 (3), 452–474.

Ceniceros, H.D., Roma, A.M., 2007. A nonstiff, adaptive mesh refinement-based method for the Cahn-Hilliard equation. J. Comput. Phys. 225 (2), 1849–1862.

Ceniceros, H.D., Nós, R.L., Roma, A.M., 2010. Three-dimensional, fully adaptive simulations of phase-field fluid models. J. Comput. Phys. 229 (17), 6135–6155.

Chan, T.F., Shen, J.J., 2005. Image Processing and Analysis: Variational, PDE, Wavelet, and Stochastic Methods. vol. 94. SIAM.

Chaudhri, A., Bell, J.B., Garcia, A.L., Donev, A., 2014. Modeling multiphase flow using fluctuating hydrodynamics. Phys. Rev. E 90 (3), 033014.

Chen, X., 1992. Generation and propagation of interfaces for reaction-diffusion equations. J. Differ. Equ. 96 (1), 116–141. ISSN 0022-0396.

Chen, X., 1994. Spectrum for the Allen-Cahn, Cahn-Hilliard, and phase-field equations for generic interfaces. Comm. Partial Differ. Equ. 19 (7–8), 1371–1395. ISSN 0360-5302.

Chen, X., 1996. Global asymptotic limit of solutions of the Cahn-Hilliard equation. J. Differ. Geom. 44 (2), 262–311. ISSN 0022-040X.

Chen, L.-Q., 2002. Phase-field models for microstructure evolution. Ann. Rev. Mater. Res. 32 (1), 113–140.

Chen, L.Q., Shen, J., 1998. Applications of semi-implicit Fourier-spectral method to phase field equations. Comput. Phys. Commun. 108 (2–3), 147–158.

Chen, Y., Shen, J., 2016. Efficient, adaptive energy stable schemes for the incompressible Cahn-Hilliard-Navier-Stokes phase-field models. J. Comput. Phys. 308, 40–56.

Chen, Y.G., Giga, Y., Goto, S., 1991. Uniqueness and existence of viscosity solutions of generalized mean curvature flow equations. J. Differ. Geom. 33 (3), 749–786. ISSN 0022-040X.

Chen, X., Elliott, C.M., Gardiner, A., Zhao, J., 1998. Convergence of numerical solutions to the Allen-Cahn equation. Appl. Anal. 69 (1–2), 47–56.

Chen, Z., Nochetto, R.H., Schmidt, A., 2000. Error control and adaptivity for a phase relaxation model. M2AN Math. Model. Numer. Anal. 34 (4), 775–797. ISSN 0764-583X.

Chen, X., Wang, X., Xu, X., 2014. Analysis of the Cahn-Hilliard equation with a relaxation boundary condition modeling the contact angle dynamics. Arch. Ration. Mech. Anal. 213 (1), 1–24. ISSN 0003-9527. https://doi.org/10.1007/s00205-013-0713-x.

Chen, R., Ji, G., Yang, X., Zhang, H., 2015. Decoupled energy stable schemes for phase-field vesicle membrane model. J. Comput. Phys. 302, 509–523.

Chen, W., Han, D., Wang, X., 2017. Uniquely solvable and energy stable decoupled numerical schemes for the Cahn-Hilliard-Stokes-Darcy system for two-phase flows in karstic geometry. Numer. Math. 137 (1), 229–255. ISSN 0029-599X. https://doi.org/10.1007/s00211-017-0870-1.

Cheng, Q., Shen, J., 2018. Multiple scalar auxiliary variable (MSAV) approach and its application to the phase-field vesicle membrane model. SIAM J. Sci. Comput. 40 (6), A3982–A4006. ISSN 1064-8275. https://doi.org/10.1137/18M1166961.

Cheng, K., Wang, C., Wise, S.M., Yue, X., 2016. A second-order, weakly energy-stable pseudo-spectral scheme for the Cahn-Hilliard equation and its solution by the homogeneous linear iteration method. J. Sci. Comput. 69 (3), 1083–1114.

Cheng, Q., Yang, X., Shen, J., 2017. Efficient and accurate numerical schemes for a hydro-dynamically coupled phase field diblock copolymer model. J. Comput. Phys. 341, 44–60.

Cheng, K., Feng, W., Wang, C., Wise, S.M., 2019. An energy stable fourth order finite difference scheme for the Cahn-Hlliard equation. J. Comput. Appl. Math. 362, 574–595.

Choksi, R., Ren, X., 2003. On the derivation of a density functional theory for microphase separation of diblock copolymers. J. Statist. Phys. 113 (1–2), 151–176.

Choksi, R., Peletier, M.A., Williams, J.F., 2009. On the phase diagram for microphase separation of diblock copolymers: an approach via a nonlocal Cahn-Hilliard functional. SIAM J. Appl. Math. 69 (6), 1712–1738.

Choksi, R., Maras, M., Williams, J.F., 2011. 2D phase diagram for minimizers of a Cahn-Hilliard functional with long-range interactions. SIAM J. Appl. Dyn. Syst. 10 (4), 1344–1362.

Chopp, D.L., Sethian, J.A., 1999. Motion by intrinsic Laplacian of curvature. Interfaces Free Bound. 1 (1), 107–123. ISSN 1463-9963.

Chrysafinos, K., 2019. Stability analysis and best approximation error estimates of discontinuous time-stepping schemes for the Allen-Cahn equation. ESAIM M2AN 53, 551–583.

Ciarlet, P.G., 1978. The Finite Element Method for Elliptic Problems. vol. 4. North-Holland Publishing Co., Amsterdam. ISBN: 0-444-85028-7, p. xix+530

Cockburn, B., 2003. Continuous dependence and error estimation for viscosity methods. Acta Numer. 12, 127–180. ISSN 0962-4929. https://doi.org/10.1017/S0962492902000107.

Cogswell, D.A., Szulczewski, M.L., 2017. Simulation of incompressible two-phase flow in porous media with large timesteps. J. Comput. Phys. 345, 856–865.

Colli, P., Laurençot, P., 2011. A phase-field approximation of the Willmore flow with volume constraint. Interfaces Free Bound. 13 (3), 341–351. ISSN 1463-9963. https://doi.org/10.4171/IFB/261.

Conti, S., Focardi, M., Iurlano, F., 2016. Phase field approximation of cohesive fracture models. Ann. Inst. Henri Poincar (C) Non-Linaire Analyze 33 (4), 1033–1067.

Copetti, M.I.M., Elliott, C.M., 1992. Numerical analysis of the Cahn-Hilliard equation with a logarithmic free energy. Numer. Math. 63 (1), 39–65. ISSN 0029-599X.

Cox, S.M., Matthews, P.C., 2002. Exponential time differencing for stiff systems. J. Comput. Phys. 176 (2), 430–455.

Da Prato, G., Debussche, A., 1996. Stochastic Cahn-Hilliard equation. Nonlinear Anal. 26 (2), 241–263. ISSN 0362-546X. https://doi.org/10.1016/0362-546X(94)00277-O.

Dai, S., Du, Q., 2012. Motion of interfaces governed by the Cahn-Hilliard equation with highly disparate diffusion mobility. SIAM J. Appl. Math. 72 (6), 1818–1841. ISSN 0036-1399. https://doi.org/10.1137/120862582.

Dai, S., Du, Q., 2014. Coarsening mechanism for systems governed by the Cahn-Hilliard equation with degenerate diffusion mobility. Multiscale Model. Simul. 12 (4), 1870–1889. ISSN 1540-3459. https://doi.org/10.1137/140952387.

Dai, S., Du, Q., 2016. Computational studies of coarsening rates for the Cahn-Hilliard equation with phase-dependent diffusion mobility. J. Comput. Phys. 310, 85–108.

Dai, S., Promislow, K., 2013. Geometric evolution of bilayers under the functionalized Cahn-Hilliard equation. Proc. R. Soc. Lond. Ser. A Math. Phys. Eng. Sci. 469 (2153), 2012505. ISSN 1364-5021. https://doi.org/10.1098/rspa.2012.0505.

Dai, S., Li, B., Lu, J., 2018. Convergence of phase-field free energy and boundary force for molecular solvation. Arch. Ration. Mech. Anal. 227 (1), 105–147. ISSN 0003-9527. https://doi.org/10.1007/s00205-017-1158-4.

Davis, T.A., Gartland Jr., E.C., 1998. Finite element analysis of the Landau-de Gennes minimization problem for liquid crystals. SIAM J. Numer. Anal. 35 (1), 336–362.

De Giorgi, E., 1991. Some remarks on γ-convergence and least squares method. In: Dal Maso, G., Dell'Antonio, G.F. (Eds.), Composite media and homogenization theory, Progress in Nonlinear Differential Equations and Their Applications, vol. 5. Birkhäuser, Boston, pp. 135–142.

De Giorgi, E., 1994. Barriers, boundaries, motion of manifolds. Conference held at Department of Mathematics of Pavia, March 18, 1994.

De Mottoni, P., Schatzman, M., 1995. Geometrical evolution of developed interfaces. Trans. Amer. Math. Soc. 347 (5), 1533–1589. ISSN 0002-9947.

Dean, E.J., Glowinski, R., Trevas, D.A., 1996. An approximate factorization/least squares solution method for a mixed finite element approximation of the Cahn-Hilliard equation. Jpn J. Ind. Appl. Math. 13 (3), 495–517. ISSN 0916-7005. https://doi.org/10.1007/BF03167260.

Debussche, A., Goudenège, L., 2011. Stochastic Cahn-Hilliard equation with double singular nonlinearities and two reflections. SIAM J. Math. Anal. 43 (3), 1473–1494. ISSN 0036-1410. https://doi.org/10.1137/090769636.

Debussche, A., Zambotti, L., 2007. Conservative stochastic Cahn-Hilliard equation with reflection. Ann. Probab. 35 (5), 1706–1739. ISSN 0091-1798. https://doi.org/10.1214/009117906000000773.

Dedè, L., Quarteroni, A., 2018. Isogeometric analysis of a phase field model for Darcy flows with discontinuous data. Chin. Ann. Math. Ser. B 39 (3), 487–512. ISSN 0252-9599. https://doi.org/10.1007/s11401-018-0079-3.

Dedè, L., Borden, M.J., Hughes, T.J.R., 2012. Isogeometric analysis for topology optimization with a phase field model. Arch. Comput. Methods Eng. 19 (3), 427–465. ISSN 1134-3060. https://doi.org/10.1007/s11831-012-9075-z.

Di, Y., Li, R., Tang, T., 2008. A general moving mesh framework in 3D and its application for simulating the mixture of multi-phase flows. Commun. Comput. Phys. 3 (3), 582–602. ISSN 1815-2406.

Diegel, A.E., Feng, X.H., Wise, S.M., 2015. Analysis of a mixed finite element method for a Cahn-Hilliard-Darcy-Stokes system. SIAM J. Numer. Anal. 53 (1), 127–152. ISSN 0036-1429. https://doi.org/10.1137/130950628.

Diegel, A.E., Wang, C., Wang, X., Wise, S.M., 2017a. Convergence analysis and error estimates for a second order accurate finite element method for the Cahn-Hilliard-Navier-Stokes system. Numer. Math. 137 (3), 495–534.

Dirr, N., Luckhaus, S., Novaga, M., 2001. A stochastic selection principle in case of fattening for curvature flow. Calc. Var. Partial Differ. Equ. 13 (4), 405–425. ISSN 0944-2669. https://doi.org/10.1007/s005260100080.

Dondl, P.W., Lemenant, A., Wojtowytsch, S., 2017. Phase field models for thin elastic structures with topological constraint. Arch. Ration. Mech. Anal. 223 (2), 693–736. ISSN 0003-9527. https://doi.org/10.1007/s00205-016-1043-6.

Du, Q., 2005. Numerical approximations of the Ginzburg-Landau models for superconductivity. J. Math. Phys. 46 (9), 095109.

Du, Q., 2011. Phase field calculus, curvature-dependent energies, and vesicle membranes. Philos. Mag. 91 (1), 165–181.

Du, Q., 2019. Nonlocal Modeling, Analysis and Computation. CBMS-NSF regional conference series in applied mathematics, vol. 94. SIAM.

Du, Q., Li, M., 2011. Analysis of a stochastic implicit interface model for an immersed elastic surface in a fluctuating fluid. Arch. Ration. Mech. Anal. 199 (1), 329–352.

Du, Q., Nicolaides, R.A., 1991. Numerical analysis of a continuum model of phase transition. SIAM J. Numer. Anal. 28 (5), 1310–1322. ISSN 0036-1429.

Du, Q., Yang, J., 2016. Asymptotically compatible Fourier spectral approximations of nonlocal Allen-Cahn equations. SIAM J. Numer. Anal. 54 (3), 1899–1919.

Du, Q., Yang, J., 2017. Fast and accurate implementation of Fourier spectral approximations of nonlocal diffusion operators and its applications. J. Comput. Phys. 332, 118–134. ISSN 0021-9991. https://doi.org/10.1016/j.jcp.2016.11.028.

Du, Q., Zhang, J., 2008. Adaptive finite element method for a phase field bending elasticity model of vesicle membrane deformations. SIAM J. Sci. Comput. 30 (3), 1634–1657.

Du, Q., Zhu, W., 2004. Stability analysis and applications of the exponential time differencing schemes and their contour integration modifications. J. Comput. Math. 22, 200–209.

Du, Q., Zhu, W., 2005. Analysis and applications of the exponential time differencing schemes. BIT Numer. Math. 45, 307–328.

Du, Q., Zhu, L., 2006. Analysis of a mixed finite element method for a phase field bending elasticity model of vesicle membrane deformation. J. Comput. Math. 24, 265–280.

Du, Q., Gunzburger, M.D., Peterson, J.S., 1992. Analysis and approximation of the Ginzburg-Landau model of superconductivity. SIAM Rev. 34 (1), 54–81.

Du, Q., Liu, C., Wang, X., 2004. A phase field approach in the numerical study of the elastic bending energy for vesicle membranes. J. Comput. Phys. 198 (2), 450–468.

Du, Q., Liu, C., Wang, X., 2005. Retrieving topological information for phase field models. SIAM J. Appl. Math. 65 (6), 1913–1932.

Du, Q., Liu, C., Wang, X., 2006. Simulating the deformation of vesicle membranes under elastic bending energy in three dimensions. J. Comput. Phys. 212 (2), 757–777.

Du, Q., Liu, C., Ryham, R., Wang, X., 2007. Diffuse interface energies capturing the Euler number: relaxation and renormalization. Commun. Math. Sci. 5 (1), 233–242. ISSN 1539-6746. http:/projecteuclid.org/euclid.cms/1175797629.

Du, Q., Liu, C., Ryham, R., Wang, X., 2009. Energetic variational approaches in modeling vesicle and fluid interactions. Physica D 238 (9–10), 923–930.

Du, Q., Ju, L., Tian, L., 2011. Finite element approximation of the Cahn-Hilliard equation on surfaces. Comput. Methods Appl. Mech. Eng. 200 (29–32), 2458–2470.

Du, Q., Ju, L., Li, X., Qiao, Z., 2018a. Maximum principles of semilinear parabolic equations and their exponential time differencing approximations. Preprint arXiv:1902.04998.

Du, Q., Ju, L., Li, X., Qiao, Z., 2018b. Stabilized linear semi-implicit schemes for the nonlocal Cahn-Hilliard equation. J. Comput. Phys. 363, 39–54.

Du, Q., Ju, L., Li, X., Qiao, Z., 2019. Maximum principle preserving exponential time differencing schemes for the nonlocal Allen-Cahn equation. SIAM J. Numer. Anal. 57, 875–898.

Ecker, K., 2004. Regularity Theory for Mean Curvature Flow, Progress in Nonlinear Differential Equations and their Applications. 57, Birkhauser Boston, Inc., Boston, MA. ISBN: 0-8176-3243-3, p. xiv+165. https://doi.org/10.1007/978-0-8176-8210-1.

Edwards, D.A., Brenner, H., Wasan, D.T., 1991. Interfacial Transport Processes and Rheology. Butterworth-Heinemann, Boston.

Elliott, C.M., French, D.A., 1987. Numerical studies of the Cahn-Hilliard equation for phase separation. IMA J. Appl. Math. 38 (2), 97–128. ISSN 0272-4960.

Elliott, C.M., French, D.A., 1989. A nonconforming finite-element method for the two-dimensional Cahn-Hilliard equation. SIAM J. Numer. Anal. 26 (4), 884–903. ISSN 0036-1429.

Elliott, C.M., Ranner, T., 2015. Evolving surface finite element method for the Cahn-Hilliard equation. Numer. Math. 129 (3), 483–534.

Elliott, C.M., Stinner, B., 2010a. A surface phase field model for two-phase biological membranes. SIAM J. Appl. Math. 70 (8), 2904–2928.

Elliott, C.M., Stuart, A.M., 1993. The global dynamics of discrete semilinear parabolic equations. SIAM J. Numer. Anal. 30 (6), 1622–1663.

Eriksson, K., Estep, D., Hansbo, P., Johnson, C., 1996. Computational Differential Equations. Cambridge University Press, Cambridge. ISBN 0-521-56312-7; 0-521-56738-6, p. xvi+538.

Escher, J., Mayer, U.F., Simonett, G., 1998. The surface diffusion flow for immersed hypersurfaces. SIAM J. Math. Anal. 29 (6), 1419–1433. ISSN 1095-7154 (electronic).

Esedoglu, S., Shen, J., 2002. Digital inpainting based on the Mumford-Shah-Euler image model. Eur. J. Appl. Math. 13 (4), 353–370.

Esedoglu, S., Rätz, A., Röger, M., 2014. Colliding interfaces in old and new diffuse-interface approximations of Willmore-flow. Commun. Math. Sci. 12 (1), 125–147.

Evans, L.C., Spruck, J., 1991. Motion of level sets by mean curvature. I. J. Differ. Geom. 33 (3), 635–681. ISSN 0022-040X.

Evans, L.C., Soner, H.M., Souganidis, P.E., 1992. Phase transitions and generalized motion by mean curvature. Comm. Pure Appl. Math. 45 (9), 1097–1123. ISSN 0010-3640.

Eyre, D.J., 1998. Unconditionally gradient stable time marching the Cahn-Hilliard equation. MRS Online Proc. Libr. Arch. 529.

Feng, X., 2006. Fully discrete finite element approximations of the Navier-Stokes-Cahn-Hilliard diffuse interface model for two-phase fluid flows. SIAM J. Numer. Anal. 44 (3), 1049–1072. ISSN 0036-1429. https://doi.org/10.1137/050638333.

Feng, X., Karakashian, O.A., 2007. Fully discrete dynamic mesh discontinuous Galerkin methods for the Cahn-Hilliard equation of phase transition. Math. Comp. 76 (259), 1093–1117. ISSN 0025-5718. https://doi.org/10.1090/S0025-5718-07-01985-0.

Feng, X., Li, Y., 2015. Analysis of symmetric interior penalty discontinuous Galerkin methods for the Allen-Cahn equation and the mean curvature flow. IMA J. Numer. Anal. 35 (4), 1622–1651. ISSN 0272-4979. https://doi.org/10.1093/imanum/dru058.

Feng, X., Prohl, A., 2003a. Numerical analysis of the Allen-Cahn equation and approximation for mean curvature flows. Numer. Math. 94 (1), 33–65. ISSN 0029-599X. https://doi.org/10.1007/s00211-002-0413-1.

Feng, X., Prohl, A., 2004a. Analysis of a fully discrete finite element method for the phase field model and approximation of its sharp interface limits. Math. Comp. 73 (246), 541–567. ISSN 0025-5718. https://doi.org/10.1090/S0025-5718-03-01588-6.

Feng, X., Prohl, A., 2004c. Error analysis of a mixed finite element method for the Cahn-Hilliard equation. Numer. Math. 99 (1), 47–84. ISSN 0029-599X. https://doi.org/10.1007/s00211-004-0546-5.

Feng, X., Prohl, A., 2005. Numerical analysis of the Cahn-Hilliard equation and approximation of the Hele-Shaw problem. Interfaces Free Bound. 7 (1), 1–28. ISSN 1463-9963. https://doi.org/10.4171/IFB/111.

Feng, X., Wise, S., 2012. Analysis of a Darcy-Cahn-Hilliard diffuse interface model for the Hele-Shaw flow and its fully discrete finite element approximation. SIAM J. Numer. Anal. 50 (3), 1320–1343. ISSN 0036-1429. https://doi.org/10.1137/110827119.

Feng, X., Wu, H., 2005. A posteriori error estimates and an adaptive finite element method for the Allen-Cahn equation and the mean curvature flow. J. Sci. Comput. 24 (2), 121–146. ISSN 0885-7474. https://doi.org/10.1007/s10915-004-4610-1.

Feng, X., Wu, H., 2008. A posteriori error estimates for finite element approximations of the Cahn-Hilliard equation and the Hele-Shaw flow. J. Comput. Math. 26 (6), 767–796. ISSN 0254-9409.

Feng, J.J., Liu, C., Shen, J., Yue, P., 2005a. An energetic variational formulation with phase field methods for interfacial dynamics of complex fluids: advantages and challenges. Modeling of Soft Matter, Springer, pp. 1–26.

Feng, W.M., Yu, P., Hu, S.Y., Liu, Z.-K., Du, Q., Chen, L.-Q., 2006. Spectral implementation of an adaptive moving mesh method for phase-field equations. J. Comput. Phys. 220 (1), 498–510.

Feng, X., He, Y., Liu, C., 2007a. Analysis of finite element approximations of a phase field model for two-phase fluids. Math. Comp. 76 (258), 539–571. ISSN 0025-5718. https://doi.org/10.1090/S0025-5718-06-01915-6.

Feng, W.M., Yu, P., Hu, S.Y., Liu, Z.-K., Du, Q., Chen, L.Q., 2009. A Fourier spectral moving mesh method for the Cahn-Hilliard equation with elasticity. Commun. Comput. Phys. 5, 582–599.

Feng, X., Li, Y., Prohl, A., 2014. Finite element approximations of the stochastic mean curvature flow of planar curves of graphs. Stoch. Partial Differ. Equ. Anal. Comput. 2 (1), 54–83. ISSN 2194-0401. https://doi.org/10.1007/s40072-014-0025-7.

Feng, X., Tang, T., Yang, J., 2015. Long time numerical simulations for phase-field problems using p-adaptive spectral deferred correction methods. SIAM J. Sci. Comput. 37 (1), A271–A294. ISSN 1064-8275. https://doi.org/10.1137/130928662.

Feng, X., Li, Y., Xing, Y., 2016. Analysis of mixed interior penalty discontinuous Galerkin methods for the Cahn-Hilliard equation and the Hele-Shaw flow. SIAM J. Numer. Anal. 54 (2), 825–847. ISSN 0036-1429. https://doi.org/10.1137/15M1009962.

Feng, X., Li, Y., Zhang, Y., 2017. Finite element methods for the stochastic Allen-Cahn equation with gradient-type multiplicative noise. SIAM J. Numer. Anal. 55 (1), 194–216. ISSN 0036-1429. https://doi.org/10.1137/15M1022124.

Feng, X., Li, Y., Zhang, Y., 2018. Strong convergence of a fully discrete finite element method for a class of semilinear stochastic partial differential equations with multiplicative noise. arXiv:1811.05028 [math.NA] (under review).

Feng, X., Li, Y., Zhang, Y., 2019. Fully discrete finite element methods for a stochastic Cahn-Hilliard equation with gradient-type multiplicative noise arXiv:1903.05146 [math. NA] (under review).

Fife, P.C., 1988. Dynamics of internal layers and diffusive interfaces. CBMS-NSF Regional Conference Series in Applied Mathematics, vol. 53. Society for Industrial and Applied Mathematics (SIAM), Philadelphia, PA. ISBN: 0-89871-225-4, p. vi+93.

Fife, P., 2003. Some nonclassical trends in parabolic and parabolic-like evolutions. In: Trends in Nonlinear AnalysisSpringer, pp. 153–191.

Fix, G., 1983. Phase field method for free boundary problems. In: Fasano, A., Primicerio, M. (Eds.), Free Boundary Problems. Pitman, London, pp. 580–589.

Furihata, D., 2001. A stable and conservative finite difference scheme for the Cahn-Hilliard equation. Numer. Math. 87 (4), 675–699.

Furihata, D., Kovács, M., Larsson, S., Lindgren, F., 2018. Strong convergence of a fully discrete finite element approximation of the stochastic Cahn-Hilliard equation. SIAM J. Numer. Anal. 56 (2), 708–731. ISSN 0036-1429. https://doi.org/10.1137/17M1121627.

Gao, M., Wang, X.-P., 2014. An efficient scheme for a phase field model for the moving contact line problem with variable density and viscosity. J. Comput. Phys. 272, 704–718. ISSN 0021-9991. https://doi.org/10.1016/j.jcp.2014.04.054.

Garcke, H., Novick-Cohen, A., 2000. A singular limit for a system of degenerate Cahn-Hilliard equations. Adv. Differ. Equ. 5 (4–6), 401–434. ISSN 1079-9389.

Garcke, H., Nestler, B., Stoth, B., 2000. A multiphase field concept: numerical simulations of moving phase boundaries and multiple junctions. SIAM J. Appl. Math. 60 (1), 295–315. ISSN 0036-1399. https://doi.org/10.1137/S0036139998334895.

Garcke, H., Preusser, T., Rumpf, M., Telea, A.C., Weikard, U., Van Wijk, J.J., 2001. A phase field model for continuous clustering on vector fields. IEEE Trans. Vis. Comput. Graph. 7 (3), 230–241.

Garcke, H., Nestler, B., Stinner, B., 2004. A diffuse interface model for alloys with multiple components and phases. SIAM J. Appl. Math. 64 (3), 775–799. ISSN 0036-1399. https://doi.org/10.1137/S0036139902413143.

Garcke, H., Lam, K.F., Nürnberg, R., Sitka, E., 2018. A multiphase Cahn-Hilliard-Darcy model for tumour growth with necrosis. Math. Models Methods Appl. Sci. 28 (03), 525–577.

Georgoulis, E.H., Lakkis, O., Virtanen, J.M., 2011. A posteriori error control for discontinuous Galerkin methods for parabolic problems. SIAM J. Numer. Anal. 49 (2), 427–458. ISSN 0036-1429. https://doi.org/10.1137/080722461.

Giga, Y., 2006. Surface Evolution Equations. Monographs in Mathematics, vol. 99. Birkhäuser Verlag, Basel. ISBN 978-3-7643-2430-8; 3-7643-2430-9, p. xii+264.

Gomez, H., Hughes, T.J.R., 2011. Provably unconditionally stable, second-order time-accurate, mixed variational methods for phase-field models. J. Comput. Phys. 230 (13), 5310–5327.

Gomez, H., 2018. Computational phase-field modeling. In: Encyclopedia of Computational Mechanics, second ed. Wiley Online Library, pp. 1–35.

Gómez, H., Calo, V.M., Bazilevs, Y., Hughes, T.J.R., 2008. Isogeometric analysis of the Cahn-Hilliard phase-field model. Comput. Methods Appl. Mech. Eng. 197 (49–50), 4333–4352.

Gottlieb, D., Orszag, S.A., 1983. Numerical Analysis of Spectral Methods: Theory and Applications. SIAM.

Grun, G., 2013. On convergent schemes for diffuse interface models for two-phase flow of incompressible fluids with general mass densities. SIAM J. Numer. Anal. 51 (6), 3036–3061.

Gu, R., Wang, X., Gunzburger, M., 2016. A two phase field model for tracking vesicle-vesicle adhesion. J. Math. Biol. 73 (5), 1293–1319.

Guan, Z., Lowengrub, J.S., Wang, C., Wise, S.M., 2014. Second order convex splitting schemes for periodic nonlocal Cahn-Hilliard and Allen-Cahn equations. J. Comput. Phys. 277, 48–71.

Gui, C., Zhao, M., 2015. Traveling wave solutions of Allen-Cahn equation with a fractional Laplacian. Ann. Inst. Henri Poincar (C) Non-Linaire Analyze 32 (4), 785–812.

Guillén-González, F., Tierra, G., 2013. On linear schemes for a Cahn-Hilliard diffuse interface model. J. Comput. Phys. 234, 140–171. ISSN 0021-9991. https://doi.org/10.1016/j.jcp.2012.09.020.

Guo, B.-Y., 1998. Spectral Methods and Their Applications. World Scientific.

Guo, R., Xu, Y., 2018. A high order adaptive time-stepping strategy and local discontinuous Galerkin method for the modified phase field crystal equation. Commun. Comput. Phys. 24 (1), 123–151. ISSN 1815-2406.

Guo, R., Ji, L., Xu, Y., 2016a. High order local discontinuous Galerkin methods for the Allen-Cahn equation: analysis and simulation. J. Comput. Math. 34 (2), 135–158. ISSN 0254-9409. https://doi.org/10.4208/jcm.1510-m2014-0002.

Guo, R., Ji, L., Xu, Y., 2016b. High order local discontinuous Galerkin methods for the Allen-Cahn equation: analysis and simulation. J. Comput. Math. 34 (2), 135–158. ISSN 0254-9409. https://doi.org/10.4208/jcm.1510-m2014-0002.

Hairer, M., Ryser, M.D., Weber, H., 2012. Triviality of the 2D stochastic Allen-Cahn equation. Electron. J. Probab. 17 (39), 14. ISSN 1083-6489. https://doi.org/10.1214/EJP.v17-1731.

Han, D., Wang, X., 2018. A second order in time, decoupled, unconditionally stable numerical scheme for the Cahn-Hilliard-Darcy system. J. Sci. Comput. 77 (2), 1210–1233. ISSN 0885-7474. https://doi.org/10.1007/s10915-018-0748-0.

He, Q., Glowinski, R., Wang, X.-P., 2011. A least-squares/finite element method for the numerical solution of the Navier-Stokes-Cahn-Hilliard system modeling the motion of the contact line. J. Comput. Phys. 230 (12), 4991–5009. ISSN 0021-9991. https://doi.org/10.1016/j.jcp.2011.03.022.

Heister, T., Wheeler, M.F., Wick, T., 2015. A primal-dual active set method and predictor-corrector mesh adaptivity for computing fracture propagation using a phase-field approach. Comput. Methods Appl. Mech. Eng. 290, 466–495.

Hele-Shaw, H.S., 1898b. On the motion of a viscous fluid between two parallel plates. Trans. R. Inst. Nav. Archit. Lond. 40, 218.

Heo, T.W., Zhang, L., Du, Q., Chen, L.-Q., 2010. Incorporating diffuse-interface nuclei in phase-field simulations. Scr. Mater. 63 (1), 8–11.

Hintermüller, M., Wegner, D., 2012. Distributed optimal control of the Cahn-Hilliard system including the case of a double-obstacle homogeneous free energy density. SIAM J. Control. Optim. 50 (1), 388–418.

Hintermüller, M., Hinze, M., Kahle, C., Keil, T., 2017. Fully adaptive and integrated numerical methods for the simulation and control of variable density multiphase flows governed by diffuse interface models. Transport Processes at Fluidic Interfaces, Springer, pp. 305–353.

Hochbruck, M., Ostermann, A., 2010. Exponential integrators. Acta Numer. 19, 209–286.

Hou, T., Tang, T., Yang, J., 2017. Numerical analysis of fully discretized Crank-Nicolson scheme for fractional-in-space Allen-Cahn equations. J. Sci. Comput. 72 (3), 1214–1231.

Hu, S.Y., Chen, L.Q., 2001. A phase-field model for evolving microstructures with strong elastic inhomogeneity. Acta Mater. 49 (11), 1879–1890.

Hu, J., Zhang, S., 2017. A canonical construction of H^m-nonconforming triangular finite elements. Ann. Appl. Math. 33 (3), 266–288. ISSN 2096-0174.

Hu, X., Li, R., Tang, T., 2009. A multi-mesh adaptive finite element approximation to phase field models. Commun. Comput. Phys. 5 (5), 1012–1029. ISSN 1815-2406.

Hua, J., Lin, P., Liu, C., Wang, Q., 2011. Energy law preserving C^0 finite element schemes for phase field models in two-phase flow computations. J. Comput. Phys. 230 (19), 7115–7131.

Huang, W., Russell, R.D., 2011. Adaptive Moving Mesh Methods, Applied Mathematical Sciences. vol. 174, Springer, New York. ISBN: 978-1-4419-7915-5, p. xviii+432. https://doi.org/10.1007/978-1-4419-7916-2.

Huisken, G., Ilmanen, T., 2001. The inverse mean curvature flow and the Riemannian Penrose inequality. J. Differ. Geom. 59 (3), 353–437. ISSN 0022-040X.

Humphries, A.R., Stuart, A.M., 1994. Runge-Kutta methods for dissipative and gradient dynamical systems. SIAM J. Numer. Anal. 31 (5), 1452–1485.

Ilmanen, T., 1993. Convergence of the Allen-Cahn equation to Brakke's motion by mean curvature. J. Differ. Geom. 38 (2), 417–461. ISSN 0022-040X. http://projecteuclid.org/euclid.jdg/1214454300.

Iserles, A., 2009. A First Course in the Numerical Analysis of Differential Equations. In: Cambridge Texts in Applied Mathematics, second ed. Cambridge University Press, Cambridge. ISBN: 978-0-521-73490-5, p. xx+459.

Jacqmin, D., 1999. Calculation of two-phase Navier-Stokes flows using phasefield modeling. J. Comput. Phys. 155 (1), 96–127. ISSN 0021-9991.

Jacqmin, D., 2000. Contact-line dynamics of a diffuse fluid interface. J. Fluid Mech. 402, 57–88.

Jacqmin, D., 2004. Onset of wetting failure in liquid-liquid systems. J. Fluid Mech. 517, 209–228.

Jamet, D., Misbah, C., 2007. Towards a thermodynamically consistent picture of the phase-field model of vesicles: local membrane incompressibility. Phys. Rev. E 76 (5), 051907.

Jeong, D., Kim, J., 2015. Microphase separation patterns in diblock copolymers on curved surfaces using a nonlocal Cahn-Hilliard equation. Eur. Phys. J. E 38 (11), 117.

Joshi, V., Jaiman, R.K., 2018. An adaptive variational procedure for the conservative and positivity preserving Allen-Cahn phase-field model. J. Comput. Phys. 366, 478–504. ISSN 0021-9991. https://doi.org/10.1016/j.jcp.2018.04.022.

Ju, L., Zhang, J., Du, Q., 2015a. Fast and accurate algorithms for simulating coarsening dynamics of Cahn-Hilliard equations. Comput. Mater. Sci. 108, 272–282.

Ju, L., Zhang, J., Zhu, L., Du, Q., 2015b. Fast explicit integration factor methods for semilinear parabolic equations. J. Sci. Comput. 62 (2), 431–455.

Jung, Y.M., Kang, S.H., Shen, J., 2007. Multiphase image segmentation via Modica-Mortola phase transition. SIAM J. Appl. Math. 67 (5), 1213–1232.

Kassam, A.-K., Trefethen, L.N., 2005. Fourth-order time-stepping for stiff PDEs. SIAM J. Sci. Comput. 26 (4), 1214–1233.

Kästner, M., Metsch, P., de Borst, R., 2016. Isogeometric analysis of the Cahn-Hilliard equation—a convergence study. J. Comput. Phys. 305, 360–371. ISSN 0021-9991. https://doi.org/10.1016/j.jcp.2015.10.047.

Kay, D., Styles, V., Süli, E., 2009. Discontinuous Galerkin finite element approximation of the Cahn-Hilliard equation with convection. SIAM J. Numer. Anal. 47 (4), 2660–2685. ISSN 0036-1429. https://doi.org/10.1137/080726768.

Kessler, D., Nochetto, R.H., Schmidt, A., 2004. A posteriori error control for the Allen-Cahn problem: circumventing Gronwall's inequality. M2AN Math. Model. Numer. Anal. 38 (1), 129–142. ISSN 0764-583X. https://doi.org/10.1051/m2an:2004006.

Kim, J., Lowengrub, J., 2005. Phase field modeling and simulation of three-phase flows. Interfaces Free Bound. 7 (4), 435–466.

Kim, J., Kang, K., Lowengrub, J., 2004. Conservative multigrid methods for Cahn-Hilliard fluids. J. Comput. Phys. 193 (2), 511–543.

Kohn, R.V., Sternberg, P., 1989. Local minimisers and singular perturbations. Proc. R. Soc. Edinburgh Sect. A 111 (1–2), 69–84. ISSN 0308-2105. https://doi.org/10.1017/S0308210500025026.

Kovács, M., Larsson, S., Mesforush, A., 2011. Finite element approximation of the Cahn-Hilliard-Cook equation. SIAM J. Numer. Anal. 49 (6), 2407–2429.

Kovács, M., Larsson, S., Mesforush, A., 2014. Erratum: Finite element approximation of the Cahn-Hilliard-Cook equation [mr2854602]. SIAM J. Numer. Anal. 52 (5), 2594–2597. ISSN 0036-1429. https://doi.org/10.1137/140968161.

Kovács, M., Larsson, S., Lindgren, F., 2015. On the backward Euler approximation of the stochastic Allen-Cahn equation. J. Appl. Probab. 52 (2), 323–338. ISSN 0021-9002. https://doi.org/10.1239/jap/1437658601.

Lakkis, O., Makridakis, C., 2006. Elliptic reconstruction and a posteriori error estimates for fully discrete linear parabolic problems. Math. Comp. 75 (256), 1627–1658. ISSN 0025-5718. https://doi.org/10.1090/S0025-5718-06-01858-8.

Lakkis, O., Makridakis, C., Pryer, T., 2015. A comparison of duality and energy a posteriori estimates for $l_\infty(0,t;l_2(\omega))$ in parabolic problems. Math. Comp. 84 (294), 1537–1569. ISSN 0025-5718. https://doi.org/10.1090/S0025-5718-2014-02912-8.

Lambert, J.D., 1991. Numerical Methods for Ordinary Differential Systems. John Wiley & Sons, Ltd., Chichester. ISBN: 0-471-92990-5, p. x+293

Landau, L.D., Lifshitz, E.M., 2013. Course of Theoretical Physics. Elsevier.

Langer, J.S., 1986. Models of pattern formation in first-order phase transitions. In: Directions in Condensed Matter Physics, World Science Publishers, pp. 164–186.

Lee, A.A., Münch, A., Süli, E., 2016. Sharp-interface limits of the Cahn-Hilliard equation with degenerate mobility. SIAM J. Appl. Math. 76 (2), 433–456. ISSN 0036-1399. https://doi.org/10.1137/140960189.

Li, Y., 2019. Error analysis of a fully discrete Morley finite element approximation for the Cahn-Hilliard equation. J. Sci. Comput. 78, 1862–1892.

Li, Y., Kim, J., 2011. Multiphase image segmentation using a phase-field model. Comput. Math. Appl. 62 (2), 737–745.

Li, D., Qiao, Z., 2017. On second order semi-implicit Fourier spectral methods for 2D Cahn-Hilliard equations. J. Sci. Comput. 70 (1), 301–341.

Li, B., Zhao, Y., 2013. Variational implicit solvation with solute molecular mechanics: from diffuse-interface to sharp-interface models. SIAM J. Appl. Math. 73 (1), 1–23. ISSN 0036-1399. https://doi.org/10.1137/120883426.

Li, B., Lowengrub, J., Rätz, A., Voigt, A., 2009. Geometric evolution laws for thin crystalline films: modeling and numerics. Commun. Comput. Phys. 6 (3), 433–482. ISSN 1815-2406.

Li, T., Zhang, P., Zhang, W., 2012. Numerical study for the nucleation of one-dimensional stochastic Cahn-Hilliard dynamics. Commun. Math. Sci. 10 (4), 1105–1132.

Li, T., Zhang, P., Zhang, W., 2013. Nucleation rate calculation for the phase transition of diblock copolymers under stochastic Cahn-Hilliard dynamics. Multiscale Model. Simul. 11 (1), 385–409. ISSN 1540-3459. https://doi.org/10.1137/120876307.

Li, D., Qiao, Z., Tang, T., 2016a. Characterizing the stabilization size for semi-implicit Fourier-spectral method to phase field equations. SIAM J. Numer. Anal. 54 (3), 1653–1681.

Li, D., Qiao, Z., Tang, T., 2016b. Characterizing the stabilization size for semi-implicit Fourier-spectral method to phase field equations. SIAM J. Numer. Anal. 54 (3), 1653–1681. ISSN 0036-1429. https://doi.org/10.1137/140993193.

Li, Y., Lee, H.G., Xia, B., Kim, J., 2016c. A compact fourth-order finite difference scheme for the three-dimensional Cahn-Hilliard equation. Comput. Phys. Commun. 200, 108–116.

Li, Y., Choi, Y., Kim, J., 2017. Computationally efficient adaptive time step method for the Cahn-Hilliard equation. Comput. Math. Appl. 73 (8), 1855–1864.

Lima, E.A.B.F., Oden, J.T., Almeida, R.C., 2014. A hybrid ten-species phase-field model of tumor growth. Math. Models Methods Appl. Sci. 24 (13), 2569–2599.

Lin, J.-T., 1988. The numerical analysis of a phase field model in moving boundary problems. SIAM J. Numer. Anal. 25 (5), 1015–1031.

Liu, C., Shen, J., 2003a. A phase field model for the mixture of two incompressible fluids and its approximation by a Fourier-spectral method. Physica D 179, 211–228.

Liu, C., Shen, J., 2003b. A phase field model for the mixture of two incompressible fluids and its approximation by a Fourier-spectral method. Physica D 179 (3–4), 211–228.

Löber, J., Ziebert, F., Aranson, I.S., 2014. Modeling crawling cell movement on soft engineered substrates. Soft Matter 10 (9), 1365–1373.

Lowengrub, J., Truskinovsky, L., 1998. Quasi-incompressible Cahn-Hilliard fluids and topological transitions. Proc. R. Soc. Lond. A Math. Phys. Eng. Sci. 454 (1978), 2617–2654.

Lowengrub, J.S., Frieboes, H.B., Jin, F., Chuang, Y.-L., Li, X., Macklin, P., Wise, S.M., Cristini, V., 2009a. Nonlinear modelling of cancer: bridging the gap between cells and tumours. Nonlinearity 23 (1), R1.

Lowengrub, J.S., Rätz, A., Voigt, A., 2009b. Phase-field modeling of the dynamics of multicomponent vesicles: spinodal decomposition, coarsening, budding, and fission. Phys. Rev. E 79(3), 0311926, 13, ISSN 1539–3755. https://doi.org/10.1103/PhysRevE.79.031926.

Luo, F., Tang, T., Xie, H., 2016. Parameter-free time adaptivity based on energy evolution for the Cahn-Hilliard equation. Commun. Comput. Phys. 19 (5), 1542–1563.

March, R., Dozio, M., 1997. A variational method for the recovery of smooth boundaries. Image Vis. Comput. 15 (9), 705–712.

Marques, F.C., Neves, A., 2014. Min-max theory and the Willmore conjecture. Ann. Math. 179, 683–782.

Marth, W., Voigt, A., 2014. Signaling networks and cell motility: a computational approach using a phase field description. J. Math. Biol. 69 (1), 91–112.

McFadden, G.B., 2002. Phase field models of solidification. In: Feng, X., Schulze, T.P. (Eds.), Recent Advances in Numerical Methods for Partial Differential Equations and Applications, Contemporary Mathematics, vol. 306. American Mathematical Society, Providence, RI, pp. 107–145.

Miehe, C., Hofacker, M., Welschinger, F., 2010a. A phase field model for rate-independent crack propagation: robust algorithmic implementation based on operator splits. Comput. Methods Appl. Mech. Eng. 199 (45–48), 2765–2778.

Miehe, C., Welschinger, F., Hofacker, M., 2010b. Thermodynamically consistent phase-field models of fracture: variational principles and multi-field finite element implementations. Int. J. Numer. Methods Eng. 83 (10), 1273–1311.

Miehe, C., Aldakheel, F., Raina, A., 2016. Phase field modeling of ductile fracture at finite strains: a variational gradient-extended plasticity-damage theory. Int. J. Plast. 84, 1–32.

Milovanov, A.V., Rasmussen, J.J., 2005. Fractional generalization of the Ginzburg-Landau equation: an unconventional approach to critical phenomena in complex media. Phys. Lett. A 337 (1–2), 75–80.

Morin, P., Nochetto, R.H., Siebert, K.G., 2002. Convergence of adaptive finite element methods. SIAM Rev. 44 (4), 631–658.

Moure, A., Gomez, H., 2017. Phase-field model of cellular migration: three-dimensional simulations in fibrous networks. Comput. Methods Appl. Mech. Eng. 320, 162–197.

Moure, A., Gomez, H., 2018. Three-dimensional simulation of obstacle-mediated chemotaxis. Biomech. Model. Mechanobiol. 17 (5), 12431268.

Mugnai, L., 2013. Gamma-convergence results for phase-field approximations of the 2D-Euler elastica functional. ESAIM Control Optim. Calc. Var. 19 (3), 740–753. ISSN 1292-8119. https://doi.org/10.1051/cocv/2012031.

Mullins, W.W., Sekerka, J., 1963. Morphological stability of a particle growing by diffusion or heat flow. J. Appl. Math. 34, 322–329.

Najem, S., Grant, M., 2016. Phase-field model for collective cell migration. Phys. Rev. E 93 (5), 052405.

Nochetto, R.H., Verdi, C., 1997. Convergence past singularities for a fully discrete approximation of curvature-driven interfaces. SIAM J. Numer. Anal. 34 (2), 490–512. ISSN 0036-1429.

Nochetto, R.H., Paolini, M., Verdi, C., 1994. Optimal interface error estimates for the mean curvature flow. Ann. Scuola Norm. Sup. Pisa Cl. Sci. 21 (2), 193–212.

Nochetto, R.H., Siebert, K.G., Veeser, A., 2009. Theory of adaptive finite element methods: an introduction. In: Multiscale, Nonlinear and Adaptive Approximation, Springer, Berlin, pp. 409–542.

Novick-Cohen, A., 2000. Triple-junction motion for an Allen-Cahn/Cahn-Hilliard system. Physica D 137 (1–2), 1–24. ISSN 0167-2789. https://doi.org/10.1016/S0167-2789(99)00162-1.

Ohta, T., Kawasaki, K., 1986. Equilibrium morphology of block copolymer melts. Macromolecules 19 (10), 2621–2632.

Osher, S., Fedkiw, R., 2003. Level Set Methods and Dynamic Implicit Surfaces. Applied Mathematical Sciences, vol. 153. Springer-Verlag, New York. ISBN: 0-387-95482-1, p. xiv+273.

Osher, S., Sethian, J.A., 1988. Fronts propagating with curvature-dependent speed: algorithms based on Hamilton-Jacobi formulations. J. Comput. Phys. 79 (1), 12–49. ISSN 0021-9991.

Otto, F., Weber, H., Westdickenberg, M.G., 2014. Invariant measure of the stochastic Allen-Cahn equation: the regime of small noise and large system size. Electron. J. Probab. 19, no. 23, 76. ISSN 1083-6489. https://doi.org/10.1214/EJP.v19-2813.

Pego, R.L., 1989. Front migration in the nonlinear Cahn-Hilliard equation. Proc. Roy. Soc. London Ser. A 422 (1863), 261–278. ISSN 0962-8444.

Pismen, L.M., 2001. Nonlocal diffuse interface theory of thin films and the moving contact line. Phys. Rev. E 64 (2), 021603.

Plapp, M., Karma, A., 2000. Multiscale finite-difference-diffusion-Monte-Carlo method for simulating dendritic solidification. J. Comput. Phys. 165 (2), 592–619. ISSN 0021-9991. https://doi.org/10.1006/jcph.2000.6634.

Pomeau, Y., 2002. Recent progress in the moving contact line problem: a review. C. R. Mecanique 330 (3), 207–222.

Preusser, T., Droske, M., Garbe, C.S., Telea, A., Rumpf, M., 2007. A phase field method for joint denoising, edge detection, and motion estimation in image sequence processing. SIAM J. Appl. Math. 68 (3), 599–618.

Priestly, E., 2012. Introduction to Liquid Crystals. Springer Science & Business Media.

Printems, J., 2001. On the discretization in time of parabolic stochastic partial differential equations. ESAIM: Math. Model. Numer. Anal. 35 (6), 1055–1078.

Prosperetti, A., Tryggvason, G., 2009. Computational Methods for Multiphase Flow. Cambridge University Press.

Provatas, N., Elder, K., 2011. Phase-Field Methods in Materials Science and Engineering. John Wiley & Sons.

Provatas, N., Goldenfeld, N., Dantzig, J., 1999. Adaptive mesh refinement computation of solidification microstructures using dynamic data structures. J. Comput. Phys. 148 (1), 265–290. ISSN 0021-9991. https://doi.org/10.1006/jcph.1998.6122.

Provatas, N., Greenwood, M., Athreya, B., Goldenfeld, N., Dantzig, J., 2005. Multiscale modeling of solidification: phase-field methods to adaptive mesh refinement. Int. J. Mod. Phys. B 19 (31), 4525–4565.

Qian, T., Wang, X.-P., Sheng, P., 2003. Molecular scale contact line hydrodynamics of immiscible flows. Phys. Rev. E 68 (1), 016306.

Rayleigh, J.W., 1899. Scientific Papers (Vol. 1). University Press.

Reddy, S.C., Trefethen, L.N., 1992. Stability of the method of lines. Numer. Math. 62, 235–267.

Ren, X., Wei, J., 2007. Many droplet pattern in the cylindrical phase of diblock copolymer morphology. Rev. Math. Phys. 19 (08), 879–921.

Rochery, M., Jermyn, I., Zerubia, J., 2005. Phase field models and higher-order active contours. In: Tenth IEEE International Conference on Computer Vision, 2005. ICCV 2005, vol. 2. IEEE, pp. 970–976.

Röckner, M., Yang, H., Zhu, R., 2018. Conservative stochastic 2-dimensional Cahn-Hilliard equation. arXiv:1802.04141 [math.NA].

Röger, M., Weber, H., 2013. Tightness for a stochastic Allen-Cahn equation. Stoch. Partial Differ. Equ. Anal. Comput. 1 (1), 175–203. ISSN 2194-0401. https://doi.org/10.1007/s40072-013-0004-4.

Rogers, T.M., Elder, K.R., Desai, R.C., 1988. Numerical study of the late stages of spinodal decomposition. Phys. Rev. B 37 (16), 9638.

Rosam, J., Jimack, P.K., Mullis, A.M., 2008. An adaptive, fully implicit multigrid phase-field model for the quantitative simulation of non-isothermal binary alloy solidification. Acta Mater. 56 (17), 4559–4569.

Rowlinson, J.S., 1979. Translation of J. D. van der Waals' "The thermodynamic theory of capillarity under the hypothesis of a continuous variation of density" J. Statist. Phys. 20 (2), 197–244. ISSN 0022-4715. https://doi.org/10.1007/BF01011513.

Rubinstein, J., Sternberg, P., Keller, J.B., 1989. Fast reaction, slow diffusion, and curve shortening. SIAM J. Appl. Math. 49 (1), 116–133. ISSN 0036-1399.

Samson, C., Blanc-Féraud, L., Aubert, G., Zerubia, J., 2000. A variational model for image classification and restoration. IEEE Trans. Pattern Anal. Mach. Intell. 22 (5), 460–472.

Schiesser, W.E., 1991. The Numerical Method of Lines: Integration of Partial Differential Equations. Academic Press, San Diego.

Schillinger, D., Borden, M.J., Stolarski, H.K., 2015. Isogeometric collocation for phase-field fracture models. Comput. Methods Appl. Mech. Eng. 284, 583–610.

Schwab, C., 1998. *p*- and *hp*-Finite Element Methods. Numerical Mathematics and Scientific Computation. The Clarendon Press, Oxford University Press, New York. ISBN: 0-19-850390-3, p. xii +374.

Sethian, J.A., 1999. Level Set Methods and Fast Marching Methods. In: Cambridge Monographs on Applied and Computational Mathematics, second ed. vol. 3. Cambridge University Press. Cambridge, p. xx+378. ISBN 0-521-64204-3; 0-521-64557-3.

Shao, D., Levine, H., Rappel, W.-J., 2012. Coupling actin flow, adhesion, and morphology in a computational cell motility model. Proc. Natl. Acad. Sci. 109 (18), 6851–6856.

Shen, J., Xu, J., 2018. Convergence and error analysis for the scalar auxiliary variable (SAV) schemes to gradient flows. SIAM J. Numer. Anal. 56 (5), 2895–2912. ISSN 0036-1429. https://doi.org/10.1137/17M1159968.

Shen, J., Yang, X., 2009a. An efficient moving mesh spectral method for the phase-field model of two-phase flows. J. Comput. Phys. 228 (8), 2978–2992.

Shen, J., Yang, X., 2010. Numerical approximations of Allen-Cahn and Cahn-Hilliard equations. Discrete Contin. Dyn. Syst. 28 (4), 1669–1691.

Shen, J., Yang, X., 2019. Decoupled, energy stable schemes for phase-field models of two-phase incompressible flows. SIAM Rev. 61, 474–506.

Shen, J., Xu, J., Yang, J., 2018. A new class of efficient and robust energy stable schemes for gradient flows. Preprint (to appear in SIAM Rev).

Shen, J., Tang, T., Wang, L.-L., 2011. Spectral Methods: Algorithms, Analysis and Applications. vol. 41. Springer Science & Business Media.

Shen, J., Xu, J., Yang, J., 2018. The scalar auxiliary variable (SAV) approach for gradient flows. J. Comput. Phys. 353, 407–416. ISSN 0021-9991. https://doi.org/10.1016/j.jcp.2017.10.021.

Shi, A.-C., 2019. Self-consistent field theory of inhomogeneous polymeric systems: a variational derivation. Adv. Theory Simul. 2, 1800188.

Shi, Y., Bao, K., Wang, X.-P., 2013. 3D adaptive finite element method for a phase field model for the moving contact line problems. Inverse Probl. Imaging 7 (3), 947–959. ISSN 1930-8337. https://doi.org/10.3934/ipi.2013.7.947.

Shin, J., Lee, H.G., Lee, J.-Y., 2017. Unconditionally stable methods for gradient flow using convex splitting Runge-Kutta scheme. J. Comput. Phys. 347, 367–381.

Shu, Y., Ai, X., Li, B.Q., 2007. Phase-field modelling of solidification microstructure formation using the discontinuous finite element method. Int. J. Numer. Methods Eng. 69 (6), 1194–1211. ISSN 0029-5981. https://doi.org/10.1002/nme.1805.

Simonett, G., 2001. The Willmore flow near spheres. Differ. Integral Equ. 14 (8), 1005–1014. ISSN 0893-4983.

Singer-Loginova, I., Singer, H.M., 2008. The phase field technique for modeling multiphase materials. Rep. Prog. Phys. 71 (10), 106501.

Song, H., Shu, C.-W., 2017. Unconditional energy stability analysis of a second order implicit-explicit local discontinuous Galerkin method for the Cahn-Hilliard equation. J. Sci. Comput. 73 (2-3), 1178–1203.

Song, F., Xu, C., Karniadakis, G.E., 2016. A fractional phase-field model for two-phase flows with tunable sharpness: algorithms and simulations. Comput. Methods Appl. Mech. Eng. 305, 376–404.

Souganidis, P.E., Yip, N.K., 2004. Uniqueness of motion by mean curvature perturbed by stochastic noise. Ann. Inst. Henri Poincar (C) Non-Linaire Analyze 21 (1), 1–23. ISSN 0294-1449. https://doi.org/10.1016/S0294-1449(03)00029-5.

Steinbach, I., 2009. Topical review: phase-field models in material science. Model. Simul. Mater. Sci. Eng. 17, 073001.

Stogner, R.H., Carey, G.F., Murray, B.T., 2008. Approximation of Cahn-Hilliard diffuse interface models using parallel adaptive mesh refinement and coarsening with C^1 elements. Int. J. Numer. Methods Eng. 76 (5), 636–661.

Sun, Z.Z., 1995. A second-order accurate linearized difference scheme for the two-dimensional Cahn-Hilliard equation. Math. Comput. 64 (212), 1463–1471.

Sun, H., Wen, J., Zhao, Y., Li, B., McCammon, J.A., 2015. A self-consistent phase-field approach to implicit solvation of charged molecules with Poisson-Boltzmann electrostatics. J. Chem. Phys. 143 (24), 12B602_1.

Takashi, S., Takayuki, A., Tomohiro, T., Akinori, Y., Akira, N., Toshio, E., Naoya, M., Satoshi, M., Malcolm, C., 2011. Peta-scale phase-field simulation for dendritic solidification on the TSUBAME 2.0 supercomputer. In: Proceedings of 2011 International Conference for High Performance Computing, Networking, Storage and Analysis (SC '11).

Tan, Z., Tang, T., Zhang, Z., 2006. A simple moving mesh method for one-and two-dimensional phase-field equations. J. Comput. Appl. Math. 190 (1–2), 252–269.

Tarasov, V.E., Zaslavsky, G.M., 2005. Fractional Ginzburg-Landau equation for fractal media. Phys. A Stat. Mech. Appl. 354, 249–261.

Tinkham, M., 2004. Introduction to Superconductivity. Courier Corporation.

Valdinoci, E., 2013. A fractional framework for perimeters and phase transitions. Milan J. Math. 81 (1), 1–23.

Van der Waals, J.D., 1894. Thermodynamic theory of capillarity assuming steady density change. J. Phys. Chem. 13 (1), 657–725.

Van Gennip, Y., Guillen, N., Osting, B., Bertozzi, A.L., 2014. Mean curvature, threshold dynamics, and phase field theory on finite graphs. Milan J. Math. 82 (1), 3–65.

Vilanova, G., Colominas, I., Gomez, H., 2017. A mathematical model of tumour angiogenesis: growth, regression and regrowth. J. R. Soc. Interface 14 (126), 20160918.

Wang, X., Du, Q., 2008. Modelling and simulations of multi-component lipid membranes and open membranes via diffuse interface approaches. J. Math. Biol. 56 (3), 347–371.

Wang, Y., Li, J., 2010. Phase field modeling of defects and deformation. Acta Mater. 58 (4), 1212–1235.

Wang, X., Zhang, Z., 2013. Well-posedness of the Hele-Shaw-Cahn-Hilliard system. Ann. Inst. Henri Poincar (C) Non-Linaire Analyze 30 (3), 367–384. ISSN 0294-1449. https://doi.org/10.1016/j.anihpc.2012.06.003.

Wang, X., Ju, L., Du, Q., 2016. Efficient and stable exponential time differencing Runge-Kutta methods for phase field elastic bending energy models. J. Comput. Phys. 316, 21–38.

Wang, J., Zhai, Q., Zhang, R., Zhang, S., 2019. A weak Galerkin finite element scheme for the Cahn-Hilliard equation. Math. Comp. 88 (315), 211–235. ISSN 0025-5718. https://doi.org/10.1090/mcom/3369.

Wanner, T., 2017. Computer-assisted equilibrium validation for the diblock copolymer model. Discrete Contin. Dyn. Syst. A 37 (2), 1075–1107.

Weber, H., 2010a. On the short time asymptotic of the stochastic Allen-Cahn equation. Ann. Inst. Henri Poincaré Probab. Stat. 46 (4), 965–975. ISSN 0246-0203. https://doi.org/10.1214/09-AIHP333.

Weber, H., 2010b. Sharp interface limit for invariant measures of a stochastic Allen-Cahn equation. Comm. Pure Appl. Math. 63 (8), 1071–1109. ISSN 0010-3640. https://doi.org/10.1002/cpa.20323.

Weinan, E., Liu, D., 2002. Gibbsian dynamics and invariant measures for stochastic dissipative PDEs. J. Statist. Phys. 108 (5–6), 1125–1156. ISSN 0022-4715. https://doi.org/10.1023/A:1019747716056.

Wheeler, A.A., Murray, B.T., Schaefer, R.J., 1993. Computation of dendrites using a phase field model. Physica D 66 (1–2), 243–262.

Wheeler, M.F., Wick, T., Wollner, W., 2014. An augmented-Lagrangian method for the phase-field approach for pressurized fractures. Comput. Methods Appl. Mech. Eng. 271, 69–85.

Wise, S., Kim, J., Lowengrub, J., 2007. Solving the regularized, strongly anisotropic Cahn-Hilliard equation by an adaptive nonlinear multigrid method. J. Comput. Phys. 226 (1), 414–446.

Wise, S.M., Wang, C., Lowengrub, J.S., 2009. An energy-stable and convergent finite-difference scheme for the phase field crystal equation. SIAM J. Numer. Anal. 47 (3), 2269–2288.

Wodo, O., Ganapathysubramanian, B., 2011. Computationally efficient solution to the Cahn-Hilliard equation: Adaptive implicit time schemes, mesh sensitivity analysis and the 3D isoperimetric problem. J. Comput. Phys. 230 (15), 6037–6060.

Wörner, M., 2012. Numerical modeling of multiphase flows in microfluidics and micro process engineering: a review of methods and applications. Microfluid. Nanofluid. 12 (6), 841–886.

Wu, S., Li, Y., 2018. Analysis of the Morley element for the Cahn-Hilliard equation and the Hele-Shaw flow. arXiv:1808.08581 [math.NA].

Xia, Y., Xu, Y., Shu, C.-W., 2007. Local discontinuous Galerkin methods for the Cahn-Hilliard type equations. J. Comput. Phys. 227 (1), 472–491.

Xia, Y., Xu, Y., Shu, C.-W., 2009. Application of the local discontinuous Galerkin method for the Allen-Cahn/Cahn-Hilliard system. Commun. Comput. Phys. 51, 821–835.

Xu, C., Tang, T., 2006. Stability analysis of large time-stepping methods for epitaxial growth models. SIAM J. Numer. Anal. 44 (4), 1759–1779. ISSN 0036-1429. https://doi.org/10.1137/050628143.

Xu, J., Li, Y., Wu, S., Bousquet, A., 2019. On the stability and accuracy of partially and fully implicit schemes for phase field modeling. Comput. Methods Appl. Mech. Eng. 345, 826–853.

Yan, Y., Chen, W., Wang, C., Wise, S.M., 2018. A second-order energy stable BDF numerical scheme for the Cahn-Hilliard equation. Commun. Comput. Phys. 23 (2), 572–602. ISSN 1815-2406.

Yang, X., 2016. Linear, first and second-order, unconditionally energy stable numerical schemes for the phase field model of homopolymer blends. J. Comput. Phys. 327, 294–316. ISSN 0021-9991. https://doi.org/10.1016/j.jcp.2016.09.029.

Yang, X., 2019. Efficient linear, stabilized, second-order time marching schemes for an anisotropic phase field dendritic crystal growth model. Comput. Methods Appl. Mech. Eng. 347, 316–339. ISSN 0045-7825. https://doi.org/10.1016/j.cma.2018.12.012.

Yang, X., Ju, L., 2017a. Efficient linear schemes with unconditional energy stability for the phase field elastic bending energy model. Comput. Methods Appl. Mech. Eng. 315, 691–712.

Yang, X., Ju, L., 2017b. Linear and unconditionally energy stable schemes for the binary fluid-surfactant phase field model. Comput. Methods Appl. Mech. Eng. 318, 1005–1029.

Yang, X., Zhao, J., Wang, Q., Shen, J., 2017. Numerical approximations for a three-component Cahn-Hilliard phase-field model based on the invariant energy quadratization method. Math. Models Methods Appl. Sci. 27 (11), 1993–2030.

Yang, J., Du, Q., Zhang, W., 2018. Uniform L_p-bound of the Allen-Cahn equation and its numerical discretization. Int. J. Numer. Anal. Model. 15 (1–2), 213–227. ISSN 1705-5105.

Yip, N.K., 1998. Stochastic motion by mean curvature. Arch. Rational Mech. Anal. 144 (4), 313–355. ISSN 0003-9527. https://doi.org/10.1007/s002050050120.

Yip, N.K., 2002. Stochastic curvature driven flows. In: Stochastic Partial Differential Equations and Applications (Trento, 2002). Lecture Notes in Pure and Appl. Math., vol. 227. Dekker, New York, pp. 443–460.

Yu, P., Chen, L.Q., Du, Q., 2008. Applications of moving mesh methods to the fourier spectral approximations of phase-field equations. In: Recent Advances in Computational Sciences: Selected Papers from the International Workshop on Computational Sciences and Its Education, World Scientific, pp. 80–99.

Yue, P., Feng, J.J., Liu, C., Shen, J., 2004. A diffuse-interface method for simulating two-phase flows of complex fluids. J. Fluid Mech. 515, 293–317.

Zhai, S., Weng, Z., Feng, X., 2016. Fast explicit operator splitting method and time-step adaptivity for fractional non-local Allen-Cahn model. Appl. Math. Model. 40 (2), 1315–1324.

Zhang, J., Du, Q., 2009. Numerical studies of discrete approximations to the Allen-Cahn equation in the sharp interface limit. SIAM J. Sci. Comput. 31 (4), 3042–3063.

Zhang, J., Du, Q., 2012. Shrinking dimer dynamics and its applications to saddle point search. SIAM J. Numer. Anal. 50 (4), 1899–1921.

Zhang, Z., Qiao, Z., 2012. An adaptive time-stepping strategy for the Cahn-Hilliard equation. Commun. Comput. Phys. 11 (4), 1261–1278.

Zhang, Z., Tang, H., 2007. An adaptive phase field method for the mixture of two incompressible fluids. Comput. Fluids 36 (8), 1307–1318.

Zhang, S., Wang, M., 2010. A nonconforming finite element method for the Cahn-Hilliard equation. J. Comput. Phys. 229 (19), 7361–7372.

Zhang, Q., Wang, X.-P., 2016. Phase field modeling and simulation of three-phase flow on solid surfaces. J. Comput. Phys. 319, 79–107. ISSN 0021-9991. https://doi.org/10.1016/j.jcp.2016.05.016.

Zhang, L., Chen, L.-Q., Du, Q., 2007. Morphology of critical nuclei in solid-state phase transformations. Phys. Rev. Lett. 98 (26), 265703.

Zhang, L., Chen, L.-Q., Du, Q., 2008. Diffuse-interface description of strain-dominated morphology of critical nuclei in phase transformations. Acta Mater. 56 (14), 3568–3576.

Zhang, J., Das, S., Du, Q., 2009. A phase field model for vesicle-substrate adhesion. J. Comput. Phys. 228 (20), 7837–7849.

Zhang, F., Xu, Y., Guo, R., 2015. Interior penalty discontinuous Galerkin based isogeometric analysis for Allen-Cahn equations on surfaces. Commun. Comput. Phys. 18 (5), 1380–1416. ISSN 1815-2406. https://doi.org/10.4208/cicp.010914.180315a.

Zhang, J., Zhou, C., Wang, Y., Ju, L., Du, Q., Chi, X., Xu, D., Chen, D., Liu, Y., Liu, Z., 2016a. Extreme-scale phase field simulations of coarsening dynamics on the Sunway TaihuLight supercomputer. In: Proceedings of the International Conference for High Performance Computing, Networking, Storage and Analysis, p. 4.

Zhang, L., Ren, W., Samanta, A., Du, Q., 2016b. Recent developments in computational modelling of nucleation in phase transformations. NPJ Comput. Mater. 2, 16003.

Zhang, F., Huang, W., Li, X., Zhang, S., 2018. A study on phase-field models for brittle fracture. arXiv preprint arXiv:1805.07357.

Zhao, J., Yang, X., Shen, J., Wang, Q., 2016b. A decoupled energy stable scheme for a hydrodynamic phase-field model of mixtures of nematic liquid crystals and viscous fluids. J. Comput. Phys. 305, 539–556. ISSN 0021-9991. https://doi.org/10.1016/j.jcp.2015.09.044.

Zhao, Y., Xu, B.-X., Stein, P., Gross, D., 2016c. Phase-field study of electrochemical reactions at exterior and interior interfaces in Li-ion battery electrode particles. Comput. Methods Appl. Mech. Eng. 312, 428–446. ISSN 0045-7825. https://doi.org/10.1016/j.cma.2016.04.033.

Zhao, Y., Ma, Y., Sun, H., Li, B., Du, Q., 2018. A new phase-field approach to variational implicit solvation of charged molecules with the coulomb-field approximation. Comm. Math. Sci. 16, 1203–1223.

Zheng, X., Yang, C., Cai, X.-C., Keyes, D., 2015. A parallel domain decomposition-based implicit method for the Cahn-Hilliard-Cook phase-field equation in 3D. J. Comput. Phys. 285, 55–70.

Zhou, C., Yue, P., Feng, J.J., Ollivier-Gooch, C.F., Hu, H.H., 2010. 3D phase-field simulations of interfacial dynamics in Newtonian and viscoelastic fluids. J. Comput. Phys. 229 (2), 498–511.

Zhu, X.-P., 2002. Lectures on Mean Curvature Flows. AMS/IP Studies in Advanced Mathematics, vol. 32. American Mathematical Society, Providence, RI; International Press, Somerville, MA. ISBN: 0-8218-3311-1, p. x+150.

Zhu, L., Wang, Y., Ju, L., Wang, D., 2010. A variational phase field method for curve smoothing. J. Comput. Phys. 229 (6), 2390–2400.

Ziebert, F., Aranson, I.S., 2016. Computational approaches to substrate-based cell motility. NPJ Comput. Mater. 2, 16019.

Further readings

Babuška, I., Rheinboldt, W.C., 1978. Error estimates for adaptive finite element computations. SIAM J. Numer. Anal. 15 (4), 736–754.

Babuška, I., Feistauer, M., Šolín, P., 2001. On one approach to a posteriori error estimates for evolution problems solved by the method of lines. Numer. Math. 89 (2), 225–256. ISSN 0029-599X.

Barrett, J.W., Garcke, H., Nürnberg, R., 2008. On sharp interface limits of Allen-Cahn/Cahn-Hilliard variational inequalities. Discrete Contin. Dyn. Syst. Ser. S 1 (1), 1–14. https://doi.org/10.3934/dcdss.2008.1.1.

Bronsard, L., Kohn, R.V., 1990. On the slowness of phase boundary motion in one space dimension. Comm. Pure Appl. Math. 43 (8), 983–997. ISSN 0010-3640. https://doi.org/10.1002/cpa.3160430804.

Bronsard, L., Reitich, F., 1993. On three-phase boundary motion and the singular limit of a vector-valued Ginzburg-Landau equation. Arch. Ration. Mech. Anal. 124 (4), 355–379. ISSN 0003-9527. https://doi.org/10.1007/BF00375607.

Bronsard, L., Wetton, B.T.R., 1995. A numerical method for tracking curve networks moving with curvature motion. J. Comput. Phys. 120 (1), 66–87. ISSN 0021-9991. https://doi.org/10.1006/jcph.1995.1149.

Caffarelli, L.A., Muler, N.E., 1995. An L^∞ bound for solutions of the Cahn-Hilliard equation. Arch. Rational Mech. Anal. 133 (2), 129–144. ISSN 0945-8396.

Cahn, J.W., Taylor, J.E., 1994. Surface motion by surface diffusion. Acta Met. Mat. 42, 1045–1063.

Chen, Z., Hoffmann, K.H., 1994. An error estimate for a finite-element scheme for a phase field model. IMA J. Numer. Anal. 14 (2), 243–255.

Coleman, B.D., Falk, R.S., Moakher, M., 1995. Stability of cylindrical bodies in the theory of surface diffusion. Physica D 89 (1–2), 123–135. ISSN 0167-2789.

Coleman, B.D., Falk, R.S., Moakher, M., 1996. Space-time finite element methods for surface diffusion with applications to the theory of the stability of cylinders. SIAM J. Sci. Comput. 17 (6), 1434–1448. ISSN 1064-8275.

Deckelnick, K., Dziuk, G., 2002. A fully discrete numerical scheme for weighted mean curvature flow. Numer. Math. 91 (3), 423–452. ISSN 0029-599X.

Diegel, A.E., Wang, C., Wang, X., Wise, S.M., 2017b. Convergence analysis and error estimates for a second order accurate finite element method for the Cahn-Hilliard-Navier-Stokes system. Numer. Math. 137 (3), 495–534. ISSN 0029-599X. https://doi.org/10.1007/s00211-017-0887-5.

Dziuk, G., Elliott, C.M., 2013. Finite element methods for surface PDEs. Acta Numer. 22, 289–396. ISSN 0962-4929. https://doi.org/10.1017/S0962492913000056.

Elliott, C.M., Garcke, H., 1996. On the Cahn-Hilliard equation with degenerate mobility. SIAM J. Math. Anal. 27 (2), 404–423. ISSN 0036-1410.

Elliott, C.M., Garcke, H., 1997. Existence results for diffusive surface motion laws. Adv. Math. Sci. Appl. 7 (1), 467–490. ISSN 1343-4373.

Elliott, C.M., Larsson, S., 1992. Error estimates with smooth and nonsmooth data for a finite element method for the Cahn-Hilliard equation. Math. Comp. 58 (198), 603–630, S33-S36. ISSN 0025-5718.

Elliott, C.M., Maier-Paape, S., 2001. Losing a graph with surface diffusion. Hokkaido Math. J. 30 (2), 297–305. ISSN 0385-4035.

Elliott, C.M., Songmu, Z., 1986. On the Cahn-Hilliard equation. Arch. Rational Mech. Anal. 96 (4), 339–357. ISSN 0945-8396.

Elliott, C.M., Stinner, B., 2010b. A surface phase field model for two-phase biological membranes. SIAM J. Appl. Math. 70 (8), 2904–2928. ISSN 0036-1399. https://doi.org/10.1137/090779917.

Elliott, C.M., French, D.A., Milner, F.A., 1989. A second order splitting method for the Cahn-Hilliard equation. Numer. Math. 54 (5), 575–590. ISSN 0029-599X.

Eriksson, K., Johnson, C., 1995. Adaptive finite element methods for parabolic problems. IV. Nonlinear problems. SIAM J. Numer. Anal. 32 (6), 1729–1749. ISSN 0036-1429.

Feng, X., Prohl, A., 2003a. Analysis of total variation flow and its finite element approximations. M2AN Math. Model. Numer. Anal. 37 (3), 533–556. ISSN 0764-583X. https://doi.org/10.1051/m2an:2003041.

Feng, X., Prohl, A., 2004b. Analysis of gradient flow of a regularized Mumford-Shah functional for image segmentation and image inpainting. M2AN Math. Model. Numer. Anal. 38 (2), 291–320. ISSN 0764-583X. https://doi.org/10.1051/m2an:2004014.

Feng, X., Schulze, T.P., 2002. Recent Advances in Numerical Methods for Partial Differential Equations and Applications, Contemporary Mathematics. vol. 306American Mathematical Society, Providence, RI.

Feng, X., von Oehsen, M., Prohl, A., 2005b. Rate of convergence of regularization procedures and finite element approximations for the total variation flow. Numer. Math. 100 (3), 441–456. ISSN 0029-599X. https://doi.org/10.1007/s00211-005-0585-6.

Feng, X., Neilan, M., Prohl, A., 2007b. Error analysis of finite element approximations of the inverse mean curvature flow arising from the general relativity. Numer. Math. 108 (1), 93–119. ISSN 0029-599X. https://doi.org/10.1007/s00211-007-0111-0.

Fife, P.C., 2000. Models for phase separation and their mathematics. Electronic J. of Diff. Eqns 2000 (48), 1–26.

Garcke, H., Kwak, D. J. C., 2006. On asymptotic limits of Cahn-Hilliard systems with elastic misfit. In: Analysis, modeling and simulation of multiscale problemsSpringer, Berlin, pp. 87–111.

Garcke, H., Stoth, B., Nestler, B., 1999. Anisotropy in multi-phase systems: a phase field approach. Interfaces Free Bound. 1 (2), 175–198. ISSN 1463-9963. https://doi.org/10.4171/IFB/8.

Garcke, H., Lam, K.F., Stinner, B., 2014. Diffuse interface modelling of soluble surfactants in two-phase flow. Commun. Math. Sci. 12 (8), 1475–1522. ISSN 1539-6746. https://doi.org/10.4310/CMS.2014.v12.n8.a6.

Giga, Y., Ito, K., 1998. On pinching of curves moved by surface diffusion. Commun. Appl. Anal. 2 (3), 393–405. ISSN 1083-2564.

Guillén-González, F., Tierra, G., 2013. On linear schemes for a Cahn-Hilliard diffuse interface model. J. Comput. Phys. 234, 140–171.

Guo, Z., Lin, P., 2015. A thermodynamically consistent phase-field model for two-phase flows with thermocapillary effects. J. Fluid Mech. 766, 226–271. ISSN 0022-1120. https://doi.org/10.1017/jfm.2014.696.

Guo, Z., Lin, P., Lowengrub, J.S., 2014. A numerical method for the quasi-incompressible Cahn-Hilliard-Navier-Stokes equations for variable density flows with a discrete energy law. J. Comput. Phys. 276, 486–507. ISSN 0021-9991. https://doi.org/10.1016/j.jcp.2014.07.038.

Guo, Z., Lin, P., Lowengrub, J., Wise, S.M., 2017. Mass conservative and energy stable finite difference methods for the quasi-incompressible Navier-Stokes-Cahn-Hilliard system: primitive variable and projection-type schemes. Comput. Methods Appl. Mech. Eng. 326, 144–174. ISSN 0045-7825. https://doi.org/10.1016/j.cma.2017.08.011.

Han, D., Wang, X., 2015. A second order in time, uniquely solvable, unconditionally stable numerical scheme for Cahn-Hilliard-Navier-Stokes equation. J. Comput. Phys. 290, 139–156. ISSN 0021-9991. https://doi.org/10.1016/j.jcp.2015.02.046.

Hele-Shaw, H.S., 1898a. The flow of water. Nature 58 (1489), 33–36.

Ilmanen, T., 2003. Problems in mean curvature flow. Preprint, downloadable at www.math.ethz.ch/ ilmanen/papers/pub.html.

Jiang, W., Bao, W., Thompson, C.V., Srolovitz, D.J., 2012. Phase field approach for simulating solid-state dewetting problems. Acta Mater. 60 (15), 5578–5592.

Karma, A., Kessler, D.A., Levine, H., 2001. Phase-field model of mode III dynamic fracture. Phys. Rev. Lett. 87 (4), 045501.

Kindermann, S., Osher, S., Jones, P.W., 2005. Deblurring and denoising of images by nonlocal functionals. Multiscale Model. Simul. 4 (4), 1091–1115.

Mayer, U.F., Simonett, G., 2000. Self-intersections for the surface diffusion and the volume-preserving mean curvature flow. Differ. Integral Equ. 13 (7–9), 1189–1199. ISSN 0893-4983.

Mullins, W.W., 1957. Theory of thermal grooving. J. Appl. Phys. 28, 333–339.

Mullins, W.W., Sekerka, R.F., 1964. Stability of a planer interface during solidification of a dilute binary alloy. J. Appl. Phys. 35, 444–451.

Nochetto, R.H., Savaré, G., Verdi, C., 2000. A posteriori error estimates for variable time-step discretizations of nonlinear evolution equations. Comm. Pure Appl. Math. 53 (5), 525–589. ISSN 0010-3640.

Novick-Cohen, A., 1998. The Cahn-Hilliard equation: mathematical and modeling perspectives. Adv. Math. Sci. Appl. 8 (2), 965–985. ISSN 1343-4373.

Nürnberg, R., 2009. Numerical simulations of immiscible fluid clusters. Appl. Numer. Math. 59 (7), 1612–1628. ISSN 0168-9274. https://doi.org/10.1016/j.apnum.2008.11.003.

Rosolen, A., Peco, C., Arroyo, M., 2013. An adaptive meshfree method for phase-field models of biomembranes. Part I: Approximation with maximum-entropy basis functions. J. Comput. Phys. 249, 303–319. ISSN 0021-9991. https://doi.org/10.1016/j.jcp.2013.04.046.

Schulze, T., Alexiades, V., Feng, X., 2008. Multi-scale modeling and simulation in materials science, preface. J. Sci. Comput. 37 (1), 1–2. ISSN 0885-7474. https://doi.org/10.1007/s10915-008-9211-y.

Sethian, J.A., 1990. Numerical algorithms for propagating interfaces: Hamilton-Jacobi equations and conservation laws. J. Differ. Geom. 31 (1), 131–161. ISSN 0022-040X.

Shen, J., Yang, X., 2009b. An efficient moving mesh spectral method for the phase-field model of two-phase flows. J. Comput. Phys. 228 (8), 2978–2992.

Shen, J., Yang, X., Wang, Q., 2013. Mass and volume conservation in phase field models for binary fluids. Commun. Comput. Phys. 13 (4), 1045–1065. ISSN 1815-2406. https://doi.org/10.4208/cicp.300711.160212a.

Shen, J., Tang, T., Yang, J., 2016. On the maximum principle preserving schemes for the generalized Allen-Cahn equation. Commun. Math. Sci. 14 (6), 1517–1534. ISSN 1539-6746. https://doi.org/10.4310/CMS.2016.v14.n6.a3.

Stoth, B.E.E., 1996. Convergence of the Cahn-Hilliard equation to the Mullins-Sekerka problem in spherical symmetry. J. Differ. Equ. 125 (1), 154–183. ISSN 0022-0396.

Tang, T., Yang, J., 2016. Implicit-explicit scheme for the Allen-Cahn equation preserves the maximum principle. J. Comput. Math. 34 (5), 471–481. ISSN 0254-9409. https://doi.org/10.4208/jcm.1603-m2014-0017.

Taylor, J.E., 2003. Some mathematical challenges in materials science. Bull. Amer. Math. Soc. (N.S.) 40 (1), 69–87 (electronic). ISSN 0273-0979 Mathematical challenges of the 21st century (Los Angeles, CA, 2000).

Walkington, N.J., 1996. Algorithms for computing motion by mean curvature. SIAM J. Numer. Anal. 33 (6), 2215–2238. ISSN 0036-1429.

Wang, H., Li, R., Tang, T., 2008. Efficient computation of dendritic growth with *r*-adaptive finite element methods. J. Comput. Phys. 227 (12), 5984–6000. ISSN 0021-9991. https://doi.org/10.1016/j.jcp.2008.02.016.

Xie, Z., Song, L., Feng, X., 2008. A moving boundary problem derived from heat and water transfer processes in frozen and thawed soils and its numerical simulation. Sci. China Ser. A 51 (8), 1510–1521. ISSN 1006-9283. https://doi.org/10.1007/s11425-008-0096-x.

Zhao, J., Wang, Q., Yang, X., 2016a. Numerical approximations to a new phase field model for two phase flows of complex fluids. Comput. Methods Appl. Mech. Eng. 310, 77–97. ISSN 0045-7825. https://doi.org/10.1016/j.cma.2016.06.008.

Chapter 6

A review of level set methods to model interfaces moving under complex physics: Recent challenges and advances

Robert I. Saye[a] and James A. Sethian[a,b,*]
[a]*Mathematics Group, Lawrence Berkeley National Laboratory, Berkeley, CA, United States*
[b]*Department of Mathematics, University of California, Berkeley, Berkeley, CA, United States*
[*]*Corresponding author: e-mail: sethian@berkeley.edu*

Chapter Outline

Handbook of Numerical Analysis, Vol. 21. https://doi.org/10.1016/bs.hna.2019.07.003

Abstract

Level set methods, and their descendents, have been valuable in providing robust numerical algorithms for tracking the dynamics of evolving interfaces moving under complex physics and have been instrumental in shedding light on a variety of scientific problems. Application areas include fluid mechanics, materials sciences, biology, and chemistry, as well as in engineering and industry such as semiconductor manufacturing, the design of commercial inkjet plotters, and the dynamics of industrial foams. To handle increasingly complex and intricate interface dynamics problems, a set of mathematical, algorithmic, and numerical challenges had to be overcome in order to extend the basic level set methodology, introduced in Osher and Sethian (1988). In this review article, we discuss how each of those challenges were tackled and show examples of how the resulting advances in core methodologies have been used to compute a variety of complex problems in computational physics.

Keywords: Level set methods, Propagating interfaces, Hamilton-Jacobi equations, Computational fluid mechanics, Multi-phase multiphysics

AMS Classification Codes: 35, 65, 76

1 Introduction

1.1 Overview

In a wide collection of scientific and engineering problems, moving interfaces, separating different phases or regions, evolve and contort under complex physics. Examples include the mixing of fluids, combustion fronts, plate tectonics, microjetting devices, the evolution of foams and soap bubble clusters, the evolution of surface profiles in semiconductor manufacturing, and the evolving shape of dendrites in crystal growth. The challenge in these problems is that the position and shape of the interface feeds into the solution of the equations for the driving physics, which then move the interface to a new configuration.

Level set methods, introduced in Osher and Sethian (1988), provide a general framework for solving such moving interface problems. The fundamental idea is to embed the interface as an implicitly defined hypersurface of a higher-dimensional function, and then derive an initial value time evolution partial differential equation for this time-dependent higher-dimensional function whose zero level set at any time coincides with the interface itself.

More precisely, consider a moving interface $\Gamma(t)$, which may be parameterized by $N - 1$ dimensions. For example, if the interface is an evolving curve in the plane, then the interface itself is a one-dimensional object separating two-dimensional space into an "inside" and an "outside." The essential idea of level set methods is to implicitly define the moving interface $\Gamma(t)$ propagating in N dimensions as the zero level set of an evolving level set function $\phi(x, t)$, $\phi : \mathbb{R}^N \times t \rightarrow \mathbb{R}$ which satisfies a time-dependent partial differential equation. In practice, one popular way to initialize this implicit function at time $t = 0$ is to let $\phi(x, t = 0)$ be the signed distance from the interface $\Gamma(t = 0)$. Thus, the interface itself is simply a particular hypersurface of this evolving function.

This formulation contains *two* embeddings:

- First, the interface $\Gamma(t)$ is represented by the implicit function $\phi(x, t)$.
- Second, the speed F of the interface is assumed to have meaning throughout the computational domain, not just at the interface itself. Thus, the motion of the zero level set interface corresponding to interface $\Gamma(t)$ is bracketed by the motion of nearby non-zero level sets, and hence a speed F must be constructed to move each of these neighbouring level sets.

This formulation, introduced in Osher and Sethian (1988), is mathematically and algorithmically appealing. First, geometric quantities on the interface may be easily computed from the level set function; as examples, the normal vector is given by $\mathbf{n} = \nabla\phi$ and the curvature κ is given by $\nabla \cdot \frac{\nabla\phi}{|\nabla\phi|}$. Second the evolving function ϕ is always the graph of a function, even though the zero level set can change topology, break and merge.

More advanced and sophisticated versions of this initial formulation have been instrumental in computing the solution to a host of complex problems. They have been used in process modelling to guide the manufacture of semiconductors, used to optimize the shape of nozzles in industrial inkjet printing, predict the location of combustion fronts in flame chemistry, extract complex structures from CT and MRI medical scans, and model the shape of evolving crystals in dendritic solidification: reviews may be found in Sethian (1996b, 1999), Sethian and Smereka (2003), and Osher and Fedkiw (2002).

1.2 Outline of review article: The evolution of level set methods

Some significant advances to the core level set methodology have contributed to the evolution and wide adoption of these methods. In this review, we discuss some core challenges, and how new techniques have been developed to meet these challenges and greatly extend the applicability of these methods. In particular:

- *Efficiency*: The above method requires computing the $N - 1$ dimensional evolving interface $\Gamma(t)$ by solving an initial value problem in one higher dimension, namely N-dimensional space. This is costly and inefficient.

The introduction of narrow band level set methods (Adalsteinsson and Sethian, 1995a) provides an adaptive approach which focuses computational labour solely in a neighbourhood of the interface, reducing the per-time-step operation count from $\mathcal{O}(n^N)$ to order $\mathcal{O}(kn^{N-1})$, where k is the width of this narrow band and n is the number of grid points per dimension. Virtually all modern level set methods use this approach or a variant (Section 2).

- *Accurate construction of extension velocities*: The choice of how to construct a normal speed F for the neighbouring level sets is important. These so-called extension velocities greatly influence the accuracy of the evolving interface. Poor choice of extension velocities can lead to significant mass loss. A general methodology for constructing extension velocities was presented in Adalsteinsson and Sethian (1999) (Section 4).
- *Multi-phase physics*: The initialization of the above method with the signed distance function requires that there be no more than two separate phases. At triple points (and higher-order junctions), it is impossible to use a signed distance function with a single level set function. A general methodology for doing so, namely the "Voronoi Implicit Interface Method" was introduced in Saye and Sethian (2011, 2012), which couples level set methods to computational geometry and provides a powerful approach to problems with hundreds of separate phases (Section 6).
- *Sharp interface physics*: The initial coupling of level set methods to physics problems, such as those arising from fluid mechanics, transformed jump conditions and sharp interface physics into smoothed out contributions on a background mesh, lowering the accuracy of the computed combined fluid/interface dynamics. A general mathematical and algorithmic approach that implicitly constructs sharp interfaces while calculating the underlying physics has allowed the accurate, high-order computation of highly delicate physical phenomena (Saye, 2017a,b) (Section 8).

In this review, we discuss the development of each of these key steps. We provide the motivation, background mathematical and algorithmic formulation, and provide computational examples of specific applications demonstrating the power of these advances. Our focus here is on outlining key numerical advances in algorithms for evolving moving fronts. We provide many example applications, trying to outline the fundamental physics and scientific motivations behind these examples. Due to space limitations, we do not discuss particulars of the numerical implementations in great detail, but instead highlight some of the key points. We refer the reader to the referenced literature for more details on the applications themselves.

2 Level set methods: Background and formulation

2.1 Formulation and equations of motion

Begin with a moving closed interface $\Gamma(t)$, that is, $\Gamma(t=0) : [0, \infty) \rightarrow \mathbb{R}^N$, propagating with a speed F in its normal direction. Here, we imagine that $\Gamma(t)$

gives the position of each point x in $\mathbb{R}^N$ on the interface at time t, and assume that the speed F in the normal direction is, for now, completely prescribed. The interface $\Gamma(t)$ can become highly complex as it moves, twists, and contorts, as well as break into multiple pieces or merge different parts together.

We wish to produce an Eulerian formulation for the motion of the hypersurface propagating along its normal direction with speed F, where F can be a function of various arguments, including the curvature, normal direction, etc. Let $\pm\, d(x)$ be the signed distance from the point $x \in \mathbb{R}^N$ to the interface at time $t = 0$, and define a function $\phi(x, t = 0)$ by the equation

$$\phi(x, t=0) = \pm d(x). \tag{1}$$

Our goal now is derive an initial value partial differential equation for the motion for this level set function ϕ. The requirement that the zero level set of the evolving function ϕ (see Fig. 1, left) always match the propagating hypersurface means that

$$\phi(x(t), t) = 0. \tag{2}$$

By the chain rule,

$$\phi_t + \nabla\phi(x(t), t) \cdot x'(t) = 0. \tag{3}$$

Since F supplies the speed in the outwards normal direction, then $x'(t) \cdot \mathbf{n} = F$ where $\mathbf{n} = \nabla\phi/|\nabla\phi|$ and this yields an evolution equation for ϕ, namely,

$$\phi_t + F|\nabla\phi| = 0, \quad \text{given} \quad \phi(x, t=0). \tag{4}$$

We stress that something subtle has happened: we have implicitly defined the speed function F to have meaning in all of space, and hence provide the speed in the normal direction for all the level sets, not just the zero level set corresponding to the interface.

This is the level set equation introduced by Osher and Sethian (1988), and for certain forms of the speed function F, one obtains a standard Hamilton–Jacobi equation. As discussed in Sethian (1982, 1985, 1987), propagating fronts can develop shocks and rarefactions in the slope, corresponding to corners and fans in the evolving interface, and numerical techniques designed for hyperbolic conservation laws can be exploited to construct schemes which produce the correct, physically reasonable entropy solution.

2.2 Advantages of this formulation

There are certain advantages to this implicit interface perspective.

- *Dimensionality:* First, the formulation is the same, regardless of the dimensionality of the problem. The initial value PDE given by Eq. (4) is the same in two, three, and higher dimensions. The only requirement is that the moving interface be a hypersurface in the background domain space.

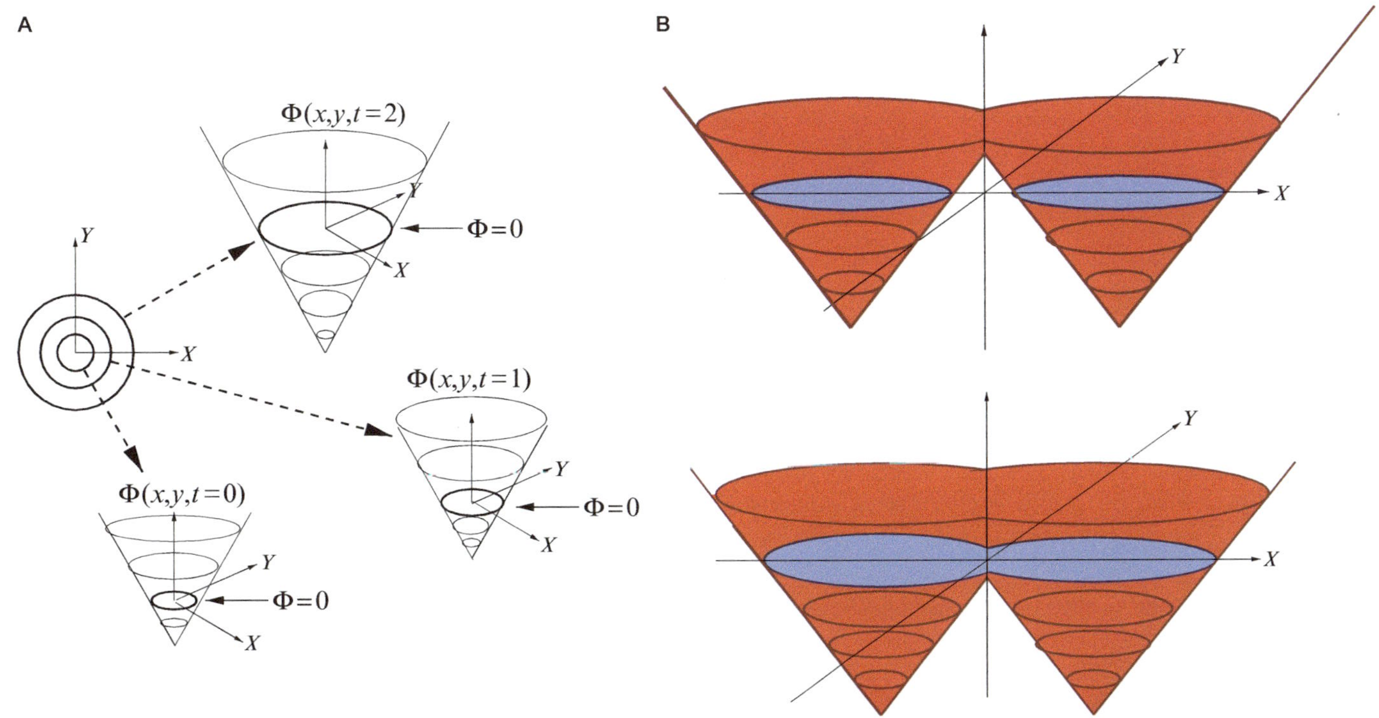

FIG. 1 *Left*: Implicit embedding of level set function. *Right*: Topological change. (A) Transformation of front motion into initial value problem. An implicitly defined surface ϕ, whose ensuing motion satisfies Eq. (4), and whose zero level set always matches the motion of the interface. (B) *Top*: The level surface ϕ in *red*. $\phi = 0$ corresponds to two separate initial fronts. *Bottom*: Later in time: the interface topology has changed, yielding a single curve as the zero level set.

- *Topological changes:* Second, topological changes in the evolving interface happen naturally; the connectivity of the zero level set may change as the interface breaks into multiple parts or merge with other regions (see Fig. 1, right).
- *Calculation of geometric quantities:* Third, geometric quantities, such as the normal direction **n** and curvature κ, are easily calculated:

$$\mathbf{n} = \nabla\phi \qquad \kappa = \nabla \cdot \frac{\nabla\phi}{|\nabla\phi|}.$$

- *Natural adaptivity:* The given equations of motion are Eulerian, in the sense that they refer to a fixed coordinate system in the physical output space. As interfaces expand, they may occupy more of the physical domain, which is naturally reflected in their representation in an Eulerian framework. This is in contrast to formulations which rely on a parameterization of the interfaces, in which the parameterization itself stretches or contracts along the propagating interfaces. This presents numerical approximation challenges, as discussed below.

There has been considerable theoretical analysis of this Eulerian partial differential equations view of interface motion and its relation to other perspectives on front propagation. The flame/entropy model from Sethian (1982) served as the basis for theoretical analysis by Barles (1985). The embedding of the front as a higher-dimensional function meant that some of the issues of topological change and corner formation could be studied in a more natural manner. Furthermore, the transformation of geometry problems into a partial differential equation setting meant that some powerful analytic techniques, including regularity of solutions, viscosity solutions of Hamilton–Jacobi equations, and tools for analyzing existence and uniqueness, could be applied.

Using the level set approach of Osher and Sethian (1988), Evans and Spruck (1991, 1992a,b, 1995), Chen et al. (1991), Giga and Goto (1992), and Giga et al. (1992) performed detailed analyses of curvature flow in a series of papers. They exploited much of the work on viscosity solutions of partial differential equations developed by Crandall and Lions (1983); see also Crandall et al. (1984, 1992). These papers examined the regularity of curvature flow equations, pathological cases, and opened up a series of investigations into further issues; we also refer the interested reader to Evans et al. (1992) and Ilmanen (1992, 1994).

2.3 Numerical approximations

A central challenge in solving the above initial value partial differential equation is that the solution need not be differentiable, even with smooth initial data. Discontinuities in the evolving front can develop. It is easy to see this from the simple case of a sinusoidal curve propagating normal to itself with unit speed: the smooth troughs narrow and develop corners. This non-differentiability is intimately connected to the notion of appropriate weak solutions. Our goal is construct numerical techniques which naturally account for this non-differentiability in the construction of accurate and efficient approximation schemes, and admit physically correct non-smooth solutions.

To illustrate, in the case where the speed $F = 1$, the level set equation given in Eq. (4) reduces to a straightforward Hamilton–Jacobi equation. Links between propagating interfaces, the formation of discontinuities in the evolving solutions and the link to hyperbolic conservation laws were analyzed in Sethian (1985, 1987), where the equation for a curve propagating normal to itself with a given speed $F - \epsilon\kappa$ was studied, where κ is the curvature and the interface remains a graph as it moves. In Sethian (1987), this was shown to be related to a hyperbolic conservation law with viscosity: the Hamilton–Jacobi equation for the interface is equivalent to a hyperbolic conservation law for the propagating slope of the interface, where curvature in the propagating front plays the role of viscosity.

Fig. 2 shows the motion of an initial cosine curve propagating normal to itself with specified speed F. On the left, $F = 1 - 0.25\kappa$ is shown. The sinusoidal wave flattens out, since the curvature term is most negative at the bottom of the trough, and hence the front moves faster there. The centre figure shows the non-entropy satisfying case in which the front crosses over itself. This solution is not physically realistic, and is not the limit as ϵ approaches zero of the case $F = 1 - \epsilon\kappa$ (for details, see Sethian, 1985, 1987). Instead, our goal is to build numerical schemes that produce the correct entropy satisfying Huygen's solution given on the right.

The key to building numerical approximations to approximate the solution of the level set equation is to build operators which approximate the gradient $|\nabla\phi|$ in an upwind fashion, and hence satisfy the entropy condition and produce the Huygens construction of sharp corners. A large body of literature has been developed to produce high-order approximations, both in finite difference and finite element settings. We refer the interested reader to

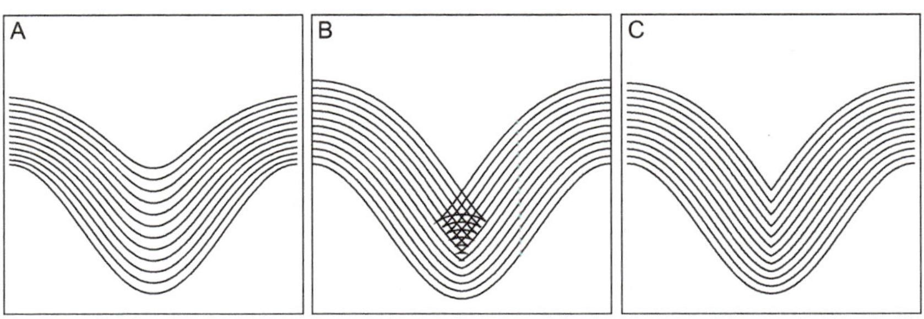

FIG. 2 Cosine curve propagating in normal direction with speed F. (A) $F = 1 - 0.25\kappa$; (B) $F = 1$: Entropy-violating; (C) $F = 1$: Entropy satisfying. *Taken from Sethian, J. A., 1996b. Level Set Methods and Fast Marching Methods. Cambridge University Press.*

Osher and Sethian (1988) and Sethian (1996b, 1999) for various schemes. Here, we simply go directly to an appropriate finite difference expression introduced in Osher and Sethian (1988), namely

$$\phi_{ijk}^{n+1} = \phi_{ijk}^{n} - \Delta t[\max(F_{ijk},0)\nabla^{+} + \min(F_{ijk},0)\nabla^{-}], \tag{5}$$

where

$$\nabla^{+} = \begin{bmatrix} \max(D_{ijk}^{-x},\ 0)^2 + \min(D_{ijk}^{+x},\ 0)^2 + \\ \max(D_{ijk}^{-y},\ 0)^2 + \min(D_{ijk}^{+y},\ 0)^2 + \\ \max(D_{ijk}^{-z},\ 0)^2 + \min(D_{ijk}^{+z},\ 0)^2 \end{bmatrix}^{1/2}$$

$$\nabla^{-} = \begin{bmatrix} \max(D_{ijk}^{+x},\ 0)^2 + \min(D_{ijk}^{-x},\ 0)^2 + \\ \max(D_{ijk}^{+y},\ 0)^2 + \min(D_{ijk}^{-y},\ 0)^2 + \\ \max(D_{ijk}^{+z},\ 0)^2 + \min(D_{ijk}^{-z},\ 0)^2 \end{bmatrix}^{1/2}$$

3 A first example: Geometry

To illustrate the level set approach and straightforward finite difference numerical algorithm we start with a purely geometric example.

Consider a simple closed curve Γ moving in the plane, with speed in the normal direction given by the negative of its curvature, that is, $F = -\kappa$. The equation of motion then becomes

$$\phi_t = \kappa|\nabla\phi| = \left[\nabla \cdot \frac{\nabla\phi}{|\nabla\phi|}\right]|\nabla\phi|, \tag{6}$$

where we have substituted $\nabla \cdot \frac{\nabla\phi}{|\nabla\phi|}$ for the curvature κ.

This problem resembles the heat equation in that the right-hand side smooths the curve as it evolves. Regions of negative curvature are pushed outwards, while regions of positive curvature move inwards. It is easy to see that an initial circle moves inwards, since the curvature and hence the speed is constant around its perimeter, collapsing into a single point.

More complicated initial curves have been the subject of much study. First, Gage (1984) and Gage and Hamilton (1986) showed that any convex curve moving under such a motion remains convex and must shrink to a point. Grayson (1987, 1989) followed this work with the remarkable result that *all* simple closed curves must shrink to a point, regardless of their initial shape.

As an example, we compute the motion of a simple closed curve moving under its curvature. To begin, we lay down a uniform background mesh across the entire computational domain. At each mesh point, we first compute the signed distance to the curve—the distance is taken as positive if the mesh point is outside the curve, negative if inside the curve, and zero if the mesh point falls exactly on the curve itself.

Armed with this initialization, the value of the level set function is updated using a straightforward finite difference approximation to Eq. (6). That is, for a mesh point i, j, using a forward time step of size Δt, we take

$$\phi_{i,j}^{n+1} = \phi_{i,j}^{n} + \Delta t \left[\frac{\phi_{xx}\phi_y^2 - 2\phi_x\phi_y\phi_{xy} + \phi_{yy}\phi_x^2}{|\phi_x^2 + \phi_y^2|} \right],$$

where central difference operators are used to approximate all the spatial derivatives. At points where the denominator vanishes, the expression is regularized by a small number in order to avoid division by zero.

In Fig. 3, we show the time evolution of geometric curvature flow, demonstrating Grayson's theorem.

4 Narrow banding and extension velocities

A major advance in the development of level set methods came through the introduction of adaptive "narrow band level set methods."

4.1 Narrow band level set methods

As initially posed, level set methods are inefficient. All the level sets throughout the entire computational domain must be advanced each time step. This can be viewed as a "full matrix approach," in that the value ϕ_{ij} at every mesh point is updated. There is little reason for this: only those near the zero level set corresponding to the interface need to be updated.

Instead, a preferable solution, introduced by Chopp (1993) and studied extensively in Adalsteinsson and Sethian (1995a), is to compute only at those computational mesh points containing level sets near the front itself, see Fig. 4.

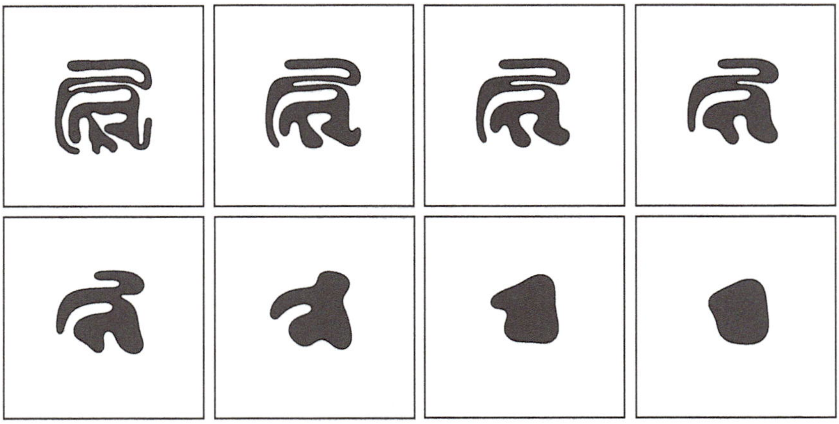

FIG. 3 Time evolution of $F(\kappa) = -\kappa$. *Taken from Sethian, J. A., 1996b. Level Set Methods and Fast Marching Methods. Cambridge University Press.*

FIG. 4 *Dark line* is zero level set corresponding to front. Grid points in *dark region* are part of the narrow band. *Taken from Sethian, J. A., 1996b. Level Set Methods and Fast Marching Methods. Cambridge University Press.*

4.1.1 Implementation of narrow band methods

To implement a narrow band level set method, several steps are required.

- Given an initial front $\Gamma(t=0)$, the first step is to tag all grid points within a distance k_1 of the interface. These are now the narrow band mesh points.
- Next, an initial level set function $\phi_{i,j}^{n=0}$ is generated by computing the signed distance from each narrow band mesh to the initial front.
- The level set equation (Eq. 5) is then updated at all narrow band mesh points.
- Once the zero level set has reached within a specified k_2 mesh points of the edge of the narrow band, the calculation is stopped. A new set of narrow band mesh points a distance k_1 from the new position of the front is built, a new signed distance function is constructed at these narrow band mesh points, and then the calculation proceeds. This procedure of rebuilding the signed distance function around a new zero level set is called "reinitialization."

Two numerical parameters are required in this implementation. The size k_1 of the narrow band determines how often the front is reinitialized. A small value means that the front will not move far before reinitialization, while a large value will require more compute time, but less reinitialization. In practice, setting k_1 to be around 10 mesh cells is a good first choice. The second parameter k_2 determines how close the evolving front may get to the edge of the narrow band before reinitialization is required. Setting k equal to three mesh cells is a good choice.

There are two additional subtleties. The first involves the boundary conditions at the edge of the narrow band. In computing quantities such as approximations to the gradient and curvature, one needs values at mesh points which, depending on the order of the approximations used, may require values outside the narrow band. In the case of outflow boundary conditions, nothing needs to be done because of the upwinding nature of the approximation, but at inflow conditions, something is required. A large amount of numerical experimentation shows that freezing the values at the mesh points at the edge of the narrow band is often sufficient: problems/instabilities stemming from these conditions will be minimized, since the calculation will be stopped and the front reinitialized before it reaches the boundary.

4.1.2 Reinitialization

The second issue involves how one performs the reinitialization. The goal is to compute a new value of the level set function ϕ such that $|\nabla\phi| = 1$ and $\phi = 0$ on the new interface, with ϕ positive on the outside and negative on inside of the interface.

- One approach is to approximate the zero level set corresponding to the front, and then recompute the signed distance function. Begin by locating all mesh cells in which the level set function changes sign, and then approximate the interface, using, for example, line segments in two dimensions or a triangulated representation in three dimensions. Then, at each of the newly designated narrow band mesh points, rebuild the level set function by evaluating the signed distance to this approximation.
- Another approach is to iterate on a pseudo-time differential equation of the form $\phi_t = switch(1 - |\nabla\phi|)$ where *switch* is a switch function depending on the location of the zero level set.

There are challenges associated with these techniques. The first method requires building an approximation to the front which can be geometrically intricate in high spatial dimensions. The second method can be inaccurate: it is easy to check that this reinitialization procedure moves the interface substantially, causing artificial mass loss or gain. While fixes have been proposed, in our experience, they produce unsatisfying results.

Instead, the preferable approach is to use an idea first introduced by Chopp (2001), as follows. Given a time-evolved level set function, the reinitialization goal is to preserve the location of zero level set and construct a new signed distance function around its location. Chopp's idea is to locate the mesh cells containing the interface, and then use the level set function at those mesh points to build a high-order polynomial approximation within that cell, typically through bi-cubic interpolation. Then, standing at each mesh point, a reinitialized level set function can be found by following the gradient of this approximation, which points normal in the direction of the interface. He then

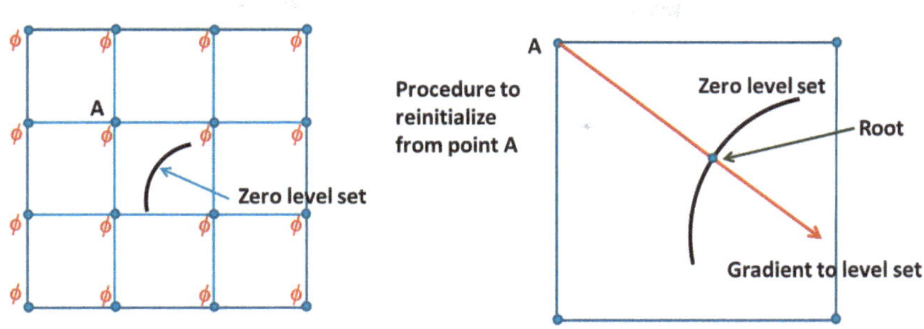

FIG. 5 Chopp's reinitialization (Chopp, 2001). *Left*: Grid points used to compute bi-cubic interpolant. *Right*: Newton solver to find intersection of gradient to level set function ϕ with level surface $\{\phi = 0\}$.

uses a quasi-Newton solver to locate where this gradient trajectory crosses the zero of the approximation, giving the signed distance. This clever and accurate technique completely avoids finding the interface (see Fig. 5).

4.1.3 Modern reinitialization techniques

In some applications of level set methods and implicit interface methods, a high-order approximation of the signed distance function is required. This is because distances or closest points to the implicitly defined interface may in some way be used to infer the geometry of the surface itself, such as when calculating normal vector fields, or curvature quantities like mean curvature or Gaussian curvature, or extension velocities discussed in the next section. Chopp's reinitialization method operates by interpolating the level set function in each grid cell via bicubic polynomials and then uses a quasi-Newton iterative method to compute distances to the zero level set of the polynomials. For sufficiently smooth problems, this procedure results in third-order accuracy. More modern and sophisticated reinitialization techniques extending this idea were introduced by Saye (2014) and allow for arbitrary order accuracy. The essential idea in this class of techniques is to fit high-order polynomials to each grid cell and apply a full Newton root finding method for increased accuracy and efficiency.

The algorithm consists of two main stages. First, in an initialization stage, a level set function ϕ defined on a mesh is piecewise approximated by high-order polynomials. In a finite element method, these polynomials may already be defined by the polynomials on the mesh; in a finite difference method, these polynomials could be constructed via piecewise bicubic polynomials (as in Chopp's approach) or Taylor polynomials of sufficient degree using compact stencils, as shown in Fig. 6 (left). These polynomials are then "sampled" by seeding points on their zero level set, with sufficient density to form a scattered cloud of points approximating the interface of ϕ. Seeding points on the zero level set of a polynomial is inexpensive and can be done via

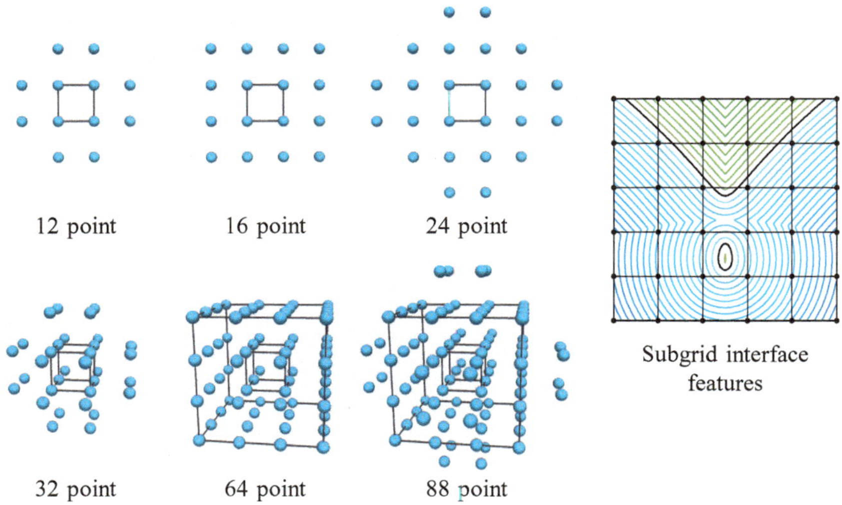

FIG. 6 *Left*: High-order accurate reinitialization methods can choose stencils of varying sizes to yield different orders of accuracy. *Right*: Subgrid details of an interface can be captured with a high-order accurate method. Here, level set values are defined only at the grid points, while the curves show the contours of the computed signed distance function—subcell features, such as the fully enclosed interface droplet, can be recovered with high-order polynomial interpolation. *Adapted from Saye, R. I., 2014. High-order methods for computing distances to implicitly defined surfaces. Commun. Appl. Math. Comput. Sci. 9, 107–141.*

a couple iterations of $x_{i+1} = x_i - p(x_i)\nabla p(x_i)/\|\nabla p(x_i)\|^2$, where p is the polynomial and x_0 is chosen somewhere in the interior of the grid cell or mesh element. In the second stage of the algorithm, given a query point x in the computational domain, the closest point in the cloud to x is found. This closest seed forms an approximation of the actual closest point to x; the approximation is then improved by using the original polynomial from which it was created, together with Newton's method for solving the minimum-distance optimization problem. The full Newton's method operates by computing the objective functional's Hessian using the gradient and Hessian of the polynomial evaluated at each iterate, and falls back to Chopp's quasi-Newton method in the very rare case that the full Hessian cannot be reliably inverted.

The combination of a discrete set of points in a cloud approximating the interface, together with Newton's method to "polish" the closest seed to find the true closest point on the interpolated interface, results in highly accurate and robust calculations. By making use of a k-d tree optimized for co-dimension one point clouds, finding the closest seed point can be made very inexpensive. Moreover, not many iterations of Newton's method are required for convergence, usually, just two or three, since the seed point is close to the true closest point. In addition, except for the initial stage of forming a piecewise polynomial high-order approximation of ϕ, the method does not rely on any computational grid and can be used to compute closest points at arbitrary locations. The results of Saye (2014) demonstrate 3rd, 4th, 5th, and 6th order accurate algorithms in 2D and 3D, capable of capturing sub-grid features (see Fig. 6, right). (C++ source code for these algorithms is available at math.lbl.gov/~saye.)

4.2 Extension velocities

Level set methods require that a velocity be established not just for the zero level set, but for the neighbouring level sets that are contained within the narrow band. Another way to say this is that each grid point in the narrow band carries a value for the evolving level set function, and a velocity must be given to transport the level set passing through that grid point.

Thus, the characterization of an interface as an embedding means that both the front and the velocity of the front are assumed to have meaning away from the interface.

4.2.1 Need for extension velocities

There are several reasons why extension velocities are required to produce accurate and robust level set methods, and wise choices for extension velocities are crucial in producing viable numerical schemes.

- In many cases, the velocity may be defined only on the front. For example, in semiconductor etching and deposition problems, see Adalsteinsson and Sethian (1995b,c, 1997) and Sethian and Adalsteinsson (1996), the interface speed has no meaning away from the etching/deposition profile. While the velocity of the evolving profile can be calculated, a velocity must be invented to move all the neighbouring level sets in the narrow band.
- Even if there is a "natural" extension velocity, it may be a poor choice. As an example, consider a fluid calculation, in which a fluid velocity is defined and updated at each grid point. This velocity may have a steep derivative in the neighbourhood of the zero level set, or even be discontinuous across the interface, causing sharp stretching and contraction of the level sets in the narrow band. This destroys the signed distance function property that $|\nabla\phi| = 1$, which then must be artificially restored periodically in order to minimize the spatial error in the update equation.
- For many interface problems, the speed of the interface delicately depends on the exact position of the front. For example, within a grid cell, the interface may be partially masked, and it is important to assign a velocity along that interface that gets correctly extended to the neighbouring grid points.

Ultimately, our goal is to build an extension velocity that equals the interface velocity on the front itself, and maintains, as best as possible, the signed distance function as the level sets evolve.

4.2.2 Constructing extension velocities

To do so, we define an extension velocity F^{ext} in the level set equation, namely

$$\phi_t + F^{ext}|\nabla\phi| = 0, \tag{7}$$

where F^{ext} is some velocity field which, at the zero level set, equals the given speed F. In other words,

$$F^{ext} = F \text{ on } \phi = 0.$$

A natural way to build an extension velocity was introduced by Malladi et al. (1995). The idea was to stand at each grid point and use the value of the speed function at the closest point on the front, thus extrapolating the velocity from the front. This was done in a purely geometric fashion: with the velocity given on a discretized version of the zero level set corresponding to the interface, the velocity at each mesh point was obtained by finding the closest point on the interface, and then simply taking that velocity as the extension velocity at the mesh point.

Formally, this approach corresponds to building an extension velocity by solving the partial differential equation

$$\nabla F_{\text{ext}} \cdot \nabla\phi = 0, \tag{8}$$

in which the velocity is transported along the characteristics: see Zhao et al. (1996). It is straightforward to show that under this velocity field, the level set function ϕ remains the signed distance function for all time, assuming that both F and ϕ are smooth.

In Adalsteinsson and Sethian (1999), a two-tiered system was used. The signed distance function is constructed around the current value of the zero level set, using the methodology described above. At the same time, an extension velocity satisfying Eq. (8) is built, which is then used to update the level set Eq. (7).

We note several advantages to this approach:

- With an accurate solution to the extension velocity equation (Eq. 8), the evolving level set function theoretically is always the signed distance function.
- There is no need to reinitialize the level set function every time step. Indeed, one can compute an extension velocity at each time step without reinitialization, greatly reducing an avoidable source of numerical reinitialization error.
- There is no need to ever "find" the front: the modern reinitialization techniques described above both reinitialize and build extension velocities without any explicit construction of the interface. For example, very high-order accurate extension velocities can be computed using the high-order closest point algorithms described in Section 4.1.3.

Many of the problems of inaccurate level set implementations, including artificial mass loss, spurious front motion, and unphysical motion of the interface during reinitialization, stem from poor choices of extension velocities. Correct choice of extension velocities, in all practicality, eliminates these issues: see the applications section of this review article for examples.

4.3 Transport and diffusion of material quantities on an evolving interface

Imagine a substance G which is spread along an interface, and both diffuses along the interface and is transported and stretched as the interface evolves. This model occurs in a variety of problems including the interplay of surfactants on a surface and in semiconductor superdeposition. The amount of material G at a particular point on the interface influences the interface evolution rate, both locally and globally.

As originally formulated, level set methods cannot handle this additional information, because all information about the local parameterization is lost by the characterization as a level set. An elegant idea which greatly extends the reach of level set methods is to embed both the interface and the function G into all of space, and then evolve both functions in time on the background fixed mesh. This was used in Adalsteinsson and Sethian (2002) to tackle some evolution geometry problems in two and three space dimensions including surface diffusion and transport.

In Sethian and Shan (2008), these ideas were used to compute problems in superconformal deposition in semiconductor manufacturing, and in Saye and Sethian (2013, 2016) they were used as part of an extensive calculation of industrial foams.

5 Applications of narrow band and extension methodologies

5.1 Application of extension velocities: Two-phase flow and industrial inkjet printing

The first application demonstrates the use of extension velocities in a fluid simulation in which the fluid velocity is computed everywhere, but an extension velocity is built from the interface. Here, we discuss work described extensively in Yu et al. (2003, 2005). We refer the reader there for more details.

5.1.1 Problem statement

Industrial microjet printing involves microfluids ejected from a reservoir through a narrow nozzle and deposited on a material. One of the most popular uses has been for industrial inkjet printing, in which commercial inkjet printers are highly functional and ubiquitous. In recent years, the applicability of microjetting printers has extended far beyond inkjets, and are now used in a range of industrial technologies, including printing of integrated circuits and tissue scaffolding.

A schematic of the basic mechanism is shown in Fig. 7. The reservoir houses the ink, which is driven through an electro-actuator mechanism through a small nozzle, which is typically axisymmetric, and then jets through the air until it lands on the material. This is a two-phase flow problem of air and ink, with boundary conditions on the solid walls. The flow is typically an incompressible Newtonian fluid for simple inks, though the introduction of coloured pigments leads to a non-Newtonian viscoelastic flow. On solid walls, both no-flow and no-slip conditions apply, with a slipping contact model needed at the air–ink–wall triple point.

There are several physical goals in building accurate simulations of this process.

- First, how is the ink ejected through the nozzle?
- How do satellite droplets, which trail the main lead ink front, form, and what causes and or inhibits their formation? As we have seen earlier, curvature acts like surface tension, and Grayson's theorem shows that surface tension in two dimensions causes all bubbles to become circular. However, in three dimensions, this is not true. The mean curvature term corresponding to surface tension causes small satellite bubbles to break off.

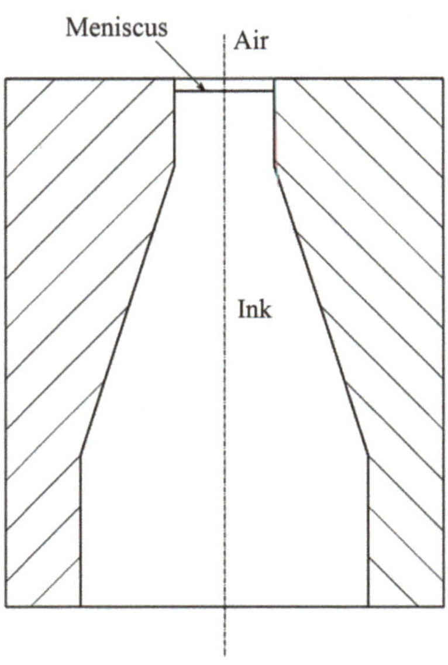

FIG. 7 Basic structure of inkjet reservoir.

- What is the effect of wetting on the nozzle walls versus no wetting? If the walls are wet, this affects the critical angles at which the air–ink–wall triple junctions switches from a sticking to a slipping condition.
- How do the nozzle shape, pressure and voltage changes, etc., affect the dynamics of droplet ejection and the integrity of the ejected droplet? How can one control the formation of satellites and droplet separation?
- How do variations in viscosity, impurities in the liquid and embedded particulates influence the fluid dynamics?

To illustrate, Fig. 8 shows the results of an experiment of a time-cycle of ink ejected from a bath.

5.1.2 Equations of motion

The equations of motion consist of an Oldroyd-B model for the visco-elastic ink, a Newtonian model for the air, and no-slip/no-flow conditions on the solid walls. Both fluids are taken as incompressible. Thus, we have

$$\begin{aligned}(Ink)\ \rho_1 \frac{D\boldsymbol{u}_1}{Dt} &= -\nabla p_1 + \nabla\cdot(2\mu_1\mathcal{D}_1) + \nabla\cdot\boldsymbol{\tau}_1, \quad \nabla\cdot\boldsymbol{u}_1 = 0,\\ \frac{D\boldsymbol{\tau}_1}{Dt} &= \boldsymbol{\tau}_1\cdot(\nabla\boldsymbol{u}_1) + (\nabla\boldsymbol{u}_1)^T\cdot\boldsymbol{\tau}_1 - \frac{1}{\lambda_1}(\boldsymbol{\tau}_1 - 2\mu_{p1}\mathcal{D}_1).\end{aligned} \tag{9}$$

$$(Air)\ \rho_2 \frac{D\boldsymbol{u}_2}{Dt} = -\nabla p_2 + \nabla\cdot(2\mu_2\mathcal{D}_2), \quad \nabla\cdot\boldsymbol{u}_2 = 0. \tag{10}$$

$$\begin{aligned}\mathcal{D}_i &= \frac{1}{2}\left[\nabla\boldsymbol{u}_i + (\nabla\boldsymbol{u}_i)^T\right], \quad i = 1,2,\\ \boldsymbol{u}_i &= u_i\boldsymbol{e}_r + v_i\boldsymbol{e}_z, \quad i = 1,2\end{aligned} \tag{11}$$

where, for the ink, $\boldsymbol{\tau}_1$ is the viscoelastic stress tensor, λ_1 is the viscoelastic relaxation time, μ_{p1} is the solute dynamic viscosity. The subscript 2 refers to the usual pressure and density for Newtonian air.

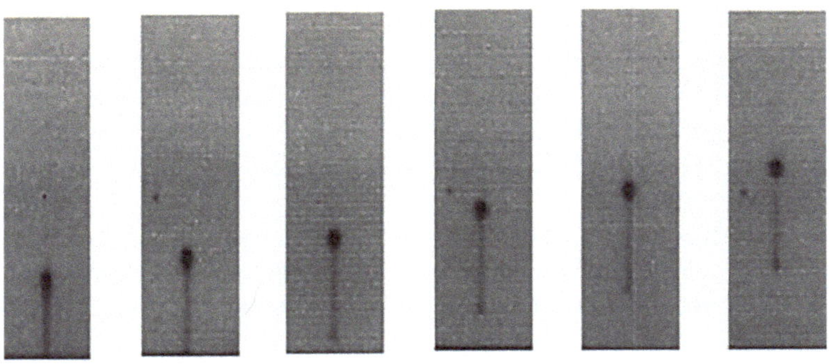

FIG. 8 Experimental profile, showing ejected ink and satellite formation. *Taken from Yu, J. D., Sakai, S., Sethian, J. A., 2005. A coupled quadrilateral grid level set projection method applied to ink jet simulation. J. Comput. Phys. 206 (1), 227–251.*

5.1.3 Algorithms and numerical implementation

To solve these equations, we use a second-order version (Bell et al., 1989) of Chorin's projection method (Chorin, 1968). More explicitly:

- We use a second-order explicit Godunov for advection and a central difference scheme for the viscous operator:
  - First, we find velocity u^* by updating the velocity in the Navier–Stokes equations.
  - Second, we then project out using a Hodge decomposition to compute $u^* = u + \nabla P$, where $\nabla \cdot u = 0$ is the incompressible component. This is done using a finite element projection for approximate projection for the Hodge decomposition, see Almgren et al. (1996).
- A level set method is used to track the air/ink boundary. The level set function is reinitialized using fast marching methods (Sethian, 1996a; Sethian and Vladimirsky, 2001) and Chopp's bicubic method (Chopp, 1993).
- Extension velocities are computed by taking the fluid velocity at the interface, and then solving Eq. (8), see Adalsteinsson and Sethian (1999).

5.1.3.1 Meshing

Some particular implementation details warrant further discussion. First, to handle the nozzle geometry, we built a body-fitted coordinate system. This was done by dividing the computational domain into three separate regions: (1) the ink chamber; (2) the nozzle geometry; and (3) free space through which the ink travels, which are then connected together. The nozzle geometry is discretized by going from the rectangular mesh into an arbitrary quadrilateral mapping which is then body-fitted to the nozzle geometry, see Fig. 9.

It is a bit of a challenge to use these arbitrary quadrilateral meshes, due to the error in the viscoelastic terms. For details, see Yu et al. (2003, 2005, 2007).

5.1.3.2 Wall contact angles, boundary conditions

A second computational issue involves building an appropriate contact model for the air/ink/wall triple point (see Fig. 9). Physically, whenever the angle made by the air/ink interface and the solid wall is too large (or too small), the contact position readjusts itself to restore an angle within the experimentally measured ranges.

The correct solution to the problem involves solving the complete lubrication/friction point equations, but this is impractical, since the spatial range required to produce an accurate solution obviates solving the large-scale incompressible Navier–Stokes equations.

Instead, we need to build an approximate, macroscale model which correctly captures the large-scale physics. There are several options:

- One option is to enforce the no-slip condition. This means that the triple point doesn't move, which is physically unreasonable.
- Another option is to restrict critical angle, and simply allow the contact point to slide. However, this violates mass conservation, since enforcing the no-slip condition is incompatible with this motion.

Instead, in Yu et al. (2003, 2005), a model is employed as follows. The no-slip condition is selectively turned off and on. It is turned off in a small neighbourhood around the triple point whenever the angle is bigger than the critical angle (or less than the receding angle). This model conserves mass, and correctly enforces both the no-slip condition and the critical angle requirement. The triple point moves towards air (that is, the no-slip condition is relaxed) if the angle is bigger than the advancing angle and the velocity is positive, and moves towards ink if the angle is less than the receding angle and the velocity is negative. Otherwise, the no-slip condition is maintained, and the triple point does not move.

5.1.3.3 Dealing with viscoelasticity

The viscoelastic term presents additional challenges. We use a mixed algorithm to integrate the viscoelastic stress equations in time, which is explicit on the upper-convected derivative terms, explicit on the solute viscosity term, and implicit on the relaxation term. In the spatial discretization of the viscoelastic stress term, we use an upwind scheme for the convection term, and central difference approximations in the rest of the upper-convected derivatives.

5.1.3.4 A comment about mass loss

The use of accurate and robust extension velocities produce a method with less than 1% mass loss over the length of the entire ejection cycle (on a 256 × 384 × 256 mesh) As an experiment, we tried using the direct fluid velocity

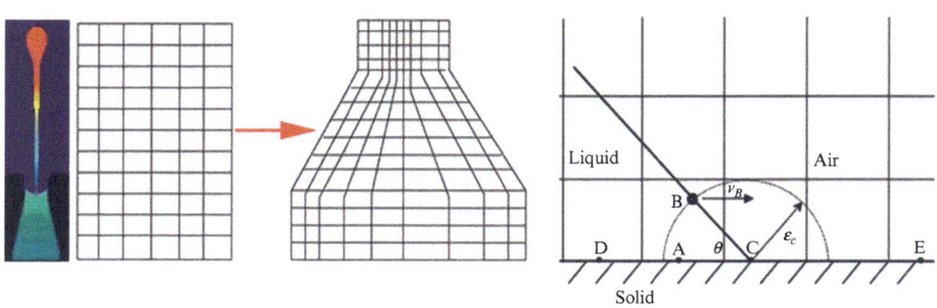

FIG. 9 *Left*: Experimental profile, showing ejected ink and satellite formation. *Middle*: Body-fitted mesh. *Right*: Moving triple point. *Taken from Yu, J. D., Sakai, S., Sethian, J. A., 2005. A coupled quadrilateral grid level set projection method applied to ink jet simulation. J. Comput. Phys. 206 (1), 227–251.*

to move the level sets: this resulted in a mass loss of over ten percent, which is unacceptable, and illustrates the advantages of using appropriate extension velocities.

5.1.3.5 Computing implementation

The code is a coarse-grained parallel MPI implementation in which the vast majority of time is spent in the projection solve.

5.1.4 *Physical parameters and results*

We used an inflow pressure from an equivalent circuit model which describes the cartridge, supply channel, vibration plate, PZT actuator, and applied voltage. The fluid is an Epson dye-based ink, with critical advancing $\theta_a = 70$ degrees and receding $\theta_r = 30$ degrees contact angle, and with $\rho_1 = 1070$ kg/m^3, $\mu_1 = 3.34 \times 10^{-3}$ kg/m s, and $\sigma = 0.032$ kg/s^2. The nozzle geometry has diameter 26 μm at opening and 65 μm at bottom.

Fig. 10 shows the results of one calculation.

5.2 Application of extension physics: Micro-bubble dynamics

We now discuss an application in which not just the velocity but several physical variables are extended away from the interface into the computational domain. Here, we summarized work on microjetting dynamics first presented in Garzon et al. (2008), see also Garzon et al. (2003, 2011, 2012a,b, 2014, 2016) and Garzon (2009).

5.2.1 *Problem statement*

Consider the problem of analyzing what happens when a thin tube of fluid pinches off. Here, we imagine a thin rotationally symmetric tube of fluid surrounded by air: for simplicity, we further assume that the problem is periodic (see Fig. 11).

Because of surface tension, the thin neck narrows, since the mean curvature drives the problem inward. This can be viewed as a blow-up version of what happens at the satellite pinch point in the previous ink jet example. At some point, the neck narrows and breaks, at which point two lobes of fluid form which quickly retract. Our goal is to understand pinch-off dynamics by modelling this process through the topological change.

5.2.2 *Equations of motion*

We consider a highly simplified model which nonetheless contains fairly complex dynamics. We begin with Euler's equation, namely

$$\nabla \cdot \mathbf{u} = 0 \text{ in } \Omega(t) \tag{12}$$

$$\mathbf{u}_t + \mathbf{u} \cdot \nabla \mathbf{u} = \frac{-\nabla p}{\rho} + \mathbf{bodyforces} \tag{13}$$

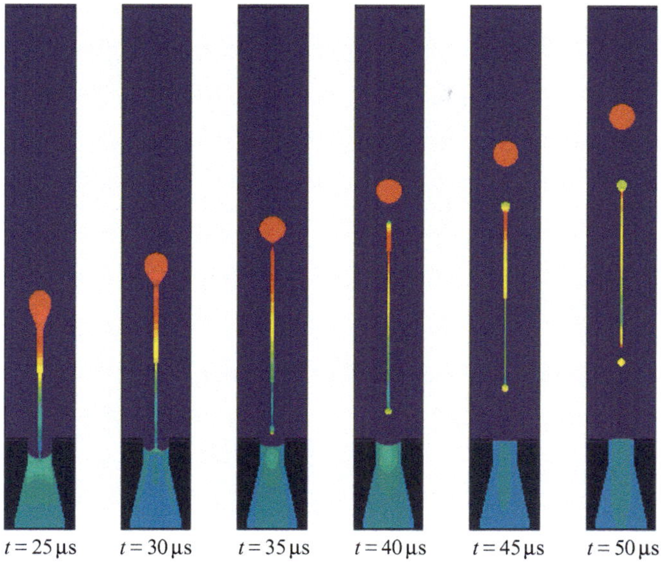

Convergence studies

Mesh number	25 × 200	50 × 400	75 × 600	100 × 800
Time to pinch off	8.0944	8.1808	8.1755	8.1738

Table 1: The time to pinch off from various meshes.

Mesh number	25 × 200	50 × 400	75 × 600	100 × 800
$t = 3.20$	1.494	1.411	1.471	1.496
$t = 4.48$	1.245	1.188	1.219	1.246
$t = 5.76$	1.047	1.012	1.065	1.093
$t = 6.72$	0.972	0.932	0.990	1.023

Table 2: Droplet head velocities from various meshes.

Mesh number	25 × 200	50 × 400	75 × 600	100 × 800
$t = 8.64$	0.5768	0.7318	0.7408	0.7440

Table 3: Droplet volumes from various meshes.

FIG. 10 Simulation of full ejection cycle. *Taken from Yu, J. D., Sakai, S., Sethian, J. A., 2005. A coupled quadrilateral grid level set projection method applied to ink jet simulation. J. Comput. Phys. 206 (1), 227–251.*

If we assume the fluid is irrotational, that is, $\nabla \times \mathbf{u} = 0$, then we can employ a fluid velocity potential $\mathbf{u} = \nabla \psi$, and write that

$$\Delta \psi = 0 \text{ in } \Omega(t) \tag{14}$$

$$\psi_t + \frac{1}{2}(\nabla \psi \cdot \nabla \psi) + \frac{p - p_a}{\rho} = 0 \quad \text{on } \Gamma_t(\mathbf{s}), \tag{15}$$

where p_a is the atmospheric pressure and ρ is the fluid density.

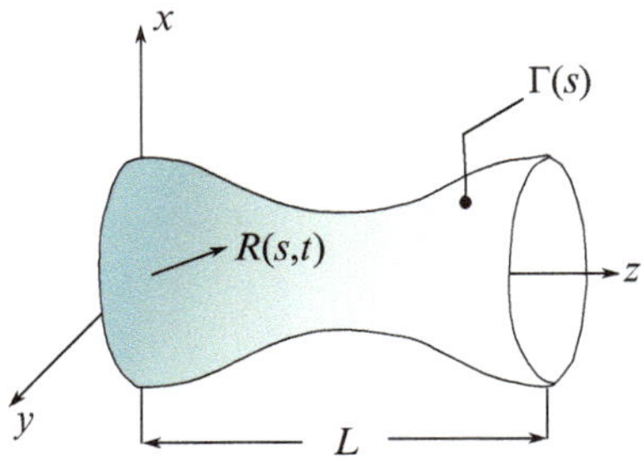

FIG. 11 Thin tube of fluid.

A calculation produces the equations of motion, namely

$$\mathbf{u} = \nabla\psi \text{ in } \Omega(t) \qquad \Delta\psi = 0 \text{ in } \Omega(t) \tag{16}$$

$$\frac{D\psi}{Dt} = \frac{1}{2}(\nabla\psi\cdot\nabla\psi) - \frac{\gamma}{\rho}\left(\frac{1}{R_1} + \frac{1}{R_2}\right) \text{ on } \Gamma_t(\mathbf{s}), \tag{17}$$

where $\Omega(t)$ is the fluid tube, Γ is the boundary of the tube, R_1 and R_2 are the principle radii of curvature, and γ is the surface tension.

We note that the potential ψ is only defined on the interface Γ. However, our goal is to compute the position of the interface as it is driven by the underlying dynamics past the point when the fluid tube separates into two regions.

The fundamental idea is to extend the relevant physics throughout the domain, so that the particular level set corresponding to the interface itself is just one of several embedded level sets. We build these extensions as follows:

- First, we smoothly extend the interface to build a level set function ϕ from the front;
- Then, we smoothly extend the velocity $\mathbf{u}_{\mathbf{ext}}$ off the front: $\mathbf{u}_{\mathbf{ext}} = \mathbf{u}$ on front, using the level set extension methodology described in Eq. (8);
- At the same time, we extend the potential $\mathbf{G}_{\mathbf{ext}}$ off the front such that $\mathbf{G}_{\mathbf{ext}} = \psi$ on the front;
- Finally, we additionally extend the right-hand side $\mathbf{f}_{\mathbf{ext}}$ off the front so that $\mathbf{f}_{\mathbf{ext}} = \frac{1}{2}(\nabla\psi\cdot\nabla\psi) - \kappa$ on the front.

This leads to a new set of equations of motion, namely

$$\begin{aligned}
&\mathbf{u} = \nabla\psi \text{ in } \Omega_d(t)\\
&\Delta\phi(r,z) = 0 \text{ in } \Omega_d(t)\\
&\Phi_t + \mathbf{u}_{\text{ext}}\cdot\nabla\Phi = 0 \text{ in } \Omega_D\\
&G_t + \mathbf{u}_{\text{ext}}\cdot\nabla G = f_{\text{ext}} \text{ in } \Omega_D\\
&\phi|\Gamma_1 = \phi|\Gamma_2
\end{aligned}$$

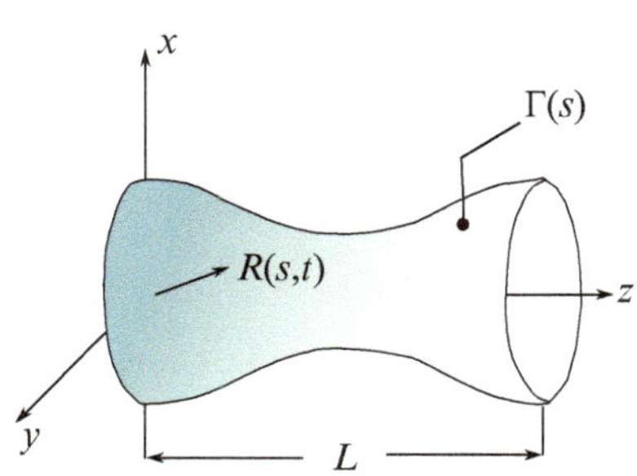

For details about the derivation of theses equations, see Garzon et al. (2003, 2008).

From a mathematical point of view, it is an open question as to the nature of these equations. Questions include issues of smoothness and well-posedness through the topological change, singularities in the evolving potential function and right-hand side, and the correctness of the sub-scale breakage model. Nonetheless, the model can be used as a basis for computation of some complex phenomena.

5.2.3 Algorithms and approximations

These equations are solved through a forward-in-time approximation on a fixed rectangular background mesh. To begin, we start with an initial curve Γ, and with initial values for the potential function ψ as well as the initial velocity u and the right-hand side f. We then build extension values for each of these quantities, and update their values on the background mesh. We solve for the potential equation at each time step by finding the zero level set, placing boundary nodes on the front, and then solving a boundary element method for the potential. These nodes are then discarded, and all the values are updated on the background mesh.

5.2.4 Results

Extensive numerical experiments are given in Garzon et al. (2003, 2008), matching experimental values for self-similar pinch-off time and computed scaling exponent. Fig. 12 shows a snapshot after pinch-off, revealing capillary surface waves on the undulating surface (left: experiment image, taken from Thorodssen (2007). Right: level set calculation of surface capillary waves, taken from Garzon et al., 2008). Fig. 13 shows the fine-scale structure of bubble dynamics after pinch-off.

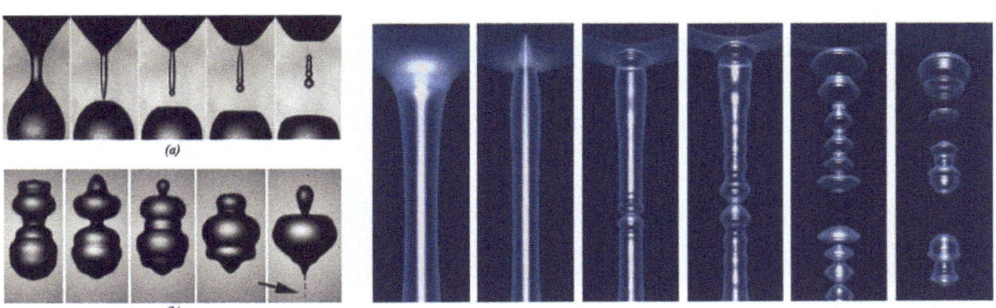

FIG. 12 Bubble dynamics. *Left*, experiment. *Right*, simulation *Left panel taken from Thorodssen, S. T., 2007. Micro-drops lets and micro-bubbles, imaging motion at small scales. Nus. Eng. Res. News 22, 1. Right panel taken from Garzon, M., Bobillo-Ares, N., Sethian, J. A., 2008. Some free boundary problems in potential flow regime using a level set method. In: Recent Advances in Fluid Mechanics. Nova Publishers.*

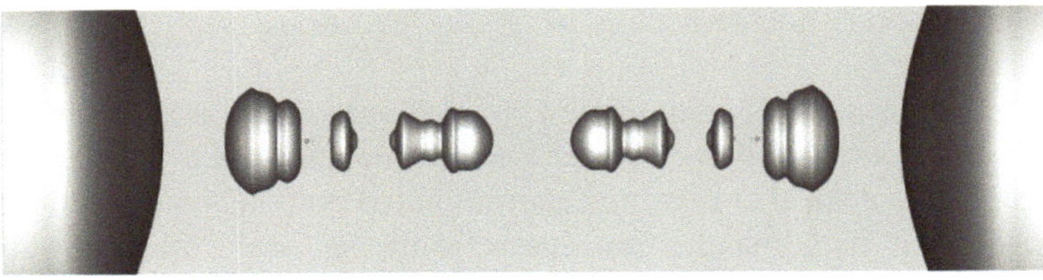

FIG. 13 Simulation: Fine-scale structure of bubble dynamics after pinch-off. *Taken from Garzon, M., Bobillo-Ares, N., Sethian, J. A., 2008. Some free boundary problems in potential flow regime using a level set method. In: Recent Advances in Fluid Mechanics. Nova Publishers.*

6 Multi-phase physics: Mathematical formulation and algorithms for tracking multiple regions

6.1 Beyond two phases

Imagine a collection of phases which share common boundaries: examples are shown in Fig. 14.

Our goal is to formulate robust, accurate, and versatile numerical methods that can track the evolution of these interfaces. These so-called multi-phase problems are challenging for several reasons:

- The interfaces advance based on geometry and physics.
- The interfaces provide boundary and jump conditions for the physics.
- Triple points where interfaces meet are singularities for curvature and are geometrically complex in three and higher spatial dimensions.
- Topological changes create strong impulse forces in the underlying physics.

A simple example shows one of the mathematical and computational challenges. Imagine an interface separating three phases, and moving under its curvature, shown in Fig. 15. It is clear that the "C" region should open up, and hence the bottom "legs" on the left should open up to form the 120 degree angles on the right. However, the curvature is zero everywhere except at the junction, where it is not defined, thus it is difficult to make sense of this problem from a simple curvature-driven front propagation point of view. This problem is far more intricate in three dimensions.

6.2 Previous algorithms to handle multi-phase evolution

Several approaches have been proposed to handle these multi-phase evolution problems. Here, we briefly summarize; for details, see the referenced work.

- *Approach 1: Lagrangian Methods*: In this approach, a Lagrangian perspective is taken in which the boundaries are discretized, either as connected line segments in two dimensions or triangles in three dimensions. These discrete representations characterize the interface, and their positions are

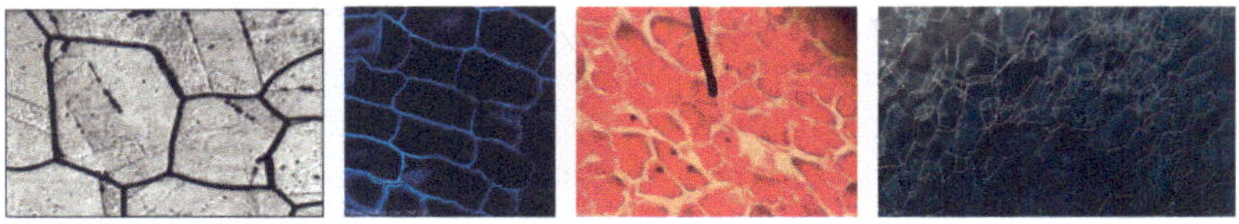

Grain metal boundaries Root structures Acinar cells Foamy fluids

FIG. 14 Examples of multi-phase problems.

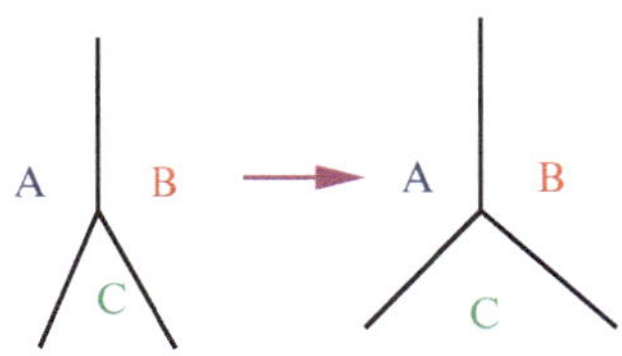

FIG. 15 Evolution of triple junction moving under curvature.

updated in time according to the underlying physics. Multiple interfaces separating multiple regions are easily handled, though care must be taken at triple points and other structures in order to establish a correct balance of forces. Because interfaces can stretch, compress and change topology, the discrete representations must add and remove elements, as well as reconnect across structures: such a process is often referred to as "surgery." Several successful versions of these approaches have been used to compute a host of interface evolution problems, we refer the interested reader to Dziuk and Elliott (2013), Chern et al. (1986), and Tryggvason et al. (2001), as well as the well-known "Surface Evolver" Brakke (1992).

- *Approach 2: Volume-of-Fluid Methods*: In volume-of-fluid methods, the domain is broken up into fixed cells, and volume fractions are assigned corresponding to the percentage of that cell filled with a particular phase. In this Eulerian view, the interface is locally reconstructed from these volume fractions, whose values are readjusted to reflect movement of the interface corresponding to the physics. First developed by Noh and Woodward (1976), they were originally used for two-phase fluid problems. They are now extensively used, see for example Lafaurie et al. (1994). Reconstructing curvature, normals, and accurate reconstruction of the interfaces from these volume fractions can be challenging, especially in three dimensions.
- *Approach 3: Diffusion-Generated Methods*: A third set of methods, mostly applicable in cases where interface motion is purely curvature-driven, comes from modelling diffusion through a convolution step followed by a reconstruction step, see, for example, Merriman et al. (1994). A sophisticated approach along these lines was introduced in Elsey et al. (2009) to compute curvature-driven motion of a large number of interconnected phases.

- *Approach 4: Level Set Methods*: Another approach is to use level set methods to track multiple interfaces. Since level set methods characterize the interface as the zero level set of a signed distance function, assigning, for example, positive distance functions outside the interface and negative inside, this by necessity implies only two phases. The approach outlined in Merriman et al. (1994) is to use multiple level set functions, and hence let each level set function delineate the boundary between that phase and the outside. In some approaches, each individual level set function is advanced, which then can cause inconsistencies and gaps/voids which are not part of any phase, and are hence filled in as part of a repair step.

6.3 The Voronoi Implicit Interface Method

We next describe in some detail the Voronoi Implicit Interface Method (VIIM) for tracking multi-phase multiphysics. As an overview, the method is a combination of level set methods and a computational geometry construction. From a mathematical point of view, some advantages of this approach include:

- *Accuracy, consistency, efficiency*: The method works in any number of dimensions, using a fixed Eulerian mesh, and a single function plus an indicator function to track the entire multi-phase system. Geometric quantities and constraints are accurately computed, and phases are coupled together in a consistent fashion, with no gaps, overlaps, or ambiguities.
- *Multiple junctions and topological change*: Multiple junctions, such as triple points, are all handled naturally and automatically, as well as breakage, merger, creation, and disappearance of phases. No special attention is paid to discontinuous topological change.
- *Coupling with time-dependent physics*: The method uses a physical time step, which then allows coupling complex physics into the interface evolution. Feedback from the physics affect the interface, and changes to the interface affects the physics.

From a numerics points of view:

- The method can be built using straightforward finite differences on fixed mesh. It also works in a finite element setting.
- Each grid point carries only the value of the evolving function, plus a phase indicator.
- The method is first-order accurate at triple points/lines, and is arbitrarily high-order away from degeneracies on smooth interfaces between neighbouring phases.
- Computational complexity is $\mathcal{O}(kN \log N)$ per time step, where N is the number of grid cells containing multi-phase boundaries, and k is the width of an adaptive narrow band.

6.4 Motivation and fundamental idea of VIIM

The starting point is to recall the definition of a Voronoi diagram. Given boundary node sets in the plane, the Voronoi diagram divides the plane into regions such that all the points in one region are closer to one boundary node set than any other boundary node set (see Fig. 16). We note that the Voronoi diagram may be obtained by solving an Eikonal equation for the distance $T(x)$ from the boundary node sets: that is, we solve

$$|\nabla T| = 1 \qquad\qquad T = 0 \quad \text{at boundary sets}$$

The points in each Voronoi region all "trace back" to the same boundary node set, hence the Voronoi diagram is the set of all points in the domain that trace back to more than one boundary node set.

The essence of the Voronoi Implicit Interface Method (VIIM) is to combine this idea of a Voronoi diagram with the unsigned distance function. Consider the triple junction shown in Fig. 17.

On the left, we construct the unsigned distance function ϕ, which is zero on the triple interface itself, and positive away from this interface. In the centre, we choose a value $\epsilon > 0$ and show the level sets corresponding to $\phi = \epsilon$. Now, given these ϵ-level sets, we can construct the Voronoi diagram on the right, which gives an approximation to the zero level set corresponding to the interface. As ϵ goes to zero, this approximation becomes exact.

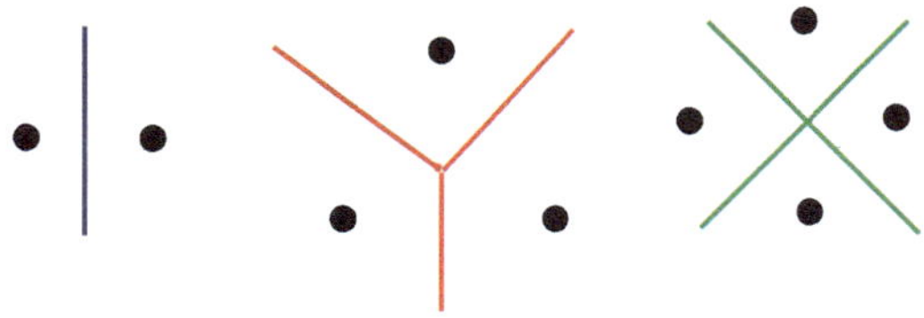

FIG. 16 Voronoi sets. *Left, blue*: Voronoi diagram between two boundary nodes. *Middle, red*: Voronoi diagram between three boundary nodes. *Right, green*: Voronoi diagram between four boundary nodes.

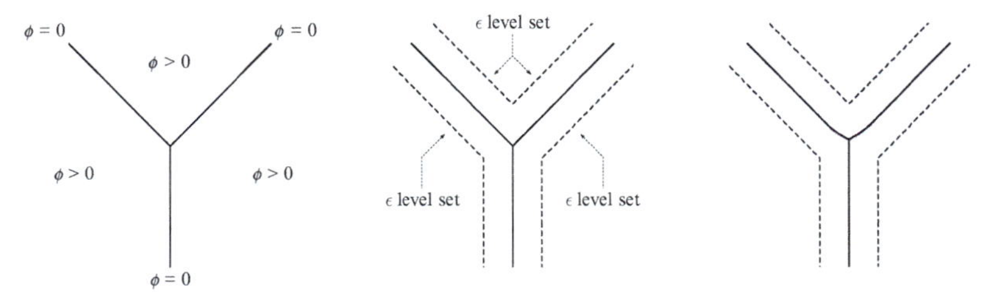

(a) Triple point of unsigned distance function (b) ϵ level sets (c) Voronoi reconstruction of interface

FIG. 17 Triple point structure. *Left*: Triple point with background level set function ϕ initialized with unsigned distance function ϕ. *Centre*: The ϵ-level sets of the function, $\epsilon > 0$. *Right*: Voronoi diagram of reconstructed interface from ϵ-level sets.

The Voronoi Implicit Interface Method builds on this geometric construction as follows. Given a collection of phases, separated by interfaces, we initialize by computing the unsigned distance to each point in the domain. Given the velocity of each point on the interface, produced by solving the appropriate physics, we then build extension velocities throughout the domain. We then evolve the unsigned level set function ϕ for some number of time steps, recomputing extension velocities as needed at each time step, after which we use the ϵ-level sets of the time-evolved unsigned distance function to compute the Voronoi diagram, giving a new position of the interfaces. Finally, this new position is used to compute a new unsigned distance function, and the calculation advances.

The key idea is that the algorithm relies on neighbouring level sets, which always exist in a "single phase," to allow reconstruction of the interface $\{\phi = 0\}$ using a Voronoi reconstruction.

6.5 Algorithm flow

Broadly, the algorithm is as follows:

Initialization:

- Choose ϵ, Δt, and a fixed regular Cartesian mesh with spacing h.
- Choose *Reinit*, which is the number of time steps before rebuilding the unsigned interface from the Voronoi reconstruction.
- Given a multi-phase problem, defined by regions in space belonging to different phases, compute the unsigned distance function ϕ from each mesh point to the closest interface boundary.
- At each mesh point, also assign a value for the integer indicator reporting the phase for that point.

Time loop:

- Use the position of the Voronoi set to compute the interface speed F depending on geometry and physics.
- Build extension velocities throughout the computational domain.
- Advance the unsigned level set function ϕ a single time step by solving $\phi_t + F|\nabla\phi| = 0$.
- After *Reinit* time steps, use the Voronoi diagram to rebuild the unsigned distance function from the ϵ-level sets.
- Loop to top.

6.6 Mathematical formulation

The above algorithm lends itself to a more formal description.

- Let $E_{\delta t}(\phi)$ be the evolution operator which evolves a given level set function ϕ for a time step δt.

- Let $V_\epsilon(\phi)$ be the operator that reconstructs the unsigned distance function from the ϵ-level sets of ϕ using the Voronoi diagram of the ϵ-level sets.
- Take the limit of $\Delta t \to 0$, followed by a limit of $\epsilon \to 0$ to define the evolution of the interface of a multi-phase system. That is, we *define* the solution at time T to be:

$$\phi(t=T) = \lim_{\epsilon\to 0}\lim_{n\to\infty}(V_\epsilon \circ E_{T/n})^n(\phi_0)$$

- (In other words:
  - For some $\epsilon > 0$, we apply n time steps, each consisting of an evolution and reconstruction.
  - We then take the limit n goes to infinity to get an ϵ-smoothed solution.
  - Taking the limit as $\epsilon \to 0^+$, we get the formal definition.)

Formally proving that this is a reasonable solution to moving triple points is not straightforward. Indeed, it is not obvious what such a solution should be under complex motions. It is possible that this definition might serve as possible way to analyze such flows, and this is an active area of research.

6.7 Implementation of Voronoi reconstruction steps

Two key steps in the algorithm are the reinitialization to build the unsigned distance function and the Voronoi reconstruction. These steps can be thought of as solving two different Eikonal equations $|\nabla T(x)| = 1$: the first time to build the unsigned function, the second time to find the Voronoi diagram characterized by points that trace back to multiple ϵ-level sets.

In fact, both of these steps can be done without ever explicitly finding either ϵ-level sets nor the Voronoi diagram. We also note that with careful programming, ϵ can be taken as zero. For details, see Saye and Sethian (2011).

7 Applications of the Voronoi Implicit Interface Method

In this section, we present some applications of the Voronoi Implicit Interface Method (VIIM).

7.1 Sintering and grain growth

In the first example, we use the VIIM to track the evolution of 93 individual phases, each sharing common interface boundaries, and moving under their curvature. The velocity of each interface is given by the local curvature: at triple point junctions, a balance of forces arguments leads to a net motion. Here, one of the advantages of the VIIM approach becomes apparent. The non-zero level sets each exist in an individual phase where the local curvature is unambiguous.

Algorithmically, we start with a background finite difference mesh, and initialize the unsigned distance function ϕ at each point as well as an indicator function specifying the phase at that mesh point. We then evolve the structure using the VIIM.

Fig. 18 shows a time evolution sequence of the ensuing flow, chosen so that the total enclosed area of each region remains constant. Here, we solve

$$\frac{\phi^{n+1} - \phi^n}{\Delta t} = \gamma\kappa^n|\nabla\phi^n| + s^n|\nabla\phi^n|$$

where $s^n(x) = \frac{V_i^0 - V_i^n}{V_i^n \Delta t}$. The first term corresponds to flow under curvature, where γ is a constant and κ is the local curvature. The second source term either inflates or deflates the ith region as a function of the change of the current area V_i^n from the baseline V_i^0.

The flow correctly demonstrates Young's law, which states that the curvature minimizes perimeter length and leads to 120 degree angles.

7.2 Variable density fluid flow

Next, we consider multi-phase fluid flow. The starting point is the variable density incompressible Navier–Stokes equations, namely

$$\begin{aligned} \rho(\boldsymbol{u}_t + (\boldsymbol{u}\cdot\nabla)\boldsymbol{u}) &= -\nabla p + \mu\Delta\boldsymbol{u} + \boldsymbol{st} + \boldsymbol{F}, \\ \nabla\cdot\boldsymbol{u} &= 0, \\ \phi_t + \boldsymbol{u}\cdot\nabla\phi &= M\sigma\kappa|\nabla\phi|, \end{aligned}$$

where ρ is the density, μ is gas viscosity, p is pressure, **st** is surface tension force, and F is the body force. Here, σ is the coefficient of surface tension, and $M > 0$ is the permeability of the individual interface.

The surface tension term for standard two-phase flow is given by $\boldsymbol{st} = -\sigma\kappa\delta(\phi)\nabla\phi$. However, for multiple junctions, the curvature is not easily defined at triple points. Hence, we take each phase separately, add up forces and normalize by two to achieve an appropriately relevant surface tension term. Thus, we take

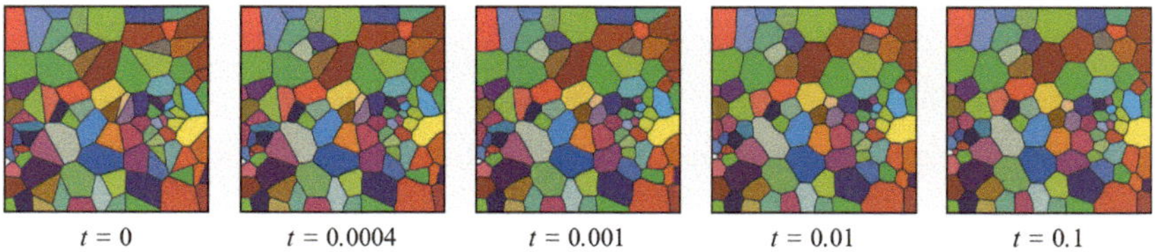

FIG. 18 Multi-phase grain growth: 93 connected phases moving under curvature. *Taken from Saye, R. I., Sethian, J. A., 2011. The Voronoi implicit interface method for computing multiphase physics. Proc. Natl. Acad. Sci. U. S. A. 108 (49), 19498–19503; Saye, R. I., Sethian, J. A., 2012. Analysis and applications of the Voronoi implicit interface method. J. Comput. Phys. 231 (18), 6051–6085.*

$$st = -\frac{\sigma}{2}\sum_{i=1}^{N} \kappa(\phi_i)\delta(\phi_i)\nabla\phi_i$$

where ϕ_i is built from the unsigned distance function ϕ and represents a level set function for each phase. Effectively, this enforces Youngs' law locally.

As was done in two-phase flow inkjet simulations, we use a second-order projection for the incompressible Navier–Stokes on a fixed Cartesian mesh, with mollifers to smooth the surface term into the right-hand side of the Navier–Stokes equations.

As application, Fig. 19 illustrates the results for a three-dimensional simulation of a variable density fluid flow, computed on a 128^3 grid with slip boundary conditions, using $\epsilon = 0^+$. The simulation starts with 15 heavy phases and approximately 100 less dense phases. The incompressible Navier–Stokes equations are solved, using a second-order projection method, and coupled to the Voronoi Implicit Interface Method. For all but the last snapshot in Fig. 19, the heavier phase is dark, while the other phases have been rendered mostly transparent, together with the triple line junctions as a network of curves. In the last snapshot, at time $t = 1.8$, we have rendered the bulk foam opaque, to make the structure of the foam more obvious.

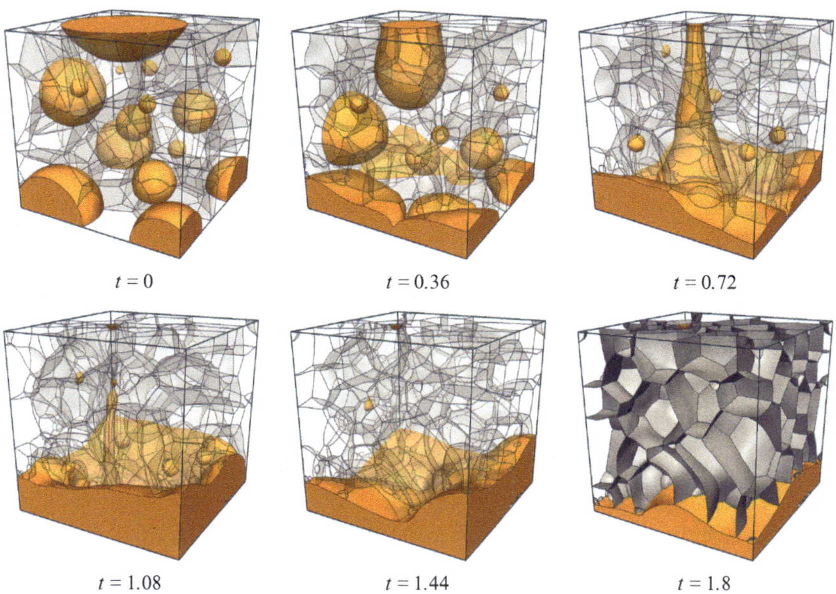

FIG. 19 Results of a fluid flow simulation in three dimensions with gravity, in which the orange coloured phase is more viscous and more dense than the other phases. The bulk foam is rendered mostly transparent except for the last frame, where it is rendered opaque to make the structure more prominent. *Taken from Saye, R. I., Sethian, J. A., 2012. Analysis and applications of the Voronoi implicit interface method. J. Comput. Phys. 231 (18), 6051–6085.*

7.3 Soap bubbles and industrial foams

Our final application is quite extensive and combines all the advances in interface methods discussed above, including core level set methods, narrow band adaptivity, extension velocities, and Voronoi Implicit Interface Methods.

7.3.1 The dynamics of foam evolution

Foams play a major role in industrial processing and manufacturing. Examples include solid foams, such as concrete, glass, and metal foams, and thin-film liquid foams:

• *Solid foams:* Metal foams, for example, are formed by mixing metallic substances such as aluminium, with a foaming agent (e.g., TiH_2). Heating above the melting temperature, hydrogen is released and dissolved in the liquid aluminium. Bubbles rise and burst, leaving a network of lightweight metal structure with a high air fraction and properties of the underlying material.

• *Liquid foams:* Soap foams are made of an ambient gas, divided into regions by a thin fluid-filled membrane consisting of multiple interconnected lamellae. Membrane fluid drains into neighbouring Plateau junctions until one ruptures, causing a macroscopic imbalance. Membranes readjust due to surface tensions forces. Neighbouring gas dynamics in each phase transports liquid in the lamella until a relatively stable equilibrium phase is reached, which is again disrupted by rupture occurring due to continued fluid drainage.

The liquid foam dynamics described above take place over six orders of magnitude in space and time. Liquid in the thin films, while only micrometres thick, drains over tens or hundreds of seconds until a membrane ruptures. Membranes burst at hundreds of centimetres per second (Bird et al., 2010), and this actuates macroscopic rearrangement of bubble topology through surface and fluid forces occurring over less than a second. Considerable mathematical analyses and experimental studies have focused on these individual stages, see, for example, Plateau (1873), von Neumann (1952), Mullins (1956), Weaire and Hutzler (1999), Polthier (2005), and Chopp (1993).

7.3.2 A scale-separation model

To gain a full view of the coupled dynamics, one approach is to model only the smallest space and time scales. However, this requires such a fine resolution that there is no practical hope of following a calculation long enough to observe the macroscale effects. Fortunately, the details at one scale are not necessarily important at another scale. By devising different models and equations at different scales, the strategy is to compute physics at different resolutions and allow these different models to communicate across the scales. In Saye and Sethian (2011), this idea of "scale separation" was used to build a realistic model that correctly couples three different regimes together:

- *Macroscopic rearrangement phase*: In this phase, the foam structure is out of macroscopic equilibrium. Surface tension at the liquid–gas interface influences the gas dynamics, which in turn evolve the network of lamellae and Plateau borders, rearranging the system of bubbles. Liquid contained in the thin films and Plateau borders is conserved and transported during this readjustment. Thus, during this phase, one needs to solve the multi-phase incompressible Navier–Stokes equations.

- *Drainage phase*: During this phase, the foam is essentially in macroscopic equilibrium. The dynamics of the gas phase are negligible, and the surface area has been locally minimized. An idealization assumes that individual lamella have constant mean curvature, suggesting a model for liquid drainage in the (fixed) network of lamellae and Plateau borders. During this phase, a version of the thin-film equations needs to be solved, with boundary conditions that couple lamella drainage to the Plateau junctions.
- *Rupture phase*: Eventually, a lamella becomes critically thin and ruptures. A tear appears, and the hole in this curved 2D sheet rapidly expands as the membrane retracts under surface tension. Depending on the scales involved (Bird et al., 2010), some of liquid breaks into small droplets, and some liquid retreats into Plateau borders.

7.3.3 A mathematical model for foam evolution

In Saye and Sethian (2013, 2016), a full mathematical model and numerical simulation framework was presented to model foam evolution. As an overview, a cluster of bubbles, initially out of equilibrium, undergoes macroscopic rearrangement, solved by a second-order projection method on a rectangular mesh, with the interface providing the right-hand side due to a mollified surface tension term, and the interface, which carries liquid, is advanced by using the Voronoi Implicit Interface Method. The liquid thickness in the lamellae is transported by a Lagrangian particle-based method of characteristics. After equilibrium is reached, the multiconnected interface is triangulated and a fourth-order PDE is solved for the thin film thickness drainage, using a finite element method for the evolving thickness. The lamella ruptures when a user-chosen minimal thickness is reached, sending the configuration into macroscopic disequilibrium. An overview of the algorithm is given in Fig. 20.

The complete dynamics are shown in Fig. 21, which shows the time evolution of a bubble cluster, starting from 26 separate bubbles end ending up in a single bubble. The bubbles colours are computed from thin film interface depending on the computed fluid thickness in the lamellae.

8 Sharp interface physics: Implicit mesh discontinuous Galerkin methods

In this section, we discuss some of the latest advances in level set methods, designed to provide high-order calculations of moving interfaces embedded in complex physics.

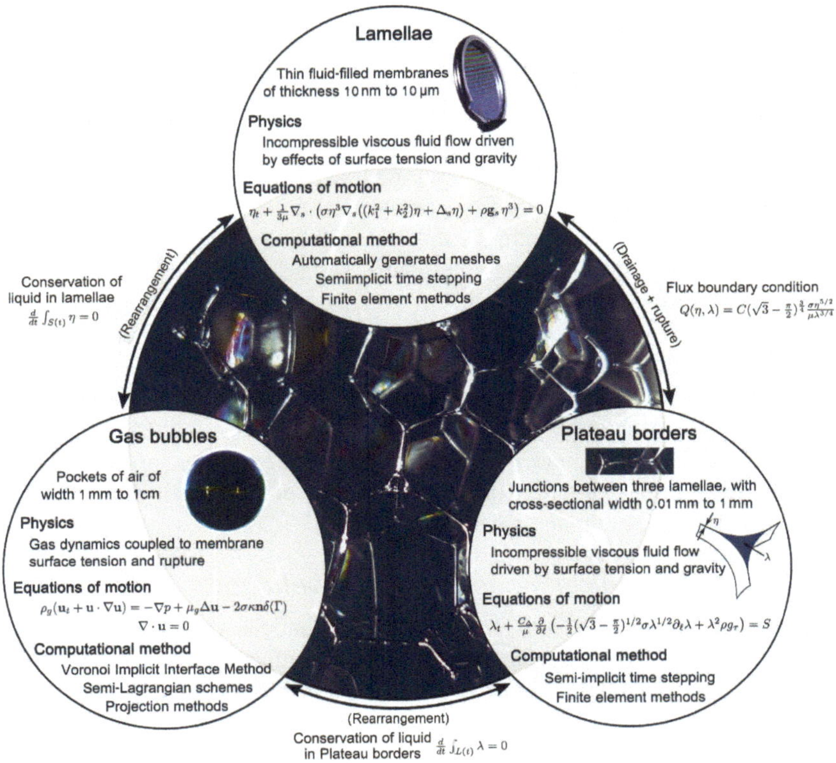

FIG. 20 A schematic of the coupled multiscale model and equations of motion. The model cycles between three stages: rearrangement, drainage, and rupture. During rearrangement, gas dynamics and surface tension-drive rearrangement of lamellae and Plateau borders, affecting their film thicknesses in the process. During drainage, these thicknesses, characterized by η and λ defined on individual membranes and junctions, change according to thin-film evolution equations. Upon rupture, liquid inside the ruptured film is conservatively redistributed, and the cycle repeats. *Taken from Saye, R. I., Sethian, J. A., 2016. Multiscale modelling of evolving foams. J. Comput. Phys. 315, 273–301.*

8.1 The challenge of sharp interface physics

As discussed earlier, a wide variety of approaches exist for computing multi-phase physics with evolving interfaces. Among them include moving mesh methods, in which the computational mesh flows and contorts along with moving interfaces or boundaries, fixed-grid methods wherein interfacial forces are transformed to forces located at grid points, and hybrids (Nochetto and Walker, 2010), such as arbitrary Lagrangian–Eulerian methods. Here, we briefly discuss these techniques.

8.1.1 *Smoothed interface methods*

The level set methods presented above are most often applied in a fixed-grid setting. For example, in multi-phase fluid flow, singular interfacial forces

FIG. 21 Collapse of a foam cluster, visualized with thin-film interference. *Taken from Saye, R. I., Sethian, J. A., 2016. Multiscale modelling of evolving foams. J. Comput. Phys. 315, 273–301.*

such as surface tension are then smoothed out onto the grid points on either side of the interface, most often via regularized Dirac delta or Heaviside functions. Prominent examples of this technique include Peskin's immersed-boundary method (Mittal and Iaccarino, 2005; Peskin, 2002) and the continuum surface tension model of Brackbill et al. (1992), Sussman et al. (1994), and Sethian and Smereka (2003). These methods have proven to be quite successful, owing to their simplicity of implementation, adaptation to modelling a wide variety of physics, and robustness.

However, the inherent smoothing of interfacial forces across several grid cells can greatly lower modelling fidelity, often to first-order accuracy or less. In cases of highly intricate physics, for example, the subtle oscillations of gas bubbles rising in a liquid, or the effect of gas boundary layers in atomizing droplet dynamics, accurate treatment is required to faithfully produce the key physics.

8.1.2 Sharp interface methods

An alternative to smoothing interfacial forces is to use a sharp interface method. Moving mesh methods are one such scheme, wherein elements move along with the interface so that elements on either side have faces precisely coinciding with the position of the interface. Then, interfacial jump conditions and boundary conditions can be imposed very accurately, e.g., through numerical fluxes in a discontinuous Galerkin framework. By expending this effort on matching the mesh geometry with the interface geometry, one can obtain very high orders of accuracy. However, moving mesh methods often break down when the interface geometry undergoes large displacements or severe distortion, and fail altogether when topological changes take place, e.g., the merger of two liquid droplets.

8.1.3 Hybrid interface methods

As a hybrid between fully static and fully dynamic meshes, an alternative approach is to have the moving interface cut through the cells of a fixed background grid to subsequently define a new mesh with changing cell shapes. Often called cut-cell, embedded-boundary, unfitted, or immersed-boundary methods, these schemes have the potential to combine the advantages of level set methods, in handling a wide variety of interface movement, with the advantages of sharp interface methods, to achieve very high precision. To illustrate, consider a uniform Cartesian grid together with an interface separating two fluids. Using the interface to cut through the cells of the grid, one obtains rectangular cells away from the interface, together with cut cell pieces on either side of the interface; these cells, cut or otherwise, form the basis of a new mesh.

In practical settings, one must be critically aware of the following property: as the interface moves across the background grid, cells cut by the interface can be arbitrarily small in size. Used as-is in a finite volume or finite element method, tiny cut cells lead to significant numerical issues, such as very ill-conditioned discretizations of PDE problems (e.g., the condition number of a discrete Laplacian operator can be arbitrarily large, akin to a finite difference method where two grid points are separated by a distance much less than the typical distance h), while time step constraints in evolutionary problems can be arbitrarily severe. Therefore, tiny cut cells must be appropriately handled.

A variety of approaches have been developed to this end, including weighting the coefficient of discrete stencils to appropriately dampen the influence of extraneous degrees of freedom associated with tiny cut cells (Bastian and Engwer, 2009; Devendran et al., 2014; Heimann et al., 2013; Muralidharan and Menon, 2016), penalty parameters to weakly impose continuity conditions (Burman et al., 2015; Nitsche, 1971), and cell merging, in which tiny cut cells are merged with their neighbours (Fröhlcke, 2013; Hunt, 2004; Johansson and Larson, 2013). In particular, this last idea of cutting cells and then gluing tiny cells to larger neighbours, provides a conceptually straightforward way to define a domain boundary- and interface-conforming mesh.

8.2 A high-order discontinuous Galerkin implicit mesh level set method for sharp interface physics

Using this idea of cell merging, in recent work, Saye (2017a,b) developed a high-order accurate implicit mesh discontinuous Galerkin framework for computing multi-phase multiphysics using the level set method. From a high level perspective, the framework uses a background grid together with a multi-phase cell merging procedure to create an implicitly defined mesh, as follows. A background reference grid is chosen, typically a uniform Cartesian grid, or an adaptively refined quadtree or octree. Curved domain boundaries and the interface between two or more fluids are specified by one or more level set functions or the Voronoi Implicit Interface Method. These implicitly defined interfaces cut through the cells of the background grid and are classified as being empty, small, large, or entire. To avoid the tiny cut cell issue, small cells are then merged with neighbouring cells to define the elements of an implicitly defined mesh, see Fig. 22. We note that:

- Except for the cell merging decisions, the geometry of the mesh is never explicitly constructed nor parameterized. Instead, the shape of curved elements is implicitly defined by the underlying level set functions. Moreover, the mesh is automatically interface-conforming, i.e., the interface is sharply represented by boundary elements belonging to different phases.
- Consequently, if the dynamics of the level set function is made high-order accurate, then this accuracy is adopted by the geometry of the implicitly defined mesh; boundary and interfacial jump conditions can then also be captured or imposed with high-order accuracy.

We remark that the resulting implicitly defined mesh (see Fig. 22, right) is somewhat unconventional. For example, the faces of some elements may overlap with more than one neighbouring element, i.e., from a finite element perspective, the mesh is non-conforming. As a result, such meshes cannot be used with a standard continuous Galerkin finite element method. Nevertheless, the mesh can be used in a discontinuous Galerkin (DG) method without problem. In a DG method, the requirement of continuity of discrete solutions is dropped, resulting in much greater flexibility in permitting non-conforming meshes, among a host of other features. Discrete solutions are piecewise

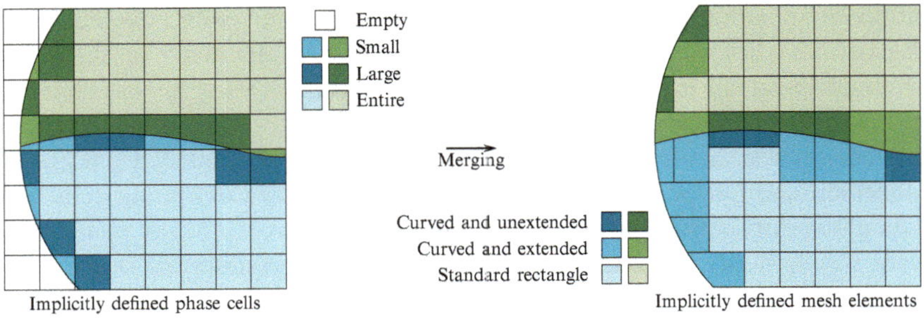

FIG. 22 Implicitly defined meshes using multi-phase cell merging. *Left*: Phase cells, defined by the intersection of each phase (*blue* and *green*) with the cells of a background Cartesian/quadtree grid, are classified according to whether they fall entirely within one phase, entirely outside the domain, or according to whether they have a small or large volume fraction. *Right*: Small cells are merged with neighbouring cells in the same phase to form a finite element mesh composed of standard rectangular elements and elements with curved, implicitly defined boundaries. *Adapted from Saye, R. I., 2017a. Implicit mesh discontinuous Galerkin methods and interfacial gauge methods for high-order accurate interface dynamics, with applications to surface tension dynamics, rigid body fluid-structure interaction, and free surface flow: Part I. J. Comput. Phys. 344, 647–682; Saye, R. I., 2017b. Implicit mesh discontinuous Galerkin methods and interfacial gauge methods for high-order accurate interface dynamics, with applications to surface tension dynamics, rigid body fluid-structure interaction, and free surface flow: Part II. J. Comput. Phys. 344, 683–723.*

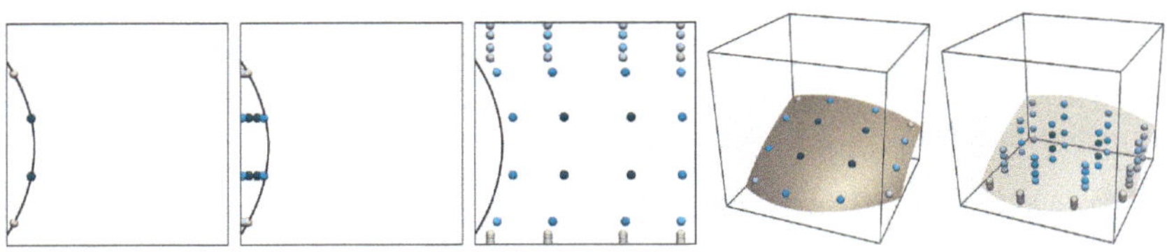

FIG. 23 Examples of quadrature schemes for implicitly defined phase cells, as constructed by the algorithms of Saye (2015). *Left three*: Integration in two dimensions, for an implicitly defined curve and area on either side. *Right two*: Integration in three dimensions, for an implicitly defined curved surface and the volume underneath. *Adapted from Saye, R. I., 2015. High-order quadrature methods for implicitly defined surfaces and volumes in hyperrectangles. SIAM J. Sci. Comput. 37 (2), A993–A1019.*

polynomial, defined by a weak or variational statement designed to discretize the model PDE. This weak formulation connects polynomials on neighbouring elements together via integral equations over individual elements and their boundaries. It follows that, in order to employ an implicitly defined mesh in a DG method, quadrature schemes must be computed for the implicitly defined elements and their boundaries.

To this end, Saye developed high-order accurate quadrature algorithms for implicitly defined surfaces and volumes in hyperrectangles (Saye, 2015) (see Fig. 23), and these have been made available in an open source C++ library (Saye, 2018). These algorithms are employed to classify the cut cells as void, small, large, or entire, for the construction of implicitly defined meshes, as well as to compute mass matrices, numerical fluxes, lifting operators, and other operators for the DG method. As the interface moves across the domain, a new implicit mesh is defined each time step and the quadrature rules need to be recomputed for elements that change shape; fortunately, the quadrature algorithms are inexpensive and have negligible cost compared to other aspects of the DG framework.

9 Application: High-order DG implicit mesh level set methods for sharp interface fluid dynamics

This implicit mesh DG framework has been applied to multi-phase incompressible fluid flow, surface tension-driven two-phase flow with phase-dependent viscosity and density, rigid body fluid–structure interaction, free surface flow, and non-Newtonian fluid modelling. Convergence analyses indicate high-order accuracy in the maximum norm, with 2nd to 10th order accurate schemes demonstrated.

Through apt choices of reinitialization schemes and extension velocities, excellent conservation of mass can be obtained: in convergence tests, mass is conserved to 10 digits of accuracy, and in very challenging high Reynolds number rapid oscillation dynamics, mass is conserved to one-hundredth of 1%.

To demonstrate one application of this framework, Fig. 24 shows a case in which high-order accuracy was used to revealed a vortex shedding phenomenon previously undetected by experiment. The problem considers a type of fluid flow behaviour which can be readily observed in a kitchen sink: when a steady

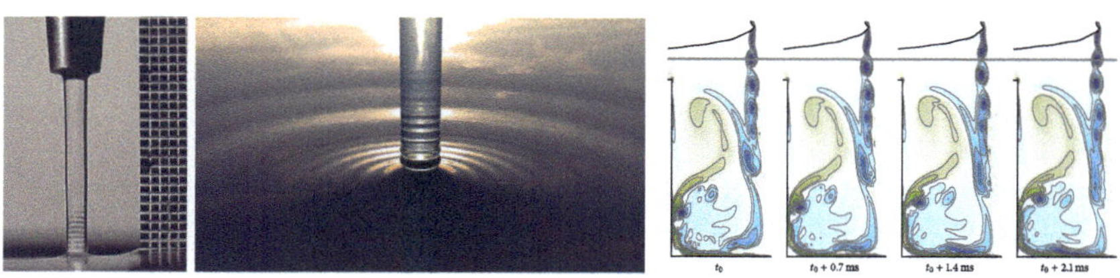

FIG. 24 Water ripples in a free surface flow. *Left*: Experimental image from Hancock and Bush Hancock and Bush (2002) showing a jet of water exiting a nozzle and entering a reservoir. Capillary waves can be seen travelling up the stream. *Middle*: Results of an axisymmetric simulation, computed using the high-order accurate implicit mesh DG framework using level set methods, with physical parameters matching those of the experiment. *Right*: Plots of the fluid vorticity, showing vortex shedding at the base of the jet stream. *Adapted from Hancock, M. J., Bush, J., 2002. Fluid pipes. J. Fluid Mech. 466, 285–304; Saye, R. I., 2017a. Implicit mesh discontinuous Galerkin methods and interfacial gauge methods for high-order accurate interface dynamics, with applications to surface tension dynamics, rigid body fluid-structure interaction, and free surface flow: Part I. J. Comput. Phys. 344, 647–682; Saye, R. I., 2017b. Implicit mesh discontinuous Galerkin methods and interfacial gauge methods for high-order accurate interface dynamics, with applications to surface tension dynamics, rigid body fluid-structure interaction, and free surface flow: Part II. J. Comput. Phys. 344, 683–723.*

stream of water of diameter 2–5 mm exits from a tap and is obstructed downstream, mildly steady ripples can be seen travelling up the stream. These ripples are caused by surface tension and the Plateau–Rayleigh instability which acts to collapse thin tubes of liquid. In this test problem, the numerical framework is compared to experimental results of Hancock and Bush (2002) and recovers the ripple behaviour with quantitative agreement to that of the experiment. The numerical results also revealed a kind of vortex shedding, in which vortex tubes are shed from the base of the ripples, which could not be seen in the experiment.

10 Summary

The introduction and subsequent development of level set methods over the years has led to algorithms that have been instrumental in computing a large number of complex physics problems across science, engineering, and industry. A large number of researchers have contributed to these developments, and this review only touches a small fraction of the large body of work that has been done and continues to be done. We refer the interested reader to a wide collection of resources, reviews, tutorials, and community software for further information.

Acknowledgement

The authors were supported in part by the Applied Mathematics Program of the U.S. DOE Office of Advanced Scientific Computing Research under contract number DE-AC02-05CH11231.

References

Adalsteinsson, D., Sethian, J.A., 1995a. A fast level set method for propagating interfaces. J. Comput. Phys. 118 (2), 269–277.

Adalsteinsson, D., Sethian, J.A., 1995b. A unified level set approach to etching, deposition and lithography I: algorithms and two-dimensional simulations. J. Comput. Phys. 120 (1), 128–144.

Adalsteinsson, D., Sethian, J.A., 1995c. A unified level set approach to etching, deposition and lithography II: three-dimensional simulations. J. Comput. Phys. 122 (2), 348–366.

Adalsteinsson, D., Sethian, J.A., 1997. A unified level set approach to etching, deposition and lithography III: complex simulations and multiple effects. J. Comput. Phys. 138 (1), 193–223.

Adalsteinsson, D., Sethian, J.A., 1999. The fast construction of extension velocities in level set methods. J. Comput. Phys. 148, 2–22.

Adalsteinsson, D., Sethian, J.A., 2002. Transport and diffusion of material quantities on propagating interfaces via level set methods. J. Comput. Phys 185 (1), 271–288.

Almgren, A.S., Bell, J.B., Szymczak, W.G., 1996. A numerical method for the incompressible Navier-Stokes equations based on an approximate projection. SIAM J. Sci. Comput. 17 (2), 358–369.

Barles, G., 1985. Remarks on a flame propagation model. INRIA Rep. 464.

Bastian, P., Engwer, C., 2009. An unfitted finite element method using discontinuous Galerkin. Int. J. Numer. Methods Eng. 79 (12), 1557–1576.

Bell, J.B., Colella, P., Glaz, H.M., 1989. A second-order projection method for the incompressible Navier-Stokes equations. J. Comput. Phys. 85, 257–283.

Bird, J.C., de Ruiter, R., Courbin, L., Stone, H.A., 2010. Daughter bubble cascades produced by folding of ruptured thin films. Nature 465, 759–762.

Brackbill, J.U., Kothe, D.B., Zemach, C., 1992. A continuum method for modeling surface tension. J. Comput. Phys. 100 (2), 335–354.

Brakke, K., 1992. The surface evolver. Exp. Math. 1 (2), 141–165. 2.

Burman, E., Claus, S.P.H., Larson, M.G., Massing, A., 2015. Cutfem: discretizing geometry and partial differential equations. Int. J. Numer. Methods Eng. 104 (7), 472–501.

Chen, Y., Giga, Y., Goto, S., 1991. Uniqueness and existence of viscosity solutions of generalized mean curvature flow equations. J. Diff. Geom 33, 749.

Chern, I.L., Glimm, J., Mcbryan, O., Plohr, B., Yaniv, S., 1986. Front tracking for gas dynamics. J. Comput. Phys. 62 (1), 83–110.

Chopp, D.L., 1993. Computing minimal surfaces via level set curvature flow. J. Comput. Phys. 106, 77–91.

Chopp, D.L., 2001. Some improvements of the fast marching method. SIAM J. Scientific Comput. 23 (1).

Chorin, A.J., 1968. Numerical solution of the Navier-Stokes equations. Math. Comput. 22, 745.

Crandall, M.G., Lions, P.L., 1983. Viscosity solutions of Hamilton-Jacobi equations. Trans. AMS 277, 1–43.

Crandall, M.G., Evans, L.C., Lions, P.L., 1984. some properties of viscosity solutions of Hamilton-Jacobi equations. Trans. AMS 282, 487–502.

Crandall, M.G., Ishii, H., Lions, P.L., 1992. User's guide to viscosity solutions of second order partial differential equations. Bull. AMS 27/1, 1–67.

Devendran, D., Graves, D.T., Johansen, H., 2014. A higher-order finite-volume discretization method for Poisson's equation in cut cell geometries. ArXiv e-prints: arXiv:1411.4283.

Dziuk, G., Elliott, C., 2013. Finite element methods for surface PDEs. Acta Numer. 22, 289–396.

Elsey, M., Esedoglu, S., Smereka, P., 2009. Diffusion generated motion for grain growth in two and three dimensions. J. Comput. Phys. 228 (21), 8015–8033.

Evans, L.C., Spruck, J., 1991a. Motion of level sets by mean curvature I. J. Diff. Geom 33, 635.

Evans, L.C., Spruck, J., 1992b. Motion of level sets by mean curvature II. Trans. Am. Math. Soc. 330 (1), 321–332.

Evans, L.C., Spruck, J., 1992. Motion of level sets by mean curvature III. J. Geom. Anal. 2, 121–150.

Evans, L.C., Spruck, J., 1995. Motion of level sets by mean curvature IV. J. Geom. Anal. 5 (1), 77–114.

Evans, L.C., Soner, H.M., Souganidis, P.E., 1992. Phase transitions and generalized motion by mean curvature. Commun. Pure Appl. Math. 45, 1097–1123.

Fröhlcke, A., 2013. A boundary conformal discontinuous Galerkin method for electromagnetic field problems on Cartesian grids. Ph.D. thesis, Technische Universität, Darmstadt. PhD thesis.

Gage, M., 1984. Curve shortening makes convex curves circular. Invent. Math. 76, 357.

Gage, M., Hamilton, R., 1986. The equation shrinking convex planes curves. J. Diff. Geom 23, 69.

Garzon, M., Gray, L.G., Sethian, J.A., 2009. Numerical simulation of non-viscous liquid pinch-off using a coupled level set-boundary integral method. J. Comput. Phys. 228 (17), 6079–6106.

Garzon, M., Gray, L.G., Sethian, J.A., 2003. Wave breaking over sloping beaches using a coupled boundary integral-level set method. Interfaces Free Bound. 7 (3), 229–239.

Garzon, M., Bobillo-Ares, N., Sethian, J.A., 2008. Some free boundary problems in potential flow regime using a level set method. In: Recent Advances in Fluid Mechanics, Nova Publishers.

Garzon, M., Gray, L.G., Sethian, J.A., 2011. Simulation of the droplet-to-bubble transition in a two-fluid system. Phys. Rev. E 83, 4.

Garzon, M., Gray, L.G., Sethian, J.A., 2012a. Axisymmetric boundary integral formulation for a two-fluid system. Int. J. Numer. Meth. Fluids 69, 1124–1134.

Garzon, M., Gray, L.G., Sethian, J.A., 2012b. Droplet and bubble pinch-off computations using level sets. J. Comput. Appl. Math. 236 (12), 3034–3041.

Garzon, M., Gray, L.G., Sethian, J.A., 2014. Numerical simulations of electrostatically driven jets from non-viscous droplets. Phys. Rev. E 89, 033011.

Garzon, M., Johansson, A., Sethian, J.A., 2016. A three-dimensional coupled Nitsche and level set method for electrohydrodynamic potential flows in moving domains. J. Comput. Phys. 309, 1–386.

Giga, Y., Goto, S., 1992. Motion of hypersurfaces and geometric equations. J. Math. Soc. Jpn 44, 99.

Giga, Y., Goto, S., Ishii, H., 1992. Global existence of weak solutions for interface equations coupled with diffusion equations. SIAM J. Math. Anal. 23 (N4), 821–835.

Grayson, M., 1987. The heat equation shrinks embedded plane curves to round points. J. Diff. Geom. 26, 285.

Grayson, M., 1989. A short note on the evolution of surfaces via mean curvatures. J. Diff. Geom. 58, 555.

Hancock, M.J., Bush, J., 2002. Fluid pipes. J. Fluid Mech. 466, 285–304.

Heimann, F., Engwer, C., Ippisch, O., Bastian, P., 2013. An unfitted interior penalty discontinuous Galerkin method for incompressible Navier-Stokes two-phase flow. Int. J. Numer. Methods Fluids 71 (3), 269–293.

Hunt, J., 2004. An adaptive 3D Cartesian approach for the parallel computation of inviscid flow about static and dynamic configurations (Ph.D. thesis). University of Michigan.

Ilmanen, T., 1992. Generalized flow of sets by mean curvature on a manifold. Indiana Univ. Math. J. 41 (3), 671–705.

Ilmanen, T., 1994. Elliptic regularization and partial regularity for motion by mean curvature. Memoirs Am. Math. Soc. 108, 520.

Johansson, A., Larson, M.G., 2013. A high order discontinuous Galerkin Nitsche method for elliptic problems with fictitious boundary. Numer. Math. 123 (4), 607–628.

Lafaurie, B., Nardone, C., Scardovelli, R., Zaleski, S., Zanetti, G., 1994. Modelling merging and fragmentation in multiphase flows with surfer. J. Comput. Phys 113 (1), 134–147.

Malladi, R., Sethian, J.A., Vemuri, B.C., 1995. Shape modeling with front propagation: a level set approach. IEEE Trans. Pattern Anal. Mach. Intell. 17 (2), 158–175.

Merriman, B., Bence, J., Osher, S.J., 1994. Motion of multiple junctions: a level set approach. J. Comput. Phys. 112 (2), 334–363.

Mittal, R., Iaccarino, G., 2005. Immersed boundary methods. Annu. Rev. Fluid Mech. 37, 239–261.

Mullins, M.M., 1956. Two-dimensional motion of idealized grain boundaries. J. Appl. Phys 27, 900–904.

Muralidharan, B., Menon, S., 2016. A high-order adaptive Cartesian cut-cell method for simulation of compressible viscous flow over immersed bodies. J. Comput. Phys. 321, 342–368.

Nitsche, J., 1971. Über ein variationsprinzip zur lösung von dirichlet-problemen bei verwendung von teilräumen, die keinen randbedingungen unterworfen sind. Abhandlungen aus dem Mathematischen Seminar der Universität Hamburg 36 (1), 9–15.

Nochetto, R.H., Walker, S.W., 2010. A hybrid variational front tracking-level set mesh generator for problems exhibiting large deformations and topological changes. J. Comput. Phys. 229, 18.

Noh, W., Woodward, P., 1976. A simple line interface calculation. Springer-Verlag.

Osher, S., Fedkiw, R., 2002. Level Set Methods and Dynamic Implicit Surfaces. Springer.

Osher, S., Sethian, J.A., 1988. Fronts propagating with curvature-dependent speed: algorithms based on Hamilton-Jacobi formulations. J. Comput. Phys. 79, 12–49.

Peskin, C.S., 2002. The immersed boundary method. Acta Numer. 11, 479–517.

Plateau, J., 1873. Statique Experimentale et Theorique des Liquides Soumis Aux Seules Forces Moleculaires. Gauthier-Villars, Trubner et cie., Paris, France.

Polthier, K., 2005. Computational aspects of discrete minimal surfaces. In: Global Theory of Minimal Surfaces, (Ed.), Proceedings of the Clay Mathematics Institute 2001 Summer School, D. Hoffman.

Saye, R.I., 2014. High-order methods for computing distances to implicitly defined surfaces. Commun. Appl. Math. Comput. Sci. 9, 107–141.

Saye, R.I., 2015. High-order quadrature methods for implicitly defined surfaces and volumes in hyperrectangles. SIAM J. Sci. Comput. 37 (2), A993–A1019.

Saye, R.I., 2017a. Implicit mesh discontinuous Galerkin methods and interfacial gauge methods for high-order accurate interface dynamics, with applications to surface tension dynamics, rigid body fluid-structure interaction, and free surface flow: Part I. J. Comput. Phys. 344, 647–682.

Saye, R.I., 2017b. Implicit mesh discontinuous Galerkin methods and interfacial gauge methods for high-order accurate interface dynamics, with applications to surface tension dynamics, rigid body fluid-structure interaction, and free surface flow: Part II. J. Comput. Phys. 344, 683–723.

Saye, R., 2018. Algoim—algorithms for implicitly defined geometry, level set methods, and Voronoi implicit interface methods. https:/algoim.github.io/.

Saye, R.I., Sethian, J.A., 2011. The Voronoi implicit interface method for computing multiphase physics. Proc. Natl. Acad. Sci. U. S. A. 108 (49), 19498–19503.

Saye, R.I., Sethian, J.A., 2012. Analysis and applications of the Voronoi implicit interface method. J. Comput. Phys. 231 (18), 6051–6085.

Saye, R.I., Sethian, J.A., 2013. Multi-scale modelling of membrane rearrangement. Drainage, and Rupture in Evolving Foams Science Magazine 340 (6133), 720–724.

Saye, R.I., Sethian, J.A., 2016. Multiscale modelling of evolving foams. J. Comput. Phys. 315, 273–301.

Sethian, J. A., 1982. An Analysis of Flame Propagation (Ph.D. dissertation). Dept. of Mathematics, University of California, Berkeley, CA.

Sethian, J.A., 1985. Curvature and the evolution of fronts. Commun. Math. Phys. 101, 487–499.

Sethian, J.A., 1987. Numerical methods for propagating fronts. In: Concus, P., Finn, R. (Eds.), Variational Methods for Free Surface Interfaces. Springer-Verlag, NY.

Sethian, J.A., 1996a. A fast marching level set method for monotonically advancing fronts. Proc. Nat. Acad. Sci. 93 (4), 1591–1595.

Sethian, J.A., 1996b. Level Set Methods and Fast Marching Methods. Cambridge University Press.

Sethian, J.A., 1999. Level Set Methods and Fast Marching Methods. Cambridge University Press.

Sethian, J.A., Adalsteinsson, D., 1996. An overview of level set methods for etching, deposition, and lithography development. IEEE Trans. Semicond. Devices 10 (1), 167–184.

Sethian, J.A., Shan, Y., 2008. Solving partial differential equations on irregular domains with moving interfaces, with applications to superconformal electrodeposition in semiconductor manufacturing. J. Comput. Phys 227, 13.

Sethian, J.A., Smereka, P., 2003. Level set methods for fluid interfaces. Annu. Rev. Fluid Mech. 35, 341–372.

Sethian, J.A., Vladimirsky, A., 2001. Ordered upwind methods for static Hamilton-Jacobi equations. Proc. Natl. Acad. Sci. 98, 11069–11074.

Sussman, M., Smereka, P., Osher, S.J., 1994. A level set approach for computing solutions to incompressible two-phase flow. J. Comput. Phys. 114 (1), 146–159.

Thorodssen, S.T., 2007. Micro-droplets and micro-bubbles, imaging motion at small scales. Nus. Eng. Res. News 22, 1.

Tryggvason, G., Bunner, B., Esmaeeli, A., Juric, D., Al-Rawahi, N., Tauber, W., Han, J., Nas, S., Jan, Y.J., 2001. A front-tracking method for the computations of multiphase flow. J. Comput. Phys. 2 (20), 708–759.

von Neumann, J., 1952. Metal interfaces. In: Herring, C. (Ed.), Cleveland. American Society for Metals, pp. 108–110.

Weaire, D., Hutzler, S., 1999. The Physics of Foams. Oxford University Press.

Yu, J.D., Sakai, S., Sethian, J.A., 2003. A coupled level set projection method applied to ink jet simulation. Interfaces Free Bound. 193 (1), 275–305.

Yu, J.D., Sakai, S., Sethian, J.A., 2005. A coupled quadrilateral grid level set projection method applied to ink jet simulation. J. Comput. Phys. 206 (1), 227–251.

Yu, J.D., Sakai, S., Sethian, J.A., 2007. Two-phase viscoelastic jetting. J. Comput. Phys. 220 (2), 568–585.

Zhao, H.K., Chan, T., Merriman, B., Osher, S., 1996. A variational level set approach to multiphase motion. J. Comput. Phys. 127, 179–195.

Chapter 7

Free boundary problems in fluids and materials

Eberhard Bänsch[a,*] and Alfred Schmidt[b]
[a]*Applied Mathematics III, Department Mathematik, Friedrich-Alexander-Universität Erlangen-Nürnberg, Erlangen, Germany*
[b]*Center for Industrial Mathematics and MAPEX Center for Materials and Processes, Universität Bremen, Bremen, Germany*
[*]*Corresponding author: e-mail: baensch@math.fau.de*

Chapter Outline

Handbook of Numerical Analysis, Vol. 21. https://doi.org/10.1016/bs.hna.2019.05.004

Abstract
Free boundary problems with geometric conditions for the interfaces are typically models for situations where interfacial energies play an important role. This happens for example in microscale models of phase transitions as well as models with fluidic interfaces in various scales, all important in technical applications. We give here an overview of related aspects of models and corresponding numerical methods, and present some representative applications and results.

Keywords: Free boundary problems, Interfaces, Free surface flow, Two-phase flow, Phase transitions, Surface tension, Level set, Finite element method, Liquid bridges, Laser welding, Dendrites

AMS Classification Codes: 35K, 35Q, 53Z, 65M, 65Z, 76D

1 Introduction to geometric free boundary problems in fluids and materials

Interfaces play a paramount role in many technical applications. Geometric conditions for interfaces come into play whenever interfacial energies are involved. This typically leads to models where the (mean) curvature (or a corresponding anisotropic quantity) of the interface enters, as curvature is directly connected to the first variation of the local surface energy. Depending on the length scale of the model considered, such effects have to be taken into account for interfaces between different phases (like solid, liquid, gas) or different materials (like different liquids).

In flow problems with two immiscible fluids (for short two-phase flow) the interface has an energy giving rise to surface tension effects, which in turn can be the dominant force, at least on small length scales or under microgravity conditions. Moreover, in case of mass transfer across the interface by for instance evaporation/condensation additional complex effects may arise.

In this contribution we focus on mathematical examples that are representative for interface problems in fluids and materials. We describe corresponding numerical methods which are based on variational formulations and finite element discretizations.

Starting several decades ago, there is a large number of publications on mathematical and numerical analysis of various aspects of free boundary models and methods. Unfortunately, the fully coupled systems which are needed and used in technical applications are typically not fully treated yet. Also for that reason, we do not want to cover most aspects of analysis here.

The rest of the chapter is organized as follows. In Section 2 we briefly outline how mathematical models are derived for two-phase flow and phase transition problems.

With these models in mind we describe the most common numerical methods for interface problems. In Section 3 this is worked out for two-phase flow and in Section 4 for phase change problems. On the one hand, we include here interface capturing methods, where an interface is represented by lower dimensional mesh facets (edges of triangles or faces of tetrahedra, for example), and a moving interface causes the motion and deformation of the underlying mesh. On the other hand, interface tracking methods represent the interface independently of the background mesh, for example as a level set, but then jumps in values or gradients of fields defined on both sides of the interface have to be approximated adequately.

Finally, in Section 4 we present several applications of the so far developed methodology. Since surface tension in two-phase flow is a crucial mechanism under microgravity conditions, we present examples along with their experiments under such conditions. These examples may not be standard and demonstrate the range of applications. Regarding phase transition problems we present examples ranging from the microscale (dendritic solidification) to the macro scale (welding). Moreover, we present applications, where different types of free boundaries have to be simultaneously taken into account.

2 Aspects of modelling

In this chapter we derive the governing equations for two-phase flow together with conditions on the free surface. The interface separating the two fluids is assumed to have a *surface energy* giving rise to surface tension effects. The case of a single phase flow subject to an outer free surface is also straightforwardly covered by neglecting the second phase. Some further consideration is necessary, when the interface is in contact to a solid wall, i.e., when there exists a *triple line* or *contact line*. The setting for two-phase flows is typically a container with two immiscible fluids (see Fig. 1A and B).

In the second part of this section we consider phase change problems. More precisely we introduce models for melting/solidification that may exhibit dendritic growth, both without and with convection in the liquid phase. Finally, we consider the combination of solid–liquid phase changes with free surface flow in the liquid phase.

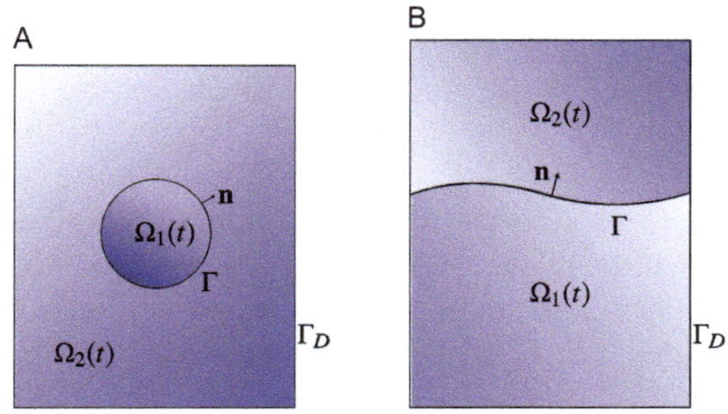

FIG. 1 Examples for two-phase flow. (A) Rising drop. (B) Container with two fluids.

2.1 Mathematical model of two-phase flow

In this section, we briefly revisit the mathematical model to describe the system under consideration. The model is derived from basic balance laws. The modelling is rather classical, see for instance (Scriven, 1960). We follow the presentation in Krahl et al. (2004), where a detailed derivation can be found.

2.1.1 Setting and generic balance equations

Sketches of typical geometries are depicted in Fig. 1A and B. The overall domain $\Omega \subseteq \mathbb{R}^n$ is divided into two domains Ω_i, $i = 1, 2$, occupied by fluids of different material properties, constant in either phase. The two phases are assumed to be separated by a smooth *sharp interface* Γ which may move over time; $\mathbf{n}$ denotes the normal vector on the interface Γ.

Our model is based on the balance equations for mass, momentum, and energy. Consider the density of some generic extensive thermodynamical quantity a. Let $V \subset \Omega$ be a test volume. The rate of change of ρa in V is determined by contributions from fluxes across the boundary and sources and sinks in V:

$$\frac{\mathrm{d}}{\mathrm{d}t}\int_V \rho a \,\mathrm{d}x = \int_S -(\rho a\mathbf{u} + \mathbf{q}_d)\cdot \mathbf{n}_S \,\mathrm{d}\mathcal{H}^{n-1} + \int_V f \,\mathrm{d}x. \tag{1}$$

Here, ρ denotes the density of the fluid, $\mathbf{u}$ the fluid velocity, $\mathbf{q}_d$ denotes the diffusive flux of a, f denotes the sources and sinks, and $\mathbf{n}_S$ the outward pointing normal of $S := \partial V$.

Assuming the test volume V to be completely contained in one of the phases Ω_1, Ω_2, applying Gauss' theorem to Eq. (1) yields

$$\int_V \partial_t(\rho a) + \nabla\cdot(\mathbf{q}_d + \rho a\mathbf{u}) - f \,\mathrm{d}x = 0. \tag{2}$$

Since Eq. (2) holds for arbitrary test volumes V it follows that

$$\partial_t(\rho a) + \nabla\cdot(\mathbf{q}_d + \rho a\mathbf{u}) = f. \tag{3}$$

We now set $a = 1$, $\mathbf{q}_d = 0$, and $f = 0$ in Eq. (3) to obtain the equation for the balance of mass:

$$\partial_t\rho + \nabla\cdot(\rho\mathbf{u}) = 0. \tag{4}$$

Combining the last two equations finally yields the generic balance equation

$$\rho(\partial_t a + \mathbf{u}\cdot\nabla a) + \nabla\cdot\mathbf{q}_d = f. \tag{5}$$

In a similar way, but working with a test volume being intersected by the interface and transported by the velocity V_Γ, one derives the balance equations for the generic quantity a on the interface Γ (we refer to Krahl et al. (2004) for a detailed derivation):

$$[\rho a(\mathbf{u}\cdot\mathbf{n} - V_\Gamma) + \mathbf{q}_d\cdot\mathbf{n}] = -f_\Gamma. \tag{6}$$

Here, $[g] := \lim_{\varepsilon \searrow 0} g(\cdot + \varepsilon \mathbf{n}) - \lim_{\varepsilon \nearrow 0} g(\cdot + \varepsilon \mathbf{n})$ denotes the jump of a generic function g across Γ and V_Γ denotes the normal velocity of the interface.

2.1.2 Incompressibility

The equation for the conservation of mass was already derived for the overall mass density of the fluid ρ, see Eq. (4). We additionally assume incompressibility, i.e.,

$$\nabla \cdot \mathbf{u} = 0. \tag{7}$$

With appropriate initial and boundary conditions this also implies constant density in either phase.

2.1.3 Conservation of momentum

Inserting $a = \mathbf{u}$ in Eq. (5) yields the equation for the conservation of momentum. According to Cauchy's Theorem, see Gurtin (1981), stresses acting on the surface of a test volume are given by $\mathbf{T}\mathbf{n}_S$, where $\mathbf{T}$ is the symmetric stress tensor and $\mathbf{n}_S$ the outward pointing normal of S. Therefore, $-\mathbf{T}$ defines the diffusive flux of momentum in Eq. (5). The stress tensor $\mathbf{T}$ can be decomposed into an isotropic part and a tracefree part:

$$\mathbf{T} = -p\mathbf{I} + \boldsymbol{\tau},$$

where p is called the pressure and $\boldsymbol{\tau}$ the viscous stress, which is, assuming the liquid to be Newtonian, given by

$$\boldsymbol{\tau} = \mu D(\mathbf{u})$$

with μ the dynamic viscosity and $D(\mathbf{u}) = \nabla \mathbf{u} + (\nabla \mathbf{u})^T$ the rate of strain tensor.

Combining all the above leads to the Navier–Stokes equations in either phase:

$$\rho_i \cdot (\partial_t \mathbf{u} + \mathbf{u} \cdot \nabla \mathbf{u}) - \mu_i \Delta \mathbf{u} + \nabla p = f \quad \text{in } \Omega_i, \quad i = 1,2 \tag{8}$$

$$\nabla \cdot \mathbf{u} = 0 \quad \text{in } \Omega_i. \tag{9}$$

Here, we have used the fact that for a solenoidal vector field, i.e., $\nabla \cdot \mathbf{u} = 0$, it holds: $\nabla \cdot D(\mathbf{u}) = \Delta \mathbf{u}$.

2.1.4 Jump condition at the interface

Set $a = \mathbf{u}$ and $\mathbf{q}_d = -\mathbf{T}$. Moreover, we assume continuity of $\mathbf{u}$ across Γ, i.e., $[\mathbf{u}] = 0$. Then from Eq. (6) we derive the jump condition across Γ:

$$[\rho \mathbf{u}(\mathbf{u} \cdot \mathbf{n} - V_\Gamma) - \mathbf{T}\mathbf{n}] = \mathbf{f}_\Gamma.$$

Denote by $\overline{a}$ the *average* of a generic quantity a across Γ. Applying the formula

$$[a_1 a_2] = \overline{a_1}[a_2] + [a_1]\overline{a_2}$$

to the above equation and assuming no mass transfer across Γ yields

$$\mathbf{f}_\Gamma = [\rho \mathbf{u}(\mathbf{u}\cdot\mathbf{n} - V_\Gamma) - \mathbf{T}\mathbf{n}] = \overline{\rho \mathbf{u}}\underbrace{[\mathbf{u}\cdot\mathbf{n} - V_\Gamma]}_{=0} + [\rho \mathbf{u}]\underbrace{\overline{(\mathbf{u}\cdot\mathbf{n} - V_\Gamma)}}_{=0} - [\mathbf{T}\mathbf{n}] = -[\mathbf{T}\mathbf{n}]. \tag{10}$$

It remains to determine $\mathbf{f}_\Gamma$. With surface tension γ, the energy of the free surface is given by

$$E_\Gamma := \int_\Gamma \gamma \, \mathrm{d}\mathcal{H}^{n-1}.$$

Assuming that Γ is closed (i.e., without boundary), a variation of the free surface energy and an application of the principle of virtual work, see Buscaglia and Ausas (2011), leads to

$$\langle \mathbf{f}_\Gamma, \mathbf{v}\rangle = -\langle E', \mathbf{v}\rangle = \int_\Gamma (\gamma\kappa\mathbf{n} + \nabla_S\gamma)\cdot \mathbf{v}\, \mathrm{d}\mathcal{H}^{n-1}$$

for arbitrary smooth $\mathbf{v} : \Gamma \to \mathbb{R}^n$. Here, ∇_S denotes the tangential gradient on Γ and κ the mean curvature (i.e., the sum of the principle curvatures) of the interface. With the help of (6) one derives the jump conditions at the interface:

$$[\mathbf{T}\mathbf{n}] = [\mu D(u)\mathbf{n} - p\mathbf{n}] = \gamma\kappa\mathbf{n} + \nabla_S\gamma. \tag{11}$$

2.2 Problem formulation of two-phase flow

Introducing appropriate characteristic values, the problem of two-phase flow without mass transfer across the interface can be cast in nondimensional form as follows, assuming γ to be constant.

Problem 1. For a given right-hand side f, boundary values $\mathbf{u}_D$, and initial data $\mathbf{u}_0$, Ω_0, for all $t \in [0, T]$ find a domain $\Omega(t)$ with fixed Dirichlet boundary Γ_D, a velocity $\mathbf{u}$, and pressure p, such that $\mathbf{u}(0) = \mathbf{u}_0$ and for all $t \in [0, T]$ fulfilling

$$\Lambda \frac{D}{Dt}\mathbf{u} - \nabla\cdot \mathbf{T}(\mathbf{u}, p) = \Lambda f \qquad \text{in } \Omega \tag{12a}$$

$$\nabla\cdot\mathbf{u} = 0 \qquad \text{in } \Omega \tag{12b}$$

$$\mathbf{u} = \mathbf{u}_D \qquad \text{on } \Gamma_D \tag{12c}$$

$$[\mathbf{T}(\mathbf{u}, p)\mathbf{n}] = -\frac{1}{\mathrm{We}}\kappa\mathbf{n} \qquad \text{on } \Gamma \tag{12d}$$

$$[\mathbf{u}] = 0 \qquad \text{on } \Gamma \tag{12e}$$

$$V_\Gamma = \mathbf{u} \cdot \mathbf{n} \qquad \text{on}\,\Gamma \tag{12f}$$

$$\Omega(0) = \Omega_0, \;\; \mathbf{u}(0, \cdot\,) = \mathbf{u}_0 \tag{12g}$$

Here, $\frac{D}{Dt} := \partial_t + \mathbf{u} \cdot \nabla$ denotes the material derivative and We the Weber number. The stress tensor reads in nondimensional form $\mathbf{T}(\mathbf{u},p) = \frac{1}{\mathrm{Re}} D[\mathbf{u}] - pI$ with $\mathrm{Re} = \frac{UL\rho}{\mu}$ the (phase-dependent) Reynolds number. Here U, L are the characteristic velocity and length scale, respectively, ρ the density and μ the kinematic viscosity. The Weber number is given by $\mathrm{We} = \rho U^2 L/\gamma$. In the two-phase flow case, Re and Λ denote the phase-dependent Reynolds number and the nondimensional densities, respectively, defined by

$$\Lambda(x) \;:= \begin{cases} \Lambda_1 := 1 & \text{if}\, x \in \Omega_1 \\ \Lambda_2 := \dfrac{\rho_1}{\rho_2} & \text{if}\, x \in \Omega_2 \end{cases}, \qquad \mathrm{Re}(x) := \begin{cases} \mathrm{Re}_1 & \text{if}\, x \in \Omega_1 \\ \mathrm{Re}_2 & \text{if}\, x \in \Omega_2 \end{cases}.$$

In the case of one-phase flow, of course no such piecewise definition is necessary.

If the surface tension γ is not constant, then from Eq. (11) one sees that there is a tangential force which induces a flow field along the surface. This effect is called the Marangoni effect. If the surface tension $\gamma = \gamma(c)$ depends on some quantity c (like for instance temperature or the density of a surfactant) then, assuming a linear dependence (for simplicity) of γ on c, Eq. (12d) has to be replaced by

$$[\mathbf{T}(\mathbf{u},p)\mathbf{n}] = -\frac{1}{\mathrm{We}}\kappa\mathbf{n} + \mathrm{Ma}\nabla_S c \tag{13}$$

with the dimensionless Marangoni number Ma.

In case of mass transfer across the interface by evaporation/condensation, an additional heat equation for the temperature ϑ has to be solved in either phase.

Taking $a = 1$, $\mathbf{q}_d = 0$, $f_\Gamma = 0$ the mass conservation Eq. (6) yields

$$[\rho(\mathbf{u} \cdot \mathbf{n} - V_\Gamma)] = 0.$$

The mass flux j across the interface then can be written in either way

$$j = \rho_1(\mathbf{u}_1 \cdot \mathbf{n} - V_\Gamma) = \rho_2(\mathbf{u}_2 \cdot \mathbf{n} - V_\Gamma)$$

and then Eqs (12e) and (12f) have to be replaced by

$$[\mathbf{u}] = -\frac{\rho_1 - \rho_2}{\rho_1\rho_2} j\mathbf{n}, \qquad [\rho\mathbf{u}] = [\rho]V_\Gamma$$

(with obvious modifications if $[\rho] = 0$). It remains to determine j. From thermodynamical considerations one deduces

$$[-\lambda\partial_{\mathbf{n}}\vartheta] = jL,$$

where λ denotes the heat conductivity and L the latent heat of evaporation.

When over/underheating at the interface is small (which is quite often the case), one may assume the temperature to equal the pressure-dependent saturation temperature

$$\vartheta = \vartheta_s(p_v),$$

thus yielding a Dirichlet condition for the temperature at the free surface. The saturation temperature in turn is given by the Clausius–Clapeyron equation, see for instance (Iribarne and Godson, 2013).

Alternatively one can use the Hertz–Knudsen law derived from kinetic theory, see Eq. (14). Here, evaporation is modelled as a result of the deviation of the partial pressure on the interface to the saturation pressure, relating the local temperature ϑ_Γ on the interface and evaporation rate j (see for instance Kennard, 1938) by

$$j = \alpha \left(\frac{m}{2\pi k_B \vartheta_\Gamma} \right)^{1/2} (p_{sat}(\vartheta_\Gamma) - p_{td}), \tag{14}$$

where m denotes the molar mass, k_B the Boltzmann constant, $p_{sat}(\vartheta_\Gamma)$ the saturation pressure corresponding to the saturation temperature ϑ_Γ via the Clausius–Clapeyron equation and α the evaporation coefficient.

2.3 The Stefan problem with surface tension and kinetic undercooling

Similar to the interface condition in two-phase flow depicted above, the mean curvature κ of the phase boundary plays an important role in microscopic models of solid–liquid phase transitions. Let $\Omega \subset \mathbb{R}^n$ denote a bounded domain containing a pure substance. For $t \geq 0$ let $\Omega_s(t) \subset \Omega$ and $\Omega_l(t) \subset \Omega$ with $\Omega_s(t) \cap \Omega_l(t) = \emptyset$, $\bar{\Omega} = \bar{\Omega}_s(t) \cup \bar{\Omega}_l(t)$ be the parts of Ω containing the solid and liquid phases of the substance at time t. The moving free boundary between solid and liquid phases will be denoted by $\Sigma(t) := \bar{\Omega}_s(t) \cap \bar{\Omega}_l(t)$.

For pure substances and temperature-induced phase transition, conservation of energy is the basis of the usual modelling. This gives a heat equation in both solid and liquid phases for the nondimensional temperature ϑ. We assume that convection with velocity $\mathbf{u}$ takes place only in the liquid subdomain Ω_l (and thus $\mathbf{u}$ is set to 0 in the solid subdomain Ω_s). With heat flux proportional to $-\nabla\vartheta$, the generic balance equation (5) leads without sources to:

$$\frac{D}{Dt}\vartheta := \frac{\partial \vartheta}{\partial t} + \mathbf{u} \cdot \nabla \vartheta = \nabla \cdot (\lambda \nabla \vartheta) \ \text{ in } \Omega_i, \quad i = s, l \tag{15}$$

with the (phase-dependent) diffusion constant $\lambda = 1/\mathrm{Pr}\sqrt{\mathrm{Gr}}$, using the Prandtl and Grashof numbers for scaling. At the free phase boundary $\Sigma(t)$, its velocity and the normal fluxes fulfil an equation known as *Stefan condition*:

$$L\, V_\Sigma \;=\; -[\lambda \nabla \vartheta \cdot \mathbf{n}_\Sigma]_\Sigma \quad \text{on } \Sigma(t), \tag{16}$$

where V_Σ is the velocity of the free boundary Σ in direction of the normal $\mathbf{n}_\Sigma$, L is the nondimensional latent heat per unit volume in the liquid phase, and as above, $[\cdot]_\Sigma$ denotes the jump at the free boundary. The left-hand side of (16) describes the rate at which energy is set free by solidification at the free boundary. The right-hand side describes the heat transport into the solid and liquid phase.

Additionally, a thermodynamical condition is needed at the free boundary. Depending on the simplicity of the model, we ask that

$$\vartheta = \vartheta_M \text{ on } \Sigma, \text{or} \tag{17}$$

$$\vartheta = \vartheta_M\left(1 - \frac{\gamma}{L}\kappa_\Sigma\right) \text{ on } \Sigma, \text{ or} \tag{18}$$

$$\vartheta = \vartheta_M\left(1 - \frac{\gamma}{L}\kappa_\Sigma - \frac{\delta}{L}V_\Sigma\right) \text{ on } \Sigma. \tag{19}$$

Here, ϑ_M is the melting temperature of the substance, and γ the surface tension between solid and liquid phase (usually depending on the direction of the normal $\mathbf{n}_\Sigma$). κ_Σ is the mean curvature of the free boundary Σ (sum of the principal curvatures). The sign of κ_Σ is taken in the way that the mean curvature for a convex solid phase Ω_s is positive. The coefficient δ depends on the direction of $\mathbf{n}_\Sigma$ in the general case, too.

Eq. (17) is just the macroscopic equation which states that the temperature at the phase boundary is equal to the melting temperature of the material. The melting temperature is typically a material-dependent constant for pure substances, while for alloys, concentrations at the interface can jump and corresponding solidus and liquidus temperatures can be different and the interface conditions are more involved, see Xie (1992). Eqs (18) and (19) are known as the Gibbs–Thomson laws. If γ or δ depend on the direction of the normal $\mathbf{n}_\Sigma$ then the coefficients are called 'anisotropic', else 'isotropic'. The term $\gamma\kappa_\Sigma$ describes the influence of surface tension, which stabilizes the motion and makes dendritic growth possible. Eq. (18) describes a situation in local thermical equilibrium (see Langer (1980) for details), while the nonequilibrium situation with a moving interface is modelled with $\delta > 0$ (compare Gurtin, 1988). In the sequel, we will only consider the nonequilibrium case.

Scaling the temperature such that the melting temperature is $\vartheta_M = 0$ and the latent heat is $L = 1$, the Stefan condition (16) and Gibbs–Thomson law (19) simplify to

$$V_\Sigma = -[\lambda\nabla\vartheta \cdot \mathbf{n}_\Sigma]_\Sigma \text{ on } \Sigma, \tag{20}$$

$$\vartheta = -\gamma\kappa_\Sigma - \delta V_\Sigma \text{ on } \Sigma. \tag{21}$$

The latter equation describes an (anisotropic) mean curvature flow with driving force $-\vartheta$ for the interface $\Sigma(t)$:

$$\delta V_\Sigma = -\gamma\kappa_\Sigma - \vartheta \text{ on } \Sigma(t),\ t > 0, \quad \Sigma(0) = \Sigma_0. \tag{22}$$

In three dimensions, it may be better to use an anisotropic surface tension to model the anisotropy. This results in some anisotropic curvature to replace the mean curvature κ_Σ. Use of this anisotropic curvature leads to weak formulations, see for example Deckelnick and Dziuk (1999) and Dziuk (1999a). For anisotropic surface energy with a direction-dependent $\gamma(\mathbf{n}_\Sigma)$, the proper formulation is given by the gradient flow of the surface energy. If $E_\gamma(\Sigma) := \int_\Sigma \gamma(\mathbf{n}_\Sigma) \mathrm{d}\mathcal{H}^{n-1}$ denotes the anisotropic surface energy, then its gradient is given in the case of one-dimensional interfaces ($n = 2$) by the functional

$$\langle E'_\gamma, \varphi \rangle = \int_\Sigma (D\gamma(\mathbf{n}_\Sigma) \cdot \nabla id)(\nabla id \cdot \nabla \varphi^\perp) + \gamma(\mathbf{n}_\Sigma) \nabla id \cdot \nabla \varphi \, \mathrm{d}\mathcal{H}^1, \tag{23}$$

which replaces the weak formulation of the curvature in the Gibbs–Thomson law.

Nonsmooth surface energies can be addressed in the context of facetted interfaces with crystalline curvature, see Roosen and Taylor (1991). Crystalline surface energies can be approximated by smooth ones, see, e.g., Barrett et al. (2012) or Schmidt (1998).

In the sequel we want to work with the simplified case, where the physical constants in the solid and liquid phases are equal. This assumption leads not only to a pure academic example; there exist real material, like *Succinonitrile*, which obey to the same physical constants in the solid and liquid phase (see Glicksman et al., 1976, table II). On the other hand, most numerical methods can be adapted to the case of different constants in both phases.

Up to here, we only considered the conservation of energy and the phase transition driven by temperature. Typically, a nonconstant temperature leads under gravity to convection in the liquid melt. Using the Boussinesq approximation, the driving force is proportional to the temperature in direction of the unit gravity vector $\mathbf{e}_g$. Thus, if the model should include the coupled flow field $\mathbf{u}$ in a fixed container Ω, the above equations are coupled to the Navier–Stokes Eqs (12a)–(12c) with additional conditions at the moving phase boundary. If the densities in liquid and solid are assumed to be the same, then a no-slip condition is appropriate for the flow velocity $\mathbf{u}$ at the interface $\Sigma(t)$. Otherwise, a jump in density would lead to a boundary condition proportional to the interface velocity. We consider here the situation without density jumps.

Altogether, we now get the following set of equations:

Problem 2. Find the temperature ϑ, velocity $\mathbf{u}$, pressure p, and a moving interface Σ solving the equations

$$\frac{D}{Dt}\vartheta - \nabla \cdot (\lambda \nabla \vartheta) = 0 \quad \text{in}\, \Omega_s \cup \Omega_l, \tag{24a}$$

$$[\lambda \nabla \vartheta \cdot \mathbf{n}_\Sigma]_\Sigma + V_\Sigma = 0 \quad \text{on}\, \Sigma, \tag{24b}$$

$$\vartheta + \gamma \kappa_\Sigma + \delta V_\Sigma = 0 \quad \text{on } \Sigma, \tag{24c}$$

$$\frac{D}{Dt}\mathbf{u} - \nabla \cdot \mathbf{T}(\mathbf{u},p) = \vartheta \mathbf{e}_g \quad \text{in } \Omega_l, \tag{24d}$$

$$\nabla \cdot \mathbf{u} = 0 \quad \text{in } \Omega_l, \tag{24e}$$

$$\mathbf{u} = \mathbf{u}_D \quad \text{on } \Gamma_D, \tag{24f}$$

$$\mathbf{u} = 0 \quad \text{on } \Sigma \tag{24g}$$

together with boundary conditions and initial values for ϑ, $\mathbf{u}$, and $\Sigma/\Omega_l/\Omega_s$.

2.4 The Stefan problem with capillary melt surface

In many applications, melt pools have a free capillary surface. Thus, solid–liquid phase transitions are combined with free surface flow in the liquid phase. The corresponding model is thus a combination of Problems 1 and 2, with an additional consideration of the triple junction where the solid–liquid interface and the capillary free boundary meet. As the size of melt pools is typically much larger than the scale of dendrites, we can use the simpler interface condition (17), which assumes that the temperature at the interface is just the constant melting temperature of the material.

If the solid–liquid interface meets the possibly free capillary surface, a triple junction exists between solid, liquid, and outer (gaseous) phase. A corresponding microscopic model of such triple junctions would typically involve angle conditions based on the various interphase surface energies, compare for example Garcke et al. (1998). Instead of this, and due to the rather large scale of the melt pool, is sufficient here to use a Dirichlet condition for the capillary surface at such triple junctions: If $\Sigma(t) \cap \Gamma(t) \neq \emptyset$, this intersection is a triple junction whose motion is given by the evolution of Σ, and we set a no-slip condition for the flow at the solid–liquid interface $\Sigma(t)$, see (25i). Altogether, we arrive at the following combined model.

Problem 3. Find the temperature ϑ, velocity $\mathbf{u}$, pressure p, moving interface Σ, and capillary boundary Γ solving the equations

$$\frac{D}{Dt}\vartheta - \nabla \cdot (\lambda \nabla \vartheta) = 0 \quad \text{in } \Omega_s \cup \Omega_l, \tag{25a}$$

$$[\lambda \nabla \vartheta \cdot \mathbf{n}_\Sigma]_\Sigma + V_\Sigma = 0 \quad \text{on } \Sigma, \tag{25b}$$

$$\vartheta = 0 \quad \text{on } \Sigma, \tag{25c}$$

$$\frac{D}{Dt}\mathbf{u} - \nabla \cdot \mathbf{T}(\mathbf{u},p) = \vartheta \mathbf{e}_g \quad \text{in } \Omega_l, \tag{25d}$$

$$\nabla \cdot \mathbf{u} = 0 \quad \text{in } \Omega_l, \tag{25e}$$

$$\mathbf{u} = \mathbf{u}_D \quad \text{on } \Gamma_D, \tag{25f}$$

$$[\mathbf{T}(\mathbf{u},p)\mathbf{n}] = -\frac{1}{\mathrm{We}}\kappa\mathbf{n} \quad \text{on } \Gamma, \tag{25g}$$

$$V_\Gamma(t) = \mathbf{u} \cdot \mathbf{n} \quad \text{on } \Gamma, \tag{25h}$$

$$\mathbf{u} = 0 \quad \text{on } \Sigma, \tag{25i}$$

$$\Omega_{s,l}(0) = \Omega_{s,l,0}, \ \mathbf{u}(0,\cdot) = \mathbf{u}_0, \ \vartheta(0,\cdot) = \vartheta_0. \tag{25j}$$

Here, we assume that the topology of the liquid subdomain Ω_l, the capillary surface Γ, and the solid–liquid interface Σ do not change during the evolution, so there is for example no nucleation of liquid or total solidification of the material. In the next subsection we describe approaches which may include topological changes.

2.5 Phase variables and phase field models

Besides the sharp interface model described above, a different approach is based on the introduction of a phase variable $\chi(x, t)$ that indicates whether a point x belongs at time t to the solid or liquid phase (or maybe to an intermediate, mushy phase). Various models can be derived in the following sections.

2.5.1 Enthalpy formulation of the Stefan problem

The phase transition with interface condition (17) can also be modelled by energy conservation with energy density (or enthalpy) e which is connected to the temperature ϑ by a nonlinear relation $\vartheta = \beta(e)$, where in the simplest nondimensional case $\beta(s) = \min(s+1,0) + \max(s-1,0)$, see Fig. 2.

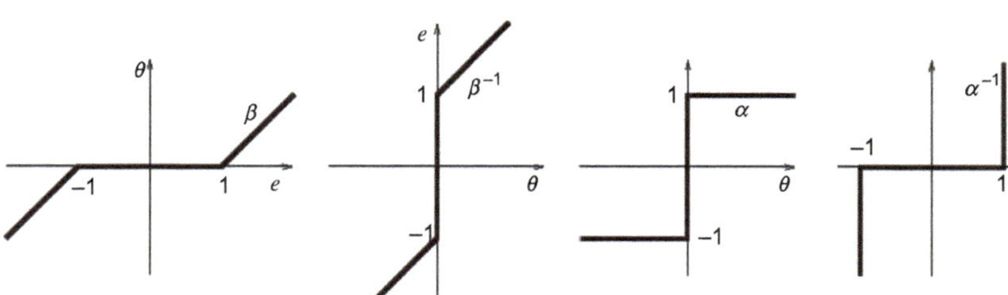

FIG. 2 Graphs of β, β^{-1}, α, and α^{-1}.

The idea behind is that the energy is affine equivalent to the temperature in both phases, but with an additional amount of latent heat $L = 2$ which is set free during phase change at temperature $\vartheta_M = 0$. The diffusive energy flux is still given by Fourier's law, so the general conservation principle without source terms leads to the equation

$$\frac{D}{Dt}e - \nabla \cdot (\lambda \nabla \vartheta) = 0, \quad \vartheta = \beta(e), \tag{26}$$

which holds for almost every time t in $H^{-1}(\Omega)$, with initial and boundary conditions. Classical solutions (with sharp interface Σ) of the Stefan problem (15), (16), (17) are also solutions of (26), but the latter permits also more general solutions, for example with nonsharp interfaces like mushy regions, or topology changes of the subdomains like nucleation of liquid or total solidification.

Using the notion of a *phase variable* χ with values in $[-1, 1]$, the energy can be written as

$$e = \vartheta + \chi$$

and thus the Stefan problem as

$$\frac{D}{Dt}(\vartheta + \chi) - \nabla \cdot (\lambda \nabla \vartheta) = 0, \quad \chi \in \alpha(\vartheta), \tag{27}$$

with the set-valued function

$$\alpha(\vartheta) = \begin{cases} -1 & \text{if } \vartheta < 0, \\ [-1, 1] & \text{if } \vartheta = 0, \\ 1 & \text{if } \vartheta > 0. \end{cases}$$

The latter inclusion can also be formulated as

$$\vartheta \in \alpha^{-1}(\chi) \tag{28}$$

with the set-valued inverse of α, the maximal monotone graph

$$\alpha^{-1}(\chi) = \begin{cases} (-\infty, 0] & \text{if } \chi = -1, \\ 0 & \text{if } \chi \in (-1, 1), \\ [0, \infty) & \text{if } \chi = 1, \end{cases} \tag{29}$$

which can be identified as the set-valued subdifferential $\partial\Psi$ of the convex generalized function Ψ

$$\Psi(s) = \begin{cases} 0 & \text{if } s \in [-1, 1], \\ \infty & \text{else}. \end{cases} \tag{30}$$

A generalization of this system, using nonconvex potentials instead of the simple Ψ above, can be used to develop diffuse interface models.

2.5.2 Allen–Cahn equation and phase field models

The idea of phase field models is the introduction of a regularized phase variable χ, which is (nearly) constant ($\pm$ 1, for example) inside the pure phases and varies smoothly in a narrow strip of width $O(\varepsilon)$ around the interface. Thus, the phase variable approximates the indicator function $-1+2\tilde{H}_{\Omega_l}$. Instead of prescribing a law of motion for the interface, like Eq. (22), a differential equation is formulated for the phase variable.

In order to push the values of χ to the constant values in the pure phases, a nonconvex potential with two minima at (or near) these values is included in the model. Phase field models can be formulated either with a double obstacle potential or a smooth double well potential, see Caginalp (1986). We want to restrict ourselves here to an obstacle formulation, as introduced by Blowey and Elliott (1993), using the potential Ψ from Eq. (30).

Given a small parameter $\varepsilon > 0$, the model describes the evolution of temperature ϑ and a *smooth* phase variable χ in a domain Ω and is a system of two parabolic equations:

$$\partial_t(\vartheta+\chi) - \nabla\cdot(\lambda\nabla\vartheta) = 0 \tag{31}$$

$$\varepsilon\,\partial_t\chi - \varepsilon\,\nabla\cdot(a(\nabla\chi)) + \frac{1}{\varepsilon}(\alpha^{-1}(\chi) - b\chi) \ni \eta\vartheta \tag{32}$$

in $(0, T)\times\Omega$ plus boundary and initial conditions. b and η are nonnegative coefficients. The function a may include anisotropic solidification parameters, and α^{-1} is the maximal monotone graph from Eq. (29). $\alpha^{-1}(\chi) - b\chi$ is the subdifferential of the double obstacle potential $\Psi(\chi) - (b/2)\chi^2$, with the effect that values of χ are in the interval $[-1, +1]$ and are pushed to ± 1, such that the phase variable χ is equal to -1 (solid) or $+1$ (liquid) everywhere but in a narrow transition region of width $O(\varepsilon)$. Eq. (32) is known as the *Allen–Cahn equation*. The second derivatives of the phase variable represent the curvature of isolevels in the transition region. So, for $\varepsilon \to 0$, the solution converges against a solution of the sharp interface model with a corresponding Gibbs–Thomson relation, see for example Paolini and Verdi (1992). For anisotropic surface tension and corresponding potentials, see Elliott and Schätzle (1995). Bellettini and Paolini (1996) get similar results in the context of Finsler geometry.

The introduction of a phase variable makes mathematical analysis of long-time solutions and numerical methods much easier. The sharp moving interface is approximated by the level set of a smooth function, thus even enabling geometrical 'catastrophes' like topology changes of the interface, nucleation, etc., without big problems in the mathematical formulation (by introducing the notion of viscosity solutions, e.g., see Evans et al. (1992) for example).

Both components of the solution, temperature, and phase variable, show a strongly different local behaviour:

- The phase variable is constant outside of a moving narrow strip of width $O(\varepsilon)$, where the phase transition occurs. Here, $|\nabla\chi| = O(\varepsilon^{-1})$.

- The temperature satisfies the heat equation outside this strip, thus it is smooth. Inside the moving strip, the gradient of temperature changes rapidly but is bounded independent of ε. In the limit $\varepsilon \searrow 0$, the Stefan condition (16) holds at the sharp interface.

3 Numerical methods for two-phase flow

This section gives an overview of the problems associated with the numerical treatment of the free boundary problems described in the previous section and corresponding solution approaches.

Typical situations for two-phase flows are sketched in Fig. 1. The difference between the two situations in Fig. 1A and B is whether the interface Γ is a closed surface or is in contact with the outer boundary (usually a solid wall).

3.1 Introduction

Most numerical methods for moving boundary problems can be categorized into two fundamentally distinct classes: *interface capturing* and *interface tracking* methods (note that authors may use different phrasings for these methods). In *interface capturing* methods, the interface is described implicitly by an additional scalar function. Representatives of this method are VOF (see Hirt and Nichols, 1981; Scardovelli and Zaleski, 1999), phase field approaches (see Jacqmin, 1999; Eck et al., 2009; Abels et al., 2012) and level set methods (see Groß and Reusken, 2011; Osher and Sethian, 1988; Sethian and Smereka, 2003; Sussman et al., 1994). Level set is the by far most common approach in two-phase flow. One of the reasons is that it computes the signed distance function to the interface, which has several appealing features. For a thorough presentation of finite element level set methods for multiphase flow see for instance the monograph by Groß and Reusken (2011).

Extensions of these methods in order to incorporate evaporation/condensation are treated for instance in Son and Dhir (1998), Yap et al. (2005), Gibou et al. (2007), and Tanguy et al. (2007) for the level set method, in Welch and Wilson (2000) and Schlottke and Weigand (2008) for VOF, and for the phase field method in Pismen and Pomeau (2000) and Jamet et al. (2001), and in Bänsch et al. (2015) for interface tracking.

In contrast to interface capturing methods, in *interface tracking* the phase boundary is tracked explicitly. We distinguish between two main classes: *mesh moving* and *front tracking*.

In view of the above discussion, interface capturing compared to tracking methods come with their specific pros and cons, which we list below.

- PROS: Interface capturing provides a very general tool for solving two-phase problem. There are no problems with changes of the topology. Moreover, level set seems to become *the* standard tool for multiphase problems.

- CONS: Computing curvature accurately requires some theoretical as well as computational effort, see for instance Groß and Reusken (2007b). Handling of discontinuities causes problems. In particular, resolving the pressure jump across the interface requires substantial effort. Otherwise this could lead to serious *spurious velocities*, see Remark 1. Thus the use of XFEM is necessary at least in a simplified fashion, see Barrett et al. (2013, 2015). Also the conservation of the volume of the individual phases on the discrete level may be an issue.

On the other hand, for interface tracking/mesh moving the picture is more or less complementary.

- CONS: Interface tracking/mesh moving is usually limited to moderate interface deformations (≈50%). Dealing with topological changes is possible in general, but very cumbersome and not really practical.
- PROS: However, there are also advantages for this class of methods. From a computational point of view, mesh moving is the simplest and cheapest, at the same time most accurate method, since the interface coincides with element boundaries. This also implies that there exist no problems computing geometric quantities like the curvature to very high accuracy. Also the resolution of discontinuities poses no problem.

In *mesh moving* the computational mesh deforms according to the interface motion, see for example Hirt et al. (1974), Hughes et al. (1981), Donea et al. (1982), and Bänsch (2001).

More precisely, let $\Omega_1(t)$, $\Omega_2(t)$ denote for instance two phases that are separated by the sharp and smooth interface $\Gamma(t)$. Computationally, we have $\Omega_{1,h}(t_k)$, $\Omega_{2,h}(t_k)$ at discrete time instants t_k. Now mesh moving means that the two computational phases are separated by $\Gamma_h(t_k)$, which is the boundary of the triangulations both of $\Omega_{1,h}(t_k)$ and $\Omega_{2,h}(t_k)$ that is $\Gamma_h(t_k) = \partial\Omega_{1,h}(t_k) \cap \partial\Omega_{2,h}(t_k)$, see Fig. 3. The joint triangulation is a conforming triangulation of the overall domain Ω. The discrete interface $\Gamma_h(t_k)$ is thus a conforming $(n-1)$-dimensional triangulation approximating the exact interface.

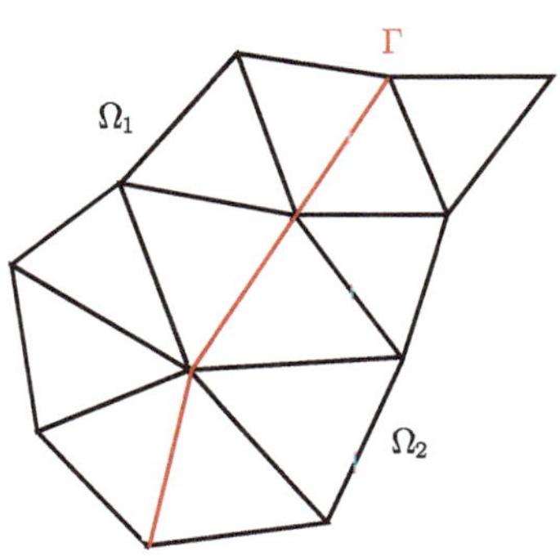

FIG. 3 Mesh moving method.

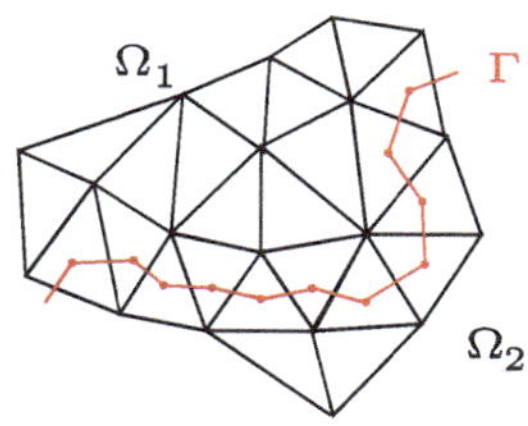

FIG. 4 Front-tracking method.

While usually limited to situations in which only moderate deformations and no topological changes of the interface occur, the explicit representation of the interface allows for a very accurate treatment of surface tension.

In *front tracking* the *background mesh* triangulating Ω is fixed. The computational interface $\Gamma_h(t_k)$ is an independent $(n-1)$-dimensional triangulation that moves through Ω according to its evolution law, see Fig. 4. The "exchange of information" between bulk mesh and interface is realized by some interface integrals.

Compared to mesh moving, front tracking allows for much larger deformations than mesh moving (think of a rising bubble in a surrounding liquid). On the other hand, there is a loss of accuracy compared to mesh moving, especially in handling discontinuities.

Maybe the first work on front tracking is the seminal article (Unverdi and Tryggvason, 1992), see also many follow-up papers by these authors.

In a sense, phase field methods are different than the other methods mentioned in that they regularize the problem and make all the variables smooth, compare Section 2.5.2. This, of course has several advantages, both from an analytical point of view as well as for numerics. Moreover, interface conditions in phase field methods are incorporated via an energy functional and thus allow for a relatively easy incorporation of additional physical phenomena.

The drawback of phase field methods is the demand to sufficiently resolve the smeared interface layer of order ε, the regularization parameter. Depending on the problem or application at hand this may result in prohibitively fine meshes. For certain applications, see Aland et al. (2013) for an example, this could mean orders of magnitude more degrees of freedom (DOF) than for instance with a sharp interface/mesh moving method. Nevertheless, if for the problem to be solved this is not the key issue, then phase field methods can be very strong numerical tools.

Since there is no explicit free boundary and all variables are smooth, the numerical approximation of a phase field problem consists in approximating a system of (second or fourth order in space) PDEs. Time discretization can take advantage of the convex–concave structure of the energy functional: convex parts are solved implicitly while the nasty concave parts are treated explicitly. For some more details, see Section 4.3.

Remark 1 (Spurious velocities).
In incompressible flow problems all methods except for mesh moving have difficulties with discontinuous pressures. To understand this problem, we consider a *static* situation, $\mathbf{u} \equiv 0$, of a two-phase flow problem, where Γ is a sphere, see Fig. 5. In this case, the Navier–Stokes equations and the interface condition reduce to

$$\nabla p = 0 \quad \text{in } \Omega_1, \Omega_2, \qquad [p] := p_2 - p_1 = \frac{1}{\text{We}}\kappa = \frac{1}{\text{We}}\frac{n-1}{R} \quad \text{on } \Gamma,$$

R the radius of the sphere. This means the pressure is piecewise constant but experiences a (possibly huge) jump.

If the numerical method does not allow for such a pressure jump (for instance, because the interface intersects an element), the numerical pressure p_h will fulfil $\nabla p_h \neq 0$ in Ω_1, Ω_2 in general. Then in turn, $\mathbf{u}_h \neq 0$, see Fig. 6. This effect has to be cured by some means, for instance XFEM, see below.

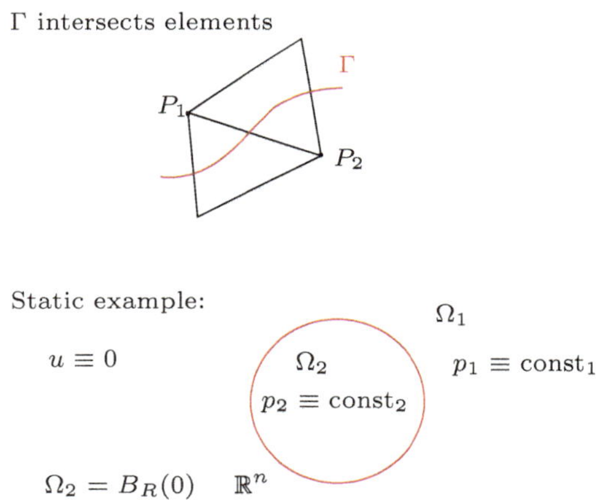

FIG. 5 Static example.

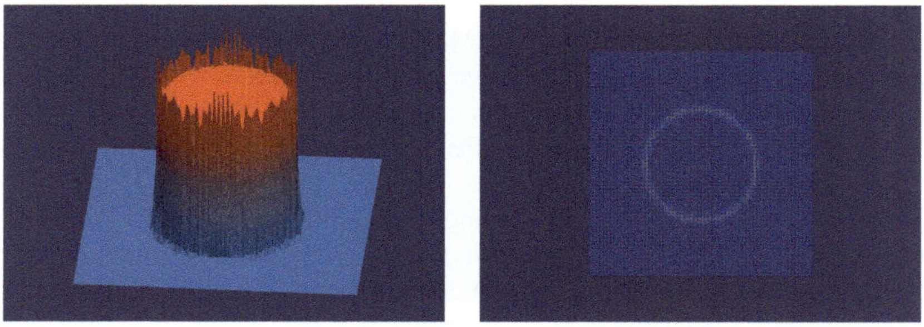

FIG. 6 Static example as above computed with the Taylor–Hood element (piecewise quadratics for the velocity, piecewise linears for the pressure) without XFEM. *Left*: pressure; *right*: magnitude of spurious velocity.

3.2 Numerics for multiphase flow: Some general considerations and problems

From the discussion above it follows that in terms of mathematical analysis, but even more importantly for the numerical methods, the most fundamental design decision that has to be made when solving a multiphase flow problem is how to represent the interface. The options introduced above are interface capturing (level set, VOF, phase field) versus interface tracking (mesh-moving, front-tracking) methods. Both of them have the pros and cons as discussed in more detail above. Generally, one is faced with the following difficulties:

- Discontinuities at the interface: Typically, quantities of interest like pressure, concentration, or gradients of velocity jump across the interface. If the underlying triangulation is not aligned with the interface, these jumps require certain measures to yield reasonable results. For instance, as seen above, the pressure may encounter a huge jump (depending on the surface tension), which can lead to spurious velocities and this in turn can spoil the computational results completely.
- Geometry (curvature, volume conservation): An accurate capturing of the curvature is indispensable for reliable computational results. However, curvature is a second order operator on the interface and thus requires special attention. Moreover, most of the computational methods are not exactly volume conservative with respect to the individual phases. For long term computations this can lead to severe imbalance of the volume and to disastrous numerical results (for instance a droplet vanishing after a certain time).
- Marangoni flow: Further physical mechanisms like Marangoni forces usually depend in a complex way on quantities like temperature, concentration of some surfactants, etc. If these quantities are not sufficiently resolved, computations may give completely wrong solutions. For instance, diffusion coefficients for surfactants are typically very small and therefore the concentration field is very hard to compute for in an accurate way. The corresponding nondimensional number is the Schmidt number Sc, which is typically of the order $1e - 5$ or smaller. There is no obvious way how to overcome this problem with nowadays methods.
- Stable time discretization: Quite often the computation of the flow field and the evolution of the interface are decoupled for convenience. This, however, leads to an overall convergence in time that can be best of first order only. Even more importantly, for stability reasons in this case one is faced with a capillary CFL condition of the form

  $$\Delta t \lesssim h^{3/2}\sqrt{\mathrm{We}}$$

  between the time step and the (local) mesh size with We the Weber number, which is inverse proportional to the coefficient of surface tension (see also Section 3.3.4).

- Finally, after having devised an appropriate discretization one has to solve the arising system of nonlinear equations in each time step. Since the systems are typically large and for accuracy reasons usually many time steps have to be computed, efficient solution techniques are mandatory. This is in conflict to the previous item: Fully implicit strategies are preferable from the point of view of stability, but of course are much harder to solve compared to decoupled subproblems.

Despite a long history of numerical methods for multiphase flow and their relative maturity, there are still persisting problems like:

- In order to avoid spurious velocities in the context of level set methods, the pressure space has to be enriched by 'aligned', truncated basis functions leading to the method of *extended finite elements* (XFEM). Implementation of XFEM is not trivial and there are still open theoretical as well as practical questions.
- In interface capturing one usually has a "conflict" between accurate volume conservation versus accuracy. This means that measures to ensure volume conservation may disturb accuracy of the solution.
- Higher order methods in time are advisable as first order methods are not only ineffective but also usually rather dissipative. However, there is only little work in this direction, especially for interface capturing.
- Time-dependent meshes: It seems natural to use adaptivity close to the interface. Also XFEM requires the adaptation of the pressure space in each time step. However, this can lead to serious pressure oscillations due to the *discrete solenoidal condition*, see Besier and Wollner (2012) and Brenner et al. (2014). This widely overlooked problem is still open.

Upshot: The appropriate choice of which method to use depends very much on the specific problem to be solved!

In what follows we will address the most prominent representatives of each method, i.e., mesh moving for interface tracking and the level set method for interface capturing in some detail.

3.3 Mesh moving

As already outlined, in mesh moving the discrete interface is explicitly tracked by a (discrete) evolution law, see Fig. 7 for a sketch. The shift of the position of the DOFs at the interface in each time step is (in the simplest case) a first order in time discretization of the kinematic boundary condition Eq. (12f). Higher order in time versions are possible.

To avoid strong mesh distortion, this movement has to be extended and smoothed out to the bulk meshes for either phase $\Omega_{1,h}$, $\Omega_{2,h}$ by an extension operator E. Hereby the mesh topology is preserved. As extension operator one can use the discrete Laplace operator or more sophisticated operators like linear or nonlinear elasticity, see Section 3.3.3.

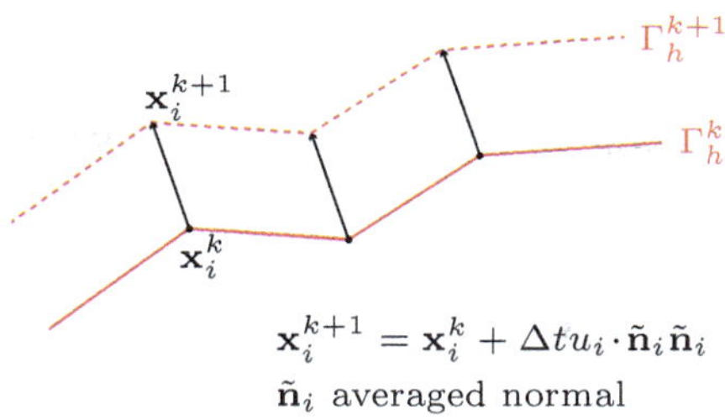

FIG. 7 Mesh moving method: defining the new position of the interface.

Since the location of the DOFs (for instance the vertices) carries the information of finite element functions, this information is also moved in turn. To compensate for this movement of information the corresponding *mesh velocity* has to be added to the transport equations. This methodology is called *arbitrary Lagrangian Eulerian coordinates* (ALE), see for instance Hughes et al. (1981), Donea et al. (1982), and Bänsch (2001). We will explain this concept next.

3.3.1 ALE formulation of the model problem

Consider a reference domain $\hat{\Omega}_1 \cup \hat{\Omega}_2 = \hat{\Omega} \subset \mathbb{R}^n$ and the corresponding reference mapping $\boldsymbol{\xi} : \mathbb{R} \times \mathbb{R}^n \to \mathbb{R}^n$ such that $\boldsymbol{\xi}(t, \hat{\Omega}_i) = \Omega_i(t)$ and $\boldsymbol{\xi}(t, \hat{\Gamma}) = \Gamma(t)$ for the (fixed) reference interface $\hat{\Gamma}$, ruling out the occurrence of topological changes. We assume $\boldsymbol{\xi}$ to be sufficiently smooth and to be a homeomorphism for each time instant t. The domain velocity is defined by

$$\mathbf{w}(t, \cdot) : \hat{\Omega} \to \mathbb{R}^n, \quad \mathbf{w}(t, \cdot) = \partial_t \boldsymbol{\xi}(t, \boldsymbol{\xi}(t, \cdot)^{-1}). \tag{33}$$

For any sufficiently smooth function $F : [0, T] \times \mathbb{R}^n \to \mathbb{R}$ the ALE time derivative of F is defined as

$$\hat{\partial}_t F(t, \mathbf{x}) := \frac{\mathrm{d}}{\mathrm{d}t} F(t, \boldsymbol{\xi}(t, \hat{\mathbf{x}})) = \partial_t F(t, \mathbf{x}) + \mathbf{w}(t, \mathbf{x}) \cdot \nabla F(t, \mathbf{x}) \tag{34}$$

for $\mathbf{x} = \boldsymbol{\xi}(t, \hat{\mathbf{x}})$, $\hat{\mathbf{x}} \in \hat{\Omega}$.

In what follows, occurring time derivatives in the balance equations are replaced by the ALE time derivative Eq. (34). We will comment on the construction of the mapping $\boldsymbol{\xi}$ in Section 3.3.3.

3.3.2 Variational treatment of interface forces

One of the crucial points of both space and time discretization is the treatment of the capillary boundary condition Eq. (12d). This condition can be integrated into the weak formulation of the momentum equation (Eq. 12a) rendering an explicit evaluation of the curvature of Γ unnecessary.

In order to simplify notation, we restrict the presentation to the isothermal case and assume Dirichlet boundary conditions for the velocity on $\partial\Omega \setminus \Gamma$.

The variational formulation of the problem then reads: find $\mathbf{u} \in X$ such that

$$\int_\Omega (\hat{\partial}_t \mathbf{u} + (\mathbf{u} - \mathbf{w}) \cdot \nabla \mathbf{u} - \nabla \cdot \mathbf{T}) \cdot \boldsymbol{\varphi} \, \mathrm{d}\mathbf{x} = \int_\Omega \mathbf{f} \cdot \boldsymbol{\varphi} \, \mathrm{d}\mathbf{x} \qquad \forall \boldsymbol{\varphi} \in X, \tag{35}$$

$$\int_\Omega (\nabla \cdot \mathbf{u}) \psi \, \mathrm{d}\mathbf{x} = 0 \qquad \forall \psi \in Y. \tag{36}$$

Hereby, X and Y denote suitable function spaces for velocity and pressure. Integrating by parts the stress term yields

$$\int_\Omega -(\nabla \cdot \mathbf{T}) \cdot \boldsymbol{\varphi} \, \mathrm{d}\mathbf{x} = \int_\Omega \mathbf{T} : (\nabla \boldsymbol{\varphi}) \, \mathrm{d}\mathbf{x} - \int_\Gamma [(\mathbf{T}\boldsymbol{\varphi}) \cdot \mathbf{n}] \, \mathrm{d}\mathcal{H}^{n-1}, \tag{37}$$

since $\boldsymbol{\varphi}$ vanishes on the Dirichlet boundary. Assume for the moment that Γ is a closed surface without boundary. Since the stress tensor $\mathbf{T}$ is symmetric, inserting boundary condition Eq. (12d) into Eq. (37) and using an integration by parts formula on manifolds (see for instance Buscaglia and Ausas, 2011) yields

$$-\int_\Gamma (\mathbf{T}\mathbf{n}) \cdot \boldsymbol{\varphi} \, \mathrm{d}\mathcal{H}^{n-1} = \frac{1}{\mathrm{We}} \int_\Gamma (\kappa \mathbf{n}) \cdot \boldsymbol{\varphi} \, \mathrm{d}\mathcal{H}^{n-1} = \frac{1}{\mathrm{We}} \int_\Gamma \nabla_S \cdot \boldsymbol{\varphi} \, \mathrm{d}\mathcal{H}^{n-1}. \tag{38}$$

Here, $\nabla_S \cdot$ denotes the tangential divergence.

Alternatively, following an idea by Dziuk (1991), one can rewrite the above equation using the identity

$$\kappa \mathbf{n} = -\Delta_S \boldsymbol{\xi} = -\nabla_S \cdot \nabla_S \boldsymbol{\xi} \quad \text{on} \quad \Gamma, \tag{39}$$

where $\Delta_S = \nabla_S \cdot \nabla_S$ denotes the Laplace–Beltrami operator. With again the help of integration by parts on manifolds one computes

$$\int_\Gamma \kappa \boldsymbol{\varphi} \cdot \mathbf{n} \, \mathrm{d}\mathcal{H}^{n-1} = -\int_\Gamma (\nabla_S \cdot \nabla_S \boldsymbol{\xi}) \cdot \boldsymbol{\varphi} \, \mathrm{d}\mathcal{H}^{n-1} = \int_\Gamma \nabla_S \boldsymbol{\xi} : \nabla_S \boldsymbol{\varphi} \, \mathrm{d}\mathcal{H}^{n-1}. \tag{40}$$

Accordingly, we define a linear form representing the surface tension based on either of the above formulae, i.e., define

$$e(\boldsymbol{\varphi}) := \frac{1}{\mathrm{We}} \int_\Gamma \nabla_S \cdot \boldsymbol{\varphi} \, \mathrm{d}\mathcal{H}^{n-1} \tag{41}$$

or

$$e(\boldsymbol{\varphi}) := \frac{1}{\mathrm{We}} \int_\Gamma \nabla_S \boldsymbol{\xi} : \nabla_S \boldsymbol{\varphi} \, \mathrm{d}\mathcal{H}^{n-1} \tag{42}$$

for $\boldsymbol{\varphi} \in H^1(\Gamma)$. The advantage of the variational strategy arises in the right-hand side of Eqs (38) or (40) yielding a term which is well defined on polygonally bounded domains for which a point-wise evaluation of the curvature does not make any sense. Note that for a (globally continuous) finite element function $\boldsymbol{\varphi}_h$ it holds $\boldsymbol{\varphi}_h|_\Gamma \in H^1(\Gamma)$.

In case Γ has a boundary corresponding to the situation depicted in Fig. 1B, integration by parts on the manifold yields an additional $(n-2)$-dimensional term:

$$\int_\Gamma \kappa\boldsymbol{\varphi}\cdot\mathbf{n}\,\mathrm{d}\,\mathcal{H}^{n-1} = \int_\Gamma (\nabla_S\boldsymbol{\xi}):(\nabla_S\boldsymbol{\varphi})\,\mathrm{d}\,\mathcal{H}^{n-1} + \int_{\partial\Gamma} \mathbf{n}_S\cdot\boldsymbol{\varphi}\,\mathrm{d}\,\mathcal{H}^{n-2}.$$

Here, $\partial\Gamma$ denotes the *contact line* between interface and wall, $\mathbf{n}_S$ denotes the outward pointing normal of the interface, i.e., the unit vector of the interface's tangent plane perpendicular to the contact line. With $\boldsymbol{\tau}$ denoting the unit vector in the wall's tangent plane which is perpendicular to the interface, one defines the *contact angle* δ by

$$\mathbf{n}_S\cdot\boldsymbol{\varphi} = \cos(\delta)\,\boldsymbol{\tau}\cdot\boldsymbol{\varphi}. \tag{43}$$

In our applications, δ will be a fixed, static contact angle, but dynamical contact angle conditions may also be treated in the same way. See for instance Gerstmann (2004) for an assessment of models of dynamical contact angles. Furthermore, in order to allow for a movement of the interface, a *slip boundary condition* for the velocity in a small vicinity around the contact line is prescribed.

3.3.3 Mesh deformation

In each time step, the solution of the flow problem defines the new position of the interface through a discretization of boundary condition Eq. (12f) (here given for an Euler discretization): Once $\mathbf{u}^{(k+1)}$ is computed, we have (abusing notations a bit)

$$V_\Gamma^{(k+1)} = \mathbf{u}^{(k+1)}\cdot\mathbf{n}, \quad \text{and} \tag{44}$$

$$\Gamma^{(k+1)} = \Gamma^{(k)} + \Delta t\, V_\Gamma^{(k+1)}\mathbf{n} \tag{45}$$

defines the new position of the interface.

For a discretization with Lagrange elements this means that the nodes located on the discrete interface are shifted using Eq. (45) to approximate $\Gamma^{(k+1)}$. In order to prevent the computational mesh from degenerating, this movement has to be extended to the whole computational domain. This can be done in different ways, see for instance Tezduyar and Benney (2003) and Wick (2011) for studies on different extension techniques for problems with rather large deformations. For not too severe distortions one can use one of the following two options:

- *Harmonic extension*: for the coordinates $\boldsymbol{\xi}$ of the DOFs of the computational grid, a Laplace problem

$$-\Delta\boldsymbol{\xi} = 0 \tag{46}$$

is solved and Dirichlet values for the nodes on the interface, i.e., $\boldsymbol{\xi} = \mathbf{x}$ for $\mathbf{x} \in \Gamma^{(k+1)}$, are prescribed. Replacing the Laplace operator by the operator of linear elasticity is superior in many cases, see for instance Matthies (2002).

- An extension operator based on a *nonlinear variational approach*, derived from principles of nonlinear elasticity theory, was introduced and analyzed in Rumpf (1996). The main idea is to find minimizers of a certain class of functionals, which measure the quality of mesh deformation.

While the first option is preferable, when deformations of the interface and the mesh can be expected to be small, some of our applications require the treatment of large deformations. In this case, the nonlinear variational approach is used.

3.3.4 Moving mesh method for the two-phase problem: Overall method

Now we have all the pieces together to define the generic form of the mesh moving method. It is given here for the backward Euler method for simplicity, higher order methods are strongly recommended though.

Algorithm 1 (Mesh moving for two-phase flow).
In each discrete time step $(t_k, t_{k+1}]$ do

1. Find $\mathbf{u}_h^{k+1}$, p_h^{k+1} fulfilling

$$\left(\Lambda \frac{\mathbf{u}_h^{k+1} - \tilde{\mathbf{u}}_h^k}{\Delta t}, \boldsymbol{\varphi}_h\right)_{\Omega_h^k} + \left(\left(\mathbf{u}_h^{k+1} - \mathbf{w}_h^k\right) \cdot \nabla \mathbf{u}_h^{k+1}\right), \boldsymbol{\varphi}_h\Big)_{\Omega_h^k} - (p, \operatorname{div} \boldsymbol{\varphi}_h)_{\Omega_h^k}$$
$$+ \left(\frac{\mathrm{Re}}{2} D(\mathbf{u}_h^{k+1}), D(\boldsymbol{\varphi}_h)\right)_{\Omega_h^k} + e_*(\boldsymbol{\varphi}_h) = 0 \quad \forall \boldsymbol{\varphi}_h \in X_h^k$$
$$(\operatorname{div} \mathbf{u}_h^{k+1}, q_h) = 0 \quad \forall q_h \in Y_h^k$$

2. Update the free boundary

$$X^{k+1} := \mathrm{Id}_{\Gamma_h^k} + \Delta t \left(\mathbf{u}_{h\,|\Gamma_h^k}^{k+1} \cdot \tilde{\mathbf{n}}\right) \tilde{\mathbf{n}},$$

where $\tilde{\mathbf{n}}$ is an averaged normal.

3. Extend X^{k+1} to the interior of Ω: $\Upsilon^{k+1} := E(X^{k+1})$ and

$$\Omega_h^{k+1} = \Upsilon^{k+1}(\Omega_h^k).$$

The DOFs of the computational grid are moved accordingly.

4. Compute the grid velocity $\mathbf{w}_h^{k+1}$ by

$$\mathbf{w}_h^{k+1} := \frac{\Upsilon^{k+1} - \mathrm{Id}}{\Delta t}.$$

The form $e_*(\cdot)$ is one of either definitions Eq. (41) or Eq. (42) evaluated at Γ_h^*, for instance

$$e_*(\boldsymbol{\varphi}) = \frac{1}{\mathrm{We}} \int_{\Gamma_h^*} \nabla_S \cdot \boldsymbol{\varphi}_h \, \mathrm{d}\mathcal{H}^{n-1},$$

where $\Gamma_h{}^*$ either denotes Γ_h^k or Γ_h^{k+1}, see the discussion below.

Remark 2.

- In the above algorithm $\tilde{\mathbf{u}}_h^k$ is the old velocity $\mathbf{u}_h^k$ on Ω_h^{k-1} but lifted to Ω_h^k:

$$\tilde{\mathbf{u}}_h^k = \mathbf{u}_h^k \circ \Upsilon^k.$$

  Numerically there is nothing to do, since we work with the same DOFs in the old and new mesh.
- Y_h^k is defined in such a way that pressure nodes at the interface are virtually *doubled*, i.e., there exist two copies of a pressure node at the interface, one belonging to $\Omega_{1,h}$ the other one to $\Omega_{2,h}$. This allows the pressure to exhibit a jump across the interface and therefore no spurious velocities occur.
- For higher order in time discretization see Weller (2008), Weller (2015) and Weller and Bänsch (2017).

It remains to determine '$*$' in the form e_*. If we take $* = k$, then all geometry terms in the algorithm are treated explicitly. Thus computation of the flow field and the geometry is completely decoupled. However, as mentioned above, in this case one faces a severe CFL-like condition for stability

$$\Delta t \lesssim h^{3/2}\sqrt{\mathrm{We}},$$

see for instance Bänsch (2001) or Brackbill et al. (1992). The choice $* = k + 1$ results in an unconditionally stable scheme, but now the geometry and the flow field are coupled (we need Γ_h^{k+1} to compute $\mathbf{u}_h^{k+1}$).

There is an alternative, decoupling geometry and flow field, but being unconditionally stable at the same time. Consider the semiimplicit integral

$$e_{semi}(\boldsymbol{\varphi}_h) = \frac{1}{\mathrm{We}} \int_{\Gamma_h^k} \nabla_S \boldsymbol{\xi}^{k+1} : \nabla_S \boldsymbol{\varphi}_h \, \mathrm{d}\mathcal{H}^{n-1},$$

where $\boldsymbol{\xi}^{k+1}$ is given by

$$\boldsymbol{\xi}^{k+1}(z) := z + \Delta t \mathbf{u}_h^{k+1}(z) \quad \text{for } z \in \Gamma_h^k.$$

This leads to

$$e_{semi}(\boldsymbol{\varphi}_h) = \frac{1}{\mathrm{We}} \int_{\Gamma_h^k} \nabla_S(\mathrm{Id} + \Delta t \mathbf{u}_h^{k+1}) : \nabla_S \boldsymbol{\varphi}_h \, \mathrm{d}\mathcal{H}^{n-1}.$$

This scheme is semiimplicit in the curvature, since in the integrand the new position of the interface is used, while the integration is performed over the current interface. There is an implicit contribution to the computation of $\mathbf{u}_h^{k+1}$, however, the geometry is still explicit. Unconditional stability of the scheme (in the context of a space-time finite element discretization) was proved in Bänsch (2001).

The semiimplicit treatment of the curvature only yields additional contributions in the system matrices and right-hand sides for the computation of $\mathbf{u}^{k+1}$ (which, due to the time dependence of the integration domain, have to be re-assembled in each time step). The main benefit lies in the decoupling of the geometry update and computation of the fluid velocity, enabling an efficient solution strategy.

3.4 Level set method for two-phase flow

In this section we briefly address the level set method and the subproblems that have to be solved in order to get an effective method. For a detailed presentation see the monograph (Groß and Reusken, 2011). Let $\Omega_1(t)$, $\Omega_2(t)$ denote the two phases separated by the sharp interface $\Gamma(t)$. In the level set method Γ is given as the zero level of the level set function ϕ:

$$\Gamma(t) := \{\mathbf{x} \in \Omega \mid \phi(t,\mathbf{x}) = 0\}.$$

Additionally we assume

$$\begin{aligned} \phi(t,\cdot) &< 0 \quad \text{in } \Omega_1(t), \\ \phi(t,\cdot) &> 0 \quad \text{in } \Omega_2(t). \end{aligned}$$

We want to find an evolution law for ϕ. Assume $\mathbf{u}$ to be smooth. We exploit the fact that Γ is a *material* boundary (in case there is no mass transfer across Γ), i.e., each phase is transported by the flow field $\mathbf{u}$. We construct ϕ in such a way that it is constant on *characteristic curves* $\xi = \xi(t,\overline{\mathbf{x}})$, that is:

$$\begin{aligned} \frac{d}{dt}\xi(t) &= \mathbf{u}(t,\xi(t)) \quad \text{for } t > 0, \\ \xi(0) &= \overline{\mathbf{x}} \in \Omega. \end{aligned}$$

On the characteristic curves the level set function is constant:

$$\frac{d}{dt}\phi(t,\xi(t,\overline{\mathbf{x}})) = 0.$$

Using the chain rule and the definition of the characteristic curve, this is equivalent to the following transport equation:

$$\partial_t\phi(t,\mathbf{x}) + \mathbf{u}(t,\mathbf{x}) \cdot \nabla\phi(t,\mathbf{x}) = 0 \tag{47}$$

for $\mathbf{x} = \xi(t,\overline{\mathbf{x}})$. If the characteristic curves cover all of Ω, then the above equation holds for all $\mathbf{x} \in \Omega$.

To close the description of ϕ, initial as well as boundary conditions on the inflow part of the boundary are needed:

$$\phi(0,\cdot)=\phi_0 \quad \text{in } \Omega,$$
$$\phi=\phi_D \quad \text{on } \partial_{in}\Omega_T$$

with ϕ_0, ϕ_D given and $\partial_{in}\Omega_T$ the inflow boundary:

$$\partial_{in}\Omega_T := \{(t,\mathbf{x}) \in [0,T] \times \partial\Omega \mid \mathbf{u}(t,\mathbf{x}) \cdot \mathbf{n}(\mathbf{x}) < 0\}.$$

It turns out that computations tend to break down, when $|\nabla\phi|$ becomes too large or close to 0. Thus we ideally want ϕ to fulfil

$$|\nabla\phi| = 1$$

at least close to the interface Γ. Interestingly, this leads to
Proposition 1. *Let* $\mathbf{u}$, ϕ *and* Γ *be smooth and* $\Gamma(t) = \{\mathbf{x} \in \Omega \mid \phi(t,\mathbf{x}) = 0\}$ *be the zero level set of* ϕ. *Then we have*

$$|\nabla\phi(t,\mathbf{x})| = 1 \ \textit{for all } t,\mathbf{x} \ \Leftrightarrow \ \phi \textit{ is the signed distance function to } \Gamma,$$

i.e., $|\phi(\mathbf{x})| = dist(\mathbf{x}, \Gamma)$ *for* $\mathbf{x}$ *close to* Γ *and where the sign of* ϕ *indicates whether* $\mathbf{x} \in \Omega_1$ *or* $\mathbf{x} \in \Omega_2$.

Even if ϕ is initially a signed distance function, this property is lost during time evolution. Thus to regain this property, one usually incorporates a step in the algorithm that tries to make ϕ a distance function again each time step or every few time steps. This step is called "redistancing" and may take a significant amount of computing time. There are many different methods for redistancing, see for instance Sethian (1996) or Groß and Reusken (2011).

Redistancing usually leads to another problem on the discrete level: The level set function is changed and then in turn also the numerical interface changes slightly, which may yield an error in the volume of either phase. This effect can accumulate over time to an unacceptable amount. Thus methods for correcting the volume have to be used, see for example Ausas et al. (2011) and Jahn and Klock (2016).

Next we will describe how to incorporate the surface force

$$-\frac{1}{\mathrm{We}}\int_\Gamma \kappa\boldsymbol{\varphi}\cdot\mathbf{n}\,\mathrm{d}\mathcal{H}^{n-1}$$

for test functions $\boldsymbol{\varphi}$ in the velocity space.

As for interface tracking one can use (assuming for the moment that Γ is closed) one of the options Eq. (41) or Eq. (42),

$$e(\boldsymbol{\varphi}) = \frac{1}{\mathrm{We}}\int_\Gamma \nabla_S\cdot\boldsymbol{\varphi}\,\mathrm{d}\mathcal{H}^{n-1}$$

or

$$e(\boldsymbol{\varphi}) = \frac{1}{\mathrm{We}} \int_{\Gamma} \nabla_{S} \boldsymbol{\xi} : \nabla_{S} \boldsymbol{\varphi}_h \, \mathrm{d}\mathcal{H}^{n-1}.$$

As pointed out above, both formulae require only the evaluation of first order derivatives on the test function $\boldsymbol{\varphi}$ and in the second case on the identity mapping on Γ and are thus well defined also in the discrete case.

Let us have a look how this can be done for a computational level set function ϕ_h. We first treat the case when ϕ_h is piecewise linear and $\Gamma_h = \{\mathbf{x} \in \Omega \mid \phi_h(\mathbf{x}) = 0\}$. As can be seen from Fig. 8, in this case the discrete level sets locally consist of subsimplices in 2d and subsimplices or quadrilaterals in 3d (except for pathological situations) and thus can be easily handled.

However, using $\mathcal{P}_1$ elements for ϕ_h is usually not accurate enough, see Schmidt (1996) or Groß and Reusken (2011). One should use (at least) $\mathcal{P}_2$ instead. Unfortunately, then the level sets of ϕ_h are no longer $(n-1)$ dimensional simplices but rather more complicated (Fig. 9).

The remedy is to (virtually) subdivide each element intersected by the level set in 2^n subsimplices and use a piecewise linear interpolation on this subcomplex, see Groß and Reusken (2011). Then define the discrete interface Γ_h by

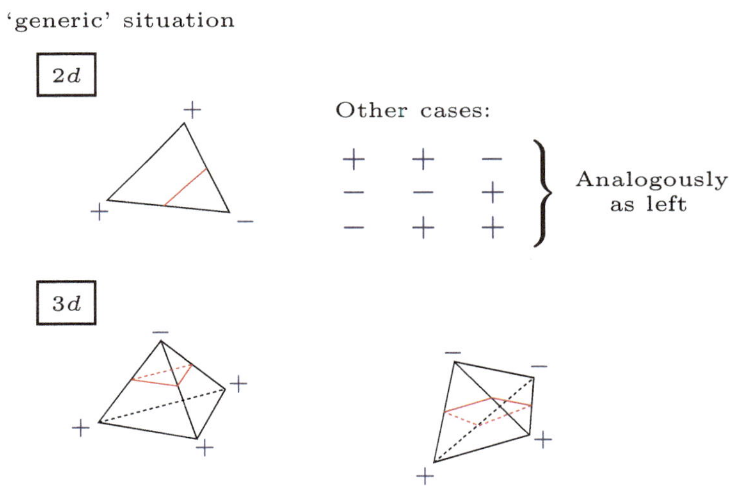

FIG. 8 Level set if ϕ_h is piecewise linear.

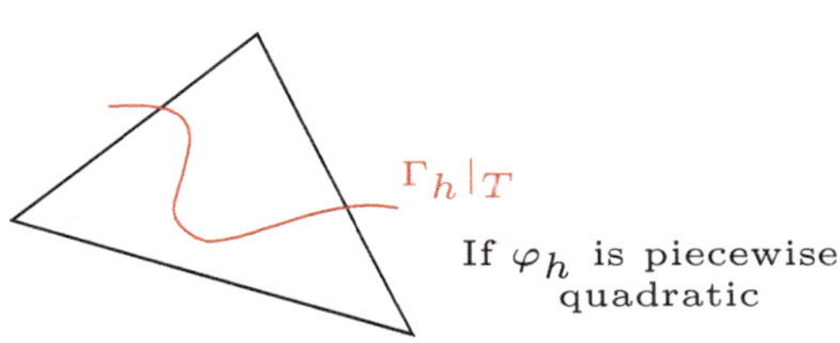

FIG. 9 Possible level set when ϕ_h is of higher order.

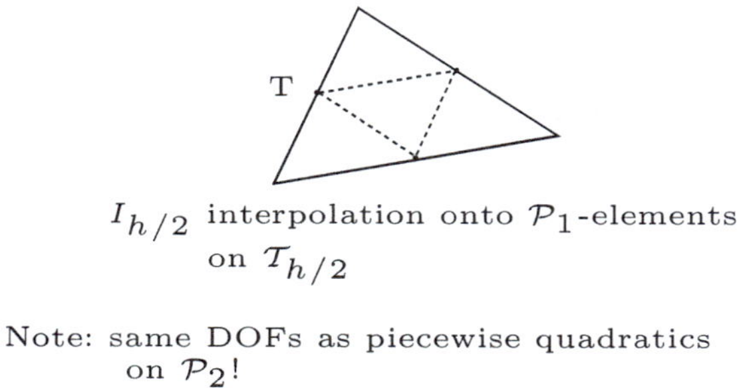

FIG. 10 Interpolation operator $I_{h/2}$ onto piecewise linears.

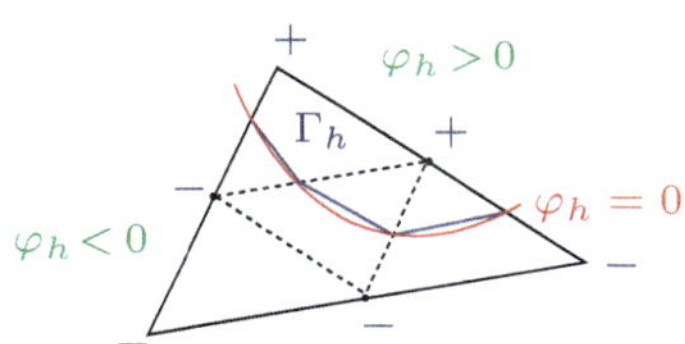

FIG. 11 Definition of Γ_h.

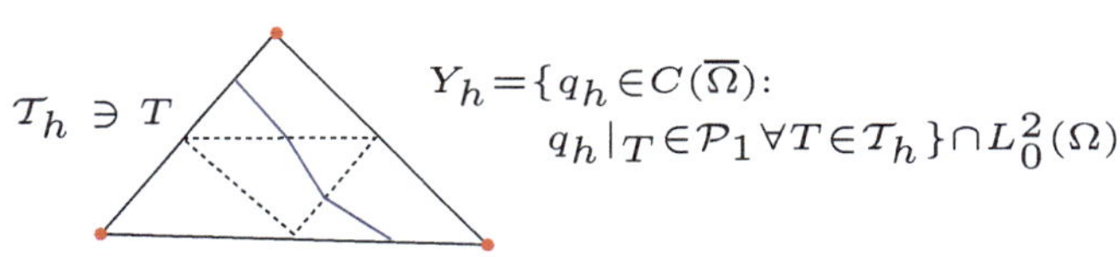

FIG. 12 Interface intersecting an element.

$$\Gamma_h := \{\mathbf{x} \in \Omega \mid I_{h/2}(\phi_h(\mathbf{x})) = 0\}$$

with $I_{h/2}$ the interpolation operator onto piecewise linears on the subcomplex, see Figs 10 and 11.

3.4.1 Pressure jump and XFEM

As outlined above, standard finite element discretization of the pressure space will inevitably lead to spurious velocities and pressure oscillations at the interface. This in turn could lead to completely wrong computational results (Figs 5, 6, and 12).

Moreover, the pressure error cannot be better than

$$\| p - p_h \| \sim \sqrt{h}.$$

The idea of *extended finite elements* (XFEM) is now to enrich the discrete pressure space locally by basis functions that are discontinuous across Γ_h such that a pressure jump can be captured by the discrete solution (see Moës et al. (1999) for a first application of XFEM to crack propagation and for instance Groß and Reusken (2007a, 2011) for XFEM in multiphase flow).

For simplicity, let the discrete pressure space Y_h consist of piecewise linears, globally continuous functions. Moreover, let $\psi_1, \ldots \psi_M$ denote the Lagrange basis functions of Y_h fulfilling

$$\psi_i(\mathbf{x}_j) = \delta_{i,j} \quad \text{for all vertices } \mathbf{x}_j, \ j = 1, \ldots M. \tag{48}$$

For a given index set $J \subseteq \{1, \ldots M\}$ define the XFEM space Y_h^{XFEM} by

$$Y_h^{XFEM} := Y_h \oplus \text{span}\{\psi_j^{XFEM} \mid j \in J\},$$

with functions ψ_j^{XFEM} discontinuous across Γ_h to be determined (Fig. 13).

Define

$$H(\mathbf{x}) = \tilde{H}(I_{h/2}(\phi_h(\mathbf{x}))) = \begin{cases} 0 & \mathbf{x} \in \Omega_{1,h} \\ 1 & \mathbf{x} \in \Omega_{2,h} \end{cases}$$

with $\tilde{H}$ the *Heavyside* function. Now for $j \in J$ we define

$$\chi_j(\mathbf{x}) := H(\mathbf{x}) - H(\mathbf{x}_j). \tag{49}$$

Note $\chi_j(\mathbf{x}) \in \{-1, 0, 1\}$.

With the help of these cut-off functions we define

$$\psi_j^{XFEM}(\mathbf{x}) = \psi_j(\mathbf{x})\chi_j(\mathbf{x}), \quad \text{for } \mathbf{x} \in \Omega,$$

see Fig. 14 for the 1d situation and Fig. 15 for XFEM in 2d.

Implementation of XFEM is demanding, see Jahn and Klock (2016) for an example using an open source toolbox. Moreover, when Γ_h is moving in time, the space Y_h^{XFEM} has to be built in each time step.

To summarize, the computational tasks for two-phase flow with the level set method consist in:

- Discretization and solution of the Navier–Stokes equation on Ω with discontinuous coefficients across Γ.
- Discretization and solution of the transport equation Eq. (47) for ϕ, with ϕ_h at least piecewise quadratic.
- Computation of the surface tension terms.

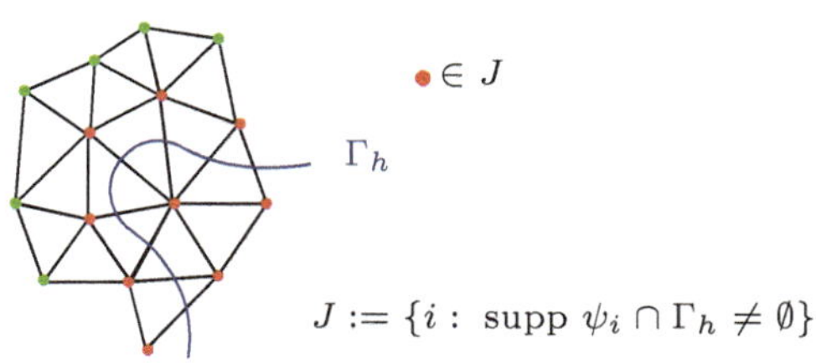

FIG. 13 Definition of the index set J.

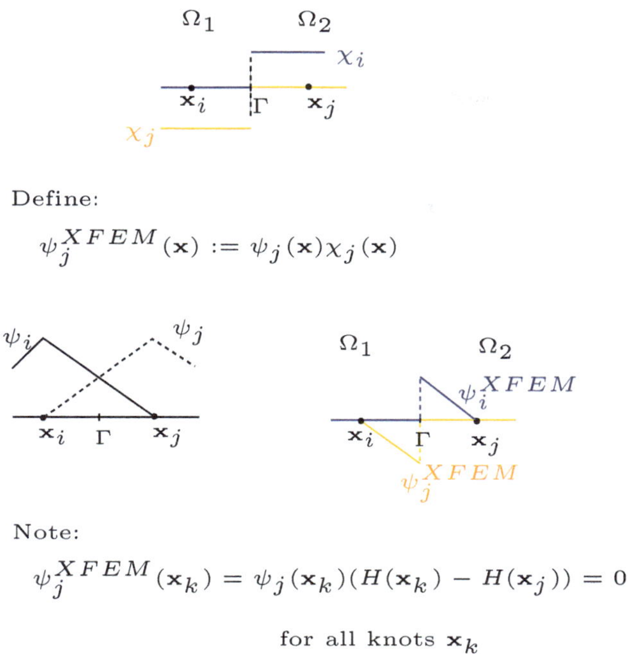

FIG. 14 ψ_j^{XFEM} in 1d.

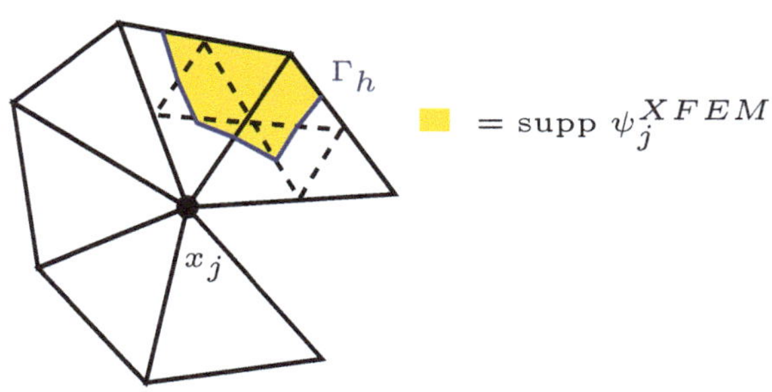

FIG. 15 ψ_j^{XFEM} in 2d.

- Redistancing.
- Volume correction.
- XFEM.
- Coupling/splitting Navier–Stokes equations and the transport equation.

4 Numerical methods for models with a parabolic interface equation

In the previous section, the geometric condition for the moving free boundary was given as a (quasi-stationary) elliptic equation. We now want to consider the case of parabolic equations for the interface motion, where the condition for the interface explicitly contains the velocity of the interface. The typical problem is motion by (anisotropic) mean curvature, like

$$\delta V_\Sigma + \gamma \kappa_\Sigma = f, \tag{50}$$

compare Eq. (22). As before, we consider explicit and implicit treatments of the interface.

4.1 Explicit treatment, parametric representation

While the interface is moved in (45) explicitly in normal direction by the velocity u_Σ, a semiimplicit time discretization of the parabolic Eq. (50), following the approach of Dziuk (1991), leads to the following equation for the parameterization X^{k+1} of Σ_h^{k+1} over the old interface Σ_h^k:

$$\int_{\Sigma_h^k} \frac{1}{\Delta t}(X^{k+1} - id)\varphi_h \, \mathrm{d}\mathcal{H}^{n-1} + \gamma \int_{\Sigma_h^k} \nabla_S X^{k+1} \cdot \nabla_S \varphi_h \, \mathrm{d}\mathcal{H}^{n-1} = \int_{\Sigma_h^k} f \varphi_h \, \mathrm{d}\mathcal{H}^{n-1}, \tag{51}$$

which gives the new interface by

$$\Sigma_h^{k+1} := X^{k+1}(\Sigma_h^k). \tag{52}$$

For anisotropic curvature with a direction-dependent surface energy density $\gamma(\mathbf{n}_\Sigma)$, the weak formulation (23) or its generalization to higher dimensions is well suited for an isoparametric finite element discretization, even for nearly crystalline surface energies, see Barrett et al. (2012) or Schmidt (1998). Direct facetted approaches are possible, too, based on generalized curvatures from geometric measure theory, see Roosen and Taylor (1991).

For the discretization of two-phase and free surface flows, the front-tracking method was already mentioned, see Section 3. Here, for the discretization of phase changes with geometric condition which may lead to the development of dendrites, the chance for large deformations of the interface is very high, so that a corresponding mesh moving approach is typically not working well. A front-tracking method with *separate* meshes for the discretization of the interface and the domain is preferred. We want to concentrate on the (lower dimensional) interface discretization first.

In the case of large and nonuniform deformations of the interface, the mesh for interface discretization would typically develop regions where mesh elements are compressed and other regions where elements would be stretched (for 2D interfaces even anisotropically). In order to prevent numerical problems, a special treatment of the meshes is needed.

One-dimensional interfaces are represented by polygons. A (local) mesh refinement (e.g., by bisection of polygon edges) can be used in order to prevent edges getting too large. In order to prevent edges getting too short, either a local coarsening might be used, or a tangential motion of vertices in order to equidistribute edge lengths. The latter can be implemented as a numerical post-processing step, or be implemented in the equation by prescribing not only the motion in normal direction but also a tangential motion of points on the interface. See Mikula and Sevcovic (2001) for examples.

For 2D interfaces, a triangular mesh may deteriorate during the evolution not only by shrinking or enlarging whole mesh elements, but also by stretching in only one direction. Typically, angles between edges would get very small or very large, destroying the shape regularity of the mesh. In order to prevent this, mesh smoothing methods (based on Laplacian or other smoothing, compare Section 3.3.3) can be combined with edge flipping, see e.g. Schmidt (1996) for more details. Like in 1D, it is possible to use formulations with a tangential motion of points on the interface also for two-dimensional interfaces, see Barrett et al. (2010) and Mikula et al. (2014).

New vertices created by (local) refinement or moved vertices should be part of the surface. As moving interfaces are typically curved, thus, a projection of points in the neighbourhood of the interface onto the (discrete) surface is needed. This may involve just a polygonal representation or a smoother, higher order approximation.

For the temperature equation, the domain should be discretized by an adaptive finite element method. Typically, steeper gradients of the temperature occur near the moving interface, but a high accuracy of temperature values at the interface is needed for the right-hand side of the geometrical evolution equation. Thus, a local mesh refinement, which moves with regions of interest and high variations, is a good way to get an accurate and efficient approximation.

The coupling between temperature and interface evolution in Problem 2 can be done by a semiexplicit time discretization: In every time step, first compute the new interface Σ^{k+1} using a parameterization over Σ^k and the old temperature there, by solving (51) with right-hand side Θ^k. This gives Σ^{k+1} by (52), and defines the discrete liquid subdomain $\Omega_{l,\ h}$ for the melt flow, if needed. The new temperature is then computed via the weak formulation of Eqs (24a) and (24b), where the jump of gradients at the interface leads to an integral $\int_{\Sigma^{k+1}} V_\Sigma \varphi \, \mathrm{d}\mathcal{H}^{n-1}$, where V_Σ can be replaced via the Gibbs–Thomson relation Eq. (19) by $\Theta^{k+1} + \gamma\kappa_\Sigma$ and thus leads to an implicit (and stabilizing) term for the temperature at the new interface.

4.2 Implicit treatment, level set

For a level set Σ of a smooth function ϕ, the velocity and the mean curvature of Σ are given by

$$V_\Sigma = \frac{\phi_t}{|\nabla\phi|}, \quad \kappa_\Sigma = -\nabla \cdot \frac{\nabla\phi}{|\nabla\phi|}.$$

Putting things together, the level set formulation for Eq. (50) with a constant γ and $\delta = 1$ leads to the system

$$\begin{aligned} \frac{\phi_t}{|\nabla\phi|} - \gamma\nabla\cdot\frac{\nabla\phi}{|\nabla\phi|} &= f \quad \text{in } \Omega, \\ \nabla\phi\cdot\mathbf{n}_\Sigma &= 0 \quad \text{on } \partial\Omega, \\ \phi &= \phi_0 \quad \text{for } t=0, \end{aligned} \tag{53}$$

where we choose homogeneous Neumann conditions on the boundary, and an initial value ϕ_0 which represents the initial interface Σ_0 as zero level set. In the case when topological changes may occur, the gradient $\nabla\phi$ may vanish, thus a regularization with $\delta > 0$ leads to the weak formulation

$$\int_\Omega \frac{\phi_t\varphi}{\sqrt{\delta^2 + |\nabla\phi|^2}} + \int_\Omega \frac{\gamma\nabla\phi\nabla\varphi}{\sqrt{\delta^2 + |\nabla\phi|^2}} = \int_\Omega f\varphi \quad \forall\varphi\in H^1(\Omega), \text{ a. e. } t\in(0,T). \tag{54}$$

Backward Euler time discretization leads to the implicit nonlinear equation

$$\int_\Omega \frac{(\phi^{k+1} - \phi^k)\varphi}{\Delta t\sqrt{\delta^2 + |\nabla\phi^{k+1}|^2}} + \int_\Omega \frac{\gamma\nabla\phi^{k+1}\nabla\varphi}{\sqrt{\delta^2 + |\nabla\phi^{k+1}|^2}} = \int_\Omega f^{k+1}\varphi \quad \forall\varphi\in H^1(\Omega), \tag{55}$$

which can be linearized by a semiexplicit treatment of the denominator, which gives

$$\int_\Omega \frac{(\phi^{k+1} - \phi^k)\varphi}{\Delta t\sqrt{\delta^2 + |\nabla\phi^{k}|^2}} + \int_\Omega \frac{\gamma\nabla\phi^{k+1}\nabla\varphi}{\sqrt{\delta^2 + |\nabla\phi^{k}|^2}} = \int_\Omega f^{k+1}\varphi \quad \forall\varphi\in H^1(\Omega). \tag{56}$$

Discretization with finite elements is straightforward. See Dziuk (1999b) for an investigation of the semiimplicit time discretization scheme for graphs. As mentioned earlier in Section 3.4, typically a redistancing of ϕ^{k+1} is added after some time steps, such that $|\nabla\phi^{k+1}| \approx 1$ near the interface Σ^{k+1}.

As we are only interested in the evolution of Σ, the zero level set of ϕ, the solution of the level set equation in not of interest when far away from Σ. In order to speedup the computation, either locally refined meshes with high resolution near Σ^k and very coarse meshes far away can be used, see Fried (2004) for example, or a narrow band technique which considers the equation only in a neighbourhood of the interface, see Peng et al. (1999). In the latter case, this band must be adjusted to the motion of the interface after some time steps, when the interface nearly reaches the boundary of the band. A corresponding restriction of the time step size is mandatory in order to keep the interface inside the current band.

4.2.1 Anisotropic motion of level sets

The motion of level sets by mean curvature is the gradient flow for the energy

$$E(\phi) := \int_\Omega |\nabla\phi|.$$

If we are given a smooth function $a : S^{n-1} \to \mathbb{R}^+$ and study the gradient flow of the anisotropic energy

$$E_a(\phi) := \int_\Omega a\left(\frac{\nabla\phi}{|\nabla\phi|}\right)|\nabla\phi|,$$

we end up with the following weak formulation:

$$\int_\Omega \frac{\phi_t\varphi}{|\nabla\phi|} + \left(a\left(\frac{\nabla\phi}{|\nabla\phi|}\right)\frac{\nabla\phi}{|\nabla\phi|} + \nabla a\left(\frac{\nabla\phi}{|\nabla\phi|}\right)\right)\cdot\nabla\varphi = f. \tag{57}$$

In two dimensions, a can be written as $a(\cos\alpha, \sin\alpha) = \tilde{a}(\alpha)$ where $\tilde{a} : [0, 2\pi] \to \mathbb{R}$. If $\tilde{a}$ is smooth and $\tilde{a} + \tilde{a}'' > 0$, this law of motion is equivalent to

$$V_\Sigma + (\tilde{a} + \tilde{a}'')\kappa_\Sigma = f.$$

Regularization and discretization of the weak formulation can be done like described above for the isotropic case.

4.3 Implicit treatment, phase field

We discuss here a time discretization of the phase field model from Section 2.5, which decouples the variables ϑ and χ, as well as the space discretization by finite elements.

First note that the inclusion Eq. (32) can be written as a variational inequality, using the convex set $\mathcal{K} := \{\phi \in H^1(\Omega) : |\phi| \le 1\}$: Find $\chi(t) \in \mathcal{K}$ such that

$$\varepsilon\langle\partial_t\chi, \chi - \phi\rangle + \varepsilon(a(\nabla\chi), \nabla(\chi - \phi)) - \frac{b}{\varepsilon}(\chi, \chi - \phi) \le \eta(\vartheta, \chi - \phi) \quad \forall\phi \in \mathcal{K}. \tag{58}$$

If M^{k+1} denotes the simplicial mesh for the next time step and V_h the corresponding piecewise linear finite element space, we define $\mathcal{K}_h := V_h \cap \mathcal{K}$ as the convex set of admissible discrete phase variables.

Let $\Theta^0 := I^0\vartheta_0$ and $X^0 := I^0\chi_0$. Then the fully discrete finite element approximation reads as follows: Given X^k and Θ^k, compute $X^{k+1} \in \mathcal{K}_h$ and $\Theta^{k+1} \in V_h$ by solving

$$\begin{aligned}\varepsilon\left\langle\frac{X^{k+1} - I^{k+1}X^k}{\Delta t}, X^{k+1} - \varphi_h\right\rangle^{k+1} &+ \alpha\varepsilon\big(a(\nabla X^{k+1}), \nabla(X^{k+1} - \varphi_h)\big) \\ -\frac{b}{\varepsilon}\langle X^{k+1}, X^{k+1} - \varphi\rangle^{k+1} &\le \eta\langle I^{k+1}\Theta^k, X^{k+1} - \varphi\rangle^{k+1} \quad \forall\ \varphi \in \mathcal{K}_h,\end{aligned} \tag{59}$$

$$\left\langle \frac{(\Theta+X)^{k+1} - I^{k+1}(\Theta+X)^k}{\Delta t}, \varphi \right\rangle^{k+1} + \left(\lambda \nabla \Theta^{k+1}, \nabla \varphi\right) = \langle I^{k+1} f, \varphi \rangle^{k+1} \quad \forall \ \varphi \in V_h. \tag{60}$$

Here, the *lumping* scalar product

$$\langle \varphi, \psi \rangle^{k+1} := \sum_{S \in M^{k+1}} \int_S I^{k+1}(\varphi \psi) \mathrm{d}\,\mathbf{x},$$

where I^{k+1} denotes the nodal interpolation on M^{k+1}, gives an easy to evaluate vertex quadrature rule.

Note that Eq. (59) is an elliptic variational inequality with double obstacle ± 1 which can be solved by using projected Gauss-Seidel or, with optimal complexity, a monotone multigrid method, see Kornhuber (1994). Once X^{k+1} is known, the linear Eq. (60) for Θ^{k+1} can be solved directly, for instance by conjugate gradient method with BPX preconditioner.

As mentioned before, the local behaviour of temperature and phase variable differ a lot, as the phase variable is piecewise constant in the bulk and varies in the narrow interface strip, while the temperature is smooth with a sharp variation of its gradient in the strip. Local mesh refinement for the discretization of both the temperature and the phase variable is mandatory. This should be based on a posteriori error indicators, see Kessler et al. (2004) for an investigation of the Allen Cahn equation without using Gronwall's inequality. An even better performance can be reached, when separate meshes are used for the temperature and the phase variable. Now, the phase variable mesh can be as coarse as possible outside of the narrow strip and very fine in the strip where gradients are of order $1/\varepsilon$, and the temperature mesh can be coarser in the strip as only the second derivatives get large. The transfer of data between meshes can be implemented quite efficiently, if both meshes are different local refinements from the same macro triangulation. A corresponding adaptive method was presented in Schmidt (2003b) for 2D and in Schmidt (2003a) for 3D phase field simulations.

5 Model problems, examples, and applications

We present here some examples of applications considered by the authors, based on the models and numerical methods discussed above. Most of the applications were examined in joint projects with scientific partners from natural sciences or engineering.

5.1 Uniaxial extensional flow in liquid bridges

In this section we present an example of flows under microgravity conditions, namely the dynamic behaviour of stretched liquid bridges, where surface tension effects play a dominant role. The physical background is to generate flow fields that are close to be linear in space at each time instant. The presentation is a condensed version of the detailed description in Bänsch et al. (2004).

5.1.1 Problem setting

Linear flow fields are commonly used for rheological studies, e.g., to measure the fluid viscosity or the deformation behaviour of the whole sample or components in it. In view of the experimental demands of rheological tests, the ideal uniaxial extensional flow field is characterized by a constant strain rate distribution, both in space and time which is given by

$$\begin{aligned} u_r &= -0.5\dot{\epsilon}_0 r, \\ u_\theta &= 0, \\ u_z &= \dot{\epsilon}_0 z, \end{aligned} \tag{61}$$

in a cylindrical coordinate system (r, θ, z) with the flow field $\mathbf{u}$ and the constant strain rate $\dot{\epsilon}_0$. For an initially cylindrical fluid volume with length L_0 and radius R_0 this specific form of $\mathbf{u}$ implies

$$L(t) = L_0 \exp(\dot{\epsilon}_0 t), \tag{62}$$

$$R(t) = R_0 \exp(-0.5\,\dot{\epsilon}_0 t), \tag{63}$$

with $L(t)$ the time-dependent length of the fluid volume and $R(t)$ its radius. In particular the liquid volume is cylindrical in shape at all times.

However, the experimental realization of such a flow field causes severe difficulties in practice. In the present example, this is achieved by stretching an almost cylindrical *liquid bridge* under microgravity that is solely kept together by surface tension. One key issue is the adjustability of the disk diameters, necessary for maintaining ideal boundary conditions. This method gives rise to much weaker end-effects than the commonly used method with unchangeable disks.

Experimentally, the stretching of the liquid bridge was performed under microgravity at the drop tower of Bremen. The microgravity environment during stretching was necessary to eliminate forces created by the hydrostatic pressure. The overall time for an experimental run under microgravity was determined by the height of the drop tower, resulting in 4.7 *s* for the Bremen experiments. The principle setting is sketched in Fig. 16.

Even under microgravity (μg) conditions and with adjustment of the disk diameters the ideal extensional flow is not totally achievable; this is due to the dynamic pressure gradient (induced by acceleration and capillary forces) deforming the bridge contour, see Berg et al. (1999). Instead, the real stretched liquid bridge is deformed in the direction of the accelerated support due to inertia and surface tension forces and will eventually pinch off.

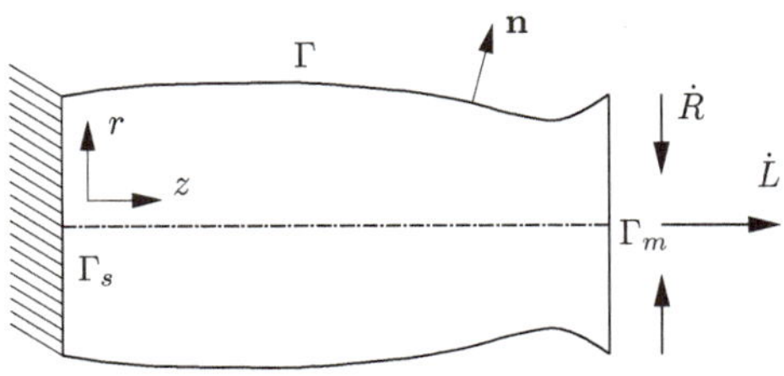

FIG. 16 Setting and basic notation. The device is dropped in the *z*-direction.

The goal of this study is to investigate this behaviour with respect to the flow parameters Ca = We/Re (capillary number) and We (Weber number) and detect those regimes with most suitable flow conditions.

To this end we first study stretching by both numerical and experimental means. The numerical results are compared with the experimental data resulting in very good agreement. The numerical method is then used to study the behaviour of liquid bridges for a fairly large range of Ca and We values.

For the numerical simulation of the problem we use a finite element mesh moving method as described in Section 3. Note that the problem constitutes a single phase flow problem with a free capillary surface, the mantle surface of the liquid bridge (Fig. 17).

Results from a typical experimental run, using castor oil in this example, is presented in Fig. 18. Picture (A) shows the liquid bridge before the onset of stretching, where the bridge is held between the membranes under microgravity conditions. The bridge is almost cylindrical before stretching and two immiscible emulsion droplets were placed inside the bridge to investigate their deformations in extensional flow in this particular case.

In the subfigures (B)–(D), the device on the right-hand side stretches the bridge for 3.5*s* with exponentially increasing velocity according to Eq. (62). Simultaneously, the membranes decrease their radii exponentially fast so that the fluid boundaries adapt to the actual bridge length.

During stretching, a small amount of necking close to the accelerated device can be seen, which is typical and caused mainly by inertia forces in this parameter range (Ca = 0.179, We = 0.0325).

For a precise comparison between experiment and numerical simulation, the real, opposed to ideal, experimental boundary conditions were taken into account.

5.1.2 Comparison of experimental and numerical results

As a first step in our investigation we compared the experimental results with corresponding numerical simulations. The results are shown in Fig. 19.

The transient bridge deformation during stretching can be subdivided into two parts. In the first time period the capillary pressure tries to stabilize the bridge contour to a cylindrical shape of constant mean curvature. A first rough estimate of this time interval is determined by the stability limit of a

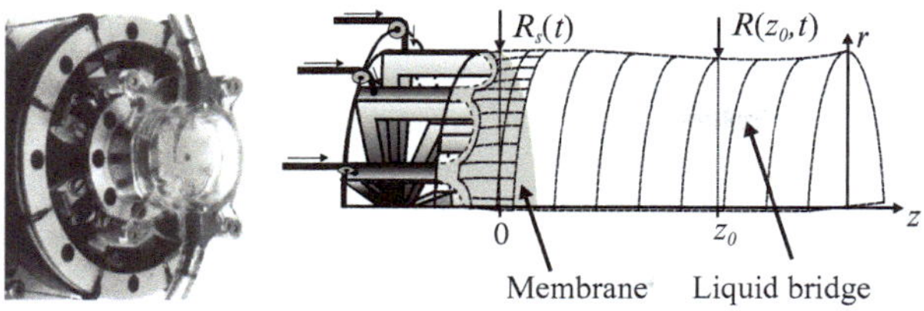

FIG. 17 Membrane with time-dependent radius to adjust to the actual bridge length; apparatus (*left*) and sketch of the device (*right*). The accelerated end moves in the z-direction, i.e., the direction where the device is dropped.

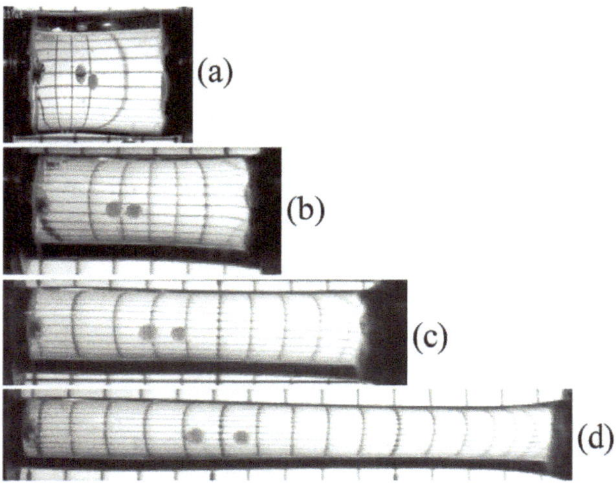

FIG. 18 Initial and stretched bridge of castor oil at $t = 0.0$ (A), $t = 0.444$ (B), $t = 0.888$ (C), $t = 1.344$ (D), respectively; time in dimensionless units.

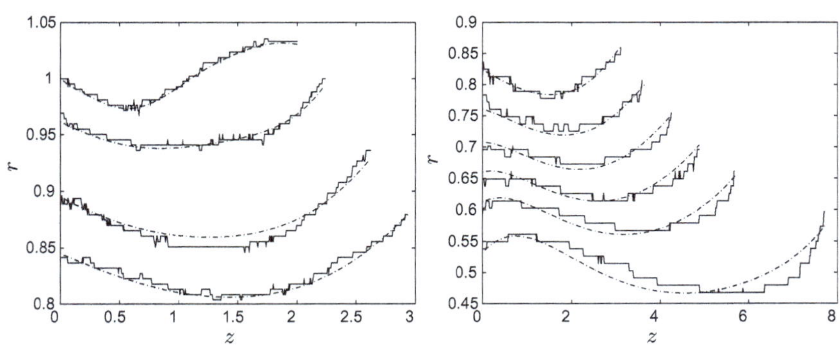

FIG. 19 Comparison experiment versus simulation: bridge shape at time instants $t = 0.000$, $t = 0.144$, $t = 0.288$, $t = 0.396$ *(left)* and $t = 0.444$, $t = 0.600$, $t = 0.744$, $t = 0.888$, $t = 1.04$, $t = 1.34$ (*right*), from top to bottom, respectively; time in nondimensional units; experiment (*solid*) and simulation (*dash-dotted*).

stationary liquid bridge, i.e., the aspect ratio $\Lambda(t) = L(t)/R(t) \leq 2\pi$, which was already studied in Plateau (1863) and Lord Rayleigh (1878).

During this period, the initial deformation reorientates and relaxes, delayed by viscous forces. The deformation changes from a concave–convex shape to a purely concave shape. Moreover, capillary forces stabilize the bridge contour, counteracting the dynamic pressure difference caused by the acceleration of the flow.

Beyond the stability limit, in the remaining stretching period, the capillary and the dynamic pressures deform the bridge contour. A significant convex–concave deformation owing to the accelerated moving membrane arises. This type of bridge contour appears during stretching because of the negative pressure gradient in the vicinity of the accelerated membrane. The typical convex–concave shape is clearly visible in Fig. 19. Eventually, instability leads to a breakup of the bridge.

The numerical simulations are capable of reliably describing the transient deformation during the bridge stretching. Fig. 19 shows that the numerical results are in fairly good agreement with the experimental contours. More precisely, the experimental uncertainty of the recording device is 2 pixels and the agreement of the numerical results is of the same order of magnitude over most parts of the bridge and at almost all times.

5.1.3 Parameter variations and bridge shapes

After this successful validation test the numerical method was then used to investigate bridge stretching and the corresponding flow quality for a wide variety of capillary and Weber numbers, see also Fig. 22. The parameters were chosen to represent the bridge stretching at realistic fluid properties and bridge dimensions.

Physically, the parameter field can be interpreted in the following way: fixing the initial bridge radius to $R_0 = 15$mm, the stretching rates are varied between 0.1/s and 10/s. The surface tension and viscosity are varied in the ranges 20–70mN/m and 0.001–1000 Pa·s, respectively, while keeping the density constant at about 1000 kg/m^3.

A typical run is shown in Figs 20 and 21. There the respective strain and shear rate distributions are shown at different time instants for a bridge with $\text{Ca} = 7.50 \cdot 10^{-1}$ and $\text{We} = 1.69 \cdot 10^{-1}$. For this example, almost no homogeneous flow exists inside the bridge for the main part of the stretching period. However, since the flow variations are weak in the convex part, a local shear-free strain flow exists for some time, which may be used for strain experiments.

Anticipating the result of our study, the parameter field can be subdivided into three major regimes:

- *capillary-dominated flow*, characterized by $\text{Ca} \ll 1$, $\text{We} \ll 1$,
- *viscous-dominated flow*, $\text{Ca} > \mathcal{O}(0.1)$, $\text{Re} < \mathcal{O}(0.1)$,
- *inertia-dominated flow*, $\text{We} > \mathcal{O}(0.1)$, $\text{Re} > \mathcal{O}(0.1)$.

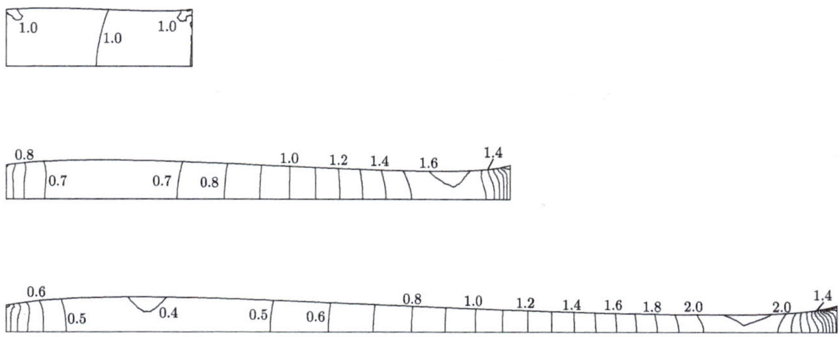

FIG. 20 Bridge with Ca = $7.50 \cdot 10^{-1}$ and We = $1.69 \cdot 10^{-1}$: strain rate distribution at $t = 0.205$, $t = 1.205$ and $t = 1.705$, from top to bottom, respectively; difference between successive isolines is 0.1.

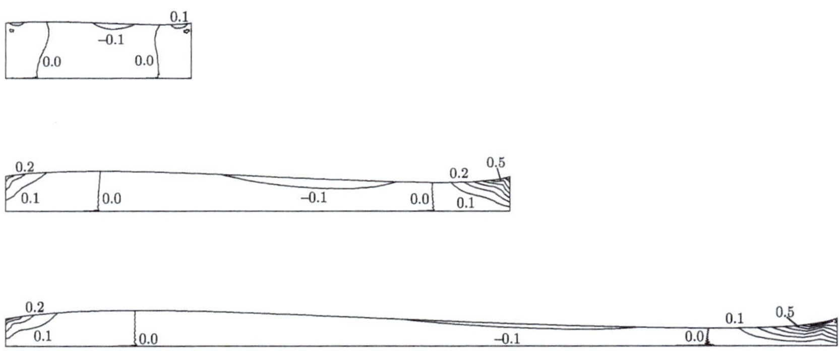

FIG. 21 Bridge with Ca = $7.50 \cdot 10^{-1}$ and We = $1.69 \cdot 10^{-1}$: shear rate distribution at t = 0.205, $t = 1.205$ and $t = 1.705$, from top to bottom respectively; difference between successive isolines is 0.1.

5.1.4 Strain and shear

The evaluation of the strain and shear rate distribution inside the bridge provides a means of assessing the quality of the uniaxial extensional flows. In cylindrical coordinates, the strain rate $\dot{\epsilon}$ and the shear rate $\dot{\gamma}$ are given by

$$\dot{\epsilon} = \frac{2}{3}(\partial_z u_z - \partial_r u_r), \quad \dot{\gamma} = \partial_r u_z + \partial_z u_r.$$

We are interested in those subregions of $\Omega(t)$ that are homogeneous with respect to the strain rate and shear rate. To this end we define the instantaneous homogeneous part Ω_{hom} of Ω, where the deviation from the desired strain and shear rate distribution of the ideal bridge is less than 5%:

$$\Omega_{hom}(t) := \{x \in \Omega(t) \mid \max\{|\dot{\epsilon}(t,x) - 1.0|, |\dot{\gamma}(t,x)|\} < 0.05\}, \tag{64}$$

as well as the fraction of the homogeneous part

$$\text{hom}(t) := \frac{|\Omega_{hom}(t)|}{|\Omega(t)|}. \tag{65}$$

The latter expression is a direct measure of the quality of a stretched bridge.

In order to get an even more condensed measure for the bridge qualities, we define the scalar quantity *qual*,

$$\text{qual} := \int_0^{t_{breakup}} \text{hom}(t)\, \mathrm{d}\, t. \tag{66}$$

This measure demonstrates increasing flow quality at decreasing Weber numbers and increasing capillary numbers, cf. Fig. 22. Large values of *qual* indicate the desirable situation of a large homogeneous flow fraction during a long-time period.

Interestingly, *qual* remains almost constant at constant Reynolds numbers in the We-Ca parameter field. This behaviour is most clearly visible for higher capillary number, where capillary forces are negligible. In this range, *qual* is a decreasing function of the Reynolds number only.

In turns out that the most suitable parameters for homogeneous extensional flow are in the range of *viscous-dominated flow*. The high viscous forces in comparison to inertia and capillary forces prevent strong local pressure variations inside the bridge. Therefore, the bridge remains cylindrical and the flow homogeneous and constant during a long stretching period. Moreover, in this range the flow quality is a decreasing function of the Reynolds number only.

5.2 Two-phase flow under microgravity without mass transfer

Mass transport across an interface liquid/gas is a fundamental process in many technical applications. Coupled to fluid dynamics in either phase and then in

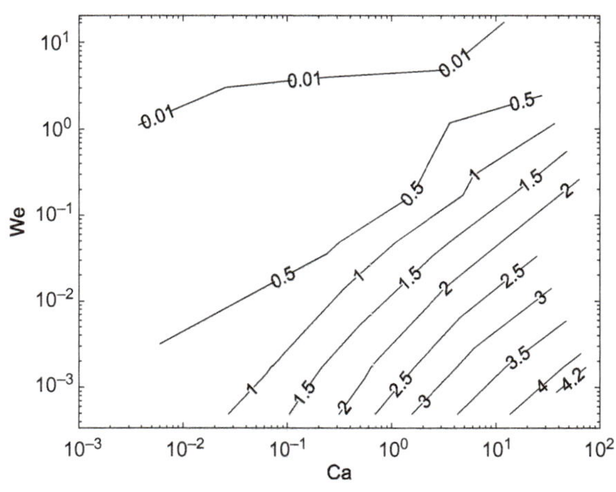

FIG. 22 Contour plot of the quality measure qual, defined by Eq. (66).

turn also to an evolution of the interface leads to a very complex system. Thus numerical simulations play an important role in understanding the relative influence of the underlying different physical mechanisms as well as predicting the behaviour of the system over time.

The application motivating the study presented here originates from space technology, more precisely from basic questions of propellants and their management during different flight phases (propelled/ballistic) of launchers.

In particular, knowing the location of the free surface liquid/gas and the heat transfer from the wall towards the fluid is of paramount importance in the design and the optimization of cryogenic upper stage tanks for launchers with ballistic phases, where residual accelerations are smaller by up to four orders of magnitude compared to the gravity acceleration on earth. This changes the driving forces drastically: free surfaces become capillary dominated and natural or free convection is replaced by thermocapillary convection, i.e., a flow driven by Marangoni effects.

This thermocapillary convection creates a velocity field directed away from the hot wall towards the colder liquid and then again back towards the wall. A deformation of the free surface resulting in an apparent contact angle is observed. The thermocapillary flow convects the heat from the wall to the liquid and significantly increases the heat transfer compared to pure conduction.

The presented results report about a long term collaboration with the Fluid Dynamics Group at ZARM, Bremen (Germany), headed by M. Dreyer. We gratefully acknowledge this stimulating collaboration, providing excellent experimental devices and data.

Our mathematical model and the numerical method take into account two-phase nonisothermal flow of an incompressible liquid and a gas phase, capillary forces at the interface of both fluids, Marangoni effects due to temperature variation of the interface and in the second experiment also mass transport across the interface by evaporation/condensation. A more detailed presentation can be found in Bänsch et al. (2015).

The numerical procedure here is again based on a mesh moving method, augmented by means to incorporate mass transfer across the interface by evaporation/condensation.

5.2.1 Surface deformation by thermocapillary convection

Here, we briefly report about the numerical simulation of an experiment which was conducted aboard a sounding rocket, the so called SOURCE experiment (SOUnding Rocket Compere Experiment). A detailed description of this experiment can be found in Fuhrmann and Dreyer (2008) and Fuhrmann and Dreyer (2009), and in the following we provide an outline of the setting for our numerical simulations.

Our numerical simulations were concerned with the evolution of the interface position during one stage of the SOURCE experiment. In this stage,

FIG. 23 Temperature distribution in the liquid phase (*left*) and triangulation in the vicinity of the contact line (*right*) for the SOURCE-I simulation.

which takes place under microgravity conditions, a cylindrical tank is partly filled with the experimental liquid, and the gaseous phase consists of a mixture of inert gas and the liquid's vapour.

The container wall is heated by a ring heat foil mounted on top of the cylindrical experiment cell, creating an axial temperature gradient towards the contact line. An approximation to this profile is depicted in Fig. 23 (left). As described above, the temperature dependency of the surface tension leads to a tangential stress which, in turn, creates a capillary (Marangoni) convection, transporting warmer liquid from the wall towards the symmetry axis.

A boundary condition in terms of the contact angle δ has to be prescribed. To prevent a simulation break down due to degenerating mesh elements, this contact angle has to be bounded somewhat from 0 from below. In the case of the SOURCE experiment, a static contact angle of 5 degrees (the experimental liquid had a static contact angle of 0 degrees) was used.

Also, initial values for the temperature distribution in the cylinder wall were modelled according to measurements obtained from the experiment. Fig. 23 (left) shows a three-dimensional representation of the simulated axially symmetric temperature distribution and a zoom of a locally adapted triangulation (right) needed to capture the viscous and thermal boundary layers of the problem.

In our simulations, which are described in detail in Fuhrmann et al. (2016) and Bänsch et al. (2015), we were interested in the influence of a

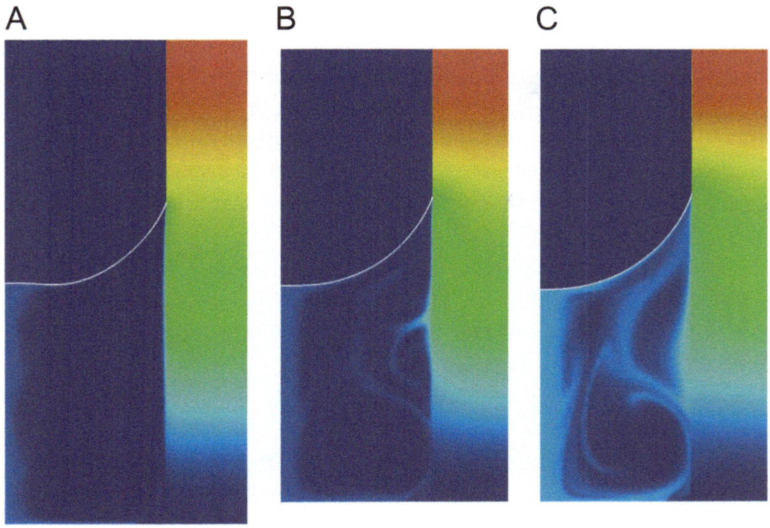

FIG. 24 Temperature distribution in the SOURCE simulation at three different time instants. (A) $t = 2$ s; (B) $t = 25$ s; (C) $t = 100$ s.

characteristic temperature difference between the liquid close to the wall and the wall itself (the driving force for the Marangoni convection).

In Fig. 24A–C, the temperature distribution at three different time instants is shown, corresponding to the initial state ($t = 2$ s), after $t = 25$ s and at the final time $t = 100$ s. Clearly, thermocapillary convection is "stirring" the fluid, leading to enhanced heat transfer towards the liquid compared to a purely conductive scenario, and a clearly visible change in the interface position.

5.2.2 *Reorientation behaviour of cryogenic liquids*

In the second experiment, we are interested in the *reorientation behaviour* of cryogenic liquids, i.e., the behaviour of the system undergoing a step reduction of gravity from $1 \times g$ to $0 \times g$ in the presence of nonisothermal boundary conditions. The corresponding experiment was performed at the drop tower facilities at ZARM, University of Bremen. A detailed description of the experiment can be found in Kulev and Dreyer (2010). In the following a brief outline is given.

As in the SOURCE experiment, a partly filled cylindrical container is considered. The container holds the test liquid, liquid argon (LAr), and is housed within a cryostat to maintain cryogenic conditions. In contrast to the SOURCE experiment, no additional inert gas is present in the gaseous phase, making it a single species system. In this case, the temperature at the liquid/gaseous interface corresponds to the saturation temperature and is constant. As a consequence, thermocapillary convection is not expected for this experiment.

At the beginning of the experiment, the experiment capsule is subject to normal gravity conditions and the interface obtains its $1 \times g$ equilibrium configuration. During a pre-heating phase before the release of the capsule, axial wall temperature gradients of different magnitude are established in the quartz wall of the container. Once the capsule is released (dropped) and subject to approximately 4.7s of microgravity conditions, these nonisothermal conditions influence the capillary driven reorientation behaviour of the liquid.

In the experiments, the frequency and dampening of the interface oscillations as well as the evolution of thermodynamic pressure were found to correlate strongly (see Kulev and Dreyer, 2010; Kulev et al., 2014). Our goal was to verify this correlation numerically.

The simulation presented here corresponds to an axial wall gradient of 1.73K/mm in the experiments presented in Kulev and Dreyer (2010). The initial $1g$ position of the interface was computed using the Young–Laplace equation, see for instance Finn (1986). Although liquid argon is a completely wetting liquid ($\delta = 0$ degree), a static contact angle of $\delta = 10$ degrees had to be prescribed to prevent the triangulation from degenerating in the vicinity of the contact point.

The simulation was started at rest ($\mathbf{u}_1 = \mathbf{u}_2 = 0$) shortly before the step reduction of gravity and an initial temperature profile corresponding to measurements from the experiment was imposed. It is worth mentioning that an accurate approximation of the initial temperature distribution in the vicinity of the interface was essential to obtain a physically realistic evaporation rate. Initially, the interface had a constant temperature corresponding to the saturation temperature of 90.8 K defined by the thermodynamic pressure of 1044 hPa. Within a layer of thickness δ_Γ parallel to Γ, the temperature drops to the subcooled temperature of the liquid bulk, $\vartheta = 85.0$ K, see Fig. 25A. An estimate for the thickness δ_Γ can be found in Das and Hopfinger (2009), and within this layer we assumed to have a linear temperature profile.

At $t = 0$ s, gravity is "turned off" numerically, corresponding to the release of the experiment capsule in the drop tower. The temperature and velocity profiles during the reorientation are depicted in Fig. 25A–D at four different time instances. Fig. 25B depicts the moment of the largest deflection of the z-coordinate of the wall contact line. We make the following observations: the initial distribution in the temperature layer below the interface is immediately distorted by the capillary driven movement of the interface. Furthermore, this movement induces a forced convection in the gaseous phase transporting "cold" gas at saturation temperature from the interface into the gaseous phase. Also a vortex forming underneath the interface is observed, see Fig. 25D.

We now compare different characteristic quantities of the simulation with experimental data: Fig. 26A shows the evolution of the z-coordinates of the centre point z_{cp}, i.e., the z-coordinate of the interface at the symmetry axis and the evolution of the z-coordinate of the wall contact point z_{wp}. While

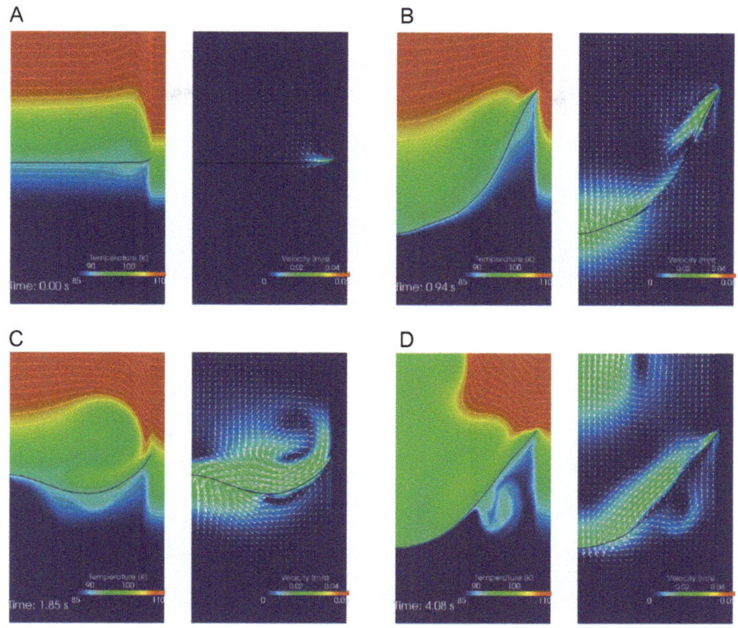

FIG. 25 Temperature distribution (*left*) and velocity profile (*right*) at four different time instants during reorientation. (A) $t = 0$ s; (B) $t = 0.94$ s; (C) $t = 1.85$ s; (D) $t = 4.08$ s.

perfect quantitative agreement between numerical and experimental results cannot be expected due to our numerical restrictions (for instance, the static contact angle $\delta = 10$ degrees), qualitative agreement is clearly visible: the number and frequency of oscillations roughly agrees. Fig. 26A also depicts the evolution of the normalized thermodynamic pressure over time. Again, qualitative agreement between numerics and experiment can be observed. Furthermore, there is a clear correlation between interface motion and pressure evolution.

To explain this behaviour, Fig. 26B shows the evolution of the total mass flux across the interface over time. As can be seen, this mass flux strongly correlates with the position of the interface: as soon as the largest deflection of the interface in z-direction occurs (at approximately $t = 0.9$ s), the interface is in contact with the hot wall and surrounding hot gas. Consequently, the evaporation rate in the vicinity of the contact point should be strongly magnified. That this is indeed the case is shown by Fig. 26C. Once the contact point recedes and reaches its minimum z-deflection, the evaporation rate also decreases. We also make the observation that although large local evaporation rates are observed close to the contact line, condensation occurs in a region close to the symmetry axis. Our numerical simulations also suggest that the region separating condensation and evaporation roughly follows the evolution of the forming vortex, compare Figs 26D and 25D.

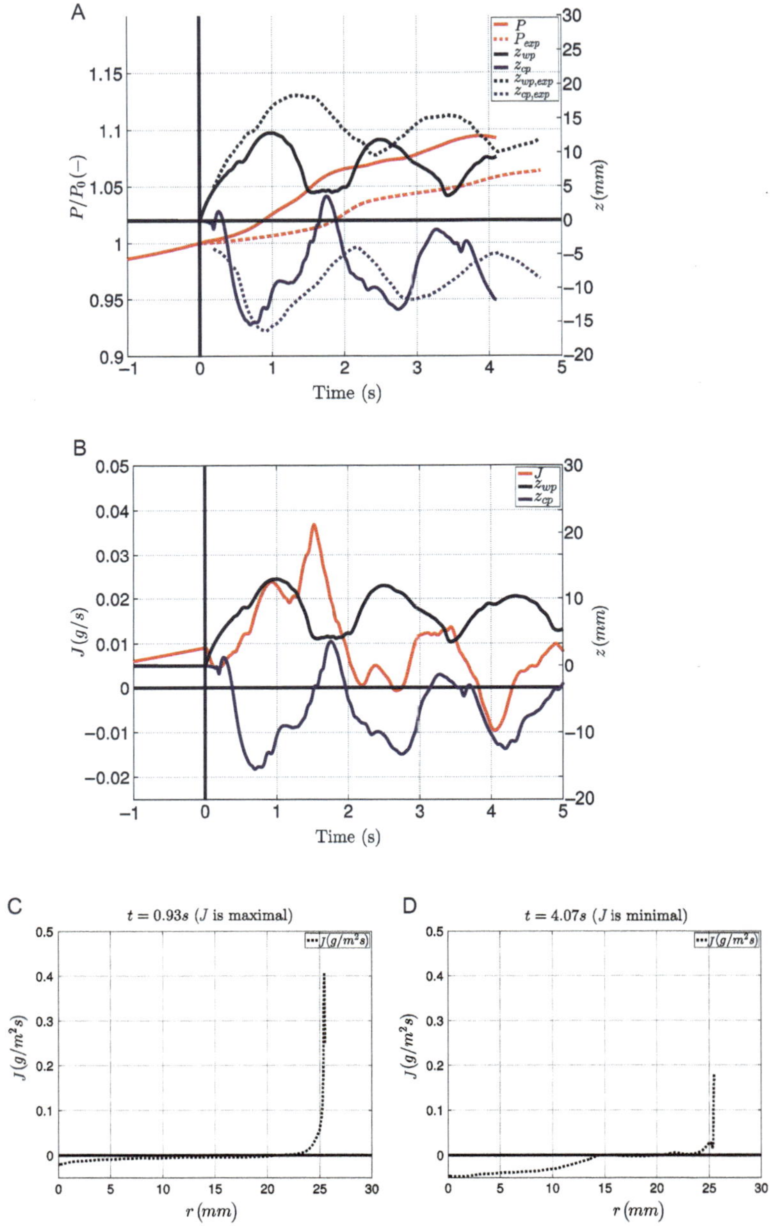

FIG. 26 Some details of the simulation of experiment 2. (A) Evolution of the normalized thermodynamic pressure, centre point z_{cp} and wall contact point z_{wp} (experiment = dotted line; simulation = solid line) over time. (B) Evolution of the evaporation rate J as well as the wall contact point z_{wp} and the centre point z_{cp} over time. (C) Distribution of the mass flux j over cylinder radius at $t = 0.93$ s. (D) Distribution of the mass flux j over cylinder radius at $t = 4.07$ s.

5.3 Material accumulation by melting and solidification

In advanced engineering, the importance of micro components has increased significantly during the last decade. For the production of such micro components, methods used in macro scale are sometimes no longer applicable to very small work pieces, so the need for new methods and processes arises. For the forming of micro components, like valves on a thin supporting rod, enough material has to be accumulated which can be formed into the final shape. In macro scale, material is usually accumulated by a repeated upsetting. In microscale, due to much stronger bending effects, etc., the applicable upset ratio is not large enough, so other accumulation methods must be used.

One newly investigated possibility is the accumulation by selective melting using a laser for heating. Due to the dominant surface tension over gravity effects, rather large melt pool drops can be generated in micro productions without separation from the supporting solid. After cooling and solidification of the melt, the generated preform is then calibrated in a subsequent forming step in a open die. By using the two-level cold forming process, upset ratios $s := \frac{l_0}{d_0} > 200$ can be reached while the conventional process is limited by the value $s = 2.1$ and decreased, if d_0 does. Here, l_0 denotes the length of accumulated (here: molten) material from a rod of diameter d_0.

A model for the accumulation process, which combines phase transition and free surface flow of the melt, was given in Section 2.4. The additional complexity of triple junctions between solid–liquid interface and capillary surface was already mentioned there.

In case of a simple and slowly varying topology of the melt Ω_l, an ALE approach can be used not only for the moving capillary surface, but also for the moving solid–liquid interface. This is easily applicable in 2D or a rotationally symmetric situation. Here, a moving sharp solid–liquid interface can be represented by moving edges of a triangular mesh.

For more complicated situations, especially with topology changes or complicated geometries, the representation of the interface by moving edges is not applicable. Thus, a different numerical approach is needed. The implicit definition of phases and the moving phase boundary on a fixed mesh via an energy/enthalpy formulation of the phase transition (compare Section 2.5.1) is much easier usable in complex situations.

For the material accumulation process, the initial nucleation of melt by heating the solid material includes a topology change by appearance of the liquid subdomain. This can easily be modelled by the energy approach. Additionally, after moving or switching off the heating, cooling with solidification will also lead to topology changes for the liquid, and maybe also the solid, subdomains. Thus, the energy approach is favourable during the cooling, too. After generation of an initial melt pool, there can be the situation that the continued heating will result in a rather stable situation where topology of the melt does not change and geometry varies rather slowly by melting additional material. Here, the ALE approach may be used efficiently.

A disadvantage of the energy approach is the definition of the varying liquid subdomain $\Omega_l(t)$ needed for the discretization of the melt flow with capillary surface. A simple approach for the usual finite element discretization is to define the discrete liquid subdomain as union of all mesh elements where the temperature is above melting temperature on the whole element:

$$\Omega_{l,h}(t) := \bigcup\{S \in \mathcal{S}_h : \theta_h(t) \geq \theta_m \text{ on } S\}. \tag{67}$$

By this definitions, mesh elements stay "solid" until they are fully melted, or get "solid" again when the material is even partially solidified. This is nicely visible in Fig. 27 in the top middle and bottom right pictures. While this approach leads to a simple and easily applicable discretization of the flow, it may create discretization problems for the capillary surface. As the discrete capillary surface is made up by boundary faces of the discrete liquid

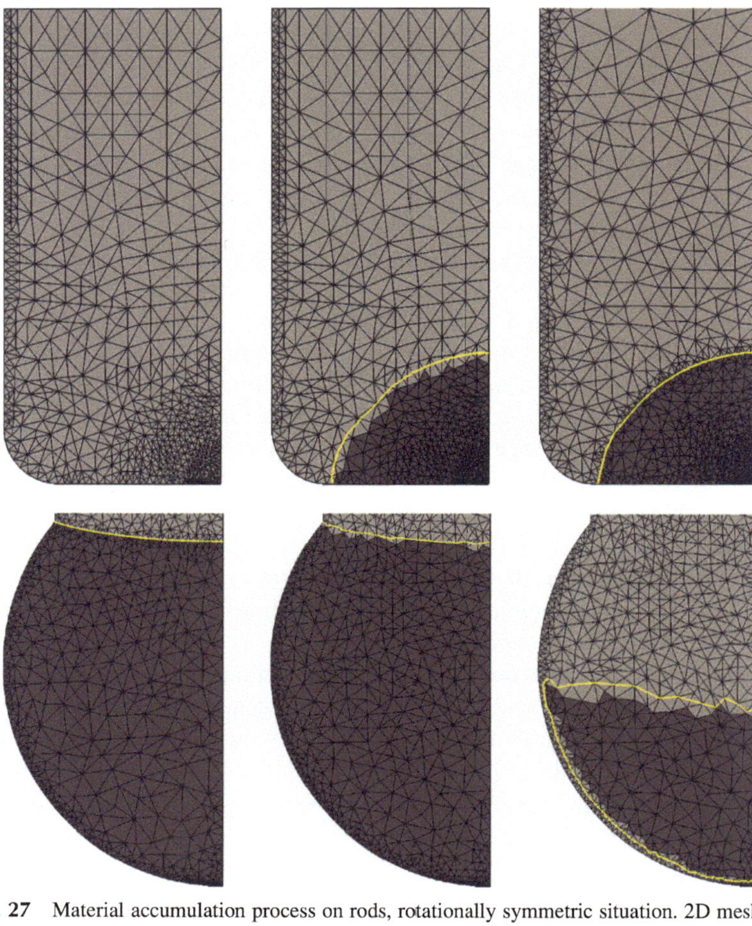

FIG. 27 Material accumulation process on rods, rotationally symmetric situation. 2D meshes at different times, together with phase boundary. Model switch from energy to ALE between pictures top middle and right, back to energy between bottom left and middle picture.

subdomain, the discretization of the capillary surface changes only (and then maybe abruptly) when new mesh elements are added to the liquid subdomain. The dominant surface tension then forces the capillary surface very fast to a sphere-like shape, which results in a rapid movement especially of the new edges belonging to the capillary surface. This introduces spurious velocities in the melt flow, leading to nonphysical waves. A (nonphysical, local, or global) enlargement of the viscosity of the fluid is able to reduce these numerical oscillations.

The ALE-approach, on the other hand, is able to implement the motion of the liquid subdomain and the capillary surface without such oscillations. But, as mentioned above, it is only applicable in simple situations. For the laser rod end melting with coaxial laser heating and a corresponding rotationally symmetric model, it is possible to use the ALE approach during the time interval after the initial melt nucleation and before the cooling.

When switching between the approaches, special care has to be taken for the phase boundary. The ALE approach needs a representation of the solid–liquid interface by mesh edges. In the energy approach, the interface is just the melting isothermal line of the temperature, which does typically not coincide with mesh edges. Thus, the domain must be remeshed when switching from the energy approach to the ALE approach. This is demonstrated in Fig. 27 for a 2D rotationally symmetric situation with symmetry axis on the right: Here, the dark grey indicates the molten and the light grey indicates the solid material. Both domains are separated by $\Sigma(t)$. In the top left picture, the initial state of the (solid) component is given. After starting the heating from below near the symmetry axis, the second and third picture visualize the first change of method from energy to ALE, and the fourth and fifth picture the second change back to energy after the laser is switched off. The bottom right picture shows typical characteristics of isothermal lines during the solidification process.

Fig. 28 shows the material accumulation at a rod end with a laterally applied, moving laser beam. Due to the nonsymmetric heating, also the geometry develops in a nonsymmetric way and the rotationally symmetric model cannot be used, but a fully 3D model is used. Here, the Energy approach is used during the whole simulation and spurious oscillations are damped by using a large viscosity.

Material accumulation at blank edges is shown in Fig. 29 for the case of a blank end, and in Fig. 30 for the case of a pierced blank, where the material is accumulated circularly around a hole.

More details of the methods and results for 2D and rotationally symmetric situations are presented in Jahn et al. (2012), Bänsch et al. (2013), and Jahn et al. (2013), for fully 3D simulations in Jahn et al. (2014). The numerical simulations were applied for control and optimization of the process, see Schmidt et al. (2017) and Rippel et al. (2018).

With transversal heating, the heat has to reach the solid–liquid interface by going through the whole liquid pool. Here, melt flow is very important for the

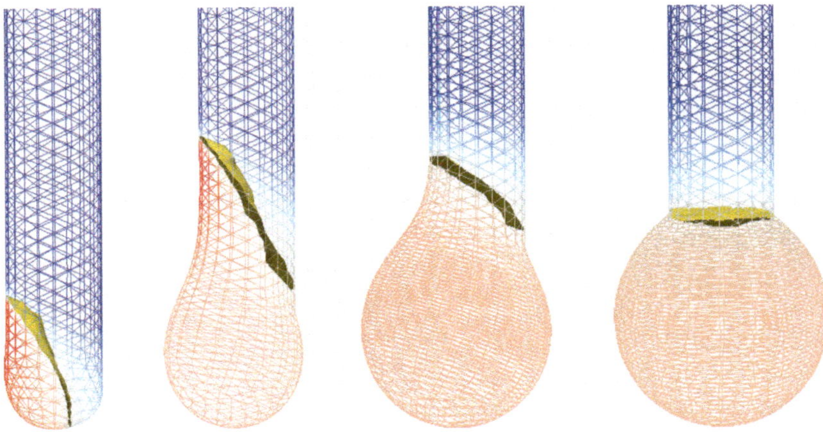

FIG. 28 Material accumulation process on rods using a laterally applied laser beam. 3D meshes (on surface) at four different times, together with phase boundary. *Colours* indicate temperature.

FIG. 29 Preform generation at blank end plains: A laterally applied laser heat source melts the material that forms a collar. 3D meshes (on surface) at two different times, together with phase boundary. *Colours* indicate temperature. *Bottom*: Final preform.

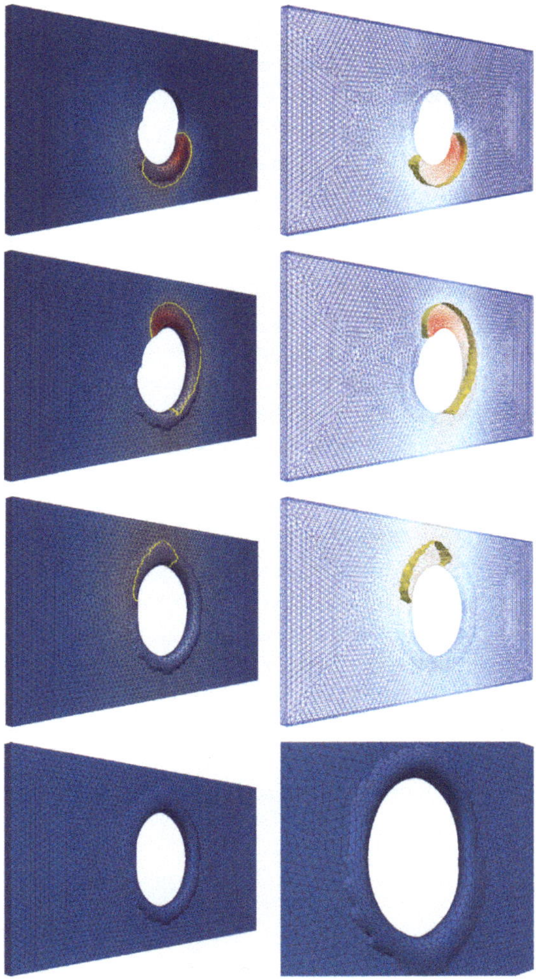

FIG. 30 Preform generation at pierced blank: A laterally applied laser heat source melts the material that forms a circular collar. 3D meshes (on surface) at three different times, together with phase boundary. *Colours* indicate temperature. *Bottom*: Final, closed thermal preform.

transport of heat. In contrast to this, using a lateral heating near the solid–liquid interface, the heat can directly affect the phase transition and melt flow is not dominant for heat transfer. Due to this fact, another possibility to avoid the spurious oscillations is a modification of the model which neglects the flow in the melt. Instead of the capillary boundary condition in the Navier–Stokes system, the surface tension could be taken into account by a quasi-stationary Young–Laplace equation which then gives the geometric shape of the capillary surface Γ at every time. The corresponding change of the subdomain Ω_l can be taken into account by a corresponding ALE method similar to the one described in Section 3.3.3.

5.4 Welded joints

The melting and solidification of material during a welding process is quite similar to the phase transitions during the material accumulation processes depicted above. Thus, the mathematical model is just the same.

In case of a hybrid joint of different materials, it is often intended that only one material (the one with lower melting temperature) melts, while the other one stays solid during the whole welding process. This is the case for hybrid joints of aluminium with another metallic material like steel or titan, as used in the production of lightweight structures. Typically, aluminium melts at a much lower temperature than steel, e.g. The molten aluminium then not only forms the welding seam geometry by capillary surface forces, but also by wetting the nonmelting material. This typically enlarges the contact area between the different materials and thus strengthens the joint. We can include this into the model by prescribing a contact angle, just as in Eq. (43).

We show here the simulation of the situation in a cut perpendicular to the welding direction. This reduces the model to a time-dependent 2D situation. Fig. 31 shows the deforming geometry of an overlap weld joint with meshes and isothermal lines at six different times during the simulation. The aluminium sheet is placed below the steel sheet, with some overlap. Heating is done by a defocused laser from above, mainly through the steel sheet.

The first four pictures show the melting phase, the fourth with nearly the maximal melt volume, and the fifth one is during the solidification. Light grey indicates triangles of the discrete liquid subdomain. Wetting of the steel sheet takes place at its lower side and at the vertical side. The final picture shows a good agreement of the computed weld seam geometry with experimental data.

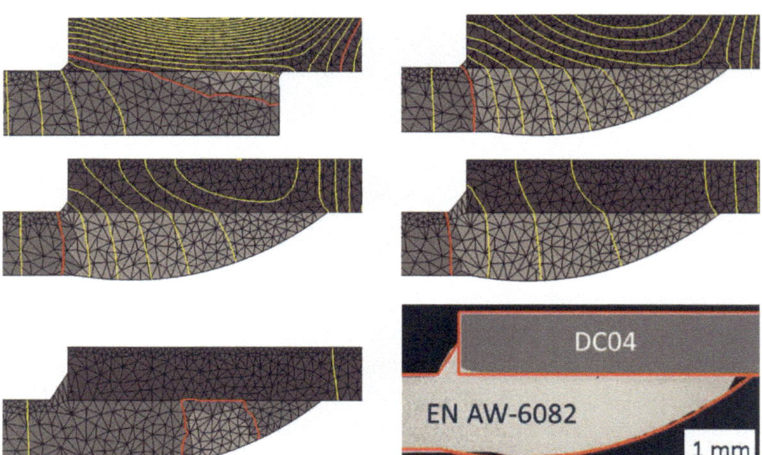

FIG. 31 Hybrid welding of aluminium with steel. Meshes and isothermal lines at five different times, and comparison with experimental data.

More details are shown in Kowalschuk et al. (2012), where more comparisons between experimental results and simulated weld seam geometries are presented and properties of the weld seam are examined. A detailed investigation of the numerical method is presented in Luttmann (2017).

5.5 Dendritic solidification with and without flow in the liquid

The surface energies of phase boundaries play an important role for the geometry and motion of interfaces when looking at a microscopic scale, where dendritic structures may grow during the solidification. Especially when the temperature in the liquid is below the melting temperature of the material, an initial solid nucleus will grow very fast and due to anisotropic surface energies (based on the possible bindings to a crystalline lattice, e.g.) there are directions in which the interface grows faster than in other ones. In alloys like steel, this happens not only in undercooled melts but even during the usual cooling and solidification in heat treatment processes.

As shown in Section 2.3, the modelling of such evolutions of phase boundaries is strongly coupled to the motion of interfaces by (anisotropic) mean curvature. Numerical simulation thus can use the various approaches depicted in Section 4. We give here some examples of front-tracking and phase field approaches.

5.5.1 Sharp interface approach for dendritic growth

First, we demonstrate results from 2D and 3D simulations of dendritic growth in undercooled liquids neglecting a melt flow. The discretization of the coupled Problem (2) combines an adaptive finite element method for the heat equation with the numerical method for interface evolution by anisotropic mean curvature from Section 4.1. The coupling of both is given by the Gibbs–Thomson relation (24c) at the solid–liquid interface, which results in a right-hand side for the interface evolution given by the temperature, and a right-hand side in the heat equation involving the curvature of the interface.

The simulation starts with a small circular (respectively, spherical) nucleus of solid inside the otherwise liquid domain. Thus, the initial discrete interface $\Sigma_h(0)$ is an approximation of the small circle resp. sphere. The initial temperature θ_0 is chosen such that it satisfies a Gibbs–Thomson relation at the interface and declines fast to the given undercooling in the liquid away from the interface.

Fig. 32 shows the interfaces at several times from a 2D simulation with sixfold symmetry, starting with the small circle in the middle and developing six nearly symmetric dendrite arms which show a nearly constant tip velocity and curvature after some initial setup time. Fig. 33 shows the graph of the temperature and a corresponding adaptively refined triangulation from a similar simulation with sixfold symmetry.

FIG. 32 Dendritic growth in 2D: Typical growth of interface for a sixfold symmetric anisotropy.

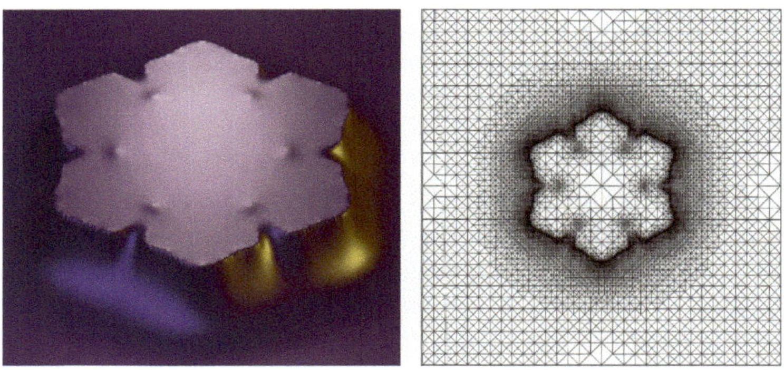

FIG. 33 Dendritic growth in 2D: Graph of temperature and corresponding adaptively refined mesh.

Results from a 3D simulation are shown in the next figures. Here, the anisotropy was chosen such that the growth in direction of the coordinate axes was preferred. Using the corresponding symmetry, computations were restricted to one octant of the cubic domain. Fig. 34 shows the growing solid at different times, starting from a small sphere to a complex dendritic shape. The final shape is shown once more in Fig. 35, together with a trace of the adaptively refined 3D mesh. Edges of (slightly shrunken) triangles from the 3D tetrahedral mesh lying on the $x_2 = 0$ coordinate plane are shown, coloured with temperature values. The temperature and corresponding mesh show the same behaviour as depicted in Fig. 33 for the 2D situation.

Fig. 36 shows not only the trace of the 3D temperature mesh, but also the 2D triangular mesh for discretization of the growing interface. In order to approximate the curvature and growth of the interface well enough, the

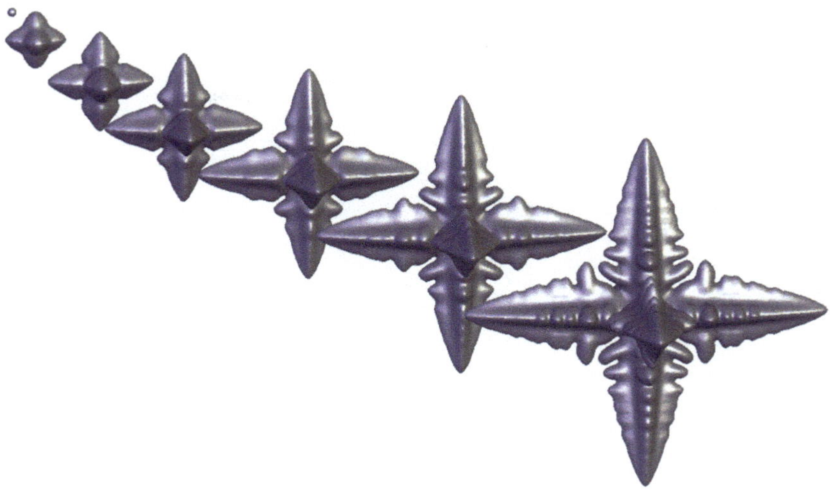

FIG. 34 Dendritic growth in 3D: Interface after 0, 40, 80, 120, 160, 200, 240 time steps.

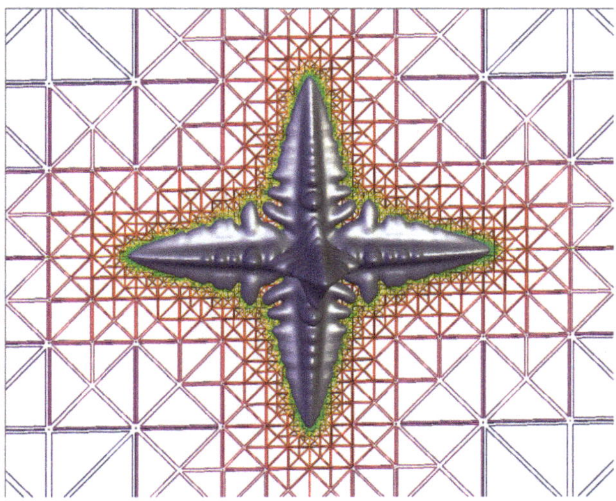

FIG. 35 Final dendrite and trace of corresponding 3D mesh, *coloured* with temperature.

interface discretization is quite fine. The 3D mesh can be chosen much coarser, with adaptive refinement near the interface based on error indicators from the heat equation.

5.5.2 Dendritic growth with convection in the melt

The simulations above show a lot of symmetry in the developing dendrites. Under the influence of convection in the melt, symmetries are reduced or even

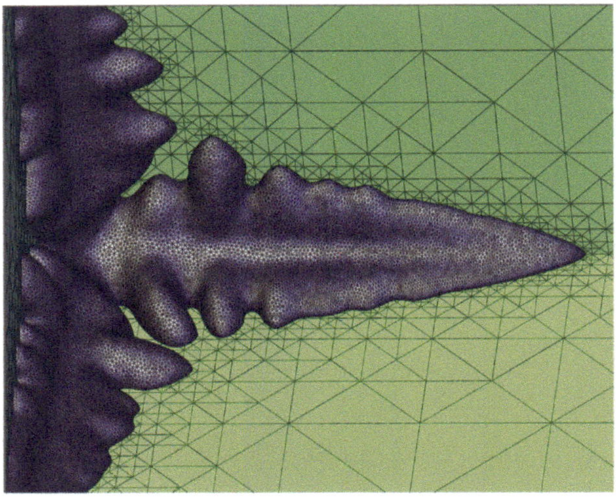

FIG. 36 2D mesh for final dendrite and traces of corresponding temperature mesh.

not visible at all. This is due to the fact that the energy set free from the latent heat of solidification is transported by the convection and thus changes the local thermal situation differently at various places near the growing interface. If we consider just thermal convection under gravity, heat is transported upwards by the convection. Thus, undercooling is stronger at lower parts of the dendrite, letting it grow faster downwards than upwards. This can be seen in Fig. 37. Here, a fourfold symmetry in the anisotropy leads to preferred growth in the diagonal directions of a square domain. By the thermal convection, the growth is amplified in the lower diagonal directions and damped in the upper ones. Temperature and flow are depicted in the left part of the picture, the corresponding mesh for temperature and flow in the right part. The liquid subdomain $\Omega_l(t)$ is defined by all mesh elements that are fully liquid at time t, similar to the specification in Eq. (67), but here not the temperature but the discrete interface $\Sigma_h(t)$ defines the boundary between solid and liquid. An additional advection from the left, given by nonzero boundary values for the flow velocity $\mathbf{u}$ at $\partial\Omega$, destroys symmetry completely, as is shown in the bottom part of Fig. 37.

More details of the numerical method are presented in Bänsch and Schmidt (2000), where also the figures were taken from.

5.5.3 Phase field approach

In phase field models, compare Section 2.5, the phase variable is constant inside of each pure phase (solid and liquid) and changes values smoothly in the narrow region between them. Thus, a good approximation needs a mesh with very high resolution in the interface region, but can be as coarse as

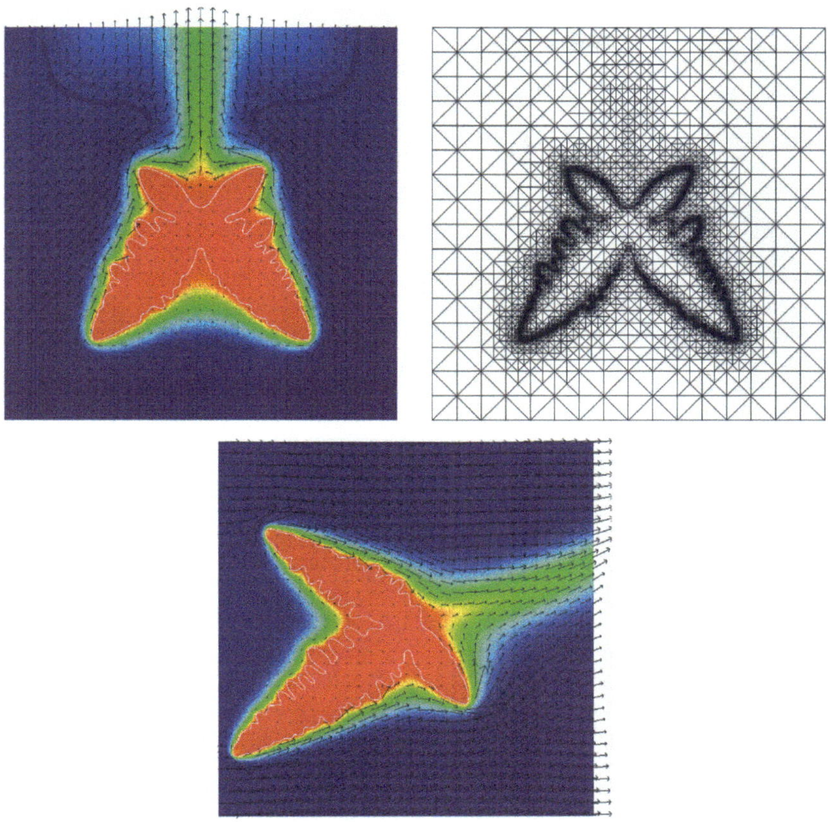

FIG. 37 Dendritic growth under influence of melt flow. *Top*: Natural thermal convection. Temperature and flow field (*left*) and corresponding mesh (*right*). *Bottom*: Additional inflow from *left*. *From Bänsch, E., Schmidt, A., 2000. Simulation of dendritic crystal growth with thermal convection. Interfaces Free Bound 2, 95–115.*

possible inside the phases. The other variable in the system, the temperature, varies inside the phases but not as abruptly as the phase variable. Thus, a coarser mesh is sufficient for the temperature in the interface region, but an (adaptively) refined mesh is needed also inside the phase regions. An optimal discretization would account for both (different) requirements by using different meshes for the two variables. Nevertheless, information must be exchanged between the finite element spaces corresponding the two meshes. Utilizing the local hierarchy between two locally refined meshes originating from the same macro triangulation can provide structures which facilitate the information exchange. A corresponding multi mesh method was presented and described for 2D meshes in Schmidt (2003b) and for 3D meshes in Schmidt (2003a). Fig. 38 shows both meshes for temperature and phase variable from a phase field simulation of dendritic growth.

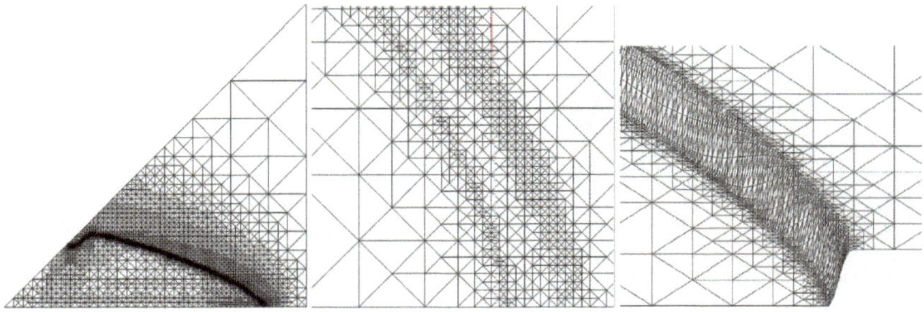

FIG. 38 Phase field simulation: Mesh for temperature (*left*), zoom into mesh for phase variable (*middle*), and graph of phase variable (*right*).

Acknowledgements

Many of the figures were created by C. Schneider from Univ. of Erlangen, whose help is gratefully acknowledged. Results in Section 5.2 were supported by DLR grants 50RL0921, 50RL0741, FKZ50JR0011 and partly also by a PAKT grant of the Leibniz Association. Results presented in Section 5.3 were supported by the German Research Foundation (DFG) via project A3 of the Collaborative Research Center 747 "Micro cold forming", results presented in Section 5.4 were supported by the German Ministry of Economics and Technology (BMWi) via AIF project IGF 3760 ZN. All fundings are gratefully acknowledged. We thank all our scientific partners for cooperation.

References

Abels, H., Garcke, H., Grün, G., 2012. Thermodynamically consistent, frame indifferent diffuse interface models for incompressible two-phase flows with different densities. Math. Models Methods Appl. Sci. 22 (3), 1150013.

Aland, S., Boden, S., Hahn, A., Klingbeil, F., Weismann, M., Weller, S., 2013. Quantitative comparison of Taylor flow simulations based on sharp-interface and diffuse-interface models. Int. J. Numer. Meth. Fluids 73 (4), 344–361.

Ausas, R.F., Dari, E.A., Buscaglia, G.C., 2011. A geometric mass-preserving redistancing scheme for the level set function. Int. J. Numer. Meth. Fluids 65, 989–1010.

Bänsch, E., 2001. Finite element discretization of the Navier-Stokes equations with a free capillary surface. Numer. Math. 88 (2), 203–235.

Bänsch, E., Schmidt, A., 2000. Simulation of dendritic crystal growth with thermal convection. Interfaces Free Bound 2, 95–115.

Bänsch, E., Berg, C.P., Ohlhoff, A., 2004. Uniaxial extensional flows in liquid bridges. J. Fluid Mech. 521, 353–379.

Bänsch, E., Paul, J., Schmidt, A., 2013. An ALE finite element method for a coupled Stefan problem and Navier-Stokes equations with free capillary surface. Int. J. Numer. Meth. Fluids 71, 1282–1296.

Bänsch, E., Basting, S., Krahl, R., 2015. Numerical simulation of two-phase flows with heat and mass transfer. Discrete Cont. Dyn. Syst. A 35 (6), 2325–2347. ISSN 1078-0947.

Barrett, J.W., Garcke, H., Nürnberg, R., 2010. On stable parametric finite element methods for the Stefan problem and the Mullins-Sekerka problem with applications to dendritic growth. J. Comput. Phys. 229, 6270–6299.

Barrett, J.W., Garcke, H., Nürnberg, R., 2012. Numerical computations of facetted pattern formation in snow crystal growth. Phys. Rev. E 86, 011604.

Barrett, J.W., Garcke, H., Nürnberg, R., 2013. Eliminating spurious velocities with a stable approximation of viscous incompressible two-phase Stokes flow. Comput. Methods Appl. Mech. Eng. 267, 511–530.

Barrett, J.W., Garcke, H., Nürnberg, R., 2015. A stable parametric finite element discretization of two-phase Navier-Stokes flow. J. Sci. Comput. 63 (1), 78–117.

Bellettini, G., Paolini, M., 1996. Anisotropic motion by mean curvature in the context of Finsler geometry. Hokkaido Math. J. 25, 537–566.

Berg, C.P., Dreyer, M., Rath, H., 1999. A large fluid-bridge device to measure the deformation of drops in uniaxial extensional flow fields. Meas. Sci. Technol. 10, 956–964.

Besier, M., Wollner, W., 2012. On the pressure approximation in nonstationary incompressible flow simulations on dynamically varying spatial meshes. Int. J. Numer. Meth. Fluids 69, 1045–1064.

Blowey, J., Elliott, C., 1993. Curvature dependent phase boundary motion andparabolic double obstacle problems. In: Wei-Ming, Ni et al., (Eds.), Degeneratediffusions, vol. 47, IMA Vol. Math. Appl, 19–60.

Brackbill, J.U., Kothe, D., Zemach, C., 1992. A continuum method for modeling surface tension. J. Comput. Phys. 100 (2), 335–354. ISSN 0021-9991.

Brenner, A., Bänsch, E., Bause, M., 2014. A-priori error analysis for finite element approximations of stokes problem on dynamic meshes. IMA J. Numer. Anal. 34, 123–146.

Buscaglia, G.C., Ausas, R.F., 2011. Variational formulations for surface tension, capillarity and wetting. Comput. Methods Appl. Mech. Eng. 200, 3011–3025.

Caginalp, G., 1986. An analysis of a phasefield model of a free boundary. Arch. Ration. Mech. Anal. 92, 205–245.

Das, S.P., Hopfinger, E.J., 2009. Mass transfer enhancement by gravity waves at a liquid-vapour interface. Int. J. Heat Mass Transf. 52 (5–6), 1400–1411. ISSN 0017-9310.

Deckelnick, K., Dziuk, G., 1999. Discrete anisotropic curvature flow of graphs. Esaim Math. Model. Numer. Anal. 33, 1203–1222.

Donea, J., Giuliani, S., Halleux, J.P., 1982. An arbitrary Lagrangian-Eulerian finite element method for transient dynamic fluid-structure interactions. Comput. Methods Appl. Mech. Eng. 33, 689–723.

Dziuk, G., 1991. An algorithm for evolutionary surfaces. Numer. Math. 58, 603–611.

Dziuk, G., 1999. Discrete anisotropic curve shortening flow. SIAM J. Numer. Anal. 36, 1808–1830.

Dziuk, G., 1999. Numerical schemes for the mean curvature flow of graphs. In: Argoul, P., Frémond, M., Nguyen, Q.S. (Eds.), IUTAM Symposium on Variations of Domain and Free-Boundary Problems in Solid Mechanics. Solid Mechanics and Its Applications, vol. 66. Springer, Netherlands, pp. 63–70.

Eck, C., Fontelos, M., Grün, G., Klingbeil, F., Vantzos, O., 2009. On a phase-field model for electrowetting. Interfaces Free Bound. 11 (2), 259–290.

Elliott, C. M., Schätzle, R., 1995. The limit of the anisotropic double-obstacle Allen-Cahn equation. CMAIA Report 95-05, Univ. of Sussex.

Evans, L.C., Soner, H.M., Souganidis, P.E., 1992. Phase transitions and generalized motion by mean curvature. Commun. Pure Appl. Math. 45, 1097–1123.

Finn, R., 1986. Equilibrium Capillary Surfaces. Grundlehren der Mathematischen Wissenschaften [Fundamental Principles of Mathematical Sciences], vol. 284. Springer-Verlag.

Fried, M., 2004. A level set based finite element algorithm for the simulation of dendritic growth. Comput. Vis. Sci. 7, 97–110.

Fuhrmann, E., Dreyer, M., 2008. Description of the sounding rocket experiment–SOURCE. Microgravity Sci. Technol. 20 (3–4), 205–212. ISSN 0938-0108.

Fuhrmann, E., Dreyer, M., 2009. Heat transfer by thermo-capillary convection. Microgravity Sci. Technol. 21 (1), 87–93. ISSN 0938-0108.

Fuhrmann, E., Dreyer, M., Basting, S., Bänsch, E., 2016. Free surface deformation and heat transfer by thermocapillary convection. Heat Mass Transf. 52 (4), 855–876.

Garcke, H., Nestler, B., Stoth, B., 1998. On anisotropic order parameter models for multi-phase systems and their sharp interface limits. Phys. D 115, 87–108.

Gerstmann, J., 2004. Numerische Untersuchung zur Schwingung freier Flüssigkeitsoberflächen. Fortschritt-Berichte VDI, Reihe 7: Strömungsmechanik. VDI-Verlag, Düsseldorf, 464.

Gibou, F., Chen, L., Nguyen, D., Banerjee, S., 2007. A level set based sharp interface method for the multiphase incompressible Navier-Stokes equations with phase change. J. Comput. Phys. 222 (2), 536–555.

Glicksman, M.E., Schaefer, R.J., Ayers, J.D., 1976. Dendritic growth—a test of theory. Metal. Trans. A 7A, 1747–1759.

Groß, S., Reusken, A., 2007. An extended pressure finite element space for two-phase incompressible flows with surface tension. J. Comput. Phys. 224 (1), 40–58.

Groß, S., Reusken, A., 2007. Finite element discretization error analysis of a surface tension force in two-phase incompressible flows. SIAM J. Numer. Anal. 45 (4), 1679–1700.

Groß, S., Reusken, A., 2011. Numerical Methods for Two-phase Incompressible Flows. Springer Series in Computational Mathematics, vol. 40. Springer, Berlin, Heidelberg. ISBN: 978-3-642-19685-0.

Gurtin, M.E., 1981. An Introduction to Continuum Mechanics. Academic Press.

Gurtin, M.E., 1988. Toward a nonequilibrium thermodynamics of two-phase materials. Arch. Ration. Mech. Anal. 100, 275–312.

Hirt, C.W., Nichols, B.D., 1981. Volume of fluid (VOF) method for the dynamics of free boundaries. J. Comput. Phys. 39 (1), 201–225.

Hirt, C.W., Amsden, A.A., Cook, J.L., 1974. An arbitrary Lagrangian-Eulerian computing method for all flow speeds. J. Comput. Phys. 14 (3), 227–253.

Hughes, T.J.R., Liu, W.K., Zimmermann, T.K., 1981. Lagrangian-Eulerian finite element formulation for incompressible viscous flows. Comput. Methods Appl. Mech. Eng. 29 (3), 329–349.

Iribarne, J.V., Godson, W.L., 2013. Atmospheric Thermodynamics. Springer Science & Business Media.

Jacqmin, D., 1999. Calculation of two-phase Navier-Stokes flows using phase-field modeling. J. Comput. Phys. 155 (1), 96–127.

Jahn, M., Klock, T., 2016. A level set toolbox including reinitialization and mass correction algorithms for FEniCS. ZeTeM Report 16-01, University of Bremen.

Jahn, M., Luttmann, A., Schmidt, A., Paul, J., 2012. Finite element methods for problems with solid-liquid-solid phase transitions and free melt surface. PAMM 12, 402–404.

Jahn, M., Luttmann, A., Schmidt, A., 2013. Finite element simulation for material accumulation and welding processes including a free melt surface. PAMM 13, 235–236.

Jahn, M., Schmidt, A., Bänsch, E., 2014. 3d finite element simulation of a material accumulation process including phase transitions and a capillary surface. ZeTeM Report 14-03, University of Bremen.

Jamet, D., Lebaigue, O., Coutris, N., Delhaye, J.M., 2001. The second gradient method for the direct numerical simulation of liquid-vapor flows with phase change. J. Comput. Phys 169 (2), 624–651.

Kennard, E.H., 1938. Kinetic Theory of Gases: With an Introduction to Statistical Mechanics. International Series in Pure and Applied Physics. McGraw-Hill.

Kessler, D., Nochetto, R.H., Schmidt, A., 2004. A posteriori error control for the Allen-Cahn problem: circumventing Gronwall's inequality. Model. Math. Anal. Num. 38, 129–142.

Kornhuber, R., 1994. Monotone multigrid methods for elliptic variational inequalities I. Numer. Math. 69, 167–184.

Kowalschuk, M., Luttmann, A., Jahn, M., Barr, A., Schmidt, A., von Hehl, A., Vollertsen, F., 2012. Challenges in simulation of welded hybrid joints. Proc. 3rd Fatigue Symposium, Montanuniversität Leoben, Austria, pp. 85–112.

Krahl, R., Adamov, M., Lozano Avilés, M., Bänsch, E., 2004. A model for two phase flow with evaporation. ISBN 88-467-1075-4. In: Celata, G.P., Di Marco, P., Mariani, A., Shah, R.K. (Eds.), Two-Phase Flow Modelling and Experimentation 2004 (3rd International Symposium on Two-Phase Flow Modelling and Experimentation, Pisa, September 2004). Edizioni ETS; Pisa S. 2381–2387. ISBN 88-467-1075-4.

Kulev, N., Dreyer, M., 2010. Drop tower experiments on non-isothermal reorientation of cryogenic liquids. Microgravity Sci. Technol. 22 (4), 463–474.

Kulev, N., Basting, S., Bänsch, E., Dreyer, M., 2014. Interface reorientation of cryogenic liquids under non-isothermal boundary conditions. Cryogenics 62, 48–59. ISSN 0011-2275.

Langer, J.S., 1980. Instabilities and pattern formation in crystal growth. Rev. Mod. Phys. 52, 1–28.

Rayleigh, Lord, 1878. On the instability of jets. Proc. London Math. Soc. 4, 207–285.

Luttmann, A., 2017. Modellierung und Simulation von Prozessen mit fest-flüssig Phasenübergang und freiem Kapillarrand. Ph.D. thesis, Dissertation, University of Bremen.

Matthies, G., 2002. Finite Element Methods for free Boundary Value Problems With Capillary Surfaces. Shaker.

Mikula, K., Sevcovic, D., 2001. Evolution of plane curves driven by a nonlinear function of curvature and anisotropy. SIAM J. Appl. Math. 61, 1473–1501.

Mikula, K., Remesikova, M., Sarkoci, P., Sevcovic, D., 2014. Manifold evolution with tangential redistribution of points. SIAM J. Sci. Comput. 36, A1384–A1414.

Moës, N., Dolbow, J., Belytschko, T., 1999. A finite element method for crack growth without remeshing. Int. J. Numer. Meth. Eng. 46 (1), 131–150.

Osher, S., Sethian, J.A., 1988. Fronts propagating with curvature-dependent speed: algorithms based on Hamilton-Jacobi formulations. J. Comput. Phys 79 (1), 12–49.

Paolini, M., Verdi, C., 1992. Asymptotic and numerical analyses of the mean curvature flow with a space-dependent relaxation parameter. Asymptotic Anal. 5, 553–574.

Peng, D., Merriman, B., Osher, S., Zhao, H., Kang, M., 1999. A PDE-based fast local level set method. J. Comput. Phys 155, 410–438.

Pismen, L.M., Pomeau, Y., 2000. Disjoining potential and spreading of thin liquid layers in the diffuse-interface model coupled to hydrodynamics. Phys. Rev. E 62 (2), 2480–2492.

Plateau, J., 1863. Experimental and theoretical researches on the figures of equilibrium of a liquid mass withdrawn from the action of gravity, etc. In: Annual Report of the Board of Regents of the Smithsonian Institution, Pt. 1, House of Representatives Misc. Doc, vol. 83, pp. 207–285.

Rippel, D., Schattmann, C., Jahn, M., Lütjen, M., Schmidt, A., 2018. Application of cause-effect-networks for the process planning in laser rod end melting. In: MATEC Web of Conferences, vol. 190, p. 15005.

Roosen, A., Taylor, J.E., 1991. Simulation of crystal growth with facetted interfaces. In: MRS Proc, vol. 237. Cambridge University Press, p. 25.

Rumpf, M., 1996. A variational approach to optimal meshes. Numer. Math. 72, 523–540. ISSN 0029-599X.

Scardovelli, R., Zaleski, S., 1999. Direct numerical simulation of free-surface and interfacial flow. Annu. Rev. Fluid Mech. 31, 567–603.

Schlottke, J., Weigand, B., 2008. Direct numerical simulation of evaporating droplets. J. Comput. Phys. 227 (10), 5215–5237.

Schmidt, A., 1996. Computation of three dimensional dendrites with finite elements. J. Comput. Phys. 125, 293–312.

Schmidt, A., 1998. Approximation of crystalline dendrite growth in two space dimensions. Acta Math. Univ. Comenianae 67, 57–68.

Schmidt, A., 2003. A multi-mesh finite element method for 3D phase field simulations. In: Colli, P., Verdi, C., Visintin, A. (Eds.), Free Boundary Problems, Int. Ser. Numer. Math, vol. 147. Birkhäuser, Basel, pp. 293–301.

Schmidt, A., 2003. A multi-mesh finite element method for phase field simulations. In: Emmerich, H., Nestler, B., Schreckenberg, M. (Eds.), Interface and Transport Dynamics–Computational Modelling, Lecture Notes in Computational Science and Engineering, vol. 32. Springer, Berlin, pp. 208–217.

Schmidt, A., Bänsch, E., Jahn, M., Luttmann, A., Niebuhr, C., Vehmeyer, J., 2017. Optimization of engineering processes including heating in time-dependent domains. In: Bociu, L., Desideri, J.A., Habbal, A. (Eds.), System Modeling and Optimization, 27th IFIP TC 7 Conference, CSMO 2015, IFIP AICT Series, vol. 494. Springer, Berlin, pp. 452–461.

Scriven, L.E., 1960. Dynamics of a fluid interface equation of motion for Newtonian surface fluids. Chem. Eng. Sci. 12 (2), 98–108.

Sethian, J.A., 1996. Level Set Methods. Evolving Interfaces in Geometry, Fluid Mechanics, Computer Vision, and Materials Science. Cambridge Monograph on Applied and Computational Mathematics. Cambridge University Press.

Sethian, J.A., Smereka, P., 2003. Level set methods for fluid interfaces. Annu. Rev. Fluid Mech. 35, 341–372.

Son, G., Dhir, V.K., 1998. Numerical simulation of film boiling near critical pressures with a level set method. J. Heat Trans. 120 (1), 183–192.

Sussman, M., Smereka, P., Osher, S., 1994. A level set approach for computing solutions to incompressible two-phase flow. J. Comput. Phys. 114 (1), 146–159.

Tanguy, S., Ménard, T., Berlemont, A., 2007. A level set method for vaporizing two-phase flows. J. Comput. Phys. 221 (2), 837–853.

Tezduyar, T., Benney, R., 2003. Mesh moving techniques for fluid-structure interactions with large displacements. J. Appl. Mech. 70 (1), 58–63.

Unverdi, S.O., Tryggvason, G., 1992. A front-tracking method for viscous, incompressible, multi-fluid flows. J. Comput. Phys. 100 (1), 25–37.

Welch, S.W.J., Wilson, J., 2000. A volume of fluid based method for fluid flows with phase change. J. Comput. Phys 160 (2), 662–682.

Weller, S., 2008. Higher Order Time Discretization for Free Surface Flows. Ph.D. thesis. Diploma thesis, University of Erlangen.

Weller, S., 2015. Time Discretization for Capillary Problems. PhD thesis, Dissertation, University of Erlangen.

Weller, S., Bänsch, E., 2017. Time discretization for capillary flow: beyond backward euler. In: Bothe, D., Reusken, A. (Eds.), Transport Processes at Fluidic interfaces. Springer, pp. 121–143.

Wick, T., 2011. Fluid-structure interactions using different mesh motion techniques. Comput. Struct. 89 (13–14), 1456–1467.

Xie, W., 1992. Phase transition in binary systems and the binary alloy problem. Math. Meth. Appl. Sci. 15, 205–221.

Yap, Y.F., Chai, J.C., Toh, K.C., Wong, T.N., Lam, Y.C., 2005. Numerical modeling of unidirectional stratified flow with and without phase change. Int. J. Heat Mass Transf. 48 (3–4), 477–486.

Chapter 8

Discrete Riemannian calculus on shell space

Behrend Heeren[a], Martin Rumpf[a], Max Wardetzky[b,*] and Benedikt Wirth[c]

[a]*Institute for Numerical Simulation, University of Bonn, Bonn, Germany*

[b]*Institute of Numerical and Applied Mathematics, University of Göttingen, Göttingen, Germany*

[c]*Institute for Analysis and Numerics, University of Münster, Münster, Germany*

[*]*Corresponding author: e-mail: wardetzky@math.uni-goettingen.de*

Chapter Outline

Handbook of Numerical Analysis, Vol. 21. https://doi.org/10.1016/bs.hna.2019.06.005

Abstract
This article discusses a Riemannian calculus on the space of discrete shells, represented by triangular meshes, and an effective as well as efficient discretization thereof. This calculus comprises shape morphing via geodesic curves, shape extrapolation via the Riemannian exponential map, the transfer of geometric detail via Riemannian parallel transport, and smooth interpolation of multiple key frames using Riemannian splines. The underlying Riemannian structure is derived from the Hessian of an elastic deformation energy. The discretization is based on an approximation of the Riemannian path energy via a sum of deformation energies defined on pairs of triangular shapes along a finite sequence of discrete shells. Furthermore, convergence results of all components of the discrete calculus to their continuous counterparts are provided, and a variety of surface processing applications are discussed.

Keywords: Riemannian shape spaces, Geometry processing, Discrete shells, Geodesic curves, Riemannian splines

AMS Classification Codes: 53B21, 68U05, 74K25

1 Introduction

Computer animated video sequences and movies are nowadays omnipresent. Artists in animation studios design sophisticated characters of increasing complexity and authenticity. Character models are typically represented by triangle meshes consisting of tens or even hundreds of thousands of degrees of freedom. Moreover, animators are able to generate movements of these characters that are almost indistinguishable from natural motions. The development of flexible and effective tools supporting artists in creating such authentic animations is linked to the mathematics of shape spaces. In particular, a comprehensive understanding of both the geometry and the physics of natural deformations and motion paths of complex shapes is essential for the creation of realistic models and efficient algorithms.

In this article we treat shapes as discrete shell surfaces, represented as triangle meshes. We view shapes as objects in a certain Riemannian shape manifold. Building on this perspective, we discuss discrete versions of fundamental tools in Riemannian calculus and show how to apply these tools for processing and animating discrete surfaces. Typical tasks in geometry processing include:

- *Morphing*. Given two poses of a complex shape one aims at computing a one-parameter family of shapes, which continuously transforms the first pose into the second pose in a natural and visually appealing manner. We show that geodesic paths in the space of shapes are promising candidates for such morphing tasks, cf. Fig. 1.
- *Extrapolation*. Given a particular shape one seeks to identify small (ideally infinitesimal) variations that represent natural movements and extrapolate these small variations to create animated versions of a given shape, cf. Fig. 2. Here, the Riemannian exponential map generates motions that correspond to given infinitesimal variations.

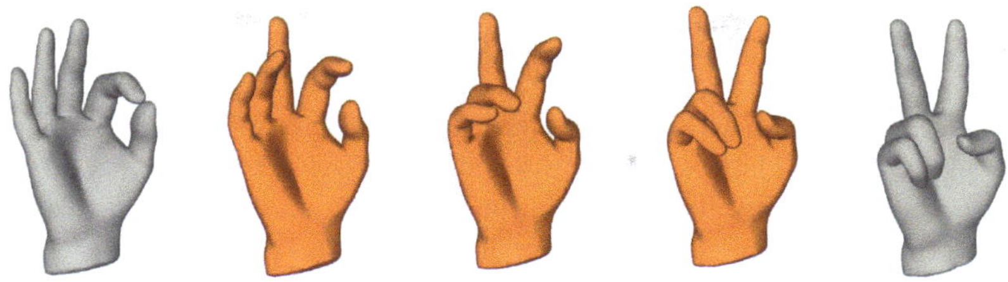

FIG. 1 An interpolating morph (*orange*) in the space of discrete shells between two different poses (*grey*).

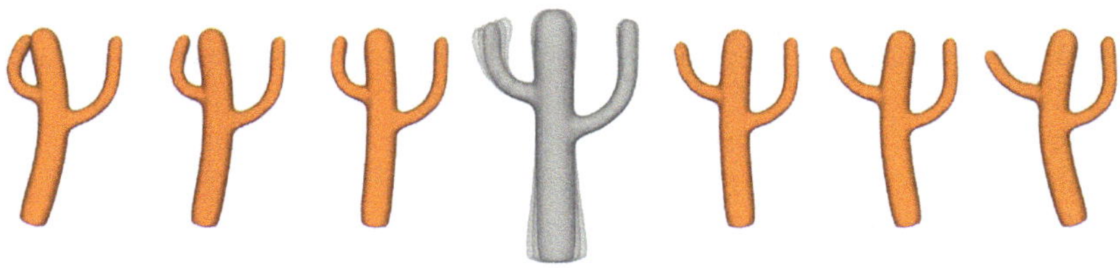

FIG. 2 Shape modelling: we determine plausible, infinitesimal variations of a static cactus pose (*grey*) and compute extrapolations to create realistic motion paths (*orange*).

FIG. 3 A keyframe interpolation via Riemannian elastic splines (*orange*) allows for a temporally smooth interpolation of a given set of keyframe poses (*grey*, *left*)—in contrast to a consecutive morphing technique (*green*).

- *Transfer of variations*. Once an animation of some shape has been generated, one would like to obtain similar animations for other shapes that differ by certain geometric details from the first shape. To this end, we consider Riemannian parallel transport of a shape variation along an animation path. This way, local (fine scale) details, e.g., skin wrinkles, can be superimposed onto an existing deformation sequence. Our discrete framework not only allows for transporting small geometric detail, but for more global shape modulations.
- *Smooth keyframe animation*. Given a sequence of character poses (keyframes), one aims at computing a smooth, realistic deformation path meeting all of the keyframes. Such deformation paths can be generated using Riemannian splines that interpolate the keyframes. In general, such smooth interpolation paths *cannot* be realized by straightforward morphing between pairs of consecutive keyframes, cf. Fig. 3.

From Hessian structure to Riemannian calculus. Our point of departure is a physically sound elastic energy that takes into account membrane and bending distortion as fundamental modes of shape variability. We then apply Rayleigh's paradigm and derive a viscous dissipation model from the elastic energy, replacing elastic strains by strain rates. The rate of viscous dissipation caused by an infinitesimal variation of one particular shape corresponds to the Hessian of the elastic energy. This Hessian structure defines a Riemannian metric on the space of shell surfaces by acting on infinitesimal shape variations, which are considered as tangent vectors to the shape manifold at a particular shape. On this basis we offer a comprehensive *geometric calculus* that induces consistent tools for navigating the space of shells. In particular, we present a model that is invariant with respect to rigid body motions. We discuss a variational time discretization of the resulting geodesic calculus.

Benefits of the Riemannian approach. It is due to Kendall (1984) that complex shapes, e.g., curves, images, or solid materials, are considered as individual elements, or points, in a high or even infinite dimensional *shape space*. Initially, this space is just a collection of shapes without any mathematical structure. Most shape spaces cannot be considered as linear vector spaces. Nevertheless, one is interested in performing mathematical analysis on the set of shapes, such as computing the distance between two shapes or shortest paths connecting them (cf. Fig. 1). Equipping the space of (discrete) shells with a Riemannian metric provides a framework for a variety of geometry processing tools with a rigorous geometric foundation. Even on very detailed and thus high-dimensional shell models the Riemannian approach allows for an appropriate geometric interpretation of complex processing tasks.

An effective variational time discretization. In a Riemannian setup, shortest paths, i.e., geodesics, are defined as *time-continuous* minimizers of the *time-continuous* path energy. In order to introduce a consistent time-discrete counterpart we make use of *a variational time discretization*, developed in a series of papers, cf. e.g., Wirth (2009), Wirth et al. (2011), Rumpf and Wirth (2013). This discrete calculus is based on a time discretization of the path energy; consequently, minimizers of this *discrete path energy* will be denoted as *discrete geodesics*. Besides the notion of discrete geodesics, the framework additionally provides (time-)discrete analogues for several basic but crucial geometric operators, e.g., for the exponential map and for parallel transport. The entire variational time discretization is based on a local approximation of the squared Riemannian distance, which is naturally linked to Rayleigh's paradigm.

Discrete and continuous Riemannian calculus differ substantially with respect to the way they are derived. In classical Riemannian calculus, one first introduces the covariant derivative. Then geodesics are defined as solutions of an ODE initial value problem—geodesics are curves whose velocity has vanishing covariant derivative. The exponential map takes an initial velocity vector to the position at time 1 under the associated geodesic flow, and the logarithm is obtained as the corresponding inverse mapping. In contrast, in discrete Riemannian calculus one first defines discrete geodesics as the solution of a variational problem. Then a notion of discrete logarithm is introduced, and the discrete exponential map is defined as the corresponding inverse. Discrete exponential map and discrete logarithm allow for the construction of discrete parallel transport. Based on this discrete transport, a discrete covariant derivative can be derived. Under certain assumptions on the underlying manifold, one can show consistency and convergence of time-discrete operators to their corresponding continuous counterparts (Rumpf and Wirth, 2015).

Finally, the discrete Riemannian calculus can be extended to *Riemannian splines*, which solve a *multinodal interpolation problem*, i.e., the computation of a smooth path interpolating more than two keyframe shapes, cf. Fig. 3. In contrast to piecewise geodesic interpolation, splines do not exhibit discontinuities in the velocity field at keyframes. In Euclidean space, cubic splines minimize the total squared second time derivative among all curves that pass through a given set of interpolation points—a property that is related to the minimization of bending energy. We build on this observation and introduce a generalized *Riemannian spline energy* based on the integration of the squared covariant derivative of the velocity field along the curve. Furthermore, we present a corresponding time-discrete functional that fits into our discrete geodesic calculus, and we discuss the consistency of this discretization based on Heeren et al. (2019).

Organization. This article is organized as follows. We discuss related work in Section 2. We then investigate the space of triangle meshes in terms of a Riemannian manifold whose metric is derived from the Hessian of deformation energy in Section 3. Then, in Section 4, we develop a discrete geodesic calculus based on a time-discrete path energy. Section 5 is dedicated to a variational time discretization of Riemannian splines. Finally, in Section 6 we discuss applications for the processing of discrete shells.

Disclaimer. This article is based on Heeren et al. (2012), Heeren et al. (2014), Rumpf and Wirth (2015), Heeren et al. (2016), and Heeren et al. (2019) and resumes the mathematical modelling, the algorithmic details, and the convergence results presented in these papers. Different from Rumpf and Wirth (2015) and Heeren et al. (2019), however, we only consider the finite dimensional case, where shapes are represented as triangle meshes. All figures have been published in Heeren et al. (2012), Heeren et al. (2014), and Heeren et al. (2016) and are used here with permission.

2 Related work

During the past decades, the notion of *shape spaces* has had an increasing impact on the development of new methods in computer vision, graphics, and imaging—ranging from shape morphing and modelling to shape statistics and computational anatomy. A variety of different shape space concepts have been investigated in the literature; some of them are finite dimensional and consider polygonal curves or triangulated surfaces as shapes, others deal with infinite dimensional spaces of shapes. In this section, we provide a summary of related work on shape spaces, with a particular focus on those shape spaces that can be considered as Riemannian manifolds.

2.1 Shape spaces

The classical treatment of shape space is due to Kendall (1984), who considered shapes as k-tuples of points in $\mathbb{R}^d$, which can, e.g., be interpreted as discretized curves or nodes of triangulated surfaces, equipped with a quotient metric that is given by the Euclidean metric modulo translation, rotation, and scaling. Linear vector spaces can be considered as shape spaces as well. However, they are usually not invariant with respect to translation or rotation, i.e., shape alignment is necessary as a preprocessing step. Examples include the vector space of landmark positions (Cootes et al., 1995; Perperidis et al., 2005; Söhn et al., 2005) or Lebesgue spaces (Dambreville et al., 2006; Leventon et al., 2000; Tsai et al., 2003). Further classical shape spaces are based on considering Hausdorff or Gromov–Hausdorff distances, which have been used to perform shape statistics (Charpiat et al., 2006), shape clustering (Mémoli and Sapiro, 2005), and classification (Bronstein et al., 2008). Gromov–Hausdorff distance quantifies deviation from isometry in a global manner, which hampers its utility for examining local deviations from isometry.

For planar *curves*, different Riemannian metrics have been devised. In their seminal work, Michor and Mumford (2006) examined Riemannian metrics on the manifold of closed regular curves. They showed that the L^2-metric is pathologic in the sense that it leads to arbitrarily short geodesic paths (Michor and Mumford, 2005). Michor and Mumford showed that the vanishing geodesic distance phenomenon for the L^2-metric also occurs in more general shape spaces. To overcome this issue they employed a curvature-weighted L^2-metric instead, see Michor and Mumford (2007). For the same reason, Mennucci et al. (2008) used Sobolev metrics in the tangent space of planar curves, with applications in image segmentation via active contours (Sundaramoorthi et al., 2007). Klassen et al. (2004) proposed a framework for geodesics in the space of arc-length parameterized curves and implemented a shooting method in order to compute them. Schmidt et al. (2006) presented an alternative variational approach for the computation of these geodesics.

Srivastava et al. (2006, 2011) utilized an elastic string representation where curves can bend and locally stretch. To this end, they assigned different weights to the L^2-metric on stretching and bending variations and obtained an elastic model of curves, which allowed them to define geodesics and distances between curves in a way that is invariant under reparameterization. In a sequence of papers, Bruveris and coworkers investigated properties of the spaces of parameterized and unparameterized curves in $\mathbb{R}^d$, equipped with Sobolev metrics (Bauer et al., 2014; Bauer et al., 2016; Bruveris, 2015; Bruveris et al., 2014). For example, in Bauer et al. (2016), the set of Sobolev metrics was divided into three groups: the L^2-metric, which is simple and reparameterization invariant, but unsuitable for shape analysis (cf. Michor and Mumford, 2006), the H^1-metric, which was shown to be well-suited for numerical computations and therefore useful in applications, and finally higher order Sobolev metrics, whose theoretical properties make them good candidates for shape analysis. In particular, the space of closed planar curves equipped with a second-order Sobolev metric is geodesically complete. Moreover, a second-order Sobolev term can be considered as a bending term that takes into account curvature changes. This circumstance has been exploited in physical simulations of viscous and elastic rods, whose centerline is mathematically described by a curve (cf. Bergou et al., 2008; Rumpf and Wirth, 2015).

With regard to the space of *surfaces*, Kurtek et al. (2010) studied L^2-metrics on special representations of parameterized surfaces. To this end, they considered geodesic paths between surfaces parameterized over the unit sphere, using local changes of the area element as a Riemannian metric. Parameterization-based metrics often lead to distances for which even isometric surfaces might be far apart from one another. Bauer et al. (2011, 2012a, b, c) generalized weighted L^2-metrics (introduced in Michor and Mumford (2007) for planar curves) to higher dimensions, i.e., to the space of surfaces described by embeddings or immersions of a given manifold. They computed geodesic equations and sectional curvatures and showed in particular that these metrics overcome the degeneracy of the L^2-metric; corresponding numerical results were shown in Bauer and Bruveris (2011).

Of particular interest in our context is the pioneering work of Kilian et al. (2007) who investigated the space of triangle meshes. They considered geodesics between consistently triangulated surfaces with respect to a Riemannian metric measuring the stretching of triangle edges. In order to compensate for the lack of a regularizing bending energy, they incorporated a supplementary regularization term. Liu et al. (2010) proposed a metric that measures resistance of an edge to both stretching and compression. In contrast to the latter examples, we make use of the regularizing effect of bending energy and stay entirely in a physical simulation framework.

Riemannian spaces have also been considered for *volumetric* shapes, where the metric imitates a physical energy dissipation induced by the deformation of a shape consisting of ductile or viscous material (Fletcher and

Whitaker, 2006; Fuchs et al., 2009; Rumpf and Wirth, 2013; Wirth et al., 2011). Using a similar paradigm, Riemannian structures have also been introduced for the space of images. The attendant *metamorphosis approach* (Holm et al., 2009; Miller and Younes, 2001; Trouvé and Younes, 2005a, b) was proposed as a generalization of the *flow of diffeomorphism* approach (Dupuis et al., 1998). In this approach, one considers temporal changes of image intensities and spatial variations thereof (controlled by the so-called material derivative) along motion paths. Physically, the underlying metric describes the viscous dissipation in a multipolar fluid model (Nečas and Šilhavý, 1991). Berkels et al. (2015) applied the variational time discretization of geodesics proposed in Wirth et al. (2011) to the Riemannian manifold induced by the metamorphosis approach. Moreover, the notion of optimal transport (Kantorovitch, 1942, 1948; Monge, 1781), in particular the formulation proposed by Benamou and Brenier (2000), can also used to define a Riemannian structure on the space of images (cf., e.g., Peyré et al., 2010, 2012; Schmitzer and Schnörr, 2013).

Only few Riemannian shape spaces allow for explicit computations of geodesic paths (see, e.g., Sundaramoorthi et al., 2011; Younes et al., 2008); in general, the system of geodesic ODEs has to be solved using numerical time stepping schemes (see, e.g., Beg et al., 2005; Klassen et al., 2004). Alternatively, geodesic paths can be approximated via the minimization of discretized path length (Schmidt et al., 2006) or path energy functionals (Fuchs et al., 2009; Wirth et al., 2011).

Instead of discretizing the underlying geodesic flow, the variational discretization proposed in Wirth et al. (2011) is based on the direct minimization of a discrete path energy subject to data given at the initial and final time. This approach turned out to be stable and robust; even very small numbers of time steps lead to qualitatively satisfactory results (Berkels et al., 2015; Maas et al., 2015; Rumpf and Wirth, 2013). Building on this variational time discretization, a comprehensive discrete geodesic calculus on finite and on certain infinite dimensional shape spaces was developed in Rumpf and Wirth (2015).

2.2 Interpolation and extrapolation

Many shape interpolation schemes in computer graphics are based on the following three-step procedure. First, select a number of geometric quantities or *shape descriptors* that determine the shape (locally). Then, based on these shape descriptors of given input shapes, compute the interpolated or extrapolated shape descriptors. Finally, reconstruct the shape or the path in shape space, respectively, that best matches the new shape descriptors. The difference between various existing methods lies in the choice of shape descriptors. Depending on whether the shape descriptors depend linearly or nonlinearly on the vertex positions, the reconstruction is a linear or nonlinear problem.

Examples of linear reconstruction schemes were given by Sumner et al. (2005) and by Lipman et al. (2005). The method proposed by Sumner et al. (2005) uses deformation gradients of the triangles as geometric quantities to be interpolated (cf. Sumner and Popović, 2004). However, the interpolation is nonlinear since the rotational components of the deformation gradients are extracted and nonlinearly blended by taking the shortest path in the rotation group. Blending of rotations is performed separately for each triangle, which might lead to undesirable interpolation results (cf. Winkler et al., 2010 for examples and a discussion of this issue). Instead of treating all triangles individually, Lipman et al. (2005) presented a method that takes the connectivity information of the triangle mesh into account and considers transformations that connect local frames in the mesh (cf. also Kircher and Garland, 2008). A key property of this ansatz is that it represents the local geometry of a mesh in a rotation-invariant way.

Examples of nonlinear reconstruction schemes were given by Winkler et al. (2010), by Fröhlich and Botsch (2011), and by Martin et al. (2011). Winkler et al. (2010) used edge lengths and dihedral angles of the triangles of a surface mesh as geometric quantities to be interpolated. To achieve an improvement in this direction, Fröhlich and Botsch (2011) introduced a fast reconstruction scheme for the interpolation based on edge lengths and dihedral angles. Their method interpolates between simplified meshes and uses deformation transfer (Botsch et al., 2006b) to map the coarse interpolated shapes to a fine mesh. We will return to their method in the context of spline interpolation.

In addition, there are numerous interpolation techniques that do not depend on a reconstruction step. A classical approach in this direction was presented by Alexa et al. (2000). They proposed a morphing technique that blends the interiors of given two- or three-dimensional shapes in an as rigid as possible manner, i.e., their method is locally least-distorting. As already mentioned above, Kilian et al. (2007) introduced a framework of geometric modelling in the space of triangle meshes that allows for interpolation, extrapolation, and even parallel transport. A more physics-based approach was considered by Chao et al. (2010). They advocated a simple geometric model for elasticity, based on penalizing the distance between the deformation differential and the rotation group. This model does not suffer from linearization artefacts, while being computationally almost as efficient as models based on linear elasticity.

von Tycowicz et al. (2015) introduced a scheme for real-time nonlinear interpolation that exploits the fact that the set of all possible interpolated shapes is a low-dimensional set in a high-dimensional shape space. To this end they constructed a reduced optimization problem that approximates its unreduced counterpart and can be solved very efficiently. We refer to the paper of von Radziewsky et al. (2016) for further applications of this ansatz and to the work by Brandt et al. (2016), where the reduction method was used for computing approximations to time-discrete geodesics in the sense of Wirth et al. (2011).

2.3 Riemannian splines

Noakes et al. (1989) introduced Riemannian cubic splines in the context of finite dimensional Riemannian manifolds. These splines are smooth curves that are stationary points of the spline energy functional, which is defined as the integrated squared covariant derivative of the velocity satisfying certain boundary conditions. A necessary condition for optimality of the resulting boundary value problem is given by the corresponding Euler–Lagrange equation, which turns out to be a fourth-order differential equation (Noakes et al., 1989). Crouch and Silva Leite (1995) considered the *multinodal interpolation problem* by minimizing the spline energy subject to interpolation constraints on the manifold.

Trouvé and Vialard (2012) presented a mathematical framework to perform interpolation on time-indexed sequences of 2D or 3D shapes with a focus on the finite dimensional case of landmarks. They developed a spline interpolation method that is related to the Riemannian cubic polynomials in Noakes et al. (1989).

Instead of dealing with the *intrinsic* formulation presented above, there are several works dealing with an *extrinsic* formulation, i.e., the minimization of integrated squared acceleration in ambient space. The restriction of the resulting curve to a submanifold is then realized as a constraint. Wallner (2004) proved existence of minimizers in this setup for finite dimensional manifolds, and Pottmann and Hofer (2005) showed that these minimizers are C^2. Additionally, Hofer and Pottmann (2004) provided a method for the computation of splines on parametric surfaces, level sets, triangle meshes, and point set surfaces. Algorithmically, they alternately computed minimizers in the tangent plane and projected them back to the manifold.

In addition to variational formulations there are numerous contributions dealing with *subdivision schemes* to produce smooth interpolating curves on manifolds (Dyn, 1992, 2002). By replacing the operation of affine averaging either by a geodesic average or by projecting the affine averages onto a constraint manifold, one generates a Riemannian extension (cf. Wallner and Dyn, 2005). More practically, Rahman et al. (2005) proposed a Deslauriers–Dubuc interpolation in the tangent space, where the mapping between tangent space and manifold is realized by means of the exponential and logarithm maps.

2.4 Discrete variational methods

Surfaces naturally exhibit curvature, and many physical problems in shell space are related to curvature functionals, which in turn lead to variational problems. Discretizing variational problems that involve curvature terms is a challenging task. A first challenge is to derive *consistent* discrete versions of curvatures on triangle meshes that converge to their smooth counterparts in the limit of refinement. Consistent discrete notions of mean curvature provide a prominent example. Mean curvature is intimately linked to Laplacians on manifolds, since the mean curvature vector of a surface embedded into Euclidean 3-space is equal to the Laplace–Beltrami operator applied to surface positions. Dziuk's seminal work (Dziuk, 1988) shows convergence of solutions to the Poisson problem for the Laplace–Beltrami operator using standard Lagrange finite elements on triangle meshes—under the condition that mesh vertices reside on the smooth limit surface. This conditions was later relaxed to meshes that are "nearby" a smooth surface in the sense that "nearness" holds for both surface positions and surface normals (Hildebrandt et al., 2006; Wardetzky, 2006). The latter analysis additionally yields convergence of mean curvature vectors in a weak sense and shows that pointwise convergence cannot be expected in general. Indeed, Cohen-Steiner and Morvan observed that convergence of positions and normals of a mesh to those of a smooth limit surface implies convergence of curvatures *in the sense of measures*, see Cohen-Steiner and Morvan (2003). Pointwise convergence can only be expected under additional assumptions, see, e.g., Meek and Walton (2000), Xu et al. (2005), Bauer et al. (2010), and Hildebrandt and Polthier (2011).

Convergence of minimizers of discrete curvature functionals is even more delicate. Perhaps one of the most prominent examples is that of *minimal surfaces*, i.e., the problem of finding a surface of least (or more precisely critical) area among all surfaces of prescribed topology spanning a prescribed boundary curve. In the smooth setting, Radó (1933) and Douglas (1931) independently solved this problem for disk-like, immersed surfaces. Douglas' existence proof is based on minimizing the Dirichlet energy of conformal surface parameterizations. Douglas' ideas have been utilized for showing convergence of discrete minimal surfaces using the finite element method (FEM) for disk-like and cylindrical surfaces (Dziuk and Hutchinson, 1999a,b; Hinze, 1996; Pozzi, 2005; Tsuchiya, 1987; Wilson, 1961). Convergence of disk-like minimal surfaces to their smooth counterparts has also been established using an entirely different approach for defining discrete minimal surfaces, based on discrete differential geometry (DDG) (Bobenko et al., 2006). Convergence of discrete minimal surfaces to smooth minimal surfaces of *arbitrary topology* has only been established recently using the notion of Kuratowski convergence (Schumacher and Wardetzky, 2019). The latter work also provides a notion of shape space that incorporates both smooth and simplicial manifolds.

Willmore energy provides a prominent example of a geometric curvature functional that is related to surface bending. The resulting Euler–Lagrange equations turn out to be a fourth-order PDE, which is demanding to treat with standard C^1-conforming finite elements on triangle meshes. An H^2-conforming discretization can be constructed using *subdivision surfaces* (Cirak and Ortiz, 2001; Cirak et al., 2000, 2002; Grinspun et al., 2002), where higher order smoothness is achieved by increasing the support of the basis functions instead of introducing additional degrees of freedom. An alternative is provided by mixed methods, where the fourth-order PDE is split into two second-order PDEs, which are approximated either by linear or by quadratic finite elements. Examples include FEM discretizations of Willmore's energy (Balzani and Rumpf, 2012; Barrett et al., 2007, 2008; Perl et al., 2014; Rusu, 2005). Yet another approach that mimics the rich structural properties of the smooth Willmore functional is provided by using methods from DDG, see, e.g., Bobenko and Schröder (2005) and Crane et al. (2013). Finally, a discretization of elastic plate problems can also be built on the *Discrete Kirchhoff Triangle* (Bartels, 2013; Batoz et al., 1980; Chapelle and Bathe, 1998).

Most approaches modelling and simulating the elastic behaviour of thin plates and shells, however, separately formulate and discretize membrane and bending contributions of a deformation. In particular the geometrically *nonlinear* treatment of bending energy, which accounts for change of curvature, is a delicate business. Terzopoulos et al. (1987) first introduced the concept of *elastically deformable models* for geometry processing. They considered a separation of an elastic energy into the sum of a membrane and bending energy measuring variations in first and second fundamental forms, respectively. Later, Grinspun et al. (2003) introduced the widespread *Discrete Shells* model, which was subsequently refined in Grinspun et al. (2006). In Bergou et al. (2006) and Wardetzky et al. (2007) a family of *discrete isometric bending models* for triangulated surfaces was derived from an axiomatic treatment of discrete Laplace operators, leading to bending energies that are *quadratic* in vertex positions for isometric deformations. Botsch et al. (2006a) proposed a novel framework for simulating and deforming 3D shapes that achieves intuitive and robust deformations by emulating physically plausible surface behaviour. Finally, there are various approaches for accelerating simulations that are based on nonlinear bending models. To this end, one might either consider differential coordinates (Kircher and Garland, 2008; Lipman et al., 2005; Sumner and Popović, 2004) or triangle descriptors, such as edge lengths and dihedral angles (Fröhlich and Botsch, 2011; Winkler et al., 2010), or alternatively make use of dimension reduction approaches and introduce efficient subspaces (Brandt et al., 2018; von Radziewsky et al., 2016).

3 Riemannian geometry and Hessian structure of shell space

A Riemannian manifold $(\mathcal{M}, g)$ is a pair of a smooth manifold $\mathcal{M}$ and a smoothly varying metric tensor g on $\mathcal{M}$. In the sequel, we only work with Hausdorff manifolds that are connected and path connected. We assume that $\mathcal{M}$ is modelled over some $\mathbb{R}^n$. In principle, the model space might also be some infinite dimensional Hilbert, Banach, or Fréchet space, in which case, however, various of the properties of the finite dimensional setting seize to hold. Therefore, we here take the comfort of working in the setting of finite dimensional manifolds—since our goal in this article is to offer methods for exploring the space of discrete shells computationally, in which case the degrees of freedom are the finite set of vertex positions.

On a Riemannian manifold $(\mathcal{M}, g)$, the metric tensor g defines an inner product $g_p(u,v)$ between tangent vectors $u,v \in T_p\mathcal{M}$ at each point $p \in \mathcal{M}$. The bilinear and symmetric operator $g_p(\cdot,\cdot)$ is required to be smoothly depending on the base point p and needs to satisfy $g_p(u,u) \geq 0$ with equality if and only if $u = 0$. In the finite dimensional setting, Riemannian manifolds have been studied for centuries, and there exists an extensive amount of literature on the subject, including various textbooks, e.g., do Carmo (1992) and Petersen (1997).

3.1 Brief recap of smooth Riemannian calculus

We briefly recapitulate the notions of Riemannian geometry that we require. Completeness of the subject is by no means our goal. Rather, our aim is to be able to contrast the smooth setting with the discrete Riemannian calculus. This has already been outlined in the introduction and will be detailed in the next section. Furthermore, we establish some useful notation.

3.1.1 Curve length and path energy

Recall that the metric tensor gives rise to the notion of length of a tangent vector $u \in T_p\mathcal{M}$ via $\|u\| = \sqrt{g(u,u)}$. Length of tangent vectors, in turn, allows for measuring lengths of smooth curves $\gamma : [0,1] \to \mathcal{M}$ via

$$\mathcal{L}[\gamma] = \int_0^1 \sqrt{g_{\gamma(t)}(\dot{\gamma}(t), \dot{\gamma}(t))}\, \mathrm{d}t = \int_0^1 \|\dot{\gamma}(t)\|\, \mathrm{d}t,$$

where $\dot{\gamma}(t) = \frac{\mathrm{d}}{\mathrm{d}t}\gamma(t) \in T_{\gamma(t)}\mathcal{M}$ denotes the curve's tangent (velocity) vector at point $\gamma(t) \in \mathcal{M}$. The curve length $\mathcal{L}[\gamma]$ is parameterization invariant, i.e., given a monotonically increasing and differentiable mapping σ: $[0,1] \to [0,1]$ with $\sigma(0) = 0$ and $\sigma(1) = 1$, we obtain by the transformation rule that $\mathcal{L}[\gamma \circ \sigma] = \mathcal{L}[\gamma]$. While this property is geometrically nice, it leads to analytical complications when dealing with the existence theory of shortest paths. It therefore turns out to be useful to consider the so-called *path energy* of curves, which is defined as

$$\mathcal{E}[\gamma] = \int_0^1 g_{\gamma(t)}(\dot{\gamma}(t), \dot{\gamma}(t))\, \mathrm{d}t = \int_0^1 \|\dot{\gamma}(t)\|^2\, \mathrm{d}t. \tag{1}$$

Notice that the path energy is *not* parameterization invariant. At first glance, this non-invariance might appear as a deficiency. From an analytical perspective, however, this turns out to be advantageous (see below). In order to relate curve length and path energy, the following observation is useful. A direct application of the Cauchy–Schwarz inequality shows that

$$\mathcal{L}[\gamma] \leq \sqrt{\mathcal{E}[\gamma]},$$

with equality if and only if $\gamma : [0,1] \to \mathcal{M}$ is parameterized by *constant speed*, i.e., $\|\dot{\gamma}(t)\| \equiv \text{const}$. Both curve length and path energy can be used in order to define the Riemannian (or geodesic) distance between two points $p,q \in \mathcal{M}$. Using curve length, distance is defined as

$$\operatorname{dist}(p,q) = \inf_{\gamma\,:\,\gamma(0)=p,\gamma(1)=q} \mathcal{L}[\gamma]. \tag{2}$$

Here the infimum is taken over all smooth curves $\gamma : [0,1] \to \mathcal{M}$. This distance turns $(\mathcal{M}, g)$ into a metric space. If this infimum is attained for some curve γ that is parameterized by constant speed, then γ is called a *minimizing geodesic*. If $\mathcal{M}$ is *finite dimensional*, then the Hopf–Rinow theorem asserts that minimizing geodesics exist for any pair of points p,q if and only if $(\mathcal{M}, g)$ is complete as a metric space. It is one of the intricacies of the infinite dimensional setting that the Hopf–Rinow theorem does in general not hold for general infinite dimensional manifolds (Atkin, 1997).

In any case, by the above discussion, we have that

$$\operatorname{dist}^2(p,q) = \inf_{\gamma\,:\,\gamma(0)=p,\gamma(1)=q} \mathcal{E}[\gamma]. \tag{3}$$

Minimizers of $\mathcal{E}[\gamma]$ not only provide minimizing geodesics but additionally provide a constant speed parameterization, since the curve's parameterization becomes part of the variational principle. Indeed, as mentioned before, it is preferable to work with $\mathcal{E}[\gamma]$ from an analytical perspective.

3.1.2 Levi-Civita connection

One crucial ingredient in the smooth setting is the notion of *covariant derivatives*. There are many ways of how to arrive at this notion. Perhaps one of the most convenient ways is an axiomatic approach, which postulates certain properties and then concludes that there exists a unique structure (namely the so-called *Levi-Civita connection*) that satisfies the desired properties. This is the approach we follow here.

The covariant derivative is a way of expressing the derivative $\nabla_v u$ of one smooth vector field u on $\mathcal{M}$ along another smooth vector field v on $\mathcal{M}$. One requires that $\nabla_v u$ is again a smooth vector field. The operator ∇ is called a *connection*. In order for $\nabla_v u$ to be the same as $D_v u$ when working in standard Euclidean space $\mathbb{R}^n$, one additionally requires that the connection be *affine* and *torsion free*. To be affine means that for every triple (u, v, w) of smooth vector fields and for any pair $(f, \tilde{f})$ of smooth scalar functions one has that, $\nabla_{fu+\tilde{f}v} w = f \nabla_u w + \tilde{f} \nabla_v w$, $\nabla_u (v + w) = \nabla_u v + \nabla_u w$, and $\nabla_u (fv) = (\nabla_u f) v + f \nabla_u v$. Notice that $\nabla_u f = df(u)$ denotes the directional derivative of a function

f along u, which is defined independently of the choice of a Riemannian metric. To be torsion free means that $\nabla_u v - \nabla_v u = [u, v]$, where $[u, v]$ denotes the *Lie bracket* between u and v, which is also defined independently of the choice of a Riemannian metric. Indeed, $[u, v]$ is the unique vector field that satisfies $\nabla_{[u,v]} f = \nabla_u(\nabla_v f) - \nabla_v(\nabla_u f)$ for all smooth functions $f : \mathcal{M} \to \mathbb{R}$. These properties alone do not suffice in order to pin down a unique connection. One additionally needs to require that the connection is compatible with the Riemannian metric in the sense that

$$\nabla_w(g(u,v)) = g(\nabla_w u, v) + g(u, \nabla_w v)$$

for all triples (u, v, w) of smooth vector fields. Indeed, the last property and the requirement to be torsion free lead to the Koszul formula

$$\begin{aligned} 2g(\nabla_u v, w) = \nabla_u(g(v,w)) + \nabla_v(g(u,w)) - \nabla_w(g(u,v)) + g([u,v],w) \\ -g([u,w],v) - g([v,w],u), \end{aligned}$$

which indeed pins down $\nabla_u v$ uniquely. The resulting connection is called the *Levi-Civita connection*.

3.1.3 Parallel transport, geodesics, exponential map, Riemann curvature tensor

With the notion of the Levi-Civita connection at hand, one proceeds by defining the notion of parallel transport. To be precise, a smooth vector field u is called parallel along a smooth curve γ in $\mathcal{M}$ if

$$\nabla_{\dot{\gamma}} u = 0,$$

where $\dot{\gamma}$ denotes the tangent vector along γ. Notice that in standard Euclidean space, a vector field is parallel along the curve, if it is constant, which justifies the name *parallel*.

A smooth curve $\gamma \subset \mathcal{M}$ is called a *geodesic* if its velocity vector is parallel along the curve:

$$\nabla_{\dot{\gamma}} \dot{\gamma} = 0. \tag{4}$$

This means that the tangent, $\dot{\gamma}$, does not turn with respect to itself along the path, i.e., γ is a straight path with respect to the Riemannian metric. Indeed, one can show that geodesics yield locally shortest path between two points in $\mathcal{M}$. In particular, (4) is the Euler–Lagrange equation for critical points of the path energy (1), hence the two definitions are consistent.

In order to define the exponential map, notice that the geodesic equation is a second-order ordinary differential equation. It therefore suffices to fix a point $\gamma(0) \in \mathcal{M}$ and a tangent vector $u = \dot{\gamma}(0)$ at that point in order to locally solve the geodesic equation. The *exponential map* is then defined as a map

$$\exp_p : U(0) \cap T_p\mathcal{M} \to \mathcal{M}$$

from an open neighbourhood $U(0)$ around 0 in the tangent space $T_p\mathcal{M}$, which maps $u \in U(0)$ to the point $\gamma(1)$ of the geodesic starting at $\gamma(0) = p$ with initial velocity $\dot{\gamma}(0) = u$. It is not difficult to see that $\exp_p$ is a radial isometry, i.e., the length of u is equal to the distance from p to $\exp_p(u)$, i.e., $\|u\| = \mathrm{dist}(p, \exp_p(u))$. This is known as Gauss' lemma. The exponential map is locally invertible. Its inverse $\log_p$ is called the *logarithm*.

Finally, the notion of the Levi-Civita connection leads to the notion of the *Riemann curvature tensor*, which associates to every pair (u, v) of smooth vector fields a mapping $R(u, v)$ from vector fields to vector fields, defined as

$$R(u,v)w = \nabla_v \nabla_u w - \nabla_u \nabla_v w + \nabla_{[u,v]} w.$$

A discretization of the Riemannian curvature tensor is beyond the scope of this article. In summary, the usual order of defining notions of smooth Riemannian calculus is

Riemannian metric ⇒ distance
Riemannian metric ⇒ Levi-Civita connection ⇒ parallel transport
⇒ geodesics ⇒ exp ⇒ log

As we will see in Section 4 this is quite different from our corresponding *discrete Riemannian calculus*.

3.2 Riemannian metric and Hessian structure

As a starting point for our discussion below, we observe that the Riemannian metric can be defined in terms of the Hessian of the squared distance function. Indeed, let

$$\mathcal{W}[p,q] = \mathrm{dist}^2(p,q)$$

denote the *squared* distance function between two points p and q in $\mathcal{M}$. Then

$$g_p(u,v) = \frac{1}{2}\mathrm{Hess}\,\mathcal{W}[p,\cdot\,]_{|_p}(u,v), \tag{5}$$

where $\mathrm{Hess}\,(f)_{|_p} = \mathrm{Hess} f[.]_{|_p}$ denotes the Riemannian Hessian of a (smooth enough) function $f: \mathcal{M} \to \mathbb{R}$ evaluated at point p.

Consequently, the path energy (1) of some smooth path $\gamma: [0,1] \to \mathcal{M}$ then takes form

$$\mathcal{E}[\gamma] = \frac{1}{2}\int_0^1 \mathrm{Hess}\,\mathcal{W}[\gamma(t),\cdot\,]_{|\gamma(t)}(\dot{\gamma}(t),\dot{\gamma}(t))\,\mathrm{d}t, \tag{6}$$

and the squared distance function can be recovered from the Hessian via

$$\mathrm{dist}^2(p,q) = \inf_{\gamma:\gamma(0)=p,\gamma(1)=q} \frac{1}{2}\int_0^1 \mathrm{Hess}\,\mathcal{W}[\gamma(t),\cdot\,]_{|\gamma(t)}(\dot{\gamma}(t),\dot{\gamma}(t))\,\mathrm{d}t. \tag{7}$$

3.2.1 Riemannian metric from Hessian

The main point of departure for our discussion below is to turn these observations around in the sense that we consider functions $\mathcal{W}$ that are meaningful *templates* for defining a Riemannian metric (and consequently a distance function) by *imposing* (5) as a definition for a Riemannian metric. For this to make sense, one is led to considering smooth enough functions $\mathcal{W} : \mathcal{M} \times \mathcal{M} \to \mathbb{R}_{\geq 0}$ that give rise to a Hessian structure, i.e., for which one has for all $p \in \mathcal{M}$ that

$$\mathcal{W}[p,p] = 0, \quad \mathcal{D}\mathcal{W}[p,\cdot]\big|_p = 0 \text{ and } \operatorname{Hess}\mathcal{W}[p,\cdot]\big|_p \text{ is positive definite.} \tag{8}$$

Notice that we do not require $\mathcal{W}$ to be symmetric.

Here a word of caution is in place: The Hessian of a function $f : \mathcal{M} \to \mathbb{R}$ on a Riemannian manifold crucially depends on the Riemannian metric itself. Indeed, Hess $(f) = \nabla(\mathrm{d}f)$, where d denotes the differential (which does not depend on the metric) and ∇ denotes the covariant derivative, which crucially depends on g. So how can the metric possibly be defined in terms of the Hessian? The answer is that at a *critical point* $p \in \mathcal{M}$ of f, the Hessian Hess $(f)\big|_p$ is *independent* of the Riemannian metric. Hence the notion $\operatorname{Hess}\mathcal{W}[p,\cdot]\big|_p$ makes sense *without* invoking a Riemannian metric first by our requirement that $\mathcal{D}\mathcal{W}[p,\cdot]\big|_p = 0$. This is precisely our point of departure for our exposition below: We *start* with a function that we regard as a *template* of a squared distance function and *then* define a metric using the Hessian of this function.

3.3 Elastic thin shell energies

Building on the discussion of the previous section, we are particularly motivated by searching for a template of a squared distance function that yields a meaningful Riemannian metric on the space of shells.

Here by a *shell* we mean a surface embedded into $\mathbb{R}^3$ that is made of some (ideally infinitesimally) thin material (e.g., paper or metal). With this in mind, a natural quantity to consider is an *elastic energy*, which measures the amount of work that is required to deform one given shell (rest state) to another shell (deformed state). Indeed, our goal is to use the Hessian of elastic energies in order to define metrics on the space of shells. Let us emphasize that we discuss deformations which map material points in the reference state to material points in the deformed state. In particular, there is no invariance with respect to reparameterization of the rest or the deformed state as in the context of purely geometric mappings between surfaces.

At first glance this may sound like a lost endeavour, since the elastic work that is required for deforming one shell into another *does not depend* on the deformation path but only on the rest and final configuration of the shell. Certainly, we cannot hope for defining a meaningful notion of distance in such a naïve manner. To this end, we consider a *viscous formulation* instead of an elastic one, where strain is replaced by strain rate and hence deformation energy (or more precisely, *viscous dissipation*) *does* depend on the deformation path.

Indeed, while nonlinear elasticity is based on the stored potential energy of reversible deformations, viscous dissipation models depend on irreversible energy conversion into heat during a deformation. Despite this conceptual difference, a viscous deformation may be seen as the limit of many infinitesimally small elastic deformations with subsequent stress relaxation (Wirth, 2009). This leads to a physical justification of deformation paths based on consecutive elastic deformations, despite the fundamental axiom of elasticity that prevents the notion of paths. In particular, the resulting formulation coincides with Rayleigh's paradigm (Rayleigh, 1896), where the dissipation density is given by the second derivative of an elastic energy density at the identity. According to this, we eventually show that the Hessian of a (discrete) elastic energy induces a well-defined Riemannian metric on the space of (discrete) shells. We start by considering the smooth case first.

In order to make things more precise, we first consider shells to be thin *three-dimensional* materials. Let δ be a scale parameter that encodes the (transversal) thickness of a smooth (mid)surface s embedded into $\mathbb{R}^3$. Let s^δ denote the corresponding material around s. Consider now a family $(\phi_t)_{0\leq t\leq 1}$ of diffeomorphisms with $\phi_0 = \mathrm{Id}$. Then ϕ_t generates a deformation path $\phi_t(s^\delta)$. The *total elastic energy* required for deforming the rest state s^δ into $\phi_t(s^\delta)$ is given by

$$\mathcal{W}[s^\delta, \phi_t(s^\delta)] = \int_{s^\delta} W(\mathcal{D}\phi_t)\,\mathrm{d}x, \tag{9}$$

where W denotes an elastic energy density acting on matrices $A \in \mathbb{R}^{3,3}$ with $\det A > 0$. A fundamental axiom of continuum mechanics is *frame indifference*, i.e., the invariance of the elastic energy with respect to rigid body motions. Hence, any coordinate transform $x \mapsto Qx+b$ for a rotation $Q \in SO(3)$ and a shift $b \in \mathbb{R}^3$ does not change the energy, i.e., $W(A) = W(QA)$ for all $Q \in SO(3)$. A direct consequence is that W only depends on the so-called right Cauchy–Green strain tensor $\mathcal{D}\phi^{\mathrm{T}}\mathcal{D}\phi$. Furthermore, we shall assume the shell to be made of an *isotropic* material, i.e., a rotation of the material before applying a deformation yields the same energy as before, i.e., $W(A) = W(AQ)$ for $Q \in SO(3)$. In fact, frame indifference and isotropy imply the identity $W(R^TAQ) = W(A)$ for all $R, Q \in SO(3)$. We further assume that the identity matrix, id, is a minimizer, implying that $W_{,A}(\mathrm{id}) = 0$, where $W_{,A}(\mathrm{id})$ denotes the derivative of W at the identity. Without loss of generality we can choose additive constants in W such that $W(\mathrm{id}) = 0$.

Suppose now that shells were made of a *viscous* material, then such a deformation would lead to viscous friction within the shell's volume. We treat viscous dissipation according to Rayleigh's analogy that derives a viscous formulation from an elastic one by replacing elastic strain by strain *rates* (Rayleigh, 1896).

According to Rayleigh's analogy, we consider the dissipation density to be given by the second derivative of elastic energy density at the identity, i.e.,

$$\mathrm{Diss}(\delta A) = \frac{1}{2}\mathrm{Hess}\, W_{|_{\mathrm{id}}}(\delta A, \delta A) \tag{10}$$

for any matrix $\delta A \in \mathbb{R}^{3,3}$. Here $\mathrm{Hess}\, W_{|_{\mathrm{id}}}(\delta A, \delta A)$ denotes the second derivative of W with respect to the variation δA, evaluated at the identity. Define the symmetrized gradient of the Eulerian flow field $v_t = \dot{\phi}_t \circ \phi_t^{-1}$ as

$$\epsilon[v_t] := \frac{1}{2}\left((Dv_t)^T + Dv_t\right).$$

Using frame indifference, the total energy dissipation associated with the *accumulated* viscous friction along the deformation path only depends on the symmetrized gradient. Hence, total energy dissipation is given by

$$\mathcal{E}[(\phi_t)_{0\le t\le 1}] = \int_0^1 \int_{\phi_t(s^\delta)} \mathrm{Diss}(\epsilon[v_t])\, \mathrm{d}x\, \mathrm{d}t\,.$$

In this view, a *best* or *shortest* path between two shells can be defined as *the path of least energy dissipation.*

Using the above definition of elastic energy, we obtain that the total energy dissipation is given by

$$\begin{aligned}
\mathcal{E}[(\phi_t)_{0\le t\le 1}] &= \int_0^1 \int_{\phi_t(s^\delta)} \frac{1}{2}\mathrm{Hess}\, W_{|_{\mathrm{id}}}(\epsilon[v_t], \epsilon[v_t])\, \mathrm{d}x\, \mathrm{d}t \\
&= \int_0^1 \frac{1}{2}\int_{\phi_t(s^\delta)} \mathrm{Hess}\, W_{|_{\mathrm{id}}}(Dv_t, Dv_t)\, \mathrm{d}x\, \mathrm{d}t \\
&= \int_0^1 \frac{1}{2}\frac{\mathrm{d}^2}{\mathrm{d}\tau^2}\int_{\phi_t(s^\delta)} W(\mathrm{id} + \tau Dv_t)\, \mathrm{d}x \Big|_{\tau=0} \mathrm{d}t \\
&= \frac{1}{2}\int_0^1 \mathrm{Hess}\, \mathcal{W}[\phi_t(s^\delta), \cdot]_{|_{\phi_t(s^\delta)}}(v_t, v_t)\, \mathrm{d}t,
\end{aligned} \tag{11}$$

where $\mathrm{Hess}\, \mathcal{W}[\phi_t(s^\delta), \cdot]_{|_{\phi_t(s^\delta)}}(v_t, v_t)$ denotes the Hessian of the elastic energy, evaluated at the current state $\phi_t(s^\delta)$ along infinitesimal deformations given by the velocity field v_t. Compare the last term to (6): This is precisely the Hessian structure on shell space that we work with in the sequel.

These observations are meant to provide motivation. On their own of course, they do not guarantee that the Hessian of an elastic energy indeed yields a positive definite metric on shell space. In Section 3.4 we show that this is indeed the case. Before doing so, however, we make precise our notion

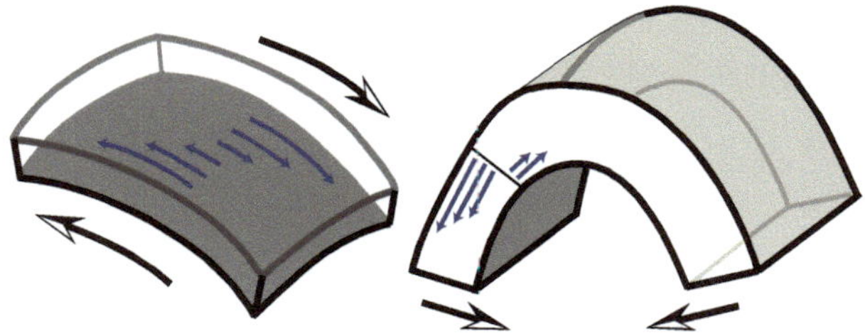

FIG. 4 Membrane (*left*) and bending (*right*) distortion, leading to viscous friction.

of deformation energies for shells. Since we view shells as *thin* material layers viscous friction is predominantly caused by two mechanisms (see Fig. 4):

(i) friction due to in-layer (tangential) shear or compression,
(ii) friction due to (transversal) shear caused by bending.

A dimension reduction approach by means of Γ-convergence (De Giorgi and Dal Maso, 1983) justifies to express the deformation energy (9) in terms of the midsurface s of the thin material s^δ and deformations $\phi : s \to \mathbb{R}^3$ thereof. Then, based on a suitable decomposition of the deformation energy, mechanism (i) can be described in terms of *membrane* strains of s, while mechanism (ii) is associated with *bending* strains of s. This splitting approach of the energy can be made precise by relying on a well-established theory for the approximation of elastic deformation energies and the geometry thereof (Friesecke et al., 2002; Le Dret and Raoult, 1995). Indeed, mechanisms (i) and (ii) are intimately related to the first and second fundamental forms of surfaces, respectively.

According to our formulation of elastic energy (9), it suffices to spell out elastic energies for *pairs* of surfaces $\bar{s}$ and $s = \phi(\bar{s})$. Here we view $\bar{s}$ and s as the undeformed and deformed surface (of zero thickness), respectively.

3.3.1 Membrane energy

The change of first fundamental forms induced by a deformation $\phi : \bar{s} \to s = \phi(\bar{s})$ can be captured by a field $A[\phi]$ of symmetric positive definite linear mappings acting on tangent vectors of $\bar{s}$, defined via

$$\bar{\mathrm{I}}(A[\phi]u, v) = \mathrm{I}(d\phi(u), d\phi(v)).$$

Here $\bar{\mathrm{I}}$ and I are the first fundamental forms on $\bar{s}$ and s, respectively, and u, v are tangent vectors on $\bar{s}$. Notice that the notion *first fundamental form* is simply another name for *Riemannian metric* (on a two-dimensional surface, not on shell space), which we use in order to underline the similarity to the second fundamental form defined below. Notice furthermore that the corresponding quadratic form $\mathrm{I}^*(u,v) = \bar{\mathrm{I}}(A[\phi]u, v)$ is the pullback of the first fundamental form (metric) of s to $\bar{s}$. Below, we provide an analogous construction for the second fundamental form.

From the viewpoint of elasticity theory $A[\phi]$ is the Cauchy–Green strain tensor field measuring tangential distortion. In fact, Le Dret and Raoult (1995) have shown in the context of Γ-convergence that to leading order, the energy of a deformation $\phi : \bar{s}^\delta \to \phi(\bar{s}^\delta)$ scales *linearly* in the thickness parameter δ and after rescaling with $\frac{1}{\delta}$ is given by the *membrane* energy

$$\mathcal{W}_{\text{mem}}[\bar{s}, \phi(\bar{s})] = \int_{\bar{s}} W_{\text{mem}}(A[\phi]) \, \mathrm{d}x \,, \tag{12}$$

where the energy density W_{mem} depends on the field $A[\phi]$. Inspired by the results in Wirth et al. (2011), we set

$$W_{\text{mem}}(A) = \frac{\mu}{2} \text{tr}A + \frac{\lambda}{4} \det A - \frac{2\mu + \lambda}{4} \log \det A \, - \mu - \frac{\lambda}{4}, \tag{13}$$

where λ and μ are the Lamé constants of a tangential Newtonian dissipation measure and where tr$A$ and det$A$ denote the trace and determinant of A. Notice that det$A$ describes area distortion, while tr$A$ measures length distortion. The polyconvex function $W_{\text{mem}}(A)$ is rigid body motion invariant, and the identity is the minimizer. The $\log \det A$ term penalizes material compression, which will be advantageous in the spatially discrete setting below in order to prevent degeneration of triangles.

3.3.2 Bending energy

Let $\overline{\text{II}}$ and II denote the second fundamental forms of $\bar{s}$ and s, respectively. Different from the first fundamental forms, which encode the *intrinsic* geometry of a surface, the second fundamental form encodes the *extrinsic* curvature of a surface embedded into $\mathbb{R}^3$. Indeed, the second fundamental form acts bilinearly and symmetrically on tangent vectors, and its two eigenvectors correspond to the two principle curvature directions at any point of a given surface. We pull back II to $\bar{s}$ via

$$\text{II}^*[\phi](u,v) := \text{II}(d\phi(u), d\phi(v)),$$

and we define linear mappings $\overline{B}$ and $B[\phi]$ corresponding to $\overline{\text{II}}$ and II*, respectively. In detail, $\overline{B}$ is the shape operator on $\bar{s}$, i.e., $\bar{\text{I}}(\overline{B}u, v) = \overline{\text{II}}(u,v)$, and $B[\phi]$ is the pulled-back shape operator on s, i.e., $\bar{\text{I}}(B[\phi]u, v) = \text{II}^*[\phi](u,v)$. The difference $Q[\phi] := \overline{B} - B[\phi]$ is a field of linear operators acting on tangent vectors of $\bar{s}$, known as the *relative Weingarten map*. Friesecke et al. (2002, 2003) demonstrated that for isometric deformations ϕ of the material midsurface (for which $\mathcal{W}_{\text{mem}}[\bar{s}, \phi(\bar{s})] = 0$) the leading order term is *cubic* in the thickness δ and after rescaling with $\frac{1}{\delta^3}$ is given by the *bending* energy

$$\mathcal{W}_{\text{bend}}[\bar{s}, \phi(\bar{s})] = \int_{\bar{s}} W_{\text{bend}}(Q[\phi]) \, \mathrm{d}x \,. \tag{14}$$

Although other choices are conceivable, we consider the simple choice $W_{\text{bend}}(Q) = \|Q\|^2 = \text{tr}(Q^T Q)$ or alternatively $W_{\text{bend}}(Q) = (\text{tr}Q)^2$.

3.3.3 Deformation energy of thin shells

Combining membrane (12) and bending (14) contributions, we formulate the elastic energy of thin shells (cf. (9)) as

$$\mathcal{W}[\bar{s},\phi(\bar{s})]=\mathcal{W}_{\text{mem}}[\bar{s},\phi(\bar{s})]+\eta\,\mathcal{W}_{\text{bend}}[\bar{s},\phi(\bar{s})]. \tag{15}$$

Accordingly, we obtain that the path energy (11) is given by

$$\mathcal{E}[(\phi_t)_{0\leq t\leq 1}]=\frac{1}{2}\int_0^1 \text{Hess}\,\mathcal{W}[\phi_t(s),\cdot]_{|\phi_t(s)}(v_t,v_t)\,dt, \tag{16}$$

which solely takes into account the two-dimensional shell (mid)surfaces. Notice that according to the above discussion, $\eta=\delta^2$; hence, $\sqrt{\eta}$ represents the thickness of the material.

3.4 Hessian structure on thin shells

We are now in the position to prove our first main result: Modulo rigid body motions of embedded surfaces, the Hessian of the thin shell elastic energy is positive definite, when evaluated at any surface that is smoothly embedded into $\mathbb{R}^3$. Hence, modulo rigid body motions, the Hessian of the elastic energy induces a Riemannian metric on "shell space." At this point we hasten to carefully stress, however, that we have not defined any rigorous notion "shell space" so far, and in fact we will not do so in the spatially continuous (or smooth), infinite dimensional setting. The reason is twofold. Firstly, we are ultimately interested in the spatially discrete, finite dimensional setting, where positive definiteness of Hessians certainly suffices for a rigorous notion of Riemannian metrics, since the intricacies of infinite dimensional manifolds are by construction avoided. Secondly, to the best of our knowledge a rigorous notion of a infinite dimensional Riemannian shell manifold on which the path energy can be shown to be lower semicontinuous and for which geodesic paths exist is still unknown.

Theorem 1 (Nondegeneracy of smooth Hessian (Heeren et al., 2014, Theorem 1)).

Let $\mathcal{S}$ denote the space of smooth orientable surfaces embedded into $\mathbb{R}^3$. Let v be a tangent vector at a point $s\in\mathcal{S}$, corresponding to a vector field $v:s\rightarrow\mathbb{R}^3$ that smoothly depends on the base point. Then $\text{Hess}\,\mathcal{W}[s,\cdot]_{|_s}(v,v)=0$ if and only if v induces an infinitesimal rigid motion. Consequently, modulo rigid body motions, $g_s(\cdot,\cdot)=\frac{1}{2}\text{Hess}\,\mathcal{W}[s,\cdot]_{|_s}(\cdot,\cdot)$ indeed induces a Riemannian metric on the space of smooth shells.

Proof. Since the energy $\mathcal{W}_s$ is nonnegative and minimized by the identity deformation, it follows that Hess $\mathcal{W}_s$, when evaluated tat the point s, is positive semidefinite. It remains to show that it is indeed *definite* on the complement of rigid transformations of s. By positive semidefiniteness, Hess $\mathcal{W}[s,\cdot]_{|_s}(v,v)=0$

if and only if Hess $\mathcal{W}_{\text{mem}}(v,v)=0$ and Hess $\mathcal{W}_{\text{mem}}(v,v)=0$, where we abbreviate Hess $\mathcal{W}_{\text{mem}}=$ Hess $\mathcal{W}_{\text{mem}}[s,\cdot]|_s$ and Hess $\mathcal{W}_{\text{bend}}=$ Hess $\mathcal{W}_{\text{bend}}[s,\cdot]|_s$. Hence it suffices to study the null spaces of Hess $\mathcal{W}_{\text{mem}}$ and Hess $\mathcal{W}_{\text{bend}}$. In order to simplify the exposition, we confine ourselves to the case where membrane and bending energy densities are given by the squared Frobenius norm. The general case is completely analogous.

In order to study variations, we consider a smooth one-parameter family (ϕ_t) of diffeomorphisms of $\mathbb{R}^3$ with $\phi_0=\text{id}$. In order to abbreviate notation, let I_t and II_t, respectively, denote the first and second fundamental forms induced by ϕ_t and pulled back to s, i.e.,

$$\text{I}_t(v,w)=\text{I}_{\phi_t(s)}(\text{d}\phi_t(v),\,\text{d}\phi_t(w)),$$
$$\text{II}_t(v,w)=\text{II}_{\phi_t(s)}(\text{d}\phi_t(v),\,\text{d}\phi_t(w))$$

for vectors v and w tangent to s.

In what follows, a superscript *dot* denotes differentiation with respect to the t variable at the point $t=0$. If the deformations ϕ_t are such that $\dot{\phi}$ is in the kernel of Hess $\mathcal{W}_s$, then $\dot{\phi}$ is in the kernel of both Hess $\mathcal{W}_{\text{mem}}$ and Hess $\mathcal{W}_{\text{bend}}$. By Lemma 1 (see below), if $\dot{\phi}$ is in the kernel of Hess $\mathcal{W}_{\text{mem}}$, then $\dot{\phi}$ induces an infinitesimal isometry, i.e., $\text{d}\dot{\phi}$ is an infinitesimal rotation at every point on s. Such an infinitesimal rotation can at each point be represented by a skew-symmetric matrix or, equivalently, by the cross product with a vector $a\in\mathbb{R}^3$ that is parallel to the rotation axis. Since $\text{d}\phi_0$ acts as the identity on tangent vectors and $\text{d}\dot{\phi}$ is an infinitesimal rotation, there exists a (unique) $a:s\to\mathbb{R}^3$ such that $\text{d}\dot{\phi}=a\times\text{d}\phi_0$ at every point of s. We now show that $a:s\to\mathbb{R}^3$ is a constant map.

By Lemma 2 (see below), if $\dot{\phi}$ is in the kernel of both Hess $\mathcal{W}_{\text{mem}}$ and Hess $\mathcal{W}_{\text{bend}}$, then we have that $\text{hess}_s(\dot{\phi})\cdot n=0$. Here n denotes the unit normal field on s and hess_s denotes the Hessian that is intrinsic to the surface s, not to be confused with Hess. Since for any (smooth) mapping $f:s\to\mathbb{R}$ one has $\text{hess}_s(f)=\nabla(\,\text{d}f)$, where ∇ denotes covariant differentiation, we obtain that

$$\begin{aligned}0&=(\nabla\,\text{d}\dot{\phi})\cdot n=\nabla(a\times\,\text{d}\phi_0)\cdot n\\&=((\nabla a)\times\,\text{d}\phi_0+a\times(\nabla\,\text{d}\phi_0))\cdot n=((\nabla a)\times\,\text{d}\phi_0)\cdot n,\end{aligned}$$

where the last equality follows from the fact that $\nabla\ \text{d}\phi_0=\text{hess}_s\phi_0$ takes values in the normal bundle of s. Indeed, one has $\text{II}(v,w)=\text{hess}_s(\phi_0)(v,w)\cdot n$ for any pair (v,w) of tangent vectors.

Additionally, observe that the symmetry of Hessians and the identity

$$\text{hess}_s(\dot{\phi})=\nabla\,\text{d}\dot{\phi}=(\nabla a)\times\,\text{d}\phi_0+a\times(\nabla\,\text{d}\phi_0)=(\nabla a)\times\,\text{d}\phi_0+a\times\text{hess}_s(\phi_0)$$

implies that $(\nabla a) \times \mathrm{d}\phi_0$ is symmetric, i.e., $(\nabla_v a) \times w = (\nabla_w a) \times v$ for all tangent vector fields v,w. In particular, fix a tangential field v that does not vanish in some open subset $U \subset s$ and (locally) choose w such that $v \times w = n$ on U. Then

$$0 = ((\nabla_v a) \times w) \cdot w = ((\nabla_w a) \times v) \cdot w = n \cdot (\nabla_w a)$$

on U. Therefore, since U and v are arbitrary, $\nabla_w a$ is a tangential field for all tangent vector fields w; hence, $(\nabla a) \times \mathrm{d}\phi_0$ takes values in the normal bundle, and from $((\nabla a) \times \mathrm{d}\phi_0) \cdot n = 0$ we deduce that $\nabla a = 0$. Hence, a is constant and therefore $\dot{\phi} = a \times \phi_0 + b$ for some constants $a, b \in \mathbb{R}^3$ is an infinitesimal rigid body motion in $\mathbb{R}^3$, which proves the claim. □

It remains to prove the two auxiliary lemmas. We keep using the notation from the proof of Theorem 1.

Lemma 1. *Hess* $\mathcal{W}_{mem}(\dot{\phi},\dot{\phi}) = 0$ *if and only if* ϕ_t *induces an infinitesimally isometric deformation of the surface* s.

Proof. Let $\mathrm{I} = \mathrm{I}_0$. Then at $t = 0$ we get

$$\text{Hess}\, \mathcal{W}_{\text{mem}}(\dot{\phi},\dot{\phi}) = \left.\frac{\mathrm{d}^2 \mathcal{W}_{\text{mem}}[s,\phi_t(s)]}{\mathrm{d}t^2}\right|_{t=0} = \left.\frac{\mathrm{d}^2}{\mathrm{d}t^2}\int_s \|\mathrm{I} - \mathrm{I}_t\|^2\, \mathrm{d}x\right|_{t=0}$$

$$= 2\frac{\mathrm{d}}{\mathrm{d}t}\left.\int_s (\mathrm{I}_t - \mathrm{I}) : \dot{\mathrm{I}}_t\, \mathrm{d}x\right|_{t=0} = 2\int_s \|\dot{\mathrm{I}}_t\|^2\, \mathrm{d}x.$$

The last term vanishes if and only if $\dot{\mathrm{I}} = 0$ a.e. (everywhere due to smoothness). Hence, ϕ_t indeed induces an infinitesimally isometric deformation. □

Lemma 2. *If the deformation* ϕ_t *induces an infinitesimally isometric deformation of* s*, i.e., if* $\dot{I} = 0$*, then*

$$Hess\, \mathcal{W}_{bend}(\dot{\phi},\dot{\phi}) = 2\int_s \|hess_s(\dot{\phi}) \cdot n\|^2\, dx,$$

where n *denotes the unit normal vector field on* s.

Proof. We abbreviate notation by referring to n_t and H_t as the unit normal field and the Hessian operator of the surface $\phi_t(s)$ pulled back to s, respectively, i.e., $n_t = n_{\phi_t(s)} \circ \phi_t$, and for a function $f : s \to \mathbb{R}$ we have

$$H_t(f)(v,w) = \text{hess}_{\phi_t(s)}(f \circ \phi_t^{-1})(\,\mathrm{d}\phi_t(v),\, \mathrm{d}\phi_t(w)),$$

where v and w are tangent to s. As mentioned before, one has $\mathrm{II}_t = H_t(\phi_t) \cdot n_t$. Thus at $t = 0$ we have

$$
\begin{aligned}
\operatorname{Hess} \mathcal{W}_{\text{bend}}(\dot{\phi},\dot{\phi}) &= \frac{\mathrm{d}^2}{\mathrm{d}t^2} \mathcal{W}_{\text{bend}}[s,\phi_t(s)]\bigg|_{t=0} \\
&= \frac{\mathrm{d}^2}{\mathrm{d}t^2} \int_s \|H_t(\phi_t)\cdot n_t - \operatorname{hess}_s(\phi_0)\cdot n\|^2 \, \mathrm{d}x \bigg|_{t=0} \\
&= 2\int_s \left\| \frac{\mathrm{d}}{\mathrm{d}t}(H_t(\phi_t)\cdot n_t) \right\|^2 \mathrm{d}x \bigg|_{t=0} .
\end{aligned}
$$

Expanding gives

$$
\frac{\mathrm{d}(H_t(\phi_t)\cdot n_t)}{\mathrm{d}t}\bigg|_{t=0} = \frac{\mathrm{d}H_t(\phi_0)}{\mathrm{d}t}\bigg|_{t=0} \cdot n + \operatorname{hess}_s(\dot{\phi})\cdot n + \operatorname{hess}_s(\phi_0)\cdot \dot{n}.
$$

In order to prove the claim, it suffices to show that both $\frac{\mathrm{d}}{\mathrm{d}t}(H_t)$ and $\operatorname{hess}_s(\phi_0)\cdot\dot{n}$ vanish at $t = 0$. To see that the first term vanishes, recall that the Hessian of a smooth function f on a Riemannian manifold can be expressed in a local coordinate chart (q^1, q^2) as

$$
\operatorname{hess}_s(f) = \nabla \, \mathrm{d}f = \sum_{i,j,k} \left(\frac{\partial^2 f}{\partial q^i \partial q^j} - \frac{\partial f}{\partial q^k} \Gamma^k_{ij} \right) dq^i \otimes dq^j,
$$

where the Christoffel symbols of the Levi-Civita connection are given by $\Gamma^k_{ij} = \frac{1}{2} g^{kl}(g_{il,j} + g_{jl,i} - g_{ij,l})$. Here $g_{il,j}$ denotes the partial derivative of the (i, l)-entry of the metric tensor g along the jth coordinate and g^{kl} denotes the (k, l)-entry of the inverse metric tensor. By assumption we have $\dot{\mathrm{I}} = 0$. Hence the Riemannian metrics $g_t = \mathrm{I}_t$ satisfy $\dot{g}_{ij} = \dot{g}^{kl} = 0$ and hence $\dot{\Gamma}^k_{ij} = 0$. Hence, $\frac{\mathrm{d}}{\mathrm{d}t}(H_t) = 0$ at $t = 0$. It remains to show that $\operatorname{hess}_s(\phi_0)\cdot\dot{n} = 0$, which immediately follows from the fact that $\operatorname{hess}_s(\phi_0)$ takes values in the normal bundle of s and from $\dot{n}\cdot n = 0$. □

3.5 Hessian structure on spatially discrete shells

Mimicking our discussion of the smooth setting, we require discrete versions of membrane and bending energies for *pairs* of triangle meshes $\bar{s}$ and s in order to compute the elastic energy (15) of a deformation. With slight abuse of notation we use the same symbols for smooth shell surfaces and for their discrete, triangulated counterparts. As before, we view $\bar{s}$ as the undeformed and s as the deformed two-dimensional surface. Since every material point has a well-defined position, we viewed the associated deformation ϕ as a priori information when computing an optimal path between two shells. In the discrete setting we adopt the exact same perspective by working with triangle meshes of a *fixed* connectivity, so that ϕ is indeed always well-defined. In particular, we now assume that $\bar{s}$ and s are triangle meshes that have the *same* connectivity, to the effect that a continuous and piecewise linear mapping ϕ between these surfaces is readily available. Following our exposition in the smooth case, we require discrete notions of first and second fundamental forms, for which various versions can be found in the literature. Here we lay out our specific choices.

3.5.1 Membrane energy

We consider the standard constant strain triangle formulation for representing the membrane energy needed to deform a triangle $\overline{T}_f$ into a triangle T_f for some face index $f \in \mathcal{F}$. In order to obtain compact expressions for the requisite first fundamental forms $\bar{\mathrm{I}}$ and I, notice that in dimension two, every quadratic form is uniquely determined by its action on three different vectors. Associating these vectors with the triangle edges e_1, e_2, e_3 of a triangle T_f of our mesh, we define $\mathrm{I}_i := \|e_i\|^2$ as the squared length of the ith edge of triangle T_f. Analogously to the smooth setting, we define the pullback of the first fundamental form from s to $\bar{s}$ by a piecewise linear mapping $A[\phi]$ via

$$\bar{\mathrm{I}}(A[\phi]\cdot, \cdot) = \mathrm{I}(d\phi\cdot, d\phi\cdot),$$

which can be expressed *per face* $f \in \mathcal{F}$ by a matrix $A_f \in \mathbb{R}^{3,3}$ given as

$$A_f = \frac{1}{8\bar{a}_f^2}\sum_{i=1}^{3}(\mathrm{I}_j + \mathrm{I}_k - \mathrm{I}_i)\,\bar{t}_i \otimes \bar{t}_i, \tag{17}$$

where the indices $j = i + 1 \pmod 3$ and $k = i + 2 \pmod 3$ refer to the cyclic ordering of edges of f, $\otimes$ denotes the outer product, $\bar{t}_i$ is the result of clockwise rotating edge $\bar{e}_i$ by $\pi/2$ in the plane of T_f, and $\bar{a}_f = |\overline{T}_f|$ is the area of f.

Notice that A_f has a null space corresponding to the normal direction; hence we also consider a corresponding two-dimensional representation—denoted by $A_f^{2\times 2} \in \mathbb{R}^{2,2}$. For details, see Heeren (2016). The associated discrete membrane energy takes the form

$$\mathcal{W}_{\mathrm{mem}}[\bar{s}, s] = \sum_{f \in \mathcal{F}} W_{\mathrm{mem}}(A_f^{2\times 2})\bar{a}_f\,, \tag{18}$$

where we sum over all faces of $\bar{s}$ (weighted by area $\bar{a}_f$) and we use the *exact same* energy density (13) as in the smooth case.

3.5.2 Bending energy

In order to be consistent with the above assumptions that first fundamental forms are *constant* per triangle, we work with second fundamental forms that share this property. Since second fundamental forms are quadratic forms that account for the change of normals, we require three normals per triangle, which we associate with edge midpoints, by adopting and adapting ideas from Grinspun et al. (2006). For every edge $e = T_1 \cap T_2$ between two faces f_1 and f_2, we define N_e as the normalized sum of the unit normals belonging to the

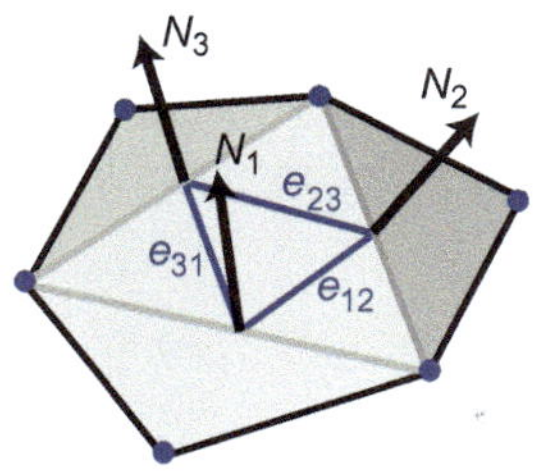

FIG. 5 Edge normals N_i on edges e_i and connecting segments $2e_{ij} = -e_k$, for i, j, k cyclic indices in $\{0, 1, 2\}$.

elements T_1 and T_2. For boundary edges, we consider the respective triangle normal. With normals associated to edge midpoints, the (discrete) 1-form dN acts on line segments connecting these midpoints. Indeed, fixing a triangle T_f with edges e_1, e_2, e_3 and corresponding edge normals N_1, N_2, N_3 we find that $dN(e_{ij}) = \int_{e_{ij}} dN = N_j - N_i$, where e_{ij} is the line segment connecting the midpoint of e_i with that of e_j, cf. Fig. 5. Using the vector identity $e_k = -2e_{ij}$, where k is the complementary index to i and j in T_f, we accordingly define $\mathrm{II}_k := \mathrm{II}(e_k, e_k) := dN(e_k) \cdot e_k = 2(N_i - N_j) \cdot e_k$ as the action of the second fundamental form associated with a triangle T_f along the direction e_k. Hence, II_k can be seen as the integrated normal curvature along e_{ij}. We make analogous definitions for the second fundamental form of the undeformed surface, i.e., $\overline{\mathrm{II}}_k := 2(\overline{N}_i - \overline{N}_j) \cdot \bar{e}_k$.

Again, the quadratic forms $\overline{\mathrm{II}}_{\overline{T}}$ and II_T are uniquely determined by $\overline{\mathrm{II}}_i$ and II_i, with $i \in \{1, 2, 3\}$, respectively. Analogously to the smooth setting, we define a discrete shape operator $\overline{B}$ and a discrete pulled-back shape operator B, respectively, via

$$\bar{\mathrm{I}}(\overline{B}\,\cdot, \cdot) = \overline{\mathrm{II}}(\,\cdot, \cdot) \quad \text{and} \quad \bar{\mathrm{I}}(B\,\cdot, \cdot) = \mathrm{II}(\mathrm{d}\phi\,\cdot, \mathrm{d}\phi\,\cdot).$$

Similarly to our treatment of first fundamental forms, *per triangle* we obtain

$$B_f = \frac{1}{8\bar{a}_f^2} \sum_{i=1}^{3} (\mathrm{II}_j + \mathrm{II}_k - \mathrm{II}_i)\, \bar{t}_i \otimes \bar{t}_i, \tag{19}$$

and analogously for $\overline{B}_f$ by replacing II_i by $\overline{\mathrm{II}}_i$ for $i \in \{1, 2, 3\}$. As for the discrete Cauchy-Green tensor (17), we make use of a corresponding two-dimensional representation—denoted by $B_f^{2\times 2} \in \mathbb{R}^{2,2}$. For details, see Heeren (2016). We define a two-dimensional representation of the discrete relative Weingarten map as $Q_f^{2\times 2} := B_f^{2\times 2} - \overline{B}_f^{2\times 2}$ and obtain the discrete bending energy via

$$\mathcal{W}_{\text{bend}}[\bar{s}, s] = \sum_{f \in \mathcal{F}} W_{\text{bend}}(Q_f^{2\times 2})\bar{a}_f, \tag{20}$$

where again we use area weights $\bar{a}_f$ and we use the exact *same* energy density $W_{\text{bend}}(Q)$ as in the smooth case.

3.5.3 Hessian of discrete elastic energy induces metric

As in the smooth case, we combine discrete membrane (18) and discrete bending (20) contributions in order to formulate the discrete elastic energy of discrete thin shells as

$$\mathcal{W}[\bar{s},s] = \mathcal{W}_{\text{mem}}[\bar{s},s] + \eta \mathcal{W}_{\text{bend}}[\bar{s},s]. \tag{21}$$

Equipped with this energy, our main result is that the space of discrete shells carries a Riemannian structure induced by the Hessian of elastic energy:

Theorem 2 (Nondegeneracy of discrete Hessian (Heeren et al., 2014, Theorem 2)).

Let $\mathcal{S} = \mathbb{R}^{3n}$ denote the space of discrete orientable triangle meshes with a fixed connectivity and with a fixed number n of vertices embedded into $\mathbb{R}^3$. Let v be a tangent vector at a point $s \in \mathcal{S}$, corresponding to a vector field that associates a 3-vector to every vertex of s. Then Hess $\mathcal{W}[s,\cdot]|_{s}(v,v) = 0$ *if and only if v induces an infinitesimal rigid motion. Consequently, modulo rigid body motions, $g_s(\cdot,\cdot) = \frac{1}{2}\text{Hess}\,\mathcal{W}[s,\cdot]|_{s}(\cdot,\cdot)$ indeed induces a Riemannian metric on the space of discrete shells.*

Proof. The proof is significantly simpler than its smooth counterpart. Suppose that Hess $\mathcal{W}[s,\cdot]|_{s}(v,v) = 0$ for some tangent vector v that corresponds to a vector field that associates a 3-vector to every vertex of s. Then v is both in the kernel of the Hessian of discrete membrane energy and the Hessian of discrete bending energy since both are positive semidefinite operators. Consider two adjacent triangles f_1 and f_2. After subtracting global translations induced by the vector field, for v to be in the kernel of the Hessian of membrane energy implies that v induces an infinitesimal rotation of f_1 since the edge lengths of f_1 must not change. The same holds for f_2. Since v is also in the kernel of the Hessian of bending energy, it follows that the corresponding vector field must not induce a change of the dihedral angle between f_1 and f_2. Hence the vector field induces a single infinitesimal rotation of the hinge $f_1 \cup f_2$. Iterating this argument over the entire mesh proves the claim. □

4 Discrete Riemannian calculus

In this section, we provide a survey of the time-discrete Riemannian calculus and summarize important convergence results. In Section 3.1 we discussed the natural order for defining notions in smooth Riemannian calculus. In *discrete* Riemannian calculus, the order is quite different, as illustrated below:

Hessian structure ⇒ Riemannian metric ⇒ local approximation of squared distance ⇒ geodesics ⇒ log ⇒ exp ⇒ parallel transport ⇒ Levi-Civita connection

In particular, the notion of discrete geodesics will serve as the core ingredient of the entire discrete calculus.[a] Hence, the discrete Riemannian calculus is also referred to as *discrete geodesic calculus*. This section summarizes the results of Wirth (2009), Wirth et al. (2009), and Rumpf and Wirth (2013, 2015). The resulting time-discrete Riemannian calculus has been applied to several Riemannian manifolds or shape spaces, e.g., in Heeren et al. (2012, 2014), Berkels et al. (2015), Maas et al. (2015), and Huber et al. (2017).

4.1 Time-discrete geodesic paths

In the continuous setting the starting point of a geometric calculus on a Riemannian manifold is usually the definition of a Riemannian metric. In contrast, discrete geodesic calculus is solely based on the notion of a (squared) Riemannian distance, resp., a local approximation thereof. What ties these two different views together is our above discussion of Hessian structures. Recall that a Hessian structure is given whenever $\mathcal{M}$ is a smooth manifold and $\mathcal{W}:\mathcal{M}\times\mathcal{M}\to\mathbb{R}_{\geq 0}$ is a sufficiently smooth mapping that satisfies

$$\mathcal{W}[s,s]=0,\quad \mathcal{W}_{,2}[s,s]=0 \;\text{ and }\; \mathcal{W}_{,22}[s,s] \text{ is positive definite} \tag{22}$$

for all $s\in\mathcal{M}$. Here and in the following $\mathcal{W}_{,i}$ denotes the partial derivative with respect to the ith argument of $\mathcal{W}$. We assume that $\mathcal{W}$ is three times continuously differentiable. Note, however, that $\mathcal{W}$ is not required to be symmetric. We discussed in Section 3 that $\text{Hess}\,\mathcal{W}[s,\cdot\,]_{|s}=\mathcal{W}_{,22}[s,s]$ defines the Riemannian metric on $\mathcal{M}$ and the squared distance function associated with this metric is obtained via (7). It has been shown in Rumpf and Wirth (2015, Lemma 4.6) that (22) along with the smoothness assumption on $\mathcal{W}$ implies

$$\mathcal{W}[\bar{s},s]=\text{dist}^2(\bar{s},s)+O(\text{dist}^3(\bar{s},s)). \tag{23}$$

A natural *time-discrete* approximation of the resulting squared distance function is then given by

$$\text{dist}^2(\bar{s},s)\approx \inf_{\substack{(s_0,\ldots,s_K)\\ s_0=\bar{s},\,s_K=s}} K\sum_{k=1}^{K}\mathcal{W}[s_{k-1},s_k],$$

where $K\in\mathbb{N}$ is given. This choice is additionally motivated by the following observation:

Lemma 3. *Let* $(\mathcal{M},g)$ *be a smooth Riemannian manifold, and let* $\bar{s},s\in\mathcal{M}$. *Suppose that there exists a smooth minimizing geodesic* $\gamma:[0,1]\to\mathcal{M}$ *with* $\gamma(0)=\bar{s}$ *and* $\gamma(1)=s$. *Then a minimizer of*

---

[a]We will often omit the prefix "time" when referring to the time-discrete notion.

$$\hat{E}^K(s_0,\ldots,s_K):=K\sum_{k=1}^{K}\mathrm{dist}^2(s_{k-1},s_k)$$

subject to $s_0=\overline{s}$ *and* $s_K = s$ *is given by* $s_k = \gamma(k/K)$. *In particular,* $\mathrm{dist}^2(\overline{s},s)=K\sum_{k=1}^{K}\mathrm{dist}^2(s_{k+1},s_k)$.

Proof. For a piecewise geodesic path consisting of segments $\gamma:[\frac{k-1}{K},\frac{k}{K}]\to\mathcal{M}$ with $\gamma(\frac{k-1}{K})=s_{k-1}$ and $\gamma(\frac{k}{K})=s_k$ we have by the Cauchy–Schwarz inequality

$$K\mathrm{dist}^2(s_{k-1},s_k) \leq \int_{\frac{k-1}{K}}^{\frac{k}{K}} g_{\gamma(t)}(\dot{\gamma}(t),\dot{\gamma}(t))\,\mathrm{d}t.$$

This inequality turns into an equality if and only if $g_{\gamma(t)}(\dot{\gamma}(t),\dot{\gamma}(t))$ is constant on each interval (t_{k-1}, t_k). The right-hand side is the path energy of the segment $\gamma:[\frac{k-1}{K},\frac{k}{K}]\to\mathcal{M}$, and the sum of these terms for $k = 1, \ldots, K$ is the total path energy. Thus, a piecewise geodesic path which corresponds to a minimizer of the sum over the left-hand side for $k = 1, \ldots, K$ cannot have a smaller path energy than the minimal path energy attained by the minimizing geodesic path which minimizes the sum over the terms on the right-hand side. □

In order to fix notation, we denote an ordered set of points in $\mathcal{M}$ as a time-discrete K-path $S^K = (s_0, \ldots, s_K)$. Often we interpret this discrete path as a uniform sampling of a smooth curve $s:[0,1]\to\mathcal{M}$, i.e., we have $s_k = s(t_k)$ with $t_k = k\tau$ for $k = 0, \ldots, K$ where $\tau = K^{-1}$ and $K\in\mathbb{N}$ denotes the sample size.

Definition 1 (Discrete path energy).
For some $\mathcal{W}:\mathcal{M}\times\mathcal{M}\to\mathbb{R}_{\geq 0}$ and a discrete K-path $S^K = (s_0, \ldots, s_K)$ with $s_k\in\mathcal{M}$ for $k = 0, \ldots, K$ we define the *discrete energy* E^K by

$$\mathrm{E}^K[S^K]=K\sum_{k=1}^{K}\mathcal{W}[s_{k-1},s_k]\,. \tag{24}$$

Then a *discrete geodesic* (K-geodesic) is defined as a minimizer S^K of $\mathrm{E}^K[\cdot]$ for fixed end points s_0,s_K.

For our discussion below, we require the following property of the mapping $\mathcal{W}$:

$$\text{(H)}\quad\begin{cases}\mathcal{W}\text{ is three times continuously differentiable in both arguments.}\\ \mathcal{W}\text{ is coercive in the sense }\mathcal{W}[\overline{s},s]\geq\gamma(\mathrm{dist}(\overline{s},s))\\ \text{for an increasing, continuous function }\gamma\text{ with }\gamma(0)=0\text{ and }\lim_{d\to\infty}\gamma(d)=\infty.\end{cases}$$

Notice that (H) implies (23) by Taylor expansion. As a direct consequence, discrete geodesics are well-defined:

Theorem 3 (Existence of discrete geodesics).

Given s_A, $s_B \in \mathcal{M}$, there is a discrete geodesic path $(s_0, \ldots, s_K)$ that minimizes the discrete energy E^K over all discrete paths $(\tilde{s}_0, \ldots, \tilde{s}_K)$ with $\tilde{s}_0 = s_A$ and $\tilde{s}_K = s_B$.

Proof. Let $(s_0^j, \ldots, s_K^j)_j$ be a minimizing sequence. Without loss of generality the energy on this sequence is bounded by

$$K(\mathcal{W}[s_A, s_B] + (K-1)\mathcal{W}[s_B, s_B]) = K\mathcal{W}[s_A, s_B] < \infty.$$

Due to the coercivity of $\mathcal{W}$ the minimizing sequence is bounded in any norm on $(\mathbb{R}^d)^{K+1}$. Hence, by the Heine–Borel theorem there is a subsequence converging to some discrete path $(s_0, \ldots, s_K)$. Then the claim follows from the continuity of $\mathcal{W}$ in both arguments. □

In the sequel, let $\mathring{\mathcal{M}}$ denote the interior of the manifold $\mathcal{M}$ whenever $\mathcal{M}$ has non empty boundary; otherwise, $\mathring{\mathcal{M}} = \mathcal{M}$. The next result asserts that the minimizer is unique if the two endpoints are sufficiently close:

Theorem 4 (Uniqueness of discrete geodesics (Rumpf and Wirth, 2015, Theorem 4.7)).

Let (H) hold. Then for all $s_A \in \mathring{\mathcal{M}}$ and $K \in \mathbb{N}$ there exists $\varepsilon > 0$ such that there exists a unique discrete geodesic $(s_0, \ldots, s_K)$ with $s_0 = s_A$ and $s_K = s_B$ for all s_B with dist$(s_A, s_B) < \varepsilon$.

The following theorem is a discrete version of the constant speed property of continuous geodesics:

Theorem 5 (Equidistribution of points along discrete geodesics (Rumpf and Wirth, 2015, Theorem 4.5)).

Under hypothesis (H) there exists $\delta > 0$ such that if $\mathrm{dist}(s_A, s_B) < \sqrt{K}\delta$, *then discrete geodesics satisfy* $\mathrm{dist}(s_{k-1}, s_k) \leq CK^{-1}$ *for all $k = 1, \ldots, K$ with the constant $C > 0$ only depending on dist(s_A, s_B).*

Furthermore, the optimal discrete path energy is a valid approximation of the squared Riemannian distance:

Theorem 6 (Bounds on discrete path energy (Rumpf and Wirth, 2015, Theorem 4.4)).

Under hypothesis (H) there exists $\delta > 0$ such that $\mathrm{dist}(s_A, s_B) < \sqrt{K}\delta$ *implies*

$$\left| \min_{(s_0, \ldots, s_K) s_0 = s_A, s_K = s_B} E^K[(s_0, \ldots, s_K)] - dist^2(s_A, s_B) \right| = O(1/K)$$

Finally, we state that sequences of successively refined discrete geodesic paths converge to a continuous geodesic path. To this end, one considers continuous paths $(s(t))_{t\in[0,\,1]}$ on $\mathcal{M}$ which are composed of shortest geodesic segments like in the proof of Lemma 3. This means, $(s(t))_{t\in\left[\frac{k-1}{K},\frac{k}{K}\right]}$ is a (possibly nonunique) shortest geodesic path connecting $s(\frac{k-1}{K})$ and $s(\frac{k}{K})$ for all $k = 1, \ldots, K$. We define an energy $\tilde{\mathrm{E}}^K : L^2((0,1);\mathcal{M}) \to \mathbb{R}$ via

$$\tilde{\mathrm{E}}^K[(s(t))_{t\in[0,1]}] = \begin{cases} \mathrm{E}^K\left[\left(s(0), s(\frac{1}{K}), \ldots, s(\frac{K-1}{K}), s(1)\right)\right], & \text{if } (s(t))_{t\in[0,1]} \text{ is a pw. geodesic path} \\ \infty, & \text{else} \end{cases}$$

where E^K is the discrete path energy in (24). Then one obtains the following convergence result:

Theorem 7 (Γ-convergence of the discrete energy (Rumpf and Wirth, 2015, Theorem 4.8)).

Assuming (H), the Γ-limit of $\tilde{\mathrm{E}}^K$ for $K \to \infty$ in the L^2-topology is the continuous path energy (1).

We refer to De Giorgi and Dal Maso (1983) and Braides (2002) for an introduction to the concept of Γ-convergence. It is a fundamental implication of the Γ-convergence $\tilde{\mathrm{E}}^K \to \mathcal{E}$ that minimizers of $\tilde{\mathrm{E}}^K$ converge to minimizers of the path energy (1). Indeed, one can show equi-mild coercivity of the discrete energies $\tilde{\mathrm{E}}^K$ to finally obtain convergence of discrete geodesics to continuous geodesics:

Corollary 1 (Convergence of discrete geodesics (Rumpf and Wirth, 2015, Corollary 4.9)).

Under (H), any sequence of minimizers of $\tilde{\mathrm{E}}^K$ contains a $C^0([0,1];\mathcal{M})$-convergent subsequence, and the limit is a minimizer of $\mathcal{E}$.

Note that a similar argument can be given for the piecewise *linear* interpolation of discrete geodesics in the definition of $\tilde{\mathrm{E}}^K$. The following theorem finally states that the convergence in C^0 ensured by the above Γ-convergence result is actually much stronger with velocities converging in L^2:

Theorem 8 (Path convergence in $W^{1,2}((0,1);\mathcal{M})$ *(Rumpf and Wirth, 2015, Theorem 4.10)*).

Suppose (H) holds and $\mathcal{W}$ is $C^4(\mathcal{M}\times\mathcal{M};\mathbb{R})$-smooth with bounded derivatives. If the interpolated discrete geodesics converge for $K \to \infty$ to the classical geodesic in L^2, then the sequence already converges in $W^{1,2}((0,1);\mathcal{M})$.

Because $W^{1,2}$ embeds continuously in $C^{0,\frac{1}{2}}$, this theorem implies in particular uniform convergence.

4.2 Time-discrete geodesic calculus

In the sequel we additionally assume that $\mathcal{M} \subset \mathbb{R}^d$ is a submanifold of $\mathbb{R}^d$, which is a natural assumption for the case where $\mathcal{M}$ is the space of triangle meshes with *fixed connectivity* (in which case $\mathbb{R}^d = \mathbb{R}^{3|V|}$, where $|V|$ denotes the number of vertices). Let $\bar{s}, s \in \mathcal{M}$ such that there is a unique geodesic $s : [0,1] \to \mathcal{M}$ with $s(0) = \bar{s}$ and $s(1) = s$. Then the logarithm of s with respect to $\bar{s}$ is the initial velocity $\dot{s}(0) \in T_p\mathcal{M}$, i.e., $\log_{\bar{s}}(s) = \dot{s}(0)$, which can be approximated by a difference quotient in time,

$$\dot{s}(0) = \frac{s(\tau) - s(0)}{\tau} + O(\tau).$$

Thus, we obtain $\tau \log_{\bar{s}}(s) = s(\tau) - s(0) + O(\tau^2)$, which gives rise to a notion of a time-discrete logarithm with underlying discrete time step size $\tau = \frac{1}{K}$:

Definition 2 (Discrete logarithm**).**
Suppose the discrete geodesic $(s_0, \ldots, s_K)$ is the unique minimizer of the discrete path energy (24) with $s_0 = \bar{s}$ and $s_K = s$. Then we define the *discrete logarithm* $(\frac{1}{K}\mathrm{LOG})_{\bar{s}}(s) = s_1 - s_0$. Note that $\frac{1}{K}$ is part of the symbol and not a factor.

We consider the difference $s_1 - s_0$ as a discrete tangent vector at $\bar{s} = s_0$. In the special case $K = 1$ we have $(\frac{1}{1}\mathrm{LOG})_{\bar{s}}(s) = s - \bar{s}$. Let us emphasize that subtracting is here defined as the subtraction in $\mathbb{R}^d$ and thus simply a vector of differences of vertex positions. As in the continuous case, the discrete logarithm can be considered as a representation of the nonlinear variation s of $\bar{s}$ in the (linear) tangent space of *displacements* in $T_{\bar{s}}\mathcal{M}$.

In the continuous setting, the exponential map $\exp_{\bar{s}}$ maps tangent vectors $v \in T_{\bar{s}}\mathcal{M}$ onto the end point $s(1)$ of the unique geodesic $(s(t))_{t \in [0,1]}$ with $s(0) = \bar{s}$ and $\dot{s}(0) = v$. That means, we have $\exp_{\bar{s}}(v) = s(1)$ and, via a simple scaling argument, $\exp_{\bar{s}}(t_k v) = s(t_k)$, for $k = 0, \ldots, K$, where $t_k = k\tau$ and $\tau = K^{-1}$. Let us again consider a discrete geodesic $(s_0, \ldots, s_K)$ with $s_0 = \bar{s}$ and $s_K = s$. Since $v = (\frac{1}{K}\mathrm{LOG})_{\bar{s}}(s) = s_1 - s_0$ is the discrete logarithm in the tangent space $T_{\bar{s}}\mathcal{M}$, we aim at defining a discrete *power k exponential map* $\mathrm{EXP}^{k}_{\bar{s}}$ such that

$$\mathrm{EXP}^{k}_{\bar{s}}(v) = s_k.$$

This notation is motivated by the observation that $\exp(ks) = \exp^k(s)$ on $\mathbb{R}$ or more general matrix groups. Furthermore, we would like to have a discrete counterpart of the following recursive property for $k = 0, \ldots, K$ and $t_k = \frac{k}{K}$ in the continuous setup:

$$s(t_k) = \exp_{\bar{s}}(kv) = \exp_{s(t_{k-2})}(2v_{k-1}), \quad v_{k-1} := \log_{s(t_{k-2})} s(t_{k-1}), \quad k \geq 2. \quad (25)$$

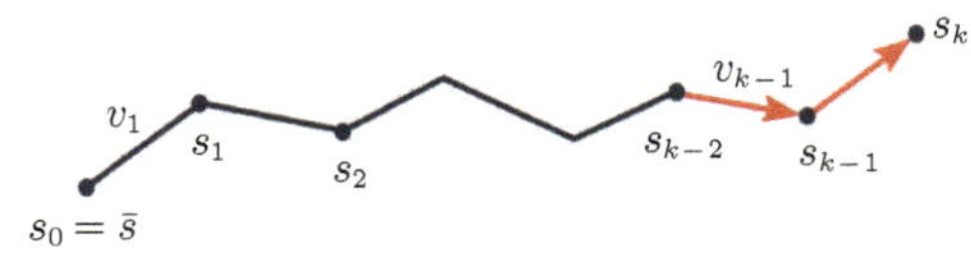

FIG. 6 A sketch of the polygonal path associated with the computation of $\mathrm{EXP}^k_{\bar{s}}(v_1)$.

That means, once we have defined a discrete version $\mathrm{EXP}^2_{\bar{s}}$ corresponding to $\exp_{\bar{s}}(2\,\cdot\,)$, we can use the recursive relation (25) to define $\mathrm{EXP}^k_{\bar{s}}$ for $k \geq 2$ by

$$s_k = \mathrm{EXP}^k_{\bar{s}}(v_1) = \mathrm{EXP}^2_{s_{k-2}}(v_{k-1}), \quad v_{k-1} = s_{k-1} - s_{k-2}, \tag{26}$$

for given $s_0 = \bar{s}$ and $s_1 = s_0 + v_1$, as shown in Fig. 6. Note that the discrete analogue of v_{k-1} in (25) is exactly $(\frac{1}{1}\mathrm{LOG})_{s_{k-2}} s_{k-1} = s_{k-1} - s_{k-2}$.

It remains to define the discrete map $\mathrm{EXP}^2_{\bar{s}}$ as an approximation of $\exp_{\bar{s}}(2\,\cdot\,)$. Formally, we have the identity

$$\frac{1}{2}\log_{\bar{s}}(\exp_{\bar{s}}(2v)) = v,$$

i.e., we can define $\mathrm{EXP}^2_{s_0}(s_1 - s_0)$ as the root of the function

$$z \mapsto (\tfrac{1}{2}\mathrm{LOG})_{s_0}(z) - (s_1 - s_0)$$

for given $s_0, s_1 \in \mathcal{M}$. In fact, we are seeking for a third point $s_2 \in \mathcal{M}$, such that $(s_0,\ s_1,\ s_2)$ is a time-discrete geodesic for $K = 2$. Using Definition 1, a necessary condition for this is given by $\mathcal{W}_{,2}[s_0,s_1] + \mathcal{W}_{,1}[s_1,s_2] = 0$, which is the system of Euler-Lagrange equation for a minimizer of the discrete path energy $2(\mathcal{W}_{,2}[s_0,s_1] + \mathcal{W}_{,1}[s_1,s_2])$ and s_1 as the degree of freedom. Thus, we define

Definition 3 (Discrete exponential map).
For given points $s_0, s_1 \in \mathcal{M}$, $v_1 = s_1 - s_0$, we define $\mathrm{EXP}^2_{s_0}(v_1)$ as the solution of

$$\mathcal{W}_{,2}[s_0,s_1] + \mathcal{W}_{,1}[s_1,s] = 0\,,$$

and hence $s_k := \mathrm{EXP}^k_{s_0}(v_1) := \mathrm{EXP}^2_{s_{k-2}}(v_{k-1})$ for $v_{k-1} = s_{k-1} - s_{k-2}$ and $k \geq 2$.

It is straightforward to verify that $\mathrm{EXP}^K_{\bar{s}} = (\frac{1}{K}\mathrm{LOG})^{-1}_{\bar{s}}$ as long as the discrete logarithm is invertible. In fact, the Euler–Lagrange equations for $(s_0,\ \ldots,\ s_K)$ being a discrete geodesic with fixed end points s_0 and s_K are given by the $K - 1$ nonlinear equations

$$0 = \mathcal{W}_{,2}[s_{k-1},s_k] + \mathcal{W}_{,1}[s_k,s_{k+1}]\,, \quad k = 1,\ldots,K-1\,, \tag{27}$$

which have to be solved *simultaneously*. On the other hand, if we compute $\mathrm{EXP}^k_{s_0}(s_1 - s_0)$ for given $s_0, s_1 \in \mathcal{M}$ and $k = 2,\ \ldots,\ K$, we get exactly the same system (27). However, in this case the system can be solved *sequentially*.

In summary, as long as a variationally computed discrete geodesic is unique we can reproduce it by discrete exponential shooting.

Next, we introduce a time-discrete notion of parallel transport along a discrete path as proposed in Rumpf and Wirth (2015). In the continuous setting, given a path $s:[0,1]\to\mathcal{M}$ and a vector $v_0 \in T_{s(0)}\mathcal{M}$, parallel transport $\mathrm{P}_{s(0)\to s(\tau)}v_0$ of v_0 along the path s is defined as the solution of the initial value problem $\nabla_{\dot{s}}v(t)=0$ for $t\in[0,\tau]$ and $v(0)=v_0$.

There is a well-known first-order approximation of parallel transport called Schild's ladder (cf. Ehlers et al., 1972; Kheyfets et al., 2000), which is based on the construction of a sequence of so-called *geodesic parallelograms*; this method has been used, e.g., by Lorenzi et al. (2011) to perform parallel transport of deformations along time series of images (see also Pennec and Lorenzi (2011)). We once more use the notation $s_k = s(t_k)$, $t_k = k\tau$, for samples of the path $s:[0,1]\to\mathcal{M}$. Given a tangent vector $v_{k-1}\in T_{s_{k-1}}\mathcal{M}$, the approximation $v_k\in T_{s_k}\mathcal{M}$ of the parallelly transported vector $\mathrm{P}_{s_{k-1}\to s_k}v_{k-1}$ via a geodesic parallelogram is illustrated in Fig. 7.

The scheme in Fig. 7 can be easily transferred to the time-discrete setup by replacing s by a discrete path $(s_0, \ldots, s_K)$ and the geodesics that define the geodesic parallelogram by time-discrete geodesics, e.g., of length 3. Conceptually, we will again replace tangent or velocity vectors by displacements of points. In fact, let $s_0,s_1,s_2\in\mathcal{M}$ and define $\hat{s}=\hat{s}(s_0,s_1,s_2)$ such that (s_0,s^c,s_2) and $(s_1,s^c,\hat{s})$ are discrete geodesics for some $s^c\in\mathcal{M}$. Then $(s_0,s_1,s_2,\hat{s})$ defines a *discrete geodesic parallelogram*, and s^c is referred to as centre point of the parallelogram, see Fig. 8.

Definition 4 (Discrete parallel transport).

Let $(s_0, \ldots, s_K)$ be a discrete path in $\mathring{\mathcal{M}}$ with $s_k - s_{k-1}$ sufficiently small for $k = 1, \ldots, K$ and v_0 a sufficiently small displacement of s_0, given as $s_0^p = s_0 + v_0$. Then the *discrete parallel transport* of v_0 along $(s_0, \ldots, s_K)$ is defined for $k = 1, \ldots, K$ via the iteration where $v_k = s_k^p - s_k$ is the transported displacement at s_k. We define

$$\mathbf{P}_{s_K,\ldots,s_0}(s_0^p - s_0) = s_K^p - s_K.$$

The notation is chosen such that $\mathbf{P}_{s_K,\ldots,s_0}\mathbf{P}_{\tilde{s}_K,\ldots,\tilde{s}_0} = \mathbf{P}_{s_K,\ldots,s_0,\tilde{s}_K,\ldots,\tilde{s}_0}$.

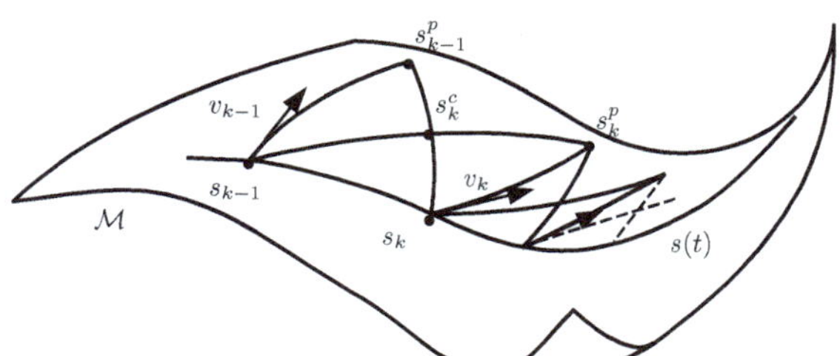

$$s_{k-1}^p = \exp_{s_{k-1}}(v_{k-1})$$

$$s_k^c = \exp_{s_{k-1}^p}\left(\tfrac{1}{2}\log_{s_{k-1}^p}(s_k)\right)$$

$$s_k^p = \exp_{s_{k-1}}\left(2\log_{s_{k-1}}(s_k^c)\right)$$

$$v_k = \log_{s_k}(s_k^p)$$

FIG. 7 A sketch of the parallel transport of $v_{k-1}\in T_{s_{k-1}}\mathcal{M}$ from s_{k-1} to s_k along s via Schild's ladder. Here, s_k^c is the midpoint of the two diagonals of a geodesic parallelogram, i.e., (s_{k-1}^p, s_k^c, s_k) and (s_{k-1}, s_k^c, s_k^p) are both discrete 2-geodesics.

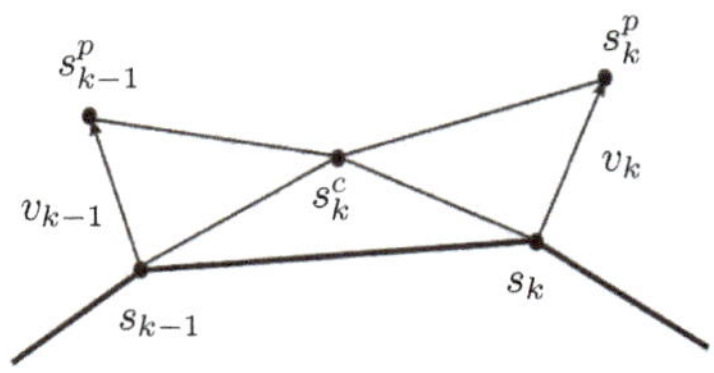

$$s_k^c = s_{k-1}^p + \left(\left(\tfrac{1}{2}\mathrm{LOG}\right)_{s_{k-1}^p}(s_k)\right),$$

$$s_k^p = \mathrm{EXP}^2_{s_{k-1}}\left(s_k^c - s_{k-1}\right),$$

FIG. 8 Discrete geodesic parallelogram associated with the computation of discrete parallel transport.

Given a discrete path $(s_0, \ldots, s_K)$, the kth step of discrete parallel transport for determining s_k^c and $s_k^p = s_k + v_k$ for given $s_{k-1}^p = s_{k-1} + v_{k-1}$ is given by solving the Euler–Lagrange equations

$$\mathcal{W}_{,2}[s_{k-1}^p, s_k^c] + \mathcal{W}_{,1}[s_k^c, s_k] = 0,$$

$$\mathcal{W}_{,2}[s_{k-1}, s_k^c] + \mathcal{W}_{,1}[s_k^c, s_k^p] = 0.$$

If $\mathcal{W}$ is symmetric, these conditions are the same as the Euler–Lagrange equations for inverse parallel transport, so that $\mathbf{P}^{-1}_{s_K,\ldots,s_0} = \mathbf{P}_{s_0,\ldots,s_K}$. However, if $\mathcal{W}$ is not symmetric this is not true in general. Finally, with a notion of (inverse) parallel transport at hand one may define a discrete connection as follows:

Definition 5 (Discrete connection).
For $s \in \mathcal{M}$, sufficiently small $\xi \in T_s\mathcal{M}$ and vectors η_0 attached to s and η_1 attached to $s + \xi$ (representing a discrete vector field), a discrete connection is defined as

$$\nabla_\xi(\eta_0, \eta_1) := \mathbf{P}^{-1}_{y+\xi,y}\eta_1 - \eta_0.$$

In the remainder of this subsection, we gather convergence statements from Rumpf and Wirth (2015) as the time step size $\tau = \frac{1}{K}$ tends to 0. The following smoothness hypothesis is required (in addition to (H)):

The energy $\mathcal{W}$ is $C^4(\mathcal{M} \times \mathcal{M}; \mathbb{R})$ – smooth with bounded derivatives. (H')

Note that $\mathcal{W} \in C^4(\mathcal{M} \times \mathcal{M}; \mathbb{R})$ implies that the metric g is C^2-smooth. In Rumpf and Wirth (2015) it is shown that under hypotheses (H) and (H') one can expect local uniqueness of $(\frac{1}{2}\mathrm{LOG})$ and local existence of (EXP^2). Moreover, it is proven that all time-discrete geometric objects introduced above converge to their continuous counterparts:

Theorem 9 (Convergence of discrete logarithm (Rumpf and Wirth, 2015, Theorem 5.1)).

Given $\overline{s}, s \in \overset{\circ}{\mathcal{M}}$, assume that hypotheses (H) and (H') hold, that the continuous and discrete geodesics between $\overline{s}, s$ are unique, and that the continuous geodesic lies in $\overset{\circ}{\mathcal{M}}$. Then $K(\frac{1}{K}\mathrm{LOG})_s\tilde{s} \to \log_s \tilde{s}$ as $K \to \infty$.

Theorem 10 (Existence and convergence of discrete exponential (Rumpf and Wirth, 2015, Theorem 5.10)).
Let $s : [0, 1] \to \overset{\circ}{\mathcal{M}}$ be a smooth geodesic. Under the hypotheses (H) and (H'), $\mathrm{EXP}^K_{s(0)}(\frac{\dot{s}(0)}{K})$ exists for K large enough, and one obtains

$$\left\| EXP^K_{s(0)}\left(\frac{\dot{s}(0)}{K}\right) - s(1)\right\| = O(K^{-1}).$$

Theorem 11 (Convergence of discrete parallel transport (Rumpf and Wirth, 2015, Theorem 5.11/5.12)).
Let $s : [0, 1] \to \mathcal{M}$ be a smooth path and v a parallel vector field along s, i.e., $v(t) \in T_{s(t)}\mathcal{M}$ for $t \in$ [0, 1]. For $K \in \mathbb{N}$ we set $s_k = s\left(\frac{k}{K}\right)$, $k = 0, \ldots, K$. Then under the hypotheses (H) and (H'), we have

$$\left\| K\mathbf{P}_{s_K,\ldots,s_0}\left(\frac{v(0)}{K}\right) - v(1)\right\| = O(K^{-1}).$$

More generally, if the sequence $(s_k)_k$ only satisfies $\| s_k - s\left(\frac{k}{K}\right)\| \leq \epsilon$, $k = 0, \ldots, K$, we still get

$$\left\| K\mathbf{P}_{s_K,\ldots,s_0}\left(\frac{v(0)}{K}\right) - v(1)\right\| = O(K^{-1} + \epsilon).$$

Finally, the convergence of the (inverse) parallel transport also implies a corresponding first-order convergence of the discrete connection defined in Definition 5 (cf. Rumpf and Wirth, 2015, Theorem 5.13).

5 Discrete Riemannian splines

In this section we consider splines on Riemannian manifolds, their effective discretization using the variational time discretization framework, and the application in the context of smooth key pose interpolation. This concept has been introduced in Heeren et al. (2019). As before, we focus on the application in the context of *finite* dimensional spaces.

5.1 Continuous Riemannian splines

We briefly recapitulate the theory of splines on smooth Riemannian manifolds as a generalization of cubic spline in Euclidean spaces. For two points

$s_A, s_B \in \mathcal{M}$ a smooth interpolation $s:[0,1] \to \mathcal{M}$ with $s(0) = s_A$ and $s(1) = s_B$ is given by the connecting geodesic path. However, for a sequence $0 = t_1 < t_2 < \cdots < t_J = 1$ and corresponding points $s_1, \ldots, s_J \in \mathcal{M}$, there is in general no geodesic curve $s:[0,1] \to \mathcal{M}$ that fulfills the *interpolation constraints*

$$s(t_j) = s^j, \quad j = 1, \ldots, J. \tag{28}$$

In particular, a curve s satisfying the interpolation constraints does in general not comply with the geodesic equation $\nabla_{\dot{s}}\dot{s} = 0$. For example, a piecewise geodesic curve connecting $s_1, \ldots, s_J$ fulfills $\nabla_{\dot{s}}\dot{s} = 0$ on each segment (t_j, t_{j+1}), $j = 1, \ldots, J-1$, but exhibits discontinuities in $\dot{s}$ at the interpolation points. Nevertheless, if one is interested in a curve that on the one hand satisfies the interpolation constraints *exactly* and on the other hand is *as smooth as possible*, one might consider the geodesic equation as a *penalty term*. Instead of requiring the interpolating curve to satisfy $\nabla_{\dot{s}}\dot{s} = 0$ exactly, we can penalize a deviation from this constraint by introducing the energy

$$\mathcal{F}[s] = \int_0^1 g_{s(t)}(\nabla_{\dot{s}}\dot{s}(t), \nabla_{\dot{s}}\dot{s}(t))\, \mathrm{d}t\,. \tag{29}$$

In addition to the interpolation constraints, we may optionally impose one of the two boundary conditions

$$\begin{aligned} &\dot{s}(0) = v_0, \quad \dot{s}(1) = v_1 \quad \text{for given } v_0, v_1 \in T\mathcal{M}, && \text{(Hermite b. c.)} \\ &s(0) = s(1), \quad \dot{s}(0) = \dot{s}(1), && \text{(periodic b. c.)} \end{aligned}$$

where $T\mathcal{M}$ denotes the tangent bundle on $\mathcal{M}$. The case without additional conditions is referred to as natural boundary condition.

For a better intuition and to motivate our terminology it is helpful to consider the Euclidean setting $\mathcal{M} = \mathbb{R}^d$, in which $\nabla_{\dot{s}}\dot{s}(t) = \ddot{s}(t)$ and $\mathcal{F}[s] = \int_0^1 \| \ddot{s}(t)\|^2 \,\mathrm{d}t$. In that setting a result by de Boor (1963) states that there is a unique minimizer s of $\mathcal{F}[s]$ with (28) and natural, Hermite, or periodic boundary conditions. Furthermore, s is given by the unique cubic spline satisfying (28) and the boundary conditions. Hence, we generalize cubic spline interpolation to Riemannian manifolds. Accordingly, we denote (29) to be the *spline energy* and a corresponding minimizer under the interpolation constraint (28) as a *Riemannian spline*.

Although there is a unique minimizer of (29) in the Euclidean setting, existence of minimizers of (29) is not guaranteed on general manifolds:

Lemma 4 (Nonexistence of Riemannian splines (Heeren et al., 2019, Lemma 2.15)).

Let $\mathcal{M}$ be any manifold with a closed geodesic curve C and a point $\overline{s}_1 \in C$ such that any locally geodesic curve connecting $\overline{s}_1$ with a point on C lies inside C. Then, minimizers of (29) under the interpolation constraints (28) do not exist in general.

Sketch of roof. It suffices to provide a counterexample. Consider the cylinder on the right with unit perimeter, where C is given by the red circle. Let $t_1 = 0$, $t_2 = r \in (0,1)\backslash\mathbb{Q}$, $t_3 = 1$ and $s_1, s_2, s_3 \in C$ such that $s_1 = s_3$ and s_2 is the opposite point, see embedded figure. Now we aim at defining $s: [0,1] \to \mathcal{M}$ with $s(t_i) = s_i$ for $i = 1, 2, 3$. For $m, n \in \mathbb{N}$ let $x: [0,1] \to \mathbb{R}$ be a cubic spline with $x(t_1) = 0$, $x(t_2) = m + \frac{1}{2}$, $x(t_3) = n$. Let $\zeta: \mathbb{R} \to \mathcal{M}$ be an arc-length parameterization of C, in particular, ζ is 1-periodic with $\zeta(0) = s_1$. Then $s = \zeta \circ x: [0,1] \to \mathcal{M}$ fulfills $s(t_i) = s_i$ for $i = 1, 2, 3$ and one obtains $\mathcal{F}[s] = c\,(m + \frac{1}{2} - r \cdot n)^2$, see Heeren et al. (2019, Lemma 2.15) for details. By Dirichlet's approximation theorem there exist $m, n \in \mathbb{N}$ that make this expression arbitrary small, i.e., $\inf \mathcal{F} = 0$. On the other hand, $\mathcal{F}[s] = 0$ implies $\nabla_{\dot{s}}\dot{s} = 0$, which on the cylinder results in a regular helix with constant speed. Since $s(0) = s(1)$, the helix degenerates and winds round the red circle C at constant speed so that necessarily $s(t) = \zeta(kt)$ for some $k \in \mathbb{N}$. Hence, the preimage of s_2 under s is a rational number, i.e., $r \notin s^{-1}(s_2)$, which is a contradiction. □

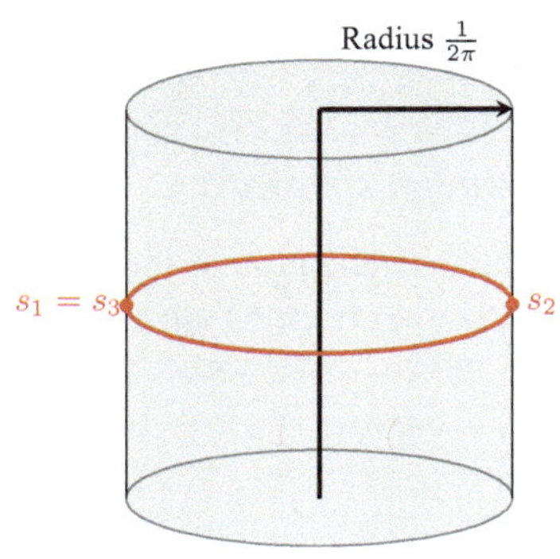

The main issue is the lack of coercivity of (29) with respect to the $W^{2,2}$-norm, i.e., a curve with zero spline energy might have unbounded path energy (1). To compensate for this it suffices to consider the regularized spline energy instead, i.e., for $\sigma > 0$ we define

$$\mathcal{F}^\sigma[s] = \mathcal{F}[s] + \sigma\mathcal{E}[s]. \tag{30}$$

Indeed, this modification is sufficient to obtain existence and regularity of spline interpolations:

Theorem 12 (Existence of spline interpolations (Heeren et al., 2019, Theorem 2.19)).

For $\sigma > 0$ there exists a minimizer of $\mathcal{F}^\sigma$ subject to (28) as well as natural, Hermite, or periodic boundary conditions in the Sobolev space $W^{2,2}((0,1); \mathcal{M})$.

Remark. The covariant derivative $\nabla_{\dot{s}} w$ of a tangential vector field w along a path s can be rewritten as $\nabla_{\dot{s}} w = \dot{w} + \Gamma_s(w, \dot{s})$, where the Christoffel operator $\Gamma_s : T_s\mathcal{M} \times T_s\mathcal{M} \to T_s\mathcal{M}$ is implicitly defined as

$$2g_s(\Gamma_s(u,v),z) = (D_s g_s)(v)(u,z) - (D_s g_s)(z)(u,v) + (D_s g_s)(u)(v,z), \quad u,v,z \in T_s\mathcal{M}.$$

Here, $D_s g$ denotes the derivative of the metric with respect to s. This allows to rewrite

$$\mathcal{F}[s] = \int_0^1 g_s(\ddot{s},\ddot{s}) + 2g_s(\ddot{s},\Gamma_s(\dot{s},\dot{s})) + g_s(\Gamma_s(\dot{s},\dot{s}),\Gamma_s(\dot{s},\dot{s}))\,\mathrm{d}t\,.$$

In order to prove the theorem one first shows a continuity property of the Christoffel operator (Heeren et al., 2019, Lemma 2.17), which implies that $\mathcal{F}^\sigma$ is continuous under convergence in $W^{2,2}$ (cf. Heeren et al., 2019, Lemma 2.18). In finite dimensions this holds with no extra condition on the structure of the manifold, whereas in infinite dimensions this requires additional assumptions on the Riemannian metric. Furthermore, an a priori bound for the minimization sequence is needed. This is not obvious because $\Gamma_s(\dot{s},\dot{s})$ does not have a sign. In Heeren et al. (2019) a monotone rearrangement argument is used to establish an a priori bound in $W^{2,2}$. The claim is then obtained by the direct method, where coercivity in $W^{2,2}$ is ensured due to bounded spline energy *and* bounded path energy.

5.2 Time-discrete Riemannian splines

We introduce a consistent time discretization of the spline energy (29) picking up the variational time discretization for the path energy (1). As a motivation, first consider discrete splines in Euclidean space, where the covariant derivative of the velocity field of a curve $s : [0,1] \to \mathbb{R}^d$ coincides with $\ddot{s}$. For a uniform sampling $s_k = s(t_k)$ for $t_k = k\tau$, $k = 0, \ldots, K$, and time step $\tau = 1/K$ one obtains the standard second-order difference quotient approximation

$$\|\ddot{s}(t_k)\|^2 \approx \left\|\frac{2s(t_k) - s(t_{k-1}) - s(t_{k+1})}{\tau^2}\right\|^2 = 4K^4 \left\|s_k - \frac{s_{k-1} + s_{k+1}}{2}\right\|^2. \tag{31}$$

For the choice $\mathcal{W}[s,\tilde{s}] = \| s - \tilde{s}\|^2$, the term on the right-hand side simplifies to $4K^4\,\mathcal{W}[s_k, \frac{1}{2}(s_{k+1} + s_{k-1})]$. Using the simple numerical quadrature $\int_0^1 f(t)\,\mathrm{d}t \approx \tau \sum_{k=1}^{K-1} f(t_k)$ yields

$$\int_0^1 \| \ddot{s}(t)\|^2\,\mathrm{d}t \approx 4K^3 \sum_{k=1}^{K-1} \mathcal{W}[s_k, \frac{1}{2}(s_{k+1} + s_{k-1})].$$

Notice that this formulation implicitly incorporates (time-discrete) natural boundary conditions. The local average $\frac{1}{2}(s_{k-1} + s_{k+1})$ can be considered as the midpoint of a straight line, i.e., a geodesic, connecting s_{k-1} and s_{k+1}.

This observation leads to a translation of the above formulation to Riemannian manifolds. Indeed, given a discrete path (s_0, …, s_K) in a *general* Riemannian manifold $\mathcal{M}$ with an approximation functional $\mathcal{W}$ as before, we may replace the local average in (31) by the midpoint of a (shortest) geodesic. Accordingly, we define the *time-discrete spline energy* by

$$\mathrm{F}^K[s_0,\ldots,s_K] = 4K^3 \sum_{k=1}^{K-1} \mathcal{W}[s_k, \tilde{s}_k], \tag{32}$$

for a discrete path (s_0, …, s_K). Here, $\tilde{s}_k$ is defined by requiring that $(s_{k-1}, \tilde{s}_k, s_{k+1})$ is a time-discrete geodesic connecting s_{k-1} and s_{k+1} for $k = 1, \ldots, K-1$, i.e.,

$$\tilde{s}_k = \arg\min_{s \in \mathcal{M}} \left(\mathcal{W}[s_{k-1}, s] + \mathcal{W}[s, s_{k+1}] \right). \tag{33}$$

In order to compute a spline curve in shell space, one has to minimize (32) among all discrete curves (s_0, …, s_K) subject to the interpolation conditions $s_{k_j} = s^j$ for a subset of indices $j = 1, \ldots, J$ and the additional built-in constraints (33). Notice that we have implicitly assumed that the interpolation times are multiples of the time step $\tau = K^{-1}$. This restriction can easily be removed using varying time step sizes and suitable adaptations of F^K.

However, as in the continuous case one considers a regularized discrete spline energy $\mathrm{F}^K + \sigma \mathrm{E}^K$ for $\sigma > 0$, as discussed in Heeren et al. (2019, Theorem 3.10). Fortunately, the interpolation constraints encountered in practice allow for a minimizer even if $\sigma = 0$. Finally, in Heeren et al. (2019) a rigorous consistency analysis was performed for the (regularized) discrete spline energy and the underlying (regularized) continuous spline energy (30) in terms of Γ-convergence (cf. also Theorem 7). In particular, the equicoercivity result (Heeren et al., 2019, Theorem 4.9) yields convergence of discrete spline interpolations to continuous ones.

Remark. Instead of mimicking the Euclidean case (as motivated by Eq. (31)) one can also start by using the discrete Levi-Civita connection as defined in Definition 5. If $\hat{s}^k$ denotes the fourth corner of a discrete geodesic parallelogram defined by s_{k-1}, s_k, s_{k+1}, then $K^2(\hat{s}_k - s_k)$ is a first-order consistent approximation of $\nabla_{\dot{s}} \dot{s}(t_k)$ (cf. Heeren, 2016). In particular, $\tilde{s}_k$ as defined in (33) is not only midpoint of the geodesic $(s_{k-1}, \tilde{s}_k, s_{k+1})$ but also midpoint of the geodesic $(s_k, \tilde{s}_k, \hat{s}_k)$. Then a scaling argument shows that $g_{s_k}(\nabla_{\dot{s}} \dot{s}(t_k), \nabla_{\dot{s}} \dot{s}(t_k)) = K^4 \mathcal{W}[s_k, \hat{s}_k] + O(\tau) = 4K^4 \mathcal{W}[s_k, \tilde{s}_k] + O(\tau)$, i.e., the simple numerical quadrature mentioned above leads again exactly to (32).

6 Discrete Riemannian calculus in discrete shell space

We presented a comprehensive discrete Riemannian calculus for generic Riemannian manifolds in Section 4. This calculus is built on an elastic dissimilarity measure which leads to a Hessian structure that finally induces the underlying Riemannian metric. In particular, we have deduced a physically sound dissimilarity measure for discrete shells given by the combination of a discrete *membrane* and *bending* energy in Section 3. In this chapter we apply the discrete calculus as well as the notion of discrete Riemannian splines as presented in Section 5 to the space of discrete shells, i.e., triangle meshes with fixed connectivity.

This section is based on previous works by Heeren et al. (2012, 2014, 2016); in particular, all figures are taken from the original articles. For details on the implementation, we further refer to Heeren (2016).

6.1 Interpolation in discrete shell space

Given two discrete shells $\bar{s}, s \in \mathbb{R}^{3n}$ with the same connectivity, we define a *discrete deformation energy* by

$$\mathcal{W}[\bar{s}, s] = \sum_{f \in \mathcal{F}} \mathcal{W}_{\mathrm{mem}}(A_f^{2\times 2})\bar{a}_f + \eta \sum_{e \in \mathcal{E}} \frac{(\bar{\theta}_e - \theta_e)^2}{\bar{d}_e} \bar{l}_e^2, \tag{34}$$

where $\eta = \delta^2$ represents the squared thickness of the shell. The first term representing the membrane energy has been defined in (18), whereas the bending term is a simplification of (20). Here, θ_e denotes the dihedral angle across that edge, $\bar{d}_e$ is one third of the sum of the areas of the two neighbouring triangles of edge e and $\bar{l}_e$ is the length of e. Note that the discrete bending energy coincides exactly with the *Discrete shells* bending model (Grinspun et al., 2003) which has also been used in Heeren et al. (2014), for further details on the relation to (20) we refer to Heeren (2016). Inserting (34) into Definition 1 we obtain immediately a notion of *time-discrete geodesics* in discrete shell space. The variational definition leads to the following necessary optimality conditions:

$$\begin{aligned} & 0 = \partial_{s_k} \mathrm{E}^K[s_0, \ldots, s_K], \quad k = 1, \ldots, K-1, \\ \Longleftrightarrow\ & 0 = \mathcal{W}_{,2}[s_{k-1}, s_k] + \mathcal{W}_{,1}[s_k, s_{k+1}], \quad k = 1, \ldots, K-1. \end{aligned} \tag{35}$$

To compute time-discrete geodesics we have to solve the system of nonlinear equations (35) simultaneously, where we fix the two end shapes s_0 and s_K.

6.1.1 Natural regularization and physical tuning

The choice of physics-based deformation energies offers several advantages. Due to everyday experience, physical models come close to human intuition so that our framework tends to lead to intuitive discrete geodesic paths. In Fig. 9, for example, the actual dissipation is concentrated at the joints—the place where the actual physical work is done. Furthermore, the material parameters in our model allow a physically based tuning of "best" paths between shells. Obviously, the bending weight parameter η in (34) has a physical interpretation and can be adjusted to change the typical behaviour of shells.

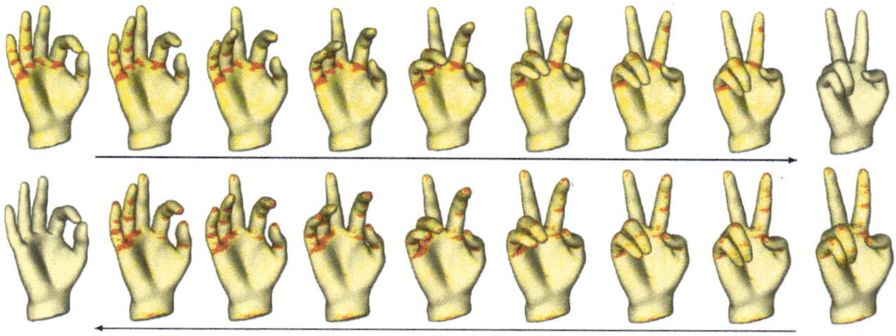

FIG. 9 Discrete geodesics $A \rightarrow B$ (*top*) vs $B \rightarrow A$ (*bottom*) between two hand poses. The underlying dissipation is represented by the two energy densities, i.e., W_{mem} (*top row*, 0 0.01) and W_{bend} (*bottom row*, 0 10).

FIG. 10 Almost isometric geodesic between two hemispheres with fixed rim ($\eta = 10^{-4}$).

FIG. 11 The bending term serves as a natural regularization; there is no need for artificial regularizers as in Kilian et al. (2007) (*top row*, $\eta = 0.8$). Decreasing η leads to the occurrence of creases until the underlying matching deformations are finally only as-isometric-as-possible (*bottom*, *left* to *right*: intermediate shape of a geodesic with $\eta = 0.8, 0.01, 0.0001$).

In detail, it represents the approximate squared thickness δ^2 of the shell, a measure of resistance against bending: On the one hand geodesics with low η allow for crumpling (cf. Fig. 10), and if the initial and final shell are near-isometric, this property is approximately maintained for intermediate shells during the deformation.

On the other hand, a larger value of η prevents crumpling (Fig. 11), which is especially desired in applications where one expects smooth intermediate shapes if the end poses are smooth. Hence the parameter η may also be seen as the *natural regularization* parameter, and there is no need for artificial regularizers as it was used in Kilian et al. (2007). However, the tuning of the physical parameters λ and μ in (13) is rather of theoretical interest, hence we will set $\lambda = \mu = 1$ in all subsequent experiments.

We have seen that bending effects play an important role for near-isometric deformations—a ubiquitous case in applications. Here, the energy dissipation due to bending serves as the proper path selection mechanism. To this end, the bending parameter η is not only a regularization weight but can also be used to manipulate the visual appearance of the deformation path. In particular, it can be used to simulate different physical materials, such as paper, aluminium foil, rubber, cloth or metal, and corresponding optimal deformation paths, i.e., geodesics, are perfectly consistent with the intuitively optimal way to deform such a material. As already discussed above, a low bending parameter η induces almost isometric deformation paths. In Fig. 12 (top) this leads to the effect that the handle is folded in order to be pushed through the outer frame. On the other hand, a larger η prevents folding at the cost of involving a fair amount of tangential stretching of the outer frame, resp., compression of the handle (Fig. 12, bottom). Hence the physical behaviour of the deformation is reminiscent of either a thin metal sheet (Fig. 12, top) or of a material made of rubber (Fig. 12, bottom).

Another example is given in Fig. 13, where a low bending weight η essentially represents a flat sheet of paper or aluminium foil that is rolled up in two orthogonal directions (top). Again, this deformation path is almost isometric as tangential distortion is highly expensive. On the other hand, if η is increased the optimal path no longer stays in the submanifold of isometric deformations but rather prefers a "short-cut" by introducing a little bit of tangential distortion (Fig. 13, bottom). Note that in both examples in Figs. 12 and 13 a simple rotation of the end poses is not allowed due to

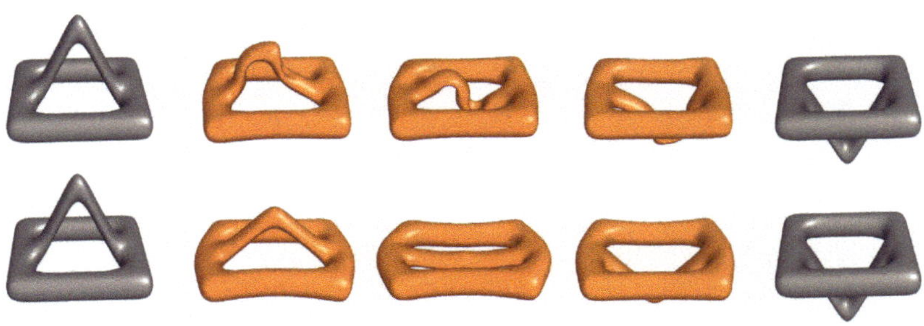

FIG. 12 Discrete geodesic for low (*top*, $\eta = 0.005$) and high (*bottom*, $\eta = 0.1$) bending resistance. Stiff materials prefer the handle being folded (inducing bending dissipation) whereas soft materials stretch the outer frame (i.e., induce membrane distortion) to avoid folding.

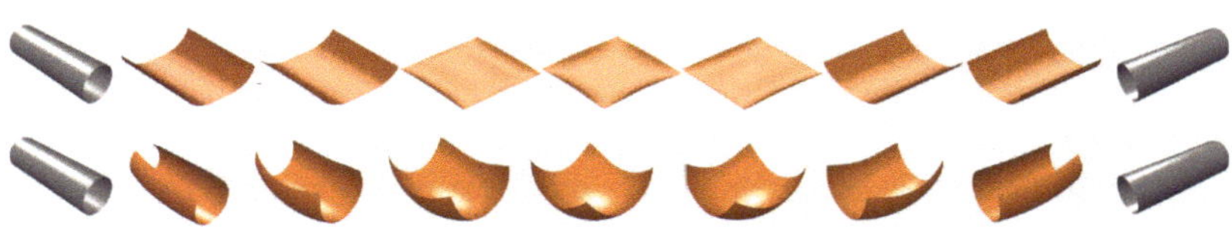

FIG. 13 Discrete geodesic for low (*top*, $\eta = 10^{-6}$) and high (*bottom*, $\eta = 10^{-1}$) bending resistance.

the dense correspondence assumption. For instance, if the four boundary edges of the plate in Fig. 13 are labelled clockwise by A, B, C, and D, then in the first pose A is supposed to be aligned with C, whereas in the last pose B is supposed to be aligned with D, which cannot be resolved by a global rotation.

6.1.2 *Lack of symmetry*

Time-discrete geodesics are in general not symmetric that means a path from A to B will be (slightly) different from a path from B to A. In particular, the approximation of the Riemannian distance given by the square root of the optimal path energy E^K is not symmetric either. This is due to the fact that the dissimilarity measure $\mathcal{W}$ in (34) is not symmetric in its arguments, as the density is integrated over the *undeformed* shell. However, the lack of symmetry of $\min \mathrm{E}^K$ decreases with increasing K, and already for a very low number K the lack of symmetry in the time-discrete geodesic is visually not perceptible: in Fig. 9 we show a time-discrete geodesic between two hand poses A and B, where the top row shows results from computations of $A \rightarrow B$ and the bottom row shows (reversed) results from $B \rightarrow A$. The relative difference between path energies of the two sequences is 3%. The deformation energy $\mathcal{W}$ between each two corresponding shapes in the two sequences is less than 0.01% of the deformation energy from the first to the final shell. The issue of missing symmetry in our model has also been investigated in Brandt et al. (2016). The authors conclude that a straightforward symmetrization of the deformation energy by using $\mathcal{W}[A,B] + \mathcal{W}[B,A]$ does not lead to significantly different results but slows down the numerical simulation instead.

6.1.3 *Nonuniqueness*

Geodesic paths need not be unique, and in particular buckling modes might lead to multiple (shortest) geodesic paths between two input shells (cf. Fig. 14). The flat plate as an intermediate shape is avoided since it would induce high tangential distortion. Therefore the bumps flip to the other side successively, while the order of flipping does not effect the accumulated dissipation. Different paths are induced by different initializations. Note that the example shown in Fig. 12 (top) also leads to a nonunique geodesic, as the left part of the handle could have been pushed through the frame first as well.

FIG. 14 Symmetric setups can lead to nonunique geodesic paths.

6.2 Extrapolation in discrete shell space

Previously, we have seen that time-discrete geodesics provide a natural *interpolation* tool in discrete shell space. Now we apply the notion of a time-discrete exponential map as introduced in Section 4 to the space of discrete shells in order to generate a consistent *extrapolation* tool. Recall that the time-discrete exponential map is built on an iterative scheme computing single *shooting steps* sequentially. i.e., we solve

$$0 = \mathcal{W}_{,2}[s_{k-2}, s_{k-1}] + \mathcal{W}_{,1}[s_{k-1}, s_k], \quad k = 2, \ldots, K, \tag{36}$$

sequentially for s_k, where the initial shapes s_0 and s_1 are given. Note that (36) coincides exactly with the $K - 1$ necessary conditions (35) for $(s_0, \ldots, s_K)$ to be a discrete geodesic. In particular, the interpolated geodesic $(s_0, \ldots, s_K)$ and the geodesic extrapolation from s_0 in direction $s_1 - s_0$ coincide because the defining system of Euler–Lagrange equations is the same and the minimizer of the variational problem is unique. For discrete shells s_0, s_K close to each other, the latter is guaranteed due to the nondegeneracy of the Hessian, which implies local strict convexity.

In Fig. 15 we show an example of geodesic extrapolation (based on initial data taken from Fröhlich and Botsch (2011)). The generation of the strongly twisted helix demonstrates the robustness of the time-discrete geodesic extrapolation, both for large time steps and large nonlinear initial variation (Fig. 15, top) as well as for many small time steps and small initial offset (Fig. 15, bottom).

Geodesic inter- and extrapolation can naturally be combined by, e.g., first interpolating between two given end poses and then extrapolating the motion in both directions. This is illustrated in Fig. 16, where the shapes at times $t = 0$ and $t = 1$ were given as input shapes. Another striking feature of computing (approximative) geodesic paths are constant dissipation rates along the path (cf. Fig. 17, bottom), which corresponds geometrically to the property $g_s(\dot{s}, \dot{s}) =$ const that holds for geodesic paths.

In animation applications geodesic shooting can further be used to generate natural motions of three-dimensional models without much user interaction. Given a particular shell s, one can compute eigenmodes of the Hessian

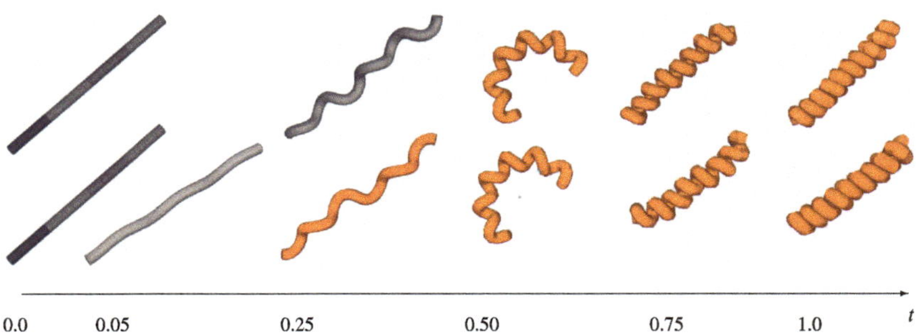

FIG. 15 Shape extrapolation is applied to generate a strong twisting of a helix.

FIG. 16 Geodesic interpolation and extrapolation of the two *grey* poses at $t = 0$ and $t = 1.0$.

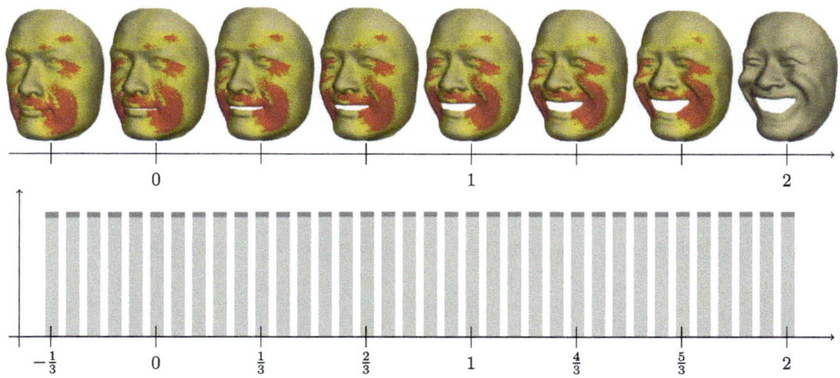

FIG. 17 *Top*: Geodesic interpolation for $0 \leq t \leq 1$ of given shells at time 0 and 1, geodesic extrapolation for $t < 0$ and $t > 1$ (local dissipation rate is *colour coded* as 0 1; bending weight $\eta = 1$). *Bottom*: The total dissipation rate (membrane contribution in *light grey*, bending contribution in *dark grey*) stays constant along the path.

of the elastic deformation energy $\mathcal{W}_{,22}[s,s]$ corresponding to the lowest eigenvalues, e.g., by a standard inverse power iteration. Obviously, the first six eigenvalues are trivial, representing the translations and rotations, respectively, in 3D space. Computing eigenvectors of the Hessian of an elastic energy to launch natural animations has also been done to perform *physical simulations* based on Newton's law of motion (Hildebrandt et al., 2011; von Tycowicz et al., 2013). Slightly different to these physical simulations, moving s in the direction of eigenmodes—corresponding to low but nontrivial eigenvalues—by means of geodesic shooting represents a deformation of s in a direction with *least energy dissipation*. Hence the motion along these so-called *dissipation modes* induces energetically preferred deformations that look very natural. In particular, the dissipation rates are again constant, which is not the case for physical simulations as in Hildebrandt et al. (2011) and von Tycowicz et al. (2013). In Fig. 18 we demonstrate geodesic extrapolation of a shell s along two different dissipation modes, given by the second and fourth nontrivial eigenmode of $\mathcal{W}_{,22}[s,s]$, respectively. Shown are the initial shell with the corresponding eigenmode scaled by $\pm$ 5 and the extrapolated shells based on 10, 20, and 30 iterations of the time-discrete extrapolation. As mentioned above, the dissipation rates are constant up to numerical errors (Fig. 18, bottom).

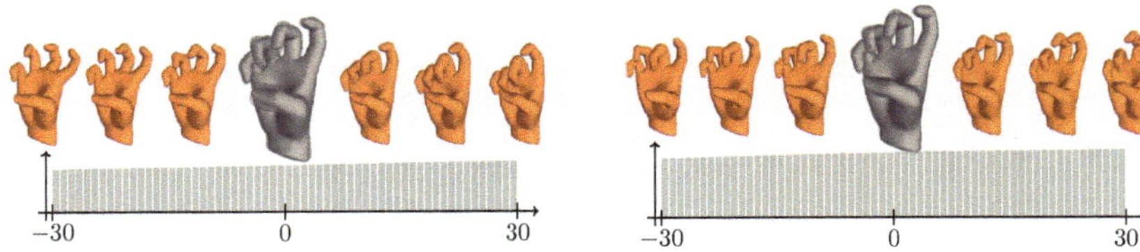

FIG. 18 Geodesic extrapolation along low dissipation modes.

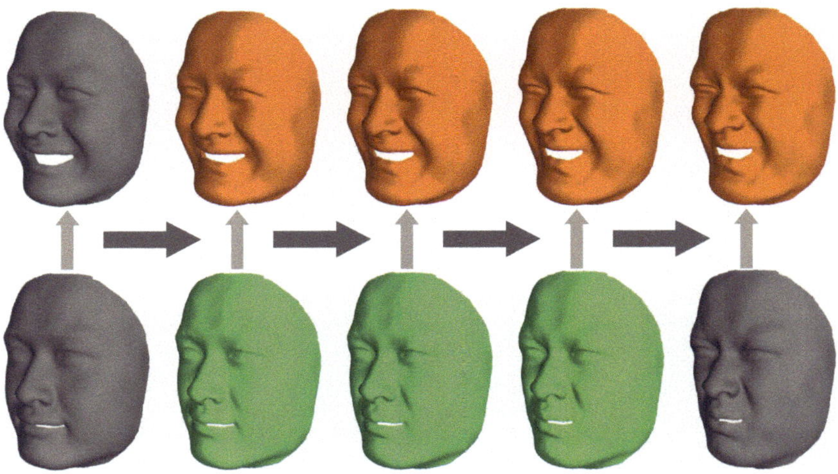

FIG. 19 Time-discrete parallel transport of the difference between a smiling (*upper left*) and a neutral face (*lower left*) along a path (*bottom row*) towards a disgusted face (*bottom right*), resulting in a smile with a frown (*upper right*).

6.3 Parallel transport in discrete shell space

In Section 4 we have introduced a well-known first-order approximation of continuous parallel transport named Schild's ladder. It is based on a geodesic parallelogram construction (cf. Fig. 7), whose building elements are local (discrete) geodesic interpolation and extrapolation operations. To obtain a discrete transport along a discrete path one computes a sequence of such discrete geodesic parallelograms iteratively. Fig. 19 uses such a concatenation of geodesic parallelograms to transport a smile along a path from a neutral to a disgusted facial expression.

6.4 Spline interpolation in discrete shell space

Interpolating curves in the space of triangular meshes are useful for many applications in computer graphics such as keyframe based animation. In particular, globally smooth curves are more desirable than piecewise smooth curves, cf. Fig. 20. Using splines for smoothly interpolating keyframe poses of animated characters in Euclidean space goes back at least to the pioneering

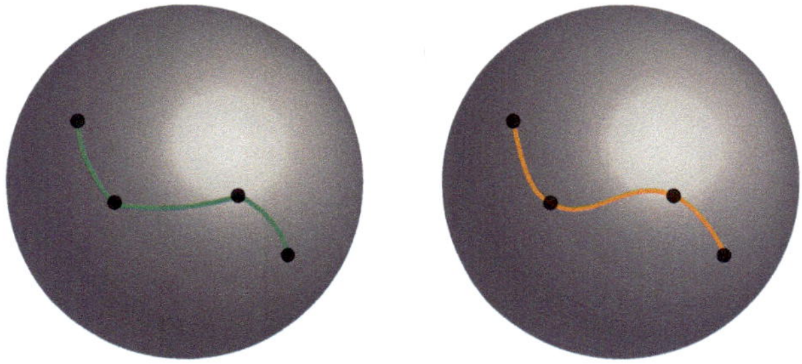

FIG. 20 Piecewise geodesic (*left*) vs spline (*right*) interpolation on Riemannian manifolds (sphere for illustration). Notice C^1-discontinuities for piecewise geodesic interpolation.

works of Kochanek and Bartels (1984) and Lakes (1987) in the 1980s. However, realizing shape interpolation through curves in the space of shells—which is considered as a Riemannian manifold—is relatively recent. Using splines for solving the multinodal interpolation problem in (discrete) shell space comes at a price, since it requires to solve highly nonlinear optimization problems with additional nonlinear constraints. Hence the computational complexity must be reduced to yield practical algorithms. To address this challenge we switch from vertex positions as primary variables to the so-called *LΘA-representation* in the spirit of Winkler et al. (2010) and Fröhlich and Botsch (2011). In this formulation, the primary variables are the edge lengths of triangles (L), dihedral angles between adjacent triangles (Θ), and triangle areas (A). To this end, discrete splines in $L\Theta A$-space can be evaluated through solving a simple linear system. However, the resulting curve in $L\Theta A$-space will in general not be realizable as a sequence of triangle meshes in 3D Euclidean space. Instead, we project to the closest curve of realizable triangle meshes in a least squares sense. Overall, this formulation greatly outperforms nonlinear optimization for computing splines based on vertex positions.

6.4.1 Splines in vertex space

To compute (time-discrete) interpolating splines as discussed in Section 4 in the discrete shell space, we simply insert the definition of a discrete deformation energy (34) into the definition of the discrete spline energy (32) and the corresponding constraints (33). For example, Fig. 21 depicts a discrete spline curve for three different keyframe poses of a cactus-type discrete shell and compares it to the piecewise geodesic interpolation ($K = 10$). The boundary keyframe poses s_0 and s_{10} are given as deformations of the cactus rest pose s_5 in two orthogonal directions, which leads to a sharp corner at s_5 when performing piecewise geodesic interpolation.

6.4.2 LΘA approximation

Although the interpolating spline curves in discrete shell space look very nice, a big limitation is computational time. To remedy the problem of high computational cost, we introduce a change of variables in order to turn the

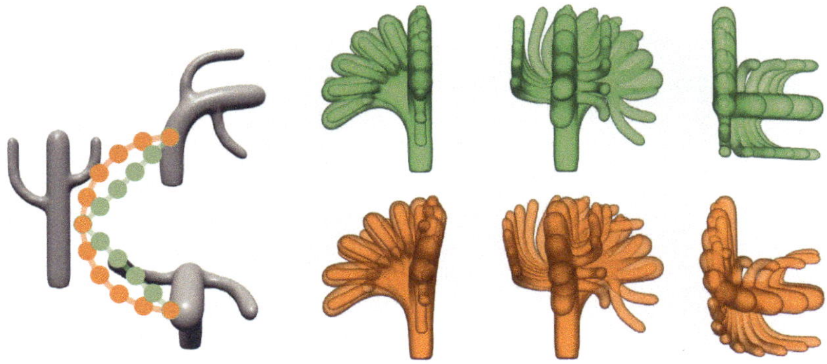

FIG. 21 *Left*: Three fixed keyframe poses (*grey*). *Right*: Different views of piecewise geodesic (*green*) and spline interpolation (*orange*), respectively.

nonlinear optimization problem for computing splines in shell space into a linear one. For this change of coordinates we heavily build on the two-step approximation scheme proposed by Fröhlich and Botsch (2011). In detail, for a discrete shell $s \in \mathbb{R}^{3n}$ we consider the vectors of edge lengths $L = L[s]$, dihedral angles $\Theta = \Theta[s]$, and triangle areas $A = A[s]$ as primary degrees of freedom. The key observation is that with these degrees of freedom, one can define a suitable approximation of the discrete deformation energy (34) which is *quadratic* in these new variables. Collecting all new primary variables in one variable $z = (L, \Theta, A)$ living in the $L\Theta A$-configuration space $\mathcal{Z} := \mathbb{R}^{|\mathcal{E}|} \times \mathbb{R}^{|\mathcal{E}|} \times \mathbb{R}^{|\mathcal{F}|}$ we set

$$\mathcal{W}_{L\Theta A}[\bar{z}, z] = \mu \sum_{e \in \mathcal{E}} \alpha_e (\bar{l}_e - l_e)^2 + \lambda \sum_{f \in \mathcal{T}} \beta_f (\bar{a}_f - a_f)^2 + \eta \sum_{e \in \mathcal{E}} \gamma_e (\bar{\theta}_e - \theta_e)^2, \tag{37}$$

with integration weights $\alpha \in \mathbb{R}^{|\mathcal{E}|}$, $\beta \in \mathbb{R}^{|\mathcal{T}|}$, and $\gamma \in \mathbb{R}^{|\mathcal{E}|}$. One possible choice is $\alpha_e = \hat{d}_e \hat{l}_e^{-2}$, $\beta_f = \hat{a}_f^{-1}$ and $\gamma_e = \hat{d}_e^{-1} \hat{l}_e^2$ computed on some representative reference shape $\hat{s}$. The discrete interpolation problem can now be reformulated as follows. We construct a spline curve in the $L\Theta A$-space defined as a minimizer of

$$\mathrm{F}^K_{L\Theta A}[z_0, \ldots, z_K] = 4K^3 \sum_{k=1}^{K-1} \mathcal{W}_{L\Theta A}[z_k, \tilde{z}_k], \tag{38}$$

where $(z_{k-1}, \tilde{z}_k, z_{k+1})$ is a discrete geodesic in the $L\Theta A$-space, i.e., $\tilde{z}_k = \mathrm{argmin}_{z \in \mathcal{Z}} (\mathcal{W}_{L\Theta A}[z_{k-1}, z] + \mathcal{W}_{L\Theta A}[z, z_{k+1}])$. Since $\mathcal{W}_{L\Theta A}$ is *quadratic*, one obtains the explicit solution $\tilde{z}_k$ as a linear combination of z_{k-1} and z_{k+1}. In particular, if we make use of the same reference mesh for all $k = 1, \ldots, K-1$, then $\tilde{z}_k = \frac{1}{2}(z_{k-1} + z_{k+1})$. This can be inserted into (38) so that we end up with an *unconstrained* optimization problem. Hence a minimizer of (38) is a (weighted) cubic spline in the linear space $\mathcal{Z} = \mathbb{R}^{|\mathcal{E}|} \times \mathbb{R}^{|\mathcal{E}|} \times \mathbb{R}^{|\mathcal{F}|}$.

Note that there is no interaction between edge lengths, triangle areas and dihedral angles in a minimizer $(z_0, \ldots, z_K)$ of (38). Moreover, there is no *spatial* coupling between any two different edge lengths, i.e., an edge length l_e^k of the kth pose interacts only with lengths l_e^j of the same edge e and poses $j \neq k$. The same applies for dihedral angles and triangle areas. As a consequence, the Euler-Lagrange equation for (38) splits into numerous independent $(K + 1)$-dimensional linear systems, i.e., one for each edge length, dihedral angle, and triangle area, which can be solved efficiently and in parallel.

However, intermediate values for edge lengths, dihedral angles, and triangle areas, provided by a minimizer $(z_0, \ldots, z_K)$ of (38), are generally not realizable as an embedded triangle mesh. Hence we consider a *reconstruction* in a least squares sense, similar to Fröhlich and Botsch (2011). For given optimal values $z_k = (L_k, \Theta_k, A_k)$ we define s_k as the minimizer of the nonlinear mapping $s \mapsto \mathcal{W}_{L\Theta A}(z[s], z_k)$, where $\mathcal{W}_{L\Theta A}$ is defined as in (37). We find the minimizer via the Gauss–Newton method (see Section 6 in Fröhlich and Botsch (2011) for details, the "target" values are given by z_k). The reconstruction can be seen as a projection of the point $z \in \mathcal{Z}$ onto the submanifold which is given by all sets of edge lengths, dihedral angles, and triangle areas that are actually realizable as an embedded triangle mesh. Computationally, the reconstruction is the hardest part in the $L\Theta A$-space approximation method, however, it can easily be parallelized.

6.4.3 Discussion

If we compare the spline interpolation to piecewise geodesic interpolation, there are two striking differences: (i) The trajectory of the spline curve exhibits a time-continuous acceleration $(\frac{D}{dt}\dot{s})$ and thus is visually smooth, whereas a piecewise geodesic curve suffers from corners at the keyframe poses (cf. Fig. 21). The continuous acceleration can in particular be seen in the energy plots in Figs. 22 and 23. (ii) The spline curve balances acceleration along the path leading to a continuous variation in speed (see Fig. 24), whereas the piecewise geodesic approximately has (piecewise) constant speed. As in the Euclidean case, a particular feature of spline interpolation is the so-called overshooting—a consequence of the smoothness requirement. Fig. 25 shows an example of this effect for the case of (discrete) Riemannian splines in shell space (orange shapes).

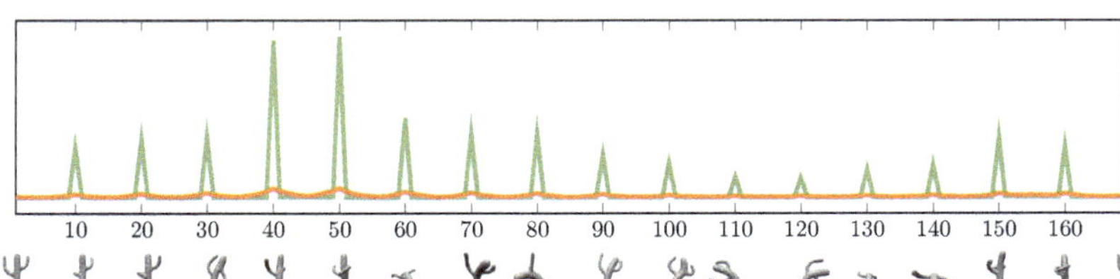

FIG. 22 Energy plot $k \mapsto \mathcal{W}_{L\Theta A}[z_k, \tilde{z}_k]$ as approximation of $g_s(\nabla_{\dot{s}}\dot{s}, \nabla_{\dot{s}}\dot{s})$ for a spline (*orange*, continuous) and piecewise geodesic (*green*, piecewise vanishing but infinite at keyframes) interpolation between 17 keyframes (*grey*).

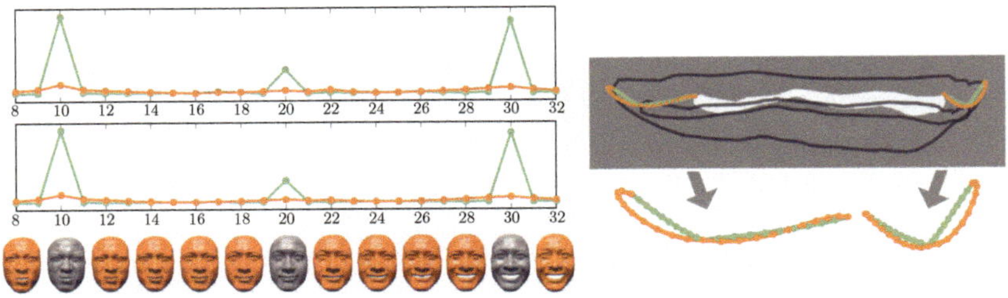

FIG. 23 *Left*: Extract from a spline curve through five fixed keyframe poses (at $k = 0$, 10, 20, 30, 40, three of which are shown in *grey*) computed via the $L\Theta A$-scheme. The energy plots show the $L\Theta A$-space energy $k \mapsto \mathcal{W}_{L\Theta A}[z_k, \tilde{z}_k]$ (*top row*) and the nonlinear energy $k \mapsto \mathcal{W}[s_k, \tilde{s}_k]$ (*second row*) for the piecewise geodesic (*green*) and the spline (*orange*) interpolation. *Right*: Zoomed contours of the mouth in keyframe pose s_{10}, s_{20}, and s_{30} with trajectories of the corners of the mouth.

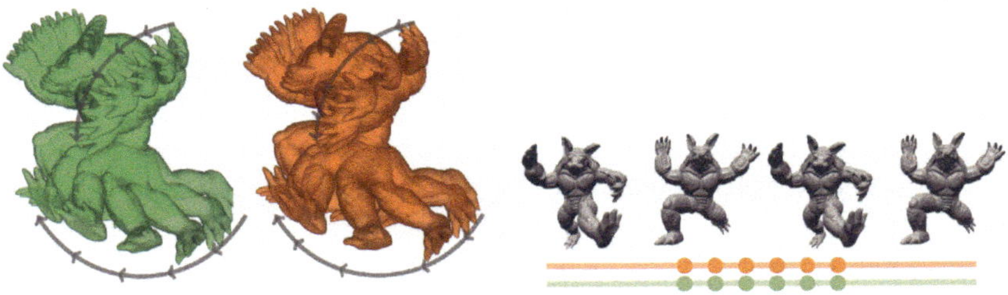

FIG. 24 Piecewise geodesic (*green*) and spline curve interpolation (*orange*) induce similar trajectories but differences in speed as indicated by varying lengths of velocity arrows.

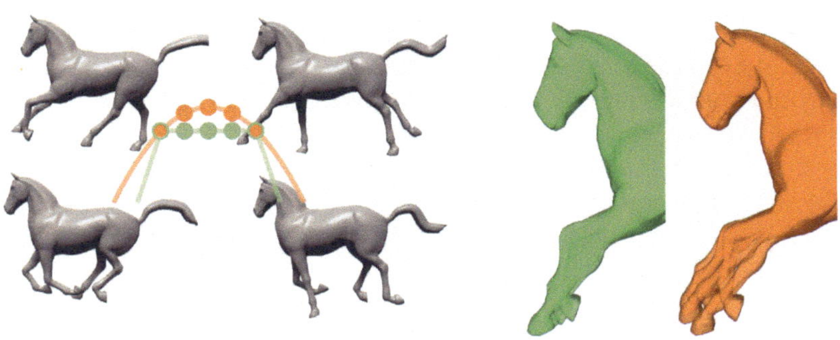

FIG. 25 Overshooting of the left front leg in spline interpolation (*orange*) compared to the piecewise geodesic interpolation (*green*) between two keyframe poses (*upper grey shapes*) taken from a sequence of running horses (Sumner and Popović, 2004).

References

Alexa, M., Cohen-Or, D., Levin, D., 2000. As-rigid-as-possible shape interpolation. In: Annual Conference Series, Proceedings of SIGGRAPH 00, pp. 157–164.

Atkin, C.J., 1997. Geodesic and metric completeness in infinite dimensions. Hokkaido Math. J. 26 (1), 1–61.

Balzani, N., Rumpf, M., 2012. A nested variational time discretization for parametric Willmore flow. Interfaces Free Bound. 14 (4), 431–454. https://doi.org/10.4171/IFB/287.

Barrett, J.W., Garcke, H., Nürnberg, R., 2007. A parametric finite element method for fourth order geometric evolution equations. J. Comp. Phys. 222, 441–467.

Barrett, J.W., Garcke, H., Nürnberg, R., 2008. Parametric approximation of Willmore flow and related geometric evolution equations. SIAM J. Sci. Comput. 31, 225–253. https://doi.org/10.1137/070700231.

Bartels, S., 2013. Approximation of large bending isometries with discrete Kirchhoff triangles. SIAM J. Numer. Anal. 51 (1), 516–525. ISSN 0036-1429. https://doi.org/10.1137/110855405.

Batoz, J.-L., Bathe, K.-J., Ho, L.-W., 1980. A study of three-node triangular plate bending elements. Int. J. Numer. Methods Eng. 15 (12), 1771–1812.

Bauer, M., Bruveris, M., 2011. A new Riemannian setting for surface registration. In: Proceedings of MICCAI Workshop on Mathematical Foundations of Computational Anatomy, pp. 182–194.

Bauer, U., Polthier, K., Wardetzky, M., 2010. Uniform convergence of discrete curvatures from nets of curvature lines. Discret. Comput. Geom. 43 (4), 798–823. ISSN 1432-0444. https://doi.org/10.1007/s00454-009-9237-4.

Bauer, M., Harms, P., Michor, P.W., 2011. Sobolev metrics on shape space of surfaces. J. Geom. Mech. 3 (4), 389–438. ISSN 1941-4889.

Bauer, M., Harms, P., Michor, P.W., 2012. Almost local metrics on shape space of hypersurfaces in n-space. SIAM J. Imaging Sci. 5, 244–310.

Bauer, M., Harms, P., Michor, P.W., 2012. Curvature weighted metrics on shape space of hypersurfaces in n-space. Differential Geom. Appl. 30 (1), 33–41.

Bauer, M., Harms, P., Michor, P.W., 2012. Sobolev metrics on shape space, II: Weighted Sobolev metrics and almost local metrics. J. Geom. Mech. 4 (4), 365–383.

Bauer, M., Bruveris, M., Marsland, S., Michor, P.W., 2014. Constructing reparametrization invariant metrics on spaces of plane curves. Differential Geom. Appl. 34, 139–165.

Bauer, M., Bruveris, M., Michor, P.W., 2016. Why use Sobolev metrics on the space of curves. In: Turaga, P., Srivastava, A. (Eds.), Riemannian Computing in Computer Vision. Springer, pp. 223–255.

Beg, M.F., Miller, M.I., Trouvé, A., Younes, L., 2005. Computing large deformation metric mappings via geodesic flows of diffeomorphisms. Int. J. Comput. Vision 61 (2), 139–157.

Benamou, J.-D., Brenier, Y., 2000. A computational fluid mechanics solution to the Monge-Kantorovich mass transfer problem. Numer. Math. 84 (3), 375–393.

Bergou, M., Wardetzky, M., Harmon, D., Zorin, D., Grinspun, E., 2006. A quadratic bending model for inextensible surfaces. In: Proc. of Eurographics Symposium on Geometry Processing.

Bergou, M., Wardetzky, M., Robinson, S., Audoly, B., Grinspun, E., 2008. Discrete elastic rods. ACM Trans. Graph. 27 (3), 63:1–63:12.

Berkels, B., Effland, A., Rumpf, M., 2015. Time discrete geodesic paths in the space of images. SIAM J. Imaging Sci. 8 (3), 1457–1488. https://doi.org/10.1137/140970719.

Bobenko, A.I., Schröder, P., 2005. Discrete Willmore flow. In: Siggraph/Eurographics Sympos. Geom. Processing, pp. 101–110.

Bobenko, A.I., Hoffmann, T., Springborn, B.A., 2006. Minimal surfaces from circle patterns: geometry from combinatorics. Ann. Math. 164 (1), 231–264.

Botsch, M., Pauly, M., Gross, M., Kobbelt, L., 2006. PriMo: coupled prisms for intuitive surface modeling. In: Proc. of Eurographics Symposium on Geometry Processing, pp. 11–20.

Botsch, M., Sumner, R., Pauly, M., Gross, M., 2006. Deformation transfer for detail-preserving surface editing. In: Proc. of Vision, Modeling & Visualization, pp. 357–364.

Braides, A., 2002. Γ-Convergence for Beginners. Oxford Lecture Series in Mathematics and Its Applications, vol. 22. Oxford University Press, Oxford.

Brandt, C., von Tycowicz, C., Hildebrandt, K., 2016. Geometric flows of curves in shape space for processing motion of deformable objects. Comput. Graph. Forum 35 (2), 295–305.

Brandt, C., Eisemann, E., Hildebrandt, K., 2018. Hyper-reduced projective dynamics. ACM Trans. Graph. 37 (4), 80:1–80:13.

Bronstein, A.M., Bronstein, M.M., Kimmel, R., 2008. Numerical Geometry of Non-Rigid Shapes. Monographs in Computer Science, Springer.

Bruveris, M., 2015. Completeness properties of Sobolev metrics on the space of curves. J. Geom. Mech. 7 (2), 125–150.

Bruveris, M., Michor, P.W., Mumford, D., 2014. Geodesic completeness for Sobolev metrics on the space of immersed plane curves. Forum Math. Sigma 2, e19.

Chao, I., Pinkall, U., Sanan, P., Schröder, P., 2010. A simple geometric model for elastic deformations. ACM Trans. Graph. 29 (4), 38:1–38:6.

Chapelle, D., Bathe, K.-J., 1998. Fundamental considerations for the finite element analysis of shell structures. Comput. Struct. 66, 19–36.

Charpiat, G., Faugeras, O., Keriven, R., Maurel, P., 2006. Distance-based shape statistics. In: Proc. of IEEE International Conference on Acoustics Speech and Signal Processing, vol. 5.

Cirak, F., Ortiz, M., 2001. Fully C1-conforming subdivision elements for finite deformation thin-shell analysis. Int. J. Numer. Methods Eng. 51 (7), 813–833.

Cirak, F., Ortiz, M., Schröder, P., 2000. Subdivision surfaces: a new paradigm for thin-shell finite-element analysis. Int. J. Numer. Methods Eng. 47 (12), 2039–2072.

Cirak, F., Scott, M.J., Antonsson, E.K., Ortiz, M., Schröder, P., 2002. Integrated modeling, finite-element analysis and engineering design for thin-shell structures using subdivision. Comput. Aided Des. 34 (2), 137–148.

Cohen-Steiner, D., Morvan, J.-M., 2003. Restricted Delaunay triangulations and normal cycle. Proc. of Symposium on Computational Geometry, pp. 312–321.

Cootes, T.F., Taylor, C.J., Cooper, D.H., Graham, J., 1995. Active shape models–their training and application. Comput. Vis. Image Underst. 61 (1), 38–59.

Crane, K., Pinkall, U., Schröder, P., 2013. Robust fairing via conformal curvature flow. ACM Trans. Graph. 32 (4), 61:1–61:10.

Crouch, P., Silva Leite, F., 1995. The dynamic interpolation problem: on Riemannian manifold, Lie groups and symmetric spaces. J. Dyn. Control. Syst. 1 (2), 177–202.

Dambreville, S., Rathi, Y., Tannenbaum, A., 2006. A shape-based approach to robust image segmentation. In: Lecture Notes in Computer Science. Proc. of Conference on Image Analysis and Recognition, vol. 4141, pp. 173–183.

de Boor, C., 1963. Best approximation properties of spline functions of odd degree. J. Math. Mech. 12, 747–749.

De Giorgi, E., Dal Maso, G., 1983. Γ-convergence and calculus of variations. In: Mathematical theories of optimization (Genova, 1981). Lecture Notes in Math., vol. 979. Springer, Berlin, pp. 121–143.

do Carmo, M. P., 1992. Riemannian Geometry. Birkhäuser, Boston.

Douglas, J., 1931. Solution of the problem of Plateau. Trans. Am. Math. Soc. 33 (1), 263–321. ISSN 0002-9947. https://doi.org/10.2307/1989472.

Dupuis, P., Grenander, U., Miller, M.I., 1998. Variational problems on flows of diffeomorphisms for image matching. Q. Appl. Math. 56, 587–600.

Dyn, N., 1992. Subdivision schemes in computer-aided geometric design. In: Advances in Numerical Analysis, vol. 2. Oxford Univ. Press, New York, pp. 36–104.

Dyn, N., 2002. Interpolatory subdivision schemes. In: Armin, I., Ewald, Q., Floater, M.S. (Eds.), Tutorials on Multiresolution in Geometric Modelling. Springer, Berlin, pp. 25–50.

Dziuk, G., 1988. Finite elements for the Beltrami operator on arbitrary surfaces. In: Hildebrandt, S., Leis, R. (Eds.), Partial Differential Equations and Calculus of Variations. Lecture Notes in Math., vol. 1357. Springer, Berlin, pp. 142–155.

Dziuk, G., Hutchinson, J.E., 1999. The discrete Plateau problem: algorithm and numerics. Math. Comput. 68 (225), 1–23. ISSN 0025-5718. https://doi.org/10.1090/S0025-5718-99-01025-X.

Dziuk, G., Hutchinson, J.E., 1999. The discrete Plateau problem: convergence results. Math. Comput. 68 (226), 519–546. ISSN 0025-5718. https://doi.org/10.1090/S0025-5718-99-01026-1.

Ehlers, J., Pirani, F. A. E., Schild, A., 1972. The geometry of free fall and light propagation. In: General Relativity (Papers in Honour of J. L. Synge), Clarendon Press, Oxford, pp. 63–84.

Fletcher, P.T., Whitaker, R.T., 2006. Riemannian metrics on the space of solid shapes. In: MICCAI Workshop on Mathematical Foundations of Computational Anatomy47–57.

Friesecke, G., James, R.D., Müller, S., 2002. A theorem on geometric rigidity and the derivation of nonlinear plate theory from three-dimensional elasticity. Comm. Pure Appl. Math. 55 (11), 1461–1506.

Friesecke, G., James, R.D., Mora, M.G., Müller, S., 2003. Derivation of nonlinear bending theory for shells from three-dimensional nonlinear elasticity by Gamma-convergence. C. R. Math. Acad. Sci. Paris 336 (8), 697–702.

Fröhlich, S., Botsch, M., 2011. Example-driven deformations based on discrete shells. Comput. Graph. Forum 30 (8), 2246–2257.

Fuchs, M., Jüttler, B., Scherzer, O., Yang, H., 2009. Shape metrics based on elastic deformations. J. Math. Imaging Vision 35 (1), 86–102.

Grinspun, E., Krysl, P., Schröder, P., 2002. CHARMS: a simple framework for adaptive simulation. ACM Trans. Graph. 21 (3), 281–290.

Grinspun, E., Hirani, A.N., Desbrun, M., Schröder, P., 2003. Discrete shells. Proc. of ACM SIGGRAPH/Eurographics Symposium on Computer animation, pp. 62–67.

Grinspun, E., Gingold, Y., Reisman, J., Zorin, D., 2006. Computing discrete shape operators on general meshes. Comput. Graph. Forum 25 (3), 547–556.

Heeren, B., 2016. Numerical Methods in Shape Spaces and Optimal Branching Patterns (Ph.D. thesis). University of Bonn.

Heeren, B., Rumpf, M., Wardetzky, M., Wirth, B., 2012. Time-discrete geodesics in the space of shells. Comput. Graph. Forum 31 (5), 1755–1764. https://doi.org/10.1111/j.1467-8659.2012.03180.x.

Heeren, B., Rumpf, M., Schröder, P., Wardetzky, M., Wirth, B., 2014. Exploring the geometry of the space of shells. Comput. Graph. Forum 33 (5), 247–256.

Heeren, B., Rumpf, M., Schröder, P., Wardetzky, M., Wirth, B., 2016. Splines in the space of shells. Comput. Graph. Forum 35 (5), 111–120.

Heeren, B., Rumpf, M., Wirth, B., 2019. Variational time discretization of Riemannian splines. IMA J. Numer. Anal. 39 (1), 61–104.

Hildebrandt, K., Polthier, K., 2011. On approximation of the Laplace-Beltrami operator and the Willmore energy of surfaces. Comput. Graphics Forum 30 (5), 1513–1520. ISSN 1467-8659. https://doi.org/10.1111/j.1467-8659.2011.02025.x.

Hildebrandt, K., Polthier, K., Wardetzky, M., 2006. On the convergence of metric and geometric properties of polyhedral surfaces. Geom. Dedicata 123, 89–112.

Hildebrandt, K., Schulz, C., Tycowicz, C.V., Polthier, K., 2011. Interactive surface modeling using modal analysis. ACM Trans. Graph. 30 (5), 119:1–11.

Hinze, M., 1996. On the numerical approximation of unstable minimal surfaces with polygonal boundaries. Numer. Math. 73 (1), 95–118. ISSN 0029-599X. https://doi.org/10.1007/s002110050186.

Hofer, M., Pottmann, H., 2004. Energy-minimizing splines in manifolds. ACM Trans. Graph. 23 (3), 284–293.

Holm, D., Trouvé, A., Younes, L., 2009. The Euler-Poincaré theory of metamorphosis. Q. Appl. Math. 67, 661–685.

Huber, P., Perl, R., Rumpf, M., 2017. Smooth interpolation of key frames in a Riemannian shell space. Comput. Aided Geom. Design 52–53, 313–328. ISSN 0167-8396. https://doi.org/10.1016/j.cagd.2017.02.008.

Kantorovitch, L.V., 1942. On the translocation of masses. C. R. (Doklady) Acad. Sci. URSS (N. S.) 37, 199–201.

Kantorovitch, L.V., 1948. On a problem of Monge. Uspekhi Mat. Nauk 3, 225–226.

Kendall, D.G., 1984. Shape manifolds, Procrustean metrics, and complex projective spaces. Bull. London Math. Soc. 16 (2), 81–121.

Kheyfets, A., Miller, W.A., Newton, G.A., 2000. Schild's ladder parallel transport procedure for an arbitrary connection. Int. J. Theoret. Phys. 39 (12), 2891–2898.

Kilian, M., Mitra, N.J., Pottmann, H., 2007. Geometric modeling in shape space. ACM Trans. Graph. 26 (64), 1–8.

Kircher, S., Garland, M., 2008. Free-form motion processing. ACM Trans. Graph. 27(2).

Klassen, E., Srivastava, A., Mio, W., Joshi, S.H., 2004. Analysis of planar shapes using geodesic paths on shape spaces. IEEE Trans. Pattern Anal. Mach. Intell. 26 (3), 372–383.

Kochanek, D.H.U., Bartels, R.H., 1984. Interpolating splines with local tension, continuity, and bias control. Comput. Graph. 18 (3), 33–41.

Kurtek, S., Klassen, E., Ding, Z., Srivastava, A., 2010. A novel Riemannian framework for shape analysis of 3D objects. In: Proc. of IEEE Conference on Computer Vision and Pattern Recognition, pp. 1625–1632.

Lakes, R., 1987. Foam structures with a negative Poisson's ratio. Science 235 (4792), 1038–1040. https://doi.org/10.1126/science.235.4792.1038.

Le Dret, H., Raoult, A., 1995. The nonlinear membrane model as variational limit of nonlinear three-dimensional elasticity. J. Math. Pures Appl. 74 (6), 549–578.

Leventon, M.E., Grimson, W. E. L., Faugeras, O., 2000. Statistical shape influence in geodesic active contours. In: Proc. of IEEE Conference on Computer Vision and Pattern Recognition.

Lipman, Y., Sorkine, O., Levin, D., Cohen-Or, D., 2005. Linear rotation-invariant coordinates for meshes. ACM Trans. Graph. 24 (3), 479–487.

Liu, X., Shi, Y., Dinov, I., Mio, W., 2010. A computational model of multidimensional shape. Int. J. Comput. Vis. 89 (1), 69–83. https://doi.org/10.1007/s11263-010-0323-0.

Lorenzi, M., Ayache, N., Pennec, X., 2011. Schild's ladder for the parallel transport of deformations in time series of images. In: Lecture Notes in Computer Science. Proc. of International Conference on Information Processing in Medical Imaging, vol. 6801, pp. 463–474.

Maas, J., Rumpf, M., Schönlieb, C., Simon, S., 2015. A generalized model for optimal transport of images including dissipation and density modulation. ESAIM Math. Model. Numer. Anal. 49 (6), 1745–1769. https://doi.org/10.1051/m2an/2015043.

Martin, S., Thomaszewski, B., Grinspun, E., Gross, M., 2011. Example-based elastic materials. ACM Trans. Graph. 30 (4), 72:1–72:8.

Meek, D.S., Walton, D.J., 2000. On surface normal and Gaussian curvature approximations given data sampled from a smooth surface. Comput. Aided Geom. Des. 17 (6), 521–543.

Mémoli, F., Sapiro, G., 2005. A theoretical and computational framework for isometry invariant recognition of point cloud data. Found. Comput. Math. 5 (3), 313–347.

Mennucci, A., Yezzi, A., Sundaramoorthi, G., 2008. Properties of Sobolev-type metrics in the space of curves. Interfaces Free Bound. 10 (4), 423–445.

Michor, P.W., Mumford, D., 2005. Vanishing geodesic distance on spaces of submanifolds and diffeomorphisms. Doc. Math. 10, 217–245.

Michor, P.W., Mumford, D., 2006. Riemannian geometries on spaces of plane curves. J. Eur. Math. Soc. 8 (1), 1–48.

Michor, P.W., Mumford, D., 2007. An overview of the Riemannian metrics on spaces of curves using the Hamiltonian approach. Appl. Comput. Harmon. Anal. 23 (1), 74–113.

Miller, M.I., Younes, L., 2001. Group actions, homeomorphisms, and matching: a general framework. Int. J. Comput. Vis. 41 (1–2), 61–84.

Monge, G., 1781. Mémoire sur la théorie des déblais et des remblais. De l'Imprimerie Royale.

Nečas, J., Šilhavý, M., 1991. Multipolar viscous fluids. Q. Appl. Math. 49 (2), 247–265.

Noakes, L., Heinzinger, G., Paden, B., 1989. Cubic splines on curved spaces. IMA J. Math. Control Inform. 6 (4), 465–473.

Pennec, X., Lorenzi, M., 2011. Which parallel transport for the statistical analysis of longitudinal deformations? In: Proc. of Colloque GRETSI.

Perl, R., Pozzi, P., Rumpf, M., 2014. A nested variational time discretization for parametric anisotropic Willmore flow. In: Griebel, M. (Ed.), Singular Phenomena and Scaling in Mathematical Models. Springer, Cham, pp. 221–241.

Perperidis, D., Mohiaddin, R., Rueckert, D., 2005. Construction of a 4D statistical atlas of the cardiac anatomy and its use in classification. In: Duncan, J., Gerig, G. (Eds.), Lecture Notes in Computer Science. Medical Image Computing and Computer Assisted Intervention, vol. 3750, pp. 402–410.

Petersen, P., 1997. Riemannian Geometry, second ed. Graduate Texts in Mathematics. Springer.

Peyré, G., Péchaud, M., Keriven, R., Cohen, L.D., 2010. Geodesic methods in computer vision and graphics. Found. Trends Comput. Graph. Vis. 5 (3–4), 197–397.

Peyré, G., Fadili, J., Rabin, J., 2012. Wasserstein active contours. Proc. of IEEE International Conference on Image Processing, pp. 2541–2544.

Pottmann, H., Hofer, M., 2005. A variational approach to spline curves on surface. Comput. Aided Geom. Design 22 (7), 693–709.

Pozzi, P., 2005. The discrete Douglas problem: convergence results. IMA J. Numer. Anal. 25 (2), 337–378. ISSN 0272-4979. https://doi.org/10.1093/imanum/drh019.

Radó, T., 1933. On the Problem of Plateau. vol. 2. Springer-Verlag, Berlin.

Rahman, I.U., Drori, I., Stodden, V.C., Donoho, D.L., Schröder, P., 2005. Multiscale representations for manifold-valued data. Multiscale Model. Sim. 4 (4), 1201–1232.

Rayleigh, J. W. S. B., 1896. The Theory of Sound. vol. 2Macmillan.

Rumpf, M., Wirth, B., 2013. Discrete geodesic calculus in shape space and applications in the space of viscous fluidic objects. SIAM J. Imaging Sci. 6 (4), 2581–2602.

Rumpf, M., Wirth, B., 2015. Variational time discretization of geodesic calculus. IMA J. Numer. Anal. 35 (3), 1011–1046. https://doi.org/10.1093/imanum/dru027.

Rusu, R.E., 2005. An algorithm for the elastic flow of surfaces. Interfaces Free Bound. 7 (3), 229–239.

Schmidt, F.R., Clausen, M., Cremers, D., 2006. Shape matching by variational computation of geodesics on a manifold. In: Lecture Notes in Computer Science. Proc. of Conference on Pattern Recognition, vol. 4174. Springer, pp. 142–151.

Schmitzer, B., Schnörr, C., 2013. A hierarchical approach to optimal transport. In: Lecture Notes in Computer Science. Proc. of International Conference on Scale Space and Variational Methods in Computer Vision, vol. 7893, pp. 452–464.

Schumacher, H., Wardetzky, M., 2019. Variational convergence of discrete minimal surfaces. Numer. Math. 141 (1), 173–213. ISSN 0945-3245. https://doi.org/10.1007/s00211-018-0993-z.

Söhn, M., Birkner, M., Yan, D., Alber, M., 2005. Modelling individual geometric variation based on dominant eigenmodes of organ deformation: implementation and evaluation. Phys. Med. Biol. 50 (24), 5893–5908.

Srivastava, A., Jain, A., Joshi, S.H., Kaziska, D., 2006. Statistical shape models using elastic-string representations. In: Lecture Notes in Computer ScienceProc. of Asian Conference on Computer Vision, vol. 3851, pp. 612–621.

Srivastava, A., Klassen, E., Joshi, S.H., Jermyn, I.H., 2011. Shape analysis of elastic curves in euclidean spaces. IEEE Trans. Pattern Anal. Mach. Intell. 33 (7), 1415–1428. ISSN 0162-8828. https://doi.org/10.1109/TPAMI.2010.184.

Sumner, R.W., Popović, J., 2004. Deformation transfer for triangle meshes. ACM Trans. Graph. 23 (3), 399–405.

Sumner, R.W., Zwicker, M., Gotsman, C., Popović, J., 2005. Mesh-based inverse kinematics. ACM Trans. Graph. 24 (3), 488–495.

Sundaramoorthi, G., Yezzi, A., Mennucci, A., 2007. Sobolev active contours. Int. J. Comput. Vis. 73 (3), 345–366.

Sundaramoorthi, G., Mennucci, A., Soatto, S., Yezzi, A., 2011. A new geometric metric in the space of curves, and applications to tracking deforming objects by prediction and filtering. SIAM J. Imaging Sci. 4 (1), 109–145.

Terzopoulos, D., Platt, J., Barr, A., Fleischer, K., 1987. Elastically deformable models. Comput. Graph. 21 (4), 205–214.

Trouvé, A., Vialard, F.-X., 2012. Shape splines and stochastic shape evolutions: a second order point of view. Q. Appl. Math. 70 (2), 219–251.

Trouvé, A., Younes, L., 2005. Local geometry of deformable templates. SIAM J. Math. Anal. 37 (1), 17–59.

Trouvé, A., Younes, L., 2005. Metamorphoses through Lie group action. Found. Comput. Math. 5 (2), 173–198.

Tsai, A., Yezzi, A., Wells, W., Tempany, C., Tucker, D., Fan, A., Grimson, W.E., Willsky, A., 2003. A shape-based approach to the segmentation of medical imagery using level sets. IEEE Trans. Med. Imaging 22 (2), 137–154.

Tsuchiya, T., 1987. Discrete solution of the Plateau problem and its convergence. Math. Comp. 49 (179), 157–165. ISSN 0025-5718. https://doi.org/10.2307/2008255.

von Radziewsky, P., Eisemann, E., Seidel, H.-P., Hildebrandt, K., 2016. Optimized subspaces for deformation-based modeling and shape interpolation. Comput. Graph. 58 (C), 128–138.

von Tycowicz, C., Schulz, C., Seidel, H.-P., Hildebrandt, K., 2013. An efficient construction of reduced deformable objects. ACM Trans. Graph. 32 (6), 213:1–213:10. ISSN 0730-0301. https://doi.org/10.1145/2508363.2508392.

von Tycowicz, C., Schulz, C., Seidel, H.-P., Hildebrandt, K., 2015. Real-time nonlinear shape interpolation. ACM Trans. Graph. 34 (3), 34:1–34:10.

Wallner, J., 2004. Existence of set-interpolating and energy-minimizing curves. Comput. Aided Geom. Design 21 (9), 883–892.

Wallner, J., Dyn, N., 2005. Convergence and C^1 analysis of subdivision schemes on manifolds by proximity. Comput. Aided Geom. Design 22 (7), 593–622.

Wardetzky, M., 2006. Discrete Differential Operators on Polyhedral Surfaces—Convergence and Approximation (Ph.D. thesis). Freie Universität Berlin.

Wardetzky, M., Bergou, M., Harmon, D., Zorin, D., Grinspun, E., 2007. Discrete quadratic curvature energies. Comput. Aided Geom. Design 24 (8–9), 499–518.

Wilson Jr., W.L., 1961. On discrete Dirichlet and Plateau problems. Numer. Math. 3, 359–373. ISSN 0029-599X.

Winkler, T., Drieseberg, J., Alexa, M., Hormann, K., 2010. Multi-scale geometry interpolation. Comput. Graph. Forum 29 (2), 309–318 (Proceedings of Eurographics).

Wirth, B., 2009. Variational Methods in Shape Space (Dissertation). University Bonn. http:/hss.ulb.uni-bonn.de/2010/2195/2195.htm.

Wirth, B., Bar, L., Rumpf, M., Sapiro, G., 2009. Geodesics in shape space via variational time discretization. In: Lecture Notes in Computer Science. Proc. of International Conference on Energy Minimization Methods in Computer Vision and Pattern Recognition, vol. 5681, pp. 288–302.

Wirth, B., Bar, L., Rumpf, M., Sapiro, G., 2011. A continuum mechanical approach to geodesics in shape space. Int. J. Comput. Vis. 93 (3), 293–318. https://doi.org/10.1007/s11263-010-0416-9.

Xu, Z., Xu, G., Sun, J.-G., 2005. Convergence analysis of discrete differential geometry operators over surfaces. In: IMA Conference on the Mathematics of Surfaces448–457.

Younes, L., Michor, P.W., Shah, J., Mumford, D., 2008. A metric on shape space with explicit geodesics. Atti Accad. Naz. Lincei Cl. Sci. Fis. Mat. Natur. Rend. Lincei (9) Mat. Appl. 19 (1), 25–57. ISSN 1120-6330. https://doi.org/10.4171/RLM/506.

Index

Note: Page numbers followed by "*f*" indicate figures.

D

G

H

I

J

K

L

M

U

V

W

X

Y

Kolibri 360 GmbH
Ferdinand-Jühlke-Straße 7
99095 Erfurt, Deutschland
produktsicherheit@kolibri360.de